Neurorehabilitation Technology

AF377609

David J. Reinkensmeyer ·
Laura Marchal-Crespo · Volker Dietz
Editors

Neurorehabilitation Technology

Third Edition

Editors
David J. Reinkensmeyer
Department of Mechanical
Aerospace Engineering
and Department of Anatomy
and Neurobiology
University of California
Irvine, CA, USA

Laura Marchal-Crespo
Delft University of Technology
Delft, The Netherlands

Volker Dietz
Spinal Cord Injury Center
University Hospital Balgrist
Zürich, Switzerland

ISBN 978-3-031-08997-8   ISBN 978-3-031-08995-4   (eBook)
https://doi.org/10.1007/978-3-031-08995-4

1st edition: © Springer-Verlag London Limited 2012
2nd edition: © Springer International Publishing 2016
3rd edition © The Editor(s) (if applicable) and The Author(s), under exclusive license
to Springer Nature Switzerland AG 2022, corrected publication 2023
Chapter "Spinal Cord Stimulation to Enable Leg Motor Control and Walking in People with
Spinal Cord Injury" is licensed under the terms of the Creative Commons Attribution 4.0
International License (http://creativecommons.org/licenses/by/4.0/). For further details see license information in the chapter.
This work is subject to copyright. All rights are solely and exclusively licensed by the Publisher,
whether the whole or part of the material is concerned, specifically the rights of translation,
reprinting, reuse of illustrations, recitation, broadcasting, reproduction on microfilms or in any
other physical way, and transmission or information storage and retrieval, electronic adaptation,
computer software, or by similar or dissimilar methodology now known or hereafter developed.
The use of general descriptive names, registered names, trademarks, service marks, etc. in this
publication does not imply, even in the absence of a specific statement, that such names are
exempt from the relevant protective laws and regulations and therefore free for general use.
The publisher, the authors, and the editors are safe to assume that the advice and information in
this book are believed to be true and accurate at the date of publication. Neither the publisher nor
the authors or the editors give a warranty, expressed or implied, with respect to the material
contained herein or for any errors or omissions that may have been made. The publisher remains
neutral with regard to jurisdictional claims in published maps and institutional affiliations.

This Springer imprint is published by the registered company Springer Nature Switzerland AG
The registered company address is: Gewerbestrasse 11, 6330 Cham, Switzerland

Introduction to the Third Edition

> *When I want to discover something, I begin by reading up everything that has been done along that line in the past—that's what all these books in the library are for. I see what has been accomplished at great labor and expense in the past. I gather data of many thousands of experiments as a starting point, and then I make thousands more.*
>
> *Attributed to Thomas Edison*

The aim of this book is to provide a comprehensive overview of the ongoing revolution in neurorehabilitation technology. World leaders have taken the time to step back from their work, evaluate the state of the art in their field, and trace the development of their own work in creating this state of the art. We wish to provide a cutting-edge resource for those seeking to use, evaluate, and improve these technologies.

There are four unique features of the book. First, we have attempted to ground the discussion of neurorehabilitation technology on neurorehabilitation science. Thus, you will find less information about the details of mechanical design or low-level machine controllers than information about the physiology of sensorimotor impairments, strategies for human-machine interaction, and the results of clinical testing.

Second, we have chosen to emphasize movement rehabilitation after stroke and spinal cord injury, thereby focusing on the leading causes of disability and the largest user groups of neurorehabilitation technology. However, we note that many of the design principles discussed can transfer to a broader user group and that several chapters cover applications for people with neuromuscular disease, cognitive impairment, and cerebral palsy as well.

Third, we have chosen to dedicate a greater amount of attention to robotic therapy than other approaches. This is because robotics therapy technologies have experienced the greatest growth over the past 40 years. Yet we recognize that other technologies—including neural stimulation, sensor-based devices, passive mechanical devices, exoskeletons, virtual reality, and mobile devices —show promise and are increasingly playing greater roles in the clinical delivery of rehabilitation, supplanting "traditional" robotic therapy in some cases. Therefore, we have expanded this third edition to include a greater amount of discussion of other emerging technologies.

Finally, we have focused on therapeutic rather than assistive technology. That is, the emphasis here is on technologies that assist the motor system in improving its intrinsic capacity to respond through training. We note however that the line between therapeutic and assistive devices is blurring because of emerging technologies like legged exoskeletons and spinal stimulation systems.

We have organized the book into seven sections, which can roughly be divided into two halves. The first half of the book (Sects. 1–4) focuses on the design and implementation of neurorehabilitation technology. Section 1 contains three chapters that explain the relevant principles of neuroplasticity, motor learning, and sensorimotor recovery that can be used to inform neurorehabilitation technology development. Section 2 contains a set of chapters that exemplify how neurorehabilitation technology can be implemented to treat specific aspects of movement pathophysiology. Section 3 overviews principles of interactive rehabilitation technology—including issues of optimal challenge, psychophysical interaction, error manipulation, haptic interactions, expectations of end-users, and device implementation in a pediatric context. Section 4 contains three chapters on assessment technology and predictive modeling, all of which suggest working toward a precision medicine approach in rehabilitation.

The second half of the book concentrates on specific technologies. Section 5 surveys a broad scope of neurorehabilitation technologies: spinal cord stimulation, functional electrical stimulation, virtual reality, wearable sensors, brain-computer interfaces, passive devices, mobile technologies for cognitive rehabilitation, and telerehabilitation. Sections 6 and 7 then provide detailed overviews of upper extremity and lower extremity robotics technologies. The book concludes with an Epilogue in the form of a debate over the current efficacy and ongoing potential of neurorehabilitation robotics.

New chapters selected for the third edition include the neurophysiological basis of rehabilitation, the role of challenge for optimizing motor performance, the role of haptic interactions in promoting motor learning, computational neurorehabilitation, precision rehabilitation, spinal cord stimulation to enable walking, wearable sensors for stroke rehabilitation, mobile technologies for cognitive rehabilitation, telerehabilitation, body weight support devices, and the debate on the current and future value of rehabilitation robotics. Other chapters published in the third edition have also been substantially updated and reorganized to reflect the ongoing revolution.

We hope that this book will inspire the next generation of innovators—clinicians, neuroscientists, and engineers—to move neurorehabilitation technology forward, thus benefiting the next generation of people with a neurologic impairment.

Irvine, USA   David J. Reinkensmeyer
Delft, The Netherlands   Laura Marchal-Crespo
Zürich, Switzerland   Volker Dietz

Contents

Part I

Basic Framework: Motor Recovery, Learning, and Neural Impairment

Learning in the Damaged Brain/Spinal Cord: Neuroplasticity

1

Andreas Luft, Amy J. Bastian, and Volker Dietz

Abstract

Neuroplasticity refers to the ability of the central nervous system (CNS) to undergo persistent or lasting modifications to the function or structure of its elements. Neuroplasticity is a CNS mechanism that enables successful learning. Likely, it is also the mechanism by which recovery after CNS lesioning is possible. The chapter gives an overview of the phenomena that constitute plasticity and the cellular events leading to them. Evidence for neural plasticity in different regions of the brain and the spinal cord is summarized in the contexts of learning, recovery, and rehabilitation therapy.

Keywords

Recovery · Rehabilitation · Stroke · Spinal cord injury · Brain lesion · Plasticity

1.1 Learning in the CNS

Rehabilitation technologies that support movement recovery make use of different brain and body mechanisms, one of which is the brain's ability to learn. Likely, the learning in the lesioned brain that mediates functional recovery is not identical to learning in the healthy state. Nevertheless, there are certain mechanisms on the cellular and the systems level, that can be broadly called *neuroplasticity mechanisms*, which are shared by healthy learning and recovery. Clearly, the main behavioral determinants of healthy learning of novel movements, activity, and repetition are also important in recovery.

Hence, movement recovery may depend in part on motor learning. Motor learning is a general term that encompasses many different processes. Distinct behavioral and neural mechanisms are engaged depending on the level of complexity of the movement to be learned and the stimulus driving learning. A few different forms of motor learning are briefly reviewed.

Motor adaptation is a type of motor learning that acts on a time scale of minutes to hours in

A. Luft (✉)
Department of Neurology, University of Zurich, Frauenkliniksrasse 26, 8091 Zurich, Switzerland
e-mail: andreas.luft@usz.ch

A. J. Bastian
Department of Neuroscience, The Johns Hopkins School of Medicine, Kennedy Krieger Institute, 707 N. Broadway, Baltimore, MD 21205, USA
e-mail: bastian@kennedykrieger.org

V. Dietz
Spinal Cord Injury Center, University Hospital Balgrist, Forchstr. 340, 8008 Zurich, Switzerland
e-mail: vdietz@paralab.balgrist.ch

© The Author(s), under exclusive license to Springer Nature Switzerland AG 2022
D. J. Reinkensmeyer et al. (eds.), *Neurorehabilitation Technology*,
https://doi.org/10.1007/978-3-031-08995-4_1

order to account for predictable perturbations to a movement. Adaptation occurs on a trial-by-trial basis, correcting a given movement from one trial to the next. It is driven by sensory prediction errors, which represent the difference between the brain's estimate of the sensory consequences of movement and the actual sensory feedback [1]. Once a movement has been adapted, it can be de-adapted when the predictable perturbation is reversed or removed. Discontinuation of training also leads to "forgetting" the adaptations over relatively short periods of time [2].

Associative learning can also occur on a time scale of minutes to hours. Classical conditioning is perhaps the most studied form of associative learning. It links two previously unrelated phenomena to improve behavior. For example, in eye-blink conditioning, a "conditioned" stimulus like a sound or tone can be repeatedly paired with a second, slightly delayed "unconditioned" stimulus like a puff of air to the eye [3]. Early in the learning process, the eye blinks in response to the puff of air (i.e., unconditioned response). However, with repeated exposure, the eye begins to blink when the tone is presented, therefore anticipating the air puff by closing the eye (i.e., conditioned response). This type of conditioning can be used to make associations between many types of behaviors.

Motor learning can also be driven by feedback. Feedback can be given in close temporal association with the movement (knowledge of performance) and is used to adjust movement parameters on a trial-to-trial basis. It may be given later, e.g., after several repetitions, and can reflect the average outcome (knowledge of result) [4]. Performance feedback may be associated with a rewarding or discouraging element thereby inducing reinforcement learning [5] or avoidance learning [6], respectively. These learning processes can occur on short- or long time scales depending on the type and complexity of the movement. The rewarding or discouraging feedback can be explicit (e.g., earning money) or implicit [7], i.e., small improvements after repeating a novel movement, e.g., when learning to play a piano piece, are often not obvious or consciously perceived. Unconscious rewarding feedback may play a role. The conscious reward of playing the piece well typically comes late and temporally unrelated to the movement (e.g., the audience applauds). Thus, implicit motor learning may be mediated through use-dependent or Hebbian-like plasticity rather than reinforcement mechanisms.

All these forms of motor learning rely on networks of neural structures rather than single areas, but there are some key regions that seem to play especially important roles in each. Adaptation is known to be cerebellum-dependent [8]. Classical conditioning can involve the cerebellum and hippocampus depending on the specific timing between stimuli [3]. Reward and avoidance learning are dependent on basal ganglia circuitry [9]. Use-dependent learning likely occurs at many levels of the nervous system, including spinal cord, brain stem, and cerebral (cortical and subcortical) structures. Complex motor skill learning induces plasticity in the motor cortex, especially during the consolidation of the learned movement [9–11]. Importantly, all forms of motor learning are dependent on cellular mechanisms of plasticity including gene/protein expression, synaptic and fiber growth, and functional changes such as long-term potentiation and long-term depression. As such, these mechanisms are reviewed below.

1.2 Mechanisms of Neuroplasticity in Learning and After Lesions

1.2.1 Gene Expression

Learning a motor skill requires gene expression in the primary motor cortex (M1) [10, 11]. If this expression is pharmacologically blocked, learning is inhibited. Gene and subsequent protein expression is a common requirement of various learning processes [12, 13] as well as for cellular equivalents of learning, i.e., the changes in neuronal structure [14] and synaptic strength in the form of long-term potentiation (LTP) and depression (LTD) [15]. For motor skill learning, proteins are not only expressed during training but also thereafter while the subject is resting [10]. This delayed synthesis can be regarded as reflecting intersession consolidation processes [16]. The genes induced by learning are manifold, including immediate early genes (IEG, transcription factors). Expression of the IEG Arc in M1 was shown to occur specifically during skill learning but not during movement without learning [17]. Dopamine-related gene expression has been shown to be related to motor skill learning in mice [18].

Gene expression is induced by ischemia, especially in the peri-infarct cortex [19]. Some of these genes could also promote cellular plasticity offering the potential for stroke-induced plasticity as self-healing mechanism of the brain. Genes and proteins induced by ischemia include axonal growth stimulators while growth inhibitors are suppressed [20, 21].

1.2.2 Cellular Plasticity

Long-term potentiation (LTP) and depression (LTD) are commonly seen as cellular equivalents of the brain's learning abilities [22]. Either by repetitive stimulation, seen as the equivalent to repetitive training, or by synchronizing two signals that converge at one neuron, potentially reflecting associative learning phenomena, an increase in synaptic strength is induced that lasts from hours to days, termed LTP [23]. LTD is induced by low-frequency stimulation and leads to a lasting reduction in synaptic strength [22]. Both LTP and LTD have been described in various brain regions including the primary motor cortex (M1) [24]. The observation that the ability of M1 neurons to undergo LTP and LTD is reduced after training provides indirect evidence for the hypothesis that primary motor cortex LTP/LTD is consumed by training, and hence can be considered as one candidate mechanism of motor skill learning [25]. Two months after a skill has been learned in a 2 week training period and is well remembered, the synaptic strengthening that is observed in M1 shortly after training persists. But, the ability to

undergo LTP has recovered and is now expressed on a higher level of synaptic strength [24].

In addition to changes in synaptic strength, the structure of neuronal networks is reorganized in association with motor skill learning. Apical and basal dendrites expand in association with skill training [26, 27]. This expansion is specific to the neurons involved in the control of the muscles used in the trained movement but not in the other musculature [28]. It remains open whether these changes are permanent or reflect a temporary expansion of M1 connectivity. Changes in dendritic spines, in contrast, were shown to be temporary and return to baseline 1 week after training has ended [29].

Microglia also play a role in plasticity by promoting synapse formation via a BDNF (brain-derived neurotrophy factor)-mediated mechanism [30].

In the context of recovery after a brain or spinal cord injury, the role of LTP and LTD is unclear. LTP is facilitated in the peri-infarct cortex [31]. This result may be incompatible with the hypothesis that LTP is consumed during recovery as it is after healthy skill learning; hence, LTP would be reduced in the peri-infarct cortex not facilitated. But, the study lacks information about the recovery of function or lesion size, so a valid comparison to healthy learning is impossible, and the issue of LTP utilization during recovery is left unanswered. In the hippocampus, short-term ischemia leads to a disruption of LTP formation [32]. In humans, preliminary evidence indicates that LTP-like phenomena elicited in M1 of the lesioned hemisphere (cortical or subcortical lesions) by repetitive transcranial magnetic stimulation (TMS) predict good recovery in 6 months [33]. Paired associative stimulation (peripheral muscle and TMS stimulation of M1)—a potential human equivalent of associative LTP—can be elicited in the affected hemisphere M1, especially in those patients with limited deficits [34]. LTP-like phenomena are enhanced by serotonin [35] possibly explaining the beneficial effect of serotonin reuptake inhibitors in stroke recovery [36]. Hence, the ability of the lesioned cortex to undergo LTP may contribute to recovery.

1.2.3 Systems Plasticity in the Brain

Plasticity phenomena not only exist on the level of single neurons or networks but also in distinct functional systems of the brain. The input–output organization and the somatotopy of M1 undergo persistent changes during motor skill learning. Skill learning leads to an expansion of the cortical representation of the trained limb [37, 38]. Longitudinal motor cortex mapping experiments in rats show that this expansion is transient and is reversed after training ends although the skill is maintained [39]. In humans who continuously train new motor skills, e.g., professional pianists, task-related activation is smaller in the area and more focused [40, 41]. Musicians also have enlarged gray matter volumes in various areas of the cortex including the motor cortices [42]. The M1 of musicians contains memory traces of practiced skills that can be probed by TMS [43].

Representations in the primary motor cortex are also modified while recovering from a stroke. Initially, large areas of motor and adjacent cortices are recruited in the attempt to accomplish a movement as detected by functional magnetic resonance imaging (fMRI) [44, 45]. If M1 itself is lesioned, expanded activation is found in the peri-infarct cortex [46] or the premotor cortex [47]. As subjects recover, this hyper-activation is reduced, and movement-related activity focuses on the ipsilesional hemisphere contralateral to the moving limb [48–50]. If recovery is unsuccessful, cortices remain hyper-activated in the lesioned as well as the non-lesioned hemisphere which has been interpreted as a sign of a continuous attempt to initiate recovery [47]. But recovery is not only accompanied by cortical activation changes. Larger activation in the cerebellum ipsilateral to the moving limb [48] and smaller activation in the contralateral cerebellum are associated with better recovery [49].

Connectivity between different cortical regions in the brain is impaired after stroke, not only in areas in the vicinity of the lesion but also in the intact hemisphere [51]. There is reduced interhemispheric connectivity after stroke, especially between primary motor cortices [52].

While movement-related activation observed with functional imaging methods demonstrates the brain areas that are involved in the control of the movement performed during imaging, TMS can directly assess the output efficacy and the viability of descending pathways in the lesioned hemisphere. Larger motor evoked potentials in response to TMS and the absence of ipsilateral responses to stimulation of the intact hemisphere are correlated with good functional recovery [53, 54].

Studies in animals and humans emphasized the importance of the GABAergic inhibition and the balance between excitation and inhibition in the peri-infarct cortex. Although the evidence from different studies in animals and humans using different methodologies is difficult to unify, it is assumed that inhibition is abnormally high early after stroke, then is reduced below the levels of healthy controls and returns to normal afterwards [55]. In addition, the interhemispheric (transcallosal) interactions of contralesional and ipsilesional motor cortices are altered after stroke and change during recovery [56], but it is yet unclear whether these observations are a cause or a consequence of (un)successful recovery [57].

1.2.4 Plasticity in Spinal Cord

There is convincing evidence in cats with a transected spinal cord that use-dependent plasticity of neuronal circuits within the spinal cord exists [58, 59]. When stepping is practiced in spinal cat, this task can be performed more successfully than when it is not practiced, but the standing duration is not improved and vice versa, i.e., training of standing has only an effect on this task [60, 61]. The training effects of any motor task critically depend on the provision of sufficient and appropriate proprioceptive feedback information to initiate a reorganization of neural networks within the spinal cord. This is usually achieved by functional training. In contrast, following a complete SCI, when locomotor networks are no longer used, a neuronal dysfunction develops below the level of the spinal cord lesion in humans [62] and rodents [46].

1.2.4.1 Spinal Reflex Plasticity

The isolated spinal cord can exhibit some neuronal plasticity. Evidence for such plasticity at a spinal level has been obtained for the relatively simple monosynaptic reflex arc [62]. Monkeys could either be trained to voluntarily increase or decrease the amplitude of the monosynaptic stretch reflex in response to an imposed muscle lengthening [62], as well as to its analogue, the H-reflex [46]. The fact that the training effects persist after spinal cord transection [60] indicates that learning by spinal neuronal circuits is possible. Similarly, humans can be trained to change the gain of the monosynaptic stretch reflex ([63]; for review, see [64]).

The idea that neuronal circuits within the spinal cord can learn is also supported by studies of spinal reflex conditioning. Simple hind limb motor responses to cutaneous or electrical stimulation are enhanced in animals with a transected spinal cord by reflex conditioning (i.e., pairing the stimulus with another stimulus that evokes a stronger motor response) [65]. These reflex responses become enhanced within minutes of conditioning indicating that synaptic efficacy along the reflex arc has changed, possibly through long-term potentiation [65].

1.2.4.2 Task-Specific Neural Plasticity

Today, it is obvious that there is also considerable task-specific plasticity of the sensorimotor networks of the adult mammalian lumbosacral spinal cord (for review, see [58, 59, 66]). The detailed assessment of the modifiability of neuronal network function is reflected in the research on central pattern generators (CPGs) underlying stepping movements [65, 67–69]. Observations made in spinal cats indicate that the lumbosacral spinal cord obviously can execute stepping or standing more successfully if that specific task is practiced. If the training of a motor task is discontinued and no similar task is subsequently trained, then the performance of the task previously trained is degraded [58]. Consequently, plasticity can be exploited for rehabilitative purposes using specific training approaches following a spinal injury.

In the cat, recovery of locomotor function following spinal cord transection can be improved using regular training, even in adult animals [70, 71]. The provision of an adequate proprioceptive input to the spinal cord during training is essential to achieve an optimal output of the spinal neuronal circuitry with the consequence of an improved function. This aspect of training could meanwhile also be demonstrated for the locomotor training of subjects suffering an SCI [72]. Furthermore, in association with hind limb exercise, the reflex activity becomes normalized in adult rats following spinal cord transection [73]. Exercise obviously helps to normalize the excitability of spinal reflexes.

Several neurotransmitter systems within the spinal cord (glycinergic and GABAergic systems) are suggested to be involved in the mediation of plastic changes following repetitive task performance [58]. In cats with a spinal cord transection, stepping can be induced by the administration of the noradrenergic agonist clonidine, which enhances the activity in spinal neuronal circuits that generate locomotor activity [74–76]. However, the application of Dopamine in patients with an SCI has no effect on the outcome of function [77]. Furthermore, serotonin seems to be involved in the production of locomotor rhythms [78].

Training paradigms of stepping and standing can modify the efficacy of the inhibitory neurotransmitter, glycine [58]. For example, when glycine is applied to a chronic spinal cat that has acquired the ability to step successfully, there is little change in its locomotor capability. If it is administered to a stand-trained cat, it becomes able to successfully step with body support [58, 66]. These findings suggest that the effect of glycine is so far specific in its action as it enables spinal networks to integrate proprioceptive input by reducing inhibition [75, 76].

1.2.5 Subcortical Contributions to Movement Learning

The cerebellum is thought to use adaptive learning mechanisms to calibrate internal models for predictive control of movement. Such models are needed because sensory feedback is too slow for movements that need to be both fast and accurate; without internal models, corrections would be issued too late. Instead, the brain generates motor commands based on internal predictions of how the command would move the body [79]. This feedforward control requires stored knowledge (i.e., models) of the body's dynamics, the environment, and any object to be manipulated. For example, recent work has demonstrated that cerebellar damage causes a bias in the brain's representation of limb inertia relative to actual inertia, which results in characteristic patterns of reaching dysmetria, i.e., over- or under-shooting targets [80]. This specific deficit was confirmed in simulation and in behavioral studies of control subjects reaching with their limb while inertia is unexpectedly changed via an exoskeleton robot. Perhaps most importantly, this work also demonstrated a way of correcting this mismatch using cerebellar patient-specific compensations rendered by an exoskeleton robot. This suggests that there may be ways to compensate for biases in internal model representations using robotics. Unfortunately, cerebellar patients cannot learn to correct their internal model biases due to a loss of a cerebellum-dependent learning process often referred to as adaptation.

Many studies have shown that the cerebellum is essential for adapting a motor behavior through repeated practice—it uses error information from one trial to improve performance in subsequent trials. It is important to note that cerebellum-dependent motor learning is driven by errors directly occurring during the movement (knowledge of performance), rather than knowledge of results after the movement is completed (e.g., hit or miss). Studies have suggested that the type of error that drives cerebellum-dependent learning is not the target error (i.e., "How far am I from the desired target?"), but instead what has been referred to as a sensory prediction error (i.e., "How far am I from where I predicted I would be?") [1]. Damage to the cerebellum impairs the ability to adapt many types of movements, including reaching [81], walking

[82], balance [83], and eye movements [84]. To date, there has been no systematic way to substitute or compensate for deficits in this form of learning.

The microcircuit involved in cerebellar adaptation was first proposed by Marr [85], Albus [86], and Ito [87]. These works continue to provide the basis for many of the current theories of cerebellar function. Central to the idea of cerebellar involvement in learning was the discovery that Purkinje cell output can be radically altered by climbing fiber induction of long-term depression (LTD) of the parallel fiber-Purkinje cell synapse [88]. Hence, climbing fiber inputs onto Purkinje cells can be viewed as providing a unique type of teaching or error signal to the cerebellum. It has been shown that the climbing fiber may not simply be an all-or-none signal indicating error [89]. Instead, the duration of climbing fiber bursts is predictive of the magnitude of plasticity and learning, making it a graded instructive signal for adaptation. In addition to the climbing fiber-dependent LTD, there are many other sites of plasticity in the cerebellar cortex and deep nuclei that involve LTP and non-synaptic plasticity (for review, see [90]). Thus, there are multiple avenues for activity-dependent plasticity to occur within the cerebellum over relatively short time scales. It is presumed that the plastic changes in cerebellar output are responsible for changing motor behavior during the process of adaptation which is processes that occur on relatively fast time scales. Purkinje cells are organized in microzones that either increase or decrease the output activity during learning (for review, see [91]). These zones interact with widespread regions of the cortex, thereby influencing learning processes across different domains of cognition and motor behavior.

While the cerebellum operates on faster time scales, the brain's reward system encodes the outcome of a behavior, i.e., the knowledge of the result, and can thereby influence learning (reinforcement learning) (for review, see [92]). Dopaminergic neurons in the substantia nigra (SN)/ventral tegmental areas (VTA) complex encode reward prediction errors and feed this information to the cortex. Especially neurons in

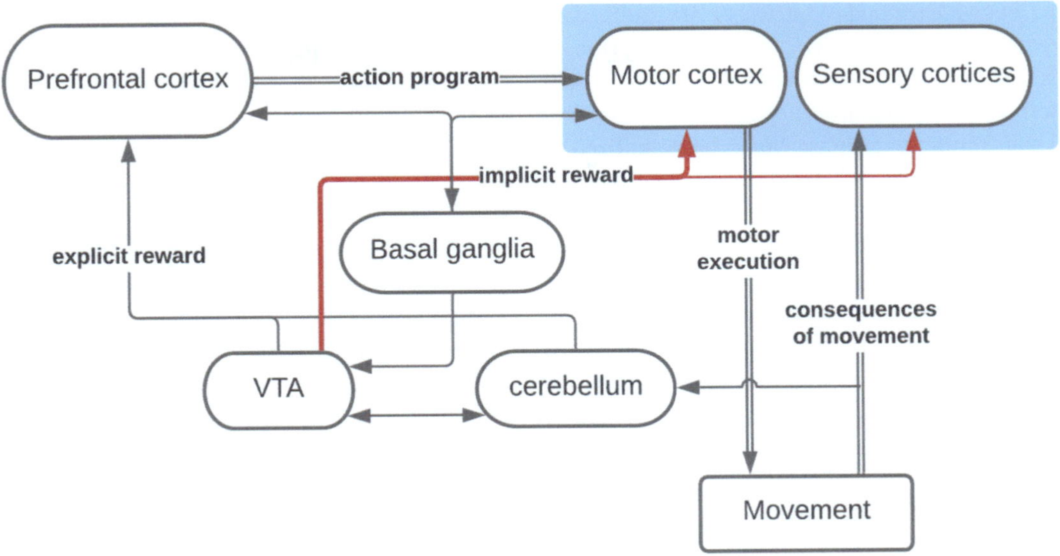

Fig. 1.1 Schematic representation of the integration of reward circuits in the sensorimotor network. Via a dopamine (DA) signal encoding a form of immediate "implicit" reward, this signal can directly modulate synaptic plasticity in sensorimotor cortex synapses enabling motor skill learning

the VTA project to the primary motor cortex and are involved in motor skill learning. Ipsilateral dopaminergic projections from VTA to M1 [93] are specifically necessary for acquiring but not for performing a skill once acquired. Elimination of dopaminergic terminals in M1 [94] or destruction of dopaminergic neurons in VTA impairs the acquisition of a reaching skill in rats [95]. Dopamine modulates the excitability of M1 [96] and S1 [97] and, more importantly, is necessary for the formation of LTP in layer II/III synapses [94] that link M1 and S1 via horizontal connections. Plasticity at these synapses is involved in skill learning—as evidenced by the fact that the capacity of these synapses undergoing LTP is reduced after skill learning [17]. It seems plausible that the VTA-to-M1 projection relays signals of the same nature as compared to those that activate dopaminergic neurons from VTA to nucleus accumbens and prefrontal cortex, i.e., information about reward value and expectedness [98]. In the complex interplay of these circuits that include the basal ganglia (esp. the ventral striatum) and the cerebellum, it may be that dopaminergic neurons signal two forms of reward, one that reflects the outcome of

an action or goal attainment (explicit reward), the second a more immediate result of a single movement component [99]. This latter form is not consciously experienced but rather causes a "good feeling" about the ongoing movement and may be termed "implicit reward" (Fig. 1.1).

1.3 Learning and Plasticity During Rehabilitation Therapy

1.3.1 Lesions of Cortex and Descending Pathways

Rehabilitative training is associated with neurophysiological alterations that are related to the improvement in motor function observed in individual stroke survivors [100]. Although correlation is not proof of causation, these studies provide an argument for neuroplasticity being one possible mechanism by which rehabilitative training can operate effectively. While bilateral arm training was associated with an increase in premotor cortex activation in both hemispheres that correlated with functional improvement in the Fugl-Meyer and Wolf tests [101, 102], conventional physical therapy (based on Bobath exercises) did not show altered brain activation despite being equally effective [102]. Conventional physical exercise may have utilized a mechanism other than those detectable by fMRI, e.g., by inducing changes in muscle, peripheral nerves, or spinal cord.

Lower extremity repetitive exercises in the form of aerobic treadmill training likely utilize yet another form of brain reorganization to improve gait. As compared with stretching exercises, improvements by treadmill training were related to increased activation of cerebellum and brainstem as detected with fMRI of paretic knee movement [103]. Interestingly, the areas recruited in the cerebellum and brainstem corresponded to regions that control spinal pattern generators (cerebellar and midbrain locomotor region). These regions may have compensated for the loss of corticospinal projections that were injured by the stroke. It has also been shown that individuals with a cerebral stroke can improve walking symmetry using adaptive mechanisms of learning on a split-belt treadmill [104–106]. Repeated split-belt training over a 1 month time resulted in improvements in step length symmetry in chronic stroke survivors [107]. Importantly, the split-belt treadmill was used to augment the step asymmetry errors that the stroke survivors produced. This was done to drive a cerebellum-dependent learning process that would correct their error. Stated simply, making their error bigger drove the nervous system to learn how to correct it. After training, when they walked over ground, they had learned to correct their step length asymmetry. Training over 4 weeks led to improvements that lasted (and even improved further) for 3 months post training. Here again, the hypothesis is that intact cerebellar mechanisms are responsible for this form of motor learning. Hence, subcortical reorganization may be the mechanism to target in lower extremity, and particularly walking, rehabilitation.

The availability of treatments that operate through distinct mechanisms may provide the rehabilitation clinician with many tools to individualize therapy for the patient. It seems likely that different patients with different brain injury and lesion profiles will require different therapeutic approaches.

1.3.2 Cerebellar Lesions

Recovery from a first ischemic cerebellar stroke is often very good, with minimal to no residual deficits in up to 83% of patients [108–110]. On the other hand, individuals with degenerative cerebellar disorders tend to have persistent or progressively worsening clinical signs and symptoms [111]. One study has shown that people with damage to the deep nuclei do not recover as well as those with damage to only the cerebellar cortex and white matter [112]. Thus, the etiology of the lesion and the extent of damage are major indicators in recovery.

There is a growing body of literature on the effectiveness of rehabilitation interventions for individuals with primary cerebellar damage. There are studies on the effects of rehabilitation interventions in this patient population, but most are noncontrolled small series (e.g., [113]) or case studies (e.g., [114]). Most work has been done on walking rehabilitation with common interventions including combinations of exercises targeting gaze, static stance, dynamic stance, gait, and complex gait activities [113, 114]. Dynamic balance activities in sitting, kneeling, and quadruped have also been advocated [113]. Individuals with acute cerebellar stroke seem to recover similarly regardless of whether they participated in a 2 week treadmill training intervention [115]. Further, individuals with superior cerebellar artery infarcts tend to show more severe ataxia than those with posterior inferior cerebellar artery infarcts early on, but both groups tend to recover to the same extent after 3 months. People with degenerative disorders tend to benefit more from rehabilitation training. Ilg and coauthors found that 4 weeks of an intensive coordination training followed by 8 weeks of home exercise could improve walking coordination and static and dynamic balance scores. It has also been shown that a 6 week home balance exercise program can improve balance and walking measures in people with cerebellar degeneration [116]. In that study, it was shown that the difficulty of the balance exercise is what predicted the best outcomes, with more challenging balance activities resulting in the greatest improvement. It was also shown that the effects of home exercise lasted for a month after therapy. In all these studies, it is not known whether such changes translate to improved real-world function.

Locomotor training over ground and on treadmills, and with and without body weight support, has also been used with some success in single case examples [117, 118]. It is not clear how imbalance is corrected in the body weight support environment, however. With all gait and balance activities, it seems critical that the exercise be sufficiently and increasingly challenging, to facilitate plasticity in other intact areas of the nervous system [119, 120].

Two controlled trials investigated the use of cerebellar direct current stimulation on ataxia and found a positive effect in patients with neurodegenerative forms of ataxias (spinocerebellar ataxia, Friedreich's ataxia, and cerebellar form of multiple system atrophy) [121, 122].

1.3.3 Spinal Lesions

1.3.3.1 Plasticity of Spinal Neuronal Circuits: Rehabilitation Issues

Based on the knowledge gained from animal experiments, the aim of rehabilitation after stroke or SCI should be focused on the improvement of function by taking advantage of the plasticity of spinal and supraspinal neuronal circuits and should less be directed to the correction of isolated clinical signs, such as the reflex excitability and muscle tone. For the monitoring of outcome and the assessment of the effectiveness of any interventional therapy, standardized functional tests should be applied.

1.3.3.2 Functional Training in Persons with a Spinal Cord Injury

The coordination of human gait seems to be controlled in much the same way as in other mammals [123]. Even arm swing during locomotion represents a residual function of quadrupedal coordination of bipedal gait [124]. Therefore, it is not surprising that in persons with complete or incomplete paraplegia, due to a spinal cord injury, locomotor EMG activity and movements can be both elicited and trained similarly as in the cat. This is achieved by partially unloading (up to 80%) the patients who are standing on a moving treadmill ([72, 125, 126] for review, see [127]). In severely affected patients, the leg movements usually must be assisted externally, especially during the transmission from stance to swing. In addition, leg flexor activation can be enhanced by flexor reflex stimulation of the peroneal nerve during the swing phase [128]. The timing of the pattern of leg muscle EMG activity recorded in such a condition is like that seen in healthy subjects. However, the amplitude of leg muscle EMG is considerably reduced, corresponding to the paresis, and is less well-modulated. This makes the body unloading necessary for locomotor training. There are several reports about the beneficial effect of locomotor training in incomplete paraplegic patients (for review, see [71, 129, 130]), and patients who undergo locomotor training have greater mobility compared to a control group without training [131]. The neuronal networks below the level of an SCI can be activated to generate locomotor activity even in the absence of supraspinal input [75, 76, 132].

The analysis of the locomotor pattern induced in complete paraplegic patients indicates that it is unlikely to be due to rhythmic stretches of the leg muscle because leg muscle EMG activity is, as in healthy subjects, equally distributed during muscle lengthening and shortening phases [133]. In addition, recent observations indicate that locomotor movements induced by a robotic device in patients who are completely unloaded do not lead to a significant leg muscle activation

[134]. This implies that the generation of the leg muscle EMG pattern in these patients is programmed at a spinal level and requires appropriate proprioceptive input from load and hip joint signaling receptors [123].

During daily locomotor training, the amplitude of the EMG in the leg extensor muscles increases and becomes better modulated during the stance phase, and inappropriate leg flexor activity decreases. Such training effects are seen both in complete and incomplete paraplegic patients [135]. The training effects lead to a greater weight-bearing function of the leg extensors, i.e., body unloading during treadmill locomotion can be reduced during training. This indicates that even the isolated human spinal cord has the capacity not only to generate a locomotor pattern but also to show some neuroplasticity which can be exploited by a functional training [136–139]. However, only persons with incomplete paraplegia benefit from the training program in so far as they can learn to perform unsupported stepping movements on solid ground [135]. Neuroplastic changes also occur in elderly SCI subjects. This becomes reflected in a recovery of neurological deficits like that of young subjects. However, the translation of this improvement into a better function is significantly worse in elderly subjects [140]. Therefore, it is required to develop and apply specific and focused rehabilitation procedures in elderly subjects.

In complete paraplegic patients, the training effects on leg muscle activation become lost after the training has been stopped [132]. Furthermore, after about 1 year after injury, complete paraplegic patients develop a neuronal dysfunction below the level of injury, especially in the long flexor muscles [141]. According to rodent experiments, this dysfunction is thought to be due to undirected sprouting within neuronal circuits [46].

1.3.3.3 Prerequisites for a Successful Training

The spinal pattern generator must be activated by the provision of an appropriate proprioceptive feedback that leads to a meaningful muscle activation associated with plastic neuronal changes and consequently to an improvement of function [134].

Proprioceptive input from receptors signaling contact forces during the stance phase of gait is essential for the activation of spinal locomotor centers [134, 142–145] and is important to achieve training effects in paraplegic patients [135] (Fig. 1.2). Furthermore, hip joint-related afferent input seems to be required to generate a locomotor pattern [134, 146]. In addition, for a successful training program for stroke and SCI subjects, spastic muscle tone must be present as partial compensation for paresis [147].

Only in patients with moderately impaired motor function, a close relationship between motor scores (clinical assessment of voluntary muscle contraction: ASIA motor score) and locomotor ability exists. More severely affected

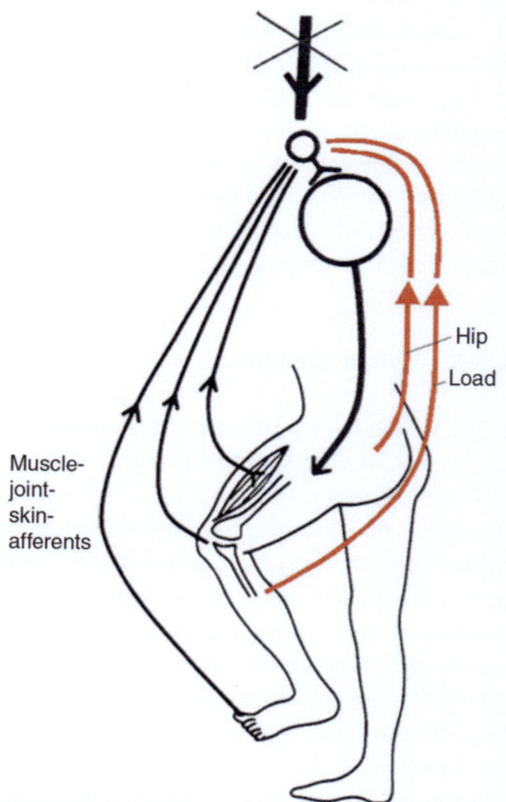

Fig. 1.2 Schematic demonstration of proprioceptive input during locomotor training in SCI subjects. The input from load and hip joint afferents was shown to be essential to achieving training effects

SCI subjects require a threshold motor score which allows performing stepping movements. During the course of a locomotor training, subjects can achieve an improved locomotor function without or with little change in motor scores [138, 148, 149]. In these cases, a relatively low voluntary force level in the leg muscles associated with an automatic synergistic muscle activation leads to an improved ability to walk.

A considerable degree of locomotor recovery can be attributed to a reorganization of spared neural pathways ([150]; for review, see [151]). It has been estimated that if as little as 10–15% of the descending spinal tracts are spared, some locomotor function can recover [152, 153]. In addition, by a training approach with the provision of appropriate proprioceptive input, directed, meaningful sprouting within neural circuits takes place below the level of lesion with the consequence of an improved recovery of function in the rat [46].

The improvement of locomotor activity might be attributed to a spontaneous recovery of spinal cord function that can occur over several months following a spinal cord injury [151, 154]. However, several observations indicate that the increase of leg extensor EMG activity also occurs independently of the spontaneous recovery of spinal cord function, as assessed by clinical and electrophysiological means [126, 137, 149, 151, 155]. Thus, functional training effects on spinal locomotor centers most likely contribute to an improvement of locomotor function in incomplete SCI subjects [126, 155], even in a chronic stage [149]. However, part of the recovery in locomotion corresponding to observations in the rat [58] might also be attributed to changes in muscle properties that occur during the training period.

1.4 Conclusion

Neuroplasticity mechanisms and training methods can improve function in patients with cerebral, cerebellar, and spinal cord injuries. However, patients with complete or almost complete hemi- or paraplegia do not, yet, profit from training because they cannot actively train.

In the future, in these patients a combination of regeneration-inducing therapy and exploitation of neuronal plasticity possibly by using novel training devices could have a beneficial effect on the recovery of function. In this aspect, the research in spinal cord regeneration appears to be quite encouraging (for review, see [156]). Novel training devices (often referred to as rehabilitation robots) become increasingly important and popular in clinical and rehabilitation settings for functional training and standardized assessments if neurophysiological requisites are met [146]. Such devices allow a prolonged training duration, increased number of repetitions of movements, improved patient safety, and less physical demands for the therapists. Supportive therapies that enhance the brain's potential to undergo plastic changes could supplement the training itself. For all these developments, testing in clinical trials will be required to prove efficacy and optimize the treatment for various disease and lesion types.

Acknowledgements This work was supported by the P&K Pühringer Foundation.

References

1. Tseng Y, Diedrichsen J, Krakauer JW, Shadmehr R, Bastian AJ. Sensory prediction errors drive cerebellum-dependent adaptation of reaching. J Neurophysiol. 2007;98(1):54–62.
2. Shmuelof L, Huang VS, Haith AM, Delnicki RJ, Mazzoni P, Krakauer JW. Overcoming motor "Forgetting" through reinforcement of learned actions. J Neurosci. 2012;32(42):14617–21.
3. Christian KM, Thompson RF. Neural substrates of eyeblink conditioning: acquisition and retention. Learn Mem. 2003;10(6):427–55.
4. Sharma DA, Chevidikunnan MF, Khan FR, Gaowgzeh RA. Effectiveness of knowledge of result and knowledge of performance in the learning of a skilled motor activity by healthy young adults. J Phys Ther Sci. 2016;28(5):1482–6.
5. Schultz W, Apicella P, Ljungberg T. Responses of monkey dopamine neurons to reward and conditioned stimuli during successive steps of learning a delayed response task. J Neurosci. 1993;13(3):900–13.
6. Iordanova MD. Dopaminergic modulation of appetitive and aversive predictive learning. Rev Neurosci. 2009;20(5–6):383–404.

7. Tranel D, Damasio AR, Damasio H, Brandt JP. Sensorimotor skill learning in amnesia: additional evidence for the neural basis of nondeclarative memory. Learn Mem. 1994;1(3):165–79.

8. Galea JM, Vazquez A, Pasricha N, de Xivry J-JO, Celnik P. Dissociating the roles of the cerebellum and motor cortex during adaptive learning: the motor cortex retains what the cerebellum learns. Cereb Cortex. 2010;21(8):1761–70.

9. O'Doherty J, Dayan P, Schultz J, Deichmann R, Friston K, Dolan RJ. Dissociable roles of ventral and dorsal striatum in instrumental conditioning. Science. 2004;304(5669):452–4.

10. Luft AR, Buitrago MM, Ringer T, Dichgans J, Schulz JB. Motor skill learning depends on protein synthesis in motor cortex after training. J Neurosci. 2004;24(29):6515–20.

11. Luft AR, Buitrago MM, Kaelin-Lang A, Dichgans J, Schulz JB. Protein synthesis inhibition blocks consolidation of an acrobatic motor skill. Learn Mem. 2004;11(4):379–82.

12. D'Agata V, Cavallaro S. Gene expression profiles —a new dynamic and functional dimension to the exploration of learning and memory. Rev Neurosci. 2002;13(3):209–20.

13. D'Agata V, Cavallaro S. Hippocampal gene expression profiles in passive avoidance conditioning. Eur J Neurosci. 2003;18(10):2835–41.

14. Steward O, Schuman EM. Protein synthesis at synaptic sites on dendrites. Annu Rev Neurosci. 2001;24(1):299–325.

15. Miyamoto E. Molecular mechanism of neuronal plasticity: induction and maintenance of long-term potentiation in the hippocampus. J Pharmacol Sci. 2006;100(5):433–42.

16. Davis HP, Squire LR. Protein synthesis and memory: a review. Psychol Bull. 1984;96(3):518–59.

17. Hosp JA, Mann S, Wegenast-Braun BM, Calhoun ME, Luft AR. Region and task-specific activation of arc in primary motor cortex of rats following motor skill learning. Neuroscience. 2013;250:557–64.

18. Qian Y, Chen M, Forssberg H, Heijtz RD. Genetic variation in dopamine-related gene expression influences motor skill learning in mice. Genes Brain Behav. 2013;12(6):604–14.

19. Carmichael ST. Plasticity of cortical projections after stroke. Neurosci. 2003;9(1):64–75.

20. Li S, Carmichael ST. Growth-associated gene and protein expression in the region of axonal sprouting in the aged brain after stroke. Neurobiol Dis. 2006;23(2):362–73.

21. Carmichael ST. Cellular and molecular mechanisms of neural repair after stroke: making waves. Ann Neurol. 2006;59(5):735–42.

22. Malenka RC, Bear MF. LTP and LTD. Neuron. 2004;44(1):5–21.

23. Bennett MR. The concept of long term potentiation of transmission at synapses. Prog Neurobiol. 2000;60(2):109–37.

24. Rioult-Pedotti M-S, Donoghue JP, Dunaevsky A. Plasticity of the synaptic modification range. J Neurophysiol. 2007;98(6):3688–95.

25. Rioult-Pedotti M-S, Friedman D, Donoghue JP. Learning-induced LTP in neocortex. Science. 2000;290(5491):533–6.

26. Greenough WT, Larson JR, Withers GS. Effects of unilateral and bilateral training in a reaching task on dendritic branching of neurons in the rat motor-sensory forelimb cortex. Behav Neural Biol. 1985;44(2):301–14.

27. Kolb B, Cioe J, Comeau W. Contrasting effects of motor and visual spatial learning tasks on dendritic arborization and spine density in rats. Neurobiol Learn Mem. 2008;90(2):295–300.

28. Wang L, Conner JM, Rickert J, Tuszynski MH. Structural plasticity within highly specific neuronal populations identifies a unique parcellation of motor learning in the adult brain. Proc Natl Acad Sci. 2011;108(6):2545–50.

29. Xu T, Yu X, Perlik AJ, Tobin WF, Zweig JA, Tennant K, et al. Rapid formation and selective stabilization of synapses for enduring motor memories. Nature. 2009;462(7275):915–9.

30. Parkhurst CN, Yang G, Ninan I, Savas JN, Yates JR, Lafaille JJ, et al. Microglia promote learning-dependent synapse formation through brain-derived neurotrophic factor. Cell. 2013;155 (7):1596–609.

31. Hagemann G, Redecker C, Neumann-Haefelin T, Freund H-J, Witte OW. Increased long-term potentiation in the surround of experimentally induced focal cortical infarction. Ann Neurol. 1998;44 (2):255–8.

32. Gasparova Z, Jariabka P, Stolc S. Effect of transient ischemia on long-term potentiation of synaptic transmission in rat hippocampal slices. Neuroendocr Lett. 2008;29(5):702–5.

33. Lazzaro VD, Profice P, Pilato F, Capone F, Ranieri F, Pasqualetti P, et al. Motor cortex plasticity predicts recovery in acute stroke. Cereb Cortex. 2009;20(7):1523–8.

34. Castel-Lacanal E, Marque P, Tardy J, de Boissezon X, Guiraud V, Chollet F, et al. Induction of cortical plastic changes in wrist muscles by paired associative stimulation in the recovery phase of stroke patients. Neurorehabil Neural Repair. 2008;23(4):366–72.

35. Batsikadze G, Paulus W, Kuo M-F, Nitsche MA. Effect of serotonin on paired associative stimulation-induced plasticity in the human motor cortex. Neuropsychopharmacology. 2013;38 (11):2260–7.

36. Chollet F, Tardy J, Albucher J-F, Thalamas C, Berard E, Lamy C, et al. Fluoxetine for motor recovery after acute ischaemic stroke (FLAME): a randomised placebo-controlled trial. Lancet Neurol. 2011;10(2):123–30.

37. Nudo R, Milliken G, Jenkins W, Merzenich M. Use-dependent alterations of movement

representations in primary motor cortex of adult squirrel monkeys. J Neurosci. 1996;16(2):785–807.

38. Kleim JA, Barbay S, Nudo RJ. Functional reorganization of the rat motor cortex following motor skill learning. J Neurophysiol. 1998;80(6):3321–5.

39. Molina-Luna K, Hertler B, Buitrago MM, Luft AR. Motor learning transiently changes cortical somatotopy. Neuroimage. 2008;40(4):1748–54.

40. Lotze M, Scheler G, Tan H-RM, Braun C, Birbaumer N. The musician's brain: functional imaging of amateurs and professionals during performance and imagery. Neuroimage. 2003;20 (3):1817–29.

41. Jäncke L, Shah NJ, Peters M. Cortical activations in primary and secondary motor areas for complex bimanual movements in professional pianists. Cogn Brain Res. 2000;10(1–2):177–83.

42. Gaser C, Schlaug G. Brain structures differ between musicians and non-musician. Neuroimage. 2001;13 (6):1168.

43. Gentner R, Gorges S, Weise D, Kampe K aufm, Buttmann M, Classen J. Encoding of motor skill in the corticomuscular system of musicians. Curr Biol. 2010;20(20):1869–74.

44. Marshall RS, Perera GM, Lazar RM, Krakauer JW, Constantine RC, DeLaPaz RL. Evolution of cortical activation during recovery from corticospinal tract infarction. Stroke. 2000;31(3):656–61.

45. Calautti C, Leroy F, Guincestre J-Y, Marié R-M, Baron J-C. Sequential activation brain mapping after subcortical stroke: changes in hemispheric balance and recovery. NeuroReport. 2001;12 (18):3883–6.

46. Beauparlant J, van den Brand R, Barraud Q, Friedli L, Musienko P, Dietz V, et al. Undirected compensatory plasticity contributes to neuronal dysfunction after severe spinal cord injury. Brain. 2013;136(11):3347–61. http://www.ncbi.nlm.nih. gov/pubmed/24080153.

47. Ward NS, Brown MM, Thompson AJ, Frackowiak RSJ. Neural correlates of motor recovery after stroke: a longitudinal fMRI study. Brain. 2003;126(11):2476–96.

48. Cramer SC, Nelles G, Benson RR, Kaplan JD, Parker RA, Kwong KK, et al. A functional MRI study of subjects recovered from hemiparetic stroke. Stroke. 1997;28(12):2518–27.

49. Seitz RJ, Höflich P, Binkofski F, Tellmann L, Herzog H, Freund H-J. Role of the premotor cortex in recovery from middle cerebral artery infarction. Arch Neurol (Chicago). 1998;55(8):1081–8.

50. Small SL, Hlustik P, Noll DC, Genovese C, Solodkin A. Cerebellar hemispheric activation ipsilateral to the paretic hand correlates with functional recovery after stroke. Brain. 2002;125(7):1544–57.

51. Rehme AK, Grefkes C. Cerebral network disorders after stroke: evidence from imaging-based connectivity analyses of active and resting brain states in humans. J Physiol. 2013;591(1):17–31.

52. Carter AR, Astafiev SV, Lang CE, Connor LT, Rengachary J, Strube MJ, et al. Resting state interhemispheric fMRI connectivity predicts performance after stroke. Ann Neurol. 2009;67(3).

53. Turton A, Wroe S, Trepte N, Fraser C, Lemon RN. Contralateral and ipsilateral EMG responses to transcranial magnetic stimulation during recovery of arm and hand function after stroke. Electroencephalogr Clin Neurophysiol Electromyogr Mot Control. 1996;101(4):316–28.

54. Escudero JV, Sancho J, Bautista D, Escudero M, López-Trigo J. Prognostic value of motor evoked potential obtained by transcranial magnetic brain stimulation in motor function recovery in patients with acute ischemic stroke. Stroke. 1998;29 (9):1854–9.

55. Grigoras I-F, Stagg CJ. Recent advances in the role of excitation–inhibition balance in motor recovery post-stroke. Fac Rev. 2021;10:58.

56. Murase N, Duque J, Mazzocchio R, Cohen LG. Influence of interhemispheric interactions on motor function in chronic stroke. Ann Neurol. 2004;55 (3):400–9.

57. Cirillo J, Mooney RA, Ackerley SJ, Barber PA, Borges VM, Clarkson AN, et al. Neurochemical balance and inhibition at the subacute stage after stroke. J Neurophysiol. 2020;123(5):1775–90.

58. Edgerton VR, de Leon RD, Tillakaratne N, Recktenwald MR, Hodgson JA, Roy RR. Use-dependent plasticity in spinal stepping and standing. Adv Neurol. 1997;72:233–47.

59. Pearson KG. Plasticity of neuronal networks in the spinal cord: modifications in response to altered sensory input. Prog Brain Res. 2000;128:61–70.

60. Lovely RG, Gregor RJ, Roy RR, Edgerton VR. Effects of training on the recovery of full-weightbearing stepping in the adult spinal cat. Exp Neurol. 1986;92(2):421–35.

61. Lovely RS, Falls LA, Al-Mondhiry HA, Chambers CE, Sexton GJ, Ni H, et al. Association of gammaA/gamma' fibrinogen levels and coronary artery disease. Thromb Haemost. 2002;88(1):26–31.

62. Dietz V, Macauda G, Schrafl-Altermatt M, Wirz M, Kloter E, Michels L. Neural coupling of cooperative hand movements: a reflex and fMRI study. Cereb Cortex. 2015;25(4):948–58.

63. Wolpaw JR. Operant conditioning of primate spinal reflexes: the H-reflex. J Neurophysiol. 1987;57 (2):443–59.

64. Wolpaw JR, Seegal RF, O'Keefe JA. Adaptive plasticity in primate spinal stretch reflex: behavior of synergist and antagonist muscles. J Neurophysiol. 1983;50(6):1312–9.

65. Harris-Warrick RM, Marder E. Modulation of neural networks for behavior. Annu Rev Neurosci. 1991;14(1):39–57.

66. Edgerton VR, Roy RR, Hodgson JA, Prober RJ, de Guzman CP, de Leon R. Potential of adult

mammalian lumbosacral spinal cord to execute and acquire improved locomotion in the absence of supraspinal input. J Neurotrauma. 1992;9(Suppl 1):S119-28.

67. Dickinson PS. Interactions among neural networks for behavior. Curr Opin Neurobiol. 1995;5(6):792–8.

68. Katz PS. Intrinsic and extrinsic neuromodulation of motor circuits. Curr Opin Neurobiol. 1995;5(6):799–808.

69. Pearson KG, Misiaszek JE. Use-dependent gain change in the reflex contribution to extensor activity in walking cats. Brain Res. 2000;883(1):131–4.

70. Barbeau H, Rossignol S. Recovery of locomotion after chronic spinalization in the adult cat. Brain Res. 1987;412(1):84–95.

71. Barbeau H, Rossignol S. Enhancement of locomotor recovery following spinal cord injury. Curr Opin Neurol. 1994;7(6):517–24.

72. Dietz V, Fouad K. Restoration of sensorimotor functions after spinal cord injury. Brain. 2014;137(3):654–67.

73. Skinner RD, Houle JD, Reese NB, Berry CL, Garcia-Rill E. Effects of exercise and fetal spinal cord implants on the H-reflex in chronically spinalized adult rats. Brain Res. 1996;729(1):127–31.

74. Chau C, Barbeau H, Rossignol S. Effects of intrathecal alpha1- and alpha2-noradrenergic agonists and norepinephrine on locomotion in chronic spinal cats. J Neurophysiol. 1998;79(6):2941–63.

75. de Leon RD, Hodgson JA, Roy RR, Edgerton VR. Locomotor capacity attributable to step training versus spontaneous recovery after spinalization in adult cats. J Neurophysiol. 1998;79(3):1329–40.

76. Leon RDD, Hodgson JA, Roy RR, Edgerton VR. Full weight-bearing hindlimb standing following stand training in the adult spinal cat. J Neurophysiol. 1998;80(1):83–91.

77. Maric O, Zörner B, Dietz V. Levodopa therapy in incomplete spinal cord injury. J Neurotrauma. 2008;25(11):1303–7. http://www.ncbi.nlm.nih.gov/pubmed/19061374.

78. Schmidt BJ, Jordan LM. The role of serotonin in reflex modulation and locomotor rhythm production in the mammalian spinal cord. Brain Res Bull. 2000;53(5):689–710.

79. Shadmehr R, Krakauer JW. A computational neuroanatomy for motor control. Exp Brain Res. 2008;185(3):359–81.

80. Bhanpuri NH, Okamura AM, Bastian AJ. Predicting and correcting ataxia using a model of cerebellar function. Brain. 2014;137(7):1931–44.

81. Smith MA, Shadmehr R. Intact ability to learn internal models of arm dynamics in Huntington's disease but not cerebellar degeneration. J Neurophysiol. 2005;93(5):2809–21.

82. Morton SM, Bastian AJ. Cerebellar contributions to locomotor adaptations during splitbelt treadmill walking. J Neurosci. 2006;26(36):9107–16.

83. Horak FB, Diener HC. Cerebellar control of postural scaling and central set in stance. J Neurophysiol. 1994;72(2):479–93.

84. Yagi T, Shimizu M, Sekine S, Kamio T, Suzuki J. A new neurotological test for detecting cerebellar dysfunction. Ann N Y Acad Sci. 1981;374(1):526–31.

85. Marr D. A theory of cerebellar cortex. J Physiol. 1969;202(2):437–70.

86. Albus JS. A theory of cerebellar function. Math Biosci. 1971;10(1–2):25–61.

87. Ito M. Long-term depression. Annu Rev Neurosci. 1989;12(1):85–102.

88. Ito M, Kano M. Long-lasting depression of parallel fiber-Purkinje cell transmission induced by conjunctive stimulation of parallel fibers and climbing fibers in the cerebellar cortex. Neurosci Lett. 1982;33(3):253–8.

89. Yang Y, Lisberger SG. Purkinje-cell plasticity and cerebellar motor learning are graded by complex-spike duration. Nature. 2014;510(7506):529–32. http://www.ncbi.nlm.nih.gov/pubmed/24814344.

90. Hansel C, Linden DJ, D'Angelo E. Beyond parallel fiber LTD: the diversity of synaptic and non-synaptic plasticity in the cerebellum. Nat Neurosci. 2001;4(5):467–75.

91. Zeeuw CID. Bidirectional learning in upbound and downbound microzones of the cerebellum. Nat Rev Neurosci. 2021;22(2):92–110.

92. Lerner TN, Holloway AL, Seiler JL. Dopamine, updated: reward prediction error and beyond. Curr Opin Neurobiol. 2021;67:123–30.

93. Luft AR, Schwarz S. Dopaminergic signals in primary motor cortex. Int J Dev Neurosci. 2009;27(5):415–21.

94. Molina-Luna K, Pekanovic A, Röhrich S, Hertler B, Schubring-Giese M, Rioult-Pedotti M-S, et al. Dopamine in motor cortex is necessary for skill learning and synaptic plasticity. PLoS ONE. 2009;4(9):e7082.

95. Hosp JA, Pekanovic A, Rioult-Pedotti MS, Luft AR. Dopaminergic projections from midbrain to primary motor cortex mediate motor skill learning. J Neurosci. 2011;31(7):2481–7.

96. Hosp JA, Molina-Luna K, Hertler B, Atiemo CO, Luft AR. Dopaminergic modulation of motor maps in rat motor cortex: an in vivo study. Neuroscience. 2009;159(2):692–700.

97. Hosp JA, Hertler B, Atiemo CO, Luft AR. Dopaminergic modulation of receptive fields in rat sensorimotor cortex. Neuroimage. 2011;54(1):154–60.

98. Schultz W. Behavioral dopamine signals. Trends Neurosci. 2007;30(5):203–10.

99. Ramnani N. Chapter 10 automatic and controlled processing in the corticocerebellar system. Prog Brain Res. 2014;210:255–85.

100. Schaechter JD. Motor rehabilitation and brain plasticity after hemiparetic stroke. Prog Neurobiol. 2004;73(1):61–72.

101. Luft AR, McCombe-Waller S, Whitall J, Forrester LW, Macko R, Sorkin JD, et al. Repetitive bilateral arm training and motor cortex activation in chronic stroke: a randomized controlled trial. JAMA. 2004;292(15):1853.

102. Whitall J, Waller SM, Sorkin JD, Forrester LW, Macko RF, Hanley DF, et al. Bilateral and unilateral arm training improve motor function through differing neuroplastic mechanisms: a single-blinded randomized controlled trial. Neurorehabil Neural Repair. 2010;25(2):118–29.

103. Luft AR, Macko RF, Forrester LW, Villagra F, Ivey F, Sorkin JD, et al. Treadmill exercise activates subcortical neural networks and improves walking after stroke: a randomized controlled trial. Stroke. 2008;39(12):3341–50.

104. Reisman DS, Wityk R, Silver K, Bastian AJ. Locomotor adaptation on a split-belt treadmill can improve walking symmetry post-stroke. Brain. 2007;130(7):1861–72.

105. Reisman DS, McLean H, Bastian AJ. Split-belt treadmill training poststroke: a case study. J Neurol Phys Ther. 2010;34(4):202–7.

106. Reisman DS, Wityk R, Silver K, Bastian AJ. Split-belt treadmill adaptation transfers to overground walking in persons poststroke. Neurorehabil Neural Repair. 2009;23(7):735–44.

107. Reisman DS, McLean H, Keller J, Danks KA, Bastian AJ. Repeated split-belt treadmill training improves poststroke step length asymmetry. Neurorehabil Neural Repair. 2013;27(5):460–8.

108. Jauss M, Krieger D, Hornig C, Schramm J, Busse O. Surgical and medical management of patients with massive cerebellar infarctions: results of the German–Austrian cerebellar infarction study. J Neurol. 1999;246(4):257–64.

109. Tohgi H, Takahashi S, Chiba K, Hirata Y. Cerebellar infarction. Clinical and neuroimaging analysis in 293 patients. The Tohoku Cerebellar Infarction Study Group. Stroke. 2018;24 (11):1697–701.

110. Kelly PJ, Stein J, Shafqat S, Eskey C, Doherty D, Chang Y, et al. Functional recovery after rehabilitation for cerebellar stroke. Stroke. 2001;32 (2):530–4.

111. Morton SM, Tseng Y-W, Zackowski KM, Daline JR, Bastian AJ. Longitudinal tracking of gait and balance impairments in cerebellar disease: longitudinal tracking of cerebellar disease. Mov Disord. 2010;25(12):1944–52.

112. Schoch B, Dimitrova A, Gizewski ER, Timmann D. Functional localization in the human cerebellum based on voxelwise statistical analysis: a study of 90 patients. Neuroimage. 2006;30(1):36–51.

113. Ilg W, Synofzik M, Brotz D, Burkard S, Giese MA, Schols L. Intensive coordinative training improves motor performance in degenerative cerebellar disease. Neurology. 2009;73(22):1823–30.

114. Gill-Body KM, Popat RA, Parker SW, Krebs DE. Rehabilitation of balance in two patients with cerebellar dysfunction. Phys Ther. 1997;77 (5):534–52.

115. Bultmann U, Pierscianek D, Gizewski ER, Schoch B, Fritsche N, Timmann D, et al. Functional recovery and rehabilitation of postural impairment and gait ataxia in patients with acute cerebellar stroke. Gait Posture. 2014;39(1):563–9.

116. Keller JL, Bastian AJ. A home balance exercise program improves walking in people with cerebellar ataxia. Neurorehabil Neural Repair. 2014;28 (8):770–8.

117. Vaz DV, de Schettino RC, de Castro TRR, Teixeira VR, Furtado SRC, de Figueiredo EM. Treadmill training for ataxic patients: a single-subject experimental design. Clin Rehabil. 2008;22(3):234–41.

118. Cernak K, Stevens V, Price R, Shumway-Cook A. Locomotor training using body-weight support on a treadmill in conjunction with ongoing physical therapy in a child with severe cerebellar ataxia. Phys Ther. 2008;88(1):88–97.

119. Kleim JA, Swain RA, Armstrong KA, Napper RMA, Jones TA, Greenough WT. Selective synaptic plasticity within the cerebellar cortex following complex motor skill learning. Neurobiol Learn Mem. 1998;69(3):274–89.

120. Kleim JA, Pipitone MA, Czerlanis C, Greenough WT. Structural stability within the lateral cerebellar nucleus of the rat following complex motor learning. Neurobiol Learn Mem. 1998;69 (3):290–306.

121. Benussi A, Koch G, Cotelli M, Padovani A, Borroni B. Cerebellar transcranial direct current stimulation in patients with ataxia: a double-blind, randomized, sham-controlled study. Mov Disord. 2015;30(12):1701–5.

122. Benussi A, Dell'Era V, Cantoni V, Bonetta E, Grasso R, Manenti R, et al. Cerebello-spinal tDCS in ataxia. Neurology. 2018;91(12):e1090-101.

123. Dietz V. Proprioception and locomotor disorders. Nat Rev Neurosci. 2002;3(10):781–90.

124. Dietz V. Do human bipeds use quadrupedal coordination? Trends Neurosci. 2002;25(9):462–7.

125. Jenson L, Colombo G, Dietz V. Locomotor activity in spinal man. Gait Posture. 1995;3(3):180–1.

126. Dietz V, Wirz M, Colombo G, Curt A. Locomotor capacity and recovery of spinal cord function in paraplegic patients: a clinical and electrophysiological evaluation. Electroencephalogr Clin Neurophysiol Electromyogr Mot Control. 1998;109 (2):140–53.

127. Dietz V. Neurophysiology of gait disorders: present and future applications. Electroencephalogr Clin Neurophysiol. 1997;103(3):333–55.

128. Popovic MR, Curt A, Keller T, Dietz V. Functional electrical stimulation for grasping and walking: indications and limitations. Spinal Cord. 2001;39(8):403–12.

129. Fung J, Stewart JE, Barbeau H. The combined effects of clonidine and cyproheptadine with inter-active training on the modulation of locomotion in spinal cord injured subjects. J Neurol Sci. 1990;100(1–2):85–93.

130. Wernig A, Müller S. Laufband locomotion with body weight support improved walking in persons with severe spinal cord injuries. Spinal Cord. 1992;30(4):229–38.

131. Wernig A, Müller S, Nanassy A, Cagol E. Laufband therapy based on "rules of spinal locomotion" is effective in spinal cord injured persons. Eur J Neurosci. 1995;7(4):823–9.

132. Wirz M, Colombo G, Dietz V. Long term effects of locomotor training in spinal humans. J Neurol Neurosurg Psychiatry. 2001;71(1):93.

133. Dietz V, Colombo G, Jensen L, Baumgartner L. Locomotor capacity of spinal cord in paraplegic patients. Ann Neurol. 1995;37(5):574–82.

134. Dietz V, Müller R, Colombo G. Locomotor activity in spinal man: significance of afferent input from joint and load receptors. Brain. 2002;125(12):2626–34.

135. Dietz V, Colombo G, Jensen L. Locomotor activity in spinal man. Lancet. 1994;344(8932):1260–3.

136. Cramer SC, Riley JD. Neuroplasticity and brain repair after stroke. Curr Opin Neurol. 2008;21(1):76–82.

137. Dietz V, Harkema SJ. Locomotor activity in spinal cord-injured persons. J Appl Physiol. 2004;96(5):1954–60.

138. Martino G. How the brain repairs itself: new therapeutic strategies in inflammatory and degenerative CNS disorders. Lancet Neurol. 2004;3(6):372–8.

139. Dietz V. Body weight supported gait training: from laboratory to clinical setting. Brain Res Bull. 2009;78(1):I–VI.

140. Wirz M, Dietz V, Network EMS of SCI (EMSCI). Recovery of sensorimotor function and activities of daily living after cervical spinal cord injury: the influence of age. J Neurotrauma. 2015;32(3):194–9. http://www.ncbi.nlm.nih.gov/pubmed/24963966.

141. Dietz V, Grillner S, Trepp A, Hubli M, Bolliger M. Changes in spinal reflex and locomotor activity after a complete spinal cord injury: a common mechanism? Brain. 2009;132(8):2196–205.

142. Dietz V. Body weight supported gait training: from laboratory to clinical setting. Brain Res Bull. 2008;76(5):459–63.

143. Dietz V, Müller R. Degradation of neuronal function following a spinal cord injury: mechanisms and countermeasures. Brain. 2004;127(10):2221–31.

144. Dobkin BH, Harkema S, Requejo P, Edgerton VR. Modulation of locomotor-like EMG activity in subjects with complete and incomplete spinal cord injury. J Neurol Rehabil. 1995;9(4):183–90.

145. Harkema SJ, Hurley SL, Patel UK, Requejo PS, Dobkin BH, Edgerton VR. Human lumbosacral spinal cord interprets loading during stepping. J Neurophysiol. 1997;77(2):797–811.

146. Gassert R, Dietz V. Rehabilitation robots for the treatment of sensorimotor deficits: a neurophysiological perspective. J Neuroeng Rehabil. 2018;15(1):46.

147. Dietz V, Sinkjaer T. Spastic movement disorder: impaired reflex function and altered muscle mechanics. Lancet Neurol. 2007;6(8):725–33.

148. Moseley AM, Stark A, Cameron ID, Pollock A. Treadmill training and body weight support for walking after stroke. Cochrane Database Syst Rev. 2003;(3):CD002840.

149. Wirz M, Zemon DH, Rupp R, Scheel A, Colombo G, Dietz V, et al. Effectiveness of automated locomotor training in patients with chronic incomplete spinal cord injury: a multicenter trial. Arch Phys Med Rehabil. 2005;86(4):672–80.

150. Curt A, Dietz V. Ambulatory capacity in spinal cord injury: significance of somatosensory evoked potentials and ASIA protocol in predicting outcome. Arch Phys Med Rehabil. 1997;78(1):39–43.

151. Curt A, Keck ME, Dietz V. Functional outcome following spinal cord injury: significance of motor-evoked potentials and ASIA scores. Arch Phys Med Rehabil. 1998;79(1):81–6.

152. Basso DM. Neuroanatomical substrates of functional recovery after experimental spinal cord injury: implications of basic science research for human spinal cord injury. Phys Ther. 2000;80(8):808–17.

153. Metz GAS, Curt A, van de Meent H, Klusman I, Schwab ME, Dietz V. Validation of the weight-drop contusion model in rats: a comparative study of human spinal cord injury. J Neurotrauma. 2000;17(1):1–17.

154. Katoh S, el Masry W. Neurological recovery after conservative treatment of cervical cord injuries. J Bone Jt Surg Br Vol. 1994;76B(2):225–8.

155. Dietz V, Wirz M, Curt A, Colombo G. Locomotor pattern in paraplegic patients: training effects and recovery of spinal cord function. Spinal Cord. 1998;36(6):380–90.

156. Schwab ME, Bartholdi D. Degeneration and regeneration of axons in the lesioned spinal cord. Physiol Rev. 1996;76(2):319–70.

Robert L. Sainburg and Pratik K. Mutha

Abstract

Research into the neural control of movement has elucidated important principles that can provide guidelines to rehabilitation professionals for enhancing the recovery of motor function in stroke patients. In this chapter, we elaborate on principles that have been derived from research on neural control of movement, including optimal control, impedance control, motor lateralization, and principles of motor learning. Research on optimal control has indicated that two major categories of cost contribute to motor planning: explicit task level costs, such as movement accuracy and speed, and implicit costs, such as energy and movement variability. Impedance control refers to neural mechanisms that modulate rapid sensorimotor circuits, such as reflexes, in order to impede perturbations that cannot be anticipated prior to movement. Research on motor lateralization has indicated that different aspects of motor control have been specialized to the two cerebral hemispheres. This organization leads to hemisphere-specific motor deficits in both the ipsilesional and contralesional arms of stroke patients. Ipsilesional deficits increase with the severity of contralesional impairment level and have a substantial effect on functional independence. Finally, motor learning research has indicated that different neural mechanisms underlie different aspects of motor learning, such as adaptation vs skill learning, and that learning different aspects of tasks can generalize across different coordinates. In this chapter, we discuss the neurobiological basis of these principles and elaborate on the implications for designing and implementing occupational and physical therapy treatment for movement deficits in stroke patients.

Keywords

Rehabilitation · Motor control · Motor learning · Motor lateralization

R. L. Sainburg (✉)
Kinesiology and Neurology, Penn State University, University Park, PA 16801, United States
e-mail: rls45@psu.edu

P. K. Mutha
Department of Biological Engineering and Centre for Cognitive Science, Indian Institute of Technology Gandhinagar, Ahmedabad, Gujarat 382424, India
e-mail: pm@iitgn.ac.in

2.1 Introduction

Deficits that result from strokes in sensory and motor regions of the brain represent a major impediment to the recovery of function in activities of daily living for stroke survivors. Such deficits most commonly include hemiparesis, a syndrome encompassing unilateral motor

© The Author(s), under exclusive license to Springer Nature Switzerland AG 2022
D. J. Reinkensmeyer et al. (eds.), *Neurorehabilitation Technology*,
https://doi.org/10.1007/978-3-031-08995-4_2

dysfunction on the side of the body opposite to the brain lesion, and spasticity, characterized by abnormally high muscle tone and atypical expression of reflexes. Occupational and physical therapy interventions often focus on reducing motor impairment, following stroke, by exposing patients to a range of movement activities, with a major focus on repetitive experience or practice. In general, the amount of practice corresponds to improvements in motor function, as measured by a variety of scales [1]. Unfortunately, gains made during therapy often show the limited translation to activities of daily living (ADLs) and carry over to the home environment.

Over the past decade, rehabilitation approaches have incorporated technological innovations that can provide more cost-effective means of achieving higher intensity practice over longer periods of time. These computer-based and robotic technologies [2–5] have been shown to match, or even exceed the efficacy of traditional therapy in promoting improvements in motor performance [6]. However, these interventions hold greater promise than simply replicating traditional therapy, by providing therapists with an unprecedented ability to specify and measure movement features such as speed, direction, amplitude, as well as joint coordination patterns. As these technologies become more readily available in the clinic, the most pressing question is how therapists can best utilize them to accelerate recovery of function. In this chapter, we will discuss principles that have been derived from research in motor control and learning that could be applied to training strategies using computer-based movement interventions.

2.2 Principle 1: Optimal Control

While most therapists recognize that practice and repetition of motor activities lead to improvements in motor performance, a systematic identification of which movements should be practiced is often lacking. This is partly because the question of what defines a desirable movement has yielded no clear answer. Traditionally, a common guiding principle employed in occupational and physical therapy

has been to make movements more "normal". Thus, the goal is to develop movement patterns that are similar to those exhibited by non-impaired individuals. This idea emerged from the observation that certain characteristics of movements made by healthy individuals are fairly similar within a given task, and even across tasks. For example, when reaching for an object in space, movement trajectories across healthy individuals appear fairly straight and smooth [7]. Such reliability of motor behavior is particularly interesting because of the abundance of possible solutions to most movement tasks and the variety of environments we move in. For example, when reaching for a cup of coffee in front of us, we have the choice of using one arm or both arms, standing up or remaining seated, leaning the trunk forward or reaching further with the selected arm(s), twisting our trunk to require shoulder abduction, or keeping it straight to require shoulder flexion, among other options (see Fig. 2.1). In addition, the relative motions between our body segments can produce a wide variety of curved trajectories of the hand, in order to procure the cup. Each possible motion can also be achieved at a variety of speeds, as well as a variety of possible muscle activation patterns. There are literally infinite solutions to this simple task.

Regardless of these vast possibilities, people tend to display movement patterns that are consistent across different instances of the same movement or even across different movements, whether made by the same or different individuals. These similarities are often referred to as "invariant characteristics" of movement. Many studies have shown that when different people make reaching movements, invariant characteristics include approximate straightness of the hand trajectory and smooth bell-shaped velocity profiles (see Fig. 2.2) [7–11]. How do different people arrive at similar solutions within and across tasks despite the extensive redundancy in the musculoskeletal system and the diversity and uncertainty of the environments we move in? One way to arrive at the "best" solution when confronted with many different options is to employ optimization strategies when planning the movement. Optimization procedures have

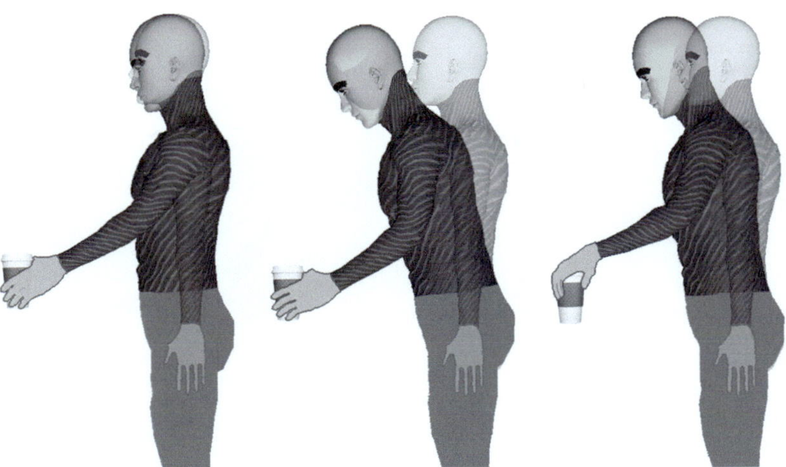

Fig. 2.1 Different ways of picking up a coffee cup starting from the same initial posture. The left pose involves shoulder flexion and elbow extension. The middle pose involves flexion of the trunk, slightly less shoulder flexion, and more elbow extension. The right pose shows some trunk flexion, shoulder abduction, elbow flexion, and forearm pronation

been developed for use in engineering applications, and seek the minimum or maximum for a given "cost function", subject to a set of constraints. For example, we can find the minimum price of a pound of coffee (function) for all the stores within a 10-mile radius of our house (constraint). Whereas this particular problem may be quite trivial, optimization routines are typically employed to find values for more complex problems, such as might be applied to human movement. Researchers have tested various cost functions that make sense heuristically and have shown that optimization of these costs reproduces many invariant characteristics observed in human motion. For example, Flash and Hogan [9] tested the idea that the smoothness of hand trajectories might reflect an important cost in the planning of reaching movements and proposed a model that minimized the jerkiness of the hand trajectory (mathematically defined as the derivative of acceleration with respect to time). Their simulation predicted straight movements with symmetrical, single-peaked, bell-shaped velocity profiles.

However, under several experimental conditions, minimum jerk trajectories and experimentally observed hand paths diverged, which led researchers to examine other plausible cost functions. For example, some researchers speculated that mechanical aspects of movements might reflect important costs for planning movements. Such cost functions have included mean squared torque change [10], peak work [12], or muscle energy [13, 14] among others. These models accounted for some experimental observations that could not be accounted for by optimizations based on kinematic parameters [11]. While minimization of cost functions such as smoothness or torque change accurately predicted average behavior, Wolpert and colleagues [8] also accounted for the small, yet important trial-to-trial variability seen during repetitions of the same task. They proposed that motor commands are corrupted with variability-inducing noise, and in the presence of such noise, the CNS seeks to minimize the variance of the final arm position. This model also predicted many observed invariant characteristics of movements such as trajectory smoothness and the tradeoff between movement accuracy and speed.

Two important inferences can be drawn from studies that have attempted to explain movement patterns based on optimization principles: (1) the nature of the costs associated with different tasks are often different and (2) costs such as endpoint variability and mechanical energy do not reflect

Fig. 2.2 Some "invariant characteristics" of point-to-point movements. The top panel shows fairly straight hand trajectories for multiple movements. The bottom panel shows fairly similar bell-shaped velocity profiles for six different movements at different speeds. When time-normalized, the bell-shaped velocity profiles closely overlap. (Adapted from Atkeson C.G. and Hollerbach J.M. Kinematic Features of Unrestrained Vertical Arm Movements. Journal of Neuroscience 1985, Voll. 5, No. 9. pp. 2318–2330. Copyright 1985 Society for Neuroscience)

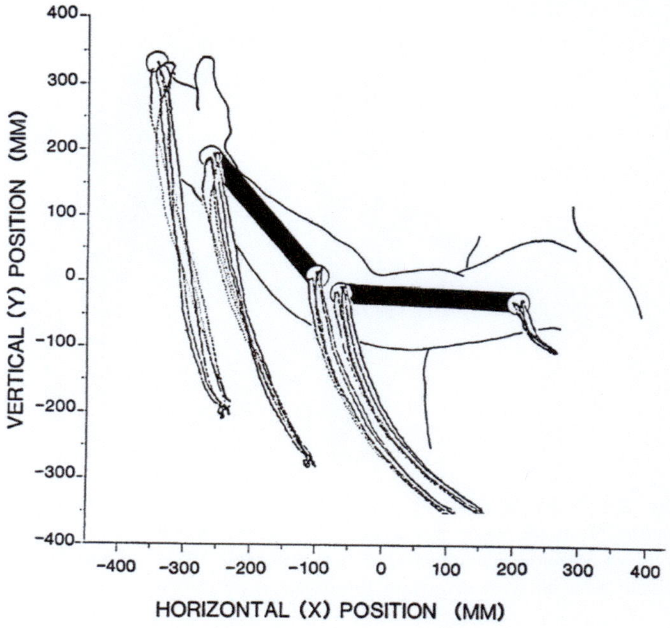

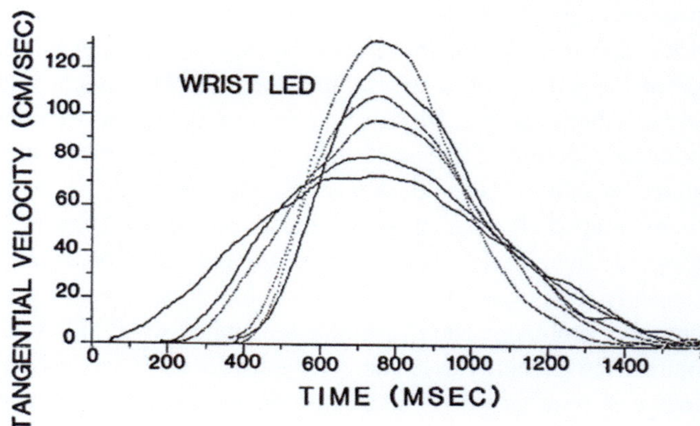

variables that we tend to have conscious awareness of, yet they appear to be accounted for during the process of motor planning. In other words, the planning of movements not only entails explicit performance criteria that are associated with successful task performance, such as getting hold of a cup of coffee, but also entails implicit criteria that we don't consciously consider, such as making energetically efficient and reliable movements.

An important aspect of the models discussed above is that optimization of a single cost function yields a desired trajectory that is then simply executed in an open-loop manner, once it is planned. The role of sensory feedback mechanisms in these models is simply to correct deviations from the planned or desired trajectory, regardless of whether these deviations resist or assist in task completion. Thus, the output of feedback circuits is not incorporated in the optimization phase. More recently, the idea that the determination of an optimal "control policy" incorporates knowledge about the "state" of the body and the environment, as relayed by feedback circuits and mechanisms that predict sensory consequences of motor commands, has gained prominence. According to this idea, the optimal solution is the best possible transformation from the current state to the motor commands that aid in achieving the task goal [15]. Not too surprisingly then, this *optimal feedback control* scheme yields task-specific cost functions that often represent a hybrid mix of explicit task-level variables that relate to performance goals, such as movement precision, as well as implicit mechanically related costs that correspond to muscle force or effort. For example, in a task that examined corrections to target displacements that occurred late in the movement, Liu and Todorov [16] showed that subject performance could be best described using a composite cost function that optimized for movement duration, accuracy, endpoint stability, and energy consumption. More importantly, subjects implicitly changed the relative contribution of these costs as the accuracy and stability requirements of the task were changed. Thus, rather than adopt a fixed policy across task conditions, subjects were able to flexibly adapt their control strategy in order to ensure maximum task success. These ideas of flexible control strategies and hybrid cost functions that include task-related and intrinsic biomechanical variables have important implications for designing therapy regimes.

Implications for Rehabilitation. It is important to recognize that damage to the CNS from stroke and the associated secondary changes in the musculoskeletal system could induce changes in the set of possible solutions as well as the costs associated with any given task. Therefore, patients may arrive at solutions to a motor task that may not look "normal", but may be "optimal" given physiological and biomechanical pathologies [17] Thus, rather than simply attempting to make movements look more "normal", it is important to understand the biomechanical costs associated with different tasks. Most importantly, if movements of the hemiparetic arm elicit energetic costs that are substantially higher than those of the ipsilesional arm, it is very unlikely that the hemiparetic arm use will be spontaneously integrated into activities of daily living. As the technologies discussed in this volume become available in the clinic, assessment of biomechanical variables, such as

joint power will also become available. While most clinical assessments of function include either the ability to perform certain ADL tasks (Functional Independence Measure—FIM [18, 19], or the ability to perform simulated ADL tasks in particular times (Jebsen-Taylor Hand Function Test—[20] we suggest that direct analysis of biomechanical costs may provide an important supplement to these tests, as an indicator of energetic efficiency. It may also be important to assess one's subjective sense of effort, which does not always accurately reflect measures of biomechanical cost [21] This should provide a valuable addition to therapeutic assessment because even when ADLs are completed independently if they are not performed within reasonable energetic costs, one might expect minimal carry over into the patient's spontaneous behavior.

It should be stressed that the role of task-level costs is also important for determining optimal control strategies for a given task. Such costs might include the accuracy and duration of movements. Computer-based technologies allow therapists to modify feedback to stress particular performance criteria, so as to emphasize certain costs. For example, in a targeted reaching task, one could provide reward based on duration, when focusing on improving movement time. However, if movement direction and straightness need to be stressed, visual feedback can be modified to amplify errors perpendicular to the desired trajectory while reducing errors in the direction of the desired movement. Such changes would penalize deviations from the desired movement path while allowing errors in the direction of movement. This approach would assign different costs to errors that contribute to task success versus those that don't. In fact, Ballester et al. [22] recently reported exactly this manipulation, using a virtual reality environment to train reaching hemiparetic stroke patients. The movements of a virtual representation of the patients' paretic limb were amplified in only the dimension parallel to the target direction. Following virtual reality training, the authors reported that the probability of using the paretic limb during a subsequent real-world task was increased by the reinforcing experience of seeing the virtual limb reach the target during training. These types of capabilities are now becoming available in the clinic, due to the increasing availability of computer-based robotic and virtual reality technologies.

2.3 Principle 2: Impedance Control

Optimal feedback control theory emphasizes that the derivation of the optimal control signal incorporates knowledge about the state of the body and the environment. If the state changes unexpectedly due to an external perturbation, or random noise, what should its influence be on the control strategy? For example, when a passenger in a vehicle drinks a cup of coffee, what should the control system do when the movement of the cup is unexpectedly perturbed by a bump in the road? Ideally, the components of the perturbing forces that assist in bringing the cup to the mouth smoothly should not be impeded. However, the components of the forces that resist in the achievement of the task goal, such as accelerating the cup too rapidly, or in the outward or downward directions, should be compensated. According to the principle of minimal intervention proposed by optimal feedback control, the central nervous system "intervenes" only when errors are detrimental to goal achievement. Such a selective compensation of errors might explain why people allow slight variability in their performance as long as the overall goals of the task are satisfied.

This type of selective modulation of feedback gains is consistent with evidence that even the simplest feedback circuits, reflexes, can be modulated based on task demands. The stretch reflex represents the simplest and most ubiquitous feedback circuit in the mammalian system. The typical response to a stretch of a muscle includes a characteristic three-phase response [23, 24], measured in the electromyogram (EMG) as shown in Fig. 2.3: the shortest latency response often referred to as M1 occurs within some 20–50 ms following perturbation onset, and reflects circuitry contained within the spinal

cord. Following this, a medium latency response, M2, is observed some 60–80 ms following the perturbation onset and is thought to reflect longer latency spinal as well as transcortical circuits. This is followed by M3, a longer latency reaction that is thought to reflect a voluntary corrective process. Studies examining how these responses are modulated have shown differential effects of different task conditions on the early and later phases of the reflex.

Early studies in which subjects were instructed to *resist* or to *not resist* a perturbation showed that M1 was not modified by such commands, while M2 could be greatly attenuated by the instruction to not resist, and M3 could actually be completely eliminated by this instruction [25]. More recent studies have shown that M2 can be modulated by spatial conditions in a task, such as when subjects are told to allow their hand to displace toward a particular target: when the arm is pushed toward the target, the later phases (M2, M3) of the stretch reflex that resist the perturbation are reduced. However, when the arm is pushed away from the target, the gains of these responses are increased. More importantly, this modulation varies with both the direction and the distance of the target [18]. This demonstrates that feedback circuits such as reflexes can be modulated in accord with task goals through implicit mechanisms. In fact modulation of reflexes appears to be a fundamental mechanism that our nervous system employs to control limb impedance and thus resist perturbations. An elegant example of such reflex modulation was provided by Lacquaniti and colleagues for a ball-catching task [26]. This study demonstrated not only modulation but also reversal of the stretch reflex, in response to ball impact. Both the amplitude and expression of the stretch reflex were modulated in a systematic way as the ball dropped toward the hand. The result of this reflex modulation was to generate impedance to the forces imposed by ball impact, thereby generating a smooth and effective catching response.

Why is active impedance control through reflex modulation important for motor performance? During everyday tasks, many environmental perturbations cannot be predicted prior to

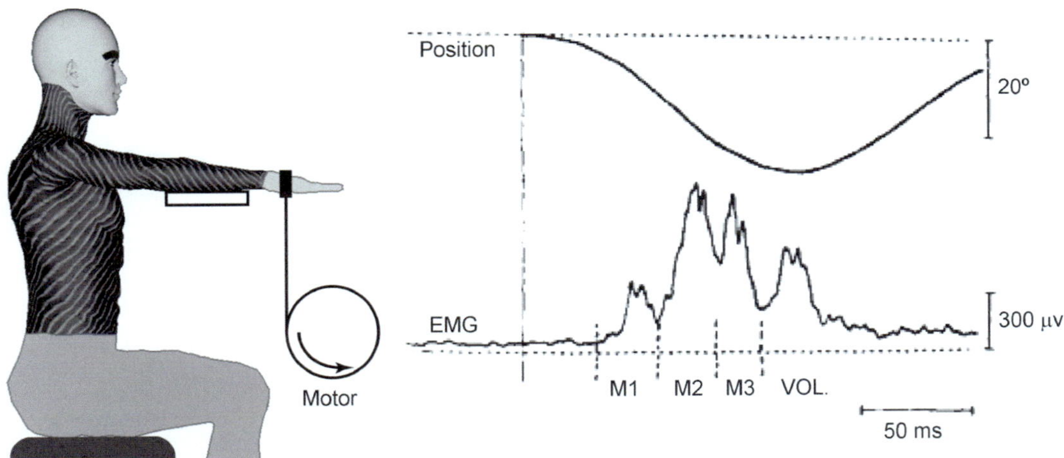

Fig. 2.3 Typical reflex response to muscle stretch. An example of the wrist extensor being stretched using a motor is shown on the left. The right panel shows the typical components of the electromyographic response to muscle stretch: the short-latency component M1 and the longer latency components M2 and M3. (Adapted from Matthews PB. The human stretch reflex and the motor cortex. Trends Neurosci. 1991 Mar;14(3):87–91.)

movement. In the example of a passenger drinking coffee in a moving vehicle, changes in vehicle acceleration due to bumps and breaking can rarely be anticipated. One can increase overall arm stiffness by co-activating muscles, but this uses a great deal of metabolic energy and interferes with the ability to bring the cup to the mouth. Franklin [27] and colleagues directly tested how subjects might selectively modify impedance without interfering with the coordination of the intended movement. In this study, subjects performed reaching movements with the arm attached to a robotic manipulandum that imposed unstable force fields that had components directed perpendicular to the required movement (see Fig. 2.4a). With practice, the participants were able to adapt to the novel dynamics and produce straight trajectories. They achieved this adaptation by selectively increasing stiffness in the direction of the instability, but not along the movement direction (see Fig. 2.4b). Remarkably, at the joint level, this impedance modification was achieved without changing baseline force and torque profiles (see Fig. 2.4c): the coordination strategy remained kinetically efficient, even though subjects were also able to effectively impede the imposed perturbations. These authors concluded that the nervous system is able to simultaneously maintain stability through impedance control and coordinate movements in a manner consistent with optimized energy expenditure.

We recently showed that such selective modification of limb impedance occurs through continuous modulation of short- and long-latency reflexes [28]. In our study, participants reached a visual target that occasionally jumped to a new location during movement initiation, thus changing the task goal during the course of motion. Unpredictable mechanical perturbations were occasionally applied, 100 ms after the target jump. Our results showed that reflex responses were tuned to the direction of the target jump: response amplitudes were increased or decreased depending on whether the perturbation opposed or assisted the achievement of the new task goal, respectively. We also showed that this reflex modulation resulted in changes in limb impedance to the perturbations. However, under conditions in which the movements were not mechanically perturbed, no changes in EMG or joint torque occurred at reflex latency relative to movements made with mechanical perturbations. These findings supported those of Franklin and colleagues by confirming that limb impedance is controlled without interfering with optimal

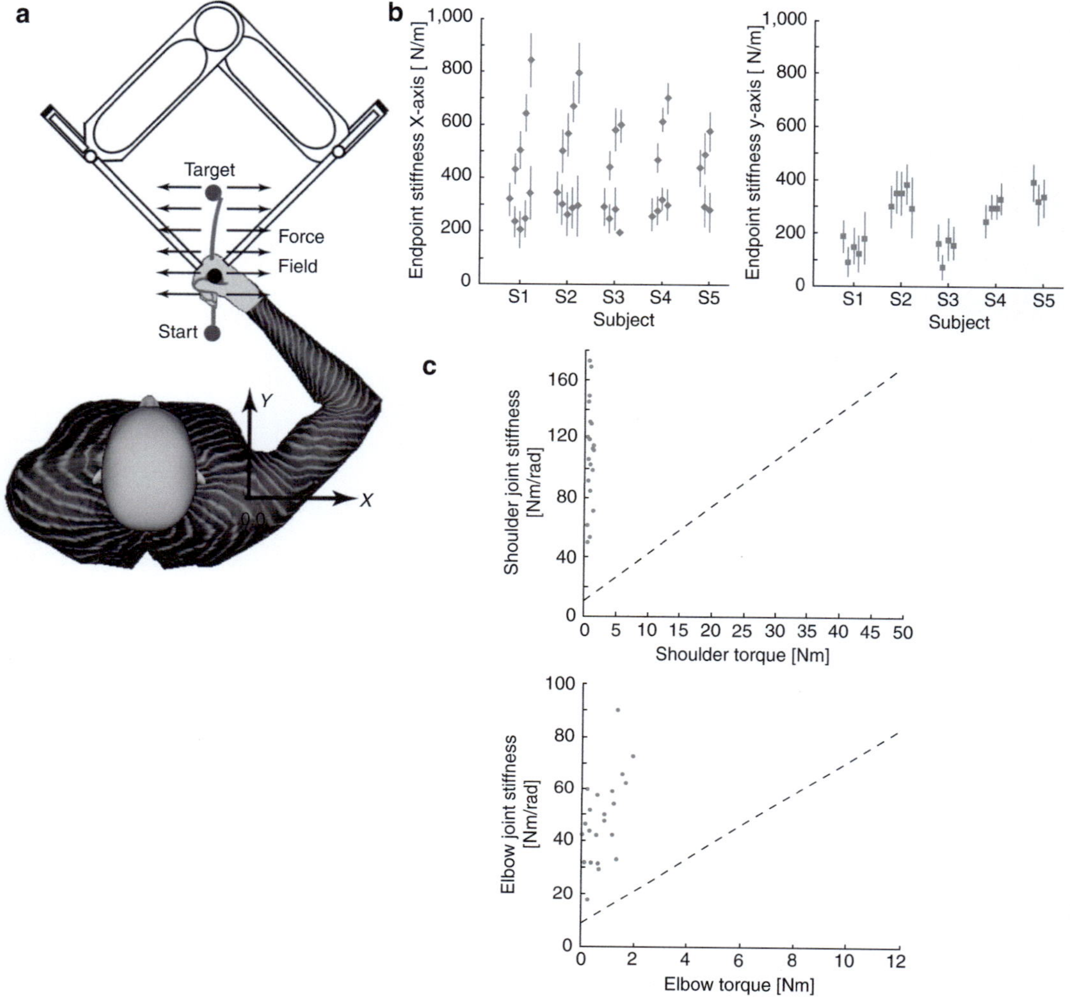

Fig. 2.4 Modulation of limb impedance. **a.** The typical setup and the perturbing force field. The field acts to push the arm perpendicular (along X-axis) to the direction of motion (Y-axis). **b.** An increase in limb stiffness along the X-, but not Y-axis for all subjects. **c.** Shoulder and elbow joint stiffnesses were independent of the respective joint torques. Adapted from Franklin DW, So U, Kawato M, Milner TE. Impedance control balances stability with metabolically costly muscle activation. J Neurophysiol. 2004 Nov;92(5):3097–105

coordination, by selectively modulating the expression of short- and long-latency reflex responses.

The studies discussed above point to the remarkable ability of the nervous system to determine optimal responses to unpredictable situations. Such control policies appear to mediate the modulation of limb impedance through the regulation of feedback circuits such as reflexes to ensure that unexpected perturbations are countered in a task-specific manner. Reflexive resistance to a perturbation is increased when it is inconsistent with the task goal but decreased when the perturbation is congruent with the goal of the task. These findings agree with the "minimum intervention principle" within the optimal feedback control framework. Thus, controlling limb impedance in a task-specific manner appears to be an integral component of the motor control process.

Implications for Rehabilitation. The research summarized above indicates that the central nervous system invokes at least two aspects of control to achieve coordinated movements. First, the commands are specified that result in optimal coordination patterns that satisfy both costs associated with task performance and energetic costs. In addition, the nervous system appears to set control policies that modulate sensorimotor circuits such as reflexes, to account for perturbations from unexpected changes in environmental or internal conditions. The importance of recognizing both of these features of control in clinical environments is fundamentally important because brain damage due to stroke can have differential effects on these two aspects of coordination. For example, Beer et al. [29] showed that hemiparesis disrupts optimal intersegmental coordination, resulting in inefficient coordination that fails to account for the dynamic interactions between the segments. This deficit does not appear to depend on the extent of hemiparesis.

Traditional therapeutic strategies, as well as more recent robot-aided rehabilitation strategies, tend to target the optimal control process by practicing fairly consistent patterns of coordination, and reinforcing task success. While this type of practice is critical for improving coordination and voluntary control, focusing on repetitive movements under consistent environmental conditions should only be the first step in rehabilitation training. In itself, this training may improve voluntary control of optimal coordination patterns, but is unlikely to train impedance control mechanisms. Because of this, patients may become adept at the training protocols, but show limited transfer to activities of daily living. We suggest that as patients improve their movement patterns under predictable conditions, training protocols should progressively incorporate unpredictable conditions. Such conditions might include random changes in target positions and varying force perturbations, thereby training patients to impede variations in environmental conditions that interfere with task performance.

2.4 Principle 3: Motor Lateralization

As discussed thus far, both optimal control and impedance control are component mechanisms underlying the control of voluntary movements. Our recent work has suggested that these two mechanisms are lateralized to the left and right brain hemispheres, respectively. The seminal research of Sperry and Gazzaniga [30] on disconnection syndrome in split-brain patients first established neural lateralization as a fundamental principle of the cerebral organization. Gazzaniga proposed that distributing different neural processes across the hemispheres was a natural consequence of developing complex functions during the course of evolution. His research provided elegant support for this view of cerebral lateralization as a neural optimization process.

Interestingly, early research on hemispheric lateralization was largely limited to cognitive and perceptual processes, with little attention to the motor systems. We introduced the dynamic dominance hypothesis of motor lateralization [31], based on left- and right-arm advantages in reaching performance in healthy adults, and expanded this hypothesis based on computational modeling studies [32, 33] and studies in patients with unilateral brain lesions [34–39]. The dynamic dominance model proposes that the left hemisphere, in right-handers, is specialized for predictive processes that specify smooth and efficient movement trajectories under mechanically stable environmental circumstances, while the right hemisphere is specialized for impedance control mechanisms that confer robustness to movements performed under unpredictable and mechanically unstable environmental conditions. In fact, this type of division of labor between the two sides of the brain appears to predate humans by half a billion years [40]. Rogers and colleagues have proposed a single organizing principle that might account for the large array of emotional, language, perceptual, and cognitive asymmetries that have been described across the evolutionary spectrum of vertebrates. While the left hemisphere appears "specialized for control of well-established patterns of behavior, under ordinary and familiar circumstances", the right hemisphere is specialized for "detecting and responding to unexpected stimuli in the environment" [41]. The dynamic dominance model provides the movement analog to Roger's model, and thus places handedness in the context of a larger array of neurobehavioral asymmetries across the animal kingdom [42].

An important feature of these models is that both hemispheres are recruited for their complimentary contributions to integrated functional activities. Thus, during the movement of a single arm, both hemispheres contribute their specific aspects of control.[43]. Because each hemisphere contributes specialized processes to control each arm, unilateral brain damage actually produces hemisphere-specific movement deficits in the non-paretic, ipsilesional arm, as well as the contralesional arm. Remarkably, this is the arm that is usually considered unaffected by unilateral brain damage. The idea that each hemisphere contributes to motor coordination of both arms is an important implication of ipsilesional, non-paretic arm motor deficits. While the role of contralateral motor areas in controlling limb movements is well-understood [44] the role of the ipsilateral hemisphere has more recently been implicated by the robust occurrence of ipsilesional motor deficits in both animal models of unilateral brain damage [45–47] as well as human stroke survivors [34, 36, 39, 48–58]. In addition, both electrophysiological and neural imaging studies have shown that unilateral arm and hand movements recruit motor-related areas in both cerebral hemispheres [43, 59–61]. Thus, it is the loss of the contributions of the ipsilateral hemisphere to movement control that gives rise to motor deficits in the non-paretic arm of stroke patients. Most importantly, these deficits can substantially limit functional performance [51, 54], a particularly concerning phenomenon, given that patients with severe contralesional paresis depend on the ipsilesional arm for the majority of their activities of daily living.

Our recent studies have examined the specific nature of the ipsilesional movement deficits that result from left or right brain damage, shedding light on motor lateralization. These studies have confirmed that right and left sensorimotor strokes produce predictable deficits in impedance control or optimal control, respectively [51, 62]. For example, Schaefer et al. [51] compared reaching movements in the ipsilesional arm of hemisphere-damaged patients with those of healthy control subjects matched for age and other demographic factors. Subjects performed targeted reaching movements in different directions within a workspace to the same side of the midline as their reaching arm. The left hemisphere-damaged group showed deficits in controlling the arm's trajectory due to impaired interjoint coordination but showed no deficits in achieving accurate final positions. In contrast, the right hemisphere-damaged group showed deficits in final position accuracy but not in interjoint coordination. These findings are exemplified in the hand paths shown in Fig. 2.5a. While control subjects made relatively straight and accurate movements, patients with left hemisphere damage made movements that were very curved, but nevertheless were accurate in the final position. In contrast, patients with right hemisphere damage made straight movements with poor final position accuracies. This double dissociation between the type of error (trajectory or final position) and the side of hemisphere damage (right or left) is emphasized in Fig. 2.5b, which shows the variance in hand positions during the initial trajectory phase (cross), or the final position phase (circle) of the movement. The ratio of errors at these two points in movement (peak velocity, movement termination) is quantified across subjects in the bar graphs, revealing that RHD patients had the greatest variance in final position, while LHD patients had the greatest variance in trajectory. Thus, these results indicate the distinct lateralization of optimal trajectory control and impedance-mediated final position control to the left and right hemispheres, respectively. It should be emphasized that these errors were associated with functional impairments in the ipsilesional arm, as measured by the Jebsen–Taylor Hand Function Test (JHFT). Thus, motor lateralization leads to deficits that depend on the side of the stroke and can lead to

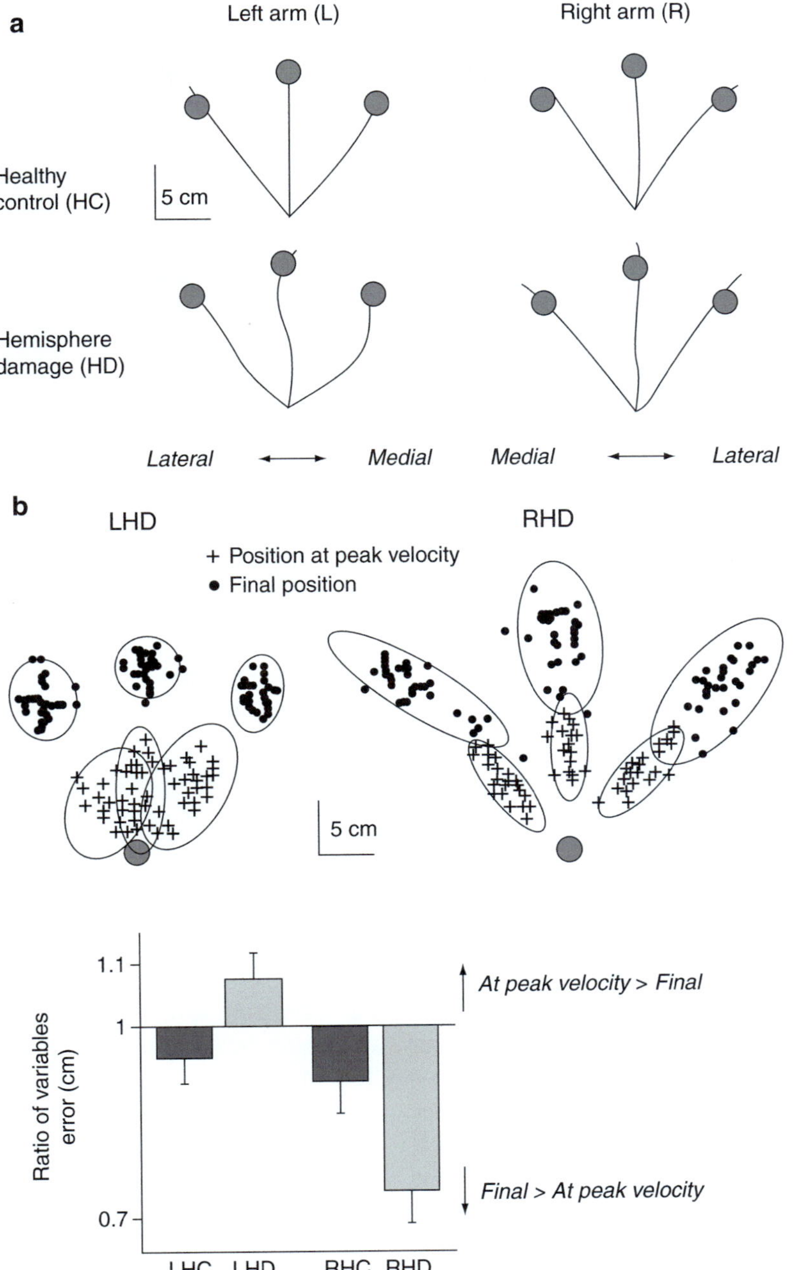

Fig. 2.5 Lateralization of motor deficits after stroke. 6A shows typical hand paths for healthy control subjects performing with their right or left arm (top panel) and left and right hemisphere-damaged stroke patients performing with their ipsilesional arm (bottom panel). 6B shows hand locations at peak velocity (crosses) and movement end (circles) for a typical left and right hemisphere-damaged stroke patient (top panel). Ellipses represent 95% confidence intervals. The bottom panel shows the mean ratio of variable errors at peak velocity to variable error at movement end across all subjects for the control and stroke groups. (Adapted from Schaefer SY, Haaland KY, Sainburg RL. Hemispheric specialization and functional impact of ipsilesional deficits in movement coordination and accuracy. Neuropsychologia. 2009 Nov;**47** (13):2953–66.)

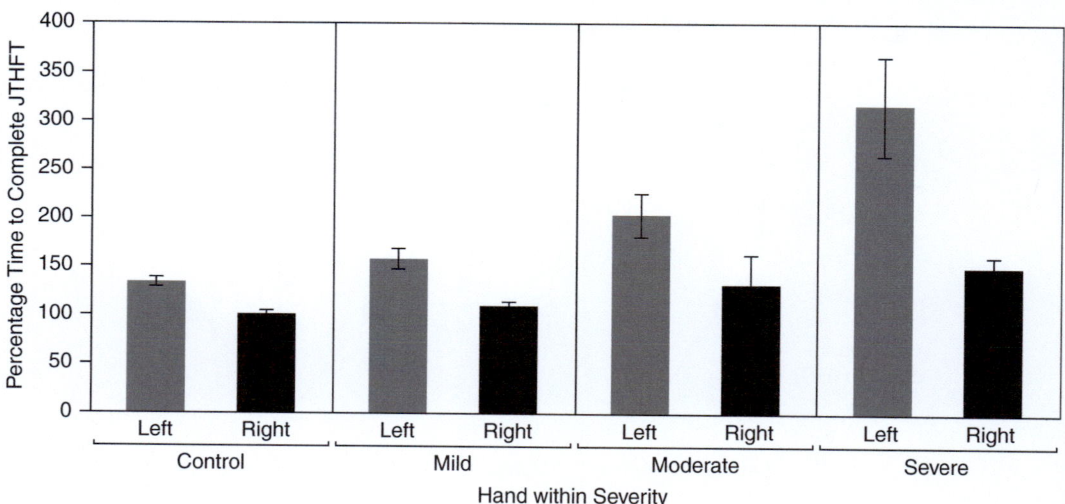

Fig. 2.6 JTHFT score, normalized to control group right-hand score. Scores for non-paretic arm of stroke survivors. Control subjects were matched to gender and age distribution of each stroke survivor group. The two control groups were comprised of those that used their left or right hands and included 18 participants each. The left hemisphere-damaged group comprised 22 stroke survivors, whereas the right hemisphere-damaged group comprised 29 stroke survivors. On the X-axis, these groups are stratified by severity of contralesional arm paresis

significant deficits, as tested with clinical assessments, such as the JHFT.

Figure 2.6 shows data from 72 age and gender-matched control subjects, 22 left hemisphere-damaged stroke survivors, and 29 right hemisphere-damaged stroke survivors. The Y-axis represents the JHFT score, taken as a percentage of the right dominant arm function in our control group. Thus, 100% is the mean for the right hand of 36 of the control subjects (those who used their right hand). The JHFT is a rather thorough assessment of unilateral arm function that includes a large range of tasks that elicit the coordination requirements of functional daily activities, such as writing, turning pages, placing large and small objects on a table, stacking checkers, and feeding. The left column (control) shows the difference between healthy subjects performing with the left arm and right arm.

The data are stratified on the X-axis by both hands (right/left: in the case of stroke survivors this is only the ipsilesional arm) and severity of contralesional paresis, as measured by the upper limb component of the Fugl–Meyer [63] assessment of motor impairment (mild >= 55, moderate > 35, Severe <= 35). In healthy subjects, the left non-dominant arm takes, on average, 33% longer than the right arm to carry out these tasks. For reference, this reflects the frustration a typical adult would experience when trying to get through their day with only the non-dominant arm, for example, due to a broken dominant arm. In our stroke survivors, there is a substantial effect of both the severity of impairment in the paretic arm, as well as the side of the brain lesion on JTHF performance with the non-paretic arm. First, the more severe the contralesional paresis, the greater the impairment in the non-paretic arm. This effect is potentiated by the side of lesion, such that left hemisphere-damaged survivors who have severe paresis in their contralesional arm, take 216% longer to complete the JTHF than the dominant arm of control subjects, whereas, right hemisphere-damaged survivors with severe contralesional paresis, take 51% longer than do control subjects. Functionally, this effect is concerning for two reasons: First, the finding that the extent of ipsilesional deficit varies with the extent of contralesional paresis indicates that the survivors who must depend most on the ipsilesional arm for function have the greatest impairments in that arm. Second, these stroke survivors were tested, on average 1.8 years following their stroke, suggesting that these deficits do not spontaneously change over time. Even right hemisphere-damaged patients with severe paresis take nearly 52% longer than aged-matched control subjects to complete the JTHF, regardless of the "forced use" of the ipsilesional arm imposed by severe contralesional paresis. This introduces the questions of whether focused remedial therapy might improve function by increasing the speed and dexterity of the non-paretic arm in patients with moderate to severe contralesional paresis.

Implications for Rehabilitation. While most robotic rehabilitation devices have been focused on training movements in the contralesional arm, the research discussed above provides compelling evidence that ipsilesional practice should also be encouraged. In fact, for many patients, the ipsilesional arm will become the primary manipulator, thus efficient coordination of this arm and hand should be critical for the effective performance of activities of daily living [64].

It is, thus likely that intensive training of the ipsilesional, non-paretic arm could substantially improve functional independence in patients with hemiparesis. However, it should be noted that *remediation of the non-paretic arm is so novel that little empirical evidence exists as to whether such intervention might lead to positive effects on motor performance and functional independence*. One recent pilot intervention study compared a group of patients who received therapy that included training of the non-paretic arm to another group who only received traditional therapy, without non-paretic arm training [65]. The results indicated that when traditional therapy was combined with non-paretic arm training, the speed and accuracy of non-paretic arm movements improved, as did the impairment level of the paretic arm when compared to patients who received traditional therapy alone. This suggests that focused non-paretic arm training might produce both improvements in non-paretic arm motor performance and modest improvements in paretic arm function, both of which should facilitate improvements in functional independence. However, some caution is indicated because of the phenomenon of learned non-use of the paretic arm, an effect that has been

successfully addressed by constraining the non-paretic arm in patients with moderate to mild paresis [66–69]. While the pilot results cited above suggest positive effects of non-paretic arm training on paretic arm function, there currently is no conclusive evidence to predict whether non-paretic arm training will influence paretic arm function, either positively or negatively. *This is an important area for future research in rehabilitation intervention for stroke patients.*

In contrast to focused non-paretic arm training, bilateral training has a long history in rehabilitation research and practice and should represent a critical component of therapeutic intervention in unilateral stroke. In fact, most activities of daily living are performed with both hands contributing to different aspects of the activity [54, 64]. For example, when buttoning a shirt, the non-dominant arm tends to stabilize the buttonhole, while the dominant arm manipulates the button through the hole. Bilateral training is not only important to facilitate remediation in the ipsilesional arm but also because unilateral training may not automatically carry over to spontaneous bilateral performance. In fact, recent research has indicated that learning novel kinetic and visuomotor environments with a single arm transfers only partially to bilateral movements, in which the same arm experiences the imposed environments [70, 71]. It is, therefore critical that rehabilitation focus not only on unilateral performance but that training be extended to bilateral movements. While some robotic devices are designed for bilateral movements [72], unilateral robotic training can be followed by bilateral training, even in the absence of bilateral robotic systems. In fact, bilateral training has a long history in Occupational Therapy treatment, where manipulation of dowels and rolling pins has often been used to encourage bilateral arm use.

More importantly is the question of whether remediation focused on the non-paretic arm might improve stroke survivors' participation in daily activities, for those patients who rely on this arm as their sole or primary manipulator and have substantial ipsilesional motor deficits. Currently, the usual standard of care in rehabilitation for patients with *low-moderate to severe paresis* tends to focus on task training in essential ADL activities, rather than on intensive remediation. We suggest that the combination of moderate to severe paresis with persistent motor deficits in the non-paretic arm limits the performance and participation in activities of daily living. We, thus, predict that intense rehabilitation, sequentially focused on each arm should provide a durable and substantial improvement in functional performance. However, this approach must be addressed with some caution because while sequential arm training has never been studied in human stroke survivors, Jones et al. [73] showed, in an acute model of stroke in rats, that initial training of ipsilesional forelimb reaches can limit the subsequent response to training in the contralesional forelimb (2010). On the other hand, interlimb transfer of motor learning often shows a positive effect in healthy individuals [74–77], and mirror training has shown positive transfer between the arms in stroke patients [78–80]. It is critical to carry out studies of ipsilesional arm intervention in survivors with moderate to severe contralesional paresis to determine whether such training can positively affect functional outcomes and participation in human stroke survivors.

2.5 Principle 4: Motor Learning

The discussion so far noted that rehabilitation should focus on improving both optimal control as well as impedance control while bearing in mind that these control mechanisms are likely lateralized to different brain hemispheres. However, rehabilitation itself rests on the assumption that patients can re-learn such control with repeated practice. As such, knowledge of how motor learning occurs, how it is retained, and how it generalizes to other conditions that haven't been practiced, is central to the development of effective rehabilitation strategies.

Motor learning is used as an umbrella term to incorporate any practice-related improvement in motor performance. The primary paradigm used in recent motor learning research has been focused on fairly short-term motor adaptation, where researchers have explored adjustments in movement patterns to various kinds of altered environments. Typically, subjects are exposed to novel task conditions such as when a cursor, representing the location of the hand on a screen, deviates from the actual hand location, or when the hand is pushed from its intended trajectory using force perturbations. Under such conditions, subjects readily adapt to the new environment, a process that appears to occur, at least in part, through changes in predictive control, or in other words, movement planning [81, 82]. The predictive nature of such adaptation is reflected by the occurrence of "aftereffects" following removal of the imposed environmental perturbation. Such after-effects tend to mirror image the movement patterns seen on early exposure to the imposed perturbation, and are based on the subject's expectation that they will continue to experience the novel environment. In other words, the effects of the perturbation are predicted and accounted for, and the motor output is appropriately modified [83, 84]. Computationally, such adaptation can be modeled as an iterative update of a forward model, defined as a transformation from movement commands to their desired sensory consequences. In this scheme, sensory prediction errors, or the difference between the intended and actual sensory feedback should drive the process of improving the accuracy of the forward model so that the predicted sensory consequences of motor commands coincide with the actual sensory feedback. This process has been shown to occur implicitly [83], although new research suggests that adaptation may also involve explicit or declarative strategies [85] as well as reinforcement mechanisms that are driven by task success [86].

In order to examine how motor learning might be represented in the nervous system, many studies have examined conditions to which the learning generalizes. Interestingly, these studies have generally suggested that the generalization of visuomotor adaption is different from the generalization of adaptation to novel dynamic conditions such as force fields. For example, Krakauer et al. [87] examined the generalization of visuomotor adaptation and found that subjects generalized to movements that were made in the same direction, but from a different starting configuration of the arm. We have also shown that such adaptation can transfer between the limbs [31]. These results are consistent with other studies that have suggested adaptation to errors introduced at the extrinsic task-level transfers along with the same coordinates [88, 89]. Generalization of adaptation to dynamic conditions such as novel force fields in contrast has been shown to occur along with intrinsic or joint coordinates [77, 90]. Malfait et al. [91] in fact showed that learning of novel force fields transferred to movements made in different regions of the workspace, but that required similar joint excursions, but poorly to movements in which joint excursions changed. Thus, the representation of the applied force field appeared to be linked to joint motions, or intrinsic coordinates. Mussa-Ivaldi and colleagues [92] have proposed that the generalization of learning novel mechanical conditions is tightly linked to the dynamic state of the arm, indicated by the velocity and positions of the arm experienced during learning. In support of this idea, when novel dynamics are learned with the dominant arm, they appear to transfer to the non-dominant arm along with intrinsic coordinates [77, 90]. Thus, while learning of novel visual-motor conditions appears to generalize in extrinsic coordinates, learning of novel dynamic conditions appears to transfer along with intrinsic coordinates.

To explore the neural basis of adaptation, which has important implications for rehabilitation post Stroke, we recently examined the impact of different brain lesions on the ability to adapt to novel visuomotor conditions. In general, we have found that left hemisphere damage, particularly to posterior parietal regions, impairs visuomotor adaptation [39]. Our results significantly expanded on prior studies that focused on the cerebellum as the neural substrate critical for adaptation [93–96]. Our results also agreed with Tanaka et al. [97] who showed that experimentally observed visuomotor adaptation and generalization patterns could be reproduced using a population-coding model in which adaptation induced changes in the synaptic weights between narrowly tuned, parietal-like neurons and units in the motor cortex. Importantly, models that utilized tuning properties of motor cortical or cerebellar neurons could not reproduce behavioral data. Thus, more recent findings have strongly implicated posterior parietal regions for adaptation, particularly under conditions in which visuomotor errors are imposed. The neural substrates critical for dynamic adaptation are less clear.

In contrast to adaptation, which requires improvement in performance in response to environmentally induced errors, learning in the absence of such errors has not been as extensively studied. Newer studies term such learning in the absence of sensory prediction errors as "skill learning", and it is thought that mechanisms that drive learning of new skills are different from those that drive adaptation [98]. Behaviorally, adaptation only focuses on return to baseline level of performance in the presence of error-inducing perturbations, progresses rapidly, is short-lived, and shows limited generalization. In contrast, skill learning occurs over much slower time scales and learned skills are rarely forgotten (86). Research suggests that learning skills may recruit reinforcement-like processes, where a successful action is found through trial and error and is then repeated since it leads to a rewarding outcome. However, this needs to be explored further. Neurophysiologically, skill learning has been mapped onto substrates that appear to be different from adaptation. For instance, the primary motor cortex and basal ganglia are believed to be crucial for learning new skills, but not for adaptation. For example, transcranial magnetic stimulation applied over M1 does not appear to impair adaptation [99], but facilitation of M1 via anodal direct current stimulation enhances skill learning [99], suggesting that M1 might play a different role in these two processes.

Despite these differences, however, there is good reason to believe that adaptation and skill learning processes interact during the learning of real-life tasks. For instance, recent results suggest that even in what would otherwise be classified as a pure adaptation task, reinforcement mechanisms are recruited [86]. Under certain conditions, adaptation to errors can in fact be driven completely by reward-based reinforcement mechanisms [100]. Other mechanisms, including the use of explicit strategies [101, 102] and declarative memory [85] have also been suggested to contribute significantly during motor learning.

Implications for Rehabilitation. The array of findings on motor learning and its underlying neural substrates have several potential implications for rehabilitation. First, it is critical to recognize that multiple mechanisms, presumably dependent on distinct neural substrates, contribute to an improvement in motor performance with practice. Loss of a particular component process because of focal lesions in different regions of the brain therefore does not automatically imply a complete loss of learning capacity. Different processes and alternate "routes" can be exploited for improvement in motor function. For instance, for a patient with an injury to parietal regions, which might affect his/her capacity to adapt to a novel environment, reinforcement mechanisms could be exploited for learning in the same environment. Second, given that adaptation and skill might recruit different neural resources, rehabilitation approaches must focus on training or facilitating both these processes, possibly along with other mechanisms such as the use of explicit strategies and declarative memory processes. Third, the fact that learning might occur and generalize in different coordinate systems must be taken into account. While learning in environments that perturb performance in the extrinsic, task space allows adaptation to task constraints, such as improving the accuracy and precision, learning in dynamic environments allows the central nervous system to optimize intrinsic coordination and mechanical energy. It is therefore important for therapists to consider both intrinsic and extrinsic aspects of task performance. It is typical to consider the similarities between two tasks in terms of extrinsic, task-related coordinates because one can readily determine whether the task is in the same region of space, is oriented similarly, and is performed at similar speeds as the task or tasks that are targeted for transfer. For example, one can practice stacking cones on a surface and expect that this might transfer to the task of procuring a glass from the cupboard (target ADL skill). However, one must also consider the dynamic requirements of the two tasks, in terms of both postural and limb movement requirements. Whether the two tasks are similar in terms of joint torques, or joint power might depend on subtle differences in body configurations, and relative segment motions. This would be difficult to determine for any given target task, let alone a large range of ADL activities. It is, therefore, important to provide a great deal of variation in dynamic experience when practicing a given task, particularly as patients become proficient at a given set of movement patterns. Robotic and technology-aided rehabilitation, which have the capacity to provide a large range of interactive visual and dynamic environments along with the capacity for high-intensity and high-dose practice, hold great promise in this regard.

2.6 Summary and Conclusions

As the technology-based intervention tools discussed in this volume enter the clinic, they will provide rehabilitation professionals with the ability to prescribe and monitor movement experiences with unprecedented precision. This introduces the question of what specific aspects of movement should be practiced and monitored with these tools. In this chapter, we presented four tenets derived from research in movement neuroscience that have an impact on this question, and that have been derived from literature on the neural control of movement. These tenets are optimal control, impedance control, motor lateralization, and motor learning. We will review these principles and the implications for rehabilitation below:

Optimal control theory has examined plausible costs that might be considered by the nervous during motor planning, and that might account for the reliable, or "invariant," features of movements that occur across tasks and individuals. This line of research has indicated two major categories of cost that contribute to motor planning: explicit task level costs, such as movement accuracy and speed, and implicit costs, such as energy and movement variability. When designing movement practice for patients, it is important to consider both types of costs, when grading the difficulty of the task. We also suggest that it is critical to consider biomechanical variables related to energetic efficiency when evaluating patients' progress. While many clinical tests assess the ability to perform ADLs, as well as the time of such performance, a critical factor that should determine carryover into spontaneous daily activities is whether the movement can be performed at a reasonable energy cost. As the technologies discussed in this volume become available in the clinic, many of the devices will allow measures of mechanically related variables, such as work, power, and torque. Such variables can be exploited to monitor progress in making not only accurate and rapid but also energetically efficient movements.

Impedance control refers to neural mechanisms that modulate rapid sensorimotor circuits, such as stretch reflexes, in order to impede perturbations that cannot be anticipated during motor planning. These include forces that arise from the environment, such as inertial forces that result from braking and acceleration of a vehicle, or even inaccurate movements of one's own body, such as the effect on the upper body and arms of stepping on an uneven surface while holding a cup. Robot-aided and virtual reality technologies allow the introduction of "perturbations into patients' movement training experience. While it is currently most common to practice repetitive patterns under stereotyped conditions, introducing unpredictable perturbations should consolidate this learning and prepare patients for movement under natural environmental conditions.

Motor lateralization research has indicated that different aspects of motor control have been specialized to the different cerebral hemispheres. The hypothesis that both hemispheres are normally recruited for each respective control mechanism, optimal trajectory control, and impedance control, predicts that damage to a single hemisphere should produce deficits in the ipsilesional arm, often considered the unaffected arm in stroke patients. Recent research has verified this prediction, demonstrating deficits in trajectory control following left hemisphere damage and deficits in achieving accurate steady-state positions following right hemisphere damage. The implications for rehabilitation are substantial: patients with persistent hemiparesis will need to use the ipsilesional arm as the lead, or often the sole, manipulator for activities of daily living. Thus, efficient performance of ADL will require well-coordinated movements of this arm. This is particularly important for patients who have severe contralesional paresis, which tends to be associated with substantial ipsilesional motor deficits. Intensive training focused on the ipsilesional arm can improve coordination, but research is needed to determine whether this will impact function either positively or negatively, of the contralesional arm. Because most ADL tasks require some degree of bilateral coordination, we recommend that following sequential unilateral training with each arm, both arms be trained simultaneously using bilateral tasks. Virtual reality environments provide an excellent paradigm to manipulate task conditions during bilateral arm training, such as requiring both arms to coordinate with each other for goal achievement and manipulating virtual objects.

Motor learning research has shown that multiple brain regions represent distinct motor learning processes. These processes include skill learning, in which one develops new sensorimotor patterns that were not previously learned, and adaptation, in which one learns to compensate for an environmental or sensory disturbance in order to perform a previously well-practiced task, such as reaching a force field, or under the influence of altered visuomotor feedback. It should be stressed that as stroke survivors learn to adapt to their new sensory and motor conditions, both of these forms of learning should be required. Even well-learned tasks, such as brushing one's teeth, may require substantially new skill development, given altered motor capacities. Similarly, distortions in sensory feedback including visual field deficits, and proprioceptive and tactile deficits can require adaptation to recover old skills. Generalization is also an aspect of motor learning with particular application to neurorehabilitation. It should be stressed that one cannot assume a particular pattern of motor generalization, following training. This is because some aspects of learning transfer along with different coordinates than others. For example, task dynamics seems to be learned and transferred in intrinsic coordinates, whereas visuomotor distortions are transferred across extrinsic coordinates. Since it is not simple, or even possible, to segregate these aspects of learning in a clinical environment, it is important to provide a range of training experiences that can ensure generalization across a range of tasks. Task-specific training, of course, should be done for key activities of daily living, but limiting training to specific tasks severely limits the potential of physical rehabilitation. We therefore strongly recommend providing a range of dynamic and kinematic training experiences that include the requirement for variability, and response to unpredictable perturbations.

References

1. Teasell R, Meyer MJ, McClure A, et al. Stroke rehabilitation: an international perspective. Topics in stroke rehabilitation. 2009;16(1):44–56.
2. Krebs HI, Hogan N, Aisen ML, Volpe BT. Robot-aided neurorehabilitation. IEEE Trans Rehabil Eng. 1998;6(1):75–87.
3. Krebs HI, Hogan N, Volpe BT, Aisen ML, Edelstein L, Diels C. Overview of clinical trials with MIT-MANUS: a robot-aided neuro-rehabilitation facility. Technol Health Care Offic J Eur Soc Eng Med. 1999;7(6):419–23.
4. Volpe BT, Krebs HI, Hogan N, Edelsteinn L, Diels CM, Aisen ML. Robot training enhanced motor outcome in patients with stroke maintained over 3 years. Neurology. 1999;53(8):1874–6.
5. Krebs HI, Volpe BT, Aisen ML, Hogan N. Increasing productivity and quality of care: robot-aided neuro-rehabilitation. J Rehabil Res Dev. 2000 Nov-Dec;37(6):639–52.
6. Fazekas G, Horvath M, Troznai T, Toth A. Robot-mediated upper limb physiotherapy for patients with spastic hemiparesis: a preliminary study. J Rehabil Med. 2007;39(7):580–2.
7. Morasso P. Spatial control of arm movements. Exp Brain Res. 1981;42(2):223–7.
8. Harris CM, Wolpert DM. Signal-dependent noise determines motor planning. Nature. 1998;394(6695):780–4.
9. Flash T, Hogan N. The coordination of arm movements: an experimentally confirmed mathematical model. J Neurosci. 1985;5(7):1688–703.
10. Uno Y, Kawato M, Suzuki R. Formation and control of optimal trajectory in human multijoint arm movement. Minimum torque-change model. Biol Cybern. 1989;61(2):89–101.
11. Atkeson CG, Hollerbach JM. Kinematic features of unrestrained vertical arm movements. J Neurosci. 1985;5(9):2318–30.
12. Soechting JF, Buneo CA, Herrmann U, Flanders M. Moving effortlessly in three dimensions: does Donders' law apply to arm movement? J Neurosci. 1995;15(9):6271–80.
13. Goble DJ, Brown SH. Task-dependent asymmetries in the utilization of proprioceptive feedback for goal-directed movement. Exp Brain Res. 2007 Jul;180(4):693–704.
14. Goble DJ, Brown SH. The biological and behavioral basis of upper limb asymmetries in sensorimotor performance. Neurosci Biobehav Rev. 2008;32(3):598–610.
15. Todorov E, Jordan MI. Optimal feedback control as a theory of motor coordination. Nat Neurosci. 2002;5(11):1226–35.
16. Liu D, Todorov E. Evidence for the flexible sensorimotor strategies predicted by optimal feedback control. J Neurosci. 2007;27(35):9354–68.
17. Latash ML. What are "normal movements" in atypical populations? Behav Brain Sci. 1996;19:55–68.
18. Pruszynski JA, Kurtzer I, Scott SH. Rapid motor responses are appropriately tuned to the metrics of a visuospatial task. J Neurophysiol. 2008;100(1):224–38.
19. Wright PJ, Fortinsky RH, Covinsky KE, Anderson PA, Landefeld CS. Delivery of preventive services to older black patients using neighborhood health centers. J Am Geriatr Soc. 2000;48(2):124–30.
20. Jebsen RH, Taylor N, Trieschmann RB, Trotter MJ, Howard LA. An objective and standardized test of hand function. Arch Phys Med Rehabil. 1969;50(6):311–9.
21. Burgess PR, Cooper TA, Gottlieb GL, Latash ML. The sense of effort and two models of single-joint motor control. Somatosens Mot Res. 1995;12(3–4):343–58.
22. Ballester BR, Nirme J, Duarte E, et al. The visual amplification of goal-oriented movements counteracts acquired non-use in hemiparetic stroke patients. J Neuroeng Rehabil. 2015;12:50.
23. Tatton WG, Forner SD, Gerstein GL, Chambers WW, Liu CN. The effect of postcentral cortical lesions on motor responses to sudden upper limb displacements in monkeys. Brain Res. 1975;96(1):108–13.
24. Matthews PB. The human stretch reflex and the motor cortex. Trends Neurosci. 1991;14(3):87–91.
25. Hammond PH. The influence of prior instruction to the subject on an apparently involuntary neuromuscular response. J Physiol. 1956;132(1):17–8.
26. Lacquaniti F, Borghese NA, Carrozzo M. Transient reversal of the stretch reflex in human arm muscles. J Neurophysiol. 1991;66(3):939–54.

27. Franklin DW, So U, Kawato M, Milner TE. Impedance control balances stability with metabolically costly muscle activation. J Neurophysiol. 2004;92(5):3097–105.

28. Mutha PK, Boulinguez P, Sainburg RL. Visual modulation of proprioceptive reflexes during movement. Brain Res. 2008;30(1246):54–69.

29. Beer RF, Dewald JP, Rymer WZ. Deficits in the coordination of multijoint arm movements in patients with hemiparesis: evidence for disturbed control of limb dynamics. Exp Brain Res. 2000;131 (3):305–19.

30. Gazzaniga MS. The split brain revisited. Sci Am. 1998;279(1):50–5.

31. Sainburg RL. Evidence for a dynamic-dominance hypothesis of handedness. Exp Brain Res. 2002;142 (2):241–58.

32. Yadav V, Sainburg RL. Motor lateralization is characterized by a serial hybrid control scheme. Neuroscience. 2011;24(196):153–67.

33. Yadav V, Sainburg RL. Handedness can be explained by a serial hybrid control scheme. Neuroscience. 2014;10(278):385–96.

34. Schaefer SY, Haaland KY, Sainburg RL. Ipsilesional motor deficits following stroke reflect hemispheric specializations for movement control. Brain. 2007;130(Pt 8):2146–58.

35. Mutha PK, Haaland KY, Sainburg RL. Rethinking motor lateralization: specialized but complementary mechanisms for motor control of each arm. PLoS ONE. 2013;8(3): e58582.

36. Mutha PK, Haaland KY, Sainburg RL. The effects of brain lateralization on motor control and adaptation. J Mot Behav. 2012;44(6):455–69.

37. Mutha PK, Sainburg RL. Independent control of velocity and position in single joint movements. Soc Neurosci Abstracts. 2005;99:9.

38. Mutha PK, Sainburg RL. Shared bimanual tasks elicit bimanual reflexes during movement. J Neurophysiol. 2009;102(6):3142–55.

39. Schaefer SY, Mutha PK, Haaland KY, Sainburg RL. Hemispheric specialization for movement control produces dissociable differences in online corrections after stroke. Cereb Cortex. 2012;22(6):1407–19.

40. Bisazza A, Rogers LJ, Vallortigara G. The origins of cerebral asymmetry: A review of evidence of behavioural and brain lateralization in fishes, reptiles and amphibians. Neurosci Biobehav Rev. 1998;22(3):411–26.

41. MacNeilage PF, Rogers LJ, Vallortigara G. Origins of the left & right brain. Sci Am. 2009;301(1):60–7.

42. Sainburg RL. Convergent models of handedness and brain lateralization. Front Psychol. 2014;5:1092.

43. Kawashima R, Roland PE, O'Sullivan BT. Activity in the human primary motor cortex related to ipsilateral hand movements. Brain Res. 1994;663 (2):251–6.

44. Kuypers HG. A new look at the organization of the motor system. Prog Brain Res. 1982;57:381–403.

45. Gonzalez CL, Gharbawie OA, Williams PT, Kleim JA, Kolb B, Whishaw IQ. Evidence for bilateral control of skilled movements: ipsilateral skilled forelimb reaching deficits and functional recovery in rats follow motor cortex and lateral frontal cortex lesions. Eur J Neurosci. 2004;20 (12):3442–52.

46. Grabowski M, Brundin P, Johansson BB. Paw-reaching, sensorimotor, and rotational behavior after brain infarction in rats. Stroke. 1993;24 (6):889–95.

47. Grabowski M, Nordborg C, Johansson BB. Sensorimotor performance and rotation correlate to lesion size in right but not left hemisphere brain infarcts in the spontaneously hypertensive rat. Brain Res. 1991;547(2):249–57.

48. Mutha PK, Sainburg RL, Haaland KY. Left parietal regions are critical for adaptive visuomotor control. J Neurosci. 2011;31(19):6972–81.

49. Mutha PK, Sainburg RL, Haaland KY. Critical neural substrates for correcting unexpected trajectory errors and learning from them. Brain. 2011;134 (Pt 12):3647–61.

50. Haaland KY, Schaefer SY, Knight RT, et al. Ipsilesional trajectory control is related to contralesional arm paralysis after left hemisphere damage. Exp Brain Res. 2009;196(2):195–204.

51. Schaefer SY, Haaland KY, Sainburg RL. Hemispheric specialization and functional impact of ipsilesional deficits in movement coordination and accuracy. Neuropsychologia. 2009;47(13):2953–66.

52. Chestnut C, Haaland KY. Functional significance of ipsilesional motor deficits after unilateral stroke. Arch Phys Med Rehabil. 2008;89(1):62–8.

53. Haaland KY. Left hemisphere dominance for movement. Clin Neuropsychol. 2006;20(4):609–22.

54. Wetter S, Poole JL, Haaland KY. Functional implications of ipsilesional motor deficits after unilateral stroke. Arch Phys Med Rehabil. 2005;86(4):776–81.

55. Haaland KY, Prestopnik JL, Knight RT, Lee RR. Hemispheric asymmetries for kinematic and positional aspects of reaching. Brain. 2004;127(Pt 5):1145–58.

56. Prestopnik J, Haaland K, Knight R, Lee R. Hemispheric dominance for open and closed loop movements. Soc Neuroscience Abstr. 2003;**30**(1).

57. Winstein CJ, Pohl PS. Effects of unilateral brain damage on the control of goal-directed hand movements. Exp Brain Res. 1995;105(1):163–74.

58. Wyke M. The effect of brain lesions in the performance of an arm-hand task. Neuropsychologia. 1967;6:125–34.

59. Kawashima R, Yamada K, Kinomura S, et al. Regional cerebral blood flow changes of cortical motor areas and prefrontal areas in humans related to ipsilateral and contralateral hand movement. Brain Res. 1993;623(1):33–40.

60. Kim SG, Ashe J, Hendrich K, et al. Functional magnetic resonance imaging of motor cortex:

hemispheric asymmetry and handedness. Science. 1993;261(5121):615–7.

61. Tanji J, Okano K, Sato KC. Neuronal activity in cortical motor areas related to ipsilateral, contralateral, and bilateral digit movements of the monkey. J Neurophysiol. 1988;60(1):325–43.

62. Schaefer SY, Haaland KY, Sainburg RL. Dissociation of initial trajectory and final position errors during visuomotor adaptation following unilateral stroke. Brain Res. 2009;17(1298):78–91.

63. Fugl-Meyer AR, Jaasko L, Leyman I, Olsson S, Steglind S. The post-stroke hemiplegic patient. 1. A method for evaluation of physical performance. Scand J Rehabil Med. 1975;7(1):13–31.

64. Haaland KY, Mutha PK, Rinehart JK, Daniels M, Cushnyr B, Adair JC. Relationship between arm usage and instrumental activities of daily living after unilateral stroke. Arch Phys Med Rehabil. 2012;93(11):1957–62.

65. Pandian S, Arya KN, Kumar D. Effect of motor training involving the less-affected side (MTLA) in post-stroke subjects: a pilot randomized controlled trial. Topics in stroke rehabilitation. 2015 Apr 28:1074935714Z0000000022.

66. Wolf SL, Blanton S, Baer H, Breshears J, Butler AJ. Repetitive task practice: a critical review of constraint-induced movement therapy in stroke. Neurologist. 2002;8(6):325–38.

67. Wolf SL, Lecraw DE, Barton LA, Jann BB. Forced use of hemiplegic upper extremities to reverse the effect of learned nonuse among chronic stroke and head-injured patients. Exp Neurol. 1989;104 (2):125–32.

68. Wolf SL, Thompson PA, Morris DM, et al. The EXCITE trial: attributes of the Wolf Motor Function Test in patients with subacute stroke. Neurorehabil Neural Repair. 2005;19(3):194–205.

69. Wolf SL, Winstein CJ, Miller JP, et al. Effect of constraint-induced movement therapy on upper extremity function 3 to 9 months after stroke: the EXCITE randomized clinical trial. JAMA. 2006;296(17):2095–104.

70. Nozaki D, Kurtzer I, Scott S. Limited transfer of learning between unimanual and bimanual skills within the same limb. Nat Neurosci. 2006;2006 (9):3.

71. Wang J, Sainburg RL. Generalization of visuomotor learning between bilateral and unilateral conditions. J Neurophysiol. 2009;102(5):2790–9.

72. Lum PS, Burgar CG, Van der Loos M, Shor PC, Majmundar M, Yap R. MIME robotic device for upper-limb neurorehabilitation in subacute stroke subjects: A follow-up study. J Rehabil Res Dev. 2006 Aug-Sep;43(5):631–42.

73. Jones RD, Donaldson IM, Parkin PJ. Impairment and recovery of ipsilateral sensory-motor function following unilateral cerebral infarction. Brain. 1989;112(Pt 1):113–32.

74. Wang J, Sainburg RL. Interlimb transfer of visuomotor rotations depends on handedness. Exp Brain Res. 2006;175(2):223–30.

75. Wang J, Sainburg RL. The symmetry of interlimb transfer depends on workspace locations. Exp Brain Res. 2006;170(4):464–71.

76. Wang J, Sainburg RL. Limitations in interlimb transfer of visuomotor rotations. Exp Brain Res. 2004;155(1):1–8.

77. Wang J, Sainburg RL. Interlimb transfer of novel inertial dynamics is asymmetrical. J Neurophysiol. 2004;92(1):349–60.

78. Oujamaa L, Relave I, Froger J, Mottet D, Pelissier JY. Rehabilitation of arm function after stroke. Literature review. Ann Phys Rehab Med. 2009 Apr;52(3):269–93.

79. Selles RW, Michielsen ME, Bussmann JB, et al. Effects of a mirror-induced visual illusion on a reaching task in stroke patients: implications for mirror therapy training. Neurorehabil Neural Repair. 2014;28(7):652–9.

80. Stevens JA, Stoykov ME. Simulation of bilateral movement training through mirror reflection: a case report demonstrating an occupational therapy technique for hemiparesis. Topics in stroke rehabilitation. 2004 Winter;11(1):59–66.

81. Shadmehr R, Mussa-Ivaldi FA. Adaptive representation of dynamics during learning of a motor task. J Neurosci. 1994;14(5 Pt 2):3208–24.

82. Krakauer JW, Mazzoni P, Ghazizadeh A, Ravindran R, Shadmehr R. Generalization of motor learning depends on the history of prior action. PLoS Biol. 2006;4(10): e316.

83. Mazzoni P, Krakauer JW. An implicit plan overrides an explicit strategy during visuomotor adaptation. J Neurosci. 2006;26(14):3642–5.

84. Sainburg RL, Ghez C, Kalakanis D. Intersegmental dynamics are controlled by sequential anticipatory, error correction, and postural mechanisms. J Neurophysiol. 1999;81(3):1045–56.

85. Keisler A, Shadmehr R. A shared resource between declarative memory and motor memory. J Neurosci. 2010;30(44):14817–23.

86. Huang VS, Haith A, Mazzoni P, Krakauer JW. Rethinking motor learning and savings in adaptation paradigms: model-free memory for successful actions combines with internal models. Neuron. 2011;70(4):787–801.

87. Krakauer JW, Pine ZM, Ghilardi MF, Ghez C. Learning of visuomotor transformations for vectorial planning of reaching trajectories. J Neurosci. 2000;20(23):8916–24.

88. Krakauer JW, Ghilardi MF, Ghez C. Independent learning of internal models for kinematic and dynamic control of reaching. Nat Neurosci. 1999;2(11):1026–31.

89. Wolpert DH, Wolf DR. Estimating functions of probability distributions from a finite set of samples.

Phys Rev E: Stat Phys Plasmas Fluids. 1995;52 (6):6841–54.

90. Criscimagna-Hemminger SE, Donchin O, Gazzaniga MS, Shadmehr R. Learned dynamics of reaching movements generalize from dominant to nondominant arm. J Neurophysiol. 2003;89(1):168–76.

91. Malfait N, Shiller DM, Ostry DJ. Transfer of motor learning across arm configurations. J Neurosci. 2002;22(22):9656–60.

92. Conditt MA, Gandolfo F, Mussa-Ivaldi FA. The motor system does not learn the dynamics of the arm by rote memorization of past experience. J Neurophysiol. 1997;78(1):554–60.

93. Bastian AJ. Learning to predict the future: the cerebellum adapts feedforward movement control. Curr Opin Neurobiol. 2006;16(6):645–9.

94. Bastian AJ, Thach WT. Cerebellar outflow lesions: a comparison of movement deficits resulting from lesions at the levels of the cerebellum and thalamus. Ann Neurol. 1995;38(6):881–92.

95. Thach WT, Bastian AJ. Role of the cerebellum in the control and adaptation of gait in health and disease. Prog Brain Res. 2004;143:353–66.

96. Tseng YW, Diedrichsen J, Krakauer JW, Shadmehr R, Bastian AJ. Sensory prediction errors drive cerebellum-dependent adaptation of reaching. J Neurophysiol. 2007;98(1):54–62.

97. Tanaka H, Sejnowski TJ, Krakauer JW. Adaptation to visuomotor rotation through interaction between posterior parietal and motor cortical areas. J Neurophysiol. 2009;102(5):2921–32.

98. Krakauer JW, Mazzoni P. Human sensorimotor learning: adaptation, skill, and beyond. Curr Opin Neurobiol. 2011;21(4):636–44.

99. Reis J, Schambra HM, Cohen LG, et al. Noninvasive cortical stimulation enhances motor skill acquisition over multiple days through an effect on consolidation. Proc Natl Acad Sci U S A. 2009;106(5):1590–5.

100. Izawa J, Shadmehr R. Learning from sensory and reward prediction errors during motor adaptation. PLoS Comput Biol. 2011;7(3): e1002012.

101. Taylor JA, Klemfuss NM, Ivry RB. An explicit strategy prevails when the cerebellum fails to compute movement errors. Cerebellum. 2010;9 (4):580–6.

102. Taylor JA, Krakauer JW, Ivry RB. Explicit and implicit contributions to learning in a sensorimotor adaptation task. J Neurosci. 2014;34(8):3023–32.

Recovery of Sensorimotor Functions After Stroke and SCI: Neurophysiological Basis of Rehabilitation Technology

Volker Dietz, Laura Marchal-Crespo, and David Reinkensmeyer

Abstract

After a stroke or spinal cord injury (SCI), there exists an inherent individual capacity for recovery of function that depends on factors such as location and severity of central nervous system (CNS) damage. This capacity can be determined early after the incident by clinical, electrophysiological, and imaging examinations. These measures can also be used as prognostic factors and, consequently, for an early selection of appropriate rehabilitation procedures. Recovery of function after a stroke mainly depends on the tract damaged and the amount of damage, e.g., recovery of hand/finger function is particularly poor after extensive lesioning of the corticospinal tract. In cervical SCI, the combination of peripheral and central nervous system damage limits recovery. As the recovery of function usually remains incomplete, an integral part of rehabilitation should be directed to compensate for the remaining motor deficit by customized assistive devices that promote independence in daily life activities. The capacity for the recovery of function can be exploited by a repetitive execution of functional movements, physically supported as far as required. This approach encourages participation by the patient and promotes appropriate proprioceptive input from limb muscles, tendons, skin, and joints under physiological movement conditions. The consequence of this knowledge is that robotic assistance has to be adapted to the actual condition and requirements of the individual patient. Furthermore, intensive training (i.e., a high number of movement repetitions and long training duration) can lead to an additional gain in function compared to low-dose conventional training. However, this gain is small compared to the spontaneous recovery of function and is often transient, due to the fact that patients will not regularly use these functions in daily life, thereby maintaining them. Finally, other promising adjuvant approaches could contribute to improving motor function in the future, such as epidural or deep brain stimulation as well as CNS repair. However, they are still in an early clinical or in a translational stage.

V. Dietz (✉)
Spinal Cord Injury Center, University Hospital Balgrist, Forchstrasse 340, 8008 Zürich, CH, Switzerland
e-mail: volker.dietz@balgrist.ch

L. Marchal-Crespo
Mechanical, Maritime and Materials Engineering, Department of Cognitive Robotics, TU Delft/Faculty 3mE, Mekelweg 2, 2628 CD Delft, The Netherlands
e-mail: L.MarchalCrespo@tudelft.nl

D. Reinkensmeyer
Department of Mechanical and Aerospace Engineering, Department of Anatomy and Neurobiology, University of California at Irvine, 4200 Engineering Gateway, Irvine, CA 92697-3975, USA
e-mail: dreinken@ucl.edu

© The Author(s), under exclusive license to Springer Nature Switzerland AG 2022
D. J. Reinkensmeyer et al. (eds.), *Neurorehabilitation Technology*,
https://doi.org/10.1007/978-3-031-08995-4_3

Keywords

Neurophysiology · Stroke · Spinal cord injury (SCI) · Recovery of function · Rehabilitation technology

3.1 Introduction

The aim of neurorehabilitation is to improve function after damage to the nervous system. This chapter focuses on the neurophysiological aspects that determine the recovery of function after stroke and SCI. Our premise is that insights into these neurophysiological aspects should influence the design of rehabilitation technologies, such as robotic devices, that are to be applied in neurorehabilitation. We specifically seek to address the following questions: What are the limits of the recovery of function? And, taking into account these limits: What are the neurophysiological aspects that can be leveraged to optimize the effectiveness of neurorehabilitation approaches for restoring upper and lower limb function in stroke and SCI? Based on the answer to these questions, we propose the following principles for promoting recovery: (1) Where the potential for biological recovery is substantially limited, relevant aspects of the residual neurophysiology should be leveraged to promote compensation; (2) Where the potential for biological recovery is high, limb muscle activation and proprioceptive input should be promoted as much as possible in a physiological manner during training to promote restoration of function; and (3) Sufficient dosage of physiologically appropriate training should be delivered to overcome an apparently nonlinear dose–response relationship. We discuss the implications of these principles for the design of rehabilitation technology.

1. Where the potential for biological recovery is limited, residual neural circuits should be leveraged.

There are fundamental constraints to recovery after stroke and SCI [1]. Much of the recovery of function in stroke [2] and SCI [3] during the first three months is due to resolving neurapraxia that occurs in parallel to the smaller effects of the rehabilitative treatments that exploit neuroplasticity [4]. For example, after stroke, most patients reach a seeming plateau after recovering approximately 70–80% of the initial proximal arm muscle function impairment [5–7].

The severity and localization of the CNS damage determine the specific range of an individual patient's achievable function, independently of the rehabilitation training [8, 9]. For example, after brain damage that includes substantial lesions of pyramidal tract connections to hands and fingers, the motor deficit can only partially be compensated by the activation of other non-damaged tracts/brain areas. Such compensation typically only restores gross hand flexion but not individual finger dexterity [5, 6, 10]. Consequently, substantial damage to the pyramidal tract means that only modest recovery of hand and finger function can usually be expected [7, 11] (Fig. 3.1a). In contrast, a more favorable recovery of upper extremity function can be expected following damage to other brain areas [5, 6, 10].

Similarly, after spinal cord damage, the improvement of upper limb function depends on the level and extent of the lesion [3]. In cervical cord injuries, a combined damage to the central (spinal tracts) and peripheral nerval structures (motoneurons and roots to arm, hand, and finger muscles) occurs. This results in an arm/hand/finger paresis associated with a mixture of spastic and flaccid muscle tone [12]. The peripheral part of nervous system damage can account for up to 50% of paresis [13] which has little potential to recover. After a sensorimotor complete cervical SCI, functionally meaningful recovery of upper extremity function cannot occur [14] (Fig. 3.1a).

Demographic and sociological factors may limit recovery as well. For example, while the age of patients has little influence on the recovery of the neurological deficit post-stroke [17] and SCI [18], a young person with a SCI can better translate the recovery of motor system deficits into movement capabilities that support daily life activities compared to elderly subjects [18].

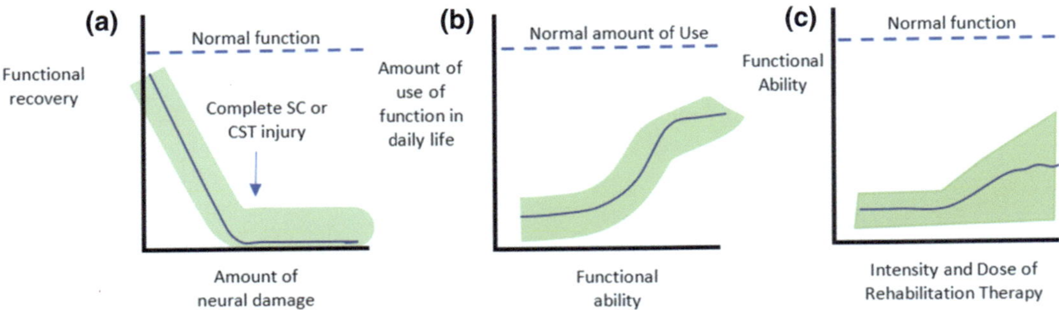

Fig. 3.1 Nonlinear relationships in neurorehabilitation recovery. **a** The amount of functional recovery that can be expected declines with more severe neuroanatomical damage, with a complete injury to the spinal cord (SC) or corticospinal tract (CST) profoundly limiting functional recovery. The green band denotes the variability in this relationship due to a range of factors, including the location of the damage, patient demographics, and intensity of rehabilitation. **b** Another nonlinearity describes the *usefulness to the patient* of any recovered functional movement ability. A patient will not regularly use a function throughout the day (such as hand grasp or walking) if the functional ability does not exceed a threshold. This graph is based on figures in [15, 16]. The green band again denotes a relatively high variability across patients. **c** The relationship of functional ability to the intensity and dose of rehabilitation may also be nonlinear, requiring a relatively high threshold of therapy to be achieved before a dose–response relationship can be identified, compared to what is usually delivered in clinical and research settings. Again, the green band denotes the effect of rehabilitation depends on other factors, including lesion location and size, patient demographics, and timing of therapy

Finally, it is important to note that moderate recovery of movement ability may not always be useful to patients. For example, a recent study used a magnetic, wearable finger movement sensor to quantify the movement of the fingers and wrist in daily life after a stroke [15]. It was observed that people with a stroke must recover a substantial amount of hand function ($\sim$50% of normal hand function as measured on the Box and Blocks Test), in order to begin to actually use their paretic hand in the home environment. This is consistent with the "Threshold Hypothesis" proposed by Schweighofer et al. [19]. Thus, even substantial recovery of hand function can have limited pragmatic usefulness to a patient in daily life (Fig. 3.2b). A similar phenomenon has been observed for walking after stroke, where individuals with a score up to 50% normal on the Berg Balance Scale had low daily walking amounts, as measured with a wearable sensor [16]. Furthermore, after a SCI, a threshold of force recovery of leg muscles (i.e., lower extremity motor score) is required for the performance of stepping movements [20]. Nevertheless, the "threshold" of usefulness might be dependent on the potential use of compensatory strategies and assistive technologies, e.g., during locomotion in SCI subjects.

For functions where severe anatomical damage limits recovery, the guiding principle is to leverage the relevant changes in the sensorimotor nervous system. For example, spasticity can contribute to the compensation of movement deficits [21–23], thereby contributing to the restoration of function. After a stroke or an incomplete SCI, a loss of supraspinal drive leads to limb paresis. Concerning lower limb function, the inability to perform stepping movements due to muscle paresis leads to reduced mobility. However, with the development of spastic muscle tone, the stiff leg can support the body during stance, i.e., the deficit becomes partially compensated. Functional movements, such as stepping, can be executed on a lower level of muscle tone regulation, i.e., the contribution of spastic muscle tone to support the body during movement performance [24]. Therefore, most post-stroke subjects can regain walking function by using the spastic-paretic leg more or less stick-like supporting the body in the stance phase and

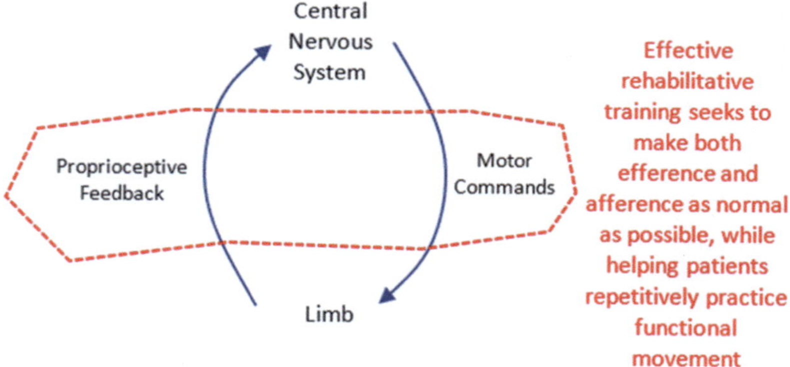

Fig. 3.2 Conceptual approach to optimizing return of function through use-dependent plasticity mechanisms. The improvement of function within the residual capacity depends on the appropriate activation of motoneuron pools and proprioceptive sensors of limb muscles under physiological movement conditions

circumducting the leg during swing (due to reduced knee flexion). However, the normal push-off at the end of the stance phase is lost. Over time, little change in biomechanical and muscle activation characteristics of the spastic-paretic leg takes place [14, 25]. In this scenario, the limited improvement of walking ability achieved over the course of rehabilitation after a stroke is due to adaptational changes (compensation) rather than due to a restoration of the "normal" stepping function.

Concerning upper limb function, early after stroke, flaccid arm muscle paresis prevails, i.e., the limbs are weak and do not resist passive displacement. In the weeks following stroke, spastic muscle tone usually becomes more pronounced in the wrist and finger flexors than in the extensor muscles, as the antigravity muscles have more muscle mass [24, 26]. Thus, with the development of some spastic muscle tone, rudimentary grips can be performed and the training of residual muscle function can be initiated [14]. In this stage, the focus of training should be directed to enable the execution of simple reach and grasp functional tasks, as well as self-care movements. Patients suffering a cervical SCI (e.g., C6/7) or severe stroke can make use of spastic muscle tone to perform simple grasp movement (the so-called tenodesis grasp). Furthermore, spastic proximal arm muscles can provide some passive gravity support to carry an object from one to another spot.

The pragmatic goal of rehabilitation therapists is not primarily to re-establish "normal" movement performance, but usually focuses on enabling compensatory movement control, which typically involves "simpler", less well-organized movements that help maximize independence during activities of daily living (ADL) for the individual patient [27]. As we have described, this may be aided by changes in muscle mechanical properties related to spastic muscle tone that develops during the spontaneous recovery of function [23]. Thus, a therapist may choose to focus on learning to use abnormal motor control and biomechanics for the performance of activities of daily living or to bypass physiological function with assistive technology (e.g., a wheelchair), rather than on restoring normal muscle control. However, it is important to note that, while the compensatory approach may enable the patient to perform needed upper and lower limb functions to regain independence, it may also establish a "local minimum" of recovery that is less than the theoretical maximum possible, specifically in cases where neuroanatomical resources for neuroplasticity and motor learning are available but not utilized.

2. Where the potential for recovery is high, limb muscle activation and proprioceptive input should be promoted in a physiological way during training.

When sufficient neuroanatomical resources remain, relearning of physiological movements can be optimized by encouraging limb muscle activation and by providing appropriate proprioceptive input to the spinal and supraspinal neural centers with the goal to activate preserved neural circuits in as normal a way as possible (Fig. 3.2). Within this framework, the level of physiological limb muscle activation and normal proprioceptive information serves as a marker for predicting the achievement of training effects.

We begin with the lower extremity. In the early nineties, functional locomotor training with body-weight unloading of para- and tetraparetic SCI subjects was introduced. This was based on the observation that key aspects of locomotor function in cat SCI models recover quite well during treadmill training with body-weight unloading (BWSSTT) [28].

In severely paralyzed patients due to an SCI, automatic stepping movements can be induced, associated with a physiological leg muscle activation (i.e., close-to-normal timing of electromyography [EMG] patterns with smaller than normal amplitude), when patients stand on a moving treadmill with the body unloaded up to 80% [29, 30]. This leg muscle activation during the stance phase of gait is triggered by load receptor input from the ground reaction forces [31]. Furthermore, as observed in studies in cats [32], sufficient hip extension at the end of the stance phase is essential for the initiation of the swing phase and contribution to the generated EMG signal pattern during stepping [30]. The consequence is a physiological limb muscle activation induced by inputs from load and hip-joint-related proprioceptive receptors that represents the prerequisite for the improvement of locomotor function in rodents [33] and patients with a stroke or SCI (for review see [14]). With a gain of strength in the proximal leg muscles, body-weight unloading can then be reduced and self-induced stepping movements become possible. This is associated with an increase in the strength of leg muscle activation. Thus, during the course of training, body un-/reloading can be adapted to the subject's actual degree of paresis.

In completely paralyzed patients suffering an SCI who do not undergo functional locomotor training, such as body-weight-supported treadmill training (BWSTT), spinal neural circuits underlying stepping movements undergo degenerative changes associated with a loss of neural activity, i.e., neurons become silent even when appropriate proprioceptive input is provided. In the long term, the lack of locomotor training results in a neuronal dysfunction below the level of the lesion in both rodents [34] and patients with SCI [35].

In incompletely paralyzed SCI patients, BWSTT has been shown to result in a similar outcome of stepping function compared to a conventional rehabilitation provided by therapists approach. Nevertheless, BWSTT reduces the physical burden for the therapist [36]. Gait speed during locomotor training represents another factor that influences the locomotor ability outcome. In ambulatory stroke patients, a successive increase of treadmill speed through a physiological range up to 20% increase of initial speed during a 4-week training period results in a better walking ability than conventional gait training [37].

Most of the recovery of function occurs during the first three to four months after CNS damage. Nevertheless, some gain in gait velocity, endurance, and performance can be achieved by automated locomotor training with a driven gait orthosis in chronic patients with an incomplete SCI and stroke [38]. However, it should be noted that passively induced leg movements by rigid robotic assistance [39] during locomotor training results in reduced therapeutic efficacy [40]. Further improvement of locomotor function after damage to the CNS is associated with minor changes in the leg muscle activity pattern and relies on a better coordination between the legs and an adapted spastic muscle tone, as shown after stroke [25] and SCI [22].

Turning to the upper extremity, evidence for the importance of generating physiological limb muscle activation during training is less direct but the evidence is growing. For the lower extremity, as described above, physiological limb activation can be seen as generating a set of necessary sensory pre-conditions for triggering and/or facilitating cyclic locomotor activity, which is then reinforced through repetitive practice. For the upper extremities, physiological limb movement activation appears to facilitate motor learning processes that contribute to the restoration of function.

Simply moving the passive upper limb is not sufficient to stimulate these learning processes. For example, no motor recovery was observed in chronic stroke patients when the paretic wrist was moved passively by a robotic therapy device for several hours per week over several weeks (except for small reductions in muscle tone) [41]. However, when the subjects were required to initiate wrist movements—measured with EMG —in order to receive movement assistance, recovery occurred. Thus, physiological self-activation of the upper limb muscles was a pre-condition for producing a training effect. A key neurophysiological mechanism that appears to be at play during passive training is "slacking", which is the algorithmic tendency of the motor system to reduce its effort and output when the kinematic error is small [42, 43].

A widely accepted approach to rehabilitate hand function after stroke is constraint-induced movement therapy (CIMT). This approach is based on the idea of enhancing recovery of function by reducing interhemispheric inhibition of the stroke hemisphere [44]. By immobilizing the non-affected hand, the patient is forced to use the paretic hand/arm for the performance of ADLs [45]. In an analysis of the Excite CIMT clinical trial, proprioceptive integrity was the strongest predictor of treatment effect from CIMT [46]. On average, patients suffering from impaired proprioception had a 20% probability of achieving a clinically meaningful outcome compared to those without clinically detected loss of proprioception.

A study of robotic-assisted finger training found that patients with impaired finger proprioception had a smaller functional benefit from robotic finger training compared to those with intact proprioception [7]. Proprioceptive integrity was quantified robotically at baseline by asking subjects to indicate when their index and middle fingers crossed each other as they were driven by a robotic exoskeleton. Proprioceptive error (i.e., the magnitude of the error in estimated finger-crossing angle) predicted 40% of the variance of the functional training effect. Further, neural injury to and abnormal activation of the somatosensory system were the strongest predictors of functional benefit from robotic hand therapy, chosen from an array of over 40 measures that included both motor and sensory variables related to anatomy, neurophysiology, and behavioral outcomes [47].

Other research increasingly implicates the importance of proprioception as a biomarker for predicting rehabilitation response of the upper extremity. For example, one study found that, for patients with chronic stroke, deficits in proprioception predicted motor learning associated with finger tracking training [48]. Clinical assessments of proprioception after stroke have shown value for predicting motor recovery [49, 50]. Lack of somatosensory evoked potentials early following stroke also predicts poor motor recovery [51, 52]. Theoretical models of recovery suggest that the reason that proprioception plays such a key role in predicting recovery is that accurate sensing of limb muscle force and movement is needed as a "teaching signal" to guide practice-driven changes in cortical activation patterns [53].

Finally, a common physiological mode of use of the upper extremities is bimanual control. Correspondingly, bilateral hand training for reaching and grasping tasks in stroke patients has been suggested to be more effective in improving unilateral execution of these tasks with the affected arm than unilateral training alone [54]. This might be a result of stronger recruitment of the contralesional hemisphere through bilateral compared to unilateral training [55]. Such bilateral hand movements, e.g., opening a bottle, are based on a task-specific control by a "neural coupling" of the hands. This is achieved either by a coupling of the hemispheres, i.e., both ipsi- and contralateral hemispheres become involved in bilateral hand movement tasks. Alternatively and more likely, each of the two hands becomes controlled by the cortico-reticulo-spinal tract during the automatically performed cooperative hand movements [56] (cf. Chap. 6). Consequently, in post-stroke patients during the training of cooperative hand tasks, the unaffected hemisphere supports movements of the paretic hand and arm [57].

3. Dosage of training required to overcome a nonlinear dose–response relationship.

Several studies indicate that more intensive training—i.e., a high number of movement repetitions per hour (intensity) and long training duration (dose)—results in an additional gain in the upper and lower limbs function. This effect was reported for post-stroke subjects [58–65], as well as for subacute [66] and chronic [38] SCI subjects. By applying a very high dose of 300 h, clinically meaningful gains in arm/hand function were described in a chronic stroke population [67]. A positive effect of training intensity on the outcome of ambulation in stroke subjects [62] was also recently confirmed for subjects with SCI [66]. The intensity and dose of physical therapy were thought to have a positive effect on outcome in both animal [33] and human [68–70] studies. These observations suggest that the intensity provided in the standard of care, which is notably low (tens of movements per practice session over a limited number of sessions [17]), might not be sufficiently large enough. Timing of delivery of training may also be an important factor in establishing its effectiveness, with higher potential in the subacute window, as was recently shown for upper extremity rehabilitation after stroke [71].

Other studies have not found an effect of movement dose on functional recovery. Several studies of upper limb function of chronic post-stroke subjects found no evidence for a dose–response effect of training intensity on functional recovery [72, 17]. For lower limb function, the improvement of outcome achieved by a more intensive training was small and/or transient (cf.

Fig. 3 of [73]) in relation to the gain in function achieved by a standard training in post-stroke [73] and SCI [66] subjects. In a large group (200 adults) of moderately to severely impaired subacute post-stroke subjects, a more intensive locomotor training was related to the improvement of stepping function. However, the BWSTT was not superior to relaxation sessions (of the same duration and in addition to standard therapy) in respect of walking speed and activities of daily living [74]. Finally, in incomplete SCI subjects, doubling the daily locomotor training time had only small effects on walking ability [66]. Therefore, questions concerning the additional gain of function achieved in relation to the spontaneous recovery of function, and whether this gain represents a lasting effect, remain open [9].

A reconciliation of these conflicting results may be possible if the dose–response relationship for movement training is nonlinear in nature (Fig. 3.1c). This has been suggested by a meta-analysis of experiments with a rodent model of upper extremity rehabilitation after ischemic stroke, which found that the dose–response relationship takes a curvilinear form [75]. This means that there is a range of levels of intensity for which changes in intensity have no effect on recovery. It may be that negative clinical trials have been in this range. Beyond this range, one might expect an increasing benefit from intensifying the training, which could account for the small cluster of successful high-dose studies. However, factors such as the neural tracts affected, the amount of damage, the level of SCI (e.g., cervical), the timing of rehabilitation, and the individual capacity for recovery are suggested to essentially determine the extent of functional recovery [9]. The appropriate range of training intensity and dose is expected to relate to the number of movements usually performed during daily life activities. Further, the neurophysiological mechanisms of this putative nonlinearity remain unclear and are an important topic for future study.

3.2 Implications for Rehabilitation Technology Design

Where the potential for recovery is limited due to the nature and extent of the anatomical damage, we have suggested that relevant aspects of the neurophysiology, such as abnormal muscle tone, should be leveraged. Several widely used non-actuated assistive devices are already based on this principle. Examples include wrist-driven tenodesis orthoses that support grip and ankle-foot orthoses that effectively further increase the tone of the ankle muscles to support walking. Several powered exoskeletons have also been developed to support overground walking (see review [76]) and hand/arm function [77, 78]). Determining how to best work with the relevant aspects of neurophysiology to maximize function is an important consideration for guiding future robotic design. We anticipate that compensatory movement strategies, strengthened with technological supports, will, for now, remain as important tools to mitigate motor deficits and promote independence [79].

When the potential for recovery is substantial due to sufficient anatomical resources, we have suggested that proprioceptive input and limb self-generated muscle activation should be promoted as much as possible in a physiological way during training to enhance the restoration of function. Special attention should be paid during the design of robotic therapy to prevent that the robot "over assists" the patient, as too much assistance might cause the patient to reduce the physiological contribution of efferents to training. One strategy is to provide limbs gravity support to allow patients to perform functional movements by their own (limited) effort [80]. Another strategy is to provide robotic assistance only following sensed self-initiated movement by the patient [81]. Another compatible approach is to keep the movement support provided by the therapist or device to a minimum in order to make the training optimally challenging and maximize the patient's contribution throughout the practiced movements (for reviews see [14, 81, 82]). An array of assist-as-needed algorithms

have already been developed and can be used as resources in robotic therapy device design [81]. Conversely, error-augmentation algorithms that seek to amplify the movement error, promote movement variability (and thus, task exploration), and/or maximize patients' effort also exhibit potential [82–84].

Implementing such a variety of robotic training algorithms requires clever engineering design so that the device can achieve the wide range of impedances needed for these algorithms, ranging from complete mechanical transparency to full assistance. This has not yet been fully achieved for wearable and untethered robotics and remains a holy grail.

Physical interfaces and controllers should also be designed so as to minimize the alteration of sensory flow during training, taking into account the tactile stimulation the robot provides via its physical interface with the patient's limbs. Enhancing the congruency of sensory information—i.e., tactile, proprioception, vision, and auditory information—might not only enhance performance during training but also promote the transfer of the acquired skills during training to activities of daily living [85]. New technological developments, such as head-mounted displays and tactile actuators, could be incorporated into current robotic solutions to allow for a more naturalistic visualization of the patients' movements within the virtual environments [85, 86] and more realistic interactions with virtual tangible objects [87]. Finally, in the case where proprioception is impaired, robots can potentially play a key role in retraining proprioception. For example, with robotics, proprioceptive training can be gamified by using the robot to "display" game elements proprioceptively by driving the patient's limb [88]. Providing meaningful and easy-to-use tools to therapists for making an impact on proprioception could open novel avenues for treatment, given that accurate proprioception seems to serve as a gateway for motor learning as described above.

We described above the possibility that for some conditions (i.e., hand function) sufficient dosage of training must be delivered to overcome an apparently nonlinear dose–response relationship. The introduction of rehabilitation robots was based on the widely accepted assumption that the recovery of function depends on the intensity of training. It now seems that this assumption may be correct only for a specific range of dose of training, i.e., there must be a sufficient dose, and the threshold dose is relatively high compared to standard clinical practice. Thus, studies of robotic therapy may have suffered from providing too low training dose, an ironic situation given that robotic therapy devices were specifically developed to allow longer training times and more repetitions. In many cases, then, the failure of a robot therapy device to prove useful may not be with the robot itself, but in the way it was applied —that is, it simply wasn't applied enough! Research is required on the institutional, structural, and pragmatic factors that limit the rehabilitation therapy dose that is typically achieved with or without rehabilitation technology.

Nevertheless, robot-assisted therapy can provide a number of other advantages besides increasing therapy dose, including a standardized training environment, adaptable support to the patient's specific needs, automatic monitoring of functional measures, and reduction of the physical burden on therapists. Rehabilitation robots are thus an ideal means to complement conventional therapy in rehabilitation centers if they are designed and applied on the basis of neurophysiological insights underlying the recovery of sensorimotor functions, as we have outlined.

3.3 Conclusion

It is concluded that there is an inherent and relatively fixed individual capacity for recovery of function after a stroke or SCI that depends on factors such as location and severity of CNS damage. This capacity can be determined early after CNS damage by clinical, electrophysiological [14], and imaging [5] examinations. These measures can also be used as prognostic factors and, consequently, for the selection of appropriate rehabilitation procedures early after CNS damage. Recovery of function after a stroke or SCI usually remains incomplete. Therefore, an integral part of rehabilitation should be directed to compensate for the remaining motor deficit by refined assistive devices that promote independence by working with the relevant aspects of residual physiology where possible.

The individual capacity for recovery of function, where it exists in a sufficient amount, can be exploited by a repetitive execution of functional movements, supported as far as required. The improvement of function within this capacity depends on the appropriate activation of descending systems as well as motoneuron pools by the input from proprioceptive sensors of limb muscles, tendons, skin, and joints under physiological movement conditions. The consequence of this knowledge is that robotic assistance has to be adapted to the actual condition and requirements of the individual patient, in such a way as to promote normal efferent and afferent physiological activation. Finally, a more intensive training can lead to an additional gain in function in relation to a standard training, if this training dose exceeds a specific threshold, although this gain is sometimes transient.

Finally, adjuvant approaches might help restore motor function in the future, such as epidural [89] or deep brain [90] stimulation as well as CNS repair, but they are still in an early clinical or in a translational stage. Their success will also likely depend on the generation of physiological patterns of limb muscle activation.

Acknowledgements DJR supported by NIH grant R01HD062744.

Disclosure David Reinkensmeyer has a financial interest in Hocoma AG and Flint Rehabilitation devices, makers of rehabilitation equipment. The terms of these arrangements have been reviewed and approved by the University of California, Irvine, in accordance with its conflict of interest policies.

References

1. Jorgensen HS, Reith J, Nakayama H, Kammersgaard LP, Raaschou HO, Olsen TS. What determines good recovery in patients with the most severe strokes? Cph Stroke Study. 1999;30(10):2008–12.
2. Zarahn E, Alon L, Ryan SL, Lazar RM, Vry MS, Weiller C, et al. Prediction of motor recovery using initial impairment and fMRI 48 h poststroke. Cereb Cortex. 2011;21(12):2712–21.
3. Curt A, Van Hedel HJ, Klaus D, Dietz V. Recovery from a spinal cord injury: significance of compensation, neural plasticity, and repair. J Neurotrauma. 2008;25(6):677–85.
4. Zeiler SR, Krakauer JW. The interaction between training and plasticity in the poststroke brain. Curr Opin Neurol. 2013;26(6):609–16.
5. Prabhakaran S, Zarahn E, Riley C, Speizer A, Chong JY, Lazar RM, et al. Inter-individual variability in the capacity for motor recovery after ischemic stroke. Neurorehabil Neural Repair. 2008;22:64–71.
6. Winters C, van Wegen EE, Daffertshofer A, Kwakkel G. Generalizability of the proportional recovery model for the upper extremity after an ischemic stroke. Neurorehabil Neural Repair. 2015;29(7):614–22.
7. Rowe JB, Chan V, Ingemanson ML, Cramer SC, Wolbrecht ET, Reinkensmeyer DJ. Robotic Assistance for training finger movement using a Hebbian model: a randomized controlled trial. Neurorehabil Neural Repair. 2017;31(8):769–80.
8. Byblow WD, Stinear CM, Barber PA, Petoe MA, Ackerley SJ. Proportional recovery after stroke depends on corticomotor integrity. Ann Neurol. 2015;78(6):848–59.
9. Dietz V. Restoration of motor function after CNS damage: is there a potential beyond spontaneous recovery? Brain Commun. 2021;3(3):fcab171.
10. Kwakkel G, Kollen B, Lindeman E. Understanding the pattern of functional recovery after stroke: facts and theories. Restor Neurol Neurosci. 2004;22(3–5):281–99.
11. Waddell KJ, Strube MJ, Bailey RR, Klaesner JW, Birkenmeier RL, Dromerick AW, et al. Does task-specific training improve upper limb performance in daily life poststroke? Neurorehabil Neural Repair. 2017;31(3):290–300.
12. Dietz V, Curt A. Neurological aspects of spinal-cord repair: promises and challenges. Lancet Neurol. 2006;5(8):688–94.
13. Curt A, Dietz V. Neurographic assessment of intramedullary motoneurone lesions in cervical spinal cord injury: consequences for hand function. Spinal Cord. 1996;34(6):326–32.
14. Dietz V, Fouad K. Restoration of sensorimotor functions after spinal cord injury. Brain. 2014;137 (Pt 3):654–67.
15. Schwerz de Lucena D, Rowe J, Chan V, Reinkensmeyer DJ. Magnetically counting hand movements: validation of a calibration-free algorithm and application to testing the threshold hypothesis of real-world hand use after stroke. Sensors (Basel). 2021;21 (4).
16. Handlery R, Regan EW, Stewart JC, Pellegrini C, Monroe C, Hainline G, et al. Predictors of daily steps at 1-Year poststroke: a secondary analysis of a randomized controlled trial. Stroke. 2021;52(5):1768–77.
17. Lang CE, Macdonald JR, Reisman DS, Boyd L, Jacobson Kimberley T, Schindler-Ivens SM, et al. Observation of amounts of movement practice provided during stroke rehabilitation. Arch Phys Med Rehabil. 2009;90(10):1692–8.
18. Wirz M, Dietz V. Recovery of sensorimotor function and activities of daily living after cervical spinal cord injury: the influence of age. J Neurotrauma. 2015;32 (3):194–9.
19. Schweighofer N, Han CE, Wolf SL, Arbib MA, Winstein CJ. A functional threshold for long-term use of hand and arm function can be determined: predictions from a computational model and supporting data from the extremity constraint-induced therapy evaluation (EXCITE) Trial. Phys Ther. 2009;89(12):1327–36.
20. Wirz M, van Hedel HJ, Rupp R, Curt A, Dietz V. Muscle force and gait performance: relationships after spinal cord injury. Arch Phys Med Rehabil. 2006;87(9):1218–22.
21. O'Dwyer NJ, Ada L, Neilson PD. Spasticity and muscle contracture following stroke. Brain. 1996;119 1737–49.
22. Dietz V, Sinkjaer T. Secondary changes after CNS damage: the significance of spastic muscle tone in rehabilitation. In: Dietz V, Ward N, editors. Oxford textbook of neurorehabilitation. Oxford: Oxford University Press; 2015. p. 76–88.
23. Katz RT, Rymer WZ. Spastic hypertonia: mechanisms and measurement. Arch Phys Med Rehabil. 1989;70(2):144–55.
24. Dietz V, Sinkjaer T. Spastic movement disorder: impaired reflex function and altered muscle mechanics. Lancet Neurol. 2007;6(8):725–33.
25. Den Otter AR, Geurts AC, Mulder T, Duysens J. Gait recovery is not associated with changes in the temporal patterning of muscle activity during treadmill walking in patients with post-stroke hemiparesis. Clin Neurophysiol. 2006;117(1):4–15.
26. Marciniak C, Rader L, Gagnon C. The use of botulinum toxin for spasticity after spinal cord injury. American journal of physical medicine and rehabilitation. Assoc Acad Physiatr. 2008;87(4):312–7; quiz 8–20, 29.
27. Latash M, Anson J. What are "normal movements" in atypic populations? Behav Brain Sci. 1996;19:55–106.
28. Barbeau H, Wainberg M, Finch L. Description and application of a system for locomotor rehabilitation. Med Biol Eng Comput. 1987;25(3):341–4.
29. Dietz V, Colombo G, Jensen L. Locomotor activity in spinal man. Lancet. 1994;344(8932):1260–3.

30. Dietz V, Colombo G, Jensen L, Baumgartner L. Locomotor capacity of spinal cord in paraplegic patients. Ann Neurol. 1995;37(5):574–82.

31. Dietz V, Muller R, Colombo G. Locomotor activity in spinal man: significance of afferent input from joint and load receptors. Brain. 2002;125(Pt 12):2626–34.

32. Pearson KG. Neural adaptation in the generation of rhythmic behavior. Annu Rev Physiol. 2000;62:723–53.

33. Edgerton VR, Tillakaratne NJ, Bigbee AJ, de Leon RD, Roy RR. Plasticity of the spinal neural circuitry after injury. Annu Rev Neurosci. 2004;27:145–67.

34. Beauparlant J, van den Brand R, Barraud Q, Friedli L, Musienko P, Dietz V, et al. Undirected compensatory plasticity contributes to neuronal dysfunction after severe spinal cord injury. Brain. 2013;136(Pt 11):3347–61.

35. Dietz V, Grillner S, Trepp A, Hubli M, Bolliger M. Changes in spinal reflex and locomotor activity after a complete spinal cord injury: a common mechanism? Brain. 2009;132(Pt 8):2196–205.

36. Dietz V. Body weight supported gait training: from laboratory to clinical setting. Brain Res Bull. 2009;78 (1):I–VI.

37. Pohl M, Mehrholz J, Ritschel C, Ruckriem S. Speed-dependent treadmill training in ambulatory hemiparetic stroke patients: a randomized controlled trial. Stroke. 2002;33(2):553–8.

38. Wirz M, Zemon DH, Rupp R, Scheel A, Colombo G, Dietz V, et al. Effectiveness of automated locomotor training in patients with chronic incomplete spinal cord injury: a multicenter trial. Arch Phys Med Rehabil. 2005;86(4):672–80.

39. Israel JF, Campbell DD, Kahn JH, Hornby TG. Metabolic costs and muscle activity patterns during robotic- and therapist-assisted treadmill walking in individuals with incomplete spinal cord injury. Phys Ther. 2006;86(11):1466–78.

40. Hidler J, Nichols D, Pelliccio M, Brady K, Campbell DD, Kahn JH, et al. Multicenter randomized clinical trial evaluating the effectiveness of the Lokomat in subacute stroke. Neurorehabil Neural Repair. 2009;23(1):5–13.

41. Hu XL, Tong KY, Song R, Zheng XJ, Leung WW. A comparison between electromyography-driven robot and passive motion device on wrist rehabilitation for chronic stroke. Neurorehabil Neural Repair. 2009;23(8):837–46.

42. Smith BW, Rowe JB, Reinkensmeyer DJ. Real-time slacking as a default mode of grip force control: implications for force minimization and personal grip force variation. J Neurophysiol. 2018;120(4):2107–20.

43. Wolbrecht ET, Chan V, Reinkensmeyer DJ, Bobrow JE. Optimizing compliant, model-based robotic assistance to promote neurorehabilitation. IEEE Trans Neural Syst Rehabil Eng. 2008;16 (3):286–97.

44. Di Pino G, Pellegrino G, Assenza G, Capone F, Ferreri F, Formica D, et al. Modulation of brain plasticity in stroke: a novel model for neurorehabilitation. Nat Rev Neurol. 2014;10(10):597–608.

45. Taub E, Uswatte G, Pidikiti R. Constraint-induced movement therapy: a new family of techniques with broad application to physical rehabilitation–a clinical review. J Rehabil Res Dev. 1999;36(3):237–51.

46. Park SW, Wolf SL, Blanton S, Winstein C, Nichols-Larsen DS. The EXCITE trial: predicting a clinically meaningful motor activity log outcome. Neurorehabil Neural Repair. 2008;22(5):486–93.

47. Ingemanson ML, Rowe JR, Chan V, Wolbrecht ET, Reinkensmeyer DJ, Cramer SC. Somatosensory system integrity explains differences in treatment response after stroke. Neurology. 2019;92(10): e1098–108.

48. Vidoni ED, Boyd LA. Preserved motor learning after stroke is related to the degree of proprioceptive deficit. Behav Brain Funct. 2009;5:36.

49. Carey LM, Lamp G, Turville M. The state-of-the-science on somatosensory function and its impact on daily life in adults and older adults, and following stroke: a scoping review. OTJR (Thorofare N J). 2016;36(2 Suppl):27S–41S.

50. Meyer S, Karttunen AH, Thijs V, Feys H, Verheyden G. How do somatosensory deficits in the arm and hand relate to upper limb impairment, activity, and participation problems after stroke? A Syst Rev Phys Ther. 2014;94(9):1220–31.

51. Lee SY, Lim JY, Kang EK, Han MK, Bae HJ, Paik NJ. Prediction of good functional recovery after stroke based on combined motor and somatosensory evoked potential findings. J Rehabil Med. 2010;42 (1):16–20.

52. Keren O, Ring H, Solzi P, Pratt H, Groswasser Z. Upper limb somatosensory evoked potentials as a predictor of rehabilitation progress in dominant hemisphere stroke patients. Stroke. 1993;24 (12):1789–93.

53. Reinkensmeyer DJ, Guigon E, Maier MA. A computational model of use-dependent motor recovery following a stroke: optimizing corticospinal activations via reinforcement learning can explain residual capacity and other strength recovery dynamics. Neural Netw. 2012;29–30:60–9.

54. Mudie MH, Matyas TA. Can simultaneous bilateral movement involve the undamaged hemisphere in reconstruction of neural networks damaged by stroke? Disabil Rehabil. 2000;22(1–2):23–37.

55. Luft AR, McCombe-Waller S, Whitall J, Forrester LW, Macko R, Sorkin JD, et al. Repetitive bilateral arm training and motor cortex activation in chronic stroke: a randomized controlled trial. JAMA. 2004;292(15):1853–61.

56. Dietz V, Macauda G, Schrafl-Altermatt M, Wirz M, Kloter E, Michels L. Neural coupling of cooperative hand movements: a reflex and FMRI study. Cereb Cortex. 2015;25(4):948–58.

57. Schrafl-Altermatt M, Dietz V. Cooperative hand movements in stroke patients: neural reorganization. Clin Neurophysiol. 2016;127(1):748–54.

58. Dancause N, Nudo RJ. Shaping plasticity to enhance recovery after injury. Prog Brain Res. 2011;192:273–95.
59. Khan F, Amatya B, Galea MP, Gonzenbach R, Kesselring J. Neurorehabilitation: applied neuroplasticity. J Neurol. 2017;264(3):603–15.
60. Cramer SC, Sur M, Dobkin BH, O'Brien C, Sanger TD, Trojanowski JQ, et al. Harnessing neuroplasticity for clinical applications. Brain. 2011;134(Pt 6):1591–609.
61. Kwakkel G, van Peppen R, Wagenaar RC, Wood Dauphinee S, Richards C, Ashburn A, et al. Effects of augmented exercise therapy time after stroke: a meta-analysis. Stroke. 2004;35(11):2529–39.
62. Kwakkel G, Wagenaar RC, Twisk JW, Lankhorst GJ, Koetsier JC. Intensity of leg and arm training after primary middle-cerebral-artery stroke: a randomised trial. Lancet. 1999;354(9174):191–6.
63. Duncan P, Studenski S, Richards L, Gollub S, Lai SM, Reker D, et al. Randomized clinical trial of therapeutic exercise in subacute stroke. Stroke. 2003;34(9):2173–80.
64. Langhorne P, Coupar F, Pollock A. Motor recovery after stroke: a systematic review. Lancet Neurol. 2009;8(8):741–54.
65. Globas C, Becker C, Cerny J, Lam JM, Lindemann U, Forrester LW, et al. Chronic stroke survivors benefit from high-intensity aerobic treadmill exercise: a randomized control trial. Neurorehabil Neural Repair. 2012;26(1):85–95.
66. Wirz M, Mach O, Maier D, Benito-Penalva J, Taylor J, Esclarin A, et al. Effectiveness of automated locomotor training in patients with acute incomplete spinal cord injury: a randomized controlled multicenter trial. J Neurotrauma. 2017;34:1891–6.
67. McCabe J, Monkiewicz M, Holcomb J, Pundik S, Daly JJ. Comparison of robotics, functional electrical stimulation, and motor learning methods for treatment of persistent upper extremity dysfunction after stroke: a randomized controlled trial. Arch Phys Med Rehabil. 2015;96(6):981–90.
68. Lohse KR, Lang CE, Boyd LA. Is more better? Using metadata to explore dose-response relationships in stroke rehabilitation. Stroke. 2014;45(7):2053–8.
69. Wu X, Guarino P, Lo AC, Peduzzi P, Wininger M. Long-term effectiveness of intensive therapy in chronic stroke. Neurorehabil Neural Repair. 2016;30(6):583–90.
70. Veerbeek JM, van Wegen E, van Peppen R, van der Wees PJ, Hendriks E, Rietberg M, et al. What is the evidence for physical therapy poststroke? A systematic review and meta-analysis. PLoS ONE. 2014;9(2): e87987.
71. Dromerick AW, Geed S, Barth J, Brady K, Giannetti ML, Mitchell A, et al. Critical Period After Stroke Study (CPASS): a phase II clinical trial testing an optimal time for motor recovery after stroke in humans. Proc Natl Acad Sci USA. 2021;118(39).
72. Lang CE, Strube MJ, Bland MD, Waddell KJ, Cherry-Allen KM, Nudo RJ, et al. Dose response of task-specific upper limb training in people at least 6 months poststroke: a phase II, single-blind, randomized, controlled trial. Ann Neurol. 2016;80(3):342–54.
73. Hubli M, Dietz V, Bolliger M. Spinal reflex activity: a marker for neuronal functionality after spinal cord injury. Neurorehabil Neural Repair. 2012;26(2):188–96.
74. Nave AH, Rackoll T, Grittner U, Blasing H, Gorsler A, Nabavi DG, et al. Physical fitness training in patients with subacute stroke (PHYS-STROKE): multicentre, randomised controlled, endpoint blinded trial. BMJ. 2019;366: l5101.
75. Jeffers MS, Karthikeyan S, Gomez-Smith M, Gasinzigwa S, Achenbach J, Feiten A, et al. Does stroke rehabilitation really matter? Part B: an algorithm for prescribing an effective intensity of rehabilitation. Neurorehabil Neural Repair. 2018;32(1):73–83.
76. Rodriguez-Fernandez A, Lobo-Prat J, Font-Llagunes JM. Systematic review on wearable lower-limb exoskeletons for gait training in neuromuscular impairments. J Neuroeng Rehabil. 2021;18(1):22.
77. Butzer T, Lambercy O, Arata J, Gassert R. Fully wearable actuated soft exoskeleton for grasping assistance in everyday activities. Soft Robot. 2021;8(2):128–43.
78. Xiloyannis M, Annese E, Canesi M, Kodiyan A, Bicchi A, Micera S, et al. Design and validation of a modular one-to-many actuator for a soft wearable exosuit. Front Neurorobot. 2019;13:39.
79. Huang VS, Krakauer JW. Robotic neurorehabilitation: a computational motor learning perspective. J Neuroeng Rehabil. 2009;6:5.
80. Beer RF, Ellis MD, Holubar BG, Dewald JP. Impact of gravity loading on post-stroke reaching and its relationship to weakness. Muscle Nerve. 2007;36(2):242–50.
81. Marchal-Crespo L, Reinkensmeyer DJ. Review of control strategies for robotic movement training after neurologic injury. J Neuroeng Rehabil. 2009;6:20.
82. Basalp E, Wolf P, Marchal-Crespo L. Haptic training: which types facilitate (re)learning of which motor task and for whom answers by a review. IEEE Trans Haptics. 2021;PP.
83. Liu LY, Li Y, Lamontagne A. The effects of error-augmentation versus error-reduction paradigms in robotic therapy to enhance upper extremity performance and recovery post-stroke: a systematic review. J Neuroeng Rehabil. 2018;15(1):65.
84. Patton JL, Stoykov ME, Kovic M, Mussa-Ivaldi FA. Evaluation of robotic training forces that either enhance or reduce error in chronic hemiparetic stroke survivors. Exp Brain Res. 2006;168(3):368–83.
85. Levac DE, Huber ME, Sternad D. Learning and transfer of complex motor skills in virtual reality: a perspective review. J Neuroeng Rehabil. 2019;16(1):121.
86. Wenk N, Penalver-Andres J, Buetler KA, Nef T, Müri RM, Marchal-Crespo L. Effect of immersive visualization technologies on cognitive load, motivation, usability, and embodiment. Virtual Rity. 2021.

87. Ranzani R, Lambercy O, Metzger JC, Califfi A, Regazzi S, Dinacci D, et al. Neurocognitive robot-assisted rehabilitation of hand function: a randomized control trial on motor recovery in subacute stroke. J Neuroeng Rehabil. 2020;17(1):115.

88. Reinsdorf D, Mahan E, Reinkensmeyer D. Proprioceptive gaming: making finger sensation training intense and engaging with the P-Pong game and PINKIE Robot. In: Proceedings of 43rd international engineering in medicine and biology conference (EMBC). 2021.

89. Wagner FB, Mignardot JB, Le Goff-Mignardot Le CG, Demesmaeker R, Komi S, Capogrosso M, et al. Targeted neurotechnology restores walking in humans with spinal cord injury. Nature. 2018;563 (7729):65–71.

90. Bachmann LC, Matis A, Lindau NT, Felder P, Gullo M, Schwab ME. Deep brain stimulation of the midbrain locomotor region improves paretic hindlimb function after spinal cord injury in rats. Sci Transl Med. 2013;5(208):208ra146.

Part II

From Movement Physiology to Technology Application

Use of Technology in the Assessment and Rehabilitation of the Upper Limb After Cervical Spinal Cord Injury

José Zariffa, Michelle Starkey, Armin Curt, and Sukhvinder Kalsi-Ryan

Abstract

The impairment of upper limb function is a central factor limiting independence and quality of life after cervical spinal cord injury. Improved approaches are needed both to comprehensively assess upper limb function throughout the continuum of care, and to enhance recovery. New technologies are expected to play a key role in both assessment and rehabilitation. For assessment, techniques ranging from established electrophysiological methods to novel wearable sensors can be used alongside clinical outcome assessments and provide a comprehensive suite of tools to assess body functions and structures, activity and participation. For rehabilitation, the variety of promising neuromodulation strategies is increasing, alongside work on robotic or interactive platforms that can support higher doses of rehabilitation. As increased evidence becomes available to support the use of these neurorehabilitation technologies after spinal cord injury, the expected benefits include improved therapeutic outcomes, new outcome measures to assess impact in clinical trials, and the possibility of large-scale automated data gathering to strengthen the evidence base for rehabilitation practices.

J. Zariffa (✉) · S. Kalsi-Ryan
KITE, Toronto Rehabilitation Institute, University Health Network, Toronto, Canada
e-mail: jose.zariffa@utoronto.ca

S. Kalsi-Ryan
e-mail: Sukhvinder.Kalsi-Ryan@uhn.ca

J. Zariffa
Institute of Biomedical Engineering, University of Toronto, Toronto, Canada

J. Zariffa · S. Kalsi-Ryan
Rehabilitation Sciences Institute, University of Toronto, Toronto, Canada

J. Zariffa
Edward S. Rogers Sr Department of Electrical and Computer Engineering, University of Toronto, Toronto, Canada

M. Starkey
Bedford, UK

A. Curt
Spinal Cord Injury Center, Balgrist University Hospital, Zurich, Switzerland
e-mail: Armin.Curt@balgrist.ch

S. Kalsi-Ryan
Department of Physical Therapy, University of Toronto, Toronto, Canada

Keywords

Spinal cord injury · Upper limb · Assessment · Rehabilitation · Wearable technology · Rehabilitation robotics · Neuromodulation

4.1 Introduction

Cervical spinal cord injury (SCI) results in complete or incomplete paralysis of the upper and lower limbs due to either a total or partial

© The Author(s), under exclusive license to Springer Nature Switzerland AG 2022
D. J. Reinkensmeyer et al. (eds.), *Neurorehabilitation Technology*,
https://doi.org/10.1007/978-3-031-08995-4_4

loss of motor and sensory function below the level of lesion. The functional impairment depends on which spinal cord segments are affected as well as the pathophysiology of the injury (e.g. traumatic or non-traumatic) [1, 2]. Due to the somatotopic organization of the spinal cord, i.e. the segmental innervation of sensory and motor nerves, the impairment of upper limb function following SCI can be accurately predicted once the location and extent of the injury are known (Fig. 4.1). As brain function is not usually affected after a SCI (unlike in stroke), motor planning and other functions related to movement initiation and control remain intact at the cortical level. Despite this, after a cervical SCI, movement control is affected by the impairment of efferent output from the spinal cord as well as the disruption of afferent input. Therefore, in order to appropriately address the specific needs for recovery of arm and hand (upper extremity) function for individuals with tetraplegia, the exact impairment must be accurately assessed and the underlying pathophysiology must be considered.

In most cases cervical SCI leads to a bilateral impairment of upper extremity function, meaning that both unimanual and bimanual tasks, such as opening a jar, are affected. As such, an individual with tetraplegia is not able to compensate for the loss of function in one limb with the less affected limb and therefore they are dependent on recovery of upper extremity function. This is in contrast to people suffering a stroke, peripheral nerve damage, and to a variable extent multiple sclerosis where impairments are either focal (in stroke unilateral) or can affect multiple areas (in MS) of the CNS where one limb often remains, at best, functionally intact or at worst less affected than the other. Consequently, following a cervical SCI, one of the highest priorities for affected individuals is recovery of upper extremity function [3, 4]. This is because useful function of the arms and hands is one of the main determinants of independence in activities of daily living (ADL) [5] which has a significant impact on quality of life. Understanding the clinical presentation of the disease through assessment is the precursor for meaningful

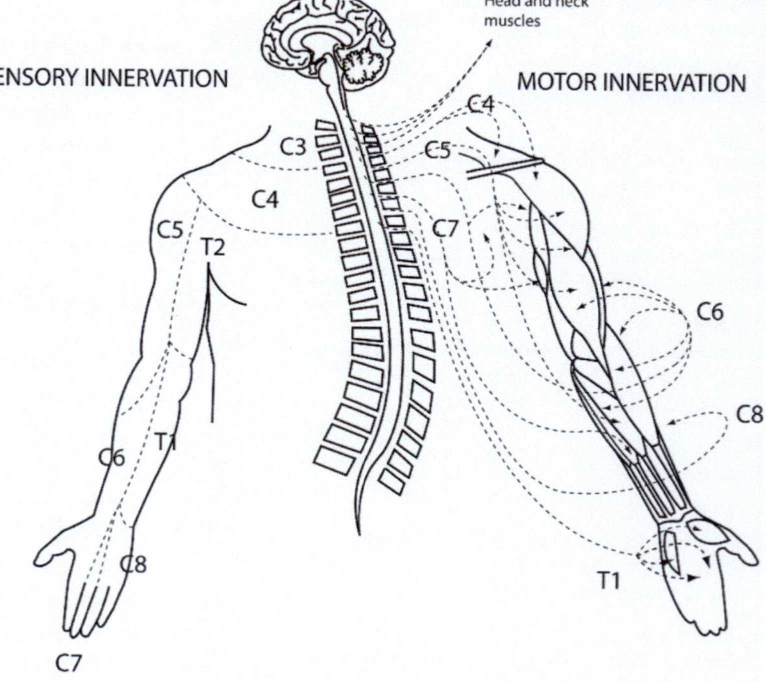

Fig. 4.1 Dermatomes (left) and myotomes (right) of the upper limbs are specifically innervated producing typical maps of sensory and motor functions and deficits, respectively

therapeutic prescription and future development of novel interventions. Therefore, the development and validation of suitable assessments and rehabilitation methods for the upper extremities following cervical SCI remain a highly relevant clinic goal.

To address this need, considerable effort has been, and remains to be, focused on the development of novel devices that can firstly assess functional loss of the upper extremity following SCI and secondly aid rehabilitation in persons with tetraplegia. In the SCI field the most notable are non-invasive electrical stimulation coupled with intensive therapy, such as functional electrical stimulation (FES), electrophysiological diagnostic approaches, robotics, passive workstations and sensor-based technology systems that will be discussed in detail below. The body of knowledge to support the clinical significance of these devices for evaluating and assessing recovery as well as their value in SCI rehabilitation is increasing.

This chapter is organized as follows: Sect. 4.2 summarizes the importance of upper limb function and how upper limb impairment is manifested in SCI; Sect. 4.3 introduces assessment frameworks and summarizes good measurement qualities; Sects. 4.4–4.6 review the available assessments that address impairment, activity and participation specific to tetraplegia; Sect. 4.7 reviews novel therapeutic approaches currently being used and/or developed for the rehabilitation or neuro-restoration of upper extremity function following spinal cord injury; Sect. 4.8 provides a discussion of trends and directions in this field.

body required for survival. For individuals with tetraplegia, not only do the upper limbs perform functions for which unimpaired individuals use their hands, but they replace the functions of other parts of the body that may no longer have even partial function (e.g. lower extremity; locomotion replaced by wheelchair propulsion). The upper limbs of an individual with tetraplegia represent all self-care activities including locomotion, bowel and bladder function, recreational activities and vocational activities. Improvement in upper limb function after cervical SCI is one of the most significant factors in improving quality of life according to individuals with tetraplegia [3, 4]. Therefore, the more normal and precise recovered upper limb function is after tetraplegia, the more functional the individual will be. The field of upper limb restoration research is relatively new; interventions have only been studied with some emphasis since the 1960s, simply because the survival of high tetraplegics was poor prior to this era [6].

The upper limb is by no means an easy limb to assess. Actions of upper extremity function are highly complex, incorporating activity in the trunk, shoulder, elbow, wrist and hand joints, musculature and nerves [7, 8]. Voluntary actions of the upper limb consist of cortical and sub-cortical control; afferent and efferent pathway involvement; motor and sensory activity as well as feedback and feedforward systems. Therefore, assessment of upper limb impairment requires a measure that acknowledges the components (sensory and motor) as well as function (integration). Not only is the upper limb an important element in the management of tetraplegia, assessing this limb requires effort and attention from the clinician or researcher.

4.2 Significance of the Upper Limb and Distinctions of the Tetraplegic Upper Extremity

4.2.1 Importance of the Upper Limb

The upper limb is the primary mode with which unimpaired humans typically interact with their environments, making it an integral part of the

4.2.2 Distinctions of the Tetraplegic Upper Limb

Sollerman characterized normal hand posturing and defined the hand postures that are used most commonly during ADLs [9]. Figure 4.2 depicts the hand postures defined by Sollerman, which have led to assessment protocols for several populations. The SCI field is no exception to using this historical literature to define how the hand is assessed after injury. Figure 4.2 also defines common hand postures that can be used to assess impairment of the hand by evaluating how far from normal hand function presents. It is important to note that there are some specific nuances in the tetraplegic hand that define how it should be assessed. An example of this is the "tenodesis grip" which is a typical tetraplegic hand posture that allows one to perform tasks with an adapted method (not one of the normal hand postures), using wrist extension to produce passive finger flexion. Therefore, one may develop the ability to perform a function using a tenodesis grasp, without demonstrating a sensory or motor gain. Therefore, assessing one's ability to perform a task is not often sufficient information to understand whether neurological status is changing.

The severity of injury along with the specific spinal cord structures (i.e. tracts/nuclei) affected also contribute to the varying degrees of presentation of impairment and potential for recovery. Therefore, interventions for the improvement of upper limb function not only need to incorporate compensatory therapeutic strategies but also restorative interventions that can reduce impairment. The potential to restore lost upper limb function in SCI is influenced by three factors: (1) an increased rate of incomplete SCI [10, 11]; (2) development of intensive restorative therapies that are applied at the periphery to affect the sensorimotor system [12, 13] and (3) development of treatments that are applied to the central nervous system such as pharmacological agents and biologics with the potential for neural repair, neuro-protection and regeneration [14, 15].

4.3 Assessment Frameworks and Psychometric Properties

4.3.1 Outcome Measure Frameworks

The primary component that defines the selection of an outcome measure is the research or clinical objective. An outcome measure should be selected so that a response to the objective can be derived from the data collected. The International Classification of Functioning, Disability and

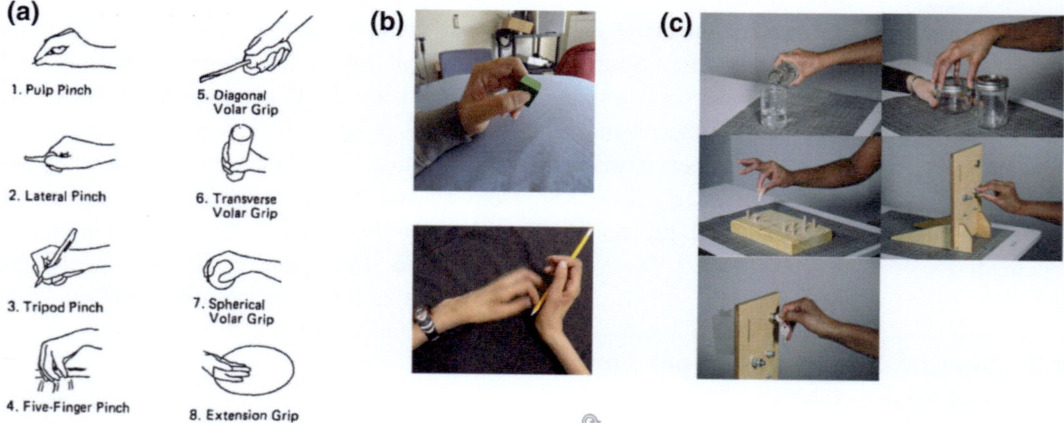

Fig. 4.2 Hand Postures and Distinctions of the Tetraplegic Hand. Legend: **a**—The eight most common grip patterns used by normal individuals. Pulp pinch and lateral pinch are used in 20% of interactions in activities of daily living; five finger pinch, diagonal volar grip, transverse volar grip 15% each; tripod pinch 10%; spherical grip and extension grip less than 5% (reproduced from [9], © Acta Chirurgica Scandinavica Society, reprinted by permission of Taylor & Francis Ltd, http://www.tandfonline.com on behalf of Acta Chirurgica Scandinavica Society). **b**—The most common and adaptive hand grip for individuals with tetraplegia, the tenodesis grip. **c**—The five grip patterns that are used to determine similarity to normal and individuals' hand function according to the GRASSP version 1 [176] (photos used with permission from Neural Outcomes Consulting Inc.)

Health (ICF) [16] and the Clinical Outcomes Assessment Compendium [17] are two measurement frameworks that can assist the researcher or clinician in understanding the domain within which their objective falls.

The ICF is an overarching framework that defines "domains of measurement" that can be considered after cervical SCI [16]. The three domains identify the construct of measurement. The objective of measurement will guide the selection of assessment. For example, where a clinician would like to determine if a therapeutic application is neurologically restorative, a measure that falls within the "body functions and structures" domain should be implemented. Table 4.1 defines the definitions of the domains in the ICF model and other terms that should have a standard meaning when used in the context of the ICF. Fig. 4.3 defines the ICF model and how it integrates with additional considerations for assessment selection. In addition to the ICF, the Clinical Outcomes Assessment Compendium (COAC) defines types of measures and further assists in selecting appropriate measures. This nomenclature has been developed by the Food and Drug Agency (FDA) and aims to facilitate the selection of outcomes assessments that measure the patient's experience and reflect the change that is being studied. This compendium aims to assist researchers in understanding how to develop outcome measures so that they are reliable patient reported, clinician reported, observer reported or performance reported outcomes [17]. Table 4.2 defines the types of outcomes. Both the ICF and COA definitions are frameworks that drive the decision-making regarding the use of neurological assessment. In Fig. 4.3, we illustrate how these

Table 4.1 ICF model domains and definitions: definitions in the context of health (from [16])

Functioning	Umbrella term for body functions, body structures, activities and participation. It denotes the positive aspects of the interaction between an individual (with a health condition) and that individual's contextual factors (environmental and personal factors)
Disability	Umbrella term for impairments, activity limitations and participation restrictions. It denotes the negative aspects of the interaction between an individual (with a health condition) and that individual's contextual factors (environmental and personal factors)
Body functions	The physiological functions of body systems (including psychological functions)
Body structures	Anatomical parts of the body such as organs, limbs and their components
Impairments	Problems in body function and structure such as significant deviation or loss
Activity	The execution of a task or action by an individual
Activity limitations	Difficulties an individual may have in executing activities
Participation	Involvement in a life situation
Participation restrictions	Problems an individual may experience in involvement in life situations
Environmental factors	The physical, social and attitudinal environment in which people live and conduct their lives
Performance	The Performance qualifier describes what an individual does in his or her current environment. Since the current environment always includes the overall societal context, performance can also be understood as "involvement in a life situation" or "the lived experience" of people in their actual context. (The current environment will be understood to include assistive devices or personal assistance, whenever the individual actually uses them to perform actions or tasks.)
Capacity	The Capacity qualifier describes an individual's ability to execute a task or an action. This construct indicates the highest probable level of functioning of a person in a given domain at a given moment

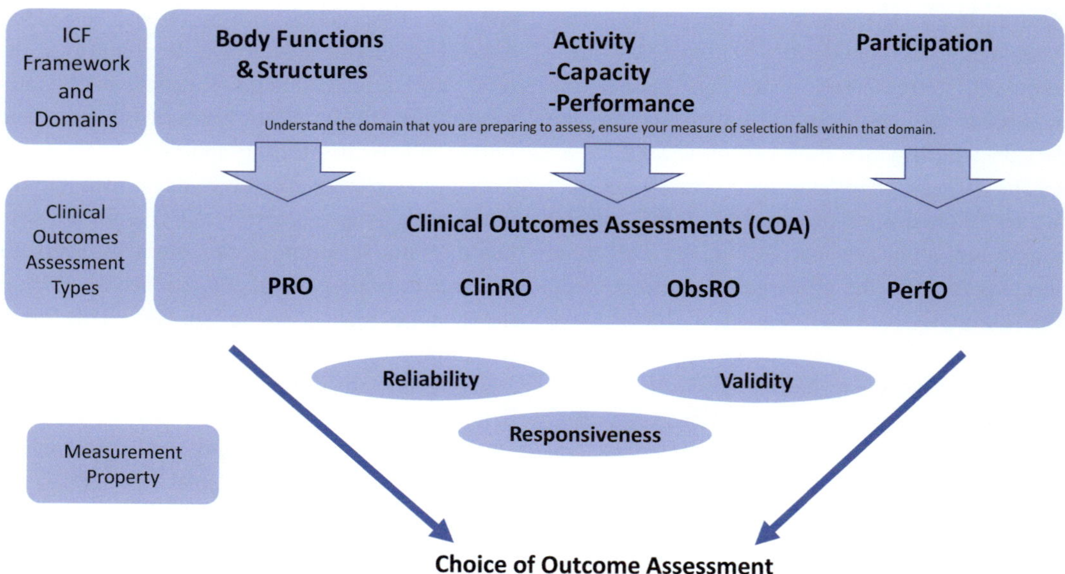

Fig. 4.3 Summary of the outcomes/assessment frameworks discussed in the text. This image provides a flow of thought on how to select the most appropriate measure when assessing the tetraplegic upper extremity

Table 4.2 FDA compendium—clinical outcome assessments—definitions of key terms. Adapted from the BEST Glossary [17]

Clinical Outcome Assessment (COA)	A COA is defined as an assessment of a clinical outcome that can be made through report by a clinician, a patient, a non-clinician observer or through a performance-based assessment
Clinician Reported Outcome (ClinRO)	A measurement based on a report that comes from a trained health-care professional after observation of a patient's health condition
Patient Reported Outcome (PRO)	A measurement based on a report that comes directly from the patient (i.e. study subject) about the status of a patient's health condition without amendment or interpretation of the patient's response by a clinician or anyone else
Observer Reported Outcome (ObsRO)	A measurement based on a report of observable signs, events or behaviours related to a patient's health condition by someone other than the patient or a health professional
Performance Outcome (PerfO)	A measurement based on standardized task(s) actively undertaken by a patient according to a set of instructions that is administered by a health care professional. PerfO assessments require patient cooperation and motivation

two frameworks and measurement properties determine assessment use in research and clinic settings.

4.3.2 Measurement Properties

The COSMIN taxonomy for measurement properties is the resource we follow for this chapter [18]. The taxonomy defines three primary domains which are reliability, validity and responsiveness. Each domain includes one or more measurement property. Reliability is "the degree to which the measurement is free from measurement error", and demonstrates internal consistency. Validity refers to "the degree to which an outcome measure measures the construct it purports to measure" and includes content validity, construct validity and criterion validity. Responsiveness "refers to the ability of an outcome to detect change over time in the construct to be measured".

Typically, reliability and validity are required measurement properties and responsiveness is important if assessment of change is the primary objective of the study. Of additional note is the minimal clinically important difference (MCID), defined as the smallest change in a measurement that signifies an important improvement from the patients' and/or clinicians' perspective [19]. MCID can be a challenging concept to measure in practice. While there are very few tools that have an MCID established for use in cervical SCI, minimal detectable difference (MDD) is a statistical concept that determines the smallest real change in an outcome which is beyond measurement error. A valid MCID therefore cannot be less than the MDD.

4.4 Measures of Body Functions and Structures After Cervical Spinal Cord Injury

In this section, we review assessments that are specific to defining impairment and that typically fall into the domain of "body functions and structures" of the ICF. Some measures assess more than one domain.

Clinically, an SCI is characterized by a combination of the neurological sensory and motor level, and the American Spinal Injury Association (ASIA) impairment scale (AIS), which provides information about the completeness of the lesion. These elements are reflected in the International Standards for the Neurological Classification of Spinal Cord Injury (ISNCSCI) that are approved by the ASIA and the International Spinal Cord Injury Society (ISCoS) [20]. The motor components of the ISNCSCI assessment rely on manual muscle testing (MMT) at a set of 10 myotomes bilaterally, whereas the sensory testing uses light touch and pinprick stimuli at 28 dermatomes bilaterally. The ISNCSCI is a classification measure that defines the severity of SCI and does so based on the global presentation of the individual. However, elements of the test quantify the impairment of the upper limb focally and are therefore used most commonly to define upper limb impairment in SCI. This assessment is not as sensitive as other measures, however does establish a solid foundation in defining the severity of the injury or disease.

Electrophysiological techniques can supplement the clinical evaluation and provide quantitative information about the extent and location of neurological damage. These assessments tend to be carried out soon after the injury and throughout rehabilitation to document any changes in sensory and motor function that occur. Likewise, biomechanical measurements can describe the motor impairment in a more detailed and quantitative manner than the ISNCSCI evaluation. The most common assessments are discussed below.

4.4.1 Electrophysiology

Electrophysiological measures used to assess an SCI include somatosensory evoked potentials (SSEP), contact heat evoked potentials (CHEP), motor evoked potentials (MEP), nerve conduction study (NCS) and electromyogram (EMG) recordings. SSEPs, CHEPs and MEPs are electrical potentials recorded from predefined locations (i.e. scalp and muscles) following the stimulation of a sensory or motor nerve and reflect conditions within the peripheral and central nervous system, whereas NCS specifically reflects the condition of peripheral nerves. EMG reflects the central and peripheral pathways involved in volitional motor control, as well as the condition of the muscles.

The application of electrophysiological assessment techniques is growing. For instance, EMG and NCS are currently heavily used to determine surgical peripheral nerve transfer candidacy. There is an increasing use of surface EMG to determine FES therapy candidacy. These techniques are currently being used as a part of clinical assessment batteries to guide clinical decision-making.

4.4.1.1 Somatosensory Evoked Potentials (SSEP)

SSEP are elicited by an electrical stimulus of a peripheral sensory or mixed nerve [21, 22]. The stimulus is applied to the skin and the evoked potential is recorded from the subject's scalp. The time taken for the sensory nerve fibres to transmit the stimulus to the sensory areas of the brain is measured. These recordings can be used to assess the integrity of the spinal cord because when the nerve pathway is damaged, the signals from the peripheral nerve to the brain become either slowed, in the case of an incomplete lesion, or completely abolished, in the case of a complete lesion. Hence, during the course of rehabilitation, changes in the latency or amplitudes of the signal can be used to indicate changes in spinal cord and brain function.

4.4.1.2 Contact Heat Evoked Potential (CHEP)

The mechanisms underlying the generation of pain in pathogenic conditions following an SCI can be studied using a CHEP stimulator [23–25]. In CHEPs the stimulus is applied with a thermode that is placed directly on the skin to stimulate the thermal pain sensory receptors on Aδ (delta) and C fibres. The pulses of heat are delivered rapidly, with adjustable peak temperatures, to elicit the different warm/heat thresholds of the receptors. The resulting evoked potentials can be measured using scalp electrodes. CHEP is used to assess the condition of the spinothalamic pathways (thermal and nociceptive sensation) and their relation to pain.

4.4.1.3 Motor Evoked Potential (MEP)

MEPs are elicited by the direct stimulation of the exposed motor cortex (during surgery) or by the transcranial stimulation of the motor cortex [26]. Transcranial electrical stimulation (TES) is applied through cutaneous electrodes, whereas transcranial magnetic stimulation (TMS) is generated with a magnetic field. In comparison to TMS, the main limitation of TES is that the electrical currents applied to the scalp can cause local discomfort. Either way the stimulus results in the contraction of a muscle contralateral to

where the MEPs were applied and are recorded with surface electrodes. TMS is used as a diagnostic and follow-up tool for neurological disorders where the impairment and eventual recovery of the corticospinal tract are assessed. It should be noted that the presence of lower motor neuron damage can be a confounding factor in the interpretation of the results.

4.4.1.4 Nerve Conduction Study (NCS)

In motor NCS, an electrical stimulus is elicited over a peripheral motor nerve and the electrical potential generated in the corresponding muscle is recorded [27]. In sensory NCS, the electrical stimulation is applied to a sensory peripheral nerve and electrical potentials are recorded at a point further along that nerve. The F-wave and H-reflex are examples of NCS and represent different reflex responses within peripheral nerves and spinal segments, respectively. Although NCS is mainly used to diagnose peripheral nerve dysfunction (such as carpal tunnel and Guillain-Barré syndromes) and muscle disorders (such as muscle atrophy), it also provides useful information on spinal cord function, specifically when damage of alpha-motoneurons (traumatic or non-traumatic) results in an alteration of motor (but not sensory) NCS resulting in reduced or abolished compound muscle action potentials.

4.4.1.5 Electromyography (EMG)

EMG uses changes in the electrical potentials of muscle cells for diagnostic purposes [27]. In surface EMG, electrodes placed on the skin are used to record signals from superficial muscles, whereas in intramuscular EMG, needles are introduced into the muscle to receive the signals from deep muscles or localized muscle activity. Surface EMGs are used to assess gross muscle activation, whereas needle EMGs assess single muscle fibres. EMG is used to diagnose neurological and muscular disorders. In the context of upper limb impairment after SCI, EMG can help to characterize the lesion in more detail, for example, by detecting remaining innervation even when voluntary contraction is not detectable through MMT [28]. Surface EMG can also be applied across multiple muscles simultaneously to provide information about motor control.

Advanced signal processing techniques can further be used to decompose a surface EMG signal to recover the activation patterns of individual motor units [29, 30]. Tools to facilitate this process are becoming more widespread, which will increase the amount of neurological information that can be extracted from surface EMG signals after SCI in the short term [31, 32].

4.4.2 Biomechanical (Kinetic, Kinematic) Measures

Changes in biomechanical parameters of upper extremity function, such as muscle strength, joint angles, ROM and movement trajectories, can be measured with specific techniques, such as robotic and motion capture systems as summarized in more detail below.

4.4.2.1 Assessment of Muscle Strength

Digital-palmar prehensile strength can be measured using a Jamar dynamometer [33], a vigorimeter (a manometer with tubing and rubber ball) [34] or another type of manometer; for review see [35]. These devices use units of force or pressure to measure strength during specific grasp or pinch postures [33, 36, 37].

4.4.2.2 Assessment of Angles, Range of Motion (ROM) and Trajectories

Upper limb passive and active joint flexion as well as ROM can be measured using traditional goniometry. However, simultaneous recordings of dynamic changes in joint angles and movement trajectories require the use of motion capture systems. The variety of available motion capture technologies has increased in recent years, and can be broadly categorized into marker-based optical methods, markerless optical methods, body-worn sensors and robotics.

Optical Motion Capture

Marker-based optical motion tracking systems use multiple infrared cameras to track reflective markers and determine their 3D position. As a result, upper limb kinematics can be quantified during tasks of interest. These systems require dedicated space and complex instrumentation and are therefore typically limited to lab-based environments, but are highly accurate [38]. Common systems of this type that have been used to describe upper limb kinematics include Vicon [39] and Qualisys [40].

More recently, markerless motion capture has become more widespread. The move to markerless systems was first popularized by novel sensors able to capture depth information in video data, notably the Microsoft Kinect [41–43]. The depth information substantially facilitates the pose estimation task. More recently, however, deep learning techniques have led to impressive performance in estimating pose from ordinary video (without a depth channel) [44, 45]. This last category estimates pose in two dimensions rather than three. However, depth information can be estimated either using stereoscopic setups [46] or recovered from the 2D pose using dedicated neural networks [47, 48]. Because these systems do not require the user to wear markers of any kind and involve a simpler camera setup, they have the major advantage of being usable in a greater variety of environments. On the other hand, their accuracy is generally lower than that of marker-based systems (on the order of centimetres versus sub-millimetre), and less thoroughly validated, with existing work on upper limb assessment focusing on stroke rather than SCI [49–54]. In one SCI-specific study, an arrangement of 3 Kinects was used to characterize upper limb kinematics during wheelchair propulsion [55]. Despite these open questions, markerless motion capture represents a substantial opportunity to describe upper limb impairment after SCI in a wide range of environment and activity contexts.

Body-Worn Sensors

Wearable approaches to kinematic tracking most commonly rely on inertial measurement units (IMUs). With appropriate sensor positioning, calibration and signal processing from multiple units, IMU systems can be used to estimate body posture and describe movement kinematics [56]. In SCI research to date, however, these types of devices have more commonly been used to assess activity rather than body function and structure, as detailed below in Sect. 4.5.1. Another wearable option relevant to upper limb function after SCI is instrumented gloves. These devices, often in combination with virtual reality, offer a relatively low-cost solution for tracking motion of the hand and fingers and, once validated, may be used in clinical research as novel assessment tools [57]. For example, custom-designed gloves equipped with force and position transducers have been used to evaluate grasping in individuals with tetraplegia [58], whereas others have been used to measure specific motor tasks performed with the hands during behavioural and functional magnetic resonance imaging studies [59]. These devices may have particular benefits in situations where subtle variations in hand movements are of interest, such as in individuals with cervical myelopathy [60].

Robotic Systems

Rehabilitation robotic systems as well as unactuated mechatronic platforms typically include a number of sensors necessary for controlling the robot and interface. These sensors additionally provide the opportunity to quantify the user's movements through kinematic and/or force measurements, and therefore to be used as assessment devices. Using robotics as novel assessment devices in the stroke field has been an area of research for a number of years. There are a broad range of devices used and outcomes assessed, for example [61–65]. However, this line of work has been less common in the SCI field. In this context, the ArmeoSpring®, a commercial gravity compensated device based on the T-WREX upper extremity rehabilitation system (see below for further details), was tested to determine whether measurements taken by the device are able to predict clinical scores in individuals with SCI [66]. The authors showed that kinematic and grip strength measurements taken with the device were able to provide relevant predictions for a number of clinical scores

and that the results were in line with previous similar studies with stroke survivors. In a separate significant advance, the specific assessment associated with the passive workstation ReJoyce® (Fig. 4.4), the ReJoyce Arm and Hand Function Test (RAHFT), was evaluated against standardized tests of arm function in individuals with stroke and SCI and was shown to be valid. This means that the RAHFT can provide a standardized assessment of arm and dexterous hand function either performed in the clinic or home environment by being administered remotely via the Internet [67], a first in this field. The ARMin exoskeleton robot has also been used to extract a range of kinetic, kinematic and timing parameters, several of which were found to be reliable and to correlate with clinical assessments [68]. Recognizing the potential of robotic technologies for assessment, subsequent studies have developed mechatronic platforms intended specifically for assessment of upper limb function [69].

Linking information collected by robotics with traditional clinical scores may allow more precise and perhaps also quicker assessments. Perhaps the most significant advantage of using robots for both treatment and assessment is that measurements could be obtained with much greater frequency than clinical assessments (at every session, instead of at intervals of several weeks). A greater frequency of assessments can be beneficial for tracking recovery and guiding therapy progression [70]. Longitudinal tracking data must however be interpreted carefully to take into account measurement error using appropriate psychometric validation, for example using the smallest real difference [71].

4.5 Measures of Activity After Cervical Spinal Cord Injury

Assessments of body structure and function are essential for describing neurological impairment and recovery. The previous section summarized a variety of clinically accepted methods as well as new technologies that can provide complementary information in this regard. Sensor-based

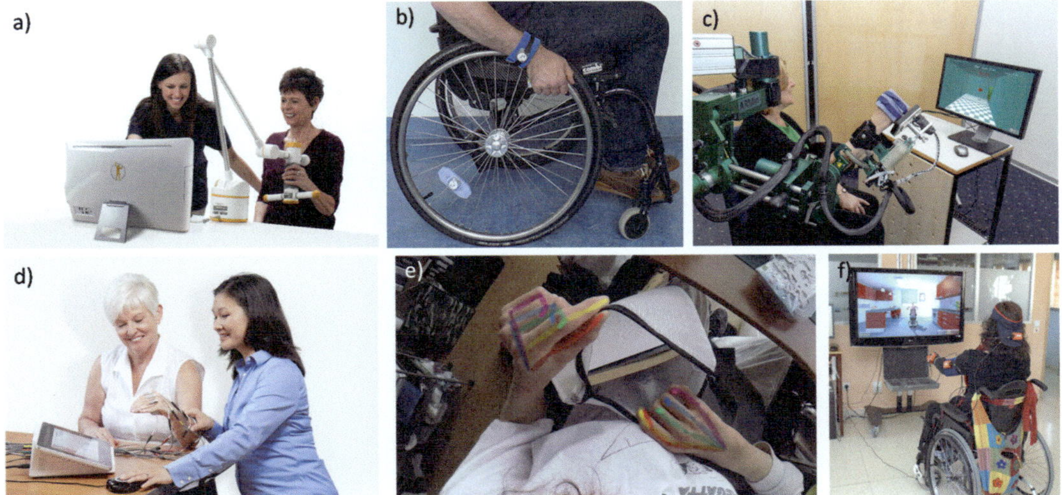

Fig. 4.4 Examples of technological approaches to the assessment and rehabilitation of upper extremity function after cervical SCI. **a** ReJoyce rehabilitation and assessment platform (photo used with permission from Rehabtronics Inc.). **b** ReSense inertial measurement units (from [177]). **c** ARMin robot (from [68]). **d** MyndMove functional electrical stimulation therapy system (photo used with permission from MyndTec Inc.). **e** Egocentric video with hand pose estimation using OpenPose [178]. **f** Training platform based on wearable devices and virtual reality (from [133])

assessment can be particularly beneficial by making it possible to collect data at more time-points and in more environments than possible with clinician-delivered evaluations, as long as appropriate psychometric properties are demonstrated. In order to fully describe the impact of injury and interventions on the affected individuals, measures of body structure and function must be complemented by measures of activity, in other words the ability to carry out activities of daily living. This section reviews clinical and technological approaches to this problem.

Of particular relevance is the Spinal Cord Independence Measure (SCIM), an objective measure that quantifies independence by assessing global functioning. Although it is a global measure, the SCIM has been used to validate many upper limb assessments. The SCIM is routinely used in the tetraplegic population to determine to what extent the upper limbs contribute to the individual's independence. SCIM is a 17 item assessment with three domains: self-care, respiratory and sphincter control and mobility [72]. The SCIM is a measure of independence and defines how much gain in global function one makes. It can be administered by clinicians' interview or by having the individual perform the tasks, thus making it a ClinRO or PerfO depending on how it is administered. The SCIM assesses the activity domain of the ICF model. The reliability ICC is above 0.95 and Cronbach's alpha values for all items range between 0.833 and 0.835. The concurrent validity when compared with the FIM is 0.84 [73].

Despite the SCIM's widespread use as a global measure of independence, interventions that specifically target upper limb function call for assessments that can provide a finer level of detail about the upper limb. Table 4.3 summarizes a range of upper limb measures that are specific to cervical spinal cord injury, or can be used to quantify cervical SCI. The methods used to rate upper extremity function are most often specific movements and/or ADL tasks carried out with a single hand or bimanually. The tasks can either be basic, such as grasping an object and transporting it from one place to another, or more complex, such as grooming. The table describes each measure and provides relevant information about the quality of the measures. Table 4.3

Table 4.3 Clinical outcome assessments developed to evaluate upper limb function in individuals with SCI

Instrument	Purpose/ICF/COAC	Proximal arm	Distal arm/hand	Time to complete	Method used	Parameter measured	Scoring	Reliability	Validity	Advantages	Disadvantages
GRASSP (Graded and redefined Assessment of Strength, sensibility, and Prehension), 2008 [176, 179, 180]	Clinical and research tool, upper limb function in C0–T1 patients Body Functions & Structure and Activity PerfO	Shoulder, elbow	Forearm, wrist, hand (grasp)	30–45 min (both sides)	10 manual muscle tests, SW filaments test, 3 grasp movements, 6 single-handed basic ADL tasks	Muscle contraction/ROM, pressure sensibility, grasp pattern, capacity to complete task	0–5 0–4 0–4 0–5	$n = 72$ Inter-rater ICC = 0.84–0.96 Test–retest ICC = 0.86–0.98 $p < 0.0001$	$n = 72$ SCIM III $r = 0.57$–0.68 SCIM III-SS $r = 0.74$–0.79 CUE $r = 0.77$–0.83 $p < 0.0001$	Applicable in acute and chronic stages of SCI Responsiveness and MDD Available	Single-handed specific; test kit has to be purchased
VLT-SV (Short Version of the Van Lieshout Test for Arm/Hand Function), 2006 [181]	Clinical and research tool, upper limb function in C5–T1 patients Activity PerfO	Shoulder, elbow	Forearm, wrist, hand grasp (pinch, grip)	25–35 min (both sides)	10 single-handed/bimanual-specific movements and basic ADL tasks	Capacity to perform task	0–5	$n = 12$ Inter-rater ICC = 0.98 and 0.99	$n = 55$ GRT $r = 0.87$ and 0.90	Scoring based on principle that different tasks have different hierarchical levels	Cannot be used in acute phase
MCS (Motor Capacities Scale), 2004 [182]	Clinical tool, upper limb function in C4–C7 patients, surgery evaluation Activity PerfO	Shoulder, elbow	Forearm, wrist, hand grasp	20–50 min	31 single-handed/bimanual-specific movements and basic ADL tasks	Independence degree, capacity to perform task, grasp pattern	1–5 1–2 1–4	$n = 52$ Inter-rater ICC = 0.99	$n = 52$ Sollerman $r = 0.96$ ASIA-MS $r = 0.74$	Measures different important parameters	Specific for surgery evaluation in C4–C7 patients, patient assumed in wheelchair
THAQ (Tetraplegia Hand Activity Questionnaire), 2004 [183]	Clinical tool, upper limb function, surgery and FES evaluation Activity PerfO	Shoulder, elbow	Forearm, wrist, hand grasp	30–45 min	Questionnaire regarding 153 self-administrated single-handed and bimanual ADL tasks	Self-perceived capacity to perform task, independence degree, task importance	0–3 0–3 0–2	Not available	Not available	Telephone interview possible	Neither tested for validity nor reliability, test self-administrated at home (no control), possible bias to please interviewer, specific for surgery and FES evaluation

(continued)

Table 4.3 (continued)

Instrument	Purpose/ICF/COAC	Proximal arm	Distal arm/hand	Time to complete	Method used	Parameter measured	Scoring	Reliability	Validity	Advantages	Disadvantages
Thorsen's Functional Test, 1999 [184]	Research tool, hand function, FES evaluation Activity PerfO	Not assessed	Forearm, wrist, hand grasp (pinch, grip)	<90 min (including preparation)	8 basic single-handed ADL tasks	Capacity to perform task	0–2	Not available	Not available	Easy administration	Neither tested for validity nor reliability, proximal arm not assessed, specific for FES evaluation, single-handed specific
CUE-Q (Capabilities of Upper Extremity Instrument), 1998 [185]	Research tool, upper limb function in C5–T1 patients Activity-Capacity PRO	Shoulder, elbow	Forearm, wrist, hand grasp	~30 min (both sides)	Questionnaire regarding 32 single-handed/bimanual self-administrated specific movements and basic ADL tasks	Self-perceived capacity to perform task	1–7	$n = 154$, Test–retest ICC = 0.94	$n = 154$ FIM-SM $r = 0.738$ $\rho = 0.798$ UEMS $r = 0.782$ $\rho = 0.798$	Telephone interview possible	Test self-administrated at home (no control), possible bias to please interviewer
CUE-T (Capabilities of Upper Extremity Instrument), 1998 [186, 187]	Clinical Tool, hand function Activity-Capacity PerfO	Not assessed	Forearm, wrist, hand grasp (pinch, grip)	20 min (both hands)	20 single-handed/bimanual basic ADL tasks	Capacity to perform task, time to complete task, grip pattern	0–4	$n = 6$ Inter-rater $r = 0.98$ Test–retest $r = 0.98$	$n = 59$ ICSHT $r = 0.76$	Performance Test MCID	Proximal arm not assessed; test kit has to be purchased
GRT (Grasp and Release Test), 1994 [188, 189]	Research tool, hand function in C5–C6 patients, neuroprosthesis evaluation Activity-Capacity PerfO	Not assessed	Forearm, wrist, hand grasp	90–150 min (both with/without neuroprosthesis)	6 single-handed basic ADL tasks	Ability to pass or fail task within a given time	Number of passed/failed tasks	$n = 19$ Test–retest ICC = 0.87–1 $p < 0.01$	$n = 19$ Partly validated FIM (in general low ρ)	Objective	Only partly validated and in general low correlation, too long, proximal arm not assessed, specific for neuroprosthesis evaluation in C5–C6 patients, single-handed specific, based only on time not on movement quality

(continued)

Table 4.3 (continued)

Instrument	Purpose/ICF/COAC	Proximal arm	Distal arm/hand	Time to complete	Method used	Parameter measured	Scoring	Reliability	Validity	Advantages	Disadvantages
Vanden Berghe Hand Function Test, 1991 [190]	Clinical tool, hand function, surgery evaluation Activity PerfO	Not assessed	Forearm, wrist, hand grasp	30–195 min (operated hand)	9 single-handed basic ADL tasks, 6 single-handed/bimanual ADL tasks	Time to complete task, independence degree	Time(s), 0–3	Not available	Not available	Easy administration	Neither tested for validity nor reliability, proximal arm not assessed, specific for surgery evaluation
RLAH (Ranchos Los Amigos Hospital Functional Activities Test), 1980 [191]	Research tool, upper limb function in traumatic C4–C7 patients, use of learned skills and orthotic device after discharge evaluation Activity ClinRO	Shoulder, elbow	Forearm, wrist, hand grasp	–	Questionnaire regarding 45 single-handed/bimanual self-care ADL tasks	Self-perceived independence degree	1–3	Not available	Not available	Telephone interview possible	Neither tested for validity nor reliability, specific for use of learned skills and orthotic device after discharge evaluation in C4–C7 Traumatic patients, patient assumed in wheelchair
Toronto Rehab Institute—Hand Function Test [192]	Research and clinical hand function in C4-C8 patients, FES evaluation Activity-Capacity PerfO	Shoulder, trunk	Forearm, wrist, hand grasp	Less than 30 min	1. Manipulation of objects 2. Strength of Pinches/Grasps	Capacity to perform a task Strength with a gauge provided	Each manipulation item is scored on a scale from 0 to 7	Inter rater reliability = 0.98 (ICC)	Concurrent Validity with FIM and SCIM respectively 0.73 and 0.61	Duration to administer is less than 30 min Self-training is sufficient for use	Test kit required

emphasizes measures that are robust and have been widely used among the SCI clinical and research fields. Prior to the development of SCI-specific upper limb measures, measures that were developed to quantify upper limb function in other populations, notably after stroke, were used in SCI studies. A selection of relevant measures is defined in Table 4.4. With the more established upper limb measures specific to tetraplegia, generic upper limb measures have been used less and less as primary measures in SCI, but may be used as ancillary tools.

4.5.1 Role of Wearable Technology in Measuring Performance

The tests reviewed in Table 4.3 predominantly measure capacity rather than performance. Advances in wearable technology provide an exciting opportunity to fill this gap and gain insights into a person's activities outside of direct clinical observation. Although the complexity of human upper extremity function creates significant challenges for automated measurement, a number of recent studies have demonstrated interesting results.

The most frequently used approach to date has been inertial measurement units (IMUs) or accelerometers, mounted on the wrists. The method has been extensively used with individuals who have experienced a stroke [74–76]; however, a few studies have also applied it to individuals living with SCI. These efforts demonstrated that the sensors could be used to measure wheeling activities [77], laterality of injury [78], physical activity [79], as well as provide metrics of upper extremity activity that correlated with aspects of the ISNCSCI and GRASSP assessments [80]. Wrist-worn sensors most strongly reflect the amount of arm movement, although one study showed that some compensatory movements could be detected, thus providing insight into the type or quality of movement [81]. Nonetheless, in order to get more detailed information about the performance of upper extremity tasks, it has been argued that wrist-worn sensors may need to be complement with sensors that can directly detect the use of the hand.

Two types of hand-worn sensors have been reported for the purpose of detecting functional hand use in unconstrained environments. In one, accelerometers were placed directly on the fingers, in order to complement the information provided by wrist-worn devices [82]. In the other, magnetometry was used to capture the movements of a finger-worn ring with respect to a sensor on the wrist [83]. Both of these approaches are promising because they capture the movements of both the arm and the fingers, which provides interesting opportunities to improve our understanding of their relative patterns of movement during functional activities. To date, however, these approaches have been used with individuals with strokes but not SCI.

A third type of approach has also been proposed to directly capture information about functional hand use, and therefore serve as the basis for measures of performance. Egocentric video is video data obtained from a wearable head-worn camera, and captures the user's point of view. Computer vision and machine learning algorithms can then be used to extract information about hand use and hand function from these videos (Fig. 4.4). The distinguishing feature of egocentric video compared to the other approaches discussed above is that it can capture contextual information in addition to body movement information. In other words, while other sensors can detect movements of the arms and/or hands, it can be difficult to associate these movements with specific functional activities. Egocentric video provides information about the tasks and objects manipulated, but at the cost of a significant increase in the complexity of data processing. In SCI, this platform has been used to demonstrate the detection of functional interactions [84, 85] and compensatory postures [86]. The recording of egocentric video in the homes of individuals with SCI has been shown to be feasible [87, 88] and acceptable [89].

Table 4.4 Clinical outcome assessment relevant to upper limb function and developed for groups other than individuals with SCI

Instrument	Purpose	Proximal arm	Distal arm/hand	Time to complete	Method used	Parameter measured	Scoring	Reliability	Validity	Advantages	Disadvantages
WMFT (Wolf Motor Function Test), 2001 [193, 194]	Research tool, upper limb function, forced nonuse of non-affected arm therapy evaluation Activity PerfO	Shoulder, elbow	Forearm, wrist, hand grasp	40 min (both sides)	21 single-handed specific movements and basic ADL tasks, 3 force measure tasks	Time, quality of movement and force	Time (s), weight (lbs)	$n = 19$, inter-rater ICC = 0.95–0.99 $P < 0.0001$	$n = 19$, FMA $\rho = 0.61$ $P < 0.02$	Norm values	Low validation correlation, specific for forced nonuse of non-affected arm therapy evaluation, single-handed specific, movement quality scoring reported to be described in non-existing reference
ARAT (Action Research Arm Test), 1981 [195]	Occupational therapy practice, clinical and research tool, upper limb function Activity PerfO	Shoulder, elbow	Forearm, wrist, hand grasp (pinch, grip)	∼40 min (both sides)	16 single-handed basic ADL tasks, 4 single-handed specific movements	Capacity to perform task or movement	0–3	$n = 20$, inter-rater $r = 0.99$, test–retest $r = 0.98$	$n = 50$, UEFT $r = 0.97$	Items arranged in hierarchical order of difficulty	Validation evaluation based on patient data from UEFT study, single-handed specific
QuickDASH (Quick Disabilities of the Arm, Shoulder, and Hand) [196, 197]	Clinical and research tool, upper limb function in patients with musculoskeletalof the upper limb and upper limb deficits due to cervical spine pathology Activity PRO	Shoulder, elbow	Forearm, wrist, hand grasp	10 min	Country-specific questionnaire regarding ∼30–40 single-handed/bimanual self-administrated ADL tasks	Self-perceived symptoms and capacity to perform task	Five point Likert scale	Test–retest reliability 0.90	Concurrent validity with NPRS: r = 0.74 MCID—8pts	Internationally validated, very robust PRO	Test self-perceived measure of disability, lacks objective data

(continued)

Table 4.4 (continued)

Instrument	Purpose	Proximal arm	Distal arm/hand	Time to complete	Method used	Parameter measured	Scoring	Reliability	Validity	Advantages	Disadvantages
COPM (Canadian Occupational Performance Measure), 1991 [198, 199]	Occupational therapy practice and clinical tool, occupational performance in patients with various disabilities Activity PRO	Shoulder, elbow	Forearm, wrist, hand grasp	20–40 min	Questionnaire regarding performance in self-care/productivity/leisure areas	Identification of problem areas and importance in daily functioning, self-perceived ability and self-satisfaction with performance	1–10 1–10 1–10	$n = 26$, test–retest item pool 56% $\rho = 0.88$–0.89 $p < 0.001$	$n = 26$, discriminant validation, BI, FAI, SASIP-30, EQ-5D, RS	International (35 countries), patient centred, not diagnosis specific, crosses developmental stages	Not upper limb specific, scores between patients not comparable
Box and Block test, 1960 [200–202]	Occupational therapy practice tool, upper limb function in patients with cerebral palsy Activity PerfO	Not assessed	Forearm, wrist, hand grasp	–	1 single-handed basic ADL task	Ability to pass or fail task within a given time	Number of passed/failed tasks	$n = 69$, test-retest ICC = 0.89–0.97	$n = 69$, ARAT $r = 0.80$–0.82, SMAF $r = 0.42$–0.54	Objective, norm values, rapid	Single-handed specific, proximal arm not assessed, based only on time not on movement quality

SCI spinal cord injury, *SW* Semmes-Weinstein, *ADL* activities of daily living, *ROM* range of motion, *ICC* intraclass correlation coefficient, *SCIM III* spinal cord independence measure version III, *SCIM III-SS* spinal cord independence measure version III-self-care subscore, *ASIA-MS* Neurological Classification of Spinal Cord Injury approved by the American Spinal Cord Association and the International Spinal Cord Injury Society-motor score, *FES* functional electrical stimulation, *FIM-SM* Functional Independence Measure-motor score, *UEMS* upper extremity motor score, *ICSHT* International Classification for Surgery of the Hand in Tetraplegia, *FIM* Functional Independence Measure, *FMA* Fugl-Meyer motor assessment, *UEFT* Upper Extremity Function Test, *BI* Barthel Index, *FAI* Frenchay Activities Index, *SASIP-30* Stroke-Adapted Sickness Impact Profile-30, *EQ-5D* EuroQol 5D, *RS* Rankin Scale, *MHQ* Michigan Hand Outcomes Questionnaire, *AUC* area under the curve, *AMA* American Medical Association, *SMAF* Système de mesure de l'autonomie fonctionnelle

4.6 Measures of Participation After Cervical Spinal Cord Injury

Assessment of participation is essential to determine the extent to which an individual is able to have "involvement in a life situation" [16]. Restrictions to participation would also define the problems an individual experiences with life situations. Most of the measures that are used in this domain are health-related quality of life measures, and therefore, not focal to the upper limb. However, they are useful in determining how an individual has an overall experience with their life situation. In conjunction with other measures specific to the upper limb, this information makes it possible to draw relationships between changes in upper limb function and quality of life. Because measures of participation are not specific to the upper limb, they are not reviewed here in detail, and the reader is referred to previous reviews on the topic [90, 91]. The Participation and Quality of life (PAR-QoL) Tool-Kit is an available resource intended to assist investigators and clinicians in selecting measures of participation and quality of life when working with individuals living with SCI [92].

4.7 Therapeutic Approaches for the Rehabilitation of Upper Extremity Function

The discussion above has focused on the use of technology for assessment. In many cases, related technologies can be used both to measure or to alter the neurological or functional state of an individual living with SCI. The remainder of this chapter will therefore focus on neurorehabilitation technologies that have been proposed for upper extremity rehabilitation after SCI. Both preclinical and clinical studies have shown the benefits of activity-based therapy on the recovery of upper extremity function [93–100], suggesting that repetitive, engaged, demanding and task-oriented training can promote neurorecovery below the level of the lesion [101, 102]. Studies carried out in rats with partial SCIs affecting the upper limb have demonstrated that training and enhanced activity increase neural plasticity and thereby improve motor recovery [95, 97, 99, 103]. Accordingly, after an incomplete cervical SCI, diverse training therapy of the upper limb is required to avoid muscular atrophy of the remaining (active) motor functions and recover, to variable extents, lost neuromotor functions [98]. Given the difficulty of delivering high doses of activity-based therapy in the context of current clinical constraints, a number of technological approaches have been suggested. These include methods to deliver therapy in the clinic (e.g. robotic rehabilitation devices), enhance the recovery trajectory (e.g. neuromodulation) or more effectively deliver rehabilitation in the home so as to extend the effective training time available (e.g. wearable, robotic, virtual reality or game-based systems). These approaches are reviewed below.

4.7.1 Robotic Systems for Upper Limb Training

Robot-assisted training has demonstrated some benefits for upper extremity function in neurorehabiltiation [104–106], and there are clear advantages to technology-based therapies. For example, robot-supported training can be more intensive, of longer duration and more repetitive than manual arm training. Additionally, the motivation of the user to perform repeated training exercises can be enhanced if they are embedded within entertaining computer games. For example, in a study comparing technological-based therapy (T-WREX) with conventional therapy, the participants reported a preference for training with the T-WREX [107]. Hence, a number of robots have been developed to train upper extremity function following neurological damage. Several studies have focused on the effectiveness of these technologies after cervical SCI, although most with relatively small samples [108–115]. Recent work with the ArmeoPower® exoskeleton (derived from the ARMin III device

[104]) and the Amadeo hand training robot found benefits comparable to those of conventional occupational therapy [116, 117]. Other robotic upper extremity training devices that have been investigated in SCI include the RiceWrist, Haptic Master, MAHI Exo-II, ReoGo and the InMotion 3.0 Wrist robot, as reviewed in [118].

The general trend of robot-assisted training providing benefits but not exceeding those offered by dose-matched conventional therapy are consistent with the findings of larger studies in stroke. Of note here is the MIT-Manus robot, which has been the focus of several large randomized controlled trials. This end-effector robot contains modules for training the arm, wrist and hand with video-guided exercises [119, 120]. The robot can move, guide or perturb the movement of a user's upper extremity while recording measures such as position, velocity and force. A multicentre randomized controlled study comparing intensive robot-assisted therapy, using the MIT-Manus, with intensive conventional therapy and usual care after stroke showed that after 12 weeks of training intensive robot-assisted therapy did not significantly improve motor function compared to either of the other two therapies. In fact, in comparison to those receiving intensive conventional therapy, the participants using the robot did worse. The participants using the robot did do better than those receiving usual care; however, the results were not significant at the 12-week time point [121]. After 36 weeks of training robot-assisted therapy, robot-trained participants showed significant improvements compared to usual care but not when compared with those receiving intensive therapy [121]. Findings were similar in the more recent Robot Assisted Training for the Upper Limb after Stroke (RATULS) trial. This study allocated 770 participants to robot-assisted training, enhanced upper limb therapy (EULT), or usual care. The groups were not found to be different in terms of the primary outcome, upper limb function success at 3 months defined based on the ARAT. Robot-assisted training showed improvement over usual care on the FMA motor subscale at 3 and 6 months, and no differences in other scales. EULT showed improvements over usual care on several measures. Robot-assisted training was found to be the least cost-effective of the three modalities studied [122]. The authors concluded that the evidence did not support this form of robot-assisted training in routine clinical practice, and that further innovations to improve efficacy and cost-effectiveness were warranted. Overall, considering that the current generation of rehabilitation robots seems able to perform comparably but not better than therapists, future efforts could focus on the development of devices that are sufficiently low-cost to be deployed at home, which would enable a much more significant increase in training dose and potentially lead to better outcomes.

An alternative direction for robotics in the context of upper limb function after SCI is the use of soft robotic gloves. While these efforts have primarily focused on using the devices as orthoses for an immediate improvement in ADL task performance [123–125], early evidence also exists for a therapeutic effect [126]. Being wearable and lightweight, this form of rehabilitation robotics is well suited for deployment into home and community environments, making it a promising avenue for research.

4.7.2 Passive Workstations and Interactive Platforms for Upper Limb Training

A variety of non-actuated interactive platforms have been used to support rehabilitation. In contrast to robotic training, these approaches may integrate passive weight support, forms of motion capture and/or interactive exercises, but do not move the user's limbs. The T-WREX, a forerunner of the ArmeoSpring®, was initially developed to enable stroke survivors with chronic hemiparesis to practice arm movements without the continuous supervision of a therapist. It consists of an orthosis that assists arm movement, a grip sensor that senses hand grip pressure and software that simulates functional activities. The exoskeleton has five degrees of freedom and passively counterbalances the weight of the arm against gravity by means of elastic bands [107].

A study comparing motor training with T-WREX versus conventional training with a table top for gravity support in chronic stroke survivors showed that all participants significantly improved motor function [127]. In addition, rehabilitation therapy with T-WREX was associated with modest maintenance of progress at the 6-month follow-up as compared with conventional therapy [127]. Later the T-WREX was commercialized as the ArmeoSpring®. The gravity compensated ArmeoSpring® robot, an example of passive workstations, has been tested with individuals with SCI. Study participants with cervical SCI completed five weeks of training with the device. While there were no statistically significant differences between the robot-trained and the control arm for any outcome measured in the overall group, individuals with some preserved hand function demonstrated increased scores in the GRASSP-Sensibility component [96]. More recently, a large randomized controlled trial compared the efficacy of self-rehabilitation sessions using the Armeo-Spring® with self-rehabilitation sessions using conventional exercises in the sub-acute stage after stroke and found no differences between groups [128].

The ReJoyce® platform can assess hand function and provide upper limb rehabilitation training for individuals with stroke as well as those with SCI. The apparatus consists of a four degrees-of-freedom spring-loaded arm (joystick), attached to a table or desk. The automated exercises are incorporated into games that comprise ADL tasks played by manipulating attachments on the device. The joystick has integrated sensors that provide quantitative information on displacement of the manipulated attachments and prehension force. A study comparing FES and ReJoyce-based therapy with FES and conventional exercises in SCI participants showed that FES together with ReJoyce-based therapy resulted in (statistically and clinically) greater improvements than those obtained with the more conventional protocol [93].

Apart from the ArmeoSpring® and ReJoyce®, other passive devices have mostly been tested in stroke survivors; for example, the HandSOME device, another passive workstation, was shown to increase finger range of motion when used for training with participants with stroke [129].

A variety of other technologies have been proposed that combine forms of movement tracking with interactive game-like training interfaces. This category includes systems based on wearable devices [130–134], virtual reality [135], and mass-market video-game systems or computer interfaces with motion capture sensors [136–138]. Most of these reports were either uncontrolled or pilot studies, and therefore the body of evidence regarding the effectiveness of these approaches in SCI remains limited. On the other hand, the rapidly growing literature describing their effects after stroke [139–141] indicates that greater translation to SCI is likely in the near future.

4.7.3 Neuromodulation Systems for Upper Limb Training

Several types of neuromodulation interventions have shown considerable promise to enhance neurorecovery after SCI. These approaches use electrical stimulation to modulate the activity of the nervous system in either peripheral or central locations. The most widely used modality to date has been functional electrical stimulation (FES), in which surface stimulation is applied to paralyzed or paretic muscles while the patient attempts to perform functional movements (for more information, refer to Chap. 19 in this book). In particular, FES-based neuroprosthesis devices for grasping, such as the Compex Motion-based neuroprosthesis [142, 143], the ETHZ-Paracare [143], the Freehand [144], the Ness Handmaster [145, 146], the Bionic Glove [147, 148] and the MyndMove [149] (Fig. 4.4) all incorporate FES and are designed to restore or improve grasping function. While FES was initially used as a dynamic orthosis to improve the ability to manipulate objects with active stimulation [145, 147, 148, 150, 151], the intent of FES therapy is to promote neurorecovery. Series of FES therapy sessions have been shown to produce significant improvements in reaching and grasping function after SCI [142, 152–154]. A noteworthy variant of these approaches is BCI-controlled FES, which has been suggested to further enhance the effectiveness of the stimulation [155–157]. Despite these promising findings, in order to better determine which types of SCI patients benefit the most from FES-based therapies and why, detailed investigations will be required. Of note, implantable FES systems for restoring upper extremity function have a long history of investigation, but are intended as assistive devices rather than therapeutic interventions, and therefore not discussed here [158, 159].

Recently, a wider variety of neuromodulation approaches have come under investigation [160]. For example, both epidural [161] and transcutaneous [162, 163] spinal cord stimulation applied at the cervical level have demonstrated promising early results after chronic SCI. Transcranial direct current stimulation is also appealing for its simplicity and has shown some indication that it may help to improve upper extremity function after SCI [164–167]. Likewise, peripheral nerve somatosensory stimulation and repetitive transcranial magnetic stimulation (TMS) have both been suggested as a means to enhance the benefits of repetitive functional task practice [168, 169]. TMS may likewise be applied in combination with peripheral nerve stimulation in a paired associative stimulation protocol intended to modulate corticospinal-motoneural synaptic transmission [170]. Another intriguing modality is the use of implanted vagus nerve stimulators to promote neuroplasticity and enhance rehabilitation outcomes. This approach has been demonstrated to date in humans after stroke [171] and in animal models of SCI [172]. Yet others have the potential to benefit individuals with SCI but have not yet been applied in this population, for example, translingual neural stimulation [173]. This surge in interest and variety of neuromodulation strategies holds great promise to improve upper extremity rehabilitation after SCI in the next few years.

4.8 Discussion

Upper extremity impairment has long been recognized as highly detrimental to independence and quality of life after cervical SCI, yet there remains a great need for interventions able to promote recovery and gains in function. Neurorehabilitation technologies have a dual role to play towards this goal. On the assessment side, technology can provide information that is complementary to current clinical assessments, including quantitative measurements, determining candidacy, more detailed longitudinal tracking and the ability to evaluate therapeutic benefits in the context of home and community environments. Accurate and sensitive measurements are valuable in the context of clinical trials, which have historically been a challenging endeavour in the field of SCI as a result of often small and heterogeneous samples. The capabilities provided by these technologies also coincide with a growing interest and progress in telerehabilitation that will increasingly call for effective strategies to measure function remotely. On the therapeutic side, robotic, wearable and other interactive platforms can increase dose and access to task-based training, while neuromodulation can directly alter the activity of neural circuitry to enhance neurorecovery. These approaches, either individually or in combinations, hold real promise of meaningful improvements in outcomes after cervical SCI.

The ICF framework can be helpful to contextualize and describe the role of technology in neurorehabilitation assessment. A wide range of clinical upper extremity capacity measures have been used in SCI studies. While many of these were originally developed for other populations (e.g. stroke), several measures have become available that are developed specifically for SCI (GRASSP, CUE-T). Measures of upper extremity performance, on the other hand, remain more limited in number and suffer from a reliance on self-report. Several types of wearable technologies are being developed in an effort to fill this gap. With proper validation, these tools may achieve the goal of accurate and sensitive longitudinal data collection in unrestricted environments. Improving our ability to assess performance is critical again from the perspective of clinical trials, because it will enable us to accurately describe the true impact of an intervention on the daily life of individuals living with SCI, and therefore help to provide concrete data showing which improvements are meaningful.

For therapeutic effectiveness, several neuromodulation approaches are showing promising results, including FES therapy and spinal cord stimulation. Robotics, passive support systems and interactive platforms have consistently demonstrated an ability to match the benefits of equivalent doses of conventional therapy, though generally not to surpass them. Their greatest promise may lie in improving access to rehabilitation in home and community environments, provided that this proves to be economically feasible. Once again, many neurorehabiltiation technologies were not developed with SCI as the primary intended indication, and may not be ideally suited to this population. For example, devices developed with a focus on hemiparesis after stroke may provide only unilateral training, rather than the bimanual training that would be required for individuals with SCI. The clinical value of these technologies in SCI and a thorough evaluation of their specific advantages/disadvantages over conventional therapies continue to be needed.

As the adoption of neurorehabilitation technology increases, the use of tools that can provide both therapy and assessment is potentially a major benefit. Wearable and robotic systems can both give guidance to a patient during exercises and provide longitudinal tracking of movement quality. Electrophysiological diagnostics could be integrated within neuromodulation systems to quantify neurorecovery. These strategies may lead to data-driven therapy progression and, by combining automated interventional and outcomes data, generate valuable practice-based evidence. Sensor-driven therapies can also allow more effective management of situations that involve subtle variations in function. In particular, the changing demographics of SCI are characterized by an increase in older individuals with non-traumatic injuries (e.g. cervical myelopathies) [2, 174]. These conditions can manifest in loss of hand dexterity, calling for more sensitive assessments to tailor therapy and determine which treatments are effective.

Realizing the benefits of novel neurorehabilitation technologies will require a concerted effort towards translation of research to the clinic. While this is a challenging task for any new clinical assessment or intervention, technological approaches can often face additional hurdles related to the cost and complexity of the devices. The financial aspects may vary across healthcare systems, but there will be a common need to create technologies that are perceived as providing clear benefits by all stakeholders, are cost-effective, easy to use, quick to set up, robust and can integrate smoothly into existing clinical workflows and information systems (e.g. electronic health records). All of these needs should be considered early on and integrated on an ongoing basis into the technology development and validation process, for example using a staged version of the FAME framework (Feasibility, Appropriateness, Meaningfulness, Effectiveness, Economic Evidence) [175]. In this manner, we can ensure that research activities lead to improvements in the quality of life of individuals living with cervical SCI.

4.9 Conclusion

Upper limb impairment is a major driver of reduced independence and quality of life after cervical SCI. New interventional strategies are needed to help restore upper limb function, and their validation should be supported by comprehensive, multi-modal assessment protocols. Neurorehabilitation technologies have a key role to play in both regards.

For assessment, the benefits of technological approaches can be summarized in two points.

First, sensors provide the opportunity to measure a broader range of parameters than what can be accomplished through clinical observation alone. This notion has long been appreciated, for example, in the form of well-established electrophysiological assessments. It also finds new embodiments in the form of wearable, robotic or optical devices measuring kinematic and kinetic parameters. Second, by enabling us to collect data in more locations and at more timepoints than previously possible, new technologies can act as valuable complements to clinical outcome assessments. For example, wearable sensors are beginning to yield significant new data about the performance domain of the ICF, which was previously difficult to reliably assess.

Similarly, therapeutic technologies are most beneficial when they go beyond what could otherwise be accomplished in the clinic. Robotic rehabilitation systems could in the future increase training doses in both clinical and home environments, a crucial consideration given the amounts of massed practice needed to support neurorecovery. Neuromodulation technologies go further by directly altering the activity of neural systems, and have demonstrated undeniable benefits across a range of protocols and applications.

As the technologies reviewed in this chapter proceed to definitive validation in well-powered studies, they have considerable potential to improve neurorecovery outcomes for individuals living with cervical SCI. New tools also enable the field to progress in new directions, including dealing with the shifting demographics of SCI and the need to create a stronger continuum of care from rehabilitation centres to the home and community.

References

1. Kalsi-Ryan S, Karadimas SK, Fehlings MG. Cervical spondylotic myelopathy: the clinical phenomenon and the current pathobiology of an increasingly prevalent and devastating disorder. Neuroscientist. 2013;19(4):409–21.
2. Nouri A, Tetreault L, Singh A, Karadimas SK, Fehlings MG. Degenerative cervical myelopathy: epidemiology, genetics, and pathogenesis. Spine. 2015;40(12):E675–93.
3. Anderson KD. Targeting recovery: priorities of the spinal cord-injured population. J Neurotrauma. 2004;21(10):1371–83.
4. Snoek GJ, IJzerman MJ, Hermens HJ, Maxwell D, Biering-Sorensen F. Survey of the needs of patients with spinal cord injury: impact and priority for improvement in hand function in tetraplegics. Spinal Cord. 2004;42(9):526–32.
5. Consortium for Spinal Cord Medicine. Outcomes following traumatic spinal cord injury: clinical practice guidelines for health-care professionals. Paralyzed Veterans of America; 1999.
6. Lamb DW, Chan KM. Surgical reconstruction of the upper limb in traumatic tetraplegia. A review of 41 patients. J Bone Joint Surg Br. 1983;65(3):291–8.
7. Brand PW, Hollister AM. Clinical mechanics of the hand. Mosby Inc.; 1999.
8. Shumway-Cook A, Woollacott MH. Motor control: translating research into clinical practice. Lippincott Williams & Wilkins; 2007.
9. Sollerman C, Ejeskar A. Sollerman hand function test. A standardised method and its use in tetraplegic patients. Scand J Plast Reconstr Surg Hand Surg. 1995;29(2):167–76.
10. Marino RJ, Ditunno JF Jr, Donovan WH, Maynard F Jr. Neurologic recovery after traumatic spinal cord injury: data from the model spinal cord injury systems. Arch Phys Med Rehabil. 1999;80 (11):1391–6.
11. Sekhon LH, Fehlings MG. Epidemiology, demographics, and pathophysiology of acute spinal cord injury. Spine. 2001;26(24S):S2–12.
12. Popovic MR, Thrasher TA, Adams ME, Takes V, Zivanovic V, Tonack MI. Functional electrical therapy: retraining grasping in spinal cord injury. Spinal Cord. 2006;44(3):143–51.
13. Beekhuizen KS, Field-Fote EC. Massed practice versus massed practice with stimulation: effects on upper extremity function and cortical plasticity in individuals with incomplete cervical spinal cord injury. Neurorehabil Neural Repair. 2005;19 (1):33–45.
14. Schwab JM, Brechtel K, Mueller C, Failli V, Kaps H, Tuli SK, et al. Experimental strategies to promote spinal cord regeneration—an integrative perspective. Prog Neurobiol. 2006;78(2):91–116.
15. Baptiste DC, Fehlings MG. Pharmacological approaches to repair the injured spinal cord. J Neurotrauma. 2006;23(3–4):318–34.
16. World Health Organization. International classification of functioning, disability and health (ICF). World Health Organization; 2001.
17. Food and Drug Administration. No title. Clinical outcome assessment compendium 2016.
18. Mokkink LB, Terwee CB, Patrick DL, Alonso J, Stratford PW, Knol DL, et al. The COSMIN study reached international consensus on taxonomy, terminology, and definitions of measurement

properties for health-related patient-reported outcomes. J Clin Epidemiol. 2010;63(7):737–45.

19. Kvien TK, Heiberg T, Hagen KB. Minimal clinically important improvement/difference (MCII/MCID) and patient acceptable symptom state (PASS): what do these concepts mean? Ann Rheum Dis. 2007;66(suppl 3):iii40–1.

20. Kirshblum SC, Burns SP, Biering-Sorensen F, Donovan W, Graves DE, Jha A, et al. International standards for neurological classification of spinal cord injury (revised 2011). J Spinal Cord Med. 2011;34(6):535–46.

21. Spiess M, Schubert M, Kliesch U, Halder P, EM-SCI study group. Evolution of tibial SSEP after traumatic spinal cord injury: baseline for clinical trials. Clin Neurophysiol. 2008;119(5):1051–61.

22. Kramer JL, Moss AJ, Taylor P, Curt A. Assessment of posterior spinal cord function with electrical perception threshold in spinal cord injury. J Neurotrauma. 2008;25(8):1019–26.

23. Chen AC, Niddam DM, Arendt-Nielsen L. Contact heat evoked potentials as a valid means to study nociceptive pathways in human subjects. Neurosci Lett. 2001;316(2):79–82.

24. Kramer J, Haefeli J, Curt A, Steeves JD. Increased baseline temperature improves the acquisition of contact heat evoked potentials after spinal cord injury. Clin Neurophysiol. 2012;123(3):582–9.

25. Ulrich A, Haefeli J, Blum J, Min K, Curt A. Improved diagnosis of spinal cord disorders with contact heat evoked potentials. Neurology. 2013;80 (15):1393–9.

26. Curt A, Keck ME, Dietz V. Functional outcome following spinal cord injury: significance of motor-evoked potentials and ASIA scores. Arch Phys Med Rehabil. 1998;79(1):81–6.

27. Kimura J. Electrodiagnosis in diseases of nerve and muscle: principles and practice. 2013.

28. Heald E, Hart R, Kilgore K, Peckham PH. Characterization of volitional electromyographic signals in the lower extremity after motor complete spinal cord injury. Neurorehabil Neural Repair. 2017;31 (6):583–91.

29. Del Vecchio A, Holobar A, Falla D, Felici F, Enoka RM, Farina D. Tutorial: Analysis of motor unit discharge characteristics from high-density surface EMG signals. J Electromyogr Kinesiol. 2020;53:102426.

30. Nawab SH, Chang S, De Luca CJ. High-yield decomposition of surface EMG signals. Clin Neurophysiol. 2010;121(10):1602–15.

31. A wearable neural interface for detecting and decoding attempted hand movements in a person with tetraplegia. In: 2019 41st annual international conference of the IEEE engineering in medicine and biology society (EMBC). IEEE; 2019.

32. Zhang X, Li X, Tang X, Chen X, Chen X, Zhou P. Spatial filtering for enhanced high-density surface electromyographic examination of neuromuscular changes and its application to spinal cord injury. J Neuroeng Rehabil. 2020;17(1):1–14.

33. Mathiowetz V, Weber K, Volland G, Kashman N. Reliability and validity of grip and pinch strength evaluations. J Hand Surg. 1984;9(2):222–6.

34. Jaber R, Hewson DJ, Duchêne J. Design and validation of the Grip-ball for measurement of hand grip strength. Med Eng Phys. 2012;34(9):1356–61.

35. Sisto SA, Dyson-Hudson T. Dynamometry testing in spinal cord injury. J Rehabil Res & Dev. 2007;44 (1).

36. Critical review of the evaluation of the results of upper limb functional surgery in tetraplegia since 50 years. Annales de readaptation et de medecine physique: revue scientifique de la Societe francaise de reeducation fonctionnelle de readaptation et de medecine physique. 2004.

37. Shin H, Moon SW, Kim G, Park JD, Kim JH, Jung MJ, et al. Reliability of the pinch strength with digitalized pinch dynamometer. Ann Rehabil Med. 2012;36(3):394.

38. Windolf M, Götzen N, Morlock M. Systematic accuracy and precision analysis of video motion capturing systems—exemplified on the Vicon-460 system. J Biomech. 2008;41(12):2776–80.

39. Hingtgen B, McGuire JR, Wang M, Harris GF. An upper extremity kinematic model for evaluation of hemiparetic stroke. J Biomech. 2006;39(4):681–8.

40. Murphy MA, Willn C, Sunnerhagen KS. Kinematic variables quantifying upper-extremity performance after stroke during reaching and drinking from a glass. Neurorehabil Neural Repair. 2011;25(1):71–80.

41. Kurillo G, Chen A, Bajcsy R, Han JJ. Evaluation of upper extremity reachable workspace using Kinect camera. Technol Health Care. 2013;21(6):641–56.

42. Mobini A, Behzadipour S, Saadat M. Test–retest reliability of Kinect's measurements for the evaluation of upper body recovery of stroke patients. Biomed Eng Online. 2015;14(1):1.

43. Kuster RP, Heinlein B, Bauer CM, Graf ES. Accuracy of KinectOne to quantify kinematics of the upper body. Gait Posture. 2016;47:80–5.

44. Cao Z, Martinez GH, Simon T, Wei S, Sheikh YA. OpenPose: realtime multi-person 2D pose estimation using part affinity fields. IEEE Trans Pattern Anal Mach Intell. 2019.

45. RMPE: Regional multi-person pose estimation. In: Proceedings of the IEEE international conference on computer vision; 2017.

46. Arac A, Zhao P, Dobkin BH, Carmichael ST, Golshani P. DeepBehavior: a deep learning toolbox for automated analysis of animal and human behavior imaging data. Front Syst Neurosci. 2019;13:20.

47. Mehta D, Sridhar S, Sotnychenko O, Rhodin H, Shafiei M, Seidel H, et al. Vnect: Real-time 3d human pose estimation with a single RGB camera. ACM Trans Graph (TOG). 2017;36(4):1–14.

48. Using a single rgb frame for real time 3d hand pose estimation in the wild. In: 2018 IEEE winter conference on applications of computer vision (WACV). IEEE; 2018.

49. Lee YM, Lee S, Uhm KE, Kurillo G, Han JJ, Lee J. Upper Limb Three-dimensional reachable workspace analysis using the kinect sensor in hemiplegic stroke patients: a cross-sectional observational study. Am J Phys Med Rehabil. 2020;99 (5):397–403.

50. Scano A, Molteni F, Molinari TL. Low-cost tracking systems allow fine biomechanical evaluation of upper-limb daily-life gestures in healthy people and post-stroke patients. Sensors. 2019;19 (5):1224.

51. Bakhti K, Laffont I, Muthalib M, Froger J, Mottet D. Kinect-based assessment of proximal arm non-use after a stroke. J Neuroeng Rehabil. 2018;15(1):1–12.

52. Bonnechère B, Sholukha V, Omelina L, Van Sint JS, Jansen B. 3D analysis of upper limbs motion during rehabilitation exercises using the KinectTM sensor: development, laboratory validation and clinical application. Sensors. 2018;18 (7):2216.

53. Lee S, Lee Y, Kim J. Automated evaluation of upper-limb motor function impairment using Fugl-Meyer assessment. IEEE Trans Neural Syst Rehabil Eng. 2017;26(1):125–34.

54. Kim W, Cho S, Baek D, Bang H, Paik N. Upper extremity functional evaluation by Fugl-Meyer assessment scoring using depth-sensing camera in hemiplegic stroke patients. PLoS ONE. 2016;11(7): e0158640.

55. Colombo Zefinetti F, Vitali A, Regazzoni D, Rizzi C, Molinero G. Tracking and characterization of spinal cord-injured patients by means of RGB-D sensors. Sensors. 2020;20(21):6273.

56. Gil-Agudo Á, de Los R-G, Dimbwadyo-Terrer I, Peñasco-Martín B, Bernal-Sahún A, López-Monteagudo P, et al. A novel motion tracking system for evaluation of functional rehabilitation of the upper limbs. Neural Regen Res. 2013;8 (19):1773.

57. Oess NP, Wanek J, Curt A. Design and evaluation of a low-cost instrumented glove for hand function assessment. J Neuroeng Rehabil. 2012;9(1):1.

58. de Castro MF, Cliquet A Jr. An artificial grasping evaluation system for the paralysed hand. Med Biol Eng Comput. 2000;38(3):275–80.

59. Vanello N, Hartwig V, Tesconi M, Ricciardi E, Tognetti A, Zupone G, et al. Sensing glove for brain studies: design and assessment of its compatibility for fMRI with a robust test. IEEE/ASME Trans Mechatron. 2008;13(3):345–54.

60. Su X, Hou C, Shen B, Zhang W, Wu D, Li Q, et al. Clinical application of a new assessment tool for myelopathy hand using virtual reality. Spine. 2020;45(24):E1645–52.

61. Krebs HI, Krams M, Agrafiotis DK, DiBernardo A, Chavez JC, Littman GS, et al. Robotic measurement of arm movements after stroke establishes biomarkers of motor recovery. Stroke. 2014;45(1):200–4.

62. Mostafavi SM, Mousavi P, Dukelow SP, Scott SH. Robot-based assessment of motor and proprioceptive function identifies biomarkers for prediction of functional independence measures. J Neuroeng Rehabil. 2015;12(1):1.

63. Bosecker C, Dipietro L, Volpe B, Krebs HI. Kinematic robot-based evaluation scales and clinical counterparts to measure upper limb motor performance in patients with chronic stroke. Neurorehabil Neural Repair. 2010;24(1):62–9.

64. Balasubramanian S, Colombo R, Sterpi I, Sanguineti V, Burdet E. Robotic assessment of upper limb motor function after stroke. Am J Phys Med Rehabil. 2012;91(11 Suppl 3):255.

65. McKenzie A, Dodakian L, See J, Le V, Quinlan EB, Bridgford C, et al. Validity of robot-based assessments of upper extremity function. Arch Phys Med Rehabil. 2017;98(10):1969–76. e2.

66. Zariffa J, Kapadia N, Kramer JL, Taylor P, Alizadeh-Meghrazi M, Zivanovic V, et al. Relationship between clinical assessments of function and measurements from an upper-limb robotic rehabilitation device in cervical spinal cord injury. IEEE Trans Neural Syst Rehabil Eng. 2012;20 (3):341–50.

67. Prochazka A, Kowalczewski J. A fully automated, quantitative test of upper limb function. J Mot Behav. 2015;47(1):19–28.

68. Keller U, Schölch S, Albisser U, Rudhe C, Curt A, Riener R, et al. Robot-assisted arm assessments in spinal cord injured patients: a consideration of concept study. PLoS ONE. 2015;10(5): e0126948.

69. Grasse KM, Hays SA, Rahebi KC, Warren VS, Garcia EA, Wigginton JG, et al. A suite of automated tools to quantify hand and wrist motor function after cervical spinal cord injury. J Neuroeng Rehabil. 2019;16(1):1–12.

70. Zariffa J. Improving neurorehabilitation of the upper limb through big data. Signal Process Mach Learn Biomed Big Data. CRC Press; 2018. p. 533–50.

71. Zariffa J, Myers M, Coahran M, Wang RH. Smallest real differences for robotic measures of upper extremity function after stroke: Implications for tracking recovery. J Rehabil Assist Tech Eng. 2018;5:2055668318788036.

72. Catz A, Greenberg E, Itzkovich M, Bluvshtein V, Ronen J, Gelernter I. A new instrument for outcome assessment in rehabilitation medicine: spinal cord injury ability realization measurement index. Arch Phys Med Rehabil. 2004;85(3):399–404.

73. Bluvshtein V, Front L, Itzkovich M, Aidinoff E, Gelernter I, Hart J, et al. SCIM III is reliable and valid in a separate analysis for traumatic spinal cord lesions. Spinal Cord. 2011;49(2):292–6.

74. Noorkoiv M, Rodgers H, Price CI. Accelerometer measurement of upper extremity movement after stroke: a systematic review of clinical studies. J Neuroeng Rehabil. 2014;11:144.

75. Waddell KJ, Strube MJ, Bailey RR, Klaesner JW, Birkenmeier RL, Dromerick AW, et al. Does task-specific training improve upper limb performance in daily life poststroke? Neurorehabil Neural Repair. 2016;31(3):290–300.

76. Lum PS, Shu L, Bochniewicz EM, Tran T, Chang L, Barth J, et al. Improving accelerometry-based measurement of functional use of the upper extremity after stroke: machine learning versus counts threshold method. Neurorehabil Neural Repair. 2020;34(12):1078–87.

77. Popp WL, Brogioli M, Leuenberger K, Albisser U, Frotzler A, Curt A, et al. A novel algorithm for detecting active propulsion in wheelchair users following spinal cord injury. Med Eng Phys. 2016;38(3):267–74.

78. Brogioli M, Popp WL, Albisser U, Brust AK, Frotzler A, Gassert R, et al. Novel sensor technology to assess independence and limb-use laterality in cervical spinal cord injury. J Neurotrauma. 2016.

79. Popp WL, Schneider S, Bär J, Bösch P, Spengler CM, Gassert R, et al. Wearable sensors in ambulatory individuals with a spinal cord injury: from energy expenditure estimation to activity recommendations. Front Neurol. 2019;10:1092.

80. Brogioli M, Popp WL, Schneider S, Albisser U, Brust AK, Frotzler A, et al. Multi-day recordings of wearable sensors are valid and sensitive measures of function and independence in human spinal cord injury. J Neurotrauma. 2017;34 (6):1141–8.

81. Predicting upper limb compensation during prehension tasks in tetraplegic spinal cord injured patients using a single wearable sensor. In: 2019 IEEE 16th international conference on rehabilitation robotics (ICORR). IEEE; 2019.

82. Lee SI, Liu X, Rajan S, Ramasarma N, Choe EK, Bonato P. A novel upper-limb function measure derived from finger-worn sensor data collected in a free-living setting. PLoS ONE. 2019;14(3): e0212484.

83. Schwerz de Lucena D, Rowe J, Chan V, Reinkensmeyer DJ. Magnetically counting hand movements: validation of a calibration-free algorithm and application to testing the threshold hypothesis of real-world hand use after stroke. Sensors. 2021;21 (4):1502.

84. Likitlersuang J, Sumitro ER, Cao T, Visée RJ, Kalsi-Ryan S, Zariffa J. Egocentric video: a new tool for capturing hand use of individuals with spinal cord injury at home. J Neuroeng Rehabil. 2019;16(1):83.

85. A wearable vision-based system for detecting hand-object interactions in individuals with cervical spinal cord injury: First results in the home environment. In: 42nd annual international conference of the IEEE engineering in medicine & biology society (EMBC). IEEE; 2020 July.

86. Dousty M, Zariffa J. Tenodesis grasp detection in egocentric video. JBHI. 2021;25(5):1463–70.

87. Likitlersuang J, Visée RJ, Kalsi-Ryan S, Zariffa J. Capturing hand use of individuals with spinal cord injury at home using egocentric video: a feasibility study. Spinal Cord Ser Cases. 2021;7 (1):17.

88. Tsai M, Bandini A, Wang RH, Zariffa J. Capturing Representative Hand Use at Home Using Egocentric Video in Individuals with Upper Limb Impairment. J Vis Exp JoVE. 2020;(166).

89. Bandini A, Kalsi-Ryan S, Craven BC, Zariffa J, Hitzig SL. Perspectives and recommendations of individuals with tetraplegia regarding wearable cameras for monitoring hand function at home: insights from a community-based study. J Spinal Cord Med. 2021:1–12.

90. Noonan VK, Kopec JA, Noreau L, Singer J, Dvorak MF. A review of participation instruments based on the international classification of functioning, disability and health. Disabil Rehabil. 2009;31 (23):1883–901.

91. Eyssen IC, Steultjens MP, Dekker J, Terwee CB. A systematic review of instruments assessing participation: challenges in defining participation. Arch Phys Med Rehabil. 2011;92(6):983–97.

92. Hitzig S, Routhier F, Noreau L, Kairy D, Lynda A, Craven BC. The Participation and Quality of Life (PAR-QoL) tool-kit: outcomes and next steps. Arch Phys Med Rehabil. 2014;95(10): e83.

93. Kowalczewski J, Chong S, Galea M, Prochazka A. In-home tele-rehabilitation improves tetraplegic hand function. Neurorehabil Neural Repair. 2011;25(5):412–22.

94. Hoffman H, Sierro T, Niu T, Sarino ME, Sarrafzadeh M, McArthur D, et al. Rehabilitation of hand function after spinal cord injury using a novel handgrip device: a pilot study. J Neuroeng Rehabil. 2017;14(1):1–7.

95. Girgis J, Merrett D, Kirkland S, Metz G, Verge V, Fouad K. Reaching training in rats with spinal cord injury promotes plasticity and task specific recovery. Brain. 2007;130(11):2993–3003.

96. Zariffa J, Kapadia N, Kramer JL, Taylor P, Alizadeh-Meghrazi M, Zivanovic V, et al. Feasibility and efficacy of upper limb robotic rehabilitation in a subacute cervical spinal cord injury population. Spinal Cord. 2012;50(3):220–6.

97. Starkey ML, Bleul C, Maier IC, Schwab ME. Rehabilitative training following unilateral pyramidotomy in adult rats improves forelimb function in a non-task-specific way. Exp Neurol. 2011;232 (1):81–9.

98. Starkey ML, Bleul C, Kasper H, Mosberger AC, Zörner B, Giger S, et al. High-impact, self-motivated training within an enriched environment with single animal tracking dose-dependently promotes motor skill acquisition and functional recovery. Neurorehabil Neural Repair. 2014;28(6):594–605.

99. Maier IC, Baumann K, Thallmair M, Weinmann O, Scholl J, Schwab ME. Constraint-induced movement therapy in the adult rat after unilateral

corticospinal tract injury. J Neurosci. 2008;28 (38):9386–403.

100. Krajacic A, Weishaupt N, Girgis J, Tetzlaff W, Fouad K. Training-induced plasticity in rats with cervical spinal cord injury: effects and side effects. Behav Brain Res. 2010;214(2):323–31.

101. Quel de Oliveira C, Refshauge K, Middleton J, de Jong L, Davis GM. Effects of activity-based therapy interventions on mobility, independence, and quality of life for people with spinal cord injuries: a systematic review and meta-analysis. J Neurotrauma. 2017;34(9):1726–ss43.

102. Behrman AL, Ardolino EM, Harkema SJ. Activity-based therapy: from basic science to clinical application for recovery after spinal cord injury. J Neurol Phys Ther JNPT. 2017;41(Suppl 3 IV STEP Spec Iss):S39.

103. Carmel JB, Berrol LJ, Brus-Ramer M, Martin JH. Chronic electrical stimulation of the intact corticospinal system after unilateral injury restores skilled locomotor control and promotes spinal axon outgrowth. J Neurosci. 2010;30(32):10918–26.

104. Klamroth-Marganska V, Blanco J, Campen K, Curt A, Dietz V, Ettlin T, et al. Three-dimensional, task-specific robot therapy of the arm after stroke: a multicentre, parallel-group randomised trial. Lancet Neurol. 2014;13(2):159–66.

105. Staubli P, Nef T, Klamroth-Marganska V, Riener R. Effects of intensive arm training with the rehabilitation robot ARMin II in chronic stroke patients: four single-cases. J Neuroeng Rehabil. 2009;6:46.

106. Brokaw EB, Nichols D, Holley RJ, Lum PS. Robotic therapy provides a stimulus for upper limb motor recovery after stroke that is complementary to and distinct from conventional therapy. Neurorehabil Neural Repair. 2014;28(4):367–76.

107. Sanchez RJ, Liu J, Rao S, Shah P, Smith R, Rahman T, et al. Automating arm movement training following severe stroke: functional exercises with quantitative feedback in a gravity-reduced environment. IEEE Trans Neural Syst Rehabil Eng. 2006;14(3):378–89.

108. Vanmulken D, Spooren A, Bongers H, Seelen H. Robot-assisted task-oriented upper extremity skill training in cervical spinal cord injury: a feasibility study. Spinal Cord. 2015;53(7):547–51.

109. Francisco GE, Yozbatiran N, Berliner J, O'Malley MK, Pehlivan AU, Kadivar Z, et al. Robot-assisted training of arm and hand movement shows functional improvements for incomplete cervical spinal cord injury. Am J Phys Med Rehabil. 2017;96(10): S171–7.

110. Cortes M, Elder J, Rykman A, Murray L, Avedissian M, Stampas A, et al. Improved motor performance in chronic spinal cord injury following upper-limb robotic training. NeuroRehabilitation. 2013;33(1):57–65.

111. Sledziewski L, Schaaf RC, Mount J. Use of robotics in spinal cord injury: a case report. Am J Occup Ther. 2012;66(1):51–8.

112. Yozbatiran N, Berliner J, O'Malley MK, Pehlivan AU, Kadivar Z, Boake C, et al. Robotic training and clinical assessment of upper extremity movements after spinal cord injury: a single case report. J Rehabil Med. 2012;44(2):186–8.

113. Kadivar Z, Sullivan JL, Eng DP, Pehlivan AU, O'Malley MK, Yozbatiran N, et al. RiceWrist robotic device for upper limb training: feasibility study and case report of two tetraplegic persons with spinal cord injury. Int J Biol Eng. 2012;2 (4):27–38.

114. Pehlivan AU, Sergi F, Erwin A, Yozbatiran N, Francisco GE, O'Malley MK. Design and validation of the RiceWrist-S exoskeleton for robotic rehabilitation after incomplete spinal cord injury. Robotica. 2014;32(8):1415.

115. A robotic exoskeleton for rehabilitation and assessment of the upper limb following incomplete spinal cord injury. In: 2015 IEEE international conference on robotics and automation (ICRA). IEEE; 2015.

116. Jung JH, Lee HJ, Cho DY, Lim J, Lee BS, Kwon SH, et al. Effects of combined upper limb robotic therapy in patients with tetraplegic spinal cord injury. Ann Rehabil Med. 2019;43(4):445.

117. Kim J, Lee BS, Lee H, Kim H, Cho D, Lim J, et al. Clinical efficacy of upper limb robotic therapy in people with tetraplegia: a pilot randomized controlled trial. Spinal Cord. 2019;57(1):49–57.

118. Singh H, Unger J, Zariffa J, Pakosh M, Jaglal S, Craven BC, et al. Robot-assisted upper extremity rehabilitation for cervical spinal cord injuries: a systematic scoping review. Disabil Rehabil Assist Technol. 2018;13(7):704–15.

119. Krebs HI, Hogan N, Aisen ML, Volpe BT. Robot-aided neurorehabilitation. IEEE Trans Rehabil Eng. 1998;6(1):75–87.

120. Masia L, Krebs HI, Cappa P, Hogan N. Design and characterization of hand module for whole-arm rehabilitation following stroke. IEEE/ASME Trans Mechatron. 2007;12(4):399–407.

121. Lo AC, Guarino PD, Richards LG, Haselkorn JK, Wittenberg GF, Federman DG, et al. Robot-assisted therapy for long-term upper-limb impairment after stroke. N Engl J Med. 2010;362(19):1772–83.

122. Rodgers H, Bosomworth H, Krebs HI, van Wijck F, Howel D, Wilson N, et al. Robot assisted training for the upper limb after stroke (RATULS): a multicentre randomised controlled trial. The Lancet. 2019;394(10192):51–62.

123. Cappello L, Meyer JT, Galloway KC, Peisner JD, Granberry R, Wagner DA, et al. Assisting hand function after spinal cord injury with a fabric-based soft robotic glove. J Neuroeng Rehabil. 2018;15 (1):1–10.

124. Bützer T, Lambercy O, Arata J, Gassert R. Fully wearable actuated soft exoskeleton for grasping assistance in everyday activities. Soft Rob. 2021;8 (2):128–43.

125. Correia C, Nuckols K, Wagner D, Zhou YM, Clarke M, Orzel D, et al. Improving grasp function

after spinal cord injury with a soft robotic glove. IEEE Trans Neural Syst Rehabil Eng. 2020;28 (6):1407–15.

126. Osuagwu BA, Timms S, Peachment R, Dowie S, Thrussell H, Cross S, et al. Home-based rehabilitation using a soft robotic hand glove device leads to improvement in hand function in people with chronic spinal cord injury: a pilot study. J Neuroeng Rehabil. 2020;17(1):1–15.

127. Housman SJ, Scott KM, Reinkensmeyer DJ. A randomized controlled trial of gravity-supported, computer-enhanced arm exercise for individuals with severe hemiparesis. Neurorehabil Neural Repair. 2009;23(5):505–14.

128. Rémy-Néris O, Le Jeannic A, Dion A, Médée B, Nowak E, Poiroux É, et al. Additional, mechanized upper limb self-rehabilitation in patients with subacute stroke: the REM-AVC randomized trial. Stroke. 2021;52(6):1938–47.

129. Brokaw EB, Black I, Holley RJ, Lum PS. Hand Spring Operated Movement Enhancer (HandSOME): a portable, passive hand exoskeleton for stroke rehabilitation. IEEE Trans Neural Syst Rehabil Eng. 2011;19(4):391–9.

130. Wittmann F, Held JP, Lambercy O, Starkey ML, Curt A, Hver R, et al. Self-directed arm therapy at home after stroke with a sensor-based virtual reality training system. J Neuroeng Rehabil. 2016;13(1):75.

131. Assessment-driven arm therapy at home using an IMU-based virtual reality system. In: 2015 IEEE international conference on rehabilitation robotics (ICORR). IEEE; 2015.

132. Pierella C, Galofaro E, De Luca A, Losio L, Gamba S, Massone A, et al. Recovery of distal arm movements in spinal cord injured patients with a body-machine interface: a proof-of-concept study. Sensors. 2021;21(6):2243.

133. Dimbwadyo-Terrer I, Gil-Agudo A, Segura-Fragoso A, De Los Reyes-Guzmán A, Trincado-Alonso F, Piazza S, et al. Effectiveness of the virtual reality system Toyra on upper limb function in people with tetraplegia: a pilot randomized clinical trial. BioMed Res Int. 2016;2016.

134. Dimbwadyo-Terrer I, Trincado-Alonso F, de Los R-G, Aznar MA, Alcubilla C, Pérez-Nombela S, et al. Upper limb rehabilitation after spinal cord injury: a treatment based on a data glove and an immersive virtual reality environment. Disabil Rehabil Assist Technol. 2016;11(6):462–7.

135. Lim DY, Hwang DM, Cho KH, Moon CW, Ahn SY. A fully immersive virtual reality method for upper limb rehabilitation in spinal cord injury. Ann Rehabil Med. 2020;44(4):311.

136. de Los Reyes-Guzmán A, Lozano-Berrio V, Alvarez-Rodríguez M, López-Dolado E, Ceruelo-Abajo S, Talavera-Díaz F, et al. RehabHand: oriented-tasks serious games for upper limb rehabilitation by using leap motion controller and target population in spinal cord injury. NeuroRehabilitation. 2021(Preprint):1–9.

137. Prasad S, Aikat R, Labani S, Khanna N. Efficacy of virtual reality in upper limb rehabilitation in patients with spinal cord injury: a pilot randomized controlled trial. Asian Spine J. 2018;12(5):927.

138. Jaramillo JP, Johanson ME, Kiratli BJ. Upper limb muscle activation during sports video gaming of persons with spinal cord injury. J Spinal Cord Med. 2019;42(1):77–85.

139. Laver KE, Lange B, George S, Deutsch JE, Saposnik G, Crotty M. Virtual reality for stroke rehabilitation. Cochrane Datab Syst Rev. 2017(11).

140. Mekbib DB, Han J, Zhang L, Fang S, Jiang H, Zhu J, et al. Virtual reality therapy for upper limb rehabilitation in patients with stroke: a meta-analysis of randomized clinical trials. Brain Inj. 2020;34(4):456–65.

141. Domínguez-Téllez P, Moral-Muñoz JA, Salazar A, Casado-Fernández E, Lucena-Antón D. Game-based virtual reality interventions to improve upper limb motor function and quality of life after stroke: Systematic review and meta-analysis. Games Health J. 2020;9(1):1–10.

142. Popovic MR, Kapadia N, Zivanovic V, Furlan JC, Craven BC, McGillivray C. Functional electrical stimulation therapy of voluntary grasping versus only conventional rehabilitation for patients with subacute incomplete tetraplegia: a randomized clinical trial. Neurorehabil Neural Repair. 2011;25 (5):433–42.

143. Popovic MR, Keller T. Modular transcutaneous functional electrical stimulation system. Med Eng Phys. 2005;27(1):81–92.

144. Taylor P, Esnouf J, Hobby J. The functional impact of the freehand system on tetraplegic hand function. Clin Results Spinal Cord. 2002;40(11):560–6.

145. Snoek GJ, IJzerman MJ, Stoffers TS, Zilvold G. Use of the NESS handmaster to restore handfunction in tetraplegia: clinical experiences in ten patients. Spinal Cord. 2000;38(4):244.

146. Alon G, McBride K. Persons with C5 or C6 tetraplegia achieve selected functional gains using a neuroprosthesis. Arch Phys Med Rehabil. 2003;84(1):119–24.

147. Popović D, Stojanović A, Pjanović A, Radosavljević S, Popović M, Jović S, et al. Clinical evaluation of the bionic glove. Arch Phys Med Rehabil. 1999;80(3):299–304.

148. Prochazka A, Gauthier M, Wieler M, Kenwell Z. The bionic glove: an electrical stimulator garment that provides controlled grasp and hand opening in quadriplegia. Arch Phys Med Rehabil. 1997;78 (6):608–14.

149. Anderson KD, Wilson JR, Korupolu R, Pierce J, Bowen JM, O'Reilly D, et al. Multicentre, single-blind randomised controlled trial comparing Mynd-Move neuromodulation therapy with conventional therapy in traumatic spinal cord injury: a protocol study. BMJ Open. 2020;10(9): e039650.

150. Ijzerman MJ, Stoffers TS, Groen F, Klatte M, Snoek GJ, Vorsteveld J, et al. The NESS Handmaster orthosis. J Rehabil Sci. 1996;9(3):86–9.

151. Mangold S, Keller T, Curt A, Dietz V. Transcutaneous functional electrical stimulation for grasping in subjects with cervical spinal cord injury. Spinal Cord. 2005;43(1):1–13.

152. Martin R, Johnston K, Sadowsky C. Neuromuscular electrical stimulation–assisted grasp training and restoration of function in the tetraplegic hand: a case series. Am J Occup Ther. 2012;66(4):471–7.

153. Kapadia NM, Bagher S, Popovic MR. Influence of different rehabilitation therapy models on patient outcomes: Hand function therapy in individuals with incomplete SCI. J Spinal Cord Med. 2014;37 (6):734–43.

154. Kapadia N, Zivanovic V, Popovic MR. Restoring voluntary grasping function in individuals with incomplete chronic spinal cord injury: pilot study. Top Spinal Cord Inj Rehabil. 2013;19(4):279–87.

155. Jovanovic LI, Kapadia N, Zivanovic V, Rademeyer HJ, Alavinia M, McGillivray C, et al. Brain–computer interface-triggered functional electrical stimulation therapy for rehabilitation of reaching and grasping after spinal cord injury: a feasibility study. Spinal Cord Ser Cases. 2021;7(1):1–11.

156. Biasiucci A, Leeb R, Iturrate I, Perdikis S, Al-Khodairy A, Corbet T, et al. Brain-actuated functional electrical stimulation elicits lasting arm motor recovery after stroke. Nat Commun. 2018;9(1):1–13.

157. Marquez-Chin C, Kapadia-Desai N, Kalsi-Ryan S. Brain-computer interfaces: neurorehabilitation of voluntary movement after stroke and spinal cord injury. Synth Lect Assist Rehabil Health-Preserv Technol. 2021;10(2):i–133.

158. Ajiboye AB, Willett FR, Young DR, Memberg WD, Murphy BA, Miller JP, et al. Restoration of reaching and grasping movements through brain-controlled muscle stimulation in a person with tetraplegia: a proof-of-concept demonstration. The Lancet. 2017;389(10081):1821–30.

159. Peckham PH, Keith MW, Kilgore KL, Grill JH, Wuolle KS, Thrope GB, et al. Efficacy of an implanted neuroprosthesis for restoring hand grasp in tetraplegia: a multicenter study. Arch Phys Med Rehabil. 2001;82(10):1380–8.

160. James ND, McMahon SB, Field-Fote EC, Bradbury EJ. Neuromodulation in the restoration of function after spinal cord injury. Lancet Neurol. 2018;17(10):905–17.

161. Lu DC, Edgerton VR, Modaber M, AuYong N, Morikawa E, Zdunowski S, et al. Engaging cervical spinal cord networks to reenable volitional control of hand function in tetraplegic patients. Neurorehabil Neural Repair. 2016;30(10):951–62.

162. Inanici F, Brighton LN, Samejima S, Hofstetter CP, Moritz CT. Transcutaneous spinal cord stimulation restores hand and arm function after spinal cord injury. TNSRE. 2021;29:310–9.

163. Freyvert Y, Yong NA, Morikawa E, Zdunowski S, Sarino ME, Gerasimenko Y, et al. Engaging cervical spinal circuitry with non-invasive spinal stimulation and buspirone to restore hand function in chronic motor complete patients. Sci Rep. 2018;8 (1):15546–10.

164. Potter-Baker KA, Janini DP, Lin Y, Sankarasubramanian V, Cunningham DA, Varnerin NM, et al. Transcranial direct current stimulation (tDCS) paired with massed practice training to promote adaptive plasticity and motor recovery in chronic incomplete tetraplegia: a pilot study. J Spinal Cord Med. 2018;41(5):503–17.

165. Cortes M, Medeiros AH, Gandhi A, Lee P, Krebs HI, Thickbroom G, et al. Improved grasp function with transcranial direct current stimulation in chronic spinal cord injury. NeuroRehabilitation (Reading, Mass.) 2017;41(1):51–9.

166. Murray LM, PhD, Edwards DJ, PhD, Ruffini G, PhD, Labar, Douglas, MD, PhD, Stampas A, MD, Pascual-Leone, Alvaro, MD, PhD, et al. Intensity dependent effects of transcranial direct current stimulation on corticospinal excitability in chronic spinal cord injury. Arch Phys Med Rehabil. 2015;96 (4):S114–21.

167. Yozbatiran N, Keser Z, Hasan K, Stampas A, Korupolu R, Kim S, et al. White matter changes in corticospinal tract associated with improvement in arm and hand functions in incomplete cervical spinal cord injury: pilot case series. Spinal Cord Ser Cases. 2017;3(1):17028.

168. Gomes-Osman J, Tibbett JA, Poe BP, Field-Fote EC. Priming for improved hand strength in persons with chronic tetraplegia: a comparison of priming-augmented functional task practice, priming alone, and conventional exercise training. Front Neurol. 2016;7:242.

169. Gomes-Osman J, Field-Fote EC. Improvements in hand function in adults with chronic tetraplegia following a multi-day 10Hz rTMS intervention combined with repetitive task practice. J Neurol Phys Ther. 2015;39(1):23–30.

170. Bunday K, Perez M. Motor recovery after spinal cord injury enhanced by strengthening corticospinal synaptic transmission. Curr Biol. 2012;22 (24):2355–61.

171. Kimberley T, Pierce D, Prudente C, Francisco G, Yozbatiran N, Smith P, et al. Vagus nerve stimulation paired with upper limb rehabilitation after chronic stroke: a blinded randomized pilot study. Stroke (1970). 2018;49(11):2789–92.

172. Darrow MJ, Torres M, Sosa MJ, Danaphongse TT, Haider Z, Rennaker RL, et al. Vagus nerve stimulation paired with rehabilitative training enhances motor recovery after bilateral spinal cord injury to cervical forelimb motor pools. Neurorehabil Neural Repair. 2020;34(3):200–9.

173. Hou J, Kulkarni A, Tellapragada N, Nair V, Danilov Y, Kaczmarek K, et al. Translingual neural stimulation with the portable neuromodulation stimulator (PoNS®) induces structural changes leading to functional recovery in patients with mild-to-moderate traumatic brain injury. EMJ Radiol. 2020;1(1):64–71.

174. Badhiwala JH, Ahuja CS, Akbar MA, Witiw CD, Nassiri F, Furlan JC, et al. Degenerative cervical myelopathy—update and future directions. Nat Rev Neurol. 2020;16(2):108–24.

175. Musselman KE, Shah M, Zariffa J. Rehabilitation technologies and interventions for individuals with spinal cord injury: translational potential of current trends. J Neuroeng Rehabil. 2018;15(1):40.

176. Kalsi-Ryan S, Beaton D, Curt A, Duff S, Popovic MR, Rudhe C, et al. The graded redefined assessment of strength sensibility and prehension: reliability and validity. J Neurotrauma. 2012;29(5):905–14.

177. Schneider S, Popp WL, Brogioli M, Albisser U, Demkó L, Debecker I, et al. Reliability of wearable-sensor-derived measures of physical activity in wheelchair-dependent spinal cord injured patients. Front Neurol. 2018;9:1039.

178. Hand keypoint detection in single images using multiview bootstrapping. In: Proceedings of the IEEE conference on computer vision and pattern recognition; 2017.

179. Kalsi-Ryan S, Curt A, Fehlings MG, Verrier MC. Assessment of the hand in tetraplegia using the Graded Redefined Assessment of Strength, Sensibility and Prehension (GRASSP): impairment versus function. Top Spinal Cord Inj Rehabil. 2009;14(4):34–46.

180. Kalsi-Ryan S, Beaton D, Ahn H, Askes H, Drew B, Curt A, et al. Responsiveness, sensitivity, and minimally detectable difference of the graded and redefined assessment of strength, sensibility, and prehension, version 1.0. J Neurotrauma. 2016;33(3):307–14.

181. Post M, Van Lieshout G, Seelen H, Snoek GJ, IJzerman MJ, Pons C. Measurement properties of the short version of the Van Lieshout test for arm/hand function of persons with tetraplegia after spinal cord injury. Spinal Cord. 2006;44(12):763–71.

182. Fattal C. Motor capacities of upper limbs in tetraplegics: a new scale for the assessment of the results of functional surgery on upper limbs. Spinal Cord. 2004;42(2):80–90.

183. Land NE, Odding E, Duivenvoorden HJ, Bergen MP, Stam HJ. Tetraplegia Hand Activity Questionnaire (THAQ): the development, assessment of arm–hand function-related activities in tetraplegic patients with a spinal cord injury. Spinal Cord. 2004;42(5):294–301.

184. Thorsen R, Ferrarin M, Spadone R, Frigo C. Functional control of the hand in tetraplegics based on residual synergistic EMG activity. Artif Organs. 1999;23(5):470–3.

185. Marino RJ, Shea JA, Stineman MG. The capabilities of upper extremity instrument: reliability and validity of a measure of functional limitation in tetraplegia. Arch Phys Med Rehabil. 1998;79(12):1512–21.

186. Marino RJ, Sinko R, Bryden A, Backus D, Chen D, Nemunaitis GA, et al. Comparison of responsiveness and minimal clinically important difference of the Capabilities of Upper Extremity Test (CUE-T) and the Graded Redefined Assessment of Strength, Sensibility and Prehension (GRASSP). Topics in spinal cord injury rehabilitation. 2018;24(3):227–38.

187. Marino RJ, Patrick M, Albright W, Leiby BE, Mulcahey M, Schmidt-Read M, et al. Development of an objective test of upper-limb function in tetraplegia: the capabilities of upper extremity test. Am J Phys Med Rehabil. 2012;91(6):478–86.

188. Wuolle KS, Van Doren CL, Thrope GB, Keith MW, Peckham PH. Development of a quantitative hand grasp and release test for patients with tetraplegia using a hand neuroprosthesis. J Hand Surg Am. 1994;19(2):209–18.

189. Mulcahey MJ, Smith BT, Betz RR. Psychometric rigor of the grasp and release test for measuring functional limitation of persons with tetraplegia: a preliminary analysis. J Spinal Cord Med. 2004;27(1):41–6.

190. Berghe AV, Van Laere M, Hellings S, Vercauteren M. Reconstruction of the upper extremity in tetraplegia: functional assessment, surgical procedures and rehabilitation. Spinal Cord. 1991;29(2):103–12.

191. Rogers JC, Figone JJ. Traumatic quadriplegia: follow-up study of self-care skills. Arch Phys Med Rehabil. 1980;61(7):316–21.

192. Kapadia N, Zivanovic V, Verrier M, Popovic M. Toronto rehabilitation institute-hand function test: assessment of gross motor function in individuals with spinal cord injury. Topics in spinal cord injury rehabilitation. 2012;18(2):167–86.

193. Wolf SL, Lecraw DE, Barton LA, Jann BB. Forced use of hemiplegic upper extremities to reverse the effect of learned nonuse among chronic stroke and head-injured patients. Exp Neurol. 1989;104(2):125–32.

194. Wolf SL, Catlin PA, Ellis M, Archer AL, Morgan B, Piacentino A. Assessing wolf motor function test as outcome measure for research in patients after stroke. Stroke. 2001;32(7):1635–9.

195. Lyle RC. A performance test for assessment of upper limb function in physical rehabilitation treatment and research. Int J Rehabil Res. 1981;4(4):483–92.

196. Gummesson C, Ward MM, Atroshi I. The shortened disabilities of the arm, shoulder and hand questionnaire (Quick DASH): validity and reliability based on responses within the full-length DASH. BMC Musculoskelet Disord. 2006;7(1):1–7.

197. Matheson LN, Melhorn JM, Mayer TG, Theodore BR, Gatchel RJ. Reliability of a visual

analog version of the QuickDASH. JBJS. 2006;88(8):1782–7.

198. Law M, Baptiste S, McColl M, Opzoomer A, Polatajko H, Pollock N. The Canadian occupational performance measure: an outcome measure for occupational therapy. Can J Occup Ther. 1990;57(2):82–7.

199. Cup EH, Scholte op Reimer W, Thijssen MC, van Kuyk-Minis M. Reliability and validity of the Canadian occupational performance measure in stroke patients. Clin Rehabil. 2003;17(4):402–9.

200. Holser P, Fuchs E. Box and block test. Occupational therapists manual for basic skills assessment: primary prevocational evaluation. Pasadena, California: Fair Oaks Printing Company; 1960. p. 29–31.

201. Desrosiers J, Bravo G, Hébert R, Dutil É, Mercier L. Validation of the box and block test as a measure of dexterity of elderly people: reliability, validity, and norms studies. Arch Phys Med Rehabil. 1994;75(7):751–5.

202. Mathiowetz V, Volland G, Kashman N, Weber K. Adult norms for the box and block test of manual dexterity. Am J Occup Ther. 1985;39(6):386–91.

Implementation of Impairment-Based Neurorehabilitation Devices and Technologies Following Brain Injury

5

Julius P. A. Dewald, Michael D. Ellis,
Ana Maria Acosta, M. Hongchul Sohn,
and Thomas A. M. Plaisier

Abstract

The implementation of electromechanical devices for the quantification and treatment of movement impairments (abnormal muscle synergies resulting in a loss of independent joint control, hypertonia, and associated spasticity and paresis) stemming from brain injury is the main topic in this chapter. The specific requirements for the use of robotic and sensing devices to quantify these impairments as well as treat them effectively both in the clinic and at home (Telerehabilitation) are discussed. A case is made that these devices not only allow the clinician to quantitatively control task practice and dosage but more importantly, allow for direct targeting of specific impairments, such as the loss of independent joint control (Dewald et al. in Top Stroke Rehabil 8(1):1–12, 2001), as well as the monitoring of the expression of such impairments during activities of daily living in the home setting. Acceptance of these new technologies is dependent on proof of their effectiveness in the reduction of movement impairments and activity limitations, as opposed to compensation, and ultimately on the carryover of benefits to activities of daily living and quality of life. Furthermore, the need of a concerted effort to simplify these new technologies, once essential treatment ingredients have been determined, is seen as being a key component for their acceptance in the clinic on a large scale. Finally, it is crucial that we demonstrate that electromechanical and sensing technologies augment existing rehabilitative care and serve to reduce treatment time and costs while maintaining, and even improving, functional outcomes. This is a requirement for future technology development, especially in a health care environment where rehabilitation services have become less accessible and telerehabilitation from home may be the only option for some.

J. P. A. Dewald (✉)
Physical Therapy & Human Movement Sciences;
Biomedical Engineering; Physical Medicine &
Rehabilitation, Northwestern University, Evanston,
IL, USA
e-mail: j-dewald@northwestern.edu

M. D. Ellis
Physical Therapy & Human Movement Sciences;
Physical Medicine & Rehabilitation, Northwestern
University, Evanston, IL, USA
e-mail: m-ellis@northwestern.edu

A. M. Acosta · M. H. Sohn
Physical Therapy & Human Movement Sciences,
Northwestern University, Evanston, IL, USA
e-mail: a-acosta@northwestern.edu

M. H. Sohn
e-mail: hongchul.sohn@northwestern.edu

T. A. M. Plaisier
Biomedical Engineering, Physical Therapy &
Human Movement Sciences, Evanston, Illinois, USA
e-mail: ThomasPlaisier2023@u.northwestern.edu

© The Author(s), under exclusive license to Springer Nature Switzerland AG 2022
D. J. Reinkensmeyer et al. (eds.), *Neurorehabilitation Technology*,
https://doi.org/10.1007/978-3-031-08995-4_5

Keywords

Stroke · Hemiparesis · Arm · Movement · Impairment · Function · Robotics · Rehabilitation Robotics · Telerehabilitation

5.1 Introduction

Sensorimotor deficits and restricted mobility are among the more prevalent problems encountered by individuals following brain injury such as stroke. Work in the past 15 years, in great part facilitated by the use of robotic technologies, quantified the impact of stereotypical muscle synergies resulting in a loss of independent joint control, hypertonia associated with spasticity and paresis, all common to many forms of brain injury, on movement and subsequent function. Robotic technologies have also been used extensively to probe the amenability of these movement impairments to restorative interventions. An important distinction in the employment of robotics in rehabilitation has been whether the application attempts to more efficiently replicate aspects of conventional care such as repetitive functional task practice or if it directly targets specific impairments with a goal of movement restoration through the amelioration of impairment. For example, some rehabilitation robotic therapies aim to more efficiently deliver conventional approaches such as practicing functional tasks but with the added benefit of a greater number of repetitions [2–5], whereas others attempt to ameliorate specific impairments such as loss of independent joint control/inter-joint coordination [6, 7, respectively] or general motor impairment through a multifaceted approach [8–11]. Computational motor learning principles suggest that the optimal rehabilitation robotics strategy may be to employ both approaches by first directly targeting impairments during early recovery to maximize restoration of impairment and then progress toward functional task practice to maximize restoration of activity limitations [12].

In this chapter, we focus on the first component of Huang and Krakauer's suggestion of an optimal rehabilitation strategy; that is, targeting impairment restoration. Since robotic devices are superior at quantifying movement impairments such as loss of independent joint control [13–19], weakness [20, 21], and spasticity [22–33], clinical decision-making regarding the response to and progression of interventions can be quantitatively driven, optimizing implementation of the rehabilitation strategy. Although some work has been done determining losses in independent joint control in the lower limb [34, 35], here we only discuss evidence for the effective use of robotics to provide high-resolution measures of motor impairment in the upper limb of individuals with stroke. This evidence has led to advances in our understanding of the neurophysiological mechanisms underlying these motor impairments. In addition, we will present results from the novel robot-mediated interventions that can complement conventional neurotherapeutic interventions in chronic stroke and introduce the application of these interventions in acute stroke. In short, we will show that new robotic and sensing technologies are ideal for the delivery of impairment-based therapeutic interventions as well as monitoring their effects in current rehabilitation clinics to augment conventional rehabilitation. Considerations for a successful transition to clinical practice, both in rehabilitation clinics and at home, will be highlighted including methods to expand the application of new technologies to telerehabilitation in the home setting.

5.2 Quantification of Impairment

5.2.1 Quantification of Abnormal Synergies and Weakness Using Instrumented Devices

A central impairment resulting from unilateral hemispheric brain injury is the loss of independent control of joint movement that is evident in the form of stereotypic movement patterns [36–38]. It is believed that these stereotypic movement patterns are an expression of abnormal muscle coactivation patterns or muscle synergies. We have presented quantitative evidence for the existence of abnormal muscle coactivation patterns using EMGs from elbow and shoulder muscles in the paretic arm of individuals with stroke during static force exertions at the shoulder and elbow in varying directions and magnitudes [39]. Using mechanical kinetic measurements during static or isometric tasks, we were able to improve the quantification of abnormal muscle coactivation patterns with a 6-degree of freedom load cell [40, 41]. Using this approach, we studied the expression of isometric elbow and shoulder torque patterns during the generation of maximum voluntary torques one direction at a time. During the execution of this single task protocol in a primary direction, we observed relative weakness in the paretic limb compared to the contralateral limb and more importantly, we found strong abnormal coupling between elbow flexion and shoulder abduction/extension/external rotation and elbow extension and shoulder adduction/internal rotation in the paretic limb of individuals with stroke [1, 41]. Conversely, individuals with no neurological injury, and individuals with stroke in their non-paretic arm, only generated nominal torques in secondary degrees of freedom. In subsequent studies, we measured maximum voluntary elbow torques under three different conditions; in combination with 10 and 50% of maximum shoulder abduction (SABD) torque and in combination with 10% of maximum shoulder adduction (SADD) torque [40]. The torque combinations most affected were those that required participants to deviate from the abnormal torque patterns observed during the single-task paradigm. Specifically, individuals with stroke exhibited an impaired ability to generate elbow extension torque with the paretic limb when increasing shoulder abduction (i.e., the 50% SABD level). The opposite trend was observed for elbow flexion torque. Individuals with stroke exhibited an enhanced ability to generate elbow flexion torque in the paretic limb with increasing levels of SABD torque. The most detailed isometric study quantifying the expression of the flexion and extension synergy during the generation of SABD or SADD torques (at 17–100% of max levels) shows a link between SABD and elbow, wrist, and finger flexion and between SADD and elbow extension, and wrist and finger flexion [18]. This is close to what Brunnstrom reported based on observations of synergies post hemiparetic stroke, with the exception of finding wrist flexion as opposed to extension as part of the extension synergy [36]. These results demonstrate the existence of a strong and abnormal linkage in the paretic limb between shoulder abduction and elbow/wrist/finger flexion and between shoulder adduction and elbow extension and wrist/finger flexion. Precise quantification of this fundamental impairment was only possible through the implementation of multi-degree of freedom force/torque sensing technologies as opposed to conventional clinical evaluation with the Fugl–Meyer Motor Assessment [42] that is limited by its ordinal scale of measurement and reliance upon subjective observational movement analysis. The application of these new technologies set the stage for the execution of dynamic experiments using robotics and sensing technologies.

Our first robotic studies investigated the effect of synergies on planar reaching and retrieval movements as a function of support condition (the paretic upper limb being either supported, partially supported or unsupported at limb weight or heavier) using an admittance controlled HapticMASTER robot (Moog Inc., The Netherlands). The robotic device was specially customized with the addition of a gimbal instrumented with position sensors and a 6-degree of freedom load cell attached to its end effector. The individual's forearm and hand are attached to the gimbal using a hand–forearm orthosis (Fig. 5.1). The modified HapticMASTER robot was integrated with a Biodex experimental chair (Biodex Medical Systems, Shirley, NY) to form the first-generation Arm Coordination Training 3D (ACT-3D) device shown in Fig. 5.1. This unique combination of technologies allows for the application of specific SABD loading forces and measurement of induced shoulder and elbow coupling during reaching expressed as a reduced work area. The

ACT-3D is a sophisticated and powerful quantification tool that can be used to characterize movement disabilities in individuals who have had brain injury resulting from a stroke. The advantage of this system is that it incorporates the ability to control the level of shoulder abduction/adduction loading while measuring movement performance in the 3D workspace, features unavailable in the early isometric and dynamic studies [1, 40, 41, 43]. In an unprecedented way, the ACT-3D has allowed us to investigate the progressive debilitating impact of SABD loading on reaching range of motion. When quantifying the effect of SABD loading on the work area of the hand, individuals with stroke and control participants were asked to slowly trace with their hands the largest possible envelope on a horizontal plane (at shoulder level) by moving their arm several times in a clockwise and counter-clockwise direction. The largest work area for each level of abduction loading was calculated from multiple trials. Participants performed the reaching movements while sliding over a haptically rendered table or under conditions where the virtual effect of gravity was enhanced or reduced by providing forces along the vertical axis of the ACT-3D and aligned with gravity. The direction of these forces dictated the amount of resulting SABD loading and was varied from 100% of limb support to 100% or more of limb weight added to the shoulder load.

An example of work area results from a single participant with moderate to severe impairments (Fugl-Meyer upper extremity score 23/66, and Chedoke–McMaster Arm Scale 3/7) is shown in Fig. 5.2. The different color traces correspond to the maximum hand work area generated while lifting the arm against varying percentages of limb weight. These ranged from 0% where the robot was compensating for the entire weight of

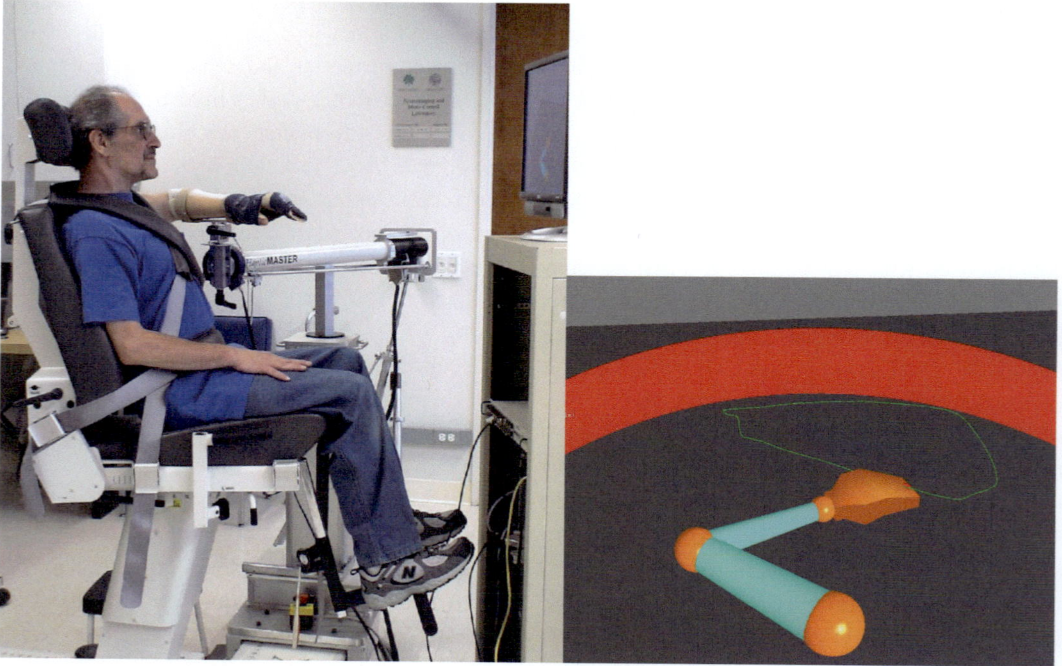

Fig. 5.1 (Left) Illustrating ACT^{3D} robot with gimbal and orthosis. (Right) Example of the visual feedback. The haptic table is shown by the *darker gray*, which the arm is resting on. In the envelope protocol (see measurement of work area below), subjects will use the *red arc* as their goal, with the *green trace* shown to give them a reference to their performance in previous circles. From Sukal TM, Ellis MD, Dewald JP (2007) Shoulder abduction-induced reductions in reaching work area following hemiparetic stroke: neuroscientific implications. Exp Brain Res 183:215–223; with kind permission from Springer Science + Business Media

the limb to 200% where the participant had to generate abduction torques twice the size of those required to lift the limb against gravity. The left panel in Fig. 5.2 shows the reduction in work area in the paretic limb (left arm in this subject) with the greatest work area reduction toward the ipsilateral and forward-reaching portion of the envelope; this area coincides with the direction requiring primarily elbow extension (the upper left portion of the envelopes). This is consistent with the expression of the flexion synergy that dictates the presence of greater coupling with elbow flexion torque for increasing levels of shoulder abduction. The reduction in work area for the same participant is displayed as a function of mean area versus percentage of active limb support. These results are in stark contrast to the non-paretic side, where no change or effect of abduction level related to shoulder and elbow range of motion is observed (see Fig. 5.2). The reductions in upper limb workspace as a function

of shoulder abduction load have been shown to exist in individuals with moderate to severe motor impairments following hemiparetic stroke [13]. This is a result of the abnormal coupling between shoulder abduction and elbow flexion or flexion synergy. This synergy has been reported to also include more distal joints of the paretic arm, namely the wrist and fingers as shown for the isometric experiments [18] discussed earlier and while using robotics [17].

The paretic wrist and fingers have also been the focus of extensive research [44–46], however, they have been examined most frequently in isolation from the rest of the upper limb, without consideration for the effect of the flexion synergy. The addition of a wrist/finger force sensing device (47—Fig. 5.3, top) to the ACT-3D robot, has allowed us to study the effect of shoulder abduction loading on wrist and finger forces in both adults and children with spastic hemiparesis. As can be appreciated from the

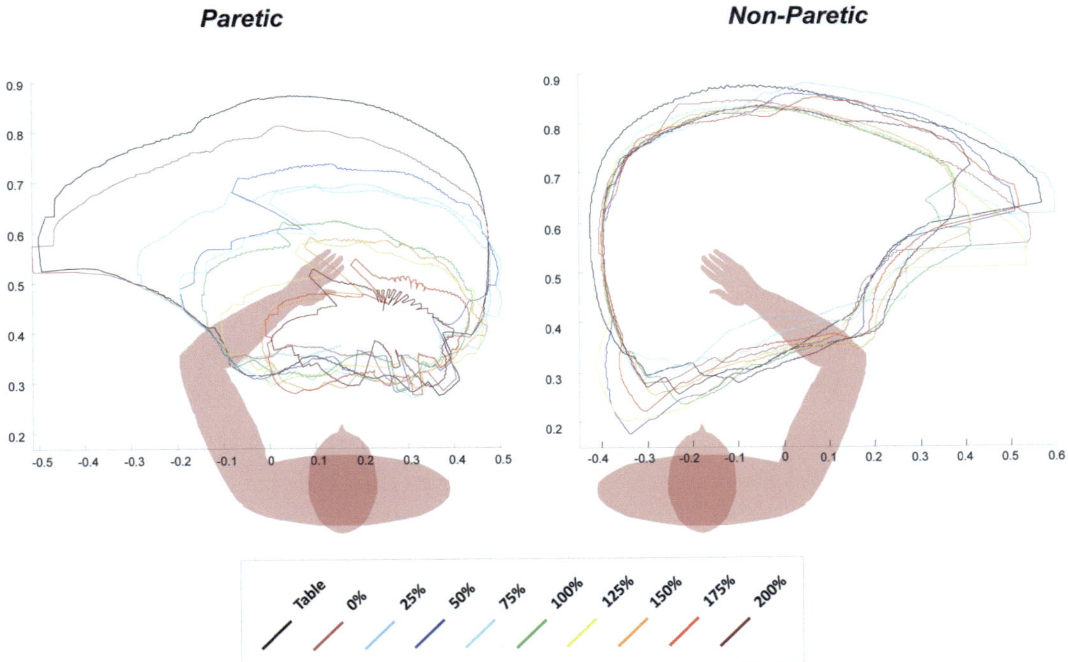

Fig. 5.2 Envelope traces consisting of shoulder/elbow flexion/extension combinations during various levels of limb support in the paretic limb (left arm) of a single participant [13]. Conditions listed in the legend are percentages of limb weight. Note the significant reduction in work area for increasing levels of shoulder abduction/external rotation. Axes units are in meters. From Sukal TM, Ellis MD, Dewald JP (2007) Shoulder abduction-induced reductions in reaching work area following hemiparetic stroke: neuroscientific implications. Exp Brain Res 183:215–223; with kind permission from Springer Science + Business Media

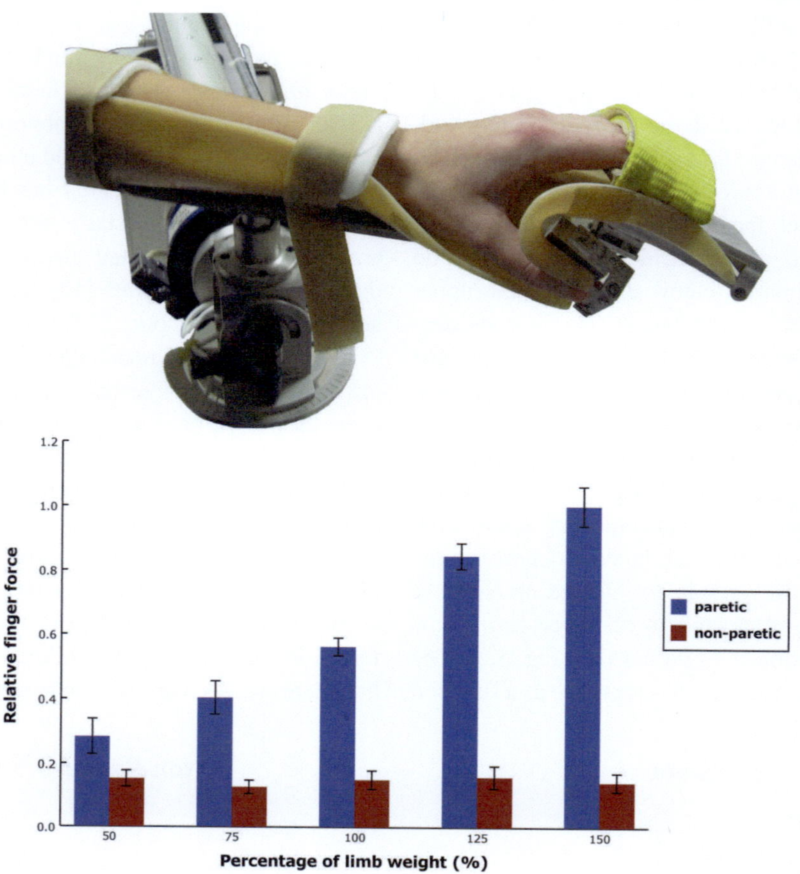

Fig. 5.3 Top: instrumented hand finger orthosis. From Miller LC, Ruiz-Torres R, Stienen AH, Dewald JP (2009) A wrist and finger force sensor module for use during movements of the upper limb in chronic hemiparetic stroke. IEEE Trans Biomed Eng 56:2312–2317; used with permission. Bottom: Relative level of finger force (normalized for each subject by the largest forces measured over the 5 shoulder abduction loading conditions) generated for increasing levels of load as percentage of maximum shoulder abduction (SABD) torque. From Miller LC, Dewald JP. Involuntary paretic wrist/finger flexion forces and EMG increase with shoulder abduction load in individuals with chronic stroke. Clin Neurophysiol 2012; 23:1216–25; used with permission. This demonstrates that increasing levels of shoulder abduction generates involuntary increases in finger flexion in the paretic hand. The error bars represent inter-subject standard errors

results shown in Fig. 5.3 (bottom), secondary finger/wrist forces increase as shoulder abduction loads increase in individuals with adult-onset stroke [16]. More recent research using a cylinder instrumented with a pressure sensor mat (Pressure Profile System Inc, Hawthorne, CA), to measure grasp forces, and 2 OptoTrak motion capture systems (Optotrak 3020 and Optotrak Certus, Northern Digital Inc, Waterloo, Ontario, Canada) to measure the kinematics of hand opening (base on the hand pentagon area, defined as the area formed by the tips of the thumb and fingers as shown in Fig. 5.4 top) allowed further characterization of the flexion synergy-induced coupling at the hand during reaching under various loading conditions applied by the ACT-3D robot [17]. Progressive SABD loading resulted in increases in grasp force in more impaired individuals (Fig. 5.4 bottom panel) whereby SABD loading reduces the ability to open the hand in moderately impaired individuals with stroke (Fig. 5.4 middle panel). Note that severely

Fig. 5.4 Top: pentagon area measure defined as the area circumscribed by the tip of the four digits and thumb.
a Pentagon area measured in stroke participants with severe ($n = 12$) and moderate ($n = 10$) impairments and in control participants ($n = 10$).
b Grasp forces during hand opening for severe ($n = 12$), moderate ($n = 11$), and control groups ($n = 10$).
$*P < 0.05$, $**P < 0.01$, $***P < 0.001$. From Lan, Y., Yao, J. & Dewald, J. P. A. The Impact of Shoulder Abduction Loading on Volitional Hand Opening and Grasping in Chronic Hemiparetic Stroke. *Neurorehabilitation and neural repair* **31**, 521–529, https://doi.org/10.1177/1545968317697033 (2017); (used with permission.)

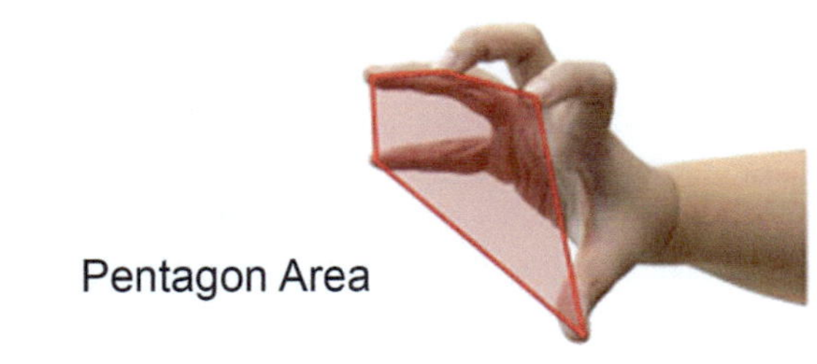

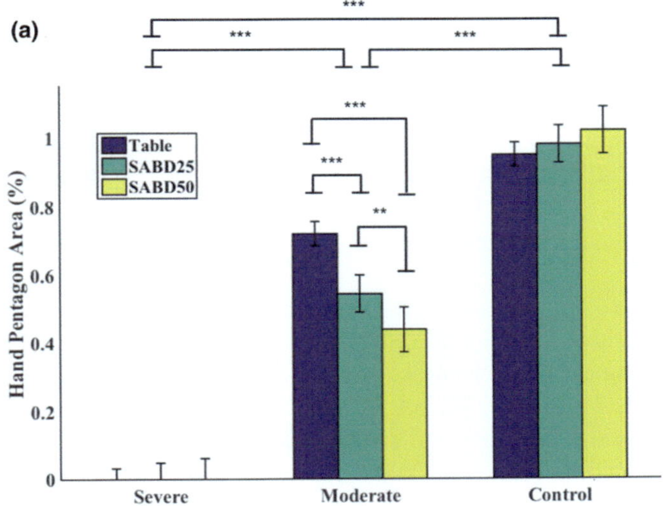

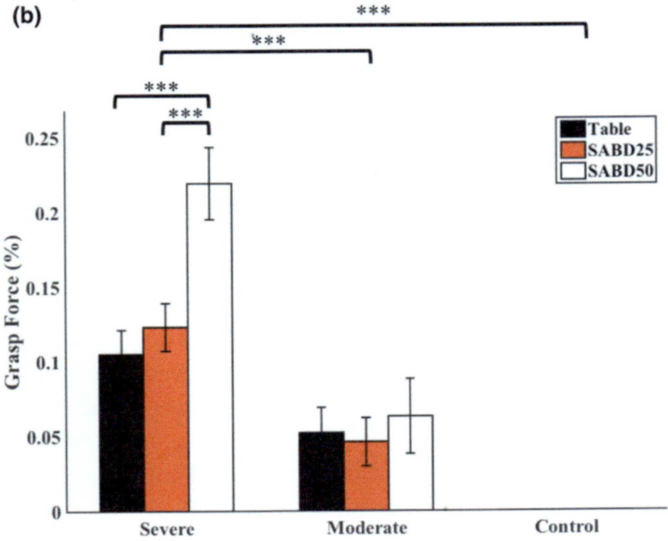

impaired individuals are not able to open their paretic hands independent of shoulder abduction loading. Even with the paretic arm fully supported, and thus the effect of the flexion synergy removed, these individuals cannot open their paretic hand. The integration of functional electrical stimulation of wrist/finger flexors/extensors can also be investigated using this setup as it allows for the measurement of finger flexion forces and hand opening kinematics generated by various electrical stimulation parameters and with various shoulder abduction loads encountered during activities of daily living.

5.2.2 Quantification of Spasticity Using Robotic Technologies

Spasticity, defined as a hyperactive velocity-sensitive stretch reflex [48], has been studied using electromechanical devices for four decades [23, 25, 27, 28, 32, 33, 49–53]. Using robotic devices, spasticity or stretch reflex hyperexcitability, associated with hypertonia due to hyperactive motoneurons, has primarily been studied in resting limbs, yet its clinical management has been directed mainly at an assumed impact on active movement. Current directions in the treatment of spasticity include stretching, serial casting, and the use of antispastic agents such as botulinum toxin and baclofen to reduce overactive muscle activity or hypertonia. The rationale for this approach is that by reducing spasticity, movement performance will improve. This conventional approach persists despite the lack of evidence demonstrating that reflex hyperexcitability (measured in a resting limb) actually impacts active movement. Numerous studies in resting limbs have reported increased mechanical resistance (reflex torques) and augmented stretch reflexes during passive joint rotation imposed by a single degree of freedom robotic devices, particularly after stroke [22–27, 49–53]. Under passive or resting conditions, the presence of spasticity in the paretic limb of individuals with stroke can be clearly identified in response to slow stretches that would

otherwise generally fail to elicit significant levels of stretch reflex activity in healthy individuals [54, 55].

Relatively little is known of spasticity in actively contracting muscles despite its obvious relevance to active movement and subsequent treatment. Even a small voluntary background contraction leads to prominent reflex activity and increased passive resistance in control limbs [53, 56]. Additionally, there is no clear demonstration that reflex EMG and torque magnitude are significantly higher in spastic limbs under analogous background activation conditions [22, 27, 49, 50, 57–59]. **Hence, it has been unclear how, or if, spasticity contributes to the movement disorder in the affected limbs. It is possible that the defining features of spasticity are a phenomenon largely confined to resting limbs. Recent findings about the effect of spasticity on unperturbed reaching movements using robotic technologies of the spastic upper limb have shown its impact to be limited.**

Most of the spasticity quantification literature to date considers hyperactive stretch reflex activity at the single joint level with the subject relaxed and does not consider its potential effects on multi-joint movements such as reaching or retrieval motions. Even if we hypothesize that spasticity expresses itself as a hyperactive stretch reflex during passive conditions only (i.e., with the subject relaxed) and stretch reflex activity during active (i.e., movement) conditions is not changed [22] then multi-joint movements may still be affected. This may be especially true during multi-joint reaching where elbow extension is mostly the result of coupling or interaction torques generated during shoulder flexion movement and not due to elbow extensor muscle activation [60]. It is likely that under such conditions, abnormal hyperactive stretch reflex activity of 'relaxed' elbow flexors (which are not reciprocally inhibited by triceps activity because coupling torques reduce the need for triceps activation) could limit the upper extremity workspace, especially at higher movement velocities and while the weight of the arm is supported by the robot [31]. In addition to the role that spasticity may play when joint movement is driven by coupling or interaction torques, as occur during multi-joint movements, stretch reflex hyperexcitability may also be affected by the expression of the flexion synergy (see section above). This is not addressed in spasticity quantification studies at the elbow where the weight of the paretic limb is supported by the measurement system [22, 23, 25, 27, 58]. Results from studies that investigate the influence of proximal joint demands (shoulder) on reflex excitability of the elbow flexors during passive single-joint elbow rotations suggest an interaction between synergy-related activation and reflex-related activation of elbow flexors. For example, stretch reflex excitability in elbow flexors is shown to be modulated by abductor activation for a single SABD load level [28] and as a function of SABD loading [32]. The next step in this work is investigating the interplay of spasticity and abnormal flexion synergy during a dynamic multi-joint reaching task already underway in our lab. Recent studies show that elbow flexor activation during an outward ballistic reach under various abduction loads reflects the negligible contribution of reflex-related flexor activation superimposed upon much larger synergy-related flexor activation in most cases. The only exception is when the arm is fully supported by the robot-generated haptic table thus eliminating the contribution of the flexion synergy to reaching and therefore allowing for greater elbow extension angular velocities sufficient to elicit the stretch reflex [31]. These data support an interaction between reflex- and synergy-related flexor activation and suggest a dominant and deleterious contribution of synergy-related flexor activation to impaired reaching function. In fact, the greater the elicitation of flexion synergy, as experienced during the movement against gravity, the greater the reduction in elbow extension velocity, minimizing any possible contribution of stretch reflex-related flexor activation. State-of-the-art robotic technologies, some of which are currently under development in our laboratory, will be required to fully elucidate the interaction between stretch reflex hyperexcitability/spasticity and impairments such as abnormal synergies during various phases of reaching and retrieval movements under a variety of SABD loading conditions similar to those experienced during functional arm activities.

5.2.3 Development of Robotic Devices for Impairment Quantification

Depending on the specific application, robotic devices used for the quantification of impairments such as abnormal muscle synergies and spasticity must possess certain key design characteristics. First, these devices must be capable of rendering haptic environments within which users can generate and experience desired forces. For example, to investigate flexion synergy, robotic devices must be capable of providing forces that result in abduction loading and unloading of shoulder muscles while measuring the effect on reaching behavior. Second, an important consideration for robotic devices seeking to capture functional movements is their functional workspace volume. If, for instance, the desired task is a center-out reaching task in multiple directions, it may be necessary to permit full extension of the arm, which will require both shoulder flexion and elbow extension in a large workspace. If, however, the goal is only elbow extension, a smaller workspace volume may be acceptable. Third, robotic devices seeking to measure the relationship between stretch reflex excitability and abnormal muscle coactivation patterns must possess an adequate number of degrees of freedom to capture functional behaviors and elicit abnormal patterns. For planar movements of the upper limb, this translates to at least 3 degrees of freedom: two for the elbow and shoulder to perform planar movements and a third perpendicular to the plane to modulate the expression of the flexion synergy through shoulder abduction loading. Finally, for spasticity-related studies, these devices must also be capable of switching between compliant and stiff operating modes, enabling low impedance movements throughout the workspace while simultaneously being able to apply precise position or speed-controlled perturbations to the user to elicit reflexes under various conditions.

An example of novel device development based on the above characteristics is the NACT-3D (New Arm Coordination Training in 3D, Fig. 5.8), a robotic device currently under development in our lab that leverages knowledge gained with the ACT-3D. This new robot combines the four principles outlined above to deliver an unprecedented combination of work area and movement control. Users interact with the device through a 6-degree of freedom force/torque sensor mounted at the endpoint, similar to the ACT-3D. However, the device drive train mimics the two segments of the human arm and allows for a considerably larger work area than was possible with previous devices. The robot is fitted with four degrees of freedom: two rotational joints in the plane of the robotic arm are analogous to the human elbow and shoulder, one translational joint at the shoulder to generate movement perpendicular to the plane of the robotic arm (e.g., for shoulder abduction loading), and finally a rotational joint to change the orientation of the robotic arm. An admittance control loop is used to create a haptic environment for the user to move freely within, as is the case with the ACT-3D described earlier. To elicit stretch reflexes, the admittance-controlled environment can be instantly deactivated and replaced by a velocity/position-controlled mode which allows for brief and rapid manipulation of the endpoint of the limb over time scales as short as tens of milliseconds. Preliminary data confirms that this device can open up novel methods of investigating stretch reflex hyperexcitability during functional movement, and its interaction with task-related muscle activation in healthy controls and individuals with stroke. Careful design considerations and working knowledge of the relevant neurophysiology are key factors in the design and implementation of novel robotic devices that allow investigators to answer specific questions about the mechanisms underlying movement impairments. In addition, simplified robotic devices can be used for the delivery of effective rehabilitation interventions that complement conventional neuro-rehabilitation approaches.

5.2.4 Development of SENSING Technologies for Home-Based Quantification of Arm Motor Deficits During Activities of Daily Living

In the past 20 years, work in our laboratories has contributed to elucidating the mechanisms underlying motor impairments in stroke through controlled experiments largely relying on advanced robotics and sensing technologies as presented here. However, only recently, we expanded our research to include quantification of arm motor deficits during activities of daily living in the home setting. This is based on recent data suggesting that (1) hypertonia due to stroke may be directly linked to the expression of spasticity and (2) constant exposure to hypertonia may have a neuroprotective effect by preventing structural adaptations due to disuse of the paretic limb. The functional implications of such altered neural drive, however, can only be captured by monitoring motor deficits in the paretic arm in a real-world context where such deficits are experienced and manifested.

Current measurement methods are only feasible in the laboratory or clinic and are often used in a functionally irrelevant context that does not correspond to real-world conditions [61, 62]. On the other hand, wearable sensors such as inertial measurement units (IMU) have been recently adopted to examine motor behaviors of neurologically injured arms in the real-world setting [63, 64]. Wearable technology, especially when combined with artificial intelligence (e.g., machine learning) techniques, has shown the potential to measure outcomes [65, 66], provide feedback [67], and engage users outside the clinic or laboratory [68]. However, we currently lack the methods to routinely gather and convey relevant and objective information containing both physical and neural activity that constantly determine the practical function of the paretic arm at every moment during daily life.

To address this gap, recent work in our laboratories has focused on the development of user-friendly technology to track physical and neural activity, integrating signal analysis techniques, artificial intelligence, and computational simulation, with wearable movement and electromyographical sensors. Our initial efforts have been focused on overcoming the major barriers in using wearable sensors, e.g., IMU + EMG to monitor motor deficits in neurologically injured individuals in the home setting. Specifically, to circumvent the challenges associated with magnitude-based measures of surface EMG (e.g., non-stationarity, noise; [69], we developed an

alternative, time-based approach to reliably quantify hypertonia expressed in paretic arm muscles during daily activities [70]. To overcome the limitations of current applications using wearable sensors and artificial intelligence (AI), whose performance inherently suffers from variations in input data and thus requires laboratory-based solutions such as careful, time-consuming control of the recordings or periodic recalibration [71], we have implemented an unsupervised transfer learning algorithm that can generalize any deep learning model to account for altered input data, i.e., different from that used for training [72]. These approaches can be readily implemented in consumer-grade, easy-to- "don and doff" wearable devices (Fig. 5.5) for seamless technology acceptance and usability by individuals with stroke [73, 74], and used to

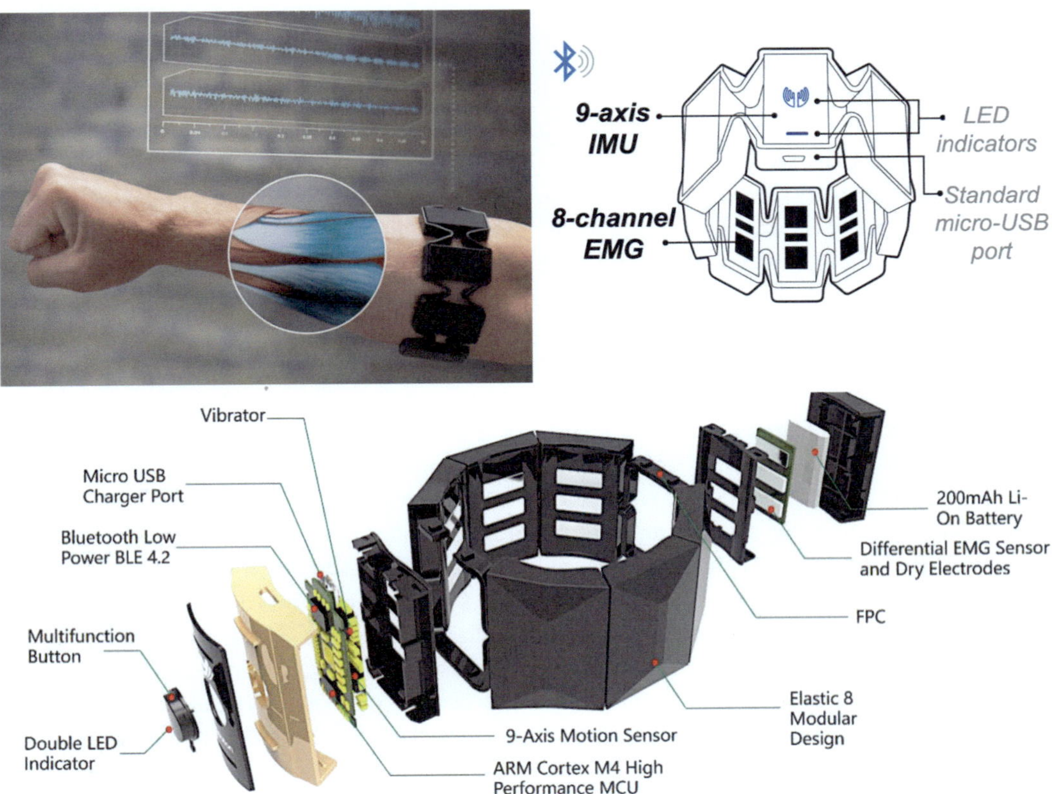

Fig. 5.5 Consumer-grade wearable sensors. Top. Typical location on forearm (left) and components (right) of Myo Armband (Thalmic Labs, Kitchener, Canada), with 9-axis IMU and 8-channel EMG sampled at 200 Hz, communicated via Bluetooth Low Energy (allows for 24 h of continuous use). Bottom. Detailed components of gForcePro + Armband (OYMotion, Shanghai, China), with identical functionality. Such easy-to- "don and doff" wearable devices (also available for laboratory-fabrication; [127]) will allow seamless technology acceptance and usability by individuals with stroke [73, 74]

quantify hypertonia and arm movement (e.g., using IMU + AI-based kinematic estimation; [75, 76]) during activities of daily living. The main requirement for this technology is for an individual with stroke to be able to place the sensors on their paretic upper arm and forearm with minimal assistance.

Taken together, methods that can objectively and comprehensively quantify motor deficits with concurrent, high-resolution information about both the physical and neural activity within the paretic arm at an individual's home will allow for the study of the daily impact of abnormal neural drive, its underlying mechanisms, and guide clinical decisions by addressing the daily impact of stroke, and ultimately enable telerehabilitation that mitigates the cost and risks involved in making visits to clinics or laboratories.

5.3 Impairment-Based Robotic Interventions

5.3.1 Introduction to a Scientifically Underpinned Concept

The strongest available evidence (Class I, Level A) in upper extremity stroke rehabilitation exists for function task practice with graded progression and activities of daily living (ADL) training [77]. However, in individuals with severe motor impairment novel interventions, perhaps directly targeting the cardinal impairments, are greatly in need. Such an innovative solution is capable of ameliorating or attenuating the development of impairments like abnormal synergies and paresis that profoundly limit functional arm use in individuals with moderate to severe stroke. Specific to abnormal synergies, basic science research discussed above has demonstrated that unavoidable and debilitating elbow/wrist and finger flexion occurs during progressively greater SABD loads in individuals with moderate to severe stroke [13, 15, 16, 32, 78, 79]. This phenomenon is attributed to abnormal co-activation of groups of muscles and results in the loss of independent joint control making it

impossible to complete functional upper extremity tasks such as reaching out to pick up a glass of water. The application of robotic technologies makes it possible to design an intervention that directly targets this impairment. Directly targeting loss of independent joint control with an impairment-based intervention such as progressive SABD loading therapy [80, 81] is the most likely avenue for optimizing functional outcomes in this population. This impairment-based approach represents a scientifically underpinned rehabilitation strategy since the neural mechanism of the impairment is well investigated and its relationship to functional movement is known. Recent evidence from our laboratory supporting this approach will be discussed below and appears to elevate the prognosis of even the most severely impaired individuals with stroke.

5.3.2 Targeting the Loss of Independent Joint Control Using Robotic Technologies

Our original intervention work targeting the flexion synergy and associated loss of independent joint control in individuals with moderate to severe stroke sought to identify a key ingredient to therapeutic reaching practice [81]. The intervention, progressive SABD loading therapy, was a randomized controlled trial comparing an intervention where reaching was practiced on a virtual/haptic support surface to reaching performed under progressively increased abduction loading. The intervention relied on the ACT-3D for systematic administration and progression. Utilization of the ACT-3D robotic device allowed us to target the flexion synergy and associated loss of independent joint control through the implementation of a dynamic multi-joint coordination task that did not involve a resistive element. In a randomized controlled design, 14 participants were assigned to one of two intervention groups. While both groups practiced reaching the ACT-3D over 8 weeks emulating traditional therapy, only the experimental group was required to actively support the arm at specified submaximal SABD (vertical) loads. The control group practiced the same reaching tasks but was fully supported on a horizontal haptic table. Therefore, only the experimental group was practicing movement outside of or against the abnormal flexion synergy. Participants in the experimental group were required to support greater percentages of arm weight (corresponding to greater SABD loads) as reaching ability improved beyond standardized kinematic performance thresholds. For example, if a participant could reach 80% of the distance to the practiced target for 8 out of 11 trials in one set for a given abduction load, the load was increased by 25% of limb weight. The same procedure was followed independently for all five of the targets that spanned the reaching work area of each participant based on standardized joint angles (Fig. 5.6). The progressive approach was implemented to maintain continued therapeutic challenge as reaching range of motion improved with a goal of progressing toward movement against gravity (100% limb weight), or more functionally, beyond limb weight as would be experienced when transporting an object. The primary outcome utilized to demonstrate effectiveness was total reaching work area

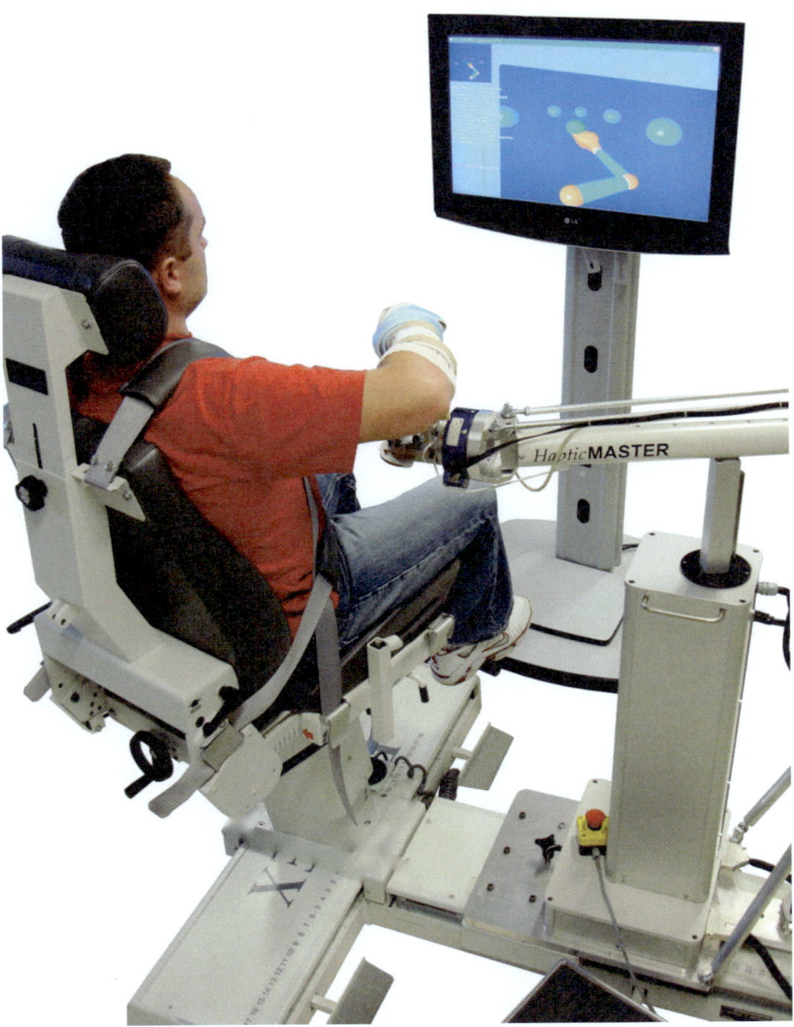

Fig. 5.6 Example of a research participant positioned with the ACT-3D showing the five reaching targets. From Ellis MD, Sukal-Moulton TM, Dewald JP (2009b) Impairment-Based 3-D Robotic Intervention Improves Upper Extremity Work Area in Chronic Stroke: Targeting Abnormal Joint Torque Coupling With Progressive Shoulder Abduction Loading. IEEE Trans Robot 25:549–555; used with permission

as a function of abduction loading, as measured with the ACT-3D, and the secondary outcome was isometric single-joint strength.

We found significantly greater increases in work area for the experimental group. Importantly, the greatest improvements in total reaching work area were at SABD loading levels equivalent to and beyond limb weight such as experienced during the transport of an object during a functional task. The results of the secondary outcome measure of strength were important to the interpretation of why improvements were observed in work area as a function of SABD loading. We found that there was no improvement in single-joint maximum strength indicating that a reduction of flexion synergy and associated increase in multi-joint coordination must have occurred. This research indicated that the SABD loading element was effective in improving arm function. Most importantly, it demonstrated the capacity of a scientifically underpinned impairment-based approach to achieve gains in individuals with chronic severe stroke whom conventional care had failed to address. Next steps will be to anchor changes in reaching function to a more global measure of meaningfulness to individuals with stroke to establish the Minimal Clinically Important Difference or MCID of reaching work area and reaching distance [78].

In a subsequent and larger research clinical trial (RCT), we further investigated progressive SABD loading therapy in an effort to determine if the effects of the intervention could be augmented [80]. Specifically, a resistive element was added to further challenge outward reaching. The resistive element was a horizontal viscous field that was only implemented in the horizontal plane so that there was no interference with abduction loading but the reaching movement felt like pushing through molasses. The comparison group received traditional progressive abduction loading therapy and the experimental group received progressive SABD loading therapy plus viscous resistance. Ultimately, both groups improved and were retained at 3 months demonstrating that progressive SABD loading is effective at improving reaching function in individuals with chronic moderate to severe stroke and is not augmented with other forms of resistance.

5.4 Successful Translation to Clinical Practice

5.4.1 Device Design that Facilitates Successful Translation

Advances in robotic and sensing technologies have given rise to multiple systems for upper extremity rehabilitation in stroke [13, 82–89]. Such systems combine robotics with computer graphics for the delivery of a rehabilitation protocol. Systematic reviews of the effect of robotic-based therapy on upper limb recovery following stroke [2–4, 90] suggest improvement in motor function of the paretic upper limb but are less conclusive on the improvement of functional abilities or activities of daily living. Recent studies have perpetuated the equivocal evidence by investigating multifaceted approaches targeting both impairment and activities of daily living with robotics. A randomized controlled trial following a multifaceted robotic intervention in comparison with conventional care found improvement in motor impairment as measured by Fugl–Meyer motor score; however, clinical meaningfulness of the gains was in question [8]. In contrast, two high-quality large-scale randomized controlled trials with sample sizes of $n = 127$ [9] and $n = 770$ [91] employing robot-assisted therapy in comparison with enhanced/dosage-matched therapy and usual care did not show significant differences between groups. Furthermore, a robotic study specifically investigating the difference between motor function [impairment] gains and clinical functional outcomes found that improvements only occurred for impairments of motor impairment [92]. Building evidence suggests that robot-assisted therapies may impact patients at the impairment level but the translation to clinically meaningful gains in activity limitation are less clear. This raises the question of what attributes of robotic training are most relevant to successful impairment restoration and what impairments will most likely translate to functional gains. What is clear though is that for robotic systems to successfully translate to practice, a clear advantage in terms of enhanced clinical outcome, both patient-reported and directly measured, beyond conventional neuro-rehabilitation therapies must be realized. Rehabilitation specialists are training to apply "best evidence" to clinical decision-making and the highest caliber of evidence remains to be randomized controlled trials that populate systematic reviews and inform clinical practice guidelines.

Many rehabilitation systems are based on traditional therapeutic approaches of functional task practice similar to conventional hands-on rehabilitation. For example, implementing a task-oriented approach where subjects complete a pick-and-place or grasp and release in a virtual task [3, 93–103] is similar to conventional therapeutic strategies [104–107]. A few groups have implemented systems based on a more constrained approach where the reaching movement or task is guided by a pre-defined trajectory or set of rules [108–110], again, similar to traditional interventions where the movement is guided by the therapist(s). The sole focus on functional task practice may explain the equivocal evidence for benefits over that of conventional care. On the other hand, some systems provide robotic assistance to the task or movement being performed either by smartly assisting the arm in a programmed endpoint or joint-space trajectory and/or by supporting the weight of the limb [98, 102, 111–117]. This approach leverages the unique features of the device to address motor impairments during functional task practice which may be difficult to replicate by a therapist in a clinical setting. However, a recent larger randomized controlled trial investigating additional self-rehabilitation using robot-assisted therapy in comparison to self-stretching and basic exercises found no benefit in additional reaching exercises when performed with the arm fully supported against gravity in the device [118].

Perhaps device design has been overly driven by a focus on conventional functional or goal-directed task practice with limited emphasis on targeting and reducing impairments. Even with a multifaceted functional and impairment-based robotic approach, only limited success is possible [9]. Therefore, we believe a sole focus on ameliorating loss of independent joint control will be most effective especially in more moderately to severely impaired individuals where function task practice is not possible. Our laboratories have taken the approach of shifting the sole focus of the robotic training intervention to reducing the most prominent impairment, loss of independent joint control, based on years of research on the mechanisms underlying upper extremity movement impairment in individuals with brain injury. Based on results from previous studies [1, 21, 39–41, 43, 60, 112], we have designed robotic systems to directly target the loss of independent joint control believed to most strongly impact upper extremity function.

Such a strategy may be more effective in improving arm function during activities of daily living in individuals with moderate to severe hemiparetic stroke that struggle to benefit from functional task practice due to the severe abnormal muscle synergies throughout the arm and hand. The ACT-3D [13, 15] was designed with this strategy in mind, to allow adjustable shoulder abduction loading, a required attribute to directly target the flexion synergy impairment. Our previous [7] and more recent [80] clinical trials have demonstrated the effectiveness of targeting the flexion synergy impairment with the ACT-3D and increasing both work area and reaching distance of the upper limb at greater shoulder abduction loads (see previous section—7, 81). Other systems like the T-WREX, PneuWREX, ARMin, L-EXOS, Freebal, ArmeoPower, and BONES [13, 84, 88, 96], have adjustable limb weight support ability and are capable of progressive shoulder abduction loading. The ability to target inter-joint coordination through progressive shoulder abduction loading is a key component for therapeutic interventions attempting to improve arm function during activities of daily living because it is the loss of independent joint control that is the most detrimental impairment in moderate to severe stroke and it is correlated with upper extremity function [15].

Based on the promising results obtained with the ACT-3D, our laboratories have continued to design robotic devices that target specific impairments present in individuals with brain injury such as weakness, synergy, and spasticity. A new device, the ACT-2D, was designed to further our understanding of spasticity during movement in stroke (see Fig. 5.7 and 32). This device allowed investigation of the interaction between abnormal flexion synergy and spasticity, through its ability to provide various shoulder abduction loads while stretching the elbow flexors without volitional activation of elbow muscles. Concurrently, a new version of the ACT-3D, the NACT-3D, was designed to augment its capabilities both in workspace and strength to allow not only implementation of impairment-based interventions but also investigations of the complex interactions between weakness, synergy, and spasticity during multi-joint dynamic conditions in order to better understand the mechanisms underlying movement dysfunction in this population, as described previously (see Figs. 5.6, 5.7, and 5.8).

Translating our technology into the clinical space has required the development of a portable and cost-effective system, the PACT-3D shown in Fig. 5.9. Presently, the PACT-3D has been deployed in an acute care stroke unit, in-patient rehabilitation facility, day rehab center, and an outpatient academic laboratory as part of ongoing clinical trials investigating progressive SABD loading therapy and the development of flexion synergy. The device was designed with portability in mind and can mount to a wheelchair, table, or stand to be useful in a small clinic setting or home. The PACT-3D retains the capacity to provide full SABD loading from weightless to multiple times limb weight in increments tied to a precise physiological measure of abduction strength. It is also fully instrumented to conduct our kinematic and kinetic (in the vertical direction) measures of reaching function and synergy expression. It is our goal that the combination of reduced expense, precision evaluation, and intervention progression, and evidence for effective attenuation of flexion synergy will facilitate the translation of the PACT-3D to clinical care and home environments.

5.4.2 Acceptance by the Rehabilitation Specialist

Despite exciting advancements in rehabilitation robotics regarding the precision evaluation of movement impairments and impairment-based interventions, translation to clinical practice has

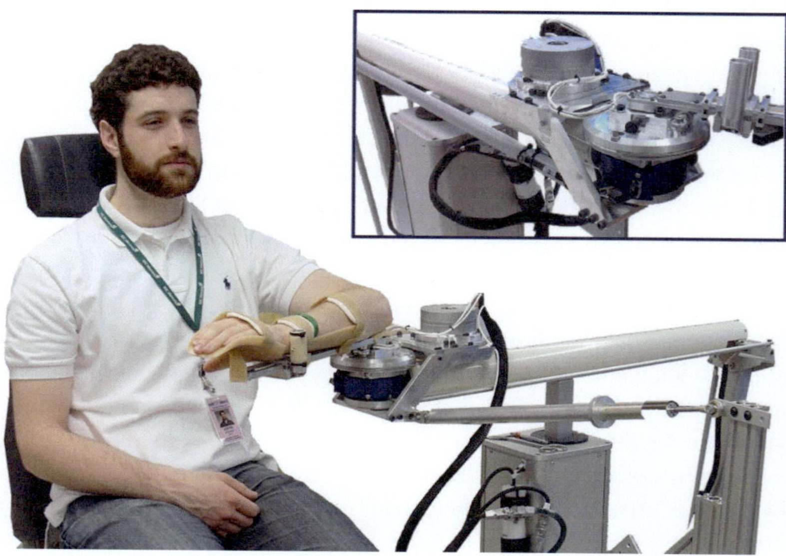

Fig. 5.7 The ACT-2D robotic device allows for single joint perturbations at the elbow combined with adjustable shoulder abduction loading to study the relationship between synergies and abnormal stretch reflex or spasticity following brain injury

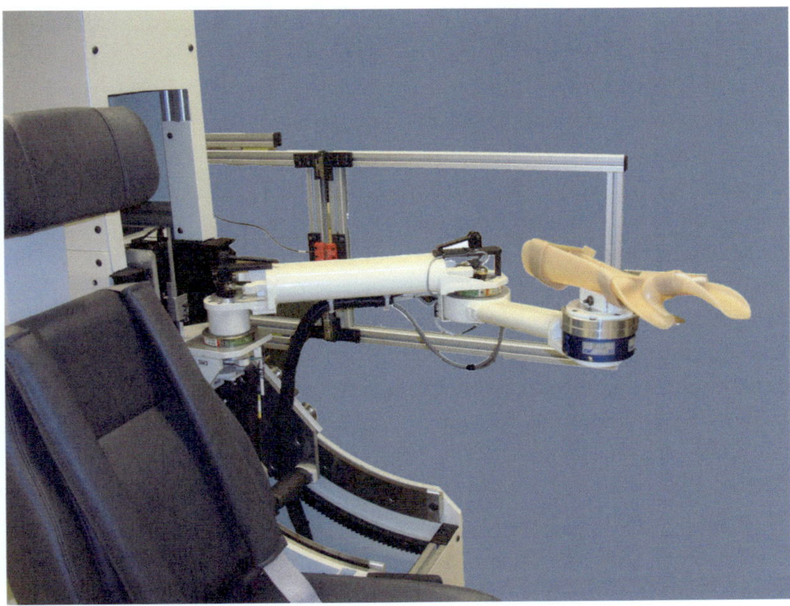

Fig. 5.8 New version of the ACT-3D, the NACT-3D, is designed to allow greater workspace measurements as well as for the application of multi-joint perturbations in the plane of movement

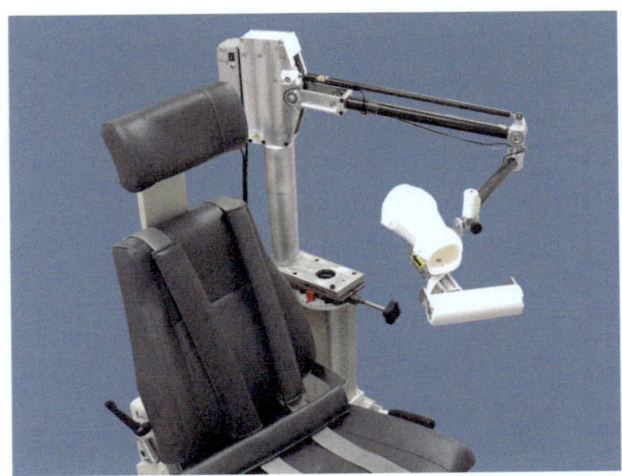

Fig. 5.9 The new PACT-3D is designed to allow safe and efficient use in the clinic and home for stroke upper limb rehabilitation and flexion synergy impairment measures

been slow and incremental. The rate of translation can be improved by increasing the quality of evidence made available to practicing clinicians. The field of rehabilitation is more likely to accept new technologies, such as the impairment-based approach utilizing novel robotic and sensing technologies, if strong and high-quality evidence of impairment reduction is provided. Evidence from our lab supports an impairment-based approach showing that amelioration of flexion synergy and improvement in reaching function is possible in individuals with chronic stroke with moderate to severe impairments [80]. Important for translation is that this study represents a very high-quality RCT as rated on the PEDRO scale. However, more high-quality RCTs must be completed at a much larger scale to facilitate translation. Our current Phase IIb RCT (ClinicalTrials.gov Identifier: NCT04118998) studying progressive abduction loading therapy in an inpatient rehabilitation facility hopes to meet this demand. Implementing progressive abduction loading therapy in the subacute phase of recovery is hypothesized to attenuate the development of flexion synergy optimizing the recovery of reaching function. This process of leveraging robotic and sensing technologies to study and then develop a novel therapy represents a scientifically underpinned strategy for achieving optimized recovery that is in stark contrast to the conventional approach of practicing functional tasks which may only be appropriate once impairment restoration has been optimized [12].

Educating clinicians on the benefits of this approach will need to go beyond marketing tutorials describing bells and whistles of robotic devices and include evidence of how the device is grounded in rehabilitation science both in concept, design, and implementation.

Convincing evidence from large-scale clinical trials is necessary to demonstrate that an impairment-based robotic intervention is superior to conventional care not just in improving function but in restoring normal movement through impairment reduction. Additionally, improvements observed should be explained by the underlying neurophysiological mechanisms. Our laboratory has recently made substantial efforts to merge quantitative evaluation of movement with high-resolution neuroimaging to evaluate intervention-related experience-dependent neuroplasticity addressing this requirement [119]. In addition, significant gains have been made in understanding the underlying neuromechanisms of flexion synergy perhaps highlighting new biomarkers representative of recovery processes [32, 120]. Future work should also seek to evaluate how other aspects of motor learning [121], beyond our current employment of optimal practice scheduling, task specificity, and augmented feedback, can be brought to bear when targeting loss of independent joint control. With convincing quantitative evidence and sound scientific underpinning, the rehabilitation specialist will readily accept the impairment-based approach catalyzing the translation to clinical practice.

5.4.3 Motivation, Ease of Use, Practical Implications, and Translation into Rehabilitation Clinics

Patient motivation is one principle of motor learning that can be readily brought to bear in targeting impairment restoration in rehabilitation robotics by integrating with video game platforms. Combining impairment-based interventions with a game has the potential to motivate patients to participate in therapy sessions and push themselves to greater performances. Advances in robotic and video game technology gave rise to multiple systems for upper extremity rehabilitation in stroke [13, 82–89, 122]. Such systems combine robotics with computer graphics for the delivery of a rehabilitation protocol.

An increasingly common approach is the use of virtual reality (VR) games that allow interaction with a three-dimensional environment simulated in a computer and integrated with haptic feedback. Reviews on the effectiveness of virtual reality programs for stroke rehabilitation [123–125] support its application albeit with limited evidence. All of these reviews recognize the potential for these therapeutic modalities, encouraging further research to establish their validity and provide evidence of their advantages over conventional therapy. The lack of directly targeting specific impairments in current gaming approaches may explain the limited improvements in arm function during activities of daily living. Preliminary results from our laboratory suggest that the combination of video games and robotics to create a haptic interface should emphasize the design of games that include specific reaching targets in the workspace compromised by the expression of the loss of independent joint control following stroke [126]. Therefore, the ultimate goal will be to develop video games that, in combination with state-of-the-art robotic devices and sensing technologies, directly address movement impairments while providing a fun and challenging experience. The combination of increased motivation and improved outcomes will facilitate successful translation to practice.

Another important element that needs to be considered for the successful translation of robotics to clinical practice, and possibly to the home environment, is its ease of use. Once the necessary ingredients have been determined to measure and reduce movement impairments resulting from brain injury, simple actuated or possibly passive devices should be developed. Setup time for use of such devices should be fast and measurement and treatment approaches, incorporating gaming, should provide intuitive interfaces that can be ultimately utilized by the individual receiving therapy. Our PACT-3D provides an example of a robotic device that requires minimal setup, is instrumented to provide measurements of flexion synergy in real time, and can be used in combination with video games to create an engaging rehabilitation environment in the clinic or home. An easy-to-use interface currently limited to rehabilitation professionals, that provides a thorough examination of the flexion synergy impairment in the paretic limb of individuals with stroke is in use in a series of clinical trials. Further development will expand to video games for use in the stroke and pediatric hemiplegia populations.

Finally, to facilitate the translation of impairment-based robotic and sensing technologies to clinical practice, they should offer evaluation and treatment approaches that are not readily reproducible by rehabilitation specialists. These technologies must allow precise quantitative evaluation of movement impairments resulting from brain injury such as the loss of independent joint control, weakness, and spasticity. Furthermore, devices must utilize standard quantitative measurements of impairment to initiate and progress the intervention. With these attributes, clinicians will be better informed of the impairments causing movement dysfunction and the response of the patient to rehabilitation.

5.5 Conclusion

This chapter discusses the use of impairment-based rehabilitation technologies and provides examples of device development that allows both the evaluation and treatment of movement impairments. Evidence is provided demonstrating that robotic and sensing technologies have the unique ability to measure the loss of independent joint control, weakness, hypertonia, and associated spasticity following brain injury. In addition to the quantification and study of mechanisms underlying the expression of these impairments, evidence was also provided demonstrating the effectiveness of specifically targeting fundamental impairments in order to improve arm function during activities of daily living. A shift in focus to impairment restoration was suggested in contrast with the repeated application of robotics that focuses on greater intensity of existing rehabilitation approaches

and multifaceted approaches of impairment-based and functional-based task practice. Finally, successful translation to clinical practice was discussed pointing to several key attributes that will facilitate both clinician and patient acceptance. With this chapter, we hope to have demonstrated that new robotic and sensing technologies are ideal for the delivery of novel therapeutic interventions grounded in a body of scientific evidence. And, that robotic interventions can be implemented in current rehabilitation clinics as well as provide a tool for clinicians to better evaluate and treat patients in a more controlled fashion with greater specificity and intensity than is currently possible with conventional rehabilitation.

The successful application of impairment-based rehabilitation technologies will depend on two factors. First, robotic devices must be proven to provide a quantitative evaluation that precisely defines movement impairments that can serve both as indicators for prognosis and response to rehabilitation. Wielding powerful diagnostic and prognostic tools, rehabilitation specialists will make more informed clinical decisions and achieve better clinical outcomes. Second, the future of rehabilitation robotics lies in our ability to demonstrate the effectiveness of robotic devices in delivering interventions that result not only in amelioration of impairments but also in clear gains in arm function during activities of daily living. This will require implementation of large-sample Phase III and IV Clinical Trials that encompass controlled impairment-based rehabilitation robotic interventions and conventional care. These trials will have the statistical power necessary to detect significant clinical effects utilizing outcomes measuring the activity of daily living that are unavoidably limited by low-resolution ordinal scales of measurement. Additionally, it is with these large Phase III and IV Clinical Trials that cost–benefit analyses can be completed demonstrating the fiscal utility of these exciting new impairment-based technologies in a changing healthcare environment (Table 5.1).

Table 5.1 Upper limb synergies in hemiparetic stroke [36]

Flexor synergy	Extensor synergy
Flexion of the wrist and fingers	Extension of the wrist and flexion of fingers
Flexion of the elbow	Extension of the elbow
Supination of the forearm	Pronation of the forearm
Abduction of the shoulder	Adduction of the arm in front of the body
External rotation of the shoulder	Internal rotation of the shoulder
Shoulder girdle retraction and/or elevation	shoulder girdle protraction

From Brunnstrom S (1970) Movement therapy in hemiplegia: a neurophysiological approach. New York: Medical Dept. Harper & Row

References

1. Dewald JP, Sheshadri V, Dawson ML, Beer RF. Upper-limb discoordination in hemiparetic stroke: implications for neurorehabilitation. Top Stroke Rehabil. 2001 Spring;8(1):1–12.
2. Kwakkel G, Kollen BJ, Krebs HI. Effects of robot-assisted therapy on upper limb recovery after stroke: a systematic review. Neurorehabil Neural Repair. 2008;22(2):111–21.
3. Mehrholz J, Platz T, Kugler J, Pohl M. Electromechanical and robot-assisted arm training for improving arm function and activities of daily living after stroke. Stroke. 2009.
4. Prange GB, Jannink MJ, Groothuis-Oudshoorn CG, Hermens HJ, Ijzerman MJ. Systematic review of the effect of robot-aided therapy on recovery of the hemiparetic arm after stroke. J Rehabil Res Dev. 2006;43(2):171–84.
5. Duret C, Grosmaire AG, Krebs HI. Robot-assisted therapy in upper extremity hemiparesis: overview of an evidence-based approach. Front Neurol. 2019;10:412.
6. Brokaw EB, Nichols D, Holley RJ, Lum PS. Robotic therapy provides a stimulus for upper limb motor recovery after stroke that is complementary to and distinct from conventional therapy. Neurorehabil Neural Repair. 2014;28(4):367–76.
7. Ellis MD, Sukal-Moulton TM, Dewald JP. Impairment-based 3-D robotic intervention improves upper extremity work area in chronic stroke: targeting abnormal joint torque coupling with progressive shoulder abduction loading. IEEE Trans Robot. 2009;25(3):549–55.
8. Klamroth-Marganska V, Blanco J, Campen K, Curt A, Dietz V, Ettlin T, et al. Three-dimensional, task-specific robot therapy of the arm after stroke: a multicentre, parallel-group randomised trial. Lancet Neurol. 2014;13(2):159–66.
9. Lo AC, Guarino PD, Richards LG, Haselkorn JK, Wittenberg GF, Federman DG, et al. Robot-assisted therapy for long-term upper-limb impairment after stroke. N Engl J Med. 2010;362(19):1772–83.
10. Milot MH, Spencer SJ, Chan V, Allington JP, Klein J, Chou C, et al. A crossover pilot study evaluating the functional outcomes of two different types of robotic movement training in chronic stroke survivors using the arm exoskeleton BONES. J Neuroeng Rehabil. 2013;10:112.
11. Gassert R, Dietz V. Rehabilitation robots for the treatment of sensorimotor deficits: a neurophysiological perspective. J Neuroeng Rehabil. 2018;15(1):46.
12. Huang VS, Krakauer JW. Robotic neurorehabilitation: a computational motor learning perspective. J Neuroeng Rehabil. 2009;6:5.
13. Sukal TM, Ellis MD, Dewald JP. Shoulder abduction-induced reductions in reaching work area following hemiparetic stroke: neuroscientific implications. Exp Brain Res. 2007;183(2):215–23.
14. Brokaw EB, Murray T, Nef T, Lum PS. Retraining of interjoint arm coordination after stroke using robot-assisted time-independent functional training. J Rehabil Res Dev. 2011;48(4):299–316.
15. Ellis MD, Sukal T, DeMott T, Dewald JP. Augmenting clinical evaluation of hemiparetic arm movement with a laboratory-based quantitative measurement of kinematics as a function of limb loading. Neurorehabil Neural Repair. 2008;22(4):321–9.
16. Miller LC, Dewald JP. Involuntary paretic wrist/finger flexion forces and EMG increase with shoulder abduction load in individuals with chronic stroke. Clin Neurophysiol. 2012;123(6):1216–25.
17. Lan Y, Yao J, Dewald JPA. The impact of shoulder abduction loading on volitional hand opening and grasping in chronic hemiparetic stroke. Neurorehabil Neural Repair. 2017;31(6):521–9.
18. McPherson LM, Dewald JPA. Differences between flexion and extension synergy-driven coupling at the elbow, wrist, and fingers of individuals with chronic hemiparetic stroke. Clin Neurophysiol. 2019;130(4):454–68.
19. Yao J, Dewald JPA. The increase in overlap of cortical activity preceding static elbow/shoulder motor tasks is associated with limb synergies in severe stroke. Neurorehabil Neural Repair. 2018;32(6–7):624–34.

20. Koo TK, Mak AF, Hung L, Dewald JP. Joint position dependence of weakness during maximum isometric voluntary contractions in subjects with hemiparesis. Arch Phys Med Rehabil. 2003;84 (9):1380–6.

21. Beer RF, Ellis MD, Holubar BG, Dewald JP. Impact of gravity loading on post-stroke reaching and its relationship to weakness. Muscle Nerve. 2007;36(2):242–50.

22. Burne JA, Carleton VL, O'Dwyer NJ. The spasticity paradox: movement disorder or disorder of resting limbs? J Neurol Neurosurg Psychiatry. 2005;76(1):47–54.

23. Schmit BD. Mechanical measures of spasticity in stroke. Top Stroke Rehabil. 2001 Spring;8(1):13–26.

24. Schmit BD, Dhaher Y, Dewald JP, Rymer WZ. Reflex torque response to movement of the spastic elbow: theoretical analyses and implications for quantification of spasticity. Ann Biomed Eng. 1999;27(6):815–29.

25. Schmit BD, Dewald JP, Rymer WZ. Stretch reflex adaptation in elbow flexors during repeated passive movements in unilateral brain-injured patients. Arch Phys Med Rehabil. 2000;81(3):269–78.

26. Powers RK, Marder-Meyer J, Rymer WZ. Quantitative relations between hypertonia and stretch reflex threshold in spastic hemiparesis. Ann Neurol. 1988;23(2):115–24.

27. Powers RK, Campbell DL, Rymer WZ. Stretch reflex dynamics in spastic elbow flexor muscles. Ann Neurol. 1989;25(1):32–42.

28. Trumbower RD, Ravichandran VJ, Krutky MA, Perreault EJ. Contributions of altered stretch reflex coordination to arm impairments following stroke. J Neurophysiol. 2010;104(6):3612–24.

29. de Vlugt E, de Groot JH, Schenkeveld KE, Arendzen JH, van der Helm FC, Meskers CG. The relation between neuromechanical parameters and Ashworth score in stroke patients. J Neuroeng Rehabil. 2010;7:35.

30. Meskers CG, Schouten AC, de Groot JH, de Vlugt E, van Hilten BJ, van der Helm FC, et al. Muscle weakness and lack of reflex gain adaptation predominate during post-stroke posture control of the wrist. J Neuroeng Rehabil. 2009;6:29.

31. Ellis MD, Schut I, Dewald JPA. Flexion synergy overshadows flexor spasticity during reaching in chronic moderate to severe hemiparetic stroke. Clin Neurophysiol. 2017;128(7):1308–14.

32. McPherson JG, Stienen AH, Drogos JM, Dewald JP. Modification of Spastic Stretch Reflexes at the Elbow by Flexion Synergy Expression in Individuals With Chronic Hemiparetic Stroke. Arch Phys Med Rehabil. 2018;99(3):491–500.

33. McPherson JG, Stienen AHA, Schmit BD, Dewald JPA. Biomechanical parameters of the elbow stretch reflex in chronic hemiparetic stroke. Exp Brain Res. 2019;237(1):121–35.

34. Sanchez N, Acosta AM, Lopez-Rosado R, Dewald JPA. Neural constraints affect the ability to generate hip abduction torques when combined with hip extension or ankle plantarflexion in chronic hemiparetic stroke. Front Neurol. 2018;9:564.

35. Sanchez N, Acosta AM, Lopez-Rosado R, Stienen AHA, Dewald JPA. Lower extremity motor impairments in ambulatory chronic hemiparetic stroke: evidence for lower extremity weakness and abnormal muscle and joint torque coupling patterns. Neurorehabil Neural Repair. 2017;31(9):814–26.

36. Brunnstrom S. Movement therapy in hemiplegia: a neurophysiological approach. [1st] ed. New York: Medical Dept. Harper & Row; 1970.

37. Foester O. Motorische Felder und Bahnen. In: Bumke O, Foester O, editors. Handbuch der Neurologie. Berlin: Springer; 1933.

38. Twitchell TE. The restoration of motor function following hemiplegia in man. Brain. 1951;74 (4):443–80.

39. Dewald JP, Pope PS, Given JD, Buchanan TS, Rymer WZ. Abnormal muscle coactivation patterns during isometric torque generation at the elbow and shoulder in hemiparetic subjects. Brain. 1995;118 (Pt 2):495–510.

40. Beer R, Dewald J. Task-Dependent Weakness at the Elbow in Patients with Hemiparesis. Arch Phys Med Rehabil. 1999;80(7):766–72.

41. Dewald JP, Beer RF. Abnormal joint torque patterns in the paretic upper limb of subjects with hemiparesis. Muscle Nerve. 2001;24(2):273–83.

42. Fugl-Meyer AR, Jaasko L, Leyman I, Olsson S, Steglind S. The post-stroke hemiplegic patient. 1. A method for evaluation of physical performance. Scand J Rehabil Med. 1975;7(1):13–31.

43. Beer RF, Dewald JP, Dawson ML, Rymer WZ. Target-dependent differences between free and constrained arm movements in chronic hemiparesis. Exp Brain Res. 2004;156(4):458–70.

44. Colebatch JG, Gandevia SC. The distribution of muscular weakness in upper motor neuron lesions affecting the arm. Brain. 1989;112(Pt 3):749–63.

45. Lang CE, Wagner JM, Edwards DF, Sahrmann SA, Dromerick AW. Recovery of grasp versus reach in people with hemiparesis poststroke. Neurorehabil Neural Repair. 2006;20(4):444–54.

46. Turton A, Wroe S, Trepte N, Fraser C, Lemon RN. Contralateral and ipsilateral EMG responses to transcranial magnetic stimulation during recovery of arm and hand function after stroke. Electroencephalogr Clin Neurophysiol. 1996;101(4):316–28.

47. Miller LC, Ruiz-Torres R, Stienen AH, Dewald JP. A wrist and finger force sensor module for use during movements of the upper limb in chronic hemiparetic stroke. IEEE Trans Biomed Eng. 2009;56(9):2312–7.

48. Lance JW. The control of muscle tone, reflexes, and movement: Robert Wartenberg Lecture. Neurology. 1980;30(12):1303–13.

49. Ada L, Vattanasilp W, O'Dwyer NJ, Crosbie J. Does spasticity contribute to walking dysfunction after stroke? J Neurol Neurosurg Psychiatry. 1998;64 (5):628–35.

50. Ibrahim IK, Berger W, Trippel M, Dietz V. Stretch-induced electromyographic activity and torque in spastic elbow muscles. Differential modulation of reflex activity in passive and active motor tasks. Brain. 1993;116 (Pt 4):971–89.

51. Meyer M, Adorjani C. Tonic stretch reflex for quantification of pathological muscle tone. In: Feldman RG YR, Koella WP, CIBA-GEIGY Corporation, editor. Spasticity: disordered motor control. Chicago: Year Book Medical Publishers, Miami, FL; 1980. p. 310–30.

52. Meyer M, Adorjani C. Quantification of the effects of muscle relaxant drugs in man by tonic stretch reflex. Adv Neurol. 1983;39:997–1011.

53. Neilson PD. Interaction between voluntary contraction and tonic stretch reflex transmission in normal and spastic patients. J Neurol Neurosurg Psychiatry. 1972;35(6):853–60.

54. Brown TI, Rack PM, Ross HF. Electromyographic responses to imposed sinusoidal movement of the human thumb. J Physiol. 1982;332:87–99.

55. Yeo W, Ada L, O'Dwyer NJ, Neilson PD. Tonic stretch reflexes in older able-bodied people. Electromyogr Clin Neurophysiol. 1998;38(5):273–8.

56. Gottlieb GL, Agarwal GC. Dependence of human ankle compliance on joint angle. J Biomech. 1978;11(4):177–81.

57. Dietz V, Trippel M, Berger W. Reflex activity and muscle tone during elbow movements in patients with spastic paresis. Ann Neurol. 1991;30(6):767–79.

58. Lee WA, Boughton A, Rymer WZ. Absence of stretch reflex gain enhancement in voluntarily activated spastic muscle. Exp Neurol. 1987;98 (2):317–35.

59. Sinkjaer T, Toft E, Larsen K, Andreassen S, Hansen HJ. Non-reflex and reflex mediated ankle joint stiffness in multiple sclerosis patients with spasticity. Muscle Nerve. 1993;16(1):69–76.

60. Beer RF, Dewald JP, Rymer WZ. Deficits in the coordination of multijoint arm movements in patients with hemiparesis: evidence for disturbed control of limb dynamics. Exp Brain Res. 2000;131 (3):305–19.

61. Lechner HE, Frotzler A, Eser P. Relationship between self- and clinically rated spasticity in spinal cord injury. Arch Phys Med Rehabil. 2006;87(1):15–9.

62. Burridge JH, Wood DE, Hermens HJ, Voerman GE, Johnson GR, van Wijck F, et al. Theoretical and methodological considerations in the measurement of spasticity. Disabil Rehabil. 2005;27(1–2):69–80.

63. Held JPO, Klaassen B, Eenhoorn A, van Beijnum BF, Buurke JH, Veltink PH, et al. Inertial sensor measurements of upper-limb kinematics in stroke patients in clinic and home environment. Front Bioeng Biotechnol. 2018;6:27.

64. van Meulen FB, Klaassen B, Held J, Reenalda J, Buurke JH, van Beijnum BJ, et al. Objective evaluation of the quality of movement in daily life after stroke. Front Bioeng Biotechnol. 2015;3:210.

65. Patel S, Hughes R, Hester T, Stein J, Akay M, Dy J, et al. Tracking motor recovery in stroke survivors undergoing rehabilitation using wearable technology. Annu Int Conf IEEE Eng Med Biol Soc. 2010;2010:6858–61.

66. Parnandi A, Wade E, Mataric M. Motor function assessment using wearable inertial sensors. Annu Int Conf IEEE Eng Med Biol Soc. 2010;2010:86–9.

67. Wittmann F, Held JP, Lambercy O, Starkey ML, Curt A, Hover R, et al. Self-directed arm therapy at home after stroke with a sensor-based virtual reality training system. J Neuroeng Rehabil. 2016;13(1):75.

68. Maceira-Elvira P, Popa T, Schmid AC, Hummel FC. Wearable technology in stroke rehabilitation: towards improved diagnosis and treatment of upper-limb motor impairment. J Neuroeng Rehabil. 2019;16(1):142.

69. Roy SH, De Luca G, Cheng MS, Johansson A, Gilmore LD, De Luca CJ. Electro-mechanical stability of surface EMG sensors. Med Biol Eng Comput. 2007;45(5):447–57.

70. Sohn MH, Deol J, Dewald JPA. A novel time-based surface EMG measure for quantifying hypertonia in paretic arm muscles during daily activities after hemiparetic stroke. medRxiv. 2022:2022.01.06.22268857.

71. Ketyko I, Kovacs F, Varga KZ. Domain adaptation for sEMG-based gesture recognition with recurrent neural networks 2019: IEEE.

72. Sohn MH, Lai SY, Elwin ML, Dewald JPA. Feasibility of Using Wearable EMG Armbands combined with Unsupervised Transfer Learning for Seamless Myoelectric Control. bioRxiv. 2022:2022.01.06.475232.

73. Feldner HA, Papazian C, Peters K, Steele KM. It's all sort of cool and interesting...but what do i do with it? A qualitative study of stroke survivors' perceptions of surface electromyography. Front Neurol. 2020;11:1037.

74. Tran VT, Riveros C, Ravaud P. Patients' views of wearable devices and AI in healthcare: findings from the ComPaRe e-cohort. NPJ Digit Med. 2019;2:53.

75. Rapp E, Shin S, Thomsen W, Ferber R, Halilaj E. Estimation of kinematics from inertial measurement units using a combined deep learning and optimization framework. J Biomech. 2021;12(116): 110229.

76. Senanayake D, Halgamuge S, Ackland DC. Real-time conversion of inertial measurement unit data to ankle joint angles using deep neural networks. J Biomech. 2021;26(125): 110552.

77. Winstein CJ, Wolf SL, Dromerick AW, Lane CJ, Nelsen MA, Lewthwaite R, et al. Effect of a task-oriented rehabilitation program on upper extremity recovery following motor stroke: the ICARE randomized clinical trial. JAMA. 2016;315(6):571–81.

78. Ellis MD, Lan Y, Yao J, Dewald JP. Robotic quantification of upper extremity loss of independent joint control or flexion synergy in individuals with hemiparetic stroke: a review of paradigms addressing the effects of shoulder abduction loading. J Neuroeng Rehabil. 2016;13(1):95.

79. Ellis MD, Hoak JM, Ellis BW, Brown JA, Sit TL, Wilkinson CA, et al. quantitative real-time PCR analysis of individual flue-cured tobacco seeds and seedlings reveals seed transmission of tobacco mosaic virus. Phytopathology. 2020;110(1):194–205.

80. Ellis MD, Carmona C, Drogos J, Dewald JPA. Progressive Abduction loading therapy with horizontal-plane viscous resistance targeting weakness and flexion synergy to treat upper limb function in chronic hemiparetic stroke: a randomized clinical trial. Front Neurol. 2018;9:71.

81. Ellis MD, Sukal-Moulton T, Dewald JP. Progressive shoulder abduction loading is a crucial element of arm rehabilitation in chronic stroke. Neurorehabil Neural Repair. 2009;23(8):862–9.

82. Burgar CG, Lum PS, Shor PC, Machiel Van der Loos HF. Development of robots for rehabilitation therapy: the Palo Alto VA/Stanford experience. Journal of Rehabilitation Research & Development. 2000;37(6):663–73.

83. Fazekas G, Horvath M, Toth A. A novel robot training system designed to supplement upper limb physiotherapy of patients with spastic hemiparesis. Int J Rehabil Res. 2006;29(3):251–4.

84. Frisoli A, Borelli L, Montagner A, Marcheschi S, Procopio C, Salsedo F, et al. Arm rehabilitation with a robotic exoskeleton in virtual reality. IEEE 10th International Conference on Rehabilitation Robotics, 2007 ICORR 2007 2007; Noordwijk, Netherlands.

85. Hesse S, Schulte-Tigges G, Konrad M, Bardeleben A, Werner C. Robot-assisted arm trainer for the passive and active practice of bilateral forearm and wrist movements in hemiparetic subjects. Arch Phys Med & Rehabil. 2003;84(6):915–20.

86. Krebs HI, Hogan N, Volpe BT, Aisen ML, Edelstein L, Diels C. Overview of clinical trials with MIT-MANUS: a robot-aided neuro-rehabilitation facility. Technol Health Care. 1999;7(6):419–23.

87. Masiero S, Celia A, Armani M, Rosati G. A novel robot device in rehabilitation of post-stroke hemiplegic upper limbs. Aging Clin Exp Res. 2006;18(6):531–5.

88. Nef T, Mihelj M, Riener R. ARMin: a robot for patient-cooperative arm therapy. Med Biol Eng Comput. 2007;45(9):887–900.

89. Reinkensmeyer DJ, Kahn LE, Averbuch M, McKenna-Cole A, Schmit BD, Rymer WZ. Understanding and treating arm movement impairment after chronic brain injury: progress with the ARM guide. J Rehabil Res Dev. 2000;37(6):653–62.

90. Mehrholz J, Pohl M, Platz T, Kugler J, Elsner B. Electromechanical and robot-assisted arm training for improving activities of daily living, arm function, and arm muscle strength after stroke. Cochrane Database Syst Rev. 2015 Nov 7(11):CD006876.

91. Rodgers H, Bosomworth H, Krebs HI, van Wijck F, Howel D, Wilson N, et al. Robot assisted training for the upper limb after stroke (RATULS): a multicentre randomised controlled trial. Lancet. 2019;394(10192):51–62.

92. Kitago T, Marshall RS. Strategies for early stroke recovery: what lies ahead? Curr Treat Options Cardiovasc Med. 2015;17(1):356.

93. Broeren J, Claesson L, Goude D, Rydmark M, Sunnerhagen KS. Virtual rehabilitation in an activity centre for community-dwelling persons with stroke. The possibilities of 3-dimensional computer games. Cerebrovasc Dis. 2008;26(3):289–96.

94. Fischer HC, Stubblefield K, Kline T, Luo X, Kenyon RV, Kamper DG. Hand rehabilitation following stroke: a pilot study of assisted finger extension training in a virtual environment. Top Stroke Rehabil. 2007 Jan-Feb;14(1):1–12.

95. Holden MK, Dyar TA, Dayan-Cimadoro L. Telerehabilitation using a virtual environment improves upper extremity function in patients with stroke. IEEE Trans Neural Syst Rehabil Eng. 2007;15(1):36–42.

96. Housman SJ, Scott KM, Reinkensmeyer DJ. A randomized controlled trial of gravity-supported, computer-enhanced arm exercise for individuals with severe hemiparesis. Neurorehabil Neural Repair. 2009;23(5):505–14.

97. Kuttuva M, Boian R, Merians A, Burdea G, Bouzit M, Lewis J, et al. The Rutgers Arm, a rehabilitation system in virtual reality: a pilot study. Cyberpsychol Behav. 2006;9(2):148–51.

98. Nef T, Quinter G, Muller R, Riener R. Effects of arm training with the robotic device ARMin I in chronic stroke: three single cases. Neurodegener Dis. 2009;6(5–6):240–51.

99. Reinkensmeyer DJ, Pang CT, Nessler JA, Painter CC. Web-based telerehabilitation for the upper extremity after stroke. IEEE Trans Neural Syst Rehabil Eng. 2002;10(2):102–8.

100. Sivak M, Mavroidis C, Holden MK. Design of a low cost multiple user virtual environment for rehabilitation (MUVER) of patients with stroke. Stud Health Technol Inform. 2009;142:319–24.

101. Stewart JC, Yeh SC, Jung Y, Yoon H, Whitford M, Chen SY, et al. Intervention to enhance skilled arm and hand movements after stroke: A feasibility study using a new virtual reality system. J Neuroeng Rehabil. 2007;4:21.

102. van der Kooij H, Prange GB, Krabben T, Renzenbrink GJ, de Boer J, Hermens HJ, et al. Preliminary

results of training with gravity compensation of the arm in chronic stroke survivors. Conf Proc IEEE Eng Med Biol Soc. 2009;1:2426–9.

103. Broeren J, Rydmark M, Sunnerhagen KS. Virtual reality and haptics as a training device for movement rehabilitation after stroke: a single-case study. Arch Phys Med Rehabil. 2004;85(8):1247–50.

104. van Peppen RP, Hendriks HJ, van Meeteren NL, Helders PJ, Kwakkel G. The development of a clinical practice stroke guideline for physiotherapists in The Netherlands: a systematic review of available evidence. Disabil Rehabil. 2007;29 (10):767–83.

105. Van Peppen RP, Kwakkel G, Wood-Dauphinee S, Hendriks HJ, Van der Wees PJ, Dekker J. The impact of physical therapy on functional outcomes after stroke: what's the evidence? Clin Rehabil. 2004;18:833–62.

106. Dobkin BH. Strategies for stroke rehabilitation. Lancet Neurol. 2004;3(9):528–36.

107. Woldag H, Hummelsheim H. Evidence-based physiotherapeutic concepts for improving arm and hand function in stroke patients: a review. J Neurol. 2002;249(5):518–28.

108. Adamovich SV, Merians AS, Boian R, Tremaine M, Burdea GS, Recce M, et al. A virtual reality based exercise system for hand rehabilitation post-stroke: transfer to function. Engineering in Medicine and Biology Society, 2004 IEMBS '04 26th Annual International Conference of the IEEE; 2004.

109. Merians AS, Poizner H, Boian R, Burdea G, Adamovich S. Sensorimotor training in a virtual reality environment: does it improve functional recovery poststroke? Neurorehabil Neural Repair. 2006;20(2):252–67.

110. Piron L, Turolla A, Agostini M, Zucconi C, Cortese F, Zampolini M, et al. Exercises for paretic upper limb after stroke: a combined virtual-reality and telemedicine approach. J Rehabil Med. 2009;41 (12):1016–102.

111. Frisoli A, Bergamasco M, Carboncini MC, Rossi B. Robotic assisted rehabilitation in Virtual Reality with the L-EXOS. Stud Health Technol Inform. 2009;145:40–54.

112. Hesse S, Werner C, Pohl M, Rueckriem S, Mehrholz J, Lingnau ML. Computerized arm training improves the motor control of the severely affected arm after stroke: a single-blinded randomized trial in two centers. Stroke. 2005;36(9):1960–6.

113. Kahn LE, Zygman ML, Rymer WZ, Reinkensmeyer DJ. Robot-assisted reaching exercise promotes arm movement recovery in chronic hemiparetic stroke: a randomized controlled pilot study. J Neuroeng Rehabil. 2006;3:12.

114. Lum PS, Burgar CG, Shor PC, Majmundar M, Van der Loos M. Robot-assisted movement training compared with conventional therapy techniques for the rehabilitation of upper-limb motor function after stroke. Arch Phys Med Rehabil. 2002;83(7):952–9.

115. Macclellan LR, Bradham DD, Whitall J, Volpe B, Wilson PD, Ohlhoff J, et al. Robotic upper-limb neurorehabilitation in chronic stroke patients. J Rehabil Res Dev. 2005 Nov-Dec;42(6):717–22.

116. Staubli P, Nef T, Klamroth-Marganska V, Riener R. Effects of intensive arm training with the rehabilitation robot ARMin II in chronic stroke patients: four single-cases. J Neuroeng Rehabil. 2009;6:46.

117. Housman S, Le V, Rahman T, Sanchez R, Reinkensmeyer D. Arm-Training with T-WREX After Chronic Stroke: Preliminary Results of a Randomized Controlled Trial2007; Noordwijk, Netherlands.

118. Remy-Neris O, Le Jeannic A, Dion A, Medee B, Nowak E, Poiroux E, et al. Additional, mechanized upper limb self-rehabilitation in patients with subacute stroke: the REM-AVC randomized trial. Stroke. 2021;52(6):1938–47.

119. Yao J, Chen A, Carmona C, Dewald JP. Cortical overlap of joint representations contributes to the loss of independent joint control following stroke. Neuroimage. 2009;45(2):490–9.

120. Karbasforoushan H, Cohen-Adad J, Dewald JPA. Brainstem and spinal cord MRI identifies altered sensorimotor pathways post-stroke. Nat Commun. 2019;10(1):3524.

121. Muratori LM, Lamberg EM, Quinn L, Duff SV. Applying principles of motor learning and control to upper extremity rehabilitation. J Hand Ther. 2013;26(2):94–102; quiz 3.

122. Krakauer JW, Kitago T, Goldsmith J, Ahmad O, Roy P, Stein J, et al. Comparing a novel neuroanimation experience to conventional therapy for high-dose intensive upper-limb training in subacute stroke: the SMARTS2 randomized trial. Neurorehabil Neural Repair. 2021;35(5):393–405.

123. Henderson A, Korner-Bitensky N, Levin M. Virtual reality in stroke rehabilitation: a systematic review of its effectiveness for upper limb motor recovery. Top Stroke Rehabil. 2007;14(2):52–61.

124. Lucca LF. Virtual reality and motor rehabilitation of the upper limb after stroke: a generation of progress? J Rehabil Med. 2009;41(12):1003–100.

125. Mumford N, Wilson PH. Virtual reality in acquired brain injury upper limb rehabilitation: evidence-based evaluation of clinical research. Brain Inj. 2009;23(3):179–91.

126. Acosta AM, Dewald HA, Dewald JP. Pilot study to test effectiveness of video game on reaching performance in stroke. J Rehabil Res Dev. 2011;48(4):431–44.

127. Cote-Allard U, Gagnon-Turcotte G, Laviolette F, Gosselin B. A low-cost, wireless, 3-D-printed custom armband for sEMG hand gesture recognition. Sensors (Basel). 2019;19(12).

The Hand After Stroke and SCI: Restoration of Function with Technology

6

Mohammad Ghassemi
and Derek G. Kamper

Abstract

Neurological injury, such as that resulting from stroke or spinal cord injury, often leads to impairment of the hand. As the hand is critical to performance of so many functional activities, diminished sensorimotor control of the distal upper extremity can profoundly impact quality of life. This is readily apparent in many stroke survivors and individuals with spinal cord injury. Technological advances have afforded promise that equipment could be developed to facilitate restoration of function. The last 30 years have seen exponential growth in robotic and mechatronic devices targeting impairment arising from neurological injury. Earlier efforts focused largely on external devices such as robotic arms that could either be used to perform tasks for the user or to facilitate movement practice. This technology was often intended to be purely assistive, helping the user to perform a specific task without addressing the underlying pathophysiology, or purely therapeutic, creating forces or motions that would help the user to regain sensorimotor control of their own limb through practice but not helping with performance of a functional task. Recent improvements in materials and actuators have spurred the development of devices that can be worn on the body. These wearable devices afford the possibility of seamlessly shifting between roles as a therapeutic or assistive device depending on the needs of the user, but potentially introduce disadvantages in terms of added mass and bulk. The optimal device remains dependent upon the needs and preferences of the specific user. For an individual with complete spinal cord injury, a robotic arm connected to their wheelchair may provide the best rehabilitation, while for a stroke survivor with good use of the proximal arm, a wearable hand exoskeleton may give the greatest benefit. This chapter describes hand physiology and pathophysiology and wearable and non-wearable robotic devices that have been developed to improve function.

Keywords

Hand function · Stroke · Spinal cord injury · Assistive technology · Therapeutic technology

M. Ghassemi · D. G. Kamper (✉)
Joint Department of Biomedical Engineering at University of North Carolina at Chapel Hill and North Carolina State University, Raleigh/Chapel Hill, NC, USA
e-mail: dgkamper@ncsu.edu

M. Ghassemi
e-mail: mghasse@ncsu.edu

© The Author(s), under exclusive license to Springer Nature Switzerland AG 2022
D. J. Reinkensmeyer et al. (eds.), *Neurorehabilitation Technology*,
https://doi.org/10.1007/978-3-031-08995-4_6

6.1 Hand Neuromechanics

The hand is a wonderfully versatile instrument. We use our hands to communicate; to express ourselves through art, music, and writing; and to manipulate objects. Our hands constitute our primary means of interacting with our environment. Human evolution is closely linked with evolution of the hand. Indeed, one of the earliest species within our genus Homo was labeled Homo habilis, the "handy man," for the presumed use of stone tools [1]. Features of the hand have facilitated this tool use, which is tightly intertwined with human existence and development. In modern humans, the thumb is longer, compared to the other digits, than in any other primate [2, 3]. Increased thumb length coupled with the saddle shape that has evolved for the carpometacarpal (CMC) joint [4] affords the human thumb the greatest range of motion among all animals, thereby facilitating opposition with the fingers for grasping objects.

The neuromechanical complexity of the hand drives its dexterity. The hand, distal to the wrist, is comprised of 19 bones connected through joints that provide more than 21 degrees of freedom (DOF) or more than three times the total of the effective DOF for the rest of the upper limb. The thumb contains five DOF, and each finger has another four, in addition to the DOF at the finger CMC joints. Rotation at the CMC joints of the ring and little fingers can be significant and enables formation of the palmar arch [5]. The rotational axes of some of these consecutive DOF run at oblique angles to each other and are offset. This arrangement facilitates certain movements, such as thumb opposition [6].

A total of 27 muscles control these DOF. Three of these muscles, flexor digitorum profundus (FDP), flexor digitorum superficialis (FDS), and extensor digitorum communis (EDC), are each comprised of multiple compartments, which give rise to separate tendons for each finger. These and the other extrinsic muscles, which provide most of the power to the hand, in accordance with their size, originate proximal to the wrist. This arrangement reduces muscle mass in the hand, thereby minimizing weight and inertia. This minimization is important as the hand, as the most distal portion of the upper extremity, can create large torque requirements for the shoulder.

The intrinsic muscles, such as the lumbricals and interossei, have both their origins and insertions within the hand. The muscle bellies for all the hand musculature, however, reside proximal to the metacarpophalangeal (MCP) joints in the digits. This arrangement permits increased joint rotation. Unlike the situation at the elbow or knee, where the contracting muscles (e.g., biceps brachii or biceps femoris) limit joint rotation, the fingers can curl completely into the palm, with rotation of the MCP, proximal interphalangeal (PIP), and distal interphalangeal (DIP) joints unimpeded by the muscles actuating them. Additionally, with the abduction/adduction DOF at the MCP joint, digits can actually overlap each other to create functional postures or gestures (e.g., crossing one's fingers). Current robotic hands are not capable of such a rich repertoire of movements. The large active range of motion is achieved without sacrificing power. Voluntary forces at the index fingertip can exceed 60 N, and thumb tip forces can exceed 100 N. Joint rotational velocity can reach 1,200 °/s.

While hand biomechanics affords considerable flexibility, it does increase motor control complexity. Each musculotendon unit influences multiple DOF simultaneously, and most tendons cross multiple joints. Many tendons interact with anatomical structures, such as the annular ligaments serving as anatomical pulleys for the flexor tendons and the extensor aponeuroses connecting up to 5 tendons per finger to the phalanges. These interactions between tendons and soft tissue impact the mapping of musculotendon force to force at the digit tip. As these interactions change substantially with joint posture, muscle activation patterns, which are often complex, may need to be adjusted to accommodate different finger postures. Significant activation of all seven muscles actuating the index finger may be required to create even a seemingly simple isometric flexion force at the fingertip, and this pattern changes with finger orientation [7].

Thus, substantial resources in the central nervous system are devoted to the hand. Disproportionately large regions of the motor and sensory cortices and the corticospinal and dorsal column pathways are associated with the hand [8]. Monosynaptic corticomotoneuronal pathways, affording direct connection between the cortex and spinal neurons, project predominantly to hand motoneurons [9]. Seemingly similar muscles for the same digit, such as EDC and extensor indicis, may be selectivity excited for different movements [10], and different compartments of even the same muscle may be activated independently [11]. Researchers have identified that coordination among a number of brain areas, such as anterior inferior parietal cortex, prefrontal cortex, and sensorimotor cortex, is involved in the planning and execution of hand movements [12, 13].

6.2 Pathophysiology

With its heavy reliance on cortical projections for both motor commands and sensory feedback, hand function may be especially impacted by injuries to the central nervous system, such as those produced by stroke or spinal cord injury. Diminished capacity to control the hand greatly reduces functionality of the entire upper extremity. As a testament of its importance, loss of the hand, such as through amputation, is considered to result in a 90% reduction in the functionality of the entire upper extremity [14]. The resulting loss of motor control can have a profound impact on self-care, employment, and leisure activities. Surveys of individuals with tetraplegia have underscored the importance of hand function, restoration of which was rated as one of the top priorities, well above restoration of walking [15, 16].

Individuals with hand and arm impairment arising from damage to the central nervous system often have concomitant impairment of one or more of the lower extremities [17]. Lower extremity impairment can exacerbate functional deficits in the hand and vice versa. Hand impairment may impact mobility in individuals with lower extremity deficits by reducing the ability to use mobility aids. Large assistive devices worn on the arm may also impact gait. Conversely, the need to control mobility aids, such as a cane, with the non-paretic hand can limit overall upper extremity function. It may become impossible to transport objects, like a glass of water, while ambulating.

6.2.1 Stroke

Stroke, produced by either hemorrhage or occlusion of blood vessels in the brain, is the leading cause of major long-term disability within the United States. Almost 800,000 Americans will incur a stroke each year, thereby leading to a population of more than 7 million stroke survivors [18]. Worldwide, an estimated 17 million strokes occur each year [18]. In fact, the World Stroke Organization reports that 1 in 4 individuals over the age of 25 will experience a stroke during their lifetime [19]. Thus, the long-term management of people with stroke-related problems is a major therapeutic, rehabilitation, and social challenge. While stroke is often considered a problem of the elderly, approximately 28% of strokes in the U.S. occur in people under the age of 65 and about 4–5% occur in individuals younger than 45 [20]. The Greater Cincinnati/ Northern Kentucky Stroke Study showed that within 10 years, the proportion of all strokes occurring in those under 55 years old rose from 13 to 19% [21]. Unfortunately, recent evidence suggests that the rate of stroke even among adolescents and young adults has been increasing [22]. Thus, a growing number of adults experience a stroke that will affect their prime working years. This contributes substantially to the enormous financial impact of stroke, with associated medical and disability costs estimated at $103.5 billion in 2016 [23]. This number is expected to exceed $129 billion by 2035 [18].

Stroke potentially impacts a number of bodily functions, including speech, vision, and sensorimotor control of the limbs. Hemiparesis affecting both the upper and lower extremities is typical.

Roughly 50% of older stroke survivors (onset of stroke at age of 65+) will have chronic hemiparesis, involving the hand in particular [24]. Deficits in voluntary digit extension are especially common [25].

The severity of hand impairment in stroke survivors can range widely, from a flaccid paralysis to trouble with finger individuation. A typical presentation has the wrist and fingers flexed with preferential weakness of extension. The deficits arise from a variety of primarily neurological sources, including flexor hypertonicity, reduced and aberrant muscle activation, and somatosensory loss.

Hypertonic activation may manifest as spasticity, excessive coactivation, and/or prolonged relaxation time. Spasticity, an abnormal velocity-dependent reflex response to imposed stretch [26], is predominantly observed in the long finger flexors, such as FDS and FDP (Fig. 6.1a). Interestingly, spasticity is largely absent in the long thumb flexor (flexor pollicus longus), even in individuals with spasticity in the finger muscles [27]. A variety of factors, such as decreased reciprocal inhibition, afferent disinhibition, and altered postactivation depression [28], may contribute to the spastic reflex. Additionally, the motoneurons of a spastic muscle may have an

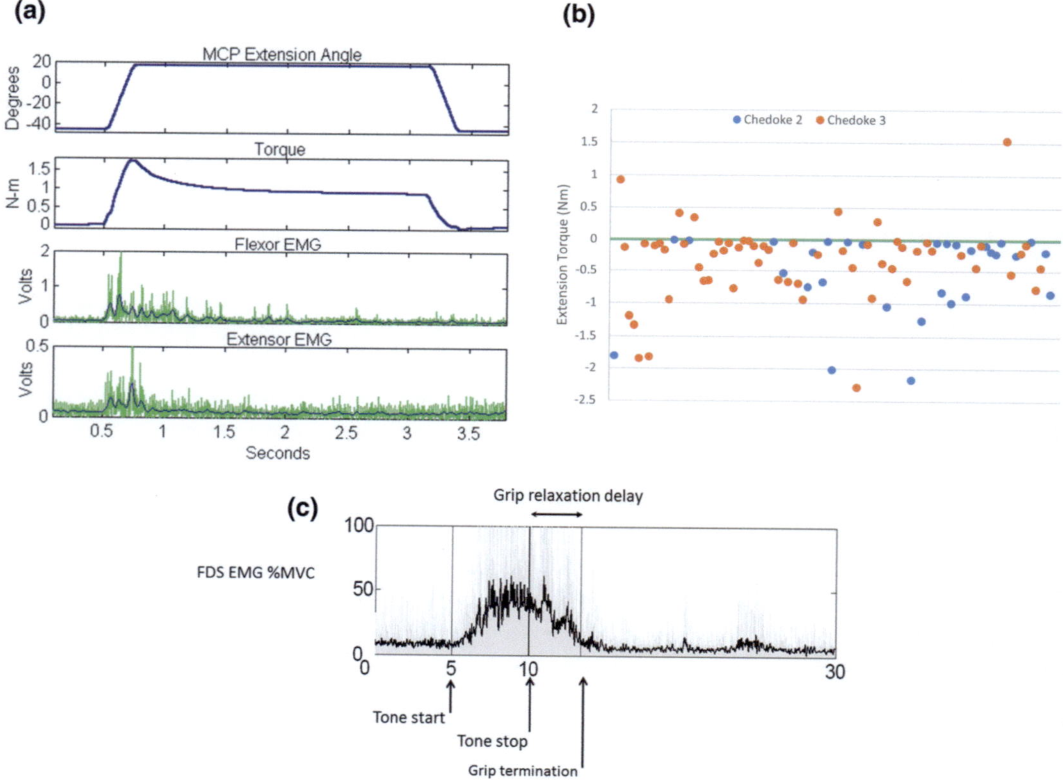

Fig. 6.1 Examples of hypertonicity in long finger flexors in stroke survivors. **a** Spastic stretch reflex evoked by rapid extension rotation (300 °/s) of the MCP joints stretches. Note increases in flexor EMG and torque. From Kamper DG, Rymer WZ. Impairment of voluntary control of finger motion following stroke: role of inappropriate muscle coactivation. Muscle Nerve 2001;24:673–81; used with permission; **b** Excessive flexor coactivation results in net flexion torque (shown as negative torque) during attempted production of voluntary extension torque about MCP joints. **c** Prolonged relaxation of long finger flexor (FDS) following generation of voluntary grasp force. From Seo NJ, Rymer WZ, Kamper DG. Delays in grip initiation and termination in persons with stroke: effects of arm support and active muscle stretch exercise. J Neurophys 2009;101:3108–15; used with permission

elevated resting potential, increasing firing probability. Indeed, spontaneous discharge of motor units is much more frequent in spastic muscle [29] and the spastic reflex response is dependent upon absolute muscle fiber length in addition to stretch velocity and magnitude [30]. It should be emphasized, however, that the degree to which spasticity itself impacts voluntary movement is open to question [28].

Excessive coactivation of nominally agonist and antagonist muscles may limit desired movement or force production, or even result in unintended movement in the opposite direction. In a recent study with 95 stroke survivors with severe hand impairment, we found that the vast majority of subjects produced a net finger *flexion* torque when trying to create a maximum finger *extension* torque [31]. This involuntary flexion torque could be substantial, exceeding 2 N-m at the MCP joints across the four fingers for some subjects (Fig. 6.1b). Thus, attempts to open the hand to position it for grasp of an object may actually result in closing of the hand, thereby precluding grasp.

Object release may also be affected, as deactivation of the finger flexors may be abnormal (Fig. 6.1c). Stroke survivors have been shown to have prolonged relaxation time in FDS following a grasp, both for the impaired and less impaired sides [32]. Relaxation time shortened following administration of cyproheptadine hydrochloride, an antiserotonergic agent, possibly suggesting a role for monoamines in prolonging activation by increasing the probability of firing within the motoneuron pool (Fig. 6.1c). In an unpublished study, we administered daily doses of either cyproheptadine hydrochloride or placebo over 3 weeks to the aforementioned group of stroke survivors with chronic, severe hand impairment. Within 1 week, the group receiving cyproheptadine significantly reduced their relaxation time (mean reduction = 45%) to a greater extent than the group receiving placebo.

Paradoxically, involuntary hyperexcitability is often accompanied by voluntary hypoexcitability. Stroke survivors may have profound activation deficits. Even in moderately impaired subjects, grip strength in the impaired hand is only 50% of that of the ipsilesional hand. In the aforementioned study by Barry et al., we observed that stroke survivors with severe impairment produced only 31% of finger flexion torque compared to neurologically intact individuals [31]. Relative weakness is even more pronounced in the extensor muscles. Only 6% of the participants in the study by Barry et al. could generate any net extension torque. These data correspond well to data we previously collected, showing that stroke survivors with severe hand impairment could generate only 27% of normal flexion force and 9% of normal extension force [33]. While excessive coactivation contributes to the net weakness, we have found that deficits in voluntary muscle activation play a primary role in muscle weakness. Using electrical stimulation techniques, we showed that activation deficits during maximum voluntary contraction closely mirror strength deficits in both flexion and extension, with activation deficits reaching 90% for EDC and 50% for FDS in stroke survivors with severe hand impairment [34]. In these subjects, almost normal levels of force could be generated by applying electrical stimulation to these muscles, even though the individual could not produce this level of activation voluntarily (Fig. 6.2a).

In addition to difficulties with fully activating muscles, stroke survivors may have difficulty modulating muscle activation appropriately with task. The range of activation levels used across a variety of tasks, even when normalized to the voluntary excitation range of the participant, is dramatically reduced for stroke survivors. We conducted a study examining muscle activation patterns during force creation with the thumb in six different directions. While neurologically intact subjects varied activation of a given muscle from 10 to 60% of maximum voluntary contraction (MVC) across force directions, activation of the same muscle in stroke survivors with severe hand impairment ranged only from 30 to 40% [35]. Stroke may also impact the ability to modulate EMG patterns across multiple

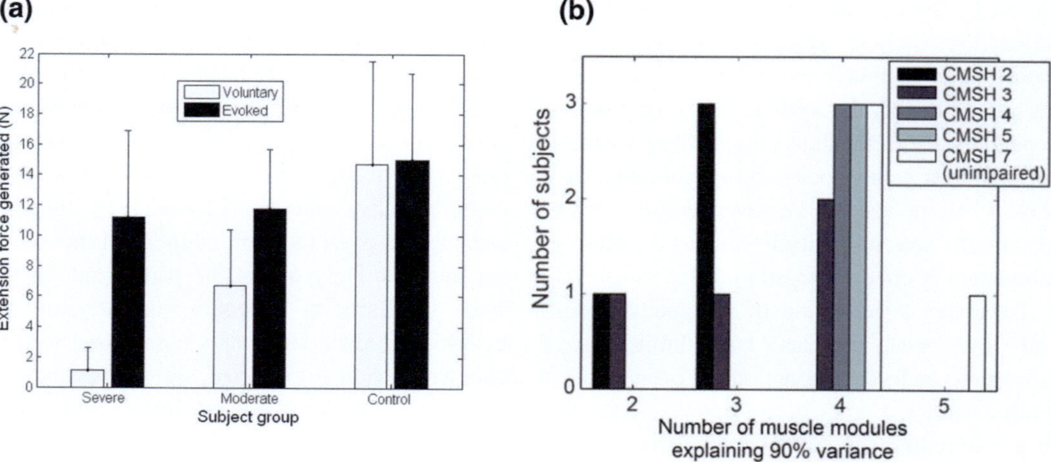

Fig. 6.2 Activation deficits following stroke. **a** Maximum extension force created either entirely voluntarily (white bars) or with the addition of external stimulation (black bars) or at the tip of the middle finger. Activation deficits are readily apparent in stroke survivors. **b** Number of muscle modules needed to explain the variance in activation patterns across six different tasks. CMSH: Chedoke-McMaster Stroke Assessment for Stage of Hand

muscles with task. One study showed that stroke survivors produced fewer distinct activation patterns across a specified set of tasks than neurologically intact individuals [36] (Fig. 6.2b).

Somatosensory data is integral to proper hand function. Accordingly, the hand is richly innervated with sensory nerves. It has been estimated that 17,000 cutaneous mechanoreceptors are present in the glabrous skin alone of the hand [37]. Proprioceptive acuity, especially in the thumb, is superior to other body segments, such as the toes [38]. To support this sensory precision, a disproportionately large portion of somatosensory cortex is devoted to the hand [39]. Unfortunately, this sense acuity is often compromised following stroke. Close to 40% of stroke survivors are thought to have chronic sensory impairment [40].

While stroke survivors typically experience hemiparesis, the non-paretic hand, ipsilateral to the brain lesion, may also be impacted after stroke, especially in stroke survivors with more severe impairment in their paretic hand [41]. We noted degradation in fine motor control even when the non-paretic hand was the dominant hand prior to the stroke [140].

6.2.2 Spinal Cord Injury

Spinal cord injury (SCI) is one of the leading causes of chronic disability in the young. Around 260,000 individuals in the U.S. have SCI, with 12,000 new cases added each year [42, 43]. The mean age at incidence is 40.2 years, and life expectancy is an additional 34 years for an injury occurring at that age. Interestingly, the increasing prevalence of SCI due to falls, primarily in the elderly, has led to a bimodal distribution of SCI incidence disproportionally skewed toward the young and the old. Falls are now the second most common cause of SCI, after automobile accidents [44].

The resulting functional impairments are dependent upon the location and extent of damage to the spinal cord. Compression, blunt trauma, and shearing, in addition to severing, of the cord are all potential mechanisms of SCI. Injury within the cervical region of the cord leads to tetraplegia, involving impairment of all four limbs. An estimated 55% of new cases of SCI will result in tetraplegia, while the other 45% of individuals will experience paraplegia due to injury below the cervical level. As acute treatment has improved, the number of incomplete spinal cord injuries has risen. With an incomplete injury, some of the neural tracts traversing the level of injury remain viable, such that some sensation and/or motor function is preserved [45]. Fifty percent or more of new SCI cases involve incomplete injury [46].

Tetraplegia typically involves the hands. Loss of descending drive can lead to flaccidity in the hand muscles, especially for high tetraplegia (C1–C4), although some muscle tone may be present, especially for low tetraplegia (C5–C8). Extensor muscle tone, however, seems to be as prevalent as flexor muscle tone, unlike the situation in stroke survivors. Abnormal interlimb reflexes, in which stimulation of lower limb nerves can produce excitation of hand muscles, may be present [47].

Hand function can be adversely impacted by a variety of means. As motoneurons and nerve roots are often damaged at the spinal level of injury, and even multiple segments below the level of the injury, flaccid paresis, and muscle atrophy are common [48]. This greatly limits functional recovery. Up to 70% of the paresis observed for C5–C7 lesions can be attributed to destruction of the nerve roots and motoneurons [49]. One study observed up to a 90% loss of motor units in the thenar muscles of the thumb in subjects at the C4–C5 level [50]. Joint movement can be further restricted by edema resulting from limited venous return, contracture arising from muscle shortening, and connective tissue formation around tendons and joint capsules.

Hand impairment mechanisms in individuals with incomplete SCI are not well described and warrant further study. Substantial atrophy in more proximal muscles has been reported [51], as have reductions in nerve conduction velocities, diminished tetanic force production, and elongated twitch times [52].

6.3 Rehabilitation Technology

Advances in sensors, actuators, and electronics have driven efforts to develop devices to restore the functional capabilities of the upper extremities that are lost due to stroke or SCI. One means of improving function involves having a device perform the activities for the user. This assistance compensates for the deficits, similar to the benefits derived from glasses or a wheelchair. While the underlying pathology remains, the ability of the user to perform desired tasks is enhanced. Assistive devices are especially employed in cases of individuals with severe impairment that is not likely to improve with treatment, such as complete SCI.

Alternatively, therapeutic interventions may improve voluntary sensorimotor control of the hand, particularly in individuals with incomplete SCI or stroke. Research has shown that the central nervous system exhibits much greater plasticity than once imagined. Even the mature nervous system is constantly changing and adapting to new circumstances. While neurogenesis is rare, synaptogenesis is constantly occurring. For example, repeated practice of hand movements, such as performed by musicians, can lead either to seemingly beneficial cortical changes in sensorimotor representation and processing [53, 54] or to harmful changes, such as in focal dystonia [55]. Experimental evidence suggests that intensive repetitive training of new motor tasks is required to induce long-term brain plasticity [56]. This finding seems to be applicable to motor relearning after brain injury, such as from stroke, as well. In animal models of brain injury, practice appears to be the primary factor leading to synaptogenesis and brain plasticity [57–59]. Similarly, in humans, repetitive practice has been shown to lead to functional improvement following stroke [60–62]. Devices which encourage and direct this therapeutic practice may benefit individuals, as long as they address specific impairment mechanisms.

Increasingly, devices focusing on assistance, therapy, or a combination of both can be separated into two classes: wearable and non-wearable. Here, a wearable device is defined as one that connects directly to the user and could be carried by the user as they move about the environment. A non-wearable device may act without contacting the user, such as a robotic arm, or be somehow constrained in space, such as by connection to an external computer, actuator, or power supply.

6.3.1 Non-wearable Devices

Without the limitations inherent to technology worn on the body, non-wearable devices can exploit a variety of commercial robots, actuators, and controllers. The intended use of the device dictates its structure. Here, we will segregate non-wearable devices into three categories: independent robots, devices contacting the user through an endpoint, and devices contacting the user at multiple points but with an actuator, power source, or controller that is fixed in space and, thus, not portable.

6.3.1.1 Independent Robots

Many of the earliest assistive devices employed commercial robots or robotic arms that served to perform tasks that the user could not do on their own. Many of these assistive devices were designed specifically to meet the needs of individuals with complete tetraplegia, where no improvement in sensorimotor control of the limbs was expected.

One of the first successful assistive robots was the Handy 1 [63], a robot workstation that could be used for eating, drinking, grooming, and even art projects (Fig. 6.3). The Handy 1 employed a Cyber 310 robotic arm, which had five DOF in addition to a gripper end-effector. The user could operate the device through a single switch. Newer robots have been incorporated into updated feeding assistants, such as My Spoon (SECOM Co., Ltd., Tokyo, Japan), a feeding robot designed explicitly for Korean food [64], and the Assistive Dexterous Arm that can work with previously unseen food [65]. These devices are more compact than their predecessors and offer greater control options for the user. Robotic workstations have also been designed to perform other services. For example, the Desktop Vocational Assistant Robot (DeVAR) was created to provide assistance within an office environment. It consisted of a commercial PUMA-260 robot

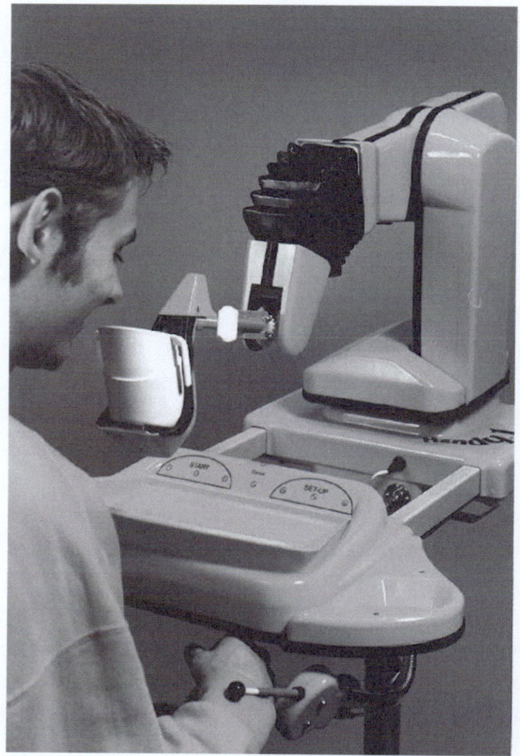

Fig. 6.3 The Handy 1 workstation, intended to help users with eating, drinking, and grooming. First developed by Mike Topping at Staffordshire University (Reprinted with permission from: Topping M. Flexibot —a multi-functional general-purpose service robot. Ind Robot 2001;28:395–401. © Emerald Group Publishing Limited; all rights reserved)

coupled to a Griefer prosthetic hand from Otto Bock Healthcare (Duderstadt, Germany).

To increase the range of tasks and situations in which they could be employed, robotic systems were developed that could be mounted directly to a wheelchair. The KARES II system, created at the Korea Advanced Institute of Science and Technology (KAIST), had six DOF in its robotic arm and a gripper at its end, all attached to a mobile platform [66]. KARES II could perform procedures such as grasping objects and turning off and on light switches under direction from the user. The Raptor Wheelchair Robot System was developed by the Rehabilitation Technologies Division of Applied Resources Corp. (RTD-ARC) expressly as an assistive device; the system was sold commercially beginning in 2000 [67].

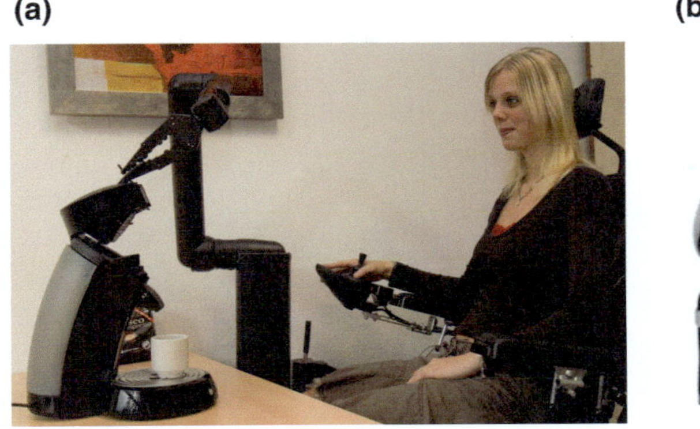

Fig. 6.4 Wheelchair-mounted assistive robots. **a** The iARM wheelchair-mounted assistive robot seen here assisting a user to make a cup of coffee (photo courtesy of Assistive Innovations, Didam, the Netherlands). **b** The JAC02 robotic arm has three fingers for grasping objects such as a cup of water (photo courtesy of Kinova Robotics, Boisbriand, Canada)

Hardware and software improvements led to widespread implementation of compliant control, such as in the KUKA LBR iiwa (KUKA, Augsburg, Germany), that facilitated safe human–robot interactions. This has enabled humans and robots to share the same space. One of the most commercially successful wheelchair-mounted devices has been the MANUS, which has evolved into the iARM (Assistive Innovations, Didam, the Netherlands). The iARM provides six DOF and a gripper end-effector and can be powered from a wheelchair battery [68]. Designed for close interaction with the user (see Fig. 6.4a), the iARM supports a wide variety of control options to accommodate the capabilities and preferences of the user. The JAC02 robotic arm (Kinova Robotics) is a lightweight six-axis robotic arm with three fingers for gripping (Fig. 6.4b). It can be mounted to a wheelchair or a tabletop.

Attempts have also been made to provide mobile robotic assistants which could move independently from the wheelchair. The MoVAR device, developed at Stanford University and the Rehabilitation Research and Development Center at the VA Palo Alto Health Care System, consists of a PUMA robot arm affixed to a powered omnidirectional base [69]. Autonomous mobile robots, intended for a number of possible applications, could also provide valuable functions for individuals with tetraplegia. For example, the assistant Care-O-bot 4 (Fraunhofer IPA) has the potential to benefit those with tetraplegia or severe stroke by retrieving and transporting objects and performing household tasks (Fig. 6.5a). The Home Exploring Robot Butler (HERB) was developed at Carnegie Mellon University for assisting individuals with household tasks [70], while the Baxter robot (Rethink Robotics, Inc.) has been proposed as an assistant to individuals with disability (Fig. 6.5b).

These external robots can perform a number of functional tasks even when the user is unable to move their extremities. The range of environments over which they can provide assistance, however, may be limited unless they can be mounted to a user's wheelchair or otherwise move with the user.

6.3.1.2 Hand-End-Effector Coupling

A number of devices have been designed to interact with the hand by coupling with the end-effector of the robot. This arrangement limits applicability for assistance, but offers benefits for use of the device as an instrument to facilitate rehabilitation therapy. The Amadeo System (Tyromotion, GmbH, Graz, Austria) uses an approach in which the fingertips are attached to linear tracks which directly control fingertip location. Each digit is attached to a separate track and can therefore be moved independently (Fig. 6.6a). The InMotion® Arm/Hand (Bionik

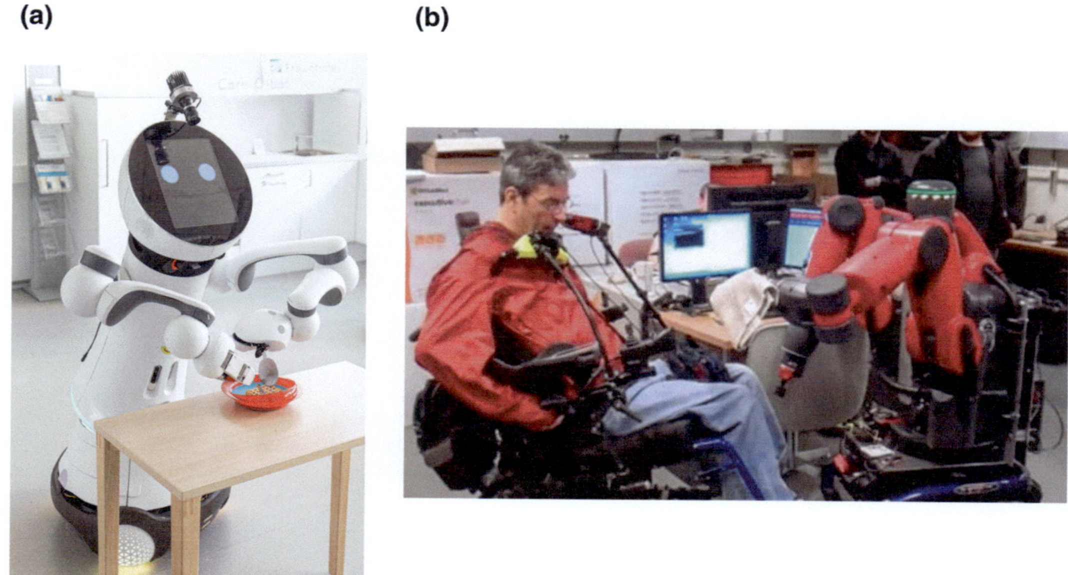

Fig. 6.5 Mobile robotic assistants. **a** Care-o-bot 4 (Fraunhofer IPA) (photo courtesy of Fraunhofer IPA: Rainer Bez). **b** BAXTER robot employed as an assistant to an individual with paralysis (photo courtesy of Rethink Robotics, Inc.)

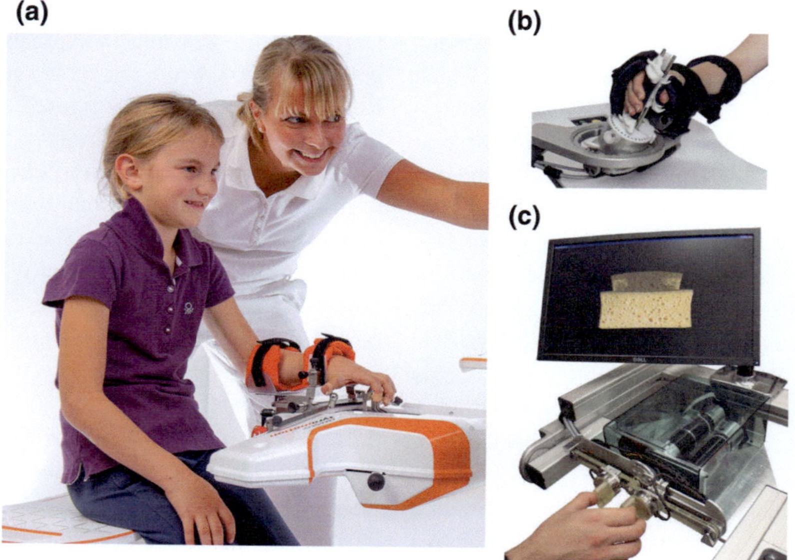

Fig. 6.6 Examples of non-wearable robots that connect with the hand through the robotic end-effector. **a** For the Amadeo (Tyromotion, Inc.), each fingertip has a separate actuated linear slide (photo courtesy of Tyromotion, Inc.). **b** End-effector expands to open the hand for the InMotion Arm/Hand (photo courtesy of Bionik Laboratories Corp., Toronto, Canada). **c** The index fingertip and thumb tip contact rigid pads that are connected to linear slides of the Haptic Knob (RELab, ETH Zurich) (photo courtesy of RELab ETH Zurich)

Laboratories Corp., Toronto, Canada) has an end-effector consisting of two curved pieces, one attached to the thumb and one to the four fingers (Fig. 6.6b). Translation of these two pieces produces gross opening and closing of the hand. The ReHapticKnob, created by the RELab at ETH Zürich, couples to the index fingertip and thumb tip through rigid pads connected to a linear slide [71]. By moving the actuated pads closer or further apart on this slide, the finger-thumb aperture can be manipulated. Additionally, this arrangement allows for the use of precise load cells and encoders that enable admittance control for rendering haptic boundaries to simulate contact with real objects. Objects of different size and stiffness can be readily simulated for therapeutic training of sensation such as touch and proprioception (Fig. 6.6c).

The end-effector coupling greatly facilitates donning and doffing of the device, as the coupling only involves the fingertip. This is especially important for a clinical therapy session with its tightly controlled duration. Potential limitations include difficulties with implementing joint-level control or practice of tasks involving object manipulation or significant arm movement.

6.3.1.3   Fixed Base

To gain more precise control over the hand and individual joints, other devices contact the hand at multiple locations. The actuators, power supply, or control, however, may still be located external to the body in order to minimize mass or bulk added to the user. These devices are typically used in therapy contexts.

The Cable Actuated Dexterous (CADEX) glove, derived from the BiomHED glove [72], employs cables routed along the digits in a manner mimicking the pathways of individual tendons in the hand [73]. Thus, assistance can be provided to facilitate training of specific activation patterns (Fig. 6.7a). The tendons run through tendon sheaths from actuators located on the tabletop to the digits. The FINGER robot can independently move the index and middle fingers through sets of eight-bar linkages (Fig. 6.7b). The linkages permit precise control of finger flexion and extension from a single actuator (one per finger). This device has been used to study training paradigms in stroke survivors [74]. Similarly, the Maestro hand exoskeleton uses linkages to actuate the digits [75]. In this case, the linkages are driven by a cable system; the cables are driven by actuators located in a fixed base on a table (Fig. 6.7c). The GloreHa system (Idrogenet S.R.L., Lumezzane, Brescia, Italy) also employs fixed actuators, but these are coupled to the digits through push–pull cables. Control of individual joints is diminished in comparison with the designs using mechanical linkages, but bulk is reduced to facilitate interaction with objects. In my laboratory, we have created a soft,

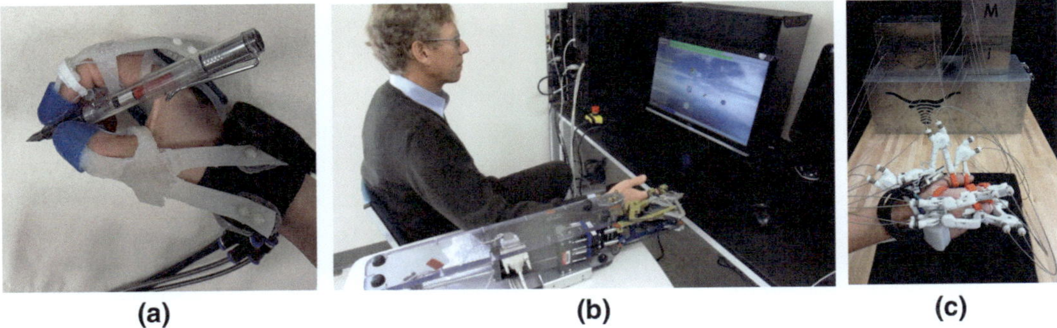

(a)   (b)   (c)

Fig. 6.7 Non-wearable robots that contact the digits at multiple points to directly control applied joint torques. **a** Cable Actuated Dexterous (CADEX) glove employs exotendons with routing pathways based on hand physiology. (photo courtesy of Dr. HS Park, KAIST University); **b** FINGER robot actuates the middle and index fingers through sets of eight-bar linkages (photo courtesy of Dr. D. Reinkensmeyer, University of California at Irvine); **c** Cable-driven exoskeleton employing multi-bar linkage transmission to actuate the digits (photo courtesy of Dr. A. Deshpande, University of Texas at Austin)

pneumatically driven glove, the PneuGlove [76], which we have coupled to a virtual keyboard. The user must flex and extend specified digits in order to "play" the virtual keys dictated by the computer program. The PneuGlove is extremely lightweight and compliant, but requires connections to an air compressor and electropneumatic servo valves located on a cart [77].

These fixed-base devices may control position or torque at multiple joints within a digit. With external support of actuators and external power supplies, more powerful actuators can be employed to provide needed assistance than would be possible with actuators worn on the body. The workspaces of these devices, however, may be quite limited.

6.3.2 Wearable Hand Robotics

The recent trend in exoskeletons, including for the hand, is a push toward wearable devices. For a number of stroke survivors and individuals with SCI, control of the arm may be relatively spared in relation to the hand. These individuals could benefit from a device which assists hand tasks while allowing free arm movement. Potential drawbacks to this approach include the weight and size of the components added to the hand.

To overcome these limitations, some designs have focused on using passive elements. One approach involves creating a set of adaptive tools that can insert into a splint worn on the wrist. These tools include modified utensils, brushes, and electric razors. In this manner, the hand is no longer required for grasping these tools; basic activities of daily living, such as feeding and grooming, can be performed with residual control of the arm. While this adaptive equipment can be very effective, the tools can only support a limited number of activities. Additionally, the set of tools has to be carried by the user and may require assistance from another individual to change from one tool to another.

Other designs involve assisting movement of the fingers themselves, but through passive actuation. The SaeboFlex and SaeboGlove (Saebo Inc., Charlotte, NC, US), for example,

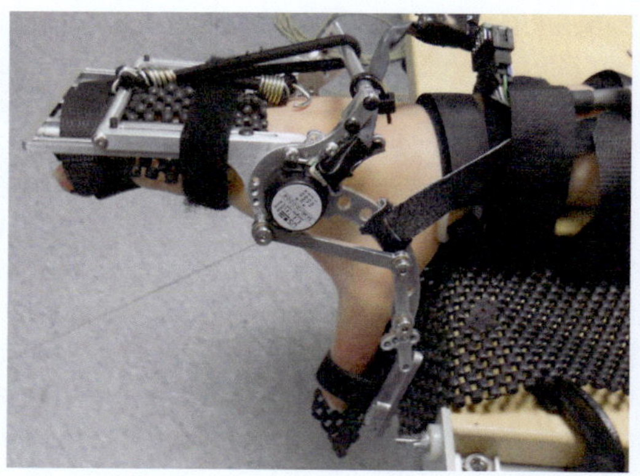

Fig. 6.8 Passive wearable exoskeleton. The HandSOME exoskeleton (Catholic University, Washington, D.C.) employs passive springs to provide nonlinear extension compensation (photo courtesy of Dr. P. Lum, Catholic University)

use passive springs and elastic bands to assist finger extension. These devices are targeted toward stroke survivors, who typically have greater trouble creating finger extension than finger flexion. HandSOME, which can provide substantial extension assistance, employs a nonlinear spring arrangement to vary extension assistance with joint angle [78] (Fig. 6.8).

A primary challenge with using passive elements is the noted profound weakness in both flexion and extension. The amount of passive extension assistance that can be provided is limited by the strength of the finger flexors to overcome this bias when finger closing is desired. The bias force required to open the hand may be substantial due to involuntary flexor coactivation [79]. Additionally, the amount of needed assistance may vary with time; we have observed that unwanted finger flexor coactivation may increase with repeated trials [80]. Thus, wearable devices are being developed to provide active assistance. Some of these devices create active extension, some produce active flexion, and a few devices provide active flexion and extension.

6.3.2.1 Active Extension Assistance

For some stroke survivors and individuals with incomplete SCI, voluntary finger flexion, while impaired, may still be sufficient to perform a number of activities of daily living. Profound

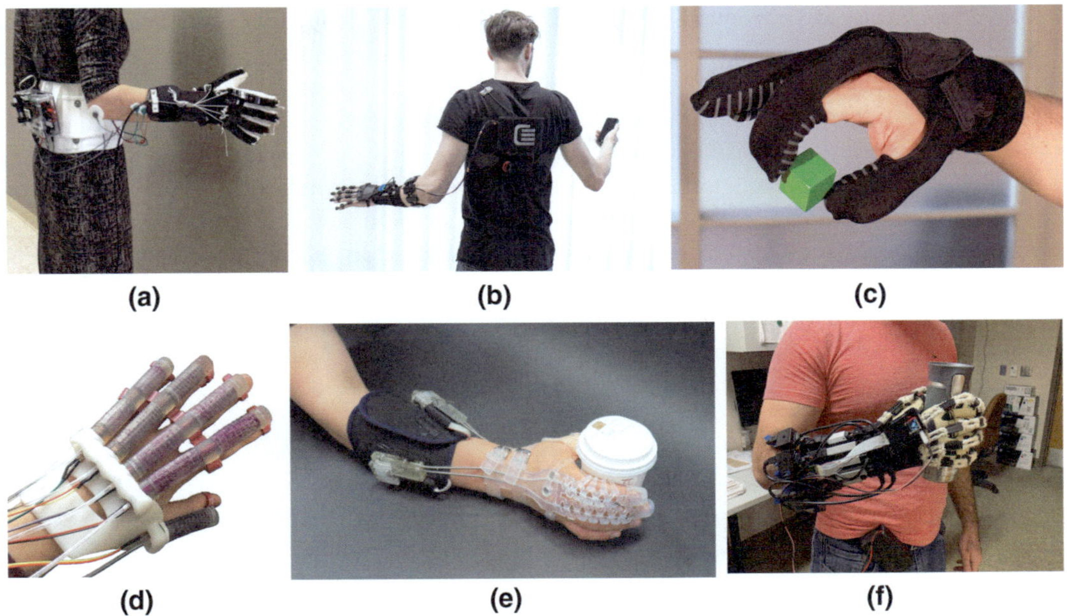

Fig. 6.9 Wearable hand devices that actively actuate digits. **a** The Voice and Electromyography-Driven Actuated (VAEDA) Glove provides simultaneous extension assistance to all digits from a servomotor that can be worn on the waist. **b** Sliding spring actuators are used in the RELab tenexo system to actively flex and extend the fingers simultaneously (photo courtesy of RELab ETH Zurich). **c** Soft glove design in which the constructed material properties of the chamber control shape, and thus finger movement, under pressurization. (photo courtesy of Dr. C. Walsh, Harvard University). **d** Pneumatic actuators actively flex digits while embedded material provides passive extension (photo courtesy of Dr. R. Tong, Chinese University of Hong Kong). **e** Exo-Glove Poly II employs a novel transmission to enable one actuator to pull both the flexion and extension cables to actively assist opening and closing (photo courtesy of Dr. KJ Cho, Seoul National University). **f** Bidirectionally Actuated Cable (BAC) Glove drives push–pull cables with linear actuators located on the forearm (photo courtesy of Dr. M. Ghassemi, North Carolina State University)

deficits in extension, however, may prevent the hand opening needed for functional task performance. Some robotic devices address this situation by providing active support of extension [81–83], while allowing the user to voluntarily flex the digits. These designs typically create extension by transmitting actuator force to structures traversing the dorsal side of the digit. The HERO Glove extends the fingers using zip ties actuated by a single linear actuator [82], while the X-Glove uses fishing line as exotendons that pass through 3D-printed guides [81]. The Voice and Electromyography-Driven Actuated (VAEDA) Glove [83] provides extension assistance to all digits simultaneously through cables connected to a single servomotor that can be worn at the waist and controlled by the user through voice and EMG signals (Fig. 6.9a).

The applied extension force can be used to not only assist hand opening, but also to produce more functional hand closing kinematics and force production. Researchers have noted that, for certain finger postures, significant contributions from the long extrinsic extensor muscles are important to creation of flexion force in the palmar direction [7, 84]. While these devices address a primary limitation for many potential users, these individuals may ultimately require assistance of flexion or greater control of posture than may be afforded by devices focusing on extension assistance.

6.3.2.2 Active Flexion Assistance

While finger flexion force is typically spared to a greater extent than extension in stroke survivors, substantial deficits are common. Flexion deficits are also prevalent in individuals with SCI. Functional deficits may arise not only from limited absolute force generation [31], but also from impaired ability to properly direct fingertip force to manipulate objects; the misdirection of fingertip forces can cause objects to slip from the grasp [85]. Some wearable devices have been developed expressly for the hand to facilitate grasp of objects [86]. For example, the Soft Extra Muscle (SEM) Glove (Bioservo Technologies, Isafjordsgatan, Sweden) helps individuals with incomplete tetraplegia by amplifying their grasping force [87] through cables running through the palmar side of a worn glove. Other devices use a similar architecture, with proximally located actuators pulling on cables running along the palmar side of the hand [88].

Some of these exoskeletons provide concomitant assistance to extension through passive elements [89, 90], but the level of assistance may be insufficient due to the large and variable flexor coactivation. Also, the additional compressive force in such devices also can be a source of discomfort for their users. To counteract the compressive forces, Kim et al. employed a flexible frame on the sides of the finger; the frame still allows the finger joints to rotate freely [91].

Arrangement of the flexion cables must be carefully considered to produce the desired finger movement and force production without interfering with object grasp. To avoid this complication, Arata et al. created a glove that assists flexion by pushing from the dorsal side of the hand rather than pulling from the palmar side. A novel actuation unit comprised of three layers of sliding springs provides active flexion torque, while the passive flat springs within the unit provide restoring extension force [92]. Nycz et al. built upon this design by moving the actuator from the back of the hand to a small backpack using a Bowden cable, in order to reduce the weight of the glove on the hand [93]. Based on the sliding spring concept of Arata et al., the RELab tenoexo also uses Bowden cables to simultaneously actuate the springs to produce gross flexion of all five digits [94] (Fig. 6.9b).

To further reduce mass, a number of soft exoskeletons, using pneumatic or hydraulic actuators, have been developed [95–97] to provide active assistance of digit flexion. Soft actuators have advantages in terms of being lightweight, safe, and able to conform to the shape of the digits. Polygerinos et al. developed an exoskeleton utilizing soft fiber-reinforced hydraulic actuators attached to the dorsal side of the fingers [95] (Fig. 6.9c). Strain-limiting fabric layers and directional fiber reinforcements of the chambers help to control flexion shape under pressurization. Heung et al. used a similar actuator design in combination with a torque-compensating layer located between the chamber and the finger to actively flex and passively extend the fingers [97] (Fig. 6.9d).

6.3.2.3 Active Assistance of Both Flexion and Extension

Other wearable hand exoskeletons actively assist both flexion and extension of the fingers. One option for achieving this involves use of a rigid structure with joint locations aligned with the anatomical finger joints. The MyoPro (Myomo, MA) simultaneously actuates the MCP joints of the index and middle fingers to form a three-jaw chuck grip with the fixed thumb [98]. Similarly, Gasser et al. developed a single DOF robot that flexes and extends the MCP and PIP joints of the four fingers simultaneously while the thumb is fixed [99]. The device additionally allows the user to passively adjust the pronation/supination of wrist to position the hand [99].

The weight and size of a rigid structure, especially one directly actuated, may limit the number of DOF that can be controlled independently; as the inertia of the finger segments is quite small, added mass can substantially impact the total inertia and control of the fingers. A number of devices seek to decrease inertia by locating the motors proximally and then transmitting the force out to the fingers through cables running along both sides (palmar and dorsal) of the digit. The routing of these cables can be modified in order to change the distribution of torque among the finger joints or to reduce the number of required actuators [100, 101].

Similarly, the Exo-Glove Poly II employs a single flexion cable routed around the index and middle fingers and two extension cables routed dorsally to the fingers and fixed to thimbles at the fingertips, which are actuated together [102] (Fig. 6.9e). The Exo-Glove Poly II also employs a novel dual-slack-enabling transmission to allow a single actuator to pull both the extension cable and the flexion cable, rather than requiring two motors [103–106]. The slack-enabling mechanism counters the cable slack around the spool that might otherwise arise from this design [102, 107, 108].

Alternatively, other investigators have employed push–pull arrangements that run entirely along the dorsal side of the hand. This arrangement minimizes interference with grasp and sensation, while actively assisting both flexion and extension with a single actuator. The Mano utilizes Bowden cables to transmit force from motors worn on the chest to push–pull cables traversing through guides secured to the digits with rings [109]. To avoid the frictional losses and potential catch-hazard associated with Bowen cables, our Bidirectionally Actuated Cable (BAC) Glove (Fig. 6.9f) employs linear actuators that directly drive push–pull cables independently for each digit. The actuators are secured to a forearm splint, which additionally serves to keep the wrist in a functional posture [110]. This arrangement involves the trade-off of greater mass applied to the forearm.

The wearable devices providing active flexion and extension have the greatest potential for assisting a variety of tasks in a variety of environments. They also offer the greatest flexibility in terms of being used for assistance, therapy, or a combination, but they are often constrained in the amount of assistance they can provide and the DOF they can independently control.

6.3.3 Control

One of the primary limitations in using robotic technology in the upper extremity is controlling the robot based on user intent. The hands perform such a wide variety of tasks that discerning intent among all of these options may not be feasible. Continuous control for even a subset may be challenging. For users with some control of muscle activation, electromyography (EMG) may provide a viable means for control and one that may also provide therapeutic benefits. Gross opening and closing of the hand can be performed by stroke survivors with only a couple of EMG electrodes [83, 111]. High-density electrode arrays and novel algorithms for processing these signals may provide an intuitive means for controlling more DOF with surface electrodes [112]. As some users may not possess sufficient EMG control, voice commands can be used additionally or solely [83, 108]. For users able to create some force in the desired direction, detection of this force could be used to drive the device.

For individuals with more severe impairment, brain–machine interface (BMI) is a means for providing facile control of multiple DOF robotic devices. Electrical signals from the brain can be decoded to determine which task the user wants to perform and even details (e.g., velocity) of the intended movement. While electroencephalograms have been used to control devices [113], finer control requires the use of indwelling electrodes such as intracortical arrays or the less invasive electrocorticographic electrodes. Human participants with implanted intracortical arrays have successfully controlled a DLR Light-Weight Robot III arm (German Aerospace Center, Oberpfaffenhofen, Germany) [114] and a modular prosthetic limb from the Johns Hopkins Applied Physics Laboratory to grasp and retrieve objects (Fig. 6.10) [115]. These BMI-controlled robots restore motor function, but do not provide any sensation to the user. Researchers are investigating how to provide sensory information, such as cutaneous sensation and proprioception, to the user. Techniques may involve stimulation of peripheral nerves [116] or somatosensory cortex [117].

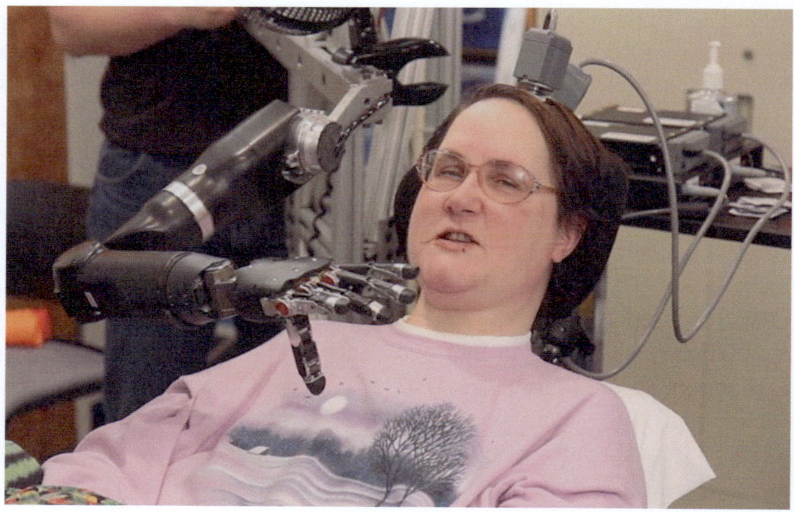

Fig. 6.10 BMI-controlled robotic limbs. Woman with tetraplegia uses modular prosthetic arm to grasp and manipulate objects (photo courtesy of Motorlab, University of Pittsburgh)

6.4 Current State of Technology and Needs for Future Development

Despite the great need for hand rehabilitation, uptake of technology into the clinic or home remains quite limited. This is partly attributable to the nature of the impairments that arise after stroke or spinal cord injury where the entire upper limb, the trunk, and the lower limb are likely to be impacted. Furthermore, activity in the proximal arm [118] or lower limb [119] can produce aberrant neural activity in the hand that hinders function. Independent robots or robotic arms are able to perform both arm and hand tasks, but their range is currently restricted unless mounted to a wheelchair. Challenges with user control may also dissuade use. Continued improvement in the dexterity of robotic hands and the development of shared human–robot control strategies may increase user acceptance, especially for individuals with tetraplegia.

The utilization of wearable hand exoskeletons to improve hand function is enticing as the devices could be omnipresent, but the aforementioned issues with the proximal arm may need to be addressed. In general, greater quantification of the overall assistance that hand exoskeletons can provide is needed. Correia et al. did find that finger force and movement and the time required to perform the Jebsen–Taylor Test of Hand Function improved with the assistance of a pneumatic hand exoskeleton in individuals with SCI [120]. In a pilot study, Yap et al. showed that stroke survivors could grasp–lift–release empty water bottles and cans more rapidly when wearing their soft pneumatic glove [96]. Use of the HERO Grip Glove improved performance of certain tasks, such as fork manipulation and the Box and Blocks, for some stroke survivors but not for others, particularly those with a little more residual function [121]. While repeated training sessions may lead to more beneficial utilization of wearable devices [122, 123], assistance with forearm pronation/supination may be critical to improving device efficacy by properly orienting the hand to perform a task. Unfortunately, present wearable options are limited in number and function, although some investigators have explored the use of pneumatic actuators [124, 125] or exo-tendons [126–128]. Functionality may also be limited by the lack of thumb actuation. Current devices mainly actuate only extension/flexion of thumb joints, although some cable designs [73, 105] and pneumatically actuated devices [129] address facets of abduction/adduction.

More dexterous control of the thumb could greatly expand the range of tasks that could be performed with the hand device.

Uptake of wearable devices is further hampered by difficulties in donning and doffing. The biomechanics of the hand, with its many DOF within a confined area and limited attachment sites, make physical connections between the device and hand difficult to implement. The gloves or Velcro straps typically utilized may be very difficult to don, particularly for stroke survivors with finger flexor hypertonicity, and the reliance on frictional coupling can be uncomfortable or result in slippage. Ideally, attachment would occur entirely on the dorsal side of the hand without requiring digit extension.

Non-wearable options seem best suited for therapy. A number of studies have been conducted in the last 10 years on the therapeutic effects of training with robotic technology for the hand. On a positive note, these studies have included more participants than seen previously. Training with the devices has generally led to improvement in cohorts of stroke survivors either at the chronic [130, 131] or subacute [132, 133] phase of recovery and individuals with incomplete tetraplegia [134], but the level of improvement is typically modest and similar to that of conventional therapy. In a study involving 50 stroke survivors with chronic impairment, Calabro et al. were able to show that training with the Amadeo device did lead to greater improvement in scores on the Fugl-Meyer assessment of motor recovery after stroke for the upper extremity and the nine-hole peg test than an equivalent duration of conventional occupational therapy [135]. Outcomes may be improved by combining robotic therapy with other treatment modalities, such as neuromuscular electrical stimulation [136, 137].

6.5 Conclusions

The need for rehabilitative technology for the hand remains great. The number of stroke survivors worldwide continues to grow. While improvements in acute care have greatly reduced the mortality rate and individuals are living longer after the stroke, a majority of stroke survivors experience chronic impairments that can greatly impact quality of life. The number of individuals with SCI continues to increase as the population increases [138]. A number of other muscular and neurological conditions, such as traumatic brain injury, brachial plexus injury, and ALS, may profoundly impact hand function. The urgency for development of new devices to meet this need is underscored by a reported study of individuals with brachial plexus who chose to undergo amputation of their hands in order to be fit with hand prostheses [139].

References

1. Dunsworth H. Origin of the genus Homo. Evo Edu Outreach. 2010;3:353–66.
2. Napier JR. Studies of the hands of living primates. Proc Zool Soc Lond. 1960;134:647–57.
3. Young RW. Evolution of the human hand: the role of throwing and clubbing. J Anat. 2003;202 (1):165–74.
4. Marzke MW, Marzke RF. Evolution of the human hand: approaches to acquiring, analysing and interpreting the anatomical evidence. J Anat. 2000;197 (Pt 1):121–40.
5. Buffi JH, Crisco JJ, Murray WM. A method for defining carpometacarpal joint kinematics from three-dimensional rotations of the metacarpal bones captured in vivo using computed tomography. J Biomech. 2014;46(12):2104–8.
6. Brand PW, Hollister AM. Clinical mechanics of the hand. 3rd ed. St. Louis, MO: Mosby; 1999. p. 369.
7. Valero-Cuevas FJ, Zajac FE, Burgar CG. Large index-fingertip forces are produced by subject-independent patterns of muscle excitation. J Biomech. 1998;31(8):693–703.
8. Penfield W, Rasmussen, T. The cerebral cortex of man: a clinical study of localization of function. New York, NY: The Macmillan Company; 1950. p. 248.
9. Rathelot JA, Strick PL. Subdivisions of primary motor cortex based on cortico-motoneuronal cells. Proc Natl Acad Sci USA. 2009;106(3):918–23.
10. Darling WG, Cole KJ, Miller GF. Coordination of index finger movements. J Biomech. 1994;27 (4):479–91.
11. Keen DA, Fuglevand AJ. Common input to motor neurons innervating the same and different compartments of the human extensor digitorum muscle. J Neurophysiol. 2004;91(1):57–62.
12. Klaes C, Kellis S, Aflalo T, Lee B, Pejsa K, Shanfield K, et al. Hand shape representations in the

human posterior parietal cortex. J Neurosci. 2015;35(46):15466–76.

13. Schaffelhofer S, Agudelo-Toro A, Scherberger H. Decoding a wide range of hand configurations from macaque motor, premotor, and parietal cortices. J Neurosci. 2015;35(3):1068–81.

14. Association AM. Guides to the evaluation of permanent impairment. Chicago: American Medical Association; 1990.

15. Hanson RW, Franklin MR. Sexual loss in relation to other functional losses for spinal cord injured males. Arch Phys Med Rehabil. 1976;57(6):291–3.

16. Snoek GJ, MJ IJ, Hermens HJ, Maxwell D, Biering-Sorensen F. Survey of the needs of patients with spinal cord injury: impact and priority for improvement in hand function in tetraplegics. Spinal Cord. 2004;42(9):526–32.

17. Tyson SF, Chillala J, Hanley M, Selley AB, Tallis RC. Distribution of weakness in the upper and lower limbs post-stroke. Disabil Rehabil. 2006;28(11):715–9.

18. Virani SS, Alonso A, Aparicio HJ, Benjamin EJ, Bittencourt MS, Callaway CW, et al. Heart disease and stroke statistics-2021 update: a report from the American heart association. Circulation. 2021;143 (8):e254–743.

19. Collaborators GLRoS. Global, regional, and country-specific lifetime risks of stroke, 1990 and 2016. N Engl J Med. 2018;379(25):2429–37.

20. Roth EJ, Lovell L. Employment after stroke: report of a state of the science symposium. Top Stroke Rehabil. 21 Suppl 1:S75–86.

21. Kissela BM, Khoury JC, Alwell K, Moomaw CJ, Woo D, Adeoye O, et al. Age at stroke: temporal trends in stroke incidence in a large, biracial population. Neurology. 2012;79(17):1781–7.

22. George MG, Tong X, Kuklina EV, Labarthe DR. Trends in stroke hospitalizations and associated risk factors among children and young adults, 1995–2008. Ann Neurol. 2011;70(5):713–21.

23. Girotra T, Lekoubou A, Bishu KG, Ovbiagele B. A contemporary and comprehensive analysis of the costs of stroke in the United States. J Neurol Sci. 2020;410: 116643.

24. Kelly-Hayes M, Beiser A, Kase CS, Scaramucci A, D'Agostino RB, Wolf PA. The influence of gender and age on disability following ischemic stroke: the Framingham study. J Stroke Cerebrovasc Dis. 2003;12(3):119–26.

25. Trombly CA. Stroke. In: Trombly CA, editor. Occupational therapy for physical dysfunction. Baltimore: Williams and Wilkins; 1989. p. 454–71.

26. Lance JW. Symposium synopsis. In: Feldman RG, Young RR, Koella WP, editors. Spasticity: disordered motor control. Chicago: Year Book Medical Publishers; 1980. p. 485–95.

27. Towles JD, Kamper DG, Rymer WZ. Lack of hypertonia in thumb muscles after stroke. J Neurophysiol. 2010.

28. Dietz V, Sinkjaer T. Spastic movement disorder: impaired reflex function and altered muscle mechanics. Lancet Neurol. 2007;6(8):725–33.

29. Mottram CJ, Suresh NL, Heckman CJ, Gorassini MA, Rymer WZ. Origins of abnormal excitability in biceps brachii motoneurons of spastic-paretic stroke survivors. J Neurophysiol. 2009;102(4):2026–38.

30. Kamper DG, Schmit BS, Rymer WZ. Effect of muscle biomechanics on the quantification of spasticity. Ann Biomed Eng. 2001;29:1122–34.

31. Barry AJ, Kamper DG, Stoykov ME, Triandafilou K, Roth E. Characteristics of the severely impaired hand in survivors of stroke with chronic impairments. Top Stroke Rehabil. 2021:1–11.

32. Seo NJ, Rymer WZ, Kamper DG. Delays in grip initiation and termination in persons with stroke: effects of arm support and active muscle stretch exercise. J Neurophysiol. 2009;101(6):3108–15.

33. Cruz EG, Waldinger HC, Kamper DG. Kinetic and kinematic workspaces of the index finger following stroke. Brain. 2005;128(Pt 5):1112–21.

34. Hoffmann G, Conrad MO, Qiu D, Kamper DG. Contributions of voluntary activation deficits to hand weakness after stroke. Top Stroke Rehabil. 2016;23(6):384–92.

35. Triandafilou KM, Fischer HC, Towles JD, Kamper DG, Rymer WZ. Diminished capacity to modulate motor activation patterns according to task contributes to thumb deficits following stroke. J Neurophysiol. 106(4):1644–51.

36. Lee SW, Triandafilou K, Lock BA, Kamper DG. Impairment in task-specific modulation of muscle coordination correlates with the severity of hand impairment following stroke. PLoS ONE. 2013;8 (7): e68745.

37. Johansson RS, Vallbo AB. Tactile sensibility in the human hand: relative and absolute densities of four types of mechanoreceptive units in glabrous skin. J Physiol. 1979;286:283–300.

38. Refshauge KM, Kilbreath SL, Gandevia SC. Movement detection at the distal joint of the human thumb and fingers. Exp Brain Res. 1998;122 (1):85–92.

39. Nakamura A, Yamada T, Goto A, Kato T, Ito K, Abe Y, et al. Somatosensory homunculus as drawn by MEG. Neuroimage. 1998;7(4 Pt 1):377–86.

40. Kessner SS, Schlemm E, Cheng B, Bingel U, Fiehler J, Gerloff C, et al. Somatosensory deficits after ischemic stroke. Stroke. 2019;50(5):1116–23.

41. Sainburg R, Good D, Przybyla A. Bilateral synergy: a framework for post-stroke rehabilitation. J Neurol Transl Neurosci. 2013;1(3).

42. The National Spinal Cord Injury Statistical Center [Internet]. https://www.nscisc.uab.edu/ (2021).

43. Lasfargues JE, Custis D, Morrone F, Carswell J, Nguyen T. A model for estimating spinal cord injury prevalence in the United States. Paraplegia. 1995;33(2):62–8.

44. Center NSCIS. www.nscisc.uab.edu.

45. Waters RL, Adkins RH, Yakura JS. Definition of complete spinal cord injury. Paraplegia. 1991;29(9):573–81.

46. Wyndaele M, Wyndaele JJ. Incidence, prevalence and epidemiology of spinal cord injury: what learns a worldwide literature survey? Spinal cord. 2006;44(9):523–9.

47. Calancie B, Molano MR, Broton JG. Interlimb reflexes and synaptic plasticity become evident months after human spinal cord injury. Brain J Neurol. 2002;125(Pt 5):1150–61.

48. Dietz V, Curt A. Neurological aspects of spinal-cord repair: promises and challenges. Lancet Neurol. 2006;5(8):688–94.

49. Dietz V, Fouad K. Restoration of sensorimotor functions after spinal cord injury. Brain J Neurol. 2014;137(Pt 3):654–67.

50. Yang JF, Stein RB, Jhamandas J, Gordon T. Motor unit numbers and contractile properties after spinal cord injury. Ann Neurol. 1990;28(4):496–502.

51. Thomas CK, Zaidner EY, Calancie B, Broton JG, Bigland-Ritchie BR. Muscle weakness, paralysis, and atrophy after human cervical spinal cord injury. Exp Neurol. 1997;148(2):414–23.

52. Hager-Ross CK, Klein CS, Thomas CK. Twitch and tetanic properties of human thenar motor units paralyzed by chronic spinal cord injury. J Neurophysiol. 2006;96(1):165–74.

53. Elbert T, Pantev C, Wienbruch C, Rockstroh B, Taub E. Increased cortical representation of the fingers of the left hand in string players. Science. 1995;270(5234):305–7.

54. Gaser C, Schlaug G. Brain structures differ between musicians and non-musicians. J Neurosci Off J Soc Neurosci. 2003;23(27):9240–5.

55. Altenmuller E, Jabusch HC. Focal dystonia in musicians: phenomenology, pathophysiology and triggering factors. Eur J Neurol. 2010;17(Suppl 1):31–6.

56. Plautz EJ, Milliken GW, Nudo RJ. Effects of repetitive motor training on movement representations in adult squirrel monkeys: role of use versus learning. Neurobiol Learn Mem. 2000;74(1):27–55.

57. Jones TA, Chu CJ, Grande LA, Gregory AD. Motor skills training enhances lesion-induced structural plasticity in the motor cortex of adult rats. J Neurosci Off J Soc Neurosci. 1999;19(22):10153–63.

58. Kleim JA, Jones TA, Schallert T. Motor enrichment and the induction of plasticity before or after brain injury. Neurochem Res. 2003;28(11):1757–69.

59. Jones TA, Schallert T. Use-dependent growth of pyramidal neurons after neocortical damage. J Neurosci Off J Soc Neurosci. 1994;14(4):2140–52.

60. Wolf SL, Winstein CJ, Miller JP, Taub E, Uswatte G, Morris D, et al. Effect of constraint-induced movement therapy on upper extremity function 3 to 9 months after stroke: the EXCITE randomized clinical trial. JAMA. 2006;296(17):2095–104.

61. Liepert J, Miltner WH, Bauder H, Sommer M, Dettmers C, Taub E, et al. Motor cortex plasticity during constraint-induced movement therapy in stroke patients. Neurosci Lett. 1998;250(1):5–8.

62. Liepert J, Bauder H, Wolfgang HR, Miltner WH, Taub E, Weiller C. Treatment-induced cortical reorganization after stroke in humans. J Cereb Circ. 2000;31(6):1210–6.

63. Topping M. An overview of the development of handy 1, a rehabilitation robot to assist the severely disabled. J Intell Rob Syst. 2002;34(3):253–63.

64. Song W, Kim J, An K, Lee I, Song W, Lee B, Hwang S, Son M, Lee E, editors. Design of novel feeding robot for Korean food. In: International conference on smart homes and health telematics; 2010; Seoul, Korea.

65. Gordon EK, Meng X, Bhattacharjee T, Barnes M, Srinivasa SS. Adaptive robot-assisted feeding: an online learning framework for acquiring previously unseen food items. In: IEEE/RSJ international conference on intelligent robots and systems (IROS); 2020. p. 9659–66.

66. Bien Z, Park K, Chung MJ. Mobile platform-based assistive robot systems. In: Helal A, Mokhtari M, Abdulrazak B, editors. The engineering handbook of smart technology for aging, disability and independence; 2008; Wiley.

67. Mahoney RM. The raptor wheelchair robot system. In: Mokhtari M, editor. Integration of Assistive Technology in the information age. IOS Press; 2001. p. 135–41.

68. Driessen BJF, Evers HG, Woerden JA. MANUS—a wheelchair-mounted rehabilitation robot. Proc Inst Mech Eng Part H: J Eng Med. 2001;215(3).

69. Van der Loos M, Michalowski S, Leifer L. Design of an omnidirectional mobile robot as a manipulation aid for the severely disabled. Foulds R, editor. New York: World Rehabilitation Fund; 1986.

70. Srinivasa S, Ferguson D, Helfrich C, Berneson D, Collet A, Diankov R, et al. HERB: a home exploring robotic butler. Auton Robot. 2010;28:5–20.

71. Metzger J-C, Lambercy O, Califfi A, Conti FM, Gassert R. Neurocognitive robot-assisted therapy of hand function. IEEE Trans Haptics. 2013;7:140–9.

72. Lee SW, Landers KA, Park HS. Development of a biomimetic hand exotendon device (BiomHED) for restoration of functional hand movement post-stroke. IEEE Trans Neural Syst Rehabil Eng. 2014;22(4):886–98.

73. Kim DH, Park HS. Cable actuated dexterous (CADEX) glove for effective rehabilitation of the hand for patients with neurological diseases. IEEE Int C Int Robot. 2018:2305–10.

74. Taheri H, Rowe JB, Gardner D, Chan V, Gray K, Bower C, et al. Design and preliminary evaluation of the FINGER rehabilitation robot: controlling challenge and quantifying finger individuation during musical computer game play. J Neuroeng Rehabil. 2014;11:10.

75. Yun Y, Dancausse S, Esmatloo P, Serrato A, Merring CA, Agarwal P, et al. Maestro: an EMG-driven assistive hand exoskeleton for spinal cord injury patients. IEEE Int Conf Robot Autom (ICRA). 2017.

76. Connelly L, Jia Y, Toro ML, Stoykov ME, Kenyon RV, Kamper DG. A pneumatic glove and immersive virtual reality environment for hand rehabilitative training after stroke. IEEE Trans Neural Syst Rehabil Eng. 2010;18(5):551–9.

77. Thielbar KO, Lord TJ, Fischer HC, Lazzaro EC, Barth KC, Stoykov ME, et al. Training finger individuation with a mechatronic-virtual reality system leads to improved fine motor control post-stroke. J Neuroeng Rehabil. 2014;11:171.

78. Brokaw EB, Holley RJ, Lum PS. Hand spring operated movement enhancer (HandSOME) device for hand rehabilitation after stroke. Conf Proc IEEE Eng Med Biol Soc. 2010;2010:5867–70.

79. Kamper DG, Fischer HC, Cruz EG, Rymer WZ. Weakness is the primary contributor to finger impairment in chronic stroke. Arch Phys Med Rehabil. 2006;87(9):1262–9.

80. Kamper DG, Rymer WZ. Impairment of voluntary control of finger motion following stroke: role of inappropriate muscle coactivation. Muscle Nerve. 2001;24(5):673–81.

81. Fischer HC, Triandafilou KM, Thielbar KO, Ochoa JM, Lazzaro ED, Pacholski KA, et al. Use of a portable assistive glove to facilitate rehabilitation in stroke survivors with severe hand impairment. IEEE Trans Neural Syst Rehabil Eng. 2016;24(3):344–51.

82. Yurkewich A, Hebert D, Wang RH, Mihailidis A. Hand extension robot orthosis (HERO) glove: development and testing with stroke survivors with severe hand impairment. IEEE Trans Neural Syst Rehabil Eng. 2019;27(5):916–26.

83. Thielbar KO, Triandafilou KM, Fischer HC, O'Toole JM, Corrigan ML, Ochoa JM, et al. Benefits of using a voice and emg-driven actuated glove to support occupational therapy for stroke survivors. IEEE Trans Neural Syst Rehabil Eng. 2017;25(3):297–305.

84. Lee SW, Qiu D, Fischer HC, Conrad MO, Kamper DG. Modulation of finger muscle activation patterns across postures is coordinated across all muscle groups. J Neurophysiol. 2020;124(2):330–41.

85. Seo NJ, Enders LR, Motawar B, Kosmopoulos ML, Fathi-Firoozabad M. The extent of altered digit force direction correlates with clinical upper extremity impairment in chronic stroke survivors. J Biomech. 2015;48(2):383–7.

86. Lucas L, DiCicco M, Matsuoka Y. An EMG-controlled hand exoskeleton for natural pinching. J Robot Mechatrons. 2004;16:1–7.

87. Nilsson M, Ingvast J, Wikander J, von Holst H, editors. The soft extra muscle system for improving the grasping capability in neurological rehabilitation. IEEE Eng Med Biol. 2012.

88. Diftler MA, Bridgwater LB, Rogers JM, Laske EA, Ensley KG, Lee JH, et al., editors. RoboGlove–a grasp assist device for Earth and space. In: 45th international conference on environmental systems; 2015

89. In H, Cho KJ, Kim K, Lee B. Jointless structure and under-actuation mechanism for compact hand exoskeleton. IEEE Int Conf Rehabil Robot. 2011;2011:5975394.

90. Biggar S, Yao W. Design and evaluation of a soft and wearable robotic glove for hand rehabilitation. IEEE Trans Neural Syst Rehabil Eng. 2016;24 (10):1071–80.

91. Kim YJ, Jeong YJ, Jeon HS, Lee DW, Kim JI. Development of a soft robotic glove with high gripping force using force distributing compliant structures. IEEE/RSJ Int Conf Intell Robot Syst (Iros). 2017;2017:3883–90.

92. Arata J, Ohmoto K, Gassert R, Lambercy O, Fujimoto H, Wada I. A new hand exoskeleton device for rehabilitation using a three-layered sliding spring mechanism. IEEE Int Conf Robot Autom (Icra). 2013;2013:3902–7.

93. Nycz CJ, Butzer T, Lambercy O, Arata J, Fischer GS, Gassert R. Design and characterization of a lightweight and fully portable remote actuation system for use with a hand exoskeleton. IEEE Robot Autom Lett. 2016;1(2):976–83.

94. Butzer T, Lambercy O, Arata J, Gassert R. Fully wearable actuated soft exoskeleton for grasping assistance in everyday activities. Soft Robot. 2021;8 (2):128–43.

95. Polygerinos P, Wang Z, Galloway KC, Wood RJ, Walsh CJ. Soft robotic glove for combined assistance and at-home rehabilitation. Robot Auton Syst. 2015;73:135–43.

96. Yap HK, Lim JH, Nasrallah F, Yeow CH. Design and preliminary feasibility study of a soft robotic glove for hand function assistance in stroke survivors. Front Neurosci. 2017;11:547.

97. Heung KHL, Tong RKY, Lau ATH, Li Z. Robotic glove with soft-elastic composite actuators for assisting activities of daily living. Soft Robot. 2019;6(2):289–304.

98. Dunaway S, Dezsi DB, Perkins J, Tran D, Naft J. Case report on the use of a custom myoelectric elbow-wrist-hand orthosis for the remediation of upper extremity paresis and loss of function in chronic stroke. Mil Med. 2017;182(7): e1963–8.

99. Gasser BW, Martinez A, Sasso-Lance E, Kandilakis C, Durrough CM, Goldfarb M. preliminary assessment of a hand and arm exoskeleton for enabling bimanual tasks for individuals with hemiparesis. IEEE Trans Neural Syst Rehabil Eng. 2020;28(10):2214–23.

100. Toochinda S, Wannasuphoprasit W, editors. Design and development of an assistive hand device for enhancing compatibility and comfortability. In: 2nd international conference on engineering innovation (ICEI); 2018; IEEE.

101. Mohammadi A, Lavranos J, Choong P, Oetomo D. Flexo-glove: a 3D printed soft exoskeleton robotic glove for impaired hand rehabilitation and assistance. Conf Proc IEEE Eng Med Biol Soc. 2018;2018:2120–3.

102. Kang BB, Choi H, Lee H, Cho KJ. Exo-glove poly II: a polymer-based soft wearable robot for the hand with a tendon-driven actuation system. Soft Robot. 2019;6(2):214–27.

103. Jian EK, Gouwanda D, Kheng TK, editors. Wearable hand exoskeleton for activities of daily living. In: IEEE-EMBS conference on biomedical engineering and sciences (IECBES); 2018; IEEE.

104. In H, Kang BB, Sin M, Cho K-J. Exo-Glove: a wearable robot for the hand with a soft tendon routing system. IEEE Robot Autom Mag. 2015;22(1):97–105.

105. Rose CG, O'Malley MK. Hybrid rigid-soft hand exoskeleton to assist functional dexterity. IEEE Robot Autom Lett. 2019;4(1):73–80.

106. Rose CG, O'Malley MK. Design of an assistive, glove-based exoskeleton. Int Symp Wearable Robot Rehabil (Werob). 2017;2017:3–4.

107. Xiloyannis M, Cappello L, Binh KD, Antuvan CW, Masia L. Preliminary design and control of a soft exosuit for assisting elbow movements and hand grasping in activities of daily living. J Rehabil Assist Technol Eng. 2017;4:2055668316680315.

108. Wang XF, Tran P, Callahan SM, Wolf SL, Desai JP. Towards the development of a voice-controlled exoskeleton system for restoring hand function. In: International symposium on medical robotics (Ismr). 2019.

109. Randazzo L, Iturrate I, Perdikis S, Millan JdR. Mano: a wearable hand exoskeleton for activities of daily living and neurorehabilitation. IEEE Robot Autom Lett. 2018;3(1):500–7.

110. Ghassemi M, Kamper D, editors. A hand exoskeleton for stroke survivors' activities of daily life. In: 43rd annual international conference of the IEEE engineering in medicine and biology society (EMBC); 2021; IEEE.

111. Hu XL, Tong KY, Wei XJ, Rong W, Susanto EA, Ho SK. The effects of post-stroke upper-limb training with an electromyography (EMG)-driven hand robot. J Electromyogr Kinesiol. 2013;23(5):1065–74.

112. Dai C, Cao Y, Hu X. Prediction of individual finger forces based on decoded motoneuron activities. Ann Biomed Eng. 2019;47(6):1357–68.

113. Pfurtscheller G, Guger C, Muller G, Krausz G, Neuper C. Brain oscillations control hand orthosis in a tetraplegic. Neurosci Lett. 2000;292(3):211–4.

114. Hochberg LR, Bacher D, Jarosiewicz B, Masse NY, Simeral JD, Vogel J, et al. Reach and grasp by people with tetraplegia using a neurally controlled robotic arm. Nature. 485(7398):372–5.

115. Collinger JL, Wodlinger B, Downey JE, Wang W, Tyler-Kabara EC, Weber DJ, et al. High-performance neuroprosthetic control by an individual with tetraplegia. Lancet (London, England). 381(9866):557–64.

116. Tan DW, Schiefer MA, Keith MW, Anderson JR, Tyler J, Tyler DJ. A neural interface provides long-term stable natural touch perception. Sci Transl Med. 6(257):257ra138.

117. Tabot GA, Dammann JF, Berg JA, Tenore FV, Boback JL, Vogelstein RJ, et al. Restoring the sense of touch with a prosthetic hand through a brain interface. Proc Natl Acad Sci USA. 110(45):18279–84.

118. Miller LC, Dewald JP. Involuntary paretic wrist/finger flexion forces and EMG increase with shoulder abduction load in individuals with chronic stroke. Clin Neurophysiol. 2012;123(6):1216–25.

119. Kline TL, Schmit BD, Kamper DG. Exaggerated interlimb neural coupling following stroke. Brain. 2007;130(Pt 1):159–69.

120. Correia C, Nuckols K, Wagner D, Zhou YM, Clarke M, Orzel D, et al. Improving grasp function after spinal cord injury with a soft robotic glove. IEEE Trans Neural Syst Rehabil Eng. 2020;28(6):1407–15.

121. Yurkewich A, Kozak IJ, Hebert D, Wang RH, Mihailidis A. Hand extension robot orthosis (HERO) grip glove: enabling independence amongst persons with severe hand impairments after stroke. J Neuroeng Rehabil. 2020;17(1):33.

122. Radder B, Prange-Lasonder G, Kottink AIR, Melendez-Calderon A, Buurke JH, Rietman JS. Feasibility of a wearable soft-robotic glove to support impaired hand function in stroke patients. J Rehabil Med. 2018;50(7):598–606.

123. Van Ommeren A, Radder B, Buurke JH, Kottink AI, Holmberg J, Sletta K, et al., editors. The effect of prolonged use of a wearable soft-robotic glove post stroke-a proof-of-principle. In: 7th IEEE international conference on biomedical robotics and biomechatronics (Biorob); 2018; IEEE.

124. Realmuto J, Sanger T. A robotic forearm orthosis using soft fabric-based helical actuators. In: 2nd IEEE international conference on soft robotics (RoboSoft); 2019. p. 591–6.

125. Park SH, Yi J, Kim D, Lee Y, Koo HS, Park YL. A lightweight, soft wearable sleeve for rehabilitation of forearm pronation and supination. In: 2nd IEEE international conference on soft robotics (Robosoft 2019); 2019. p. 636–41.

126. Lessard S, Pansodtee P, Robbins A, Baltaxe-Admony LB, Trombadore JM, Teodorescu M, et al. CRUX: a compliant robotic upper-extremity exosuit for lightweight, portable, multi-joint muscular augmentation. IEEE Int Conf Rehabil Robot. 2017;2017:1633–8.

127. Li N, Yang T, Yu P, Zhao L, Chang JL, Xi N, et al. Force point transfer method to solve the structure of soft exoskeleton robot deformation due to the driving force. In: Proceedings of 2018 IEEE international conference on real-time computing and robotics (IEEE Rcar). 2018:236–41.

128. Ismail R, Ariyanto M, Perkasa IA, Adirianto R, Putri FT, Glowacz A, et al. Soft elbow exoskeleton for upper limb assistance incorporating dual motor-tendon actuator. Electronics. 2019;8(10).

129. Shiota K, Kokubu S, Tarvainen TVJ, Sekine M, Kita K, Huang SY, et al. Enhanced Kapandji test evaluation of a soft robotic thumb rehabilitation device by developing a fiber-reinforced elastomer-actuator based 5-digit assist system. Robot Auton Syst. 2019;111:20–30.

130. Rowe JB, Chan V, Ingemanson ML, Cramer SC, Wolbrecht ET, Reinkensmeyer DJ. Robotic assistance for training finger movement using a hebbian model: a randomized controlled trial. Neurorehabil Neural Repair. 2017;31(8):769–80.

131. Nijenhuis SM, Prange-Lasonder GB, Stienen AH, Rietman JS, Buurke JH. Effects of training with a passive hand orthosis and games at home in chronic stroke: a pilot randomised controlled trial. Clin Rehabil. 2017;31(2):207–16.

132. Ranzani R, Lambercy O, Metzger JC, Califfi A, Regazzi S, Dinacci D, et al. Neurocognitive robot-assisted rehabilitation of hand function: a randomized control trial on motor recovery in subacute stroke. J Neuroeng Rehabil. 2020;17(1):115.

133. Aprile I, Germanotta M, Cruciani A, Loreti S, Pecchioli C, Cecchi F, et al. Upper limb robotic rehabilitation after stroke: a multicenter, randomized clinical trial. J Neurol Phys Ther. 2020;44 (1):3–14.

134. Jung JH, Lee HJ, Cho DY, Lim JE, Lee BS, Kwon SH, et al. Effects of combined upper limb robotic therapy in patients with tetraplegic spinal cord injury. Ann Rehabil Med. 2019;43(4):445–57.

135. Calabro RS, Accorinti M, Porcari B, Carioti L, Ciatto L, Billeri L, et al. Does hand robotic rehabilitation improve motor function by rebalancing interhemispheric connectivity after chronic stroke? Encouraging data from a randomised-clinical-trial. Clin Neurophysiol. 2019;130(5):767–80.

136. Huang Y, Nam C, Li W, Rong W, Xie Y, Liu Y, et al. A comparison of the rehabilitation effectiveness of neuromuscular electrical stimulation robotic hand training and pure robotic hand training after stroke: a randomized controlled trial. Biomed Signal Process Control. 2020;56.

137. Qian Q, Nam C, Guo Z, Huang Y, Hu X, Ng SC, et al. Distal versus proximal—an investigation on different supportive strategies by robots for upper limb rehabilitation after stroke: a randomized controlled trial. J Neuroeng Rehabil. 2019;16(1):64.

138. Jain NB, Ayers GD, Peterson EN, Harris MB, Morse L, O'Connor KC, et al. Traumatic spinal cord injury in the United States, 1993–2012. JAMA. 2015;313(22):2236–43.

139. Maldonado AA, Kircher MF, Spinner RJ, Bishop AT, Shin AY. The role of elective amputation in patients with traumatic brachial plexus injury. J Plast Reconstr Aesthet Surg. 2016;69 (3):311–7.

140. Barry AJ, Triandafilou KM, Stoykov ME, Bansal N, Roth EJ, Kamper DG. Survivors of chronic stroke experience continued impairment of dexterity but not strength in the nonparetic upper limb. Arch Phys Med Rehabil. 2020;101(7):1170–5.

Neural Coordination of Cooperative Hand Movements: Implications for Rehabilitation Technology

7

Volker Dietz and Miriam Schrafl-Altermatt

Abstract

The coordination of cooperative hand movements, required for many activities of daily living (e.g. opening a bottle), is achieved by a neural coupling. This neural coupling is manifested in electromyographic reflex responses in the forearm muscles of both sides to unilateral arm nerve stimulation. During such cooperative movements fMRI recordings show a bilateral task-specific activation and functional coupling of the secondary somatosensory cortical areas (S2). This activation is suggested to reflect processing of shared cutaneous afferent input from the hands. The efferent/executive part of bilateral hand coordination on the basis of animal evidence is assumed to be mediated by the cortico-reticulo-spinal pathway. In chronic post-stroke patients, arm nerve stimulation of the unaffected arm during bimanual movements leads to a bilateral response pattern, i.e. also in paretic forearm muscles, similar to those seen in healthy subjects. In contrast, arm nerve stimulation of the affected side is followed only by ipsilateral reflex responses, suggesting an impaired processing of afferent input from the affected side leading to a defective neural coupling. According to these results, hand rehabilitation of stroke patients, currently focused on reach and grasp movements of the affected side, should be supplemented by a training of bilateral hand movements required during activities of daily living. According to these clinical/neurophysiological aspects, robotic devices designed to assist arm/hand therapy should include cooperative hand training opportunities. They might be combined with an artificial sensory input applied to the unaffected side in order to achieve an additional activation of paretic arm muscles.

Keywords

Stroke · Hand function · Rehabilitation · Neural limb coupling · Reflex activity · fMRI recordings

7.1 Introduction

In contrast to lower limbs, a great variety of functional movements are performed by hands/arms during activities of daily living (ADL). During such movements, hands and fingers have to interact in different ways. This requires a task-specific neural coordination. There exist a number of studies concerning the neural

V. Dietz (✉) · M. Schrafl-Altermatt
Spinal Cord Injury Center, University Hospital
Balgrist, Forchstrasse 340, 8008 Zürich, Switzerland
e-mail: volker.dietz@balgrist.ch

M. Schrafl-Altermatt
e-mail: miriam.altermatt@hest.ethz.ch

© The Author(s), under exclusive license to Springer Nature Switzerland AG 2022
D. J. Reinkensmeyer et al. (eds.), *Neurorehabilitation Technology*,
https://doi.org/10.1007/978-3-031-08995-4_7

control of uni-and bilateral arm/hand movements. In non-human primates [1–3] and humans [4–6] several studies suggest a common neural control of hand and finger movements by direct cortico-motoneuronal (CM) connections. In monkeys, it is suggested that the supplementary motor area (SMA) of one hemisphere influences the motor outflow of both hemispheres [7, 8]. Furthermore, the primary [9, 10] and non-primary motor cortex [11], as well as the prefrontal cortex [12] are assumed to play an essential role in the execution of bimanual tasks. Alternatively, it is assumed that distributed neural networks, including cortical and subcortical areas [9, 13–17] control interlimb coordination during bimanual tasks. Alongside these general control mechanisms, task-specificity of neural control seems to exist for various bimanual movements (for review cf. [18]). It has been shown that interhemispheric connections between the primary motor cortical areas are involved in the control of uni- and bilateral in-phase movements, while connections between the premotor cortex and the contralateral primary motor cortex regulate bimanual anti-phase movements [19]. In bimanual tasks, rapid grip force adjustments of one hand are generated by sensory input from contralateral hand and fingers [20, 21]. Similarly, during a two hand grasp, bimanual reflex responses occur following a unilateral mechanical hand perturbation [22].

For upper limbs one has to take into account that during unilateral as well as bilateral movements hand/finger interactions differ between power grip tasks, i.e. when fingers act as part of the hand, and precision grip movements, i.e. when fingers move locally independently of the hand, with the consequence of a task-specific neural control [23]. Bilateral precision finger movements, such as playing piano, are achieved by a strong, task-specific CM control. In contrast, on the basis of animal experiments, the hands are suggested to be coupled by the reticular system [24].

During cooperative hand movements, both hands act in synchrony, in order to accomplish the task, i.e. the action of one hand is supported by the other one, for example when opening a bottle, or when lifting and balancing a tray. In such tasks a 'neural coupling' coordinates bilateral power grip hand movements [25]. Nevertheless, the neural coupling mechanism underlies not only cooperative hand movements but automatically coordinates all bilateral, synchronously performed hand movements [25]. When subjects are forced to perform bilateral asynchronous, i.e. independent hand movements, a visually guided voluntary control dominates movement performance. This is associated with larger movement errors and, consecutively, corrective movements [25].

This chapter gives an overview of actual research on the neural coupling mechanism underlying the neural coordination of bilateral hand movements and its function and dysfunction in healthy and post-stroke subjects. Various aspects of normal and impaired hand movement control have to be taken into account for rehabilitation of hand function in post-stroke subjects, e.g. in how far a damage of the corticospinal tract limits recovery of hand/finger function. Consequently, also the design of technology applied in rehabilitation should be based on this knowledge. In the first part of this chapter, specific aspects of neural control of cooperative hand movements are described. The second part deals with the neural adaptations of impaired hand movement control in relation to the defective neural coupling mechanism in post-stroke subjects. In the last part, the consequences of the first and second parts for the rehabilitation of hand function and the implementation in an appropriate technology become established.

7.2 Coordination of Bilateral Hand Movements by Neural Coupling

Recent studies indicate a differential neural control between power and precision grip motor tasks, not only for unilateral, but also for bilateral hand and finger movements [23]. During bilateral precision finger movements, such as playing piano, hands and fingers have a different function compared to a power grip task such as opening a bottle. For piano playing, independent finger movements are needed for the execution of quavers by the fingers of one and triplets by those of the other hand (e.g. piano pieces of Debussy or Chopin). Therefore, during playing piano fingers have to move independently on both sides, while the associated hand movements appear to be coupled [23] in order to uphold the rhythm. The fingers obviously are under the same task-specific cortico-motoneuronal control as it is the case during unilateral skilled finger movements [23, 24].

In contrast, bilateral power grips include cooperative hand movements. In these motor tasks, fingers act as part of the hands to achieve a specific functional goal, such as opening a bottle. These movements are coordinated by the neural coupling mechanism. Many daily life activities require cooperative hand movements, but little is known about their neural control [26]. Previous electrophysiological research on hand function has focused on the execution of unilateral hand movements. During such movements, a task-dependent amplitude modulation of EMG responses in forearm muscles, with larger amplitudes during a dynamic compared with a static muscle contraction, is present [27–29]. Similarly, during synchronously performed pro-/supination movements of both hands only ipsilateral EMG responses to arm nerve stimulations are recorded [17, 29].

In contrast, during cooperative hand movements, a different reflex behavior, reflecting the neural coupling, has been described [29]. During such movements (e.g. opening a bottle, lifting/balancing a tray), a bilateral EMG response pattern (N2–P2 complex) appears in forearm muscles of both sides with approximately the same latency (80 ms) to unilateral arm nerve stimulation (Fig. 7.1). This observation reflects the neural coupling mechanism underlying the coordination of cooperative hand movements [29].

A corresponding observation of a task-specific neural coupling, i.e. the appearance of bilateral arm muscle responses to unilateral leg nerve stimulation during locomotion in healthy subjects [30] has been described. It was suggested to reflect the residual quadrupedal coordination of arms and legs during stepping [30]. Although locomotor function differs basically from

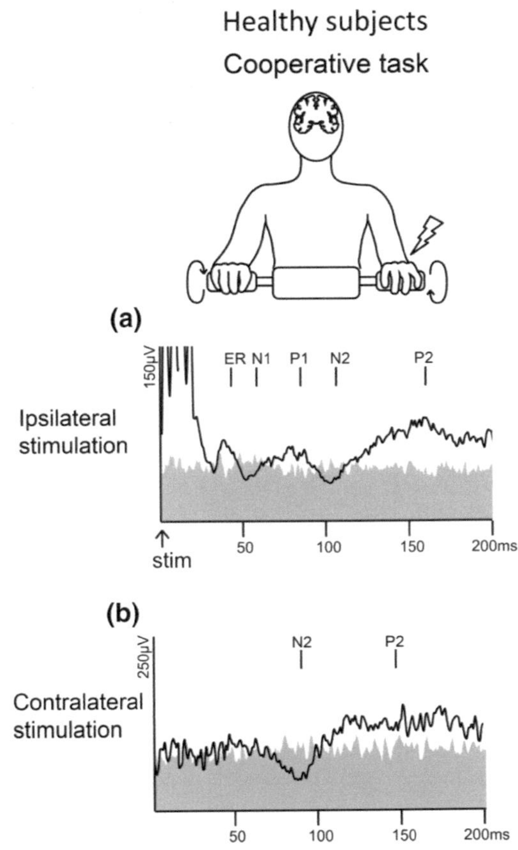

Fig. 7.1 Reflex response pattern during cooperative hand movements of a healthy subjects. Grand averages (n = 24) of the EMG reflex responses following electrical unilateral ulnar nerve stimulation in the forearm muscles ipsilateral (**a**) and contralateral (**b**) to the stimulation site. The ipsilateral response pattern (A) consists of an early reflex response (ER) followed by a first component composed of a first negativity (N1) and a first positivity (P1) followed by a late component (N2 and P2). On the contralateral side, the reflex response consists only of the late components, i.e. N2 and P2. Upper part: schematic drawings of the movement task performed and stimulation site. Shaded areas represent the level of background EMG. Vertical arrow in A indicates the onset of electrical stimulation. Note the different calibrations

cooperative hand movements, the underlying mechanism of a task-dependent neural coupling of limbs might be achieved in a similar way [31].

Afferent pathway of neural coupling: Role of S2 cortical areas.

fMRI recordings performed during cooperative hand movements [29] show a task-specific activation of secondary somatosensory (S2) cortical areas which suggests that these brain areas are involved in the afferent control of the neural coupling underlying these movements. This assumption is supported by observations in humans [32] and nonhuman primates [33] showing that each S2 cortical area receives afferent input from receptor fields of both hands during cooperative hand movements.

The S2 cortical area is suggested to be involved in the exchange and integration of information from both sides of the body [34]. S2 is thought to have a role in combining somatosensory information from both sides of the body to allow its interhemispheric unification [35]. Several observations further support the assumption that the S2 cortical area is engaged in the inter-hemispheric processing of afferent input during cooperative hand movements: First, the spatial extent of fMRI activation of the S2 cortical areas in humans are larger for bilateral than for unilateral hand stimulation [32]. Second, a connectivity analysis indicates that the left and right S2 areas are functionally connected during cooperative hand movements [29]. These findings support the view that the interaction of S2 cortical areas represents the afferent part of the coordination of cooperative hand movements.

Efferent pathway of neural coupling: Role of the reticular system.

Different executive, i.e. efferent, pathways might be involved in the neural coupling, i.e. the generation of the bilateral response pattern to unilateral arm nerve stimulation. Both ipsi- and contralateral M1 cortical areas might become activated in the cooperative movement condition, mediated by the corpus callosum. Alternatively, for the coordination of bilateral hand movements direct ipsi- as well as contralateral cortico-spinal pathways arising from one M1 cortical area might lead to the bilateral response pattern to unilateral arm nerve stimulation. However, both mechanisms seem to be rather unlike candidates responsible for the neural coupling: The ipsilateral cortico-spinal pathway is suggested to play a functionally rather minor role in motor control in both monkey [36] and human beings [37] for mediating the quite robust mechanism of neural coupling.

A likely mechanism generating the bilateral response pattern is based on animal evidence: the cortico-reticulo-spinal pathway is known to distribute a unilateral cortical command signal to both body sites [38, 39]. The reticular system represents a phylogenetically old motor system with an important functional role in the coordination of forelimb movements throughout the mammalian species, most probably including human beings [37]. It allows an automatic coordination of upper limb movements, i.e. exactly what is happening during cooperative hand movements. It is, therefore, suggested to reflect the executory part of neural coupling.

7.3 Defective Neural Coupling in Post-Stroke Subjects

In post-stroke subjects, it is known that both anticipatory postural adjustments [40] and bimanual coordination are impaired, most probably due to a defective processing of somatosensory information [41]. Also the neural coupling of cooperative hand movements was shown to be defective after a stroke [42]. In post-stroke patients, electrical arm nerve stimulation of the unaffected arm during cooperative hand movements leads to bilateral EMG responses in forearm muscles, similar to those obtained in healthy subjects. In contrast, nerve stimulation of the affected arm in most patients elicits only ipsilateral EMG responses [42].

This indicates an impaired processing of afferent input by the cortico-spinal tract at the paretic side after a stroke [2] (Fig. 7.2), although in the clinical testing of post-stroke patients light touch perception is only slightly impaired on the affected side, while the paresis dominates. In contrast, the largely preserved reflex responses in both the affected and unaffected forearm muscles following unaffected arm nerve stimulation indicate a preserved reflex transmission to both sides, most probably mediated by the cortico-reticular-spinal tract, arising from the non-damaged hemisphere.

In fact, an involvement of the ipsilateral cortico-spinal pathway in the compensation of

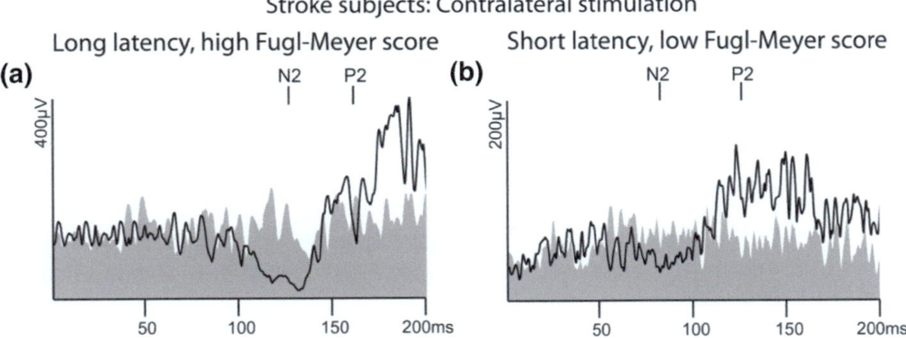

Fig. 7.2 Contralateral reflex responses to ipsilateral arm nerve stimulation in post-stroke subjects during cooperative movements. Averages of EMG recordings from the affected side to 15 ulnar nerve stimulations of the unaffected side of a patient with a high Fugl-Meyer score and a longer N2 latency (compared to healthy subjects) (**a**) and a patient with a low Fugl-Meyer score and shortened N2 latency (**b**). The N2 latency seen in A is somewhat longer than that observed in healthy volunteers, indicating a slower processing of afferent input. Shaded areas represent the level of background EMG. Note the different calibrations

motor deficit after a stroke cannot be excluded. In healthy subjects these fibers can be activated under specific conditions [43, 44]. It is also assumed that ipsilateral tract fibers contribute to movement performance after stroke for a partial compensation of the deficit [45] and defective inter-hemispheric interactions [46]. Ipsilateral fibers of the corticospinal tract may, in fact, contribute to stroke recovery, especially in patients suffering a severe lesion (for a review see [47]). In this context, one has to be aware that in such cases motor deficits and functional impairments concern both ipsi- and contralesional arms [48, 49].

Nevertheless, ipsilateral cortico-spinal connections are weak [36] which favors a dominant role of the cortico-reticulo-spinal pathway in the preservation of the EMG response pattern in both the unaffected and paretic arm muscles following unaffected arm nerve stimulation [50]. The pathways that are suggested to be involved in cooperative hand movements in healthy subjects and post-stroke subjects are displayed in Fig. 7.3.

The finding of an impaired neural coordination of hand movements is in line with a defective task-specific neural coupling of arms and legs during locomotion in post-stroke subjects [51]. In this condition, stimulation of the tibial nerve of the unaffected leg during stepping in

most cases produces normal EMG responses in proximal arm muscles of both sides. In contrast, nerve stimulation of the paretic leg elicits neither ipsilateral nor contralateral reflex responses in the arms. This is also assumed to be due to an impaired processing of afferent input by the damaged cortico-spinal tract.

These observations made in post-stroke subjects are important for hand rehabilitation in so far that after CNS damage improvement in function depends on the training of motor tasks required in activities of daily life [52]. Rehabilitation of hand function is currently mainly focused on unimanual reach and grasp movements of the affected arm and hand. According to the insights established here, bilateral/cooperative hand movements should be included in training approaches.

7.4 Consequences for Conventional and Robotic Assisted Therapy

After stroke, the impairment of the affected limb is usually compensated by more intensive utilizing of the unaffected limb in tackling activities of daily living (ADL), leading to the non-use phenomenon [53]. To avoid this, constraint induced movement therapy (CIMT) [54] became established in neurorehabilitation. This approach

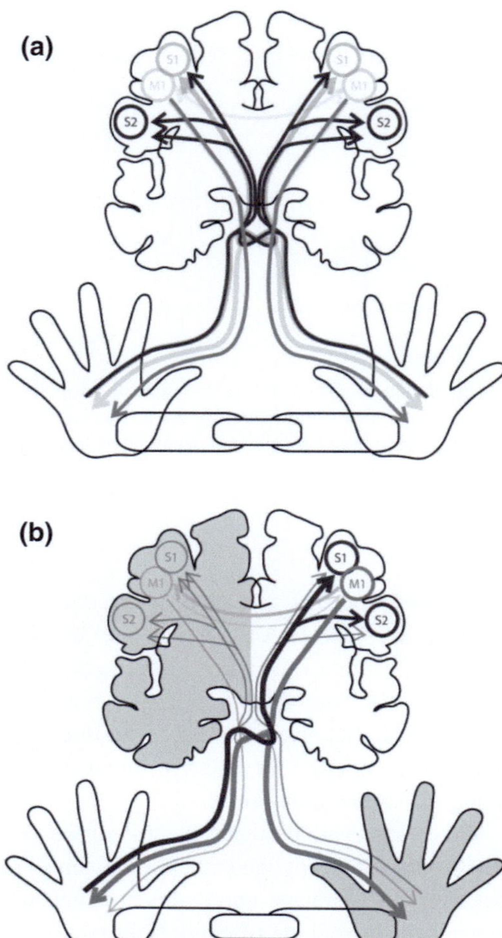

Fig. 7.3 Schematic illustration of pathways involved in the control of cooperative hand movements of healthy and post-stroke subjects. During cooperative movements specific ascending and descending (e.g. cortico-reticulo-spinal) pathways are suggested to be involved in the generation of bilateral reflex responses to unilateral arm nerve stimulation in healthy (**a**) and post-stroke (**b**) subjects

intensifies exclusive training of the affected arm/hand. No clear evidence of superiority has been demonstrated when CIMT becomes compared with a bimanual training approach [55–57]. Nevertheless, bimanual training of reaching and grasping tasks in stroke patients has been shown to be more effective in improving execution of ADL tasks than unilateral training of the affected arm/hand alone [58].

There is also some evidence in post-stroke patients that the activation of the neural coupling mechanism by the performance of bimanual movements improves function of the affected limb: Stroke patients perform a simple tapping task faster when they use both arms/hands compared to execution of the task only by the paretic arm/hand [59]. This is in line with the idea of a contribution of the unaffected hemisphere to task performance of the paretic arm/hand, probably by using the cortico-reticulo-spinal pathway for movement execution. Furthermore, children with cerebral palsy can perform mirrored movements to accomplish a task better as compared with the more affected arm alone [60, 61].

Based on experiments in rodents, improvement of function after a CNS damage depends on the specific task and its underlying neural control to be trained (for reviews [52, 62]). Therefore, current approaches to exploit neuroplasticity after a stroke are directed at training motor tasks required during ADL. This should include a training of bilateral hand movements in the rehabilitation of hand function after a stroke as many ADL require cooperation of the hands. Nevertheless, the question concerning the best training approaches to improve hand function after a stroke remains open and needs further experimental and clinical studies [63].

A wide variety of rehabilitation technology for upper limb training after a stroke is available today (for review cf. [64]). Robot assisted training has been shown to lead to a better or about equal effect on functional improvements in post-stroke patients when compared to conventional therapy [65, 66]. Currently available robotic devices provide training for the affected hand, e.g. Amadeo [67], arm, e.g. Armeo [68] or MIT Manus [69], for bimanual training, e.g. Bi-Manu Track [70], or mirror movement training, e.g. MIME [71] (for review cf. [72]). Nevertheless, only few of these devices support the training of cooperative hand movement tasks, i.e. activation of the neural coupling mechanism. It is suggested that robot-assisted therapy should be supplemented by a technology that allows training of bilateral/cooperative hand and arm movements covering a great range of upper limb tasks needed in ADL. Such a training might be combined with additional afferent input (e.g. electrical nerve or vibratory stimulations of the unaffected limb) in order to co-activate paretic arm muscles.

References

1. Lemon RN, Griffiths J. Comparing the function of the corticospinal system in different species: organizational differences for motor specialization? Muscle Nerve. 2005;32(3):261–79.
2. Lemon R. Descending pathways in motor control. Annu Rev Neurosci. 2008;31:195–218.
3. Maier MA, Bennett KM, Hepp-Reymond MC, Lemon RN. Contribution of the monkey corticomotoneuronal system to the control of force in precision grip. J Neurophysiol. 1993;69(3):772–85.
4. Palmer E, Ashby P. Corticospinal projections to upper limb motoneurones in humans. J Physiol. 1992;448:397–412.
5. Baldissera F, Cavallari P. Short-latency subliminal effects of transcranial magnetic stimulation on forearm motoneurones. Exp Brain Res. 1993;96(3):513–8.
6. de Noordhout AM, Rapisarda G, Bogacz D, Gerard P, De Pasqua V, Pennisi G, et al. Corticomotoneuronal synaptic connections in normal man: an electrophysiological study. Brain. 1999;122(Pt 7):1327–40.
7. Jenny AB. Commissural projections of the cortical hand motor area in monkeys. J Comp Neurol. 1979;188(1):137–45.
8. Rouiller EM, Babalian A, Kazennikov O, Moret V, Yu XH, Wiesendanger M. Transcallosal connections of the distal forelimb representations of the primary and supplementary motor cortical areas in macaque monkeys. Exp Brain Res. 1994;102(2):227–43.
9. Donchin O, Gribova A, Steinberg O, Bergman H, Vaadia E. Primary motor cortex is involved in bimanual coordination. Nature. 1998;395(6699):274–8.
10. Kermadi I, Liu Y, Tempini A, Calciati E, Rouiller EM. Neuronal activity in the primate supplementary motor area and the primary motor cortex in relation to spatio-temporal bimanual coordination. Somatosens Mot Res. 1998;15(4):287–308.
11. Tanji J, Okano K, Sato KC. Relation of neurons in the nonprimary motor cortex to bilateral hand movement. Nature. 1987;327(6123):618–20.
12. Theorin A, Johansson RS. Selection of prime actor in humans during bimanual object manipulation. J Neurosci. 2010;30(31):10448–59.
13. Debaere F, Swinnen SP, Beatse E, Sunaert S, Van Hecke P, Duysens J. Brain areas involved in interlimb coordination: a distributed network. Neuroimage. 2001;14:947–58.
14. Kazennikov O, Hyland B, Corboz M, Babalian A, Rouiller EM, Wiesendanger M. Neural activity of supplementary and primary motor areas in monkeys and its relation to bimanual and unimanual movement sequences. Neuroscience. 1999;89:661–74.
15. Kermadi I, Liu Y, Rouiller EM. Do bimanual motor actions involve the dorsal premotor (PMd), cingulate (CMA) and posterior parietal (PPC) cortices? Comparison with primary and supplementary motor cortical areas. Somatosens Mot Res. 2000;17(3):255–71.
16. Stephan KM, Binkofski F, Halsband U, Dohle C, Wunderlich G, Schnitzler A, et al. The role of ventral medial wall motor areas in bimanual co-ordination. A combined lesion and activation study. Brain. 1999;122 (Pt 2):351–68.
17. Swinnen SP. Intermanual coordination: from behavioural principles to neural-network interactions. Nat Rev Neurosci. 2002;3:348–59.
18. Wiesendanger M, Serrien DJ. The quest to understand bimanual coordination. Prog Brain Res. 2004;143:491–505.
19. Liuzzi G, Horniss V, Zimerman M, Gerloff C, Hummel FC. Coordination of uncoupled bimanual movements by strictly timed interhemispheric connectivity. J Neurosci. 2011;31(25):9111–7.
20. Ohki Y, Johansson RS. Sensorimotor interactions between pairs of fingers in bimanual and unimanual manipulative tasks. Exp Brain Res. 1999;127(1):43–53.
21. White O, Dowling N, Bracewell RM, Diedrichsen J. Hand interactions in rapid grip force adjustments are independent of object dynamics. J Neurophysiol. 2008;100:2738–45.
22. Lum PS, Reinkensmeyer DJ, Lehman SL. A bimanual reflex during two hand grasp. Conf Proc IEEE. 1993; San Diego, CA: Med Biol Soc:1163–4.
23. Caldelari P, Lemon R, Dietz V. Differential neural coordination of bilateral hand and finger movements. Physiol Rep. 2020;8(6): e14393.
24. Dietz V. Neural coordination of bilateral power and precision finger movements. Eur J Neurosci. 2020.
25. Kochli S, Scharfenberger T, Dietz V. Coordination of bilateral synchronous and asynchronous hand movements. Neurosci Lett. 2020;720: 134757.
26. Obhi SS. Bimanual coordination: an unbalanced field of research. Mot Control. 2004;8(2):111–20.
27. Datta AK, Harrison LM, Stephens JA. Task-dependent changes in the size of response to magnetic brain stimulation in human first dorsal interosseous muscle. J Physiol. 1989;418:13–23.
28. Zehr EP, Kido A. Neural control of rhythmic, cyclical human arm movement: task dependency, nerve specificity and phase modulation of cutaneous reflexes. J Physiol. 2001;537(Pt 3):1033–45.

29. Dietz V, Macauda G, Schrafl-Altermatt M, Wirz M, Kloter E, Michels L. Neural coupling of cooperative hand movements: a reflex and FMRI study. Cereb Cortex. 2015;25(4):948–58.

30. Michel J, van Hedel HJ, Dietz V. Obstacle stepping involves spinal anticipatory activity associated with quadrupedal limb coordination. Eur J Neurosci. 2008;27(7):1867–75.

31. Dietz V, Fouad K, Bastiaanse CM. Neuronal coordination of arm and leg movements during human locomotion. Eur J Neurosci. 2001;14(11):1906–14.

32. Disbrow E, Roberts T, Poeppel D, Krubitzer L. Evidence for interhemispheric processing of inputs from the hands in human S2 and PV. J Neurophysiol. 2001;85(5):2236–44.

33. Whitsel BL, Petrucelli LM, Werner G. Symmetry and connectivity in the map of the body surface in somatosensory area II of primates. J Neurophysiol. 1969;32(2):170–83.

34. Lin YY, Forss N. Functional characterization of human second somatosensory cortex by magnetoencephalography. Behav Brain Res. 2002;135(1–2):141–5.

35. Hari R, Hanninen R, Makinen T, Jousmaki V, Forss N, Seppa M, et al. Three hands: fragmentation of human bodily awareness. Neurosci Lett. 1998;240 (3):131–4.

36. Soteropoulos DS, Edgley SA, Baker SN. Lack of evidence for direct corticospinal contributions to control of the ipsilateral forelimb in monkey. J Neurosci. 2011;31(31):11208–19.

37. Ziemann U, Ishii K, Borgheresi A, Yaseen Z, Battaglia F, Hallett M, et al. Dissociation of the pathways mediating ipsilateral and contralateral motor-evoked potentials in human hand and arm muscles. J Physiol. 1999;518(Pt 3):895–906.

38. Brocard F, Ryczko D, Fenelon K, Hatem R, Gonzales D, Auclair F, et al. The transformation of a unilateral locomotor command into a symmetrical bilateral activation in the brainstem. J Neurosci. 2010;30(2):523–33.

39. Fisher KM, Zaaimi B, Baker SN. Reticular formation responses to magnetic brain stimulation of primary motor cortex. J Physiol. 2012;590(16):4045–60.

40. Massion J, Ioffe M, Schmitz C, Viallet F, Gantcheva R. Acquisition of anticipatory postural adjustments in a bimanual load-lifting task: normal and pathological aspects. Exp Brain Res. 1999;128 (1–2):229–35.

41. Torre K, Hammami N, Metrot J, van Dokkum L, Coroian F, Mottet D, et al. Somatosensory-related limitations for bimanual coordination after stroke. Neurorehabil Neural Repair. 2013;27(6):507–15.

42. Schrafl-Altermatt M, Dietz V. Cooperative hand movements in stroke patients: neural reorganization. Clin Neurophysiol. 2016;127(1):748–54.

43. Bradnam LV, Stinear CM, Byblow WD. Theta burst stimulation of human primary motor cortex degrades selective muscle activation in the ipsilateral arm. J Neurophysiol. 2010;104(5):2594–602.

44. Howatson G, Taylor MB, Rider P, Motawar BR, McNally MP, Solnik S, et al. Ipsilateral motor cortical responses to TMS during lengthening and shortening of the contralateral wrist flexors. Eur J Neurosci. 2011;33(5):978–90.

45. Baker SN. The primate reticulospinal tract, hand function and functional recovery. J Physiol. 2011;589(Pt 23):5603–12.

46. Edwards JD, Meehan SK, Linsdell MA, Borich MR, Anbarani K, Jones PW, et al. Changes in thresholds for intracortical excitability in chronic stroke: more than just altered intracortical inhibition. Restor Neurol Neurosci. 2013.

47. Teasell R, Bayona NA, Bitensky J. Plasticity and reorganization of the brain post stroke. Top Stroke Rehabil. 2005;12(3):11–26.

48. Schaefer SY, Haaland KY, Sainburg RL. Ipsilesional motor deficits following stroke reflect hemispheric specializations for movement control. Brain. 2007;130(Pt 8):2146–58.

49. Sainburg RL, Duff SV. Does motor lateralization have implications for stroke rehabilitation? J Rehabil Res Dev. 2006;43(3):311–22.

50. Senesh MR, Barragan K, Reinkensmeyer DJ. Rudimentary dexterity corresponds with reduced ability to move in synergy after stroke: evidence of competition between corticoreticulospinal and corticospinal tracts? Neurorehabil Neural Repair. 2020;34 (10):904–14.

51. Kloter E, Wirz M, Dietz V. Locomotion in stroke subjects: interactions between unaffected and affected sides. Brain. 2011;134(Pt 3):721–31.

52. Edgerton VR, Tillakaratne NJ, Bigbee AJ, de Leon RD, Roy RR. Plasticity of the spinal neural circuitry after injury. Annu Rev Neurosci. 2004;27:145–67.

53. Taub E, Uswatte G, Mark VW, Morris DM. The learned nonuse phenomenon: implications for rehabilitation. Eura Medicophys. 2006;42:241–56.

54. Taub E, Uswatte G, Pidikiti R. Constraint-induced movement therapy: a new family of techniques with broad application to physical rehabilitation–a clinical review. J Rehabil Res Dev. 1999;36(3):237–51.

55. Stoykov ME, Lewis GN, Corcos DM. Comparison of bilateral and unilateral training for upper extremity hemiparesis in stroke. Neurorehabil Neural Repair. 2009;23:945–53.

56. van Delden AL, Peper CL, Beek PJ, Kwakkel G. Match and mismatch between objective and subjective improvements in upper limb function after stroke. Disabil Rehabil. 2013.

57. Lin KC, Chen YA, Chen CL, Wu CY, Chang YF. The effects of bilateral arm training on motor control and functional performance in chronic stroke: a randomized controlled study. Neurorehabil Neural Repair. 2010;24(1):42–51.

58. Mudie MH, Matyas TA. Can simultaneous bilateral movement involve the undamaged hemisphere in reconstruction of neural networks damaged by stroke? Disabil Rehabil. 2000;22(1–2):23–37.

59. Jung R, Dietz V. Delayed initiation of voluntary movements after pyramidal lesions in man (author's transl). Arch Psychiatr Nervenkr. 1975;221(2):87–109.

60. Carr LJ, Harrison LM, Evans AL, Stephens JA. Patterns of central motor reorganization in hemiplegic cerebral palsy. Brain. 1993;116(Pt 5):1223–47.

61. Mayston MJ, Harrison LM, Stephens JA. A neurophysiological study of mirror movements in adults and children. Ann Neurol. 1999;45(5):583–94.

62. Buma F, Kwakkel G, Ramsey N. Understanding upper limb recovery after stroke. Restor Neurol Neurosci. 2013;31(6):707–22.

63. Sleimen-Malkoun R, Temprado JJ, Berton E. A dynamic systems approach to bimanual coordination in stroke: implications for rehabilitation and research. Medicina. 2010;46(6):374–81.

64. Laffont I, Bakhti K, Coroian F, van Dokkum L, Mottet D, Schweighofer N, et al. Innovative technologies applied to sensorimotor rehabilitation after stroke. Ann Phys Rehabil Med. 2014;57(8):543–51.

65. Lum PS, Burgar CG, Shor PC, Majmundar M, Van der Loos M. Robot-assisted movement training compared with conventional therapy techniques for the rehabilitation of upper-limb motor function after stroke. Arch Phys Med Rehabil. 2002;83:952–9.

66. Klamroth-Marganska V, Blanco J, Campen K, Curt A, Dietz V, Ettlin T, et al. Three-dimensional, task-specific robot therapy of the arm after stroke: a multicentre, parallel-group randomised trial. Lancet Neurol. 2014;13(2):159–66.

67. Sale P, Lombardi V, Franceschini M. Hand robotics rehabilitation: feasibility and preliminary results of a robotic treatment in patients with hemiparesis. Stroke Res Treat. 2012;2012: 820931.

68. Guidali M, Duschau-Wicke A, Broggi S, Klamroth-Marganska V, Nef T, Riener R. A robotic system to train activities of daily living in a virtual environment. Med Biol Eng Comput. 2011;49(10):1213–23.

69. Ang KK, Chua KS, Phua KS, Wang C, Chin ZY, Kuah CW, et al. A Randomized Controlled Trial of EEG-Based Motor Imagery Brain-Computer Interface Robotic Rehabilitation for Stroke. Clin EEG Neurosci. 2014.

70. Formaggio E, Storti SF, Boscolo Galazzo I, Gandolfi M, Geroin C, Smania N, et al. Modulation of event-related desynchronization in robot-assisted hand performance: brain oscillatory changes in active, passive and imagined movements. J Neuroeng Rehabil. 2013;10:24.

71. Lum PS, Burgar CG, Shor PC. Evidence for improved muscle activation patterns after retraining of reaching movements with the MIME robotic system in subjects with post-stroke hemiparesis. IEEE Trans Neural Syst Rehabil Eng. 2004;12(2):186–94.

72. Basteris A, Nijenhuis SM, Stienen AH, Buurke JH, Prange GB, Amirabdollahian F. Training modalities in robot-mediated upper limb rehabilitation in stroke: a framework for classification based on a systematic review. J Neuroeng Rehabil. 2014;11:111.

Robotic Gait Training in Specific Neurological Conditions: Rationale and Application

8

Markus Wirz, Jens Bansi,
Marianne Capecci, Alberto Esquenazi,
Liliana Paredes, Candy Tefertiller,
and Hubertus J. A. van Hedel

Abstract

This chapter focuses on robotic gait training. As a basis, it summarizes the neurophysiological rationale for such training. These neurophysiological findings are mostly based on animal studies. The observations from these studies led to the development of theories such as the spinal central pattern generator (CPG). In a deductive manner, studies have then also been performed on human participants showing similar phenomena. Based on the neurophysiological mechanisms, robot-assisted locomotor training is justified even in patients with severe functional limitations. Those patients would not be able to maintain an upright posture while performing stepping movements in a conventional training condition. The clinical application of robotic gait training is, therefore, another focus of this chapter. Different robotic devices and their relevant characteristics are introduced. Not all devices are designed to purely assist locomotor training by maximally exploiting the neuroplastic potential of the central nervous system. Some devices can also be considered assistive technologies which support patients in their daily life mobility. General aspects which are relevant for robotic gait training during rehabilitation are summarized. Then the chapter takes the translation of the fundamental principles one step further and addresses the application of robotic gait training in specific neurological conditions, that

M. Wirz (✉)
School of Health Sciences, Institute of
Physiotherapy, ZHAW Zurich University of Applied
Sciences, Katharina-Sulzer-Platz 9, 8401 Winterthur,
Switzerland
e-mail: Markus.Wirz@zhaw.ch

J. Bansi
Rehabilitation Centre Valens, Taminaplatz 1, 7317
Valens, Switzerland
e-mail: Jens.Bansi@kliniken-valens.ch

M. Capecci
Department of Experimental and Clinical Medicine,
Neurorehabilitation Clinic, University Hospital
Riuniti Di Ancona, "Politecnica Delle Marche"
University, Ancona, Italy
e-mail: M.Capecci@staff.univpm.it

A. Esquenazi
Department of Physical Medicine and Rehabilitation,
MossRehab, Elkins Park, PA, USA
e-mail: Aesquena@einstein.edu

L. Paredes
Center for Neurological Rehabilitation, Hauptstrasse
2, 8588 Zihlschlacht, Switzerland
e-mail: L.Paredes@rehaklinik-zihlschlacht.ch

C. Tefertiller
Department of Physical Therapy, Craig Hospital,
Englewood, CO, USA
e-mail: CTefertiller@craighospital.org

H. J. A. van Hedel
University-Children's Hospital Zurich, Swiss
Children's Rehab, Muehlebergstrasse 104, CH8910
Affoltern am Albis, Switzerland
e-mail: Hubertus.vanhedel@kispi.uzh.ch

© The Author(s), under exclusive license to Springer Nature Switzerland AG 2022
D. J. Reinkensmeyer et al. (eds.), *Neurorehabilitation Technology*,
https://doi.org/10.1007/978-3-031-08995-4_8

145

is, in stroke, traumatic brain injury (TBI), spinal cord injury (SCI), multiple sclerosis (MS), and Parkinson's disease. These parts are written by experts in their respective fields. Every section follows the same structure and informs the reader about the condition in general, specific gait limitations, and the rehabilitation thereof. If available, the latter is supported by recent high-level evidence. In the other cases, relevant primary studies are summarized. Because specific clinical guidelines for the application of robotic gait training are largely missing, the experts provide information about the most important clinical aspects from their perspective. This information encompasses indications, devices, training parameters and duration, assessments, and potential adverse events. Therefore, this chapter will help clinicians who consider introducing robotic gait training how to shape a training and assessment program for their patients.

Keywords

Stroke · Spinal cord injury · Traumatic brain injury · Multiple sclerosis · Parkinson's disease · Central pattern generator · Locomotor training · Rehabilitation robotics · Robot-assisted gait training · Physiological prerequisites

8.1 General Introduction

After a damage to the central nervous system (CNS), such as stroke or spinal cord injury (SCI), walking function can be severely impaired, which can limit independence in daily life activities and restrict participation in the community. Indeed, the loss of the ability to walk represents a major disability for subjects with SCI or stroke [1, 2]. Almost two-thirds of all stroke survivors cannot walk without assistance in the acute phase following the incident [3].

Robot-assisted gait training (RAGT) has been introduced during the last decades to retrain gait complementing the conventional active physiotherapeutic program. RAGT evolved from manually assisted bodyweight-supported treadmill training (BWSTT). Both have in common that patients walk on a treadmill while being partially bodyweight-supported [4–9]. Depending on the ability to perform repetitive stepping on the treadmill, during manually assisted BWSTT, one or two therapists assisted in performing the leg movements throughout the gait cycle. Although an improvement in locomotor function was achieved following manually assisted treadmill training, its practical implementation in the clinical setting was limited by the labor-intensive nature of the approach. In patients with SCI, usually two therapists had to assist in leg movements, i.e., one therapist per leg [10]. Particularly, in individuals with severe motor deficits and/or a high degree of spasticity, appropriate manual assistance proved difficult to provide over longer training times, as it was very strenuous for the therapists. Therefore, a major limitation of manual-assisted BWSTT was that the duration and intensity of the training sessions were limited by the physical abilities of the therapist (rather than the patient), which could negatively impact the effectiveness of the therapeutic approach. The success and promise of BWSTT and the limitations and resource constraints in the therapeutic settings have inspired the design and development of robotic devices to improve the rehabilitation of ambulation in patients following stroke, SCI, traumatic brain injury (TBI), multiple sclerosis (MS), and M. Parkinson [11, 12], which will be discussed later in this chapter.

Interestingly, BWSTT and the consecutive RAGT interventions are based on findings derived from years of animal and subsequently human research investigating the neurophysiological basis of locomotion. These interventions are excellent examples of how findings initially derived from basic animal research have been translated to the human condition (i.e., from "bench" to "bedside"). Nowadays, these therapy options are widely accepted and have become evidence-based therapeutic applications in the multidisciplinary neurorehabilitation programs that such patients undergo.

The current chapter focuses on the neurorehabilitation of locomotor function. First, it summarizes the neuroscientific rationale for RAGT, followed by a general introduction to the application of rehabilitation technologies for locomotion. Consecutively, this chapter provides more detailed diagnosis-specific paragraphs on the clinical application written by clinical experts who provide these interventions on a daily basis. These paragraphs include a description of the clinical presentation including epidemiological numbers, specific diagnosis-related walking impairments, the time-course of recovery, the scientific evidence available to date on the effectiveness of RAGT, and finally, a more practically relevant paragraph on diagnosis-specific issues concerning the practical application.

8.2 The Neurophysiology Underlying Locomotion

8.2.1 Introduction

The control of locomotion is distributed throughout the CNS and includes structures from within the spinal cord up to the cerebellum and cortex. The very basis of the locomotor system, however, is located within the spinal cord, where inter-neuronal spinal circuits can generate a basic flexor–extensor stepping output pattern. For the lower limb, these neural circuits are distributed along the entire lumbosacral spinal cord. The general pattern of motor coordination during walking is ancient and seems to be shared not only by rodents, cats, and humans but also birds, which suggests that it had evolved already in reptiles [13]. Thomas Graham Brown (1882–1965) was the first who proposed the view that a stepping output pattern, whose timing does not depend upon descending or sensory inputs, can be generated in mammals by the intrinsic capability of the spinal cord (see [14]). Based on his experiments in cats and guinea pigs, Brown proposed mutually inhibitory connections between a pair of intrinsically active flexor and extensor "half-centers" on each side of the spinal cord. Nowadays, we name this locomotor network of neurons coordinating the pattern of muscle activation in each step cycle, a central pattern generator network (CPG). The term "fictive locomotion" refers to the stepping-like rhythmic activity recorded in ventral roots or peripheral nerves in paralyzed and deafferented animals, i.e., in the absence of sensory information. The models for the CPG have become more complex over time (for recent reviews, see, e.g., [15, 16]). This is understandable because (human) locomotion is very complex to control. On the one hand, locomotion can be varied in many ways, for example, walking forwards, backwards, and sideways, and it is likely that these variations are controlled by the same CPG (see, e.g., [17]). On the other hand, the CPG needs to control different muscle groups in a very precise manner.

So, modulation of CPG activity is needed, for example, to adapt the walking pattern to different environmental conditions. The CPG receives inputs from peripheral mechanoreceptors located in muscles, tendons, joints, and skin, collectively termed somatosensory feedback, as well as from various supraspinal structures [18]. The sensory inputs to the CPG are of critical importance for normal locomotion [16, 19]. Particularly, the information from hip joint and load receptors [20, 21] has received attention. Afferent information from the hip flexors and ankle extensors is important for the transition from the stance to the swing phase in cats [22]. Furthermore, skin mechanoreceptors of the paw provide sensory input needed to position the paws, which is particularly important during complex high-precision locomotor tasks, such as ladder walking in the cat [23]. Cutaneous input from the plantar surface of the paw reinforces extensor activity in decerebrated cats walking on a treadmill [24]. Based on findings from the literature, it is assumed that group II hip flexor afferents, group I ankle extensor afferents, and low-threshold cutaneous afferents from the cat paw have direct access to spinal locomotor circuitries and can reset or entrain fictive locomotion in adult decerebrated cats [25].

While CPG investigations have merely been performed in animal models, there are strong indications that the CPG is also involved in human locomotion (see, e.g., [26, 27]). These observations include stepping in newborns [28] and prenatal coordinated whole-body movements [29], as well as various observations in individuals with a motor-complete SCI, for example, sleep-related rhythmic leg movements (e.g., [30]), spinal cord stimulation-induced coordinated leg movements [31], or vibration-induced air stepping [32]. Unfortunately, the amplitude of leg muscle EMG activity in human patients with a motor-complete SCI is small compared to healthy individuals, although an increase in amplitude during the course of locomotor training can be observed [20]. This reduction in EMG amplitudes is probably caused by a loss of input from descending noradrenergic pathways to spinal locomotor centers [3].

Indeed, especially in humans, supraspinal input is required for walking. Earlier investigations in the cat had already shown that substances mimicking the action of long descending pathways such as clonidine resulted in distinct and consistent alternating bursts of electromyographic activity inducing spinal stepping [33]. Both the CPG and the reflex mechanisms that mediate afferent input to the spinal cord are under the control of the brainstem ([34]; for review see [20]). While the cerebellum fine-tunes the movements according to the needs of the task and may modulate the step cycle to alter step patterns, the frontal cortex and basal ganglia are expected to play a role in controlling gait during rapid changes in environmental conditions. Indeed, imaging studies in humans showed that various parts of the cortex become activated during walking. In their excellent systematic review, Hamacher et al. [35] summarize the findings regarding brain activity during gait. In line with La Fougere et al. [36] and Zwergal et al. [37], they distinguish between a direct locomotor pathway that guides locomotion via the primary motor cortex (M1), cerebellum, and spinal cord, and an indirect pathway including the prefrontal cortex, premotor cortex, and supplementary motor area, and basal ganglia. They report that goal-directed locomotion is associated with activations within the indirect locomotor pathway. Also, performing more complex locomotor tasks requires higher activities in certain (sub-) cortical such as the motor cortex, parietal lobule, thalamus, and basal ganglia [38] as well as the frontal and occipital lobule [39–41].

8.2.2 Training the Spinal Circuitry in Animals and Humans

There is convincing evidence from research on spinalized animals that a use-dependent plasticity of the spinal cord exists [42, 43]. When stepping is practiced in a spinal cat, this task can be performed more successfully than when it is not practiced [44, 45]. Thus recovery of locomotor function following spinal cord transection can be improved using regular training even in adult animals [4]. In contrast, the cat loses the ability to step spontaneously if it is not regularly performed. During such a locomotor training, the animal is supported (i.e., comparable to the partial bodyweight unloading during BWSTT or RAGT in human patients). Locomotor movements of the hind limbs are induced by a moving treadmill while the forelimbs stand on a platform. With ongoing training, body support can be decreased, associated with improving locomotor abilities. Later on, the cat can completely take over its body weight and perform well-coordinated stepping movements [45]. Furthermore, after hind limb exercise in adult rats, the excitability of spinal reflexes becomes normalized [46]. Stepping movements can also be released in a monkey after transection of the spinal cord, suggesting that the isolated primate spinal cord is capable of generating hind limb stepping movements [47, 48]. To successfully train the spinal circuitry in conditions of lacking or drastically reduced supraspinal input, two factors from peripheral receptors appear particularly important.

8.2.2.1 Role of Cyclic Body Unloading and Loading

The spinal central pattern generator must be activated by providing appropriate afferent input and proprioceptive feedback to induce plastic neuronal changes [20, 49]. Body unloading and reloading are considered crucial to inducing training effects on the spinal locomotor centers because the afferent input from receptors signaling contact forces during the stance phase (corresponding to the initiation of newborn stepping by foot sole contact, see above) is essential to activate spinal neuronal circuits underlying locomotion [50]. Therefore, a cyclic loading is important for achieving training effects in cats [51] and humans [20, 52]. Overall, observations of healthy subjects [51, 53], small children [54], and patients with paraplegia [55, 56] indicate that afferent input from load receptors essentially contributes to the activation pattern of leg muscles during locomotion. This suggests that proprioceptive inputs from extensor muscles and probably also from mechanoreceptors in the foot sole provide load information [57]. In addition, similar to the cat [58], afferent input from muscles that act around the hip is important for the leg muscle activation during locomotion [20]. This afferent activity from load and hip joint receptors is required to shape the locomotor pattern, control phase transitions, and reinforce ongoing activity (Fig. 8.1). Short-latency stretch and cutaneous reflexes may be involved in the compensation of small irregularities and in the adaptation to the actual ground conditions.

8.2.3 Spastic Muscle Tone

In individuals with stroke and particularly SCI, spastic muscle tone must be present as a partial compensation for paresis [60]. For a patient with spasticity, a low-amplitude, tonic activation of the lower limb muscles takes place during locomotor movements, i.e., a normal modulation of

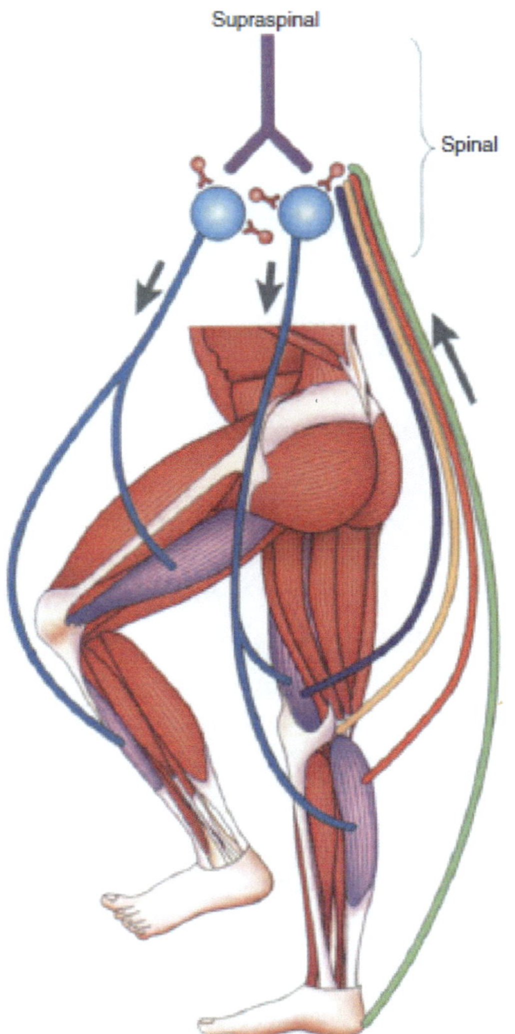

Fig. 8.1 Schematic drawing of the neuronal mechanisms involved in human gait. Leg muscles become activated by a programmed pattern that is generated in spinal neuronal circuits. This pattern is modulated by multisensory afferent input that adapts the pattern to meet existing requirements. Both the programmed pattern and the reflex mechanisms are normally under supraspinal control. In addition, there is differential neuronal control of leg extensor and flexor muscles. While extensors are primarily activated by proprioceptive feedback, the flexors are predominantly under supraspinal control (adapted with kind permission from [59]). To evoke a locomotor pattern in individuals with a complete spinal cord injury, where supraspinal control is interrupted, load and hip joint-related afferent input seem crucially important

EMG activity is lacking while the timing of muscle activity is preserved [61, 62]. The amplitude reduction of leg muscle activity is suggested to be due to a diminished excitatory drive from supraspinal centers and an attenuated activity of polysynaptic (or long-latency) reflexes [59, 63]. Polysynaptic reflexes are known to modulate leg extensor muscle activity [59] and thereby adapt the movement pattern to the environmental requirements. In contrast, short-latency reflexes neither in healthy subjects nor in patients with spasticity contribute significantly to muscle activity during natural movements [60]. These observations indicate that the muscle tone required during a movement (e.g., to support the body during the stance phase of stepping) after CNS damage develops on a lower level of organization [60].

Initially, after a severe stroke or SCI, the muscle force produced by the leg muscle activation (small EMG amplitude) is likely insufficient to support the body during walking. Therefore, partial bodyweight unloading is necessary to enable stable stepping movements. During the course of daily locomotor training, the amplitude of leg extensor EMG activity increases during the stance phase, while an inappropriate tibialis anterior activation decreases [20, 52]. This is associated with a greater weight-bearing function of the leg extensors, i.e., body unloading during treadmill locomotion can be reduced. These training effects are seen in both patients with incomplete and complete paraplegia. However, only individuals with incomplete paraplegia benefit from the training program insofar as they learn to perform stepping movements on solid ground.

In summary, a locomotor pattern can be induced and trained in individuals without supraspinal control of spinal circuits by having patients walk with partial bodyweight on a treadmill while their leg movements are assisted by a robotic device. A considerable degree of locomotor recovery in mammals with SCI can be attributed to a reorganization of spared neural pathways ([64, 65]; for review, see [66]). It has been estimated that sparing of 25% of the lateral or ventral white matter in non-human primates permitted walking [67] and even less needs to be preserved in rats [68, 69] to recover some locomotor function. It can be concluded that assisted training represents an important factor in the recovery of locomotor function by exploiting the neural plasticity of various structures within the CNS, including the spinal cord. Furthermore, these finding increases the probability that one day, experimental interventions that might be able to regenerate a (small) proportion of damaged descending axons could result in an improved stepping capacity in individuals with disturbed supraspinal control of spinal circuits [70].

8.3 Clinical Application of Robotics and Technology in the Restoration of Walking

The following sections comprise general aspects of the clinical application of robots for the training of locomotor function. It starts with an overview of the designs of robotic devices for locomotor training or mobility. More specifically, some of the devices become applied either for therapeutic purposes, that is for a temporary period, e.g., during rehabilitation with the aim to improve motor function, or as assistive devices, which support the user in his or her daily mobility by compensating for the loss of function. For some devices, the purpose is not as clearly discernible. The subsequent text addresses aspects that are valid irrespective of the underlying pathology.

8.3.1 Robotic Devices

8.3.1.1 For Locomotor Training

Robotic devices for locomotor training in rehabilitation centers must be adaptable to cover patients showing a wide range of anthropometrics and abilities regarding voluntary motor control. Typically, these robots are either designed as exoskeletons or as end-effector devices. This book contains chapters specifically dedicated to the design of rehabilitation robots. In short, exoskeletons comprise segments and joints corresponding to the human body. Each joint is equipped with movement and force sensors and actuated separately. This allows accurate measurements of the mechanical forces acting in each joint, e.g., when patients actively perform walking movements. It also allows to individually gauge the supporting or resistance forces produced by the actuators. With exoskeletons, it takes longer to adapt the device to an individual patient.

End-effector devices comprise mainly support for the most distal segment, i.e., for walking devices, the foot. This footplate moves the foot along a defined movement trajectory. This means that throughout the gait cycle, the footplate acts as a manipulator of the foot during the swing phase and as the floor during the stance phase. It seems that the unloading during the swing phase is not as clear as with exoskeletons. Unloading and loading are known to be pivotal for the activation of central pattern generators. However, the literature indicates that both types of robotic assistance are equally effective from a clinical perspective [71]. An advantage of end-effector devices is that the set-up time is shorter. However, there is the disadvantage that adjacent joints are not controllable. For example, in an end-effector device, the knee joints might be bent during the stance phase, not allowing proper weight-bearing. Exoskeletons and end-effector devices used in rehabilitation clinics are stationary. The so-called tethered exoskeletons are typically mounted over a treadmill. End-effector devices do not need a treadmill, as the footplates simulate the level floor. In addition, both types of devices have mechanisms to partially unload the bodyweight of the patients. Another common feature of both types of devices is a form of augmented feedback. Training performance is provided to the patient and therapist by means of visual, acoustic, or haptic information.

Recently, mobile devices also become increasingly applied during inpatient rehabilitation [72]. Typically, these devices are designed as exoskeletons either supporting the trunk and legs or only single body segments, e.g., the ankle.

Although clear application regimens are still lacking, robotic devices have become established in neurorehabilitation, at least in highly resourced countries [73]. In some clinics, supporting interventions become applied concomitant with the robotic training. These add-ons cannot be considered standard but are still part of experimental evaluations. Examples are functional electrical stimulation (FES) of leg muscles, repetitive transcranial magnetic stimulation of the brain (rTMS), spinal cord stimulation, or virtual or augmented reality environments (VR, AR). Details about some of these applications can be found in the respective chapters of this book.

8.3.1.2 For Daily Life Mobility

As opposed to devices used to train locomotor function within a certain period of rehabilitation, there are also devices designed to assist daily life mobility. Ultimately, these devices strive to allow bipedal locomotion in patients experiencing severe functional limitations and thus replace wheelchairs. However, as of today, the advantage of the mechanically simple but effective construction of a wheelchair still prevails [74, 75]. Also, in the light that the built environment takes wheeled mobility more and more into account. That is the number of places that are accessible with a wheelchair increases. Nonetheless, there seems to be more value to be mobile on two legs than can be explained by pure accessibility [76, 77]. Powered exoskeletons created the opportunity to become mobile outside a clinical setting. This development of exoskeletons for personal use can be observed especially for persons with SCI [78]. The design of the devices regards the extent of the functional limitation. Rigid-powered exoskeletons allow bipedal mobility for persons with complete paraplegic SCI [79]. Soft exoskeletons are appropriate for persons with residual but limited walking function [80]. However, the use of overground exoskeletons is still limited to a subset of persons with SCI. In summary, overground exoskeletons seem to fulfill two main purposes, mobility, i.e., to move from one location to another, and regular physical activity in order to prevent inactivity-related conditions [81].

8.3.2 Why Robotic Locomotor Training?

The prognosis to regain ambulatory function after a neurological event is quite substantial but varies depending on the underlying pathology. Diagnosis-specific information can be found in the following sections of this chapter. Accordingly, walking is one of the major goals of neurological rehabilitation. The crucial meaning of the achievable mobility status is the fact that it is strongly correlated with the discharge destination, independence, and general health [82]. However, whether a patient will regain ambulatory function is not always clear shortly after a neurologic event. Hence, valid and reliable assessments, biomarkers, or algorithms for the prognosis are important to set appropriate rehabilitation goals, especially for those cases where the outcome is not obvious. In these situations, the functional limitation is more pronounced in the early phase of rehabilitation and resolves as a function of time and training [83]. For those patients, locomotor training with the use of robotic devices is most appropriate because these devices allow training of a physiological gait pattern which wouldn't be possible otherwise because of the high amount of required support. It is, however, important to adapt the amount of support throughout the training series to the current stage of functional abilities of the patient performance. Otherwise, patients might not be challenged sufficiently because too much support does not facilitate self-produced movements. Studies in the stroke population have also shown that a combination of robotic training and conventional methods is more effective [71].

8.3.3 Evolution of Motor Abilities During Rehabilitation

Rehabilitation after a neurological event includes the training of gross motor functions. The term refers to global movements of the whole body by coordinated activations of large muscle groups. Mastering gross motor functions is essential for the reestablishment of functional task execution, including standing and walking. The hierarchical sequence of the gross motor functions is mapped in functional tests, e.g., the Motor Assessment Scale (MAS) [84]. According to the MAS, task complexity starts from "supine to side-lying" followed by "supine to sitting over the edge of a bed", "balanced sitting", "sitting to standing", "walking", "upper-arm function", "hand movements", and "advanced hand activities". Within these stages, multiple tasks must be trained to become skilled in that given stage, e.g., walking initiation or stopping. The training of locomotor function with the use of robotic devices has to be meaningfully embedded into this evolution of motor abilities.

8.3.4 Effects of Locomotor Training

8.3.4.1 Specific Effects

Obviously, locomotor training aims specifically at improving ambulatory function. Effects can be achieved at various levels, including the neuromotor and musculoskeletal systems, as well as circulatory tolerance for the upright position. Clinically, ambulatory performance can be measured regarding the required assistance, e.g., using the Functional Ambulation Categories (FAC) [85], or capacity, e.g., with timed walking tests such as the Ten-Meter Walk Test or the Six-Minute Walk Test [86]. It is advisable to use these measures for goal setting and to collect them on a regular basis.

8.3.4.2 Unspecific Effects

Besides the specific effects of robot-assisted locomotor training, participants of such training also report unspecific effects, i.e., not directly related to walking function. These outcomes encompass, for example, improved cardiovascular function [87], or in persons with SCI, improved bowel function [88], and less pain and spasticity [89]. Specifically, persons who depend on a wheelchair for daily mobility and are thus forced to a sedentary lifestyle might profit from a powered exoskeleton to become physically active on a regular basis [90]. The health-enhancing effects of regular physical activity are very well established and resulted in the World Health Organizations' guideline on physical activity and sedentary behavior [91].

8.4 Specific Neurological Conditions

8.4.1 Introduction

After the introduction to the general aspects of locomotor training, in the subsequent sections, experts provide diagnosis-specific information pertaining to robotic-assisted gait training. The respective neurological conditions encompass stroke, traumatic brain injury (TBI), spinal cord injury (SCI), multiple sclerosis (MS), and Parkinson's disease.

All diagnosis-specific sections are organized along the following structure and should support clinicians in configuring their robotic-assisted interventions.

Introduction related to the condition: In this paragraph the neurological conditions are described pertaining to their etiology, clinical presentation, and prognosis. It also comprises the most important epidemiological numbers.

Walking-related introduction: Information about the diagnoses-specific impairments affecting the walking function of the respective neurological condition, including their frequency, extent, and prognosis, as well as the usual treatment or rehabilitation path, are introduced in this part.

Evidence for robotic locomotor training: This paragraph focuses on published evidence regarding the clinical effectiveness of robotic locomotor training. Priority is given to clinical guidelines, systematic reviews/meta-analyses, or primary studies if no source with a higher evidence grade was available at the time.

Practical application of robotic locomotor training: Because the practical application is often not described in detail, the experts share their clinical experience [73]. These parts can hence be considered as best-practice examples.

The start of robotic-assisted locomotor training should be based on a reliable indication in conjunction with achievable goals. To allow comparisons across the different neurological conditions, all experts characterized indications and goals. Before robotic training starts, a structured examination should be performed in order to rule out contraindications and to obtain medical clearance.

Robotic training is typically applied along with other interventions during inpatient rehabilitation. The section about the practical application provides information about other means of locomotor training which optimally complement the robotic intervention.

Information will be provided about the planning and duration of training series, i.e., when to start the training in relation to the neurologic event or in relation to rehabilitation milestones and the number of weekly training sessions.

The section will then focus on single training sessions. This will be repeated for all devices which are typically applied in each condition. Practical information will be provided, e.g., training parameters like duration, bodyweight support, walking speed, guidance force, or how to include feedback modalities. In addition, the required number of staff and the mode of supervision and monitoring of the training will be described.

Since robotic training typically lasts longer than a conventional therapy session, motivational aspects will also be explained. The challenge is to shape the training in a way that not the movement is repeated but the movement task.

Another aspect of the practical application is the assessment of effectiveness. The experts state when certain assessment instruments are applied.

Although robotic locomotor training is associated with only minor risks, it is important to consider these adverse events (AE). Depending on the condition, different AEs might be more frequent, e.g., pathological muscle tone, lesions to the skin, pain, over-strain, etc. The section will include how to assess AE, criteria to stop a single training, pause, or even stop the training series.

8.4.2 Stroke

8.4.2.1 Introduction Stroke

Stroke is "classically characterized as a neurological deficit attributed to an acute focal injury of the central nervous system (CNS) by a vascular cause, including cerebral infarction, intracerebral hemorrhage (ICH), and subarachnoid hemorrhage (SAH), and is a major cause of disability and death worldwide" [92]. Ischemic stroke can be classified according to the TOAST classification system in (1) large-artery atherosclerosis, (2) cardioembolism, (3) small-artery occlusion (lacune), (4) stroke of other determined etiology such as hypercoagulable states, hematological disorders, and (5) stroke of undetermined etiology (no cause can be found) [93]. In Caucasian populations, approximately 80% of all strokes are ischemic, 10–15% intracerebral hemorrhage (ICH), 5% subarachnoid hemorrhage (SAH), and the rest are due to other causes of stroke [94]. Asian studies report a higher incidence of ICH compared to the numbers in Caucasians, with approximately 20–30% being hemorrhagic.

A combination of large-artery diseases and cardioembolic strokes constitutes the most significant proportion of ischemic strokes. The average age of patients affected by stroke is 70 years in men and 75 years in women. The overall case-fatality within one month of stroke onset is about 23%. The case-fatality is higher for intracerebral hemorrhage (42%) and subarachnoid hemorrhage (32%) than for ischemic stroke (16%). While most studies on stroke incidence, case-fatality, and prevalence included Caucasian people, the WHO has reported a considerable geographical variation in both mortality and case-fatality [95].

The location of the brain lesion can be identified by brain CT or MRI and grouped by anatomical locations as done by Pan and colleagues. They classified the regions into the basal ganglion (BG), the corona radiata (CR), the region of the anterior cerebral artery (ACA), the region of the middle cerebral artery (MCA), the brain stem (BS), the region of the posterior cerebral artery (PCA), and the thalamus (THA). They found that infarct in the PCA territory may result in less motor deficit and better outcome in activities of daily living (ADL) than infarct in other sites. The PCA territory is not in the vicinity of the motor cortex. If brain injuries are involved in BG or internal capsule area, patients may be more likely to have severe sensorimotor deficits since the major sensorimotor neural pathways (e.g., corticospinal tract) congregate in this location. It might also be that subcortical infarct lesions have a lower potential to reorganize than cortical lesions [96, 97]. The type of cerebral infarction can also be classified according to the Oxford Community Stroke Project (OCSP) as total anterior circulation infarcts (TACI), partial anterior circulation infarcts (PACI), posterior circulation infarcts (POCI), and lacunar anterior circulation infarcts (LACI). An anterior circulation infarct refers to a cortical stroke occurring in both the deep and superficial areas of the MCA/ACA. Posterior circulation infarction corresponds to any infarction occurring within the vertebrobasilar vascular territory, which includes the brainstem, cerebellum, midbrain, thalami, and areas of temporal and occipital lobes. Lacunar infarcts are infarcts confined to the region of the deep small perforating arteries [98]. A larger infarct size indicates more significant neural damage, leading to a greater neurological deficit and worse outcome [96]. Bamford et al. [98] found that patients in the TACI group had a negligible chance of good functional outcome and mortality was high. Patients in the POCI group were at greater risk of a recurrent stroke later in the first year after the index event but had the best chance of a good functional outcome. Despite the small anatomical size of the infarcts in the LACI group, many patients remained with substantial functional limitations.

Multiple prospective epidemiological studies have found that approx. 60% of all individuals with stroke will regain independence in basic activities of daily living (ADL) within six months post-injury [99]. Approximately 14% of these stroke survivors achieve full recovery in their basic ADLs, between 25 and 50% require at least some assistance in ADLs, and approximately half experience severe long-term dependency [99, 100].

8.4.2.2 Gait in Persons with Stroke

Multiple longitudinal cohort studies have shown that approx. 60–80% of patients with stroke are able to walk independently at six months post-stroke [101]. The prognosis for the recovery of walking is therefore favorable. Although recovery of walking function mainly occurs within the first six months after stroke [101, 102], about 10% of the patients will continue to show significant functional changes after this period [103, 104]. Three months after stroke, approximately 85% of the patients walk at a reduced gait speed and capacity [105]. Although the majority of stroke survivors learn to walk independently by six months after stroke, gait disabilities persist through the chronic stages. Walking endurance, as measured by the distance walked in 6 min (Six-Minute Walk Test), remains the most striking area of difficulty among individuals with chronic stroke [105].

For patients who can initially not walk independently after a first-ever anterior circulation stroke, recovery of independent walking ability can be predicted during the first two to nine days after stroke. Valuable predictors include the early assessment of sitting balance with the Trunk Control Test and motor function of the paretic leg with the Motricity Index (MI) or Fugl-Meyer lower extremity motor score. Patients assessed within the first 72 h post-stroke who were able to sit independently on the edge of the bed with the feet off the ground for at least 30 s and had a MI leg score of 25 points or more or 19 points or more on the motor part of the Fugl-Meyer Assessment for the lower extremity had about a 98% chance of regaining independent gait within six months. In comparison, patients who were unable to regain sitting balance and were not or hardly able to contract the muscles of the paretic lower limb within 72 h post-stroke had a probability of about 27% of achieving independent gait [101]. The probability of regaining independent gait dropped to 23% on day five and 10% on day nine post-stroke if sitting balance and lower limb strength did not recover [101].

Further factors such as initially reasonable activities of daily living (ADL) skills, younger age, absence of homonymous hemianopia, visuospatial attention, urinary continence, level of consciousness at admission, and the absence of premorbid limitations of walking ability and ADLs can also influence regaining of independent walking ability [101, 106]. Therefore, in patients with a less favorable prognosis of independent walking ability, it is recommended to administer these determinants weekly during the first month, thereafter monthly for up to six months, and biannually if recovery after six months is still incomplete. These evaluations are needed to monitor progress, inform clients and relatives properly, redefine realistic treatment goals multidisciplinary, facilitate the planning of resources in inpatient and outpatient settings, and anticipate home and community adjustments [101].

Ambulatory individuals, after rehabilitation discharge, considered the ability to walk in the community to be essential or very important [107]. Although most patients with stroke are able to walk one-year post-stroke, many walk slowly, i.e., at speeds of 0.38–0.80 m/s, and hesitantly or dependently. At the very least, this means that they cannot walk fast enough to cross the road, while at worst, they are unable to leave the house [108]. Not being able to walk at a speed necessary to walk safely outside hinders daily life independence, social (re)integration, and the ability to participate in society.

The debilitating motor consequences of a stroke can markedly decrease mechanical efficiency and increase the energy expenditure of walking up to two times the expenditure of healthy persons. Housekeeping tasks such as making the bed or vacuuming require substantially more energy expenditure in individuals with stroke than among their healthy counterparts [100]. Objective activity monitoring of stroke survivors has shown that they spent more than 80% of their time sedentary in the first year post-stroke [109] and even stroke survivors with mild motor impairments do not meet recommended levels of physical activity [110]. Indeed, long-term gait impairments frequently lead to the sedentary behavior in stroke survivors [111]. This is alarming given that physical inactivity after stroke contributes to cardiovascular and metabolic de-conditioning, muscle weakness, gait impairment, and related declines in physical and social functioning, and an increased risk of cardiovascular problems. In addition, patterns of non-use/inactivity negatively affect brain activation and recovery. Moreover, approximately 30% of individuals with stroke are at risk of developing a second stroke, and cardiovascular diseases are the leading cause of death in patients with a chronic stroke [100].

Comprehensive programs with adequate resources, therapy dose, and duration are essential in post-acute care and rehabilitation. Stroke rehabilitation requires a sustained and coordinated effort from a large interdisciplinary team, including physicians and therapists taking into account the patient and his/her goals, family, friends, and other caregivers. Isolated efforts are unlikely to achieve the full patient's recovery potential [112]. In the outpatient setting, at home, or in the community, activities that promote mobility and physical activity are necessary to improve quality of life and prevent further complications such as falls resulting in fracture or cardiovascular problems due to sedentary behavior [113].

Restoration of walking ability and gait rehabilitation are highly relevant for people who are unable to walk independently after stroke as well as for family members. The brain has the capability to reorganize its structure, functions, and connections due to tissue damage after stroke through plasticity. This ability can be magnified by intensive, repetitive, active, and task-oriented training. The amount, challenge, and timing of training determine its effectiveness [114, 115]. This was one of the reasons why (manual assisted) bodyweight-supported treadmill training was introduced to complement the conventional gait rehabilitation program. Patients step with or without partial bodyweight unloading on a treadmill, while for those unable to move their legs themselves, therapists provide assistance, which is relatively strenuous.

In a recent systematic review update [71], Mehrholz et al. found that the use of treadmill training with or without BWS did not increase the chance of becoming an independent walker compared with people after stroke receiving other physiotherapy interventions without treadmill training. The ability to walk was measured with the Functional Ambulatory Category (FAC). Walking speed and capacity measured with the 10 MWT (Ten-Meter Walk Test) and Six-Minute Walk Test improved significantly for people in the first three months after stroke who were able to walk independently at training onset (FAC > 2), only the improvement in walking capacity was clinically relevant. Patients after stroke who were initially not able to walk nearly independently showed no additional benefit of treadmill training. For people treated after three months post-stroke, the effects were lower (and not clinically relevant). While treadmill training with or without BWS was safe and acceptable for most patients, it seemed that training more intensively (five times per week versus less than three times per week) might produce greater effects, but the difference was statistically not significant. The beneficial effect for the group of independent walkers was not persistent at follow-up as measured by studies between three weeks and 12 months after the end of the intervention. The available data was very heterogeneous.

8.4.2.3 Evidence for Robotic Gait Training in Persons with Stroke

To overcome some of the limitations of conventional treadmill training, rehabilitation robotic devices were introduced that provide a repetitive and task-specific training of walking. The "Gait Trainer" was introduced in 1999. It is an electromechanical end-effector device with two driven footplates that simulates the phases of gait [12]. The "Lokomat" was introduced in 2000. It is a robotic exoskeleton [11] combined with a treadmill. Patients wear a harness and are partially unloaded with a bodyweight support system (BWS). The main difference from the treadmill training is that these electromechanical-assisted devices guide the patient's legs according to a pre-programmed gait pattern. Various electromechanical devices have been developed after the introduction of the Lokomat and Gait Trainer: The Exowalk, the Haptic Walker, the

Anklebot, the LOPES, GE-O, among others [71, 116]. In addition, new powered mobile solutions have been developed, and their functional benefits have been described: ReWalk, Ekso NR, Indego [116], SMA [117], HAL [118], and ReStore Exo-Suit [119], among others.

Electromechanical-assisted devices can be used for non-ambulatory patients with severe impairments after stroke. The advantage of these devices compared with treadmill training is the reduced physical effort required by the therapists since they neither have to move the paretic limbs manually nor control weight shift. According to the latest Cochrane systematic review [71] and national treatment guidelines (e.g., [106]) on electromechanical- and robot-assisted gait training (RAGT) for walking after stroke, RAGT interventions plus physiotherapy can help more people (age: 18–80 year) walk independently than only physiotherapy or usual care alone. Mehrholz et al. [71] calculated the "apparent" effectiveness of RAGT plus physiotherapy for walking after stroke to be 45%, while the "apparent" effectiveness of the conventional gait therapy control group was 29%. The quality of the evidence was high. The authors recommended this combined approach for people in the first three months after stroke (subacute phase) who were initially unable to walk (quality of evidence: moderate). The effect of electromechanical devices and physiotherapy on walking distance or velocity was less strong. Reviewers were moderately confident about the effects of this combined approach on the distance walked in six minutes and less confident about the effects measured with the 10MWT. RAGT interventions plus physiotherapy did not increase walking capacity more than usual care alone. Studies that included non-ambulatory patients and dependent walkers treated with this combined approach had the greatest effect in improving the walking velocity by 0.09 m/s (value below the minimal clinically important difference, MCID = 0.15–0.25 m/s). For independent walkers, there was neither an improvement nor a difference between the combined approach and usual care alone. Still, some questions remain open. Training duration varied largely (most interventions lasted three to four weeks, but durations actually varied between ten days and eight weeks), making it difficult to recommend the optimal frequency, duration, and optimal timing of RAGT. In addition, while the use of end-effector and exoskeleton devices seems safe and acceptable to most participants (quality of evidence: moderate), it is still unclear whether the type of device could play a role in improving walking capacity after stroke. Finally, no definitive conclusion could be drawn with respect to a longer-lasting effect of the use of RAGT in combination with physiotherapy (mean follow-up between 18 and 22 weeks) due to the lack of follow-up and heterogeneity of the studies.

8.4.2.4 Practical Application of Robotic Gait Training in Persons with Stroke

The medical physician and responsible therapist should check the contraindications for each device type. The indication for locomotor training should consider the training goal and the phase after stroke. We recommend 60 min therapy sessions when using a robotic orthosis (Fig. 8.2) or mobile exoskeleton (Fig. 8.3) and 30–45 min therapy sessions when applying a treadmill or BWS system. According to our experience, the frequency of the training should range from two to five times a week depending on the rehabilitation goal and the patient's compliance and motivation. For robotic devices, the therapist should carefully adjust the settings of the device in the first session and let the patient walk for at least five minutes to verify that the settings are correct and to allow the patient a first experience in walking with the technology. When training with a treadmill or a BWS system, the first therapy can already be more intense. Across sessions, the therapists should increase the challenge of the therapy gradually by adjusting the training parameters, for example, by decreasing the bodyweight support, increasing the walking velocity and therapy duration, and decreasing the guidance force in the case of the electromechanical systems (Table 8.1).

A therapy session should be stopped when the patient communicates that he/she is very tired

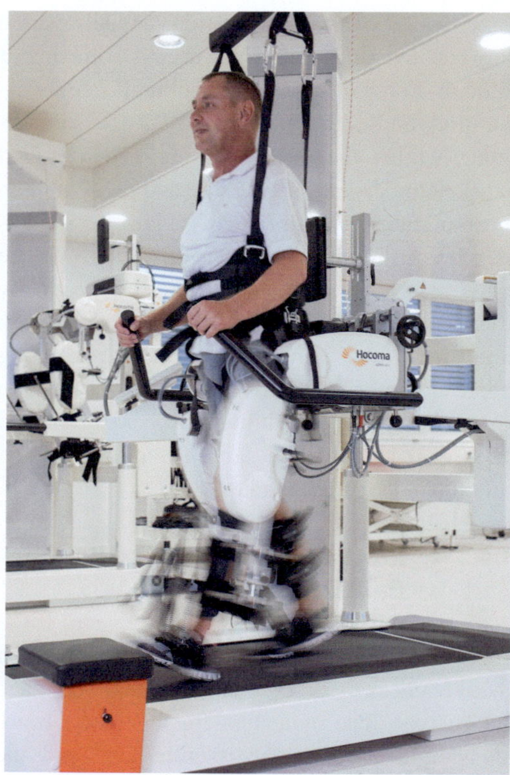

Fig. 8.2 Tethered exo: Lokomat

and cannot actively walk any longer or if the patient's respiratory and physical capacity decreases, leading to dizziness, cold sweat, or turning pale in cases where patients cannot properly communicate how they are feeling. The treatment (training series) should immediately be interrupted if the patient is not compliant or in case of any contraindication (Table 8.1).

For improving respiratory and physical capacity in non-ambulatory patients, robotic orthosis can be used, and the patient does not necessarily need to be active. As therapy progresses, it is very important that the patient can actively participate in the robotic or treadmill therapy and walk bearing an appropriate amount of his/her body weight. The therapist can motivate the patient by integrating various features in training such as augmented feedback, gaming, and dual tasking (Table 8.1).

To evaluate the neurological and functional progress of patients after stroke, we recommend the use of the following assessments: FAC for measuring walking ability, 10MWT for walking speed, Six-Minute Walk Test for walking endurance, Trunk Control Test for assessing trunk movement, Motricity Index for measuring limb strength, Fugl-Meyer Assessment for measuring motor function of the lower extremities, de Morton Mobility Index (DEMMI) for measuring changes in mobility, and the Barthel Index for measuring functional disability.

8.4.3 Traumatic Brain Injury

8.4.3.1 Introduction Traumatic Brain Injury

A blow or jolt to the head from blunt or penetrating trauma can result in a traumatic brain injury (TBI). Every year, an estimated 2.8 million people in the United States visit an emergency department, are hospitalized, or die as a result of a TBI. More than 5.3 million people live with a TBI-related disability in the United States. A recent systematic review estimated that 69 million (95% CI 64–74) individuals worldwide sustain a TBI each year [120]. The proportion of TBIs resulting from traffic accidents was greatest in Africa and Southeast Asia (both 56%) and lowest in North America (25%). One in five adults experienced TBIs of sufficient severity to cause loss of consciousness (LOC) and almost 10% experienced their first TBI with LOC before the age of 15 years [120].

While the risk of having a TBI is substantial among all age groups, this risk is highest among adolescents, young adults, and persons older than 75 years. The risk of TBI among males is twice that of females. The major causes of TBI are motor vehicle accidents, violence caused by suicidal behavior and assaults that involve firearms. For the elderly, falls are the leading cause of TBI.

The severity of a TBI can be initially classified according to the Glasgow Coma Scale (GCS) score at admission, which ranges from mild (GCS 13–15) over moderate (GCS 9–12) to severe (GCS $\leq$ 8). The risk of dying from TBI

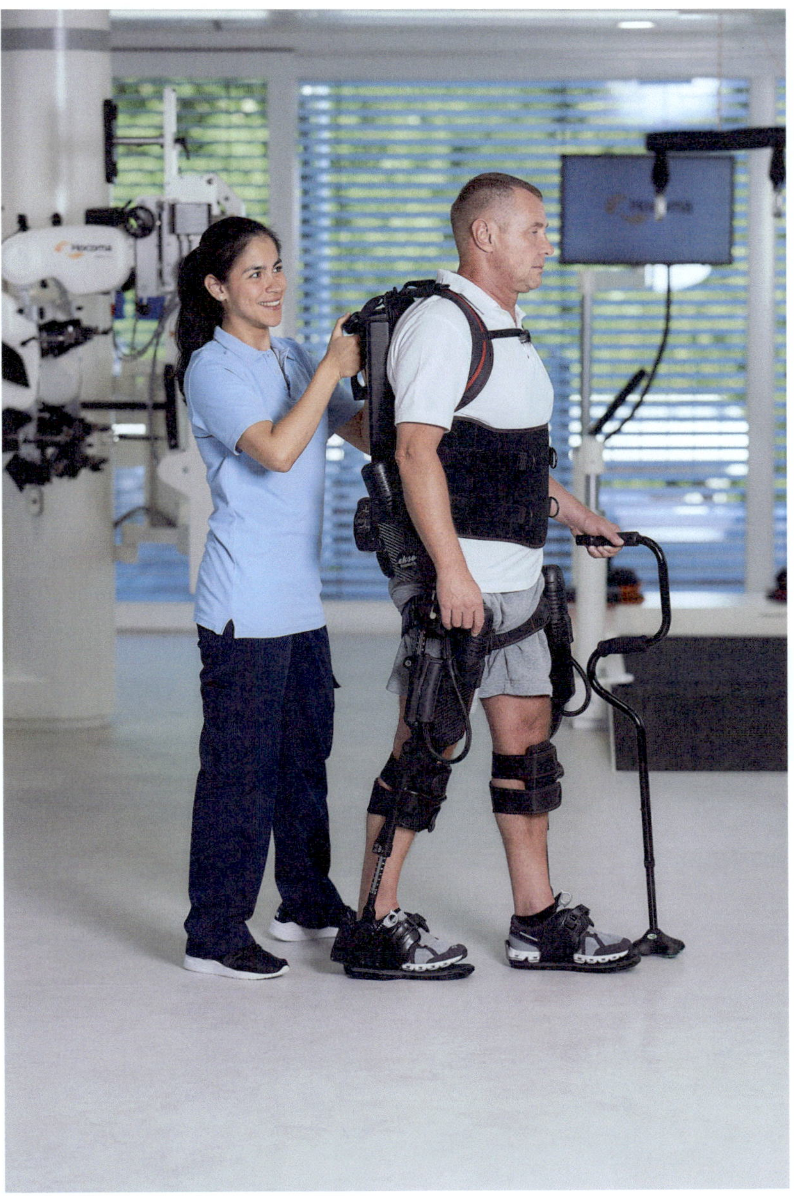

Fig. 8.3 Mobile exo

is low after mild ($\sim 1\%$), intermediate after moderate (up to 15%), and high (up to 40%) after severe TBI [121]. Most brain injuries are diffuse with widely distributed damage to axons, diffuse vascular injury, hypoxic-ischemic injury, and brain swelling (edema). The main injury mechanism is rapid acceleration–deceleration of the head, such as seen after high-speed motor-vehicle accidents. Impact forces act on the cranium at the site of impact (coup) or of tissue oppositely to the impact (contre-coup) and the soft brain may move against the rigid encasement provided by the bones.

Patients may present with initial loss of consciousness (LOS) or coma, varying from seconds up to weeks. Patients with a severe TBI may go through a stage with disorder of consciousness which may be followed by a period of disorientation, memory disorders, and behavioral disturbances. This period can last minutes to months.

Table 8.1 Practical application of robotic and treadmill locomotor training

Device	Training goal	Contrain-dications	Start of training	Training parameters	Number of staff	Mode of supervision	Motivational aspects
Lokomat—robotic orthosis	Increase respiratory and physical capacity for non-ambulatory patients. Improve walking ability for patients with FAC 0–1	1–9	Highly recommended for patients in the first three months post-stroke. The effects reduce afterwards markedly	Increment training duration and velocity, reduce BWS and guidance force for each leg as soon as possible	1–2 therapists or in an efficient setting (e.g. 1 therapist for 2 patients)	Continuous to little supervision. Only under supervision for patients with stable cardiac fitness who can communicate	Dual tasking, augmented feedback, combination with treadmill training with BWS for patients beginning to generate steps, add sit-to-stand exercises
Ekso NR—powered mobile solution	Increase respiratory and physical capacity for non-ambulatory patients. Improve walking ability for patients with FAC 0–1	1–5, 7–11	Highly recommended for patients in the first three months post-stroke. The effects reduce afterwards markedly	Increment training duration, adjust assistance mode for each leg individually (full guidance of motors, assist-as-needed or free in space), adjust amount of weight shifting individually	1 to 2 therapists	At least 1 therapist giving continuous to intermediate physical support	Acoustic and visual feedback for weight shifting, exploration training through clinical routes
Treadmill training with or without BWS (e.g., treadmill of Lokomat, C-Mill, locomotion 150/50 DE med)	Improve walking speed and walking capacity for patients with FAC $\geq$ 2. Improve walking quality (gait symmetry parameters such as step length, weight shifting)	1, 2, 4, 7	Recommended use within the first three months post-stroke. Effects get reduced for patients in the chronic phase	Increase velocity, decrease BWS	1 therapist for moving the paretic leg, group therapy (e.g., 1 therapist for 2–4 patients)	Continuous to little supervision. Only under supervision for patients with good walking ability who do not stumble	Dual tasking, augmented feedback, projections on the treadmill, gaming, add interval training

(continued)

Table 8.1 (continued)

Device	Training goal	Contrain-dications	Start of training	Training parameters	Number of staff	Mode of supervision	Motivational aspects
BWS systems without treadmill (e.g., Rysen)	Improvement of balance, improvement of gait (in combination with stairs) for patients with FAC $\geq$ 2	1, 2, 4, 7	No existing recommendation from Cochrane reviews	Decrease BWS	1 therapist for physical support, group therapy	Continuous to little supervision	Free movement in space and safely, augmented feedback, dual tasking (large floor projections, gaming), practicing stairs and obstacles, playing basketball

1. Patients cannot tolerate an upright standing position for 15 min
2. Skeletal instability (non-consolidated fractures, instability of the vertebral column, osteopenia, osteoporosis)
3. Severe vascular disturbances in lower limbs (uncured deep vein thrombosis)
4. Any medical condition preventing active rehabilitation (e.g., respiratory disease, orthostatic dysregulation, infectious or inflammatory disorders, osteomyelitis)
5. Fix joint contractures that limit the range of motion of the orthosis
6. Not suitable body dimensions for the Lokomat orthoses and harness system: Weight >135 or <10 kg, height >2 m, upper leg length <35 or >47 cm
7. The orthoses or the harness cannot be fitted appropriately (e.g., deformities, colostomy bags, skin lesions)
8. Pregnancy
9. Amputation
10. Not suitable body dimensions for the Ekso orthoses: Weight >100 kg. Upper leg difference of more than 1.27 and 1.9 cm for the lower leg
11. Limited joint mobility: Hip flexion <110°, Knee flexion contractures >12°, Inability to achieve a 0° neutral ankle dorsal extension with accompanying knee flexion up to 12°

TBI can present significant residual challenges to the individual their family, and society. An injured person may experience a wide range of physical, cognitive, emotional, and behavioral changes that affect everyday function. Functional problems caused by a TBI may include paralysis, cognitive and speech changes, and impaired motor control and dexterity, as well as abnormal muscle activity that includes spasticity, clonus, dystonia, co-contraction, associated reactions, and flexor and extensor spasms frequently seen in the upper motor neuron syndrome [122].

8.4.3.2 Gait in Persons with Traumatic Brain Injury

In patients with TBI, gait disturbance is a manifestation of a primary problem that alters neural control of ambulation. The altered gait pattern that emerges is actually the body's attempt to achieve the goal of walking utilizing the remaining resources [123, 124]. Just as normal gait has stereotypic characteristics with individual differences, there are general patterns of gait disturbance associated with common central nervous system (CNS) disorders. Commonly these patterns do not present in isolation but usually present in association with others of the lower limb or even patterns affecting the upper limb [122, 125, 126]. Muscle overactivity can result in multiple patterns of clinical motor dysfunction affecting the lower limb (e.g., equinovarus, stiff knee, striatal toe, adducted thighs, flexed hip), which interfere with ambulation [125].

Due to the diffuse nature of the brain damage with the possibility of involvement of other subcortical structures such as the basal ganglia and cerebellum, individuals with TBI show a very variable presentation of the gait pattern. Generally, individuals with TBI walk slower compared to healthy individuals. In TBI, the reduced walking speed was found to be associated with reduced ankle power generation at push-off [127]. When walking at similar speeds compared to healthy controls, people with TBI seem to use a strategy of increased hip power generation both in early stance and in pre-swing

to account for the reduced ankle power generation. The impairments in ankle joint power generation at push-off are critical, as these were found to be the strongest predictor of mobility outcome in ambulant people with TBI after six months of rehabilitation [128].

Unilateral involvement: Studies have shown that muscle weakness of the hip flexors, knee extensors, and ankle plantar flexors on the affected side are key factors in contributing to the decreased speed of ambulation, and also limit the capacity to increase speed [123, 124, 126, 129]. Initial contact at the forefoot and decreased ankle dorsiflexion are typically observed [126]. The affected ankle predominantly remains in plantar flexion caused by muscle overactivity and/or contracture. Excessive knee extension or knee flexion may also occur. These deviations result in impaired limb stability that correlates with a decreased stance time on the affected limb, increased double support duration, and shorter step length for the unaffected limb [124, 126]. Overall, this pattern provides for the preservation of stability since increased time in double support and increased time weight-bearing on the sound limb, resulting in decreased time weight-bearing through the affected limb. The initiation of the swing phase is delayed and more effortful and accompanied by an increased swing time on the affected side. From a biomechanical standpoint, this is due to an abnormal force transfer from the hindfoot to the forefoot and reduced or absent push-off in the terminal stance [130]. Indeed, during the swing phase, decreased hip and knee flexion and ankle equinovarus can lead to impaired clearance on the affected side with compensatory hip hiking or circumduction.

Bilateral involvement: In individuals with TBI with bilateral involvement, a scissoring gait pattern can be observed, leading to a reduced base of support and stability [126]. However, using the hip adductors may also be a compensatory strategy to assist with hip flexor weakness. This differentiation becomes particularly important when considering treatment, as elimination of hip adduction may render a paraparetic patient unable to walk [123, 131]. The knees can be flexed or can be hyperextended with the ankles in equinus, further impairing clearance and stability [126].

8.4.3.3 Evidence for Robotic Gait Training in Persons with Traumatic Brain Injury

Recently, a clinical practice guideline to improve locomotor function following chronic stroke, incomplete spinal cord injury, and brain injury was published [115]. The systematic review underlying the guideline had the primary purpose of providing clinicians with recommendations for optimizing rehabilitation outcomes to improve walking speed and distance. Their findings suggest that large amounts of locomotor practice performed at higher cardiovascular intensities or with augmented feedback may be critical for improving walking function. In contrast, lower intensity walking interventions or impairment-based training strategies demonstrated equivocal or limited efficacy. It should be noted, though, that these recommendations were mainly based on studies performed on patients after stroke and not TBI or incomplete spinal cord injury.

There are indeed only a few randomized controlled clinical trials (RCTs) investigating the effectiveness of rehabilitation therapy robots in individuals with TBI to improve gait. Esquenazi et al. investigated manual-assisted versus robotic-assisted BWSTT with the exoskeleton Lokomat in 16 persons with TBI [132]. They noted no significant in-between group differences after 18 therapy sessions (three sessions per week) of 45 min each. Interestingly, the improvements in self-selected walking speed and step-length asymmetry ratio improved higher in the robotic group (mean 50% and 33%, respectively) compared to the manually assisted group (31 and 9%). In contrast, maximal walking velocity and distance walked during 6 min increased more in the manually-assisted group (31% and 19%, respectively), compared to the robotic group (15% and 12%, respectively). Less staffing and effort were needed when using the robotic technology [132].

The same group also evaluated differences between partial bodyweight-supported treadmill, an exoskeleton (Lokomat), and an end-effector (GE-O) training in 22 individuals with chronic hemiparetic TBI [133]. Again, the individuals participated in 18 sessions of 45 min. The interventions resulted in a significantly increased self-selected walking speed for all three groups and an increased maximal velocity for manual and Lokomat-supported training. Unlike the previous study, the authors did not note significant changes in gait symmetry. Staffing was, again, least for the Lokomat compared to the other groups [133].

In a retrospective analysis of individuals with an acquired brain injury, training with the exoskeleton Lokomat (n = 28) was compared to training with a very similar device, the Walkbot [134] (n = 34). Both interventions were combined with conventional physiotherapy. The median number of sessions was 14. Of the 62 individuals, who were initially not able to walk, 15 had TBI. Both groups improved in various gait and gait-related outcomes, and the improvements were comparable.

There are now also studies that have evaluated the effects of training with over-ground exoskeleton devices in individuals with an acquired brain lesion, i.e., including both stroke and TBI. In an uncontrolled study, seven individuals aged 14–27 years with chronic acquired brain injury (four TBI, all hemiparetic) participated in twelve 45 min sessions with the over-ground exoskeleton device Ekso (Ekso Bionics, Inc., Richmond, CA, USA) [135]. The pre–post evaluation of loading/unloading and spatial-temporal characteristics showed that step length, speed, and an overall progression toward healthy bilateral loading tended to improve, with linearity of loading showing a statistically significant effect. In a retrospective analysis, the authors tried to identify predictors for the independence in daily-life motor performance (motor score of the Functional Independence Measure) at discharge of 36 patients with acquired brain injury (nine TBI), who had participated in Ekso training [136]. The authors found that admission motor

FIM score and the total number of robotic steps to be statistically significant predictors. Interestingly, for every 1,000 steps taken in the robotic exoskeleton, the discharge motor FIM score increased by three points.

8.4.3.4 Practical Application of Robotic Gait Training in Persons with Traumatic Brain Injury

The overall goal of rehabilitation, also in individuals with TBI, is to achieve the highest level of independence in daily-life activities. In rehabilitation, this means that rehabilitation therapy technologies are always combined with conventional treatment, particularly for those patients with a Functional Ambulation Category (FAC) less or equal to 3. Individuals with a FAC of 4 or 5 rarely receive robotic therapy; they are mainly trained conventionally, i.e., without robotic devices.

In MossRehab, patients can be admitted to the rehabilitation program if they are able to tolerate 3 h of therapy per day. Patients receive about 1.5–2 h of physiotherapy (including robotic locomotor therapy) per day, besides one hour of occupational therapy and one hour of speech and language therapy. Therapy is provided seven days a week, but robotic therapy is rarely applied on the weekend. Conventional and robotic therapy follow the functional progress of the patient. A major advantage of including goal-directed and task-specific robotic therapy in the multidisciplinary program is to improve the consistency of the training for the patient without taxing the therapist too much. While initially, two staff members (a physiotherapist and aid) might be available to support the strongly impaired patient, the use of robotics allows reducing staff over time. It should be noted, though, that patients with TBI should be strongly supervised throughout the initial rehabilitation program. On the one hand, supervision is needed due to cognitive and behavioral issues. On the other hand, therapists should regularly monitor vital signs because autonomic dysfunctions can present.

Appropriate use of rehabilitation robots depends on the clinician's knowledge of the different robotic devices and technical features, thereby allowing patients to benefit from robot-aided gait training throughout the rehabilitation continuum with the ultimate goal of returning to safe and efficient over-ground walking [132]. As a general rule of thumb, therapists select a technology that is considered best concerning the motor impairments of the patient. For example, patients with little leg and trunk muscle strength and difficulties in controlling multi-joint movements might benefit from therapy in an exoskeleton device like the Lokomat that guides the leg through a physiological gait pattern. If patients have achieved better lower limb control for standing, an end-effector device can be utilized. Few studies have attempted to directly compare the different robots in a comprehensive manner. At MossRehab, the clinical setup places a Lokomat and a GE-O next to each other. Taking advantage of this, we were able to set our 3D gait kinematic recording system and obtain sequential data from patients with traumatic brain injury using the two systems and compare that to bodyweight-supported manual-assisted therapy on a treadmill. Our data confirms our clinical paradigm that a more controlled and repetitive gait pattern when using a Lokomat with a gait pattern that is more similar to that of over-ground walking. The GE-O provides a gait pattern that has more variability of motion for the hips and knees with slightly reduced knee motion and the gait pattern differs slightly from that observed during over-ground walking. Finally, the gait pattern achieved during manually assisted treadmill bodyweight-supported therapy was most variable with a lack of symmetry of movement and timing [137, 138] (Fig. 8.4).

Also, in the field of TBI, therapists use various rules of thumb. First, training intensity can be increased by reducing bodyweight unloading, assistance from the robotic device, increased walking velocity, and active therapy duration. However, there are currently no evidence- or practice-based standards informing therapists about a preferred order of adjusting these

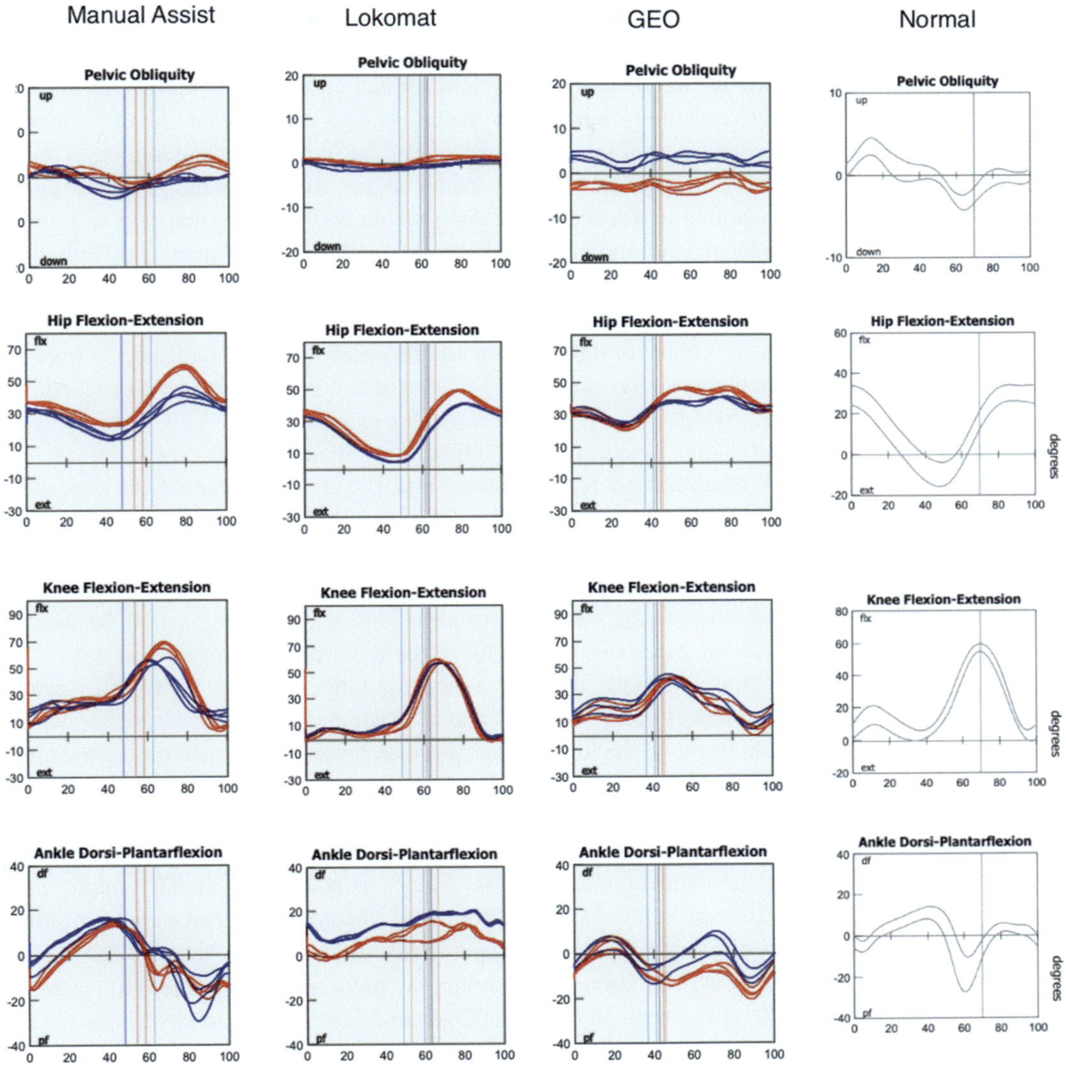

Fig. 8.4 A representative data set looking at the knee kinematics in a 32-year-old man with left spastic hemiparesis. Red represents the right knee and blue represents the left knee over several walking cycles. Data recorded with CODA CX1 optoelectronic sensors

parameters to optimize the training intensity for the patient with TBI. Second, as the overall goal is to achieve independence in daily-life activities, therapists prioritize improving function first (e.g., being able to walk independently) before focusing on the movement quality (e.g., the physiological gait pattern).

However, like no other field within adult neurorehabilitation, therapists might frequently deviate from these rules of thumb, as the behavioral and cognitive impairments of the patient can have a huge impact on the rehabilitation goals and program. Some of these impairments can be more challenging to treat than motor impairments. This makes it initially very challenging to make an early prognosis of therapy outcome when performing gait rehabilitation in patients with TBI. Therapists, physicians, and other members of the multidisciplinary rehab team sit weekly together, discuss the progress of the patient, and formulate the consecutive short-term functional goals. As the behavioral and cognitive impairments might also accompany the individual with TBI after discharge, the rehab team needs to search for the most appropriate solution for community reintegration. For some patients, e.g., with aggressive behavior, insufficient safety awareness, or cortico-visual impairments resulting in a reduced visual field, achieving independent mobility might actually create a potentially dangerous situation. In such a case, the rehab team will work toward a solution where sufficient supervision is provided when re-entering the community, e.g., through assisted living.

The behavioral and cognitive impairments can also influence the choice of rehabilitation technology. For example, based on the motor impairments of the patient, an exoskeleton device like the Lokomat might be the most appropriate device to train the patient early during the rehab program. However, if the patient perceives the harness and cuffs as too confining, reports discomfort in the device, or experiences orthostatic hypotension before being able to move the legs, the therapist might select another less constrictive technology.

Besides the selection of the technology, the behavioral and cognitive impairments also impact the therapy session. Therapists should be aware that patients with TBI might not perceive their environment as we do, which can strongly affect their participation in therapy. Events in the environment can distract the patient due to neurocognitive impairments, e.g., in focusing attention or inhibiting irrelevant visual or acoustic stimuli. When individuals with TBI train in a device, therapists should be aware that patients might have difficulties expressing discomfort or cannot precisely formulate what they experience. Slight dizziness might be indicative of a drop in blood pressure requiring immediate action. Indeed, therapists will initially repeatedly monitor vital signs during therapy, i.e., heart rate and blood pressure. As communication might be impaired, the therapist must know the patient well, can interpret non-verbal signs indicating discomfort, and communicate in an empathic way. Knowing the patient well is also advantageous to know the internal motivators of the

patient to have him/her participate actively in the robotic therapy. While some patients respond better to visual cues, others might better respond to auditory cues (e.g., their favorite music). Particularly, early during rehabilitation, it might be better to not use exergaming, as the games might provide distracting visual and/or auditory information. When, over time, the individual with TBI can integrate more inputs simultaneously, exergames complementing the rehabilitation technologies might be indicated to improve motivation and increase the number of repetitions in a playful manner.

Of course, there are also neurological and motor impairments that influence the session. In cases of high levels of spasticity, it could be that robotic training can be performed more effectively if anti-spastic medication is given in advance. Casts and splinting might be used to achieve the required range of motion.

When training patients with TBI, certain (relative) contraindications need to be considered, besides those device-specific contraindications formulated in the device approval. Adequate control of seizures, but it should be noted that epilepsy could also develop later after TBI. Some individuals with TBI might have experienced disorders of consciousness for a longer time. After a long time of no weight bearing, the physician might decide to perform a DEXA-scan to determine sufficient bone density before robotic therapy can be started. Cardiovascular stability is needed until the patient can start moving his/her legs. Furthermore, it should be noted that some patients might have feeding tubes or catheters. Therapists should ensure that during the robotic training, these tubes do not become blocked or cause pressure on the underlying skin. Importantly, if patients are unable to reliably communicate (verbally or non-verbally) discomfort or pain, robotic therapy is best avoided.

Reasons for pausing a therapy session are sudden complaints of pain or discomfort, altered cognitive status, significant changes in vital signs, changed behavior (e.g., the patient becomes vocally loud or starts cursing), or changes in the environment that make further training difficult. Of course, any adverse events like a sudden strong increase in pathological muscle tone or skin breakdown requiring nursing should interrupt the session, and perhaps temporarily the training series. One reason to pause or stop the training series is when the amount of setup outweighs the use of the robotic device (e.g., you need 40 min to prepare the patient in the device and have only 10 min left to train the patient).

Functional progress is evaluated by various standardized assessments. These include the 2 or Six-Minute Walk Test, the five times sit-to-stand test, and the timed up and go. In addition, for patients who progressed well, additional tests like the Berg Balance Scale, the Dynamic Gait Index, and the High-Level Mobility Assessment Tool (HiMAT) are used.

In summary, task-oriented, high-repetition gait training can improve, among other things, muscular strength, motor control, and movement coordination in patients with neurologic impairments [139, 140], help reduce muscle atrophy, osteoporosis, joint stiffness, and muscle and soft tissue shortening, and promote the reduction of spasticity, among other benefits [137, 141].

However, particularly when treating individuals with TBI, the behavioral and cognitive components are important in shaping patient-tailored sessions.

8.4.4 Spinal Cord Injury

8.4.4.1 Introduction Spinal Cord Injury

Damage to the spinal cord often results in lasting sensory and motor impairments that can significantly impact an individual's mobility and independence, ultimately leading to reduced life satisfaction [142–144]. Each year, 250,000–500,000 individuals worldwide experience a traumatic spinal cord injury (SCI), with an annual global incidence of 40–80 cases per million population [145]. In the United States, there are approximately 18,000 new SCI cases each year, with an estimated 296,000 individuals living with SCI. Less than 1% of individuals with SCI achieve full neurologic recovery before discharge from the hospital, and many are left with permanent functional limitations, leaving them reliant on rehabilitation services and technology to maximize their quality of life and community participation [146].

8.4.4.2 Gait in Persons with Spinal Cord Injury

Return to walking is a priority for individuals in rehabilitation after SCI [142, 147, 148]. The severity of injury remains the primary predictor for walking recovery after SCI. Individuals diagnosed with complete SCI ASIA (American Spinal Cord Injury Association) Impairment Scale (AIS) A are not expected to regain functional ambulation unless heavily reliant on assistive devices, equipment, and/or technology [149]. Studies have shown that individuals who recover the ability to accurately detect pain (using pinprick sensation) have a better prognosis to recover independent ambulation [150–153]. Recovery of lower extremity motor function at one-month post-injury has also been reported as a significant predictor of walking recovery. Waters et al. [154] reported that the majority of individuals with SCI who achieve a score of at least 10 points on a lower extremity strength test by one month post-injury will be community walkers at one year post-injury. Zörner et al. [155] reported that individuals who achieve 25 points on a lower extremity strength test by one month are able to walk at limited community speeds (0.6 m/s) by six months post-injury, and that initial lower extremity strength score is the most significant predictor of independent walking. Walking prediction algorithms have been published, but predicting who will recover walking, especially in the middle of the severity spectrum (AIS B and C), remains challenging for health care professionals [156, 157].

A number of therapeutic interventions have been developed that concentrate on walking recovery after SCI. Locomotor training (LT)—the repetition of stepping-like patterning to promote walking recovery in either a bodyweight-supported (BWS) or a non-bodyweight-supported condition and with or without robotic assistance—has been a focus of research to date.

There is a growing evidence-base to support that various forms of LT may be effective in promoting walking recovery in individuals with motor incomplete SCI [50, 158–165]. Published literature supports that participation in intensive walking programs for this patient population may result in improved walking function, including but not limited to speed, endurance, and independence [161, 163, 164, 166–168].

Recent technological advancements have resulted in the development of robotic exoskeletons with the goal of providing an energy-efficient avenue for individuals to regain the ability to walk over-ground after neurologic injury. Multiple versions of robotic exoskeleton systems have been developed, which rely on hip and knee joint motors, a computerized control system, and rechargeable batteries to promote walking for individuals who are otherwise unable to mobilize in an upright manner.

8.4.4.3 Evidence for Robotic Gait Training in Persons with Spinal Cord Injury

Current literature on exoskeletons includes a systematic review with meta-analysis published by Miller et al. [169] in 2016, which included 14 studies representing 111 individuals with SCI. Reported training programs were generally conducted 3×/week, 60–120 min/session for 1–24 weeks. The majority of studies (10) focused on training on indoor surfaces, while four studies incorporated more complex walking tasks that include outdoor walking, navigating obstacles, and climbing/descending stairs. Following training, 76% of individuals were able to ambulate with no physical assistance. The weighted mean distance for the Six-Minute Walk Test was 98 m, and the physiologic demand of powered exoskeleton-assisted walking was 3.3 metabolic equivalents, while the rating of perceived exertion was 10 on the Borg 6–20 scale. Improvements in spasticity and bowel movement regularity were reported in 38% and 61% of individuals, respectively. No serious adverse events occurred in the 14 studies reviewed. However, a 3.4% incidence of bone fracture was reported. Walking speeds of 0.03–0.51 m/s have been reported in case series and pilot studies [75, 170, 171].

8.4.4.4 Practical Application of Robotic Gait Training in Persons with Spinal Cord Injury

In general, individuals participate in exoskeleton-assisted walking with the goal of compensating for lost walking function or improving walking recovery. Exoskeleton-assisted walking can begin in the inpatient rehabilitation setting as soon as individuals with SCI are medically stable and cleared for full weight-bearing and upright tolerance. Training may extend through outpatient therapy and may also take place in community health and wellness centers. During inpatient and outpatient rehabilitation, exoskeletons are often used to improve walking outcomes for those individuals with motor incomplete SCI who have the potential to regain functional ambulation. Evaluating and training individuals with SCI and their caregivers who purchase a device for home use may also be the focus of an outpatient therapy program. Community health and wellness centers also offer fee-for-service opportunities for individuals with SCI to use exoskeletons, with the focus primarily being on improving overall health.

Currently, there are two exoskeletons approved by the FDA for use in the home/community with the assistance of a trained companion as well as in rehabilitation facilities with trained therapists, the ReWalk™ and the Indego.™ A third exoskeleton, the Ekso,™ is only approved for use in healthcare facilities with trained therapists.

The ARGO ReWalk™ orthotically fits to the lower limbs and part of the upper body and is intended to enable individuals with SCI at levels T7 to L5 to ambulate with the supervision of a specially trained companion in accordance with the user assessment and training certification program. The device is also intended to enable individuals with SCI at levels T4 to T6 to ambulate in rehabilitation facilities in conjunction with the user assessment and training certification program. The ReWalk is not intended for sports or stair climbing. Individuals must be between 160 and 188 cm tall to use the device, and it has a maximum weight capacity of 100 kg. The ReWalk weighs 25 kg, and individuals use a wrist band to activate movement. When in the "walk" mode, forward flexion of the upper body is detected by the tilt sensor and triggers a step. This results in a three-point gait pattern, advancing one step at a time. There are four additional modes: sit-stand, stand-sit, up steps, and down steps. The maximal walking velocity is 0.6 m/s (2.2 km/h) [172]. Forearm crutches are necessary to walk with the device.

The Ekso™ by Ekso Bionics is intended to support ambulation for individuals with SCI in rehabilitation facilities under the supervision of a trained physical therapist. It is intended for use with individuals who have functional upper extremity (UE) strength and SCI levels T4-L5 as well as C7-T3 (AIS D). Therapists must complete a training program prior to using the device. The Ekso GT™ provides the option of variable assistance, which allows the assistance in either or both lower extremities to be reduced when individuals have motor activity in their lower extremities to assist with walking. The Ekso weighs ~22 kg and is intended for use with individuals who are 162–189 cm in height and have a maximum weight of 100 kg. Walking can be activated by achieving a threshold of torso tilt or using an external controller to manually trigger stepping. Sit-stand and stand-sit are activated using an external controller. Individuals use crutches or a walker for balance support while walking.

The Indego™ exoskeleton is intended for use with individuals with SCI at C7 and lower injury levels in rehabilitation facilities and T3 and lower injury levels for use in the home and community settings. Walking is activated when the individual moves his or her center of pressure (COP) in an anterior direction, which signals the controller to initiate walking and sit-stand or in a posterior direction to initiate stopping or stand-sit. The Indego offers a variable assistance mode which allows for the reduction of assistance on one or

both lower extremities for individuals who have preserved motor activity to assist with ambulation. The Indego weighs 12 kg and consists of five modular components: a hip segment, a right and left upper leg segment, and a right and left lower leg segment. It is intended for use in individuals with a height of 162–189 cm and a maximum weight of 100 kg. Individuals use crutches or a walker for balance support while walking.

The use of an exoskeleton requires medical clearance for full weight-bearing. During inpatient rehabilitation, exoskeleton-assisted walking often requires the assistance of two staff (one physical therapist and one aide) to provide appropriate physical support while also adjusting software parameters. Once hardware and software parameters have been established for an individual and appropriate training has been completed, it generally only requires one trained physical therapist or family member to provide support for walking using an exoskeleton. As previously discussed, individuals with SCI prioritize regaining walking ability and are motivated to use exoskeletons throughout the continuum of rehabilitation.

Primary contraindications for using an exoskeleton are poor bone density, active pressure injuries, severe, uncontrolled spasticity in the lower extremities, and lower extremity joint contractures that prevent functional gait. An initial evaluation that includes strength and range of motion testing by a trained physical therapist is required before using an exoskeleton. Once the evaluation is completed, and the hardware is set to fit the individual with SCI, they transfer into the device (with or without assistance), and the device is secured to their torso and lower extremities. The various control mechanisms previously described for each system will be triggered to bring the individual into an upright standing position using an assistive device for balance. After successful standing has been achieved, the individual will learn how to trigger stepping with the device (i.e., postural tilt, weight shift, manual trigger) and walk using a reciprocal gait pattern with the support of crutches or a walker. The initial walking session after the evaluation may only last 5–10 min with the goal of increasing walking time, speed, endurance, and independence with every session. With practice and experience, individuals can achieve up to 40–45 min of walking in a 1-h session. When individuals are using the Indego or the Ekso, therapists will evaluate the ability to reduce assistance on one or both lower extremities (variable assistance) based on the individuals' residual motor function. To evaluate the efficacy of these devices to improve walking recovery or to compensate for lost walking function, outcome measures evaluating speed (Ten-Meter walk test), endurance (Six-Minute Walk Test), and independence (assistance level) are often used.

Potential adverse events from exoskeleton-assisted walking include joint swelling, muscle strains, lower extremity fractures, and skin breakdown. Skin checks should be completed before, during, and after each session with any abnormalities clearly documented. Training should be stopped with the presence of adverse events and should not be started again until medically appropriate.

8.4.5 Multiple Sclerosis

8.4.5.1 Introduction Multiple Sclerosis

Multiple sclerosis (MS) is a chronic neurodegenerative disease of the central nervous system that affects approximately 2.8 million people worldwide [173]. MS is characterized by central nervous system demyelination and axonal loss that causes a variety of symptoms such as spasticity, paresis, walking difficulties, bladder dysfunction, and cognitive abnormalities [174]. Decreased exercise capacity, excessive (post-exercise) fatigue, and reduced muscle contractile function are frequently observed and further impede mobility [175]. Together these symptoms impact the active design and participation of persons with MS in everyday life and lower quality of life [176].

8.4.5.2 Gait in Persons with Multiple Sclerosis

Limitations of walking are common in persons with MS but the spectrum of severity is high: while 85% report gait disturbances with mainly mild symptoms (atactic pattern), 40% report direct restrictions on their gait quality [177]. Often, these restrictions are not due to one symptom, but an interaction of several factors like muscle weakness, reduced motor coordination, spasticity, balance disorders, and fatigue. The disturbances not only increase the risk and frequency of falling but also increase the fear and uncertainty to fall [177]. Therefore, persons with MS prioritize safety during walking and significantly reduce their walking speed [178]. In general, no walking aids are necessary for the early MS stages (Expanded Disability status scale, EDSS of 1.0–5.5) where the mobility and gait restrictions are low [179]. The EDSS is a method for quantifying and monitoring the changes in disability in MS over time. The EDSS is a 10-point Likert scale evaluated by a neurologist reaching from 0 (no MS) to 10 (Death by MS) in 0.5 increments with higher units representing higher levels of disability [180].

For the treatment and management of movement disorders, the physical and energy management approaches show the best evidence in these stages [181]. Especially concerning physical fitness and disease-associated biomarkers, studies showed that high-intensity interval training (HIIT) leads to greater improvements than classical moderate continuous training (MCT) [182]. In the early phases, the rehabilitative strategies for improving gait are key supportive treatment options that aim to allow persons with MS to participate in activities of daily living and maintain their health-related quality of life (HR-QoL).

As the disease progresses and motor limitations intensify (EDSS 6.0–7.5), physical counter retention through exercise becomes increasingly difficult. In these phases, aids like walking sticks, walkers, and wheelchairs are increasingly used and over-ground walking training becomes more difficult if not impossible [183]. Thus, reducing the physical effort through bodyweight-supported training (BWST) or with technological/robotic devices is in these stages useful in the severe MS stages. Especially for individuals with balance disorders severe spasticity, and/or fatigue BWST enables the patient to train safer and longer [184].

8.4.5.3 Evidence of Robotic Gait Training in Persons with Multiple Sclerosis

To date, MS research shows no strong and convincing evidence for superiority of robotic gait training or BWST compared to conventional training for improving gait disturbances (e.g., walking speed). Recently, two major RCTs evaluated the impact of robotic locomotor training on treating gait and balance disturbances in MS [185]. Both studies reported improvements on the Six-Minute Walking Test and in the balance after robotic gait training but not in gait speed measured by the Ten-Meter Walk Test. However, both studies were heterogeneous in their methods as they used different devices, adopted different training protocols (12 versus 15 sessions over 3 versus 6 weeks), and included participants with different MS phenotypes representing a wide range of gait disabilities (EDSS between 3.0 and 7.5). In addition, sample sizes were considerably small [185].

Systematic reviews that include other neurological diagnoses (e.g., stroke or Parkinson's disease) show that the advantages of robotic gait training may lie in other (softer) dimensions such as safety, reduced fear of falling, repeatability, or influencing the motor-paradigm-induced fatigue of the participants.

8.4.5.4 Practical Application of Robotic Gait Training in Persons with Multiple Sclerosis

Robotic devices in MS mainly provide functional movement therapy for restoring gait and, balance by lower limb rehabilitation. Devices include various rehabilitation robots such as the Lokomat (Hocoma AG, Volketswil, Switzerland), assistive orthoses, or exoskeletons that support patients' movements. Newer devices combine robotics with wearable sensors shifting toward more smart and adaptive applications. These devices mostly use targeted epidural electrical stimulation for restoring motor functions [186].

Independent of the patient's disease phase, the applied training strategies must follow specific training principles that comprise specificity, overload, progression, initial values, reversibility, and diminishing returns [187]. We summarized these in Table 8.2. These principles represent fundamental components for developing exercise programs in order to respect physiological aspects of performance and to address all opportunities for improvements [187]. To date, no specific training guidelines for robotic gait training exist. The application depends therefore on the experience of the therapists. Disregarding these principles may lead to an inadequate dosage of the prescribed exercise sessions.

8.4.6 Parkinson's Disease

8.4.6.1 Introduction Parkinson's Disease

Parkinson's disease (PD) is the second most frequent age-related neurodegenerative disorder after Alzheimer's disease [188], characterized by a progression of global disability, despite the best pharmacological therapy, and with a high human and economic burden for individuals and society. More than 6 million people worldwide are diagnosed with PD, and both the incidence and prevalence of the disease are increasing faster than other neurological disorders [189].

PD is a heterogeneous and complex pathology, considering the concomitant presence of both motor and non-motor disorders and different kinds of disability progression [190]. Diagnosis is based on history and examination. History can include prodromal features, i.e., rapid eye movement sleep behavior disorder, hyposmia, constipation, and characteristic emergent movement difficulties. The examination has to demonstrate the presence of bradykinesia and rigidity, while rest tremor is considered a typical but not a mandatory diagnostic feature.

Table 8.2 Principles for locomotor training in patients with MS

Principle	Example
Specificity: Training adaptations are specific to the organ system or muscles trained with exercise	Aerobic exercise such as brisk walking is more appropriate for a session aimed at increasing cardiovascular fitness than strength training
Progression: Over time, the body adapts to exercise. For continued improvement, the volume or intensity of training must be increased	Increase duration of walking program by 5% every two weeks depending on exercise tolerance
Overload: For an intervention to improve fitness, the training volume must exceed current habitual physical activity and/or training levels	Prescribing intensity for a resistance training program based on % of measured and/or estimated 1-repetition maximum
Initial values: Improvements in the outcome of interest will be greatest in those with lower initial values	Perform baseline measurements before participating in an aerobic/strength training program to increase cardiovascular fitness and reduce fatigue
Reversibility: Once a training stimulus is removed, fitness levels will eventually return to baseline	Participants who maintain their training after a supervised exercise program preserve strength whereas those who stop exercising return to baseline
Diminishing returns: The expected degree of improvement in fitness decreases as individuals become more fit, thereby increasing the effort required for further improvements. Also known as the 'ceiling effect'	Gains in muscle strength are greatest in the first half of a training program unless the training stimulus continually increases

Imbalance is the fourth pathognomonic feature that appears later in the disease. In 1967, Margaret Hoehn and Melvin Yahr described how the motor symptoms of PD progress in the scale that was named after them. The original scale included stages 1 through 5 [190], and was modified by Goetz et al. in 2004 [191]. However, reducing the complexity of Parkinson-related disability to these five or seven stages is difficult because of the heterogeneity of disease progression and symptoms (Table 8.3).

Between motor symptoms, gait problems that worsen as the disease progresses are major disease burden and markedly affect independence and quality of life [192]. From the onset of the disease, non-motor symptoms may be present. These symptoms can include psychological or cognitive problems (e.g., anxiety, insomnia, depression, and mild cognitive impairment). However, the evolution of dysphagia, gait disorders, balance problems, and other non-motor symptoms, such as orthostatic hypotension,

Table 8.3 Hoehn and Yahr stage

Stage		Hoehn and Yahr Scale	Modified Hoehn and Yahr Scale
Early	1	Unilateral involvement only usually with minimal or no functional disability	Unilateral involvement only
Early	1.5	-	Unilateral and axial involvement
Moderate	2	Bilateral or midline involvement without impairment of balance	Bilateral involvement without impairment of balance
Moderate	2.5	-	Mild bilateral disease with recovery on pull test
Moderate / Advanced	3	Bilateral disease: mild to moderate disability with impaired postural reflexes; physically independent	Mild to moderate bilateral disease; some postural instability; physically independent
Advanced	4	Severely disabling disease; still able to walk or stand unassisted	Severe disability; still able to walk or stand unassisted
Terminal	5	Confinement to bed or wheelchair unless aided	Wheelchair bound or bedridden unless aided

dementia, and psychosis, complicate the advanced stage of the disease that is associated with severe global disability. Moreover, postural disorders—either static or dynamic—are complex, not entirely understood, and contribute to PD motor-related disability. Stooped posture, with flexion of the hip and knees and rounding of the shoulders, is the most recognized static deformity, evident shortly after the onset of the illness. More severe abnormalities of static posture disrupting spinal alignment and leading to significant disability include camptocormia, antecollis, Pisa syndrome, and scoliosis [193]. A complex integration of multisensory inputs results in a final posture and a motor adjustment process. All or some of the components of this system may be dysfunctional in people with PD, rendering postural instability one of the most disabling features of PD. Balance control is critical for moving safely in and adapting to the environment. PD induces a multilevel impairment of this function, therefore worsening the patients' physical and psychosocial disability, compromising gait quality and safety [193].

PD has multiple disease variants with different prognoses: prominent early motor (severe rigidity or bradykinesia and prominent gait problem) and non-motor symptoms (rapid eye movement sleep behavior disorder, mild cognitive impairment, and orthostatic hypotension) identify subjects with a diffuse malignant subtype (9–16%) that have a poor response to medication and faster disease progression. Individuals with mild motor-predominant PD (49–53%) are younger at the onset, present mild symptoms, and show a good response to dopaminergic medications (e.g., carbidopa-levodopa, dopamine agonists) and slower disease progression. Frequently, they have a tremor-dominant form of PD and lower severity of gait disorder at the onset. Other individuals have an intermediate subtype. The treatment of PD is symptomatic, focused on improving motor (i.e., tremor, rigidity, bradykinesia, gait, and postural disorders) and non-motor (genitourinary disturbances, orthostatic hypotension, constipation, cognition, mood,

sleep) signs and symptoms, prevalently, through the replacement of dopaminergic stimulation [194, 195].

8.4.6.2 Gait in Persons with Parkinson's Disease

The pathognomonic motor manifestations of PD, i.e., bradykinesia, rigidity, and reduced amplitude and automaticity of movement affect the gait patterns of people with PD since the disease onset [196].

In the early stages of PD, the gait is slow and smooth with short steps, compared with those of age-matched healthy adults, with a reduced amplitude of arm swing and increased inter-limb asymmetry, which is more specific to PD. These are often the first motor symptoms. In the early stages of the disease, symptoms are often unilateral, corresponding to asymmetrical basal ganglia neuropathology. Both inter-limb and intra-limb movements, such as the timing of swing duration, are also impaired. Hip, knee, and ankle range of motion progressively reduce during walking, especially during the late-stance phase of the gait cycle [196–198]. Changes in posture and range of motion further affect the magnitude of movement, for example, by contributing to a reduction in step length [193]. Gait variability is larger than seen in age-matched healthy controls, and performance on complex locomotor tasks is also impaired (e.g., reduced angular velocity of turning). Additionally, as ambulation in this stage becomes less automatic, many gait alterations increase during dual tasks [196].

In the mild-to-moderate stage of the disease, many of the spatiotemporal features altered in the early stages progress bilaterally so that inter-limb asymmetry decreases and movement becomes more bradykinetic with disease progression [196]. Shuffling steps, in which the feet slide forward instead of being lifted up off the floor, may appear, while gait cadence generally decreases. The magnitude of arm swing is reduced bilaterally with accompanied reduction of axial rotation [196–198]. Postural changes, such as stooped posture, might further deteriorate gait by altering gait kinematics [193]. Motor automaticity becomes further impaired, resulting in fragmented motor function, such as defragmentation of turns (turning en-block), and problems with gait initiation [199, 200]. Freezing of gait and festination might appear, and patients have an increased risk of falling in this stage [196].

Freezing of Gait (FoG) is defined as the sudden inability to start or continue walking despite the intent to maintain locomotion. It is common in advanced PD and other primary or secondary parkinsonian syndromes [201]. FoG is one of the most disabling and difficult to assess impairments of gait, as it is episodic and variable by nature [202]. It usually lasts a couple of seconds, but episodes can occasionally exceed 30 s. It includes episodes of gait "starting hesitation", arrests in forward progression during walking, mainly during "turning" or when the subject reaches the destination. It can appear as episodes of shuffling forward with steps that proceed for millimeters in length, complete akinesia with foot or toe not leaving the ground; alternate trembling of the legs in place at a frequency of 3–8 Hz. FoG can be asymmetrical, affecting mainly one foot or being elicited more easily by turning in one direction [201]. A longer disease duration, gait and postural disturbances at the disease onset, concomitance of mood depression, long-term treatment with dopamine agonists (excluding levodopa) [201, 203], mental-loading or dual-tasking, disturbances of executive functions [204] were related to a higher risk of developing FoG. Gait analysis studies highlighted that FoG is characterized by higher step timing variability, reduction of bilateral coordination, stride amplitude, and postural stability, whereas a profound and incremental decrease in stride length, highly reduced joint ranges, disordered temporal control of gait cycle, and high-frequencies alternate trembling-like leg movement (3–8 Hz) often precedes FoG [202, 203, 205].

In the advanced stage, gait worsens. The postural reactions begin to be impaired [194, 196, 198] or inadequate [194], and patients manifest abnormal dynamic postural control (i.e., postural instability) [193]. Additionally, blocks in motor function (e.g., FoG) become more frequent, accompanied by reduced balance and postural control and a severe risk of falling [196, 201, 202]. Finally, motor fluctuations and dyskinesias are present in most patients and negatively impact gait. Endurance and muscle force further decline, resulting in a reduced motor capacity and the need for assistive devices or wheelchair use.

Management of Parkinson's Disease: The practical indication and the available guidelines for PD treatment [194, 206–210] underline the need for a multidisciplinary and personalized approach. Different symptomatic, pharmacological, functional, and neuro-surgical therapies are proposed according to the phases. A continuous rehabilitation, from the disease onset to the end of life, guarantees a tailored approach to the different needs emerging with the disease progression [208].

The most common treatment is dopaminergic medication [206]. However, there is increasing evidence on the efficacy of non-pharmacological, i.e., exercise interventions to improve motor function and gait and to reduce the risk of falling [196, 208].

Speed and stride length of gait, turning speed, and various types of FoG improved with levodopa treatment [197, 206–210]. In contrast, temporal parameters, such as cadence and double support, or coordination and step variability, did not improve [211, 212].

Long-term use of levodopa may lead to motor response fluctuations and dyskinesia, which affects gait and promotes falls [194, 196]. Exercise can improve gait through both central nervous system mechanisms, such as re-balancing sensory-motor networks and promoting attentional strategies, and peripheral mechanisms via better fitness, strength, and balance [196, 206–210, 213–225]. Mainly repetitive intensive task-oriented training was demonstrated to be effective in improving gait and balance in PD [206–212, 215, 224]. Vigorous-intensity aerobic exercise has not shown statistically significant improvements in motor and non-motor impairments in people with PD as compared to moderate/low-intensity aerobic exercise [217].

Conventional physiotherapy significantly improved motor symptoms, gait, and quality of life. Resistance training improved gait. Strategy training improved balance and gait. Dance, Nordic walking, balance and gait training, and martial arts improved motor symptoms, balance, and gait. Exergaming improved balance and quality of life. Hydrotherapy improved balance. Dual-task training did not significantly improve any of the outcomes studied [225]. The use of external auditory or visual cues and cue-augmented training improved both spatial and temporal measures (i.e., velocity, stride length, and cadence), various aspects of gait variability (i.e., length and time of step), and reduced the duration of turning and freezing of gait [196, 212–215]. Integrative approaches with neuroplasticity potentiation approaches (i.e., motor imagery, action observation) and interventions that address motor-cognitive interactions (e.g., dual-tasking) improved gait speed. However, the current evidence is too limited to allow recommendations for clinical practice [221–223].

Several RCTs demonstrated that treadmill training interventions can effectively improve gait, balance, and significantly reduce falls in people with PD. Moreover, treadmill training could be a useful approach for improving FoG symptoms, possible because treadmill-induced changes in gait parameters (i.e., speed and stride length) might indirectly reduce FoG episodes or also because often treadmill training had been combined with auditory and visual cues [224].

The European Physiotherapy Guideline stresses the importance of accurate referrals to skilled physiotherapists as a key tool for improving the quality of care in PD and offers referral criteria as to when and why physicians can consider referring patients for physiotherapy [210]. These recommendations are in line with those of the American Academy of Neurology [226] and of the English National Institute for Health and Care Excellence [227], who recommend physicians to discuss the potential of physiotherapy with persons with PD at least once a year (Fig. 8.5).

8.4.6.3 Evidence for Robotic Gait Training in Persons with Parkinson's Disease

Emerging technology-driven approaches, such as virtual reality, robotics, exergaming (i.e., gaming platforms that incorporate physical exercise), and transcranial direct-current stimulation, show some benefit in improving measures of gait, such as velocity, distance walked, step and stride length, and falls [196]. However, the current guidelines do not include rehabilitation technologies as recommended therapy options due to limited evidence [213–227].

Still, the interest in robotic locomotor training to address the motor impairment associated with gait and posture disturbances in PD is growing. Notwithstanding, no guidelines for non-pharmacological management of PD list robotic locomotor training as a recommended approach because the evidence is just about to emerge and has to be confirmed in larger studies [228–230]. In fact, few randomized controlled trials [212, 231–239] with a sample size ranging from 20 [237] to 96 [212] are available online, while the follow-up is short, often limited to the immediate post-training [228, 230].

Robotic locomotor training showed some benefits on the Unified Parkinson's Disease Rating Scale (UPDRS) [232–234, 236, 239–242], walking speed (Ten-Meter Walk Test) [212, 231–235, 242–244], the Timed Up and Go [212, 231, 233, 234, 236, 242, 244], balance (Berg Balance Scale or BBS) [233–236, 244], confidence in balance (Activities-specific Balance Confidence Scale or ABC) [233, 234, 236, 244], and spatio-temporal parameters [233–235, 237–241, 245]. However, robotic training did not seem superior to other conventional physical therapy interventions [228–230].

All enrolled trials showed robotic-training-induced improvements in outcomes measures, both specific for postural instability (i.e., BBS or Tinetti scores) and more generic gait-related measures [227–229]. Randomized controlled trials showed a possibly better effect of robotic training with respect to over-ground physiotherapy not specific for balance and gait problems at the end of the treatment [233–236] and at follow-

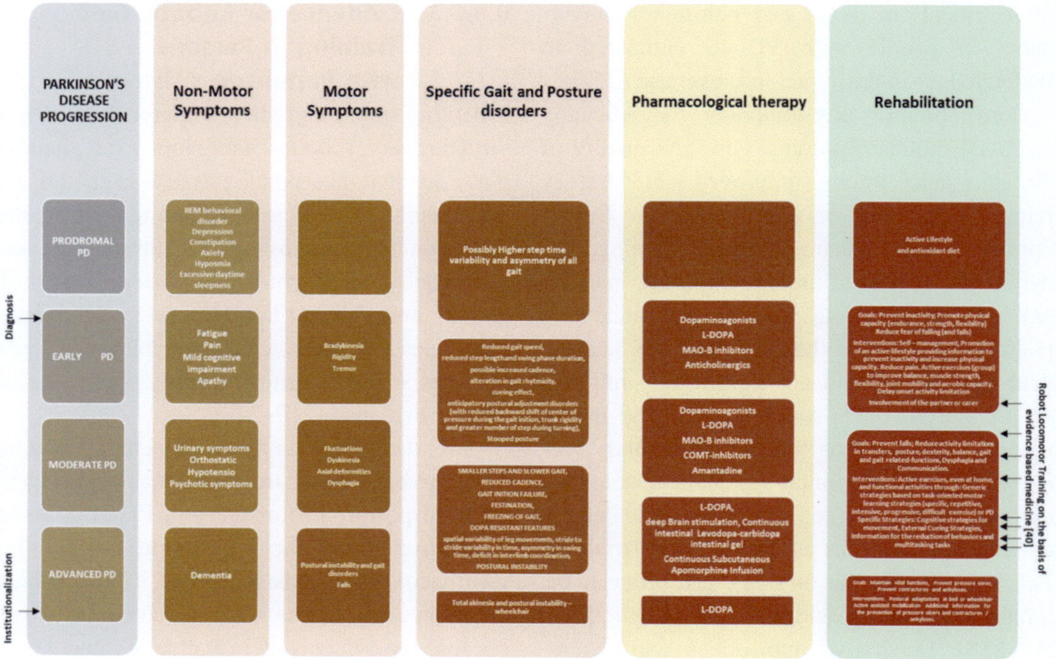

Fig. 8.5 Management of PD

up [232]. Conversely, robot-assisted gait training seems not superior to treadmill training [212, 235–238], possibly except for balance [229], and some spatial and temporal measures acquired by 3D quantitative gait analysis, i.e., step length, gait velocity, cadence, and pelvic kinematics in the frontal plane [238]. One review [229], an RCT [212], and some case series [239–241, 244] addressed the specific problem of the effect of robotic locomotor training on FoG as the main study objective. Alwardat et al. [229] included in a recent review four non-controlled studies. They noted improvements when using robotic gait training in FoG, as confirmed by an RCT [212]. Post-hoc analysis showed that people with PD who were dependent walkers benefited more than independent walkers from any gait training, whereas freezers gained more from robot-assisted than treadmill training in terms of FoG reduction.

Data about the effect and effectiveness of robotic gait training are available mainly at a short-term follow-up: at the end of the treatment [212, 237, 238, 240, 241, 244] or after the 1st [233, 234, 239, 243, 245] and 3rd [232, 235, 236, 242] month after the end of the training. At the moment, only Carda et al. assessed outcomes at six months follow-up [231].

The quality of the RCTs ranged from a score of 6–8 on the PEDro scale; concealed allocation, lack of blinding the patient or the assessor, and lack an adequate follow-up were the most often omitted study features [230].

There is, currently, no explicit study on the safety of robotic locomotor training. Few articles reported data on side effects and drop-outs, denying that they occurred during the study [212, 236–238]. The most frequently reported exclusion criteria, for being enrolled in a trial using robotic locomotor training, are suffering from severe dyskinesias or "on–off" fluctuations, deficits of somatic sensation involving the lower limbs (assessed by means of a physical and neurological examination), vestibular disorders or paroxysmal vertigo, other neurological or orthopedic conditions involving the lower limbs (musculoskeletal diseases, severe osteoarthritis, peripheral neuropathy, joint replacement), cardiovascular comorbidity (recent myocardial infarction, heart failure, uncontrolled hypertension, orthostatic hypotension) were excluded from the studies.

8.4.6.4 Practical Application of Robotic Gait Training in Persons with Parkinson's Disease

We can learn from the literature how robotic locomotor training in PD is being practiced. Both exoskeleton and end-effector devices are used: the former (Lokomat, Hocoma, Volketswil, Switzerland; Walkbot_S, P&S Mechanics, Seoul, Korea) [231, 232, 239–243, 245] more frequently than the latter (Robotic Gait Trainer—RGT Rehastim, Berlin; Gang-Trainer GT1, Reha-Stim, Berlin, Germany; G-EO System Evolution, Reha Technology, Olten, Switzerland) [212, 233–238, 244]. Device selection generally depends on the availability and not on a clinical indication. Individuals at an early–moderate (Hoehn and Yahr $\leq$ 2) [212, 231–233, 239, 242] and moderate–advanced (Hoehn and Yahr $\geq$ 3) [212, 232, 234–239, 241–245] severity level are being trained with these technologies. Carda et al. [231] enrolled only people with early stage of disease, while Nardo et al. studied the effect of robotic training on people with PD with deep brain stimulation [245].

Robotic training series (both with exoskeleton and end-effector devices) lasted mostly between 4 and 5 weeks (range: 2–5 weeks). The number of therapy sessions ranged from 10 to 20 with a frequency that could be two, three, or five times a week. The duration of each treatment session was between 20 and 45 min [230]. The training protocols are generally progressive with an aerobic program in which gait velocity was increased from 1.3 to 1.5 km/h to 2.2–3.0 km/h monitoring heart rate [230]. Most of the studies used the bodyweight support integrated into the robotic device [212, 231–243, 245]. When used, the bodyweight support was progressively reduced during the single treatment session and/or along the treatment series according to the growing confidence with the robotic device.

Some studies used robot-assisted gait training as part of more comprehensive and sophisticated training protocols, which lasted from 1 to 3 h per session [232, 237, 238]. Treatments were inpatients in two cases [232, 245]. A physiotherapy protocol performed over-ground was provided as control treatment in five randomized controlled trials [232–234, 236]. Five randomized controlled trials proposed treadmill training as comparative treatment [231, 235–237].

With regard to the treatment outcomes, both gait and balance disorders and specific scales for the assessment as well as more global scales were used (including issues related to instrumental gait analysis and axial control in real-life conditions). The most addressed robotic locomotor training outcomes and measures were:

- the overall motor symptoms and global disability of people with PD measured through a disease-specific rating scale, the Unified Parkinson's Disease Rating Scale (UPDRS) [191]
- gait velocity measured by the Ten-Meter Walk Test (10 MWT) or Timed Up And Go Test (TUG)
- walking endurance through the 6MWT
- gait spatial–temporal parameters (i.e., gait velocity, step length, step width, coefficient of variation of stride time, cadence, single and double support duration)
- balance through TUG, ABC, and the BBS [47–50, 57, 59], as well as the Fear of Falling Efficacy Scale (FFES) and the Korean version of the Falls Efficacy Scale-International (KFES).

Finally, other less frequently addressed outcomes were: the confidence in axial movements, using the Activities-Specific Balance Confidence Scale (ABC), and the quality of life, using the Parkinson's Disease Quality of life Questionnaire (PDQ 8 or 39 items).

In conclusion, robot-assisted gait training is a task-oriented forced intensive training, generally aerobic, that showed, in the management of axial-related disorders in PD, to be superior to conventional over-ground training, and more similar to treadmill training, from which it may possibly differ when evaluating balance and FoG. This difference may be related to the training approach: both robotic locomotor training and treadmill training are considered a "forced-use therapy", the former more intensive than the latter, from which robotic training also differs because individuals are constrained to produce gait cycles at greater speed symmetry and stride length than people with PD would automatically select during over-ground walking [246, 247]. Aerobic physical activity has been shown to inhibit the pro-oxidative pro-inflammatory state that increases dopaminergic neuron vulnerability and the risk of developing PD with aging while intensive task-oriented exercise promotes neuroplasticity improving learning and reducing disability progression in diagnosed people with PD [248].

8.5   Conclusion

It can be concluded that the rationale for intensive and functional gait training using robotic devices has not changed in recent years. The foundation for this rationale comprises neurophysiological phenomena and principles of motor learning. Neuronal centers at the spinal cord can be activated and trained with adequate peripheral stimuli. Such training improves EMG activity regarding modulation and amplitude. Clinically, this is relevant in situations where voluntary muscle activation is not sufficient for standing or walking. The main principles of motor learning which can be pursued with robot-assisted training are the amount of practice, distributed practice, and movement specificity. Robotic gait training has become established in the rehabilitation of patients with neurological conditions, at least in high-resourced countries. However, rehabilitation robots are not effective per se. They should rather be considered as tools that open training possibilities that wouldn't be possible without those devices. A successful application of rehabilitation robots requires the skill of therapists to meaningfully integrate robotic training into the rehabilitation program and to shape the training according to the underlying principles taking the actual state of

the patients into account. Also, the selection of patients might be crucial for achieving beneficial effects. It seems that robotic gait training is only suitable for a subgroup of patients with functional limitations and a favorable prognosis to regain ambulatory capacity. This complexity might be the reason why effectiveness trials still don't result in clear conclusions. However, systematic reviews indicate that the combined approach is most successful. The complexity might also be the reason why widely accepted clinical guidelines are still missing. Clinical studies indicate that the principles of locomotor training seem to be generally valid. However, according to the experts and the body of literature, the underlying neurological condition must be regarded when robots are applied for gait training in specific patient populations.

In recent years, there haven't been major developments in devices that are designed for locomotor training in a rehabilitation setting, such as tethered exoskeletons or end-effector devices. In contrast, the development of mobile exoskeletons has achieved significant steps. The number of commercially available mobile exoskeletons is increasing. These devices serve two purposes, to enable gait training and to allow for bi-pedal daily mobility. It can be expected that the functionality for the latter will be further refined and that in the near future, those exoskeletons will allow patients to independently move to places, that are not accessible with wheelchairs. Another observation is the growing body of studies that investigate the effects of robotic gait training combined with other means of therapeutic intervention such as functional electrical stimulation (FES) of leg muscles, repetitive transcranial magnetic brain stimulation (rTMS), or transcutaneous or implanted spinal cord stimulation. These approaches are still experimental and didn't find their way into the clinical routine so far.

Acknowledgments The coordinating authors (MW and HVH) thank Theresa Toczylowski for sharing her valuable experience in the area of robotic gait training in patients with TBI.

References

1. Kelly-Hayes M, Robertson JT, Broderick JP, Duncan PW, Hershey LA, Roth EJ, Thies WH, Trombly CA. The American heart association stroke outcome classification. Stroke. 1998;29:1274–80.

2. Waters RL, Adkins R, Yakura J, Sie I. Donal Munro lecture: functional and neurologic recovery following acute SCI. J Spinal Cord Med. 1998;21:195–9.

3. Jorgensen HS, Nakayama H, Raaschou HO, Olsen TS. Recovery of walking function in stroke patients: the Copenhagen stroke study. Arch Phys Med Rehabil. 1995;76:27–32.

4. Barbeau H, Wainberg M, Finch L. Description and application of a system for locomotor rehabilitation. Med Biol Eng Comput. 1987;25:341–4.

5. Dietz V, Wirz M, Colombo G, Curt A. Locomotor capacity and recovery of spinal cord function in paraplegic patients: a clinical and electrophysiological evaluation. Electroencephalogr Clin Neurophysiol. 1998;109:140–53.

6. Hesse S, Werner C. Partial body weight supported treadmill training for gait recovery following stroke. Adv Neurol. 2003;92:423–8.

7. Teasell RW, Bhogal SK, Foley NC, Speechley MR. Gait retraining post stroke. Top Stroke Rehabil. 2003;10:34–65.

8. Wernig A, Müller S. Laufband locomotion with body weight support improved walking in persons with severe spinal cord injuries. Paraplegia. 1992;30:229–38.

9. Wernig A, Nanassy A, Müller S. Laufband (treadmill) therapy in incomplete paraplegia and tetraplegia. J Neurotrauma. 1999;16:719–26.

10. Morrison SA, Backus D. Locomotor training: is translating evidence into practice financially feasible? J Neurol Phys Ther JNPT. 2007;31:50–4.

11. Colombo G, Joerg M, Schreier R, Dietz V. Treadmill training of paraplegic patients using a robotic orthosis. J Rehabil Res Dev. 2000;37:693–700.

12. Hesse S, Uhlenbrock D. A mechanized gait trainer for restoration of gait. J Rehabil Res Dev. 2000;37:701–8.

13. Dominici N, Ivanenko YP, Cappellini G, et al. Locomotor primitives in newborn babies and their development. Science. 2011;334:997–9.

14. Stuart DG, Hultborn H. Thomas Graham Brown (1882–1965), Anders Lundberg (1920–), and the neural control of stepping. Brain Res Rev. 2008;59:74–95.

15. Grillner S, Kozlov A. The CPGs for limbed locomotion-facts and fiction. Int J Mol Sci. 2021;22:5882.

16. Grillner S, El Manira A. Current principles of motor control, with special reference to vertebrate locomotion. Physiol Rev. 2020;100:271–320.

17. Harnie J, Audet J, Klishko AN, Doelman A, Prilutsky BI, Frigon A. The spinal control of backward locomotion. J Neurosci Off J Soc Neurosci. 2021;41:630–47.

18. Merlet AN, Harnie J, Frigon A. Inhibition and facilitation of the spinal locomotor central pattern generator and reflex circuits by somatosensory feedback from the lumbar and perineal regions after spinal cord injury. Front Neurosci. 2021;15:720542.

19. Rossignol S, Dubuc R, Gossard J-P. Dynamic sensorimotor interactions in locomotion. Physiol Rev. 2006;86:89–154.

20. Dietz V, Muller R, Colombo G. Locomotor activity in spinal man: significance of afferent input from joint and load receptors. Brain. 2002;125:2626–34.

21. Dietz V, Duysens J. Significance of load receptor input during locomotion: a review. Gait Posture. 2000;11:102–10.

22. Pearson KG. Role of sensory feedback in the control of stance duration in walking cats. Brain Res Rev. 2008;57:222–7.

23. Bouyer LJG, Rossignol S. Contribution of cutaneous inputs from the hindpaw to the control of locomotion. I. Intact cats. J Neurophysiol. 2003;90:3625–39.

24. Duysens J, Pearson KG. The role of cutaneous afferents from the distal hindlimb in the regulation of the step cycle of thalamic cats. Exp Brain Res. 1976;24:245–55.

25. Frigon A, Sirois J, Gossard J-P. Effects of ankle and hip muscle afferent inputs on rhythm generation during fictive locomotion. J Neurophysiol. 2010;103:1591–605.

26. Dietz V. Central pattern generator. Paraplegia. 1995;33:739.

27. Duysens J, van Wezel BM, van de Crommert HW, Faist M, Kooloos JG. The role of afferent feedback in the control of hamstrings activity during human gait. Eur J Morphol. 1998;36:293–9.

28. Forssberg H. Ontogeny of human locomotor control. I. Infant stepping, supported locomotion and transition to independent locomotion. Exp Brain Res. 1985;57:480–93.

29. Rayburn WF. Fetal movement monitoring. Clin Obstet Gynecol. 1995;38:59–67.

30. Lee MS, Choi YC, Lee SH, Lee SB. Sleep-related periodic leg movements associated with spinal cord lesions. Mov Disord Off J Mov Disord Soc. 1996;11:719–22.

31. Rosenfeld J, McKay W, Halter J, Pollo F, Dimitrijevic M. Evidence of a pattern generator in paralyzed subjects with spinal cord injury during spinal cord stimulation. Soc Neurosci Abstr. 1995;21.

32. Gurfinkel VS, Levik YS, Kazennikov OV, Selionov VA. Locomotor-like movements evoked by leg muscle vibration in humans. Eur J Neurosci. 1998;10:1608–12.

33. Edgerton VR, Roy RR, Hodgson JA, Prober RJ, de Guzman CP, de Leon R. Potential of adult mammalian lumbosacral spinal cord to execute and acquire improved locomotion in the absence of

supraspinal input. J Neurotrauma. 1992;9(Suppl 1): S119–28.

34. Jankowska E, Lundberg A. Interneurones in the spinal cord. Trends Neurosci. 1981;4:230–3.

35. Hamacher D, Herold F, Wiegel P, Hamacher D, Schega L. Brain activity during walking: a systematic review. Neurosci Biobehav Rev. 2015;57:310–27.

36. la Fougère C, Zwergal A, Rominger A, Förster S, Fesl G, Dieterich M, Brandt T, Strupp M, Bartenstein P, Jahn K. Real versus imagined locomotion: a [18F]-FDG PET-fMRI comparison. Neuroimage. 2010;50:1589–98.

37. Zwergal A, Linn J, Xiong G, Brandt T, Strupp M, Jahn K. Aging of human supraspinal locomotor and postural control in fMRI. Neurobiol Aging. 2012;33:1073–84.

38. Godde B, Voelcker-Rehage C. More automation and less cognitive control of imagined walking movements in high- versus low-fit older adults. Front Aging Neurosci. 2010;2:139.

39. Lacourse MG, Orr ELR, Cramer SC, Cohen MJ. Brain activation during execution and motor imagery of novel and skilled sequential hand movements. Neuroimage. 2005;27:505–19.

40. Malouin F, Richards CL, Jackson PL, Dumas F, Doyon J. Brain activations during motor imagery of locomotor-related tasks: a PET study. Hum Brain Mapp. 2003;19:47–62.

41. Wai Y-Y, Wang J-J, Weng Y-H, Lin W-Y, Ma H-K, Ng S-H, Wan Y-L, Wang C-H. Cortical involvement in a gait-related imagery task: comparison between Parkinson's disease and normal aging. Parkinsonism Relat Disord. 2012;18:537–42.

42. Edgerton VR, de Leon RD, Tillakaratne N, Recktenwald MR, Hodgson JA, Roy RR. Use-dependent plasticity in spinal stepping and standing. Adv Neurol. 1997;72:233–47.

43. Pearson KG. Neural adaptation in the generation of rhythmic behavior. Annu Rev Physiol. 2000;62:723–53.

44. Lovely RG, Gregor RJ, Roy RR, Edgerton VR. Effects of training on the recovery of full-weight-bearing stepping in the adult spinal cat. Exp Neurol. 1986;92:421–35.

45. Barbeau H, Rossignol S. Enhancement of locomotor recovery following spinal cord injury. Curr Opin Neurol. 1994;7:517–24.

46. Skinner RD, Houle JD, Reese NB, Berry CL, Garcia-Rill E. Effects of exercise and fetal spinal cord implants on the H-reflex in chronically spinalized adult rats. Brain Res. 1996;729:127–31.

47. Vilensky JA, O'Connor BL. Stepping in nonhuman primates with a complete spinal cord transection: old and new data, and implications for humans. Ann N Y Acad Sci. 1998;860:528–30.

48. Hubli M, Dietz V. The physiological basis of neurorehabilitation—locomotor training after spinal cord injury. J NeuroEngineering Rehabil. 2013;10:5.

49. Dietz V, Fouad K. Restoration of sensorimotor functions after spinal cord injury. Brain J Neurol. 2014;137:654–67.

50. Harkema SJ, Ferreira CK, van den Brand RJ, Krassioukov AV. Improvements in orthostatic instability with stand locomotor training in individuals with spinal cord injury. J Neurotrauma. 2008;25:1467–75.

51. Pearson KG, Collins DF. Reversal of the influence of group Ib afferents from plantaris on activity in medial gastrocnemius muscle during locomotor activity. J Neurophysiol. 1993;70:1009–17.

52. Dietz V. Human neuronal control of automatic functional movements: interaction between central programs and afferent input. Physiol Rev. 1992;72:33–69.

53. Harkema SJ, Hurley SL, Patel UK, Requejo PS, Dobkin BH, Edgerton VR. Human lumbosacral spinal cord interprets loading during stepping. J Neurophysiol. 1997;77:797–811.

54. Banala SK, Kim SH, Agrawal SK, Scholz JP. Robot assisted gait training with active leg exoskeleton (ALEX). IEEE Trans Neural Syst Rehabil Eng. 2009;17:2–8.

55. Wirz M, Zemon DH, Rupp R, Scheel A, Colombo G, Dietz V, Hornby TG. Effectiveness of automated locomotor training in patients with chronic incomplete spinal cord injury: a multicenter trial. Arch Phys Med Rehabil. 2005;86:672–80.

56. Katoh S, el Masry WS. Neurological recovery after conservative treatment of cervical cord injuries. J Bone Joint Surg Br. 1994;76:225–8.

57. Dietz V. Body weight supported gait training: from laboratory to clinical setting. Brain Res Bull. 2008;76:459–63.

58. Grillner S, Rossignol S. On the initiation of the swing phase of locomotion in chronic spinal cats. Brain Res. 1978;146:269–77.

59. Dietz V. Proprioception and locomotor disorders. Nat Rev Neurosci. 2002;3:781–90.

60. Dietz V, Sinkjaer T. Spastic movement disorder: impaired reflex function and altered muscle mechanics. Lancet Neurol. 2007;6:725–33.

61. Ibrahim IK, Berger W, Trippel M, Dietz V. Stretch-induced electromyographic activity and torque in spastic elbow muscles. Differential modulation of reflex activity in passive and active motor tasks. Brain J Neurol. 1993;116(Pt 4):971–89.

62. Dietz V, Quintern J, Berger W. Electrophysiological studies of gait in spasticity and rigidity. Evidence that altered mechanical properties of muscle contribute to hypertonia. Brain J Neurol. 1981;104:431–49.

63. Kloter E, Wirz M, Dietz V. Locomotion in stroke subjects: interactions between unaffected and affected sides. Brain J Neurol. 2011;134:721–31.

64. Curt A, Dietz V. Ambulatory capacity in spinal cord injury: significance of somatosensory evoked potentials and ASIA protocol in predicting outcome. Arch Phys Med Rehabil. 1997;78:39–43.

65. Curt A, Keck ME, Dietz V. Functional outcome following spinal cord injury: significance of motor-evoked potentials and ASIA scores. Arch Phys Med Rehabil. 1998;79:81–6.

66. Curt A, Dietz V. Electrophysiological recordings in patients with spinal cord injury: significance for predicting outcome. Spinal Cord. 1999;37:157–65.

67. Dobkin BH, Havton LA. Basic advances and new avenues in therapy of spinal cord injury. Annu Rev Med. 2004;55:255–82.

68. Basso DM. Neuroanatomical substrates of functional recovery after experimental spinal cord injury: implications of basic science research for human spinal cord injury. Phys Ther. 2000;80:808–17.

69. Metz GA, Curt A, van de Meent H, Klusman I, Schwab ME, Dietz V. Validation of the weight-drop contusion model in rats: a comparative study of human spinal cord injury. J Neurotrauma. 2000;17:1–17.

70. Chen K, Marsh BC, Cowan M, Al'Joboori YD, Gigout S, Smith CC, Messenger N, Gamper N, Schwab ME, Ichiyama RM. Sequential therapy of anti-Nogo-A antibody treatment and treadmill training leads to cumulative improvements after spinal cord injury in rats. Exp Neurol. 2017;292:135–44.

71. Mehrholz J, Thomas S, Kugler J, Pohl M, Elsner B, Mehrholz J, Thomas S, Kugler J, Pohl M, Elsner B. Electromechanical-assisted training for walking a er stroke (Review). 2020. https://doi.org/10.1002/14651858.CD006185.pub5.www.cochranelibrary.com

72. Tan K, Koyama S, Sakurai H, Teranishi T, Kanada Y, Tanabe S. Wearable robotic exoskeleton for gait reconstruction in patients with spinal cord injury: a literature review. J Orthop Transl. 2021;28:55–64.

73. Calabrò RS, Sorrentino G, Cassio A, et al. Robotic-assisted gait rehabilitation following stroke: a systematic review of current guidelines and practical clinical recommendations. Eur J Phys Rehabil Med. 2021;57:460–71.

74. Lajeunesse V, Vincent C, Routhier F, Careau E, Michaud F. Exoskeletons' design and usefulness evidence according to a systematic review of lower limb exoskeletons used for functional mobility by people with spinal cord injury. Disabil Rehabil Assist Technol. 2016;11:535–47.

75. Esquenazi A, Talaty M, Jayaraman A. Powered exoskeletons for walking assistance in persons with central nervous system injuries: a narrative review. PM&R. 2017;9:46–62.

76. Gagnon DH, Vermette M, Duclos C, Aubertin-Leheudre M, Ahmed S, Kairy D. Satisfaction and perceptions of long-term manual wheelchair users with a spinal cord injury upon completion of a locomotor training program with an overground robotic exoskeleton. Disabil Rehabil Assist Technol. 2019;14:138–45.

77. Thomassen G-KK, Jørgensen V, Normann B. "Back at the same level as everyone else"—user perspectives on walking with an exoskeleton, a qualitative study. Spinal Cord Ser Cases. 2019;5:103.

78. Kandilakis C, Sasso-Lance E. Exoskeletons for personal use after spinal cord injury. Arch Phys Med Rehabil. 2021;102:331–7.

79. Gagnon DH, Escalona MJ, Vermette M, Carvalho LP, Karelis AD, Duclos C, Aubertin-Leheudre M. Locomotor training using an overground robotic exoskeleton in long-term manual wheelchair users with a chronic spinal cord injury living in the community: lessons learned from a feasibility study in terms of recruitment, attendance, learnability, performance and safety. J Neuroeng Rehabil. 2018;15:12.

80. Haufe FL, Schmidt K, Duarte JE, Wolf P, Riener R, Xiloyannis M. Activity-based training with the Myosuit: a safety and feasibility study across diverse gait disorders. J Neuroeng Rehabil. 2020;17:135.

81. Hicks AL. Locomotor training in people with spinal cord injury: is this exercise? Spinal Cord. 2021;59:9–16.

82. Selves C, Stoquart G, Lejeune T. Gait rehabilitation after stroke: review of the evidence of predictors, clinical outcomes and timing for interventions. Acta Neurol Belg. 2020;120:783–90.

83. Langhorne P, Bernhardt J, Kwakkel G. Stroke rehabilitation. Lancet Lond Engl. 2011;377:1693–702.

84. Carr JH, Shepherd RB, Nordholm L, Lynne D. Investigation of a new motor assessment scale for stroke patients. Phys Ther. 1985;65:175–80.

85. Holden MK, Gill KM, Magliozzi MR. Gait assessment for neurologically impaired patients. Standards for outcome assessment. Phys Ther. 1986;66:1530–9.

86. Lam T, Noonan VK, Eng JJ, SCIRE Research Team. A systematic review of functional ambulation outcome measures in spinal cord injury. Spinal Cord. 2008;46:246–54.

87. Duddy D, Doherty R, Connolly J, McNally S, Loughrey J, Faulkner M. The effects of powered exoskeleton gait training on cardiovascular function and gait performance: a systematic review. Sensors. 2021;21:3207.

88. Chun A, Asselin PK, Knezevic S, Kornfeld S, Bauman WA, Korsten MA, Harel NY, Huang V, Spungen AM. Changes in bowel function following exoskeletal-assisted walking in persons with spinal cord injury: an observational pilot study. Spinal Cord. 2020;58:459–66.

89. Juszczak M, Gallo E, Bushnik T. Examining the effects of a powered exoskeleton on quality of life and secondary impairments in people living with spinal cord injury. Top Spinal Cord Inj Rehabil. 2018;24:336–42.

90. Escalona MJ, Brosseau R, Vermette M, Comtois AS, Duclos C, Aubertin-Leheudre M, Gagnon DH. Cardiorespiratory demand and rate of

perceived exertion during overground walking with a robotic exoskeleton in long-term manual wheelchair users with chronic spinal cord injury: a cross-sectional study. Ann Phys Rehabil Med. 2018;61:215–23.

91. World Health Organization. WHO guidelines on physical activity and sedentary behaviour. Geneva: World Health Organization; 2020.

92. Sacco RL, Kasner SE, Broderick JP, et al. An updated definition of stroke for the 21st century: a statement for healthcare professionals from the American Heart Association/American Stroke Association. Stroke. 2013;44:2064–89.

93. Adams HP, Bendixen BH, Kappelle LJ, Biller J, Love BB, Gordon DL, Marsh EE. Classification of subtype of acute ischemic stroke. Definitions for use in a multicenter clinical trial. TOAST. Trial of Org 10172 in acute stroke treatment. Stroke. 1993;24:35–41.

94. Truelsen T, Begg S, Mathers C. The global burden of cerebrovascular disease. Glob Burd Dis. 2000:1–67

95. Feigin VL, Lawes CMM, Bennett DA, Anderson CS. Stroke epidemiology: a review of population-based studies of incidence, prevalence, and case-fatality in the late 20th century. Lancet Neurol. 2003;2:43–53.

96. Pan S-L, Wu S-C, Wu T-H, Lee T-K, Hsiu T, Chen H, Wu C, Chen TH-H. Location and size of infarct on functional outcome of noncardioembolic ischemic stroke. Disabil Rehabil. (2017). https://doi.org/10.1080/09638280500404438.

97. Weimar C, Ziegler A, König IR, Diener HC. Predicting functional outcome and survival after acute ischemic stroke. J Neurol. 2002;249:888–95.

98. Bamford J, Sandercock P, Dennis M, Warlow C, Burn J. Classification and natural history of clinically identifiable subtypes of cerebral infarction. Lancet. 1991;337:1521–6.

99. Kwakkel G, Veerbeek JM, Harmeling-Van Der Wel BC, Van Wegen E, Kollen BJ. Diagnostic accuracy of the barthel index for measuring activities of daily living outcome after ischemic hemispheric stroke: does early poststroke timing of assessment matter? Stroke. 2011;42:342–6.

100. Gordon NF, Gulanick M, Costa F, Fletcher G, Franklin BA, Roth EJ, Shephard T. Physical activity and exercise recommendations for stroke survivors: an American Heart Association scientific statement from the council on clinical cardiology, subcommittee on exercise, cardiac rehabilitation, and prevention; the council on cardiovascular nursing; the council on nutrition, physical activity, and metabolism; and the stroke council. Stroke. 2004;35:1230–40.

101. Veerbeek JM, Van Wegen EEH, Harmeling Van Der Wel BC, Kwakkel G. Is accurate prediction of gait in nonambulatory stroke patients possible within 72 hours poststroke? The EPOS study. Neurorehabil Neural Repair. 2011;25:268–74.

102. Jørgensen HS, Nakayama H, Raaschou HO, Olsen TS. Recovery of walking function in stroke patients: the copenhagen stroke study. Arch Phys Med Rehabil. 1995;76:27–32.

103. Kwakkel G, Kollen BJ, Wagenaar RC. Long term effects of intensity of upper and lower limb training after stroke: a randomised trial. J Neurol Neurosurg Psychiatry. 2002;72:473–9.

104. Van De Port IGL, Kwakkel G, Schepers VPM, Lindeman E. Predicting mobility outcome one year after stroke: a prospective cohort study. J Rehabil Med. 2006;38:218–23.

105. Mayo Nancy E, Wood-Dauphinee S, Ahmed S, Carron G, Higgins J, Mcewen S, Salbach N. Disablement following stroke. Disabil Rehabil. 1999;21:258–68.

106. Royal Dutch Society for Physical Therapy. KNGF clinical practice guideline for physical therapy in patients with stroke. Pract Guidel. 2014;04:1–67.

107. Lord SE, McPherson K, McNaughton HK, Rochester L, Weatherall M. Community ambulation after stroke: how important and obtainable is it and what measures appear predictive? Arch Phys Med Rehabil. 2004;85:234–9.

108. Ada L, Dean CM, Hall JM, Bampton J, Crompton S. A treadmill and overground walking program improves walking in persons residing in the community after stroke: a placebo-controlled, randomized trial. Arch Phys Med Rehabil. 2003;84:1486–91.

109. Tieges Z, Mead G, Allerhand M, Duncan F, Van Wijck F, Fitzsimons C, Greig C, Chastin S. Sedentary behavior in the first year after stroke: a longitudinal cohort study with objective measures. Arch Phys Med Rehabil. 2015;96:15–23.

110. Rand D, Eng JJ, Tang PF, Jeng JS, Hung C. How active are people with stroke? use of accelerometers to assess physical activity. Stroke J Cereb Circ. 2009;40:163–8.

111. Michael KM, Allen JK, MacKo RF. Reduced ambulatory activity after stroke: the role of balance, gait, and cardiovascular fitness. Arch Phys Med Rehabil. 2005;86:1552–6.

112. Winstein CJ, Stein J, Arena R, et al. AHA/ASA guideline guidelines for adult stroke rehabilitation and recovery. 2016. https://doi.org/10.1161/STR.000000000000098.

113. Rand D, Eng JJ, Tang PF, Hung C, Jeng JS. Daily physical activity and its contribution to the health-related quality of life of ambulatory individuals with chronic stroke. Health Qual Life Outcomes. 2010;8:1–8.

114. Pollock A, Baer G, Pomeroy V, Langhorne P. Physiotherapy treatment approaches for the recovery of postural control and lower limb function following stroke. Cochrane Database Syst Rev. 2007. https://doi.org/10.1002/14651858.CD001920.pub2.

115. Hornby TG, Reisman DS, Ward IG, et al. Clinical practice guideline to improve locomotor function following chronic stroke, incomplete spinal cord

injury, and brain injury. J Neurol Phys Ther. 2020. https://doi.org/10.1097/NPT.0000000000000303.

116. Morone G, Paolucci S, Cherubini A, De Angelis D, Venturiero V, Coiro P, Iosa M. Robot-assisted gait training for stroke patients: current state of the art and perspectives of robotics. Neuropsychiatr Dis Treat. 2017;13:1303–11.

117. Buesing C, Fisch G, O'Donnell M, Shahidi I, Thomas L, Mummidisetty CK, Williams KJ, Takahashi H, Rymer WZ, Jayaraman A. Effects of a wearable exoskeleton stride management assist system (SMA®) on spatiotemporal gait characteristics in individuals after stroke: a randomized controlled trial. J Neuroeng Rehabil. 2015;12:1–14.

118. Watanabe H, Tanaka N, Inuta T, Saitou H, Yanagi H. Locomotion improvement using a hybrid assistive limb in recovery phase stroke patients: a randomized controlled pilot study. Arch Phys Med Rehabil. 2014;95:2006–12.

119. Awad LN, Bae J, O'Donnell K, et al. A soft robotic exosuit improves walking after stroke. Sci Transl Med. 2017. In Press.

120. Taylor CA, Bell JM, Breiding MJ, Xu L. Traumatic brain injury-related emergency department visits, hospitalizations, and deaths—United States, 2007 and 2013. MMWR Surveill Summ. 2017;66:1–16.

121. Andriessen TMJC, Jacobs B, Vos PE. Clinical characteristics and pathophysiological mechanisms of focal and diffuse traumatic brain injury. J Cell Mol Med. 2010;14:2381–92.

122. Mayer NH, Esquenazi A, Childers MK. Common patterns of clinical motor dysfunction. Muscle Nerve Suppl. 1997;6:S21-35.

123. Esquenazi A. Evaluation and management of spastic gait in patients with traumatic brain injury. J Head Trauma Rehabil. 2004;19:109–18.

124. Ochi F, Esquenazi A, Hirai B, Talaty M. Temporal-spatial feature of gait after traumatic brain injury. J Head Trauma Rehabil. 1999;14:105–15.

125. Mayer N, Esquenazi A, Keenan M. Assessing and Treating muscle overactivity in the upper motor neuron syndrome. In: Zasler N, Katz D, Zafonte R, editors. Brain Inj. Med. Princ. Pract. New York, NY, USA: DEMOS;2006. pp. 615–53.

126. Moon D, Esquenazi A. Instrumented gait analysis: a tool in the treatment of spastic gait dysfunction. JBJS Rev. 2016. https://doi.org/10.2106/JBJS.RVW.15.00076.

127. Williams G, Morris ME, Schache A, McCrory PR. People preferentially increase hip joint power generation to walk faster following traumatic brain injury. Neurorehabil Neural Repair. 2010;24:550–8.

128. Williams GP, Schache AG, Morris ME. Mobility after traumatic brain injury: relationships with ankle joint power generation and motor skill level. J Head Trauma Rehabil. 2013;28:371–8.

129. Eng JJ, Tang P-F. Gait training strategies to optimize walking ability in people with stroke: a synthesis of the evidence. Expert Rev Neurother. 2007;7:1417–36.

130. Jonkers I, Delp S, Patten C. Capacity to increase walking speed is limited by impaired hip and ankle power generation in lower functioning persons post-stroke. Gait Posture. 2009;29:129–37.

131. Vachranukunkiet T, Esquenazi A. Pathophysiology of gait disturbance in neurologic disorders and clinical presentations. Phys Med Rehabil Clin N Am. 2013;24:233–46.

132. Esquenazi A, Lee S, Packel AT, Braitman L. A randomized comparative study of manually assisted versus robotic-assisted body weight supported treadmill training in persons with a traumatic brain injury. PM R. 2013;5:280–90.

133. Esquenazi A, Lee S, Wikoff A, Packel A, Toczylowski T, Feeley J. A Comparison of locomotor therapy interventions: partial-body weight—supported treadmill, Lokomat, and G-EO training in people with traumatic brain injury. PM&R. 2017;9:839–46.

134. Lee HY, Park JH, Kim T-W. Comparisons between Locomat and Walkbot robotic gait training regarding balance and lower extremity function among non-ambulatory chronic acquired brain injury survivors. Medicine (Baltimore). 2021;100:e25125.

135. Karunakaran KK, Ehrenberg N, Cheng J, Bentley K, Nolan KJ. Kinetic gait changes after robotic exoskeleton training in adolescents and young adults with acquired brain injury. Appl Bionics Biomech. 2020;2020:1–10.

136. Treviño LR, Roberge P, Auer ME, Morales A, Torres-Reveron A. Predictors of functional outcome in a cohort of hispanic patients using exoskeleton rehabilitation for cerebrovascular accidents and traumatic brain injury. Front Neurorobotics. 2021;15:682156.

137. Esquenazi A, Lee S, Wikoff A, Packel A, Toczylowski T, Feeley J. A Randomized comparison of locomotor therapy interventions: partial body weight supported treadmill, Lokomat® and G-Eo® training in traumatic brain injury. PM R. 2016;8:S154.

138. Esquenazi A, Maier IC, Schuler TA, et al. Clinical application of robotics and technology in the restoration of walking. In: Reinkensmeyer DJ, Dietz V, editors., et al., Neurorehabilitation Technology. Cham: Springer International Publishing; 2016. p. 223–48.

139. Esquenazi A, Packel A. Robotic-assisted gait training and restoration. Am J Phys Med Rehabil. 2012;91:S217–27; quiz S228–31.

140. Reinkensmeyer DJ, Emken JL, Cramer SC. Robotics, motor learning, and neurologic recovery. Annu Rev Biomed Eng. 2004;6:497–525.

141. Dobkin BH, Plummer-D'Amato P, Elashoff R, Lee J, SIRROWS Group. International randomized clinical trial, stroke inpatient rehabilitation with reinforcement of walking speed (SIRROWS), improves outcomes. Neurorehabil Neural Repair. 2010;24:235–42.

142. Anderson KD. Targeting recovery: priorities of the spinal cord-injured population. J Neurotrauma. 2004;21:1371–83.

143. Post MW, Van Dijk AJ, Van Asbeck FW, Schrijvers AJ. Life satisfaction of persons with spinal cord injury compared to a population group. Scand J Rehabil Med. 1998;30:23–30.

144. Putzke JD, Richards JS, Hicken BL, DeVivo MJ. Predictors of life satisfaction: a spinal cord injury cohort study. Arch Phys Med Rehabil. 2002;83:555–61.

145. Spinal cord injury. https://www.who.int/newsroom/fact-sheets/detail/spinal-cord-injury. Accessed 7 Dec 2021.

146. National Spinal Cord Injury Statistical Center. https://www.nscisc.uab.edu/. Accessed 16 Dec 2021.

147. Scivoletto G, Tamburella F, Laurenza L, Torre M, Molinari M. Who is going to walk? A review of the factors influencing walking recovery after spinal cord injury. Front Hum Neurosci. 2014;8:141.

148. Ditunno PL, Patrick M, Stineman M, Ditunno JF. Who wants to walk? Preferences for recovery after SCI: a longitudinal and cross-sectional study. Spinal Cord. 2008;46:500–6.

149. Kirshblum S, Millis S, McKinley W, Tulsky D. Late neurologic recovery after traumatic spinal cord injury. Arch Phys Med Rehabil. 2004;85:1811–7.

150. Crozier KS, Graziani V, Ditunno JF, Herbison GJ. Spinal cord injury: prognosis for ambulation based on sensory examination in patients who are initially motor complete. Arch Phys Med Rehabil. 1991;72:119–21.

151. Waters RL, Adkins RH, Yakura JS, Sie I. Motor and sensory recovery following incomplete paraplegia. Arch Phys Med Rehabil. 1994;75:67–72.

152. Waters RL, Adkins RH, Yakura JS, Sie I. Motor and sensory recovery following incomplete tetraplegia. Arch Phys Med Rehabil. 1994;75:306–11.

153. Poynton AR, O'Farrell DA, Shannon F, Murray P, McManus F, Walsh MG. Sparing of sensation to pin prick predicts recovery of a motor segment after injury to the spinal cord. J Bone Joint Surg Br. 1997;79:952–4.

154. Waters RL, Adkins R, Yakura J, Vigil D. Prediction of ambulatory performance based on motor scores derived from standards of the American Spinal Injury Association. Arch Phys Med Rehabil. 1994;75:756–60.

155. Zörner B, Blanckenhorn WU, Dietz V, EM-SCI Study Group, Curt A. Clinical algorithm for improved prediction of ambulation and patient stratification after incomplete spinal cord injury. J Neurotrauma. 2010;27:241–52.

156. Phan P, Budhram B, Zhang Q, et al. Highlighting discrepancies in walking prediction accuracy for patients with traumatic spinal cord injury: an evaluation of validated prediction models using a Canadian multicenter spinal cord injury registry. Spine J Off J North Am Spine Soc. 2019;19:703–10.

157. Engel-Haber E, Zeilig G, Haber S, Worobey L, Kirshblum S. The effect of age and injury severity on clinical prediction rules for ambulation among individuals with spinal cord injury. Spine J Off J North Am Spine Soc. 2020;20:1666–75.

158. Spiess MR, Jaramillo JP, Behrman AL, Teraoka JK, Patten C. Unexpected recovery after robotic locomotor training at physiologic stepping speed: a single-case design. Arch Phys Med Rehabil. 2012;93:1476–84.

159. Dobkin B, Apple D, Barbeau H, et al. Weight-supported treadmill versus over-ground training for walking after acute incomplete SCI. Neurology. 2006;66:484–93.

160. Dobkin B, Barbeau H, Deforge D, et al. The evolution of walking-related outcomes over the first 12 weeks of rehabilitation for incomplete traumatic spinal cord injury: the multicenter randomized spinal cord injury locomotor trial. Neurorehabil Neural Repair. 2007;21:25–35.

161. Field-Fote EC, Yang JF, Basso DM, Gorassini MA. Supraspinal control predicts locomotor function and forecasts responsiveness to training after spinal cord injury. J Neurotrauma. 2017;34:1813–25.

162. Behrman AL, Harkema SJ. Locomotor training after human spinal cord injury: a series of case studies. Phys Ther. 2000;80:688–700.

163. Manella KJ, Torres J, Field-Fote EC. Restoration of walking function in an individual with chronic complete (AIS A) spinal cord injury. J Rehabil Med. 2010;42:795–8.

164. Nooijen CFJ, Ter Hoeve N, Field-Fote EC. Gait quality is improved by locomotor training in individuals with SCI regardless of training approach. J Neuroeng Rehabil. 2009;6:36.

165. Winchester P, McColl R, Querry R, Foreman N, Mosby J, Tansey K, Williamson J. Changes in supraspinal activation patterns following robotic locomotor therapy in motor-incomplete spinal cord injury. Neurorehabil Neural Repair. 2005;19:313–24.

166. Field-Fote E, Ness LL, Ionno M. Vibration elicits involuntary, step-like behavior in individuals with spinal cord injury. Neurorehabil Neural Repair. 2012;26:861–9.

167. Hoffman LR, Field-Fote EC. Functional and corticomotor changes in individuals with tetraplegia following unimanual or bimanual massed practice training with somatosensory stimulation: a pilot study. J Neurol Phys Ther JNPT. 2010;34:193–201.

168. Ness LL, Field-Fote EC. Whole-body vibration improves walking function in individuals with spinal cord injury: a pilot study. Gait Posture. 2009;30:436–40.

169. Miller LE, Zimmermann AK, Herbert WG. Clinical effectiveness and safety of powered exoskeleton-assisted walking in patients with spinal cord injury: systematic review with meta-analysis. Med Devices Auckl NZ. 2016;9:455–66.

170. Tefertiller C, Hays K, Jones J, Jayaraman A, Hartigan C, Bushnik T, Forrest GF. Initial outcomes from a multicenter study utilizing the Indego

powered exoskeleton in spinal cord injury. Top Spinal Cord Inj Rehabil. 2018;24:78–85.

171. Yang A, Asselin P, Knezevic S, Kornfeld S, Spungen AM. Assessment of in-hospital walking velocity and level of assistance in a powered exoskeleton in persons with spinal cord injury. Top Spinal Cord Inj Rehabil. 2015;21:100–9.

172. Zeilig G, Weingarden H, Zwecker M, Dudkiewicz I, Bloch A, Esquenazi A. Safety and tolerance of the ReWalk™ exoskeleton suit for ambulation by people with complete spinal cord injury: a pilot study. J Spinal Cord Med. 2012;35:96–101.

173. Atlas-3rd-Edition-Epidemiology-report-EN-updated-30-9-20.pdf.

174. Desai RA, Davies AL, Tachrount M, Kasti M, Laulund F, Golay X, Smith KJ. Cause and prevention of demyelination in a model multiple sclerosis lesion. Ann Neurol. 2016;79:591–604.

175. Dalgas U, Stenager E, Jakobsen J, Petersen T, Overgaard K, Ingemann-Hansen T. Muscle fiber size increases following resistance training in multiple sclerosis. Mult Scler Houndmills Basingstoke Engl. 2010;16:1367–76.

176. Dalgas U, Stenager E, Jakobsen J, Petersen T, Hansen HJ, Knudsen C, Overgaard K, Ingemann-Hansen T. Fatigue, mood and quality of life improve in MS patients after progressive resistance training. Mult Scler Houndmills Basingstoke Engl. 2010;16:480–90.

177. Kalron A, Allali G. Gait and cognitive impairments in multiple sclerosis: the specific contribution of falls and fear of falling. J Neural Transm Vienna Austria. 2017;124:1407–16.

178. Ahmadi MA, Arastoo DAA, Nikbakht DM. The effects of a treadmill training programme on balance, speed and endurance walking, fatigue and quality of life in people with multiple sclerosis. 9.

179. Broekmans T, Gijbels D, Eijnde BO, Alders G, Lamers I, Roelants M, Feys P. The relationship between upper leg muscle strength and walking capacity in persons with multiple sclerosis. Mult Scler Houndmills Basingstoke Engl. 2013;19:112–9.

180. Kurtzke JF. Rating neurologic impairment in multiple sclerosis: an expanded disability status scale (EDSS). Neurology. 1983;33:1444–52.

181. Khan F, Amatya B. Rehabilitation in multiple sclerosis: a systematic review of systematic reviews. Arch Phys Med Rehabil. 2017;98:353–67.

182. Zimmer P, Bloch W, Schenk A, Oberste M, Riedel S, Kool J, Langdon D, Dalgas U, Kesselring J, Bansi J. High-intensity interval exercise improves cognitive performance and reduces matrix metalloproteinases-2 serum levels in persons with multiple sclerosis: a randomized controlled trial. Mult Scler Houndmills Basingstoke Engl. 2018;24:1635–44.

183. Callesen J, Cattaneo D, Brincks J, Dalgas U. How does strength training and balance training affect gait and fatigue in patients with multiple sclerosis?

184. Pilutti LA, Paulseth JE, Dove C, Jiang S, Rathbone MP, Hicks AL. Exercise training in progressive multiple sclerosis: a comparison of recumbent stepping and body weight-supported treadmill training. Int J MS Care. 2016;18:221–9.

185. Gandolfi M, Geroin C, Picelli A, Munari D, Waldner A, Tamburin S, Marchioretto F, Smania N. Robot-assisted versus sensory integration training in treating gait and balance dysfunctions in patients with multiple sclerosis: a randomized controlled trial. Front Hum Neurosci. 2014;8:318.

186. Capogrosso M, Milekovic T, Borton D, et al. A brain-spine interface alleviating gait deficits after spinal cord injury in primates. Nature. 2016;539:284–8.

187. Hoffman J. Physiological aspects of sport training and performance. 2nd ed. Champaign, IL: Human Kinetics; 2014.

188. Dorsey ER, Bloem BR. The Parkinson pandemic—a call to action. JAMA Neurol. 2018;75:9–10.

189. GBD 2016 Neurology Collaborators. Global, regional, and national burden of neurological disorders, 1990–2016: a systematic analysis for the Global Burden of Disease Study 2016. Lancet Neurol. 2019;18:459–80.

190. Hoehn MM, Yahr MD. Parkinsonism: onset, progression and mortality. Neurology. 1967;17:427–42.

191. Goetz CG, Poewe W, Rascol O, et al. Movement disorder society task force report on the Hoehn and Yahr staging scale: status and recommendations. Mov Disord Off J Mov Disord Soc. 2004;19:1020–8.

192. van Uem JMT, Marinus J, Canning C, van Lummel R, Dodel R, Liepelt-Scarfone I, Berg D, Morris ME, Maetzler W. Health-related quality of life in patients with Parkinson's disease—a systematic review based on the ICF model. Neurosci Biobehav Rev. 2016;61:26–34.

193. Rinalduzzi S, Trompetto C, Marinelli L, Alibardi A, Missori P, Fattapposta F, Pierelli F, Currà A. Balance dysfunction in Parkinson's disease. BioMed Res Int. 2015;2015:434683.

194. Armstrong MJ, Okun MS. Diagnosis and treatment of Parkinson disease: a review. JAMA. 2020;323:548–60.

195. Fereshtehnejad S-M, Zeighami Y, Dagher A, Postuma RB. Clinical criteria for subtyping Parkinson's disease: biomarkers and longitudinal progression. Brain. 2017;140:1959–76.

196. Mirelman A, Bonato P, Camicioli R, Ellis TD, Giladi N, Hamilton JL, Hass CJ, Hausdorff JM, Pelosin E, Almeida QJ. Gait impairments in Parkinson's disease. Lancet Neurol. 2019;18:697–708.

197. Galna B, Lord S, Burn DJ, Rochester L. Progression of gait dysfunction in incident Parkinson's disease:

impact of medication and phenotype. Mov Disord Off J Mov Disord Soc. 2015;30:359–67.

198. Pistacchi M, Gioulis M, Sanson F, De Giovannini E, Filippi G, Rossetto F, Zambito Marsala S. Gait analysis and clinical correlations in early Parkinson's disease. Funct Neurol. 2017;32:28–34.

199. Mellone S, Mancini M, King LA, Horak FB, Chiari L. The quality of turning in Parkinson's disease: a compensatory strategy to prevent postural instability? J Neuroeng Rehabil. 2016;13:39.

200. Son M, Youm C, Cheon S, Kim J, Lee M, Kim Y, Kim J, Sung H. Evaluation of the turning characteristics according to the severity of Parkinson disease during the timed up and go test. Aging Clin Exp Res. 2017;29:1191–9.

201. Nutt JG, Bloem BR, Giladi N, Hallett M, Horak FB, Nieuwboer A. Freezing of gait: moving forward on a mysterious clinical phenomenon. Lancet Neurol. 2011;10:734–44.

202. Capecci M, Pepa L, Verdini F, Ceravolo MG. A smartphone-based architecture to detect and quantify freezing of gait in Parkinson's disease. Gait Posture. 2016;50:28–33.

203. Zhang F, Shi J, Duan Y, Cheng J, Li H, Xuan T, Lv Y, Wang P, Li H. Clinical features and related factors of freezing of gait in patients with Parkinson's disease. Brain Behav. 2021;11:e2359.

204. Amboni M, Barone P, Picillo M, Cozzolino A, Longo K, Erro R, Iavarone A. A two-year follow-up study of executive dysfunctions in parkinsonian patients with freezing of gait at on-state. Mov Disord Off J Mov Disord Soc. 2010;25:800–2.

205. Palmerini L, Rocchi L, Mazilu S, Gazit E, Hausdorff JM, Chiari L. Identification of characteristic motor patterns preceding freezing of gait in Parkinson's disease using wearable sensors. Front Neurol. 2017;8:394.

206. Okun MS. Management of Parkinson disease in 2017: personalized approaches for patient-specific needs. JAMA. 2017;318:791–2.

207. Bouça-Machado R, Rosário A, Caldeira D, Castro Caldas A, Guerreiro D, Venturelli M, Tinazzi M, Schena F, Ferreira JJ. Physical activity, exercise, and physiotherapy in Parkinson's disease: defining the concepts. Mov Disord Clin Pract. 2020;7:7–15.

208. Martignon C, Pedrinolla A, Ruzzante F, Giuriato G, Laginestra FG, Bouça-Machado R, Ferreira JJ, Tinazzi M, Schena F, Venturelli M. Guidelines on exercise testing and prescription for patients at different stages of Parkinson's disease. Aging Clin Exp Res. 2021;33:221–46.

209. Kim Y, Lai B, Mehta T, Thirumalai M, Padalabalanarayanan S, Rimmer JH, Motl RW. Exercise training guidelines for multiple sclerosis, stroke, and Parkinson disease: rapid review and synthesis. Am J Phys Med Rehabil. 2019;98:613–21.

210. Domingos J, Keus SHJ, Dean J, de Vries NM, Ferreira JJ, Bloem BR. The European physiotherapy guideline for Parkinson's disease: implications for neurologists. J Park Dis. 2018;8:499–502.

211. Curtze C, Nutt JG, Carlson-Kuhta P, Mancini M, Horak FB. Levodopa is a double-edged sword for balance and gait in people with Parkinson's disease. Mov Disord Off J Mov Disord Soc. 2015;30:1361–70.

212. Capecci M, Pournajaf S, Galafate D, et al. Clinical effects of robot-assisted gait training and treadmill training for Parkinson's disease. A randomized controlled trial. Ann Phys Rehabil Med. 2019;62:303–12.

213. Cosentino C, Baccini M, Putzolu M, Ristori D, Avanzino L, Pelosin E. Effectiveness of physiotherapy on freezing of gait in Parkinson's disease: a systematic review and meta-analyses. Mov Disord Off J Mov Disord Soc. 2020;35:523–36.

214. Okada Y, Ohtsuka H, Kamata N, et al. Effectiveness of long-term physiotherapy in Parkinson's disease: a systematic review and meta-analysis. J Park Dis. 2021;11:1619–30.

215. Lorenzo-García P, Cavero-Redondo I, Torres-Costoso AI, Guzmán-Pavón MJ, Núñez de Arenas-Arroyo S, Álvarez-Bueno C. Body weight support gait training for patients with Parkinson disease: a systematic review and meta-analyses. Arch Phys Med Rehabil. 2021;102:2012–21.

216. van de Wetering-van Dongen VA, Kalf JG, van der Wees PJ, Bloem BR, Nijkrake MJ. The effects of respiratory training in Parkinson's disease: a systematic review. J Park Dis. 2020;10:1315–33.

217. Rodríguez MÁ, Albillos-Almaraz L, López-Aguado I, Crespo I, Del Valle M, Olmedillas H. Vigorous aerobic exercise in the management of parkinson disease: a systematic review. PM R. 2021;13:890–900.

218. Choi H-Y, Cho K-H, Jin C, et al. Exercise therapies for Parkinson's disease: a systematic review and meta-analysis. Park Dis. 2020;2020:2565320.

219. Abou L, Alluri A, Fliflet A, Du Y, Rice LA. Effectiveness of physical therapy interventions in reducing fear of falling among individuals with neurologic diseases: a systematic review and meta-analysis. Arch Phys Med Rehabil. 2021;102:132–54.

220. de Almeida HS, Porto F, Porretti M, Lopes G, Fiorot D, Bunn PDS, da Silva EB. Effect of dance on postural control in people with Parkinson's disease: a meta-analysis review. J Aging Phys Act. 2020;29:130–41.

221. Johansson H, Hagströmer M, Grooten WJA, Franzén E. Exercise-induced neuroplasticity in Parkinson's disease: a metasynthesis of the literature. Neural Plast. 2020;2020:8961493.

222. Perrochon A, Borel B, Istrate D, Compagnat M, Daviet J-C. Exercise-based games interventions at home in individuals with a neurological disease: a systematic review and meta-analysis. Ann Phys Rehabil Med. 2019;62:366–78.

223. Wang B, Shen M, Wang Y-X, He Z-W, Chi S-Q, Yang Z-H. Effect of virtual reality on balance and gait ability in patients with Parkinson's disease: a

223. systematic review and meta-analysis. Clin Rehabil. 2019;33:1130–8.

224. Robinson AG, Dennett AM, Snowdon DA. Treadmill training may be an effective form of task-specific training for improving mobility in people with Parkinson's disease and multiple sclerosis: a systematic review and meta-analysis. Physiotherapy. 2019;105:174–86.

225. Radder DLM, Silva L, de Lima A, Domingos J, Keus SHJ, van Nimwegen M, Bloem BR, de Vries NM. Physiotherapy in Parkinson's disease: a meta-analysis of present treatment modalities. Neurorehabil Neural Repair. 2020;34:871–80.

226. Cheng EM, Tonn S, Swain-Eng R, Factor SA, Weiner WJ, Bever CT, American Academy of Neurology Parkinson Disease Measure Development Panel. Quality improvement in neurology: AAN Parkinson disease quality measures: report of the quality measurement and reporting subcommittee of the American Academy of Neurology. Neurology. 2010;75:2021–7.

227. Rogers G, Davies D, Pink J, Cooper P. Parkinson's disease: summary of updated NICE guidance. BMJ. 2017;358:j1951.

228. Alwardat M, Etoom M, Al Dajah S, Schirinzi T, Di Lazzaro G, Sinibaldi Salimei P, Biagio Mercuri N, Pisani A. Effectiveness of robot-assisted gait training on motor impairments in people with Parkinson's disease: a systematic review and meta-analysis. Int J Rehabil Res Int Z Rehabil Rev Int Rech Readaptation. 2018;41:287–96.

229. Alwardat M, Etoom M. Effectiveness of robot-assisted gait training on freezing of gait in people with Parkinson disease: evidence from a literature review. J Exerc Rehabil. 2019;15:187–92.

230. Picelli A, Capecci M, Filippetti M, et al. Effects of robot-assisted gait training on postural instability in Parkinson's disease: a systematic review. Eur J Phys Rehabil Med. 2021;57:472–7.

231. Carda S, Invernizzi M, Baricich A, Comi C, Croquelois A, Cisari C. Robotic gait training is not superior to conventional treadmill training in parkinson disease: a single-blind randomized controlled trial. Neurorehabil Neural Repair. 2012;26:1027–34.

232. Furnari A, Calabrò RS, De Cola MC, Bartolo M, Castelli A, Mapelli A, Buttacchio G, Farini E, Bramanti P, Casale R. Robotic-assisted gait training in Parkinson's disease: a three-month follow-up randomized clinical trial. Int J Neurosci. 2017;127:996–1004.

233. Picelli A, Melotti C, Origano F, Waldner A, Gimigliano R, Smania N. Does robotic gait training improve balance in Parkinson's disease? A randomized controlled trial. Parkinsonism Relat Disord. 2012;18:990–3.

234. Picelli A, Melotti C, Origano F, Waldner A, Fiaschi A, Santilli V, Smania N. Robot-assisted gait training in patients with Parkinson disease: a randomized controlled trial. Neurorehabil Neural Repair. 2012;26:353–61.

235. Picelli A, Melotti C, Origano F, Neri R, Waldner A, Smania N. Robot-assisted gait training versus equal intensity treadmill training in patients with mild to moderate Parkinson's disease: a randomized controlled trial. Parkinsonism Relat Disord. 2013;19:605–10.

236. Picelli A, Melotti C, Origano F, Neri R, Verzè E, Gandolfi M, Waldner A, Smania N. Robot-assisted gait training is not superior to balance training for improving postural instability in patients with mild to moderate Parkinson's disease: a single-blind randomized controlled trial. Clin Rehabil. 2015;29:339–47.

237. Sale P, De Pandis MF, Le Pera D, Sova I, Cimolin V, Ancillao A, Albertini G, Galli M, Stocchi F, Franceschini M. Robot-assisted walking training for individuals with Parkinson's disease: a pilot randomized controlled trial. BMC Neurol. 2013;13:50.

238. Galli M, Cimolin V, De Pandis MF, Le Pera D, Sova I, Albertini G, Stocchi F, Franceschini M. Robot-assisted gait training versus treadmill training in patients with Parkinson's disease: a kinematic evaluation with gait profile score. Funct Neurol. 2016;31:163–70.

239. Barbe MT, Cepuran F, Amarell M, Schoenau E, Timmermann L. Long-term effect of robot-assisted treadmill walking reduces freezing of gait in Parkinson's disease patients: a pilot study. J Neurol. 2013;260:296–8.

240. Lo AC, Chang VC, Gianfrancesco MA, Friedman JH, Patterson TS, Benedicto DF. Reduction of freezing of gait in Parkinson's disease by repetitive robot-assisted treadmill training: a pilot study. J Neuroeng Rehabil. 2010;7:51.

241. Ustinova K, Chernikova L, Bilimenko A, Telenkov A, Epstein N. Effect of robotic locomotor training in an individual with Parkinson's disease: a case report. Disabil Rehabil Assist Technol. 2011;6:77–85.

242. Paker N, Bugdayci D, Goksenoglu G, Sen A, Kesiktas N. Effects of robotic treadmill training on functional mobility, walking capacity, motor symptoms and quality of life in ambulatory patients with Parkinson's disease: a preliminary prospective longitudinal study. NeuroRehabilitation. 2013;33:323–8.

243. Yun SJ, Lee HH, Lee WH, Lee SH, Oh B-M, Seo HG. Effect of robot-assisted gait training on gait automaticity in Parkinson disease: a prospective, open-label, single-arm, pilot study. Medicine (Baltimore). 2021;100:e24348.

244. Pilleri M, Weis L, Zabeo L, Koutsikos K, Biundo R, Facchini S, Rossi S, Masiero S, Antonini A. Overground robot assisted gait trainer for the treatment of drug-resistant freezing of gait in Parkinson disease. J Neurol Sci. 2015;355:75–8.

245. Nardo A, Anasetti F, Servello D, Porta M. Quantitative gait analysis in patients with Parkinson treated with deep brain stimulation: the effects of a robotic gait training. NeuroRehabilitation. 2014;35:779–88.
246. Pohl M, Rockstroh G, Rückriem S, Mrass G, Mehrholz J. Immediate effects of speed-dependent treadmill training on gait parameters in early Parkinson's disease. Arch Phys Med Rehabil. 2003;84:1760–6.
247. Hesse S, Schmidt H, Werner C, Bardeleben A. Upper and lower extremity robotic devices for rehabilitation and for studying motor control. Curr Opin Neurol. 2003;16:705–10.
248. Petzinger GM, Fisher BE, McEwen S, Beeler JA, Walsh JP, Jakowec MW. Exercise-enhanced neuroplasticity targeting motor and cognitive circuitry in Parkinson's disease. Lancet Neurol. 2013;12:716–26.

Part III

Principles for Interactive Rehabilitation Technology

Designing User-Centered Technologies for Rehabilitation Challenge that Optimize Walking and Balance Performance

David A. Brown, Kelli L. LaCroix, Saleh M. Alhirsan, Carmen E. Capo-Lugo, Rebecca W. Hennessy, and Christopher P. Hurt

Abstract

The purpose of this chapter is to provide the reader with motivation to design technology solutions that provide effective levels of challenge during walking and balance training activities for people with motor disabilities. When clinicians choose to use certain technologies for training, they are investing time and effort toward an activity that they expect will result in greater abilities and skills as outcomes. During training sessions, there are many factors at play that determine how well a person participates and benefits from the time that is spent relearning skills and increasing abilities. In addition to the usual considered physiological factors such as strength, and skill, a client must overcome psychological barriers during the training session. Our focus is on using rehabilitation robotics and virtual reality to provide safety and assurance, along with motivation and enjoyment, which is necessary to encourage high levels of balance and walking training performance so that clients get the most benefit during the training session. First, we describe some exemplar frameworks of challenge (Regulatory Focus Theory and OPTIMAL Theory) and their relation to high walking and balance training performance, including overcoming psycho-behavioral barriers to achieving high training performance. Second, we explore device considerations when designing for high training performance, including mechanisms to provide safety from falls, allowance for a large variety of ecologically valid walking and balance tasks with graded difficulty, motivational experiences with sustained attention, and added resistance to increase walking workload. We conclude by providing recommendations for future device development with an introductory guide to a design process that is user-centered from the start.

D. A. Brown (✉) · K. L. LaCroix
University of Texas Medical Branch, Galveston, TX, USA
e-mail: davibrow@utmb.edu

K. L. LaCroix
e-mail: klfaaiti@UTMB.EDU

S. M. Alhirsan · C. E. Capo-Lugo · R. W. Hennessy · C. P. Hurt
University of Alabama at Birmingham, Birmingham, AL, USA
e-mail: alhirsan@uab.edu

C. E. Capo-Lugo
e-mail: capolugo@uab.edu

R. W. Hennessy
e-mail: rwhite31@uab.edu

C. P. Hurt
e-mail: cphurt@uab.edu

Keywords

Walking and balance training performance · Rehabilitation technology · Physical challenge · Motivation · Attention · Augmented feedback · Psychomotor performance · User-centered design

© The Author(s), under exclusive license to Springer Nature Switzerland AG 2022
D. J. Reinkensmeyer et al. (eds.), *Neurorehabilitation Technology*,
https://doi.org/10.1007/978-3-031-08995-4_9

9.1 Frameworks of Promotion, Prevention, Motivation, and Attention and Their Relation to High Performance

After brain injury, with stroke and other acquired neurological conditions, recovery of walking and balance is a tricky business. In addition to the acquired loss of motor control, there are long-term, secondary conditions, ranging from muscle atrophy and cardiovascular deconditioning to psycho-behavioral maladaptation, such as fear of falling, malaise, and apathy, that make the recovery process more difficult. Clinical solutions require a multi-pronged approach to addressing the requirements for successful walking and balance recovery (Table 9.1). One major clinical objective is to create an environment that allows clients to perform at their highest level so that they can most benefit from the training regime.

While the process of physical rehabilitation is usually conducted by trained professional clinicians, clients may perceive that their safety is at risk when practicing challenging walking and balance tasks. The very act of standing up out of a wheelchair can be fraught with anxiety and dread unless proper safety precautions are taken, and the client is reassured that no harm will come to them if

Table 9.1 Requirements for successful walking and balance recovery

1. Generate and modulate muscle force
2. Coordinate muscle activity (intra- and inter-limb)
3. Respond to perturbations
4. Adapt to environmental changes
5. Modulate and control the speed
6. Process and integrate sensory information
7. Maintain psychological and cognitive function
8. Develop cardiovascular and aerobic capacity
9. Develop muscular endurance
10. Maintain musculoskeletal integrity and viability
11. Strategize and problem-solve
12. Strive for social interaction and personal independence

they participate in the upcoming training activities. When designing technological solutions, one must understand these factors and seek to find ways to assure safety and protection against harm. Several frameworks seek to explain the psycho-behavioral influencers toward fully engaged and motivated performance during training under conditions where fear of failure and/or harm are present. We will discuss two such frameworks (The Regulatory Focus Theory and Optimizing Performance through Intrinsic Motivation and Attention for Learning (OPTIMAL Theory)) with the aim of presenting ways that they can be applied when designing technological solutions to enhance the performance of clients during training sessions.

9.1.1 The Regulatory Focus Theory

A framework that can provide insight into the decision-making process and perception of the individual involved in recovery is the *Regulatory Focus Theory* [1]. This framework has been used in a variety of disciplines, including marketing [2], public safety [3], exercise physiology [4], and education [5]. E. Tory Higgins proposed the *Regulatory Focus Theory* in 1997 and 1998, which is based on an expansion of the hedonistic principle (i.e., approach of pleasure and avoidance of pain) [6]. It has since been used as a framework to study the impact of human motivation stemming from two self-regulatory strategies: "promotion-focus" and "prevention-focus". A promotion-focus individual tends to regulate behavior based on goal-oriented ideas and aspirations (pursuing positive outcomes). A prevention-focused individual, however, tends to base behavior on security, duty, and responsibility (avoiding negative outcomes) [7, 8]. This can also translate to responses to non-verbal cues as promotion-focus people prefer eager engagement strategies as opposed to prevention-focus people who prefer vigilant engagement with tasks [9]. For example, a promotion-focused person will benefit from interactions with audience members who applaud and cheer risk-taking, while a prevention-focused person will benefit from close proximity and handling by the clinician, with a clinician body stance that belies readiness to catch a person who loses balance.

Regulatory Fit implies that individuals use preferred engagement strategies "fit" their current regulatory orientation (i.e., either promotion or prevention) when performing certain tasks. The "fit" applies to the strategy that is most closely oriented to the focus of the person performing the task [1, 10]. It is important to note that the Regulatory Fit can be applied to an individual in any given circumstance—no one regulatory focus is superior to the other. It has been shown that high vs. low levels of stressors can differentially impact the performance of those who are promotion vs. preventions focus, which highlights the advantage of individuals who can modify their motivational focus based on fluctuating situations (11).

Clinicians try to maximize rehabilitation outcomes through individualized client-centered care but encouraging full engagement in the rehabilitation training efforts is one of many objectives of rehabilitation efforts. It is therefore essential to create an appropriate level of rehabilitation challenge to motivate individuals to attain higher levels of performance while not causing undue discouragement if they are unable to complete the requested tasks. It has been shown that the effective regulation of challenges can coincide with promotion-focused success as well as emotional well-being [12]. Franks et al. showed that having too many or too few challenges can result in poor mental health. It was concluded that challenge *regulation*, not necessarily minimization, is needed for optimal well-being [12]. Pfeffer et al. found evidence that having a regulatory fit within these challenges can lead to stronger intentions to perform physical activities. In the same study, it was suggested that matching printed messages (handouts that emphasized either "possible gains" or "costs avoided") that match the regulatory focus of the individual is useful to enhance motivation for physical activity [4]. Maximizing the appropriate level of rehabilitation efforts also means maximizing communication and client motivation. We propose that there is value in knowing a client's motivational focus and matching that focus with rehabilitation materials and

instruction that will better motivate and engage them. An individual who is motivated in a way that fits their focus will feel more at ease and is more likely to perceive the event as positive and of value.

It is proposed that the Regulatory Fit Theory can be used as a general framework for understanding how the clinician and client communication impacts client participation during rehabilitation (Fig. 9.1). This promotion vs. prevention model addresses motivational aspects of client decision-making to participate and perception of rehabilitation value, thus highlighting the importance of tailoring efforts to the individual in a way that maximizes motivational fit between the client and clinician. It is hypothesized that creating a rehabilitation environment meant to cultivate motivational fit for clients will increase client effort (observed as increased strenuous physical or mental exertion). Ideally, new technologies have the flexibility to allow for each focus. For example, technological solutions may be designed to allow a prevention-focused, safety-enhanced technology meant to combat decreased walking performance due to fall avoidance (a prevention-focused behavior), which acts as a barrier to achieving maximum walking capacity. It is expected that a safety-enhanced environment will not only allow individuals to perform at a higher level than during standard overground tests that are performed with minimal safeguards, but it would also decrease the clinician's burden of ensuring client fall safety. In addition, safety constraints may be

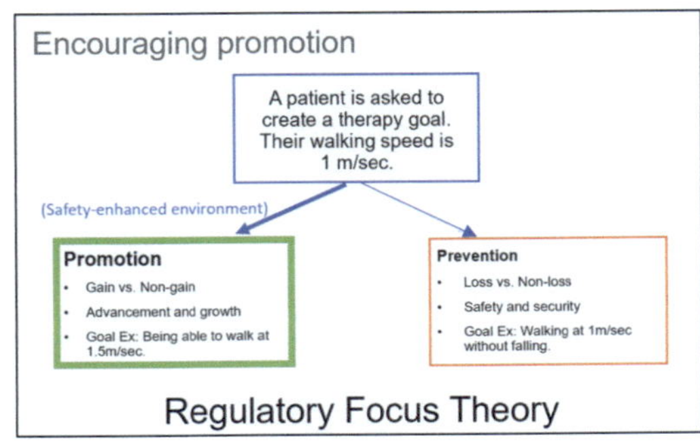

Fig. 9.1 Diagram of the application of the Regulatory Focus Theory to balance/walking goals

relaxed to allow more challenge. Technological solutions may be designed to allow flexibility to be used for a promotion-focused, safety challenge element meant to encourage individuals to overcome situations where a new strategic approach to balance and walking will result in greater confidence and self-reliance in dealing with real-world situations. As a result, when clients present themselves with a prevention-based or with a promotion-based mindset, the technology can be switched into modes that provide the correct fit.

9.1.2 Optimizing Performance through Intrinsic Motivation and Attention for Learning Theory

Another theoretical framework to consider when designing technological solutions to enhance training performance during walking and balance recovery involves amplifying both intrinsic motivation and attention. Intrinsic motivation is conceptualized as the act of engaging in a task or activity due to the inherent enjoyment or challenge involved rather than to seek external rewards or avoid punishments [13]. The opposite of intrinsic motivation is extrinsic motivation, where the focus is on external rewards or avoidance of punishment [13]. In poststroke rehabilitation, one study found that most patients are extrinsically motivated during early rehabilitation, and their intrinsic motivation slightly increased toward the final phase of rehabilitation. This increase in intrinsic motivation seemed to result in improvement in their ability to perform activities of daily living [14]. In this case, intrinsic motivation may be regarded as an essential construct that reflects recovery and adaptation [13, 14].

The Optimizing Performance through Intrinsic Motivation and Attention for Learning (OPTI-MAL) Theory suggests that both motivational and attentional constructs contribute to performance and learning by strengthening the connection between goals and actions [15]. Based on this theory, several other constructs also influence performance and learning, including enhancing expectancies, autonomy, and an external focus of attention. Expectancies refer to a range of forward-directed anticipatory or predictive cognitions or beliefs about what is to occur. The type of feedback received (i.e., positive or negative) enhances expectancies and reflects people's implicit (non-conscious) knowledge. For example, if you are in a racing competition with other competitors, you can estimate your ability to win the race based on where you are in the race. The racers behind you indicate that you are walking faster than they are while the racers in front of you indicate that you are walking slower than they are. Thus, you can expect that if you increase your walking speed, you might catch up to the racer in front of you while you might lose your current position to the race behind you if you slow down. Autonomy is the sense of volition people have during task performance (i.e., self-control). Finally, the external focus of attention refers to the ability to concentrate on the task at hand and not the movement of individual body segments by filtering out information when there are distractions [15].

The OPTIMAL Theory classifies autonomy and enhanced expectancies as motivational constructs. It also classifies external focus as an attentional construct. Both motivational and attentional constructs impact motor performance and motor learning via enhancing focus on task goals and minimizing self-focus [15]. Thus, it is expected that individuals with high motivation (desire to achieve some result) might perform motor tasks better than those with low or lack of motivation.

Performance of motor tasks at high intensities is one of the most effective methods to enhance function after a stroke [16]. People with hemiparesis can perform movements faster than they typically choose to, and when they do, movement quality is improved [17]. For example, they can walk faster than their self-selected walking speed when provided with an appropriate environment [17]. During gait training sessions, clinicians provide feedback in various forms, including verbal remarks and demonstrations [18]. These methods allow patients to become aware of how they move and to correct their compensatory movement strategies.

There are mainly two types of performance feedback [19]: (1) intrinsic feedback, which refers to the sensory-perceptual information that each of the sensory systems provides during a task and (2) augmented (external) feedback, described as the addition to or improvement of the body's intrinsic feedback mechanism. Augmented feedback can be provided in several forms and times (i.e., before, during, and after movement). Before the movement, the instructor or coach might provide information (instructions) to augment how the performer thinks about or conceptualizes approaching the task. During movement, information might be provided when a person performs a continuous task (i.e., walking). This type of feedback is often termed concurrent feedback because the instructor provides it during an ongoing movement with the aim to alter the movement.

Augmented feedback after a movement provides a result of the movement itself. After movement, augmented feedback can be provided as knowledge of performance or knowledge of results. Knowledge of performance provides information about the movement that has just been made and sometimes is referred to as kinematic feedback. Knowledge of results provides information related to the outcome; it is based on achieving the goal or task [20]. Augmented feedback plays a significant role in enhancing the performer's motivation by (1) directing their focus of attention and (2) providing information about errors and corrective actions. These, taken together, enhance performance and encourage more attempts in a motor task.

Augmented feedback provided during the movement in a virtual reality (VR) system provided during the movement (i.e., knowledge of performance) allows for better movement patterns and motor learning than augmented feedback given after completing the motor task (i.e., knowledge of results) in individuals after stroke [21]. It has been shown that treatments where feedback is provided by innovative technology are the most effective in improving gait parameters and functional recovery of patients with gait dysfunction [22].

For rehabilitation purposes, VR has been delivered and used in various forms (i.e., non-game-based VR applications and game-based VR applications). Clinical practice guidelines suggest that high-intensity gait training with virtual reality (VR) is an effective way to enhance walking after six months of stroke onset [18]. The non-game-based VR applications generally use VR features (i.e., immersion or presence and real-time interaction) to mimic and simulate the real-world training conditions in the VR environment without game scoring systems (e.g., experience constraints with walking in daily life and stepping over virtual objects with vibrotactile stimulus) [23, 24]. These systems usually use haptic sensations, sounds, and different movement visualizations and optical flow to provide real-time movement feedback [20, 23, 24].

Game-based VR applications in rehabilitation, usually referred to as VR exergames (a combination of exercise and video games), have been used as adjuncts to gait training to enhance enjoyment [25, 26]. VR exergames implement game features into activities and tasks that increase task enjoyment and motivation by making tasks feel more play-like [25, 26]. The Theory of Work Gamification suggests that gamification provides motivational and performance benefits through two main pathways, informational and affective [25]. The informational pathway includes access to visual and immediate feedback while the affective pathway includes task enjoyment. Feedback, challenge, and rewards are the fundamental mechanisms by which exergames provide enjoyable training environments [27]. For example, interacting with a game that uses simple visual feedback results in improved accuracy of a movement (i.e., performing the right and correct motor task) compared to performing an exercise from memory or with limited feedback (e.g., instructional video or demonstration) [26]. Also, VR exergames can provide objective feedback for both therapists and patients. For example, it can provide feedback on the quality of movement, allowing the patient to adjust their movements and focus on their treatment [26].

To provide a framework that outlines the influential pathways affecting motor performance, we developed a conceptual framework that describes the connection between the OPTIMAL Theory and the Theory of Work Gamification. We developed The Enhanced OPTIMAL Theory by combining the OPTIMAL Theory and the Theory of Work Gamification (Fig. 9.2) to provide possible relevant constructs associated with motivation and motor performance inducement [28]. The Enhanced OPTIMAL Theory details four pathways that could be manipulated to induce immediate changes to motor performance

(Fig. 9.2). The pathways are (1) motivation, (2) attention, (3) informational, and (4) affective. The affective pathway and the informational pathway are both pathways that enhance or reduce motivation. The attention and motivation pathways induce changes in locomotor training session performance through an increased focus on the task. A non-motivated, unengaged, bored, inattentive client who is working with a clinician will not learn much from the training session and the technology will be relatively useless. The clinician makes the decision to engage the client in a meaningful training task that is meant to help a person acquire a skill that they can learn to use in the real world.

Within the context of walking and balance training, the environment would be set up to provide elements of feedback about task effectiveness and elements of enjoyment with attention to accomplishing the task and discovering self-derived solutions that enhance the sense of autonomy. One example of an exergame scenario that can incorporate all elements of The Enhanced OPTIMAL framework might be a stepping task that incorporates a virtual environment containing balloons of different sizes that are strewn across the path. As a person walks, they are instructed to kick as many balloons as possible to clear the path (autonomy).

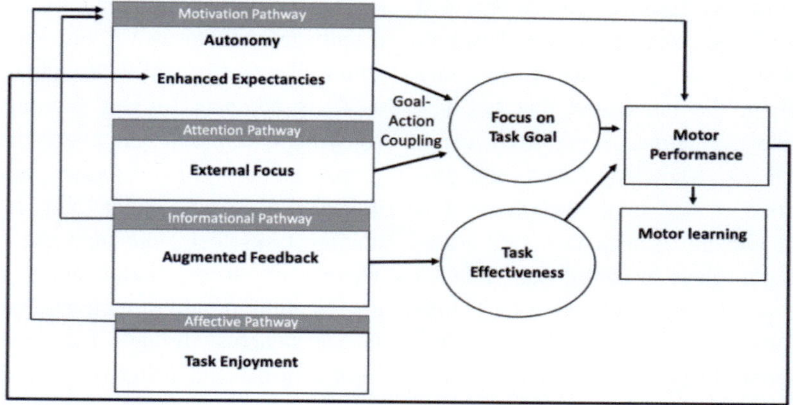

Fig. 9.2 The enhanced OPTIMAL Theory—a conceptual framework [28] adapted from the combination of OPTIMAL Theory (providing elements of Goal-Action coupling) and A Theory of Work Gamification (adding the element of Task Enjoyment to enhance Motivation) (from *licensed under CC-BY 4.0*)

The more balloons that they kick, the more points that they acquire (enhanced expectancies). The larger balloons may be worth more points and can motivate the gamer to preferentially aim for them while walking. The balloons can also be strategically placed so that the gamer will need to reach further with the kicking leg, outside the straight and narrow path, in order to kick them successfully (external focus). To add further enjoyment, the balloons may "pop" and streamers can be propelled into the air (augmented feedback with task enjoyment).

In summary, the frameworks discussed in this section provide guideposts for creating training environments that possess:

1. elements of prevention-focus for individuals that prefer enhanced safety and reduced anxiety during intensive training;
2. elements of promotion-focus for individuals that prefer challenges that are goal-oriented and allow a person to feel a sense of accomplishment;
3. provide autonomy and augmented feedback with enhanced expectancies that demonstrate a direct relationship between a person's actions and the task achievement or failure;
4. provide an experience of enjoyment that engages a person's attention over long periods of high-intensity training and motivates them to improve performance over several sessions.

9.2 Technology Considerations When Designing for High Performance

The above discussion about theoretical frameworks establishes an argument for designing new technologies that can be used to enhance the training environment to enable the best performance for clients recovering from walking and balance disorders. This section describes design elements that can be targeted to assure that training task goals can be achieved at a high level of intensity and performance.

A. **Providing safety from falls:** Walking and balance training puts a client at risk for falls and other bodily stresses and strains. An experience with falls during a training session can greatly reduce a person's confidence and trust in the rehabilitation process, and the clinician becomes overly cautious with clients, whether warranted or not. Technological solutions can protect individuals from falls. The clinician must work with great sensitivity toward understanding the client's fear of falling and concerns about training sessions that involve fall risk.

Technological solutions can protect individuals from falls. Designs might include a catch mechanism, such as a well-fitted harness, at the very least. The catch must occur prior to a height loss or device configuration that would result in contact with the ground or other surrounding objects. Further, the catch mechanism might include a compliant end-feel so that the person does not encounter a rigid constraint by the harness system. In addition, a catch mechanism that restores the person's upright orientation soon after the drop event can allow a person to get physically and psychologically ready for the next walking and balance task trial. Designing a catch mechanism that is located at a person's center of mass may allow the client to restore their upright posture without having to counter extra-rotational disequilibrium.

Ideally, falls during a challenging balance and walking training session would be part of the learning process, where the drop or tripping event triggers a client's desire to learn how to regulate balance so that a drop event can be avoided. For those clients with a more prevention-focused approach, drop events can be introduced slowly and with gradual drop intensities. Constraints can be introduced so that, other than the drop in body position, very few other body movements, such as trunk rotations and limb reaching can be limited. In the authors' experience, graded exposure, coupled with many repetitions of self-induced, autonomously controlled, dropping trials, can help a client to feel comfortable with the safety of the device and will establish trust in the clinician. Finally, the clinician's response to a drop can greatly influence a client's confidence in the technology. If the clinician responds with concern or fear or appears to be anxious about the event, then the client will likely amplify these responses. The clinician can spend time deliberately practicing drops with the technology so that they feel comfortable with putting their clients in the same situations.

B. **Allowing for a large variety of ecologically valid tasks with graded difficulty:** Ecologically, valid tasks assure that the client is focused on meaningful challenges that are important for the person in their daily living experience (Table 9.2). A key component of presenting ecologically valid challenge opportunities is making sure that the technology remains a degree of transparency of movement and error tolerance so that the device does not interfere with the success of the individual. When the client is presented with different tasks they will need an opportunity to learn, at a high level of challenge and without fear of bodily harm, whether the task results in success or failure. With success, the individual may discover a new and improved way to accomplish a meaningful task. With failure, the individual is driven to try new approaches to movement control that can eventually result in success. Tasks must have a clear success vs failure outcome immediately after the person attempts the task, with the eventual goal being the elimination of failure. Tasks must be gradable from a minimal level of difficulty to an extraordinarily high level so that clients at different levels of motor recovery can find a task that is anchored to their ability. Then as the task becomes more difficult, movement control strategy must be compelled toward an optimal solution. One recent study describes the use of this type of approach to treadmill training [29]. When compared with a hands-free treadmill environment as a comparison group, both the graded challenge and hands-free group showed significant improvements in walking speed, but the change was equal in magnitude.

C. **Motivational virtual experiences with sustained attention:** In many cases, a person who has been injured and acquired a brain

Table 9.2 Examples of ecologically valid walking and balance tasks with graded difficulty

Task description	Graded difficulty progression with increments
• Step onto raised step	• Adding two-inch risers
• Step over hurdle	• Adding two-inch heights
• Targeted long step	• Adding two-inch target distance
• Forward reach	• Adding two-inch target distance
• Step onto foam surface	• Adding two-inch heights
• Forward push	• Increasing distance and/or velocity of push
• Backward push	• Increasing distance and/or velocity of push
• Sit to stand	• Lowering two-inch seat height
• Backward stepping	• Increasing target stepping cadence by 10 steps/minute
• Narrow base stepping	• Decreasing step width target by two inches
• Variable speed	• Increasing range of speeds by ±0.2 meters/seconds
• Sprint stepping	• Increasing fastest treadmill speed by 0.2 meters/seconds
• Side-stepping	• Increasing target stepping cadence by 10 steps/minute

injury may move toward a more prevention-focused regulatory focus. One common example of how movement performance can fundamentally change a person's focus is when a person experiences pain-related fear of movement. For example, negative pain-related beliefs can interfere with recovery from acute injury. The consequential prevention-focused behavior promotes a cycle of physical deterioration, depression, and ultimately disability, further reinforcing the pain experience and impeding movement. Using the Regulatory Fit Theory to approach and communicate with clients, the focus might be on helping to change the person's orientation from prevention to promotion. Graded exposure (GEXP) is one such approach, in which individuals progressively confront fearful expectations through carefully selected activities [30].

Virtual reality (VR) is a tool that can generate GEXP environments not otherwise possible in a clinical or laboratory setting. Specifically, VR may enhance GEXP by offering enhanced training based on patient goals, reduced clinician workload, and improved user monitoring through movement tracking [31]. VR can generate an unlimited number of elements within uniquely designed environments that can enhance user interaction and improve the intrinsic motivation of GEXP therapy. Graded exposure in VR has been used to provide individuals with progressively more difficult physical challenges in a fun and enjoyable environment. For example, we have developed an engaging, locomotion-enabled GEXP VR application for pain-related fear of walking and reaching [32] (Fig. 9.3). The VR application, Lucid, consists of six 3-min modules that challenge participants to complete engaging walking, bending, and reaching tasks in VR that employ the enjoyment of fighting monsters with a sword and shield (Table 9.3).

D. **Adding resistance to increase workload:** Treadmill-based methods progressively increase the treadmill belt speed and grade or slope of the treadmill surface [33] to increase the mechanical work that is performed by the person on their center of mass to maintain their position on the treadmill. Concern of falling off the back of the treadmill may limit effort or compel the use of handrails while walking at faster speeds on a progressively increasing slope. Application of posteriorly directed resistive force, either through mechanically applied resistance or software designed to require more effort to drive the treadmill belt, at an individual's center of mass while walking on a treadmill offers an alternative to increasing the slope and speed of the treadmill. Recently, we have used posteriorly directed resistive forces to study responses to the self-selected walking speed of individuals poststroke and nonimpaired

Fig. 9.3 Virtual reality technology used to present graded rehabilitation challenges (Products | From the Future (ftfvr.com))

Table 9.3 Movement requirements and activity goals for the virtual reality GEXP modules. (Modified from [32] *licensed under CC-BY 4.0*)

Session	Movement requirements	VR activity goal
1 (low-intensity challenge)	Walking: any pace Reaching: requires one hand Bending: no Carry weights: no	"Walk at your own pace and rid the realm of monsters. Swing your sword to damage foes and block their attacks with your shield"
1 (low-intensity challenge)	Walking: walking quickly Reaching: requires one hand Bending: no Carry weights: no	"Walk at an increased pace to save as many animals as you can. Monsters have started to prey on the wildlife, and it's up to you to save the animals before the monster consumes them"
2 (medium-intensity challenge)	Walking: any pace Reaching: requires both hands Bending: no Carry weights: no	"The monsters have desolated the land, and it's up to you to collect food and coins for the realm. You are given two swords to reach both your foes and your items in all directions"
2 (medium-intensity challenge)	Walking: any pace Reaching: requires one hand Bending: yes Carry weights: no	"Crouch under trees and tunnels to explore more of the realm. You'll want to make sure you avoid limbs and the ceiling, or you'll bring your journey to an end"
3 (high-intensity challenge)	Walking: any pace Reaching: requires both hands Bending: yes Carry weights: yes	"Wield a weighted sword and shield while you crouch under trees and tunnels to explore more of the realm. You'll want to make sure you avoid limbs and the ceiling, or you'll bring your journey to an end"
3 (high-intensity challenge)	Walking: walking quickly Reaching: requires both hands Bending: yes Carry weights: yes	"Wield a weighted sword and shield to defeat your enemies"

control participants. We found that all individuals tolerated walking against relatively high posteriorly directed resistive forces without external aids to stabilize themselves like handrails [34]. We developed a method to perform graded exercises that use posteriorly directed resistive forces. This system increases externally applied work rates by increasing posteriorly directed resistive forces while keeping the treadmill belt speed constant (Fig. 9.4). We validated the system in comparison to a traditional incline-based treadmill in a cohort of nonimpaired individuals [35]. We found that the system

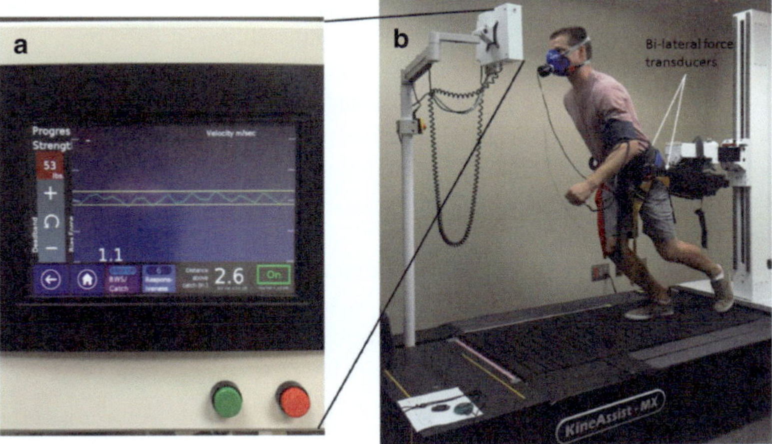

Fig. 9.4 Combination of (A) visual feedback of walking speed target, and (B) intent-driven treadmill system that can apply posteriorly directed resistance at progressive levels. (From [35] *licensed under CC-BY 4.0*)

generated VO2peak values that were comparable to incline-based treadmill walking, and further, that there was a significant correlation between values at different stages of work rates. These resistive forces could also be integrated into a VR exergame to simulate walking up hills or through viscous environments like walking in water, to create graded challenges.

In summary, the technology designer has many ways to design deliberate elements that allow clients to experience high-intensity balance and walking learning training sessions without undue stress and anxiety. The suggested considerations that are described in this section can enable designers to purposefully plan for adding progressive challenges to clients while they are recovering from walking and balance disabilities, including:

A. providing safety from falls by allowing a transparent physical interaction with an event-triggered method for preventing a large drop and easy recovery back to an upright position;
B. allowing for a large variety of ecologically valid tasks with graded difficulty so that individuals, regardless of recovery level, feel engaged, focused, and appropriately challenges as they recover walking and balance function;
C. designing virtual reality exergames that provide graded exposure to motivational experiences and require sustained attention to remain engaged with an enjoyable task;
D. adding resistance to increase workload so that the task involves an appropriate level of effort and challenge that results in cardiac, respiratory, and musculoskeletal benefits.

9.3 A User-Centered Approach for Future Technology Development

When considering the above device development factors, it is important to remember that the end-users of rehabilitative technology, both the clinician and individual seeking treatment, are motivated to solve problems and optimize recovery during walking and balance rehabilitation. Technologists who design devices without input from users are destined to feel disappointment when they receive feedback about the impracticality of the device and the lack of understanding of the realities of the busy clinical environment.

Entire books have been published, NSF, NIH, and other agencies have sponsored workshops, private companies have been created, and educational institutions have adopted opportunities to educate technologists on how to approach problem-solving from a variety of user-centered development standpoints. The following section is merely a summary of key principles that can be used to ensure some amount of success with the eventual adoption of new technologies that are designed to enhance performance during walking and balance training sessions.

A. *Identify the stakeholders:* The first step in the user-centered design process is to identify the key individuals, organizations, and communities that have a stake in the final use of the new technology. With respect to rehabilitation, these stakeholders may be classified within five basic categories: 1. Health care professionals; 2. Health care clients; 3. Payment agencies; 4. Key purchase decision-makers; 5. Industry partners.

1. Health care professionals (such as physicians, rehab clinicians, and allied health professionals) are typically the gatekeepers on whether a particular client group will benefit from access to the technology. If these stakeholders are not involved in the design process and do not understand how the device will be used, then they are not likely to recommend the technology to their clients.

2. Health care clients (such as people with disability or people recovering from injury or disease) are the target users of the device. If clients do not have a satisfactory experience with the device, then they are likely to refuse this as part of their care plan.

3. Payment agencies (such as insurance companies, advocacy organizations, or personal funds) must be willing to provide funds for the acquisition and use of the technology. If payment agencies will not pay, then the underlying expenses associated with marketing, selling, and distributing the technology will not be supported.

4. Key purchase decision-makers (such as purchasing agents, rehab facility administrators, or technology review committees) will evaluate the cost vs benefits of the technology for making purchasing decisions. A review process is often used with standards and requirements that must be met in order to be considered for purchase.

5. Industry partners are key for marketing, sales, and distribution of the technology. If they cannot clearly identify a customer base and subsequent positive profit margin, they are not likely to support the technology as a product that they might offer.

B. *Observe the stakeholders as they tackle everyday problems:* The ecosystem of technology use can be very complex, but must be understood to assure that the designed device will fit into the stakeholder context. Observing stakeholders requires careful attention to capturing the realities of the motivation and environment in which the stakeholders operate. If successful, connections will emerge between and amongst the various stakeholders that will help to build a relatively valid understanding of the ecosystem. For example, careful observation will reveal the motivation behind clinicians using a technology (e.g., "I value safety more than anything during walking training"). Industry partners may be driven by a desire to maintain a profit margin while also helping people recover from disability. Payment agencies may wish to speed recovery and/or prevent further complications that may be costly.

C. *Understand the problem(s) that the stakeholders need help to solve. What is the pain and what is the gain?* In the end, the new technology must solve an existing and important problem. One way to recognize whether or not a problem is important is to identify what is the "pain point" that can be relieved and what is the potential gain to be made by using the technology. For example, if a client wants to be able to spend time with friends while walking outside in a community setting while being safe from falls and other hazards, then technology should allow this level of safety, while also providing the pleasurable gain associated with the experience of being with friends. If the device is safe, yet requires complex equipment and interferes with socialization, then the client is likely to be dissatisfied and refuse to use the device.

D. ***Make prototypes that allow the stakeholders to give you feedback:*** Next, the prototyping process can allow for relatively frequent displays of concepts and design ideas that are subject to stakeholder feedback. Frequently, designers will try to shortcut this process by moving rapidly to well-developed prototypes with many built-in features. The danger with this shortcut is that it does not allow for feedback to be used to change, remove, or add other features that will be more responsive to the stakeholders. Ideally, each new iteration of features would test a hypothesis about a concept and its suitability for providing a particular needed solution for the stakeholders.

E. Narrow down the final solution only when major stakeholders are satisfied and feel excited about eventually using the technology: A final solution is achieved once all stakeholders express satisfaction with the technology and can visualize the device to be used in their particular context. A client might express a feeling of safety and comfort, without embarrassment when using the device. A health care professional might express that they feel the device will add to success with goals and outcomes of therapy. A payment agency might evaluate whether the technology merits financial support. Key industry partners will consider whether the device is a valuable addition to their product line.

With the user-centered design process, outlined above, a designer will be in a strong position to succeed in creating a technology that will be accepted and used for the important purpose of helping people to recover needed abilities. We've provided only a brief introduction to this process that may help designers to seek deeper

understanding and skills in executing this approach along with their strong engineering skills.

In summary, user-centered design approaches are very important to allow success with an eventual uptake of the technology by stakeholders involved with client care. Many different approaches exist, but the basic elements to consider are:

A. identifying stakeholders who will be involved with the acquisition and use of the technology;

B. observing stakeholders in their natural environment so as to establish the context of the ecosystem within which the technology will be used;

C. understanding the problems that require solutions, especially with the goal to reduce pain and provide gain;

D. providing experience prototypes to allow stakeholder feedback and a sense of buy-in with the direction of the design;

E. establishing stakeholder satisfaction with the final product.

9.4 Final Thoughts

We have presented an array of suggestions for the technology designer who intends to create new solutions for walking and balance recovery within the rehabilitation setting. The approach that we have provided considers the "person in the loop" of design. These suggestions provide a starting point for designers to consider matching their design elements with psycho-behavioral constructs that will enable clinicians and clients to benefit from motivational and enjoyment opportunities as a person recovers walking and balance function after injury or disease. While new materials, electronics, sophisticated artificial intelligence algorithms, and other advanced engineering approaches may be readily available, our hope is that the designer will focus their attention on providing technological solutions that will be preferred by stakeholders who care deeply about recovering walking and balance function so that they can continue to enjoy a high quality of life with mobility function that allows them to go places they never thought possible.

References

1. Higgins ET. Making a good decision: value from fit. Am Psychol. 2000;55(11).
2. Paulssen M, Bagozzi RP. A self-regulatory model of consideration set formation. Psychol Mark. 2005;22 (10).
3. Pierro A, Giacomantonio M, Pica G, Giannini AM, Kruglanski AW, Higgins ET. Persuading drivers to refrain from speeding: effects of message sidedness and regulatory fit. Accid Anal Prev. 2013;50.
4. Pfeffer I. Regulatory fit messages and physical activity motivation. J Sport Exerc Psychol. 2013;35 (2).
5. Toyama M, Nagamine M, Tang L, Miwa S, Kurozumi R, Aikawa A. Can regulatory fit improve performance? Effect of types of regulatory fit and performance. Shinrigaku Kenkyu. 2017;88(3).
6. Higgins ET. Promotion and prevention: regulatory focus as a motivational principle. Adv Exp Soc Psychol. 1998;30(C).
7. Higgins ET. Beyond pleasure and pain. Am Psychol. 1997;52(12).
8. Higgins ET, Silberman I. Development of regulatory focus: promotion and prevention as ways of living. In: Motivation and self-regulation across the life span. 2009.
9. Cesario J, Higgins ET. Making message recipients "feel right": how nonverbal cues can increase persuasion. Psychol Sci. 2008;19(5).
10. Higgins ET. Value from regulatory fit. Curr Dir Psychol Sci. 2005;14(4).
11. Byron K, Peterson SJ, Zhang Z, LePine JA. Realizing challenges and guarding against threats: interactive effects of regulatory focus and stress on performance. J Manag. 2018;44(8).
12. Franks B, Chen C, Manley K, Higgins ET. Effective challenge regulation coincides with promotion focus-related success and emotional well-being. J Happiness Stud. 2016;17(3).
13. Ryan RM, Deci EL. Intrinsic and extrinsic motivations: classic definitions and new directions. Contemp Educ Psychol. 2000;25(1).
14. Ryan RM, Rigby CS, Przybylski A. The motivational pull of video games: A self-determination theory approach. Motiv Emot. 2006;30(4).
15. Lewthwaite R, Wulf G. Optimizing motivation and attention for motor performance and learning. Curr Opin Psychol. 2017;16
16. Madhavan S, Lim H, Sivaramakrishnan A, Iyer P. Effects of high intensity speed-based treadmill training on ambulatory function in people with chronic stroke: a preliminary study with long-term follow-up. Sci Rep. 2019;9(1).
17. Dejong SL, Schaefer SY, Lang CE. Need for speed: Better movement quality during faster task performance after stroke. Neurorehabilitation Neural Repair. 2012;26(4).
18. Hornby TG, Reisman DS, Ward IG, Scheets PL, Miller A, Haddad D, et al. Clinical Practice Guideline to Improve Locomotor Function Following Chronic Stroke, Incomplete Spinal Cord Injury, and Brain Injury. J Neurol Phys Ther. 2020; 44
19. Schmidt RA, Lee TD. Motor learning and performance: from principles to application (5th éd.). Champaign, IL: Human Kinetics. 2014;
20. Ferreira dos Santos L, Christ O, Mate K, Schmidt H, Krüger J, Dohle C. Movement visualisation in virtual reality rehabilitation of the lower limb: a systematic review. BioMedical Engineering Online. 2016; 5
21. Arlindo Soares MA, Gatinho Bonuzzi GM, Coelho DB, Torriani-Pasin C. Effects of extrinsic feedback on the motor learning after stroke. Motriz Revista de Educacao Fisica. 2019;25(1).
22. Chamorro-Moriana G, Moreno AJ, Sevillano JL. Technology-based feedback and its efficacy in improving gait parameters in patients with abnormal gait: a systematic review. Vol. 18, Sensors (Switzerland). 2018.
23. Borrego A, Latorre J, Llorens R, Alcañiz M, Noé E. Feasibility of a walking virtual reality system for rehabilitation: Objective and subjective parameters. J NeuroEng Rehabil. 2016;13(1).
24. Jaffe DL, Brown DA, Pierson-Carey CD, Buckley EL, Lew HL. Stepping over obstacles to improve walking in individuals with poststroke hemiplegia. J Rehabil Res Dev. 2004;41(3 A).
25. Cardador MT, Northcraft GB, Whicker J. A theory of work gamification: Something old, something new, something borrowed, something cool? Hum Resour Manag Rev. 2017;27(2).
26. Doyle J, Kelly D, Patterson M, Caulfield B. The effects of visual feedback in therapeutic exergaming on motor task accuracy. In: 2011 international conference on virtual rehabilitation, ICVR 2011. 2011.
27. Lyons EJ. Cultivating engagement and enjoyment in exergames using feedback, challenge, and rewards. Games Health J. 2015.
28. Alhirsan SM, Capó-Lugo CE, Brown DA. Effects of different types of augmented feedback on intrinsic motivation and walking speed performance in poststroke: a study protocol. Contemp Clin Trials Commun. 2021 24:100863.
29. Graham SA, Roth EJ, Brown DA. Walking and balance outcomes for stroke survivors: A randomized clinical trial comparing body-weight-supported treadmill training with versus without challenging mobility skills. J NeuroEng Rehabil. 2018;15(1).
30. George SZ, Wittmer VT, Fillingim RB, Robinson ME. Comparison of graded exercise and graded exposure

clinical outcomes for patients with chronic low back pain. J Orthop Sport Phys Ther. 2010;40(11).

31. Trost Z, Zielke M, Guck A, Nowlin L, Zakhidov D, France CR, et al. The promise and challenge of virtual gaming technologies for chronic pain: the case of graded exposure for low back pain. Pain Manag. 2015; 5.

32. Hennessy RW, Rumble D, Christian M, Brown DA, Trost Z. A graded exposure, locomotion-enabled virtual reality app during walking and reaching for individuals with chronic low back pain: Cohort gaming design. JMIR Serious Games. 2020;8(3).

33. Fletcher GF, Ades PA, Kligfield P, Arena R, Balady GJ, Bittner VA, et al. Exercise standards for testing and training: a scientific statement from the American heart association. Circulation. 2013;128(8).

34. Hurt CP, Wang J, Capo-Lugo CE, Brown DA. Effect of progressive horizontal resistive force on the comfortable walking speed of individuals post-stroke. J NeuroEng Rehabil. 2015;12(1).

35. Hurt CP, Bamman M, Naidu A, Brown DA. Comparison of resistance-based walking cardiorespiratory test to the Bruce protocol. J Strength Cond Res [Internet]. 2017 Dec 11 [cited 2018 Mar 1]; http://www.ncbi.nlm.nih.gov/pubmed/29239992

Psychophysiological Integration of Humans and Machines for Rehabilitation

10

Vesna D. Novak, Alexander C. Koenig, and Robert Riener

Abstract

In conventional man–machine interfaces for motor rehabilitation, the primary goal is to control the biomechanical interaction between the human and the machine or environment. However, integrating the human into the loop can be considered not only from a biomechanical view, but also with regard to physiological and psychological aspects. Such psychophysiological integration involves recording and controlling the patient's physiological responses so that the patient receives appropriate stimuli and is challenged in a moderate but engaging way without causing undue stress or harm. In this chapter, we present examples first of physiological integration (without taking psychological aspects into account) and then of full psychophysiological integration where the patient's cognitive workload is automatically estimated from physiological data. Examples are given both for gait rehabilitation and arm rehabilitation.

Keywords

Human in the loop · Rehabilitation · Rehabilitation robotics · Biocooperative control · Affective computing · Psychophysiology · Physiological computing · Lokomat · Gait training

10.1 Introduction: Multimodal Human–Machine Interaction in Rehabilitation

A rehabilitation device can only be effective if it can efficiently guide the patient and react to their needs and desires. In the past, many rehabilitation systems that included novel actuation and digital processing capabilities worked in a "master–slave" relationship: they forced the user only to follow predetermined reference trajectories without taking into account individual properties, spontaneous intentions, or voluntary efforts of that particular person. For example, many early actuated orthoses applied predetermined motion patterns to the patient's legs but did not react to the patient's voluntary effort.

In contrast, modern rehabilitation technologies place the human "in the loop", making them more than just the sender of a command to a

V. D. Novak (✉)
Department of Electrical Engineering and Computer Science, University of Cincinnati, 2600 Clifton Ave, Cincinnati, OH 45221, USA
e-mail: novakdn@ucmail.uc.edu

A. C. Koenig
Reactive Robotics, Landsbergerstrasse 234, 80687 Munich, Germany
e-mail: koenig@reactive-robotics.com

R. Riener
Sensory-Motor Systems Lab, ETH Zurich, Tannenstrasse 1, C8092 Zurich, Switzerland
e-mail: robert.riener@hest.ethz.ch

© The Author(s), under exclusive license to Springer Nature Switzerland AG 2022
D. J. Reinkensmeyer et al. (eds.), *Neurorehabilitation Technology*,
https://doi.org/10.1007/978-3-031-08995-4_10

device or the passive receiver of a device action. Instead, the human closes the loop by feeding information to a computer processing unit, which then takes into account the user's properties, intentions, actions, and environmental factors. The interaction thus becomes bidirectional, optimizing the user experience and thus the rehabilitation outcome.

In the first decade of this century, rehabilitation robots were developed that worked with patients in a "patient-cooperative" [1] or "assist-as-needed" [2] fashion: adapting themselves to the user's preferred movements and providing only as much assistance needed to successfully complete the exercise. This provided integration of the human "into the loop" in a biomechanical sense, making the rehabilitation system safe, ergonomically acceptable, and compliant [2–5]. However, there are further possible levels of integration: physiological and even psychological (see Fig. 10.1). Such psychophysiological integration involves recording and controlling the patient's physiological reactions so that they receive appropriate stimuli and are challenged in a moderate but engaging and motivating way without causing undue stress or harm [6].

In the following sections, we will discuss how patients during rehabilitation can be integrated into the control loop. We first present examples of purely physiological integration (with no consideration of the psychological state) and then full psychophysiological integration.

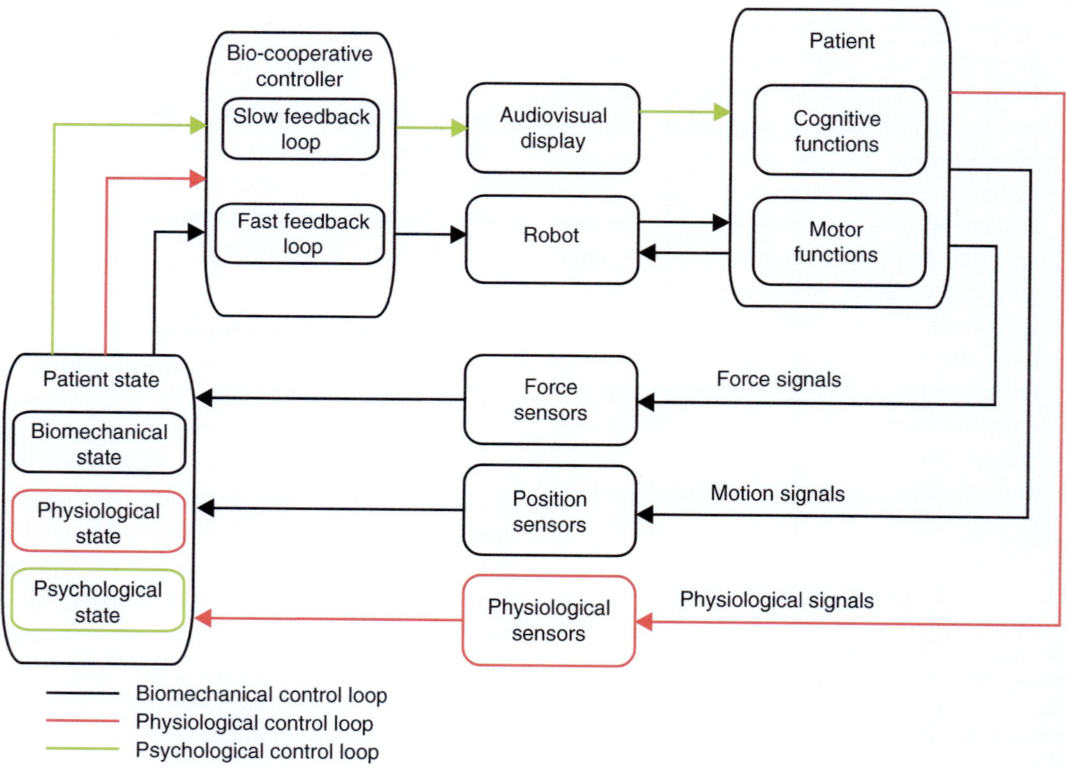

Fig. 10.1 The human is in the loop with respect to biomechanical (black), physiological (red), and psychological aspects (green). A fast feedback loop with update frequencies in the millisecond range controls the robot. A slow feedback loop adapts robot and audiovisual display with an updating frequency of several seconds. Figure created for book chapter to describe work of Riener and Munih [6]

10.2 Physiological Integration of Humans and Rehabilitation Technologies

10.2.1 Rationale

Neurological patients in need of motor rehabilitation can greatly benefit from cardiovascular training, i.e., performing exercises in which their heart rate (HR) is controlled to a desired level [7]. In the absence of advanced technologies, such training is performed either on treadmills or on stationary bicycles; bicycles are commonly used in patients with more severe motor impairments, who would not be able to maintain balance on treadmills. However, gait robots such as the Lokomat [8] and the LOPES [9] allow even nonambulatory patients to exercise walking by guiding the legs on a walking trajectory while maintaining proper balance, with consequent significant improvement of gait function after neurological injury [10, 11]. Integration of cardiovascular therapy into gait rehabilitation robotics could thus combine the benefits of both training types and consequently improve rehabilitation outcomes.

Three major challenges of cardiovascular training with a gait robot compared to standard treadmill walking need to be considered. First, for patient safety, treadmill speed during robot-assisted rehabilitation is typically limited to very slow walking speeds and does not allow fast walking or running. The Lokomat gait orthosis, for example, is limited to 3.2 km/h, which is low compared to non-robotic cardiovascular training approaches that involve walking speeds greater than 3.6 km/h [12]. Second, for facilitation of stance, the patient can be body weight supported, which causes HR to decrease as body weight support (BWS) increases [13]. Finally, all gait robots use actuators to provide supportive guidance forces in order to enable walking movements in patients with little leg force or little coordinative capabilities. This guidance force can be expected to alter HR as it decreases the energy required by the subject to perform the walking movement.

10.2.2 Model-Based Heart Rate Control

The Lokomat gait orthosis (Hocoma AG, Switzerland) is a popular gait rehabilitation robot that consists of a lower limb exoskeleton attached to a treadmill in order to help patients train gait with robotic support (Chap. 8). The robot can apply large forces to the patient's legs to guide them along a reference trajectory, which has a major effect on HR. At high guidance forces with a stiff impedance controller, patients can either walk actively (i.e., push into the orthosis with high forces) or behave passively, letting the robot move their legs. Since changing different parameters of the exercise (e.g., guidance force, treadmill speed) may have different effects on HR, it would be useful to have a model that predicts changes in HR before they occur. Such a model could be used, for example, to predict situations that may become harmful to the patient. Additionally, it could be used as a basis for controlling HR to a desired level by modifying the settings of the gait robot.

In an experimental study with eight chronic stroke patients, co-authors Koenig and Riener evaluated the effects of BWS, treadmill speed, and guidance force on the patient's HR [14]. Changes in guidance force did not significantly alter HR; however, both BWS and treadmill speed had a major impact on HR. Increasing BWS reduces the loading the patient has to carry during walking, and appropriate loading of the patient during treadmill training was previously shown to be key for successful rehabilitation outcome [15]. Similarly, treadmill speed has previously been shown to linearly increase HR in healthy participants [16]. Thus, BWS and treadmill speed were used as control variables in a model-based HR controller for the Lokomat.

The HR control model developed based on the above experimental data included four subject-specific parameters and six subject-independent parameters [14]. For example, the linear relationship between treadmill speed and HR was interpreted as a low-pass reaction to a sudden increase in oxygen demand and modeled as a first-order delay element. Treadmill acceleration and deceleration resulted, respectively, in an overshoot and undershoot of HR before steady state was reached [17, 18], which was modeled as a second-order derivative element. The power expenditure of the human, which has also been reported to correlate linearly with HR, was taken as a linear input parameter modeled as a first-order delay element. As a fourth example, the fatigue effect that results in increased resting HR after longer training durations [17, 19] was modeled as a first-order low-pass element.

When fitted to the eight chronic stroke patients, the model achieved a coefficient of determination r^2 of 79%. It was expected in advance to be more applicable to robot-aided gait rehabilitation than other control techniques (e.g., proportional–integrative–derivative control) since it explicitly models important factors such as human power expenditure, a key driver of changes in HR (Fig. 10.2).

After fitting, the model was evaluated in a real-time fashion with three healthy subjects as well as with three stroke patients by controlling HR to 70, 80, and 90 beats/min. In healthy subjects, the controller could stabilize HR within 1 bpm $\pm$ 3 bpm. To mimic the training situation in which patients exercise, we limited the treadmill speed of the Lokomat to 3 km/h. When trying to control the subjects' HR to 90 bpm, treadmill speed saturated. In patients, the effectiveness of HR control depended on the baseline HR during standing, as resting HR of stroke survivors has been shown to be higher than the

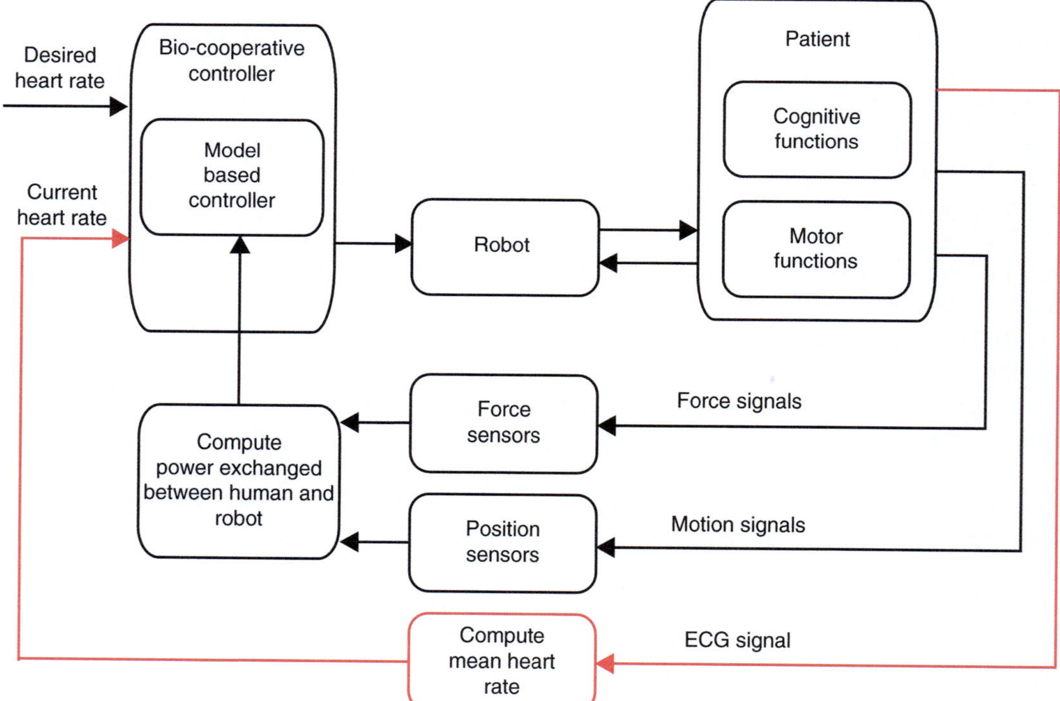

Fig. 10.2 Model-based control of heart rate [14]. Heart rate control is performed based on a model by taking power exchange between the patient and robot into account. Figure created for book chapter to describe work of Koenig et al. [14]

resting HR of healthy subjects [20]. However, it was possible to control HR of stroke patients in a range between resting HR and plus 10 beats/min.

10.2.3 Heart Rate Control Using Treadmill Speed and Visual Stimuli

As an alternative to the above model-based approach, co-authors Koenig and Riener also investigated HR control with a simpler approach: extracting mean HR in real time, comparing it to a desired value, and feeding the error into a proportional–integrative controller [21]. In this study, two approaches to influencing heart rate were tested: changing the treadmill speed (as in the previous study) and changing the visual feedback provided to the patient. For the visual feedback, patients were provided with a virtual scenario consisting of a dog walking in front of the patient (Fig. 10.3). The dog corresponded to desired effort while a white dot on the ground corresponded to actual effort, thus encouraging patients to modify their walking speed (and consequently heart rate) to bring the dot underneath the dog.

Fig. 10.3 Visual feedback for heart rate control in the Lokomat [21]. The patient's actual effort is shown as a white dot on the screen while the dog represents the desired effort, thus encouraging patients to modify their walking speed. Figure originally from Koenig et al. [21], licensed under CC-BY 2.0

In an experimental evaluation, patients first walked for 10 min at a gait speed of 1.5 km/h in order to familiarize themselves with the Lokomat and to obtain baseline HR values. HR was then controlled to a desired temporal profile that corresponded to four levels of patient activity: low, intermediate, high and very high, each for three minutes. Both treadmill speed and visual feedback were able to successfully control patient activity quantified by HR to a desired level, and the setup was adaptable to the specific cognitive and biomechanical needs of each patient.

10.2.4 Further Examples of Physiological Integration

Though we covered two examples of physiological integration via heart rate control in the Lokomat in detail, similar approaches have been used in other areas of rehabilitation robotics. For example, co-author Riener's group used a nonlinear model predictive control approach to automatically control HR and blood pressure in bed rest patients via automated leg mobilization and body tilting [22], with mean values differing on average less than 1 bpm from desired HR and less than 2.5 mmHg from desired blood pressure values. In follow-up studies, further models were developed to incorporate additional procedures such as functional electrical stimulation of leg muscles [23]. As cardiovascular and endocrine systems experience significant deconditioning during bed rest, leading to secondary problems such as muscle atrophy and pneumonia, such automated control has the potential to reduce complications during bed rest [24] and represents an application of similar HR control principles to a different area of rehabilitation.

As another example, co-author Novak's team used a simple proportional controller to automatically control respiration rate and arm muscle activation in a competitive robot-aided arm rehabilitation scenario via automated changes in visual feedback and robot resistance [25]. While only tested with participants without neurological injuries, the proposed control approach was able to influence respiration rate and muscle activation, thus automatically modulating the intensity of a competitive arm rehabilitation exercise. Other groups have also recently begun exploring the use of physiological responses for adaptation of competitive exercises [26]. As such competitive exercises (involving either two patients or a patient and an unimpaired loved one) have the potential to generally increase patient motivation and exercise intensity compared to solo exercise [27, 28], two-person physiological integration could thus enhance long-term rehabilitation outcome.

10.2.5 Implementing Physiological Integration into Daily Clinical Routine

The major obstacle to utilizing HR control and other physiological control schemes in daily clinical routines has so far been the necessity of using sensors that require constant body contact to record the subject's physiology. While electrocardiography electrodes, chest belts, or even HR-measuring wristwatches are commonly available, they all need to be attached to the patient and later disinfected, which represents a time-consuming process in the already time-consuming clinical routine of physical therapy staff.

In the future, this obstacle could be addressed via the use of noncontact sensors. For example, state-of-the-art image processing algorithms allow quantification of HR through changes in blood flow that are invisible to the human eye [29]. As each heartbeat pumps blood through the veins and, therefore, through the head, small changes in the red color channel of the head occur. By magnifying these color changes, an algorithm can evaluate the frequency at which these changes occur and extract HR from it. This could, in the future, allow therapists to simply direct a webcam toward the face of the patient and start the training. As another example, studies have shown that HR and respiration rate can be extracted from fabric-based sensors built into seat cushions [30]. This would not be

suitable for gait rehabilitation but could potentially be used with arm rehabilitation robots, which are commonly used while the patient is seated.

Alternatively, physiological sensors could be built directly into the rehabilitation robot. For example, one study built HR, skin conductance, and skin temperature sensors directly into the handle of an arm rehabilitation robot and found that they were able to measure physiological responses during a physical task with reasonably good accuracy [31]. While this would need to consider, e.g., different patient sizes and grasping capabilities, it may be more practical for robot-aided rehabilitation since therapists and patients would not need to work with any separate sensors such as cameras.

Finally, another challenge of physiological integration is the fact that patients in rehabilitation exhibit different physiological responses from unimpaired controls. For example, after a stroke, abnormalities in sweating and heart rate variability can persist for months or years [20, 32] Thus, technologies developed on unimpaired participants cannot easily be transferred to patients. Furthermore, as patients exhibit significant intersubject variability, control strategies developed on one patient may not transfer to another, and control strategies developed on one patient population (e.g., stroke survivors) may not transfer to another population (e.g., spinal cord injury survivors). This variability greatly complicates the development and evaluation of rehabilitation technologies, which must be developed with input from a large patient sample and/or include sufficient calibration and adaptation capabilities to handle previously unseen physiological behavior.

10.3 Psychophysiological Integration

10.3.1 Rationale

In motor rehabilitation, it is known that high motivation and active participation during a difficult but feasible task can enhance motor learning and this improves the rehabilitation outcome [33]. In robot-aided rehabilitation, motivation and participation are encouraged via the use of diverse virtual environments [34]; these simulate activities that range from everyday tasks such as cooking and cleaning [35] to fictional tasks such as exploring tropical islands [36]. However, as it is often unclear what elements of virtual environments actually contribute to motivation and to what degree [34], it would be useful to have a measure of the patient's cognitive and emotional state. This could then be used both to evaluate existing technologies and to adapt rehabilitation tasks to the patient's psychological state.

Patient psychological states were first measured in technology-aided rehabilitation using questionnaires such as the Intrinsic Motivation Inventory [37]. However, such questionnaires interrupt the rehabilitation process whenever they are used and thus cannot be used in real time. Furthermore, patients with neurological injuries, particularly stroke survivors, may exhibit cognitive deficits such as aphasia or limited self-perception capabilities, making the use of questionnaires difficult. Thus, an objective measurement of psychological state that is unobtrusive and does not rely on the patient's active participation would be more useful for rehabilitation.

Automated recognition of human psychological states is part of the field of affective computing and involves applying machine learning algorithms to diverse measurements such as speech and facial expressions. In motor rehabilitation, particular attention has been given to psychophysiological responses: signals such as heart rate and skin conductance that also contain information about psychological states. For example, a common psychophysiological response is increased sweating (measured as skin conductance) due to stress. Psychophysiological measurements were popular as indicators of stress in applications such as pilot monitoring in the 1990s [38], and began to see increased attention in the early twenty-first century when classification algorithms were applied to them in order to automate the psychological state recognition process [39]. In the last 20 years, psychophysiological measurements have been used to estimate stress, cognitive workload, and emotional states in diverse applications such as computer games [40], driver monitoring [41], and education [42].

Psychological state recognition is commonly done with supervised machine learning algorithms that learn from labeled "training" data: examples of psychophysiological measurements in known psychological states. The training data are obtained by externally inducing desired states (e.g., making the participant play a boring videogame or watch a scary movie) and using either self-report questionnaires or external observers to verify that the desired state has been correctly induced [43]. To evaluate the performance of the machine learning algorithms, participants are exposed to new example scenarios, and psychophysiological measurements and ground-truth data (self-report questionnaires or external observation) are collected again. The accuracy of the algorithm is then defined as the ability to correctly reproduce the ground-truth information on these new "test" data. While several alternatives to this procedure (e.g., unsupervised algorithms [43]) have been suggested as an alternative, supervised machine learning with self-report ground-truth information remains the dominant approach.

In motor rehabilitation, psychophysiological measurements of the autonomic nervous system have the advantage that the sensors can be applied relatively quickly compared to bulkier measurements such as electroencephalography. Additionally, other measurement modalities are not highly useful, for example, speech emotion analysis is not relevant for physical exercise, and facial expressions may be suppressed or altered in some patients with neurological injuries. Thus, most studies on psychophysiological integration in motor rehabilitation have been performed with measurements of autonomic nervous system responses.

10.3.2 Psychophysiological Integration in Arm Rehabilitation in the MIMICS Project

The European-Union-funded MIMICS project [6] carried out the first major work on psychophysiological integration in motor rehabilitation (though pilot studies had been done earlier [44]). We first present an implementation in arm rehabilitation, followed by application to gait rehabilitation in the next section.

The arm rehabilitation setup involved the Haptic Master (Moog FCS, Netherlands) haptic robot and a virtual environment in which patients could perform reaching and grasping tasks. Four psychophysiological signals were measured: HR (via the electrocardiogram), skin conductance (using electrodes on the fingers of the nondominant hand), respiration rate (using a thermistor-based sensor underneath the nose), and peripheral skin temperature (using a sensor on the nondominant hand). While the overall goal was to extract psychological information, these physiological responses are also all affected by physical workload. A study was thus first carried out to determine whether psychological information is obscured by physiological effects of physical workload [45]. It involved 30 subjects with no physical or cognitive impairments, who performed a visual-motor task (balancing a virtual pendulum) at three levels of cognitive load (corresponding to different degrees of pendulum instability) and two levels of physical load (corresponding to different degrees of physical resistance provided by the Haptic Master). While it found that several physiological responses (e.g., HR) are significantly affected by physical load, effects of cognitive load still cause significant physiological changes at a stable physical load level. A follow-up exploratory study in a dual-task scenario found similar results [46].

Following these exploratory studies, a "biocooperative" control system was developed to automatically adapt the difficulty of an arm rehabilitation task based on the level of patient workload. Patients used the Haptic Master to interact with a virtual environment with a sloped

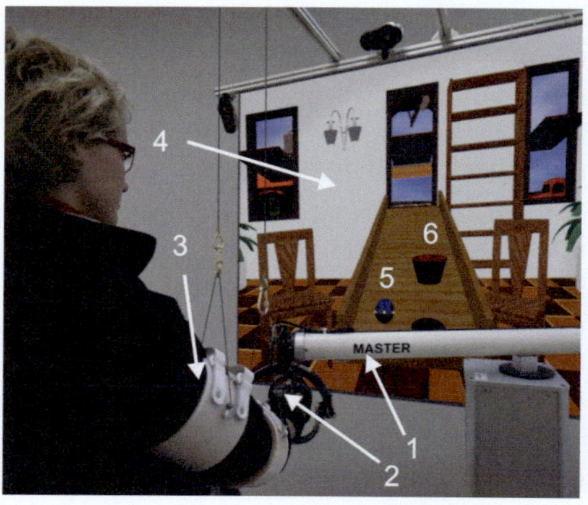

Fig. 10.4 A human participant interacting with a biocooperative arm rehabilitation system [48]. The participant performs the task using the robot (1) and grasping device (2) while their arm is gravity compensated (3). The screen (4) shows a sloped table, a ball (5) and a basket (6). Figure originally from Ziherl et al. [48], licensed under CC-BY 2.0

table; balls periodically rolled across the table toward the patient, and the patient had to catch each ball and place it in a basket to the side of the table [47] (Fig. 10.4). The robot provided assistance with catching the ball, but the speed of the incoming balls was increased or decreased every two minutes based on the patient's cognitive workload. This matches the general design of a biocooperative system seen in Fig. 10.1: the robot provides immediate assistance (fast feedback loop) while psychological state integration provides slower adaptation of the task.

To determine the level of cognitive load, the same four psychophysiological responses (HR, skin conductance, respiration, skin temperature) were measured and multiple features such as mean HR and metrics of HR variability [49] were extracted from consecutive 2-min task periods. These features are known to correspond to cognitive load outside rehabilitation, for example, both temporal and spectral metrics of HR variability are significantly affected by psychological stress [50], and the frequency of skin conductance responses ("peaks" in the skin conductance signal) is correlated with psychological arousal [51]. The features were then input into a supervised machine learning algorithm that output the level of cognitive load: either low or high. If cognitive load was low, difficulty (i.e., ball speed) was increased; conversely, if cognitive load was high, difficulty was decreased. To determine the "ground truth" for training and testing the algorithm, patients were asked how they themselves would change the difficulty. The classification accuracy was then defined as the percentage of times the algorithm's decision matches the patient's preference.

When the supervised machine learning algorithm was provided only with task score and movement data, it achieved a classification accuracy of 81.8% in a sample of 11 patients [47]. Conversely, when the algorithm was also provided with psychophysiological data, it achieved an accuracy of 89.4% in the same sample. This illustrated that the addition of psychophysiological information can improve the accuracy of machine behavior in robot-aided arm rehabilitation, though it did not demonstrate an improved user experience. Additionally, the psychophysiological measurements were used to adapt both physical and cognitive workload, as higher ball speed requires participants to both move and think faster.

10.3.3 Psychophysiological Integration in Leg Rehabilitation in the MIMICS Project

In parallel to the above arm rehabilitation system, the MIMICS project also developed a leg rehabilitation system based on the Lokomat robotic leg orthosis (previously described for physiological integration). The same four physiological measurements were taken. Similar to the arm scenario, a study was first carried out to determine whether psychological information is obscured by physiological effects of physical workload [52]. Seven participants with no physical and cognitive impairment experienced five scenarios in the Lokomat: standing, walking, walking with soccer, standing with mental arithmetic and walking with mental arithmetic. Again, several physiological responses were affected by physical workload, but effects of cognitive load were nonetheless visible, and it was decided to proceed with the development of a biocooperative controller.

For the biocooperative control system, the Lokomat was combined with a virtual environment where participants walked along a straight line through a virtual forest displayed on a screen in front of them [53]. The walking speed in the scenario was controlled via the participant's voluntary effort performed in the Lokomat: an increase in effort led to an increase in virtual walking speed while a decrease in effort led to a decrease in virtual walking speed. Virtual objects periodically appeared on the screen in front of the participant and slowly disappeared; the participant could speed up to collect the object or slow down to avoid it. In addition to these objects, questions were displayed on the screen and had to be mentally answered by participants. If the statement was correct (e.g., $1 + 1 = 2$), the participant had to collect the object before it disappeared. If the statement was false (e.g., $1 + 1 = 3$), the participant had to avoid a collision by decreasing their speed until the object disappeared. This mental challenge thus presented an additional component to the physical effort generally required by the Lokomat.

In the Lokomat system, the MIMICS team was interested in specifically focusing on modulating cognitive workload using psychophysiological responses. The biocooperative controller thus modified the difficulty of the questions presented by participants and the amount of time the questions were visible, influencing cognitive load without strongly influencing physical workload. As in the arm rehabilitation system, features were extracted from the four physiological signals and input into a supervised machine learning algorithm that output the level of cognitive load (Fig. 10.5). However, in the Lokomat system, there were four levels of cognitive load corresponding to four possible changes to the questions: they could become much harder, slightly harder, slightly easier, or much easier. Difficulty was adapted every 60 s while participants performed the task. In this scenario, ground-truth data for training and testing the

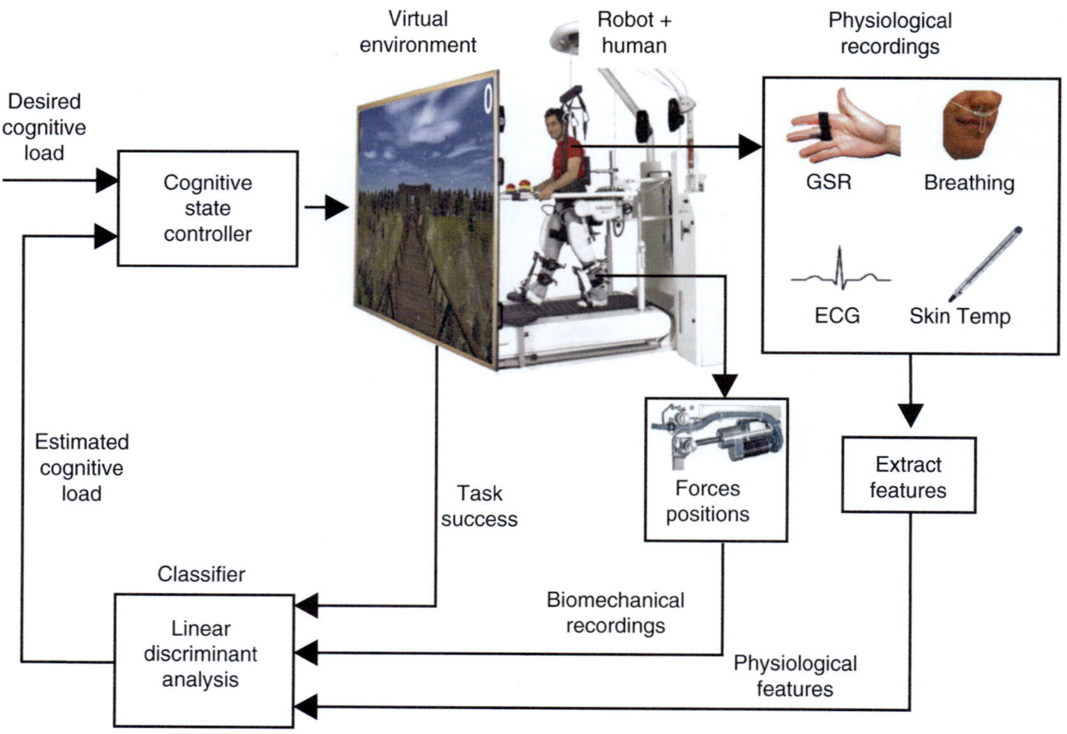

Fig. 10.5 System setup for psychophysiological integration in gait rehabilitation. Figure [53] modified from Koenig et al., © 2011 IEEE

machine learning algorithm were obtained by asking both patients and supervising occupational therapists how they would prefer to change the difficulty of the questions.

If the therapists' opinions on difficulty change were used as the ground truth, the algorithm achieved a classification accuracy of 60% based only on task score [53]. When the algorithm was also provided with psychophysiological data, it achieved a classification accuracy of 70%, and adding movement data increased it to 75%. This again demonstrated the usefulness of psychophysiological measurements in providing more accurate machine behavior and was done in a situation where primarily cognitive load was modulated by the system. However, as in the arm rehabilitation scenario, no effects on user experience were evaluated. Furthermore, when the patients' opinions were used as the ground truth instead, the algorithm was unable to achieve adequate classification accuracies.

10.3.4 Further Examples of Psychophysiological Integration

Besides the MIMICS project, several other groups have demonstrated examples of psychophysiological integration of humans and machines for rehabilitation. A prominent early example was the work of Guerrero et al. [54]. It bore some similarities to the aforementioned arm rehabilitation example: participants used an arm robot to interact with a virtual environment where droplets fell from the top of the screen and had to be caught with the robot. Three psychophysiological responses (HR, skin conductance, skin temperature) were measured and input into a controller that modified the droplet speed to suit the participant's psychophysiological state. However, unlike the classifiers used in the MIMICS project, this controller was based on fuzzy logic with expert-defined rules, avoiding the need for training data collection that was a time-consuming part of the MIMICS project [47, 53]. Guerrero reported an increase in self-reported dominance and emotional valence using their controller compared to a simpler calibration task, indicating potential positive effects on patient psychological state. A later study by Guerrero used an updated version of the system [55], and again found higher self-reported dominance and valence compared to a simpler calibration task and compared to a system with only biomechanical integration.

Another interesting example was presented by Shirzad and Van der Loos [56], who fused performance-based and physiology-based estimates of task difficulty in a robot-aided arm rehabilitation system. Their system first estimated task difficulty using performance measurements and also estimated its own confidence in the difficulty estimate. If confidence was high, it proceeded to adapt difficulty based on the performance-based difficulty estimate. If confidence was low, it estimated task difficulty using psychophysiological measurements (skin conductance, respiration, skin temperature). If confidence in that estimate was high, it adapted difficulty based on that estimate instead; if not, it chose an adaptation action randomly. Estimates were made with neural networks, another supervised machine learning approach. Participants (all people with no physical or cognitive impairment) perceived the difficulty changes as more motivating, more satisfying, and less effortful than a random difficulty adaptation schedule, again demonstrating potential benefits of the controller.

As a potentially interesting expansion of the above systems, co-author Novak's team recently presented a biocooperative controller for a competitive arm rehabilitation task [57]. It collected psychophysiological responses of both participants, and extracted both the features presented above (e.g., mean HR, skin conductance response frequency) as well as measures of physiological synchronization: the degree of similarity between two people's physiological responses, which is known to correlate with perceived competition intensity [58, 59]. The individual features and synchronization features were then input into a supervised machine learning algorithm that adapted different task parameters to try and keep the exercise intense and enjoyable for both participants simultaneously. While it achieved promising open-loop (i.e.., classification but no adaptation) results, a more extensive evaluation was interrupted by the COVID-19 pandemic in 2020. Other groups have also begun exploring the degree of synchronization between two people's brain activity as a tool for potentially enhancing motor rehabilitation [60], and such multi-user biocooperative control may thus see significant development in the near future.

Finally, besides arm rehabilitation robots and gait orthoses, biocooperative control could be applied to other technologies for people with disabilities. For example, Knaepen et al. [61] conducted a study on psychophysiological responses during gait (similar to an expanded version of the above first MIMICS leg study [52]), and suggested that psychophysiological responses could be used in a biocooperative robotic lower limb prosthesis. Such a prosthesis would be able to detect the cognitive workload of walking in amputees and adapt its own behavior to the wearer's cognitive workload and abilities. While such a prosthesis has not yet been implemented, it represents an intriguing concept that shows the possibilities of psychophysiological integration.

10.3.5 Implementing Psychophysiological Integration into Daily Clinical Routine

Much like physiological integration, psychophysiological integration is limited by sensors that require constant contact with the body. Still, this problem appears to be solvable for psychophysiological integration as well: for example, one study with unimpaired participants used a digital camera to measure HR, HR variability, and respiratory rate at a distance of up to 3 m, and used it to classify cognitive workload into two classes (low/high) with an 85% classification accuracy [62]. Thus, less obtrusive sensing is also possible for psychophysiological integration.

However, psychophysiological integration suffers from another limitation: it has not yet demonstrated significant benefits in patients. While the MIMICS project showed that the addition of psychophysiological measurements enables more accurate classification of patient state [47, 53], it did not demonstrate any improvements in patient motivation, effort or rehabilitation outcome. Follow-up studies by co-author Novak's team involving healthy participants indicated that improved classification accuracy in a psychophysiological feedback loop does tend to lead to an improved user experience, but that at least ~ 10% higher accuracy is needed for a noticeable improvement in user experience [63, 64]. Thus, psychophysiological measurements are likely to be more useful in the MIMICS gait rehabilitation system, where they improved accuracy by approximately 15% [53], rather than in the arm rehabilitation system, where they improved it by approximately 8% [47].

Other groups have shown improvements in psychological state as a result of biocooperative control [54–56], indicating potential benefits. However, two weaknesses of this work must be acknowledged. First, these studies were conducted with participants with no physical or cognitive impairment, and results are not guaranteed to transfer to patients with neurological injuries. Patients not only exhibit different psychophysiological responses than participants without neurological injuries [65], but also may have different priorities and interests during a rehabilitation task, thus exhibiting different reactions to biocooperative control. Second, these studies have mostly shown benefits of biocooperative control compared to, e.g., a simple calibration task [54] or a random task adaptation schedule [56]. For biocooperative control to be broadly adopted, it would need to demonstrate benefits compared to a system that makes intelligent adaptation decisions using fewer sensors. For example, questionnaires could be used

to measure a patient's personality and exercise preferences, and this information could thus be used to personalize robot-aided rehabilitation (by providing emotionally unstable patients with less stressful exercises) without the need for constant psychophysiological monitoring [66, 67]. Alternatively, since the self-reported opinions of patients or therapists are commonly used as the "ground-truth" data for machine learning in psychophysiological integration, patients could simply be asked to manually adapt the difficulty themselves. Psychophysiological integration is thus only likely to be adopted by clinicians if it can demonstrate higher patient motivation or higher exercise intensity than such simpler solutions.

10.4 Conclusion

In motor rehabilitation, humans can be integrated with machines at multiple levels: biomechanical, physiological, and even psychological. While biomechanical integration in the form of assist-as-needed techniques has become fairly standard in robot-aided rehabilitation, both physiological and psychophysiological integration still remain at the prototype stage due to the reliance on often inconvenient sensors and (in the case of psychophysiology) unclear clinical benefits.

Nonetheless, the development of noncontact and robot-mounted physiological sensors is likely to soon overcome one of the key barriers to practical psychophysiological integration. This will then lead to broader adoption of physiological integration, which has already shown the ability to train patients at safe, effective exercise intensities. Simultaneously, it will encourage further evaluation of biocooperative controllers with patients. If such biocooperative controllers are shown to have short-term benefits in patients, they may be able to improve robot-assisted rehabilitation by enabling clinicians to provide therapy that is tailored to each patient's specific needs and demands. This may in term lead to a better rehabilitation outcome and improved quality of life for patients with diverse neurological injuries.

Acknowledgements This work was supported in part by the U. S. National Science Foundation under grant no. 2024813.

References

1. Riener R, Lünenburger L, Jezernik S, Anderschitz M, Colombo G, Dietz V. Patient-cooperative strategies for robot-aided treadmill training: first experimental results. IEEE Trans Neural Syst Rehabil Eng. 2005;13(3):380–94.
2. Wolbrecht ET, Chan V, Reinkensmeyer DJ, Bobrow JE. Optimizing compliant, model-based robotic assistance to promote neurorehabilitation. IEEE Trans Neural Syst Rehabil Eng. 2008;16 (3):286–97.
3. Gassert R, Dietz V. Rehabilitation robots for the treatment of sensorimotor deficits: A neurophysiological perspective. J Neuroeng Rehabil. 2018;15:46.
4. Marchal-Crespo L, Reinkensmeyer DJ. Review of control strategies for robotic movement training after neurologic injury. J Neuroeng Rehabil. 2009;6(20).
5. Hobbs B, Artemiadis P. A review of robot-assisted lower-limb stroke therapy: unexplored paths and future directions in gait rehabilitation. Front Neurorobot. 2020;14:19.
6. Riener R, Munih M. Guest editorial: Special section on rehabilitation via bio-cooperative control. IEEE Trans Neural Syst Rehabil Eng. 2010;18(4):337–8.
7. Gordon NF, Gulanick M, Costa F, Fletcher G, Franklin BA, Roth EJ, et al. Physical activity and exercise recommendations for stroke survivors: an American Heart Association scientific statement from the Council on Clinical Cardiology, Subcommittee on Exercise, Cardiac Rehabilitation, and Prevention; the Council on Cardiovascula. Stroke. 2004;35 (5):1230–40.
8. Colombo G, Joerg M, Schreier R, Dietz V. Treadmill training of paraplegic patients using a robotic orthosis. J Rehabil Res Dev. 2000;37(6):693–700.
9. Veneman JF, Kruidhof R, Hekman EE, Ekkelenkamp R, van Asseldonk EH, van der Kooij H. Design and evaluation of the LOPES exoskeleton robot for interactive gait rehabilitation. IEEE Trans Neural Syst Rehabil Eng. 2007;15(3):379–86.
10. Nam KY, Kim HJ, Kwon BS, Park J-W, Lee HJ, Yoo A. Robot-assisted gait training (Lokomat) improves walking function and activity in people with spinal cord injury: a systematic review. J Neuroeng Rehabil. 2017;14:24.
11. Bruni MF, Melegari C, De Cola MC, Bramanti A, Bramanti P, Calabrò RS. What does best evidence tell us about robotic gait rehabilitation in stroke patients: a systematic review and meta-analysis. J Clin Neurosci. 2018;48:11–7.
12. Su SW, Huang S, Wang L, Celler BG, Savkin A V., Guo Y, et al. Nonparametric Hammerstein model based model predictive control for heart rate regulation. In: Proceedings of the 29th Annual International Conference of the IEEE Engineering in Medicine and Biology Society. 2007. p. 2984–7.
13. Colby SM, Kirkendall DT, Bruzga RF. Electromyographic analysis and energy expenditure of harness supported treadmill walking: implications for knee rehabilitation. Gait Posture. 1999;10(3):200–5.
14. Koenig A, Caruso A, Bolliger M, Somaini L, Omlin X, Morari M, et al. Model-based heart rate control during robot-assisted gait training. In: Proceedings of the 2011 IEEE International Conference on Robotics and Automation. 2011. p. 4151–4156.
15. Dietz V, Duysens J. Significance of load receptor input during locomotion: a review. Gait Posture. 2000;11(2):102–10.
16. Achten J, Jeukendrup AE. Heart rate monitoring: applications and limitations. Sport Med. 2003;33 (7):517–38.
17. Hájek M, Potůček J, Brodan V. Mathematical model of heart rate regulation during exercise. Automatica. 1980;16(2):191–5.
18. Baum K, Essfeld D, Leyk D, Stegemann J. Blood pressure and heart rate during rest-exercise and exercise-rest transitions. Eur J Appl Physiol Occup Physiol. 1992;64(2):134–8.
19. Fagraeus L, Linnarsson D. Autonomic origin of heart rate fluctuations at the onset of muscular exercise. J Appl Physiol. 1976;40(5):679–82.
20. Tokgözoglu SL, Batur MK, Topçuoglu MA, Saribas O, Kes S, Oto A. Effects of stroke localization on cardiac autonomic balance and sudden death. Stroke. 1999;30(7):1307–11.
21. Koenig A, Omlin X, Bergmann J, Zimmerli L, Bolliger M, Müller F, et al. Controlling patient participation during robot-assisted gait training. J Neuroeng Rehabil. 2011;8(14).
22. Wieser M, Gisler S, Sarabadani A, Ruest RM, Buetler L, Vallery H, et al. Cardiovascular control and stabilization via inclination and mobilization during bed rest. Med Biol Eng Comput. 2014;52(1):53–64.
23. Tafreshi AS, Okle J, Klamroth-Marganska V, Riener R. Modeling the effect of tilting, passive leg exercise, and functional electrical stimulation on the human cardiovascular system. Med Biol Eng Comput. 2017;55:1693–708.
24. Taveggia G, Ragusa I, Trani V, Cuva D, Angeretti C, Fontanella M, et al. Robotic tilt table reduces the occurrence of orthostatic hypotension over time in vegetative states. Int J Rehabil Res. 2015;38(2):162–6.
25. Darzi A, Goršič M, Novak D. Difficulty adaptation in a competitive arm rehabilitation game using real-time control of arm electromyogram and respiration. In: Proceedings of the 2017 IEEE International Conference on Rehabilitation Robotics. 2017. p. 857–62.
26. Catalán JM, García-Pérez JV, Blanco A, Martínez D, Lledó LD, García-Aracil N. Differences in physiological reactions due to a competitive rehabilitation game modality. Sensors. 2021;21(11):3681.

27. Goršič M, Cikajlo I, Goljar N, Novak D. A multisession evaluation of an adaptive competitive arm rehabilitation game. J Neuroeng Rehabil. 2017;14:128.
28. Pereira F, Bermúdez i Badia S, Jorge C, Cameirão MS. The use of game modes to promote engagement and social involvement in multi-user serious games: a within-person randomized trial with stroke survivors. J Neuroeng Rehabil. 2021;18:62.
29. Wu H-Y, Rubinstein M, Shih E, Guttag J, Durand F, Freeman W. Eulerian video magnification for revealing subtle changes in the world. ACM Trans Graph. 2012;31(4):1–8.
30. Dziuda L, Skibniewski FW, Krej M, Lewandowski J. Monitoring respiration and cardiac activity using fiber Bragg grating-based sensor. IEEE Trans Biomed Eng. 2012;59(7):1934–42.
31. Jakopin B, Mihelj M, Munih M. An unobtrusive measurement method for assessing physiological response in physical human–robot interaction. IEEE Trans Human-Machine Syst. 2017;47(4):474–85.
32. Korpelainen JT, Sotaniemi KA, Myllylä VV. Autonomic nervous system disorders in stroke. Clin Auton Res. 1999;9:325–33.
33. Lotze M, Braun C, Birbaumer N, Anders S, Cohen LG. Motor learning elicited by voluntary drive. Brain. 2003;126(4):866–72.
34. Laver KE, Lange B, George S, Deutsch JE, Saposnik G, Crotty M. Virtual reality for stroke rehabilitation. Cochrane Database Syst. Rev. 2017;**2017**.
35. Guidali M, Duschau-Wicke A, Broggi S, Klamroth-Marganska V, Nef T, Riener R. A robotic system to train activities of daily living in a virtual environment. Med Biol Eng Comput. 2011;49(10):1213–23.
36. Mihelj M, Novak D, Milavec M, Ziherl J, Olenšek A, Munih M. Virtual rehabilitation environment using principles of intrinsic motivation and game design. Presence Teleoperators Virtual Environ. 2012;21(1):1–15.
37. Colombo R, Pisano F, Mazzone A, Delconte C, Micera S, Carrozza MC, et al. Design strategies to improve patient motivation during robot-aided rehabilitation. J Neuroeng Rehabil. 2007 Jan;4(3).
38. Byrne EA, Parasuraman R. Psychophysiology and adaptive automation. Biol Psychol. 1996;42(3):249–68.
39. Picard RW, Vyzas E, Healey J. Toward machine emotional intelligence: Analysis of affective physiological state. IEEE Trans Pattern Anal Mach Intell. 2001;23(10):1175–91.
40. Chanel G, Rebetez C, Bétrancourt M, Pun T. Emotion assessment from physiological signals for adaptation of game difficulty. IEEE Trans Syst Man Cybern – Part A Syst Humans. 2011;41(6):1052–63.
41. Healey JA, Picard RW. Detecting stress during real-world driving tasks using physiological sensors. IEEE Trans Intell Transp Syst. 2005;6(2):156–66.
42. Gashi S, Lascio E Di, Santini S. Using students' physiological synchrony to quantify the classroom emotional climate. In: UbiComp/ISWC 2018-Adjunct Proceedings of the 2018 ACM International Joint Conference on Pervasive and Ubiquitous Computing and Proceedings of the 2018 ACM International Symposium on Wearable Computers. 2018.
43. Novak D, Mihelj M, Munih M. A survey of methods for data fusion and system adaptation using autonomic nervous system responses in physiological computing. Interact Comput. 2012;24:154–72.
44. Cameirão MS, Badia SB i, Mayank K, Guger C, Verschure PFMJ. Physiological responses during performance within a virtual scenario for the rehabilitation of motor deficits. In: Proceedings of PRESENCE 2007. Barcelona, Spain; 2007. p. 85–8.
45. Novak D, Mihelj M, Munih M. Psychophysiological responses to different levels of cognitive and physical workload in haptic interaction. Robotica. 2011;29(03):367–74.
46. Novak D, Mihelj M, Munih M. Dual-task performance in multimodal human-computer interaction: a psychophysiological perspective. Multimed Tools Appl. 2012;56(3):553–67.
47. Novak D, Mihelj M, Ziherl J, Olenšek A, Munih M. Psychophysiological measurements in a biocooperative feedback loop for upper extremity rehabilitation. IEEE Trans Neural Syst Rehabil Eng. 2011;19(4):400–10.
48. Ziherl J, Novak D, Olenšek A, Mihelj M, Munih M. Evaluation of upper extremity robot-assistances in subacute and chronic stroke subjects. J Neuroeng Rehabil. 2010;7(52).
49. Task Force of the European Society of Cardiology and the North American Society of Pacing and Electrophysiology. Heart rate variability: Standards of measurement, physiological interpretation, and clinical use. Eur Heart J. 1996;17(3):354–81.
50. Delaney JP, Brodie DA. Effects of short-term psychological stress on the time and frequency domains of heart-rate variability. Percept Mot Skills. 2000;91(2):515–24.
51. Boucsein W. Electrodermal Activity. 2nd ed. Springer; 2012.
52. Riener R, Koenig A, Bolliger M, Wieser M, Duschau-Wicke A, Vallery H. Bio-cooperative robotics: controlling mechanical, physiological and mental patient states. In: Proceedings of the 2009 IEEE 11th International Conference on Rehabilitation Robotics. 2009. p. 407–12.
53. Koenig A, Novak D, Omlin X, Pulfer M, Perreault E, Zimmerli L, et al. Real-time closed-loop control of cognitive load in neurological patients during robot-assisted gait training. IEEE Trans Neural Syst Rehabil Eng. 2011;19(4):453–64.
54. Rodriguez-Guerrero C, Fraile Marinero JC, Turiel JP, Muñoz V. Using, "human state aware" robots to enhance physical human-robot interaction in a cooperative scenario. Comput Methods Progr Bomed. 2013;19(112):250–9.
55. Rodriguez-Guerrero C, Knaepen K, Fraile-Marinero JC, Perez-Turiel J, Gonzalez-de-Garibay V,

Lefeber D. Improving challenge/skill ratio in a multimodal interface by simultaneously adapting game difficulty and haptic assistance through psychophysiological and performance feedback. Front Neurosci. 2017;11:242.

56. Shirzad N, Van der Loos HFM. Evaluating the user experience of exercising reaching motions with a robot that predicts desired movement difficulty. J Mot Behav. 2016;48(1):31–46.

57. Darzi A, Novak D. Automated affect classification and task difficulty adaptation in a competitive scenario based on physiological linkage: an exploratory study. Int J Hum Comput Stud. 2021;153: 102673.

58. Liu T, Saito G, Lin C, Saito H. Inter-brain network underlying turn-based cooperation and competition: a hyperscanning study using near-infrared spectroscopy. Sci Rep. 2017;7:8684.

59. Spapé MM, Kivikangas JM, Järvelä S, Kosunen I, Jacucci G, Ravaja N. Keep your opponents close: Social context affects EEG and fEMG linkage in a turn-based computer game. PLoS ONE. 2013;8(11): e78795.

60. Short MR, Hernandez-Pavon JC, Jones A, Pons JL. EEG hyperscanning in motor rehabilitation: a position paper. J Neuroeng Rehabil. 2021;18:98.

61. Knaepen K, Marušič U, Crea S, Guerrero CR, Vitiello N, Pattyn N, et al. Psychophysiological response to cognitive workload during symmetrical, asymmetrical and dual-task walking. Hum Mov Sci. 2015;40:248–63.

62. McDuff D, Gontarek S, Picard RW. Remote measurement of cognitive stress via heart rate variability. In: Proceedings of the 2014 36th Annual International Conference of the IEEE Engineering in Medicine and Biology Society. 2014. p. 2957–60.

63. McCrea SM, Geršak G, Novak D. Absolute and relative user perception of classification accuracy in an affective videogame. Interact Comput. 2017;29 (2):271–86.

64. Novak D, Nagle A, Riener R. Linking recognition accuracy and user experience in an affective feedback loop. IEEE Trans Affect Comput. 2014;5(2):168–72.

65. Goljar N, Javh M, Poje J, Ocepek J, Novak D, Ziherl J, et al. Psychophysiological responses to robot training in different recovery phases after stroke. In: Proceedings of the 2011 IEEE International Conference on Rehabilitation Robotics. 2011.

66. Nagle A, Wolf P, Riener R. Toward a system of customized video game mechanics based on player personality: relating the Big Five personality traits with difficulty adaptation in a first-person shooter game. Entertain Comput. 2016;13:10–24.

67. Darzi A, McCrea S, Novak D. User experience comparison between five dynamic difficulty adjustment methods for an affective computer game. JMIR Serious Games. 2021;9(2): e25771.

Sensory-Motor Interactions and the Manipulation of Movement Error

Pritesh N. Parmar, Felix C. Huang, and James L. Patton

Abstract

Brain injury often results in a partial loss of the neural resources communicating to the periphery that controls movements. Consequently, the signals that were employed prior to injury may no longer be appropriate for controlling the muscles for the intended movement. Hence, a new pattern of signals may need to be learned that appropriately uses the residual resources. The learning required in these circumstances might in fact share features with sports, music performance, surgery, teleoperation, piloting, and child development. Our lab has leveraged key findings in neural adaptation as well as established principles in engineering control theory to develop and test new interactive environments that enhance learning (or re-learning). Successful application comes from the use of robotics and video feedback technology to augment error signals. These applications test standing hypotheses about error-mediated neuroplasticity and illustrate an exciting prospect for rehabilitation environments of tomorrow. This chapter highlights our works, identifies our acquired knowledge, and outlines some of the successful pathways for restoring function to brain-injured individuals.

Keywords

Learning · Motor control · Movement · Human · Rehabilitation · Adaptation · Training · Feedforward control

P. N. Parmar
The Shirley Ryan AbilityLab, University of Illinois at Chicago, Chicago, IL, USA
e-mail: pparma2@uic.edu

F. C. Huang
Department of Physical Medicine & Rehabilitation, Northwestern University, Chicago, Illinois, USA
e-mail: f-huang@northwestern.edu

J. L. Patton (✉)
Biomedical Engineering at University of Illinois at Chicago, The Robotics Lab at the Shirley Ryan Ability Lab, The Department of Physical Medicine Rehabilitation at Northwestern University, University of Illinois at Chicago, Shirley Ryan Ability Lab, & Northwestern University, Chicago, IL, USA
e-mail: jPatton@sralab.org

11.1 Introduction

As rehabilitation research continues to provide support for extended practice on daily activities, technology has emerged as a promising modality to enhance neurorehabilitation. Enhanced training through robotic interfaces and virtual reality systems in the rehabilitation from neurological injury (e.g., stroke) could facilitate the reorganization of neural circuits to re-establish normal movement patterns [1, 2]. For many researchers, the central

© The Author(s), under exclusive license to Springer Nature Switzerland AG 2022
D. J. Reinkensmeyer et al. (eds.), *Neurorehabilitation Technology*,
https://doi.org/10.1007/978-3-031-08995-4_11

issue to be addressed is how technology can deliver therapeutic advantages over simply administering greater intensity or prolonged treatment in traditional approaches. This chapter will focus on how technology can augment error to speed up, enhance, or trigger motor relearning.

In this chapter, we first provide a review and motivation from neuroplasticity and recovery in rehabilitation. Then, we present how augmentation of feedback can be used to leverage neuroplasticity. Next, we present a key form of augmented feedback based on error (error augmentation) to enhance motor learning. Next, we discuss challenges and opportunities associated with personalizing the error augmentation technique for an individual. Finally, we share some preliminary research on novel versions of error augmentation (statistical error augmentation and functional error augmentation).

11.2 Neuroplasticity and Recovery

While the fundamental principles of neurorehabilitation are still being actively debated, nearly all clinicians and researchers that do rehabilitation agree that a key mode of recovery is the nervous system's natural capacity to adapt— neuroplasticity of neural control. Although for brain injuries such as stroke, there are many associated deficits (e.g., contractures, weakness, cognitive deficits, attentional deficits, etc.) that may not be directly involved in neural reorganization, one mode of recovery is certainly to *learn* how to perform normal functions. Hence, the capacity of neuroplasticity is believed to be one of the most powerful resources that can be leveraged to foster functional recovery through the proper conditions of training, feedback, encouragement, motivation, and time. The proper cocktail of these components leads to varying outcomes and is the subject of a large family of research projects over the years.

Early investigations of training-induced neuroplasticity were motivated by results from studies of sensorimotor adaptation in healthy individuals. Earlier studies dating back to the 1950's used a variety of more traditional motor learning tasks to understand the adaptation processes (see Schmidt and Lee [3] for an excellent review). These early studies trace back to Bernstein who studied the skills of the blacksmith [4], and a vast number of visuomotor distortion experiments [5], and pursuit tracking tasks [6]. Tasks such as reaching for a cup seem trivial but extremely difficult and frustrating to some survivors of stroke. This is because even mundane everyday actions are complex dynamic control problems that require skills acquired through learning processes. Moreover, the factors influencing arm movements include coupled nonlinear arm dynamics [7], long feedback delays [8], and slow activation times for muscle [9]. Consequently, movements (especially rapid movements) must be pre-planned using a prediction or "neural representation" of the outcomes. These representations, also called internal models, are typically acquired via a lifetime of experience [10]. Yet, research has shown that exposing individuals to tasks with altered sensorimotor relationships modifies these representations at least for a short period of time following training.

One such distortion of sensory-motor relationships is through mechanical alterations (haptics), such as the introduction of a heavy weight in one's hand, which causes reaching errors that prompt adaptation. The motivation is to recover the ability to move with one's normal pattern within a single motion [11], but complex loads (loads not so easily understood, like wielding a hammer) can take hundreds of movement attempts [12–14]. People often stiffen (i.e., cocontract their muscles) as a first strategy in response to external force perturbation [15, 16]. However, the stiffness quickly fades as they learn to counteract the forces. Such compensation for the forces results in *after-effects* when forces are unexpectedly removed (Fig. 11.1) [14, 17, 18]. It is important to note that both the adaptation and after-effects can occur implicitly with minimal conscious attention to any goal [19]. Beyond the investigation of basic motor control principles, we have shown that this type of training can be used constructively to teach a desired set of new movements [19,20]. Also, beyond effecting a

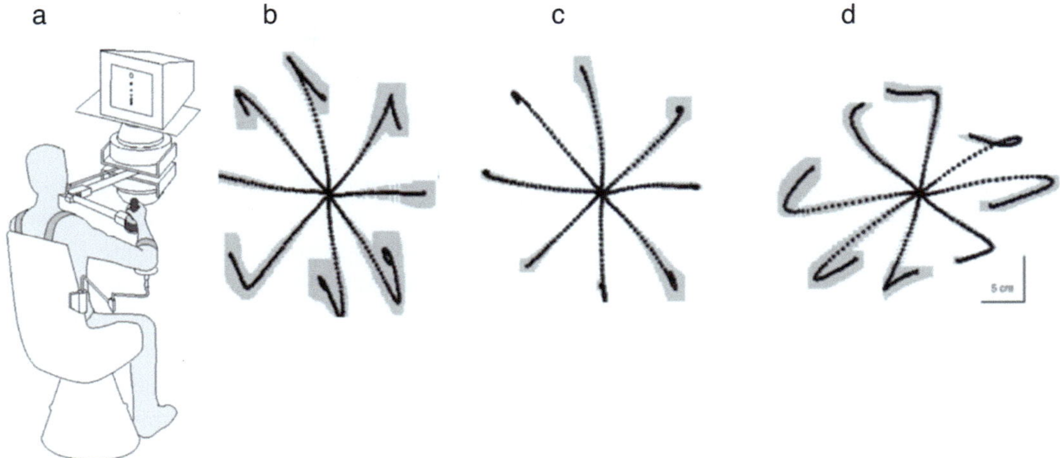

Fig. 11.1 A classic adaptation experiment in which a robot exerts a mechanical distortion (force field) on the subject's hand during reaching. The subject attempted reaching movements to targets in 8 different directions. **a** Experimental apparatus with a subject seated at the robot. **b** Reaching movements during the Initial exposure to the force field. **c** Reaching movements at the end of training. **d** Reaching movements when the force field was unexpectedly removed post training, which resulted in after-effects. Adapted from *Shadmehr R, Mussa-Ivaldi FA. Adaptive representation of dynamics during learning of a motor task. J Neurosci. 1994;14(5 Pt 2):3208–24* (14). Copyright 1994 Society for Neuroscience

change in overall force-motion relationships, robotic training can introduce haptic interactions at the point of contact [21, 22]. Such interactions can alter both the energetics and sensory information in a task, and hence could elicit additional neural responses [23].

Adaptation can also occur when one is exposed to a visual distortion. In fact, this type of motor learning is an older body of research on visuomotor adaptations, such as those induced by prisms (see Harris [24] for a review), rotations, stretches and other distortions of the conventional hand-to-screen mapping [26–28). All of these distortions appear to induce learning and can reduce sensory dysfunction such as hemispatial neglect [29).

Motor adaptation induced by interactive forces and/or visual distortion may be temporary and fragile, however. People de-adapt faster than they adapt [25], and the learning effects often does not persist across timescales that are good for use in neurorehabilitation [19, 26–28]. Furthermore, it has been shown that a subsequent training experience can interfere with the motor adaptation from previous training [29–31]. Such findings would at first seem to undermine the

prospects of successfully transferring the skills acquired in an engineered environment to functional ability in the real world. However, one key premise of robot-mediated training is that adaptation is retained if the resulting behaviors have functional utility. Our studies [19, 32, 33] and the work of others [29, 34, 35] have demonstrated some evidence for persistent effects after training in the presence of visuomotor distortions. Hence individuals de-adapt if conditions require it, and some motor memory is preserved well beyond the training phase. Further work is needed to understand what neural processes mediate the successful evolution between adaptation from a single session training and long-term retention, and it may be that the two share many common neural resources, with a continuum between short and long-term neuroplasticity.

Quite importantly, the adaptive responses can also be observed in stroke survivors, and evidence is found in the oculomotor [36] and limb motor systems [28, 37, 38] for neuroplasticity, induced by enriched interactive experiences. In fact, errors seen in reaching movements of stroke survivors reflect poor compensation of interaction torques and resemble the problems seen in healthy subjects when they are exposed to force fields [39]. At least, part of impairment has been attributed to "learned non-use" that can be reversed by encouraging individuals to practice and relearn how to move their affected arm [40]. What is clear is that the process of *learning* deserves a closer look.

11.3 Multiple Functional Forms of Neuroplasticity

Physiologically, plasticity comes in many forms across many time scales, making it difficult to fully identify all underlying mechanisms at work during each therapeutic context [41, 42]. Changes can range from very temporary shifts in neurotransmitter concentrations, facilitation or inhibition from collateral neurons, neural growth to establish synapses, or actual neurogenesis where entire neurons are established. Making matters more complicated, neuroplasticity can be described as residing within a much larger spectrum of mechanisms with overlapping times scales. These can include dendritic spine formation and synaptic strength changes across seconds, long term potentiation over minutes, permanent consolidation of learning, muscle hypertrophy, healing, degeneration of tissue, growth in development, and loss during natural aging. Finally, aspects of the nervous system's control apparatus can be seen as hierarchical agents, where people *learn to learn*, recall better or worse after training, and even learn to make decisions to learn.

Functionally, or algorithmically, or computationally, learning is typically divided in three groups. First, supervised learning involves adjustments to action commands based on error between intended and resulting outcomes. Second, the reinforcement learning (RL) allows for exploration of many options, and then rewards (or punishes) the system after multiple attempts or sequences of actions. Third, unsupervised "Hebbian" learning that only strengthens computational units used while pruning those that are unused. These three algorithmic processes are non-physiologic in their conception but are connected by some evidence to the nervous system. Supervised is related to the idea of backpropagation of error in the nervous system [43] and the cortical-cerebellar processes [44–49]. Dopaminergic neurons and cortico-Basal Ganglia networks are thought to play an important role in the reinforcement learning [50–53]. Hebbian learning is related by some to the Spinal Cord and brain stem in its retraining [54–56]. Hebbian learning may also be related to "error-free learning" [57–59]. It is also connected to the benefits seen from repetitive practice which consolidates successfully learned actions.

Supervised motor learning involves reduction of performance errors [60, 61]. Tests in experiments often use simple tasks to explore how humans respond to such errors. In reaching, these errors are presumed to be simply deviations away from the straight-line hand path to a target [62, 63]. Experiments have demonstrated that it is possible to train subjects to produce new movements, such as a prespecified (but straight) path during reaching. Such adaptive training has resulted in altered motion patterns in both the arms [28] and legs [64] by accentuating trajectory errors using robotic forces. Subjects in those studies were exposed to custom-designed force fields that promoted the learning of specific movements by exploiting short-term adaptive processes [64]. Besides reducing performance error, supervised learning framework can also explain minimization of motor effort [65–67], pain [68], uncertainty [69–73], and sensory mistrust [74–76].

Even within the scope of error-based learning, there are many ways in which the nervous system alters behavior in response to error experiences (Fig. 11.2). It is not clear *which error* is the error the nervous system is using to learn. There has been recent debate over whether kinetic and kinematic adaptations are separate and independent processes. Krakauer et al. [30] suggested that learning of kinematic distortions (a 30° rotation of visual display) and kinetic distortions (distortions of added mass) were independent processes because learning one did not interfere with the other. Basic modeling assumptions can easily show that separate error-motivated adaptation processes could arise (different red lines of Fig. 11.2). Flanagan and colleagues also showed similar results with a visuomotor rotation and a viscous force field [77]. However, Tong and colleagues argued that these studies should not be expected to show interference because the kinetic and kinematic distortions involved different variables, and the kinematic rotation depended on position while the kinetic mass depended on acceleration [31]. They demonstrated that when both the force field and the visuomotor rotation depended on position (or on acceleration), interference was observed. These results strongly suggest that kinetic and kinematic adaptation share at least some common neural resources in motor working memory. As a logical extension of this concept of shared resources, we might employ multiple environmental effects to "trick" the nervous system into learning more. One possibility is to *facilitate* (rather than interfere with) learning. Consequently mixed experience of both force and visual feedback distortions can enhance learning even further [78].

11.4 Augmentation of Feedback to Leverage Neuroplasticity

The great enlightenment philosopher George Berkeley pioneered the idea "Esse est percipi' (to be is to be perceived). Technology has recently allowed us to constructively alter motor behavior through novel distortions to perception, essentially creating a "lie" to the interacting subject in a variety of ways through augmented feedback. This approach to facilitating training offers a bright prospect, not only in the world of engineering for rehabilitation, but in many areas in which people must learn to make new actions.

Augmentations of feedback can provide the learner with information to reinforce movement patterns [79, 80]. Studies have shown that neuroplasticity can be stimulated when the visual or haptic feedback conditions are altered [70, 76, 77, 81–83], falsified [84] and amplified [64]. Subjects learning how to counteract a force disturbance in a walking study increased their rate of learning by approximately 26% when a

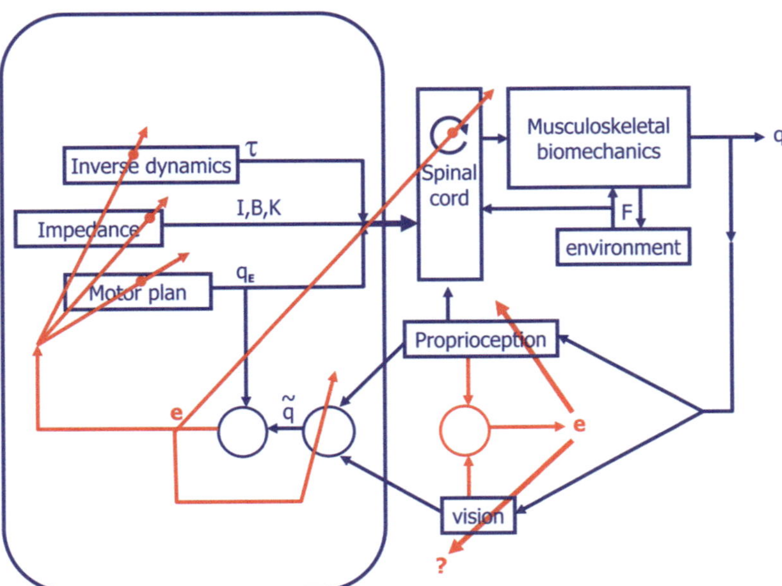

Fig. 11.2 A schematic flowchart that illustrates the believed error-mediated adaptation for the control of movement. News of outcome movements are fed back to the central nervous system to calculate errors, e, that are used for adjusting motor commands (adapting). Several known mechanisms exist that use error (red lines) to make alterations, such as recalibration of the proprioceptive system, alterations in preplanned inverse dynamics, impedance, and the intended trajectory

disturbance was transiently amplified [64]. In another study, providing feedback that was less than the actual force production caused subjects to apply larger forces to compensate [84]. Conversely, suppression of visual feedback has been shown to slow the un-learning process [19]. Nevertheless, not all kinds of augmented feedback on practice conditions have proven to be therapeutically beneficial for healthy participants and for stroke survivors [85].

11.5 Movement Augmentation

Training environments can simply amplify motor actions. Robot guided training can exaggerate movements in real-time, effectively augmenting the dynamic behavior of the arm. Robotic intervention can certainly expand human capabilities through interactive environments using a function of applied forces or speed [86, 87]. Such approaches use active impedance, such as *negative damping* (or *negative viscosity),* which constantly pushes the limb more in the same direction it is moving. Beyond altering online performance, negative damping can increase awareness of deviations from expected behavior —information critical for driving adaptation. Furthermore, a major advantage is that it allows training even when weakness limits voluntary motion. Most importantly, however, such environments facilitate training and allow generalization to out-of-training situations (e.g., activities of daily living).

To test negative damping as a supplement to training, we investigated the efficacy in a skills training experiment using a robotic interface. One key feature of our approach was to allow self-directed movement during training. While goal-directed movement typically focuses on kinematic performance, we expected that allowing training via exploratory movements would emphasize relevant force and motion relationship and provide better improvement in overall function than repetition of the same task [88, 89]. This free training paradigm also served as an excellent way to test subjects' abilities to *generalize* what they learned, since the structured

evaluations after training (making circles) differed from practice.

We found improvements in performance that persisted even when the negative viscosity portion of the forces were removed [90]. In a follow-up study with stroke survivors (Fig. 11.3), similar training with negative viscosity resulted in improved skill within a single training session, while no improvement was observed in the control group where no forces were administered [91]. It is important to emphasize that each group was evaluated in the absence of applied forces, which demonstrates that patients' training with negative viscosity transferred their learned skills to better actions in the real world.

Excitingly, these ideas were expanded to *free exploration therapy*, where the robotic forces were applied in response to individual deficits of the patient [92, 93]. The training environments were established using a probability density function (PDF), showing the distribution of free exploration movements (in the kinematic space of hand position and velocity), and the forces for training were calculated using the gradient of this probability distribution [94–96]. Comparing these probability distributions of the patients with that of the healthy individual enabled us to readily obtain estimates for the patients' underrepresented motions or motor deficits.

11.6 Guidance Versus Error Augmentation

Human–machine interactions have the extremely powerful ability to foster learning, but it is less obvious how to precisely program them to alter these interactions for therapeutic benefit. One possibility would be to have a system that *guides* one's actions to help one learn. This enables the patient to visit the positions and velocities of a task; being "shown the way" as a template. This template may offer the added benefits of keeping the joint mobile through the range of motion and preventing secondary effects such as contractures from immobility. In one study on healthy people, simply watching the robot make a template motion caused subjects to learn about as well as

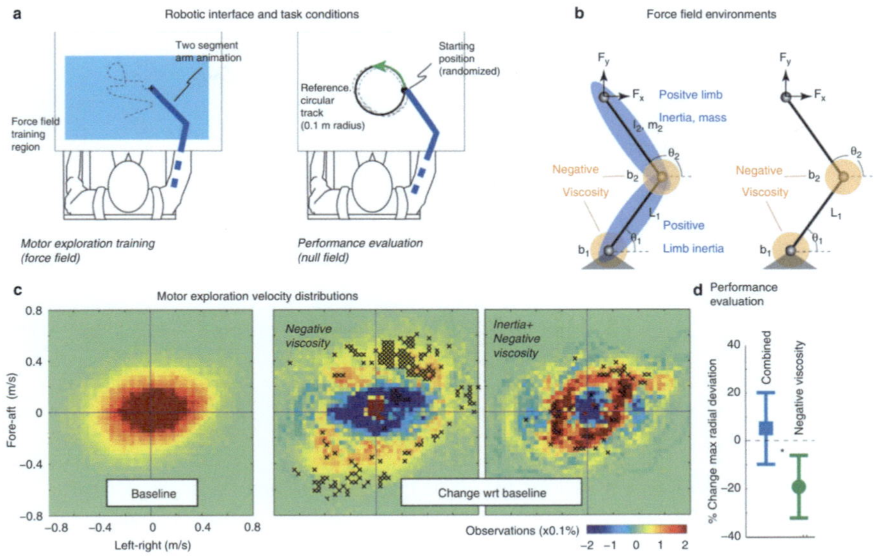

Fig. 11.3 Patients benefit from free exploration training with robot-applied negative viscosity to augment movement. **a** The robot interfaced to the arm about a free pivot at the wrist. Subjects were allowed to freely interact with each load in a "motor exploration" stage. Following exploration, subjects made counterclockwise circular movements during task performance trials at random starting locations of a 0.1 m radius circular track. **b** The virtual arm augmented the existing dynamics of the human arm with negative viscosity in the elbow and shoulder and/or positive inertia to the upper and forearm. **c** Stroke survivors (n = 10) perform motor exploration with no load, revealing average baseline distribution with evident asymmetry in range. Negative Viscosity training prompted significant increases (indicated as x) especially in elbow flexion–extension. **d** Tests of learning show error decreased (-19.1 ± 0.1%, p = 1.3e-2) from Negative Viscosity training, while no change was found from Inertia + Negative Viscosity training (+5.1 ± 16.2%, p = 4.3e-1). Adapted from *Huang FC, Patton JL. Augmented dynamics and motor exploration as training for stroke. IEEE Trans Biomed Eng. 2013;60(3):838–44* (98). RightsLink License 5,260,980,732,698 Mar 2, 2022

people when they practiced using robotic guidance [97]. While this may be an answer for people entirely paralyzed, this approach only provides the correct kinematics without the correct kinetics. There have been a few studies that have shown a benefit for haptic guidance in learning motions [98–100].

The guidance approach, however, does not ensure that the person produces all motor actions necessary for the correct motion. The guidance interventions are designed to supplement the movement deficits such that task objectives are achieved, and the trainee is not challenged to generate total compensatory motor actions. This type of supplementary intervention in effect reduces deviations away from expected behavior. Such guidance algorithms generate unnatural interactive forces unless individuals also actively participate in making the desired motions, which

renders a guiding robot unnecessary. For example, guiding forces, such as those from path control [101, 102] and haptic channel [103, 104], can improve the performance, but still require the user to move. Furthermore, the guidance interactions may encourage unwanted resistance, promote laziness [105, 106], or reduce the subject to inattention. This can remove any desire to learn, and lead the individual to simply rely on guidance like one might rely on a crutch [107]. People can fall asleep while practicing. The negative damping approach (see above), on the other hand, augments the overall movements, and if there are deviations from expected behavior (i.e., movement errors), those are augmented as well. The negative damping, however, also amplifies other aspects of the movements (such as reach extent and visits to novel kinematic configurations) that may be irrelevant to

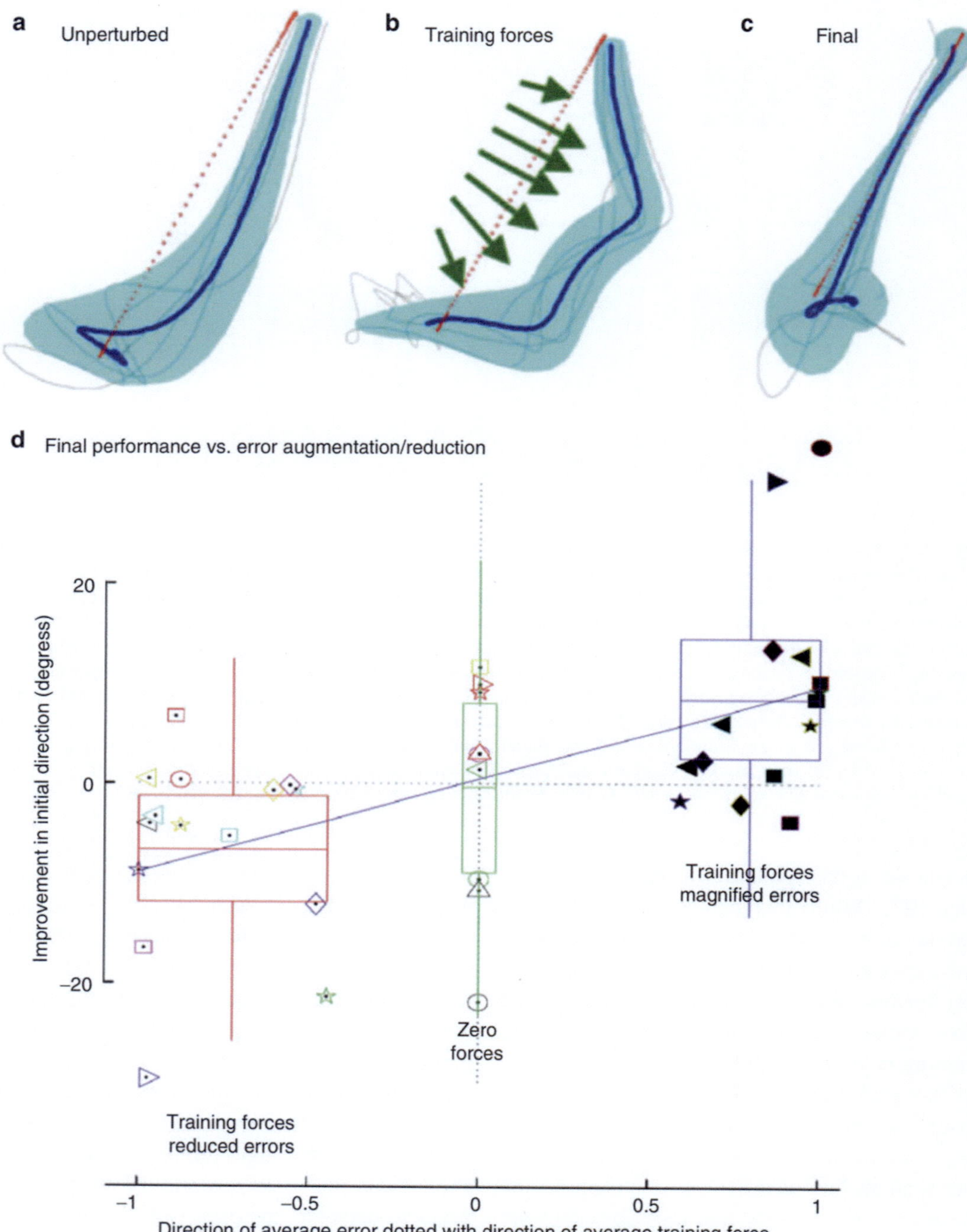

Fig. 11.4 **a** One stroke survivor's unperturbed reaching movement, showing initial movement error. **b** Response to training forces that amplify the original counterclockwise movement error. **c** The force field during training (arrows in b) resulted in a reduction of error following training that was sustained until the end of the experiment. **d** Cross plot of all stroke subjects' final performance improvements versus the amount of error magnification/reduction in training. Error magnification was determined by calculating the dot product between the average training force direction and the average movement error direction. Performance improvement was calculated by measuring the reduction initial direction error from the baseline phase to the final phase of the experiment. Boxes represent mean and 95% confidence intervals, and whiskers indicate 2-standard deviations. Adapted from *Patton JL, Stoykov ME, Kovic M, Mussa-Ivaldi FA. Evaluation of robotic training forces that either enhance or reduce error in chronic hemiparetic stroke survivors. Experimental Brain Research. 2006;168(3):368–83* (115). License 5,267,841,080,641 date Mar 14, 2022

learning and can cause a person to experience exhaustion from such confounding learning signals. A better approach is to provide a focused intervention based on error signals alone to facilitate learning.

An often-striking idea is the opposite – to make errors larger using forms of anti-guidance that we term *Error Augmentation (EA)*. In an early study of error augmentation, our group focused on the chronic stroke population and compared error-magnifying forces to error-reducing forces in a short therapy session. We exposed hemiparetic stroke survivors and healthy aged-matched controls to a pattern of disturbing forces that has been found by previous studies to induce dramatic after-effects in healthy individuals [108]. Eighteen stroke survivors made 834 movements on a manipulandum robot in the presence of a robot-generated force field. This field generated forces that were proportional to hand speed, perpendicular to movement direction —either clockwise or counterclockwise (Fig. 11.4a-c). We found significant after-effects from the stroke surviving participants, indicating the presence of a reserve capacity for neuroplasticity in these patients that has very little or nothing to do with stroke severity [108]. Importantly, significant improvements occurred only when the training forces magnified errors, and not when the training forces reduced errors or when there were no forces (Fig. 11.4d). Interestingly, adaptation during practice in stroke survivors is concurrent with anatomical and cellular evidence that the nervous system reorganizes areas of the brain to accommodate feedback [109]. These results point to a unifying concept —errors induce learning, and judicious manipulation of error feedback can lead to lasting desired changes.

11.7 Error Augmentation for Enhanced Training

Many of the feedback augmentations discussed earlier engage error-driven learning processes, which are believed to be central to skill acquisition [60, 61]. One key implementation of augmented feedback is error augmentation, where we isolate and selectively enhance the performance error during learning [110]. Neural networks (and other forms of artificial intelligence) can show that learning can be enhanced when feedback error is larger [64, 111–113]. Therefore, the error augmentation approach could enhance motor learning. However, the optimal method for error augmentation in motor learning is not yet known.

One method of error augmentation is magnification of error (EA-gain). In a preliminary error augmentation study, we conducted simple evaluation of the rate change of trial-to-trial hand path while subjects made point-to-point reaching movements under a visuomotor distortion [110, 114]. Here, error deviations from a straight-line trajectory were visually augmented with either a magnification of 2 (EA-gain 2) or a magnification of 3.1 (EA-gain 3.1). We found smaller time constants (fitting performance changes to an exponential curve) for the EA-gain 2 compared to the control (EA-gain 1), demonstrating that error augmentation could increase the rate of learning (Fig. 11.5). However, further augmentation using the EA-gain 3.1 showed no benefit. Similar result was observed in a reaching study where there was diminishing effectiveness from the presentation of larger feedback errors, causing smaller performance changes from one movement to the next [115]. Furthermore, van der Kooij, Brenner [116] also reported a faster adaptation to visuomotor rotation at moderate levels of error augmentation gains during reaching.

Another method of error augmentation might be to shift, or *offset,* feedback in the direction of the error. This same study [110] also tested another type of error augmentation via the addition of a constant "error-offset" to the performance feedback (EA-offset). This has been shown to influence the rate of learning (time constant, Fig. 11.5) and the amount of learning (total error reduction). While the error magnification amplifies error magnitude in the feedback, it also magnifies motor variability, sensor inaccuracies and other uncertainties [67, 76, 117], and these factors can confound learning. Therefore, the error magnification may be practicably limited

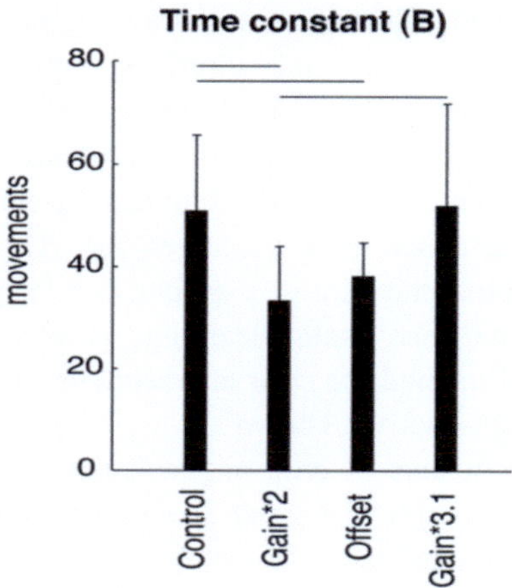

Fig. 11.5 Time constant of error decay during visuomotor learning with error augmentation in healthy subjects. Faster time constant was observed at moderate EA-gain (Gain*2), but the EA-gain effect was somehow lost for higher gains (Gain*3.1). Error bars indicate 95% confidence intervals. Horizontal lines indicate significant differences (post-hoc) between groups. [110] Adapted from *Patton JL, Wei YJ, Bajaj P, Scheidt RA. Visuomotor learning enhanced by augmenting instantaneous trajectory error feedback during reaching. PLoS One. 2013;8 (1):e46466.* Copyright: © 2013 Patton et al. open-access

to moderate gains. On the other hand, adding a constant error bias to augment feedback may be equal or more effective because such augmented feedback would be independent of noise and other confounding factors. A constant offset presents persistent errors throughout training, even as the learner improves. This technique may motivate learning longer during practice and hence cause the amount of learning to increase (total error reduction from the initial error or lower error in the steady-state performance).

While there are several neural mechanisms for how error augmentation might work, a full understanding of the sources is not known. One possible mechanism explaining why magnifying error may be effective is that it makes subjects less tolerant to smaller errors that are now made larger. Modeling and experimental systems have demonstrated better and faster learning if errors are larger [60, 111, 118]. Error bias, such as in the *offset* condition mentioned above, can motivate subjects to persistently try to reduce

fictitious errors. This can lead to learning beyond the desired goal, so such a technique may be beneficial in situations where subjects do not fully learn. Since EA-gain seems to influence the rate of learning and EA-offset seems to influence the amount of learning, a mixture of these two feedback augmentation methods is needed to enhance learning. However, optimal parameters governing such a mixture are not yet known and are likely to differ from patient to patient.

Another possible reason why error augmentation may lead to benefits is that the impaired nervous system is not as sensitive to error and hence does not react to small errors. Error augmentation might make errors more noticeable by raising signal-to-noise ratios in sensory feedback. It may heighten motivation, attention, or anxiety, which has been suggested to correlate with learning [119]. Errors that are more noticeable may trigger responses that would otherwise remain dormant.

Error perception appears to be on a continuum that is not yet understood. Error reduction appears to stifle learning [103] and suppression of visual feedback has been shown to slow down the de-adaptive process [19]. Considering the other extreme, too much error augmentation appears to dampen results. Robotically reducing kinematic errors in a golf training improved skill more for the less skilled, but increasing errors had no effect, and additionally decreased motivation [120]. The nervous system may react to excessively large error signals by decreasing learning so that there is little change in response to subsequent performance errors. Large errors thus may be regarded as outliers by a nonlinear "loss function" that governs motor adaptation [117]. These and other studies that induce sensorimotor conflict suggest that the nervous system *adapts the way it adapts*. In other words, it can re-weigh the interpretation of sensory information if it no longer is perceived reliable [76, 121]. This suggests that there is a "sweet spot" of error augmentation intensities [122]. Thus, to understand the optimal settings, we first need to better understand the landscape of combined effects of the different forms of error augmentation.

11.8 Combined Effects of the Forms of Error Augmentation

Each error augmentation approach (magnifying or biasing) has benefits and pitfalls: gain augmentation is vulnerable to feedback instability, whereas the biasing approach risks learning beyond the goal. As a high feedback gain can induce oscillations in artificial system states, a higher EA-gain may cause frequent changes in launch angles during reaching about an ideal straight-line (changes in the sign of error). Our earlier work [110] could not account for such oscillations about zero error because we used a continuous exponential model. While inter-trial variability has been associated with faster learning [123], excessive oscillations may be perceived as noise and therefore adversely affect learning. Thus, such oscillations should be properly accounted for when studying the effects of EA-gain on the rate of adaptation. Furthermore, the EA-offset can lead to *overcompensation* where there is continued learning beyond the goal, influencing the steady-state performance. Our past study [110] also could not detect such overcompensation effects because we used an unsigned error metric. Various combinations of EA-gain and EA-offset conditions might influence such inter-trial oscillations, the overall learning rate, and steady-state error.

We addressed some of these questions in our recent error augmentation study [124, 125], where we comprehensively examined combined effects of both EA-gain and EA-offset conditions on the time course of visuomotor learning process. We used intermittent "catch trials", where the visual feedback about movements was removed. These catch trials served as "proxy" of the underlying learning process and provided a reasonable measurement of the learned feedforward response that was uncontaminated by the training conditions. We also used a signed error metric to denote any overcompensation as *negative error*. While this study replicated the results of our preliminary study [110], our analysis provided new interpretations. While the preliminary study used an exponential function to estimate time course of motor learning, this newer study used a discrete first-order affine model to fit the learning curves (Fig. 11.6). We measured a significantly faster learning rate (model parameter B) when the EA-gain level was 2 compared to 1 (control) and 3 (with EA-offset level 0; $p = 0.0368$; the Wilcoxon rank sum test with the Bonferroni correction). We found a significant bias (model parameter A) towards the overcompensation in learning (negative error) when EA-gain was 1 and EA-offset was 1 compared to the control (EA-gain 1 and EA-offset 0). We also measured the inter-trial noise (σ) as a standard deviation of the model residual errors post-fit. The inter-trial noise (σ) on average was the lowest when the EA-gain was 2 and EA-offset was 0 and the highest when the EA-gain was 0 and EA-offset was either 0 or 1. The data exhibited a moderate level of inter-trial noise (σ) on average at the other EA coordinates.

Expanding further in this same study [124], we examined several other discrete model-structures that ranged from the first-order to eight-order models with constant learning rates and a few first-order models with error-dependent learning rates. All these models iteratively predicted the next trial movement error given the most recent movement error (first-order) or a history of movement errors (higher-orders). Our exhaustive cross-validation analysis found that the second-order model with constant learning rates was consistently superior in predicting how individuals respond to the error augmented visual feedback. This second-order model used current and previous trial movement errors (e_n and e_{n-1}, respectively; n is current trial number) to predict the next trial movement error (e_{n+1}). Here, we also found that the EA-gain and EA-offset combinations varied in their influence on the learning rates, steady-state performance, and inter-trial variability.

Fig. 11.6 Average parameter values for the first-order affine model across the error-augmentation (EA-gain and EA-offset) coordinates. Blue circles and black lines represent averages, and 95% confidence intervals of the model parameter values, respectively. *dz* represents the stochastic Weiner process. There was a significant bias (model parameter A) towards the overcompensation in learning (negative error) when EA-gain was 1 and EA-offset was 1 compared to the control (EA-gain 1 and EA-offset 0; $p < 0.05$ with post-hoc correction). A significantly faster learning rate (model parameter B) was observed when EA-gain was 2 and EA-offset was 0 compared to the control ($p < 0.05$ with post-hoc correction). The inter-trial noise (σ) on average was the lowest when the EA-gain was 2 and EA-offset was 0 and the highest when the EA-gain was 0 and EA-offset was either 0 or 1. The data exhibited a moderate level of intera-trial noise (σ) on average at the other EA coordinates. [124] Adapted from *Parmar PN, Patton JL. A Second-Order Process Model Best Describes Influence of Error-Augmentation on The Time Course of Motor Learning. PLoS One. 2022;(in revision).* Copyright: © 2022 Parmar et al. open-access

First-order model:

$$e_{n+1} = e_n + A + B\, e_n + \sigma\, dz$$

Model-offset parameter (a)

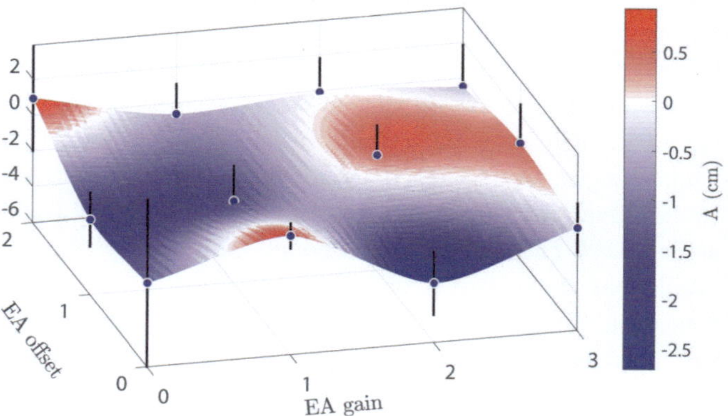

First-order learning rate (b)

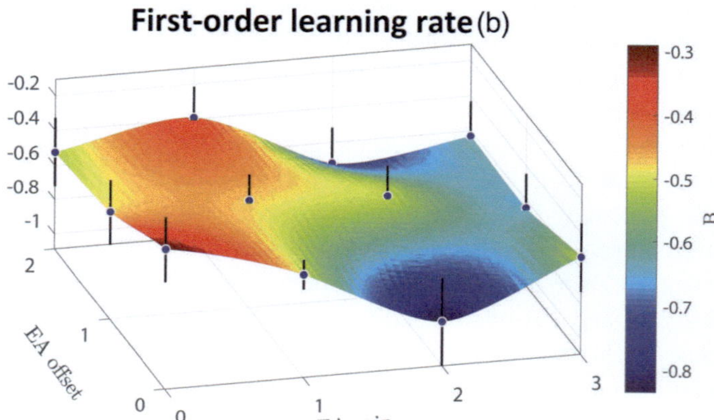

Inter-trial variability (σ)

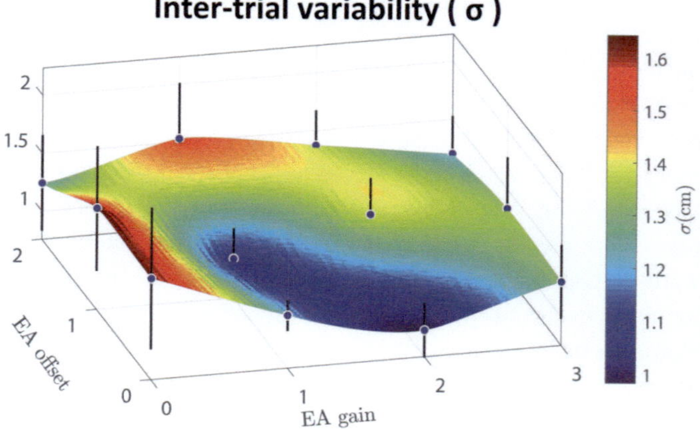

11.9 Challenges and Opportunities for Personalized Training

While the second-order model [124] presented in the previous section can predict how individuals learn in response to the error augmented visual feedback, the model does not readily prescribe training conditions that can be used in rehabilitation for an individual. A better modeling framework is required that can elucidate the training conditions and enable clinicians to design the personalized training for an individual. Such engineering of treatment, tailored to the individual, is part of the now well-known field of personalized or precision medicine. We provide a discussion on a few of these challenges below and suggest potential solutions.

The initial challenge is that data is expensive for personalization. Such data-driven models comprise the field of *system identification* [126], where data is collected from many forms of disturbing (exciting) the system and observing how it changes. These often require large amounts of data and persistent excitation, with a variety of inputs that span the space of possibilities. In the case of neurorehabilitation, each input can be a separate learning experience, and learning may be confounded by prior experience. For example, repeated testing conditions have been used to find best settings for augmented feedback [127–129]. It is sometimes difficult to avoid long experiments that obtain data as we seek patient-specific estimates and conduct personalized training. There are many situations where gathering enough data is costly, time-consuming, or otherwise difficult. Hence, it is imperative to find the minimum amount of data to be collected on a patient, obtaining a "stopping criterion" that can sufficiently characterize an individual's learning tendencies.

The solution to this sample size problem can come from a synthetic analysis. Assuming exponentially decaying transient signals, our recent study [130] simply employed Monte-Carlo methods to determine the minimum number of samples required for accurate identification from noisy transient responses. We then tracked the accuracy of recovery of synthesized data to reveal a prescription for the minimum number of samples for a robust identification of the underlying learning process, given preliminary estimates of the time constant and noise levels. Our results revealed a systematic relationship for the minimum number of samples required from transient signals that can be used as a stopping criterion for data collection [130]. We evaluated these results by using a past motor learning study and determining the minimum number of required samples (movements) to confidently estimate the learning process.

Another challenge is to optimize training *schedules*. While higher doses of motor therapy in chronic poststroke hemiparesis do not necessarily result in better outcome [131], a variety of tasks should be scheduled properly to best redevelop motor ability [132–134]. Repeated actions provide certainty while variability brings better performance through properly assigning credit to relevant factors [135–137]. *Randomized training* of multiple tasks has been shown to enhance retention and transfer capability of the acquired skill compared to the single-task training across *blocked trials* [138, 139]. Most importantly, the variety of training tasks play at least a partial role in restoring function where *generalization* (transfer) is considered a hallmark of good neurorehabilitation training [140]. Randomized practice, however, introduces "trial gaps" where *interference* and *forgetting* may occur (Fig. 11.7). Internal model updates from nonconsecutive, interfering, tasks have not been well modeled.

A solution to this challenge is a model with *memory dynamics* that can also *generalize* from one task to another. Our recent work in a preliminary study [142], included a memory for each movement direction that also shares and informs the learned controls of nearby directions (Fig. 11.7). In our subsequent study [141], we expanded our list of candidate models that considered the first-order versus higher-order effects and constant versus error-dependent learning rates. We also employed cross-validation to select models, and we found that a first-order model with a constant learning rate was the best. This model

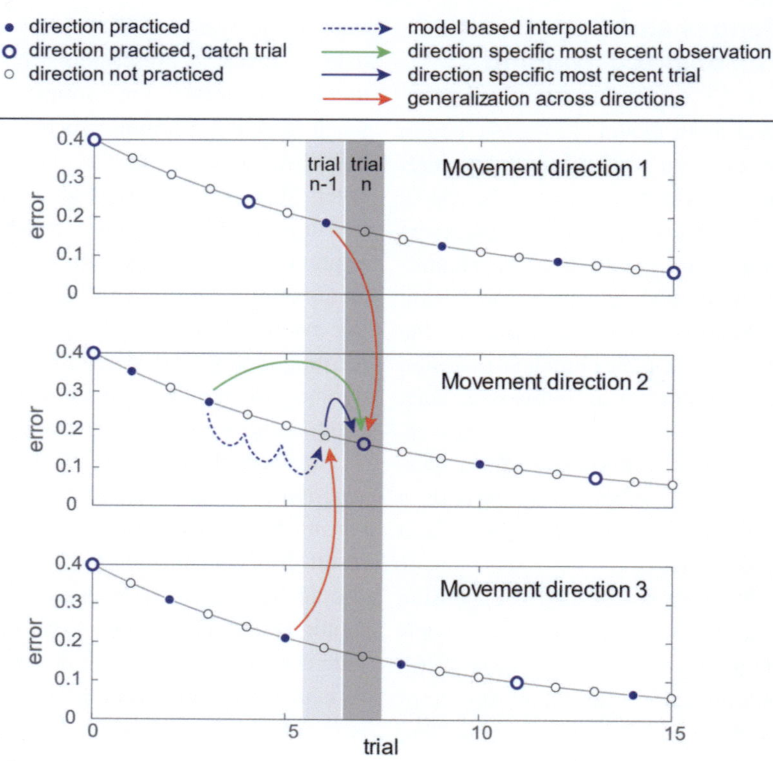

Fig. 11.7 Challenges associated with assessing the time course of motor learning. The randomized trial-to-trial training schedule across several related motor tasks (directions) provides the opportunity to exercise repetition as well as the variation of the motor commands to redevelop motor ability. The randomized practice, however, introduces trial gaps for repetition of the same task, introducing sparsity in the dataset, and interference pattern across different tasks. Empirical modeling to predict performance during a catch trial could consider various forms of interpolation across trial gaps (blue vs. green arrows) and interferences or generalizations across related tasks (red arrows). (see Parmar and Patton [141])

revealed an interaction between the learning and forgetting processes using the direction-specific memory of error. As expected, learning effects were observed at the practiced movement direction on a given trial. Forgetting effects (error increasing) were observed at the unpracticed movement directions with learning effects from generalization from the practiced movement direction.

11.9.1 Statistical Approaches to Personalize Error Augmentation

While error augmentation might be quite effective in healthy individuals, one concern is the lack of one powerful tool commonly used by clinicians—*customizing* treatment for each patient. This limitation is the reason interactive therapies have not been effective for some patients. Studies in customization have had preliminary success by assisting the patient only as much as needed [1, 105], by gradually reducing assistance [143, 144], and by using patient-customized forces [28, 145, 146]. What is missing, however, is a principled method that relates errors to intervention. The answer may lie in statistical modeling of a patient's motor deficits, which can be used to customize therapy [95].

Recent work has shown how interactive machines can inform a direct mathematical relationship between patient deficits and applied interventions [147]. This builds upon some recent and exciting aspects that consider the statistical relationship of errors to learning. Recent research has shown that the nervous system is quite clever—it takes advantage of information on error *statistics* to shape learning [115, 148]. First, spurious errors are discarded [115]. More importantly, prior experience of error alone appears to govern the amount of learning [149]. Our current paradigm ensures that only repeatable errors are augmented in regions of the *error space* where errors were part of previous experience. This greatly aligns with recent findings that learning is greatest when errors can be expected [150]. Because the learning part of the nervous system appears to hinge on error, we suggest that approaches should consider how error probabilities also change during learning and should be updated. Because the error experience plays a clear and prominent role in learning, it follows that neurorehabilitation should consider error statistics in its arsenal for recovery.

Using the statistics of an individual's error is a fairly straightforward extension of the already tested methods, which allows us to further customize training and provide an even better error augmentation that varies appropriately as needed [94]. While "offset" used the average initial error

and was a step in the right direction, our most recent efforts employ more comprehensive statistics. A statistical profile of error, created from the patient's own assessment data, is used to construct a probability density function. A typical statistical profile from a healthy individual's 100 movements is shown in the (Fig. 11.8). For this subject, the concentration of errors was centered in a small region to the side of zero error.

Next, a statistical model of error informs the design of customized therapy [151]. The innovation is that the error distribution is used to directly determine the appropriate training forces that one might experience throughout reaching. Forces magnify one's often-repeated error tendencies and leave spurious and rare errors alone. Very large errors are also not further enlarged, making the system gentler. Hence, the algorithm first learns the human, gaining an individualized probabilistic "picture" of error tendencies, which then serves as a basis for the forces that augment error—only in these error-prone regions where they were observed.

As an example, we show a chronic stroke subject training across several days (Fig. 11.9). For this direction of motion, the subject first was

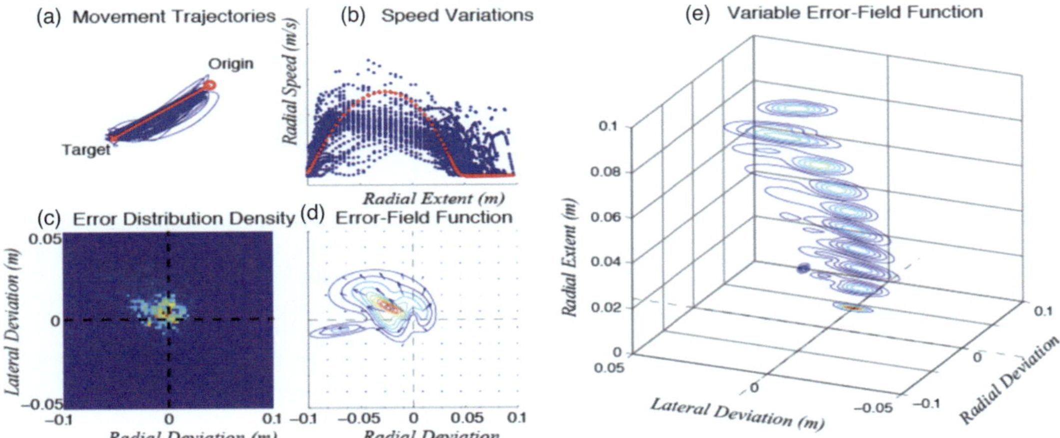

Fig. 11.8 Illustration of the construction of error fields based on the individual's patterns of error from ideal straight-line motion shown in red (**a**). Profiles of movement speed also showed large variability and deviation from a specified goal profile shown in red (**b**). A distribution of observed error patterns in a two-dimensional space is modeled as a linear combination of Gaussian distributions (**c**). An error-field is based on this statistical probability of error times the actual magnitude (**d**). This error field function is shown as probability contours in progressive slices along the extent of the movement toward the target (**e**)

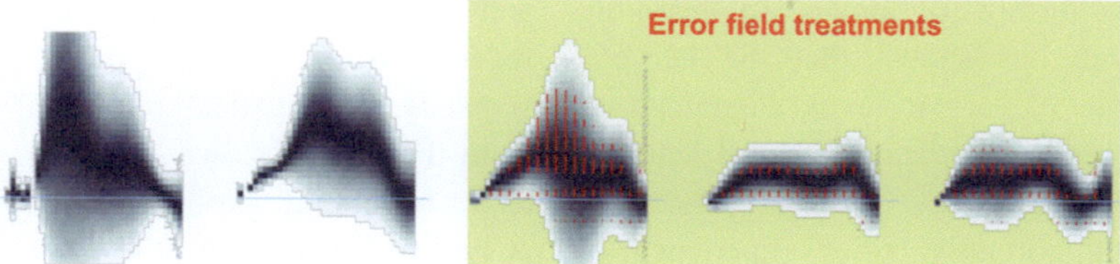

Fig. 11.9 Application of the Error Field approach on a chronic stroke survivor, with each plot showing successive days visiting the laboratory. Error versus time shows a probability (indicated by a degree of shading) to curve in the counterclockwise direction. Probabilities of error times the magnitude of error lead to a magnification of error only where errors are likely to occur (indicated by red arrows), and where they have repeatedly occurred. Error field treatments began only on visit three, where error began to reduce. Each pixel represents 5 square millimeters, and forces were tuned such that the maximum force was 15 N

assessed for two days before being treated with an error field. Error varied across the motion and did not change across the two days of assessment. Only after the error field treatment began did the errors decline.

One can speculate on other exciting applications of such techniques. Training over a broader domain on a larger variety of tasks should provide functional improvements that are better than from simple repetitions of the same task [88, 89], and will facilitate *system identification* as a part of learning [21]. Importantly, any task from simple to complicated is applicable. Because they only amplify *repeatable* negative actions but otherwise do not impede, error-fields should also benefit highly impaired patients who are often excluded from clinical trials because of their inability to move [152]. Success may be further improved because rehabilitation is typically most effective in these less impaired patients [153, 154], mainly because residual capabilities are normally required to perform therapy [152, 155].

This proposed framework should also stimulate new research on how such error distributions might be linked to specific motor pathologies (such spasticity, weakness, synergies, contractures, etc.). Once better known, the error statistics of each individual might guide therapy, and a person's error signature should provide a unique and valuable assessment of their motor deficits and how they may be resolving over time. While here our focus was on error, it is part of a family of methods for therapy that concentrate on a variety of motor deficits that first identifies the

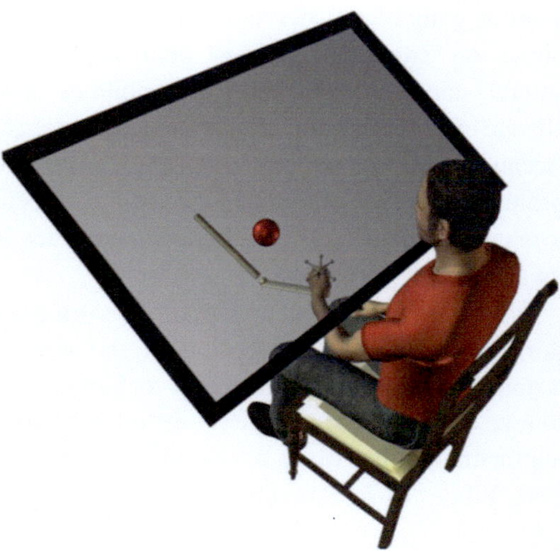

Fig. 11.10 A subject seated at a large-workspace haptic/graphic display. The system tracks the head position to render appropriate 3d graphics to the location of the eyes in spite of where they are in space. The robotics device provides haptic forces that render the desired display of the texture of the environment or any other set of forces that might be desired by the clinician. We assert that augmenting error *visually*, through distortions of the display, or *haptically*, using forces magnifying error (or both), are viable methods of heightening recovery in the neurally injured individual

statistical tendencies of deficits, then uses it to create a training environment.

11.9.2 Functional Error Augmentation

When a robotic device is coupled with a three-dimensional graphic display the sensorimotor

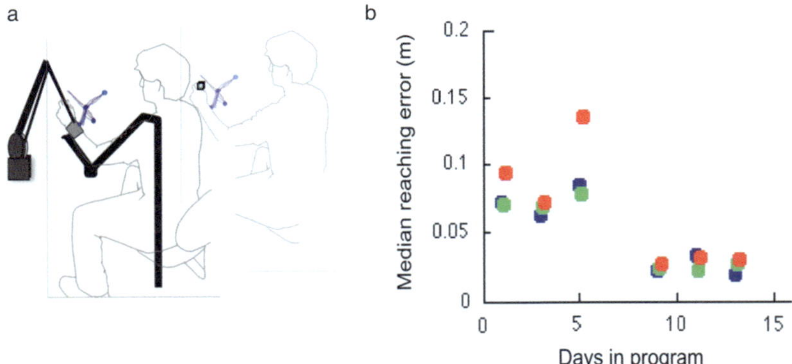

Fig. 11.11 a An error augmentation application for stroke rehabilitation where a subject and therapist work together, seated and using the large-workspace haptic/graphic display to practice movement. The Therapist provides a cue for the subject and can tailor conditioning to the needs of the patient. The robot provides forces that push the limb away from the target and the visual feedback system enhances the error of the cursor. **b** Typical chronic stroke patients improved from day to day, each dot representing the median error measured for a 2-min block of stereotypical functional movements Each color is a different patient. While the patient shows progress across the 2-week period and final benefit, this person did not always improve each day. This is [49] adapted from Sharp, I., Patton, J., Listenberger, M., Case, E. Haptic/Graphic Rehabilitation: Integrating a Robot into a Virtual Environment Library and Applying it to Stroke Therapy. J. Vis. Exp., e3007, doi:10.3791/3007 (2011)

system is able to engage all the types of visual and motor learning described above [156]. The haptic actuator is typically a specially designed robot to allow the user to easily move (back-drive) and may also exert forces that render the sense of touch. The augmented reality graphic display presents images in stereo, in first-person, and using head tracking to appropriately correspond to the current eye location (Fig. 11.10). Images can be superimposed on the real world.

Importantly, studies now must understand the retention and progressive accumulation of benefit from repeated therapy sessions. In a recent study [32], stroke survivors with chronic hemiparesis simultaneously employed the trio of patient, the therapist, and machine. Error augmentation treatment, where haptic (robotic forces) and graphic (visual display) distortions are used to enhance the feedback of error, was compared to comparable practice without such a treatment. The 6-week randomized crossover design involved approximately 60 min of daily treatment three times per week for two weeks, followed by 1- week of rest, then another two weeks of the other treatment. A therapist "teleoperated" the patient using a tracking device. Here the tracked location of the therapist's hand moved a

target (shifted sideways to a point in front of the patient) for the patient to track. The patient was encouraged to track match their hand on top of the target (Fig. 11.11a). Most master–slave robots would perform such teleoperation using forces that attract the patient to the target (i.e., guidance), but here we applied anti-assistance, or error augmentation. This error augmentation, using both haptic ($F = 100[N/m] \cdot e$) and visual ($x = 1.5 \cdot e$) exaggeration of instantaneous error, was employed for one of the 2-weeks without being disclosed explicitly to anyone (thus blinding the patient, therapist, technician-operator, and rater). Several clinical measures gauged outcome at the beginning and end of each 2-week epoch and 1 week post training.

Results showed incremental benefit across most but not all days, abrupt gains in performance (Fig. 11.11b), and most importantly, a significant increase in benefit to error augmentation training in final evaluations. This application of interactive technology may be a compelling new method for enhancing a therapist's productivity in stroke functional restoration. This was a small but significant benefit to robotic training over simple repetitive practice, with a mean 2-week gain in Fugl-Meyer UE

Fig. 11.12 Clinical score changes (Upper extremity Fugl-Meyer) from the first visit. **a** Results from Fig. 4 of Abdollahi, Case Lazarro [32]. **b** Similar measures from the follow up bimanual study [33]. Red lines show the EA treatment, blue lines show standard treatment, and dashed lines show rest periods without treatment. In (**b**), a single outlier subject from the standard treatment group was removed for this analysis. Each of these studies showed significant superior gains for EA compared to comparable practice

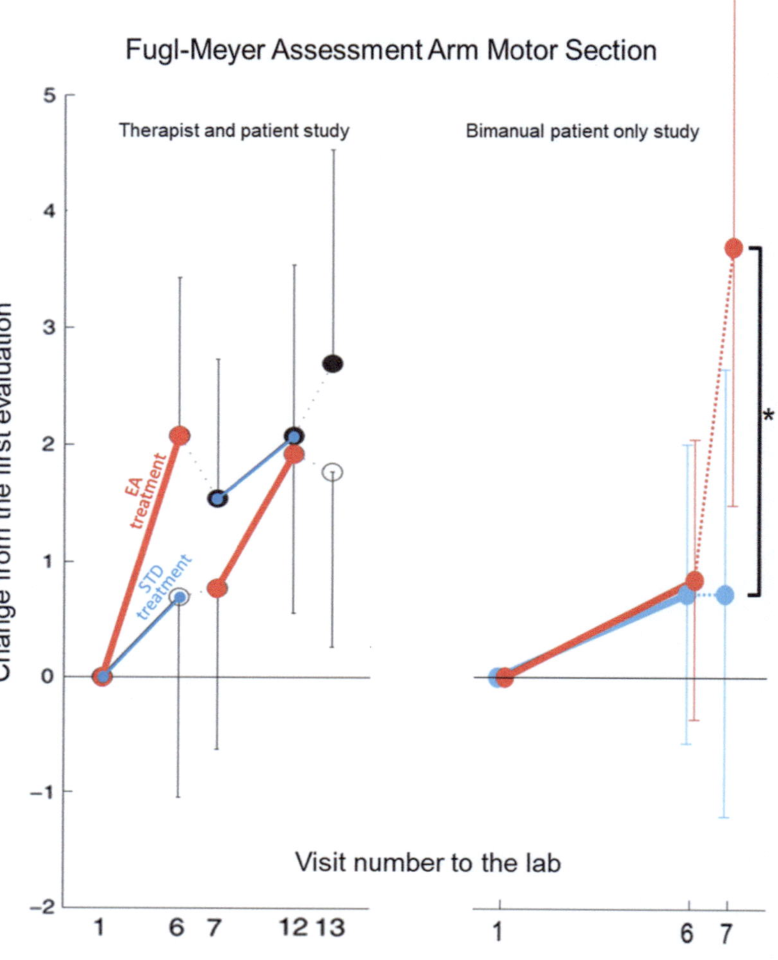

motor score of 2.08 and Wolf Motor Function Test of timed tasks of 1.48 s (Fig. 11.12a). This small amount is encouraging, however, because the interactive technology was only applied for two weeks although a significant gain was observed. Such an effect may improve more given a longer course of therapy.

More encouraging evidence comes from a follow-up pilot study, which invited stroke survivors to use both their limbs [33]. In this double-blinded randomized controlled study (N = 26), we evaluated the short-term effects of bilateral arm training to foster functional recovery of a hemiparetic arm, with half of our subjects unknowingly also receiving the same type of error augmentation as the previous study on their paretic arm. Twenty-six individuals with chronic stroke were randomly assigned to practice an equivalent amount of bimanual reaching either with or without error augmentation. Participants were instructed to coordinate both arms while reaching to two targets (one for each arm) in three 45-min treatments per week for two weeks, with a follow-up visit after one week without treatment. Subjects' 2-week gains in Fugl-Meyer score averaged 2.92, and we also observed improvements in Wolf Motor Functional Ability Scale average 0.21, and Motor Activity Log of 0.58 for quantity and 0.63 for quality-of-life scores (Fig. 11.12b). The extra benefit of error augmentation over the three weeks became apparent in Fugl-Meyer score only after removing an outlier subject from consideration. It is important to mention that these modest advantages for error augmentation were detectable over short intervals (2 or 3 weeks), suggesting further clinical trials in interactive self-rehabilitation systems that can enhance motor recovery.

Also note that differences between groups were not always evident until 1 week later at follow-up. It remains to be seen whether some neural consolidation takes place during this period, or that sufficient rest allowed subjects to perform their best, or that each group simply performs at the end differently because of the differences in effort required for these types of training.

11.9.3 Conclusion

There are a variety of compelling aspects of error augmentation that arise from the fact that we often evaluate and adjust our control based on the error of previous movements rather than the current one—we learn to walk by repeatedly falling and trying again. Such *post movement evaluations* imply that we can gain insights into the nature of the learning process from one attempt to the next. This is a compelling new prospect in many areas that include rehabilitation, where the machine encourages the patients to adapt based on their past attempts.

Regardless of the details of the mechanism, the bioengineering community is now observing successes with Error Augmentation, and the clinical research world calls for more studies to discover its optimal application [122]. Statistical techniques provide an enhanced approach that is tailored to the individual's more likely errors. The work outlined in the chapter provides early evidence. Once these approaches are even better understood, they should provide a broad approach to serve therapeutic goals and questions. Even more broadly, these approaches should provide a powerful strategy to improve capabilities for healthy and patient activities alike, covering any situation requiring repetitive motor skill training.

Acknowledgements This work was supported by American Heart Association 0330411Z, NIH R24 HD39627, NIH 5 RO1 NS 35673, NIH F32HD08658, Whitaker RG010157, NSF BES0238442, NIH R01HD053727, NIH R01NS053606, NIDILRR 90REGE0005, NIH F31NS100520, and the summer internship in neural engineering (SINE) program at the Sensory Motor Performance Program at the Rehabilitation Institute of Chicago, and the Falk Trust. For additional information see https://www.sralab.org/research/labs/Robotics-Lab.

References

1. Krebs HI, Palazzolo JJ, Dipietro L, Volpe BT, Hogan N. Rehabilitation robotics: performance-based progressive robot-assisted therapy. Auton Robot. 2003;15(1):7–20.
2. Van der Loos HM, Reinkensmeyer DJ, Guglielmelli E. Rehabilitation and health care robotics. Springer handbook of robotics: Springer; 2016. p. 1685–728.
3. Schmidt R, Lee T. Motor control and learning: a behavioral emphasis. 5 ed. Champaign, IL, USA: Human Kinetics; 2011 2011.
4. Bernshteĭn N. The co-ordination and regulation of movements. Oxford. New York: Pergamon Press; 1967.
5. Harris CS. Perceptual adaptation to inverted, reversed, and displaced vision. Psychol Rev. 1965;72:419–44.
6. Bachman JC. Influence of age and sex on the amount and rate of learning two motor tasks. Res Quart Am Assoc Health, Phys Educ Recreat. 1966;37(2):176–86.
7. Hollerbach JM, Flash T. Dynamic interactions between limb segments during planar arm movement. Biol Cybern. 1982;44(1):67–77.
8. Hogan N. Mechanical impedance of single and multi-articular systems. In: Winters JM, Woo SL-Y, editors. Multiple Muscle Systems. New York: Springer-Verlag; 1990. p. 149–64.
9. Zajac FE. Muscle and tendon: properties, models, scaling, and application to biomechanics and motor control. Crit Rev Biomed Eng. 1989;17(4):359–411.
10. Ghez C. The control of movement. In: Kandel ER, Scwartz JH, Jessel TM, editor. Principles of Neural Science1991. p. 533–47.
11. Bock O. Load compensation in human goal-directed arm movements. Behav Brain Res. 1990;41(3):167–77.
12. Lackner JR, Dizio P. Rapid adaptation to Coriolis force perturbations of arm trajectory. J Neurophysiol. 1994;72(1):299–313.
13. Sainburg RL, Ghez C, Kalakanis D. Intersegmental dynamics are controlled by sequential anticipatory,

error correction, and postural mechanisms. J Neurophysiol. 1999;81(3):1045–56.

14. Shadmehr R, Mussa-Ivaldi FA. Adaptive representation of dynamics during learning of a motor task. J Neurosci. 1994;14(5 Pt 2):3208–24.

15. Tee KP, Franklin DW, Kawato M, Milner TE, Burdet E. Concurrent adaptation of force and impedance in the redundant muscle system. Biol Cybern. 2010;102(1):31–44.

16. Franklin DW, So U, Kawato M, Milner TE. Impedance control balances stability with metabolically costly muscle activation. J Neurophysiol. 2004;92(5):3097–105.

17. Osu R, Burdet E, Franklin DW, Milner TE, Kawato M. Different mechanisms involved in adaptation to stable and unstable dynamics. J Neurophysiol. 2003;90(5):3255–69.

18. Thoroughman KA, Shadmehr R. Electromyographic correlates of learning an internal model of reaching movements. J Neurosci. 1999;19 (19):8573–88.

19. Patton JL, Mussa-Ivaldi FA. Robot-assisted adaptive training: custom force fields for teaching movement patterns. IEEE Trans Biomed Eng. 2004;51(4):636–46.

20. Mussa-Ivaldi FA, Patton JL, editors. Robots can teach people how to move their arm. IEEE International Conference on Robotics and Automation (ICRA); 2000; San Francisco, CA.

21. Huang FC, Gillespie RB, Kuo AD. Visual and haptic feedback contribute to tuning and online control during object manipulation. J Mot Behav. 2007;39(3):179–93.

22. Huang FC, Gillespie RB, Kuo AD. Human adaptation to interaction forces in visuo-motor coordination. IEEE Trans Neural Syst Rehabil Eng. 2006;14(3):390–7.

23. Lackner JR, DiZio P. Motor control and learning in altered dynamic environments. Curr Opin Neurobiol. 2005;15(6):653–9.

24. Harris CS. Perceptual adaptation to inverted, reversed, and displaced vision. Psychol Rev. 1965;72(6):419–44.

25. Smith MA, Ghazizadeh A, Shadmehr R. Interacting adaptive processes with different timescales underlie short-term motor learning. PLoS Biol. 2006;4(6): e179.

26. Reisman DS, Wityk R, Silver K, Bastian AJ. Split-belt treadmill adaptation transfers to overground walking in persons poststroke. Neurorehabil Neural Repair. 2009;23(7):735–44.

27. Finley JM, Statton MA, Bastian AJ. A novel optic flow pattern speeds split-belt locomotor adaptation. J Neurophysiol. 2014;111(5):969–76.

28. Patton JL, Kovic M, Mussa-Ivaldi FA. Custom-designed haptic training for restoring reaching ability to individuals with poststroke hemiparesis. J Rehabil Res Dev. 2006;43(5):643–56.

29. Shadmehr R, Brashers-Krug T. Functional stages in the formation of human long-term motor memory. J Neurosci. 1997;17(1):409–19.

30. Krakauer JW, Ghilardi MF, Ghez C. Independent learning of internal models for kinematic and dynamic control of reaching. Nat Neurosci. 1999;2(11):1026–31.

31. Tong C, Wolpert DM, Flanagan JR. Kinematics and dynamics are not represented independently in motor working memory: evidence from an interference study. J Neurosci. 2002;22(3):1108–13.

32. Abdollahi F, Case Lazarro ED, Listenberger M, Kenyon RV, Kovic M, Bogey RA, et al. Error augmentation enhancing arm recovery in individuals with chronic stroke: a randomized crossover design. Neurorehabil Neural Repair. 2014;28 (2):120–8.

33. Abdollahi F, Corrigan M, Lazzaro EDC, Kenyon RV, Patton JL. Error-augmented bimanual therapy for stroke survivors. NeuroRehabilitation. 2018;43(1):51–61.

34. Brashers-Krug T, Shadmehr R, Bizzi E. Consolidation in human motor memory. Nature. 1996;382 (6588):252–5.

35. Shadmehr R, Holcomb HH. Neural correlates of motor memory consolidation. Science. 1997;277 (5327):821–5.

36. Weiner MJ, Hallett M, Funkenstein HH. Adaptation to lateral displacement of vision in patients with lesions of the central nervous system. Neurology. 1983;33(6):766–72.

37. Dancause N, Ptito A, Levin MF. Error correction strategies for motor behavior after unilateral brain damage: short-term motor learning processes. Neuropsychologia. 2002;40(8):1313–23.

38. Takahashi CD, Reinkensmeyer DJ. Hemiparetic stroke impairs anticipatory control of arm movement. Exp Brain Res. 2003;149(2):131–40.

39. Beer RF, Given JD, Dewald JPA. Task-dependent weakness at the elbow in patients with hemiparesis. Arch Phys Med Rehabil. 1999;80(7):766–72.

40. Wolf SL, Lecraw DE, Barton LA, Jann BB. Forced use of hemiplegic upper extremities to reverse the effect of learned nonuse among chronic stroke and head-injured patients. Exp Neurol. 1989;104 (2):125–32.

41. Kandel ER. Cellular mechanisms of learning and the biological basis of individuality. Principles of neural science. 31991. p. 1009–31.

42. Hawkins RD, Kandel ER, Siegelbaum SA. Learning to modulate transmitter release: themes and variations in synaptic plasticity. Annu Rev Neurosci. 1993;16:625–65.

43. Lillicrap TP, Cownden D, Tweed DB, Akerman CJ. Random synaptic feedback weights support error backpropagation for deep learning. Nat Commun. 2016;7:13276.

44. Criscimagna-Hemminger SE, Bastian AJ, Shadmehr R. Size of error affects cerebellar contributions

to motor learning. J Neurophysiol. 2010;103(4):2275–84.

45. Houk JC, Buckingham JT, Barto AG. Models of the cerebellum and motor learning. Behavioral and Brain Sciences. 1996;19(3):368-+.

46. Ito M. Mechanisms of motor learning in the cerebellum. Brain Res. 2000;886(1–2):237–45.

47. Kawato M, Gomi H. A computational model of four regions of the cerebellum based on feedback-error learning. Biol Cybern. 1992;68(2):95–103.

48. Llinás R, Welsh JP. On the cerebellum and motor learning. Curr Opin Neurobiol. 1993;3(6):958–65.

49. Popa LS, Streng ML, Hewitt AL, Ebner TJ. The errors of our ways: understanding error representations in cerebellar-dependent motor learning. Cerebellum. 2016;15(2):93–103.

50. Ito M, Doya K. Multiple representations and algorithms for reinforcement learning in the cortico-basal ganglia circuit. Curr Opin Neurobiol. 2011;21(3):368–73.

51. Fee MS, Goldberg JH. A hypothesis for basal ganglia-dependent reinforcement learning in the songbird. Neuroscience. 2011;198:152–70.

52. Bar-Gad I, Morris G, Bergman H. Information processing, dimensionality reduction and reinforcement learning in the basal ganglia. Prog Neurobiol. 2003;71(6):439–73.

53. Kaplan A, Mizrahi-Kliger AD, Israel Z, Adler A, Bergman H. Dissociable roles of ventral pallidum neurons in the basal ganglia reinforcement learning network. Nat Neurosci. 2020;23(4):556–64.

54. Hebb DO. The organization of behavior: a neuropsychological theory. New York: Wiley; 1949. xix, 335 p.

55. Rowe JB, Chan V, Ingemanson ML, Cramer SC, Wolbrecht ET, Reinkensmeyer DJ. Robotic assistance for training finger movement using a Hebbian model: a randomized controlled trial. Neurorehabil Neural Repair. 2017;31(8):769–80.

56. Ingemanson ML, Rowe JR, Chan V, Wolbrecht ET, Reinkensmeyer DJ, Cramer SC. Somatosensory system integrity explains differences in treatment response after stroke. Neurology. 2019;92(10):e1098–108.

57. van Vugt FT, Ostry DJ. Early stages of sensorimotor map acquisition: learning with free exploration, without active movement or global structure. J Neurophysiol. 2019;122(4):1708–20.

58. Marchal-Crespo L, Reinkensmeyer DJ. Review of control strategies for robotic movement training after neurologic injury. J Neuroeng Rehabil. 2009;6:20.

59. Marchal-Crespo L, Schneider J, Jaeger L, Riener R. Learning a locomotor task: with or without errors? J Neuroeng Rehabil. 2014;11(1):25.

60. Kawato M. Feedback-error-learning neural network for supervised motor learning. Advanced Neural Comput. 1990;6(3):365–72.

61. Wolpert DM, Ghahramani Z, Jordan MI. An internal model for sensorimotor integration. Science. 1995;269(5232):1880–2.

62. Flanagan JR, Rao AK. Trajectory adaptation to a nonlinear visuomotor transformation: evidence of motion planning in visually perceived space. J Neurophysiol. 1995;74(5):2174–8.

63. Scheidt RA, Reinkensmeyer DJ, Conditt MA, Rymer WZ, Mussa-Ivaldi FA. Persistence of motor adaptation during constrained, multi-joint, arm movements. J Neurophysiol. 2000;84(2):853–62.

64. Scheidt RA, Dingwell JB, Mussa-Ivaldi FA. Learning to move amid uncertainty. J Neurophysiol. 2001;86(2):971–85.

65. Nelson WL. Physical principles for economies of skilled movements. Biol Cybern. 1983;46(2):135–47.

66. Todorov E. Optimality principles in sensorimotor control. Nat Neurosci. 2004;7(9):907–15.

67. Todorov E, Jordan MI. Optimal feedback control as a theory of motor coordination.[see comment]. Nature Neuroscience. 2002;5(11):1226–35.

68. Hesse S, Schulte-Tigges G, Konrad M, Bardeleben A, Werner C. Robot-assisted arm trainer for the passive and active practice of bilateral forearm and wrist movements in hemiparetic subjects. Arch Phys Med Rehabil. 2003;84(6):915–20.

69. Körding K, Tenenbaum J, Shadmehr R. Multiple timescales and uncertainty in motor adaptation. Adv Neural Inf Process Syst. 2006;19:745–52.

70. Kording KP, Wolpert DM. Bayesian integration in sensorimotor learning. Nature. 2004;427(6971):244–7.

71. Wei K, Kording K. Uncertainty of feedback and state estimation determines the speed of motor adaptation. Front Comput Neurosci. 2010;4:11.

72. Wei K, Stevenson IH, Kording KP. The uncertainty associated with visual flow fields and their influence on postural sway: Weber's law suffices to explain the nonlinearity of vection. J Vis. 2010;10(14):4.

73. Milner TE, Hinder MR, Franklin DW. How is somatosensory information used to adapt to changes in the mechanical environment? Prog Brain Res. 2007;165:363–72.

74. Hwang S, Agada P, Kiemel T, Jeka JJ. Dynamic reweighting of three modalities for sensor fusion. PLoS ONE. 2014;9(1): e88132.

75. Logan D, Kiemel T, Jeka JJ. Asymmetric sensory reweighting in human upright stance. PLoS ONE. 2014;9(6): e100418.

76. Ernst MO, Banks MS. Humans integrate visual and haptic information in a statistically optimal fashion. Nature. 2002;415(6870):429–33.

77. Flanagan JR, Nakano E, Imamizu H, Osu R, Yoshioka T, Kawato M. Composition and decomposition of internal models in motor learning under altered kinematic and dynamic environments. J Neurosci. 1999;19(20):RC34.

78. Wei Y, Patton J, editors. Forces that Supplement Visuomotor Learning: A 'Sensory Crossover' Experiment. Symposium on Haptic Interfaces, a satellite to the IEEE Conference on Virtual Reality; 2004; Chicago.

79. Schmidt RA, Wrisberg CA. Motor learning and performance: A situation-based learning approach: Human kinetics; 2008.

80. Wulf G, Shea CH. Principles derived from the study of simple skills do not generalize to complex skill learning. Psychon Bull Rev. 2002;9(2):185–211.

81. Srinivasan MA, LaMotte RH. Tactual discrimination of softness: abilities and mechanisms. Somesthesis and the Neurobiology of the Somatosensory Cortex: Springer; 1996. p. 123–35.

82. Sainburg RL, Lateiner JE, Latash ML, Bagesteiro LB. Effects of altering initial position on movement direction and extent. J Neurophysiol. 2003;89(1):401–15.

83. Brewer BR, Klatzky R, Matsuoka Y, editors. Effects of visual feedback distortion for the elderly and the motor-impaired in a robotic rehabilitation environment. IEEE International Conference on Robotics and Automation, 2004 Proceedings ICRA'04 2004; 2004: IEEE.

84. Brewer BR, Fagan M, Klatzky RL, Matsuoka Y. Perceptual limits for a robotic rehabilitation environment using visual feedback distortion. IEEE Trans Neural Syst Rehabil Eng. 2005;13(1):1–11.

85. Winstein CJ, Merians AS, Sullivan KJ. Motor learning after unilateral brain damage. Neuropsychologia. 1999;37(8):975–87.

86. Kazerooni H. The human power amplifier technology at the University of California, Berkeley. Robot Auton Syst. 1996;19(2):179–87.

87. Aguirre-Ollinger G, Colgate JE, Peshkin MA, Goswami A, editors. Active-Impedance Control of a Lower-Limb Assistive Exoskeleton. Rehabilitation Robotics, 2007 ICORR 2007 IEEE 10th International Conference on; 2007 13–15 June 2007.

88. Hanlon RE. Motor learning following unilateral stroke. Arch Phys Med Rehabil. 1996;77(8):811–5.

89. Jarus T, Gutman T. Effects of cognitive processes and task complexity on acquisition, retention, and transfer of motor skills. Can J Occup Ther. 2001;68(5):280–9.

90. Huang FC, Patton JL, Mussa-Ivaldi FA. Manual skill generalization enhanced by negative viscosity. J Neurophysiol. 2010;104(4):2008–19.

91. Huang FC, Patton JL. Augmented dynamics and motor exploration as training for stroke. IEEE Trans Biomed Eng. 2013;60(3):838–44.

92. Wright ZA, Carlsen AN, MacKinnon CD, Patton JL. Degraded expression of learned feedforward control in movements released by startle. Exp Brain Res. 2015;233(8):2291–300.

93. Wright ZA, Lazzaro E, Thielbar KO, Patton JL, Huang FC. Robot training with vector fields based on stroke survivors' individual movement statistics.

94. Fisher ME, Huang FC, Wright ZA, Patton JL. Distributions in the error space: goal-directed movements described in time and state-space representations. Annu Int Conf IEEE Eng Med Biol Soc. 2014;2014:6953–6.

95. Wright ZA, Fisher ME, Huang FC, Patton JL. Data sample size needed for prediction of movement distributions. Annu Int Conf IEEE Eng Med Biol Soc. 2014;2014:5780–3.

96. Huang FC, Patton JL. Movement distributions of stroke survivors exhibit distinct patterns that evolve with training. Journal of NeuroEngineering and Rehabilitation. 2016;13(1):23-.

97. Liu J, Cramer SC, Reinkensmeyer DJ. Learning to perform a new movement with robotic assistance: comparison of haptic guidance and visual demonstration. J Neuroeng Rehabil. 2006;3:20.

98. Chib VS, Patton JL, Lynch KM, Mussa-Ivaldi FA, editors. The effect of stiffness and curvature on the haptic identification of surfaces. First Joint Eurohaptics conference and symposium on haptic interfaces for virtual environment and teleoperator systems (IEEE-WHC); 2005 18–20 March 2005; Pisa, Italy.

99. Heuer H, Rapp K. Active error corrections enhance adaptation to a visuo-motor rotation. Exp Brain Res. 2011;211:97–108.

100. van Asseldonk EH, Wessels M, Stienen AH, van der Helm FC, van der Kooij H. Influence of haptic guidance in learning a novel visuomotor task. J Physiol Paris. 2009;103(3–5):276–85.

101. Pennycott A, Wyss D, Vallery H, Klamroth-Marganska V, Riener R, Roy R, et al. Towards more effective robotic gait training for stroke rehabilitation: a review. J NeuroEng Rehabilit 2012;9(1):65-.

102. Tucker MR, Olivier J, Pagel A, Bleuler H, Bouri M, Lambercy O, et al. Control strategies for active lower extremity prosthetics and orthotics: a review. BioMed Central Ltd.; 2015.

103. Scheidt RA, Conditt MA, Secco EL, Mussa-Ivaldi FA. Interaction of visual and proprioceptive feedback during adaptation of human reaching movements. J Neurophysiol. 2005;93(6):3200–13.

104. Smith MA, Ghazizadeh A, Shadmehr R. Interacting adaptive processes with different timescales underlie short-term motor learning. PLoS Biology. 2006;4(6):e179-e.

105. Emken JL, Benitez R, Reinkensmeyer DJ. Human-robot cooperative movement training: learning a novel sensory motor transformation during walking with robotic assistance-as-needed. J Neuroeng Rehabil. 2007;4(1):8.

106. Emken JL, Benitez R, Sideris A, Bobrow JE, Reinkensmeyer DJ. Motor adaptation as a greedy optimization of error and effort. J Neurophysiol. 2007;97(6):3997–4006.

107. Hornby TG, Reinkensmeyer DJ, Chen D. Manually-assisted versus robotic-assisted body weight-supported treadmill training in spinal cord injury: what is the role of each? PM R. 2010;2 (3):214–21.

108. Patton JL, Stoykov ME, Kovic M, Mussa-Ivaldi FA. Evaluation of robotic training forces that either enhance or reduce error in chronic hemiparetic stroke survivors. Exp Brain Res. 2006;168(3):368–83.

109. Nudo RJ, Friel KM. Cortical plasticity after stroke: implications for rehabilitation. Rev Neurol (Paris). 1999;155(9):713–7.

110. Patton JL, Wei YJ, Bajaj P, Scheidt RA. Visuomotor learning enhanced by augmenting instantaneous trajectory error feedback during reaching. PLoS ONE. 2013;8(1): e46466.

111. Rumelhart DE, Hinton GE, Williams RJ. Learning representations by back-propagating errors. Nature. 1986;323(6088):533–6.

112. Thoroughman KA, Shadmehr R. Learning of action through adaptive combination of motor primitives. Nature. 2000;407(6805):742–7.

113. Wolpert DM, Kawato M. Multiple paired forward and inverse models for motor control. Neural Netw. 1998;11(7–8):1317–29.

114. Wei Y, Bajaj P, Scheidt RA, Patton JL, editors. Visual Error Augmentation for Enhancing Motor Learning and Rehabilitative Relearning. IEEE-International Conference on Rehabilitation Robotics (ICORR); 2005; Chicago, IL, USA.

115. Wei K, Kording K. Relevance of error: what drives motor adaptation? J Neurophysiol. 2009;101 (2):655–64.

116. van der Kooij K, Brenner E, van Beers RJ, Smeets JB. Visuomotor adaptation: how forgetting keeps us conservative. PLoS ONE. 2015;10(2): e0117901.

117. Kording KP, Wolpert DM. The loss function of sensorimotor learning. Proc Natl Acad Sci USA. 2004;101(26):9839–42.

118. Lisberger SG. The neural basis for learning of simple motor skills. Science. 1988;242(4879):728–35.

119. Alleva E, Santucci D. Psychosocial vs. "physical" stress situations in rodents and humans: role of neurotrophins. Physiol Behav. 2001;73(3):313–20.

120. Duarte JE, Reinkensmeyer DJ. Effects of robotically modulating kinematic variability on motor skill learning and motivation. J Neurophysiol. 2015;113 (7):2682–91.

121. Ravaioli E, Oie KS, Kiemel T, Chiari L, Jeka JJ. Nonlinear postural control in response to visual translation. Exp Brain Res. 2005;160(4): 450–9.

122. Parmar PN, Patton JL, editors. Optimal gain schedules for visuomotor skill training using error-augmented feedback. 2015 IEEE International Conference on Robotics and Automation (ICRA); 2015: IEEE.

123. Wu HG, Miyamoto YR, Gonzalez Castro LN, Olveczky BP, Smith MA. Temporal structure of motor variability is dynamically regulated and predicts motor learning ability. Nat Neurosci. 2014;17(2):312–21.

124. Parmar PN, Patton JL. A second-order process model best describes influence of error-augmentation on the time course of motor learning. PLoS One. 2022;(in revision).

125. Parmar PN, Patton JL, editors. Sparse identification of motor learning using proxy process models. 2019 IEEE 16th International Conference on Rehabilitation Robotics (ICORR); 2019 24–28 June 2019.

126. Ljung L. System identification : theory for the user. Englewood Cliffs, NJ: Prentice-Hall; 1987. xxi, 519 p.

127. Suzuki S, Furuta K. Adaptive impedance control to enhance human skill on a haptic interface system. Journal of Control Science and Engineering; 2012. p. 1–10.

128. Kim M, Ding Y, Malcolm P, Speeckaert J, Siviy CJ, Walsh CJ, et al. Human-in-the-loop Bayesian optimization of wearable device parameters. PLoS ONE. 2017;12(9): e0184054.

129. Ding Y, Kim M, Kuindersma S, Walsh CJ. Human-in-the-loop optimization of hip assistance with a soft exosuit during walking. Sci Robot. 2018;3(15): eaar5438.

130. Parmar PN, Patton JL. Optimal design of motor learning experiments informed by Monte-carlo simulation. Intern J Exper Des Proc Optim. 2021;6(4):289–303.

131. Lang CE, Strube MJ, Bland MD, Waddell KJ, Cherry-Allen KM, Nudo RJ, et al. Dose response of task-specific upper limb training in people at least 6 months poststroke: A phase II, single-blind, randomized, controlled trial. Annals of Neurology. 2016.

132. Shea CH, Kohl RM. Specificity and variability of practice. Res Q Exerc Sport. 1990;61(2):169–77.

133. Memmert D. Long-term effects of type of practice on the learning and transfer of a complex motor skill. Percept Mot Skills. 2006;103(3):912–6.

134. Lage GM, Ugrinowitsch H, Apolinario-Souza T, Vieira MM, Albuquerque MR, Benda RN. Repetition and variation in motor practice: A review of neural correlates. Neurosci Biobehav Rev. 2015;57:132–41.

135. Cothros N, Wong J, Gribble P. Are there distinct neural representations of object and limb dynamics? Exp Brain Res. 2006;173(4):689–97.

136. Berniker M, Kording K. Estimating the sources of motor errors for adaptation and generalization. Nat Neurosci. 2008;11(12):1454–61.

137. Kluzik J, Diedrichsen J, Shadmehr R, Bastian AJ. Reach adaptation: what determines whether we learn an internal model of the tool or adapt the model of our arm? J Neurophysiol. 2008;100 (3):1455–64.

138. Wulf G, Schmidt RA. Variability of practice and implicit motor learning. J Exper Psychol Learn Memory Cogn. 1997;23(4):987–1006.

139. Lee JY, Oh Y, Kim SS, Scheidt RA, Schweighofer N. Optimal schedules in multitask motor learning. Neural Comput. 2016;28(4):667–85.

140. Schaefer SY, Patterson CB, Lang CE. Transfer of training between distinct motor tasks after stroke: implications for task-specific approaches to upper-extremity neurorehabilitation. Neurorehabilitation and neural repair. 2013;27(7).

141. Parmar PN, Patton JL. Direction-specific iterative tuning of motor commands with local generalization during randomized reaching practice across movement directions. Front Neurorobot. 2021;15: 651214.

142. Parmar PN, Patton JL, editors. Models of Motor Learning Generalization. 2018 40th Annual International Conference of the IEEE Engineering in Medicine and Biology Society (EMBC); 2018: IEEE.

143. Colombo R, Sterpi I, Mazzone A, Delconte C, Pisano F. Taking a lesson from patients' recovery strategies to optimize training during robot-aided rehabilitation. IEEE Trans Neural Syst Rehabil Eng. 2012;20(3):276–85.

144. Kahn LE, Rymer WZ, Reinkensmeyer DJ. Adaptive assistance for guided force training in chronic stroke. Conf Proc IEEE Eng Med Biol Soc. 2004;2004:2722–5.

145. Kahn LE, Lum PS, Rymer WZ, Reinkensmeyer DJ. Robot-assisted movement training for the stroke-impaired arm: does it matter what the robot does? J Rehabil Res Dev. 2006;43(5):619–30.

146. Johnson MJ, Van der Loos HF, Burgar CG, Shor P, Leifer LJ. Experimental results using force-feedback cueing in robot-assisted stroke therapy. IEEE Trans Neural Syst Rehabil Eng. 2005;13 (3):335–48.

147. Sundaram H, Chen Y, Rikakis T. A computational framework for constructing interactive feedback for assisting motor learning. Annu Int Conf IEEE Eng Med Biol Soc. 2011;2011:1399–402.

148. Herzfeld DJ, Shadmehr R. Motor variability is not noise, but grist for the learning mill. Nat Neurosci. 2014;17(2):149–50.

149. Herzfeld DJ, Vaswani PA, Marko MK, Shadmehr R. A memory of errors in sensorimotor learning. Science. 2014;345(6202):1349–53.

150. Takiyama K, Hirashima M, Nozaki D. Prospective errors determine motor learning. Nat Commun. 2015;6:5925.

151. Fisher ME, Huang FC, Klamroth-Marganska V, Riener R, Patton JL. Haptic error fields for robotic training. 2015 IEEE World Haptics Conference (WHC); Evanston, IL, USA: IEEE 2015. p. 434–9.

152. Buch ER, Modir Shanechi A, Fourkas AD, Weber C, Birbaumer N, Cohen LG. Parietofrontal integrity determines neural modulation associated with grasping imagery after stroke. Brain. 2012;135 (Pt 2):596–614.

153. Parry R, Lincoln N, Vass C. Effect of severity of arm impairment on response to additional physiotherapy early after stroke. Clin Rehabil. 1999;13 (3):187–98.

154. Winstein CJ, Rose DK, Tan SM, Lewthwaite R, Chui HC, Azen SP. A randomized controlled comparison of upper-extremity rehabilitation strategies in acute stroke: a pilot study of immediate and long-term outcomes. Arch Phys Med Rehabil. 2004;85(4):620–8.

155. Nowak DA, Grefkes C, Ameli M, Fink GR. Interhemispheric competition after stroke: brain stimulation to enhance recovery of function of the affected hand. Neurorehabil Neural Repair. 2009;23 (7):641–56.

156. Patton JL, Dawe G, Scharver C, Muss-Ivaldi FA, Kenyon R. Robotics and virtual reality: a perfect marriage for motor control research and rehabilitation. Assist Technol. 2006;18(2):181–95.

157. Imamizu H, Miyauchi S, Tamada T, Sasaki Y, Takino R, Putz B, et al. Human cerebellar activity reflecting an acquired internal model of a new tool. [see comments.]. Nature. 2000;403(6766):192–5.

158. Krakauer JW, Pine ZM, Ghilardi MF, Ghez C. Learning of visuomotor transformations for vectorial planning of reaching trajectories. J Neurosci. 2000;20(23):8916–24.

159. Rossetti Y, Rode G, Pisella L, Farne A, Li L, Boisson D, et al. Prism adaptation to a rightward optical deviation rehabilitates left hemispatial neglect. Nature. 1998;395(6698):166–9.

160. Emken JL, Reinkensmeyer DJ. Robot-enhanced motor learning: accelerating internal model formation during locomotion by transient dynamic amplification. IEEE Trans Neural Syst Rehabil Eng. 2005;13(1):33–9.

161. Eldar E, Hauser TU, Dayan P, Dolan RJ. Striatal structure and function predict individual biases in learning to avoid pain. Proc Natl Acad Sci U S A. 2016;113(17):4812–7.

162. Robles-De-La-Torre G, Hayward V. Force can overcome object geometry in the perception of shape through active touch. Nature. 2001;412 (6845):445–8.

163. Patton JL, Wei Y, Scharver C, Kenyon RV, Scheidt R, editors. Motivating rehabilitation by distorting reality. BioRob 2006: The first IEEE / RAS-EMBS International Conference on Biomedical Robotics and Biomechatronics,; 2006 February 20–22, 2006; Pisa, Tuscany, Italy.

164. Abdollahi F, Rozario S, Case E, Listenberger M, Kovic M, Kenyon R, et al. Arm control recovery enhanced by error augmentation IEEE International Conference on Rehabilitation Robotics; Zurich, Switzerland: IEEE; 2011.

The Role of Haptic Interactions with Robots for Promoting Motor Learning

12

Niek Beckers and Laura Marchal-Crespo

Abstract

Robot-assisted haptic training has the potential to facilitate motor learning and neurorehabilitation for a diverse number of motor tasks, ranging from manipulating objects with unknown dynamics to relearning walking using robotic exoskeletons. In this chapter, we provide an overview of current haptic methods evaluated in motor (re)learning studies with the goal to discuss implications for the design of rehabilitation technology. We highlight the challenge point framework as a unifying view on how to guide the design of haptic training paradigms, based on the initial skill level of the learner and the characteristics of the task to be learned. Future work on robot-aided haptic training strategies should focus on adaptive training algorithms, providing more naturalistic congruent multisensory feedback that resembles out-of-the-lab training, and conduct long-term studies to assess the efficacy of haptic training on learning not only the trained task but importantly, on skill transfer to real tasks.

Keywords

Motor learning · Neurorehabilitation · Robots · Haptic interfaces · Robot-assisted training · Augmented haptic feedback

12.1 Introduction

Humans go through a continuous process of acquiring new motor skills. Some skills are crucial to meet fundamental needs such as ambulation, nourishment, and self-care, and others involve more skilled movements that bring joy and sense to our lives, including playing sports, music, and dancing. We never stop learning new skills or brushing up on already gained skills. Despite their lower motor performance, elderly people still learn new motor skills [1] although at a slower rate [2, 3]. We might also encounter detrimental situations that demand us to relearn lost functions or learn other motor strategies to circumvent the loss of motor control through intensive neurorehabilitation, after suffering a brain injury. It is thought that motor learning and neurorehabilitation can be optimized by providing intensive meaningful movement training that

N. Beckers · L. Marchal-Crespo (✉)
Department of Cognitive Robotics, TU Delft/Faculty 3mE (Mechanical, Maritime and Materials Engineering), Mekelweg 2, 2628 CD Delft, The Netherlands
e-mail: L.MarchalCrespo@tudelft.nl

N. Beckers
e-mail: niekbeckers@gmail.com

L. Marchal-Crespo
ARTORG Center for Biomedical Engineering Research, University of Bern, Freiburgstrasse 3, 3010 Bern, Switzerland

© The Author(s), under exclusive license to Springer Nature Switzerland AG 2022
D. J. Reinkensmeyer et al. (eds.), *Neurorehabilitation Technology*,
https://doi.org/10.1007/978-3-031-08995-4_12

promotes multi-sensory input to the central neural system (see Chap. 3). Given the impact on people's lives, topics of motor learning and re-learning of lost functions, and specifically how robotics can be employed to stimulate neurorehabilitation, have been extensively studied—see reviews in [4–6].

The possibility of using robotics to stimulate neurorehabilitation and motor learning is attractive because robots can provide controllable, repeatable, and intensive training paradigms while ensuring patients' safety. Robots are generally used to assist the learner by physically guiding their limbs during movement training (haptic guidance) toward a physiological movement or task goal, thus alleviating physical strain on therapists/trainers. Alternatively, robots could also be used to challenge the learner (haptic disturbance) to improve their task performance, by for example stimulating the exploration of novel training environments or novel movement strategies [4]. In addition, robots can provide haptic feedback, combined with other sensory modalities such as auditory and visual feedback, to stimulate motor learning [4, 5, 7].

While evidence is accumulating that haptic training could benefit the motor recovery of stroke patients [8, 9], the efficacy of robot-aided motor (re)learning in particular for healthy or less-severely impaired persons is not fully established yet. Although haptic guidance is often used in motor training to provide the central nervous system with sensory input from physiological movements, there is currently little evidence that robotic guidance is more beneficial for human motor learning of healthy participants than unassisted practice [10, 11] or to a different form of guidance, such as auditory or visual feedback [4, 5]. Several studies have confirmed that only physically guiding movements to reduce performance errors does not aid motor learning and may, in fact, hamper it [12, 13]. Indeed, research in motor learning has stated that the learners' effort and performance errors are crucial elements to drive motor learning [14, 15] and neuroplasticity [16]. Therefore, new haptic training methods have been proposed that make motor tasks more challenging, suggesting that

enhancing or inducing errors, rather than reducing them, could be beneficial to some learners [17, 18].

Although the effectiveness of haptic training methods has been investigated by a myriad of motor learning studies, results remain inconclusive. A potential rationale might be due to the diversity of the selected motor tasks to be learned, study protocol designs, selected haptic training methods, and the learners' skill/disability level across studies. In this chapter, we provide an overview of current haptic training strategies and their effect on an individual's motor learning. We end with the implications for robot-aided rehabilitation paradigms and possible research avenues.

12.2 Haptic Training Methods

Williams and Carnahan categorized the different haptic training methods into two main groups: performance-enhancing (haptic guidance) and performance-degrading (haptic disturbance) methods [11]. Performance-enhancing methods are commonly explored in robot-based therapy and encompass the haptic training methods that: (i) use a robotic device to **haptically demonstrate** the desired tasks' kinesthetic characteristics while the learner remains passive (e.g., [19]); (ii) provide **haptic cueing** through tactile actuators to signal the correct time to initiate an upcoming desired movement (e.g., [20]); (iii) use a robotic device to provide **haptic assistance** to guide a learner's movements while the learner actively executes the motor task (e.g., [21]); and (iv) promote a participant's motivation (e.g., [9, 22]). Haptic assistance methods are derived from traditional physical and occupational therapy, in which therapists manually apply physical assistance to aid patients in accomplishing their intended movements.

Several different robotic controller designs can be found in the literature to provide haptic assistance, depending on which task features the experimenter is interested to teach (e.g., spatial, temporal, and/or spatiotemporal features). Among the most used controllers, we find classical feedback controllers such as *proportional and/or derivative controllers* that apply forces depending on the difference between the desired and actual position and/or velocity of the limb [23–26]. *Path controllers* are used to restrict the limb's movement to an area around the desired trajectory by correcting the movement with forces that prohibit the limb to go outside of the predefined boundaries, providing safety while allowing for free movement [27, 28]. Other controllers apply a position-dependent velocity profile [29], enforce pre-recorded force profiles (*haptic guidance in force*) [30, 31], or match the frequency of a limb's motion with that of a robotic device [32].

While haptic guidance generally aims to reduce movement errors, research on motor learning has emphasized that errors are fundamental signals that drive motor adaptation [33, 34]. Robotic strategies that deteriorate the learners' performance during task execution are likely to increase effort, energy consumption, and attention [18, 19]. Thereby, researchers have proposed **haptic disturbance** methods that apply forces to degrade the performance during training rather than enhance performance (e.g., [14, 35]). One of the first haptic disturbance methods studied in literature aimed at amplifying the learners' movements errors while they were executing the task (*error augmentation*) by applying forces to push learners away from the desired movement trajectory [17, 36, 37]. Other approaches used *haptic resistance* to the participant's limb movements during exercise, or force fields that required specific patterns of force generation [38, 39].

Not only the magnitude of the movement errors but also the history of the experienced errors drive motor learning. Variability in task execution, often assessed as variability in experienced movement errors, has also been shown to predict motor learning in unassisted reaching movements. Participants with higher task-relevant motor variability improve faster their task performance compared to participants with lower motor variability [40]. Motor variability is believed to originate from noise in our motor system, in which a distinction is made between *planning noise*, originating in the brain, and *execution noise*, emerging from the periphery (e.g., muscle activation noise, noise in neuronal transmission) [41]. Higher planning noise results in higher learning rates, while execution noise reduces learning rates; humans seem to optimally tune their learning rate to the planning and execution noise [42]. Yet, the causality of motor variability and motor learning still has to be fully established.

Importantly, it is not well understood how motor variability, particularly motor variability stemming from planning noise, can be successfully stimulated by externally applying haptic forces. Some studies applied unexpected disturbing forces (*haptic noise*) during training [18, 35]. Other approaches aimed to hinder the participants' natural motor variability as little as possible while still providing assistive forces when needed (minimal intervention), for example through using model predictive controllers [21].

In the majority of the aforementioned haptic training strategies, the controller parameters do not change during training (referred to as *fixed guidance/disturbance*). Fixed haptic training strategies often rely on manual tuning by a therapist or teacher to adjust the haptic assistance/resistance to account for interpersonal differences in skill and progress. Moreover, a learner's performance and learning state evolve over time, warranting training strategies that adapt accordingly. A learner might initially benefit from haptic assistance that ensures movement with safety boundaries while exploring the task, and haptic disturbances to improve performance in later learning stages. This modulation can be achieved by adapting the controller's parameters (e.g., the impedance) or applied haptic forces based on the online measurement of the learner's performance (*performance-based adaptive haptic training*) [43–45]. Controller parameters can also be modulated based on the progression of the training, independent of the learner's performance [28, 46, 47].

12.3 Assessing Motor Learning

Motor learning refers to permanent changes in performance in a motor task, together with a reduction of physical and mental effort, as a result of training [48]. Because the general aim of haptic training is to enhance a learner's performance in the long term, the evaluation of their learning progress should only be performed by assessing the learners' performance before (*baseline*) and after (*retention*) the haptic training, when no assistance or resistance from the robot is applied. As a general guideline, retention tests should be performed at least 24 h after training to ensure the memory consolidation of the acquired skill, i.e., to assess long-term learning [11, 49–51]. Retention tests performed right after the training only assess short-term learning.

As important as getting skilled in the trained task is to transfer the acquired skill to untrained altered versions of the trained motor task (*generalization*). This is especially important in neurorehabilitation, in which acquired or recovered skills and functionalities during robotic rehabilitation are desired to be transferred to better function of activities of daily living, beyond the tasks trained during the rehabilitation sessions (e.g., [52]). Despite the importance of skill transfer in motor learning, only a few studies on haptic training methods have evaluated long-term skill transfer using a modified albeit similar version of the trained tasks [18, 32, 53], and even fewer studies assessed the skill generalization to real-life tasks [54, 55].

Different outcome metrics can be selected to evaluate motor learning depending on the movement aspects to be mastered. Performance metrics can be based on deviation from the desired movement path (**spatial** aspect, e.g., in [56]), the timing of an action (**temporal** aspects, e.g., in [57]), or a combination of temporal and spatial aspects, such as velocity error or movement smoothness (**spatiotemporal** aspects, e.g., in [31, 58]).

A common approach to quantify learning is by comparing average task performance before and after training. However, depending on the task, average task performance can be similar between highly skilled and lowly skilled learners. Task performance variability—e.g., the standard deviation of movement errors with respect to a movement goal at the beginning of a training period and at the end of a training period—could then be indicative of motor learning. Highly skilled learners often show lower task performance variability compared to higher task variability in lower-skilled learners, e.g., in [26, 59, 60].

The listed haptic training strategies might have contrasting effects on the learning of different movement aspects. For example, several studies have shown the benefit of haptic guidance in learning to reproduce the temporal—but not the spatial—characteristics of complex spatiotemporal curves [24, 61]. Schmidt et al. also highlight the importance of measuring physical and mental effort [48], as less physical and mental effort are expected in the final stages of motor learning [62]. However, measurements of physical and mental effort are hardly conducted in motor learning experiments, probably because the objective measurement of physical effort (e.g., using electromyography [19]) and mental effort (e.g., brain activation [63]) is cumbersome.

Along with mental effort, there are other relevant psychological factors that might have an effect on motor learning. The OPTIMAL theory states that trainees' motivation and attention enhance motor learning, possibly due to the release of dopamine [64]. Motivation has been shown to have both indirect (e.g., by increasing the number of movement repetitions) and direct (e.g., improving memory consolidation) positive effects on learning [9, 65]. Other psychological factors, such as the sense of agency—i.e., the feeling of being in control over our own movements [66], or personality traits are less studied in the motor learning literature, yet might play an important role in motor learning [21, 67].

12.4   Current Evidence of the Effectiveness of Haptic Methods on Motor (Re) Learning

Several factors might play a role in the effectiveness of robot-based training, making it challenging to compare the results from studies in which similar haptic training methods were employed for different tasks or different strategies for similar tasks. It is generally accepted that the learner's initial skill level might play a crucial role in the effectiveness of robotic training methods [4]. This finding is in line with the **challenge point framework**, which states that motor learning is enhanced when the difficulty of the motor task to be learned is matched with the learner's skill level (Fig. 12.1) [68]. Skill is defined as the ability to perform a task "*with maximum certainty and minimum outlay of energy, or of time and energy*", which progresses as a result of task practice [69, 70]. This learning progression has been proposed to follow three stages: a first cognitive stage (novice), a

motor/associative stage (advanced), and a final autonomous stage (expert) [62]. The majority of studies on robot-aided motor learning have been conducted with novice learners during the cognitive stage, while the number of studies on advanced learners and experts is scarce [4].

Although task difficulty has been studied in a large number of motor learning studies (e.g., [71, 72]), a definition of the term has not yet been explicitly stated. Instead, three different but interconnected concepts are employed when talking about task difficulty: nominal task difficulty, functional task difficulty, and conditional task difficulty.

Nominal task difficulty can be defined as the objective and inherent challenge of the task to be learned due to the task's spatial, temporal, and spatiotemporal performance requirements regardless of the differences between learners' initial skill levels. In their recent review [4], Basalp et al. proposed a task classification taxonomy—an extension of the motor task organization introduced by Schmidt and Wrisberg in [70]—to categorize motor tasks depending on

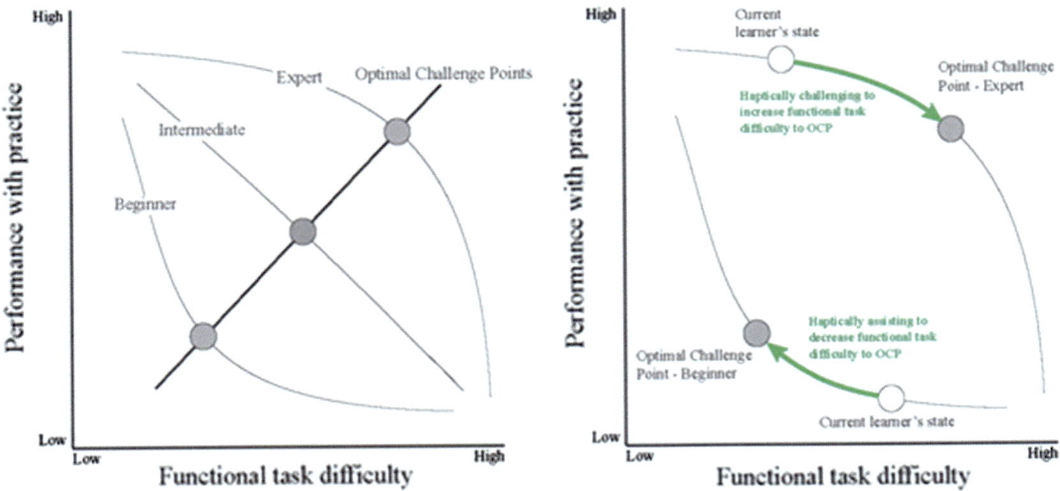

Fig. 12.1 Haptic training methods can help match the functional task difficulty to the learners' skill level. Left figure: schematic representation of the optimal challenge points (gray circles) in relation to the learner's skill level and functional task difficulty. According to the challenge point theory, motor learning is enhanced when the functional task difficulty is matched with the learner's skill level, defined as their optimal challenge points. Right figure: examples of how haptic training methods can help adapt the functional task difficulty through modulating the conditional task difficulty such that the optimal challenge point (OCP) for a certain skill level is reached. For example, beginners can benefit from haptic assistance to decrease the functional task difficulty, and experts can benefit from haptic methods that challenge the learner. Figures adapted from [68]

their continuity (discrete vs. continuous), rhythmicity (single execution vs. rhythmic), and complexity—related to several factors, e.g., demands on attention, memory, and/or processing capacity, or number of degrees of freedom, among others [7, 72]. Different motor task types, e.g., those that incorporate single task execution (e.g., pressing a key) versus rhythmic/repetitive motions (e.g., rowing or walking) have been shown to involve distinct control primitives/actions [73] and activate distinct brain areas [74]. Thus, haptic methods that support learning of one type of motor task might not be suitable to also support the learning of another task category [56, 75].

The **functional task difficulty** depends on the initial skill level of the learner. It is related to how challenging the execution of the task itself is perceived by the—novice, advanced, or expert—learners during training. Importantly, providing haptic guidance or disturbance during training may change the challenge presented to the learner by modulating the amount of task-relevant information conveyed by the haptic training strategy (Fig. 12.1), which is referred to as the **conditional task difficulty**. Robots can adapt the conditional task difficulty, for example by modulating the task environment—e.g., changing the simulated water density in a rowing task (e.g., [76])—or by haptically assisting/challenging the learners (e.g., [27]).

Current evidence supports the idea that the (lack of) effectiveness of state-of-the-art haptic methods can be explained by the challenge point framework. In particular, the effectiveness of haptic training seems to depend on: (1) the nominal task difficulty; (2) the task-relevant information conveyed by the haptic training method (conditional task difficulty); and (3) the initial skill level of the learner (functional task difficulty).

When learners face the training of tasks with low nominal difficulty, for example, simple motor tasks such as steering a virtual car without dynamics [35] or synchronizing between leg movements [19], it was observed that the learners' initial skill level was adequate to successfully learn the task. Thus, training with haptic methods did not promote motor learning in particularly simple tasks. Learning benefits of haptic methods over training without haptics were observed when learning tasks with higher nominal difficulty, e.g., steering nonholonomic vehicles [23, 28] and tracking of letters [31].

However, when haptic training was compared to training with other forms of feedback—e.g., *visual feedback* provided in virtual environments [47] or *terminal feedback* (i.e., knowledge of results and performance after the task is performed [5], no evident differences in motor learning were found between feedback modalities. Thus, for a general sample of healthy learners, providing task-relevant information (conditional task difficulty) by other sources of feedback (e.g., visual or auditory) might promote motor learning at the same level as haptic methods. Nevertheless, when other sources of feedback are not available and/or when the initial skill/disability level of the learners is too low to perform the task by themselves in a safe and motivating environment, the employment of haptic training might be effective to enhance learning.

Indeed, performance-enhancing haptic methods seem to be especially promising in promoting motor learning in initially less skilled (novice) healthy participants [57, 77], children [46, 78], and in brain-injured patients [9, 79]. Healthy novices seem to benefit from performance-enhancing haptic methods to learn the spatial (e.g., reducing the spatial error during path tracing tasks [39, 80]), temporal (e.g., timing turning in curves [23, 78]), and spatiotemporal aspects of the tasks (e.g., learning velocities [26]). This enhanced learning is probably due to the robotic assistance reducing the conditional task difficulty, and thereby, optimally challenging novices. The studies conducted with children, who usually adapt at slower rates [81], further support these findings [46]. Studies performed with neurologic patients seem to be in line with these findings, especially in the learning of the task's temporal aspects [9, 79].

Performance-degrading haptic methods, on the other hand, might provide a more optimal task challenge to advanced learners, by increasing the learner's effort and attention [18, 19] and by promoting the exploration of more advanced movements to achieve the task more efficiently [26]. Although only a few studies with rather small sample sizes have been conducted with advanced and expert learners, initial findings suggest that performance-degrading haptic methods are especially beneficial for learning spatial aspects of the tasks, but not temporal nor spatiotemporal aspects, in initially more skilled participants. The limited effectiveness of haptic disturbance methods to improve temporal and spatiotemporal aspects might be due to the design of these methods, as most error augmentation methods have been designed to only augment spatial errors [37, 55].

Haptic error augmentation and haptic noise increase movement variability and—although results are still inconclusive—could benefit advanced learners more than novices [18, 57]. Haptic error augmentation methods have also been found to be more effective than conventional repetitive training (e.g., [17, 82, 83]) and performance-enhancing haptic methods (e.g., [17, 84, 85]) for re-learning motor tasks' spatial aspects in neurologic patients. Yet, caution should be put when designing performance-degrading methods, as in some cases, the performance degradation might result in a decrease of the learners' perceived competence, hampering learners' motivation [26], and therefore, potentially hindering motor learning [64].

12.5   Implications for Rehabilitation Technology Design

12.5.1   The Personal and Temporal Nature of Motor Learning Highlights the Need for Adaptive Haptic Training Paradigms

As outlined earlier, current evidence highlights the essential role that the learner's initial skill level plays in the effectiveness of the different haptics methods on motor learning. As stated by the challenge point framework, motor learning can be maximized when the difficulty of the task to be learned matches the current learner's skill level [68]. Thus, adapting the haptic methods to adjust the task difficulty to match the learner's ongoing performance may have direct positive effects (i.e., by providing the optimal amount of information to enhance the performance and prevent slacking), and indirect effects (e.g., by enhancing learner's motivation and agency) on motor learning.

Although recent efforts have been made to develop adaptive algorithms (e.g., [43, 45, 86–88]), those have not yet been extensively investigated in motor learning studies. Most studies assess adaptive algorithms for haptic assistance over short time periods, ranging from hours to a few days, likely for practical reasons. Yet, learning is a long-term process that typically starts at a cognitive stage when the learner is still a novice and ends in an autonomous stage as a skillful performer or expert [62]. So, the requirements for haptic training paradigms depend on the learning stage over extended periods of time, in which the haptic training could be used to appropriately challenge the learners to promote their learning. Systematic studies showed that different learners need different types and levels of assistance, and adaptive paradigms need to appropriately account for these differences across individuals and time.

The majority of the adaptive haptic training paradigms focus on isolated aspects of motor learning, including cognitive and/or physical states, yet due to the interdependence of the factors governing motor learning, there is a need for holistic approaches that combine the insights gained in haptic training studies. Recent artificial intelligence (AI) approaches for therapy personalization have yielded promising results [89–91]; however, there are raising concerns about the interpretability and trustworthiness of opaque-box algorithms [92–94]. Furthermore, previous research only employed single metrics (i.e., single performance metrics, e.g., ongoing tracking error), which are, given the complexity of an individual's recovery process, inherently a poor descriptor of the overall patient characterization.

These limitations might be mitigated by developing novel therapist-in-the-loop personalization approaches that combine machine learning to learn and identify meaningful features that define the current cognitive and motor status of the patient from large amounts of high-dimensional data—e.g., biomechanical and physiological metrics—with the possibility to explicitly model the therapists' reasoning (e.g., using symbolic AI) to provide explainable, trustworthy, and interpretable therapy recommendations, such as the level of challenge for the learner, for example by adapting the task difficulty.

12.5.2 Appropriate Delivery of Task-Relevant Information Provided by Haptic Training Methods is Key to Enhance Motor Learning and Transfer

As (re)training functional motor tasks involves physical interaction with tangible objects, haptic training methods might impede motor learning if the haptic feedback hinders the learner's perception of task-relevant information. Such task-relevant information includes somatic (proprioceptive and tactile) information from the interaction with the environment (e.g., tangible objects) which is crucial for fine motor control [95, 96] and motor (re)learning [97–99].

The corrupted perception of task-relevant information during robotic training might be behind the observed (poor) transfer of learning from the virtual training environment to real-life tasks [7]. Current rehabilitation robotics does not support patients in regaining the functional movements needed to perform their activities of daily living and achieve their independence [100, 101]. Despite the crucial role that physiological sensory information plays in motor learning and neurorehabilitation (see Chap. 3), current haptic strategies rely on rather abstract visual feedback while meaningful somatic information from the interaction with virtual tangible objects/environments is neglected [98, 102]. Only a few studies have

incorporated haptic rendering—i.e., the simulation of the interaction forces between humans and tangible virtual objects/environments—into motor learning studies [21, 56, 76, 99]. This is probably due to the limitations of the used robots, especially the bulky and heavy exoskeletons employed in clinical settings, as they suffer from low transparency, which limits their capability to haptically render these informative interaction forces.

The learners' perception of the haptic rendering might also be hampered because the forces from the haptic rendering and the assistive/resistive haptic forces are provided from the same actuators. Several efforts have been made to provide these different forces in a way that the interference is minimized. For example, Power and O'Malley evaluated the effect on motor learning of separating the assisting forces from the task rendered haptic forces (a spring-damper dynamic system) spatially (i.e., using different robotic devices), or temporally (i.e., by the sequential provision of the assisting and haptic rendering forces [56]). None of these strategies was found to be effective in learning the dynamic task, which the authors attributed to the difficulty to interpret the feedback designs. More recent attempts to disentangle the assistive from the haptic rendering forces include solutions that employ robots to provide the task-relevant kinesthetic haptic rendering, while assistive guidance is provided through cutaneous skin stretch devices [103].

It is also important to take into account whether the learner perceives the haptic training forces as intended. Several studies suggest that human force perception, both magnitude and direction, is impacted by uncertainty (random errors) and systematic errors (biases). Systematic errors in force magnitude perception often manifest in incorrect force reproductions: humans typically reproduce higher forces than the presented force, indicating that we overestimate externally applied forces, such as an interaction force [104–106]. For low force levels (<10 N), humans seem to rely more on position sensory feedback than on force sensory feedback [107]. In addition, humans are inaccurate in estimating the direction of an applied force [108] and reproducing the direction and magnitude of the applied force [109]. Hence, the question remains: how accurately the learner perceives and subsequently interprets the information provided by the haptic training forces, in particular when these forces can change in direction and magnitude quickly? Also, how should this knowledge be taken into account when designing haptic training methods?

Finally, the haptic training strategy may also alter the learners' perception of the intended goal of the task to be learned. For example, in a virtual tracking task, researchers found that participants trained with error amplification—with repulsive forces that systematically pushed them to the opposite direction of the correct movement—got used to their low performance instead of trying to improve their tracking skills [36]. In addition, when the assistive forces do not align with the learner's own goal or how to reach that goal, conflicts between the learner and robot controller can occur. Interaction conflicts can impact learning and can even lead to disuse of the haptic training [110]. Therefore, when designing haptic methods, the task goals should be clearly established, communicated, and reachable.

In short, the provision of more naturalistic congruent visuo-haptic feedback might grant a more optimal training environment that might promote motor learning, and importantly, the transfer of skills gained during robotic training to real-life activities [7, 111]. Besides providing more realistic interactions with tangible virtual objects [99], providing a more naturalistic visualization of the learners' movements within the virtual environment might enhance motor learning and transfer [7]. To this date, most motor learning studies have provided a rather abstract visualization of the performed movements on computer screens, televisions, or projection systems. The reduced depth cues provided by these displays and the visuospatial transformation from the movements performed in the three-dimensional space to their two-dimensional visualization on conventional screen are far from being natural, realistic, and might enhance the trainees' cognitive load, and thus, negatively impact learning [112]. New

off-the-shelf virtual or augmented reality head-mounted displays offer the possibility to provide a more naturalistic virtual representation of the trainers' movements, for example by employing an avatar from a first-person perspective, that might reduce the cognitive load, enhance the sense of agency, and importantly, result in higher motivation [113].

12.5.3 Long-Term Effects and Generalization of Learning of Haptic Training Need More Attention

The primary goal of haptic training is to facilitate long-term learning and generalization of motor skills. However, most haptic training paradigms are only assessed on short-term learning with retention tests right after the training is finished, possibly under- or overestimating their benefits. Therefore, in future studies, researchers are encouraged to conduct long-term transfer tests, along with the delayed retention tests (at least 24 h after training is finished), for a more thorough investigation of the effectiveness of haptic training methods.

12.5.4 More Research is Needed to Understand How Haptic Trainings Could Modulate Motor Variability to Stimulate Motor Learning

Research on unassisted human motor learning found evidence that motor learning rate is positively correlated with the learner's motor variability [40], specifically the planning noise originating from the brain [42]. Some studies attempted to increase task-related motor variability through haptic forces (e.g., haptic noise or force disturbances) in order to modulate a learner's motor variability to subsequently stimulate learning [36, 114, 115]. However, it is unclear whether and how externally provided haptic forces can indeed modulate the learner's internal motor variability to facilitate learning through exploration, e.g., specifically their planning noise as hypothesized by researchers [41, 42]. Hence, despite the accumulating evidence of the impact of motor variability on motor learning in fundamental motor learning research, more research is needed before it can be used to inform the design of haptic training paradigms.

12.6 Conclusion

Current evidence from robot-aided motor (re) learning studies indicates that the effectiveness of the haptic training strategies on motor learning and neurorehabilitation could mainly be explained by the challenge point framework [68]. The functional task difficulty, nominal task difficulty, and conditional task difficulty play central roles in the effectiveness of robot-aided training. Performance-enhancing haptic training methods seem to be especially effective for novice learners and to train the temporal aspects of the task, while performance-degrading haptic methods might be more effective when training more skilled participants, especially in learning the spatial aspects of the tasks.

The findings from studies with brain-injured patients are in line with those from motor learning studies with healthy participants. This is an important observation, as the gained insights from past and future studies with healthy participants could be leveraged to improve current robotic-aided neurorehabilitation paradigms. Although haptic training was found to be as effective as training with other feedback modalities in healthy participants, brain-injured patients might still benefit from the robotic assistance when facing too difficult or frustrating tasks.

Based on the current evidence, we suggest that future research should focus on designing adaptive algorithms that can accommodate the learner's skill, progress level, and learning strategy by identifying and reducing hindrances that could impede learning, or by challenging more skilled learners. Finally, to enhance motor learning and the transfer of the gained skill during robot-aided training to real life, future research should focus on: (1) providing more naturalistic multisensory feedback that resembles out-of-the-lab training and (2) conducting long-term studies including transfer tests.

Acknowledgements We would like to thank Dr. Peter Wolf and Dr. Ekin Basalp for their support during the literature research. This work was supported in part by the Swiss National Science Foundation (SNF) through the grant PP00P2163800, the Dutch Research Council (NWO) Talent Program VIDI TTW 2020, and AiTech, TU Delft's initiative on Meaningful Human Control.

References

1. Voelcker-Rehage C. Motor-skill learning in older adults—a review of studies on age-related differences. Eur Rev Aging Phys Act. 2008;5:5–16. https://doi.org/10.1007/s11556-008-0030-9.
2. Wishart LR, Lee TD. Effects of aging and reduced relative frequency of knowledge of results on learning a motor skill. Percept Mot Skills. 1997;84:1107–22. https://doi.org/10.2466/pms. 1997.84.3.1107.
3. Swinnen SP. Age-related deficits in motor learning and differences in feedback processing during the production of a bimanual coordination pattern. Cogn Neuropsychol. 1998;15:439–66. https://doi. org/10.1080/026432998381104.
4. Basalp E, Wolf P, Marchal-Crespo L. Haptic training: Which types facilitate (re)learning of which motor task and for whom answers by a review. IEEE Trans Haptics 2021:1–1. https://doi. org/10.1109/TOH.2021.3104518.
5. Sigrist R, Rauter G, Riener R, Wolf P. Augmented visual, auditory, haptic, and multimodal feedback in motor learning: a review. Psychon Bull Rev. 2013;20:21–53. https://doi.org/10.3758/s13423-012-0333-8.
6. Marchal-Crespo L, Reinkensmeyer DJ. Review of control strategies for robotic movement training after neurologic injury. J Neuroeng Rehabil. 2009;6:20. https://doi.org/10.1186/1743-0003-6-20.
7. Levac DE, Huber ME, Sternad D. Learning and transfer of complex motor skills in virtual reality: a perspective review. J Neuroeng Rehabil. 2019;16:121. https://doi.org/10.1186/s12984-019-0587-8.
8. Dehem S, Gilliaux M, Stoquart G, Detrembleur C, Jacquemin G, Palumbo S, et al. Effectiveness of upper-limb robotic-assisted therapy in the early rehabilitation phase after stroke: a single-blind, randomised, controlled trial. Ann Phys Rehabil Med. 2019;62:313–20. https://doi.org/10.1016/j. rehab.2019.04.002.

9. Rowe JB, Chan V, Ingemanson ML, Cramer SC, Wolbrecht ET, Reinkensmeyer DJ. Robotic assistance for training finger movement using a hebbian model: a randomized controlled trial. Neurorehabil Neural Repair. 2017;31:769–80. https://doi.org/10.1177/1545968317721975.

10. Wulf G, Shea CH, Whitacre CA. Physical-guidance benefits in learning a complex motor skill. J Mot Behav. 1998;30:367–80. https://doi.org/10.1080/00222899809601351.

11. Williams CK, Carnahan H. Motor learning perspectives on haptic training for the upper extremities. IEEE Trans Haptics. 2014;7:240–50. https://doi.org/10.1109/TOH.2013.2297102.

12. Winstein CJ, Pohl PS, Lewthwaite R. Effects of physical guidance and knowledge of results on motor learning: support for the guidance hypothesis. Res Q Exerc Sport. 1994;65:316–23. https://doi.org/10.1080/02701367.1994.10607635.

13. O'Malley MK, Gupta A, Gen M, Li Y. Shared Control in Haptic Systems for Performance Enhancement and Training. J Dyn Sys, Meas, Control. 2005;128:75–85. https://doi.org/10.1115/1.2168160.

14. Emken JL, Reinkensmeyer DJ. Robot-enhanced motor learning: accelerating internal model formation during locomotion by transient dynamic amplification. IEEE Trans Neural Syst Rehabil Eng. 2005;13:33–9. https://doi.org/10.1109/TNSRE.2004.843173.

15. Hertzog MA. Considerations in determining sample size for pilot studies. Res Nurs Health. 2008;31:180–91. https://doi.org/10.1002/nur.20247.

16. Cramer SC, Sur M, Dobkin BH, O'Brien C, Sanger TD, Trojanowski JQ, et al. Harnessing neuroplasticity for clinical applications. Brain. 2011;134:1591–609. https://doi.org/10.1093/brain/awr039.

17. Patton JL, Stoykov ME, Kovic M, Mussa-Ivaldi FA. Evaluation of robotic training forces that either enhance or reduce error in chronic hemiparetic stroke survivors. Exp Brain Res. 2006;168:368–83. https://doi.org/10.1007/s00221-005-0097-8.

18. Marchal-Crespo L, Michels L, Jaeger L, Lopez-Oloriz J, Riener R. Effect of error augmentation on brain activation and motor learning of a complex locomotor task. Front Neurosci 2017;11. https://doi.org/10.3389/fnins.2017.00526.

19. Marchal-Crespo L, Schneider J, Jaeger L, Riener R. Learning a locomotor task: with or without errors? J Neuroeng Rehabil. 2014;11:25. https://doi.org/10.1186/1743-0003-11-25.

20. Sigrist R, Rauter G, Riener R, Wolf P. Terminal feedback outperforms concurrent visual, auditory, and haptic feedback in learning a complex rowing-type task. J Mot Behav. 2013;45:455–72. https://doi.org/10.1080/00222895.2013.826169.

21. Özen Ö, Buetler KA, Marchal-Crespo L. Promoting motor variability during robotic assistance enhances motor learning of dynamic tasks. Front Neurosci 2020;14. https://doi.org/10.3389/fnins.2020.600059.

22. Reinkensmeyer DJ, Housman SJ. "If I can't do it once, why do it a hundred times?": connecting volition to movement success in a virtual environment motivates people to exercise the arm after stroke. Virtual Rehabilitation. 2007;2007:44–8. https://doi.org/10.1109/ICVR.2007.4362128.

23. Marchal Crespo L, Reinkensmeyer DJ. Haptic guidance can enhance motor learning of a steering task. J Mot Behav. 2008;40:545–56. https://doi.org/10.3200/JMBR.40.6.545-557.

24. Lüttgen J, Heuer H. The influence of haptic guidance on the production of spatio-temporal patterns. Hum Mov Sci. 2012;31:519–28. https://doi.org/10.1016/j.humov.2011.07.002.

25. Lüttgen J, Heuer H. The influence of robotic guidance on different types of motor timing. J Mot Behav. 2013;45:249–58. https://doi.org/10.1080/00222895.2013.785926.

26. Duarte JE, Reinkensmeyer DJ. Effects of robotically modulating kinematic variability on motor skill learning and motivation. J Neurophysiol. 2015;113:2682–91. https://doi.org/10.1152/jn.00163.2014.

27. Rauter G, Sigrist R, Marchal-Crespo L, Vallery H, Riener R, Wolf P. Assistance or challenge? Filling a gap in user-cooperative control. In: 2011 IEEE/RSJ international conference on intelligent robots and systems; 2011. p. 3068–73. https://doi.org/10.1109/IROS.2011.6094832.

28. Chen X, Agrawal SK. Assisting versus repelling force-feedback for learning of a line following task in a wheelchair. IEEE Trans Neural Syst Rehabil Eng. 2013;21:959–68. https://doi.org/10.1109/TNSRE.2013.2245917.

29. Marchal-Crespo L, Rauter G, Wyss D, Zitzewitz J von, Riener R. Synthesis and control of an assistive robotic tennis trainer. In: 2012 4th IEEE RAS EMBS international conference on biomedical robotics and biomechatronics (BioRob); 2012. p. 355–60. https://doi.org/10.1109/BioRob.2012.6290262.

30. Srimathveeravalli G, Thenkurussi K. Motor skill training assistance using haptic attributes. In: First joint eurohaptics conference and symposium on haptic interfaces for virtual environment and teleoperator systems. World haptics conference; 2005. p. 452–7. https://doi.org/10.1109/WHC.2005.96.

31. Bluteau J, Coquillart S, Payan Y, Gentaz E. Haptic guidance improves the visuo-manual tracking of trajectories. PLoS ONE. 2008;3: e1775. https://doi.org/10.1371/journal.pone.0001775.

32. Zondervan DK, Duarte JE, Rowe JB, Reinkensmeyer DJ. Time flies when you are in a groove: using entrainment to mechanical resonance to teach a desired movement distorts the perception of the movement's timing. Exp Brain Res. 2014;232:1057–70. https://doi.org/10.1007/s00221-013-3819-3.

33. Wei K, Körding K. Relevance of error: what drives motor adaptation? J Neurophysiol. 2009;101:655–64. https://doi.org/10.1152/jn.90545.2008.

34. Shadmehr R, Smith MA, Krakauer JW. Error correction, sensory prediction, and adaptation in motor control. Annu Rev Neurosci. 2010;33:89–108. https://doi.org/10.1146/annurev-neuro-060909-153135.

35. Lee H, Choi S. Combining haptic guidance and haptic disturbance: an initial study of hybrid haptic assistance for virtual steering task. In: 2014 IEEE haptics symposium (HAPTICS); 2014. p. 159–65. https://doi.org/10.1109/HAPTICS.2014.6775449.

36. Lee J, Choi S. Effects of haptic guidance and disturbance on motor learning: Potential advantage of haptic disturbance. In: 2010 IEEE haptics symposium; 2010. p. 335–42.https://doi.org/10.1109/HAPTIC.2010.5444635.

37. Fisher ME, Huang FC, Klamroth-Marganska V, Riener R, Patton JL. Haptic error fields for robotic training. In: 2015 IEEE world haptics conference (WHC); 2015. p. 434–9. https://doi.org/10.1109/WHC.2015.7177750.

38. Morris D, Tan H, Barbagli F, Chang T, Salisbury K. Haptic feedback enhances force skill learning. In: Proceedings of the second joint EuroHaptics conference and symposium on haptic interfaces for virtual environment and teleoperator systems. Washington, DC, USA: IEEE Computer Society; 2007. p. 21–6. https://doi.org/10.1109/WHC.2007.65.

39. Rauter G, Sigrist R, Riener R, Wolf P. Learning of temporal and spatial movement aspects: a comparison of four types of haptic control and concurrent visual feedback. IEEE Trans Haptics. 2015;8:421–33. https://doi.org/10.1109/TOH.2015.2431686.

40. Wu HG, Miyamoto YR, Gonzalez Castro LN, Ölveczky BP, Smith MA. Temporal structure of motor variability is dynamically regulated and predicts motor learning ability. Nat Neurosci. 2014;17:312–21. https://doi.org/10.1038/nn.3616.

41. Cheng S, Sabes PN. Modeling sensorimotor learning with linear dynamical systems. Neural Comput. 2006;18:760–93. https://doi.org/10.1162/089976606775774651.

42. van der Vliet R, Frens MA, de Vreede L, Jonker ZD, Ribbers GM, Selles RW, et al. Individual differences in motor noise and adaptation rate are optimally related. ENeuro 2018;5: ENEURO.0170-18.2018. https://doi.org/10.1523/ENEURO.0170-18.2018.

43. Maggioni S, Reinert N, Lünenburger L, Melendez-Calderon A. An adaptive and hybrid end-point/joint impedance controller for lower limb exoskeletons. Front Robot AI. 2018;5:104. https://doi.org/10.3389/frobt.2018.00104.

44. Bernardoni F, Özen Ö, Buetler K, Marchal-Crespo L. Virtual reality environments and haptic strategies to enhance implicit learning and motivation in robot-assisted training. In: 2019 IEEE 16th international conference on rehabilitation robotics (ICORR); 2019. p. 760–5. https://doi.org/10.1109/ICORR.2019.8779420.

45. Rauter G, Gerig N, Sigrist R, Riener R, Wolf P. When a robot teaches humans: automated feedback selection accelerates motor learning. Sci Robot 2019;4. https://doi.org/10.1126/scirobotics.aav1560.

46. Palluel-Germain R, Bara F, Boisferon AH de, Hennion B, Gouagout P, Gentaz E. A visuo-haptic device - telemaque - increases kindergarten children's handwriting acquisition. In: Second joint EuroHaptics conference and symposium on haptic interfaces for virtual environment and teleoperator systems (WHC'07); 2007. p. 72–7. https://doi.org/10.1109/WHC.2007.13.

47. Marchal-Crespo L, van Raai M, Rauter G, Wolf P, Riener R. The effect of haptic guidance and visual feedback on learning a complex tennis task. Exp Brain Res. 2013;231:277–91. https://doi.org/10.1007/s00221-013-3690-2.

48. Schmidt RA, Lee TD. Motor control and learning: a behavioral emphasis. Champaign, IL, USA: Human Kinetics Publishers; 2005; 2010.

49. Shadmehr R, Holcomb HH. Neural correlates of motor memory consolidation. Science. 1997;277:821–5. https://doi.org/10.1126/science.277.5327.821.

50. McGaugh JL. Memory–a century of consolidation. Science. 2000;287:248–51. https://doi.org/10.1126/science.287.5451.248.

51. Heuer H, Lüttgen J. Robot assistance of motor learning: a neuro-cognitive perspective. Neurosci Biobehav Rev. 2015;56:222–40. https://doi.org/10.1016/j.neubiorev.2015.07.005.

52. Kitago T, Krakauer JW. Motor learning principles for neurorehabilitation. Handb Clin Neurol. 2013;110:93–103. https://doi.org/10.1016/B978-0-444-52901-5.00008-3.

53. Williams CK, Tremblay L, Carnahan H. It pays to go off-track: practicing with error-augmenting haptic feedback facilitates learning of a curve-tracing task. Front Psychol. 2016;7. https://doi.org/10.3389/fpsyg.2016.02010.

54. Zhang Z, Sternad D. Back to reality: differences in learning strategy in a simplified virtual and a real throwing task. J Neurophysiol. 2021;125:43–62. https://doi.org/10.1152/jn.00197.2020.

55. Marchal-Crespo L, Tsangaridis P, Obwegeser D, Maggioni S, Riener R. Haptic error modulation outperforms visual error amplification when learning a modified gait pattern. Front Neurosci. 2019;13. https://doi.org/10.3389/fnins.2019.00061.

56. Powell D, O'Malley MK. The task-dependent efficacy of shared-control haptic guidance paradigms. IEEE Trans Haptics. 2012;5:208–19. https://doi.org/10.1109/TOH.2012.40.

57. Milot M-H, Marchal-Crespo L, Green CS, Cramer SC, Reinkensmeyer DJ. Comparison of error-amplification and haptic-guidance training

techniques for learning of a timing-based motor task by healthy individuals. Exp Brain Res. 2010;201:119–31. https://doi.org/10.1007/s00221-009-2014-z.

58. Sigrist R, Rauter G, Marchal-Crespo L, Riener R, Wolf P. Sonification and haptic feedback in addition to visual feedback enhances complex motor task learning. Exp Brain Res. 2015;233:909–25. https://doi.org/10.1007/s00221-014-4167-7.

59. Lee TD, Ishikura T, Kegel S, Gonzalez D, Passmore S. Head-putter coordination patterns in expert and less skilled golfers. J Mot Behav. 2008;40:267–72. https://doi.org/10.3200/JMBR.40.4.267-272.

60. Su ELM, Ganesh G, Yeong CF, Teo CL, Ang WT, Burdet E. Effect of grip force and training in unstable dynamics on micromanipulation accuracy. IEEE Trans Haptics. 2011;4:167–74. https://doi.org/10.1109/TOH.2011.33.

61. Feygin D, Keehner M, Tendick R. Haptic guidance: experimental evaluation of a haptic training method for a perceptual motor skill. In: 10th symposium on haptic interfaces for virtual environment and teleoperator systems. HAPTICS 2002. Proceedings; 2002. p. 40–7. https://doi.org/10.1109/HAPTIC.2002.998939.

62. Fitts PM. Perceptual-motor skill learning. In: Melton AW, editor. Categories of human learning. Academic Press; 1964. p. 243–85. https://doi.org/10.1016/B978-1-4832-3145-7.50016-9.

63. Penalver-Andres J, Buetler KA, Koenig T, Müri RM, Marchal-Crespo L. Providing task instructions during motor training enhances performance and modulates attentional brain networks. Front Neurosci. 2021;15. https://doi.org/10.3389/fnins.2021.755721

64. Wulf G, Lewthwaite R. Optimizing performance through intrinsic motivation and attention for learning: the OPTIMAL theory of motor learning. Psychon Bull Rev. 2016;23:1382–414. https://doi.org/10.3758/s13423-015-0999-9.

65. Lohse KR, Boyd LA, Hodges NJ. Engaging environments enhance motor skill learning in a computer gaming task. J Mot Behav. 2016;48:172–82. https://doi.org/10.1080/00222895.2015.1068158.

66. Endo S, Fröhner J, Musić S, Hirche S, Beckerle P. Effect of external force on agency in physical human-machine interaction. Front Hum Neurosci. 2020;14:114. https://doi.org/10.3389/fnhum.2020.00114.

67. Darzi A, Wondra T, McCrea S, Novak D. Classification of multiple psychological dimensions in computer game players using physiology, performance, and personality characteristics. Front Neurosci. 2019;13:1278. https://doi.org/10.3389/fnins.2019.01278.

68. Guadagnoli MA, Lee TD. Challenge point: a framework for conceptualizing the effects of various practice conditions in motor learning. J Mot Behav. 2004;36:212–24. https://doi.org/10.3200/JMBR.36.2.212-224.

69. Guthrie ER. Psychology of learning. Oxford, England: Harper; 1935.

70. Schmidt RA, Wrisberg CA. Motor learning and performance: a situation-based learning approach, 4th ed. Champaign, IL, US: Human Kinetics; 2008.

71. Fleishman EA, Quaintance MK, Broedling LA. Taxonomies of human performance: the description of human tasks. San Diego, CA, US: Academic; 1984.

72. Wulf G, Shea CH. Principles derived from the study of simple skills do not generalize to complex skill learning. Psychon Bull Rev. 2002;9:185–211. https://doi.org/10.3758/BF03196276.

73. Buchanan JJ, Park J-H, Ryu YU, Shea CH. Discrete and cyclical units of action in a mixed target pair aiming task. Exp Brain Res. 2003;150:473–89. https://doi.org/10.1007/s00221-003-1471-z.

74. Schaal S, Sternad D, Osu R, Kawato M. Rhythmic arm movement is not discrete. Nat Neurosci. 2004;7:1136–43. https://doi.org/10.1038/nn1322.

75. Marchal-Crespo, Rappo N., Riener R. The effectiveness of robotic training depends on motor task characteristics. Exp Brain Res. 2017. https://doi.org/10.1007/s00221-017-5099-9.

76. Basalp E, Marchal-Crespo L, Rauter G, Riener R, Wolf P. Rowing simulator modulates water density to foster motor learning. Front Robot AI. 2019;6. https://doi.org/10.3389/frobt.2019.00074.

77. Marchal-Crespo L, McHughen S, Cramer SC, Reinkensmeyer DJ. The effect of haptic guidance, aging, and initial skill level on motor learning of a steering task. Exp Brain Res. 2010;201:209–20. https://doi.org/10.1007/s00221-009-2026-8.

78. Chen X, Ragonesi C, Galloway JC, Agrawal SK. Training toddlers seated on mobile robots to drive indoors amidst obstacles. IEEE Trans Neural Syst Rehabil Eng. 2011;19:271–9. https://doi.org/10.1109/TNSRE.2011.2114370.

79. Bouchard AE, Corriveau H, Milot M-H. A single robotic session that guides or increases movement error in survivors post-chronic stroke: which intervention is best to boost the learning of a timing task? Disabil Rehabil. 2017;39:1607–14. https://doi.org/10.1080/09638288.2016.1205151.

80. Marchal-Crespo L, Baumann T, Imobersteg M, Maassen S, Riener R. Experimental evaluation of a mixed controller that amplifies spatial errors and reduces timing errors. Front Robot AI. 2017;4. https://doi.org/10.3389/frobt.2017.00019.

81. Rossi C, Chau CW, Leech KA, Statton MA, Gonzalez AJ, Bastian AJ. The capacity to learn new motor and perceptual calibrations develops concurrently in childhood. Sci Rep. 2019;9:9322. https://doi.org/10.1038/s41598-019-45074-6.

82. Rozario SV, Housman S, Kovic M, Kenyon RV, Patton JL. Therapist-mediated post-stroke rehabilitation using haptic/graphic error augmentation. Annu Int Conf IEEE Eng Med Biol Soc. 2009;2009:1151–6. https://doi.org/10.1109/IEMBS.2009.5333875.

83. Huang FC, Patton JL. Augmented dynamics and motor exploration as training for stroke. IEEE Trans Biomed Eng. 2013;60:838–44. https://doi.org/10.1109/TBME.2012.2192116.

84. Cesqui B, Aliboni S, Mazzoleni S, Carrozza MC, Posteraro F, Micera S. On the use of divergent force fields in robot-mediated neurorehabilitation. In: 2008 2nd IEEE RAS EMBS international conference on biomedical robotics and biomechatronics; 2008. p. 854–61. https://doi.org/10.1109/BIOROB.2008.4762927.

85. Tropea P, Cesqui B, Monaco V, Aliboni S, Posteraro F, Micera S. Effects of the alternate combination of "error-enhancing" and "active assistive" robot-mediated treatments on stroke patients. IEEE J Transl Eng Health Med. 2013;1:2100109. https://doi.org/10.1109/JTEHM.2013.2271898.

86. Li Y, Huegel JC, Patoglu V, O'Malley MK. Progressive shared control for training in virtual environments. In: World haptics 2009 - third joint eurohaptics conference and symposium on haptic interfaces for virtual environment and teleoperator systems; 2009. p. 332–7. https://doi.org/10.1109/WHC.2009.4810873.

87. Li Y, Carboni G, Gonzalez F, Campolo D, Burdet E. Differential game theory for versatile physical human–robot interaction. Nat Mach Intell. 2019;1:36–43. https://doi.org/10.1038/s42256-018-0010-3.

88. Tamagnone I, Basteris A, Sanguineti V. Robot-assisted acquisition of a motor skill: evolution of performance and effort. In: 2012 4th IEEE RAS EMBS international conference on biomedical robotics and biomechatronics (BioRob); 2012. p. 1016–21. https://doi.org/10.1109/BioRob.2012.6290881.

89. Xu G, Gao X, Pan L, Chen S, Wang Q, Zhu B, et al. Anxiety detection and training task adaptation in robot-assisted active stroke rehabilitation. Int J Adv Rob Syst. 2018;15:1729881418806433. https://doi.org/10.1177/1729881418806433.

90. Giang C, Pirondini E, Kinany N, Pierella C, Panarese A, Coscia M, et al. Motor improvement estimation and task adaptation for personalized robot-aided therapy: a feasibility study. Biomed Eng Online. 2020;19:33. https://doi.org/10.1186/s12938-020-00779-y.

91. Menner M, Berntorp K, Zeilinger MN, Di Cairano S. Inverse learning for data-driven calibration of model-based statistical path planning. IEEE Trans Intell Veh. 2020:1–1. https://doi.org/10.1109/TIV.2020.3000323.

92. Doran D, Schulz S, Besold TR. What does explainable AI really mean? a new conceptualization of perspectives; 2017. arXiv:171000794 [Cs].

93. Garcez A d'Avila, Gori M, Lamb LC, Serafini L, Spranger M, Tran SN. Neural-symbolic computing: an effective methodology for principled integration of machine learning and reasoning; 2019. arXiv:190506088 [Cs].

94. Rudin C. Stop explaining black box machine learning models for high stakes decisions and use interpretable models instead. Nat Mach Intell. 2019;1:206–15. https://doi.org/10.1038/s42256-019-0048-x.

95. Sternad D, Duarte M, Katsumata H, Schaal S. Bouncing a ball: tuning into dynamic stability. J Exp Psychol Hum Percept Perform. 2001;27:1163–84. https://doi.org/10.1037/0096-1523.27.5.1163.

96. Danion F, Diamond JS, Flanagan JR. The role of haptic feedback when manipulating nonrigid objects. J Neurophysiol. 2012;107:433–41. https://doi.org/10.1152/jn.00738.2011.

97. Borich MR, Brodie SM, Gray WA, Ionta S, Boyd LA. Understanding the role of the primary somatosensory cortex: Opportunities for rehabilitation. Neuropsychologia. 2015;79:246–55. https://doi.org/10.1016/j.neuropsychologia.2015.07.007.

98. Gassert R, Dietz V. Rehabilitation robots for the treatment of sensorimotor deficits: a neurophysiological perspective. J Neuroeng Rehabil. 2018;15:46. https://doi.org/10.1186/s12984-018-0383-x.

99. Özen Ö, Buetler KA, Marchal-Crespo L. Towards functional robotic training: motor learning of dynamic tasks is enhanced by haptic rendering but hampered by arm weight support. J Neuroeng Rehabil. 2022;19:19. https://doi.org/10.1186/s12984-022-00993-w.

100. Bertani R, Melegari C, De Cola MC, Bramanti A, Bramanti P, Calabrò RS. Effects of robot-assisted upper limb rehabilitation in stroke patients: a systematic review with meta-analysis. Neurol Sci. 2017;38:1561–9. https://doi.org/10.1007/s10072-017-2995-5.

101. Veerbeek JM, Langbroek-Amersfoort AC, van Wegen EEH, Meskers CGM, Kwakkel G. Effects of robot-assisted therapy for the upper limb after stroke. Neurorehabil Neural Repair. 2017;31:107–21. https://doi.org/10.1177/1545968316666957.

102. Handelzalts S, Ballardini G, Avraham C, Pagano M, Casadio M, Nisky I. Integrating tactile feedback technologies into home-based telerehabilitation: opportunities and challenges in light of COVID-19 pandemic. Front Neurorobot. 2021;15:4. https://doi.org/10.3389/fnbot.2021.617636.

103. Pezent E, Fani S, Clark J, Bianchi M, O'Malley MK. Spatially separating haptic guidance from task dynamics through wearable devices. IEEE Trans Haptics. 2019;12:581–93. https://doi.org/10.1109/TOH.2019.2919281.

104. Shergill SS, Bays PM, Frith CD, Wolpert DM. Two eyes for an eye: the neuroscience of force escalation. Science. 2003;301:187–187. https://doi.org/10.1126/science.1085327.

105. Walsh LD, Taylor JL, Gandevia SC. Overestimation of force during matching of externally generated forces. J Physiol. 2011;589:547–57. https://doi.org/10.1113/jphysiol.2010.198689.

106. Onneweer B, Mugge W, Schouten AC. Force reproduction error depends on force level, whereas the position reproduction error does not. IEEE Trans Haptics. 2016;9:54–61. https://doi.org/10.1109/TOH.2015.2508799.

107. Mugge W, Schuurmans J, Schouten AC, van der Helm FCT. Sensory weighting of force and position feedback in human motor control tasks. J Neurosci. 2009;29:5476–82. https://doi.org/10.1523/JNEUROSCI.0116-09.2009.

108. van Beek FE, Tiest WMB, Kappers AML. Anisotropy in the haptic perception of force direction and magnitude. IEEE Trans Haptics. 2013;6:399–407. https://doi.org/10.1109/TOH.2013.37.

109. Toffin D, McIntyre J, Droulez J, Kemeny A, Berthoz A. Perception and reproduction of force direction in the horizontal plane. J Neurophysiol. 2003;90:3040–53. https://doi.org/10.1152/jn.00271.2003.

110. Abbink DA, Carlson T, Mulder M, de Winter JCF, Aminravan F, Gibo TL, et al. A topology of shared control systems—finding common ground in diversity. IEEE Trans Hum-Mach Syst. 2018;48:509–25. https://doi.org/10.1109/THMS.2018.2791570.

111. de Mello Monteiro CB, Massetti T, da Silva TD, van der Kamp J, de Abreu LC, Leone C, et al. Transfer of motor learning from virtual to natural environments in individuals with cerebral palsy. Res Dev Disabil. 2014;35:2430–7. https://doi.org/10.1016/j.ridd.2014.06.006.

112. Wenk N, Penalver-Andres J, Palma R, Buetler KA, Muri R, Nef T, et al. Reaching in several realities: motor and cognitive benefits of different visualization technologies. IEEE Int Conf Rehabil Robot. 2019;2019:1037–42. https://doi.org/10.1109/ICORR.2019.8779366.

113. Wenk N, Penalver-Andres J, Buetler KA, Nef T, Müri RM, Marchal-Crespo L. Effect of immersive visualization technologies on cognitive load, motivation, usability, and embodiment. Virtual Real. 2021. https://doi.org/10.1007/s10055-021-00565-8.

114. Brookes J, Mushtaq F, Jamieson E, Fath AJ, Bingham G, Culmer P, et al. Exploring disturbance as a force for good in motor learning. PLoS ONE. 2020;15:e0224055. https://doi.org/10.1371/journal.pone.0224055.

115. Marchal-Crespo L, López-Olóriz J, Jaeger L, Riener R. Optimizing learning of a locomotor task: amplifying errors as needed. Conf Proc IEEE Eng Med Biol Soc. 2014;2014:5304–7. https://doi.org/10.1109/EMBC.2014.6944823.

Implementation of Robots into Rehabilitation Programs: Meeting the Requirements and Expectations of Professional and End Users

13

Rüdiger Rupp and Markus Wirz

Abstract

This chapter covers the various aspects related to the practical application of robots in neurorehabilitation. For successful implementation of robotic therapy devices into motor rehabilitation programs of patients with neurological impairments, the specific requirements of users need to be met. Users in this case are patients with neurological conditions but also therapists who operate rehabilitation robots. Both claim different requirements, which need to be addressed for a robotic therapy to be accepted. Robots are valuable tools to apply intensive training in respect to number of task repetitions and task specificity. The complexity of robotic devices is mainly determined by the patients' residual functions. In patients with muscular weakness, a body weight support system might be sufficient, whereas in patients with severe paralysis, actively driven exoskeletons with multiple degrees of freedom are necessary. Robotic devices have to be adjustable to a wide range of anthropometric properties and to the severity and the characteristics of the impairments of patients. The user-friendliness and intuitiveness of the robot's human–machine interface consisting of the mechanical, the control, and the feedback interfaces represent essential determinants of a robot's clinical acceptance. An inherent advantage of the more complex training robots is their ability to measure joint-specific angles and forces. On one hand, these data can be used for objective documentation of the therapy outcome and for shaping the training. On the other hand, the data can be used to provide real-time feedback to patients with substantially impaired proprioceptive but sufficient motor functions to enable them to actively correct their pathological gait pattern. In the future, robotic devices which allow the continuation of a therapy at home will further enhance functional recovery and lead to a better long-term health status.

Keywords

User requirements · Complexity vs. usability · Practical application · End user safety · Human–machine interfaces · Robotic assessments

R. Rupp (✉)
Spinal Cord Injury Center, Heidelberg University Hospital, Schlierbacher Landstrasse 200a, 69118 Heidelberg, Germany
e-mail: ruediger.rupp@med.uni-heidelberg.de

M. Wirz
School of Health Sciences, Institute of Physiotherapy, ZHAW Zurich University of Applied Sciences, Katharina-Sulzer-Platz 9, 8401 Winterthur, Switzerland
e-mail: markus.wirz@zhaw.ch

© The Author(s), under exclusive license to Springer Nature Switzerland AG 2022

D. J. Reinkensmeyer et al. (eds.), *Neurorehabilitation Technology*,
https://doi.org/10.1007/978-3-031-08995-4_13

13.1 Introduction

Technology represents an essential part of modern medicine particularly for diagnosis of diseases and their treatment. Technical devices also play an important role in the rehabilitation of patients with central nervous system (CNS) diseases or lesions. While assistive devices such as wheelchairs are used since centuries for compensation of severely impaired functions, based on the growing knowledge about human neurobiology, highly sophisticated training devices are increasingly applied in the early stage after the CNS lesion or disease with the aim of restoration of motor functions to the highest possible degree. In the chronic phase after a CNS injury or disease, technology might support physical activities to maintain a functional status and enhance the general health condition.

This chapter focuses on the possibilities and challenges of implementing rehabilitation technologies including robotic devices, which have become increasingly available for clinical use over the last decades, into multifaceted rehabilitation programs of patients with neurological conditions, e.g., spinal cord injury (SCI) or stroke. While these robotic devices do not represent a radically new therapeutic approach compared to conventional restorative therapies, they enable (1) to start a task-specific and goal-oriented training at an early time point after the CNS injury and (2) allow to generate intensive afferent feedback with a high number of repetitions of functional movements such as walking or reaching/grasping. Robots not only perform movements repeatedly, but they allow the introduction of task variation and provide feedback thereby maintaining an adequate level of challenge during the training. All of these aspects are called principles of motor learning and are known to contribute to a better functional outcome of patients with neurological impairments [1]. The issues discussed in this chapter about rehabilitation robotics may partially also be valid for other types of rehabilitation and assistive technologies.

The starting point for developing any new assistive and training device should be the definition of the intended use and the identification of the specific requirements of users. Users in this case are end users with neurological conditions, who want to benefit in a more effective way from robotic training, meaning that they want to achieve maximum recovery of motor functions in the shortest period of time and with the least efforts possible. Users are also professional users, i.e., therapists, who, by using robotic devices, experience less physical exertion, have extended therapeutic possibilities, and can quantitatively assess the functional improvements. However, these benefits come with complex and time-consuming setup procedures, which reduce the effective training time within a therapy session. Hence, patients and therapists claim different requirements which need to be respected in a meaningful way. Those requirements should be in the focus as opposed to technical feasibility which does not always comply with rehabilitative demands [2]. It is absolutely mandatory that engineers regard patients and therapists as integral components of the therapy program who need to be involved throughout the whole process of development of a robotic device. Patients and therapists are likely to set priorities in the development of robotic therapy devices differently than engineers. The potential clinical application has to be borne in mind throughout the whole developmental process. An iterative user-centered design approach is necessary to identify needs for improvement of devices in early tests with end users. A widely adopted classification of developmental stages of technology is the technology readiness level (TRL, see Table 13.1) which has originally been established for the aerospace engineering by the National Aeronautics and Space Administration (NASA) [3].

Nowadays, this classification represents the standard of reference (ISO 16290:2013) and is widely accepted by several organizations (e.g., the European Commission [4]). The TRL has not specifically been designed for the requirements of the development of rehabilitative robotics but may serve as a structural framework for the developmental process.

Table 13.1 Technology readiness level (TRL) according to the European Horizon 2020—Work Programme 2014–2015 [4] and their translation to the development of rehabilitation robotics

TRL	Description	Translation to rehabilitation robotics
TRL 1	Basic principles observed	Proof of principle observed in animal models and in pilot human applications (e.g., on motor learning)
TRL 2	Technology concept formulated	Technical requirements and specifications including safety measures with regard to the application in patients defined. Review of these concepts by end users (patients and therapists) is recommended
TRL 3	Experimental proof of concept	Development and implementation of an experimental model
TRL 4	Technology validated in lab	First usability studies in healthy volunteers (human factor study) for the refinement of technical specifications in a user-centered design process. Development of training scenarios for the application in patients
TRL 5	Technology validated in relevant environment (industrially relevant environment in the case of key enabling technologies)	First usability studies in selected patients representing typical use cases. These studies do not focus on investigating efficacy but feasibility of both hardware and software components as well as acceptance by users. Establishment of reliability and validity of the devices' measurement capacities. Ethical approval and involvement of regulatory authorities required. Aim: certification of the product
TRL 6	Technology demonstrated in relevant environment (industrially relevant environment in the case of key enabling technologies)	Clinical trials with broader inclusion criteria and a larger number of patients to investigate efficacy. Health technology assessment
TRL 7	System prototype demonstration in operational environment	Effectiveness studies
TRL 8	System complete and qualified	Broader application, commercialization
TRL 9	Actual system proven in operational environment (competitive manufacturing in the case of key enabling technologies; or in space)	Ongoing refinement according to end user feedback and to the technological progress. Different manufacturers with comparable products

Besides the specifications which are framed by patients and therapists, there are several technological issues and principles regarding the clinical application of therapeutic robots. Both aspects will be covered in the next sections.

13.2 Patients' Requirements

The following sections focus on aspects related to patients who train with and therapists who operate technological devices. We refer to the former as primary users or end users and to the latter as secondary or professional users. In the process of a therapy session, both user groups demand different requirements.

13.2.1 Neurological Condition

The extend of the impairment varies to a high degree in patients with neurological conditions. This applies not only to individuals between diagnostic groups but also within the same diagnostic entity. For example, in persons with incomplete SCI the individual motor deficits may vary to a high degree, ranging from an isolated drop foot on one body side to an almost complete loss of motor function in both legs. In stroke survivors, an increased spastic muscle tone may restrict the successful application of a robotic training. In traumatic brain injury, cognitive restrictions may occur additionally to the physical impairments, which reduce the cooperativeness of the patient to a minimum. All these patient-related factors require an individualized shaping of the training paradigms including feedback modalities. More information about diagnosis-specific aspects of robotic gait training can be found in Chap. 8.

Patients presenting with severe functional limitations requiring technological devices for performing movement exercises are usually in a vulnerable state. Especially in the early stage after a neurological injury or damage, a poor general condition is present which needs to recover to a certain extent before intensive rehabilitation can be initiated. Also, in the early phase after an SCI or stroke, patients' stability in terms of circulation, mood, and motivation is negatively affected. Although emergency events or device-related adverse events are rare, robotic devices should account for them in such a way that subjects can be evacuated from the device within a short period of time. Fittings must be designed that they can be removed quickly, and the whole device must be removable in order to get access to the patient or to transport an unconscious patient from the device without constraints. That is, that robotic joints must not lock in the current position after an emergency stop but must be unlockable on demand or freely movable with only low friction.

13.2.2 Autonomic Nervous System

The clinical presentation of an SCI or a stroke comprises motor weakness up to a complete paralysis, partial or complete loss of sensory function, and a more or less pronounced derailment of vegetative functions [5, 6]. The latter include, among others, lack of voluntary bladder and bowel voiding function and/or lack of blood pressure adaptation as a response to upright position named orthostatic hypotension. Besides the vegetative symptoms, patients have a reduced vital capacity which may become evident in upright standing and during exercise. For robotic gait training, the most relevant aspect is that patients with SCI have a marked propensity to faint once they are elevated in an upright position. The possibility to position patients horizontally when the blood pressure starts to drop is therefore crucial. While bringing patients into a robotic device, it is advisable that one staff member focuses on the patient and observes him or her while other persons are responsible for fitting and mounting the patient into the device. The blood pressure monitoring can be realized by checking clinical symptoms or by using a measuring device. Because the risk for fainting is lower during walking compared to static standing the mounting-phase should be kept as short as possible. This can be achieved by preparing all settings and training protocols before each training session and by assigning clear roles to the different staff members. A specific complication in patients with SCI is autonomic dysreflexia (AD) which is a potentially life-threatening condition with uncontrolled increase of blood pressure triggered by afferent autonomic stimuli. In rare cases, AD might be triggered by somatosensory stimuli occurring during locomotor training, so therapists need to be aware of the typical symptoms of AD to take appropriate countermeasures [7].

13.2.3 Musculoskeletal System and Skin

In most cases, a traumatic SCI is caused by an instable fracture of the spine. In addition, spine fractures are often associated with fractures of the extremities, pelvis, or ribs. Rehabilitation therapists need to check that the musculo skeletal system is stable enough to tolerate the applied loads and forces applied by robotic devices before including patients in a robotic locomotion training program. This holds even true in cases where fractures and instabilities have been treated surgically. The partial or complete lack of sensibility has to be considered when a patient with a neurological condition is trained. After every training session, the spots where forces are exchanged between the robotic device and the patient have to be inspected visually. Optimally such an inspection is also performed at the start of the training. Any sign of skin strain must be documented and carefully controlled. Robotic devices enable intensive and long training sessions with a large number of repetitions. Some patients may react to that amount of workload with signs of overload, e.g., joint swelling, increased spasticity, or pain. In older patients with a known history of osteoporosis, the training intensity and body weight unloading have to be set very carefully. The repeated stress on bony structures may result in fatigue fractures [8].

13.2.4 Cognition

Patients who experience an impairment of their cognitive function, e.g., distorted self-perception, might not be able to cooperate with a robotic device. Even though some devices use virtual environments which are similar to the real world and the control of these environments is very intuitive, patients still require the ability to cope with the robotic intervention. In order to able to participate actively in the robotic training and to make use of numerous ways of training modalities, patients need to have no more than mild to moderate cognitive deficits. However, in those end users who are compliant to the robotic therapies not only motor but also cognitive functions might be improved [9].

There might be patients who are generally not used to working with new information technologies or other technologies and may thus be reluctant to train in a robotic device, for example, patients of advanced age. Without complete confidence in a robotic training device, the success of the intervention is endangered. It is therefore important that end users are able to acknowledge robotic training as an important component on the way to their maximum possible independence.

13.3 Therapists' Requirements

Therapists who operate robotic therapy devices have the responsibility to shape the training in a way that it is most effective while preventing any harmful effects. Indication, medical clearance, time of training initiation within the rehabilitation process, dosage, outcome measures, and the integration into the overall rehabilitation program are some aspects which have to be taken into consideration. Unfortunately, even 20 years after the beginning of the integration of rehabilitation robots into clinical therapy programs, the usage of these devices is only partially a subject in basic therapist training. The reason for this might be that the field of rehabilitation robotics is growing rapidly with a large number of new devices being developed every year and that not all clinics where students do their internship provide robotic-assisted training and the students therefore do not have the chance to gain practical experience. Another issue is the fact that the application of robotic-assisted training is handled differently depending on the center-specific rehabilitation path. This concerns the timing, frequency, duration of training sessions, the characteristics of subjects who could benefit from robotic training, and most obviously the availability of the devices. Clear and specific clinical guidelines addressing the application of robotic training for the lower and upper extremities, which are widely adopted, do still not exist [10, 11]. One source of information is the International Industry Society in Advanced Rehabilitation Technology (IISART) is formed by manufacturers of robotic devices, equipment for virtual rehabilitation and neuromuscular electrical stimulation. The society's Internet resources provide a comprehensive overview of the actual state of various devices in this fast-advancing area. IISART also provides educational material, e.g., for the rationale of robotic training and for specific devices [12].

13.3.1 Instruction of Therapists

Therapists who operate rehabilitation robots are considered secondary users of these devices. Their role is critical for the success of robotic training. Despite the fact that the level of matureness of robotic technology is quite high, its proper use is critical for the success of the training with respect to the rehabilitation outcome. For complex devices, a sufficient period of time should be scheduled for the instruction of therapists. It is important that every therapist does as many one-to-one trainings under supervision of an expert user as needed until s/he is able to apply the device accurately and safely. If a robotic training is associated with a high risk for severe adverse events (e.g., large and powerful devices, which are mounted to the whole body), it is recommended that in a given institution special safety procedures become defined. It must be ensured that every person who trains with a robotic device has been instructed properly beforehand. The emergency procedures should be trained practically. Liability issues in case of an accident must be clarified. For the use of smaller and less strong devices (e.g., where a patient remains seated) with fewer operating modes, a basic instruction can be sufficient. However, other devices require extensive training and experience in order to respond to variations and irregularities. Usually, manufacturers provide certified user and refresher courses for their devices. In some cases, there are separate courses for trainers or users. The former is targeted to a kind of super-users who become qualified to train other users, the latter for regular users.

13.3.2 Implementation of Robots into Clinical Therapy Programs

For the practical implementation of robotic devices into a clinical therapy program, it is recommended to evaluate if multiple or only few therapists are assigned to use a device. In the case of a large number of users, a single therapist will require more time to become confident with a complex device. On the other hand, when only few staff members know how to run such a device, experience can be accumulated in a shorter period of time. Additionally, knowledge exchange is easier among a smaller group of experienced users. However, this approach is associated with the risk of losing crucial know-how when such an expert leaves his or her position. There are also mixed models where an experienced user does the setup for a given patient during an initial training session. The subsequent trainings will then be performed by a therapist with less specific knowledge, usually the therapist who conducts the non-robotic, conventional training interventions. If required, the more experienced colleague provides supervision in this phase. The advantage of such a model is that a therapist who knows a patient from the conventional therapy can more easily integrate the robotic training into the other rehabilitation interventions. For example, the transfer of skills trained in the robot towards daily life scenarios is easier. Vice versa, the therapists are familiar with the specific deficits and can adapt the robotic training accordingly. However, there are other constellations conceivable, where teams of specialized therapists are exclusively responsible for the technology-assisted training applied as an adjunct to the regular therapy program. In those cases, several devices are grouped in one room, and few therapists assist the patients in the setup of the devices and the adaptation of the training parameters. This allows for supervision of multiple patients during the technology-assisted training similar to a gym with strength training equipment for non-disabled individuals. Such parallel trainings seem more appropriate in later stages of inpatient rehabilitation or in an outpatient setting.

13.4 Principles of Robotic Training

At the current stage, robots do not introduce completely new rehabilitation strategies [13], but rather enhance and amend existing approaches.

Electromechanical devices are valuable tools for motor training because they can generate and apply greater forces for a longer period of time than human therapists and follow more precisely physiological trajectories. In addition, robots can measure muscle activations during task execution and can document training progression far more accurately than human therapists and free from subjective perception (see Chap. 10) [14, 15]. However, most robotic devices usually measure forces only in one plane or degree of freedom. A human therapist is able to perceive forces acting in multiple directions, in particular, rotational forces which are relevant for successful execution of gross motor functions.

There is a range of training robots available (for a comprehensive overview see Chaps. 26–30); however, there is no device that fits all patients during all phases, e.g., of locomotor rehabilitation. As patients recover muscle strength, the amount of support by a robotic device should adapt accordingly. Robots designed for the early stages of rehabilitation can provide full body weight and movement support. An example is the Lokomat, a motorized exoskeleton mounted on a treadmill. The Lokomat is extensively described in Chap. 29 of this book. Devices capable of providing full support are typically used in rehabilitation clinics and specialized outpatient institutions and not designed for home use due to their size, weight, required staff and costs. Training devices for later stages of rehabilitation or for end users with less severe functional limitations provide support to a lesser degree. This allows downsizing of the device's actuators, which allows mobile use and also application in a home environment. The soft exoskeleton Myosuit is one example of such a device (Fig. 13.1). There are also specific approaches where a patient can train on a robotic device at home without direct supervision of a therapist [16]. In this case, the end user and therapist are connected through the Internet, allowing the therapist to monitor the progress of the patient and adapt the training parameters and procedures [17].

13.4.1 Training Parameters

The question pertaining to the principles behind robotic training is the question regarding the principles of neurological rehabilitation (for more details see Chaps. 2 and 3) [2]. Even after many years of clinical research, to date there is still no consensus about the choice of the optimal robotic intervention, the timing of its application, and the intensity required to maximally exploit the rehabilitative capacity of an end user. Thus, training variables vary considerably between studies [10, 11]. This holds not only true for robotic-assisted training but for rehabilitative interventions in general. There is preclinical evidence that suggests a strong influence of the abovementioned factors [18]. Studies addressing sequencing and intensity as well as individual tailoring of rehabilitation interventions need to be conducted in order to get a better understanding of the exposure-outcome association. In recent years, there have been many reports on the principles and strategies on which neurological rehabilitation is based upon [19–25]. Most reports which have been published regarding this topic relate to the stroke population since this is one of the most common conditions for acquired neurological disability. Nevertheless, from an empiric point of view, most of the described principles can be transferred to other groups of patients, e.g., SCI, Multiple Sclerosis, or Parkinson's disease.

13.4.2 Principles of Motor Learning

One major and persistent principle of neurological rehabilitation is that of motor learning even though applied in different ways [19, 24, 26]. During rehabilitation, patients have to re-learn motor tasks in order to overcome disability and limitations in the completion of everyday activities. These processes are initiated by task-specific trainings which support either restoration of impaired motor functions by neuroplasticity or compensation of lost motor functions by

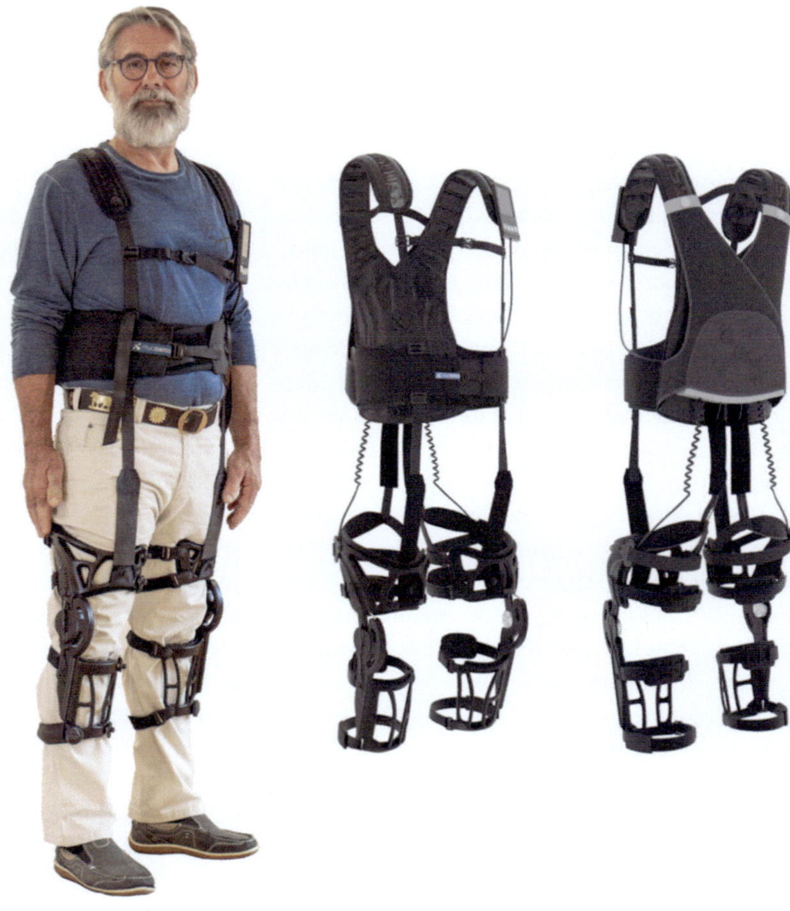

Fig. 13.1 The Myosuit is a lightweight, soft exosuit for actively supporting movements of the hip and knee initiated by an end user under everyday life conditions (Photo courtesy of MyoSwiss AG, Zuerich, Switzerland. Used with permission)

still preserved motor functions [19, 27]. Regardless the underlying mechanism, the principles of motor learning apply in both approaches [24, 28]. These principles comprise among others task specificity, goal orientation, meaningfulness, active involvement, and most importantly a high amount of practice. Rehabilitation robots allow task-specific training early after a neurological incident [29]. For example, for the training of gait function, robotic devices are applied, which support upright posture, partial weight bearing, and leg movements of patients to successfully perform the motor task of walking. At such an early stage, patients cannot stand up independently and are not or only partially able to perform leg movements on their own. Studies have shown that adequate proprioceptive afferent input is critical for training functional tasks, e.g., walking in patients with SCI [30–33]. Part I

(Chaps. 1–3) of this book provides detailed information about the neurophysiological basis of the application of rehabilitation technologies. In short and concerning locomotor training, the correct unloading and loading of the legs as well as hip extension at the end of stance phase seem to be key afferent stimuli for the appropriate facilitation of neural circuitry which are involved in the control of walking [34].

The most recent report of an incremental systematic review included 62 studies and 2,440 patients with stroke. The main conclusion was that the combination of electromechanical-assisted gait training with physiotherapy is associated with a higher likelihood of achieving independent walking than gait training without robotic assistance. This observation seems to be valid, in particular, if such a training is applied in the acute to subacute phase early after the event (i.e., within 3 months) and in patients with severe activity-related limitations. The authors calculated a number needed to treat of eight, meaning that eight patients need to be treated to prevent one patient from dependency for walking [33].

Also, robotic devices for the training of upper limb functions seem to be valuable for improvement of rehabilitation outcomes [35, 36]. These robots assist patients to follow task-specific trajectories with their arms and hands. There are upper extremity robots which are designed for the use in a very early stage when the patient still lies in bed for most of the time [37, 38]. A number of devices work in conjunction with a display, on which the patient completes meaningful tasks of daily living within a virtual environment. An advantage of such a training using virtual environments is that patients do not focus on the learning of specific movements itself but on the effects of these movements. This so-called external focus is beneficial for the learning of task automatism [39–41]. Upper extremity robots in combination with virtual environments allow for the implementation of relatively unconventional concepts of occupational therapy. Among them is the so-called error augmentation concept, where haptic (via robot-rendered forces) and graphic (via a virtual environment) distortions are used to amplify upper extremity tracking errors and thereby maximizing therapy outcome [42, 43]. Chapter 11 of this book addresses this concept in more detail.

Without the support of electromechanical devices, patients will not be able to start functional exercises (e.g., walking) at such an early stage or may get exhausted after a short while and few repetitions. Compared to the human therapist, who might get tired while providing extensive amount of support to patients with severe disabilities, robotic devices allow longer training durations and a higher number of task repetitions. Studies have shown that augmented exercise results in an improved outcome [32]. However, it seems not sufficient just to repeat a specific movement or completion of a task. Task variability improves the acquisition of that task [26]. Robotic devices which have been developed so far offer numerous ways to adapt and vary training. Systematic reviews indicate that effectiveness is independent from the design of the device [35]. The introduction of virtual environments wherein the patients take over control enables multiple ways of task variation within the same robotic setup. Further possibilities to adapt tasks are the number of degrees of freedom which are under control of the end user. The amount of support to control a given degree of freedom, e.g., hip flexion or extension, could be adapted according to the end user's abilities. Robots may not only provide assisting forces but in later stages also resisting forces. Increased resistance perpendicular to a defined trajectory helps to guide a patient through a desired movement path. The changes of movement velocity entail a different level of challenge. Walking within a robotic device allows dynamic walking at a nearly normal walking speed as opposed to walking within parallel bars or other walking aids where speed is markedly slowed down. This is, in particular, true for tethered systems which are used in combination with a body weight support system on a treadmill. Walking speed during training is considered important to warrant further improvements [44].

13.4.3   Feedback and Virtual Reality

In order to maintain physiological movement trajectories in different operating conditions and for safety reasons, robots are equipped with sensors. These sensors measure positions, velocities, and accelerations on one hand and torques and forces on the other. These signals are not only important for the internal state control of the machine, but can also be used to provide specific feedback to primary and secondary users. Feedback can be provided using various cues such as auditory, visual, or haptic. Based on the forces that patients exert on the machine, selected actions occur in the virtual environment, e.g., an avatar walks left or right or a virtual hand grasps an object (for more details see Chap. 20). In such way, robotic devices act as an interface between the real and a virtual world [45]. Although there is conflicting evidence about the additional benefit of the use of virtual reality and interactive video gaming on upper and lower limb functional recovery, the gamification of the therapy increases the compliance of patients during the training and their active participation [46–48].

In all these virtual scenarios and tasks, it is important to consider their appropriateness for the translation of the learned task to the movement demands in a real everyday life environment. The human sensorimotor system improves its skills in all tasks which are trained extensively also in those which are not appropriate. For example, exaggerated hip flexion and knee extension during swing phase required for a virtual task during robotic locomotor training is not appropriate for swing phase during walking over ground which is more characterized by a passive swing movement and an eccentric braking action of the hamstring muscles.

Another important issue of the use of virtual scenarios and tasks is that great care needs to be taken not to make patients dependent on the feedback. The goal of the robotic therapy is to achieve an independent overground ambulation and not to reach a high score in a virtual game controlled by movements! Therapists need to inform patients about the basic principle of the gaming scenarios and to implement strategies for improving patients' own body perception [49, 50].

A comparison of the sensor data of the robotic device such as joint kinematics or active forces with reference data obtained from non-disabled individuals or with previous sessions of patients supports therapists in the objective documentation of the progression within a training series. A helpful model to set the appropriate task difficulty is the theory of the optimal challenge point described by Guadagnoli and Lee [51]. The idea is that motor tasks represent different challenges for performers of different abilities. Depending on the degree of performance, the difficulty of the task should be adapted to the skill level of the performer. In neurological patients, the skill level does not only change

from day to day or within the same day, but might also occur within one training session. After all, it is the skill of the human therapist to integrate various signals and parameters of the robotic system and feedback from the patients to assess their actual skill level. Based on those findings, therapists will shape exercises and set up conditions in a way that patients are challenged and motivated without being under- or overstrained. Robotic devices represent a useful tool for therapists as well as patients to easily adapt the training parameters to a wide range without being limited by physical constraints.

13.5 Technical Aspects of Robots for Restorative Therapies

13.5.1 Complexity of Training Devices

The main goal of task-oriented neurorehabilitative training is to improve patients with neurological impairments to perform activities of daily living to the best possible degree by enhancing neuroplasticity at all level of the CNS. A key factor for the success of the training is the number of task repetitions, the generation of physiological afferent—in particular proprioceptive [52]—input and the active participation of patients [53]. With robotic devices, a high training intensity is achieved without physical burden for the therapist. The therapy intensity in particular with respect to total therapy session duration represents one of the main factors for the therapeutic success [54]. In this context, robotic devices are beneficial and valuable tools.

The complexity of electromechanical training devices is mainly determined by the residual functions of the patient group in the focus. In patients with minor to moderate impairments, passive devices may be sufficient to enable the execution of relevant tasks. This is particularly true for the upper extremity, where passive devices like the Swedish Help Arm (also known as Helparm, Swedish Sling, Deltoid Aide, or OB Help arm) or stationary and mobile versions of the ARMON orthosis (Microgravity

Products BV, Rotterdam, Netherlands) are used to reduce or eliminate the effects of gravity, thereby allowing the user to effectively use his or her weak muscles for performing everyday tasks like eating, drinking, or grooming. More complex devices such as the ArmeoSpring (Hocoma AG, Volketswil, Switzerland) have integrated sensors, in which data may be used for performing motor tasks in a virtual environment, thereby providing a patient with feedback about her or his individual performance. These devices may also help the patient retain or re-establish important proprioceptive information about the achievable workspace that the impaired limb should be able to reach as recovery progresses. Since the purely passive devices are relatively simple in their construction, they are affordable also for the patients themselves and are easy to use. The main disadvantage of these simple passive devices, which are mainly based on springs or counterweights, is that they basically provide a constant amount of weight reduction regardless of the position of the extremity. Even in positions of the arm, where low or no support is necessary, the patient is supported. Additionally, the desired movement trajectory cannot be predefined, and therefore the user may train a wrong, non-physiological movement pattern. In the worst case, the patient cannot complete the desired movement at all. To overcome this limitation, passive devices are often used during occupational therapy sessions under supervision of a therapist, who manually assist the movements to ensure that a physiological movement trajectory is achieved.

To free the therapist from this physically exhausting and mechanistic work of manually guiding the movements and to perform a therapy in a more standardized way, active robotic devices with integrated actuators have been introduced. The active components of the robots consist nowadays mainly of electric motors or pneumatically driven actuators in combination with spindles, gears, or Bowden cables.

13.5.2 End-Effector Devices Versus Exoskeletons

Within the class of active devices, there are technically more simple devices, which are mainly based on an end-effector approach, and complex exoskeleton devices, in which several degrees of freedom (DOF) of several joints are actively driven independently.

The end-effector-based systems use dedicated handles or footplates to guide the movements of the hand or foot in space [55–57] (Fig. 13.2). Their main advantage is their easy setup since no technical joints of the device have to be aligned with the anatomical joints of the human body. Furthermore, they only use one or two drives per extremity to generate a two-dimensional planar motion. However, the movements originate from the most distal segment of the extremity, and therefore—though the kinematic movement pattern of the guided body segments (e.g., the foot) looks similar to the physiological situation—the kinetics of the generated movements in the adjacent joints (e.g., knee) and the principle of weight bearing may not be perfectly physiological [58]. However, this seems to be crucial for the success of the therapy [30]. Additionally, in end-effector-based robots, only information about forces and/or position of the most distal part of the extremity is available, which may be too unspecific for control of a physiological kinetic and kinematic movement trajectory. Examples of machines based on the end-effector approach for the upper extremity are the MIT-Manus [59] or the ReJoyce (Rehabtronics, Edmonton, Canada) and for the lower extremity the Gait Trainer [57] (Reha-Stim, Berlin, Germany), the LYRA (Medica Medizintechnik, Hochdorf, Germany), the GEO (Fig. 13.2, Reha Technology AG, Olten, Switzerland), and the LEXO (Tyromotion, Graz, Austria) systems. A physiological movement of all joints of an extremity can only be achieved by the use of active drives, which support the movements of

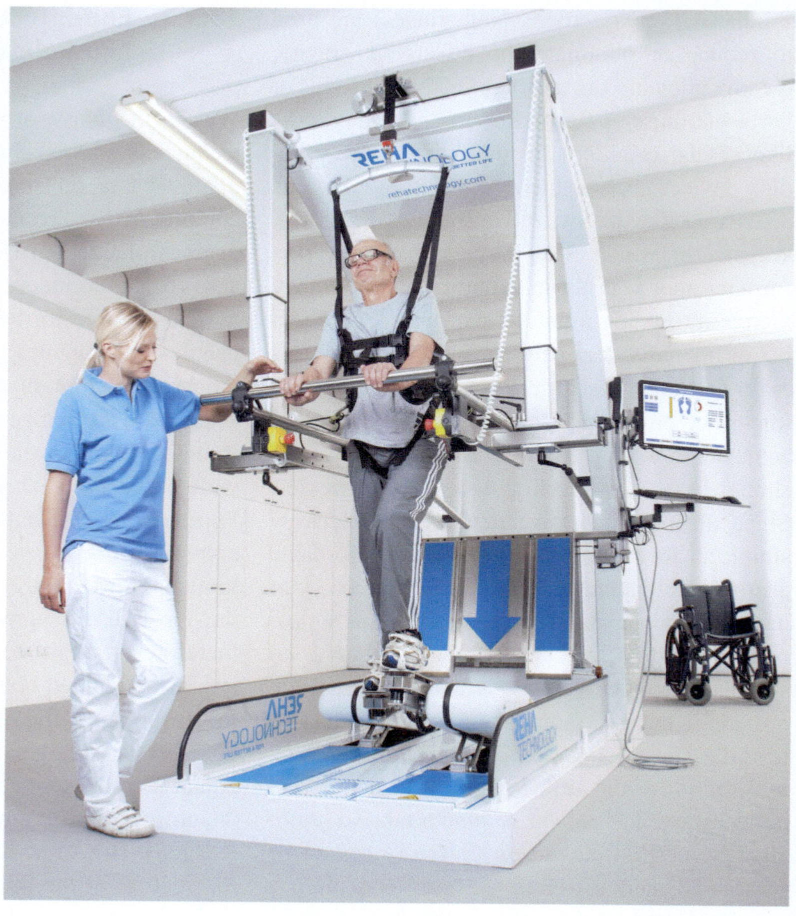

Fig. 13.2 The GEO System assists the patient during gait training using an end-effector-based approach combined with a system for partial unloading of the body weight (Photo courtesy Reha Technology AG, Olten, Switzerland. Used with permission)

every DOF of a given joint. Additionally, an individualized setup of a joint- and movement-phase-related resistance is only possible with actively driven exoskeletons. Locomotion robots for training of patients with less preserved motor functions are often constructed as actuated exoskeletons which operate either in conjunction with a system for partial body weight unloading and a moving treadmill (Fig. 13.3) [60–63]. Devices for people with a stable trunk and the ability to operate a walker or crutches might use fully mobile devices for overground walking. Since active components form the most expensive parts of a robotic device, usually a compromise between costs and functionality in terms of perfectly following a given trajectory has to be made. Therefore, most robotic locomotion training machines are generating movements only in the sagittal plane, whereas movements in the frontal or transversal plane are most often restricted to passive movements. A general challenge of the application of exoskeletons is their proper adjustment and alignment to the anatomical constraints of the different types of joints. Due to their mechanical complexity, exoskeletons are often time-consuming in their initial setup and in everyday applications. Examples for stationary actively driven exoskeletons for the lower extremity are the Lokomat and Lopes I devices [63]. The most frequently used commercially available wearable exoskeletons are the ReWalk (ReWalk Robotics, Yokneam Ilit, Israel), the Ekso (Ekso Bionics, Richmond, California, USA), the Indego (Parker Hannifin, Macedonia, Ohio, USA) [64], and the Hybrid-Assistive Limb systems (HAL, Cyberdyne, Tsukuba, Japan) [65]. The exoskeleton market is rapidly evolving and many research

Fig. 13.3 The Lokomat is an active exoskeleton which is operated in conjunction with a moving treadmill

and RUPERT devices are examples of active exoskeletons for training of upper extremity functions [66–68].

13.5.3 Body Weight Support Devices for Overground Training

It is known that the biomechanics, in particular, the muscle activation pattern is not fully identical in treadmill and overground walking [70]. While these differences might not be therapeutically relevant in patients with severe impairments, people who are basically ambulatory perceive the walking on a treadmill as non-physiological. In those individuals, locomotor training should be performed overground. However, in people with lack of coordination and/or an unstable walking pattern, overground training might be challenging because of the patients' fear of falling, early muscular fatigue, or pain due to overuse of joints. With versatile body weight support systems, safe training conditions can be achieved which result in a higher patient compliance to the training tasks. Examples are the ZeroG (Aretech, Ashburn, VA, USA) and the Vector Gait & Safety System (Bioness, Valencia, CA, USA). These systems are based on an overhead rail system

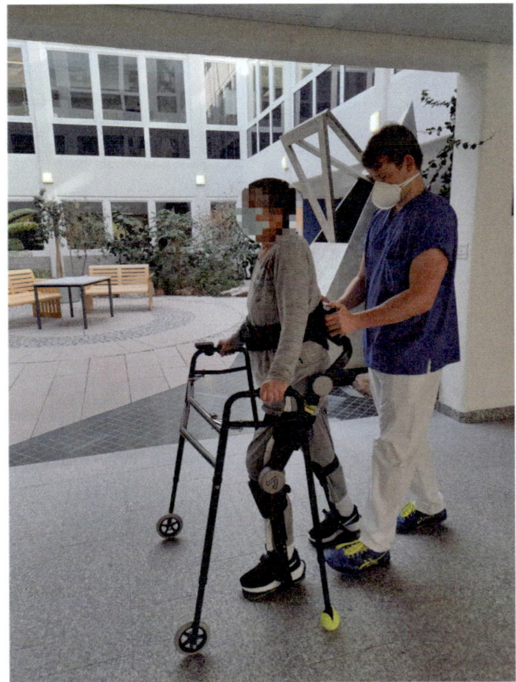

Fig. 13.4 The lightweight ABLE exoskeleton (ABLE Motion, Barcelona, Spain) with actuated knee joints for overground locomotion training [69]

prototypes are at the stage of market entry such as the ABLE exoskeleton (ABLE Motion, Barcelona, Spain) (Fig. 13.4). The ARMin (Armeo Power, Hocoma AG, Volketswil, Switzerland)

which direction an end user has to follow [71]. Body weight support systems that allow for unrestricted movements in a room are the cable-driven Float (Reha-Stim Medtec AG, Schlieren, Switzerland) and Rysen systems (Motek Medical B.V., Houten, Netherlands) [72]. However, these systems have substantial space requirements and only a single patient can train at a time. To overcome this, mobile body weight support systems that follow the route of a patient in a room have been developed, e.g., the Andago device (Hocoma, Volketswil, Switzerland). However, while these devices behave almost transparent to a user when used on straight paths, turns on the spot can be quite challenging.

13.5.4 Control Algorithms for Active Robotic Training Devices

Besides its hardware, the control algorithm represents the most important part of an active training robot. Most of the systems are operating in a position control mode, which means that the actively driven joints follow reference trajectories. This control concept has the disadvantage that even if movements in the robot look normal, the underlying muscle activation may still not be physiological. As a consequence, a position control mode should only be used in conjunction with a feedback of joint torques for informing the patient about the deviations from a physiological gait pattern. Using a position control, the patient's movements are supported even during phases, where the voluntary force of the patient would be sufficient to follow a physiological trajectory. In these cases, the robotic device does not help but hinders a patient to perform a certain movement task. Therefore, an "assist-as-needed" (AAN) control scheme seems to be the most appropriate for active devices to provide support, when and where it is needed. Hereby, patients with only moderate to minor impairments are challenged in an optimal way [73]. Although a variety of AAN-control methods have been investigated in research prototypes, in particular of locomotion robots (see Chap. 32 for more details), their translation into clinically

applicable, easy-to-use, and robust implementations is still lacking [74]. One of the reasons for this is that only devices with powerful drives providing a highly dynamic force generation are capable of realizing an AAN control. From a clinical viewpoint, in patients with increased muscle tone of, e.g., the ankle plantar flexors or clonic muscle activities, position control might be the better choice because it can prevent musculoskeletal injuries, e.g., of the ankle joint due to dragging of the foot on the treadmill.

Ideally, an active robotic device allows for small deviations from the reference trajectories, because it is known that a physiological movement does not consist of a highly reproductive movement pattern but contains some variability [75]. The nonlinear control scheme of "force fields" implemented in the T-/Pneu-WREX device [76] or an impedance control scheme of the Lokomat [77] or the Lopes II has the potential to promote the active involvement of end users in the therapy, thereby possibly leading to better outcomes. Recent studies, however, do not show a superior impact of an AAN-control scheme on the general walking ability, but rather on a joint-specific level [78].

13.5.5 Combinatory Robotic Training Approaches

Although robotic trainings contribute to the functional recovery of patients, there might be the possibility to boost the therapeutic effect by combining them with innovative neuromodulative therapies. These neuromodulation therapies interact with the CNS at different levels, namely, the brain, the spinal cord, or in addition the peripheral nervous system. The basic principle of neuromodulation therapies is to increase the activity level of neuronal networks, thereby providing a substrate for fostering neuroplasticity. An enriched proprioceptive feedback to the spinal cord can be generated by activation of (partly) paralyzed muscles by externally applied electrical current impulses. It has been shown that robotic training combined with functional electrical stimulation (FES) of the lower extremities leads to a better outcome than the robotic training alone [79–81]. However, the differences disappear when the training is stopped indicating that the effect is at least partly based on the muscle strengthening effect of the FES. Another recently introduced neuromodulation therapy is the stimulation of the thoracolumbar spinal cord either non-invasively by transcutaneous stimulation [82] or invasively by epidural electrodes [72]. It has been shown that the combination of targeted epidural spinal cord stimulation and an intensive locomotor training program in a body weight support system can substantially recover walking functions in people with chronic incomplete SCI [72].

There are several neuromodulation therapies that act on brain structures, which are vagal nerve stimulation [83] or transcranial direct current stimulation (tDCS) with partly conflicting outcome results [84–86]. Another emerging technology is non-invasive Brain–Computer Interfaces (BCIs) based on the recording of the electroencephalogram (EEG). With those systems, the user's movement intention can be measured to control the movements of a therapy robot, e.g., hand opening/closing [87] or walking/standing [88]. With this, a proprioceptive feedback is generated by the robotic movements congruent to the intention of a user, thereby expecting to result in improved motor functions (see Chap. 22).

While the listed combinatorial approaches hold promise to result in a better outcome, they impose additional burden to therapists. Invasive neuromodulative devices such as epidural spinal cord stimulators (see Chap. 18) or vagus nerve stimulators do not represent realistic solutions in the acute stage after an injury to the CNS. Non-invasive adjunct therapies, such as EEG-based BCIs, tDCS, FES of peripheral nerves and muscles, or transcutaneous spinal cord stimulation, increase the setup time of a combined robotic therapy substantially and might render these approaches as impossible in a clinical environment with typically 30 min to 1 h therapy session times.

13.5.6 User-Centered Design Process and Legal Challenges

There is a consensus that all user groups of a robotic therapy device should be included into its development cycle in the sense of an interactive user-centered design cycle. Besides, in some countries usability tests are a regulatory requirement for the certification of a device as a medical product. An overview over several methods of usability testing can be found in the book of Wiklund et al. [89]. The early involvement of users will help to refine the device specifications in the initial development stage. In later stages of the development, usability tests help to make robotic devices safer, more efficient, and easier to operate. The inclusion of the perspectives of as many user groups as possible will foster the future acceptance and implementation into routine training programs. Besides the obvious patient and therapist groups, also manufacturers, people from administration, healthcare payers, and lay caregivers can be considered as stakeholders in the decision-making process.

However, while from an exploitation viewpoint it is highly desirable and advisable to obtain structured feedback on the safety and usability of robotic devices already in the early phase of development, the legal constraints for performing tests with functional demonstrators are becoming increasingly challenging. In fact, the Medical Device Regulation (MDR), which became fully applicable in all European Union member states from 26 May 2021 renders such early feasibility experiments almost impossible at least for academic developers or small (start-up) enterprises with limited resources. On the one hand, the main aim of the MDR is to improve patient safety. To achieve this goal, it introduces new regulations for the certification of medical devices and aims to improve post-market surveillance. While the regulation is intended to guarantee medical devices' safety, performance, and clinical benefits, it contains a number of new measures including stricter requirements for the certification process and puts stern demands on notified bodies which are responsible for the final certification of a new device [90]. The MDR distinguishes between regular clinical investigations of devices (Article 62) and so-called "other clinical investigations" (Article 82), which are early feasibility trials of innovative, non-CE-marked devices. The conduction of an in-house other clinical investigation puts a substantial administrative burden and introduces high regulatory hurdles on researchers of academic institutions. By this, the new rules of the MDR result in innovative devices being withheld from end users or becoming available on the market only with a huge delay, and even prevent some devices from being developed in the first place. With this, in spite of its valid intentions with regard to patient safety, the MDR contradicts its own goals and could jeopardize patients' safety and care, because it poses major challenges for innovations [91, 92].

But even worse, the MDR imposes problems not only for highly innovative devices under development, but also for medical devices that were certified under the previous directives. These will lose their conformity under the new legislation and certificates issued under the Medical Devices Directive 93/42/EEC (MDD) will expire either on or before May 27, 2024, depending on the device type and certificate expiry date. Devices certified under the old MDD may only be put into service for one further year after this date. Despite calls from many manufacturers, there is no stock protection beyond this deadline, which will result in withdrawal of products with small sales numbers. Unfortunately, this might apply to several robotic devices used in the rehabilitation of patients with neurological impairments.

13.5.7 Individually Tailored Training

Most rehabilitation robots can only be effectively used when they are properly adjusted to the needs of the end user. If this setup is not performed carefully, the training in such a device can be harmful. In the instruction courses for secondary users, specific processes and checklists are introduced to ensure that all aspects are regarded appropriately. The following sections provide an overview over aspects which have to be considered.

13.5.8 Anthropometrics

Human beings vary to a great degree in their anthropometric data like height, weight, and body proportions such as length or widths of extremities. In order to perform the training in 95% of the population with one device, the machine has to be adjustable to a large degree and in many ways. This means that, e.g., in a locomotion exoskeleton, the length of the shank and thigh, the width of the pelvis, and the position of the trunk in all three planes must be adaptable to the anthropometrics of the individual end user. To ensure that the imposed moments of the device result in a desired joint movement the joint of the exoskeleton and the joint of the patient must be aligned properly. Any spatial mismatch may result in unphysiological strain-like skin frictions, translatory or shear movements. In some end users, the passive range of motion (ROM) may be reduced, e.g., due to spastic adaptations or pre-existing conditions. It is advisable to perform a ROM examination before starting the initial robotic training. Also, leg length asymmetries might not be evident in a patient who is seated in a wheelchair and should therefore be included in the physical examination. Differences in leg length can and should be compensated by the use of elevation insoles. For the future, the continuous increase of the body mass index of the general population of industrial countries represents a challenge for the level of adaptability of orthotic and robotic devices.

13.5.8.1 Setup Time for a Robotic Training

Since a regular therapy session is for personnel resources reasons limited to 45–60 min, every effort has to be made to keep the changeover time at a minimum. In reality, it takes one therapist about 5 min to prepare an end-effector-based robotic system to a patient and about 10–15 min in case of an exoskeleton. Much more time has to be reserved when the system is initially being set up to an individual end user.

Ideally, a machine would automatically adapt to different end users or not need any type of adjustment, since technical solutions have been provided which do not need manual adaptations. Surprisingly, up to now not a lot of effort has been made into this direction.

13.5.8.2 Task Specificity

Robotic training machines have to provide the possibility for setup of a large variety of training paradigms in order to broaden their fields of application. Most importantly, the function that is trained has to be the same as the one which should be improved. Recent developments in robotics for the lower extremities take this prerequisite into account and offer the possibility for training of, for example, stair climbing [31, 56].

Nevertheless, it has to be kept in mind that practically none of the robotic devices are able to generate a fully physiological movement since not every DOF is equipped with an actuator and therefore cannot be controlled independently.

13.6 Human–Machine Interface

The user interface is a crucial part of any robotic therapy system since it determines to a large degree whether a device is regularly integrated in the rehabilitation programs of neurological patients or not. Since the robotic systems are designed by research and development engineers, the user interfaces tend to be complicated and are not intuitive to understand for therapists. This is a general problem of the human–machine interface in almost every technical product intended to be operated by users with different technical expertise and nontechnical professional background. Therefore, the ISO 9241–210 standard, which refers to "Ergonomics of human-system interaction—Part 210: Human-centered design for interactive systems," may be a good

starting point to continuously improve the human–machine interface of a technical system [93]. The ISO 9241–210 standard defines the framework of an iterative approach to involve end users during all stages of development of a product and explicitly includes parts which are important for any type of assistive technology.

As outlined above, in rehabilitation, robotics users include primary (patients) and secondary (therapists) users. Therefore, their feedback should be addressed very carefully by developers and implemented into novel designs for increasing the acceptance.

13.6.1 Mechanical Interfaces

An extremely important component of a robotic device is the mechanical interfaces between the robot and the end user. Special attention must be paid on the design of these mechanical interfaces because skin damages up to pressure injuries represent the most frequent adverse events in robotic trainings [94]. At the points where the robot is attached to the patient, high forces are transmitted depending on the mode of operation, i.e., either a robot assists the performance of movements or applies resistance forces. Force vectors have to be in alignment to the joint axes to generate pure rotational moments and not shear forces. The fixations of the robot have to be soft and mold to fit the respective part of the body in order to prevent the occurrence of pressure lesions or abrasions of the skin. In contrast to that requirement, the interfaces must transmit the forces without loss, e.g., by deformation or loose fit. This will ensure appropriate monitoring and modeling of the forces which are exerted on the patient. This is especially important pertaining to the assessment and feedback features of robotic devices. Fixations have to be adaptable to a wide range of anthropometrics. The usage has to be unambiguous and easy. This is of importance in the case when a patient has to be removed from the device quickly, i.e., in case of an emergency.

13.6.2 Control and Feedback Interfaces

Important components of the robotic system are the control and the feedback interfaces. The control interface is used by the therapist to set and adapt the most important therapy parameters like speed, amount of support, or ROM of specific joints. The feedback interface provides the patient as well as the therapist with information about the current status and the progress of the training. The control interface has to provide a very intuitive and simple-to-use graphical user interface, which by design, e.g., an appropriate size of the control elements on the screen or on the operator panel, minimizes the risk for wrong inputs and faulty operation by the therapist during the training. A general requirement of the robotic device often demanded by therapists is a high degree of "transparency," i.e., the most relevant machine parameters and options are accessible and adjustable. However, a balance has to be found between maximal adjustability and easy handling. A possible way to meet both claims could be the common implementation of a standard and an expert mode together with the possibility for individualization of the graphical user interface.

Additional to the graphical user interface, the input device is of crucial importance, since keyboards and mice are not easy to operate while having the patient in the focus, which often results in mismatch of parameter settings. Therefore, touch-panel-based interface systems are a proper choice, in particular, if the system is operated by a patient without supervision. However, touch-panel-based input devices are prone to random and intentional entries.

13.6.3 Automated Adaptation of Training Parameters

Since most of the robotic machines are equipped with sensors, which provide feedback about the current state and performance of the patient, the implementation of an automated adaptation scheme would free the therapist from continuously adjusting the relevant parameters of the therapy. In some cases, such an adaptation scheme may allow the application of a robotic therapy without the need for continuous supervision by a therapist. However, in this condition, an adequate feedback has to be provided to the therapist and the patient, so that both are informed about the internal control state of the machine and to give them the confidence that they have the machine under control and not vice versa [95].

13.6.4 Selection of Feedback Parameters

At the current stage of knowledge, the benefit of any neurorehabilitative approach seems to be based on the enhancement of neuroplasticity in the CNS. In order to enhance neuroplasticity on a supraspinal level, the patient has to be provided with an adequate feedback of her/his current performance, in particular, in patients with sensory deficits. This is also most important for increasing motivation [96]. Comparable to the design criteria of the control interface, the number of dynamic feedback parameters presented to a patient at a time has to be carefully chosen, since a patient is only capable to influence one or two parameters simultaneously. The selection of the feedback parameters needs to be adapted to each individual according to the prominent functional deficit and the most severe impairment, respectively. In case of the lower extremities, this might be a joint angle of a dedicated gait phase like swing or stance phase. The feedback should be provided in an absolute scale so that patients are able to compare their current status to their status at the end of the last therapy session [97]. Also feedback modalities other than visual may provide a more effective way to enhance the perception of the patient [98].

13.6.5 Robotic Assessment and Therapy Documentation

Rehabilitation robots are not only equipped with actuators but also with multiple sensors. These sensor signals are not only used to safely control the operation of the robots, but can also serve as basis for providing feedback and for measuring certain biomechanical properties. While angular sensors can measure range of movement, with force or torque transducers the voluntary strength of muscle groups can be estimated.

Combined signals can assess resistance against passive movements and the phases of the movement cycle during which resistance occurs. Changes in resistance can be attributed to altered muscular tone or spasticity. Robots have also been investigated regarding their ability to assess balance during standing and walking [99]. Assessments are important to control the course of the training, to legitimate training, and to document progresses or deteriorations (see Chap. 15). Measurement results can be used to monitor the actual state of the patient and to shape the training accordingly. Some improvements may not be perceived by the patient (and the therapist) but are accessible by the high-resolution sensors [100]. Detection of any functional gains is important to foster motivation [101]. However, for any assessment, there are basic requirements which have to be met in order to be useful. Assessments have to be practical, reliable, valid, and responsive to changes. The measurement within a robotic device is easy to perform since it can be performed along with training or as a part of the training. Nevertheless, the assessment within a robotic device is restricted to that particular situation, e.g., a robot is able to measure the range of motion in the sagittal plane, but its mechanical construction does not allow measurments in the other planes. Appropriate software can record and compare the results to previous measurements or normative values. On the first sight, it seems obvious that a mechanical sensor has a higher accuracy than a human examiner. A reduction of error leads to increased reliability. Still, there are more sources for errors, e.g., the instructions given by therapists or pain may influence measurements. Few studies pertaining to this issue affirmed feasibility and reliability [102–105]. The concept of validity states that a given testing procedure aims at measuring a specified property. Regarding range of movement and voluntary muscle strength, there are no controversies as opposed to the measurement of spasticity. Even widely used tests such as the Modified Ashworth Scale (MAS) are under debate and may be improved if applied by a robot [106]. Additionally, robots allow for assessments which are hard to perform without the robot like the measurement of lower limb joint position sense with the patient in an upright position [107].

Although only few studies addressed the issue of the quality of assessments recorded by rehabilitation training robots, it can be stated that these devices provide reliable data about the performance of an end user during a dynamic motor task. Appropriate measurements whose results are transferable into an everyday setting still need to be defined.

13.6.6 Continuation of the Robotic Therapy at Home

Due to increasing economical restrictions in the healthcare system, the length of primary rehabilitation is getting shorter, e.g., in the US Model Spinal Cord Injury System the mean initial rehabilitation period of incomplete patients was 89 days in 1975, which continuously decreased to 28 days in 2005 [108]. It can be expected that this trend will continue in the future.

With the help of robotic devices, the sufficient intensity of task-oriented gait training can be sustained in the clinical setting, whereas a dramatic reduction of the quantity and quality of the training occurs after the discharge from the rehabilitation unit. This is especially true if patients return to their homes in rural areas.

Though systematic experimental investigations are missing, it may be concluded from review of the literature that long-term, moderate-intensity locomotion training over several months is more effective than the application of training protocols with high intensity for only a few weeks [108, 109]. However, up to now only a few robotic training devices exist for home-based locomotion training. A simple transfer of the existing robotic devices to the patients' homes is not possible since most of them are mainly restricted to the application in a clinical setting due to their size, weight, and price. Furthermore, most of the devices have to be operated by skilled therapist.

13.6.7 Safety of Home-Based Robotic Systems

The main challenges of therapeutic devices for application in the home environment are safety issues and the self-operation of the device by the users [110]. This is especially true for the use of locomotion training devices. Whereas in the clinical environment, the therapy is supervised by trained therapists, in the home environment a safe operation without the need for supervision has to be guaranteed.

Only a few studies exist which describe the development and application of dedicated home-based robotic training systems [17, 111, 112]. In locomotion robotics, a key method to minimize the risk of injuries is to put the user in a safe training position, like a semi-recumbent position of the body in case of the MoreGait device (Fig. 13.5).

From the available results of real home-based training, it may be concluded that a safe application without a high risk for serious adverse events is feasible and that the outcomes of the training are in the same range than of systems used in clinics [16].

Nevertheless, a certain amount of supervision is necessary to assess the current status of the patient, to individually adjust therapy parameters to the patient's progress, and to help patients in solving small hardware problems. Here, Internet-based telemonitoring methods are cheap and effective tool for transfer of sensor data and diagnostic trouble codes of the machine to a centralized location, e.g., a large rehabilitation center or an outpatient clinic [113]. Personal video conferences between therapists and end users or among end users are very valuable to keep motivation at a high level and to share experiences also in respect to troubleshooting.

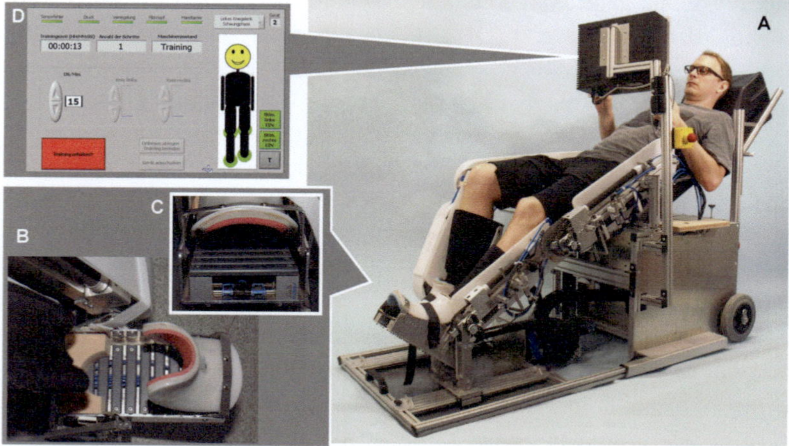

Fig. 13.5 The MoreGait is a pneumatically actuated robot for the training of walking function. The device allows the use at the patient's home: An individual during training (**A**), top (**B**), and front (**C**) view of the mediolateral bars of the stimulative shoe for mechanical stimulation of foot loading afferents, user interface, and feedback screen (**D**)

13.6.8 Conventional Gaming Consoles

For home-based therapy, especially in patients with minor motor and cognitive deficits, the use of conventional gaming consoles like Nintendo's Wii or Microsoft's Xbox, in particular, with the camera-based KINECT option represents a very attractive option for an entertaining self-training [114–116]. The latter allows for full body movement analysis and therefore a joint-specific therapy without the need for specific markers or sensors fixed to the body. The main advantage of using such type of technology is the non-limited availability and the low price. The gaming console-based training relies mainly on the feedback principles of the external focus, which is beneficial for the learning of task automatism. This form of training is motivating and provides the possibility for giving feedback about the current state of the functional impairment and the improvement over time to the user. Due to the motivating nature of games, training times can be increased which is known to contribute to a better outcome.

A recent meta-analysis of serious games for neurorehabilitation concludes that custom-made casual games that resort to the first-person perspective do not feature a visible player character, are played in single-player mode, and use non-immersive virtual reality to attain the best results in terms of positive clinical outcomes. In addition, the use of custom-made games versus commercial off-the-shelf games tends to give better clinical results, although the latter are perceived as more motivating and engaging [117]. In general, there is a need for better integration of motor learning principles into the development of future virtual reality systems [118] and for studies with comparable study protocols and sufficient sample sizes [46].

13.7 Conclusion

For the successful development, application, and integration of robotic systems into rehabilitation programs, the needs and expectations of all stakeholders including developers, clinicians, and end users have to be considered. The devices' specifications should be based on rehabilitative goals and neurobiological knowledge. The characteristics of robotic devices should comply with the demands of patients and therapists. In order to justify the costs of rehabilitation robots, they should allow for adaptation to a wide range of patients with respect to anthropometrics but also with respect to different grades of capabilities reflecting the actual state of rehabilitation. While in the beginning, supporting forces are required, in later stages a device may apply resisting forces in order to challenge patients appropriately at every skill level. The setup and operation of robots should fit in a clinical setting. Signals from sensors enable sophisticated feedback modalities and the surveillance of training progression.

Robotic devices are very useful enhancements of rehabilitation interventions, offering additional training as well as measurement options. Studies suggest that the main advantage of therapies with robotic devices compared with conventional therapies is the capability of increasing motor task repetitions without putting additional physical burden on the therapists. Robot-assistive training devices therefore represent a valuable component in therapy paradigms, which are intensive, frequent, repetitive, and comply with the principles of motor learning.

In the future, the demographic change with an increasing age of the general and patient population represents a substantial challenge for neurorehabilitation. On the one hand, less personnel will be available for hands-on trainings and the duration of inpatient rehabilitation will further decrease. On the other hand, the proportion of patients with comorbidities and cognitive restrictions in high need for individually tailored rehabilitation programs will grow. In this setting, active robotic devices and sensors may evolve to the essential component of an intense, task-specific neurorehabilitative therapy program. However, substantial efforts have to be made to improve the usability and applicability of robotic technology. First of all, there is a high need for clinically effective devices which are reduced to the technological minimum and therefore can be used in a home-based setting. Algorithms are needed for calculating outcome measures which allows for self-supervision of the machine-based training by the end user. Vendor-independent, harmonized interfaces for data-protection-compliant, cloud-based sharing of sensor, therapy, and clinical data from robotic users should be defined and implemented. By this, a large data pool can be generated which can be further analyzed with machine learning methods to identify the most successful therapy parameters on a patient-subgroup-specific level. A possible starting point would be to add sensor and robot-generated data to registries with high-quality clinical data such as the European Multicenter Study about Spinal Cord Injury [28]. All this shows that the full potential of robotic therapies is far from being fully exploited.

References

1. Maier M, Ballester BR, Verschure P. Principles of neurorehabilitation after stroke based on motor learning and brain plasticity mechanisms. Front Syst Neurosci. 2019;13:74.
2. Gassert R, Dietz V. Rehabilitation robots for the treatment of sensorimotor deficits: a neurophysiological perspective. J Neuroeng Rehabil. 2018;15 (1):46.
3. Mankins JC. Technology readiness assessments: a retrospective. Acta Astronaut. 2009;65(9–10):1216–23.
4. European Commission. Horizon 2020–Work Programme 2014–2015, Annex G: Technology Readiness Levels (TRL) (2014)
5. Kelly-Hayes M, Robertson JT, Broderick JP, Duncan PW, Hershey LA, Roth EJ, et al. The American heart association stroke outcome classification. Stroke. 1998;29(6):1274–80.
6. Wecht JM, Krassioukov AV, Alexander M, Handrakis JP, McKenna SL, Kennelly M, et al. International standards to document autonomic function following SCI (ISAFSCI): Second Edition. Top Spinal Cord Inj Rehabil. 2021;27(2):23–49.
7. Geigle PR, Frye SK, Perreault J, Scott WH, Gorman PH. Atypical autonomic dysreflexia during robotic-assisted body weight supported treadmill training in an individual with motor incomplete spinal cord injury. J Spinal Cord Med. 2013;36 (2):153–6.
8. van Herpen FHM, van Dijsseldonk RB, Rijken H, Keijsers NLW, Louwerens JWK, van Nes IJW. Case report: description of two fractures during the use of a powered exoskeleton. Spinal Cord Ser Cases. 2019;5:99.
9. Aprile I, Guardati G, Cipollini V, Papadopoulou D, Mastrorosa A, Castelli L, et al. Robotic rehabilitation: an opportunity to improve cognitive functions in subjects with stroke an explorative study. Front Neurol. 2020;11: 588285.
10. Calabro RS, Sorrentino G, Cassio A, Mazzoli D, Andrenelli E, Bizzarini E, et al. Robotic-assisted gait rehabilitation following stroke: a systematic review of current guidelines and practical clinical recommendations. Eur J Phys Rehabil Med. 2021;57(3):460–71.
11. Morone G, Palomba A, Martino Cinnera A, Agostini M, Aprile I, Arienti C, et al. Systematic review of guidelines to identify recommendations for upper limb robotic rehabilitation after stroke. Eur J Phys Rehabil Med. 2021;57(2):238–45.
12. IISART. International Industry Society in Advanced Rehabilitation Technology https://iisart.org.
13. Reinkensmeyer DJ, Maier MA, Guigon E, Chan V, Akoner O, Wolbrecht ET, et al. Do robotic and non-robotic arm movement training drive motor recovery after stroke by a common neural mechanism? Experimental evidence and a computational model. Conf Proc IEEE Eng Med Biol Soc. 2009;2009:2439–41.
14. Nordin N, Xie SQ, Wunsche B. Assessment of movement quality in robot- assisted upper limb rehabilitation after stroke: a review. J Neuroeng Rehabil. 2014;11:137.
15. Schwarz A, Kanzler CM, Lambercy O, Luft AR, Veerbeek JM. Systematic review on kinematic assessments of upper limb movements after stroke. Stroke. 2019;50(3):718–27.
16. Rupp R, Schliessmann D, Plewa H, Schuld C, Gerner HJ, Weidner N, et al. Safety and efficacy of at-home robotic locomotion therapy in individuals with chronic incomplete spinal cord injury: a prospective, pre-post intervention, proof-of-concept study. PLoS ONE. 2015;10(3): e0119167.
17. Sanchez RJ, Liu J, Rao S, Shah P, Smith R, Rahman T, et al. Automating arm movement training following severe stroke: functional exercises with quantitative feedback in a gravity-reduced environment. IEEE Trans Neural Syst Rehabil Eng. 2006;14(3):378–89.
18. Wahl AS, Schwab ME. Finding an optimal rehabilitation paradigm after stroke: enhancing fiber growth and training of the brain at the right moment. Front Hum Neurosci. 2014;8:381.
19. Dobkin BH. Strategies for stroke rehabilitation. Lancet Neurol. 2004;3(9):528–36.
20. Floel A, Cohen LG. Translational studies in neurorehabilitation: from bench to bedside. Cogn Behav Neurol. 2006;19(1):1–10.
21. Graham JV, Eustace C, Brock K, Swain E, Irwin-Carruthers S. The Bobath concept in contemporary

clinical practice. Top Stroke Rehabil. 2009;16 (1):57–68.

22. Hubbard IJ, Parsons MW, Neilson C, Carey LM. Task-specific training: evidence for and translation to clinical practice. Occup Ther Int. 2009;16(3–4):175–89.

23. Kalra L. Stroke rehabilitation 2009: old chestnuts and new insights. Stroke. 2010;41(2):e88-90.

24. Krakauer JW. Motor learning: its relevance to stroke recovery and neurorehabilitation. Curr Opin Neurol. 2006;19(1):84–90.

25. Oujamaa L, Relave I, Froger J, Mottet D, Pelissier JY. Rehabilitation of arm function after stroke literature review. Ann Phys Rehabil Med. 2009;52(3):269–93.

26. Hanlon RE. Motor learning following unilateral stroke. Arch Phys Med Rehabil. 1996;77(8):811–5.

27. Levin MF, Kleim JA, Wolf SL. What do motor "recovery" and "compensation" mean in patients following stroke? Neurorehabil Neural Repair. 2009;23(4):313–9.

28. Curt A, Van Hedel HJ, Klaus D, Dietz V. Recovery from a spinal cord injury: significance of compensation, neural plasticity, and repair. J Neurotrauma. 2008;25(6):677–85.

29. Molteni F, Gasperini G, Cannaviello G, Guanziroli E. Exoskeleton and end-effector robots for upper and lower limbs rehabilitation: narrative review. PM R. 2018;10(9 Suppl 2):S174–88.

30. Dietz V, Muller R, Colombo G. Locomotor activity in spinal man: significance of afferent input from joint and load receptors. Brain. 2002;125(Pt 12):2626–34.

31. Harkema SJ, Hurley SL, Patel UK, Requejo PS, Dobkin BH, Edgerton VR. Human lumbosacral spinal cord interprets loading during stepping. J Neurophysiol. 1997;77(2):797–811.

32. Lünenburger L, Bolliger M, Czell D, Müller R, Dietz V. Modulation of locomotor activity in complete spinal cord injury. Exp Brain Res. 2006;174(4):638–46.

33. Mehrholz J, Thomas S, Kugler J, Pohl M, Elsner B. Electromechanical-assisted training for walking after stroke. Cochrane Database Syst Rev. 2020;10:CD006185.

34. Spiess MR, Steenbrink F, Esquenazi A. Getting the best out of advanced rehabilitation technology for the lower limbs: minding motor learning principles. PMR. 2018;10(9 Suppl 2):S165–73.

35. Mehrholz J, Pollock A, Pohl M, Kugler J, Elsner B. Systematic review with network meta-analysis of randomized controlled trials of robotic-assisted arm training for improving activities of daily living and upper limb function after stroke. J Neuroeng Rehabil. 2020;17(1):83.

36. Zhang C, Li-Tsang CW, Au RK. Robotic approaches for the rehabilitation of upper limb recovery after stroke: a systematic review and meta-analysis. Int J Rehabil Res. 2017;40(1):19–28.

37. Stefano M, Patrizia P, Mario A, Ferlini G, Rizzello R, Rosati G. Robotic upper limb rehabilitation after acute stroke by NeReBot: evaluation of treatment costs. Biomed Res Int. 2014;2014: 265634.

38. Dehem S, Gilliaux M, Stoquart G, Detrembleur C, Jacquemin G, Palumbo S, et al. Effectiveness of upper-limb robotic-assisted therapy in the early rehabilitation phase after stroke: a single-blind, randomised, controlled trial. Ann Phys Rehabil Med. 2019;62(5):313–20.

39. Subramanian SK, Massie CL, Malcolm MP, Levin MF. Does provision of extrinsic feedback result in improved motor learning in the upper limb poststroke? a systematic review of the evidence. Neurorehabil Neural Repair. 2010;24(2):113–24.

40. van Vliet PM, Wulf G. Extrinsic feedback for motor learning after stroke: what is the evidence? Disabil Rehabil. 2006;28(13–14):831–40.

41. Wulf G, Shea C, Lewthwaite R. Motor skill learning and performance: a review of influential factors. Med Educ. 2010;44(1):75–84.

42. Abdollahi F, Case Lazarro ED, Listenberger M, Kenyon RV, Kovic M, Bogey RA, et al. Error augmentation enhancing arm recovery in individuals with chronic stroke: a randomized crossover design. Neurorehabil Neural Repair. 2014;28(2):120–8.

43. Abdollahi F, Corrigan M, Lazzaro EDC, Kenyon RV, Patton JL. Error-augmented bimanual therapy for stroke survivors. NeuroRehabilitation. 2018;43(1):51–61.

44. Sullivan KJ, Knowlton BJ, Dobkin BH. Step training with body weight support: effect of treadmill speed and practice paradigms on poststroke locomotor recovery. Arch Phys Med Rehabil. 2002;83(5):683–91.

45. Janeh O, Steinicke F. A review of the potential of virtual walking techniques for gait rehabilitation. Front Hum Neurosci. 2021;15: 717291.

46. Laver KE, Lange B, George S, Deutsch JE, Saposnik G, Crotty M. Virtual reality for stroke rehabilitation. Cochrane Database Syst Rev. 2017;11: CD008349.

47. Lohse KR, Hilderman CG, Cheung KL, Tatla S, Van der Loos HF. Virtual reality therapy for adults post-stroke: a systematic review and meta-analysis exploring virtual environments and commercial games in therapy. PLoS ONE. 2014;9(3): e93318.

48. de Rooij IJM, van de Port IGL, Punt M, Abbink-van Moorsel PJM, Kortsmit M, van Eijk RPA, et al. Effect of virtual reality gait training on participation in survivors of subacute stroke: a randomized controlled trial. Phys Ther. 2021;101(5).

49. Qaiser T, Eginyan G, Chan F, Lam T. The sensorimotor effects of a lower limb proprioception training intervention in individuals with a spinal cord injury. J Neurophysiol. 2019;122(6):2364–71.

50. Takeoka A. Proprioception: bottom-up directive for motor recovery after spinal cord injury. Neurosci Res. 2020;154:1–8.

51. Guadagnoli MA, Lee TD. Challenge point: a framework for conceptualizing the effects of various

practice conditions in motor learning. J Mot Behav. 2004;36(2):212–24.

52. Takeoka A, Vollenweider I, Courtine G, Arber S. Muscle spindle feedback directs locomotor recovery and circuit reorganization after spinal cord injury. Cell. 2014;159(7):1626–39.

53. Prange GB, Jannink MJ, Groothuis-Oudshoorn CG, Hermens HJ, Ijzerman MJ. Systematic review of the effect of robot-aided therapy on recovery of the hemiparetic arm after stroke. J Rehabil Res Dev. 2006;43(2):171–84.

54. Wirz M, Mach O, Maier D, Benito-Penalva J, Taylor J, Esclarin A, et al. Effectiveness of automated locomotor training in patients with acute incomplete spinal cord injury: a randomized, controlled multicenter trial. J Neurotrauma. 2017;34 (10):1891–6.

55. Freivogel S, Mehrholz J, Husak-Sotomayor T, Schmalohr D. Gait training with the newly developed 'LokoHelp'-system is feasible for non-ambulatory patients after stroke, spinal cord and brain injury a feasibility study. Brain Inj. 2008;22 (7–8):625–32.

56. Hesse S, Waldner A, Tomelleri C. Innovative gait robot for the repetitive practice of floor walking and stair climbing up and down in stroke patients. J Neuroeng Rehabil. 2010;7:30.

57. Hesse S, Werner C, Uhlenbrock D, von Frankenberg S, Bardeleben A, Brandl-Hesse B. An electromechanical gait trainer for restoration of gait in hemiparetic stroke patients: preliminary results. Neurorehabil Neural Repair. 2001;15(1):39–50.

58. Hesse S, Sarkodie-Gyan T, Uhlenbrock D. Development of an advanced mechanised gait trainer, controlling movement of the centre of mass, for restoring gait in non-ambulant subjects. Biomed Tech (Berl). 1999;44(7–8):194–201.

59. Hogan NKH, Sharon A, Charnnarong J. Inventor; Massachusetts Institute of Technology, assignee. Interactive robotic therapist. USA1995 1995.

60. Banala SK, Kim SH, Agrawal SK, Scholz JP. Robot assisted gait training with active leg exoskeleton (ALEX). IEEE Trans Neural Syst Rehabil Eng. 2009;17(1):2–8.

61. Colombo G, Joerg M, Schreier R, Dietz V. Treadmill training of paraplegic patients using a robotic orthosis. J Rehabil Res Dev. 2000;37(6):693–700.

62. Mantone J. Getting a leg up? Rehab patients get an assist from devices such as HealthSouth's AutoAmbulator, but the robots' clinical benefits are still in doubt. Mod Healthc. 2006;36(7):58–60.

63. Veneman JF, Kruidhof R, Hekman EE, Ekkelenkamp R, Van Asseldonk EH, van der Kooij H. Design and evaluation of the LOPES exoskeleton robot for interactive gait rehabilitation. IEEE Trans Neural Syst Rehabil Eng. 2007;15(3):379–86.

64. Tan K, Koyama S, Sakurai H, Teranishi T, Kanada Y, Tanabe S. Wearable robotic exoskeleton for gait reconstruction in patients with spinal cord injury: a literature review. J Orthop Translat. 2021;28:55–64.

65. Brinkemper A, Aach M, Grasmucke D, Jettkant B, Rosteius T, Dudda M, et al. Improved physiological gait in acute and chronic SCI patients after training with wearable cyborg hybrid assistive limb. Front Neurorobot. 2021;15: 723206.

66. Nef T, Mihelj M, Riener R. ARMin: a robot for patient-cooperative arm therapy. Med Biol Eng Comput. 2007;45(9):887–900.

67. Sugar TG, He J, Koeneman EJ, Koeneman JB, Herman R, Huang H, et al. Design and control of RUPERT: a device for robotic upper extremity repetitive therapy. IEEE Trans Neural Syst Rehabil Eng. 2007;15(3):336–46.

68. Klamroth-Marganska V, Blanco J, Campen K, Curt A, Dietz V, Ettlin T, et al. Three-dimensional, task-specific robot therapy of the arm after stroke: a multicentre, parallel-group randomised trial. Lancet Neurol. 2014;13(2):159–66.

69. Font-Llagunes JM, Lugris U, Clos D, Alonso FJ, Cuadrado J. Design, control, and pilot study of a lightweight and modular robotic exoskeleton for walking assistance after spinal cord injury. J Mech Robot. 2020;12(3).

70. Lee SJ, Hidler J. Biomechanics of overground vs. treadmill walking in healthy individuals. J Appl Physiol (1985). 2008;104(3):747–55.

71. Fenuta AM, Hicks AL. Metabolic demand and muscle activation during different forms of body-weight supported locomotion in men with incomplete SCI. Biomed Res Int. 2014;2014: 632765.

72. Wagner FB, Mignardot JB, Le Goff-Mignardot CG, Demesmaeker R, Komi S, Capogrosso M, et al. Targeted neurotechnology restores walking in humans with spinal cord injury. Nature. 2018;563 (7729):65–71.

73. Edgerton VR, Courtine G, Gerasimenko YP, Lavrov I, Ichiyama RM, Fong AJ, et al. Training locomotor networks. Brain Res Rev. 2008;57 (1):241–54.

74. Cao J, Xie SQ, Das R, Zhu GL. Control strategies for effective robot assisted gait rehabilitation: the state of art and future prospects. Med Eng Phys. 2014;36(12):1555–66.

75. Kruger M, Eggert T, Straube A. Joint angle variability in the time course of reaching movements. Clin Neurophysiol. 2011;122(4):759–66.

76. Wolbrecht ET, Chan V, Reinkensmeyer DJ, Bobrow JE. Optimizing compliant, model-based robotic assistance to promote neurorehabilitation. IEEE Trans Neural Syst Rehabil Eng. 2008;16 (3):286–97.

77. Duschau-Wicke A, Caprez A, Riener R. Patient-cooperative control increases active participation of individuals with SCI during robot-aided gait training. J Neuroeng Rehabil. 2010;7:43.

78. Alingh JF, Fleerkotte BM, Groen BE, Rietman JS, Weerdesteyn V, van Asseldonk EHF, et al. Effect of

assist-as-needed robotic gait training on the gait pattern post stroke: a randomized controlled trial. J Neuroeng Rehabil. 2021;18(1):26.

79. Bae YH, Ko YJ, Chang WH, Lee JH, Lee KB, Park YJ, et al. Effects of robot-assisted gait training combined with functional electrical stimulation on recovery of locomotor mobility in chronic stroke patients: a randomized controlled trial. J Phys Ther Sci. 2014;26(12):1949–53.

80. Daly JJ, Ruff RL. Construction of efficacious gait and upper limb functional interventions based on brain plasticity evidence and model-based measures for stroke patients. Sci World J. 2007;7:2031–45.

81. Hesse S, Werner C, Bardeleben A. Electromechanical gait training with functional electrical stimulation: case studies in spinal cord injury. Spinal Cord. 2004;42(6):346–52.

82. Minassian K, Hofstoetter US. Spinal cord stimulation and augmentative control strategies for leg movement after spinal paralysis in humans. CNS Neurosci Ther. 2016;22(4):262–70.

83. Dawson J, Liu CY, Francisco GE, Cramer SC, Wolf SL, Dixit A, et al. Vagus nerve stimulation paired with rehabilitation for upper limb motor function after ischaemic stroke (VNS-REHAB): a randomised, blinded, pivotal, device trial. Lancet. 2021;397(10284):1545–53.

84. Comino-Suarez N, Moreno JC, Gomez-Soriano J, Megia-Garcia A, Serrano-Munoz D, Taylor J, et al. Transcranial direct current stimulation combined with robotic therapy for upper and lower limb function after stroke: a systematic review and meta-analysis of randomized control trials. J Neuroeng Rehabil. 2021;18(1):148.

85. Leon D, Cortes M, Elder J, Kumru H, Laxe S, Edwards DJ, et al. tDCS does not enhance the effects of robot-assisted gait training in patients with subacute stroke. Restor Neurol Neurosci. 2017;35(4):377–84.

86. Simis M, Fregni F, Battistella LR. Transcranial direct current stimulation combined with robotic training in incomplete spinal cord injury: a randomized, sham-controlled clinical trial. Spinal Cord Ser Cases. 2021;7(1):87.

87. Ramos-Murguialday A, Curado MR, Broetz D, Yilmaz O, Brasil FL, Liberati G, et al. Brain-machine interface in chronic stroke: randomized trial long-term follow-up. Neurorehabil Neural Repair. 2019;33(3):188–98.

88. Donati AR, Shokur S, Morya E, Campos DS, Moioli RC, Gitti CM, et al. Long-term training with a brain-machine interface-based gait protocol induces partial neurological recovery in paraplegic patients. Sci Rep. 2016;6:30383.

89. Wiklund MEK, J.; Strochlic, A.Y. Usability Testing of Medical Devices. Massachusetts, USA Taylor & Francis Inc.; 2016.

90. Commission E. Regulation (EU) 2017/745 2017. https://www.medical-device-regulation.eu/.

91. Letourneur D, Joyce K, Chauvierre C, Bayon Y, Pandit A. Enabling MedTech translation in academia: redefining value proposition with updated regulations. Adv Healthc Mater. 2021;10(1):e2001237.

92. Maresova P, Rezny L, Peter L, Hajek L, Lefley F. Do regulatory changes seriously affect the medical devices industry? evidence from the Czech Republic. Front Public Health. 2021;9: 666453.

93. International Organisation for Standardization. ISO 9241-210: 2019 (en) Ergonomics of human-system interaction—Part 210: human-centred design for interactive systems. 2019. https://www.iso.org/obp/ui/#iso:std:iso:9241:-210:ed-2:v1:en

94. Benson I, Hart K, Tussler D, van Middendorp JJ. Lower-limb exoskeletons for individuals with chronic spinal cord injury: findings from a feasibility study. Clin Rehabil. 2016;30(1):73–84.

95. Maggioni S, Reinert N, Lunenburger L, Melendez-Calderon A. An adaptive and hybrid end-point/joint impedance controller for lower limb exoskeletons. Front Robot AI. 2018;5:104.

96. Quattrocchi G, Greenwood R, Rothwell JC, Galea JM, Bestmann S. Reward and punishment enhance motor adaptation in stroke. J Neurol Neurosurg Psychiatry. 2017;88(9):730–6.

97. Schliessmann D, Schuld C, Schneiders M, Derlien S, Glockner M, Gladow T, et al. Feasibility of visual instrumented movement feedback therapy in individuals with motor incomplete spinal cord injury walking on a treadmill. Front Hum Neurosci. 2014;8:416.

98. Koritnik T, Koenig A, Bajd T, Riener R, Munih M. Comparison of visual and haptic feedback during training of lower extremities. Gait Posture. 2010;32(4):540–6.

99. Shirota C, van Asseldonk E, Matjacic Z, Vallery H, Barralon P, Maggioni S, et al. Robot-supported assessment of balance in standing and walking. J Neuroeng Rehabil. 2017;14(1):80.

100. Maggioni S, Melendez-Calderon A, van Asseldonk E, Klamroth-Marganska V, Lunenburger L, Riener R, et al. Robot-aided assessment of lower extremity functions: a review. J Neuroeng Rehabil. 2016;13(1):72.

101. Banz R, Riener R, Lunenburger L, Bolliger M. Assessment of walking performance in robot-assisted gait training: a novel approach based on empirical data. Conf Proc IEEE Eng Med Biol Soc. 2008:1977–80.

102. Bolliger M, Banz R, Dietz V, Lunenburger L. Standardized voluntary force measurement in a lower extremity rehabilitation robot. J Neuroeng Rehabil. 2008;5:23.

103. Bosecker C, Dipietro L, Volpe B, Krebs HI. Kinematic robot-based evaluation scales and clinical counterparts to measure upper limb motor performance in patients with chronic stroke. Neurorehabil Neural Repair. 2010;24(1):62–9.

104. Keller U, Scholch S, Albisser U, Rudhe C, Curt A, Riener R, et al. Robot-assisted arm assessments in spinal cord injured patients: a consideration of concept study. PLoS ONE. 2015;10(5): e0126948.

105. Schmartz AC, Meyer-Heim AD, Muller R, Bolliger M. Measurement of muscle stiffness using robotic assisted gait orthosis in children with cerebral palsy: a proof of concept. Disabil Rehabil Assist Technol. 2010.

106. Meseguer-Henarejos AB, Sanchez-Meca J, Lopez-Pina JA, Carles-Hernandez R. Inter- and intra-rater reliability of the Modified Ashworth Scale: a systematic review and meta-analysis. Eur J Phys Rehabil Med. 2018;54(4):576–90.

107. Domingo A, Lam T. Reliability and validity of using the Lokomat to assess lower limb joint position sense in people with incomplete spinal cord injury. J Neuroeng Rehabil. 2014;11(1):167.

108. NSCISC. The 2018 Annual Statistical Report for the Model Spinal Cord Injury Care System: National SCI Statistical Center; 2018 www.uab.edu/NSCISC.

109. Wirz M, Zemon DH, Rupp R, Scheel A, Colombo G, Dietz V, et al. Effectiveness of automated locomotor training in patients with chronic incomplete spinal cord injury: a multicenter trial. Arch Phys Med Rehabil. 2005;86(4):672–80.

110. Fritz H, Patzer D, Galen SS. Robotic exoskeletons for reengaging in everyday activities: promises, pitfalls, and opportunities. Disabil Rehabil. 2019;41(5):560–3.

111. Hicks AL, Adams MM, Martin Ginis K, Giangregorio L, Latimer A, Phillips SM, et al. Long-term body-weight-supported treadmill training and subsequent follow-up in persons with chronic SCI: effects on functional walking ability and measures of subjective well-being. Spinal Cord. 2005;43(5):291–8.

112. Rupp R, Plewa H, Schuld C, Gerner HJ, Hofer EP, Knestel M. MotionTherapy@Home - First results of a clinical study with a novel robotic device for automated locomotion therapy at home. Biomed Tech (Berl). 2011;56(1):11–21.

113. Dobkin BH. A rehabilitation-internet-of-things in the home to augment motor skills and exercise training. Neurorehabil Neural Repair. 2017;31(3):217–27.

114. Anderson KR, Woodbury ML, Phillips K, Gauthier LV. Virtual reality video games to promote movement recovery in stroke rehabilitation: a guide for clinicians. Arch Phys Med Rehabil. 2015;96(5):973–6.

115. Webster D, Celik O. Systematic review of Kinect applications in elderly care and stroke rehabilitation. J Neuroeng Rehabil. 2014;11:108.

116. Perrochon A, Borel B, Istrate D, Compagnat M, Daviet JC. Exercise-based games interventions at home in individuals with a neurological disease: a systematic review and meta-analysis. Ann Phys Rehabil Med. 2019;62(5):366–78.

117. Vieira C, Ferreira da Silva Pais-Vieira C, Novais J, Perrotta A. Serious Game Design and Clinical Improvement in Physical Rehabilitation: Systematic Review. JMIR Serious Games. 2021;9(3):e20066.

118. Demers M, Fung K, Subramanian SK, Lemay M, Robert MT. Integration of motor learning principles into virtual reality interventions for individuals with cerebral palsy: systematic review. JMIR Serious Games. 2021;9(2): e23822.

Clinical Application of Rehabilitation Therapy Technologies to Children with CNS Damage

14

Hubertus J. A. van Hedel, Tabea Aurich Schuler, and Jan Lieber

Abstract

The application of rehabilitation therapy technologies in children with neurological impairments appears promising. Characteristics of these robotic-supported and computer-assisted therapies are in line with principles of motor learning. They include high numbers of repetitions, prolonged training durations, and online feedback about the patient's active participation. When applied by experienced therapists, the technologies provide a safe and simultaneously fun and motivating therapy to young patients. Furthermore, the evidence of the effectiveness for lower and upper limb applications is rising. We wrote this chapter to account for the lack of knowledge concerning the implementation and evidence-based application of such technologies. We discuss in this chapter the complementary role of rehabilitation therapy technologies and present general considerations for their implementation and application. Furthermore, we propose a conceptual ordering of categories of lower and upper limb rehabilitation therapy technologies, based on whether the systems provide weight support, physical assistance, or 'only' augmented feedback. Finally, we present some of these technologies representing the categories in more detail and summarize the research and evidence of these technologies.

Keywords

Adolescents · Robot-assisted therapy · Computer-assisted systems · ICF-CY · Virtual reality · Pediatric neurorehabilitation · Habilitation · In- and exclusion criteria · Training intensity · Clinical evidence

14.1 Introduction

The treatment of lower and upper extremity functions and activities presents particularly in children with neurological impairments a high challenge as the developmental status of the child interferes with the complexity of neurological, functional, cognitive, and motivational aspects. Symptoms that affect overall functioning are, among others, spasticity, muscle weakness, impaired balance, contractures as well as joint and bone deformities. In young patients with congenital or acquired brain lesions, co-morbidities such as epilepsy, cognitive functioning, learning difficulties, behavioral challenges, or sensory impairments can be of similar complexity to treat as the motor disabilities, and this will affect both

H. J. A. van Hedel (✉) · T. A. Schuler · J. Lieber
Swiss Children's Rehab, University Children's Hospital Zurich, Mühlebergstrasse 104, CH-8910 Affoltern am Albis, Switzerland
e-mail: hubertus.vanhedel@kispi.uzh.ch

Children's Research Center, University Children's Hospital Zurich, University of Zurich, Zurich, Switzerland

© The Author(s), under exclusive license to Springer Nature Switzerland AG 2022
D. J. Reinkensmeyer et al. (eds.), *Neurorehabilitation Technology*,
https://doi.org/10.1007/978-3-031-08995-4_14

therapy planning and execution (see for example [1]). Due to the ongoing development and growth, it remains important to monitor these children regularly, even in case of non-progressive neurological lesions. For example, gait can worsen during growth in children with cerebral palsy (CP), especially during the pubertal growth spurt, due to an increasingly disproportional relationship between leg muscle strength and body weight. Therefore, regular check-ups, including standardized assessments, are necessary to detect a deterioration of impairments early [1].

While early neurodevelopmental treatment concepts promoted passive inhibition techniques, these principles have been adapted over time, and the focus has switched towards self-activity of the child [2, 3]. Besides the conventional physical and occupational therapy, multidisciplinary rehabilitation programs nowadays also include approaches such as sports therapy, strength training [4], or functional task-orientated training [5], including activities of daily living. The intensity of therapy, repetition, and a goal-oriented and task-specific training program are considered essential to achieve successful functional outcome [5, 6].

In line with this change in therapeutic focus, rehabilitation goals become increasingly defined at the domain of activities and participation (according to the International Classification of Functioning, Disability and Health, Child & Youth version, ICF-CY) [7]. Also in children with neurological impairments, the goal of rehabilitation is to improve the independence in daily life activities. Improved independence will reduce the burden of care for the whole family and positively affect the quality of life for the young patient and the family.

14.1.1 The Complementing Role of Rehabilitation Therapy Technologies

In our opinion, it is important that rehabilitation therapy technologies should be complementing —and not replacing—conventional therapies. First, not each child can be trained with technology, e.g., due to very young age, anthropometrics not fitting to the technology, lack of cognitive understanding, or contractures limiting the applicability. Second, not each therapeutic goal can be influenced with rehabilitation therapy technology. Third, in our view, most currently available technologies improve motor function (ICF body function domain), while conventional therapies are needed to transfer the improved function to relevant activities of daily living (ICF activities performance domain) to achieve the overall rehabilitation goals. As the technology should contribute to the overall rehabilitation goal, carefully selecting the most appropriate system is needed.

However, adding rehabilitation therapy technologies to the conventional multidisciplinary therapy program can have several advantages. First, the technologies could contribute to the motor learning concept in rehabilitation. They can deliver high-dose (i.e., number of practice movements) and high-intensive (i.e., number of movements per time unit) training interventions, have accurate movement controllability, and provide immediate and precise feedback. These points are all considered important for successful neurorehabilitation [8]. Second, motivation is essential, especially in children and adolescents. Nowadays, most rehabilitation technologies are equipped with games to provide feedback and enhance motivation. Child-friendly exergames with strong, immersive qualities could distract children from monotonous, repetitive exercises. Furthermore, these games can be adapted to the developmental and functional status of the child, which can lead to more challenging training situations, thereby increasing compliance over a prolonged time (see for example [9–13]), which should lead to an improved functional outcome and active participation.

It is crucial that rehabilitation programs are tailored to the special needs of an individual child. Pediatric neurorehabilitation is personnel-intensive and costly. Limited resources can hinder the optimal dosage of rehabilitation measures and, consequently, the achievement of the desired functional outcome. Rehabilitation programs combining conventional therapies with rehabilitation therapy technologies might partly solve this situation by achieving a higher therapeutic dose in a similar amount of time.

14.1.2 Focus of This Chapter and Definitions

In this chapter, we present practice-based clinical recommendations and the current scientific body of evidence on the applicability of rehabilitation *therapy* technologies in children and adolescents with neurological impairments. We narrow the contents of this chapter down to the application of rehabilitation *therapy* devices that therapists use to improve impairments in function and limitations in activities. We do not include a discussion on rehabilitation *assistive* technologies that compensate for the loss of function during daily life activities.

Furthermore, we use terms such as *exoskeleton* and *end-effector* devices. Exoskeletons are devices that connect with the human body in a 'wearable' way and can control the movement of all joints in the training process. End-effector devices are connected to patients at one distal point, and their joints do not match the human joints. Furthermore, we discuss *exergames* (i.e., games that are controlled by human movement), which are a category of serious games (i.e., games that are not just intended for fun), or interactive computer-play (which also include games steered by a mouse or keyboard). Finally, we mostly refrain from the term *virtual reality* (VR), as VR includes various additional factors, including immersion and interaction, that might not apply to all the technologies (despite that some literature suggests otherwise).

14.2 General Considerations for Implementing Rehabilitation Technologies

The integration of rehabilitation technologies in the overall therapeutic setting depends on patient-related, infrastructural, and economic aspects, as well as the organization of the rehabilitation clinic and the healthcare system in general. Especially in pediatrics, the general therapy conditions, the hard- and software, and the tasks should enable an intensive and challenging yet, positive experience to participate repeatedly in the consecutive therapeutic sessions.

14.2.1 Technologies

The hard- and software should be robust to prevent negative experiences for the child reducing compliance. Due to the relatively large age span (5- to 21-year-old children and youths) and consecutive wide range of anthropometrics, the adjustment of especially exoskeleton devices might take longer to fit the device optimally to each patient. However, it is important that the harness and cuffs are comfortable and that the movements do not cause any unpleasant sensations. Skin redness can lead to skin lesions in case of prolonged therapy duration, and young patients will quickly lose acceptance for devices that are repeatedly uncomfortable. Furthermore, practicality issues like a short calibration and start-up time and being able to change between games quickly contribute to a better acceptance. For patients who have difficulties concentrating for a longer time, it proved useful when therapists could pause the game in-between when needed. In case of minor technical difficulties, therapists should be able to solve these issues themselves to prevent abortion of a training session. Major technical difficulties should be solved fast and adequately by the company.

Nowadays, most of our young patients are experienced in playing computer games. Therefore, they have high demands for using exergames. They appreciate high-quality graphics and lose motivation if the games are not multifaceted and challenging. Highly appreciated are games that include strategic gameplay and choice [14]. In addition, children quickly find ways to 'trick' the system, i.e., they find out how they can increase their game score while performing less active and often undesirable compensatory movements if they are not continuously reminded by the therapist. However, therapists should consider that executive functions such as short-term memory, selective or divided attention, or alertness can be impaired after a congenital or acquired brain injury, which can impact both conventional as well as technological therapy interventions. While specific computer games designed to train such executive functions playfully could improve specific impairments during game playing, therapists should consider that such impairments could also limit the selection of appropriate games, for example, because they are too difficult or provide too many distracting elements. As an additional note, a large proportion of young patients might have visual impairments. Therefore, exergames with less distractive elements and high contrasts might be preferred.

When rehabilitation specialists and management have to decide on what system to purchase, several questions appear relevant: the patient group with its specific impairments and severity (e.g., age groups, more proximal or distal or uni- or bilateral impairments, cognitive capacity), the costs (purchase and annual recurring), practicability (e.g., the time needed for donning and doffing, but also how easy it is to use for patients and therapists), and the available space. Therefore, it makes sense to ask the supplier to deliver a system and test it for a couple of months before deciding to buy.

14.2.2 Therapists

Importantly, therapists should be given time and support in getting experienced in handling and applying these technologies [15]. This is important because despite that some practice-based guidelines have been published (for example, for the Lokomat see [16]), evidence-based guidelines, including details on the application of the technology in children, are missing. Similar to working with conventional techniques, therapists should get a 'feeling' which patient requirements are needed to train with and profit from a particular technology. While companies provide initial instructions on how to use the technology

safely, they are often unable to deliver in-depth patient-specific recommendations for the clinical application. Therapists need to become confident in using the technologies. Any uncertainties will limit therapists from exploiting all the possibilities that some of the devices have. Therefore, it is advantageous when collaborations with technically well-versed people exist.

Good therapeutic quality requires that therapists apply such systems regularly. This might mean for a small center with not too many patients that only a select group of therapists should use the technologies, despite that such an organization also has some negative consequences in practice. From our experience, we would recommend building teams of therapists who work part-time with the technologies and part-time with conventional therapeutic techniques. This increases the flexibility of the therapists and allows therapists to have full programs, also in times when only a few patients fulfill the requirements to train with the available rehabilitation therapy technologies. Such an organization would also facilitate the transfer of therapy contents, particularly considering the transfer from improved motor function to patient-relevant activities. Furthermore, including experienced therapists in clinical research would be beneficial, as we need more studies investigating how the technologies work and whether they are effective.

14.2.3 Scheduling Robotic Therapies

When planning therapies, particularly for the exoskeleton devices that require more time for donning and doffing, we recommend including a preparation time in the absence of the child to set the hard- and software settings. In our center, a therapy session lasts 45 min and includes the donning and doffing of the system. The effective training time amounts, therefore, to about 30–35 min. A selection of therapeutic games is based on the individual requirements of the patient and the therapeutic goals. We recommend that training takes place 3–5 times per week. Based on early studies, we recommend at least

12 training sessions, as this can result in functional improvements [17, 18]. Training is complemented with high intensive conventional therapies, including strength training or sports.

14.2.4 Environment

The environment should consist of an adequately spaced quiet room with a pleasant atmosphere. Sufficient space should be available for children and adolescents to maneuver their wheelchairs. Quite a number of rehabilitation technologies require that the patient wears a harness or is fixated in a seat. Particularly in summer, it can become warm for patients when training intensively in such devices (i.e., due to the harness, cuffs, etc.). Therefore, appropriate measures for keeping an optimal climate are needed to accomplish an intensive therapy session.

14.2.5 Assessments

Introducing technologies could mean that therapists need to adapt the assessments to determine the initial status of the child and to monitor progress in functions that are assumed to undergo change through the application of the technology. In our lower and upper extremity teams, these assessments include, on the body function level, range of motion, the modified Ashworth Scale, and manual muscle testing. In the lower extremity robotic team, these body function tests can be complemented with the Selective Control Assessment of the Lower Extremity Scale. The therapists further assess the 10-m and 6-minute walk tests and the Gross Motor Function Measure (capacity). Performance measures include the Functional Mobility Scale, Gillette Functional Assessment Questionnaire walking scale, and the Functional Ambulation Category.

In the upper extremity robotic team, the body function tests are complemented with Jamar and pinch dynamometry test for hand and finger strength, respectively, and sometimes the Selective Control of the Upper Extremity Scale and

somatosensory testing. Further capacity tests include the evaluation of grasping, transferring, and releasing objects (e.g., with the box and block test indicating gross motor functioning), collecting coins, performing the nine-hole peg test (for quantifying fine motor functioning), opening and closing a bottle, and manipulating small objects to investigate bimanual tasks. When indicated, therapists assess the Melbourne assessment of unilateral upper limb function (capacity test) or the Assisting Hand Assessment (performance).

In both teams, we apply the Goal Attainment Scale to evaluate whether goals defined by patients and their parents are achieved at the end of the rehabilitation period.

14.3 Applying Rehabilitation Therapy Technologies to Children

Children with congenital or acquired neurological lesions can profit from periods of intensive therapeutic interventions to enhance their motor development. Rehabilitation therapy technologies can be implemented in such 'regular' programs but also be a valuable component of post-surgical rehabilitation (see also [16]), such as after multi-level surgery, selective dorsal rhizotomy, or after implanting intrathecal baclofen pumps or deep brain stimulators [19, 20]. For patients who underwent such procedures, practice-based clinical pathways were developed to standardize the process of rehabilitation and to find the best onset for robotic therapy [21].

Children can have their own relationship to rehabilitation technologies, particularly to large robotic devices [22]. Young children can be afraid or respectful of large robotic devices. Adolescents sometimes prefer to keep their distance from the therapist ('hands-off') and might like to train with such technology. As each child or adolescent has to be motivated in a very specific way, we always recommend 1:1 supervision for rehabilitation technology therapy.

14.3.1 Patient Selection

The diagnoses of the children who are trained with rehabilitation therapy technologies are various, including children with congenital (CP) or acquired brain lesions (e.g., stroke, traumatic brain injury, encephalitis, brain tumor), spinal cord injuries, or spina bifida, resulting in sensorimotor impairments. There are many contraindications, but most are relative. Contraindications depend on the device that will be applied and should always be discussed with the responsible physician. Examples of such contraindications are severe obesity (e.g., leg, arm, or trunk cannot be fitted into the orthosis), severe joint contractures, joint instabilities, fractures, osteoporosis, allergic reactions against material that is in contact with the skin, or open skin lesions. Rehabilitation therapy devices might be contraindicated for children with certain implanted technologies like baclofen pumps, shunts, or defibrillators, as the manufacturers cannot guarantee failure-free functioning. Depending on how well the device can be disinfected, patients with contagious infections should not train with such technologies. Relative contraindications can be lesions of nerves, pain reacting negatively to the intervention, or strong spontaneous movements such as seen in children with ataxia, dystonia, or myoclonic twitches. Also, unstable vital functions (e.g., pulmonary or cardiovascular), apraxia, substantial visual impairments, strong spasticity (modified Ashworth 4), or severe epilepsy can be relative contraindications. Besides, severe cognitive deficits, uncooperative or aggressive behavior, and insufficient trunk and head stability, or the inability to position the patient well in the device are considered relative contraindications. In general, several of these issues can be cleared if an initial test training is performed.

We consider it important that the young patient should have a certain understanding of the treatment situation and can respond adequately to demands that arise during robotic therapy. It is crucial that the patient (the functionally better he/she is) can understand and implement specific instructions from the therapist to progress in his/her voluntary motor skills. To us, cognitive understanding is more important than 'just' defining a minimum age. Furthermore, abnormal muscle tone expressed as spasticity, hypotonus, or dyskinetic movements needs to be considered carefully before, during, and after the training, as it might influence the training settings. For patients with strong spasticity, it might be beneficial to first optimize the dosage of antispastic medication before starting with the robotic therapy.

14.3.1.1 Congenital Versus Acquired Neurological Lesions

We apply the same rehabilitation therapy technologies to children with a congenital brain lesion and children with an acquired lesion, as we take into account the specific impairments in body functions and limitations in activities rather than the diagnosis. Nevertheless, robotic therapists note some differences when treating children with a congenital brain lesion ('habilitation') versus children with an acquired lesion ('rehabilitation' in combination with (depending on the age and developmental status of the child) 'habilitation'). First, when considering factors such as age, size, and location of the brain lesion, or severity of functional and cognitive impairment, children with acquired brain lesions seem to improve functionally better. This could partly be caused by the spontaneous neurological recovery that occurs during the (sub-)acute phase in those with an acquired lesion. This has also consequences for the application of rehabilitation technologies, as exemplified in Fig. 14.1. In addition, perhaps due to the presence of previously learned movement patterns, we are under the impression that those with an acquired brain lesion might have better chances of restoring physiological selective voluntary movements. As a consequence, it could be that a child with a congenital brain lesion might practice with one technology throughout the rehabilitation stay, while another child with an acquired lesion switches from one technology to another to account for the improvement in motor function and reduced need for physical support from the technology. Second, children with congenital

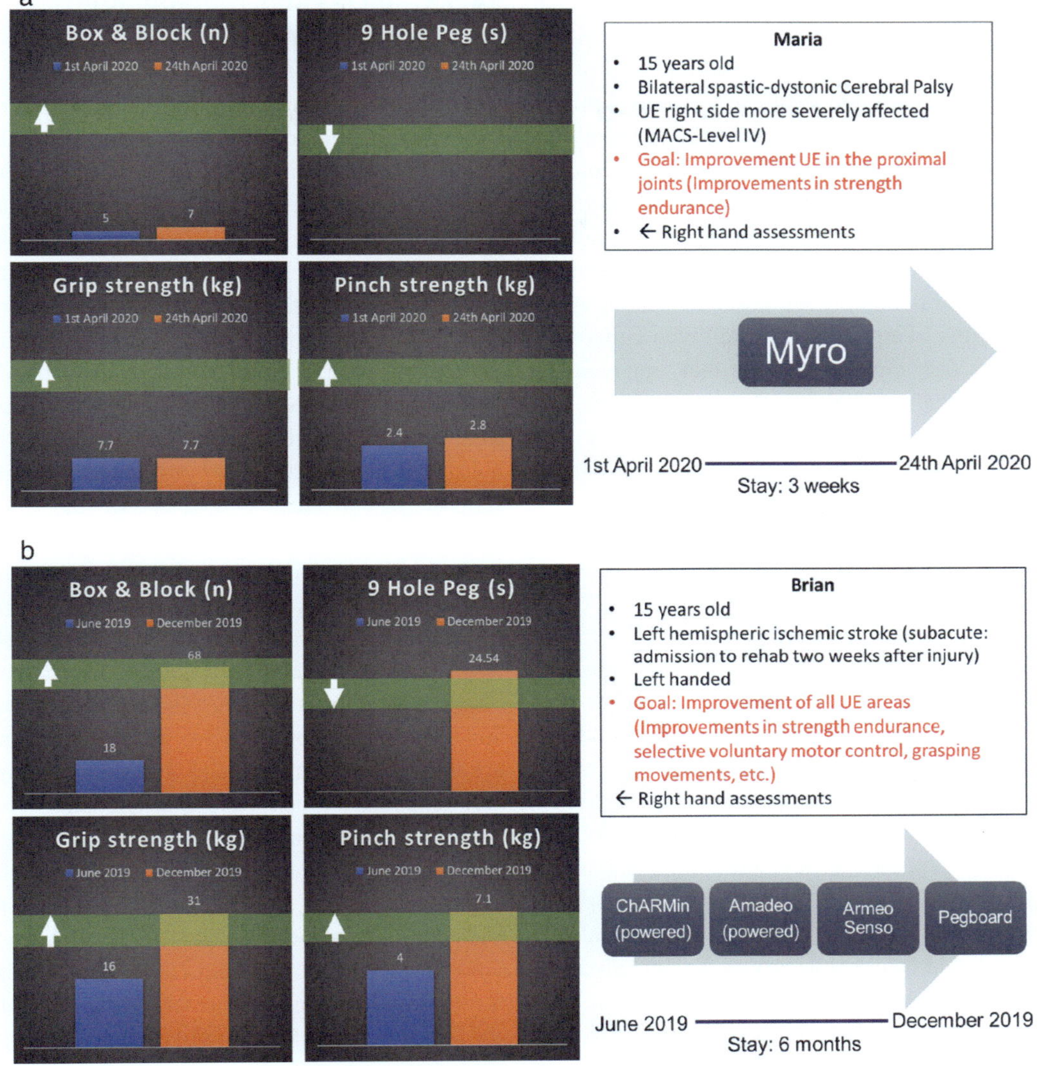

Fig. 14.1 Differences in applying therapy technologies between individuals with congenital versus acquired brain lesions. While the motor status of a patient with a congenital lesion does not change rapidly, and sometimes the goal is even to prevent deterioration, the motor status of a patient with an acquired brain Injury can change during his/her rehabilitation stay. Therefore, to train the patient and his/her functional goals optimally, different therapy devices are needed, which are adapted to the (changing) motor status of the patient in line with the motto 'as much support as needed, as little as possible.' **a,** Maria is 15 years old and has bilateral spastic-dystonic cerebral palsy. The box and block test for the right arm/hand show that she fatigues quickly during reaching, grasping, and transport movements. She cannot perform fine motor tasks with objects that are too small (Nine Hole Peg Test) and has reduced hand and finger strength. She reports that her arm also tires quickly during her daily life. Maria trains on the same device during the rehabilitation period of three weeks. Since her goal is strength endurance, the Myro, a machine without assistance from motors or weight support, was chosen. **b,** Brian is 15 years old and was admitted to our center two weeks after a left hemispheric ischemic stroke. The goal of robot-assisted therapy: improvements in all motor areas of the right upper extremity. Initially, he showed deficits in reaching, grasping, and transport movements (Box and Block Test), fine motor skills (Nine Hole Peg Test), and hand and finger strength. At the beginning of the rehabilitation, we selected therapy devices for arm and hand movements that offer powered support (i.e., ChARMin and Amadeo, see 14.5.1). In the course of time, his voluntary movement control improved, strength and strength-endurance increased, so we changed to augmented feedback technologies that provided no support requiring him to perform hand movements actively against gravity (e.g., the Armeo Senso or an electronic Smart Pegboard). Abbreviations: UE, upper extremity; MACS, Manual Ability Classification System

brain lesions might have developed functional compensatory strategies over (many) years, including learned-non-use. These strategies can be accompanied by joint contractures limiting functional training and improvement. Third, our therapists report that it is often easier for children with acquired lesions (and their parents) to define specific treatment goals because they are aware of the previous capabilities of the child. As such, a reference is missing for children with congenital lesions. (Parents of) children with congenital lesions frequently mention that maintaining (rather than improving) the current level of functionality is the rehabilitation goal, particularly for those children with major impairments.

14.3.2 Initial Consultation and Test Training

Pediatric neurorehabilitation implies working with the child *and* its parents. Therapists and physicians should keep in mind that parents might have excessive expectations and hopes when therapists apply such expensive, high-tech rehabilitation systems. Parents consider robotic training often as the next best thing and have high expectations, for example, to restore physiological walking movements [23]. Realistic goals should be defined early in the rehabilitation process, preferably in an initial consultation, and communicated clearly to the parents. In this consultation, a physician and an experienced robotic therapist should identify all medical and device-related exclusion criteria. By combining this information with the therapeutic goal, the sensorimotor impairments, and other patient-related factors, the most appropriate robotic device can be selected. A test training should confirm whether the technology is also acceptable for the young patient, as anxiety can be an issue in treating children. It is essential that the therapist selects the most appropriate device for the child, which requires therapeutic experience, and communicates with the child empathically. In some cases, it may make sense to use a different device initially to overcome the patient's fear of technology.

14.3.3 Increasing Therapy Intensity Over Time

The intensity of training with rehabilitation technologies should change in parallel to the progress of the patient over time. Like conventional therapies, therapy can be intensified by prolonging the training duration, increasing the number of movement repetitions, or making the tasks more difficult, for example, by increasing the velocity, the complexity, or the position of the patient (e.g., a less stable underground). While therapists reduce their hands-on support over time, the settings of the rehabilitation technologies can be changed to provide less weight support or physical assistance to the joint movements.

Rehabilitation is about pushing limits, again and again. However, continuously motivating children to perform at their best might be challenging, as the literature shows that children with developmental disorders generally appear less motivated and more passive in their playing behavior (less complex and less challenging) despite equal curiosity and pleasure than typically developing children [24, 25]. This might particularly apply to those children with significant impairments, as children with CP with higher IQ, better motor skills, and fewer restrictions in self-care, communication, and social skills seem better motivated than those who do not [21]. It is, therefore, important that therapists become confident and creative to play around with the technologies to find solutions that fit best to the (changing) requirements of the patient.

14.4 Technology Supported Lower Extremity Rehabilitation in Children

Being able to walk takes on a high value throughout life. At a young age, independent walking enables children to experience and expand their environment. Later, the inability to walk can lead to increased stigma in society [26]. Furthermore, mobility ranks second on the priorities of parents of children with CP, right after self-care [27]. In our pediatric neurorehabilitation center, improving walking short distances is even the most frequently mentioned rehabilitation goal [28].

14.4.1 Overview of Pediatric Lower Extremity Systems

Nowadays, there are many commercial rehabilitation therapy technologies that aim to improve standing and walking, so a complete overview cannot be given here. In Fig. 14.2, we suggest a conceptual ordering of these devices, roughly based on the motor impairment level of the patient and the ability of the technologies to provide weight support and physical assistance. The higher the level of impairment (left part), the higher the need for weight support and physical assistance to move the limbs or joints by the technology. While the categorization could assist a therapist in selecting an appropriate device for a specific patient, the clinical situation can deviate from this concept for various reasons. These reasons include the specific therapy goal, other sensorimotor and cognitive impairments, variable settings of the technology, the expertise of the therapist, the personal and environmental factors of the patient, and, of course, the availability of the technology.

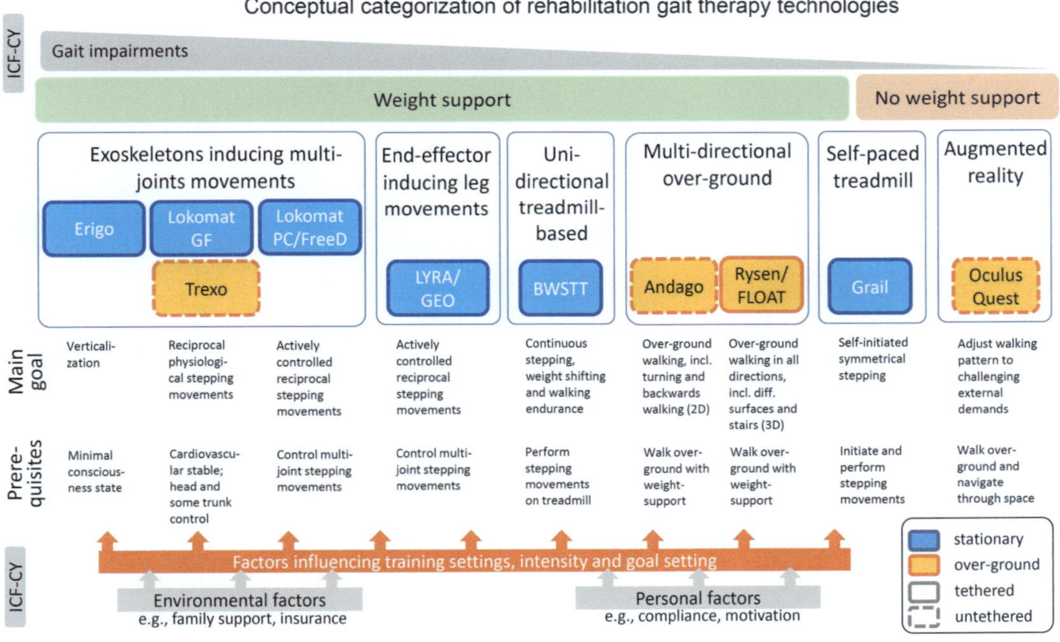

Fig. 14.2 Conceptual categorization of lower limb rehabilitation gait technologies. The technologies can be positioned on a continuum reflecting the amount of motor impairment and the assistance that the technology can provide to the patient taking into account weight support and guidance of movements or individual joints. The continuum ranges from initiating stepping movements on a dynamic tilt table over robot-assisted gait training (RAGT), including exoskeletons and end-effector robot-supported systems, various bodyweight-supported treadmill systems, and multi-directional bodyweight supporting over-ground devices to technologies that provide no weight support or assistance to joint movements at all (like augmented reality). We included in the figure the main goals and prerequisites of the patients for each technology, but more goals and prerequisites apply for each application and are mentioned in the text. Please note that some systems, like the treadmill-based systems, can also be applied without bodyweight unloading. Some patients train with multiple technologies during their rehabilitation stay to work on their various goals. When taking into account the domains of the International Classification of Functioning, Disability and Health (ICF), both environmental and personal factors play an important role in determining the individual goals, selecting the technology, and adjusting the settings. Abbreviations: GF, Guidance Force; PC, Path Control

A dynamic tilt table like the Erigo (Hocoma AG, a DIH brand, Volketswil, Switzerland) is an exoskeleton developed to train verticalization of taller children and adolescents, as there is no specific pediatric version available. The patient is positioned in a harness while the legs are attached to drives that induce leg movements. The Erigo can be applied particularly to adolescents with severe motor impairments, such as those in a minimally conscious state or early after a severe traumatic brain injury or stroke (Fig. 14.3). Besides verticalization, therapeutic goals include improving cardiovascular circulation, perception, muscle tone regulation, and initial stepping movements. There are also other motorized medical devices intended to experience assisted, guided, repetitive leg movements in a static upright position. Examples are the

Innowalk and Innowalk Pro (Made for Movement (global), Skien, Norway), where children of 80–135 cm height and maximally 35 kg weight can already train in the small version.

In general, robot-assisted gait training (RAGT) utilizing exoskeleton devices is indicated for patients with limited leg muscle strength, as these devices provide partial body-weight unloading and guide the lower limb joint movements. The first commercially widely distributed exoskeleton gait trainer was the Lokomat (Hocoma AG, a DIH brand, Volketswil, Switzerland), see Fig. 14.4. It consists of a treadmill belt, a weight-support system, and a robotic exoskeleton. The system allows full biomechanical guidance of hip and knee joints. Such an exoskeleton device can be applied to

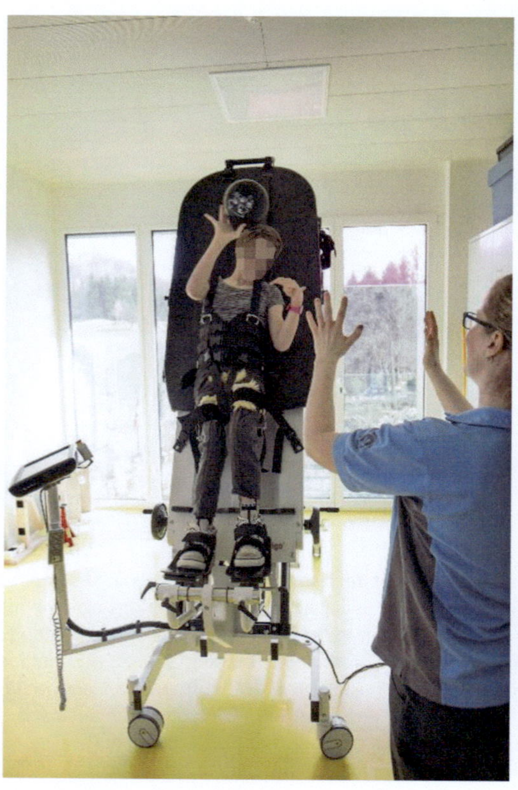

Fig. 14.3 Erigo. The dynamic tilt-table Erigo enables verticalization and supports leg movements of patients with major trunk and lower extremity motor impairments. Picture with kind permission from the University Children's Hospital Zurich

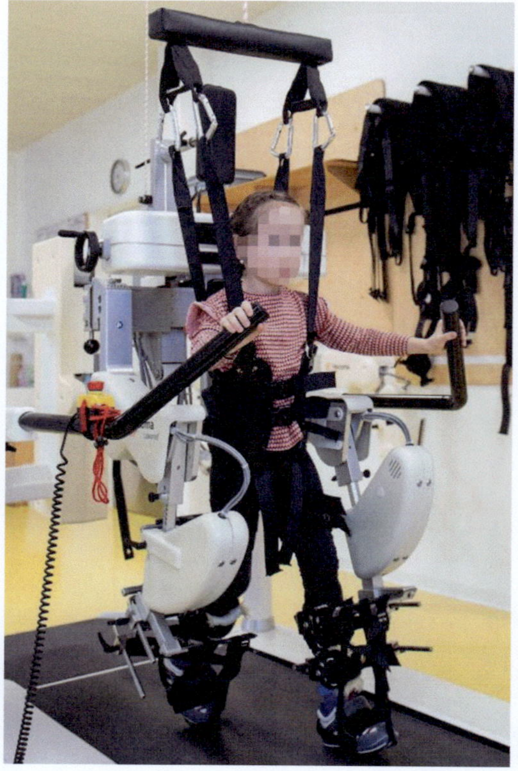

Fig. 14.4 Lokomat. The exoskeleton Lokomat consists of a treadmill, a bodyweight support system and a robotic gait orthosis that can fully guide the leg movements. Newer versions also allow more kinematic freedom, making the device applicable to patients with fewer lower limb impairments. Picture with kind permission from the University Children's Hospital Zurich

improve self-initiation of stepping and reciprocal leg movements while having some head and trunk control and being cardiovascular stable. Training can be intensified by lowering body-weight unloading or robotic support or by increasing walking velocity or training duration. Additional therapy goals can be to improve body alignment, trunk control, ankle control, range of motion, but also to regulate muscle tone, decrease body weight support and biomechanical/ kinematical guidance, or improve speed, walked distance, and gait symmetry. Many other robotic devices similar to the Lokomat have been developed in the meantime. One could consider it a limitation that these devices do not allow the patient to walk around and navigate through the environment.

A novel robotic system like the Trexo (Trexo Robotics Inc., Mississauga, Canada) overcomes these limitations and provides guided leg movements while walking over-ground (Fig. 14.5). This robotic gait trainer is currently offered to children with a weight limit of 68 kg (150lbs). It is available in small, medium, large, and x-large

Fig. 14.5 Trexo. The mobile robotic gait training device Trexo consists of a wheeled frame including a battery, robotic legs, and a tablet to control it. A seat, arm, and chest prompts can be used to support the weight and remain balanced. Picture with kind permission from Trexo Robotics Inc

sizes. Weight support can be provided by a seat (that can also be removed), and arm prompts are available to provide additional trunk stability. Features such as speed of their steps, amount of weight-bearing, and the gait pattern can be customized for each child. A unique feature of the Trexo is that the initiation percentage, indicating the level of active participation from the child, can easily be monitored through its tablet. Based on first experiences from users, walking around, particularly in the home environment of the child, is much appreciated by the children and their parents. Another device that has recently received CE clearance is the exoskeleton device ATLAS Pediatric Exo (Marsi Bionics, Madrid, Spain). While it looks similar to the adult exoskeleton devices that are used as rehabilitation *assistive* devices to walk around in daily life, children still need an additional wheeled frame to walk around with it.

Generally, exoskeleton devices have the advantage that they can guide leg movements in a physiological pattern. When patients are better able to control multi-joint lower limb movements, one could switch to a different control modus of the exoskeleton robot (like path-control or FreeD, see e.g., ([29, 30]) or use an end-effector device. Both options allow more kinematic freedom to perform leg movements. This results in higher muscle activity, inter-muscular coordination, joint control, and a more variable walking pattern, but it also requires that patients can control multi-joint leg movements better. Otherwise, the physiological walking pattern can deteriorate [29]. End-effector devices aiming at inducing repetitive step movements vary from rather simple constructed end-effector equipment that can be placed on a treadmill to complement the existing bodyweight support system (e.g., Lokohelp Pedago, LokoHelp Group, Weil am Rhein, Germany) to complete and more complex end-effector systems like the THERA-Trainer LYRA (THERA Trainer by medica Medizintechnik GmbH, Hochdorf, Germany Fig. 14.6), or the G-EO System (Reha Technology AG, Olten, Switzerland). The latter also allows practicing climbing-stair-like movements. These systems use footplates as already integrated with

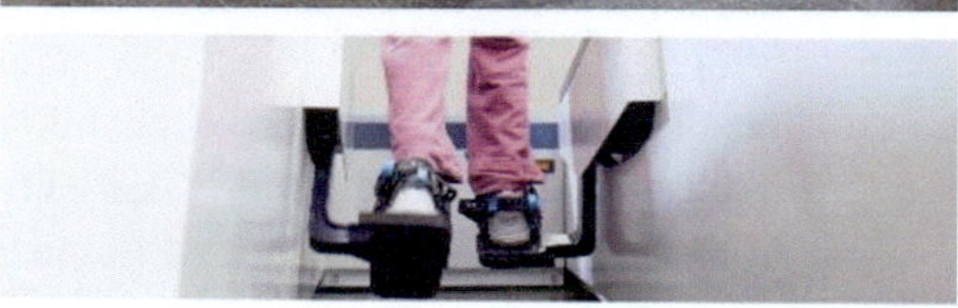

Fig. 14.6 LYRA. The end-effector THERA-trainer LYRA (THERA-Trainer by medica Medizintechnik GmbH, Hochdorf, Germany) allows to train children and youths with body heights varying between 1.00–1.95 m and a body weight up to 150 kg. The patient is positioned on footplates, and step lengths can easily be adjusted from 39 to 67 cm according to the patient's need. Foot bindings are interchangeable, taking into account different shoe sizes and different step widths. Pediatric handrails can be mounted onto the handrails supporting the younger patients, and additional hip stabilization is available in various sizes. Picture with kind permission from THERA Trainer

the Gait Trainer (GT II, Reha-Stim Medtec AG, Schlieren, Switzerland).

If patients can perform unsupported repetitive stepping movements but require some bodyweight unloading and/or manual assistance, walking on a treadmill can be indicated, particularly if the goal is to increase walking endurance. One disadvantage of walking on a treadmill is that the treadmill induces the walking speed. This has several consequences for the therapy, including a reduction in variability between the steps [31]. Children and youths with neurological impairments who walk on a treadmill might even show less stride-to-stride variability than typically developing children walking over-ground

[32]. While more sophisticated treadmill systems like the Grail (Motek, a DIH brand, Amsterdam, Netherlands) enable self-paced treadmill walking, i.e., where the treadmill adapts to the walking velocity of the child and not vice versa, these high-tech systems are often used by children who do not require bodyweight support anymore (see Fig. 14.2). While a system like the Grail allows training self-initiated (symmetrical) stepping, due to the combination with a large virtual reality screen surrounding the person walking on the treadmill, children and adolescents can receive direct feedback about their walking movements or obtain the illusion of moving around.

While, in our opinion, most of the previously presented technologies focus on improving body functions, some newer technologies that enable multi-directional walking over-ground might allow training walking abilities that might transfer relatively well to activities of daily life. Such technologies might enable us to work on therapeutic goals such as improving static and dynamic balance (especially at unstable surfaces and uneven terrains), step variations, timing of steps, stop and go, or reacting to external environmental disturbances, overcoming obstacles (stair walking), etc. One device, the Andago (Hocoma AG, a DIH brand, Volketswil, Switzerland), provides partial bodyweight unloading enabling the patient to move around indoors, independent of the room, without the risk of falling (Fig. 14.7). The focus is on the ability to balance and walk over-ground. When children experience impairments in trunk balance, they can hold on to the parallel bars. After lower limb surgery, the device can support early mobilization, as it allows making steps with reduced levels of load.

Comparable yet different technologies that allow for walking over-ground while being secured and partially bodyweight unloaded are the RYSEN (Motekmedical, a DIH brand, Amsterdam, Netherlands) and the FLOAT (Reha-Stim Medtec AG, Schlieren, Schweiz). These systems are mounted to the ceiling while four cables and a harness are connected to the patient to prevent falls and allow dynamic

Fig. 14.7 Andago. The Andago allows over-ground bodyweight supported walking on flat surfaces. It enables the therapist to focus more on the child, as the device provides partial bodyweight unloading and prevents falls. Picture with kind permission from the University Children's Hospital Zurich

bodyweight unloading. Patients can move around freely in space. The space depends on the height of the ceiling (in our center, approximately 10×2.5 m). The device supports movements in a horizontal and vertical direction, and a fall detection system ensures a safe training of the patient. The technology can be used to train balance (see Fig. 14.8a with the RYSEN) and practice activities such as standing up and sitting down, walking over hurdles or small stairs, and stop and go tasks. In our center, we can combine RYSEN therapy with large floor projections displaying funny exergames or more complex daily-life-related challenging tasks motivating patients to practice longer (Fig. 14.8b).

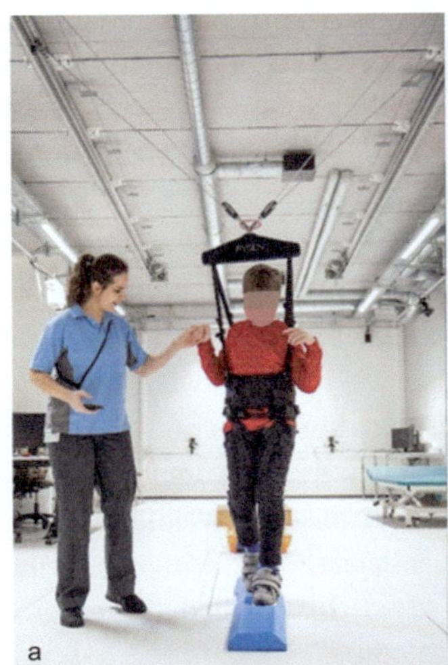

Fig. 14.8 RYSEN. a, The RYSEN allows relatively free walking within the space that is predefined by the system. It has excellent transparency, particularly in the forward–backward direction. It can be used to train various tasks like balancing, stepping over obstacles, walking up or downstairs, standing up, and sitting down. **b,** RYSEN can be combined with three beamers enabling large floor projections that can provide funny, challenging, and goal-specific exergaming. Picture with kind permission from the University Children's Hospital Zurich

As displayed in Fig. 14.2 at the right, virtual reality (VR) or augmented reality head-mounted displays could be used to practice more complex, challenging locomotor tasks in a motivating, playful environment. Particularly the VR goggles can provide a strong immersive environment that could be used to practice challenging, daily-life-relevant tasks repetitively in a playful environment. While these technologies provide no support to the patient, future studies could investigate whether it might be possible to combine these goggles with some of the technologies previously mentioned to allow patients to experience immersive environments while being bodyweight supported and/or receiving physical assistance to joint movements.

14.4.2 Clinical Evidence

In a recent systematic meta-analysis, the authors investigated the effects of mechanically-assisted walking training in children with CP aged 3–18 years [33]. This study included all kinds of mechanical systems, e.g., motorized devices such as a treadmill, gait trainers (i.e., wheeled walking aids), or robotic exoskeleton and end-effector therapy devices. The key results from this study were that for mechanically assisted walking without bodyweight support, there was a small benefit in terms of walking speed, gross motor function, and participation compared to the same amount of over-ground walking. However, for the mechanically assisted walking systems with bodyweight support, there was no benefit in walking speed, gross motor function, or participation compared to the same amount of over-ground walking. These results might be somewhat disappointing and, of course, might be criticized because of the large variability between technologies, outcomes, patients, etc. However, as the effects were not worse than those found for over-ground training, the authors' conclusion was clinically relevant because they acknowledged that mechanically-assisted walking could provide high-dose, repetitive training. They concluded that these technologies might be valuable to provide practice for younger children

with poor concentration when it is hard to apply the same dose of over-ground walking.

Here, we summarize the research and evidence for the categories of the technologies as proposed in Fig. 14.2.

Dynamic tilt table: While some studies have found promising effects of Erigo therapy in adult patients after acute stroke on lower limb motor and cognitive function [34, 35], we are unaware of any studies that have been performed in children with neurological impairments. However, two recent studies investigated the motorized Innowalk device [36, 37]. When pooling data of 31 individuals (mostly children and adolescents), the authors reported that the Innowalk training results in various improved outcomes, particularly in the passive range of motion of the hip joint in multiple directions [36]. In the other study [37], 20 non-ambulatory children with CP participated in a static and dynamic intervention (cross-over design). Better effects on spasticity and range of motion were observed for the dynamic, i.e., Innowalk, intervention.

RAGT using exoskeleton devices: Two meta-analyses concluded that the evidence regarding the use of RAGT for children with gait disorders is still weak and inconsistent [38, 39], despite some tendencies for benefits (i.e., an improvement in walking speed (10-m walking test), endurance (6-min walking test), and gross motor function (Gross Motor Function Measure or GMFM)) [39].

Most literature on children and adolescents undergoing RAGT involved the application of the Lokomat (approximately 40 published studies), whereas only a few of these were randomized trials. The first randomized controlled trial for RAGT with the pediatric Lokomat was performed by Druzbicki et al. [40]. While 26 children with spastic bilateral CP completed the RAGT, only nine from 26 finished the control intervention. The authors concluded that the children could improve their walking speed slightly, but without significant differences between the intervention and control groups. Wallard et al. investigated the effect of Lokomat therapy on kinetic parameters and dynamic equilibrium in 30 children with bilateral spastic

CP [41]. After 20 sessions, the authors reported improved balance control in the Lokomat group. Peri and colleagues investigated the effects of frequency and duration of RAGT on motor outcome in 44 children with CP [42]. The children were allocated to one of four groups: (i) 40 sessions (4 sessions/week) over 10 weeks of Task-Oriented Physiotherapy (TOP); (ii) same number of sessions of RAGT; (iii) ten weeks combined RAGT and TOP (2 + 2 sessions/week); and four weeks combined RAGT and TOP (5 + 5 sessions/week). The authors reported no relevant differences among the four protocols, although both groups with exclusive physiotherapy or RAGT obtained significant improvements in gross motor function, while the mixed approaches did not show significant changes. This stands in contrast to Ammann et al. [43], who concluded that RAGT as a single intervention was not effective in improving walking abilities in the selected children and recommended embedding RAGT in a holistic treatment approach, as it cannot cover all aspects relevant to gait. They compared in their pragmatic cross-over study with randomized treatment sequences 5 weeks of RAGT (three times per week with a maximum of 45 min walking time each) with a 5-week period of standard treatment. They had to stop the trial prematurely due to poor recruitment, while their preliminary results indicated no relevant changes. One of the latest randomized studies comparing RAGT to conventional therapy was performed by Klobucka and colleagues [44]. While their sample of 47 individuals with bilateral spastic CP was relatively old (mean age of 21.2 ± 5.33 years), they concluded that RAGT had significantly better effects on gross motor function in adolescents and young adults with bilateral spastic CP compared to conventional therapy.

When summarizing results from non-controlled studies, it appears that RAGT with the Lokomat is well accepted by patients, parents, and therapists [45]. It appears to be safe [46] and, in combination with a multidisciplinary therapy program, can improve gait speed, walking endurance, gross motor functioning [47], and balance [40]. Improvements seem to sustain over

six months [48, 49], and patients who underwent repetitive RAGT blocks could maintain and slightly increase their gross motor functions over time [50].

There does not seem to be agreement on who might profit more from RAGT. Patritti et al. [51] found that patients with Gross Motor Function Classification System (GMFCS) level II improved more in the outcome than GMFCS level III, and also Schroeder and colleagues [17] found a tendency that less severely impaired patients responded better (as well as younger patients), see [52]. Beretta et al. [53] reported improvements in children with GMFCS III (and not I-II or IV), while van Hedel et al. [54] reported improvements particularly in those with GMFCS level IV (and a dose–response relationship between the number of Lokomat therapy sessions and improved outcome in GMFCS III and IV).

When children with CP walk in a device like the Lokomat, their muscle activity is reduced (for example, in comparison to treadmill walking) and responds little to reducing the guidance force or bodyweight unloading [55]. RAGT can be combined with exergames to motivate the children while simultaneously providing feedback, which should inform about the active participation of the child and ensure higher leg muscle activations. Exergames might indeed lead to better motor outcomes [51], more challenging therapeutic situations [10], and more active participation during the therapy session [9], as also shown by increased leg muscle activations [13], without affecting the muscle activation patterns negatively [56]. Exergames were designed with a particular purpose, for example, to include periods of more and less activity within a therapy session [12] or include a dual-task condition (see also Fig. 14.9) [57]. Interestingly, while some of the participants responded as anticipated, others were not able to perform these more complex tasks. As participants were well-characterized, such studies provide preliminary evidence for identifying potential therapy responders.

A long-term criticism, particularly for exoskeleton devices like a Lokomat, was that the early control strategies did not allow for

Fig. 14.9 Innovative movement therapy in childhood. The screenshot was taken from the serious game 'Magic Castle'. To perform dual-task training early during rehabilitation, serious game designers collaborating with therapists developed several games. In Magic Castle, the avatar of the young patient is a little wizard who travels on the back of an animal. If the child is little active in the Lokomat system, this animal is a snail. If the child becomes more active, the animal switches to a turtle or, with even more activity, a sheep. To make it a dual task, the young patient holds a stick in his/her hand. In the game environment, the wizard holds this magic wand and has to point it on virtual objects. When pointing sufficiently long on an object (i.e., accurate and prolonged pointing task), the object becomes alive, and the child playing the game gets additional points. Picture with kind permission of the Zurich University of the Arts/Specialization in Game Design and University Children's Hospital Zurich

sufficient variability in task performance. Movement variability is an important prerequisite for motor learning (consider Bernstein's 'repetition without repetition' principle [58]), and technologies that just produce a pre-programmed movement pattern are not considered optimal, especially for functionally better patients [59]. The consolidation of a learned task over time seems better if kinematic variability is introduced [60], and variable practice might facilitate translation of the skill into everyday life [61]. Indeed, some studies have suggested the superiority of assist-as-needed compared to fully-assisted control strategies [62–67]. Therefore, several control strategies have been developed to allow more movement variability.

The Lokomat was initially equipped with the 'Guidance Force' control, which functions as a position control mode, where the legs are moved along a strictly defined trajectory. The newer 'Path Control' induces a patient-cooperative behavior by enabling the patient to move within a virtual tunnel instead of having to stay on the specific trajectory. The 'FreeD module' also includes a hardware module adding more kinematic variability by enabling the pelvis to translate laterally up to 4 cm per side and rotate up to 4° per side. Additionally, the leg cuffs allow a lateral movement of the child's legs. Aurich et al. showed that these novel control strategies increased the amount of muscular activity in patients with neuromotor disorders [29]. However, some young participants walked with a less physiological pattern with the FreeD module (compared to a reference curve of typically developing children), and Path Control and FreeD affected the activation patterns and step variability differently [30]. A subsequent study verified that this issue was most likely caused by the inability of the young patients to handle the large amount of kinematic freedom because healthy adults could walk with a physiological pattern in FreeD [68].

While we are unaware of any studies investigating the ATLAS Pediatric Exoskeleton, a first study on the over-ground robotic gait trainer

Trexo in a non-ambulatory child with CP showed that its regular use might have positive effects on head control, knee flexor spasticity, and the frequency and quality of bowel movements [69].

RAGT using end-effector devices: Studies that evaluated end-effector RAGT systems in children seem rare. One randomized controlled trial evaluated the effectiveness of repetitive locomotor training with the Gait Trainer GT I versus conventional training in 18 ambulatory children with diplegic and tetraplegic CP [70]. Gait velocity and endurance in children absolving RAGT were significantly improved and maintained 1 month after the treatment. The effects of the experimental treatment outweighed those of the control treatment.

Bodyweight-supported treadmill training: While we do not present here a detailed overview of all studies that investigated treadmill training and bodyweight-supported treadmill training in children and adolescents with neurological gait impairments, we refer to the recent systematic review by Novak and colleagues [71]. The authors concluded that these interventions should be considered effective health interventions for children with CP with positive effects on walking speed, endurance, and gross motor function.

Multi-directional over-ground devices: Two studies have been published on the Andago. In the first [32], the clinical utility of the device was investigated in children with (merely) neurological gait impairments. The authors concluded that the Andago is a practical and well-accepted device to train walking over-ground with bodyweight unloading in children and adolescents with gait impairments safely. The system allows individual stride-to-stride variability of temporospatial gait parameters without affecting antigravity muscle activity strongly. Some children reported that walking in Andago required more attention and appeared more similar to normal walking. Therefore, the same group investigated differences in brain activation when children and adolescents with neurological gait impairments walked in Andago and on a treadmill at the same speed and with the same level of bodyweight unloading [72]. Functional Near-Infrared Spectroscopy was used to study hemodynamic responses of the Supplementary Motor Area and the Prefrontal Cortex. In those children showing a typical hemodynamic response, which was only a small proportion of all participants, the responses in these brain regions tended to be larger when walking in Andago compared to the control condition indicating higher involvement of these areas for controlling walking.

As the RYSEN or Float are relatively new technologies, we are unaware of any studies investigating these systems in pediatric patients.

Self-paced treadmill with VR environment: The Grail system has been used to study whether ambulatory children with CP are able to online-adapt their walking pattern in response to real-time feedback [73]. The first findings were positive, indicating that ambulatory children with CP, particularly those with poorer initial gait, could improve hip and knee extension based on visual feedback of hip and knee kinematics. Furthermore, 16 children with CP who participated in 18 sessions with the Grail improved in various parameters (e.g., gait speed and endurance, stride length, gross motor function, kinematics) [74]. Interestingly, the changes were mainly predicted by age and cognitive abilities. The same group also investigated the navigation skills of children with CP and typically developing children [75]. They could show that children with CP seemed to learn and adapt to new conditions differently than their healthy peers, which could have an influence on everyday life and participation.

Augmented reality: While two recent systematic reviews suggest that VR technology can improve balance and motor skills [76] or postural control and ambulation [77] in children with CP, it should be noted that these reviews included all kinds of technologies, including those that we would refer to as exergames rather than VR.

The therapeutic application of VR in children has mainly been performed in the fields of pain management [78] or for educational purposes, e.g., to cross the street safely [79].

We have started to explore the application of head-mounted displays in youths undergoing neurorehabilitation. Based on initial interviews with adolescents with neurological gait impairments and their parents, we collected gait-related activities that they consider problematic in daily life and should be addressed in therapy. In the next step, we investigated whether children with neurological gait impairments and their therapists would prefer mixed reality head-mounted displays or virtual reality head-mounted displays [80]. The results showed that youths and therapists accepted both systems well, with advantages regarding usability, user experience, and preference for the virtual reality head-mounted display. Based on this work, we are developing several apps for the Oculus Quest (Facebook, CA, USA) and are currently performing first investigations, for example, in determining whether children and adolescents with gait performance deal with different objects in the same way when they are real or only present in the virtual environment.

14.5 Technology Supported Upper Extremity Rehabilitation in Children

For many years, walking was considered and pursued as the most important rehabilitation goal. However, if we think about our everyday life, it quickly becomes clear that we need good arm and hand function for almost every activity. Getting dressed, using a mobile phone, or eating are just a few examples. It should, therefore, not be forgotten that arm and hand functions are important requirements for many activities of daily living and thus for independence. The importance is reflected in the relatively high number of upper-limb-related rehabilitation goals, which are defined by the children with neurological impairments, their parents, and the interdisciplinary rehabilitation team [28]. As conventional training performed by occupational or physical therapists of the arm and hand is very labor-intensive, more and more systems to train the upper extremity have entered the field of rehabilitation in the past years. Not all systems were specifically designed or modified for young patients. While for some exoskeleton devices,

hardware adjustments were made to fit the system to the anthropometrics of the child, it was often forgotten to adjust the exergames to the requirements of a younger target group, and children had to play games initially developed for senior patients after stroke.

14.5.1 Overview of Pediatric Upper Extremity Systems

While we presented in Fig. 14.2 a conceptual ordering of lower limb gait-improving rehabilitation technologies, we propose a similar conceptual ordering for the various upper limb technologies that are nowadays available (Fig. 14.10). Also this overview is incomplete, and exceptions are possible. The ordering is roughly based on the level of motor impairment of the patient and the ability of the technologies to provide physical assistance and/or weight support. While powered exoskeletons can guide movements providing more control over individual joints and even lock certain joints to focus on adjacent joints, end-effector devices support movements without specific guidance. For the upper limb, it is furthermore relevant to consider whether the technology can be used unilaterally or bilaterally and whether the technology focuses on proximal or distal individual joints or includes the whole arm and hand. In addition, in contrast to the lower limbs, there are many upper limb systems that provide augmented feedback without providing additional physical and/or weight support. All these device characteristics have an influence on which goals can be pursued with rehabilitation therapy technology. For example, a device that can be used bilaterally can also be used to pursue goals such as hand-hand coordination. Furthermore, it indicates the minimum requirements for a patient to train with the therapy device (Fig. 14.10). This is directly related to the support provided by the device. If the therapy device actively guides the patient, this applies to powered devices, it is sufficient if the patient has the potential to perform a voluntary movement. In the case of weight support, the patient should be able to actively move his/her arm

Fig. 14.10 Conceptual categorization of upper limb rehabilitation therapy technologies. Upper limb rehabilitation therapy devices, their support, and the requirements to train on them. Both the functional goals and the goals of the Assisting Hand Assessment are marked with traffic light colors: Red = goal cannot be pursued; Orange = goal can be partially pursued; Green = goal can be pursued. If the device is marked with *exo, it is an exoskeleton. Below is also indicated which joints can be trained with the technology. The overview shows why it is probably not sufficient to have only one device, why children with the same goal might train on different devices, and why it is helpful to be aware of your patient population before equipping with various devices. The figure shows the selection of equipment that we have summoned in our center over the past years, always with the thought to close a gap in need with regard to our patients and their therapy goals. Results are based on the opinions of three experienced occupational therapists from Swiss Children's Rehab, who scored each system first independently and then discussed differences until they found a consensus

independently with support against gravity. If the therapy device offers neither of these, i.e., only augmented feedback, the patient must be able to move his/her arm or the target movement independently against gravity.

Upper limb therapy technologies that include *powered support* for guiding arm and hand movements are relatively rare among the therapy devices. These systems can either guide the patient's entire arm through its passive range of motion or perform particular movements, such as opening and closing the fingers. To train with these powered therapy devices, we generally select patients who are not yet able to plan their movements independently ('I want to move my arm from starting point Y to target point Z) and/or have no or too few voluntary motor skills in their extremities. Some of these technologies also allow an assistive mode, where the device assists the patient in a target movement, either through weight support or motors. In these cases, both the device and patient are active.

In our center, we apply the exoskeleton ChARMin (Fig. 14.11), which was developed in a collaboration between the Sensory Motor Systems Lab of the ETH Zurich and the Swiss Children's Rehab for guiding shoulder, elbow, forearm, and wrist movements. A pressure-sensitive bulb registers squeezing (and opening) of the hand, which is also part of the controls to steer the exergames.

There are also powered end-effector devices that enable practicing movements of the arm, such as the InMotion2, which is the

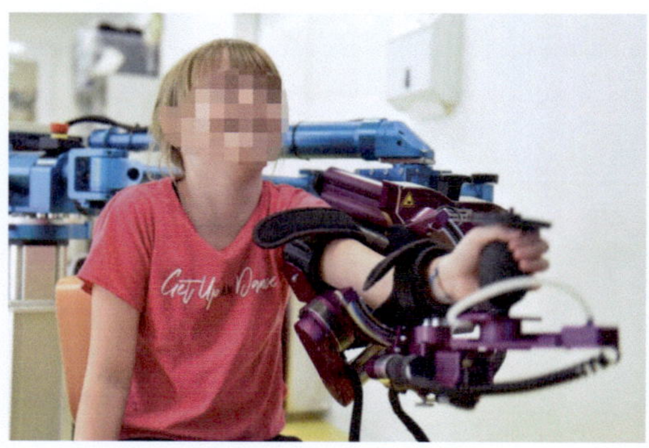

Fig. 14.11 ChARMin. The Children's Arm Mechatronic Interface (ChARMin) is an upper extremity exoskeleton robot with actuated joints. It was designed to train shoulder, elbow, forearm, wrist, and hand grasping movements of children and adolescents with major upper limb motor impairment. Picture with kind permission from the University Children's Hospital Zurich

commercially available version of the MIT-Manus (Interactive Motion Technologies, Watertown, MA, USA), or the REAPlan (Axinesis, Wavre, Belgium).

A powered device specifically developed for training active, assistive, or passive movements of the hand and fingers is the Amadeo (Tyromotion GmbH, Graz, Austria). This device focuses on hand and finger movements. Various exergames can be played by performing corresponding finger movements.

Several devices provide *weight support* of the arm to relieve the patient and enable the performance of high numbers of movement repetitions without muscle weakness limiting the therapy. This can be indicated, for example, in patients after a long period of immobilization or in patients with insufficient strength to move their arm against gravity but with sufficient motor control to perform coordinated movements when gravity is 'eliminated'.

The Diego (Tyromotion GmbH, Graz, Austria) resembles a modern help-arm (Fig. 14.12). Upper and lower arm cuffs are fixed through cables to one or both arms (i.e., bilateral training is possible). Motors can move the patient in the sagittal plane or provide weight support, which is why this system can be listed under 'powered' systems. Various exergames can be performed in

Fig. 14.12 The Diego. The end-effector device Diego provides weight support of the arms and allows both unilateral and bilateral exercises. Picture with kind permission from the University Children's Hospital Zurich

either the sagittal, frontal, horizontal, or transverse plane. The Diego can be used actively without or assistive with weight support and passively in the sagittal plane.

The pediatric Armeo Spring (Hocoma AG, a DIH brand, Volketswil, Switzerland) is a weight-supporting exoskeleton device with integrated springs (Fig. 14.13). The tension of the springs can be adjusted to support the arm's weight against gravity and train movements in three dimensions. The patient, therefore, has to perform the arm movements actively but is supported by the weight support.

Most existing upper limb systems, which are not used as supportive rehabilitation assistive technologies in everyday life, but as therapy devices, provide the patient with software-based *augmented feedback* via so-called exergames

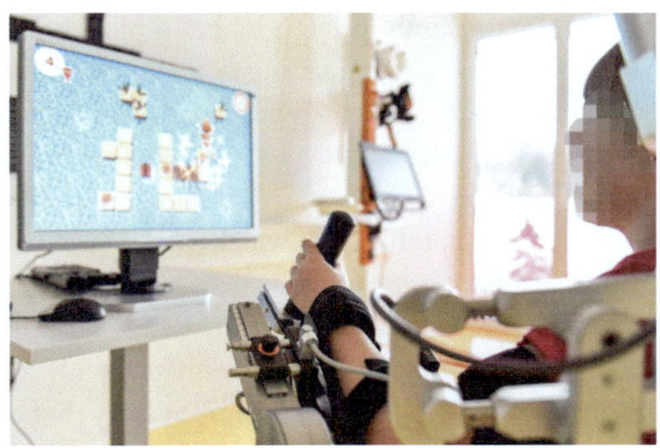

Fig. 14.13 The pediatric Armeo Spring. The exoskeleton device Armeo Spring allows weight-supported 3D movements in space. It can be switched from the left to the right side and vice versa. Picture with kind permission from the University Children's Hospital Zurich

Fig. 14.14 The Rapael smart glove. This glove-based augmented reality system includes an impressive number of exergames of good graphical quality. Picture with kind permission from the University Children's Hospital Zurich

[81]. These exergames have the aforementioned advantage that they increase the patient's motivation, but above all, they provide visual, auditory, or haptic (e.g., vibration or virtual walls) feedback as to whether the desired movement is being executed correctly. There are several systems that provide *augmented feedback* only and do not provide any physical support. Due to the gaming, children and adolescents with neurological upper limb impairments are motivated by the games to perform high numbers of goal-directed tasks in a fun and challenging way.

While our therapists still use the glove-based system YouGrabber (Reha-Stim Medtec, Schlieren, Switzerland), which contains two data gloves, an infrared camera, and a display, it seems that this system is currently not sold anymore. A novel, somewhat similar system is the Rapael smart glove (Neofect, South Korea, see Fig. 14.14). This system allows selective single-joint movements and does not support more complex multi-joint movements.

The Armeo Senso (Hocoma AG, Volketswil, Switzerland) uses similar software as the pediatric Armeo Spring (see weight-supported devices). However, the patient's movements are not weight-supported, and the system uses motion sensors on the chest and upper and lower arm to steer the various exergames. In addition, the patient can also perform grasping movements through a pressure-sensitive hand module.

A completely different approach is used by Myro (Tyromotion GmbH, Graz, Austria). This is a huge tablet (Fig. 14.15), an interactive screen, that can be used in a variety of ways. The patient can use the screen, combine it with real objects that require different grasp forms, and even practice pushing and pulling tasks on the screen. The screen can be adapted to the patient's range of motion and can be adjusted in both height and tilt.

Fig. 14.15 The Myro. The huge interactive screen enables much more exercise conditions than one would expect at first sight. Therapy difficulty can be influenced by the games, the tasks, the use of different objects requiring individual grasps (see figure), but also the positioning of the patient (i.e., sitting, sitting on a roll, standing), requiring less or more trunk control in combination with the fine-motor tasks. Picture with kind permission from the University Children's Hospital Zurich

14.5.2 Clinical Evidence

When summarizing recent systematic reviews, it seems that there is agreement that robotic therapy and augmented feedback technologies have the potential to improve upper limb function in children with neurological upper limb impairments. For example, Chen and colleagues concluded that augmented feedback seems to be an effective intervention for improving arm function in children with CP [77]. Two years before, the same group acknowledged the potential of robotic therapy to improve upper limb function, but they also mentioned that the paucity of group design studies summons the need for more rigorous research [82]. These findings seem to be in line with the results from Novak and colleagues [71]. Yet, it should be noted that others have concluded that the role of augmented feedback to improve hand function in children with CP is unclear due to limited evidence, while the use as an adjunct has some support [83]. This is in line with our concept that technological interventions should be combined with conventional therapies to achieve improvement in daily-life relevant tasks. Indeed, upper limb robotic therapies can be well integrated with camps combining the technologies with conventional interventions (e.g., [84, 85]). However, applying technologies such as augmented feedback exergames at home, without professional guidance and motivation, remains challenging, as some excellently designed trials showed [86, 87].

Below, we summarize the research and evidence for the upper limb technologies presented in 14.5.1.

Powered devices: While there are some studies on the development of the exoskeleton ChARMin (e.g., [88]), and we tried to investigate contextual interference in children with brain lesions with this device [89], no randomized trials have examined the effectiveness of this technology. Such evidence is available, though, for some upper limb end-effector devices. One of the first randomized controlled trials was performed by Gilliaux et al. [90]. They investigated the effectiveness of the REAPlan in 16 children with CP. This end-effector robot can move the patient's upper limb in the horizontal plane via a handle that the patient can grasp, while force and position sensors allow controlling lateral and longitudinal interaction forces between the patient and the robot. One group received five conventional therapies per week (for 8 weeks), while the other group received two robotic and three conventional interventions. Per session, participants performed 744 movements with the robot. The authors reported significantly better movement fluency and Box and Block Test results for the robotic/conventional group.

Concerning the Amadeo, one pilot study investigated the effects of 18 one-hour sessions in children with hemiparesis and found significant improvements in upper limb impairments as well as bimanual performance [91].

Weight-supporting technologies: While we are unaware of any pediatric studies investigating the Diego, the Armeo Spring has been investigated in several studies. Keller and van Hedel found evidence indicating the successful acquisition, transfer, and retention of upper extremity skills in children with CP [92]. Noteworthy, the young participants performed over 3000 point-to-point movements and hand-opening and closing movements while playing a game for 70 min, distributed over 3 days. Furthermore, in an uncontrolled pre-post design study, 10 children with bilateral CP undergoing 40 Armeo Spring sessions improved in upper limb coordination, fluency, and quality of movements [93]. In 2018, a randomized controlled trial comparing Armeo Spring therapy with conventional therapy (12 weeks, three times per week) concluded that the robotic therapy was significantly more effective than the conventional therapy in improving the upper limb quality of movement in children with hemiplegic CP [94]. In children with acquired brain injury, effects from the Armeo Spring were compared to effects from constrained induced movement therapy and conventional physiotherapy [95]. The authors reported that Armeo treatment delivered improvements, particularly in a vertical motor task and movement efficiency, while it reduced compensatory movements of the shoulder.

Augmented feedback technologies: The YouGrabber was previously investigated for its feasibility [96] and preliminary findings indicated the YouGrabber's ability to improve dexterity in children with CP [18]. In a recent randomized controlled trial investigating in 80 children with brain lesions the effectiveness of a combination of Rapael therapy and occupational therapy versus conventional occupational therapy alone, the authors reported significant improvements in upper-limb dexterity, activities of daily living, and forearm supination in the combined conventional/Rapael group [97]. The authors also noted that children with more severe motor impairment seemed to profit more from the intervention than those with less severe impairment. We are unaware of any randomized studies in children with upper limb neurological impairments that investigated the Armeo Senso or the Myro. However, we use the Armeo Senso (and a surface electromyography system) as an input device to steer an exergame that we developed to specifically train selective voluntary motor control [14]. In an ongoing randomized controlled trial, we are currently investigating whether we can improve selective voluntary motor control in children with upper motor neuron lesions [98].

14.6 Concerns about Using Rehabilitation Therapy Technologies

There are some concerns when applying these technologies to children. We discuss two. The first concern is one of the leading criticisms for using rehabilitation therapy devices and addresses the risk of passivity of the patient. We minimize this concern by striving to select the technology for an individual patient that provides as much assistance as needed but as little as possible (see also Figs. 14.2 and 14.10).

However, we agree that passivity can be an issue for powered exoskeleton devices that fully guide the extremity through the movement. In our experience, we rarely train the upper extremity under such conditions, so we report here on our strategy to reduce the risk of passivity in patients receiving therapy with the exoskeleton Lokomat. As previously discussed, not fully guiding control strategies such as assist-as-needed or Path Control reduce the risk of passivity of the child, as the patient needs to take over parts of or the whole movement. However, also in conditions of fully guided movements, it is possible to increase the activity of the patient. Here, it is essential that both the therapist and the patient can monitor the patient's activity within the device. This is possible using the biofeedback values derived from the human–machine interactions. Various visualizations of the biofeedback signals can be helpful. For example, therapists might appreciate accurate and timely joint and gait-phase specific information to instruct the patient when and how to improve his/her activity. However, such information, presented for each step, is too complex to present to young patients during therapy. Instead, they profit from a more generalized, motivating, and fun visualization utilizing exergames (see, for example, Fig. 14.9). Please note that while we can use biofeedback values to estimate the patient's activity level, these values are currently not sophisticated enough to indicate the quality of the movement. In addition, we have investigated other options to counteract passivity when walking in the Guidance Force condition in two unpublished studies. In one study, 14 children with congenital or acquired brain lesions, aged 13.6 ± 3.3 years, walked at three different speeds, varying levels of bodyweight unloading (minimal required level of unloading, minimal level plus 15%, and minimal level plus 30%), and with various intensity instructions from the therapist (to remain passive in the device, to walk normally, and to participate as actively as possible). We used electromyography to measure the activity of the m. tibialis anterior, m. gastrocnemius medialis, m. vastus medialis, and m. rectus femoris. Participants showed significantly higher mean leg muscle activity levels when walking fast and receiving encouraging instructions. Reducing bodyweight support only increased the vastus medialis and biceps femoris muscles' activity. In the other study, we investigated changes in the activity of the same leg muscles, heart rate, and subjective level of effort in 19 participants with similar diagnoses and aged 13.0 ± 3.4 years. We varied the level of guidance force (i.e., 100%, 80%, 60%, and 40%, uni- and bilateral, randomized order). None of the outcomes changed due to the varying levels of guidance force in these patients. While these studies provide information on how therapists can influence therapy with the Lokomat to reduce the risk of passivity of the patient, it needs to be investigated whether these findings might also apply to other technologies or adult patients.

The second concern addresses the issue that particularly adolescents with long-term disabilities devote a considerable part of their leisure time to screen-time behaviors, including playing computer games. Results from a large international survey from 2013/14 inform us that 58% of the boys with long-term disabilities spent more than 2 h per weekday playing computer games [99]. This percentage was 74% for weekend days. The percentages were smaller for girls with long-term disabilities, yet 31% played for more than 2 h each weekday and 43% each day of the weekend. Despite that exergaming includes many movement repetitions, from time to time, we hear critical voices asking how much additional exergaming is beneficial for patients that already spend a considerable amount of their time playing computer games.

14.7 Outlook

Various developments seem promising to improve therapy outcomes or independence in daily life for children and adolescents with neurological impairments. Here, we mention some that we consider relevant.

14.7.1 Training the Central Core, the Trunk

The trunk is a core element relevant to lower and upper extremity functioning. Investigations from others [100] and our group [92, 101] have shown that a considerable proportion of the variability in

gross motor function, mobility, and self-care in children with CP can be explained by the functionality of the trunk. Therefore, more playful technologies should be developed to train the trunk. While some technologies are already on the market, like the Hirob (Intelligent Motion GmbH, Wartberg/Krems Austria), less expensive and more practical alternatives (like the Wii balance board) are desirable for children and adolescents with neuromotor impairments (see [102]).

14.7.2 From Rehabilitation Therapy to Assistive Technology

We know that functional improvement can be limited for some children. At a certain point, they achieve a plateau in function and do not progress anymore, despite intensive rehabilitation. Such patients might use technologies in daily life to compensate for the loss of function, i.e., rehabilitation *assistive* technologies. We currently participate in developing assistive technology, i.e., the fully wearable pediatric whole hand exoskeleton PEXO [103, 104]. It consists of a hand exoskeleton and a backpack containing the motors and the battery, connected via a cable allowing the patient to move freely in daily life. In the long term, PEXO could assist children with severe hand impairments to compensate for the loss of grasping and holding of the affected hand and enable performing bimanual everyday-life activities, such as opening a bottle, independently. When comparing our experiences to those from our partners who work with adult patients, we notice that the children's compliance is much more influenced by the comfort, robustness, and practicability of the technology. This shows the need to include children and adolescents early in the design and development process of such technologies.

14.8 Conclusion

We presented a conceptual ordering of lower and upper limb rehabilitation therapy technologies that could help identify devices that might be appropriate for a particular child or adolescent requiring neurorehabilitation. While the 'ordering' is merely based on years of practical experience working with these technologies, some experimental studies confirm parts of our concept, particularly for those technologies that are already somewhat longer on the market. A better understanding of the technology could contribute to a more evidence-based application. We presented various devices and summarized the current level of evidence. However, due to the rapid technological developments and the vast number of devices in this field, it would be beyond the scope of this chapter to cover all of them. Even though some rehabilitation technologies have been applied in the pediatric field for about 20 years, readers of this chapter should not forget that the multidisciplinary programs nowadays still merely include conventional therapies, not each child can be treated with a rehabilitation technology, and there are rehabilitation goals that cannot be achieved with the available therapy technologies. In addition, the level of evidence of the effectiveness of the various applications appears inconclusive. On one hand, this is caused by the relatively small number of and huge variability between pediatric patients, as well as various methodological issues such as the lack of a control group and long-term follow-up, variable and poorly described training regimes, and unresponsive outcome measures. On the other hand, we should not forget that if the control intervention consists of evidence-based active goal-directed therapy interventions performed by experienced therapists, finding similarly good results with rehabilitation technologies should not be considered a negative result. The rapid development of rehabilitation technologies will make it challenging for rehabilitation specialists and researchers to keep up with clinical research. Perhaps we exaggerate somewhat, but at the time when a clinical researcher has finalized the study protocol, obtained the funding and the ethical approval, the technology's hard- and software has already been upgraded. The pediatric field should work to substantiate the evidence in the next years. Important topics will be to identify responsive patient groups, come up with objective and responsive functional outcome measures, and initiate collaborations with other centers to recruit appropriate sample sizes.

Acknowledgements We are grateful to the therapists of the lower and upper extremity robotic teams of our center, particularly Esther Zamarrón Sánchez, for sharing their long-term experience and knowledge. Furthermore, we thank the young persons, their parents, and the companies who approved that their pictures could be used. Some of the work presented in this chapter is based on scientific projects financially supported by the Mäxi-Foundation and the J&K Wonderland Foundation.

References

1. O'Shea TM. Diagnosis, treatment, and prevention of cerebral palsy. Clin Obstet Gynecol. 2008;51(4):816–28. https://doi.org/10.1097/GRF.0b013e3181870ba7.
2. Mayston M. Evidence-based physical therapy for the management of children with cerebral palsy. Dev Med Child Neurol. 2005;47(12):795. https://doi.org/10.1017/S0012162205001672.
3. Mayston MJ. People with cerebral palsy: effects of and perspectives for therapy. Neural Plast. 2001;8(1–2):51–69. https://doi.org/10.1155/NP.2001.51.
4. Verschuren O, Ada L, Maltais DB, Gorter JW, Scianni A, Ketelaar M. Muscle strengthening in children and adolescents with spastic cerebral palsy: considerations for future resistance training protocols. Phys Ther. 2011;91(7):1130–9. https://doi.org/10.2522/ptj.20100356.
5. Salem Y, Godwin EM. Effects of task-oriented training on mobility function in children with cerebral palsy. NeuroRehabilitation. 2009;24(4):307–13.
6. Damiano DL. Activity, activity, activity: Rethinking our physical therapy approach to cerebral palsy. Phys Ther. 2006;86(11):1534–40. https://doi.org/10.2522/ptj.20050397.
7. WHO. International classification of functioning, disability and health: children & youth version: ICF-CY. Geneva, Switzerland: World Health Organisation Press; 2007.
8. Holden MK. Virtual environments for motor rehabilitation: Review. Cyberpsychol Behav. 2005;8(3):187–211. https://doi.org/10.1089/cpb.2005.8.187.
9. Brutsch K, Koenig A, Zimmerli L, Merillat S, Riener R, Jancke L, et al. Virtual reality for enhancement of robot-assisted gait training in children with neurological gait disorders. J Rehabil Med. 2011;43(6):493–9. https://doi.org/10.2340/16501977-0802.
10. Brutsch K, Schuler T, Koenig A, Zimmerli L, Merillat S, Lünenburger L, et al. Influence of virtual reality soccer game on walking performance in

robotic assisted gait training for children. J Neuroeng Rehabil. 2010;7:15. https://doi.org/10.1186/1743-0003-7-15.

11. Koenig A, Wellner M, Köneke S, Meyer-Heim A, Lünenburger L, Riener R. Virtual gait training for children with cerebral palsy using the Lokomat gait orthosis. Stud Health Technol. 2008;132:204–9.

12. Labruyere R, Gerber CN, Birrer-Brutsch K, Meyer-Heim A, van Hedel HJA. Requirements for and impact of a serious game for neuro-pediatric robot-assisted gait training. Res Dev Disabil. 2013;34 (11):3906–15. https://doi.org/10.1016/j.ridd.2013.07.031.

13. Schuler T, Brutsch K, Muller R, van Hedel HJA, Meyer-Heim A. Virtual realities as motivational tools for robotic assisted gait training in children: a surface electromyography study. NeuroRehabilitation. 2011;28(4):401–11. https://doi.org/10.3233/Nre-2011-0670.

14. Fahr A, Klay A, Keller JW, van Hedel HJA. An interactive computer game for improving selective voluntary motor control in children with upper motor neuron lesions: development and preliminary feasibility study. JMIR Serious Games. 2021;9(3): e26028. https://doi.org/10.2196/26028.

15. Gilardi F, De Falco F, Casasanta D, Andellini M, Gazzellini S, Petrarca M, et al. Robotic technology in pediatric neurorehabilitation. A pilot study of human factors in an Italian pediatric hospital. Int J Environ Res Public Health. 2020;17(10):3503. https://doi.org/10.3390/ijerph17103503.

16. Aurich T, Warken B, Graser JV, Ulrich T, Borggraefe I, Heinen F, et al. Practical Recommendations for robot-assisted treadmill therapy (Lokomat) in children with cerebral palsy: indications, goal setting, and clinical implementation within the WHO-ICF framework. Neuropediatrics. 2015;46(4):248–60. https://doi.org/10.1055/s-0035-1550150.

17. Schroeder AS, Homburg M, Warken B, Auffermann H, Koerte I, Berweck S, et al. Prospective controlled cohort study to evaluate changes of function, activity and participation in patients with bilateral spastic cerebral palsy after Robot-enhanced repetitive treadmill therapy. Eur J Paediatr Neuro. 2014;18(4):502–10. https://doi.org/10.1016/j.ejpn.2014.04.012.

18. Van Hedel HJ, Wick K, Eng K, Meyer-Heim A. Improving dexterity in children with cerebral palsy —preliminary results of a randomized trial evaluating a glove based VR-system. Int Conf on Virtual Rehab.2011. p. 127–32.

19. Berker AN, Yalcin MS. Cerebral palsy: orthopedic aspects and rehabilitation. Pediatr Clin North Am. 2008;55(5):1209–25, ix. https://doi.org/10.1016/j.pcl.2008.07.011.

20. Damiano DL, Alter KE, Chambers H. New clinical and research trends in lower extremity management for ambulatory children with cerebral palsy. Phys Med Rehabil Clin N Am. 2009;20(3):469–91. https://doi.org/10.1016/j.pmr.2009.04.005.

21. Majnemer A, Shevell M, Law M, Poulin C, Rosenbaum P. Level of motivation in mastering challenging tasks in children with cerebral palsy. Dev Med Child Neurol. 2010;52(12):1120–6. https://doi.org/10.1111/j.1469-8749.2010.03732.x.

22. Phelan SK, Gibson BE, Wright FV. What is it like to walk with the help of a robot? Children's perspectives on robotic gait training technology. Disabil Rehabil. 2015;37(24):2272–81. https://doi.org/10.3109/09638288.2015.1019648.

23. Beveridge B, Feltracco D, Struyf J, Strauss E, Dang S, Phelan S, et al. "You gotta try it all": Parents' Experiences with Robotic Gait Training for their Children with Cerebral Palsy. Phys Occup Ther Pediatr. 2015;35(4):327–41. https://doi.org/10.3109/01942638.2014.990547.

24. Jennings KD, Connors RE, Stegman CE. Does a physical handicap alter the development of mastery motivation during the preschool years. J Am Acad Child Psy. 1988;27(3):312–7. https://doi.org/10.1097/00004583-198805000-00008.

25. Messier J, Ferland F, Majnemer A. Play behavior of school age children with intellectual disability: their capacities, interests and attitude. J Dev Phys Disabil. 2008;20(2):193–207. https://doi.org/10.1007/s10882-007-9089-x.

26. Gibson BE, Teachman G, Wright V, Fehlings D, Young NL, McKeever P. Children's and parents' beliefs regarding the value of walking: rehabilitation implications for children with cerebral palsy. Child Care Hlth Dev. 2012;38(1):61–9. https://doi.org/10.1111/j.1365-2214.2011.01271.x.

27. Chiarello LA, Palisano RJ, Maggs JM, Orlin MN, Almasri N, Kang LJ, et al. Family priorities for activity and participation of children and youth with cerebral palsy. Phys Ther. 2010;90(9):1254–64. https://doi.org/10.2522/ptj.20090388.

28. Rast FM, Labruyere R. ICF mobility and self-care goals of children in inpatient rehabilitation. Dev Med Child Neurol. 2020;62(4):483–8. https://doi.org/10.1111/dmcn.14471.

29. Aurich-Schuler T, Grob F, van Hedel HJA, Labruyere R. Can Lokomat therapy with children and adolescents be improved? An adaptive clinical pilot trial comparing Guidance force, Path control, and FreeD. J Neuroeng Rehabil. 2017;14(1):76. https://doi.org/10.1186/s12984-017-0287-1.

30. Aurich-Schuler T, Labruyere R. An increase in kinematic freedom in the Lokomat is related to the ability to elicit a physiological muscle activity pattern: a secondary data analysis investigating differences between guidance force, path control, and freed. Front Robot AI. 2019;6:109. https://doi.org/10.3389/frobt.2019.00109.

31. Oudenhoven LM, Booth ATC, Buizer AI, Harlaar J, van der Krogt MM. How normal is normal: consequences of stride to stride variability, treadmill walking and age when using normative paediatric gait data. Gait Posture. 2019;70:289–97. https://doi.org/10.1016/j.gaitpost.2019.03.011.

32. van Hedel HJA, Rosselli I, Baumgartner-Ricklin S. Clinical utility of the over-ground bodyweight-supporting walking system Andago in children and youths with gait impairments. J Neuroeng Rehabil. 2021;18(1):29. https://doi.org/10.1186/s12984-021-00827-1.

33. Chiu HC, Ada L, Bania TA. Mechanically assisted walking training for walking, participation, and quality of life in children with cerebral palsy. Cochrane Database Syst Rev. 2020;11:CD013114. https://doi.org/10.1002/14651858.CD013114.pub2.

34. Calabro RS, Naro A, Russo M, Leo A, Balletta T, Sacca I, et al. Do post-stroke patients benefit from robotic verticalization? A pilot-study focusing on a novel neurophysiological approach. Restor Neurol Neurosci. 2015;33(5):671–81. https://doi.org/10.3233/RNN-140475.

35. Kumar S, Yadav R, Aafreen. Comparison between Erigo tilt-table exercise and conventional physiotherapy exercises in acute stroke patients: a randomized trial. Arch Physiother. 2020;10:3. https://doi.org/10.1186/s40945-020-0075-2.

36. Schmidt-Lucke C, Kaferle J, Rydh Berner BM, Ahlborg L, Hansen HM, Skjellvik Tollefsen U et al. Effect of assisted walking-movement in patients with genetic and acquired neuromuscular disorders with the motorised Innowalk device: an international case study meta-analysis. PeerJ. 2019;7:e7098. https://doi.org/10.7717/peerj.7098.

37. Tornberg AB, Lauruschkus K. Non-ambulatory children with cerebral palsy: effects of four months of static and dynamic standing exercise on passive range of motion and spasticity in the hip. PeerJ. 2020;8:e8561. https://doi.org/10.7717/peerj.8561.

38. Lefmann S, Russo R, Hillier S. The effectiveness of robotic-assisted gait training for paediatric gait disorders: systematic review. J Neuroeng Rehabil. 2017;14(1):1. https://doi.org/10.1186/s12984-016-0214-x.

39. Carvalho I, Pinto SM, Chagas DDV, Praxedes Dos Santos JL, de Sousa Oliveira T, Batista LA. Robotic gait training for individuals with cerebral palsy: a systematic review and meta-analysis. Arch Phys Med Rehabil. 2017;98(11):2332–44. https://doi.org/10.1016/j.apmr.2017.06.018.

40. Druzbicki M, Rusek W, Snela S, Dudek J, Szczepanik M, Zak E, et al. Functional effects of robotic-assisted locomotor treadmill thearapy in children with cerebral palsy. J Rehabil Med. 2013;45(4):358–63. https://doi.org/10.2340/16501977-1114.

41. Wallard L, Dietrich G, Kerlirzin Y, Bredin J. Robotic-assisted gait training improves walking abilities in diplegic children with cerebral palsy. Eur J Paediatr Neuro. 2017;21(3):557–64. https://doi.org/10.1016/j.ejpn.2017.01.012.

42. Peri E, Turconi AC, Biffi E, Maghini C, Panzeri D, Morganti R, et al. Effects of dose and duration of robot-assisted gait training on walking ability of children affected by cerebral palsy. Technol Health Care. 2017;25(4):671–81. https://doi.org/10.3233/Thc-160668.

43. Ammann-Reiffer C, Bastiaenen CHG, Meyer-Heim AD, van Hedel HJA. Lessons learned from conducting a pragmatic, randomized, crossover trial on robot-assisted gait training in children with cerebral palsy (PeLoGAIT). J Pediatr Rehabil Med. 2020;13(2):137–48. https://doi.org/10.3233/PRM-190614.

44. Klobucka S, Klobucky R, Kollar B. Effect of robot-assisted gait training on motor functions in adolescent and young adult patients with bilateral spastic cerebral palsy: A randomized controlled trial. NeuroRehabilitation. 2020;47(4):495–508. https://doi.org/10.3233/NRE-203102.

45. Meyer-Heim A, Borggraefe I, Ammann-Reiffer C, Berweck S, Sennhauser FH, Colombo G, et al. Feasibility of robotic-assisted locomotor training in children with central gait impairment. Dev Med Child Neurol. 2007;49(12):900–6. https://doi.org/10.1111/j.1469-8749.2007.00900.x.

46. Borggraefe I, Klaiber M, Schuler T, Warken B, Schroeder SA, Heinen F, et al. Safety of robotic-assisted treadmill therapy in children and adolescents with gait impairment: a bi-centre survey. Dev Neurorehabil. 2010;13(2):114–9. https://doi.org/10.3109/17518420903321767.

47. Meyer-Heim A, Ammann-Reiffer C, Schmartz A, Schafer J, Sennhauser FH, Heinen F, et al. Improvement of walking abilities after robotic-assisted locomotion training in children with cerebral palsy. Arch Dis Child. 2009;94(8):615–20. https://doi.org/.10.1136/adc.2008.145458.

48. Borggraefe I, Kiwull L, Schaefer JS, Koerte I, Blaschek A, Meyer-Heim A, et al. Sustainability of motor performance after robotic-assisted treadmill therapy in children: an open, non-randomized baseline-treatment study. Eur J Phys Rehabil Med. 2010;46(2):125–31. https://doi.org/R33102225[pii].

49. Arellano-Martinez IT, Rodriguez-Reyes G, Quinones-Uriostegui I, Arellano-Saldana ME. [Spatial-temporal analysis and clinical findings of gait: comparison of two modalities of treatment in children with cerebral palsy-spastic hemiplegia. Preliminary report]. Cir Cir. 2013;81(1):14–20.

50. Weinberger R, Warken B, Konig H, Vill K, Gerstl L, Borggraefe I, et al. Three by three weeks of robot-enhanced repetitive gait therapy within a global rehabilitation plan improves gross motor development in children with cerebral palsy - a retrospective cohort study. Eur J Paediatr Neurol. 2019;23(4):581–8. https://doi.org/10.1016/j.ejpn.2019.05.003.

51. Patritti B, Sicari M, Deming L, Romaguera F, Pelliccio M, Benedetti MG, et al. Enhancing robotic gait training via augmented feedback. Conference proceedings : Annu Int Conf IEEE Eng Med Biol Soc. 2010;2010:2271–4. https://doi.org/10.1109/IEMBS.2010.5627707.

52. Schroeder AS, Von Kries R, Riedel C, Homburg M, Auffermann H, Blaschek A, et al. Patient-specific

determinants of responsiveness to robot-enhanced treadmill therapy in children and adolescents with cerebral palsy. Dev Med Child Neurol. 2014;56 (12):1172–9. https://doi.org/10.1111/dmcn.12564.

53. Beretta E, Storm FA, Strazzer S, Frascarelli F, Petrarca M, Colazza A, et al. Effect of robot-assisted gait training in a large population of children with motor impairment due to cerebral palsy or acquired brain injury. Arch Phys Med Rehabil. 2020;101(1):106–12. https://doi.org/10.1016/j.apmr.2019.08.479.

54. Van Hedel HJ, Meyer-Heim A, Rüsch-Bohtz C. Robot-assisted gait training might be beneficial for more severely affected children with cerebral palsy: Brief report. Dev Neurorehabil. 2016;19(6):410-415.

55. van Kammen K, Reinders-Messelink HA, Elsinghorst AL, Wesselink CF, Meeuwisse-de Vries B, van der Woude LHV et al. Amplitude and stride-to-stride variability of muscle activity during Lokomat guided walking and treadmill walking in children with cerebral palsy. Eur J Paediatr Neurol. 2020;29:108–17. https://doi.org/10.1016/j.ejpn.2020.08.003.

56. Aurich Schuler T, Muller R, van Hedel HJ. Leg surface electromyography patterns in children with neuro-orthopedic disorders walking on a treadmill unassisted and assisted by a robot with and without encouragement. J Neuroeng Rehabil. 2013;10:78. https://doi.org/10.1186/1743-0003-10-78.

57. Ricklin S, Meyer-Heim A, van Hedel HJA. Dual-task training of children with neuromotor disorders during robot-assisted gait therapy: prerequisites of patients and influence on leg muscle activity. J Neuroeng Rehabil. 2018;15(1):82.

58. Bernstein N. The coordination and regulation of movements. New York Pergamon, 1967.

59. Hornby TG, Reinkensmeyer DJ, Chen D. Manually-assisted versus robotic-assisted body weight-supported treadmill training in spinal cord injury: what is the role of each? PM R. 2010;2(3):214–21. https://doi.org/10.1016/j.pmrj.2010.02.013.

60. Winstein C, Wing AM, Whitall J. Motor control and learning principles for rehabilitation of upper limb movements after brain injury. In: Grafman J, Robertson IH, editors. Handbook of neuropsychology, 2nd ed., 2003.

61. Krakauer JW. Motor learning: its relevance to stroke recovery and neurorehabilitation. Curr Opin Neurol. 2006;19(1):84–90. https://doi.org/10.1097/01.wco.0000200544.29915.cc.

62. Cai LL, Fong AJ, Otoshi CK, Liang YQ, Burdick JW, Roy RR, et al. Implications of assist-as-needed robotic step training after a complete spinal cord injury on intrinsic strategies of motor learning. J Neurosci. 2006;26(41):10564–8. https://doi.org/10.1523/Jneurosci.2266-06.2006.

63. Duschau-Wicke A, Caprez A, Riener R. Patient-cooperative control increases active participation of individuals with SCI during robot-aided gait training. J Neuroeng Rehabil. 2010;7. Artn 43. https://doi.org/10.1186/1743-0003-7-43.

64. Duschau-Wicke A, von Zitzewitz J, Caprez A, Lunenburger L, Riener R. Path control: a method for patient-cooperative robot-aided gait rehabilitation. IEEE Trans Neural Syst Rehabil Eng. 2010;18 (1):38–48. https://doi.org/10.1109/TNSRE.2009.2033061.

65. Krishnan C, Kotsapouikis D, Dhaher YY, Rymer WZ. Reducing robotic guidance during robot-assisted gait training improves gait function: a case report on a stroke survivor. Arch Phys Med Rehab. 2013;94(6):1202–6. https://doi.org/10.1016/j.apmr.2012.11.016.

66. Lee C, Won D, Cantoria MJ, Hamlin M, de Leon RD. Robotic assistance that encourages the generation of stepping rather than fully assisting movements is best for learning to step in spinally contused rats. J Neurophysiol. 2011;105(6):2764–71. https://doi.org/10.1152/jn.01129.2010.

67. Schuck A, Labruyere R, Vallery H, Riener R, Duschau-Wicke A. Feasibility and effects of patient-cooperative robot-aided gait training applied in a 4-week pilot trial. J Neuroeng Rehabil. 2012;9:31. https://doi.org/10.1186/1743-0003-9-31.

68. Aurich-Schuler T, Gut A, Labruyere R. The FreeD module for the Lokomat facilitates a physiological movement pattern in healthy people - a proof of concept study. J Neuroeng Rehabil. 2019;16(1):26. https://doi.org/10.1186/s12984-019-0496-x.

69. Diot CM, Thomas RL, Raess L, Wrightson JG, Condliffe EG. Robotic lower extremity exoskeleton use in a non-ambulatory child with cerebral palsy: a case study. Disabil Rehabil-Assi. 2021. https://doi.org/10.1080/17483107.2021.1878296.

70. Smania N, Bonetti P, Gandolfi M, Cosentino A, Waldner A, Hesse S, et al. Improved gait after repetitive locomotor training in children with cerebral palsy. Am J Phys Med Rehabil Assoc Acad Phys. 2011;90(2):137–49. https://doi.org/10.1097/PHM.0b013e318201741e.

71. Novak I, Morgan C, Fahey M, Finch-Edmondson M, Galea C, Hines A, et al. State of the evidence traffic lights 2019: systematic review of interventions for preventing and treating children with cerebral palsy. Curr Neurol Neurosci Rep. 2020;20 (2):3. https://doi.org/10.1007/s11910-020-1022-z.

72. van Hedel HJ, Bulloni A, Gut A. Prefrontal cortex and supplementary motor area activation during robot-assisted weight-supported over-ground walking in young neurological patients: a pilot fNIRS study. Front Rehabilit Sci. 2021. https://doi.org/10.3389/fresc.2021.788087.

73. van Gelder L, Booth ATC, van de Port I, Buizer AI, Harlaar J, van der Krogt MM. Real-time feedback to improve gait in children with cerebral palsy. Gait Posture. 2017;52:76–82. https://doi.org/10.1016/j.gaitpost.2016.11.021.

74. Gagliardi C, Turconi AC, Biffi E, Maghini C, Marelli A, Cesareo A, et al. Immersive virtual

reality to improve walking abilities in cerebral palsy: a pilot study. Ann Biomed Eng. 2018;46 (9):1376–84. https://doi.org/10.1007/s10439-018-2039-1.

75. Biffi E, Gagliardi C, Maghini C, Genova C, Panzeri D, Redaelli DF, et al. Learning my way: a pilot study of navigation skills in cerebral palsy in immersive virtual reality. Front Psychol. 2020;11: 591296. https://doi.org/10.3389/fpsyg.2020.591296.

76. Ravi DK, Kumar N, Singhi P. Effectiveness of virtual reality rehabilitation for children and adolescents with cerebral palsy: an updated evidence-based systematic review. Physiotherapy. 2017;103 (3):245–58. https://doi.org/10.1016/j.physio.2016.08.004.

77. Chen Y, Fanchiang HD, Howard A. Effectiveness of virtual reality in children with cerebral palsy: a systematic review and meta-analysis of randomized controlled trials. Phys Ther. 2018;98(1):63–77. https://doi.org/10.1093/ptj/pzx107.

78. Bailey JO, Bailenson JN. Immersive virtual reality and the developing child. In: Blumberg FC, Brooks PJ, editors. Cognitive development in digital contexts. Academic Press; 2017. p. 181–200.

79. Morrongiello BA, Corbett M, Beer J, Koutsoulianos S. A pilot randomized controlled trial testing the effectiveness of a pedestrian training program that teaches children where and how to cross the street safely. J Pediatr Psychol. 2018;43(10):1147–59. https://doi.org/10.1093/jpepsy/jsy056.

80. Ammann-Reiffer C, Keller U, Klay A, Meier L, van Hedel HJA. Do youths with neuromotor disorder and their therapists prefer a mixed or virtual reality head-mounted display? J Rehabil Med Clin Commun. 2021;4:1000072. https://doi.org/10.2340/20030711-1000072.

81. van Hedel HJ, Lieber J, Ricklin S, Meyer-Heim A. Die praktische Anwendung von Exergames und virtueller Realität in der pädiatrischen Rehabilitation. Neurorehabil Neural Repair. 2017;9(1):35–40.

82. Chen YP, Howard AM. Effects of robotic therapy on upper-extremity function in children with cerebral palsy: a systematic review. Dev Neurorehabil. 2016;19(1):64–71. https://doi.org/10.3109/17518423.2014.899648.

83. Rathinam C, Mohan V, Peirson J, Skinner J, Nethaji KS, Kuhn I. Effectiveness of virtual reality in the treatment of hand function in children with cerebral palsy: a systematic review. J Hand Ther. 2019;32(4):426–34 e1. https://doi.org/10.1016/j.jht.2018.01.006.

84. Bono GLP, Achermann P, Rückriem B, Lieber J, van Hedel HJA. Goal-directed personalized upper limb intensive therapy (PULIT) for children with hemiparesis: a retrospective analysis. Am J Occup Ther 2021;In press.

85. Roberts H, Shierk A, Clegg NJ, Baldwin D, Smith L, Yeatts P, et al. Constraint induced movement therapy camp for children with hemiplegic cerebral palsy augmented by use of an exoskeleton to play games in virtual reality. Phys Occup Ther Pediatr. 2021;41(2):150–65. https://doi.org/10.1080/01942638.2020.1812790.

86. James S, Ziviani J, Ware RS, Boyd RN. Randomized controlled trial of web-based multimodal therapy for unilateral cerebral palsy to improve occupational performance. Dev Med Child Neurol. 2015;57 (6):530–8. https://doi.org/10.1111/dmcn.12705.

87. Sakzewski L, Lewis MJ, McKinlay L, Ziviani J, Boyd RN. Impact of multi-modal web-based rehabilitation on occupational performance and upper limb outcomes: pilot randomized trial in children with acquired brain injuries. Dev Med Child Neurol. 2016;58(12):1257–64. https://doi.org/10.1111/dmcn.13157.

88. Keller U, van Hedel HJA, Klamroth-Marganska V, Riener R. ChARMin: the first actuated exoskeleton robot for Pediatric Arm rehabilitation. IEEE/ASME Trans Mechatron. 2016;21(5):2201–13.

89. Graser JV, Bastiaenen CHG, Gut A, Keller U, van Hedel HJA. Contextual interference in children with brain lesions: a pilot study investigating blocked vs. random practice order of an upper limb robotic exergame. Pilot Feasibility Stud. 2021;7(1):135. https://doi.org/10.1186/s40814-021-00866-4.

90. Gilliaux M, Renders A, Dispa D, Holvoet D, Sapin J, Dehez B, et al. Upper limb robot-assisted therapy in cerebral palsy: a single-blind randomized controlled trial. Neurorehabil Neural Repair. 2015;29(2):183–92. https://doi.org/10.1177/1545968314541172.

91. Bishop L, Gordon AM, Kim H. Hand robotic therapy in children with hemiparesis: a pilot study. Am J Phys Med Rehabil. 2017;96(1):1–7. https://doi.org/10.1097/PHM.0000000000000537.

92. Keller JW, Fahr A, Lieber J, Balzer J, van Hedel HJA. Impact of upper extremity impairment and trunk control on self-care independence in children with upper motor neuron lesions. Phys Ther. 2021;101(8). https://doi.org/10.1093/ptj/pzab112.

93. Turconi AC, Biffi E, Maghini C, Peri E, Servodio Iammarone F, Gagliardi C. Can new technologies improve upper limb performance in grown-up diplegic children? Eur J Phys Rehabil Med. 2016;52(5):672–81.

94. El-Shamy SM. Efficacy of Armeo(R) robotic therapy versus conventional therapy on upper limb function in children with hemiplegic cerebral palsy. Am J Phys Med Rehabil. 2018;97(3):164–9. https://doi.org/10.1097/PHM.0000000000000852.

95. Beretta E, Cesareo A, Biffi E, Schafer C, Galbiati S, Strazzer S. Rehabilitation of upper limb in children with acquired brain injury: a preliminary comparative study. J Healthc Eng. 2018;2018:4208492. https://doi.org/10.1155/2018/4208492.

96. Wille D, Eng K, Holper L, Chevrier E, Hauser Y, Kiper D, et al. Virtual reality-based paediatric interactive therapy system (PITS) for improvement of arm and hand function in children with motor impairment–a pilot study. Dev Neurorehabil.

2009;12(1):44–52. https://doi.org/10.1080/1751842
0902773117.

97. Choi JY, Yi SH, Ao L, Tang X, Xu X, Shim D, et al. Virtual reality rehabilitation in children with brain injury: a randomized controlled trial. Dev Med Child Neurol. 2021;63(4):480–7. https://doi. org/10.1111/dmcn.14762.

98. Fahr A, Klay A, Coka LS, van Hedel HJA. Game-based training of selective voluntary motor control in children and youth with upper motor neuron lesions: protocol for a multiple baseline design study. Bmc Pediatr. 2021;21(1):505. https://doi.org/ 10.1186/s12887-021-02983-8.

99. Ng KW, Augustine L, Inchley J. Comparisons in screen-time behaviours among adolescents with and without long-term illnesses or disabilities: results from 2013/14 HBSC study. Int J Environ Res Public Health. 2018;15(10). https://doi.org/10.3390/ ijerph15102276.

100. Curtis DJ, Butler P, Saavedra S, Bencke J, Kalle-mose T, Sonne-Holm S, et al. The central role of trunk control in the gross motor function of children with cerebral palsy: a retrospective cross-sectional study. Dev Med Child Neurol. 2015;57(4):351–7. https://doi.org/10.1111/dmcn.12641.

101. Balzer J, Marsico P, Mitteregger E, van der Linden ML, Mercer TH, van Hedel HJA. Influence of trunk control and lower extremity impairments on gait capacity in children with cerebral palsy. Disabil Rehabil. 2018;40(26):3164–70. https://doi. org/10.1080/09638288.2017.1380719.

102. Park SH, Son SM, Choi JY. Effect of posture control training using virtual reality program on sitting balance and trunk stability in children with cerebral palsy. NeuroRehabilitation. 2021;48 (3):247–54. https://doi.org/10.3233/NRE-201642.

103. Butzer T, Dittli J, Lieber J, van Hedel HJA, Meyer-Heim A, Lambercy O, et al. PEXO - a pediatric whole hand exoskeleton for grasping assistance in task-oriented training. IEEE Int Conf Rehabil Robot. 2019;2019:108–14. https://doi.org/10.1109/ ICORR.2019.8779489.

104. Lieber J, Dittli J, Lambercy O, Gassert R, Meyer-Heim A, van Hedel HJA. Clinical utility of a pediatric hand exoskeleton: identifying users, prac-ticability, and acceptance, and recommendations for design improvement. J Neuroeng Rehabil. 2022;19 (1):17. https://doi.org/10.1186/s12984-022-00994-9.

Assessment Technology and Predictive Modeling

Robotic Technologies and Digital Health Metrics for Assessing Sensorimotor Disability

15

Christoph M. Kanzler, Marc Bolliger, and Olivier Lambercy

Abstract

Neurological disorders such as stroke, multiple sclerosis, traumatic brain injury, cerebral palsy, or spinal cord injury result in partial or complete sensorimotor impairments in the affected limbs. To provide an optimal and personalized rehabilitation program, a detailed assessment of the nature and degree of the sensorimotor deficits, as well as their temporal evolution, is crucial. Valid, reliable, and standardized assessments are essential to define the rehabilitation setting and adapt it over the course of a therapy. Many clinical assessments have a limited sensitivity and are not able to capture behavioral intra- and inter-participant variability, which limits their suitability as endpoints for clinical trials and for clinical decision-making. Technological solutions, such as robotics or wearable sensors, are promising approaches that can provide objective, sensitive, and reliable digital health metrics, which could help overcome the common limitations of conventional clinical assessments. This chapter focuses on the novel possibilities that robotic devices offer for assessing upper and lower limb disability and provides an overview of existing approaches. Further, we discuss how such digital health metrics can be selected and validated, and how they could be integrated into predictive computational models. We conclude that robotics and digital health metrics are excellent tools to describe sensorimotor disability that they promise novel insights into long-term recovery and provide the basis for a more data-driven and personalized clinical decision-making.

C. M. Kanzler
Future Health Technologies, Singapore-ETH Centre, Campus for Research Excellence and Technological Enterprise (CREATE), Singapore, Singapore
e-mail: christoph.kanzler@sec.ethz.ch

M. Bolliger
Spinal Cord Injury Center, University of Zurich, Balgrist University Hospital, Forchstrasse 340, 8008 Zurich, Switzerland
e-mail: marc.bolliger@balgrist.ch

O. Lambercy (✉)
Department of Health Sciences and Technology, Rehabilitation Engineering Laboratory, ETH Zurich, BAA Lengghalde 5, 8008 Zurich, Switzerland
e-mail: olivier.lambercy@hest.ethz.ch

Keywords

Neurorehabilitation · Sensorimotor impairment · Clinical assessment · Medical robotics · Rehabilitation robotics · Robot-assisted assessment · Computational models · Prediction models

© The Author(s), under exclusive license to Springer Nature Switzerland AG 2022
D. J. Reinkensmeyer et al. (eds.), *Neurorehabilitation Technology*,
https://doi.org/10.1007/978-3-031-08995-4_15

15.1 Introduction/Motivation

Neurological damage following a stroke, spinal cord injury (SCI), or other neurological disorder can result in severe impairment of sensorimotor function. A detailed assessment and understanding of the nature and level of sensorimotor deficits is crucial for neurorehabilitation in several ways. In an early phase after the neurological injury, assessments are used to diagnose the level of sensorimotor impairment. This diagnosis then serves as a basis to identify the most suitable therapy, i.e., to establish appropriate protocols tailored to the patient's needs and goals. In a subsequent phase, therapy is progressively adapted based on assessments, by tuning training parameters, e.g., type and complexity of a task, to optimally challenge and engage patients during rehabilitation. In addition, valid, reliable, and standardized assessments are needed in order to evaluate the effect of new therapeutic interventions. Finally, due to rising healthcare costs, assessments have an important socioeconomic role, as hospitals and insurance companies offer their services based on clinically meaningful thresholds on standardized assessment scales.

A *clinical assessment* can be defined as the evaluation of a patient's physical condition and prognosis based on a physical inspection by a clinician or therapist. Throughout the course of a rehabilitation therapy, clinical assessments are usually repeated only at a few stages to monitor the patient's status and progress. The quality of an assessment method is defined by its sensitivity, validity, and reliability. Validity describes how precisely a scale assesses what it intends to measure. Hence, validity cannot be described by an all-or-nothing metric, but is rather multi-dimensional and continuous. Reliability is given if results are consistent on repeated administrations of the same test [1], which can be by the same or different raters (intra- or inter-rater reliability), or at two different points in time (test–retest reliability). Sensitivity, or responsiveness, of an assessment is defined as its ability to detect and quantify real changes (i.e., recovery or decline). For use in a clinical setting, it is essential that an assessment can detect changes due to therapeutic effects of interventions [2]. Nevertheless, many clinical assessments suffer from limitations such as low intra- or inter-rater reliability, low sensitivity, poor validity, or floor/ceiling effects, which is known to introduce bias when predicting sensorimotor recovery [3–5]. Furthermore, they are often time-consuming to administer, which limits the number and frequency of assessments that can be performed, especially outside of research studies.

The field of rehabilitation technologies has seen increasing interest and development over the last decades [6–10]. Robotic devices emerged as a promising approach to complement conventional therapy. Through their embedded sensors, such rehabilitation robots or similar standalone technology-based assessments provide unique platforms for more objective and sensitive patient evaluation [11–15]. By *robotic assessments*, we understand the evaluation of the physical condition (in terms of sensorimotor function) of a patient by interpreting kinematic and kinetic data (both during active/assisted movement of the patient as well as during passive guidance or perturbations). Robotic systems offer—depending on the used technology—the possibility to precisely and objectively record movement trajectories, limb posture, completion time, task precision, etc., and measure interaction forces during well-controlled and standardized tasks. These allow extracting task-related metrics descriptive of sensorimotor function of a patient, herein referred to as *digital health metrics* [16, 17]. Additional to this observational approach, robotic devices can actively excite or perturb the patients' movement in order to investigate neuromuscular control and related dysfunctions, and even be used concurrently with neuroimaging to gain deeper insights into the impaired neural mechanisms [18].

Clinical assessments and those based on advanced technology are fundamentally different, but both aim at providing patients and therapists with a precise evaluation of sensorimotor functions or their impact on activities or participation. Both types of assessment provide valuable and complementary information, but the specific characteristics of each assessment may often preclude a direct comparison. With the International Classification of Functioning, Disability and Health (ICF), a common reference framework for describing disability has been established. The goal of the ICF is to serve as a scientific basis to describe the health status of an individual with a common language. This common language allows comparison of results between clinics all over the world. In the context of the ICF, the health condition of an individual can be described by three main components: (1) body functions and structures, (2) activities, and (3) participation. There is a dynamic interaction between these three entities: changes or intervention in one may have influence on one of the other components or both [19]. ICF further distinguishes measures of activity capacity (i.e., what an individual can do in a standardized environment, such as during a clinical evaluation) from measures of activity performance (i.e., what an individual actually does in his/her usual environment, for example, at home). In an attempt to provide a description of clinical- and technology-based sensorimotor assessments, we here group them according to whether they are time-based (e.g., 10 m-timed walking test or the nine hole peg test), observer-based (e.g., Fugl-

Meyer assessment or modified Ashworth scale), or technology-based (e.g., movement smoothness) assessments, considering various levels of complexity (Fig. 15.1).

Despite robotic assessments being promising and their integration into research studies steadily increasing, they still face fundamental challenges that prevent their widespread usage. These relate, for example, to practicalities in the application of complex technical devices in clinical environments, for example, due to the required resources to familiarize caregivers and patients with the devices and to perform the assessments. Another important challenge in robotic assessments is to translate raw measurements of physical quantities (e.g., acceleration) collected by the various sensors into clinically meaningful scales representative of sensorimotor disability (i.e., digital health metrics, Fig. 15.1). Depending on the technology, task, and the measured quantity, digital health metrics can either be directly deduced from the physical quantities or may require advanced processing algorithms. Another challenge lies in the selection and validation of specific digital health metrics for a use-case or research question, as a large variety of metrics is applied in literature and best practice methodological approaches for validating such metrics have not been described. Lastly, a major question

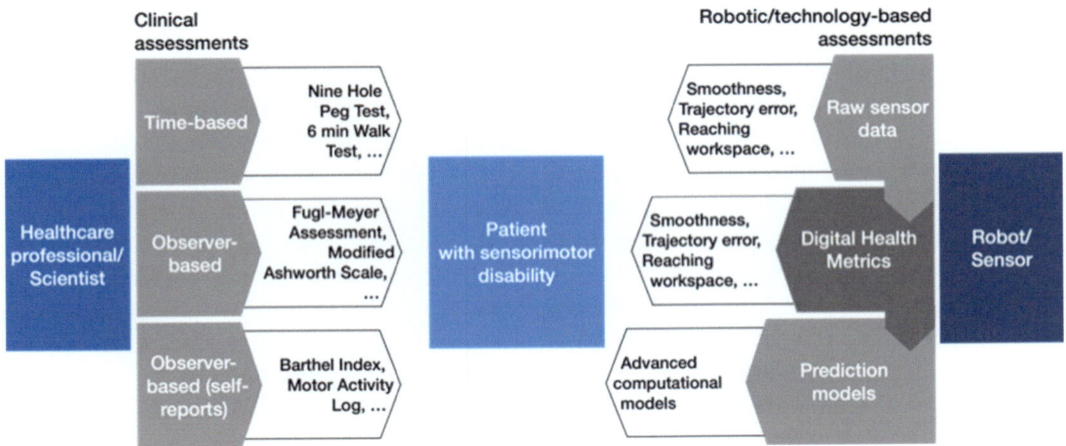

Fig. 15.1 Schematic representation of approaches for clinical and robotic/technology-based assessments used to evaluate patients' sensorimotor disability resulting from neurological injuries

is how digital health metrics can be used beyond outcome measures in clinical trials and instead influence daily clinical decision-making.

This chapter will briefly review clinical scales commonly employed after neurological injuries. It will then provide an overview of the current state of the art in robotic assessments for the upper and lower extremities, and how these can be used to automatically adapt technology-assisted therapy. We first present different strategies used to evaluate sensorimotor impairments, briefly mention examples of alternative technology-based approaches, and discuss how digital health metrics of various complexity can be selected and validated. Further, we report on the possibility of using digital health metrics in computational models for predicting sensorimotor recovery and their potential for clinical routine and research. The chapter will conclude with an outlook on the main challenges toward realizing generally accepted technology-based assessments. Thus, this chapter will serve as a concise overview of the field of robotic assessments in neurological injuries and open grand challenges, thereby aspiring to guide novice and expert neurorehabilitation researchers.

15.2 Clinical Assessments

Many clinical assessments for upper or lower limb function have been developed for use in different neurological conditions. Unfortunately, many challenges remain with regard to the systematic application of these clinical assessments. In the following, examples of clinical assessments of lower and upper limb function are given. The assessments are grouped into time-based and observer-based assessments.

Time-based assessments rate the individual abilities based on the time required to complete the assessment task. The measurements are done on an interval or ratio scale (time). Following neurological diseases that affect the lower limbs, time-based walking tests are widely performed. The time required to accomplish the test can also be used to calculate the walking speed, a parameter that has been linked to a number of

higher functions, such as cognitive function [20, 21]. A typical time-based clinical assessment is the 10 m walking test (10 mWT), in which the patient is asked to walk 10 m along a defined direction. An example for a time-based assessment for the upper extremity is the Nine Hole Peg Test (NHPT [22]). The NHPT was developed to measure finger dexterity and can be applied to patients with low to moderate impairment of upper limb function due to a variety of neurological diseases. The task consists in taking nine pegs from a container (one by one) and placing them into nine holes on a square board as fast as possible.

Observed-based assessments measure how well a patient can achieve a specific task, which is expected to describe a specific sensorimotor impairment or to be representative of a daily life activity. A common impairment after a neurological lesion is spasticity, which is characterized by disordered sensorimotor control, resulting from an upper motor neuron lesion, presenting as intermittent or sustained involuntary activation of muscles [23]. However, the role of spasticity in walking impairment is debated. More recently, a differentiated look at spasticity has been provided [24], based on the different neural and anatomical effects that are relevant in passive conditions, during non-functional and during functional movements. A clinical assessment method of spasticity in the passive condition is the Modified Ashworth Scale (MAS) [25]. The test can be applied to muscles of the lower as well as the upper limbs. The rater flexes and extends the patient's limb from maximal extension to maximal flexion or vice versa while the patient is instructed to remain passive and rates the perceived resistance on a six-point ordinal scale. The Ashworth scale and the modified Ashworth scale are the most commonly used assessment methods in clinical as well as scientific research to measure spasticity, despite being strongly criticized [26]. Impairments of the sensory pathways are also common, specifically those affecting proprioception. Clinical assessments of proprioception usually focus on detecting or replicating a movement executed by a rater. In the first case, the patient is asked to indicate the direction in which his/her limb was moved by a rater, e.g., [27]. For assessments using movement replication, the patients have to move their limbs according to previously presented positions [28]. Another example for an observer-based assessment is the Fugl-Meyer Assessment (FMA, [29]) that is primarily used in stroke patients. The FMA is well established and widely used clinically and in research studies thanks to its good validity and reliability [30]. Voluntary movement of the upper and lower limbs, balance, sensation, passive range of motion, and pain are assessed, each being scored on a three-point ordinal scale. Patient-reported outcome measures or self-reports can be considered as another form of observer-based clinical assessments (i.e., the patient is the observer). Such assessments typically attempt to describe the behavior of patients in daily life (activity performance) either through questionnaires and/or a diary. An example for this is the Barthel index [31], which consists of 10 items describing daily life activities, mobility, and self-care and participants are asked to answer based on an ordinal scale indicating the level of independence in these items.

Whereas clinical assessments are widely used to diagnose individuals after a neurological injury, they also suffer from limitations. Time-based assessments are usually fast and easy to administer, provide quantitative values, and usually have good validity and repeatability [22, 32]. However, they do not provide information on movement quality, and thus on the impairments or functional limitations underlying the time loss. Typically, time-based assessments were found to be limited when it comes to distinguishing true recovery of motor function from compensatory movements [33]. Observer-based assessments are subjective by nature as they rely on the interpretation of trained raters. Furthermore, they typically have limited sensitivity and present ceiling/flooring effects due to the ordinal scales they use [23, 25, 30, 34]. The responsiveness of these tools and consequently their usability in clinical trials to investigate new intervention therapies are limited. When considering self-reports, these additionally suffer from recollection bias, especially in persons with neurological injuries that might also be affected by cognitive decline [31]. The drawbacks of current clinical assessments could in the future be addressed by technology-based assessments. Despite still being in their infancy, such technology-based assessments have shown the possibility to quantitatively measure and record several parameters concurrently from multiple joints during well-controlled, highly repeatable tasks.

15.3 Robotic Assessments and Digital Health Metrics

The desire to quantify the effect of a specific therapy and the resulting improvements, along with the (financial) pressure on the health system to restrict reimbursement to quantifiably increased therapy outcomes, have motivated the extension of rehabilitation robots to allow performing assessment, as well as the parallel development of robotic tools dedicated to performing assessments. This is especially interesting as robotic systems are per se equipped with sensors required for the control of their multiple degrees of freedom. This can provide detailed information about the movement kinematics and kinetics (e.g., interaction force, active range of motion, movement smoothness, movement accuracy, movement velocity, motor coordination, and amount of robotic assistance), thus promising more objective assessments with higher sensitivity.

15.3.1 Robotic Assessments Based on Raw Sensor Data

The assessment of upper and lower extremity functions with robotic devices can be based on a large variety of data collected by the robot during interaction with a patient, and the main challenge is to properly interpret these data and extract information in a meaningful way mainly not only for clinicians but also for the patients themselves [12, 13]. A first approach is to use raw sensor data or apply to it simple signal processing steps. For example, a simple way to describe disability is to record the number of successful trials in a specific task the patient has to perform with the robotic system. For example, the number of successful reaching movements to a target position represented in a virtual environment during a specific amount of time can be a good general indicator of overall upper limb (shoulder, elbow, and wrist) motor function. Similarly, the time required to perform a specific task, for example, moving a virtual object from one point to another, with or without assistance from a robotic device is a commonly used measure to assess motor function [35, 36]. While easily implementable on any platform, this type of measurement does not provide any information on how well the task is performed by the subject and does not take full advantage of the measurement capabilities of a robotic system.

Training parameters can also be used to assess performance, for example, the desired walking speed (e.g., [37]) or the required amount of body weight support to evaluate gait performance. Although these parameters can be set relatively arbitrarily by the therapist during the training, maximum challenge or minimal assistance required for a patient to perform a task with the robot can be used as a simple way to reflect the sensorimotor ability. If the assistance of the device is automatically adapted through dedicated algorithms, more objective data could eventually be extracted [38, 39].

Raw sensor data can be advantageously collected by most robotic devices for therapy and assessment of upper extremity function equipped with position and force/torque sensors during specific movements with the device. This allows for objective measurement of metrics such as the range of motion (ROM) and maximum voluntary force/torque. Exoskeleton devices provide a simple means to assess joint ROM. For assessment of the passive ROM, the therapist moves the corresponding joint manually through the patient's ROM while the device records the maximal and minimal angles as measured by the integrated position sensors. This procedure was reported for the driven gait orthosis Lokomat (Hocoma, Switzerland) [40] and is generally applicable to all devices with backdrivable joints (i.e., those that can be moved by an external force). When the device is configured to compensate for its own weight and the patient actively moves his/her limb, this method also allows for assessing active ROM. In another example, here for a measurement of maximum voluntary muscle force [41], the exoskeleton system is controlled to maintain pre-defined joint positions while the patient is instructed to generate maximum voluntary force in one joint (e.g., left knee) and in one movement direction (extension or flexion). The computer instructs the movement on the screen and uses audio cues according to a pre-defined fixed sequence of joints and movement directions. The key outcome variable is the maximum torque a patient can generate. It has been shown that this method can provide an objective and reliable outcome measure to record changes in muscle strength following robot-aided gait therapy in patients with incomplete spinal cord injury [42]. Simpler devices focusing on single joints (e.g., ankle) were also used to assess isometric force and passive/active ROM [43, 44]. Similar approaches were used at the level of the upper limb using the arm exoskeleton ARMin to evaluate shoulder, elbow, forearm, and wrist ROM, as well as maximal joint torques in spinal cord injury patients [45]. As they do not follow the position of each joint, the raw position measurements of end-effector robots can be used to assess end-point workspace. For example, the ACT3D arm robot has been used to evaluate arm-reaching workspace of stroke subjects on a virtual table, as well as the effect of shoulder abduction loading [46].

From joint position readouts, attempts have been made to assess proprioception, specifically joint position sense, with robotic devices. While different methodological approaches exist, they all take advantage of the robot to move a limb segment to precise reference positions that patients are asked to either reproduce (i.e., position matching [47]) or compare to a second passive stimulus (i.e., difference threshold, see

[48] for a review). Domingo and Lam [49] evaluated a method that uses an exoskeleton robot to move the leg of a patient in an objective and repeatable manner. The robot passively moves the patient's leg to a target position and subsequently to a distractor position. Then the patient controls the robot using a joystick to replicate the target position. The angular difference between the final and the target position is the main outcome metric. This robotic assessment method was found to be reliable and valid in able-bodied subjects and subjects with incomplete spinal cord injury [49].

15.3.2 Robotic Assessments Based on Advanced Digital Health Metrics

More advanced robotic assessment techniques have been proposed with the aim of extracting additional information from the data collected by robotic platforms. In the following, we consider as advanced digital health metrics the metrics that are extracted from the raw data using dedicated algorithms, with the aim of better evaluating motor function and typical impairments [12, 13].

As example in the case of robots for lower extremity rehabilitation that enables deviation from a prescribed trajectory, such as the LOPES [50], the actual foot trajectory can be analyzed similar to motion capture data in gait analysis. Using this exoskeleton device and footswitches, Van Asseldonk et al. [51] could determine stride length, duration of stride, stance, and swing, as well as double-stance ratio kinematic parameters, to assess the subjects' gait. When no deviation from the prescribed trajectory is possible—e.g., for a high-impedance setting in an impedance controller—the trajectory does not provide any information on the quality of gait. Instead, the drive torques required to keep the patient's movement along the pre-defined trajectory are indicative of the patient's actions. One approach is to use torques measured by the device multiplied by a weighting function selected in order to provide positive values when the patient performs correct movements [52–54]. Averaging for stance and swing phases provides two values per leg and joint that can be displayed to the patient and therapist as an index of the patient's activity, as well as stored for later analysis [54].

Even though devices for lower extremity rehabilitation are mainly designed to support gait movements, they can also be used to perform specific physiological assessments, e.g., to evaluate biomechanical correlates of spasticity. Robot-assisted assessments of spasticity apply passive movements controlled by the device to a single joint while the torque is recorded and analyzed during repetitive movements [55, 56] (see [57] for a review). The addition of electromyography can help determine the actual muscle activity, but increases the complexity of the measurement. One interesting direction is the use of pseudorandom binary perturbations, which is based on system identification [58, 59]. Also, stiffness measurement in multi-joint robots has been described [40], where the mechanical stiffness was calculated offline taking into account the passive effects of the orthosis and of the patient's legs during passive movements using mathematical models. The mentioned assessment methods all focus on non-functional movements, typically isokinetic or sinusoidal patterns or passive conditions, whereas most clinical assessments like the Modified Ashworth Scale (MAS) [25] always use passive conditions. Nevertheless, joint stiffness measured by a device showed a reasonable relation to spasticity measured using the MAS [40]. In principle, robotic devices would have the capability to assess spasticity also during functional movements and could therefore inform clinicians whether treatment of clinical signs of spasticity have a positive or negative effect on functional outcome.

In upper extremity rehabilitation, movements with a robot are less stereotyped than in the case of gait rehabilitation. This requires the development of metrics assessing movement quality without relying on a pre-defined movement pattern (see [13] for a detailed review). Movement smoothness is a typical digital health metric representative of upper limb coordination that has been extensively studied using robotic devices training arm-reaching movements. In the literature, smoothness has been evaluated based on the jerk as the third derivative of position [60], the ratio of mean speed over peak speed [61], the number of zero crossings of the acceleration reflecting the number of putative submovements the movement is composed of [62], the number peaks in speed [63], or through frequency analysis of the movement speed profile [64] (see [65] for a review). Another promising metric describing movement smoothness is the spectral arc length, which describes the frequency content of the velocity signal [64, 66]. This metric has been extensively validated based on systematic reviews, simulations, and experimental data from multiple technology-based assessments and neurological injuries. Based on the excellent statistical properties of the spectral arc length, it was consequently recommended by multiple groups as the standard in assessing upper limb movement smoothness [12, 64, 66, 67]. Also, several studies with stroke patients have shown that movement smoothness improves over the course of rehabilitation [67–71], suggesting that smoothness indicators are valid measures of motor recovery [68]. During point-to-point reaching movements, the error with respect to a straight trajectory, or equivalent measures such as hand path ratio, e.g., ratio of trajectory length over straight-line length, are also used to evaluate motor control. It has been shown that neurologically impaired patients tend to deviate more from an ideal straight line, reflecting impaired inter-joint coordination in the upper limb [35, 72]. Abnormal muscle synergies can be evaluated from simultaneous force recordings at different joints of the upper limb while asking subjects to produce isolated isometric force, e.g., shoulder flexion/extension or elbow flexion/extension in different position [73, 74]. Miller et al. [75] proposed a similar approach with a robotic platform recording isometric flexion and extension forces generated by the fingers, wrist, and thumb during robot-mediated movements of the upper limb.

Digital health metrics have also been developed in an attempt to assess hand function using haptic interfaces as dedicated assessment platforms, where neurologically impaired patients perform object manipulation in a virtual environment [76, 77]. Bardorfer et al. [78] used a PHANToM haptic device (Sensable/Geomagic, USA) to create a virtual labyrinth in which subjects have to navigate. Hand and arm function was evaluated using performance metrics such as movement velocity, number of collisions with the labyrinth walls, impact duration, as well as impact force, and allowed to distinguish between patients suffering from different types of neurodegenerative diseases. Using a similar approach, Emery et al. proposed the Virtual Peg Insertion Test (VPIT, Fig. 15.2), a virtual reality assessment motivated by movement components of the conventional NHPT and the box and block test [79], where subjects have to insert nine virtual pegs into nine virtual holes by controlling the position and orientation of an instrumented handle attached to a PHANToM Omni device [80–83]. Other similar approaches using PHANToM haptic devices in combination with virtual reality to extract features representative of upper limb function have also been reported [84–87].

More advanced algorithms to evaluate upper limb proprioception have also been proposed and validated in stroke patients, using paradigms where patients mirror-match a movement presented on their impaired side by moving the unimpaired arm using the KINARM (Kinarm, Canada), a two-arm robotic apparatus specifically designed for assessments [88–93]. Other robotic assessments of proprioception focused on the evaluation of difference threshold in joint position sense in stroke patients, at the level of the arm [94], wrist [95–98], or fingers [99–101]. A promising approach for assessing finger proprioception is by relying on an exoskeleton to induce criss-cross movements of the index and middle finger. In this paradigm, participants are asked to indicate when the position of the index and middle finger is exactly overlapping, without having vision of the fingers [102, 103]. The outcome metric is the angular position deviation between index and middle finger at the instant that was indicated by the participant, which has

a Virtual Peg Insertion Test (VPIT)

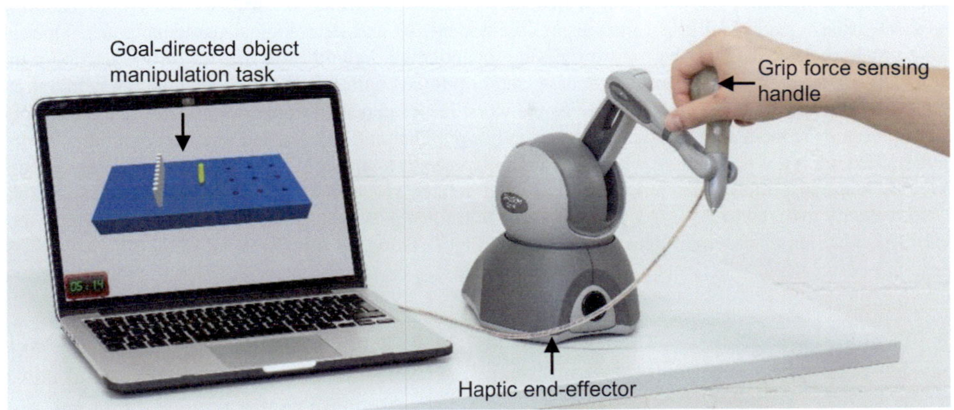

b Able-bodied participant: raw data

c Neurological participant: raw data

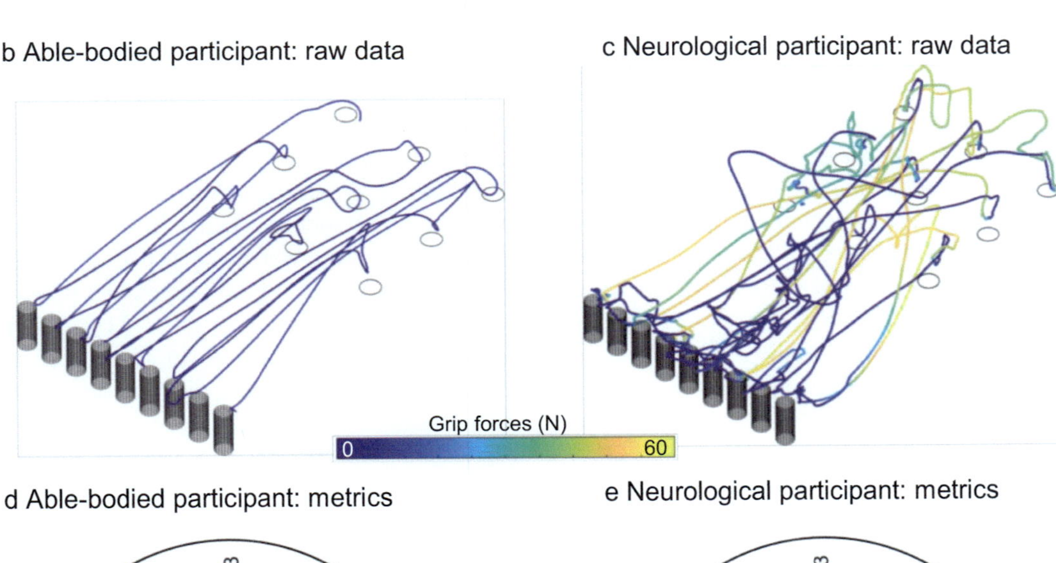

d Able-bodied participant: metrics

e Neurological participant: metrics

◀ **Fig. 15.2** Exemplary robotic assessment. The Virtual Peg Insertion Test (VPIT, **a**) features a goal-directed object manipulation task using haptic end-effector and a custom-made force sensing handle. This enables the recording of data on movement patterns and hand grip forces from able-bodied (**b**) and neurological participants (**c**). Through a data processing pipeline, these data can be transformed into quantitative digital health metrics (**d, e**) describing movement efficiency, movement speed, movement smoothness, and grip force control. In **d** and **e**, each pie segment represents one metric. The outer boundary of the circle represents the worst recorded task performance (100%). The inner boundary of the circle represents the median of an able-bodied reference population (0%). Dashed lines represent the 95th-percentile of an able-bodied reference population, which is commonly used as a threshold to identify abnormal task behavior. M1-M10: M1: log jerk transport. M2: log jerk return. M3: SAL return. M4: path length ratio transport. M5: path length ratio return. M6: velocity max return. M7: jerk peg approach. M8: force peaks transport. M9: force rate SAL transport. M10: force rate SAL hole approach. SAL: spectral arc length. Figure based on Kanzler et al. [17]

shown to capture age-related and stroke-related decline in proprioception. Another approach relies on a robotic end-effector to move the index finger to pseudorandom locations in flexion or extension direction. Without having vision of the actual finger position, the participant has to indicate his/her perceived finger position on a tablet that is overlayed on the workspace of the device. The main outcome metric of the approach is the error between actual and perceived finger position, which has shown to be a valid, reliable, and informative descriptor of proprioception in stroke patients [104, 105].

15.3.3 Non-Robotic Technology-Based Assessments

Robotic assessments are a promising approach to provide a fine-grained assessment of sensorimotor impairments in neurological injuries. However, they still rely on rather expensive and complex setups. In many cases, robotic assessments could ideally be complemented by other technology-based assessment solutions similarly relying on movement kinematics and kinetic, which will be briefly summarized in the following.

Stationary sensors, such as cameras, motion capture systems, or force plates, have been used extensively in gait labs for biomechanics studies over the past decades. With the availability of relatively low-cost alternatives, such technology has penetrated the clinics for rehabilitation and assessment applications. For example, instrumented mats such as the GAITRite™ (GAITRite, USA) have proved valid and reliable for estimating spatio-temporal gait patterns [106–108]. 3D cameras, such as the Microsoft Kinect (Microsoft, USA), are also of high interest for assessing balance [109, 110] as well as gait kinematic parameters [111, 112] (see [113] for a review), as they further allow the reconstruction of individual joint angles. They also proved useful for therapy and assessment of arm function (e.g., 3D reachable workspace [114]), providing patients with immersive and motivating training conditions using virtual reality [114–116]. For the evaluation of hand function, which requires the detection of fine movements, gloves instrumented with position sensors (e.g., CyberGlove, (Immersion, USA)) have been used in stroke or spinal cord injury patients [117–119]. These can further be complemented by objects instrumented with force sensors allowing the evaluation of grip force control during interaction with real objects, which is often affected after neurological injuries [120–124]. Thanks to such kinematic and kinetic data, it is possible to identify impairments beyond what is achievable with clinical scales.

However, the use of stationary sensors requires a dedicated space (e.g., motion capture volume) and continuous data processing, making them valuable for laboratory experiments, but difficult to translate to the evaluation of real activities of daily living (ADL) tasks, or to move them out of the research/clinical environment. In that respect, assessments that rely exclusively on wearable sensors (e.g., simple accelerometers or inertial measurement units (IMUs)) bear high potential [125, 126]. One of their main advantages is the possibility to perform assessments in functional conditions and during ADL, as they only marginally interfere with movements. Another advantage is their relatively low cost, especially compared to robotic devices or optical motion capture equipment.

The most widely used type of wearable sensors are accelerometers, typically placed at the wrist or foot, to record changes in acceleration during movements and offer the possibility to label periods where a subject is active [127]. From a clinical assessment perspective, actigraphy can provide valuable information on activity levels of neurological patients. The type and duration of certain ADL such as walking, sitting, and laying can be detected through triaxial accelerometers placed on the lower back [128–131] or on the sternum [132]. These measures can replace self-reported questionnaires that are subjective and do not provide detailed information on the intensity and frequency of ADL.

Several studies with stroke patients wearing accelerometers on both arms aimed at evaluating the amount of use of the impaired arm, or the ratio of use between impaired and unimpaired arms [133–135] (see [136] for a review). These values are expected to provide information on how patients involve their paretic arm in real-life activities, with the possibility to track patients over several hours or days, which is a real advantage over punctual clinical assessment. Using such an approach, Leuenberger et al. classified ambulatory activity in a group of 24 chronic stroke patients wearing IMUs at both wrists, both ankles and the trunk during 24 h recordings. It was possible to distinguish level walking from stair ascent/descent with high sensitivity, highlighting the potential of wearable sensors for gaining insights on patient behavior outside of the clinic [137]. More complex setups of IMUs, for example, embedded in sensor suits providing whole body kinematic information have been used to assess upper limb synergies [138] or dissect performance during specific ADL [139].

15.4 Challenges and Future Directions

Given the promises of robotic assessments, the question of why such scales are not more rapidly adopted by the clinical and research communities can be raised. Robotic assessments are currently still in their infancy and several important limitations, both theoretical and practical, must be addressed before novel scales can be adopted in clinical practice.

15.4.1 Usability and Influence of Robotic Assessment Platforms

A wider acceptance of robotic assessments is strongly dependent on the technology readiness level and availability of the platforms that are used. Technology should prove to be safe and robust enough for daily use with patients, while minimizing additional effort required from therapists to perform the assessments. Many systems are still too close to research prototypes and require the presence of engineers to properly operate them, which is not clinically viable. In parallel, it would be important to better inform clinicians and therapists about what is available and how technology can support their daily work, to help them become more confident in the interpretation of the new technology-based assessments.

A prerequisite for a valid and reliable robotic assessment is that the device does not negatively influence the voluntary movement of patients performing certain movements (e.g., through inertia that is not optimal compensated). This issue gets more challenging the more complex a robotic device is and underlines the key role of sophisticated controllers to increase transparency of the devices. In general, it appears that specific types of devices are tailored to assess specific physiological construct. For example, exoskeletons that offer the possibility to assist or perturb a joint or the entire leg while recording the active torque generated by the patient appear to be the preferred choice for assessments of lower limb sensorimotor function and impairments (e.g., spasticity). For assessing free walking, robots can be advantageously replaced by wearable IMUs providing quantitative gait metrics in daily life situations [140, 141]. End-effector, low impedance (i.e., transparent) robotic devices appear to be well suited to assess functional ability of the upper limb, as they do not constrain the complex and highly dynamic movements of the arm and hand. Additionally, factors such as usability (for patient and therapists), size, and portability should be considered when developing dedicated assessment tools. In that sense, systems based on tabletop haptic devices (such as the VPIT [81]) and IMUs, which could be complemented by cameras or instrumented objects (e.g., [142–144]), bear high potential. Also, wearable sensor technologies are unique solutions for long-term and unobtrusive monitoring, offering new ways of following patients' physical activity over extended periods of time after discharge from the hospital [145]. This has the potential to provide not only useful information to establish detailed patient profiles, but also unique data to help scientists investigate mechanisms of recovery underlying neurological disorders and their evolution over time [33].

15.4.2 Selection and Validation of Digital Health Metrics

A major challenge that researchers face when developing technology-based assessment is the question of which digital health metrics should be extracted from the recorded sensor data and how these can be validated. This is a challenge because a large variety of different metrics have been proposed but most of them are insufficiently validated, as validity is mostly evaluated through comparison with established clinical assessments (concurrent validity). This points toward a key issue when establishing robotic assessments, which are expected to be more objective and sensitive to clinical assessments that rely on subjective judgment, and that are known to be limited (e.g., ceiling/flooring effect).

These issues were highlighted in a systematic review that found that 151 different metrics are used for describing upper limb kinematics in stroke but that favorable statistical properties were reported for only eight of the metrics [12]. This underlines a clear research gap in the validation of digital health metrics, which can be seen as one of the reasons for their limited acceptance in clinical research and practice. In the following, we aspire to provide recommendations for the selection and validation of such metrics.

First, a suitable instrumented assessment task needs to be chosen that involves the relevant physiological constructs that are targeted by a specific research question. For this, it is essential to define whether proximal (e.g., for the upper limb: trunk/shoulder/elbow movements) or distal (e.g., for the upper limb: wrist/hand/finger movements) body parts should be involved. Also, the assessment task should consider the physical capabilities of the patient population, as, for example, severely affected patients will likely not be able to perform complex multidimensional movements. In addition, in case of an interventional study, the assessment task should not be equal to the therapy task, as assessment outcomes are otherwise confounded by task-related learning. The instrumentation of the assessment task should be chosen such that it has minimal influence on the performed movements.

Second, an initial set of candidate metrics should be selected based on the constructs targeted by the research questions. These constructs can include, for example, movement accuracy, efficacy, efficiency, planning, precision, smoothness, posture, speed, or workspace, and dedicated metrics for each of them have been identified previously [12]. Also, the specific metrics should be selected based on recommendations from literature, available clinimetric evidence, the frequency of use, and insights from motor control, as well as technical and clinical perspectives [12, 66, 67].

Third, given that the clinimetric properties of digital health metrics are strongly dependent on the specific assessment task and the observed patient population, the initial set of metrics should be refined into a validated core set based on a data-driven selection procedure. Ideally, such data would be gathered within a test–retest study, where participants from the target patient population are asked to repeat the technology-based assessment multiple times over a period of few days, where it can be assumed that patients are without a considerable change in physiological conditions. This permits the calculation of test–retest reliability and measurement error and evaluates potential learning effects. Based on these constructs and established cut-off values, a core set of use-case-specific digital health metrics with ideal statistical properties can then be selected [17]. As an exemplary use-case, this framework was applied to kinematic and kinetic data collected from able-bodied and neurological populations with the VPIT. In more details, a core set of 10 digital health metrics describing movement smoothness, efficiency, speed, and grip force control was established and successfully validated and refined in patients with multiple sclerosis, hereditary ataxia, or after stroke [17, 146, 147]. It should also be noted that the approach included a processing pipeline that relies on mixed effect models to remove the effect of confounding factors (e.g., age, gender, etc.) on the metrics. This allowed for an adapted z-score normalization of all metrics with respect to an age-stratified control population of 120 participants, thereby permitting a direct comparison between the metrics, which are typically expressed in different units. In addition, this enabled the identification of impairments in individual patients or patient populations, by relying on the 95th-percentile of able-bodied controls as a threshold, even when patients were defined as asymptomatic according to common clinical scales.

This three-step process, which fuses a hypothesis-driven with a data-driven approach, should help to select digital health metrics that are ideal to answer a specific research question. In case it is not possible to run a test–retest study (e.g., due to the required resources), it is an option to only rely on the first and second step to select metrics. However, in this case, the metrics and assessment results should be interpreted with appropriate caution under a potential risk of bias.

15.4.3 Interpretation of Digital Health Metrics

Since digital health metrics provide information that were not possible to obtain with existing clinical assessments (at least not in an objective and sensitive manner, such as metrics evaluating movement smoothness), many of the proposed novel metrics remain abstract values that are difficult to interpret for therapists and patients compared to the well-established and standardized clinical measures. Further studies are therefore needed to determine what these metrics represent based on concurrent physiological examinations, and how these metrics can document functional changes, predict therapy outcome, and reflect sensorimotor impairment. Some of the signals captured by technological assessments may, for example, represent a purely clinical sign (i.e., different from a healthy subject) which does not, or at least not negatively, affect function. In this context, it is important to think about how to visually present data from technology-based assessments to the different user groups, such as patients, therapists, or clinicians. Clearly, different levels of abstractions might be needed to tailor the information content to the use-case and the expertise of the target user. For patients, it might be more intuitive to show a visual representation of raw data (e.g., Fig. 15.2) and inform them about their task completion time. On the other hand, for clinicians it might be relevant to get detailed information on the digital health metrics and whether they indicate abnormal sensorimotor control.

Another challenge when interpreting digital health metrics is that the underlying mathematical constructs and signal processing methods can be applied to any robotic assessment, but the results and interpretation of the robotic measurements are specific to the device with which they were generated and therefore cannot be easily generalized across different devices.

15.4.4 Computational Models for Predicting Neurorehabilitation Outcomes

While carefully selected and validated digital health metrics can be seamlessly integrated as primary or secondary endpoints in clinical studies, it is yet another open question whether such metrics can be used to influence clinical decision-making in neurorehabilitation. One approach for this might be to integrate them into predictive computational models [148]. Essentially, those models attempt to solve the long-standing clinical challenge to establish an accurate prognosis, as this is the basis for the definition of realistic therapy goals and an optimally tailored and personalized therapy plan. This is a challenging task, because a multitude of participant-dependent and/or environmental variables is expected to influence the prognosis and these variables are often challenging to capture (e.g., motivation) or data is not available in clinical routine [149]. Thus, researchers attempted to identify the main prognostic factors determining rehabilitation outcomes and integrated those in computational models that allow to algorithmically determine outcomes for individual patients.

Historically, in stroke research, one of the first proposed computational model was the propositional recovery model [150, 151], where the upper limb impairment level at 6 months was predicted from upper limb impairment at day 3 after stroke following a simple linear equation. While several large studies demonstrated the potential of this remarkably simple model, several recent studies highlighted methodological issues suggesting caution when applying and interpreting this model [3–5].

Thus, more complex and methodologically adequate approaches are warranted. One of them is the Predict Recovery Potential (PREP and PREP2) algorithm, which predicts upper limb functional outcomes (excellent, good, limited, poor) at 3 months post-stroke with an accuracy of 75% based on data collected 3 to 7 days after stroke [152]. The required data includes two measures of upper limb disability (SAFE: shoulder abduction, finger extension score; NIHSS: National Institute of Health Stroke Scale), the age of participants, and information on the integrity of their corticospinal tract, as measured by motor evoked potentials via transcranial magnetic stimulation. More recently, additional models have been proposed that further increase accuracy, allow dynamic predictions with multiple timepoints across the time course of recovery, and identified multiple subpopulations with different recovery trajectories [153, 154]. Importantly, when providing the prognostic data from the PREP as a support for healthcare professional in clinical routine, therapist modified their treatment decisions based on the output from the algorithm and were able to significantly reduce length of hospitalization by 1 week [155].

These promising initial results motivate to explore the integration of digital health metrics into computational prediction models. The working hypothesis is that this could help improve the accuracy of such models, as digital health metrics are able to capture behavioral variability that is not picked up by clinical scales. It is believed that especially this behavioral variability could allow to identify behavioral subpopulations that exhibit different patterns of disease progression or recovery.

First preliminary work has been done in this area, attempting to predict changes over an 8-week neurorehabilitation intervention in persons with multiple sclerosis by relying on clinical routine data, digital health metrics from the VPIT, and conventional assessments [156] that were collected before the intervention. While this work is clearly limited by its small sample size, it was still able to show that machine learning models trained on such data can achieve moderate to good accuracy when predicting changes in fine and gross dexterity, respectively. Further research with more representative datasets is warranted to fully explore whether digital health metrics can provide a benefit for computational prediction models.

In the context of computational models, it is also important to mention the emerging applications of artificial intelligence and machine learning for selecting digital health metrics [157], estimating clinical scores from remotely collected wearable sensor data [158], and automatically adapting exercise settings in robot-assisted therapy [159]. Such approaches are promising, as they allow to bring heterogeneous multi-modal sensor data or metrics into a more applied (clinical) context, for example, by transforming abstract wearable sensor data (e.g., metrics derived from acceleration) that is collected remotely into an established clinical score (e.g., FMA-UE) [158]. However, one challenge is that such methods often rely on black-box models that lack interpretability and might therefore not be accepted by healthcare practitioners. Such methodological considerations should be addressed in the future, together with the need for high-quality big datasets that can unleash the full potential of machine learning models, which will allow to better judge the full potential of this promising research area.

15.4.5 Influencing Therapy Decisions with Digital Health Metrics and Prediction Models

Technology-based assessments and prediction models will only be integrated in clinical practice if they can generate actionable information that is not available with conventional means and can help to, for example, optimize the personalization of therapy. An exemplary use-case of technology-based assessments would be to use digital health metrics to identify proprioceptive impairments when providing therapy to chronic post-stroke individuals. This could help to identify persons that may benefit the most from the intervention, as the level of proprioceptive impairments has been shown to correlate to the level of therapy gains [103]. Even though these results are preliminary and need to be confirmed in a more representative population, this use-case would provide a clear benefit to healthcare practitioners and patients and cannot be achieved with clinical assessments or subjective inspection, as sensory deficits are notoriously challenging to assess.

Further, accurate prognostic information from computational prediction models building on digital health metrics might help to find the optimal trade-off between therapy that focuses on the restoration of body functions or on learning compensatory strategies to increase the spectrum of activities an individual can perform despite the available impairments. For example, for an individual where no considerable improvement in body functions can be expected through a restorative neurorehabilitation intervention, it could be discussed whether the therapy should focus more on learning compensatory strategies. While the research community has a strong focus on exploring novel technical developments, we want to emphasize that there is a strong unmet need for providing evidence for the clinical usefulness of the technologies, for example, by exploring these important clinical use-cases.

15.5 Conclusions

The promising results of recent studies using advanced technologies such as robotics demonstrate the potential of using technology not only to complement conventional therapy but also to assess sensorimotor function in a more objective, reliable, and continuous way and to use this information to guide clinical decision-making. Whether relying on basic or more sophisticated metric extraction algorithms, digital health metrics obtained from robotic systems offer new possibilities to objectively investigate sensorimotor impairments under reproducible conditions. These metrics can provide unique information on the quality of patients' performance in a defined task, which cannot be captured with conventional clinical scales. Because of the quantifiable assistance that robots can provide, robotic assessments can be administered even if the patient is not able to perform the movement without support. This can enlarge the measurable range of impairment and improve sensitivity. These robotic assessments can be complemented by other technology-based approaches, such as wearable sensors monitoring patients in daily life, in order to provide a comprehensive picture of disability. Table 15.1 aims at providing an (non-exhaustive) overview

Table 15.1 Overview of clinical assessment domains including their advantages and disadvantages, as well as opportunities for robotic and sensor-based technologies to complement these assessments. Corresponding examples of metrics typically used in studies with neurological patients are listed in an attempt to highlight their relevance. The list of examples is for illustrative purposes only and is not comprehensive

Assessment domain	Examples of clinical assessments	Pros (+) and cons (−)	Opportunities for technology-based assessments	Examples of digital health metrics
Time-based	• 10 m Walking Test (10 m WT) • 6 min Walk Test (6MWT) • Timed up and go (TUG) • Action research arm test (ARAT) • Nine hole peg test (NHPT) • Box and block test (BBT)	+ Easy and fast to administer + Reflects overall functional ability + Objective measure + No expensive equipment required − No information about movement quality − Do not distinguish recovery from compensation − Patients must be able to achieve the task	• Time can be easily and precisely measured with robotic/sensor-based systems • Allow decomposing a task into different sub-times • Robotic devices would enable application to more severely impaired patients, e.g., by providing controlled assistance while performing a task	• Time to execute a motor task [35, 36, 83] • Sub-times for phases of a task [81]
Observer-based				
− Lower limbs and balance	• Berg balance scale (BBS) • Spinal cord independence measure (SCIM) • Fugl-Meyer assessment (FMA) (lower limb) • Modified Ashworth scale (MAS)	+ Subscales providing a detailed overview of functional ability + Many standardized well-validated assessment scales − Subjective (based on perception of therapist or patient) − Require trained therapists − Usually ordinal	• Can provide objective and quantitative kinetic and kinematic data • From raw sensor measurements, linear and sensitive metrics representative of different types of impairments can be extracted, which can be used to construct models predicting recovery • Possibility to extract metrics during robot-assisted therapy and adapt therapy settings online • Wearable sensors can monitor activity levels during long-term recording or activities of daily living • Robotic technology can impose a movement or force pattern repeatedly in a	• Gait spatio-temporal parameters (speed, step length, etc.) [51, 106, 113, 160] • Gait symmetry [161] • Gait variability [162, 163]
− Upper limbs	• FMA (upper limb) • Motor assessment scale (MS) • Motricity index (MI) • Chedoke McMaster stroke assessment (CMSA)			• Movement smoothness [61, 64, 65] • Movement speed [78, 164, 165] • Trajectory error [35, 78] • Interaction force/torque [73–75]

(continued)

Table 15.1 (continued)

Assessment domain	Examples of clinical assessments	Pros (+) and cons (−)	Opportunities for technology-based assessments	Examples of digital health metrics
	• Modified Ashworth scale (MAS) • Somatosensory assessments (e.g., Nottingham sensory assessment (NSA)) • Range of motion (ROM)	scales, non-linear and with limited sensitivity − Suffer from flooring and ceiling effects	systematic way, while objectively recording the resulting force/motion. This should increase reliability and validity	• Reaching workspace [46, 114, 166, 167] • Joint stiffness [40, 168] • Sensory thresholds [48, 49, 89, 99] Passive ROM [40]
− Self-reported reports (activities of daily living)	• Barthel index (BI) • Motor activity log (MAL)			• Activity counts [127, 136] • Index of arm use in activities of daily life [133, 169]

of examples of clinical and technology-based assessments, and a summary of key opportunities where robotic assessments can advantageously complement conventional clinical scales.

It is clear that digital health metrics will, in the near future, play an increasingly prominent role in the assessment of sensorimotor function of the lower and upper extremities. The tight coupling between robotic assessments and robot-assisted therapy and the appealing possibility of achieving both on the same hardware platform is enticing. By embedding short and independent assessment modules within robotic therapy sessions, it becomes possible to track the performance of patients on a daily basis, without having to perform time-consuming clinical assessments [170]. Robotic assessments could also be performed online during therapy, offering the possibility to continuously and automatically adapt type and complexity of a therapy to the current state and principal impairment of the patient, with the aim of maximizing engagement and therapeutic effect. This approach of assessment-driven therapy has been successfully implemented in several pilot trials on robot-assisted or sensor-based stroke rehabilitation [166, 171–175].

Robotic- and sensor-based assessments should be seen as independent but complementary tools to conventional assessments. Technological assessments will never replace neurological examinations, such as reflex testing, but by combining both clinical and technological assessments, clinicians would benefit from more sensitive, reliable, and objective evaluations of different aspects of sensorimotor function/impairment, which could ultimately impact the way neurorehabilitation therapy is administered.

References

1. Finch E, Brooks D, Stratford PW, Mayo NE. Physical rehabilitation outcome measures: a guide to enhanced clinical decision making, 2nd ed. Canadian Physiotherapy Assiciation;2002.
2. Kirshner B, Guyatt G. A methodological framework for assessing health indices. J Chronic Dis. 1985;38 (1):27–36.
3. Hope TMH, Friston K, Price CJ, Leff AP, Rotshtein P, Bowman H. Recovery after stroke: not so proportional after all? Brain. 2019;142(1):15–22.
4. Lohse KR, Hawe RL, Dukelow SP, Scott SH. Statistical considerations for drawing conclusions about recovery. Neurorehabil Neural Repair. 2021;35(1):10–22.
5. Kundert R, Goldsmith J, Veerbeek JM, Krakauer JW, Luft AR. What the proportional recovery rule is (and is not): methodological and statistical considerations. Neurorehabil Neural Repair. 2019;33(11):876–87.

6. Kwakkel G, Kollen BJ, Krebs HI. Effects of robot-assisted therapy on upper limb recovery after stroke: a systematic review. Neurorehabil Neural Repair. 2008;22(2):111–21.

7. Mehrholz J, Elsner B, Werner C, Kugler J, Pohl M. Electromechanical-assisted training for walking after stroke. Cochrane Database Syst Rev. 2013;7: CD006185.

8. Mehrholz J, Hadrich A, Platz T, Kugler J, Pohl M. Electromechanical and robot-assisted arm training for improving generic activities of daily living, arm function, and arm muscle strength after stroke. Cochrane Database Syst Rev. 2012;6:CD006876.

9. Lo AC, Guarino PD, Richards LG, Haselkorn JK, Wittenberg GF, Federman DG, et al. Robot-assisted therapy for long-term upper-limb impairment after stroke. New Engl J Med. 2010;362(19):1772–83.

10. Maciejasz P, Eschweiler J, Gerlach-Hahn K, Jansen-Troy A, Leonhardt S. A survey on robotic devices for upper limb rehabilitation. J Neuroeng Rehabil. 2014;11:3.

11. Prange GB, Jannink MJ, Groothuis-Oudshoorn CG, Hermens HJ, Ijzerman MJ. Systematic review of the effect of robot-aided therapy on recovery of the hemiparetic arm after stroke. J Rehabil Res Dev. 2006;43(2):171–84.

12. Schwarz A, Kanzler CM, Lambercy O, Luft AR, Veerbeek JM. Systematic review on kinematic assessments of upper limb movements after stroke. Stroke. 2019;50(3):718–27.

13. Nordin N, Xie SQ, Wunsche B. Assessment of movement quality in robot- assisted upper limb rehabilitation after stroke: a review. J Neuroeng Rehabil. 2014;11:137.

14. Tran VD, Dario P, Mazzoleni S. Kinematic measures for upper limb robot-assisted therapy following stroke and correlations with clinical outcome measures: a review. Med Eng Phys. 2018;53:13–31.

15. Kwakkel G, Van Wegen E, Burridge JH, Winstein CJ, van Dokkum L, Alt Murphy M, et al. Standardized measurement of quality of upper limb movement after stroke: consensus-based core recommendations from the second stroke recovery and rehabilitation roundtable. Int J Stroke. 2019;14 (8):783–91.

16. Reinkensmeyer DJ, Emken JL, Cramer SC. Robotics, motor learning, and neurologic recovery. Annu Rev Biomed Eng. 2004;6:497–525.

17. Kanzler CM, Rinderknecht MD, Schwarz A, Lamers I, Gagnon C, Held JPO, et al. A data-driven framework for selecting and validating digital health metrics: use-case in neurological sensorimotor impairments. NPJ Digit Med. 2020;3:80.

18. Gassert R, Burdet E, Chinzei K. Opportunities and challenges in MR-compatible robotics: reviewing the history, mechatronic components, and future directions of this technology. IEEE Eng Med Biol Mag. 2008;27(3):15–22.

19. World Health Organization. International classification of functioning, disability and health: ICF: WHO Library Cataloguing-in-Publication Data;2001.

20. Gale CR, Allerhand M, Sayer AA, Cooper C, Deary IJ. The dynamic relationship between cognitive function and walking speed: the English longitudinal study of ageing. Age (Dordr). 2014;36(4):9682.

21. Welmer AK, Rizzuto D, Qiu C, Caracciolo B, Laukka EJ. Walking speed, processing speed, and dementia: a population-based longitudinal study. J Gerontol. 2014;69(12):1503–10.

22. Mathiowetz V, Weber K, Kashman N, Volland G. Adult norms for the nine hole peg test of finger dexterity. Occup Ther J Res. 1985;5:24–38.

23. Pandyan AD, Gregoric M, Barnes MP, Wood D, Van Wijck F, Burridge J, et al. Spasticity: clinical perceptions, neurological realities and meaningful measurement. Disabil Rehabil. 2005;27(1–2):2–6.

24. Dietz V, Sinkjaer T. Spastic movement disorder: impaired reflex function and altered muscle mechanics. Lancet Neurol. 2007;6(8):725–33.

25. Bohannon RW, Smith MB. Interrater reliability of a modified Ashworth scale of muscle spasticity. Phys Ther. 1987;67(2):206–7.

26. Fleuren JFM, Voerman GE, Erren-Wolters CV, Snoek GJ, Rietman JS, Hermens HJ, et al. Stop using the Ashworth scale for the assessment of spasticity. J Neurol Neurosurg Psychiatry. 2010;81 (1):46–52.

27. Gilman S. Joint position sense and vibration sense: anatomical organisation and assessment. J Neurol Neurosurg Psychiatry. 2002;73(5):473–7.

28. Goble DJ. Proprioceptive acuity assessment via joint position matching: from basic science to general practice. Phys Ther. 2010;90(8):1176–84.

29. Fugl-Meyer AR, Jaasko L, Leyman I, Olsson S, Steglind S. The post-stroke hemiplegic patient: a method for evaluation of physical performance. Scand J Rehabil Med. 1975;7(1):13–31.

30. Gladstone DJ, Danells CJ, Black SE. The fugl-meyer assessment of motor recovery after stroke: a critical review of its measurement properties. Neurorehabil Neural Repair. 2002;16(3):232–40.

31. Sinoff G, Ore L. The Barthel activities of daily living index: self-reporting versus actual performance in the old-old (> or =75 years). J Am Geriatr Soc. 1997;45(7):832–6.

32. Rossier P, Wade DT. Validity and reliability comparison of 4 mobility measures in patients presenting with neurologic impairment. Arch Phys Med Rehabil. 2001;82(1):9–13.

33. Kitago T, Liang J, Huang VS, Hayes S, Simon P, Tenteromano L, et al. Improvement after constraint-induced movement therapy: recovery of normal motor control or task-specific compensation? Neurorehabil Neural Repair. 2013;27(2):99–109.

34. Blackburn M, van Vliet P, Mockett SP. Reliability of measurements obtained with the modified

Ashworth scale in the lower extremities of people with stroke. Phys Ther. 2002;82(1):25–34.

35. Dovat L, Lambercy O, Salman B, Johnson V, Gassert R, Burdet E, et al. Post-stroke training of a pick and place activity in a virtual environment. Proc Int Conf Virtual Rehabil. 2008:28–34.

36. Feys P, Alders G, Gijbels D, De Boeck J, De Weyer T, Coninx K, et al. Arm training in multiple sclerosis using phantom: clinical relevance o robotic outcome measures. In: Proc of the IEEE Int Conf on Rehabilitation Robotics (ICORR);2009. p. 671–6.

37. Hesse S, Waldner A, Tomelleri C. Innovative gait robot for the repetitive practice of floor walking and stair climbing up and down in stroke patients. J Neuroeng Rehabil. 2010;7:30.

38. Emken JL, Harkema SJ, Beres-Jones JA, Ferreira CK, Reinkensmeyer DJ. Feasibility of manual teach-and-replay and continuous impedance shaping for robotic locomotor training following spinal cord injury. IEEE Trans Biomed Eng. 2008;55 (1):322–34.

39. Maggioni S, Lunenburger L, Riener R, Melendez-Calderon A. Robot-aided assessment of walking function based on an adaptive algorithm. In: Proc of the IEEE Int Conf on Rehabilitation Robotics (ICORR);2015. p. 804–9.

40. Lunenburger L, Colombo G, Riener R, Dietz V. Clinical assessments performed during robotic rehabilitation by the gait training robot Lokomat. In: Proc of the IEEE Int Conf on Rehabilitation Robotics (ICORR);2005. p. 345–8.

41. Bolliger M, Banz R, Dietz V, Lunenburger L. Standardized voluntary force measurement in a lower extremity rehabilitation robot. J Neuroeng Rehabil. 2008;5:23.

42. Galen SS, Clarke CJ, McLean AN, Allan DB, Conway BA. Isometric hip and knee torque measurements as an outcome measure in robot assisted gait training. NeuroRehabilitation. 2014;34(2):287–95.

43. Waldman G, Yang CY, Ren Y, Liu L, Guo X, Harvey RL, et al. Effects of robot-guided passive stretching and active movement training of ankle and mobility impairments in stroke. NeuroRehabilitation. 2013;32(3):625–34.

44. Zhang LQ, Chung SG, Bai Z, Xu D, van Rey EM, Rogers MW, et al. Intelligent stretching of ankle joints with contracture/spasticity. IEEE Trans Neural Syst Rehabil. 2002;10(3):149–57.

45. Keller U, Scholch S, Albisser U, Rudhe C, Curt A, Riener R, et al. Robot-assisted arm assessments in spinal cord injured patients: a consideration of concept study. PLoS ONE. 2015;10(5):e0126948.

46. Ellis MD, Sukal-Moulton T, Dewald JP. Progressive shoulder abduction loading is a crucial element of arm rehabilitation in chronic stroke. Neurorehabil Neural Repair. 2009;23(8):862–9.

47. Carey LM, Oke LE, Matyas TA. Impaired limb position sense after stroke: a quantitative test for clinical use. Arch Phys Med Rehabil. 1996;77 (12):1271–8.

48. Han J, Waddington G, Adams R, Anson J, Liu Y. Assessing proprioception: a critical review of methods. J Sport Health Sci. 2015. https://doi.org/10.1016/j.jshs.2014.10.004.

49. Domingo A, Lam T. Reliability and validity of using the Lokomat to assess lower limb joint position sense in people with incomplete spinal cord injury. J Neuroeng Rehabil. 2014;11:167.

50. Veneman JF, Kruidhof R, Hekman EE, Ekkelenkamp R, Van Asseldonk EH, van der Kooij H. Design and evaluation of the LOPES exoskeleton robot for interactive gait rehabilitation. IEEE Trans Neural Syst Rehabil Eng. 2007;15(3):379–86.

51. van Asseldonk EH, Veneman JF, Ekkelenkamp R, Buurke JH, van der Helm FC, van der Kooij H. The effects on kinematics and muscle activity of walking in a robotic gait trainer during zero-force control. IEEE Trans Neural Syst Rehabil. 2008;16(4):360–70.

52. Banz R, Bolliger M, Colombo G, Dietz V, Lunenburger L. Computerized visual feedback: an adjunct to robotic-assisted gait training. Phys Ther. 2008;88 (10):1135–45.

53. Lunenburger L, Colombo G, Riener R. Biofeedback for robotic gait rehabilitation. J Neuroeng Rehabil. 2007;4:1.

54. Lunenburger L, Colombo G, Riener R, Dietz V. Biofeedback in gait training with the robotic orthosis Lokomat. Proc IEEE Eng Med Biol Conf. 2004;7:4888–91.

55. Wood DE, Burridge JH, van Wijck FM, McFadden C, Hitchcock RA, Pandyan AD, et al. Biomechanical approaches applied to the lower and upper limb for the measurement of spasticity: a systematic review of the literature. Disabil Rehabil. 2005;27(1–2):19–32.

56. Kakebeeke TH, Lechner H, Baumberger M, Denoth J, Michel D, Knecht H. The importance of posture on the isokinetic assessment of spasticity. Spinal Cord. 2002;40(5):236–43.

57. Johnson GR. Outcome measures of spasticity. Eur J Neurol. 2002;9(Suppl 1):10–6; dicussion 53–61.

58. Mirbagheri MM, Barbeau H, Kearney RE. Intrinsic and reflex contributions to human ankle stiffness: variation with activation level and position. Exp Brain Res. 2000;135(4):423–36.

59. Mirbagheri MM, Barbeau H, Ladouceur M, Kearney RE. Intrinsic and reflex stiffness in normal and spastic, spinal cord injured subjects. Exp Brain Res. 2001;141(4):446–59.

60. Flash T, Hogan N. The coordination of arm movements—an experimentally confirmed mathematical-model. J Neurosci. 1985;5(7):1688–703.

61. Rohrer B, Fasoli S, Krebs HI, Hughes R, Volpe B, Frontera WR, et al. Movement smoothness changes during stroke recovery. J Neurosci. 2002;22 (18):8297–304.

62. Burdet E, Milner TE. Quantization of human motions and learning of accurate movements. Biol Cybern. 1998;78(4):307–18.

63. Kahn LE, Zygman ML, Rymer WZ, Reinkensmeyer DJ. Effect of robot-assisted and unassisted exercise on functional reaching in chronic hemiparesis. Proc IEEE Eng Med Biol Conf. 2001;23:1344–7.

64. Balasubramanian S, Melendez-Calderon A, Burdet E. A robust and sensitive metric for quantifying movement smoothness. IEEE Trans Biomed Eng. 2012;59(8):2126–36.

65. Hogan N, Sternad D. Sensitivity of smoothness measures to movement duration, amplitude, and arrests. J Mot Behav. 2009;41(6):529–34.

66. Balasubramanian S, Melendez-Calderon A, Roby-Brami A, Burdet E. On the analysis of movement smoothness. J Neuroeng Rehabil. 2015;12:112.

67. Saes M, Mohamed Refai MI, van Kordelaar J, Scheltinga BL, van Beijnum BF, Bussmann JBJ, et al. Smoothness metric during reach-to-grasp after stroke: part 2. Longitudinal association with motor impairment. J Neuroeng Rehabil. 2021;18(1):144.

68. Rohrer B, Fasoli S, Krebs HI, Volpe B, Frontera WR, Stein J, et al. Submovements grow larger, fewer, and more blended during stroke recovery. Mot Control. 2004;8(4):472–83.

69. Kahn LE, Zygman ML, Rymer WZ, Reinkensmeyer DJ. Robot-assisted reaching exercise promotes arm movement recovery in chronic hemiparetic stroke: a randomized controlled pilot study. J Neuroeng Rehabil. 2006;3:12.

70. Lambercy O, Dovat L, Yun H, Wee SK, Kuah CW, Chua KS, et al. Effects of a robot-assisted training of grasp and pronation/supination in chronic stroke: a pilot study. J Neuroeng Rehabil. 2011;8:63.

71. Yoo DH, Kim SY. Effects of upper limb robot-assisted therapy in the rehabilitation of stroke patients. J Phys Ther Sci. 2015;27(3):677–9.

72. Levin MF. Interjoint coordination during pointing movements is disrupted in spastic hemiparesis. Brain. 1996;119:281–93.

73. Guidali M, Schmiedeskamp M, Klamroth V, Riener R. Assessment and training of synergies with an arm rehabilitation robot. In: Proc of the IEEE Int Conf on Rehabilitation Robotics (ICORR);2009. p. 772–6.

74. Dewald JPA, Beer RF. Abnormal joint torque patterns in the paretic upper limb of subjects with hemiparesis. Muscle Nerve. 2001;24(2):273–83.

75. Miller LC, Ruiz-Torres R, Stienen AHA, Dewald JPA. A wrist and finger force sensor module for use during movements of the upper limb in chronic hemiparetic stroke. IEEE Trans Biomed Eng. 2009;56(9):2312–7.

76. Metzger JC, Lambercy O, Califfi A, Conti FM, Gassert R. Neurocognitive robot-assisted therapy of hand function. IEEE Trans Haptics. 2014;7(2):140–9.

77. Kazemi H, Kearney R, Milner T. Characterizing coordination of grasp and twist in hand function of healthy and post-stroke subjects. In: Proc of the IEEE Int Conf on Rehabilitation Robotics (ICORR), vol. 2013. 2013. p. 1–6.

78. Bardorfer A, Munih M, Zupan A, Primozic A. Upper limb motion analysis using haptic interface. IEEE-ASME Trans Mechatron. 2001;6(3):253–60.

79. Mathiowetz V, Volland G, Kashman N, Weber K. Adult norms for the box and block test of manual dexterity. Am J Occup Ther. 1985;39(6):386–91.

80. Emery C, Samur E, Lambercy O, Bleuler H, Gassert R. Haptic/VR assessment tool for fine motor control. In: Proc Haptics: generating and perceiving tangible sensations, vol. 6192. 2010. p. 186–93.

81. Fluet MC, Lambercy O, Gassert R. Upper limb assessment using a virtual peg insertion test. IEEE Int Conf Rehabil Robot. 2011;2011:5975348.

82. Lambercy O, Fluet MC, Lamers I, Kerkhofs L, Feys P, Gassert R. Assessment of upper limb motor function in patients with multiple sclerosis using the virtual peg insertion test: a pilot study. IEEE Int Conf Rehabil Robot. 2013;2013:6650494.

83. Gagnon C, Lavoie C, Lessard I, Mathieu J, Brais B, Bouchard JP, et al. The virtual peg insertion test as an assessment of upper limb coordination in ARSACS patients: a pilot study. J Neurol Sci. 2014;347(1–2):341–4.

84. Amirabdollahian F, Johnson G. Analysis of the results from use of haptic peg-in-hole task for assessment in neurorehabilitation. Appl Bionics Biomech. 2011;8(1):1–11.

85. Xydas EG, Louca LS. Upper limb assessment of people with multiple sclerosis with the use of a haptic nine-hole peg-board test. In: Proc ASME biennial conference on engineering systems design and analysis. 2008. p. 159–66.

86. Hussain N, Alt Murphy M, Sunnerhagen KS. Upper limb kinematics in stroke and healthy controls using target-to-target task in virtual reality. Front Neurol. 2018;9:300.

87. Hussain N, Sunnerhagen KS, Alt MM. End-point kinematics using virtual reality explaining upper limb impairment and activity capacity in stroke. J Neuroeng Rehabil. 2019;16(1):82.

88. Semrau JA, Herter TM, Scott SH, Dukelow SP. Robotic identification of kinesthetic deficits after stroke. Stroke. 2013;44(12):3414–21.

89. Dukelow SP, Herter TM, Moore KD, Demers MJ, Glasgow JI, Bagg SD, et al. Quantitative assessment of limb position sense following stroke. Neurorehabil Neural Repair. 2010;24(2):178–87.

90. Dukelow SP, Herter TM, Bagg SD, Scott SH. The independence of deficits in position sense and visually guided reaching following stroke. J Neuroeng Rehabil. 2012;9:72.

91. Deblock-Bellamy A, Batcho CS, Mercier C, Blanchette AK. Quantification of upper limb

position sense using an exoskeleton and a virtual reality display. J Neuroeng Rehabil. 2018;15(1):24.

92. Mochizuki G, Centen A, Resnick M, Lowrey C, Dukelow SP, Scott SH. Movement kinematics and proprioception in post-stroke spasticity: assessment using the Kinarm robotic exoskeleton. J Neuroeng Rehabil. 2019;16(1):146.

93. Kenzie JM, Semrau JA, Hill MD, Scott SH, Dukelow SP. A composite robotic-based measure of upper limb proprioception. J Neuroeng Rehabil. 2017;14(1):114.

94. Simo L, Botzer L, Ghez C, Scheidt RA. A robotic test of proprioception within the hemiparetic arm post-stroke. J Neuroeng Rehabil. 2014;11(77).

95. Cappello L, Elangovan N, Contu S, Khosravani S, Konczak J, Masia L. Robot-aided assessment of wrist proprioception. Front Hum Neurosci. 2015;9:198.

96. D'Antonio E, Galofaro E, Zenzeri J, Patane F, Konczak J, Casadio M, et al. Robotic assessment of wrist proprioception during kinaesthetic perturbations: a neuroergonomic approach. Front Neurorobot. 2021;15:640551.

97. Marini F, Squeri V, Morasso P, Campus C, Konczak J, Masia L. Robot-aided developmental assessment of wrist proprioception in children. J Neuroeng Rehabil. 2017;14(1):3.

98. Rinderknecht MD, Popp WL, Lambercy O, Gassert R. Reliable and rapid robotic assessment of wrist proprioception using a gauge position matching paradigm. Front Hum Neurosci. 2016;10:316.

99. Rinderknecht M, Popp W, Lambercy O, Gassert R. Experimental validation of a rapid, adaptive robotic assessment of the MCP joint angle difference threshold. In: Haptics: neuroscience, devices, modeling, and applications. Berlin Heidelberg: Springer;2014. p. 3–10.

100. Rinderknecht MD, Lambercy O, Raible V, Busching I, Sehle A, Liepert J, et al. Reliability, validity, and clinical feasibility of a rapid and objective assessment of post-stroke deficits in hand proprioception. J Neuroeng Rehabil. 2018;15(1):47.

101. Rinderknecht MD, Lambercy O, Raible V, Liepert J, Gassert R. Age-based model for metacarpophalangeal joint proprioception in elderly. Clin Interv Aging. 2017;12:635–43.

102. Ingemanson ML, Rowe JB, Chan V, Wolbrecht ET, Cramer SC, Reinkensmeyer DJ. Use of a robotic device to measure age-related decline in finger proprioception. Exp Brain Res. 2016;234(1):83–93.

103. Ingemanson ML, Rowe JR, Chan V, Wolbrecht ET, Reinkensmeyer DJ, Cramer SC. Somatosensory system integrity explains differences in treatment response after stroke. Neurology. 2019;92(10):e1098-108.

104. Zbytniewska M, Kanzler CM, Jordan L, Salzmann C, Liepert J, Lambercy O, et al. Reliable and valid robot-assisted assessments of hand proprioceptive, motor and sensorimotor impairments after stroke. J Neuroeng Rehabil. 2021;18(1):115.

105. Zbytniewska M, Rinderknecht MD, Lambercy O, Barnobi M, Raats J, Lamers I, et al. Design and characterization of a robotic device for the assessment of hand proprioceptive, motor, and sensorimotor impairments. IEEE Int Conf Rehabil Robot. 2019;2019:441–6.

106. Bilney B, Morris M, Webster K. Concurrent related validity of the GAITRite walkway system for quantification of the spatial and temporal parameters of gait. Gait Posture. 2003;17(1):68–74.

107. McDonough AL, Batavia M, Chen FC, Kwon S, Ziai J. The validity and reliability of the GAITRite system's measurements: a preliminary evaluation. Arch Phys Med Rehabil. 2001;82(3):419–25.

108. Menz HB, Latt MD, Tiedemann A, Mun San Kwan M, Lord SR. Reliability of the GAITRite walkway system for the quantification of temporospatial parameters of gait in young and older people. Gait Posture. 2004;20(1):20–5.

109. Clark RA, Pua YH, Fortin K, Ritchie C, Webster KE, Denehy L, et al. Validity of the microsoft kinect for assessment of postural control. Gait Posture. 2012;36(3):372–7.

110. Lange B, Chang CY, Suma E, Newman B, Rizzo AS, Bolas M. Development and evaluation of low cost game-based balance rehabilitation tool using the Microsoft Kinect sensor. In: Proc IEEE engineering in medicine and biology Conf. 2011. p. 1831–4.

111. Lee MM, Song CH, Lee KJ, Jung SW, Shin DC, Shin SH. Concurrent validity and test-retest reliability of the OPTOGait photoelectric cell system for the assessment of spatio-temporal parameters of the gait of young adults. J Phys Ther Sci. 2014;26(1):81–5.

112. Clark RA, Vernon S, Mentiplay BF, Miller KJ, McGinley JL, Pua YH, et al. Instrumenting gait assessment using the Kinect in people living with stroke: reliability and association with balance tests. J Neuroeng Rehabil. 2015;12:15.

113. Wikstrom J, Georgoulas G, Moutsopoulos T, Seferiadis A. Intelligent data analysis of instrumented gait data in stroke patients—a systematic review. Comput Biol Med. 2014;51:61–72.

114. Han JJ, Kurillo G, Abresch RT, De Bie E, Nicorici A, Bajcsy R. Upper extremity 3-dimensional reachable workspace analysis in dystrophinopathy using Kinect. Muscle Nerve. 2015;52(3):344–55.

115. Olesh EV, Yakovenko S, Gritsenko V. Automated assessment of upper extremity movement impairment due to stroke. PLoS ONE. 2014;9(8):e104487.

116. Chang YJ, Chen SF, Huang JD. A Kinect-based system for physical rehabilitation: a pilot study for young adults with motor disabilities. Res Dev Disabil. 2011;32(6):2566–70.

117. Raghavan P, Santello M, Gordon AM, Krakauer JW. Compensatory motor control after stroke: an alternative joint strategy for object-dependent shaping of hand posture. J Neurophysiol. 2010;103(6):3034–43.

118. Oess NP, Wanek J, Curt A. Design and evaluation of a low-cost instrumented glove for hand function assessment. J Neuroeng Rehabil. 2012;9:2.

119. Gentner R, Classen J. Development and evaluation of a low-cost sensor glove for assessment of human finger movements in neurophysiological settings. J Neurosci Methods. 2009;178(1):138–47.

120. Hermsdorfer J, Hagl E, Nowak DA, Marquardt C. Grip force control during object manipulation in cerebral stroke. Clin Neurophysiol. 2003;114 (5):915–29.

121. Nowak DA, Hermsdorfer J. Grip force behavior during object manipulation in neurological disorders: toward an objective evaluation of manual performance deficits. Mov Disord. 2005;20(1):11–25.

122. Nowak DA, Hermsdorfer J. Objective evaluation of manual performance deficits in neurological movement disorders. Brain Res Rev. 2006;51(1):108–24.

123. Nowak DA, Hermsdorfer J, Marquardt C, Topka H. Moving objects with clumsy fingers: how predictive is grip force control in patients with impaired manual sensibility? Clin Neurophysiol. 2003;114 (3):472–87.

124. Nowak DA, Hermsdorfer J, Topka H. Deficits of predictive grip force control during object manipulation in acute stroke. J Neurol. 2003;250(7):850–60.

125. Kim GJ, Parnandi A, Eva S, Schambra H. The use of wearable sensors to assess and treat the upper extremity after stroke: a scoping review. Disabil Rehabil. 2021;30:1–20.

126. Schwarz A, Bhagubai MMC, Wolterink G, Held JPO, Luft AR, Veltink PH. Assessment of upper limb movement impairments after stroke using wearable inertial sensing. Sensors (Basel). 2020;20(17).

127. Chen KY, Bassett DR Jr. The technology of accelerometry-based activity monitors: current and future. Med Sci Sports Exerc. 2005;37(1):S490-500.

128. Dijkstra B, Kamsma YP, Zijlstra W. Detection of gait and postures using a miniaturized triaxial accelerometer-based system: accuracy in patients with mild to moderate Parkinson's disease. Arch Phys Med Rehabil. 2010;91(8):1272–7.

129. Godfrey A, Conway R, Meagher D, ÓLaighin G. Direct measurement of human movement by accelerometry. Med Eng Phys. 2008;30(10):1364–86.

130. Zijlstra W, Hof AL. Assessment of spatio-temporal gait parameters from trunk accelerations during human walking. Gait Posture. 2003;18(2):1–10.

131. Dijkstra B, Zijlstra W, Scherder E, Kamsma Y. Detection of walking periods and number of steps in older adults and patients with Parkinson's disease: accuracy of a pedometer and an accelerometry-based method. Age Ageing. 2008;37(4):436–41.

132. Paraschiv-Ionescu A, Perruchoud C, Buchser E, Aminian K. Barcoding human physical activity to assess chronic pain conditions. PLoS ONE. 2012;7 (2):e32239.

133. Michielsen ME, Selles RW, Stam HJ, Ribbers GM, Bussmann JB. Quantifying nonuse in chronic stroke patients: a study into paretic, nonparetic, and bimanual upper-limb use in daily life. Arch Phys Med Rehabil. 2012;93(11):1975–81.

134. van der Pas SC, Verbunt JA, Breukelaar DE, van Woerden R, Seelen HA. Assessment of arm activity using triaxial accelerometry in patients with a stroke. Arch Phys Med Rehabil. 2011;92(9):1437–42.

135. Thrane G, Emaus N, Askim T, Anke A. Arm use in patients with subacute stroke monitored by accelerometry: association with motor impairment and influence on self-dependence. J Rehabil Med. 2011;43(4):299–304.

136. Noorkoiv M, Rodgers H, Price CI. Accelerometer measurement of upper extremity movement after stroke: a systematic review of clinical studies. J Neuroeng Rehabil. 2014;11.

137. Leuenberger K, Gonzenbach R, Wiedmer E, Luft A, Gassert R. Classification of stair ascent and descent in stroke patients. In: Proc International conference on wearable and implantable body sensor networks workshops. 2014. p. 11–6.

138. Schwarz A, Veerbeek JM, Held JPO, Buurke JH, Luft AR. Measures of interjoint coordination post-stroke across different upper limb movement tasks. Front Bioeng Biotechnol. 2020;8:620805.

139. Held JPO, Klaassen B, Eenhoorn A, van Beijnum BF, Buurke JH, Veltink PH, et al. Inertial sensor measurements of upper-limb kinematics in stroke patients in clinic and home environment. Front Bioeng Biotechnol. 2018;6:27.

140. Kanzler CM, Barth J, Rampp A, Schlarb H, Rott F, Klucken J, et al. Inertial sensor based and shoe size independent gait analysis including heel and toe clearance estimation. Annu Int Conf IEEE Eng Med Biol Soc. 2015;2015:5424–7.

141. Vienne A, Barrois RP, Buffat S, Ricard D, Vidal PP. Inertial sensors to assess gait quality in patients with neurological disorders: a systematic review of technical and analytical challenges. Front Psychol. 2017;8:817.

142. Carpinella I, Cattaneo D, Ferrarin M. Quantitative assessment of upper limb motor function in multiple sclerosis using an instrumented action research arm test. J Neuroeng Rehabil. 2014;11:67.

143. Campolo D, Taffoni F, Formica D, Iverson J, Sparaci L, Keller F, et al. Embedding inertial-magnetic sensors in everyday objects: assessing spatial cognition in children. J Integr Neurosci. 2012;11(1):103–16.

144. Memberg WD, Crago PE. Instrumented objects for quantitative evaluation of hand grasp. J Rehabil Res Dev. 1997;34(1):82–90.

145. Rast FM, Labruyere R. Systematic review on the application of wearable inertial sensors to quantify everyday life motor activity in people with mobility impairments. J Neuroeng Rehabil. 2020;17(1):148.

146. Kanzler CM, Schwarz A, Held JPO, Luft AR, Gassert R, Lambercy O. Technology-aided assessment of functionally relevant sensorimotor impairments in arm and hand of post-stroke individuals. J Neuroeng Rehabil. 2020;17(1):128.

147. Kanzler CM, Lessard I, Gassert R, Brais B, Gagnon C, Lambercy O. Reliability and validity of digital health metrics for assessing arm and hand impairments in an ataxic disorder. Ann Clin Transl Neurol. 2022;9(4):432–443.

148. Reinkensmeyer DJ, Burdet E, Casadio M, Krakauer JW, Kwakkel G, Lang CE, et al. Computational neurorehabilitation: modeling plasticity and learning to predict recovery. J Neuroeng Rehabil. 2016;13(1):42.

149. Stinear C. Prediction of recovery of motor function after stroke. Lancet Neurol. 2010;9(12):1228–32.

150. Prabhakaran S, Zarahn E, Riley C, Speizer A, Chong JY, Lazar RM, et al. Inter-individual variability in the capacity for motor recovery after ischemic stroke. Neurorehabil Neural Repair. 2008;22(1):64–71.

151. Winters C, van Wegen EE, Daffertshofer A, Kwakkel G. Generalizability of the proportional recovery model for the upper extremity after an ischemic stroke. Neurorehabil Neural Repair. 2015;29(7):614–22.

152. Stinear CM, Byblow WD, Ackerley SJ, Smith MC, Borges VM, Barber PA. PREP2: a biomarker-based algorithm for predicting upper limb function after stroke. Ann Clin Transl Neurol. 2017;4(11):811–20.

153. Selles RW, Andrinopoulou ER, Nijland RH, van der Vliet R, Slaman J, van Wegen EE, et al. Computerised patient-specific prediction of the recovery profile of upper limb capacity within stroke services: the next step. J Neurol Neurosurg Psychiatry. 2021.

154. van der Vliet R, Selles RW, Andrinopoulou ER, Nijland R, Ribbers GM, Frens MA, et al. Predicting upper limb motor impairment recovery after stroke: a mixture model. Ann Neurol. 2020;87(3):383–93.

155. Stinear CM, Byblow WD, Ackerley SJ, Barber PA, Smith MC. Predicting recovery potential for individual stroke patients increases rehabilitation efficiency. Stroke. 2017;48(4):1011–9.

156. Kanzler CM, Lamers I, Feys P, Gassert R, Lambercy O. Personalized prediction of rehabilitation outcomes in multiple sclerosis: a proof-of-concept using clinical data, digital health metrics, and machine learning. Med Biol Eng Comput. 2021.

157. Lu L, Tan Y, Klaic M, Galea MP, Khan F, Oliver A, et al. Evaluating rehabilitation progress using motion features identified by machine learning. IEEE Trans Biomed Eng. 2021;68(4):1417–28.

158. Adans-Dester C, Hankov N, O'Brien A, Vergara-Diaz G, Black-Schaffer R, Zafonte R, et al. Enabling precision rehabilitation interventions using wearable sensors and machine learning to track motor recovery. NPJ Digit Med. 2020;3(1).

159. Badesa FJ, Morales R, Garcia-Aracil N, Sabater JM, Casals A, Zollo L. Auto-adaptive robot-aided therapy using machine learning techniques. Comput Methods Programs Biomed. 2014;116(2):123–30.

160. Aminian K, Najafi B, Bula C, Leyvraz PF, Robert P. Spatio-temporal parameters of gait measured by an ambulatory system using miniature gyroscopes. J Biomech. 2002;35(5):689–99.

161. Allen J, Kautz S, Neptune R. Step length asymmetry is representative of compensatory mechanisms used in post-stroke hemiparetic walking. Gait Posture. 2011;33(4):538–43.

162. Hausdorff JM. Gait variability: methods, modeling and meaning. J Neuroeng Rehabil. 2005;2:19.

163. Hamacher D, Hamacher D, Taylor WR, Singh NB, Schega L. Towards clinical application: repetitive sensor position re-calibration for improved reliability of gait parameters. Gait Posture. 2014;39(4):1146–8.

164. Colombo R, Pisano F, Micera S, Mazzone A, Delconte C, Carrozza MC, et al. Robotic techniques for upper limb evaluation and rehabilitation of stroke patients. IEEE Trans Neural Syst Rehabil. 2005;13(3):311–24.

165. Celik O, O'Malley MK, Boake C, Levin HS, Yozbatiran N, Reistetter TA. Normalized movement quality measures for therapeutic robots strongly correlate with clinical motor impairment measures. IEEE Trans Neural Syst Rehabil. 2010;18(4):433–44.

166. Wittmann F, Lambercy O, Held J, Gonzenbach R, Höver R, Starkey M, et al. Assessment-driven arm therapy at home using an IMU-based virtual reality system. In: Proc of the IEEE Int Conf on Rehabilitation Robotics (ICORR). 2015. p. 707–12.

167. Rihar A, Mihelj M, Pasic J, Kolar J, Munih M. Infant trunk posture and arm movement assessment using pressure mattress, inertial and magnetic measurement units (IMUs). J Neuroeng Rehabil. 2014;11:133.

168. Lunenburger L, Oertig M, Brunschwiler A, Colombo G, Riener R, Dietz V. Assessment of spasticity with the robotic gait orthosis Lokomat. In: Proc 6th world congress on brain injury. 2005.

169. Uswatte G, Giuliani C, Winstein C, Zeringue A, Hobbs L, Wolf SL. Validity of accelerometry for monitoring real-world arm activity in patients with subacute stroke: evidence from the extremity constraint-induced therapy evaluation trial. Arch Phys Med Rehabil. 2006;87(10):1340–5.

170. Zariffa J, Kapadia N, Kramer JLK, Taylor P, Alizadeh-Meghrazi M, Zivanovic V, et al. Relationship between clinical assessments of function and measurements from an upper-limb robotic rehabilitation device in cervical spinal cord injury. IEEE Trans Neural Syst Rehabil. 2012;20(3):341–50.

171. Metzger JC, Lambercy O, Califfi A, Dinacci D, Petrillo C, Rossi P, et al. Assessment-driven selection and adaptation of exercise difficulty in robot-assisted therapy: a pilot study with a hand rehabilitation robot. J Neuroeng Rehabil. 2014;11:154.

172. Colombo R, Sterpi I, Mazzone A, Delconte C, Pisano F. Taking a lesson from patients' recovery strategies to optimize training during robot-aided rehabilitation. IEEE Trans Neural Syst Rehabil. 2012;20(3):276–85.

173. Cameirao MS, Badia SB, Oller ED, Verschure PF. Neurorehabilitation using the virtual reality based rehabilitation gaming system: methodology, design, psychometrics, usability and validation. J Neuroeng Rehabil. 2010;7:48.

174. Kan P, Huq R, Hoey J, Goetschalckx R, Mihailidis A. The development of an adaptive upper-limb stroke rehabilitation robotic system. J Neuroeng Rehabil. 2011;8:33.

175. Ranzani R, Lambercy O, Metzger JC, Califfi A, Regazzi S, Dinacci D, et al. Neurocognitive robot-assisted rehabilitation of hand function: a randomized control trial on motor recovery in subacute stroke. J Neuroeng Rehabil. 2020;17(1):115.

Computational Neurorehabilitation

16

Nicolas Schweighofer

Abstract

Computational Neurorehabilitation is an emerging field at the intersection of Neurorehabilitation, Computational Neuroscience, Motor Control and Learning, and Statistical Learning. The overarching goals of Computational Neurorehabilitation are to understand and to further improve motor recovery following neurologic injury by mathematically modeling and simulating the neural processes underlying the change in behavior due to rehabilitation (1). This chapter is organized into three main sections. First, we review the overall framework of Computational Neurorehabilitation and argue that computational neurorehabilitation models belong to the general class of dynamical system models. Second, we discuss the three categories of plastic processes that have been incorporated in previous models: unsupervised, supervised, and reinforcement learning. Third, we discuss the two main types of models in Computational Neurorehabilitation: Qualitative "biological" models whose main goal is to advance our understanding of the neural mechanisms of recovery and Quantitative "predictive" models whose main goal is to predict long-term changes in functional outcomes for individual patients. We illustrate these two types of models by briefly reviewing a number of recent relevant qualitative and quantitative models. We conclude by suggesting future directions for the field.

Keywords

Stroke · Arm and hand function · Neurorehabilitation · Neuro-computational models · Predictive models · Reinforcement learning · Supervised learning · Unsupervised learning

16.1 The Computational Neurorehabilitation Framework

As a preamble to this section, we first note that we use the terms "learning" and "plasticity," as we previously proposed in [1]: As people (with or without a neurologic injury) practice a motor task, their ability to perform the task will improve through normal skill acquisition. This process of "motor learning" is dependent on neural plasticity, both in health and disease. A caveat is that in the theoretical field of machine learning, which largely influences research in Computational Neuroscience and therefore in Computational Neurorehabilitation, the word "learning" is used to describe the change in synaptic efficacy. In

N. Schweighofer (✉)
Biokinesiology and Physical Therapy, University of Southern California, Los Angeles, CA, USA
e-mail: schweigh@usc.edu

© The Author(s), under exclusive license to Springer Nature Switzerland AG 2022

D. J. Reinkensmeyer et al. (eds.), *Neurorehabilitation Technology*,
https://doi.org/10.1007/978-3-031-08995-4_16

addition, and again as in [1], we use the term "recovery" to describe improvements in movement ability over time, resulting in improvements in impairment, performance, motor function, or activities. Recovery can occur through restitution, defined as the ability to perform movements with normative biological structures and functions. Recovery can also occur through compensation, defined as the ability to perform movement goals via the use of biological structures and/or functions differently from those originally used before the injury. In the rest of this chapter, we emphasize the rehabilitation of upper extremity functions following stroke, as this reflects the majority of the work in the field. However, many of the principles and models discussed here can be adapted for the rehabilitation of other motor functions, notably locomotion [2], and to other conditions affecting the motor system, such as Parkinson's disease [3] and spinal cord injury [2].

16.1.1 The Three Essential Characteristics of Computational Neurorehabilitation Models

Because a fundamental premise of neurorehabilitation is that sensorimotor activity can improve motor recovery via brain plastic processes, we earlier proposed that computational neurorehabilitation models contain the following three key properties [1]:

First, the models take as input u(t) quantitative descriptions of sensorimotor activity. Such activity can be the result of motor practice during in-person therapy or during semi-automated therapy (with rehabilitation robots or connected objects), or be the result of "self-practice" as a result of spontaneous use of the limbs [4–6], or both. Whereas some simpler models only consider the dose of practice as inputs [6], other models include the specific features of practice, such as movement errors or rewards (e.g., successfully reaching the target) [4, 7–9].

Second, models produce outputs variables y(t) related to measurement variables, typically related to motor outcomes. Examples of outputs for models of arm recovery post-stroke are predictions of the changes in motor function, changes in arm movement kinematics or kinetics, or daily amount of use of the arm.

Third, and crucially, computational neurorehabilitation models are "dynamical". Because plasticity means "changeability", the models must include differential equations (or difference equations if the time is discretized) to characterize changes in the state of the system due to rehabilitation. As a result, the output depends both on the input signal and on the current state of the system, introducing some type of memory. Mathematically, the models therefore include internal state variables x(t) that represent the information necessary to compute the output y(t), given input u(t), and initial conditions x(t0).

This use of differential or difference equations in computational neurorehabilitation models distinguishes these models from "static" models that have been developed for prognostic purpose, e.g., [10–12]. These latter models take as inputs various predictors of recovery, such as clinical features and baseline behavior, and/or brain imaging measurements, and then predict functional outcomes at future time points using regression techniques. We note however that the inclusion of such predictors is not limited to such regression models, as "Quantitative" computational neurorehabilitation models (see below) can also include such predictors to improve the accuracy and the precisions of the predictions.

16.1.2 Computational Neurorehabilitation Models as Dynamical System Models

Because of the three properties reviewed above, computational neurorehabilitation models, despite their vastly different implementations, all belong to the general class of dynamical systems. A convenient and compact description of a dynamical system is via the **state-space representation**, which is a mathematical model of a physical system with the input, output, and state variables composed of first-order differential equations (or difference equations if time is discretized). The system is represented by three sets of equations: the (possibly vectorial) state update equation, the (possibly vectorial) output equation, and a specification of the initial conditions:

$$x(t+dt) = f(x(t), u(t), t),$$

$$y(t) = g(x(t), u(t), t),$$

and

$$x(t0) = x0$$

where $x(t) \in \mathbb{R}^{n \times 1}$ are the internal states, $y(t) \in \mathbb{R}^{my \times 1}$ are the outputs that are related to motor outcomes, $u(t) \in \mathbb{R}^{mu \times 1}$ are the inputs that provide quantitative descriptions of sensorimotor activity, $f: \mathbb{R}^{n \times 1} \times \mathbb{R}^{mu \times 1} \to \mathbb{R}^{n \times 1}$ is the process function, and $g: \mathbb{R}^{n \times 1} \times \mathbb{R}^{mu \times 1} \to \mathbb{R}^{my \times 1}$ is the output function. In addition, not explicitly shown in the three equations above, models can include (1) noise terms in the state and output equations and (2) known (clinical, imaging, baseline behavioral, etc.) covariates. At time t, the input $u(t)$ can either be 0 (during periods of no sensorimotor activity), simply reflecting the dose of training, or carry movement error information or reward information, as we will discuss below in the different categories of learning. As noted above, what fundamentally differentiates the computational neurorehabilitation models from static models is the state update equation (the first equation).

Whenever possible, it is advantageous to linearize the model above since this simplifies analysis and understanding. The discretized system can then be rewritten as a system of first-order linear-time invariant (LTI) difference equations and output equations; assuming no noise and no covariates given by (assuming no effect of the inputs on the output):

$$x(t+dt) = Ax(t) + Bu(t),$$

$$y(t) = Cx(t),$$

$$x(t0) = x0,$$

where $A \in \mathbb{R}^{n \times n}$, $B \in \mathbb{R}^{n \times mu}$, and $C \in \mathbb{R}^{my \times n}$. The dynamics of the system (such as plastic processes) are determined by matrix A. In particular, each first-order equation in this system of equations has a unique time scale that is inversely related to the "decay rate" (see below). The matrix B controls the strength of the input, and, in some models, its elements are referred to as "learning rates".

16.1.3 Multiple Time Scales in Computational Neurorehabilitation Models

Computational neurorehabilitation models often include multiple time scales of plasticity. For instance, models that include *"activity-driven recovery"*, which is driven by thousands of movements over weeks that influence the slow reorganization of surviving neural networks, e.g., [13], include state variable(s) with time constants ranging from weeks to months—see for instance [4, 5, 8, 14]. Models that include *"spontaneous recovery"*, which is greatest in the first month and continues for $\sim$6 months [15], and involve the reduction in lesion edema, ischemic penumbra, and brain reorganization [16, 17], could include state variable(s) with long-time constants in the range of months. Models that include *"motor learning"*, which occurs at the shorter time scales within and between practice sessions, either for learning compensatory movements or for motor adaptation, include state variable(s) with time constant ranging from minutes to hours [7, 18–22].

16.2 The Three Categories of Theoretical Learning Rules in Computational Neurorehabilitation Models

As noted above, computational neurorehabilitation models account for a change in behavior driven by neuroplastic processes (see Chap. 1). From a theoretical standpoint, such plastic processes are categorized into three categories of learning: unsupervised learning, supervised learning, and reinforcement learning. These categories of learning are defined by the nature of the teaching signals that guide learning [23]: *none* in unsupervised learning, *directional error vectors* in supervised learning, and *scalar rewards* in reinforcement learning. Here, we review these categories of learning and their relevance to Computational Neurorehabilitation. We will review examples of models that include one or more categories of learning when we discuss Qualitative and Quantitative models below.

16.2.1 Learning without feedback: Unsupervised learning and homeoplastic processes

The goal of an unsupervised learning algorithm is to learn a mapping from $x \in \mathbb{R}^n$ so that the output $y \in \mathbb{R}^m$, often with $m < n$, characterizes the statistical regularities in the data. For instance, it can be shown that minimization of the reconstruction error in a linear mapping from $y = Wx$ leads to a simple form and biologically plausible form of unsupervised learning called Hebbian plasticity ("Cells that fire together, wire together" [24]), given mathematically by

$$w_{ij}(t + dt) = w_{ij}(t) + \alpha x_i y_j,$$

where $w_{ij}(t)$ is the weight of the connection from presynaptic neuron x_i to postsynaptic neuron y_j, and α a learning rate. Note that this rule above is a simplification, as some type of normalization process (such as weight decay or synaptic normalization) is necessary to keep the weights from diverging. Earlier models have shown that Hebbian plasticity allows the cerebral cortex to self-organize during development or to re-organize post-lesions [25–27] (see also [28]).

Another form of the unsupervised process is homeostatic plasticity, in which neurons regulate their own activity. Because of the death of afferent neurons due to stroke, the mean inputs of multiple neurons will change (typically decrease), which will lead to a change in mean firing. Homeostatic plastic processes will restore the neurons' excitability either via overall increases in synaptic strength (synaptic scaling) or via changes in current conductances in the axon initial segment (hillock). Thus, by renormalizing the activity of networks indirectly affected by the lesion, homeostatic plasticity may be in part responsible for the spontaneous recovery post-stroke [16, 29].

16.2.2 Learning from Errors: Supervised Learning

To learn a motor task, the learner needs feedback, external, internal, or both. Thus, in addition to unsupervised learning, supervised or reinforcement forms of learning have been incorporated into computational neurorehabilitation models. The goal of supervised learning is to learn an input–output mapping that predicts the output y for an input data x by minimizing the expected error between the desired and actual output. For example, given a linear mapping, a simple supervised learning rule is given by

$$w_{ij}(t + dt) = w_{ij}(t) + \alpha x_i(\widehat{y}_j - y_j),$$

where $\widehat{y}_j$ and y_j are desired and actual outputs, respectively. This rule shows the two factors minimally required for supervised learning: a stimulus signal and the output error.

Supervised learning is important to improve performance in motor learning and has notably been involved in adaptive motor control and motor adaptation via updates of forward and inverse internal models [30]. Accordingly, models of motor adaptation in rehabilitation, e.g., [7, 18], and models of the neuromotor recovery process [4, 8] use various forms of the supervised rule shown above.

16.2.3 Learning from Rewards: Reinforcement Learning

In a situation in which no error is given, but only scalar reward feedback such as "success/failure" is provided, reward information can be used to improve motor performance via reinforcement learning. Current human motor control research is exploring how reinforcement learning can lead to the development of (near) optimal controllers for movements such as reaching movements [31] by maximizing expected rewards (or equivalently minimizing cost) that comprise terms such as final error and effort.

Because motor errors are not directly available to the learner in reinforcement learning, exploration is needed to find motor comments that increase the expected reward. A possibility is to perturb the motor commands with noise and then update the commands in the direction that increase the reward. For instance, if the motor output is a random variable depending on the activation level, and the baseline reward is $\underline{r}$ (computed, for instance by the running average of previous rewards), the input weights to the motor neuron can be updated according to

$$w_{ij}(t+dt) = w_{ij}(t) + \alpha(r - \underline{r})x_i(y_j - u_j),$$

where r is the reward and y_j the actual (noisy) output and the u_j noiseless output. If the perturbation has improved the movement outcome by generating a reward r greater than $\underline{r}$, then the motor commands should be updated in the direction of the perturbation. If not, the motor commands should be updated in the other direction. This rule shows the three factors are required in reinforcement learning: a stimulus signal, x_i; the action produced in its presence, y_j; and the consequent evaluation, r. Such "stochastic policy gradient" reinforcement learning methods can include rules for modulation of the amplitude of the perturbations as performance improves [32, 33]. However, because of the need for exploration, reinforcement learning systems are slow, especially in high dimensions. Imitation learning, in which the learner learns target representation via supervised learning [34], can significantly speed up the learning process because it drastically reduces the size of the state–action space that needs to be explored.

Reinforcement learning can model skill acquisition in the healthy or lesioned brain, e.g., [9, 31] via learning, or re-learning, a new controller. In addition, besides its role in learning motor controllers, reinforcement learning provides a plausible framework for human adaptive decision-making, for instance to simulate the reorganization of arm use post-stroke [4].

16.3 The Two Types of Computational Neurorehabilitation Models: Qualitative and Quantitative

In the literature, two types of computational neurorehabilitation models can be distinguished: Qualitative models and quantitative models.

16.3.1 Qualitative "Biological" Models

The primary goals of qualitative models are to understand the biological entities and processes that drive recovery and to make new qualitative predictions. Much is known about various neurobiological processes important to effective rehabilitation—see [35] and Chap. 1 for instances for in-depth reviews. However, what is lacking is the integration of these processes, which operate at widely different temporal (see above) and spatial scales (neural, plastic, muscular, behavioral, etc.) [1]: Qualitative computational models of rehabilitation constrain the researcher to specify the dynamic interactions of these processes. Such models are often «large», as they may model neural systems, possibly in different brain areas, plastic processes, and often a (simplified) musculoskeletal sub-system. Typically, no data are available to estimate all the parameters at the multiple spatial levels of these models. Thus, the (free) parameters of these models are extracted from the literature when possible, or, in many other cases, are "tuned" by hand. For instance, the learning rates, the amplitudes of neural noise, the number of neurons, etc., are all parameters that need to be chosen by the researcher. Thus, a drawback of such models is that they often contain dozens of such "free" parameters, requiring careful sensitivity analyses to test the robustness of the predictions.

In Qualitative models, at least one of the three categories of theoretical learning rules, as described above, are implemented, as we illustrate in the three following examples.

The first example is a computational model of the sensory cortex, in which Bains and Schweighofer studied the effect of the relative timing of homeostatic and Hebbian plasticity on recovery following lesion [16] (Fig. 16.1). Hebbian plasticity modulated synaptic strength and homeostatic plasticity modified cell excitability to maintain the desired firing distribution. After initial training, the network was lesioned, leading to areas of hyper- and hypoactivity due to the loss of lateral synaptic connections. The network was then retrained via rehabilitative arm movements that generated proprioceptive inputs.

A first finding was that network recovery was unsuccessful in the absence of homeostatic plasticity. A second finding was that Hebbian plasticity was maladaptive and led to poor map reorganization if the network activity was not previously re-normalized via homeostatic plasticity during a period of arm inactivity (which led to low sensory input). Recent clinical data showing that the optimal timing of rehabilitation in humans is 1–3 months post-stroke support this view [36].

The second example is a model of cortical control of wrist forces, in which Reinkensmeyer et al. showed that improvements in motor function following lesions depend on the ability to activate spared portions of the damaged corticospinal system [9] (Fig. 16.2). In the model, wrist force was produced by the summed effect of corticospinal cells targeting motor neuronal pools to both flexor and extensor muscle. The model used a reinforcement learning algorithm:

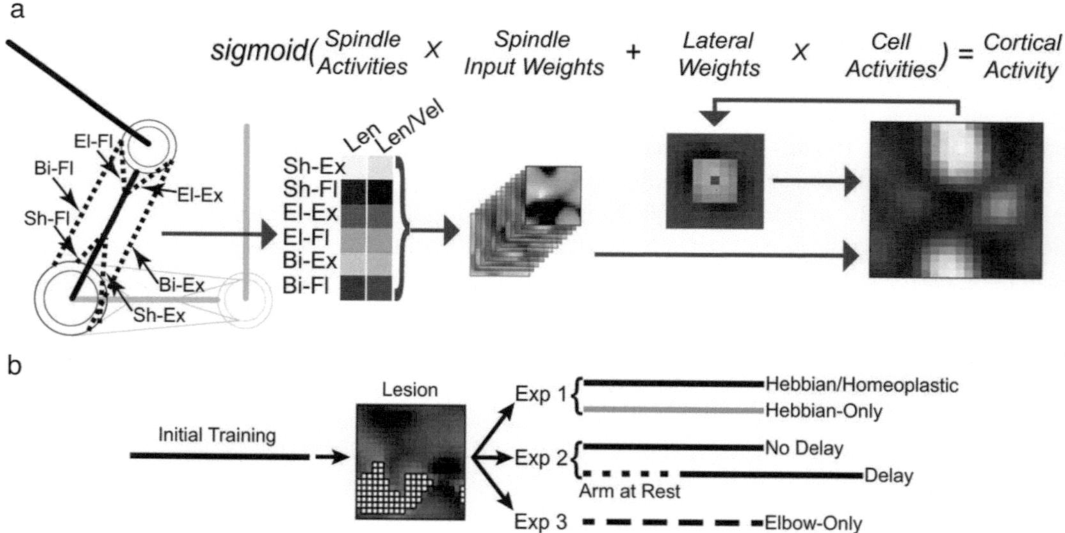

Fig. 16.1 A model of reorganization of the sensory cortex post-lesion via homeostatic and Hebbian plasticity [16]. **a** Effects of arm movement on cortical activity. *Left* to *right*: simulated two-joint, six-muscle arm during movement (background shows arm in home position for reference). Black dashed lines are muscles and include shoulder flexor and extensor (Sh-Fl and Sh-Ex), elbow flexor and extensor (El-Fl and El-Ex), and biarticular flexor and extensor (Bi-Fl and Bi-Ex). Each cortical cell received weighted spindle activities and weighted activity from neighboring cells. Homeostatic plasticity modified the cortical neurons' activity with the goal of maintaining an exponential firing rate distribution with the desired mean. Hebbian plasticity modified afferent and lateral (positive and negative) weights. **b** Schematic demonstrating time course of the initial network training period, lesion, and rehabilitation training condition tested. In the Delay condition, training starts following the "Arm at rest" phase. Figure reproduced with permission from [16]

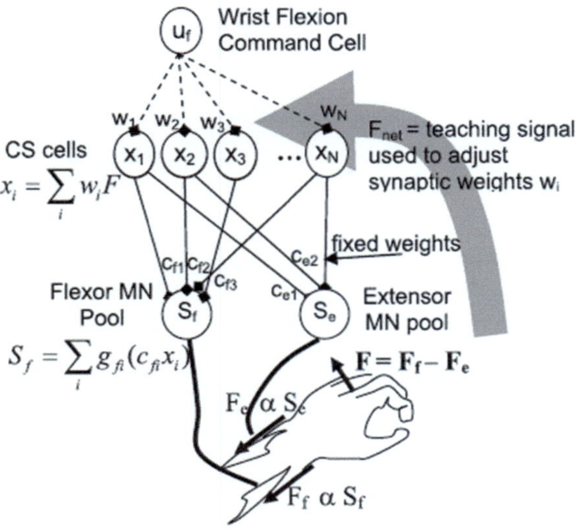

Fig. 16.2 A model of adaptive cortical control of wrist forces following lesion via reinforcement learning [9]. A reinforcement learning rule adjusts the synaptic weights wi that determine the activations xi of the N residual corticospinal CS cells given a flexion command, in order to maximize the net force output F_{net} of the wrist (which is based on measured force F generated during movements). The activities of CS cells sum in the spinal cord, activating flexor and extensor motoneuronal pools. gi is a saturation function. Figure reproduced with permission from [9]

cell firing rate changed by a random amount from trial to trial, and activation patterns that produce more force were remembered for future use. The model predicted exponential-like strength recovery curves that exhibit a residual capacity for further recovery with further movement practice. It also predicted that patients with a larger residual corticospinal network recover more and that movement-related activation in secondary motor areas increases following a stroke that damages output from the primary motor area. These predictions have received experimental support [37–39].

The third example is a model of arm choice following the cortical lesion [4], which was extended in a model of bimanual coordination post-stroke [40]. In this model, Han et al. studied the interaction between recovery of motor performance and arm choice post-stroke (Fig. 16.3). The model contains two main plastic neural processes: (1) A bilateral model of the motor cortex/cerebellum networks that generate reaching movements that is (unilaterally) lesioned by stroke and that is updated via a supervised learning processes which aim to decrease error in movements (before and after the lesion), and (2) a decision-making process, loosely based on the basal ganglia, that selects the arm to reach a given target and that is updated via a reinforcement plastic process which aims to maximize rewards (e.g., target "hits"). The model predicted that if stroke suddenly decreases motor performance, the value of the more affected UE is downregulated because of reach failures, leading to learned non-use and compensatory choice of the less affected arm. In addition, the model predicted that recovery is bistable: (Simulated) patients using the affected arm above a threshold experienced progressive amelioration of both use and performance via a virtuous cycle induced via "self-practice". In contrast, (simulated) patients who did not use their affected arm above this threshold experienced progressive deterioration of both use and performance via a vicious cycle, with rehabilitation becoming "in vain". Evidence for such a threshold has since been observed both in animal and clinical data [6, 14, 41, 42]. In addition, the model has led to a new method to promote arm use in stroke patients by boosting their confidence in their more affected UE

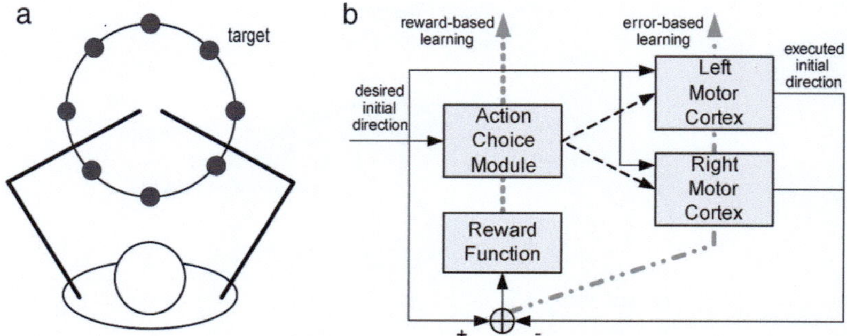

Fig. 16.3 A model of arm choice following cortical lesion [4]. **a** the simulated patient. At each trial, a target appears and the patient is either instructed to use the impaired arm (rehabilitation) or free to use one of the two arms (free choice, mimicking daily arm use). **b** The model simplified neurorehabilitation dynamics by assuming that a reinforcement learning (reward-based) mechanism determines the probabilities of choosing the impaired or unimpaired arms after stroke in one motor cortex and that a separate, supervised (error-based) learning mechanism accounts for improvements in motor control through practice (see text for further details). Figure reproduced with permission from [4]

function. Participants with hemiparesis were exposed to reduced errors while performing arm shooting movements in a non-immersive virtual-reality system [43]. Unaware of the manipulations, patients reported making internal attributions of the success they experienced through training and showed a higher probability of using their more affected arm.

16.3.2 Quantitative "Predictive" Models

Quantitative models predict long-term changes in functional outcomes after brain injury and rehabilitation. Because the parameters in these models are estimated from actual data, which in rehabilitation studies are often sparse and often include only behavioral data (and not neural data), the models are necessarily «small». Concretely, for a LTI model, the dimensions of the states n and motor commands mu and motor outputs my are small (typically 1 or 2) such that the matrices A, B, and C contain few parameters. Besides predicting individual recovery trajectories, these models can also be used to test specific hypotheses: the data can be used to compare predictions made by different models, and model comparison techniques can be used to select the best fitting model, controlling for the number of parameters (as increasing model complexity always improves the model fit). Finally, the predictions of these models can also be used to personalize motor training—see for instances [44, 45].

The first example of such models is a model of motor recovery during robotic training [8]. The model is driven by data from actual robotic training sessions and describes the trial-by-trial evolution of the recovery process induced by robotic training. This model provides insights into the role of assistive force in the recovery process, and the extent to which changes in voluntary control decay over time and transfer to subsequent training sessions. A striking result of this model is that the retention parameter predicted the percent change of the Fugl-Meyer score at the 3-month point following the end of the robotic treatment.

The second example of such models is a model of the interaction between arm use and function following rehabilitation [5]. The model largely simplified the Han et al. model described above via two coupled first-order nonlinear equations. A model with reciprocal interactions between arm function and use was the best fitting model and accounted for the virtuous and vicious cycles. Furthermore, it was found that therapy increased the parameter that modulated the effect of UE function on use. Simulations showed that increasing this parameter, which can be thought of as the confidence to use the arm for a given level of function (i.e., self-efficacy for paretic limb use), led to an increase in spontaneous use and the development of a virtuous cycle.

16.4 Concluding Remarks and Future Directions

In Quantitative computational neurorehabilitation models, accurate and precise parameter estimation is primordial. Unfortunately, available data in the form of repeated clinical tests used to fit the models have been typically sparse. As a result, previous models were necessarily simple to avoid overfitting [5, 6]. Compounding the problem is that researchers have often used "point estimation" methods such as least square or maximum likelihood estimation to fit average or individual data. However, recent developments have seen the emergence of mixed effect models to improve individual predictions when the data for each individual is sparse [6]. In addition, Bayesian modeling, e.g., [5], seamlessly generates credible intervals for these predictions and can incorporate prior information (e.g., parameter mean and variance) from previous studies to further improve individual predictions.

Richer data sets would allow us to increase the complexity of the quantitative models. Fortunately, a revolution in data availability is underway. For instance, data from rehabilitation robots or exoskeletons [46] or from sophisticated wearable sensors that monitor both arm and hand movements such as the *Manumeter* [47] could largely increase the data available for modeling. In addition, semi-automatic analysis of MRI scans post-stroke [48] could provide (summarized) neural data that could be included in the models. At the same time, new qualitative models that incorporate different types of plasticity reviewed above, both following lesion and during rehabilitation, will further predict new effective rehabilitation methods. In particular, new models that use reinforcement learning are needed to better understand the development of compensation versus restitution. We expect that new quantitative models of recovery, based on early clinical data, kinematic data both in joint and task space [49], and routine scans, and informed by new qualitative models, will soon provide the basis for future models that suggest timing, dosage, and content of therapy. Such an approach will transform neurorehabilitation by guiding clinicians, patients, and health providers in the optimization of treatments via Precision Rehabilitation (see next chapter).

Acknowledgements The author acknowledges grant NIH R21NS120274 and thanks Denis Mottet, Etienne Burdet, Carolee Winstein, Cheol Han, Jim Gordon, Jun Izawa, Michael Arbib, and David Reinkensmeyer for fruitful discussions that have led to the ideas presented in this chapter.

References

1. Reinkensmeyer DJ, Burdet E, Casadio M, Krakauer JW, Kwakkel G, Lang CE, et al. Computational neurorehabilitation: modeling plasticity and learning to predict recovery. J Neuroeng Rehabil. 2016;13(1):42.
2. Reinkensmeyer DJ, Aoyagi D, Emken JL, Galvez JA, Ichinose W, Kerdanyan G, et al. Tools for understanding and optimizing robotic gait training. J Rehabil Res Dev. 2006;43(5):657–70.
3. Frank MJ. Computational models of motivated action selection in corticostriatal circuits. Curr Opin Neurobiol. 2011;21(3):381–6.
4. Han CE, Arbib MA, Schweighofer N. Stroke rehabilitation reaches a threshold. PLoS Comput Biol. 2008;4(8): e1000133.
5. Hidaka Y, Han CE, Wolf SL, Winstein CJ, Schweighofer N. Use it and improve it or lose it: interactions between arm function and use in humans post-stroke. PLoS Comput Biol. 2012;8(2): e1002343.
6. Wang C, Winstein C, D'Argenio DZ, Schweighofer N. The efficiency, efficacy, and retention of task practice in chronic stroke. Neurorehabil Neural Repair. 2020;34(10):881–90.
7. Scheidt RA, Stoeckmann T. Reach adaptation and final position control amid environmental uncertainty after stroke. J Neurophysiol. 2007;97(4):2824–36.
8. Casadio M, Sanguineti V. Learning, retention, and slacking: a model of the dynamics of recovery in robot therapy. IEEE Trans Neural Syst Rehabil Eng. 2012;20(3):286–96.
9. Reinkensmeyer DJ, Guigon E, Maier MA. A computational model of use-dependent motor recovery following a stroke: optimizing corticospinal activations via reinforcement learning can explain residual

capacity and other strength recovery dynamics. Neural Netw. 2012;29–30:60–9.

10. Riley JD, Le V, Der-Yeghiaian L, See J, Newton JM, Ward NS, et al. Anatomy of stroke injury predicts gains from therapy. Stroke. 2011;42(2):421–6.

11. Stinear CM, Barber PA, Petoe M, Anwar S, Byblow WD. The PREP algorithm predicts potential for upper limb recovery after stroke. Brain. 2012;135 (Pt 8):2527–35.

12. Kwakkel G, Kollen B. Predicting improvement in the upper paretic limb after stroke: a longitudinal prospective study. Restor Neurol Neurosci. 2007;25 (5–6):453–60.

13. Nudo RJ, Wise BM, SiFuentes F, Milliken GW. Neural substrates for the effects of rehabilitative training on motor recovery after ischemic infarct. Science. 1996;272(5269):1791–4.

14. Schweighofer N, Han CE, Wolf SL, Arbib MA, Winstein CJ. A functional threshold for long-term use of hand and arm function can be determined: predictions from a computational model and supporting data from the extremity constraint-induced therapy evaluation (EXCITE) trial. Phys Ther. 2009;89(12):1327–36.

15. Duncan PW, Lai SM, Keighley J. Defining post-stroke recovery: implications for design and interpretation of drug trials. Neuropharmacology. 2000;39 (5):835–41.

16. Bains AS, Schweighofer N. Time-sensitive reorganization of the somatosensory cortex post-stroke depends on interaction between Hebbian plasticity and homeoplasticity: a simulation study. J Neurophysiol. 2014;jn 00433 2013.

17. Murphy TH, Corbett D. Plasticity during stroke recovery: from synapse to behaviour. Nat Rev Neurosci. 2009;10(12):861–72.

18. Schweighofer N, Lee JY, Goh HT, Choi Y, Kim SS, Stewart JC, et al. Mechanisms of the contextual interference effect in individuals poststroke. J Neurophysiol. 2011;106(5):2632–41.

19. Smith MA, Ghazizadeh A, Shadmehr R. Interacting adaptive processes with different timescales underlie short-term motor learning. PLoS Biol. 2006;4(6): e179.

20. Kording KP, Tenenbaum JB, Shadmehr R. The dynamics of memory as a consequence of optimal adaptation to a changing body. Nat Neurosci. 2007;10(6):779–86.

21. Lee JY, Schweighofer N. Dual adaptation supports a parallel architecture of motor memory. J Neurosci. 2009;29(33):10396–404.

22. Kim S, Ogawa K, Lv J, Schweighofer N, Imamizu H. Neural substrates related to motor memory with multiple timescales in sensorimotor adaptation. PLoS Biol. 2015;13(12): e1002312.

23. Doya K. What are the computations of the cerebellum, the basal ganglia and the cerebral cortex? Neural Netw. 1999;12(7–8):961–74.

24. Hebb D. The organisation of behaviour. New York: Wiley; 1949.

25. Goodall S, Reggia JA, Chen Y, Ruppin E, Whitney C. A computational model of acute focal cortical lesions. Stroke. 1997;28(1):101–9.

26. Goodall S, Reggia JA, Cho S. Modeling brain adaptation to focal damage. In: Proceedings of the annual symposium on computer application [sic] in medical care. 1994:860–4.

27. Reggia JA. Neurocomputational models of the remote effects of focal brain damage. Med Eng Phys. 2004;26(9):711–22.

28. Varier S, Kaiser M, Forsyth R. Establishing, versus maintaining, brain function: a neuro-computational model of cortical reorganization after injury to the immature brain. J Int Neuropsychol Soc. 2011;17 (6):1030–8.

29. Butz M, van Ooyen A, Worgotter F. A model for cortical rewiring following deafferentation and focal stroke. Front Comput Neurosci. 2009;3:10.

30. Kawato M. Internal models for motor control and trajectory planning. Curr Opin Neurobiol. 1999;9 (6):718–27.

31. Izawa J, Shadmehr R. Learning from sensory and reward prediction errors during motor adaptation. PLoS Comput Biol. 2011;7(3): e1002012.

32. Williams RJ. Simple statistical gradient-following algorithms for connectionist reinforcement learning. Mach Learn. 1992;8(3–4):229–56.

33. Gullapalli V. A stochastic reinforcement learning algorithm for learning real-valued functions. Neural Netw. 1990;3:671–92.

34. Schaal S. Is imitation learning the route to humanoid robots? Trends Cogn Sci. 1999;3(6):233–42.

35. Krakauer JW, Carmichael ST. Broken movement: the neurobiology of motor recovery after stroke. Cambridge: The MIT Press; 2017. xiv, p. 269.

36. Dromerick AW, Geed S, Barth J, Brady K, Giannetti ML, Mitchell A, et al. Critical Period After Stroke Study (CPASS): a phase II clinical trial testing an optimal time for motor recovery after stroke in humans. Proc Natl Acad Sci USA. 2021;118(39).

37. Page SJ, Gater DR, Bach YRP. Reconsidering the motor recovery plateau in stroke rehabilitation. Arch Phys Med Rehabil. 2004;85(8):1377–81.

38. Brouwer BJ, Schryburt-Brown K. Hand function and motor cortical output poststroke: are they related? Arch Phys Med Rehabil. 2006;87(5):627–34.

39. Ward NS, Newton JM, Swayne OB, Lee L, Frackowiak RS, Thompson AJ, et al. The relationship between brain activity and peak grip force is modulated by corticospinal system integrity after subcortical stroke. Eur J Neurosci. 2007;25(6):1865–73.

40. Takiyama K, Okada M. Recovery in stroke rehabilitation through the rotation of preferred directions induced by bimanual movements: a computational study. PLoS ONE. 2012;7(5): e37594.

41. MacLellan CL, Keough MB, Granter-Button S, Chernenko GA, Butt S, Corbett D. A critical threshold of rehabilitation involving brain-derived neurotrophic factor is required for poststroke recovery. Neurorehabil Neural Repair. 2011;25(8):740–8.

42. Schwerz de Lucena D, Rowe J, Chan V, Reinkens-meyer DJ. Magnetically counting hand movements: validation of a calibration-free algorithm and application to testing the threshold hypothesis of real-world hand use after stroke. Sensors (Basel). 2021; 21(4).

43. Ballester BR, Nirme J, Duarte E, Cuxart A, Rodriguez S, Verschure P, et al. The visual amplification of goal-oriented movements counteracts acquired non-use in hemiparetic stroke patients. J Neuroeng Rehabil. 2015;12:50.

44. Burdet E, Li Y, Kager S, Chua KSG, Hussain A, Campolo D. Interactive robot assistance for upper-limb training. Rehabilitation robotics: Academic Press; 2018. p. 137–48

45. Reinkensmeyer DJ. How to retrain movement after neurologic injury: A computational rationale for incorporating robot (or therapist) assistance. Proc Annu Int IEEE EMBS. 2003;25:1479–82.

46. Schweighofer N, Wang C, Mottet D, Laffont I, Bakhti K, Reinkensmeyer DJ, et al. Dissociating motor learning from recovery in exoskeleton training post-stroke. J Neuroeng Rehabil. 2018;15(1):89.

47. Friedman N, Rowe JB, Reinkensmeyer DJ, Bachman M. The manumeter: a wearable device for monitoring daily use of the wrist and fingers. IEEE J Biomed Health Inform. 2014;18(6):1804–12.

48. Akkus Z, Galimzianova A, Hoogi A, Rubin DL, Erickson BJ. Deep learning for brain MRI segmentation: state of the art and future directions. J Digit Imaging. 2017;30(4):449–59.

49. Nibras N, Liu C, Mottet D, Wang C, Reinkensmeyer D, Remy-Neris O, et al. Dissociating sensorimotor recovery and compensation during exoskeleton training following stroke. Front Hum Neurosci. 2021;15: 645021.

Precision Rehabilitation: Can Neurorehabilitation Technology Help Make It a Realistic Target?

17

W. Zev Rymer and D. J. Reinkensmeyer

Abstract

A major theme that has emerged in medicine over the last 10–15 years is "precision medicine", in which therapeutic interventions are tailored toward the specific biological makeup and clinical needs of a given patient. This chapter briefly introduces the concept of precision medicine, exemplified by recent successes in oncology. Then, we summarize attempts at developing precision medicine approaches for stroke rehabilitation with a focus on upper extremity movement recovery. These attempts have mainly focused on predicting long-term outcomes rather than on predicting who can benefit from rehabilitation therapy, or, more importantly, who can benefit from what type of therapy. The most promising results have come from using early voluntary movement ability as a predictor, supplemented by motor evoked potential status, rather than genetic or brain structure factors. Of all the components that could further a clinically viable precision rehabilitation, we argue that precision measurement of impairment with neurorehabilitation technology is most directly within our reach and will have the greatest impact.

Keywords

Precision medicine · Stroke · Assessment · Prediction · Rehabilitation · Therapy · Modeling

W. Z. Rymer
Shirley Ryan AbilityLab, Chicago, IL, USA
e-mail: w-rymer@northwestern.edu

Department of Physical Medicine and Rehabilitation, Northwestern University, Chicago, IL, USA

D. J. Reinkensmeyer (✉)
Department of Anatomy and Neurobiology, University of California at Irvine, 4200 Engineering Gateway, Irvine, CA 92617, USA
e-mail: dreinken@uci.edu

Department of Mechanical and Aerospace Engineering, University of California at Irvine, 4200 Engineering Gateway, Irvine, CA 92617, USA

Department of Biomedical Engineering, University of California at Irvine, 4200 Engineering Gateway, Irvine, CA 92617, USA

17.1 Introduction

There have been an enormous number of technical advances in neurological rehabilitation in the last 10–15 years as outlined in multiple chapters of this book. In spite of these advances, there are many areas of medicine in which therapeutic progress has been even more rapid. Here, we consider one of these areas—precision medicine, which tailors therapeutic interventions toward the specific biological makeup and clinical needs of a given patient or patient population. The goal of this chapter is to understand the

© The Author(s), under exclusive license to Springer Nature Switzerland AG 2022

D. J. Reinkensmeyer et al. (eds.), *Neurorehabilitation Technology*,
https://doi.org/10.1007/978-3-031-08995-4_17

potential for precision medicine techniques to be applied in neurologic rehabilitation. With some exceptions, as we detail below, we have been unable so far to develop an advanced precision framework for neurological rehabilitation. And yet, we argue, technology-enabled measurement of sensory motor impairments is the most promising avenue for improving precision rehabilitation and should be further pursued.

This chapter first briefly reviews a success story of precision medicine—oncology. Then, we briefly discuss the fledgling attempts at bringing precision medicine into upper extremity stroke rehabilitation, summarizing what is known about genetic, etiologic, brain structure, brain function, and behavioral predictors. Finally we discuss the role that technology-enabled measurement of impairments could play in achieving precision rehabilitation.

17.2 Oncology—The Exemplary Case of Precision Medicine

The most advanced applications of precision medicine have been in the areas of human oncology [1]. Currently, therapeutic interventions in many cancers are based on the detailed identification of key factors that may impact the likelihood of success for cancer treatments. So, for example, strategies for treatment of breast cancer have advanced enormously because of the systematic application of several key components of precision medicine [2]. In particular, precision oncology has achieved greater 5 years survivals in patients with metastatic breast cancer than prevailing chemotherapy techniques that were utilized before [3]. These new techniques include a detailed analysis of patient genomics, rigorous analysis and pathological staging of the breast cancer itself, the selection of a suitable, standardized intervention protocol, a systematic assessment of response to therapy and a detailed and frequent measurement of outcomes using quantitative imaging and other methods.

These "precision" approaches are also being implemented now in other areas of medicine, including cardiology, gastroenterology, and endocrinology. Other areas of medicine now seek to make this transition to precision medicine, although most have not yet been able to replicate the effectiveness of results achieved in precision oncology.

17.3 Initial Attempts at Precision Rehabilitation

The two most pressing questions for neurorehabilitation are: "What is a patient's potential for recovery?" and "What rehabilitation treatments should be applied to maximize recovery based on the patient's profile?" [4]. The vast majority of effort toward a precision rehabilitation has been applied to the first question. Here, we summarize these efforts, referring the reader to recent reviews for more details [4–6].

17.3.1 Genetic Markers

Brain plasticity supports the improvements in sensory motor function that result from rehabilitation therapy. Two genes have been hypothesized to play a particularly important role in plasticity: brain derived neurotrophic factor (BDNF) [7] and ApolipoproteinE (ApoE) [8]. These genes vary across humans, with each having a common polymorphism that affects neural function.

BDNF is a widely expressed growth factor in the neurotrophin family that supports neuron survival and encourages growth of new neurons and synapses [9]. The val66met polymorphism in the BDNF gene is a common mutation present in $\sim 30\%$ if the population. This mutation is associated with reductions in activity-dependent BDNF secretion [10], human motor plasticity [11], and motor skill acquisition [12].

ApoE is a protein that plays a role in the metabolism of fats. The ApoE ε4 allele present in 24% of the population is associated with a weakening of neural repair processes [13] and reduced microvascular flow [14]. ApoE ε4 is associated with poorer outcome after hemorrhagic stroke [15] and traumatic brain injury [16] but its influence after ischemic stroke is unclear [17].

A recent phase 3 randomized controlled trial of upper extremity stroke rehabilitation tested the potential role of these genetic variations in predicting response to therapy [8]. The Interdisciplinary Comprehensive Arm Rehabilitation Evaluation (ICARE) study enrolled 216 patients with mild and moderate impairment and randomized them to task-oriented upper extremity training, dose-equivalent occupational therapy, or standard of care. The primary outcome measure was the 12-month change in Wolf Motor Function Test. Neither the val66met BDNF nor ApoE ε4 polymorphism predicted response to rehabilitation therapy, although the BDNF val66met polymorphism predicted baseline cerebral atrophy and the ApoE ε4 polymorphism was associated with younger age at stroke onset.

Thus, while there are reasons to expect that genetic variations will impact the effectiveness of upper extremity stroke rehabilitation therapy due to their influence on brain plasticity, as yet this link has yet to be demonstrated. Therefore, including genetic measures in precision rehabilitation remains aspirational.

17.3.2 Stroke Etiology: Ischemia Versus Hemorrhage

Some studies suggest that the mechanism of injury to the brain plays a role in predicting outcome. In a retrospective study of 1064 consecutive admissions to a rehabilitation hospital over a 4-year period, patients with intracerebral hemorrhage had greater functional impairment at admission compared to those with ischemic stroke, but made greater gains, as measured by FIM score. Other studies support the finding of better prognosis after hemorrhage, although a recent study found no differences in outcomes predicted by etiology, highlighting the mixed nature of results [18]. Of relevance to neurorehabilitation technologies, a recent study compared robotic gait training outcomes between hemorrhagic and ischemic patients who were on average 28 weeks after stroke. The participants received an intervention blending conventional and robot-assisted gait training with the Lokomat [19]. Hemorrhagic participants made significantly greater gains in Functional Ambulation Category.

17.3.3 Brain Structure and Function

A recent consensus statement on biomarkers of stroke recovery suggested the viability of several measures of brain structure and function for predicting motor recovery and treatment response after stroke [4]. Potential predictors include the site of damage, lesion volume, lesion load to the corticospinal tract (CST), presence of motor evoked potential (MEP), fMRI and EEG resting state functional connectivity, fMRI measures of laterality and activation, and EEG measures of corticoneuronal oscillation amplitude. The committee concluded that CST injury measured with diffusion tensor imaging or lesion overlap, as well as the TMS measure of MEP+ or MEP−, were ready to be used in clinical trials. However, in a large review that pooled individual data from 372 people with arm impairment after stroke, the presence of a MEP was the only biomarker related to a better motor outcome, as judged by the Upper Extremity Fugl-Meyer Score [20]. Motor outcome was unrelated to stroke lesion volume, location (cortical, subcortical, mixed) or side (left versus right), as well as to corticospinal tract asymmetry.

Limited progress has been made in predicting response to rehabilitation treatment with biomarkers. Of note, several of the key studies in this area have relied on robotic and other neurorehabilitation technologies to deliver motor training. For example, the extent of CST injury predicted 40% of the variance of treatment gains from robotic hand therapy [21] and helped better identify responders in an unsuccessful epidural brain stimulation trial [22]. Prediction of the benefit of robotic hand therapy was improved to

44% variance accounted for by combining interhemispheric M1 rsFC with CST injury [23]. For chronic stroke participants, cortical connectivity with ipsilesional M1 measured with dense array EEG at baseline predicted 60% of the variance of motor gains across four weeks of intense, home-based virtual game motor training for the upper extremity [24]. Finally, a model combining total sensory system injury and sensorimotor cortical connectivity between ipsilesional primary motor and secondary somatosensory cortices explained 56% of variance in treatment-induced hand functional gains resulting from intensive robotic finger training [25].

Thus, measures of brain structure and function show potential for predicting the response to treatment with neurorehabilitation technologies. However, such markers have not yet been used to distinguish who might benefit from what treatment. A limiting factor in the application of these biomarkers is the need for the specialized equipment and data processing expertise associated with the techniques of MRI, fRMI, TMS, and EEG.

17.3.4 Behavioral Approaches and the First Clinical Precision Rehabilitation Tools

Behavioral measurements made early after stroke strongly predict long-term outcome. For example, upper extremity impairment measured within 2 weeks of stroke strongly predicts extremity impairment at 3 or 6 months, with most patients recovering 70–80% of amount they could recover [26]. However, a certain subgroup of patients, the so-called "non-fitters" to this Proportional Recovery Law, recover substantially less than predicted. These individuals lack hand MEP at baseline [27], and ultimately fail to recover hand dexterity or the ability to move the arm out of an abnormal synergistic pattern [28].

Baseline proprioceptive integrity has been shown to predict the ability to benefit from constraint induced therapy [29] or robotic hand movement training [30]. In the former study, chronic stroke participants with impaired proprioception had a 20% chance of achieving a meaningful clinical improvement in hand function in the EXCITE trial. In the latter study, finger proprioception measured robotically at baseline predicted 40% of the variance of functional response to three weeks of robotic finger training.

The strength of behavioral predictors has led to what are arguably the first two precision rehabilitation tools, although they are primitive (Figs. 17.1 and 17.2). The first is based on the shoulder-abduction-finger-extension (SAFE) score, which measures the ability to perform some voluntary shoulder abduction and some voluntary finger extension early after stroke [31]. If these movements are present in the first 72 h after stroke, there is a 98% probability of achieving some upper limb function at 6 months post-stroke as measured with the ARAT score [31]. Further, 60% of patients exhibiting these movements early recover full upper limb function. In contrast, patients without these movement abilities showed only a 25% probability of recovering some upper limb function. It was suggested to monitor finger extension for the first 12 weeks post-stroke because of its prognostic power [31, 32]. 45% of individuals without finger extension in the first week after stroke will eventually recover it, with a median time for regaining it of 4 weeks [32]. Based on these findings, one of the first primitive precision rehabilitation tools is an app that suggests different therapeutic activities based on SAFE score and other behavioral features [33] (Fig. 17.1).

The second tool, the PREP algorithm, supplements the SAFE score with MEP status when needed to resolve prediction uncertainty [34]. As noted above, persons with a low SAFE score early after stroke can end up ultimately achieving a wide variation of upper extremity function. Adding MEP status achieved a 75% predictive accuracy into four recovery classes (Complete, Notable, Limited, None, see definitions in Fig. 17.2), each with its own suggested therapy content (Fig. 17.2). PREP was further refined into a more efficient PREP2 algorithm that relies

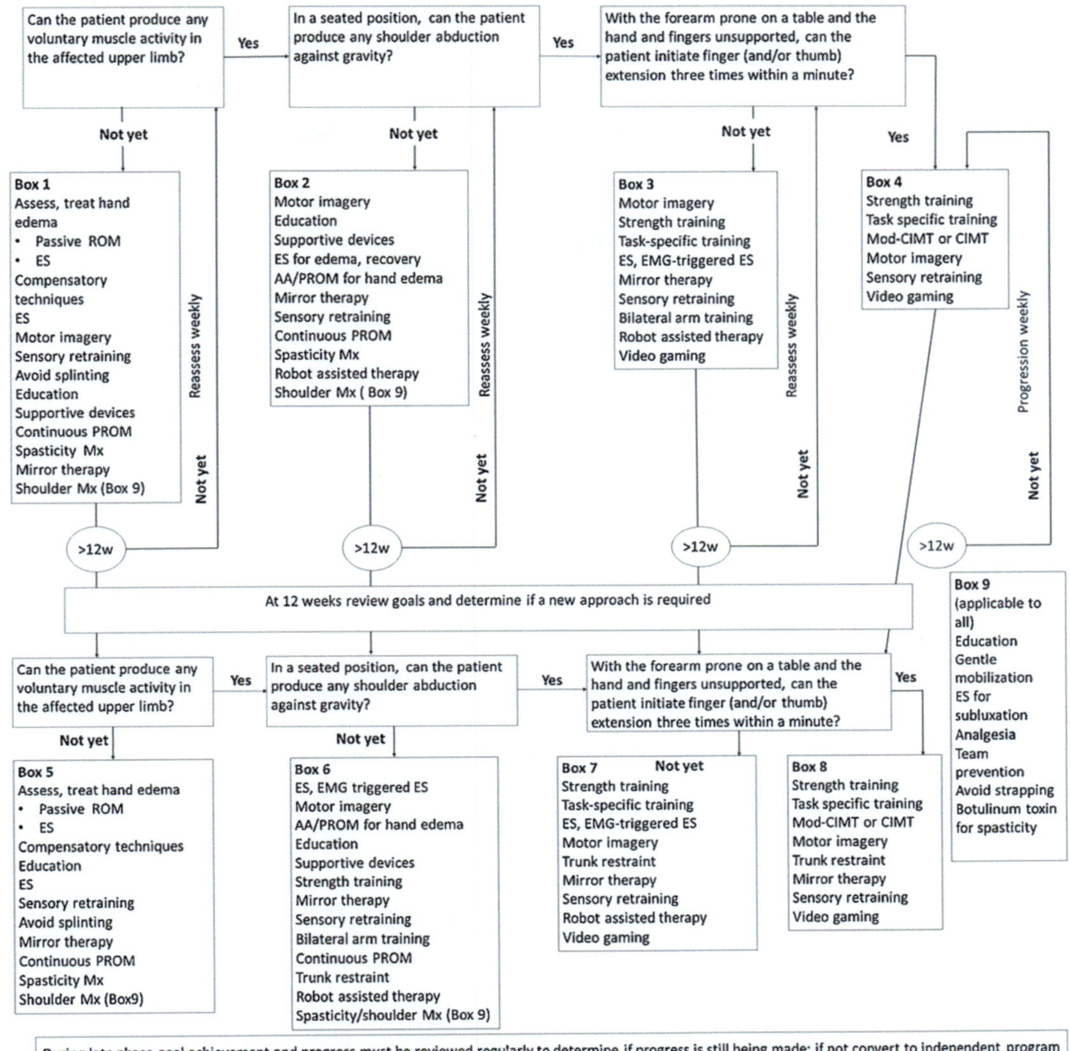

Fig. 17.1 Clinical decision-making flow diagram underlying the therapist app developed by Wolf et al. Figure used from [33] with permission

less frequently on acquiring MEP status and draws on stroke lesion load obtained from MRI or stroke severity assessed with the NIHSS score [35] (Fig. 17.2). Arguing against its robustness, a recent study found low classification accuracy with PREP2, as low as 33% for the functional category Limited [36].

Nevertheless, use of PREP in clinical practice boosted therapists' self-rating of confidence (in terms of knowing what to expect for the patient's upper limb recovery), shortened inpatient stays (perhaps by increasing therapist confidence), and altered therapeutic content selection in a rational way (by, for example, causing therapists to reduce the amount of passive upper limb exercise for patients with good recovery prognosis) [37]. Results from a recent survey in Denmark indicated that 89% of therapists believe it is important to know how upper limb function will evolve after stroke; 35% reported familiarity

Recovery	Definition	Goal
Complete	The patient has the potential to return to normal or near-normal hand and arm function within 12 weeks.	Rehabilitation could focus on task-specific therapy in order to facilitate a return to full or near-full use of the hand and arm in activities of daily living.
Notable	The patient has the potential to be using their affected hand and arm in most activities of daily living within 12 weeks, though normal function is unlikely.	Rehabilitation could focus on strength, coordination and fine motor control, in order to maximize recovery of function and minimize compensation with the other hand.
Limited	The patient has the potential to have some movement in their affected hand and arm within 12 weeks, but it is unlikely to be used functionally for activities of daily living.	Rehabilitation could focus on reducing impairment by strengthening the paretic upper limb and improving active range of motion, in order to promote adaptation and incorporation of the affected upper limb in daily activities wherever possible.
None	The patient can expect to have minimal movement in their affected hand and arm, with little improvement at 12 weeks.	Rehabilitation could focus on prevention of secondary complications, such as spasticity and shoulder instability, and reducing disability by learning to complete activities of daily living with the unaffected hand and arm.

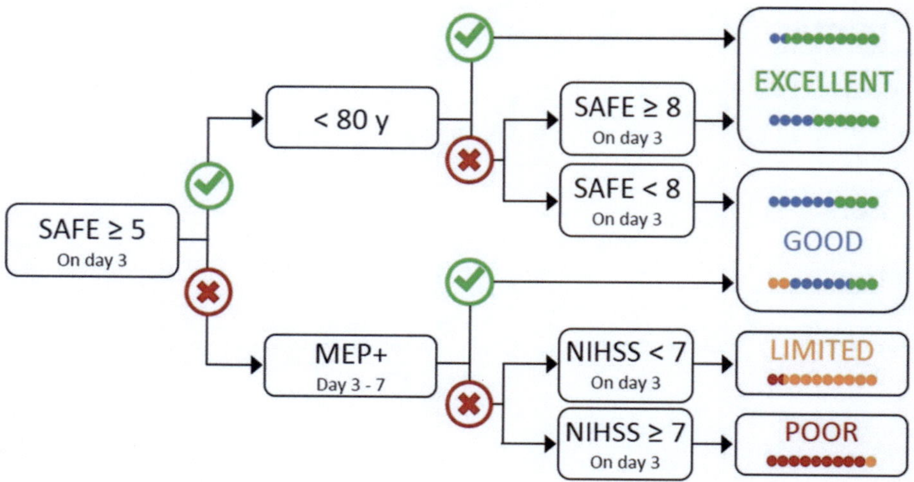

Fig. 17.2 Top: Recovery definitions and examples of feasible rehabilitation goals used in the PREP2 algorithm. Bottom: The PREP2 algorithm. SAFE is an acronym for Shoulder Abduction Finger Extension score. NIHSS refers to the NIH stroke scale. MEP+ indicates the presence of a motor evoked potential in a hand muscle, as assessed with TMS. Table and Figure used with permission from [35]

with prediction models for UL function after stroke; but only 9% confirmed the use of prediction models for UL function in clinical practice [38].

17.4 Discussion

17.4.1 Summary

In rehabilitation, where our most challenging and difficult patients are those who have sustained major neurologic disease or neurotrauma, we do not yet have useful biomarkers to help us decide whether the response to different therapies is likely to be different in particular patients or patient groups. We are somewhat able to gauge gross levels of potential for patients with hemispheric stroke, a major target for neurologic rehabilitation. Thus, rehabilitation treatments are relatively uniform and are usually not modified systematically based on objective pathological, biochemical, genetic, or radiological data. Clinical conditions such as aphasia, dysphagia, and impaired bladder control may modify the rehabilitation therapies prescribed. But these interventions are designed to deal with specific functional impairments rather than to tailor the response to rehabilitation therapy.

As a result, our interventions remain broadly generic in character, although clinical teams do make a strong effort to tailor treatments according to the specific needs of a patient. While this tailoring approach is laudable, there are very few if any clinical trials that guide a clinician in making well-informed choices about which type of therapy is best suited to the needs of a particular neurological patient or patient group.

17.4.2 What Are the Bottlenecks and What Can We Do?

For stroke survivors, it is not yet practical to undertake a full genomic analysis of every survivor, largely because there is not yet demonstrable patient benefit. Similarly, we do not yet know whether there are specific structural or pathological features of a stroke which are strong predictors of clinical outcome. We do know that there are some differences between motor recovery after thrombosis and hemorrhage, and there are clear functional implications of the amount of lesion overlap with the CST. But we do not yet have evidence that specific structural or genetic factors can impact the way we apply specific treatments. As a result, our approaches to movement training after stroke are relatively uniform and are focused on functional restoration.

It is likely that our current clinical measurement tools are inadequate to detect impairment changes unless they are quite large in scale. This uncertainty is aggravated by the current focus on functional assessments above the value of impairment scales. So, for example, a key functional/clinical target after stroke is walking capacity. Currently, we measure walking speed using standardized tests including the 10 m walk, and we can assess endurance using six-minute distance measurements. However, we make few if any other quantitative measurements of motor function. We do not routinely measure muscle strength quantitatively, even when a key objective of neurologic rehabilitation is to improve muscle strength. Our other tests of function are largely qualitative in nature—these include Fugl-Meyer, the ARAT and other clinical tests of upper extremity function. Why is there this apparent lack of interest in greater diagnostic precision? There are basically two potential reasons.

The first reason is that clinicians providing the treatment often believe that the precision of their measurements is adequate. This assertion holds even as prevailing clinical precision levels are known to be quite limited. Second, there is often a belief that clinical measures applied by different therapists at different times are consistent with each other. Unless there is specific training to promote this outcome, this is unlikely to be correct, but this belief is still widely held. So, we are left with the potentially difficult situation where therapeutic outcomes can only be recognized if they are large. While this may be useful from a practical tracking viewpoint, it limits our ability to detect change early and to track magnitude of change over time as therapy progresses.

It is our assertion that the capacity to emulate the advances achieved in areas of precision medicine is limited by the limitations in outcome measurement. Because our measurement scales are coarse, functionally focused, and applied at only one or two time points, we can only detect large changes in response to therapy. This means we cannot tell readily whether a particular patient will respond well, depending on their neuropathology. But it is also relevant at the population level where there may be major characteristics in different patient groups that allow us to determine whether one type of therapeutic intervention is potentially more useful than another. Thus, of all the components involved in precision medicine, precision measurement is most directly within our reach, and it may ultimately have the greatest impact. This is where we should focus, going forward, leveraging robotic and sensor-based technologies to obtain finer-resolution, impairment-focused signals at frequent time points that can rationally shape therapeutic content.

Acknowledgements DJR supported by NIH grant R01HD062744.

Disclosure David Reinkensmeyer has a financial interest in Hocoma AG and Flint Rehabilitation Devices, makers of rehabilitation equipment. The terms of these arrangements have been reviewed and approved by the University of California, Irvine in accordance with its conflict of interest policies.

References

1. Tsimberidou AM, Fountzilas E, Nikanjam M, Kurzrock R. Review of precision cancer medicine: evolution of the treatment paradigm. Cancer Treat Rev. 2020;86:102019.
2. Odle TG. Precision medicine in breast cancer. Radiol Technol. 2017;88(4):401M – 21.
3. Burney IA, Lakhtakia R. Precision medicine. Sultan Qaboos Univ Med J. 2017;17(3):e255-8.
4. Boyd LA, Hayward KS, Ward NS, Stinear CM, Rosso C, Fisher RJ, et al. Biomarkers of stroke recovery: consensus-based core recommendations from the stroke recovery and rehabilitation roundtable. Int J Stroke Off J Int Stroke Soc. 2017;12(5):480–93.
5. Kwah LK, Herbert RD. Prediction of walking and arm recovery after stroke: a critical review. Brain Sci. 2016;6(4):E53.
6. Stinear CM. Prediction of motor recovery after stroke: advances in biomarkers. Lancet Neurol. 2017;16(10):826–36.
7. Balkaya M, Cho S. Genetics of stroke recovery: BDNF val66met polymorphism in stroke recovery and its interaction with aging. Neurobiol Dis. 2019;126:36–46.
8. Cramer SC, See J, Liu B, Edwardson M, Wang X, Radom-Aizik S, et al. Genetic factors, brain atrophy, and response to rehabilitation therapy after stroke. Neurorehabil Neural Repair. 2022;36(2):131–9.
9. Lu B, Nagappan G, Lu Y. BDNF and synaptic plasticity, cognitive function, and dysfunction. Handb Exp Pharmacol. 2014;220:223–50.
10. Egan MF, Kojima M, Callicott JH, Goldberg TE, Kolachana BS, Bertolino A, et al. The BDNF val66met polymorphism affects activity-dependent secretion of BDNF and human memory and hippocampal function. Cell. 2003;112(2):257–69.
11. Kleim JA, Chan S, Pringle E, Schallert K, Procaccio V, Jimenez R, et al. BDNF val66met polymorphism is associated with modified experience-dependent plasticity in human motor cortex. Nat Neurosci. 2006;9(6):735–7.
12. Fritsch B, Reis J, Martinowich K, Schambra HM, Ji Y, Cohen LG, et al. Direct current stimulation promotes BDNF-dependent synaptic plasticity: potential implications for motor learning. Neuron. 2010;66(2):198–204.
13. White F, Nicoll JA, Roses AD, Horsburgh K. Impaired neuronal plasticity in transgenic mice expressing human apolipoprotein E4 compared to E3 in a model of entorhinal cortex lesion. Neurobiol Dis. 2001;8(4):611–25.
14. Koizumi K, Hattori Y, Ahn SJ, Buendia I, Ciacciarelli A, Uekawa K, et al. Apoε4 disrupts neurovascular regulation and undermines white matter integrity and cognitive function. Nat Commun. 2018;9(1):3816.
15. Alberts MJ, Graffagnino C, McClenny C, DeLong D, Strittmatter W, Saunders AM, et al. ApoE genotype and survival from intracerebral haemorrhage. Lancet Lond Engl. 1995;346(8974):575.
16. Teasdale GM, Nicoll JA, Murray G, Fiddes M. Association of apolipoprotein E polymorphism with outcome after head injury. Lancet Lond Engl. 1997;350(9084):1069–71.
17. Broderick J, Lu M, Jackson C, Pancioli A, Tilley BC, Fagan SC, et al. Apolipoprotein E phenotype and the efficacy of intravenous tissue plasminogen activator in acute ischemic stroke. Ann Neurol. 2001;49(6):736–44.
18. Perna R, Temple J. Rehabilitation outcomes: ischemic versus hemorrhagic strokes. Behav Neurol. 2015;2015:891651.
19. Dierick F, Dehas M, Isambert J-L, Injeyan S, Bouché A-F, Bleyenheuft Y, et al. Hemorrhagic versus ischemic stroke: who can best benefit from blended conventional physiotherapy with robotic-assisted gait therapy? PLoS ONE. 2017;12(6):e0178636.
20. Hayward KS, Schmidt J, Lohse KR, Peters S, Bernhardt J, Lannin NA, et al. Are we armed with the right data? Pooled individual data review of biomarkers in people with severe upper limb impairment after stroke. Neuroimage Clin. 2017;1(13):310–9.
21. Riley JD, Le V, Der-Yeghiaian L, See J, Newton JM, Ward NS, et al. Anatomy of stroke injury predicts gains from therapy. Stroke. 2011;42(2):421–6.
22. Nouri S, Cramer SC. Anatomy and physiology predict response to motor cortex stimulation after stroke. Neurology. 2011;77(11):1076–83.
23. Burke Quinlan E, Dodakian L, See J, McKenzie A, Le V, Wojnowicz M, et al. Neural function, injury, and stroke subtype predict treatment gains after stroke. Ann Neurol. 2015;77(1):132–45.
24. Wu J, Quinlan EB, Dodakian L, McKenzie A, Kathuria N, Zhou RJ, et al. Connectivity measures are robust biomarkers of cortical function and plasticity after stroke. Brain J Neurol. 2015;138(Pt 8):2359–69.
25. Ingemanson ML, Rowe JR, Chan V, Wolbrecht ET, Reinkensmeyer DJ, Cramer SC. Somatosensory system integrity explains differences in treatment response after stroke. Neurology. 2019;92(10).
26. Winters C, van Wegen EEH, Daffertshofer A, Kwakkel G. Generalizability of the proportional recovery model for the upper extremity after an ischemic stroke. Neurorehabil Neural Repair. 2015;29(7):614–22.
27. Stinear CM, Barber PA, Smale PR, Coxon JP, Fleming MK, Byblow WD. Functional potential in chronic stroke patients depends on corticospinal tract integrity. Brain. 2007;130:170–80.
28. Senesh MR, Reinkensmeyer DJ. Breaking proportional recovery after stroke. Neurorehabil Neural Repair. 2019;33(11).
29. Park S-W, Wolf SL, Blanton S, Winstein C, Nichols-Larsen DS. The EXCITE trial: predicting a clinically

meaningful motor activity log outcome. Neurorehabil Neural Repair. 2008;22(5):486–93.

30. Rowe JB, Chan V, Ingemanson ML, Cramer SC, Wolbrecht ET, Reinkensmeyer DJ. Robotic assistance for training finger movement using a hebbian model: a randomized controlled trial. Neurorehabil Neural Repair. 2017;31(8).

31. Nijland RHM, van Wegen EEH, van der Wel Harmeling BC, Kwakkel G, EPOS Investigators. Presence of finger extension and shoulder abduction within 72 hours after stroke predicts functional recovery: early prediction of functional outcome after stroke: the EPOS cohort study. Stroke. 2010;41 (4):745–50.

32. Winters C, Kwakkel G, Nijland R, van Wegen E, Consortium E. When does return of voluntary finger extension occur post-stroke? A prospective cohort study. PLoS ONE. 2016;11(8):e0160528.

33. Wolf SL, Kwakkel G, Bayley M, McDonnell MN, Upper Extremity Stroke Algorithm Working Group. Best practice for arm recovery post stroke: an international application. Physiotherapy. 2016;102 (1):1–4.

34. Stinear CM, Barber PA, Petoe M, Anwar S, Byblow WD. The PREP algorithm predicts potential for upper limb recovery after stroke. Brain. 2012;135 (8):2527–35.

35. Stinear CM, Byblow WD, Ackerley SJ, Smith M-C, Borges VM, Barber PA. PREP2: a biomarker-based algorithm for predicting upper limb function after stroke. Ann Clin Transl Neurol. 2017;4(11):811–20.

36. Lundquist CB, Nielsen JF, Arguissain FG, Brunner IC. Accuracy of the upper limb prediction algorithm PREP2 applied 2 weeks poststroke: a prospective longitudinal study. Neurorehabil Neural Repair. 2021;35(1):68–78.

37. Stinear CM, Byblow WD, Ackerley SJ, Barber PA, Smith M-C. Predicting recovery potential for individual stroke patients increases rehabilitation efficiency. Stroke. 2017;48(4):1011–9.

38. Kiær C, Lundquist CB, Brunner I. Knowledge and application of upper limb prediction models and attitude toward prognosis among physiotherapists and occupational therapists in the clinical stroke setting. Top Stroke Rehabil. 2021;28(2):135–41.

Part V

General Technological Approaches in Neurorehabilitation

Ismael Seáñez, Marco Capogrosso,
Karen Minassian, and Fabien B. Wagner

Abstract

Spinal cord injury (SCI) disrupts the communication between the brain and the spinal circuits that control movement and integrate sensory feedback, which are usually located below the lesion. The disruption of the differ-

The original version of this chapter was previously published non-open access. A correction to this chapter is available at https://doi.org/10.1007/978-3-031-08995-4_35

I. Seáñez
Biomedical Engineering, Washington University in St. Louis, St. Louis, MO, USA
e-mail: ismaelseanez@wustl.edu

Neurosurgery, Washington University School of Medicine in St. Louis, St. Louis, MO, USA

M. Capogrosso
Department of Neurological Surgery, Rehabilitation and Neural Engineering Laboratories, University of Pittsburgh, 3520, Fifth Av., Suite 300, Pittsburgh, PA 15213, USA
e-mail: MCAPO@pitt.edu

K. Minassian
Center for Medical Physics and Biomedical Engineering, Medical University of Vienna, Währinger Gürtel 18-20/4L, 1090 Vienna, Austria
e-mail: karen.minassian@meduniwien.ac.at

F. B. Wagner (✉)
Institute of Neurodegenerative Diseases (CNRS, UMR 5293), French National Centre for Scientific Research (CNRS), University of Bordeaux, 146 rue Léo Saignat, 33076 Bordeaux, France
e-mail: fabien.wagner@u-bordeaux.fr

ent anatomical sources of descending motor control and ascending sensory afferents can result in complete or partial, but permanent motor paralysis. For decades, recovery of motor function after long-standing SCI was thought impossible because of the severe and multi-modal failure of these bidirectional communication pathways. This conclusion was supported by overwhelming and disappointing empirical evidence showing poor recovery in people with chronic (>6 months post-injury), severe SCI despite intensive rehabilitation. However, a recent wave of clinical studies has reported unprecedented outcomes in people with both incomplete and complete SCI, independently demonstrating the long-term recovery of voluntary motor function in the chronic stage after SCI. These studies utilized a combination of intensive rehabilitation and electrical spinal cord stimulation (SCS), which was delivered via epidural multi-electrode arrays implanted between the vertebral bone and the dura mater of the lumbosacral spinal cord. SCS has a long history of applications in motor control, which started soon after its first applications as interventional studies in pain management. To date, SCS has been applied in thousands of individuals with neuromotor disorders ranging from multiple sclerosis to SCI. However, even though the motor-enabling effects of SCS were first observed about half a century ago, the lack of a coherent conceptual framework to

© The Author(s) 2022, corrected publication 2023

D. J. Reinkensmeyer et al. (eds.), *Neurorehabilitation Technology*,
https://doi.org/10.1007/978-3-031-08995-4_18

interpret and expand these clinical findings hindered the evolution of this technology into a clinical therapy. More importantly, it led to substantial variability in the clinical reports ranging from anecdotal to subjective descriptions of motor improvements, without standardized methods and rigorous statistical analyses. For several decades, these limitations clouded the potential of SCS to promote long-term recovery in individuals with SCI. In this chapter, we present the historical background for the development of SCS to treat motor disorders and its evolution toward current applications for neurorehabilitation in individuals with SCI (Sect. 18.1). We then provide an overview of the conjectured mechanisms of action (Sect. 18.2), and how this collective knowledge has been used to develop SCS into a promising approach to treat motor paralysis after SCI, ranging from tonic stimulation to more sophisticated spatiotemporal protocols (Sect. 18.3). Finally, we open up this review to the recent development of non-invasive methods to deliver SCS, namely transcutaneous SCS, and its comparison with epidural SCS in terms of functional effects and underlying mechanisms (Sect. 18.4).

Keywords

Spinal cord injury · Spinal cord stimulation · Neuromodulation · Locomotion · Motor control · Rehabilitation

18.1 The Origins of SCS: From Pain to Motor Control

18.1.1 The Rise of SCS for Chronic Pain Management

SCS was originally developed for the treatment of chronic, intractable pain based on neurophysiological studies suggesting the possibility to inhibit pain fiber input in the spinal cord by stimulation of the larger-diameter sensory fibers [1, 2]. The spinal cord dorsal columns, which contain the longitudinal ascending continuations

of cutaneous fibers from many spinal cord segments, appeared as an optimal target for inhibiting pain, in particular because of their easy surgical access from the dorsal aspect of the spinal cord. Subdural electrical stimulation of these structures proved successful in managing pain in cats [3], followed by the first human application of SCS in a patient with cancer [4]. The evolution from subdural to less invasive epidural electrodes, and the development of fully implantable commercial systems led to the FDA approval of epidural SCS for chronic pain management in 1989. Today, SCS accounts for about 70% of all neuromodulation treatments [5].

18.1.2 First Evidence of Improved Motor Function During SCS: From Multiple Sclerosis to SCI

The ability of SCS to restore function after neuromotor disorders was first unexpectedly observed in 1973 when a subject with motor deficits as a consequence of multiple sclerosis (MS) received SCS therapy to treat intractable pain. Investigators unexpectedly observed that the subject regained volitional control of nearly normal strength of the lower limbs, facilitation of sitting, standing, and ambulation when SCS was active [6]. Four additional implanted MS participants without pain reported a feeling of lightness of the legs during movement, increased endurance during ambulation, as well as the recovery of some voluntary motor function and improved bladder control. Cook improved the technique of electrode placement in the epidural space and implanted more than 200 additional patients with MS within the next few years [7, 8]. Following this first example of SCS-mediated motor improvements, Illis and colleagues introduced the use of SCS for motor disorders to Europe by replicating Cook's methods. Their first study demonstrated marked improvements in motor, sensory, and bladder function in two participants with MS receiving SCS [9]. The first participant, who had signs of an upper motor neuron lesion and presence of spasticity, regained the ability to walk independently with SCS within the first 24 h of continuous stimulation, while the second participant had marked improvements in sensory function.

Follow-up large-cohort studies by the same group revealed high variability in outcomes that reduced the initial enthusiasm for SCS. Indeed, although motor and sensory improvements surpassed those ever achieved by any other method at the time, it was only 5 out of 19 individuals that experienced these types of improvements [10]. A subsequent study on 90 participants confirmed that these striking improvements in motor function were infrequent and observed only in a few individuals [11]. An objective effect was only seen on bladder dysfunction and limb spasticity. Similarly, studies by other groups showed remarkable effects of SCS on altered motor function after MS, but with variability across subjects. Siegfried and colleagues tested 111 patients with MS with temporary SCS and considered only about 33% of them as responders [12]. Davis and colleagues also reported on 69 MS patients with full implantations [13]. 64% showed improvements in gait, endurance, and muscle strength. Among these patients, nine who were wheelchair users could walk again with SCS.

In spite of the variable outcomes of these initial studies, the clinical importance of the potential effects led to a cascade of studies assessing the off-label use of SCS in subjects with a wide variety of motor disorders, including amyotrophic lateral sclerosis, cerebral palsy, traumatic brain injury, dystonia, torticollis, and a few individuals with SCI. Overall, these studies reported improvements in strength, balance, walking, coordination, speech, swallowing, eye movements, and bladder function [12–16]. Initial investigations on SCI focused primarily on the management of spasticity, but also explored secondary effects on autonomic function including bowel control and sexual function, as well as motor capacity [17–21]. More specifically, Waltz and colleagues observed improved motor control in 65% of 303 participants with SCI who received SCS of the cervical region [22]. These improvements in motor control were initially ascribed to a reduction in spasticity enabled by SCS [17]. However, further studies such as the one by Barolat in 1986 revealed that voluntary

control of paralyzed muscles was strictly dependent on SCS and stopped immediately when SCS was turned off, independently of changes in spasticity [23] (Fig. 18.1). This effect is discussed in more details later (Sect. 18.3.1.1).

In his observations after treating 1336 individuals with a broad range of disorders over a quarter of a century Waltz reported marked or moderate improvements in a majority of them (Table 18.1) [24]. However, despite the initial unprecedented improvements in motor function observed through SCS, the lack of a conceptual framework kept these observations anecdotal. There was no identified physiological criterion to predict responders to SCS, and a lack of agreement on the optimal electrode implantation site, which ranged from high-cervical (C2) to low-thoracic (T10) vertebral levels. These two factors significantly contributed toward the observed inconsistency in clinical outcomes [12, 13, 25], and led to a declining interest in SCS for motor disorders during the 1990s [26].

18.1.3 Standardization of SCS Location for Leg Motor Control Following Motor Disorders

A critical contribution to current applications of SCS came from Dimitrijevic and colleagues, who observed that the optimal electrode placement to control leg spasticity was at the T11-L1 vertebral levels, which corresponds to the site of innervation of leg motoneurons in the lumbosacral spinal cord [19]. Indeed, tonic SCS, delivered with higher stimulation amplitudes and lower frequencies than those used for spasticity, could elicit rhythmic, step-like activity in paralyzed muscles of six subjects with chronic, complete SCI [27]. The multi-segmental muscle activity showing consistent rhythmicity and clear alternation between antagonistic muscles was interpreted as the most direct evidence at the time for the existence of central pattern generators (CPGs) in the human spinal cord [27, 28].

These important observations in humans were further supported by investigations using

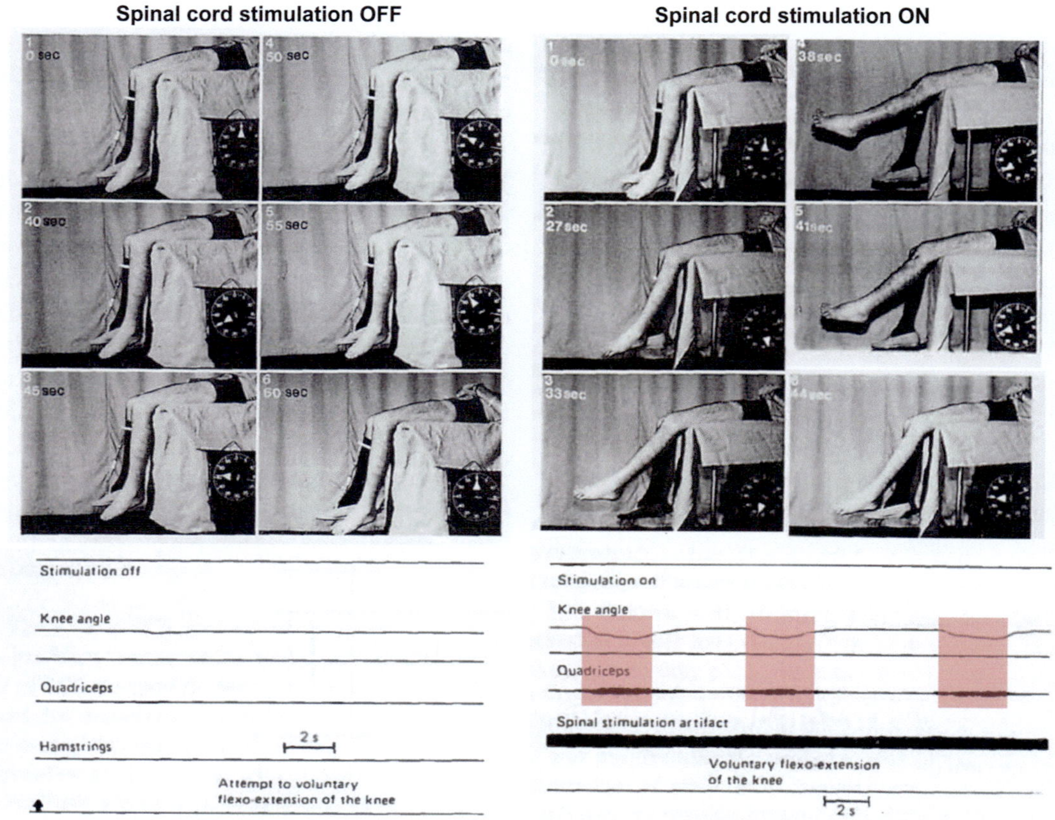

Fig. 18.1 First discovery that SCS enables voluntary movement after SCI in 1986. Pictures of one individual with SCI and with some residual motor control over the left toes and ankle, during an attempt to perform a voluntary left knee extension in the absence (left panel) or presence (right panel) of SCS. The knee angle and EMG activity of quadriceps and hamstrings muscles are represented. During SCS, EMG signals from the hamstrings are affected by stimulation artifacts. Red rectangles indicate the 3 repetitions of the flexion–extension movement, highlighting the EMG activity in the quadriceps, which is facilitated by SCS but voluntarily triggered. Adapted with permission from [23]. All rights reserved

Table 18.1 Improvements in 1336 individuals treated with SCS (from [24]). All rights reserved

Disorder	N	%
Cerebral palsy	456	80
Spinal cord injury	303	44
Dystonia	173	59
Multiple sclerosis	130	67
Posttraumatic brain injury	113	59
Torticollis	90	77
Spinocerebeller degneration	71	58

simulations of the biophysical interactions between electrical fields and neural structures, which led to the understanding that SCS recruits primarily proprioceptive afferents [29–31]. Proprioceptive afferents then convey mono and poly-synaptic excitatory potentials to spinal motoneurons. To maximize motoneuron activation, SCS electrodes must be therefore placed in positions that favor the direct activation of posterior spinal roots that carry these afferents [32–34]. In the case of lower limb movements, these correspond exactly to the optimal location observed by Dimitrijevic. Currently, all scientific investigations that aim at improving lower extremity function via SCS use the same level for epidural electrode placement: below the injury and over the posterior aspect of the lumbar and upper sacral spinal cord, regardless of the level of SCI [34–39].

18.2 Potential Mechanisms of SCS for Motor Control

18.2.1 General Principles

The generation and modulation of spinal motor activity by SCS results from biophysical phenomena that occur upon the application of electrical fields to the spinal cord, which in turn causes physiological effects in the spinal circuitry. Epidural electrodes are positioned in direct contact with the dura mater. During each stimulation pulse, most of the generated ionic currents flow between the active contacts through the dural sac that contains the spinal cord and roots, owing to the relatively high electrical conductivity of the cerebrospinal fluid [29–31, 40–42]. Depending on the current flow and the relative localization and orientation of the neural structures within the dural sac, specific subsets of neural structures are depolarized to a level where action potentials are generated according to an all-or-nothing principle. These immediate electrical effects of the stimulation were studied using computer models that are able to solve established physics in arbitrary complex geometrical structures, such as the spinal cord, to calculate the electric current flow and electric potential distribution [29, 31, 40–43]. This method allows to calculate currents and voltages but does not provide per se an estimation of the response of electrically active neural tissue to an electrical field applied externally. To this end, nerve fiber models utilizing the Hodgkin-Huxley formalism enable to calculate membrane potentials in response to external currents [44]. Such models were used to estimate the sites of maximum depolarization and activation thresholds for individual neurons or their substructures. The electrically activated neurons that lead to motor effects were also investigated by neurophysiological studies in rats and individuals with SCI [28, 29, 38, 45–47]. The physiological effects that are caused by electrically activated neurons into their post-synaptic targets involve local and potentially suprasegmental circuits. The underlying mechanisms are not completely clear and still a topic of active research.

18.2.2 Electrically Activated Neural Structures

The neuronal substructure that is the most excited by external electrical stimulation is the axon. In particular, myelinated and large-diameter axons have the lowest thresholds to stimulation [48]. The axonal depolarization is proportional to the second-order spatial derivative of the electric potential along the axonal path [49, 50]. The stimulation-induced currents within the dural sac primarily flow within the relatively well conductive cerebrospinal fluid, in which the spinal roots are bathed, with maximum current densities in the vicinity of the active electrodes and rapid attenuation when entering the spinal cord [29, 31, 42]. Because of their small size, the lack of myelin sheath, and the fact that the induced current poorly enters into the spinal cord, direct electrical activation of spinal interneurons is highly unlikely to happen with SCS [29]. On the other hand, strong depolarizations are caused in the sensory axons of the longitudinally running lumbar and upper sacral posterior roots of passage that are the closest to the active cathode location [31]. Additional low-threshold sites are created along the sensory axons at the posterior rootlet-spinal cord interface, owing to anatomical inhomogeneities, i.e., electrical conductivity boundaries and changes of the orientation of the axon paths with respect to the electric field [31,

51]. Based on their fiber diameters, electrical activation is likely limited to a subset of posterior root fibers ranging from group Ia muscle spindle afferents to group II afferents, both proprioceptors and cutaneous mechanoreceptors [42].

Upon entering the spinal cord through their posterior rootlets, the sensory axons bifurcate to rostral and caudal projections in the posterior columns. Collaterals connect to locally confined spinal neurons or to relay neurons with ascending axons. There are anatomical differences in the rostral projections between the Ia muscle spindle afferents and the cutaneous afferents that are relevant for their recruitment by SCS [38, 40, 52]. While the rostral projections of the cutaneous afferents can ascend the entire length of the posterior columns, those of the Ia muscle spindle afferents from the lower extremity largely terminate in the upper lumbar and lower thoracic segments to synapse with the relay neurons of Clarke's column, and in doing so, they occupy deep positions in the white matter [53–55]. Therefore, electrical activation of neural structures within the posterior columns—that is limited to its most superficial layers—is essentially restricted to a small population of ascending projections of cutaneous afferents [56]. On the other hand, a large proportion of the total number of the large-diameter proprioceptive afferents can be in theory recruited in the posterior roots or rootlets [31], especially in the lumbosacral and cervical enlargements where the roots cover the entire cord with rootlet projections [31, 40, 42].

All neurophysiological studies in humans and animals to date support the notion that the motor effects evoked by SCS are triggered by posterior root fiber stimulation. Responses tested by paired pulses applied epidurally with a step-wise decrease of interstimulus intervals clearly demonstrated post-stimulation depression [38, 47, 57], a hallmark of monosynaptic reflexes evoked in proprioceptive afferents [58–60]. Complete suppression of responses with repetitive stimulation also rules out the direct electrical activation of motor axons in the anterior roots [38, 40, 49, 61]. Finally, the order of recruitment of lower- and upper extremity muscles with different segmental innervation, either from

different segmental cathode positions, with different cathode–anode combinations, or with graded stimulation amplitudes, can be entirely explained by stimulation of the respective posterior roots [34, 38, 40]. Typically, moving the stimulating cathode in the rostral direction away from the segmental posterior root entries increases the response threshold for muscles innervated by motoneuron pools located in these segments, in accordance with earlier predictions by computational models [29–31, 40]. For instance, S1-innervated lower leg muscles cannot be recruited by clinically applicable stimulation amplitudes with a cathode located medially over the upper lumbar segments [38]. This can only be explained if the effects of the stimulation are mediated by the posterior roots and not by the longitudinal sensory fiber projections within the posterior columns. Overall, the notion that SCS can be configured to selectively activate proprioceptive afferents of subsets of posterior roots can be used to direct SCS effects toward specific motoneurons to define more targeted stimulation strategies [32–34, 39, 62] (Fig. 18.2).

18.2.3 Evidence for Post-synaptic Activation of Neural Circuits

The directly activated proprioceptive afferents make connections with multiple classes of interneurons and motoneurons in the spinal cord. Therefore, the direct recruitment of proprioceptive afferents will generate synchronized volleys of excitatory post-synaptic potentials into all the neurons directly connected by these fibers. This means that the effects of SCS can occur in multiple neural circuits, having effects that can propagate to both spinal and supraspinal structures. A major goal of current research in SCS is to elucidate the contribution of these activated circuits and their relevance to the effects observed in human clinical trials. In this section, we will discuss only a few of the most studied circuits underlying the effects of SCS, while acknowledging that many more may contribute to the restoration of voluntary motor control.

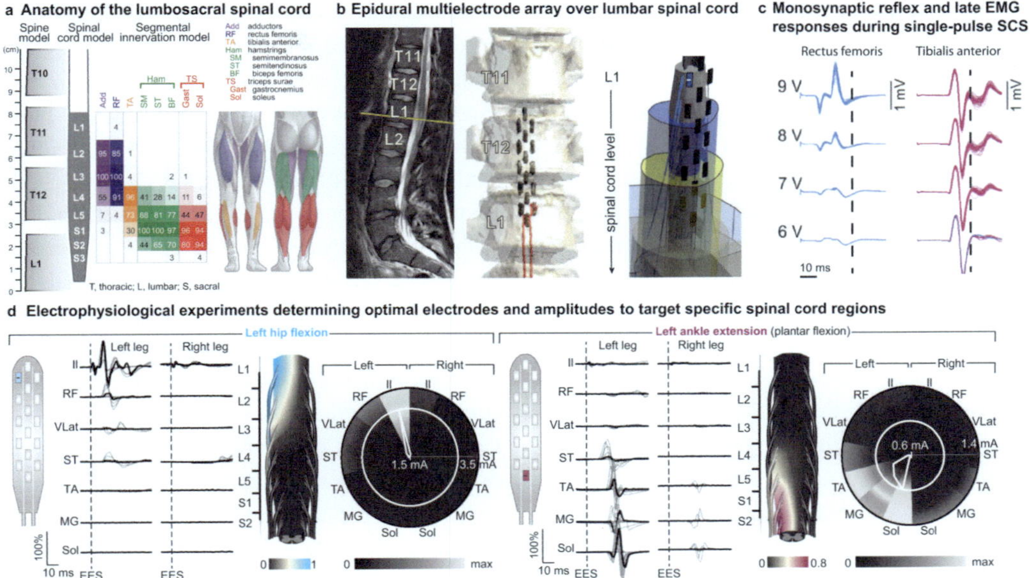

Fig. 18.2 **Segmental recruitment of posterior roots and innervated muscles during SCS**. **a** Anatomical representation of the spine, (defined by vertebral body and intervertebral disc heights), lumbosacral spinal cord (aligned with the spine), and segmental innervation probabilities (0%–100%) of lower extremity muscles, reflected by the opacity of their respective colors. These representations correspond to an average over thousands of individuals published in the literature. Adapted with permission from [38]. (Figure published under a Creative Commons license: Fig. 18.2a, licensed under CC-BY 4.0). **b** Left: mid-sagittal MRI image of the spine and spinal cord of a study participant with SCI, with indications of the vertebral levels, estimated level of the tip of the spinal cord (conus medullaris, yellow line) and electrode array location in this subject (red line). Middle: reconstruction of the location of the electrode array based on MRI and Computed Tomography (CT) scan. Right: Computational model showing the location of the epidural electrode array, epidural fat (yellow), cerebrospinal fluid (dark blue), gray and white matter (gray), and posterior roots (light blue). **c** Responses of rectus femoris and tibialis anterior to 2-Hz SCS with incremental amplitudes show the presence of late EMG components in tibialis anterior, but not in rectus femoris, with EES amplitudes greater than two times the response threshold. Adapted with permission from [66]. (Figure published under a Creative Commons license: Fig. 18.2c, licensed under CC-BY 4.0). **d** Electrophysiological recordings were used to determine optimal electrodes and amplitudes for targeting specific spinal cord regions. EMG responses when delivering single-pulse EES at increasing amplitudes are shown (gray traces). Motor neuron activation maps correspond to optimal amplitudes (black traces). Circular plots report EMG amplitude (in gray scale) at increasing amplitudes (radial axis). White circles show optimal amplitudes; polygons quantify selectivity at this amplitude. **c–d**: adapted with permission from [34]. All rights reserved

18.2.3.1 Recruitment of the Monosynaptic Reflex

In previous paragraphs, we briefly anticipated that most of the motor effects of SCS can be largely explained by assuming the direct recruitment of large proprioceptive afferents. Indeed, Ia-afferents form strong monosynaptic excitatory connections to spinal motoneurons, especially with those innervating extensor muscles. In consequence, each pulse of SCS will generate strong synchronized volleys of excitatory post-synaptic potentials on motoneuron membranes, which can either lead to their direct activation (H-reflex-like responses also known as posterior root-muscle reflexes), or to an increase in their membrane excitability depending on how many afferents are recruited. Since the number of recruited afferents is directly proportional to the strength of the electrical fields, one can argue that stimulation amplitude will then determine whether SCS is in "subthreshold" (i.e., modulating motoneuron membrane excitability) or "supra-threshold" (i.e., directly inducing action potentials in the motoneurons). Therefore, supra-threshold SCS pulses can induce single, distinct H-reflex-like responses in the muscles with a segmental recruitment order. Stimulation of the more rostral roots will mainly induce responses in rostrally innervated muscles, whereas stimulation of the more caudal roots will elicit responses in the more caudally innervated muscles [29, 33, 34, 38, 62]. This property is remarkably robust across animal species and humans as well as the spinal cord region (cervical vs lumbar). This large body of evidence demonstrates that the monosynaptic reflex is certainly involved in the generation of motor outputs during SCS and likely contributes to the facilitation of voluntary motor control by modulating motoneuron excitability.

18.2.3.2 Recruitment of Excitatory Spinal Circuits

The group I and group II afferent fibers activated by SCS have rich synaptic connections to spinal motoneurons, to first-order interneurons of spinal circuits with a pivotal role in the control of locomotion, as well as to supraspinal circuits—given that residual longitudinal connectivity exists following a trauma [60, 63, 64]. Studies so far are limited to local spinal effects that have been indirectly deduced in individuals with SCI, using trains of stimulation with various frequencies (2–100 Hz) and intensities [57]. As introduced above, the most prominent events are stimulus-triggered, short-latency responses that are predominantly monosynaptic in nature—called posterior root-muscle reflexes—and are most evident at lower stimulation frequencies (i.e., 2–10 Hz) [28, 46, 47, 57, 65]. Low-frequency SCS was also shown to evoke crossed-reflexes in thigh muscle groups [66], posterior root-muscle reflexes with superimposed, delayed-latency electromyographic (EMG) components in flexor muscles (by ~ 8 ms with respect to the monosynaptic response) [66], and complex long-latency responses (>50 ms) [66, 67] with characteristics reminiscent of the late flexion reflex observed in individuals with chronic SCI [60, 66, 68]. These observations hence suggest that SCS can activate

commissural neurons as well as interneurons specific to flexor-related oligo- and poly-synaptic pathways, i.e., interneuron types shown to be essential components of the mammalian lumbar locomotor circuitry [66, 69]. Indeed, previous studies demonstrated that tonic lumbar SCS at frequencies around 30 Hz applied in individuals with chronic, motor complete SCI examined in the supine position could generate periods (10–60 s) of rhythmic EMG activities in the paralyzed lower extremity muscles [27, 47]. The generation of rhythmicity with cycle frequencies compatible with slow to fast walking speeds and various patterns of muscle recruitment, including reciprocity between antagonistic muscles [70] by a sustained stimulation with constant parameters, were interpreted as evidence for the activation of a human CPG for locomotion [27, 71]. It should be noted that the generation of these rhythmic EMG activities required a minimum frequency of 22.5 Hz and rather high SCS amplitudes, about three times the posterior root-muscle reflex threshold of the mid-lumbar-innervated quadriceps muscle group [70]. Such stimulation likely recruits a large proportion of group I and II afferent fibers within the lumbar posterior roots. The antidromic action potentials carried along these afferents toward the periphery likely cancel a large part of the naturally generated, orthodromically traveling action potentials from proprioceptors [71, 72]. The generation of rhythmic activity independent from phasic peripheral feedback (largely canceled following antidromic collision) is essential in the demonstration of centrally generated rhythmicity and significant from a neuroscientific point of view. However, the blocking of proprioceptive feedback required for generating adaptive movements limits the applicability of such high stimulation intensities in neurorehabilitation approaches [72, 73].

18.2.3.3 Recruitment of Inhibitory Spinal Circuits

Despite their historical use in spinal spasticity [6, 21, 74], the recruitment of inhibitory spinal circuits by SCS has received less attention. A phenomenon observed with low SCS frequencies sheds some light on the activation of inhibitory interneurons [49]. Stimulation at around 16 Hz can produce specific modulation patterns of repetitively evoked posterior root-muscle reflexes, with every other response being attenuated, resulting in simple periodic patterns. The rhythmic attenuation of responses without changes in stimulation amplitude and the reciprocal organization of these patterns between antagonistic muscles suggests the recruitment of inhibitory circuits involving Renshaw cells and Ia inhibitory interneurons [49]. Indeed, similarly to monosynaptic connections to the motoneurons, Ia-inhibitory interneurons are known to have direct inputs from Ia-afferents [75, 76]. These interneurons inhibit the motoneurons that innervate antagonistic muscles. Therefore, the potentiation of these inhibitory cells, or even their direct post-synaptic activation by SCS, is likely involved in the alternation of agonist and antagonist muscle activations, which can, in turn, contribute to the production and modulation of coordinated movements.

18.2.4 Lessons from Animal Studies

18.2.4.1 Recruitment of Different Reflex Circuits by SCS in Animal Studies

In intact and spinalized rats, single-pulse SCS of the lumbosacral spinal cord evoked composite EMG responses in the hindlimb muscles with a succession of stimulus-triggered, partially overlapping potentials [29, 45, 77]. Three physiologically different components were suggested, and distinguished according to the relative latencies of their major EMG peaks: an early, middle, and late-latency response. The early response is a direct, M-wave-like response [60], elicited by direct electrical stimulation of motor axons within the ventral roots—a response type not evoked in humans with electrodes placed over the midline (see above). The middle response is well documented to be a monosynaptic reflex, likely corresponding to the

monosynaptic posterior root-muscle reflex in humans [47]. The late response occurs with an additional delay of 4–5 ms with respect to the middle response, distinguishing it as an oligosynaptic spinal reflex [38]. There is the uncertainty of the origin of the late response, but it was suggested to involve group II muscle afferents [29] or the flexor reflex afferents [45, 77].

18.2.4.2 Recruitment of Locomotor Circuits by SCS in Animal Studies

Early animal studies employing spinal cord and/or root stimulation were conducted in cats with the aim to investigate the intrinsic capability of the lumbar spinal cord to generate the rhythm and pattern underlying hindlimb locomotion. Stimulation of bilateral pairs of lumbar dorsal roots, typically with 30–50 Hz, induced rhythmic alternating activity in spinal preparations after elimination of phasic inputs from the hindlimbs (through deafferentation or curarization), and hence demonstrated the existence of a CPG [78]. Similar patterns were observed in acutely spinalized cats pretreated with L-DOPA and in chronic spinal cats acutely decerebrated but without administration of drugs. Another classical study demonstrated that subdural and epidural stimulation over the posterior aspect of the lumbar spinal cord could elicit stepping in acutely spinalized cats suspended over a moving treadmill without pharmacological manipulation [79]. Electrodes positioned over the spinal cord midline or laterally over the dorsal root entry zones were both effective. This animal study was perhaps one of the first to suggest the value of SCS for locomotor rehabilitation, as it could be suitable for *"a degree of "exercise" and perhaps provide sufficient force to propel the subject as long as postural support is provided"* [79]. Later studies in rodents and non-human primates provided further ground for the development of SCS into a tool for enabling locomotion after SCI.

18.3 The Evolution of SCS into a Neuroprosthetic Technology and a Neurorehabilitation Therapy

In the previous sections, we discussed the early history of human applications of SCS in motor disorders, how empirical observations highlighted the potential of SCS to enable voluntary motor control, and its potential mechanisms of action. In this section, we bring all this information together to discuss how the concept of SCS shifted from a neurophysiological tool to activate CPGs to a therapy capable of amplifying residual voluntary motor control [80]. Currently, the application of SCS from the epidural space located over the posterior aspect of the lumbosacral spinal cord represents a state-of-the-art neuromodulation technique for facilitating lower limb motor control after SCI. Here, we report the technological innovations that enabled this transition toward a neuroprosthetic solution for motor impairments and a neurorehabilitation therapy for long-term recovery. In particular, we discuss different approaches for delivering SCS, from tonic stimulation to more sophisticated spatiotemporal stimulation protocols, which can be either applied at a pre-defined pace, triggered by external events, or adjusted in closed loop.

18.3.1 Tonic SCS for the Recovery of Voluntary Motor Control in People with SCI

18.3.1.1 The Initial Discovery that SCS Enables Voluntary Motor Control

The very first discovery that SCS may be used as a motor-enabling neuromodulation tool after SCI is attributed to Barolat and colleagues in 1986 [23]. This case study showed that one subject with incomplete SCI regained voluntary motor control in the presence of SCS after several months of stimulation (Fig. 18.1). Specifically, SCS allowed the subject to perform a full knee extension against gravity, despite the complete absence of voluntary activity in the thigh muscles in the absence of SCS. This effect was present only when the stimulation was turned on, illustrating for the first time an immediate effect of SCS for enabling motor function after SCI. Importantly, this first observation of improved motor control did not involve the triggering of automatic stepping patterns, which was the main focus of animal research in the field. Instead, SCS enabled single-joint voluntary movements, which is still the main clinical outcome of modern clinical trials.

18.3.1.2 SCS as a Tool to Trigger Movement Primitives

Several years later, Dimitrijevic and colleagues demonstrated in six subjects with complete SCI that SCS delivered over the lumbar spinal cord (T11-L1 vertebrae) in supine position produced rhythmic EMG responses and flexion–extension leg movements resembling stepping [27]. Stimulation was applied at relatively high amplitudes (5–9 V) and for frequencies between 25 and 60 Hz. Beyond its potential prosthetic implications, this discovery provided indirect evidence for the existence of a CPG in humans, defined as a circuit within the spinal cord capable of generating rhythmic outputs in response to a tonic input (i.e., a continuous pulse train at a given frequency without rhythmic modulations).

In addition to the rhythmic movements induced by stimulation of the CPG, other simpler movements could be produced. For example, Jilge and colleagues showed in five subjects that tonic SCS at frequencies between 5 and 15 Hz induced bilateral extension of the lower limbs [81]. Minassian and colleagues next demonstrated in ten subjects that tonic SCS at different frequencies can switch the functional state of spinal circuits between distinct functional units that produce muscle synergies associated with either bilateral extension of the lower limbs (for frequencies of 5–15 Hz) or rhythmic stepping-like movements (for frequencies of 25–50 Hz) [47]. This latter study was the first one to coin the term "neuroprostheses" in the context of SCS: *"This study opens the possibility for developing neuroprostheses for activation of inherent spinal networks involved in generating functional synergistic movements using a single electrode implanted in a localized and stable region".*

However, it is important to highlight that the recruitment of CPGs or other movement primitives should not be taken as a goal to produce automatic, non-voluntary stepping (which may not be relevant as a clinical outcome), but rather as a demonstration that SCS can engage the spinal circuitry that is necessary to produce complex motor patterns such as locomotion. For this reason, subsequent studies investigated the combination of SCS with physical training as a means to improve the capacity of residual descending inputs to regain control over spared spinal circuits.

18.3.1.3 The Combination of SCS with Training and the Re-Discovery that SCS Enables Voluntary Motor Control

The first combination of SCS with partial weight-bearing locomotor training was performed by Herman and colleagues in a subject with incomplete SCI [82, 83], and later expanded to a second subject [84]. Both participants had incomplete SCI with no independent ambulatory function and were graded as AIS C with low lower extremity motor scores—following the American Spinal Injury Association (ASIA) Impairment Scale (AIS). Treadmill stepping alone improved gait performance on the treadmill but did not improve overground walking capabilities. Instead, the addition of SCS led to the immediate facilitation of walking, as well as further training-related gait improvements and a reduced sense of effort. Despite remarkable improvements in walking capability, there was no change in muscle strength or lower extremity motor scores when the stimulation was turned off, suggesting that this approach was insufficient to trigger neuroplasticity mechanisms in the tested subjects.

The next milestone in the application of SCS to neurorehabilitation came from a case study by Harkema and colleagues in 2011, which aimed at promoting standing and assisted treadmill stepping in a subject with motor-complete, sensory-incomplete SCI (AIS-B) [85]. SCS enhanced rhythmic EMG activity during assisted treadmill stepping, evoked sustained activation patterns in lower extremity muscles, and allowed independent, full weight-bearing standing after 80 sessions of intensive training. An additional major outcome of the study was the incidental re-discovery of the so-called motor-enabling effect of SCS, initially observed by Barolat in 1986. Indeed, the participant reported that SCS enabled him to perform voluntary movements of paralyzed muscles, including toe extension, ankle dorsiflexion, and leg flexion. As in 1986, this effect was present only when the stimulation was on.

The investigators focused on this motor-enabling effect in a subsequent study involving intensive training of voluntary leg movements under SCS in four participants with motor complete SCI (two graded AIS-A, and two AIS-B), including the participant from the original study [35]. All new participants were able to voluntarily induce movement when SCS was applied from the very first day and without any training, even for the two sensory and motor complete subjects (Fig. 18.3).

With training enhanced by SCS, all participants improved over time and became able to control movements based on visual or auditory cues. Three of them could generate graded levels of force in at least one leg, and two could modulate EMG activity during assisted treadmill stepping and SCS by thinking about moving the legs. Additionally, all participants became able to perform full weight-bearing standing with SCS after 80 sessions of stand training [86]. A follow-up study found that rehabilitation in the presence of SCS needs to be task-specific, and that stand and step training leads to different functional outcomes [87]. After completion of the study, one of the participants (AIS-B) was enrolled to receive additional activity-based training with

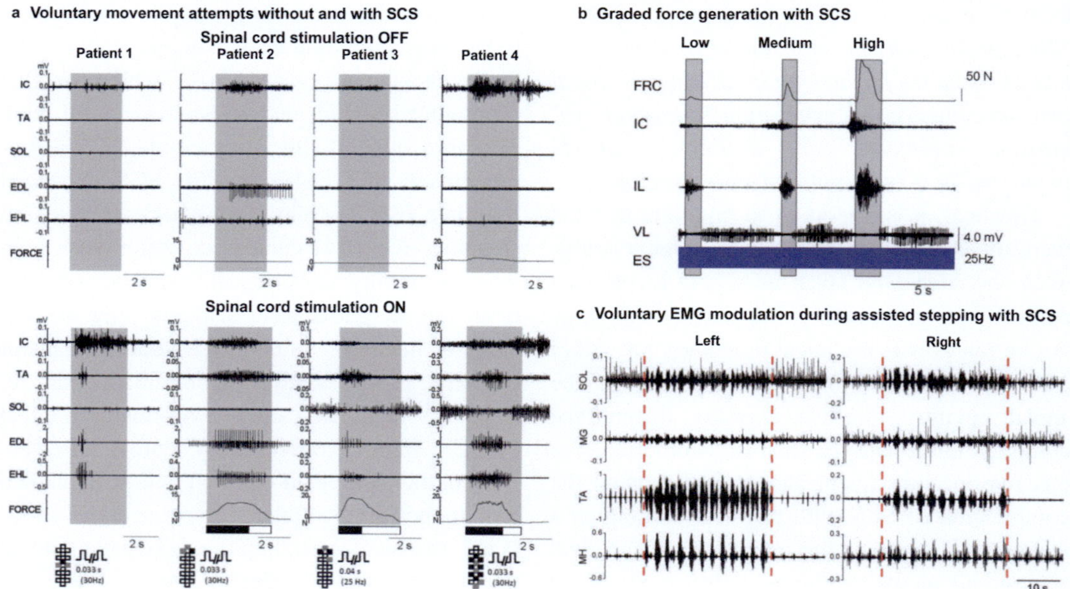

Fig. 18.3 **Tonic SCS immediately enables voluntary motor control**. **a** Lower extremity EMG activity during voluntary movement attempts (ankle dorsiflexion) without and with SCS in four individuals with clinically determined motor complete SCI. Electrode representations show cathodes (gray) and anodes (black). Stimulation frequency: 25–30 Hz. Muscles, surface EMG: intercostal sixth rib (IC), tibialis anterior (TA), soleus (SOL); fine wire EMG: iliopsoas (IL), extensor digitorum longus (EDL), extensor hallucis longus (EHL). Gray highlighted: active 'flexion/extension' period. **b** Left leg force and iliopsoas, vastus lateralis and intercostals EMG activity generated during a low (20%), medium (60%), and high (100%) effort of hip/knee flexion with SCS from patient 3. Gray shading: force duration. **c** Volitional modulation in EMG activity by an individual with motor complete SCI during manually assisted stepping (40% body weight support, 1.07 m/s) in the presence of SCS. Initial steps show EMG pattern while the subject (patient 3) is not thinking about stepping. Section within the red dashed lines show the period of steps while the subject is consciously thinking about stepping and facilitating each step (with voluntary intent). Adapted with permission from [35]. All rights reserved

SCS, and his voluntary leg motor control progressively improved throughout the 3.7 years of training, to a level such that he could produce voluntary leg movement and standing even with SCS turned off [88].

With the goal to confirm the motor-enabling effect of SCS on participants with severe SCI, an independent group at the Mayo Clinic conducted a 2-week, 8-session study on a participant with chronic, complete SCI (AIS-A) attempting volitional control of leg movements with SCS [89]. Stimulation enabled voluntary knee flexion, initiation and termination of rhythmic leg movements, full weight-bearing standing, and voluntary generation of step-like movements while stationary in an upright position with body-weight support. These abilities were present only when the stimulation was turned on.

18.3.2 Spatiotemporal SCS for Neuroprosthetics and Neurorehabilitation in Animal Models of SCI

In parallel with these investigations utilizing tonic SCS in humans, several studies in animal models of SCI were laying the groundwork for a new stimulation paradigm that would combine the ability to promote voluntary motor control and overt synergistic movements during functional tasks: spatiotemporal SCS. So far, we described applications in which SCS was applied tonically, i.e., stimulation parameters such as amplitude, frequency, pulse width, and electrode configurations (choice of anodes and cathodes) were set manually by the experimenter at the beginning of a trial and were kept constant across consecutive steps. In real-life situations, however, the fine control of gait requires the ability to modulate muscle activity and kinematic outputs depending on the environment, task requirements, or levels of fatigue, which motivated the development of spatiotemporal SCS.

18.3.2.1 Spatiotemporal SCS Controlled by Residual Kinematics

During SCS-enabled locomotion, this adjustment can be done artificially by linking the task requirements and the observed kinematics with the stimulation parameters used to control SCS in real time. In a first pioneering study, Wenger and colleagues established the first proof-of-concept closed-loop SCS in rat models of SCI [90]. They identified a linear relationship between SCS frequency and the elicited step height, which they exploited as part of a closed-loop proportional-integral (PI) controller. This controller adjusted SCS frequency in real-time and for each step based on the desired and observed step heights, which depended on the task requirements and environment. For example, climbing a staircase required a higher step height than overground walking on flat surfaces. The observed kinematics were obtained by means of a motion capture system that monitored the 3D position of infrared-reflective markers placed on the joints of the hindlimb. This technological solution shows the possibility to develop intelligent systems that can support an individual in modulating motor output by adjusting SCS parameters in real time.

In a subsequent study, the same group introduced a new concept of real-time control of SCS parameters, which selected spatially-distinct sets of electrodes during different phases of the gait cycle. This protocol aimed at independently activating specific muscle synergies such as leg flexion or extension at the appropriate time, a concept termed spatiotemporal SCS [62].

Specifically, they aimed at replicating the dynamics of motoneuron activation underlying gait in non-injured animals, which was indirectly inferred from the EMG activity of several key hindlimb muscles innervated at various segments within the lumbosacral spinal cord. The obtained spatiotemporal maps of motoneuron activation during bipedal locomotion highlighted two spatially distinct activation patterns associated with major muscle synergies, and with, respectively, flexion and extension movements of the hindlimb, hereafter referred to as "hotspots".

To reactivate these hotspots in animals with SCI, spatially specific electrode configurations were first selected to recruit the dorsal roots projecting to the corresponding spinal segments. Next, kinematic gait events corresponding to the time when the animal places the foot on the ground ("foot strike") and lifts it off ("foot off") were extracted in real-time using the motion capture system described previously. Extracting these gait events was possible thanks to the residual kinematics of the animals with incomplete SCI placed in a body-weight support system. In this new paradigm, SCS was not applied tonically to the lumbosacral spinal cord, but as short pulse trains (lasting a few hundreds of milliseconds) triggered by these extracted gait events. Specifically, electrode configurations able to activate the upper lumbar and sacral segments of the spinal cord (associated with hindlimb flexion and extension, respectively) were triggered by the "foot off" and "foot strike" gait events, respectively. In addition, SCS was delivered at an amplitude sufficiently high to elicit motor outputs through the generation of powerful spinal reflexes, which supported the animals in movement execution.

This study demonstrated that spatiotemporal SCS allows the facilitation of specific movement phases through targeted activation of the appropriate motoneuron pools through spinal circuits during gait, which can be achieved using epidural multi-electrode arrays. This approach also enables to reduce muscle co-activations observed during tonic SCS, to use higher stimulation amplitudes that can provide better weight-bearing capacity, and to vary stimulation frequency throughout the gait cycle depending on the desired functional outcomes.

18.3.2.2 Spatiotemporal SCS Controlled by Brain Signals

The principles underlying spatiotemporal SCS were then tested in non-human primates, which represent the most suitable animal model for the translation to humans because of the unique organization of the corticospinal tract in these species [91, 92]. In their study, Capogrosso and colleagues extracted motor signals from intracortical recordings in macaque monkeys with SCI to control spatiotemporal SCS, thereby pioneering the concept of a "brain-spine interface" [33]. Specifically, a 96-channel intracortical microelectrode array (Utah array) was surgically implanted into the hindlimb area of the primary motor cortex, which sends motor commands down to the spinal cord. Thanks to a state-of-the-art wireless neuronal amplifier (able to amplify and broadcast wirelessly 96 channels of neuronal data at a sampling rate of 20 kHz), they recorded the spiking activity across the 96 electrodes while animals performed treadmill and overground locomotion. This neuronal activity was used as an input to a machine learning algorithm (a discrete classifier) able to identify neural states associated with flexion or extension of the hindlimb. At the spinal cord level, the animals were implanted with an epidural multi-electrode array designed specifically to cover the lumbosacral segments of the macaque spinal cord. The spinal electrode array was in turn connected to a modified version of an implantable pulse generator (IPG) clinically approved for deep brain stimulation. Modifications to the IPG firmware provided real-time control over stimulation parameters such as electrode configurations, amplitude, and frequency.

This technological framework enabled the implementation of brain-triggered spatiotemporal SCS in freely moving non-human primates. The same principles as used in rats to optimize SCS were extended to non-human primates [32], and stimulation protocols facilitating flexion or extension of the leg were extracted. After a first proof-of-concept in intact animals, two macaque monkeys received a unilateral corticospinal tract lesion that left one hindlimb completely paralyzed. As early as a few days after the experimental lesion, the brain-spine interface enabled to reestablish both treadmill and overground locomotion, with the paralyzed hindlimb moving in coordination with the three other limbs.

This study brought important advancements, both technologically and scientifically. On the technological side, it demonstrated that it is possible to implement brain-triggered spatiotemporal SCS with currently available technologies that are ready for human use. From a scientific standpoint, it demonstrated that spatiotemporal SCS immediately restored weight-bearing locomotion as early as six days post-injury in non-human primates, which bears substantial clinical relevance. The concept of brain-controlled SCS for rehabilitation was further developed by Bonizzato and colleagues who linked cortical ensemble activity to the amplitude of SCS in rat models of SCI and showed a more pronounced and faster recovery of locomotion when training with brain-controlled SCS compared to tonic SCS [93].

In summary, this series of studies laid the technological and scientific premises for the application of spatiotemporal SCS to humans with SCI. First, spatiotemporal SCS provides a way to activate different spinal cord locations and thus different muscle synergies at different phases during the gait cycle or any other motor task. Next, these different stimulation protocols can be triggered based on residual kinematics or neuronal signals to enable a smooth integration into the ongoing locomotor activity. Finally, closed-loop control policies can be additionally used to adjust various parameters such as stimulation amplitude or frequency in real-time to adapt to task requirements and environmental constraints.

18.3.3 Tonic and Spatiotemporal SCS Combined with Intensive Rehabilitation Restore Independent Overground Walking in People with SCI

The year 2018 marked a milestone for the application of SCS in people with SCI. For the first time, three groups in parallel demonstrated in a total of six subjects with chronic, severe SCI that SCS, delivered with either tonic or spatiotemporal protocols and combined with intensive rehabilitation, could enable independent overground walking [34, 37, 94].

18.3.3.1 Tonic SCS

At the Kentucky Spinal Cord Injury Center and the University of Louisville, Angeli and colleagues enrolled four participants with motor-complete SCI (two AIS-A, two AIS-B) who performed training sessions for standing, treadmill stepping with body-weight support and manual assistance, and overground walking when possible, all in the presence of tonic SCS [94]. All four participants achieved assisted standing and improved trunk stability in the sitting position in the presence of SCS and after several weeks of training. Most importantly, the two participants with motor-complete, sensory-incomplete SCI (AIS-B) achieved the ability to walk overground with tonic SCS after 278 and 81 training sessions respectively, over a period of 85 and 15 weeks. Walking only occurred when SCS was turned on, and while the participant consciously intended to walk. After 147 sessions, the second participant was able to walk independently with a walker and with SCS, which was an unprecedented level of recovery for a person with motor-complete SCI.

In parallel, at the Mayo Clinic, Gill and colleagues enrolled an individual with chronic motor- and sensory-complete SCI (AIS-A), who had previously trained to perform step-like movements with SCS in a side-lying position [89] to receive additional motor task training with tonic SCS [37]. After 43 weeks of training and in the presence of tonic SCS, this participant was able to stand, step on a treadmill without body-weight support, and walk overground with a walker and assistance of a physiotherapist for hip stability, for the first time in a participant graded AIS-A in the chronic state of SCI. In a recent follow-up study, Gill and colleagues also showed that maximizing participants' intention to walk and minimizing body-weight support during training with tonic SCS improved independence and decreased the need for external assistance by a physiotherapist (Gill et al. 2020). Conversely to the earlier study by Rejc and colleagues [87], dynamic training combining the repetition of different motor tasks found positive effects on both stand and gait performance simultaneously.

Clinical studies by these two groups illustrated the potential of tonic SCS combined with several months of intensive rehabilitation for restoring motor function after SCI.

18.3.3.2 Spatiotemporal SCS

Meanwhile, at the Lausanne University Hospital and Ecole Polytechnique Fédérale de Lausanne (EPFL) in Switzerland, Wagner and colleagues pioneered the use of spatiotemporal SCS in three participants with chronic, incomplete but severe SCI (one AIS-D and two AIS-C, including one with motor scores of 0 in all key leg muscles but with remaining sphincter control) [34] (Fig. 18.4). They demonstrated both an immediate facilitation of body-weight-supported walking and long-term recovery of motor function even in the absence of SCS. This strategy leveraged the IPG with real-time control capabilities previously tested in non-human primates [33], which was connected to the same 16-electrode array as used in the other clinical studies cited above (Specify 5-6-5, Medtronic, clinically approved for the treatment of chronic pain). Taking inspiration from their previous methodology in rodents and non-human primates [32], they developed a stimulation protocol that alternated between the swing, weight acceptance, and propulsion phases of the right and left legs at appropriate times and amplitudes during the gait cycle. Each functionality was associated with a stimulation pattern consisting of a spatially specific set of anodes and cathodes optimized to recruit the associated posterior roots, and with a stimulation amplitude and frequency that further maximized the activation of the desired muscle synergy. For example, stimulation frequencies between 40 and 120 Hz tended to better promote a whole-leg flexion synergy (at the hip, knee, and ankle joints simultaneously), while frequencies of 20–30 Hz preferentially recruited functional knee, and ankle extensors. Finally, the alternation of stimulation patterns could either be delivered automatically at a pre-defined sequence and pace, or they could be triggered in real time by residual kinematics for people with sufficient residual control. Movement feedback used to trigger SCS was initially obtained from an infrared-based 3D motion capture system as previously shown in rodents [62], and later by wearable sensors containing inertial measurement units (IMU) placed on the subjects' feet. Such sensors, along with appropriate algorithms, enabled to extract the foot inclination angle as the participant attempted to initiate movement and to trigger a stimulation pattern that enabled flexion of the corresponding leg.

This spatiotemporal SCS paradigm, combined with a cable-based robotic body-weight support system, allowed a wide variety of locomotor tasks both on a treadmill and overground in the three participants with chronic SCI at the cervical level. One participant (AIS-C) had complete motor paralysis on the left leg but residual activity on the right, the second (AIS-D) had paralysis in the leg flexor muscles, and the third (AIS-C, based on the presence of sphincter contraction) had motor-complete paralysis in both legs. Spatiotemporal SCS immediately (i.e., without training) facilitated EMG activity underlying locomotion in otherwise inactive or poorly active leg muscles and enabled participants to walk overground with assistive devices and body-weight support. Participants could voluntarily modulate the effect of the stimulation by exaggerating step elevations, could walk at different speeds and could cover distances of up to 1.2 km on a treadmill without deterioration of kinematics or muscle activity. The first two

participants regained the ability to transition from sitting to standing and to walk independently with crutches without SCS or body-weight support. Neurological recovery, tested according to clinical standards and without SCS, was observed to different degrees in all three participants. The first participant improved from AIS-C to AIS-D and gained 16 points in his lower extremity motor scores (from 14 to 30, maximum of 50). The second participant gained 11 points (from 25 to 36). The third participant gained four points (0–4). Although he was not able to perform voluntary movements against gravity in the absence of SCS, the researchers observed an increase in the maximum isometric torques that the participant was able to produce in the presence of SCS. Although the investigators did not attempt to train participants with tonic SCS, they performed a comparison of the immediate facilitation of locomotor activity with tonic versus spatiotemporal SCS. In the three reported participants, tonic SCS created an important co-activation of antagonistic muscles preventing smooth locomotion. Additionally, it disrupted the residual proprioceptive inputs to the spinal cord and the brain, thereby blocking important feedback cues to the spinal locomotor circuitry as well as impairing the conscious perception of the lower limbs in space [72].

In a recent study by the same group, Rowald and colleagues expanded their approach to people with motor-complete SCI (two AIS-A, one AIS-B), who were implanted with a new 16-electrode array specifically designed for the rehabilitation of both leg and trunk motor function after severe SCI [39]. Their approach also involved the development of personalized computational models of the spinal cord derived from structural and functional magnetic resonance imaging (MRI). This study demonstrated that spatiotemporal SCS immediately enables (i.e., within a week of using SCS) powerful facilitation of walking even in motor-complete participants, whereas similar functional outcomes could only be achieved after several months of intensive training using tonic SCS [37, 94]. Furthermore, it provides a path forward in the refinement of neurotechnologies for delivering SCS, ranging

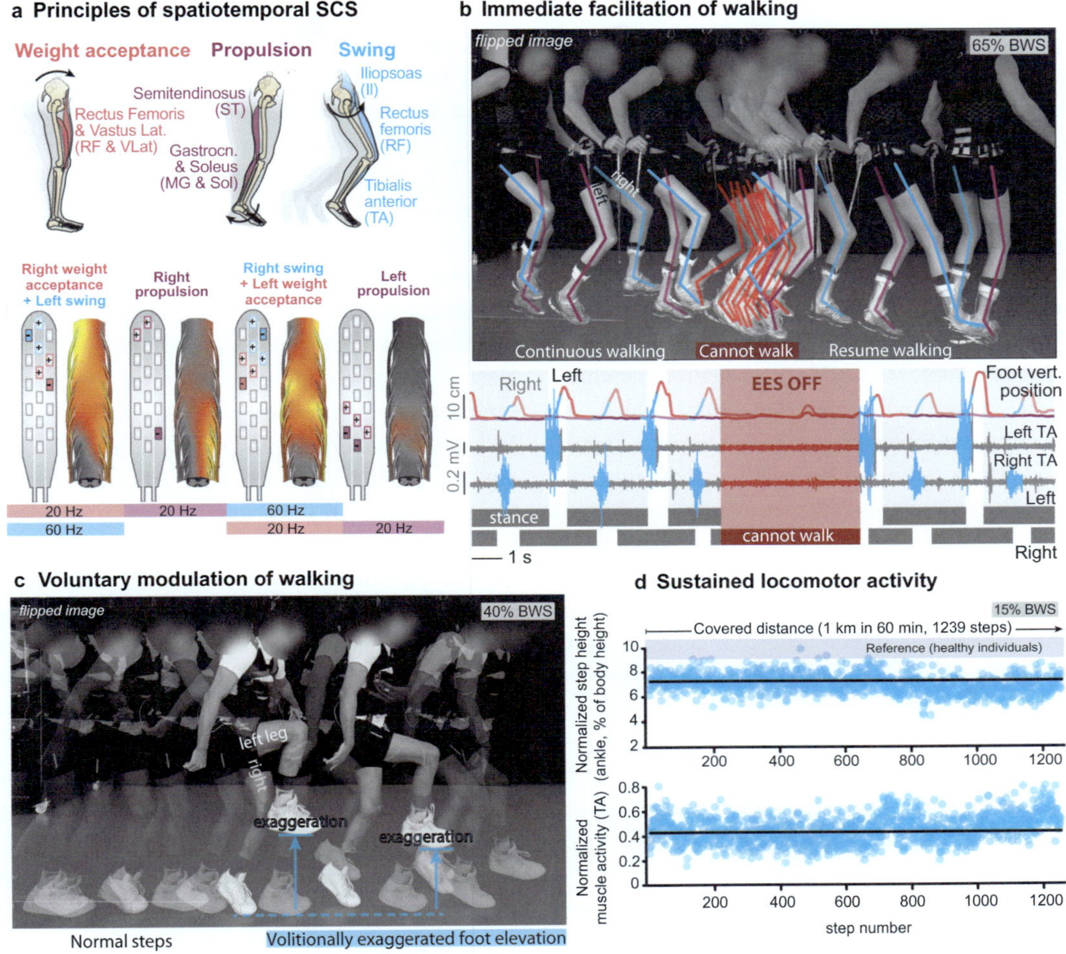

Fig. 18.4 **Spatiotemporal SCS immediately enables independent walking.** **a** Top: the three muscle synergies underlying human walking, and which can be targeted by SCS. Bottom: typical sequence of spatiotemporal SCS and associated parameters for immediate facilitation of walking after SCI. **b** Chrono-photography, tibialis anterior EMG activity and foot vertical position during overground walking with body-weight support and walking sticks while SCS is switched on, then off, then on in a subject with severe SCI. **c** Overground walking when a subject with incomplete SCI but completely paralyzed left leg is asked to perform first normal and then exaggerated step heights. **d** Consecutive values of step height and EMG activity over 60 min of walking with EES (1 km). BWS: body-weight support. Adapted with permission from [34]. All rights reserved

from new electrodes arrays and personalized computational models of the spinal cord to versatile software platforms for configuration and use of spatiotemporal SCS by non-experts. The future deployment of such technologies in widespread clinical practice will require the additional development of new implantable neurostimulators and automated pipelines for optimizing SCS parameters.

18.3.3.3 Limitations of Locomotor Rehabilitation Facilitated by SCS

Improvements in motor function mediated by SCS require high-intensity neurorehabilitation sessions, spread over a time period that is much longer than provided in current clinical practice and covered by insurance. For SCS to become a clinically accepted method for augmenting rehabilitation outcomes, the duration of the rehabilitation phase should be therefore considerably shorter. This could be achieved for example by starting neuromodulation therapies in the sub-acute phase after the injury, which would leverage the intrinsic capacity of the spinal cord to reorganize. In individuals living with a chronic SCI, combining SCS with pharmacological interventions will likely further improve and accelerate rehabilitation outcomes [95, 96]. Administration of pharmacological agents would thereby mimic the effects of neurotransmitters such as serotonin and dopamine, which are essential for locomotion. These neurotransmitters are synthesized in the brainstem and the posterior hypothalamus, but their descending axons become partially separated from the lumbar spinal cord after SCI. Finally, all subjects with severe SCI who achieved overground walking with SCS required their arms to maintain balance using either a walker or crutches. This means that their arms cannot easily serve other purposes, such as reaching for an object and carrying it from one spot to another, potentially limiting the usability of this technology in certain daily life situations. To improve dynamic balance, SCS protocols will need to target additional muscle groups involved in hip/trunk movement and stabilization, such as leg abductors and adductors, as well as the quadratus lumborum and the paraspinal muscles. In fact, there is early demonstration that multi-electrode arrays placed over the low-thoracic and lumbosacral spinal segments, combined with activity-specific stimulation programs, can augment the control of both trunk and leg movements in individuals with chronic, motor complete spinal cord injury [39].

18.3.4 Other Recent Studies of SCS for Improving Motor and Autonomic Functions After SCI

Following the three seminal studies from 2018, which focused on overground walking, several groups sought to improve a wider range of motor functions and additionally target autonomic functions.

In terms of motor functions, trunk stability turned out to be a key element to improve in motor- and sensory-complete SCI, as already shown by Angeli and colleagues in 2018 [94]. Later, Gill and colleagues also demonstrated in two participants with motor- and sensory-complete SCI that SCS could improve seated reaching distance [97]. On the technological side, Rowald and colleagues showed that a longer electrode array could target both trunk and lower limb motor functions in subjects with complete SCI and considerably improve several daily living and leisure activities that critically require trunk stability [39]. At the University of Minnesota, Darrow and colleagues showed in two female participants with chronic motor- and sensory-complete SCI (AIS-A) that tonic SCS could immediately enable volitional leg movements [98]. In a follow-up study, Pena Pino and colleagues studied the effect of long-term exposure to tonic SCS without intensive neurorehabilitation [99]. After one month of optimization of various stimulation programs for volitional motor control, spasticity, and autonomic functions, participants were allowed to use SCS at home as much as 24 h a day during their daily living activities. Out of seven participants with motor-complete SCI, four of them (all graded AIS-A) recovered the ability to perform voluntary movements even without SCS after a period ranging from 3 to 13 months. Importantly, these movements were not present at every clinical visit, showing variability over time in these motor effects. Even more importantly, higher levels of spasticity seemed to correlate positively with the recovery of voluntary movements. These results add up to the recovery without SCS observed by Rejc and colleagues [88], and independently by Wagner and colleagues [34].

In terms of autonomic functions, Darrow and colleagues showed that SCS improved bowel-bladder synergy in their two participants, with SCS, and cardiovascular function in one of the two participants who had otherwise drops in blood pressure during tilt-table tests [98]. This same participant also reported the ability to achieve orgasm during sexual intercourse when SCS was on or immediately after it was turned off. Although a thorough review is beyond the scope of this chapter, we would like to highlight that SCS has been shown to improve bladder function [21, 98, 100], body composition, and metabolism [101], and blood pressure [98, 102–104]. Targeting such autonomic functions is of tremendous importance for improving the quality of life of people with SCI.

18.3.5 Comparison Between SCS and Functional Electrical Stimulation (FES)

Functional Electrical Stimulation is an established technology that targets efferent axons innervating specific muscles to produce a desired movement, using electrical stimulation applied at the surface of the skin [105–107] or with leads implanted in the periphery [108, 109]. FES has important clinical applications in hemiplegia, used for example as a commercially available foot-drop stimulator, and for the rehabilitation of upper extremity motor function. Moreover, it has been extensively used in research applications for SCI [107, 110–112], but did not become a standard clinical practice for this condition. In this paragraph, we discuss the conceptual and practical differences between SCS and FES.

18.3.5.1 Conceptual Differences: Stimulation of Muscles Versus Spinal Circuits

Functional Electrical Stimulation aims at generating force and movement by recruiting the efferent axons that innervate muscle fibers via pulses of electrical stimulation. This direct recruitment of muscles enables a high degree of controllability because each targeted muscle can be independently stimulated. However, the stimulation patterns required to coordinate a functional movement can be extremely complex and are gravity-dependent [110]. Therefore, a specific set of parameters can only work for a pre-determined movement but can hardly be generalized [108]. This aspect significantly increases the complexity of FES systems, as they must be specifically tuned for each task.

Conceptually, SCS works very differently than FES, as SCS engages surviving spinal circuits below the lesion via their input fibers, the excitatory sensory afferents. SCS, therefore, overcomes some of the limitations of FES because it requires simple stimulation protocols that leverage existing neural architectures to perform complex movements of a whole limb [33, 34, 113]. Indeed, excitatory spinal circuits producing synergistic movements receive rich innervation from the primary afferents stimulated by SCS [62, 114] and a single Ia afferents connects to all the motoneurons of the homonymous muscle and up to 60% of synergistic motoneurons even at different joints [115].

Another key difference is that FES imposes a specific movement according to a preprogrammed pattern, irrespective of the subject's voluntary intention. On the other hand, SCS protocols are thought to synergistically act and enhance residual voluntary inputs. Motor outputs can then be modulated and naturally shaped by movement-specific feedback [75] as well as volitional contributions [34, 35].

Concerning the ability to produce large forces, FES suffers from the "inverse recruitment effect". Since large axons have a lower threshold to electrical stimulation than smaller diameter fibers [48], FES systems first recruit large motor axons instead of smaller axons [116]. Large motor axons recruit muscle fibers that generate large forces but are not resistant to fatigue. This is the opposite of what happens with a natural movement, during which larger fibers are only activated when substantial forces are required. This artificial inverse recruitment rapidly leads to the generation of fatigue, making it technically challenging to produce and sustain large forces [117, 118]. By contrast, SCS does not recruit spinal motoneurons directly. The activation of spinal motoneurons by means of pre-synaptic recruitment of primary afferents leads to a natural recruitment order that is resistant to fatigue and can produce forces capable of sustaining the whole body weight for extended periods of time [72, 90].

Finally, FES applications that rely on the stimulation of motor axons in the peripheral nerves bypass the spinal cord and consequently cannot directly lead to neuroplasticity of spinal circuits. By contrast, such neuroplasticity is believed to mediate the neurological recovery observed during neurorehabilitation facilitated by SCS [119, 120].

18.3.5.2 Practical Differences: Assistance Versus Therapy

Because of the difficulty to coordinate complex activations of muscles, both implantable and non-invasive FES systems can almost exclusively be used in controlled environments. For this reason, FES therapy is applied during laboratory or clinic sessions of physical therapy. In this sense, an FES system works to assist physical exercise with a therapeutic goal. It is not a wearable assistive system that supports daily living activities in community settings. By contrast, epidural SCS is a fully implantable system that is seamlessly integrated in patient's lives. Therefore, SCS can be used both to assist physical therapy as well as support activity of daily livings in community settings [34, 39]. Outside the laboratory or clinics, patients are able to use their fully-implanted systems similarly to what patients with Parkinson's disease do with a DBS implant. In this regard, SCS addresses needs of assistance that cannot be addressed with modern FES devices.

18.3.6 Conclusion: SCS, a Promising Neuroprosthetic Technology and Neurorehabilitation Therapy After SCI

In this section, we have described how early human studies using tonic SCS in people with SCI laid the groundwork for its subsequent use as a neurorehabilitation technique, in particular its

motor-enabling effect which dates back to 1986 [23], and the discovery of stepping-like movements in humans with SCI [27]. Next, the recent re-discovery of the motor-enabling effect of SCS in 2011 led to its integration into intensive rehabilitation programs for volitional control of joint movements and standing [85]. In parallel with these clinical studies, the development of new stimulation paradigms for delivering SCS to the spinal cord, called spatiotemporal SCS, restored locomotion in rodent and non-human primate models of SCI [33, 62]. Finally, we showed that all these scientific and technological advances converged to the demonstration that SCS combined with intensive rehabilitation can support the recovery of voluntary motor control and overground walking in participants with severe and chronic SCI [34, 37, 94]. Current studies now aim at expanding the range of motor and autonomic functions enabled by SCS (e.g., [39, 98]).

Overall, SCS has shown promises both as neuroprosthesis, i.e., an assistive technology that replaces a lost function and provides immediate relief to a particular deficit, and a rehabilitation tool, i.e., a means to train people with SCI to improve their motor functions when combined with intensive physiotherapy (Fig. 18.5). Although both tonic and spatiotemporal SCS bear great potential in terms of neurorehabilitation, spatiotemporal SCS leads to faster functional outcomes in the absence of training, hence a good indication as an effective neuroprosthesis. Another key aspect to consider is the emergence of the long-term recovery of motor functions in the absence of SCS [34, 88], which likely relies on neuroplastic mechanisms triggered by prolonged use of SCS [119, 120].

In conclusion, although clinical studies have uncovered a formidable potential of SCS both for neuroprosthetics and neurorehabilitation, technological developments and larger multicentric clinical trials are required to assess its safety and efficacy for the millions of people living with SCI worldwide.

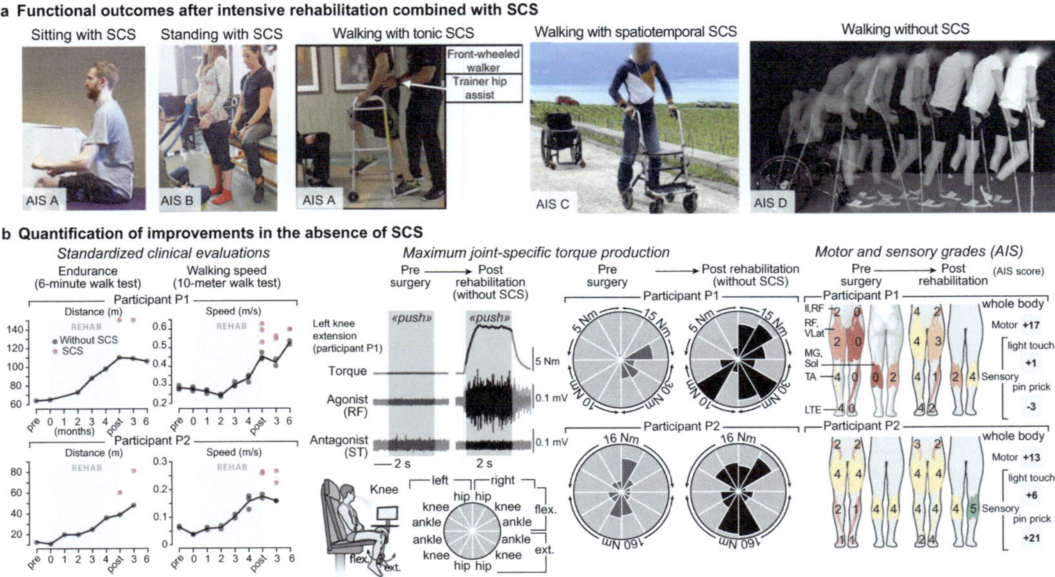

Fig. 18.5 SCS combined with rehabilitation leads to long-term motor improvements. a Functional outcomes that can be reached after tonic or spatiotemporal SCS combined with intensive rehabilitation over several months. For each picture, the severity of injury of the depicted individual is indicated. Adapted with permission from [34, 37, 94]. **b** Neurological recovery observed in two subjects with incomplete SCI after six months of intensive rehabilitation combined with SCS. Left: plots reporting changes in 6-min and 10-m walk tests. Tests were performed without body-weight support, following clinical standards. Middle: evaluations of isometric torque production for each joint, quantified before surgery and after rehabilitation without SCS. Right: changes in lower limb motor and sensory scores after rehabilitation. Changes in motor and sensory scores on abbreviated injury scale for all levels below injury are summarized (motor scores; 0: total paralysis, 1: palpable or visible contraction, 2: active movement, gravity eliminated, 3: active movement against gravity, 4: active movement against some resistance, 5: active movement against full resistance). Adapted with permission from [34]. All rights reserved

18.4 Transcutaneous SCS as a Non-Invasive Complement to Epidural SCS

18.4.1 Non-Invasive SCS: Stimulating Posterior Roots via Transcutaneous Electrodes

Fifty years after the development of epidural SCS, transcutaneous SCS was developed as a non-invasive method to activate similar neural structures as epidural SCS, i.e., the posterior roots, but from outside the skin [121]. This study was also the first to suggest that "continuous transcutaneous posterior root stimulation represents a novel, non-invasive, neuromodulative approach for individuals with different neurological disorders". Inspired by an earlier method of high-voltage percutaneous electrical stimulation of the anterior roots to assess afferent peripheral nerve conduction [122, 123], transcutaneous SCS utilizes skin-surface electrodes with one electrode over the spine at the thoracolumbar junction overlying the lumbosacral spinal cord, and a much larger return electrode placed over the lower abdomen or the iliac crests. Computational modeling of epidural SCS has predicted low-threshold sites along proprioceptive fibers at the posterior rootlet-spinal cord interface [31]. In fact, the high angles in fiber orientation and the crossing of the electrical conductivity boundary between the cerebrospinal fluid and the spinal cord make the recruitment of posterior roots possible also by skin-surface electrodes, albeit with lower root specificity than epidural SCS [30].

Because of the high external voltages required to elicit sufficiently strong voltages inside the vertebral bones, transcutaneous SCS requires external stimulators similar to those typically used for traditional functional electrical stimulation (FES) or transcutaneous electrical nerve stimulation (TENS). With this approach, it is possible to directly stimulate proprioceptive afferent fibers within the posterior roots and evoke activity in extremity muscles through spinal reflexes that resemble those elicited by epidural SCS [121, 126, 127]. The sharing of the same low-threshold sites along the posterior root afferent fibers between epidural and transcutaneous SCS found by studies in computational modeling [30, 128], combined with the near-identical reflex responses evoked by both stimulation methods in humans with SCI (Fig. 18.6 a), suggest that both epidural and transcutaneous SCS recruit common neural structures through similar mechanisms [46]. However, because of the larger distance from the spine and the reduced focality of the electrical field, the specificity in muscle recruitment is lower with transcutaneous SCS compared to its epidural counterpart [32, 34, 40, 129].

18.4.2 Transcutaneous SCS for Generating Locomotor-Like Movements

Despite reduced specificity, transcutaneous SCS remains an attractive solution for applications that do not seek to achieve highly selective muscle recruitment. Indeed, it does not require any surgical procedure, thereby significantly reducing the risks and costs of the intervention. It also provides a potentially inclusive and affordable solution to obtain, at least in part, motor improvements comparable with those achieved by epidural SCS.

Therefore, multiple studies have investigated the possibility to augment EMG activity and movements (Fig. 18.6b), as well as functional movements during active treadmill stepping in individuals with chronic, motor-incomplete SCI (AIS D) [49, 130, 131]. As a robotic gait orthosis moved the legs of individuals with clinically complete SCI (AIS A) in a walking pattern over a treadmill, a small number of muscles exhibited electromyographic responses [125]. Adding 30 Hz transcutaneous SCS increased the number of activated leg muscles, which had rhythmic activity during the different phases of gait (Fig. 18.6c). Moreover, transcutaneous SCS alone could produce rhythmic activation of leg muscles without the need for triggered stimulation. Four individuals with incomplete SCI (AIS D) were able to voluntarily modify the generated lower limb muscle activity [131, 132]. By consciously modifying their augmented muscle activity according to the gait phase, participants were able to improve the quality of their stepping kinematics, the range of hip and knee movements, and their stride length. As in epidural SCS, turning the stimulation off would immediately cause degradation of walking quality and muscle activity.

The current interpretation of these results is that mechanisms of transcutaneous SCS are at least partially similar to those of epidural SCS. In particular, this interpretation implies that transcutaneous SCS interacts with the flow of step-induced proprioceptive feedback to modulate muscle activity during locomotion through spindle feedback circuits [72]. However, the rhythmic activation without peripheral feedback suggests that transcutaneous SCS could also engage CPGs [27, 70, 71, 133] in addition to proprioceptive feedback circuits [29, 63, 75]. Moreover, a summation process between residual supraspinal inputs (via clinically silent translesional neural connections that survived the injury), and the increased spinal circuits excitability enhanced by transcutaneous SCS, may be a likely explanation for the voluntary

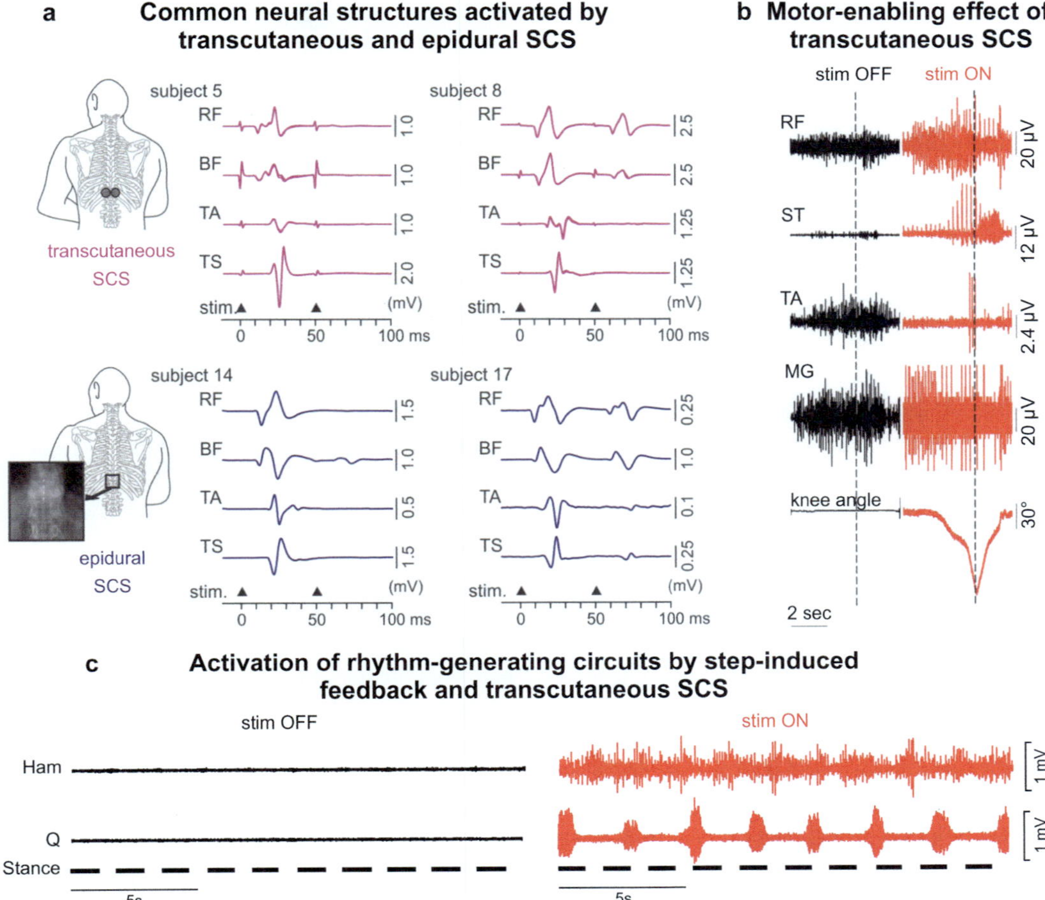

Fig. 18.6 **Transcutaneous SCS can enhance motor and locomotor function by recruiting similar structures to epidural SCS. a** Post-activation depression and similar response latencies, peak-to-peak amplitudes, and waveforms by paired-pulse transcutaneous and epidural SCS suggest the activation of common neural input structures (predominantly primary afferent fibers within multiple posterior roots) by both techniques. Modified with permission from [46]. **b** EMG activity and knee angle excursion during voluntary knee flexion attempts with and without tonic transcutaneous SCS. Modified with permission from [124] (Figure published under a Creative Commons license: Figs. 18.6a and b, licensed under CC-BY 4.0). **c** Modification of muscle activity by 30-Hz transcutaneous SCS during robotic-driven treadmill stepping. Modified with permission from [125], All rights reserved. Note that the EMG activity in quadriceps during SCS suggests an overt activation of rhythm-generating spinal circuits, because the rhythmic bursts are not properly synchronized with stepping. RF, rectus femoris; BF, biceps femoris; TA, tibialis anterior; TS, triceps surae; ST, semitendinosus; MG, medial gastrocnemius; Ham, hamstring; Q quadriceps

control of movement enabled in otherwise paralyzed muscles (Fig. 18.6b) [125].

18.4.3 Stimulation Parameters for Transcutaneous SCS

Two general methods for stimulation pulse shape configuration are currently used to deliver transcutaneous SCS: conventional and "Russian" current stimulation. Conventional currents consist of mono- or biphasic charge-balanced pulses with rectangular pulses of 1 ms width, delivered at frequencies ranging from about 5 to 100 Hz. Stimulation at 30 Hz is commonly applied to elicit muscle activity or movement, while 50 Hz pulses are used to improve spasticity [130, 134–136]. The "Russian current" is composed of 1 ms bursts filled with 10 kHz pulses [137]. Several studies have suggested that it may be possible to use these currents to reduce the potential discomfort of transcutaneous SCS directly below the stimulating surface electrodes [138, 139], while maintaining recruitment of deep neural structures to elicit comparable electrophysiological responses as conventional transcutaneous SCS. The hypothesis is that the temporal summation of graded potentials created by the rapid depolarization and repolarization of high-frequency stimulation may raise the membrane potential of larger fibers enough to be activated without depolarizing unmyelinated C-fibers [140]. Additionally, there is some neurophysiological evidence suggesting C-fibers are less likely to fire in response to high-frequency stimulation compared to large-diameter fibers [141, 142]. However, a recent study comparing the tolerance and responses elicited by the "Russian current" and conventional transcutaneous SCS protocols contradicted this view. While participants could indeed tolerate significantly higher levels of stimulation amplitude with Russian currents, both protocols produced the same amount of discomfort when the amplitudes were adjusted to obtain the same spinally-evoked muscle responses [143]. Therefore, the apparent higher comfortability may be due to the fact that it takes higher currents with 10 kHz burst to accumulate enough charge to stimulate the same deep structures in the spinal cord as with standard protocols.

18.4.4 Functional Recovery by Long-Term Transcutaneous SCS and Activity-Based Training

Long-term activity-based training with transcutaneous SCS has also been investigated as a method to induce functional recovery in the chronic phase after SCI. An 18-week training strategy involving voluntary modulation of

transcutaneous SCS-generated locomotor movements in combination with buspirone, an orally active serotonergic agonist, was tested in five individuals with chronic motor-complete, sensory-incomplete (AIS B) individuals [137]. The ability of individuals to voluntarily modulate the step-like movements induced by transcutaneous SCS improved with training, and participants gained the ability to generate voluntary movements in previously paralyzed muscles, even without stimulation. With stimulation, their motor function was equivalent to that of individuals with AIS C impairments.

In a subsequent single-case study to evaluate the contribution of buspirone [124], a 4-week training paradigm in robot-assisted overground training found that transcutaneous SCS alone or in combination with buspirone, but not buspirone alone, resulted in the lowest dependence on robotic assistance. Moreover, the participant could perform voluntary knee flexion when in the supine position after a 1-week training with transcutaneous SCS alone, but not after 1-week training with the drug alone. Sayenko and colleagues [139] subsequently demonstrated the ability of transcutaneous SCS to enable standing without previous training in 15 individuals with chronic SCI (AIS A, B, C), and a 12-session training program followed by six participants improved their upright balance control and reduced their dependence on external assistance. A post-training clinical evaluation revealed an increase in muscle tone of all participants. Notably, the results of this study are quantitatively similar to those seen in previous investigations with SCS [85, 86].

Although a single session of transcutaneous SCS can facilitate residual voluntary control of single joints and modify the excitability of cortical, corticospinal, and spinal reflexes [144–146], it is not sufficient to observe statistically significant improvements in walking performance [145]. Recent studies by several groups highlight the importance of combining SCS therapy with activity-based training to enable consistent improvements in walking [147, 148] and sit-to-stand ability [149].

18.4.5 Recent Advances to Improve Muscle Recruitment Selectivity in Transcutaneous SCS

Transcutaneous SCS presents a promising non-invasive approach to enable movement, locomotor function, and long-term recovery after SCI. However, the low specificity of transcutaneous SCS in muscle recruitment compared to its epidural counterpart may limit the type of movements and exercises that can be supported and trained during rehabilitation. Nevertheless, recent studies by Calvert, Krenn, and colleagues have demonstrated that positioning surface electrodes in lateralized (Fig. 18.7a) and rostro-caudal locations (Fig. 18.7 b) can target specific mediolateral and rostro-caudal spinal cord circuitry toward improved muscle recruitment selectivity [150, 151]. These observations suggest that future advances in electrode configurations, as well as stimulation amplitudes, frequencies, and timing, may further enhance the potential of transcutaneous SCS in non-invasive rehabilitation approaches.

18.4.6 Conclusion: Transcutaneous SCS is Less Specific Than Epidural SCS but Provides an Inclusive Access to Advanced Healthcare

In summary, transcutaneous SCS is a promising tool to investigate the combination of SCS and rehabilitation for applications that do not require high specificity. However, the high levels of current required to produce substantial muscle activity to support the body weight in people with severe motor paralysis may be uncomfortable and cause large contractions of the back and abdominal muscles. Nevertheless, we believe that transcutaneous SCS should not necessarily be seen in opposition with epidural SCS, but they can be seen as having unique advantages. While less invasive, transcutaneous SCS requires electrode application for every use and requires an

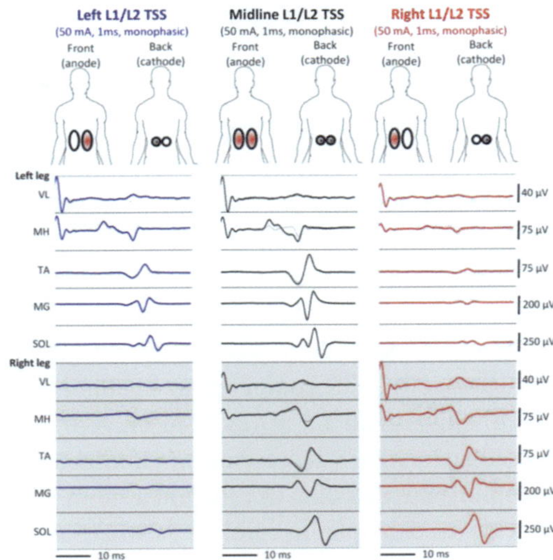

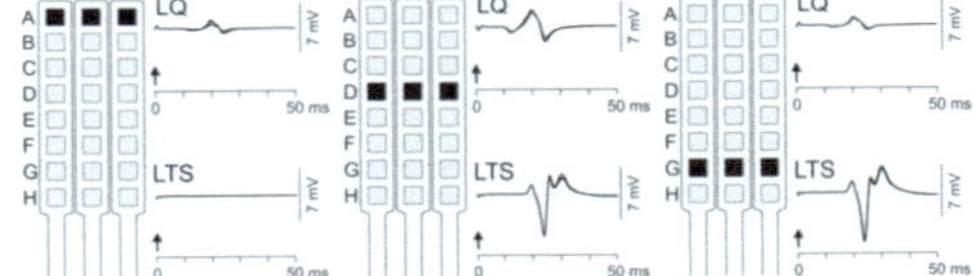

Fig. 18.7 **Limitations in muscle recruitment selectivity by transcutaneous SCS can be partially overcome by electrode configuration. a** Elicitation of spinal reflexes by lateral and midline transcutaneous SCS. Transcutaneous SCS applied ∼2 cm lateral to the midline of the lumbosacral spinal cord can selectively activate ipsilateral spinal sensorimotor circuits and thus ipsilateral lower extremity muscles. Modified with permission from [150]. **b** Elicitation of posterior root-muscle reflexes by stimulation from different rostro-caudal sites along a multi-electrode array. The multi-electrode array was 152 mm in length with row D positioned over the T11-T12 vertebra. Stimulation of rostral electrodes (row A) elicits responses in quadriceps but not in triceps surae. In contrast, stimulation of the caudal-most row elicits a large response in triceps surae and a small response in quadriceps. Modified with permission from [151]. MG, medial gastrocnemius; MH, medial hamstrings; SOL, soleus; TA, tibialis anterior; VL, vastus lateralis; L, left; Q, quadriceps; TS, triceps surae. All rights reserved

external stimulator connected to the electrodes, and may thus suffer from limited repeatability and portability. Therefore, transcutaneous SCS is very well suited to study and perform SCS-enhanced physical training and rehabilitation protocols in the hospital, and perhaps at home with appropriate guidance, but does not currently serve as a neuroprosthetic intervention that sustains motor activity in daily living. Instead, while more invasive, epidural SCS is fully implantable and therefore, by definition portable, thus sustaining motor activities for people with SCI not only in the clinic but in their daily life. These differences should be considered in the evaluation of the risk–benefit ratio, as it is possible that the interactions between descending voluntary input and SCS are limited when the stimulation is turned off. This would reduce the effectiveness of transcutaneous SCS compared to epidural SCS, which can instead be always on. Nevertheless, transcutaneous SCS represents an affordable tool to amplify the outcome of rehabilitation in many centers, for example, in rural areas where access to neurosurgery expertise and expensive invasive devices may be limited.

References

1. Melzack R, Wall PD. Pain mechanisms: a new theory. Science. 1965;150(3699):971–9.
2. Wall PD, Sweet WH. Temporary abolition of pain in man. Science. 1967;155(3758):108–9.
3. Shealy CN, Taslitz N, Mortimer JT, Becker DP. Electrical inhibition of pain: experimental evaluation. Anesth Analg. 1967;46(3):299–305.
4. Shealy CN, Mortimer JT, Reswick JB. Electrical inhibition of pain by stimulation of the dorsal columns: preliminary clinical report. Anesth Analg. 1967;46(4):489.
5. Krames ES, Peckham PH, Rezai A, Aboelsaad F. Neuromodulation. Section Introduction. Neuromodulation; 2009. p. 3–8.
6. Cook AW, Weinstein SP. Chronic dorsal column stimulation in multiple sclerosis. Preliminary report. N Y State J Med. 1973;73(24):2868–72.
7. Cook AW. Electrical stimulation in multiple sclerosis. Hosp Pract. 1976;11(4):51–8.
8. Cook AW, Taylor JK, Nidzgorski F. Functional stimulation of the spinal cord in multiple sclerosis. J Med Eng Technol. 1979;3(1):18–23.
9. Illis LS, Sedgwick EM, Oygar AE, Awadalla MA. Dorsal-column stimulation in the rehabilitation of patients with multiple sclerosis. The Lancet. 1976;307(7974):1383–6.
10. Illis LS, Sedgwick EM, Tallis RC. Spinal cord stimulation in multiple sclerosis: clinical results. J Neurol Neurosurg Psychiatry. 1980;43(1):1.
11. Illis LS, Read DJ, Sedgwick EM, Tallis RC. Spinal cord stimulation in the United Kingdom. J Neurol Neurosurg Psychiatry. 1983;46(4):299.
12. Siegfried J, Lazorthes Y, Broggi G. Electrical spinal cord stimulation for spastic movement disorders. Stereotact Funct Neurosurg. 1981;44(1–3):77–92.
13. Davis R, Gray E, Kudzma J. Beneficial augmentation following dorsal column stimulation in some neurological diseases. Stereotact Funct Neurosurg. 1981;44(1–3):37–49.
14. Dooley DM, Sharkey J. Electrostimulation of the nervous system for patients with demyelinating and degenerative diseases of the nervous system and vascular diseases of the extremities. Stereotact Funct Neurosurg. 1977;40(2–4):208–17.
15. Siegfried J, Krainick J-U, Haas H, Adorjani C, Meyer M, Thoden U. Electrical spinal cord stimulation for spastic movement disorders. Stereotact Funct Neurosurg. 1978;41(1–4):134–41.
16. Waltz JM, Reynolds LO, Riklan M. Multi-lead spinal cord stimulation for control of motor disorders. Stereotact Funct Neurosurg. 1981;44(4):244–57.
17. Barolat G, Myklebust JB, Wenninger W. Effects of spinal cord stimulation on spasticity and spasms secondary to myelopathy. Stereotact Funct Neurosurg. 1988;51(1):29–44.
18. Barolat-Romana G, Myklebust JB, Hemmy DC, Myklebust B, Wenninger W. Immediate effects of spinal cord stimulation in spinal spasticity. J Neurosurg. 1985;62(4):558–62.
19. Dimitrijevic MM, Illis LS, Nakajima K, Sharkey PC, Sherwood AM. Spinal cord stimulation for the control of spasticity in patients with chronic spinal cord injury: I. Clinical observations. Cent Nerv Syst Trauma. 1986;3(2):129–43.
20. Richardson RR, Cerullo LJ, McLone DG, Gutierrez FA, Lewis V. Percutaneous epidural neurostimulation in modulation of paraplegic spasticity. Acta Neurochir (Wien). 1979;49(3–4):235–43.
21. Richardson RR, McLone DG. Percutaneous epidural neurostimulation for paraplegic spasticity. Surg Neurol. 1978;9(3):153–5.
22. Waltz J. Chronic stimulation for motor disorders. Textb Stereotact Funct Neurosurg. 1998.
23. Barolat G, Myklebust JB, Wenninger W. Enhancement of voluntary motor function following spinal cord stimulation–case study. Appl Neurophysiol. 1986;49(6):307–14.
24. Waltz JM. Spinal cord stimulation: a quarter century of development and investigation. Stereotact Funct Neurosurg. 1997;69(1–4):288–99.
25. Bamford JA, Davis SF. Principles of neurophysiological assessment, mapping, and monitoring; 2019. p. 171–9.
26. Nagel SJ, Wilson S, Johnson MD, Machado A, Frizon L, Chardon MK, et al. Spinal cord stimulation for spasticity: historical approaches, current status, and future directions. Neuromodulation Technol Neural Interface. 2017;20(4):307–21.

27. Dimitrijevic MR, Gerasimenko Y, Pinter MM. Evidence for a spinal central pattern generator in humans. Ann N Y Acad Sci. 1998;860:360–76.

28. Minassian K, Persy I, Rattay F, Pinter MM, Kern H, Dimitrijevic MR. Human lumbar cord circuitries can be activated by extrinsic tonic input to generate locomotor-like activity. Hum Mov Sci. 2007;26 (2):275–95.

29. Capogrosso M, Wenger N, Raspopovic S, Musienko P, Beauparlant J, Bassi Luciani L, et al. A computational model for epidural electrical stimulation of spinal sensorimotor circuits. J Neurosci. 2013;33(49):19326–40.

30. Ladenbauer J, Minassian K, Hofstoetter US, Dimitrijevic MR, Rattay F. Stimulation of the human lumbar spinal cord with implanted and surface electrodes: a computer simulation study. IEEE Trans Neural Syst Rehabil Eng. 2010;18(6):637–45.

31. Rattay F, Minassian K, Dimitrijevic MR. Epidural electrical stimulation of posterior structures of the human lumbosacral cord: 2. quantitative analysis by computer modeling. Spinal Cord. 2000;38(8): 473–89.

32. Capogrosso M, Wagner FB, Gandar J, Moraud EM, Wenger N, Milekovic T, et al. Configuration of electrical spinal cord stimulation through real-time processing of gait kinematics. Nat Protoc. 2018;13 (9):2031–61.

33. Capogrosso M, Milekovic T, Borton D, Wagner F, Moraud EM, Mignardot J-B, et al. A brain-spine interface alleviating gait deficits after spinal cord injury in primates. Nature. 2016;539(7628):284–8.

34. Wagner FB, Mignardot J-B, Le Goff-Mignardot CG, Demesmaeker R, Komi S, Capogrosso M, et al. Targeted neurotechnology restores walking in humans with spinal cord injury. Nature. 2018;563 (7729):65–71.

35. Angeli CA, Edgerton VR, Gerasimenko YP, Harkema SJ. Altering spinal cord excitability enables voluntary movements after chronic complete paralysis in humans. Brain J Neurol. 2014;137 (Pt 5):1394–409.

36. Calvert JS, Grahn PJ, Strommen JA, Lavrov IA, Beck LA, Gill ML, et al. Electrophysiological guidance of epidural electrode array implantation over the human lumbosacral spinal cord to enable motor function after chronic paralysis. J Neurotrauma. 2019;36(9):1451–60.

37. Gill ML, Grahn PJ, Calvert JS, Linde MB, Lavrov IA, Strommen JA, et al. Neuromodulation of lumbosacral spinal networks enables independent stepping after complete paraplegia. Nat Med. 2018;24(11):1677–82.

38. Hofstoetter US, Perret I, Bayart A, Lackner P, Binder H, Freundl B, et al. Spinal motor mapping by epidural stimulation of lumbosacral posterior roots in humans. iScience. 2021;24(1):101930.

39. Rowald A, Komi S, Demesmaeker R, Baaklini E, Hernandez-Charpak SD, Paoles E, et al. Activity-dependent spinal cord neuromodulation rapidly restores trunk and leg motor functions after complete paralysis. Nat Med. 2022.

40. Greiner N, Barra B, Schiavone G, Lorach H, James N, Conti S, et al. Recruitment of upper-limb motoneurons with epidural electrical stimulation of the cervical spinal cord. Nat Commun. 2021;12(1):435.

41. Holsheimer J. Computer modelling of spinal cord stimulation and its contribution to therapeutic efficacy. Spinal Cord. 1998;36(8):531–40.

42. Lempka SF, Zander HJ, Anaya CJ, Wyant A, Ozinga JG, Machado AG. Patient-specific analysis of neural activation during spinal cord stimulation for pain. Neuromodulation J Int Neuromodulation Soc. 2020;23(5):572–81.

43. Coburn B, Sin WK. A theoretical study of epidural electrical stimulation of the spinal cord–Part I: Finite element analysis of stimulus fields. IEEE Trans Biomed Eng. 1985;32(11):971–7.

44. Hodgkin AL, Huxley AF. A quantitative description of membrane current and its application to conduction and excitation in nerve. J Physiol. 1952;117 (4):500–44.

45. Gerasimenko YP, Lavrov IA, Courtine G, Ichiyama RM, Dy CJ, Zhong H, et al. Spinal cord reflexes induced by epidural spinal cord stimulation in normal awake rats. J Neurosci Methods. 2006;157(2):253–63.

46. Hofstoetter US, Freundl B, Binder H, Minassian K. Common neural structures activated by epidural and transcutaneous lumbar spinal cord stimulation: Elicitation of posterior root-muscle reflexes. PLoS ONE. 2018;13(1):1–22.

47. Minassian K, Jilge B, Rattay F, Pinter MM, Binder H, Gerstenbrand F, et al. Stepping-like movements in humans with complete spinal cord injury induced by epidural stimulation of the lumbar cord: electromyographic study of compound muscle action potentials. Spinal Cord. 2004;42(7):401–16.

48. McNeal DR. Analysis of a model for excitation of myelinated nerve. IEEE Trans Biomed Eng. 1976;23(4):329–37.

49. Hofstoetter US, Danner SM, Freundl B, Binder H, Mayr W, Rattay F, et al. Periodic modulation of repetitively elicited monosynaptic reflexes of the human lumbosacral spinal cord. J Neurophysiol. 2015;114(1):400–10.

50. Rattay F. Analysis of models for external stimulation of axons. IEEE Trans Biomed Eng BME. 1986;33(10):974–7.

51. Struijk J, Holsheimer J, Boom H. Excitation of dorsal root fibers in spinal cord stimulation: a theoretical study. IEEE Trans Biomed Eng. 1993;40 (7):632–9.

52. Davidoff RA. The dorsal columns. Neurology. 1989;39(10):1377–85.

53. Lloyd DPC, McIntyre AK. Dorsal column conduction of group I muscle afferent impulses and their relay through Clarke's column. J Neurophysiol. 1950;13(1):39–54.

54. York DH. Somatosensory evoked potentials in man: differentiation of spinal pathways responsible for conduction from the forelimb vs hindlimb. Prog Neurobiol. 1985;25(1):1–25.

55. Whitsel B, Petrucelli L, Sapiro G. Modality representation in the lumbar and cervical fasciculus gracilis of squirrel monkeys. Brain Res. 1969;15 (1):67–78.

56. Holsheimer J. Which neuronal elements are activated directly by spinal cord stimulation. Neuromodulation J Int Neuromodulation Soc. 2002;5 (1):25–31.

57. Minassian K, McKay WB, Binder H, Hofstoetter US. Targeting lumbar spinal neural circuitry by epidural stimulation to restore motor function after spinal cord injury. Neurotherapeutics. 2016;13 (2):284–94.

58. Hofstoetter US, Freundl B, Binder H, Minassian K. Recovery cycles of posterior root-muscle reflexes evoked by transcutaneous spinal cord stimulation and of the H reflex in individuals with intact and injured spinal cord. PLoS ONE. 2019;14(12): e0227057.

59. Magladery J, Teasdall R, Park A, Porter W. Electrophysiological studies of nerve and reflex activity in normal man. V. Excitation and inhibition of two-neurone reflexes by afferent impulses in the same trunk. Bull Johns Hopkins Hosp. 1951;88(6): 520–37.

60. Pierrot-Deseilligny E, Burke D. The circuitry of the human spinal cord. Cambridge: Cambridge University Press; 2012.

61. Sharpe AN, Jackson A. Upper-limb muscle responses to epidural, subdural and intraspinal stimulation of the cervical spinal cord. J Neural Eng. 2014;11(1):016005.

62. Wenger N, Moraud EM, Gandar J, Musienko P, Capogrosso M, Baud L, et al. Spatiotemporal neuromodulation therapies engaging muscle synergies improve motor control after spinal cord injury. Nat Med. 2016;22(2):138–45.

63. Hofstoetter US, Knikou M, Guertin PA, Minassian K. Probing the human spinal locomotor circuits by phasic step-induced feedback and by tonic electrical and pharmacological neuromodulation. Curr Pharm Des. 2017;23(12):1805–20.

64. Jankowska E. Interneuronal relay in spinal pathways from proprioceptors. Prog Neurobiol. 1992;38 (4):335–78.

65. Minassian K, Freundl B, Hofstoetter US. The posterior root-muscle reflex. In: Neurophysiology in Neurosurgery. Elsevier; 2020. p. 239–53.

66. Hofstoetter US, Danner SM, Freundl B, Binder H, Lackner P, Minassian K. Ipsi- and contralateral oligo- and polysynaptic reflexes in humans revealed by low-frequency epidural electrical stimulation of the lumbar spinal cord. Brain Sci. 2021;11(1):112.

67. Sayenko DG, Angeli C, Harkema SJ, Edgerton VR, Gerasimenko YP. Neuromodulation of evoked muscle potentials induced by epidural spinal-cord stimulation in paralyzed individuals. J Neurophysiol. 2014;111(5):1088–99.

68. Roby-Brami A, Bussel B. Long-latency spinal reflex in man after flexor reflex afferent stimulation. Brain J Neurol. 1987;110(Pt 3):707–25.

69. Kiehn O. Decoding the organization of spinal circuits that control locomotion. Nat Rev Neurosci. 2016;17(4):224–38.

70. Danner SM, Hofstoetter US, Freundl B, Binder H, Mayr W, Rattay F, et al. Human spinal locomotor control is based on flexibly organized burst generators. Brain. 2015;138(3):577–88.

71. Minassian K, Hofstoetter US, Dzeladini F, Guertin PA, Ijspeert A. The human central pattern generator for locomotion: does it exist and contribute to walking? Neuroscientist. 2017;23(6): 649–63.

72. Formento E, Minassian K, Wagner F, Mignardot JB, Le Goff-Mignardot CG, Rowald A, et al. Electrical spinal cord stimulation must preserve proprioception to enable locomotion in humans with spinal cord injury. Nat Neurosci. 2018;21(12):1728–41.

73. Minassian K, Hofstoetter US. Spinal cord stimulation and augmentative control strategies for leg movement after spinal paralysis in humans. CNS Neurosci Ther. 2016;22(4):262–70.

74. Pinter MM, Gerstenbrand F, Dimitrijevic MR. Epidural electrical stimulation of posterior structures of the human lumbosacral cord: 3. Control of spasticity. Spinal Cord. 2000;38(9):524–31.

75. Moraud EM, Capogrosso M, Formento E, Wenger N, DiGiovanna J, Courtine G, et al. Mechanisms underlying the neuromodulation of spinal circuits for correcting gait and balance deficits after spinal cord injury. Neuron. 2016;89 (4):814–28.

76. Talpalar AE, Endo T, Löw P, Borgius L, Hägglund M, Dougherty KJ, et al. Identification of minimal neuronal networks involved in flexor-extensor alternation in the mammalian spinal cord. Neuron. 2011;71(6):1071–84.

77. Lavrov I, Gerasimenko YP, Ichiyama RM, Courtine G, Zhong H, Roy RR, et al. Plasticity of spinal cord reflexes after a complete transection in adult rats: Relationship to stepping ability. J Neurophysiol. 2006;96(4):1699–710.

78. Grillner S, Zangger P. On the central generation of locomotion in the low spinal cat. Exp Brain Res. 1979;34(2):241–61.

79. Iwahara T, Atsuta Y, Garcia-Rill E, Skinner RD. Spinal cord stimulation-induced locomotion in the adult cat. Brain Res Bull. 1992;28(1):99–105.

80. Seáñez I, Capogrosso M. Motor improvements enabled by spinal cord stimulation combined with physical training after spinal cord injury: review of experimental evidence in animals and humans. Bioelectron Med. 2021 Oct28;7(1):16.

81. Jilge B, Minassian K, Rattay F, Pinter MM, Gerstenbrand F, Binder H, et al. Initiating extension of the lower limbs in subjects with complete spinal

cord injury by epidural lumbar cord stimulation. Exp Brain Res. 2004;154(3):308–26.

82. Carhart MR, He J, Herman R, D'Luzansky S, Willis WT. Epidural spinal-cord stimulation facilitates recovery of functional walking following incomplete spinal-cord injury. IEEE Trans Neural Syst Rehabil Eng Publ IEEE Eng Med Biol Soc. 2004;12(1):32–42.

83. Herman R, He J, D'Luzansky S, Willis W, Dilli S. Spinal cord stimulation facilitates functional walking in a chronic, incomplete spinal cord injured. Spinal Cord. 2002;40(2):65–8.

84. Huang H, He J, Herman R, Carhart MR. Modulation effects of epidural spinal cord stimulation on muscle activities during walking. IEEE Trans Neural Syst Rehabil Eng Publ IEEE Eng Med Biol Soc. 2006;14(1):14–23.

85. Harkema S, Gerasimenko Y, Hodes J, Burdick J, Angeli C, Chen Y, et al. Effect of epidural stimulation of the lumbosacral spinal cord on voluntary movement, standing, and assisted stepping after motor complete paraplegia: a case study. Lancet Lond Engl. 2011;377(9781):1938–47.

86. Rejc E, Angeli C, Harkema S. Effects of lumbosacral spinal cord epidural stimulation for standing after chronic complete paralysis in humans. PLoS ONE. 2015;10(7):e0133998.

87. Rejc E, Angeli CA, Bryant N, Harkema SJ. Effects of stand and step training with epidural stimulation on motor function for standing in chronic complete paraplegics. J Neurotrauma. 2017;34(9):1787–802.

88. Rejc E, Angeli CA, Atkinson D, Harkema SJ. Motor recovery after activity-based training with spinal cord epidural stimulation in a chronic motor complete paraplegic. Sci Rep. 2017;7(1):13476.

89. Grahn PJ, Lavrov IA, Sayenko DG, Van Straaten MG, Gill ML, Strommen JA, et al. Enabling task-specific volitional motor functions via spinal cord neuromodulation in a human with paraplegia. Mayo Clin Proc. 2017;92(4):544–54.

90. Wenger N, Moraud EM, Raspopovic S, Bonizzato M, DiGiovanna J, Musienko P, et al. Closed-loop neuromodulation of spinal sensorimotor circuits controls refined locomotion after complete spinal cord injury. Sci Transl Med. 2014;6 (255):255ra133.

91. Courtine G, Bunge MB, Fawcett JW, Grossman RG, Kaas JH, Lemon R, et al. Can experiments in nonhuman primates expedite the translation of treatments for spinal cord injury in humans? Nat Med. 2007;13(5):561–6.

92. Lemon RN, Griffiths J. Comparing the function of the corticospinal system in different species: organizational differences for motor specialization? Muscle Nerve. 2005;32(3):261–79.

93. Bonizzato M, Pidpruzhnykova G, DiGiovanna J, Shkorbatova P, Pavlova N, Micera S, et al. Brain-controlled modulation of spinal circuits improves recovery from spinal cord injury. Nat Commun. 2018;9(1):3015.

94. Angeli CA, Boakye M, Morton RA, Vogt J, Benton K, Chen Y, et al. Recovery of overground walking after chronic motor complete spinal cord injury. N Engl J Med. 2018;379(13):1244–50.

95. Musienko P, van den Brand R, Märzendorfer O, Roy RR, Gerasimenko Y, Edgerton VR, et al. Controlling specific locomotor behaviors through multidimensional monoaminergic modulation of spinal circuitries. J Neurosci Off J Soc Neurosci. 2011;31(25):9264–78.

96. Radhakrishna M, Steuer I, Prince F, Roberts M, Mongeon D, Kia M, et al. Double-blind, placebo-controlled, randomized phase I/IIa study (safety and efficacy) with Buspirone/Levodopa/Carbidopa (SpinalonTM) in subjects with complete AIS A or motor-complete AIS B spinal cord injury. Curr Pharm Des. 2017;23(12):1789–804.

97. Gill ML, Linde MB, Hale RF, Lopez C, Fautsch KJ, Calvert JS, et al. Alterations of spinal epidural stimulation-enabled stepping by descending intentional motor commands and proprioceptive inputs in humans with spinal cord injury. Front Syst Neurosci. 2020;14: 590231.

98. Darrow D, Balser D, Netoff TI, Krassioukov A, Phillips A, Parr A, et al. Epidural spinal cord stimulation facilitates immediate restoration of dormant motor and autonomic supraspinal pathways after chronic neurologically complete spinal cord injury. J Neurotrauma. 2019;36(15):2325–36.

99. Peña Pino I, Hoover C, Venkatesh S, Ahmadi A, Sturtevant D, Patrick N, et al. Long-term spinal cord stimulation after chronic complete spinal cord injury enables volitional movement in the absence of stimulation. Front Syst Neurosci. 2020;30 (14):35.

100. Herrity AN, Williams CS, Angeli CA, Harkema SJ, Hubscher CH. Lumbosacral spinal cord epidural stimulation improves voiding function after human spinal cord injury. Sci Rep. 2018;8(1):8688.

101. Terson de Paleville DGL, Harkema SJ, Angeli CA. Epidural stimulation with locomotor training improves body composition in individuals with cervical or upper thoracic motor complete spinal cord injury: A series of case studies. J Spinal Cord Med. 2019;42(1):32–8.

102. Aslan SC, Legg Ditterline BE, Park MC, Angeli CA, Rejc E, Chen Y, et al. Epidural spinal cord stimulation of lumbosacral networks modulates arterial blood pressure in individuals with spinal cord injury-induced cardiovascular deficits. Front Physiol. 2018;9:565.

103. Harkema SJ, Wang S, Angeli CA, Chen Y, Boakye M, Ugiliweneza B, et al. Normalization of Blood Pressure With Spinal Cord Epidural Stimulation After Severe Spinal Cord Injury. Front Hum Neurosci. 2018;12:83.

104. Squair JW, Gautier M, Mahe L, Soriano JE, Rowald A, Bichat A, et al. Neuroprosthetic baroreflex controls haemodynamics after spinal cord injury. Nature. 2021;590(7845):308–14.

105. Ganzer PD, Colachis 4th SC, Schwemmer MA, Friedenberg DA, Dunlap CF, Swiftney CE, et al. Restoring the sense of touch using a sensorimotor demultiplexing neural interface. Cell. 2020.

106. Granat M, Heller B, Nicol D, Baxendale R, Andrews B. Improving limb flexion in FES gait using the flexion withdrawal response for the spinal cord injured person. J Biomed Eng. 1993;15(1): 51–6.

107. Peckham PH, Knutson JS. Functional electrical stimulation for neuromuscular applications. Annu Rev Biomed Eng. 2005;7:327–60.

108. Ajiboye AB, Willett FR, Young DR, Memberg WD, Murphy BA, Miller JP, et al. Restoration of reaching and grasping movements through brain-controlled muscle stimulation in a person with tetraplegia: a proof-of-concept demonstration. Lancet Lond Engl. 2017 May 6;389(10081):1821–30.

109. Kilgore KL, Hoyen HA, Bryden AM, Hart RL, Keith MW, Peckham PH. An implanted upper-extremity neuroprosthesis using myoelectric control. J Hand Surg. 2008;33(4):539–50.

110. Popovic MR, Popovic DB, Keller T. Neuroprostheses for grasping. Neurol Res. 2002;24(5): 443–52.

111. Ho CH, Triolo RJ, Elias AL, Kilgore KL, DiMarco AF, Bogie K, et al. Functional electrical stimulation and spinal cord injury. Phys Med Rehabil Clin N Am. 2014;25(3):631–54, ix.

112. Kapadia N, Masani K, Catharine Craven B, Giangregorio LM, Hitzig SL, Richards K, et al. A randomized trial of functional electrical stimulation for walking in incomplete spinal cord injury: effects on walking competency. J Spinal Cord Med. 2014;37 (5):511–24.

113. Barra B, Conti S, Perich MG, Zhuang K, Schiavone G, Fallegger F, et al. Epidural electrical stimulation of the cervical dorsal roots restores voluntary arm control in paralyzed monkeys. 2020;2020.11.13.379750.

114. Levine AJ, Hinckley CA, Hilde KL, Driscoll SP, Poon TH, Montgomery JM, et al. Identification of a cellular node for motor control pathways. Nat Neurosci. 2014;17(4):586–93.

115. Eccles JC, Eccles RM, Lundberg A. Synaptic actions on motoneurones caused by impulses in Golgi tendon organ afferents. J Physiol. 1957;138 (2):227–52.

116. Jones KE, Bawa P. Computer simulation of the responses of human motoneurons to composite 1A EPSPS: effects of background firing rate. J Neurophysiol. 1997;77(1):405–20.

117. Giat Y, Mizrahi J, Levy M. A musculotendon model of the fatigue profiles of paralyzed quadriceps muscle under FES. IEEE Trans Biomed Eng. 1993;40(7):664–74.

118. Kirsch R, Rymer W. Neural compensation for muscular fatigue: evidence for significant force regulation in man. J Neurophysiol. 1987;57 (6):1893–910.

119. Asboth L, Friedli L, Beauparlant J, Martinez-Gonzalez C, Anil S, Rey E, et al. Cortico-reticulo-spinal circuit reorganization enables functional recovery after severe spinal cord contusion. Nat Neurosci. 2018;21(4):576–88.

120. Nishimura Y, Perlmutter SI, Eaton RW, Fetz EE. Spike-timing-dependent plasticity in primate corticospinal connections induced during free behavior. Neuron. 2013;80(5):1301–9.

121. Minassian K, Persy I, Rattay F, Dimitrijevic MR, Hofer C, Kern H. Posterior root-muscle reflexes elicited by transcutaneous stimulation of the human lumbosacral cord. Muscle Nerve. 2007;35(3): 327–36.

122. de Noordhout AM, Rothwell JC, Thompson PD, Day BL, Marsden CD. Percutaneous electrical stimulation of lumbosacral roots in man. J Neurol Neurosurg Psychiatry. 1988;51(2):174.

123. Troni W, Bianco C, Moja MC, Dotta M. Improved methodology for lumbosacral nerve root stimulation. Muscle Nerve. 1996;19(5):595–604.

124. Gad P, Gerasimenko Y, Zdunowski S, Turner A, Sayenko D, Lu DC, et al. Weight bearing overground stepping in an exoskeleton with non-invasive spinal cord neuromodulation after motor complete paraplegia. Front Neurosci. 2017;11:1–8.

125. Minassian K, Hofstoetter US, Danner SM, Mayr W, Bruce JA, McKay WB, et al. Spinal rhythm generation by step-induced feedback and transcutaneous posterior root stimulation in complete spinal cord-injured individuals. Neurorehabil Neural Repair. 2016;30(3):233–43.

126. Dy CJ, Gerasimenko YP, Edgerton VR, Dyhre-Poulsen P, Courtine G, Harkema SJ. Phase-dependent modulation of percutaneously elicited multisegmental muscle responses after spinal cord injury. J Neurophysiol. 2010;103(5):2808–20.

127. Hofstoetter US, Minassian K, Hofer C, Mayr W, Rattay F, Dimitrijevic MR. Modification of reflex responses to lumbar posterior root stimulation by motor tasks in healthy subjects. Artif Organs. 2008;32(8):644–8.

128. Danner SM, Hofstoetter US, Ladenbauer J, Rattay F, Minassian K. Can the human lumbar posterior columns be stimulated by transcutaneous spinal cord stimulation? A Modeling Study Artif Organs. 2011;35(3):257–62.

129. de Freitas RM, Sasaki A, Sayenko DG, Masugi Y, Nomura T, Nakazawa K, et al. Selectivity and excitability of upper-limb muscle activation during cervical transcutaneous spinal cord stimulation in humans. J Appl Physiol Bethesda Md 1985. 2021;131(2):746–59.

130. Hofstoetter US, McKay WB, Tansey KE, Mayr W, Kern H, Minassian K. Modification of spasticity by transcutaneous spinal cord stimulation in

individuals with incomplete spinal cord injury. J Spinal Cord Med. 2013;37(2):202–11.

131. Minassian K, Hofstoetter U, Tansey K, Rattay F, Mayr W, Dimitrijevic M. Transcutaneous stimulation of the human lumbar spinal cord: facilitating locomotor output in spinal cord injury. 2010.

132. Minassian K, Hofstoetter US, Rattay F. Transcutaneous lumbar posterior root stimulation for motor control studies and modification of motor activity after spinal cord injury. In: Restorative neurology of spinal cord injury. 2011. p. 226–55.

133. Wiley J. The human central pattern generator for locomotion: does it exist and contribute to walking? (Unpublished). The Neuroscientist. 2010.

134. Estes SP, Iddings JA, Field-Fote EC. Priming neural circuits to modulate spinal reflex excitability. Front Neurol. 2017;8:17.

135. Hofstoetter US, Freundl B, Lackner P, Binder H. Transcutaneous spinal cord stimulation enhances walking performance and reduces spasticity in individuals with multiple sclerosis. Brain Sci. 2021;11(4):472.

136. Hofstoetter US, Freundl B, Danner SM, Krenn MJ, Mayr W, Binder H, et al. Transcutaneous spinal cord stimulation induces temporary attenuation of spasticity in individuals with spinal cord injury. J Neurotrauma. 2020;37(3):481–93.

137. Gerasimenko Y, Gorodnichev R, Moshonkina T, Sayenko D, Gad P, Edgerton VR. Transcutaneous electrical spinal-cord stimulation in humans. Ann Phys Rehabil Med. 2015;58(4):225–31.

138. Gad P, Lee S, Terrafranca N, Zhong H, Turner A, Gerasimenko Y, et al. Non-invasive activation of cervical spinal networks after severe paralysis. J Neurotrauma. 2018;35(18):2145–58.

139. Sayenko DG, Rath M, Ferguson AR, Burdick JW, Havton LA, Edgerton VR, et al. Self-assisted standing enabled by non-invasive spinal stimulation after spinal cord injury. J Neurotrauma. 2019;36(9):1435–50.

140. Ward AR, Chuen WLH. Lowering of sensory, motor, and pain-tolerance thresholds with burst duration using kilohertz-frequency alternating current electric stimulation: part II. Arch Phys Med Rehabil. 2009;90(9):1619–27.

141. Joseph L, Butera RJ. High-frequency stimulation selectively blocks different types of fibers in frog sciatic nerve. IEEE Trans Neural Syst Rehabil Eng. 2011;19(5):550–7.

142. Torebjörk HE, Hallin RG. Responses in human A and C fibres to repeated electrical intradermal stimulation 1. J Neurol Neurosurg Psychiatry. 1974;37(6):653.

143. Manson GA, Calvert JS, Ling J, Tychhon B, Ali A, Sayenko DG. The relationship between maximum tolerance and motor activation during transcutaneous spinal stimulation is unaffected by the carrier frequency or vibration. Physiol Rep. 2020;8(5):e14397.

144. Kumru H, Flores Á, Rodríguez-Cañón M, Edgerton VR, García L, Benito-Penalva J, et al. Cervical electrical neuromodulation effectively enhances hand motor output in healthy subjects by engaging a use-dependent intervention. J Clin Med. 2021;10(2):195.

145. Meyer C, Hofstoetter US, Hubli M, Hassani RH, Rinaldo C, Curt A, et al. Immediate effects of transcutaneous spinal cord stimulation on motor function in chronic, sensorimotor incomplete spinal cord injury. J Clin Med. 2020;9(11):3541.

146. Parhizi B, Barss TS, Mushahwar VK. Simultaneous cervical and lumbar spinal cord stimulation induces facilitation of both spinal and corticospinal circuitry in humans. Front Neurosci. 2021;15:615103.

147. Estes S, Zarkou A, Hope JM, Suri C, Field-Fote EC. Combined transcutaneous spinal stimulation and locomotor training to improve walking function and reduce spasticity in subacute spinal cord injury: a randomized study of clinical feasibility and efficacy. J Clin Med. 2021;10(6):1167.

148. Shapkova EY, Pismennaya EV, Emelyannikov DV, Ivanenko Y. Exoskeleton walk training in paralyzed individuals benefits from transcutaneous lumbar cord tonic electrical stimulation. Front Neurosci. 2020;14:416.

149. Al'joboori Y, Massey SJ, Knight SL, Donaldson N de N, Duffell LD. The effects of adding transcutaneous spinal cord stimulation (tSCS) to sit-to-stand training in people with spinal cord injury: a pilot study. J Clin Med. 2020;9(9):2765.

150. Calvert JS, Manson GA, Grahn PJ, Sayenko DG. Preferential activation of spinal sensorimotor networks via lateralized transcutaneous spinal stimulation in neurologically intact humans. J Neurophysiol. 2019;122(5):2111–8.

151. Krenn M, Hofstoetter US, Danner SM, Minassian K, Mayr W. Multi-electrode array for transcutaneous lumbar posterior root stimulation. Artif Organs. 2015;39(10):834–40.

Open Access This chapter is distributed under the terms of the Creative Commons Attribution 4.0 International License (http://creativecommons.org/licenses/by/4.0/), which permits use, duplication, adaptation, distribution and reproduction in any medium or format, as long as you give appropriate credit to the original author(s) and the source, a link is provided to the Creative Commons license and any changes made are indicated.

The images or other third party material in this chapter are included in the work's Creative Commons license, unless indicated otherwise in the credit line; if such material is not included in the work's Creative Commons license and the respective action is not permitted by statutory regulation, users will need to obtain permission from the license holder to duplicate, adapt or reproduce the material.

Functional Electrical Stimulation Therapy: Mechanisms for Recovery of Function Following Spinal Cord Injury and Stroke

19

Milos R. Popovic, Kei Masani, and Matija Milosevic

Abstract

Electrical stimulation is a tool that applies low-energy electrical pulses to artificially generate muscle contractions. If electrical stimulation is used to enable functional movements, such as walking and grasping, then this intervention is called functional electrical stimulation (FES). When FES is used as therapy instead of being used as an orthosis, it is called FES therapy or FET. In this chapter, we introduce recent findings and advances in the field of FET. The findings to date clearly show that FET for reaching and grasping is a therapeutic modality that should be imple-mented in every rehabilitation institution that is treating individuals with stroke and Spinal Cord Injury (SCI). There is also considerable evidence to support the use of FET as a therapeutic modality to treat drop-foot problem in stroke and incomplete populations. Although phase I randomized control trials have been completed with chronic SCI popu-lation using this new FET technology and preliminary findings are encouraging, further research and development are required before the multichannel FET for walking will be ready for clinical implementation. Finally, emerging evidence for the beneficial use of brain-computer interface (BCI) combined with FET (BCI-FET) for improving upper and lower limb function will also be presented.

M. R. Popovic (✉) · K. Masani
Institute of Biomedical Engineering, University of Toronto, Toronto, ON, Canada
e-mail: milos.popovic@utoronto.ca

K. Masani
e-mail: k.masani@utoronto.ca

M. R. Popovic · K. Masani
KITE Research Institute, Toronto Rehabilitation Institute – University Health Network, Toronto, ON, Canada

M. R. Popovic · K. Masani
CRANIA, University Health Network & University of Toronto, Toronto, ON, Canada

M. Milosevic
Graduate School of Engineering Science, Osaka University, Toyonaka, Osaka, Japan
e-mail: matija@ieee.org

Keywords

Functional electrical stimulation (FES) · Therapy · FES therapy · Spinal cord injury · Stroke · Rehabilitation · Neuroprosthesis · Neurorehabilitation · Neuromodulation · Restoration of voluntary function · Brain-computer interface (BCI)

19.1 Introduction

Functional electrical stimulation (FES) is a technology one can use to artificially generate body movements in individuals who have paralyzed muscles due to injury to the central nervous

© The Author(s), under exclusive license to Springer Nature Switzerland AG 2022
D. J. Reinkensmeyer et al. (eds.), *Neurorehabilitation Technology*,
https://doi.org/10.1007/978-3-031-08995-4_19

system. More specifically, FES can be used to generate functions such as grasping and walking in individuals with paralyses such as stroke and spinal cord injury (SCI). This technology was originally used to develop neuroprostheses that were implemented to permanently substitute impaired functions such as bladder voiding, grasping, and walking. In other words, a consumer would use the device each time he wanted to generate a desired function. In recent years FES technology has been used to deliver therapies to retrain voluntary motor functions such as grasping, reaching, and walking. In this embodiment, FES is used as a short-term therapy for several weeks to months, with the objective to restore voluntary function and not lifelong dependence on the FES device, hence the name FES therapy or FET. In other words, FET is used as a short-term intervention to help the central nervous system of the consumer to relearn how to execute impaired functions instead of making the consumer dependent on neuroprostheses for the rest of her/his life. In this chapter, we introduce recent findings and advances in the field of FET.

19.2 Functional Electrical Stimulation (FES)

19.2.1 Definitions

Individuals with stroke and SCI have injuries that prevent the central nervous system from generating a desired motor command and/or transmitting the desired motor command to the parts of the peripheral nervous system that innervate muscles. As a result, these individuals are frequently unable to voluntarily move different body parts and perform functions such as sitting, standing, reaching, grasping, and bladder voiding. However, as long as the peripheral nerves innervating the muscles, the muscles themselves, and the joints and soft tissues supporting the muscle-joint structures are intact, the electrical stimulation can be used to generate joint movements by contracting the muscles that actuate them. The electrical stimulation used for this

purpose is called neuromuscular electrical stimulation (NMES). An organized and patterned NMES that aims to generate coordinated limb or body movements such as grasping, standing, and walking, instead of isolated muscle contractions is called functional electrical stimulation (FES). In such a context, the FES technology is used as a prosthetic/orthotic device. In the literature, this use of FES technology is referred to as a neuroprosthesis or neuroprosthetics.

19.2.2 Physiology

In nerve cells, information is coded and transmitted as a series of electrical impulses called action potentials, which represent a brief change in cell electric potential of approximately 80–90 mV. These nerve signals are frequency modulated; that is, the number of action potentials that occur in a unit of time is proportional to the intensity of the transmitted signal. The typical frequency of action potentials is between 8 and 20 Hz [1]. An electrical stimulation can artificially elicit this action potential by changing the electric potential across nerve cell/axon membranes by inducing electrical charge in the immediate vicinity of the outer membrane of the cell (Fig. 19.1).

In most of the applications, FES activates the nerves. However, in some applications, FES can be used to directly stimulate muscle fibers if their peripheral nerves have been severely damaged (i.e., denervated muscles) [2]. The majority of the FES systems used today to stimulate the nerve trunk or the nerve ending at the neuromuscular junction. The main reason is the fact that direct muscle stimulation requires considerably more energy to generate contractions (at least three orders of magnitude more [3]), which makes these systems more challenging to implement at home and in clinical settings. Nevertheless, it should be noted that an electric stimulator that has been purposefully designed to generate contractions in denervated muscles is currently commercially available. Its name is Stimulette edition5, and it is manufactured by Dr. Schuhfried, Medical Technology, Austria (www.

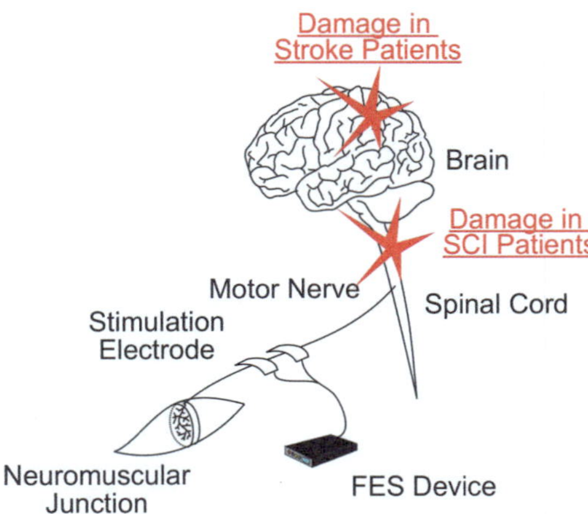

Fig. 19.1 A schematic representation of the surface functional electrical stimulation (FES) system. The FES system causes a muscle contraction by electrically stimulating the motor axons that are connected to the muscles. The electrical stimulation generates action potentials in the motor neurons, which propagate along the motor neurons toward the muscle. When the action potentials reach the muscle, they cause the muscle to contract

schuhfriedmed.at). In the remainder of this document, we will only discuss FES systems that have been developed to stimulate innervated muscles.

In some FES applications, the stimulation electrode is located on muscle bellies, while the stimulation electrode is located over superficial nerve trunk in other FES applications. While differences exist in how nerve trunk and muscle belly stimulation affect the recruitment of motor (efferent) and sensory (afferent) nerves [4], applications of FES on the muscle belly can produce selective muscle contractions to generate functional movements.

In most FES applications, the electrical charge can stimulate both motor nerves (efferent nerves —descending nerves from the central nervous system to muscles) and sensory nerves (afferent nerves—ascending nerves from sensory organs to the central nervous system). In some applications, the nerves are stimulated to generate localized muscle activity, i.e., the stimulation is aimed at generating muscle contraction orthodromically via motor nerves. In other applications, stimulation is used to activate simple or complex reflexes via sensory nerves. In other

words, the sensory nerves are stimulated to evoke a reflex, which is typically expressed as a coordinated contraction of one or more muscles. Notably, activation of the sensory system (i.e., the reflex pathway) is believed to be one of the important contributing factors for the recovery of motor function after FES [5].

When a nerve is stimulated, i.e., when sufficient electrical charge is provided to a nerve cell, a localized depolarization of the cell wall occurs resulting in an action potential that propagates toward both ends of the axon. Typically, one "wave" of action potentials will propagate along the axon toward the muscle (orthodromic propagation), and concurrently, the other "wave" of action potentials will propagate toward the cell body in the central nervous system (antidromic propagation). While the direction of propagation in case of the antidromic stimulation and the sensory nerve stimulation is the same, i.e., toward the central nervous system, their end effects are very different. The antidromic stimulus has been considered an irrelevant side effect of FES. According to Rushton [6], repeated antidromic stimulation through Hebb-type processes may over time enable sparse supraspinal commands to activate anterior motor neuron and produce desired muscle contraction(s). Typically, FES is concerned with orthodromic stimulation and uses it to generate coordinated muscle contractions.

In the case where sensory nerves are stimulated, the reflex arcs are triggered by the stimulation of sensory nerve axons at specific peripheral sites. One example of such a reflex is the flexor withdrawal reflex. The flexor withdrawal reflex occurs naturally when a sudden, painful sensation is applied to the sole of the foot. It results in flexion of the hip, knee, and ankle of the affected leg and extension of the contralateral leg in order to get the foot away from the painful stimulus as quickly as possible. The sensory nerve stimulation can be used to generate desired motor tasks, such as evoking flexor withdrawal reflex to facilitate walking in individuals following stroke, or they can be used to alter reflexes or the function of the central nervous system. In the latter case, the electrical stimulation is commonly described by the term neuromodulation.

19.2.3 Technology

Nerves can be stimulated using either surface (transcutaneous) or subcutaneous (percutaneous or implanted) electrodes. The surface electrodes are placed on the skin surface above the nerve or muscle that needs to be "activated." They are non-invasive, easy to apply, and generally inexpensive. To generate muscle contraction, the impedance, location, size, and orientation of the electrodes need to be optimized to maximize the current flow and muscle recruitment. Placing a smaller cathode electrode closer to the target nerve with the larger anode away from the cathode can be used to generate more accurate stimulation under the cathode while allowing a larger area of the skin under the anode to be used to close the electrical circuit and minimize discomfort. Empirically, we know that there are motor point locations where muscles are most sensitive to electrical stimulation. Placement of electrodes on these motor points also plays an important role in generating strong muscle contractions (for a review of upper and lower limb motor points, see Bersch et al. [7] and Botter et al. [8], respectively). Typically, smaller muscles tend to have one motor point [7], while larger muscles are known to have several (e.g., quadriceps group has seven motor points [8, 9]).

Until recently the common belief in the FES field has been that due to the electrode–skin contact impedance, skin and tissue impedance, and current dispersion during stimulation, much higher-intensity pulses are required to stimulate nerves using surface stimulation electrodes as compared to the subcutaneous electrodes. This statement is correct for all commercially available stimulators except MyndMove® stimulator (Fig. 19.2), which is manufactured by a Canadian company MyndTec (www.myndtec.com). MyndMove® has implemented a new stimulation pulse that allows the stimulator to generate muscle contractions using electrical pulses, which steady-state amplitudes are 10–15 times lower in intensity than those required by other transcutaneous electrical stimulation systems. The key aspects of this new technology are stimulation pulses that have very a fast slew rate, which is the time that a circuit needs to go from 0 to targeted amplitude within a single pulse (US Patent 20130090712), and are able to rapidly engage Aα efferent nerve fibers (i.e., descending nerves from the central nervous system to muscles) using very low stimulation amplitudes and at the same time minimize engagement of afferent Aδ and C nerve fibers responsible for the transmission of pain sensation [10]. This new technology not only reduces the intensity of stimulation, but it also reduces discomfort during stimulation, which is a common problem with commercially available transcutaneous electrical stimulation systems [10].

Typical FES systems could use different types of waveforms to stimulate the muscles. Common stimulation pulses are balanced biphasic impulses that ensure that the residual charge in the tissues is removed, and they also generate a discharge under both the anode and the cathode. Most FES systems use asymmetric balanced biphasic impulses to ensure that the muscle contractions occur only under the stimulating cathode electrode. The magnitude of muscle contractions can be varied by changing the stimulating pulse amplitude, pulse width, or pulse frequency. The pulse amplitude, or intensity, is related to the depolarizing effect, with higher amplitudes inducing a stronger depolarization. Increasing the amplitude results in the additional recruitment of smaller fibers near the electrode and larger fibers farther from the electrode [11]. The pulse width, or pulse duration, required to achieve adequate depolarization and cause muscle to contractions is typically around 200–500 µs. The frequency during FES determines the rate of action of the motor and sensory pathways. Depending on the application, a variety of frequencies can be used to generate contractions using FES, with most in the range between 20–50 Hz. Such frequencies are needed because electrical stimulation activates muscle fibers synchronously, which requires higher firing rates to generate tetanic contractions. Lower

Fig. 19.2 Use of MyndMove® to retrain upper arm voluntary movements (Photo courtesy MyndTec Inc., Toronto, ON, Canada)

frequency stimulation (<16 Hz) produces unfused contractions, while high frequency stimulation (50–80 Hz) may induce the rapid onset of muscle fatigue [12].

Since muscle belly stimulation activates localized muscle fibers around the electrodes and with a relatively high frequency, such electrical stimulation can also induce muscle fatigue [13, 14]. This is a major limitation of many electrical stimulation applications. There are multiple techniques to reduce the onset of muscle fatigue during FES, which are discussed elsewhere [14–16]. We proposed using spatially sequentially distributed electrical stimulation (SDSS), which was shown to reduce rapid muscle fatigue by 30%, compared to conventional stimulation [13, 17–19]. The SDSS method can be realized by using a simple generic adapter that can be applied to FES devices to realize an effective and low-cost solution for dealing with fatigue [20].

Another limitation of the transcutaneous electrical stimulation is that some nerves, for example, those innervating the hip flexors and the trunk or the less superficial upper extremity muscles, are too deep to be stimulated using surface electrodes. This limitation can be partly addressed by using arrays of electrodes, which can use several electrical contacts to increase selectivity [21–23].

Subcutaneous electrodes can be divided into percutaneous and implanted electrodes. The percutaneous electrodes consist of thin wires inserted through the skin and into muscular tissue close to the targeted nerve. These electrodes typically remain in place for a short period of time and are only considered for short-term FES interventions. However, it is worth mentioning that some groups, such as Cleveland FES Center, have been able to safely use percutaneous electrodes with individual patients for months and years at a time. One of the drawbacks of using the percutaneous electrodes is that they are prone to infection, and special care has to be taken to prevent such events.

The other class of subcutaneous electrodes is implanted electrodes. These are permanently implanted in the consumer's body and remain in the body for the remainder of the consumer's life. Compared to surface stimulation electrodes, implanted and percutaneous electrodes potentially have higher stimulation selectivity, which is a desired characteristics of FES systems. To achieve higher selectivity while applying lower stimulation amplitudes, it is recommended that both cathode and anode are in the vicinity of the nerve that is stimulated [24]. The drawbacks of the implanted electrodes are they require an invasive surgical procedure to install, and as is the case with every surgical intervention, there exists a possibility of infection following implantation.

19.3 FES Therapy (FET)

19.3.1 Definition

FES can be used for neuroprosthetic and therapeutic purposes. If FES is used as a neuroprosthesis, the purpose of this device is to generate a body function that the consumer is unable to perform alone, such as walking, biking, bladder voiding, grasping, etc. In these applications, the FES system needs to be worn or used each and every time the consumer needs to perform the desired function. In essence, the consumer uses the FES device as a permanent orthotic system. Good examples of these FES systems are neuroprostheses for rowing and biking. Each time the consumer wants to row or bike he needs to use the neuroprosthetic system, without which he would not be able to perform this task at all. Examples commercially available FES biking technologies are the RT300 FES bike from Restorative Therapies (www.restorative-therapies.com) and RehaMove by Hasomed (www.hasomed.de).

The implanted FES systems are primarily used as permanent neuroprostheses. However, some attempts have been made to use the BION implantable FES system for FET [25]. On the other hand, the surface FES systems have been used equally well as neuroprostheses and platforms to deliver FET. In the past, the main focus of the FES field was on developing neuroprosthetic systems, in particular those that patients had to use daily. In recent years, the advances made in the field of FET and the use of neuroprostheses for muscle strengthening and cardiovascular exercises have shifted the focus of the FES field, at least partially, toward the use of surface FES systems. As a result, a number of commercially available surface FES systems have been developed in last decade.

The use of neuroprostheses as a means of providing short-term therapeutic intervention for improving and restoring voluntary function has been termed FES therapy or FET [26]. When the FES technology is used to deliver FET, the purpose of that intervention is to restore voluntary function. In other words, FES is used only temporarily as a short-term intervention with the objective of helping the neuromuscular system relearn to execute a function impaired due to neurological injury or disorder. In this application, the ultimate goal of the FES intervention is for the consumer to recover voluntary function, as much as possible, so the consumer does not need to use the FES system for the rest of her/his life. In this application, the central nervous system essentially relearns how to control the impaired muscles and how to contract them in a temporarily appropriate manner to generate the desired body function. Since FET systems are generally non-invasive and are used to produce diverse upper or lower limb movements/therapies, FET-dedicated systems can have many more stimulation protocols (e.g., ten or more for upper limb FET) that at times target different muscle groups and can be used with a single consumer. However, the neuroprostheses that are used as permanent orthotic systems often target one set of muscles or muscle groups and have one or at best two/three consumer-specific stimulation protocols.

19.3.2 Neuroplasticity and Carry-Over Effect After FET

Since the 1970s, some researchers and practitioners in the field of FES have observed that many patients who use FES on a regular basis experience significant carry-over in function that persists even when the device is not in use. This "enigma" of "carry-over effect" has interested researchers [27], even though most of these reports were anecdotal in nature at the beginning.

One of the first papers that specifically discussed this phenomenon was an article authored by Merletti et al. in 1975 [28]. They investigated the carry-over effect of FES on hand opening and elbow extension functions for stroke patients. Three of five patients showed the carry-over effects after a 2-month training period, i.e., after the FES intervention session, functional tasks such as the shifting of an object between two specified areas on a desk were improved even without wearing the FES device. The observed carry-over effect supported the potential role of neuroprostheses as therapeutic interventions in clinical practice. Despite the fact that FES-related carry-over results were observed as early as the 1970s, a rigorous investigation of FES carry-over effect started only recently. Currently, compelling evidence exists that the therapeutic application of FES over a period of time could help individuals with neurological impairments regain some of the voluntary function after the intervention. By taking advantage of this, FES has been used to restore voluntary upper limb movements in individuals with neurological injuries through FET. Recent FET results suggest that long-term clinical benefits could be obtained after using FES systems [29, 30]. It was believed that clinical changes after FET could partially be due to the muscle training and strengthening as well as the improved flexibility and range of motion of the affected limbs. In addition to these peripheral effects, spasticity reduction after FET is believed to be one of the main reasons for clinical improvements in motor function [31–33]. Specifically, it was shown that even short duration application of FES at the levels sufficient to generate muscle contractions can inhibit spinal excitability in multiple muscles after the stimulating period [33]. In addition, compelling evidence also suggests that cortical reorganization takes place after FET [34–36] and that approximately 36–40 h of training may be required for cortical reorganization [36]. Evidence of neuroplasticity after FET implies that task-specific and repetitive FET helps guide the central nervous system after neurological injury to relearn to control muscles by creating new control centers in the brain [36].

19.4 Current Evidence of FET Effectiveness

It took almost two decades to start seriously examining the carry-over effects of FET. As described next, it was first examined with the drop-foot FES systems, where scientists explored the ability of the system to restore voluntary walking function in individuals with stroke. These studies were then followed by investigations examining the use of a neuroprosthesis for grasping and, later, neuroprostheses for reaching and grasping for restoring voluntary arm and hand functions in individuals with stroke and SCI. Finally, the neuroprosthesis for walking was used to investigate the restoration of voluntary walking function in individuals with incomplete SCI.

Initially, FET did not exist as a field on its own, and the first FET studies were essentially examining carry-over effect of the neuroprostheses. Once, it becomes clear that FET is actually helping reprogram the central nervous system and that the carry-over effect is not due to the muscle strengthening (which was initially suspected [37]) but was due to neuroplasticity, the FET field has been established and FET-dedicated systems started being developed. The systems used to test FET concept were originally neuroprostheses that were normally used as orthoses. Today we are experiencing the development of FET-dedicated systems, which design requirements are very different from the "garden variety" neuroprosthetic systems developed for orthotic applications. In a recent article, we summarized practical considerations that therapies should follow for applying the Toronto FET intervention [38], which can improve motor function after stroke and SCI.

19.4.1 FET for Restoration of Lower Limb Function Following Stroke

Among stroke patients, the drop-foot is a common symptom, characterized by a lack of dorsiflexion during the swing phase of gait, resulting in short, shuffling strides. It has been shown that the drop-foot stimulator effectively compensates for the drop-foot during the swing phase of the gait. At the moment just before a heel off phase of the gait occurs, the drop-foot stimulator induces a stimulus at the common peroneal nerve, which results in contraction of the muscles responsible for dorsiflexion (Fig. 19.3). There are a number of drop-foot stimulators, which use surface FES technology and have been FDA (US Food and Drug Administration) approved, that have been developed to date: the Odstock® Dropped Foot Stimulator (ODFS® Pace) by Odstock Medical (www.odstockmedical.com) [39], the Walk-Aide® by Innovative Neurotronics (www.walk-aide.com) [40], and the NESS L300 Go for Foot Drop by Bioness (www.bioness.com) [41]. The ActiGait® by Ottobock (www.ottobock.com) [42] and the STIMuSTEP® by Finetech Medical (www.finetech-medical.co.uk) [43] are implantable drop-foot stimulators that are also commercially available and have the CE mark in Europe. Drop-foot stimulators are one of the most successful neuroprostheses to date after cochlear implants. Overall, consumer perception of the drop-foot stimulators is they are superior to the ankle–foot orthosis [44].

There has been a great deal of evidence showing the benefits of the drop-foot FES for the lower limbs of stroke patients. In most of the studies, the effect of the drop-foot stimulator as an orthosis has been studied. Only few studies have investigated the FET effect in stroke patients with drop-foot problem (e.g., [45]). In the early phase, some studies showed a negative result with respect to the FET effect [46, 47], while other studies showed positive effect on the FET effect [39]. For example, Granat et al. [47] investigated the effect of a drop-foot stimulator on hemiplegic patients ($n = 19$) in a two-period crossover study design (4-week control period followed by 4-week FES treatment period). The results demonstrated that there was a significant orthotic effect (positive effect when the subject was using the FES system) in inversion of ankle, while the same study did not show a therapeutic effect (positive effect when the subjects were not using the FES system, i.e., FET effect). In a randomized controlled trial, Burridge et al. [46] investigated the effect of a drop-foot stimulator on individuals with stroke. The intervention group ($n = 16$) received conventional physiotherapy and FES treatment, while the control group ($n = 16$) received conventional physiotherapy alone. They demonstrated that the mean increase in walking speed was 20.5% in the intervention group when the subjects in that group used the drop-foot stimulator as an orthosis. The control group showed only a 5.2% increase in mean walking speed. The physiological cost index (PCI) was reduced 24.9% in the intervention group when they were using the drop-foot stimulator as an orthosis and was reduced 1% in the control group. However, the same study did not show any improvements in the intervention group when the drop-foot stimulator was removed. In other words, they were not able to demonstrate the drop-foot stimulator's FET effect. Taylor et al. [39] investigated the effect of a drop-foot stimulator in stroke ($n = 9$) and multiple sclerosis (MS) ($n = 2$) patients. Stroke patients showed a mean increase in walking speed of 27% and a reduction in PCI of 31% when the system was used as an orthosis. However, the same study showed a 14% increase

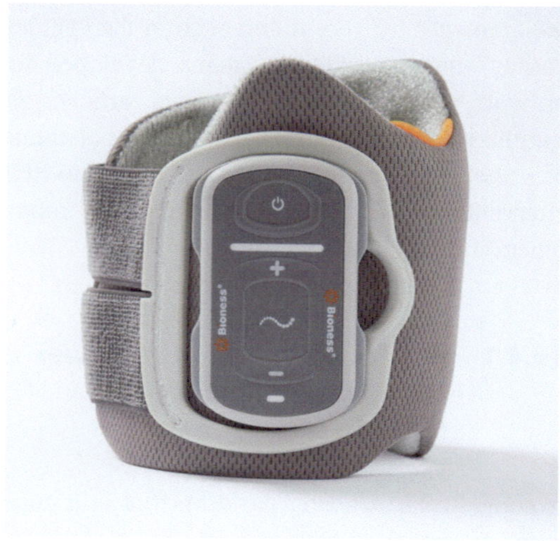

Fig. 19.3 NESS L300 Go for foot drop (Photo courtesy Bioness Inc., Valencia, CA, USA)

in walking speed and a 19% reduction in PCI, when the stimulator was removed from the patients, i.e., FET effect. The MS patients showed similar benefits when they used the drop-foot stimulator as an orthosis, with no noticeable FET effects.

In a relatively larger population study, Stein et al. [40] investigated the effect of a drop-foot stimulator in stroke ($n = 41$) and MS ($n = 32$) patients. They demonstrated that both stroke and MS patients showed increased walking speed when the system is used as a therapeutic and orthotic devices. After 3 months of drop-foot stimulator training, both groups had a similar and significant orthotic (increments of 5.0% and 5.7% for stroke and MS patients, respectively) and FET (17.8% and 9.1% for stroke and MS patients, respectively) effects on walking speed, during over ground walking. After 11 months of following the baseline, the FET effect on speed diverged between the two groups to 28.0% and 7.9% for stroke and MS patients, respectively. Overall, PCI showed a decreasing trend. They concluded that both subject groups had an orthotic benefit from FES for up to 11 months. The FET effect increased up to 11 months in stroke patients, which is a nonprogressive neurologic disorder, while in the MS patients, as expected, the therapeutic effect increased only in the first 3 months following the baseline.

In summary, there is considerable evidence that the drop-foot stimulators, if they are used to deliver FET, produce lasting positive changes in gait in individuals with stroke.

19.4.2 FET for Restoration of Lower Limb Function Following SCI

Impairment in lower limb function is a common symptom following SCI. Various FES systems have been developed to help individuals with SCI to improve walking function. In individuals with SCI, the scope of impairment is not limited to the ankle joint, as is the case with many stroke individuals, but rather affects many muscles in the legs, pelvis, and trunk. Thus, the FES technology for walking for individuals with SCI is more diverse and targets the muscles of the entire lower limb. However, it is not uncommon that in some individuals with SCI, the above-discussed drop-foot stimulators have been also used as a means to assist with gait.

As early as the 1960s, Kantrowitz demonstrated paraplegic standing by applying continuous electrical stimulation to the quadriceps and gluteus maximus muscles of a patient with complete SCI, using surface FES technology [48]. This earliest neuroprosthesis for paraplegic "gait" provided continuous stimulation to the quadriceps to produce a mode of gait similar to long leg-brace walking, by inducing stiffened legs. Later systems used alternating bilateral quad/glut stimulation (during stance phase) out of phase with peroneal nerve stimulation to induce the flexor withdrawal reflex (during swing phase) [49]. Following that, Kralj et al. described a technique for paraplegic gait using surface electrical stimulation, which remains the most popular method in use today [50]. Electrodes are placed over the quadriceps muscles and peroneal nerves bilaterally. The user controls the neuroprosthesis with two pushbuttons attached to the left and right handles of a walking frame, or on canes, or crutches. When the neuroprosthesis is turned on, both quadriceps muscles are stimulated to provide a standing posture. The left button initiates the swing phase in the left leg by briefly stopping stimulation of the left quadriceps and stimulating the peroneal nerve. This stimulation is applied suddenly, so as to trigger the flexor withdrawal reflex, resulting in simultaneous hip and knee flexion, as well as dorsiflexion. After a fixed period of time, peroneal nerve stimulation is stopped, and quadriceps stimulation is initiated, while the reflex is still active to complete the stride. Similarly, the right button initiates the swing phase in the right leg.

As microprocessor technology developed, neuroprostheses for walking became more portable and flexible. Examples of this type of neuroprosthesis are Parastep [51, 52], HAS [53], and RGO [54] and the Case Western Reserve University (CWRU)/VA neuroprosthesis [55–58]. The Parastep system is one of the most popular products and uses Kralj's technique [51, 52]. The HAS and the RGO walking neuroprostheses are devices that, in addition to FES, also apply active and passive braces, respectively. The braces were introduced to provide additional stability during standing and walking and to conserve the user's energy. CWRU/VA neuroprosthesis is an implant system [55–58]. Parastep, HAS, and RGO systems were designed for orthotic use; however, they could be potentially implemented as FET devices as well.

The above neuroprostheses for walking apply the flexor withdrawal reflex to generate stepping movement during the walking cycle. There is a disadvantage in using this approach as the flexor withdrawal reflex is highly variable and is subject to rapid habituation. The reflex also may not activate the hip in patients with very limited strength/mobility. However, there are systems that do not use the flexor withdrawal reflex, instead they stimulate muscles in a manner that is as close as possible to the physiologically correct muscle activation pattern that generates the bipedal walking cycle. Good examples of such systems are the Case Western Reserve University (CWRU)/VA neuroprosthesis [55–58], Praxis [59], and Compex Motion neuroprosthesis for walking [60, 61]. The Praxis and CWRU/VA neuroprosthesis are implantable FES device systems that have 22 and 8–16 stimulation channels, respectively. They are able to generate sit-to-stand, walking, and stand-to-sit functions and are suitable to orthotic applications. However, recently the Cleveland team tested the therapeutic effects of their implantable system in a single-subject study [55].

Complex Motion neuroprosthesis for walking is an 8–16 channel surface FES system used to restore walking in stroke and SCI individuals [60]. The system uses a push button control strategy, similar to the one used in the Parastep system, and a gate phase detection sensor [62] to trigger the FES sequences. What is unique about this FES system is that it was specifically developed for FET applications. The benefits of FES for lower limbs of individuals with incomplete SCI were discussed in a review by Bajd et al. [63]. The review concluded that there are various benefits including therapeutic effect of FES for individuals with SCI and of strength training, drop-foot stimulator, and plantar flexor stimulation during gait phase.

In addition to those studies, Wieler et al. [64] investigated, in a multicenter study, the effect of a drop-foot stimulator and a withdrawal reflex stimulator on individuals with SCI ($n = 31$) and with cerebral impairment ($n = 9$). The results showed that the walking speed increased by approximately 40% when the drop-foot stimulator was used as an orthotic device and 20% as when it was used as FET device. Similar findings have been published by Field-Fote and her team [65, 66].

Thrasher et al. [61] investigated whether patterned stimulation on individual muscles would have greater rehabilitative potential than the stimulation of flexor withdrawal reflexes. Specifically, they investigated the effect of a gait-patterned multichannel FES in five individuals with chronic, incomplete SCI. These subjects were trained for 12–18 weeks using Compex Motion multichannel neuroprosthesis for walking (Fig. 19.4). All subjects demonstrated significant improvements in walking function over the training period. Four of the subjects achieved significantly increased walking speeds, which were due to increases in both stride length and step frequency. The fifth subject experienced a significant reduction in preferred assistive devices. The results suggest that the proposed FES-based gait training regimen was effective for improving voluntary walking function in a population for whom significant functional changes are not expected and that this application of FET is viable for restoration of voluntary gait in incomplete SCI.

Inspired by Thrasher et al. [61] results, Toronto team carried out phase I randomized control trial in which they compared the gait-patterned multichannel FET against an equal dose of convectional exercise [67–69]. Patient population was incomplete chronic SCI individuals. The results of the study suggested that 40 h of exercise and 40 h of multichannel FET both generated clinically meaningful improvements in this patient population. At the same time, the

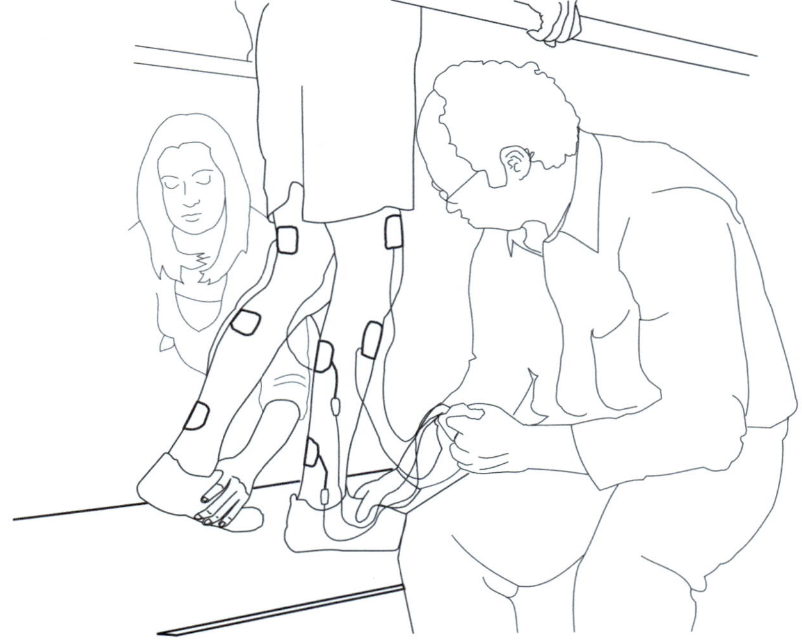

Fig. 19.4 Conceptual illustration showing a participant during the multichannel FES intervention supported by the therapist

differences between the two groups were minimal, meaning that FET in this patient population did not generate superior outcomes compared to the control group. However, it should be noted that the Spinal Cord Independence Measure (SCIM) Mobility Subscore improved in FET group significantly more than in the control group [69].

In summary, there is mounting evidence that, in individuals with incomplete SCI, neuroprostheses for walking can be used as FET devices to improve voluntarily walking function. Most of the work has been done using drop-foot stimulators. However, more complex gait-patterned multichannel FES systems have been recently tested as FET systems and have shown encouraging results with respect to improving voluntary walking function in more severely disabled individuals with SCI.

19.4.3 FET for Restoration of Upper Limb Function Following Stroke

Impaired reaching and grasping functions are common symptoms among stroke patients. Numerous neuroprostheses have been designed to compensate for lost grasping [70–81] and grasping and reaching [26, 29, 60, 80, 82] functions in stroke patients.

Some notable grasping and/or reaching neuroprostheses are the Freehand system [24], the NESS H200 for Hand Paralysis by Bioness (www.bioness.com) [74] (Fig. 19.5), the Bionic Glove [75, 80, 83], the ETHZ-ParaCare neuroprosthesis for grasping [60, 84, 85], the systems developed by Rebersek and Vodovnik [81], the Belgrade Grasping-Reaching System [86], Compex Motion neuroprosthesis for reaching and grasping [60], the percutaneous systems by Chae et al. [71, 72], and recently MyndMove® by MyndTec (www.myndtec.com) [87]. The above neuroprostheses for grasping were shown to restore the power grasp and the precision grip. The power grasp is used to hold larger and heavier objects between the palm of the hand and the four fingers. During a power grasp, the object is held in a clamp formed by partly flexed fingers and the palm counter pressure being applied by the thumb lying more or less in the plane of the palm. Precision grip is used to hold smaller and thinner objects, such as keys and paper, between the thumb and forefinger. The precision grip is generated by flexing the fingers followed by opposition of the thumb. In addition to these two grasping styles, Compex Motion neuroprosthesis and MyndMove® system offer a variety of additional grasping styles, such as pinch grasp, lumbrical grasp, tripod grasp, and proper hand opening that involves activation of the intrinsic muscles of the hand. The Belgrade Grasping-Reaching System, Freehand system, Compex Motion system, and MyndMove® also offer reaching capabilities. Of these systems, Mynd-Move® offers the largest diversity of grasping and/or reaching tasks that can be performed with a single FES system. The Freehand system is an implantable FES system designed for individuals with SCI, while the remaining devices are surface FES systems that can be used to deliver FET.

The use of FES as means of improving hand function following stroke has been intensively studied for a long time. A meta-analysis in 1996 already proved that FES is effective in the recovery of muscle strength after stroke [88]. Recent studies that have specifically examined FET have suggested positive outcomes in acute [26, 75, 78, 82] and chronic [74, 76, 77, 83] stroke patients. These were then followed by randomized control trials that confirmed the positive outcomes of FET in acute [29, 70, 89] and chronic [29, 71] stroke patients. In most of the discussed studies, surface FES technology

has been used to deliver FET, while a percutaneous FES system has been used in studies published by Chae et al. [71, 72]. In most studies the upper limb FET has been delivered in a clinical setting with the assistance of therapists. However, a self-administered FET intervention, i.e., those that were conducted at home, has been recently explored using the NESS system [23] and a new version of the Bionic Glove [75, 83, 90].

It is important to mention that, to date, most of the clinical trials conducted using FET for grasping in the stroke population targeted individuals who had partially preserved reaching and/or grasping functions. Namely, the targeted patients typically had Chedoke McMaster Stages of Motor Recovery scores 4 and 5 or Upper Extremity Fugl-Meyer Assessment Score greater than 30, which means that they were able to place the hand voluntarily within at least 20–30% of the hand/arm workspace and were able to initiate some or many wrist, hand, and finger movements. However, recently in randomized controlled trials, Popovic and colleagues [29, 82] as well as Hebert et al. [91] investigated the use of FET for reaching and grasping in severe stroke patients, i.e., stroke patients who had Chedoke McMaster Stages of Motor Recovery scores 1 and 2 or Upper Extremity Fugl-Meyer Assessment Score $\leq$ 15. These individuals were unable to initiate or execute voluntarily any component of reaching or grasping function. Popovic et al. have shown that the FET is able to improve both reaching and grasping functions in severe stroke patients [29]. The median improvement achieved in this study in the FET group was 24.5 points on the Upper Extremity Fugl-Meyer Assessment, while the median improvement in the control group (received conventional occupational therapy and physio-therapy without FES therapy) was 0 [29]. Hebert et al. [91] study has shown similar improvement, where 7.1 points on the Upper Extremity Fugl-Meyer Assessment were achieved in chronic stork patients after only 20 h of FET.

It is worth mentioning that a small study with chronic pediatric stroke patients has been carried out where FET was used to improve reaching and

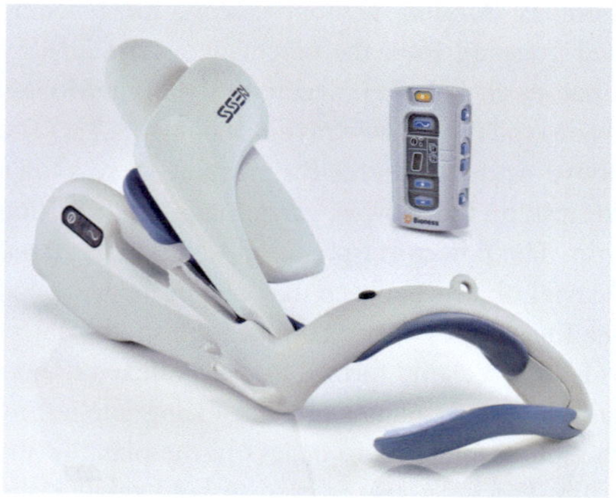

Fig. 19.5 NESS H200 for hand paralysis (Photo courtesy Bioness Inc., Valencia, CA, USA)

grasping function in this patient population [92]. Although only four individuals participated in this pilot study, the outcomes achieved were very encouraging, and they indicated that FET for upper limb could be effectively delivered in pediatric patients.

In summary, there is mounting evidence that in individuals with moderate and severe upper limb deficit, which results from stroke, FET can enable substantial improvement in their voluntary upper limb function. Also, these studies suggested that the improvements achieved are long lasting [36].

19.4.4 FET for Restoration of Upper Limb Function Following SCI

A SCI at a T1 level or above frequently results in a partial or complete loss of grasping and reaching functions. Various therapies, surgical interventions, and/or devices have been proposed to help improve those functions in individuals with SCI. Among these interventions, FES devices have shown the most promise [93]. The same neuroprostheses for grasping and reaching as discussed above have been used with the SCI population. However, almost all these devices, except for Bionic Glove, ETHZ-ParaCare neuroprosthesis, Compex Motion system, and MyndMove®, have been used with SCI subjects almost exclusively as orthotic systems and were all efficacious as orthoses.

While the benefit of FET has been intensively investigated with stroke patients, it has not been investigated as intensely with individuals who have SCI. From the above-listed FES systems that were used to deliver FET in individuals with SCI, ETHZ-ParaCare and Compex Motion systems were able to deliver both palmar and lateral grasps using the same electrode configuration. The ETHZ-ParaCare grasping neuroprosthesis was primarily used as an orthotic system. However, Mangold et al. [94] provided some evidence that a few of the SCI patients who used the device experienced a weak FET effect. A clinical trial using Bionic Grove showed that the Bionic Glove can considerably improve upper limb function in individuals with C5–C7 SCI. This study was conducted by Popovic et al. (not the author of this article) and presents the first concrete evidence that FET for grasping could be effective in SCI population [92].

In 2006, the first randomized controlled trial was carried out carefully examining the impact of FET on grasping function in individuals with traumatic C4–C7 SCI [95]. In this study, the individuals received 40 1-h FET treatments (intervention group) or 40 1-h conventional occupational therapy treatments (control group). The therapy was tested on individuals with complete and incomplete subacute (<6 months) SCI. Although this particular study was underpowered, it provided clear evidence that both individuals with complete and incomplete subacute SCI greatly benefited from the FET for grasping. This study was then followed by another phase II randomized controlled trial; FET for grasping was evaluated in individuals with incomplete, traumatic subacute C3–C7 SCI [30]. What is relevant to mention is that this was a very conservative study with respect to FET. In this study, both control and intervention groups received 1 h of conventional occupational therapy daily, as described in [95]. Then both groups were given at least a 2-h break followed by another dose of therapy where the control group got 1 h of conventional occupational therapy, and the intervention group received 1 h of FET for grasping. Both groups received therapy 5 days a week (working days) for 8 weeks (40 session days in total). At the end of the study, there were 12 subjects in the intervention group and nine in the control group. The results obtained were statistically significant and have revealed that FET dramatically improved hand function in this patient population. Also, the long-term follow-up in this study has shown that 6 months after the baseline assessment, both control and intervention groups maintained or further improved their hand function as compared to the assessments performed at discharge from the study [30]. In other words, this study suggests that the changes in the hand function produced by FET are dramatic, and they persist over time. Recently, a phase I randomized control trial study was performed using FET for grasping in chronic (>24 months) incomplete SCI individuals [96]. Forty 1-h sessions of FET (intervention group) were compared against 40 1-h sessions of conventional occupational therapy (control group). The results of the study showed that the individuals who received FET improved considerably better than the individuals who had the same dose of conventional occupational therapy.

In summary, there is mounting evidence that individuals with incomplete C3–C7 SCI, both chronic and subacute, can benefit from the FET for grasping. The existing studies also suggest that early engagement in the FET would result in better outcomes compared to later engagement. Also, a recently published study suggested that simple increase in intensity of conventional therapy is not able to match outcomes that were achieved with FET [97], further confirming that FET for grasping should be considered the new best practice with respect to incomplete SCI population. As for the complete SCI individuals, there is weak evidence that FET is beneficial for that population as well, if it is used early during subacute phase of rehabilitation.

19.5 Hybrid FET

19.5.1 Hybrid FET with Orthoses or Robotic Devices

In the past, it has been shown that FES-assisted walking has several limitations such as muscle fatigue, reduced joint torques generated using FES alone as compared to volitionally activated torques in healthy subjects, modified reflex activities, and spasticity [98]. To overcome these limitations, a combined use of FES and a mechanical brace or an orthosis has been suggested. These systems are better known as hybrid assistive systems (HAS) or hybrid orthotic systems (HOS) [53, 99, 100]. Such mechanical supports have been used mainly for safety and prevention of adverse events during standing and gait [98].

In recent years the rehabilitation robotics field has experienced rapid growth. Instead of being passive orthotic systems or braces, rehabilitation robots now have active joints and are used to help move upper and lower limbs in a physiologically correct manner, mimicking proper reaching and walking functions, respectively. Similarly, FET has been used to allow patients to execute various repetitive upper and lower limb tasks. Since both technologies have advantages and disadvantages, it was only natural to consider merging these technologies as means to overcome the disadvantages and benefit from the advantages that these two technologies offer. For example, FES systems are currently unable to generate very accurate limb movements but are able to engage flaccid and spastic muscles in task execution and generate much more significant proprioceptive and sensory feedback, which is critical for retraining the neuromuscular system. Specifically, Takeoka et al. [101] demonstrated that muscle spindle feedback is critical and probably essential for the functional recovery following SCI. On the other hand, robotic systems are very good in executing accurate limb movements, but, in general, these systems themselves do not generate muscle activations. The FES systems are able to achieve that, although not as good as the intact central nervous system does. The robotic systems, because of the nature of this technology, have neither capability to produce desired muscle tension nor are able to regulate muscle tension as a function of joint angle. In robotic systems, the more substantial afferent feedback can be produced if the consumer has tone. However, it is not clear if the afferent feedback produced under such circumstances matches the one that the intact central nervous system would naturally produce. Therefore, it has been suggested that the combination of FES with robotic devices will enhance the therapeutic effects of both interventions. A recent study by Freeman et al. [102] has proposed a robotic device for reaching movement with upper limbs that can be combined with FES. The study tested and confirmed the accuracy of the trajectory that the robotic system executed with 18 healthy subjects using FES applied to the triceps muscle. The results confirmed the efficacy of a combined robotic device and FES system and showed the feasibility of the proposed device. The same authors started to test the system with five stroke patients in treatment sessions comprised of up to 25 1-h visits. For walking, Stauffer et al. [103] developed a hybrid robotic and FES system (WalkTrainer). The robotic device consisted of leg and pelvic orthoses, active bodyweight support, and a mobile frame that allowed the user to perform walking therapy during overground walking. The system also had a closed-loop controlled FES system. This system was tested with six paraplegic patients, and its feasibility as a rehabilitation tool was confirmed.

Recently, a new hybrid robotics-FET system has been proposed for the restoration of grasping and reaching after stroke [104]. The system combines ALEX (an upper limb exoskeleton), which provides the reaching support [105], together with a FES system that uses electrode arrays to provide grasp control. Real reaching and grasping tasks can be achieved by using a satellite robot, which presents the objects to be grasped. Specific rehabilitation tasks can be implemented by taking advantage of the possibility to quantify the support needed by patients and to modulate both the mechanical and FES support over the reachable workspace.

Hybrid rehabilitation systems, consisting of a robotic device and an FES system, are not a new idea. However, this idea has become a more attractive and realistic solution in recent years. It is very likely that in the near future, we will see more devices that are combining FES and robotic technologies to develop advanced neurorehabilitation tools and interventions.

19.5.2 Comparison of FET and Robotic Therapies

To the best of our knowledge, a proper comparison of the FET and robotic therapy was not conducted to date. The only comparison that we are aware of is the one conducted by Hess et al. [106], where Bi-Many-Track system (Reha-Stim, Germany) (www.reha-stim.de) was compared to electrical stimulation of the wrist extensor muscles. The study was performed in subacute stroke individuals (between 4 and 8 weeks following stroke) patients, which Upper Extremity Fugl-Meyer scores were less than 18. Bi-Many-Track was used to deliver therapy to the wrist (flexion/extension and pronation/supination), elbow (flexion/extension), and indirectly to shoulder (flexion/extension). The electrical stimulation was delivered to wrist extensors only and was activated manually or using biofeedback approach. Although both therapies were delivered over 30 sessions that were 20 min long (10 h of therapy in total), Bi-Many-Track delivered between 12,000 and 24,000 movement repetitions (spread over different joints), and electrical stimulation delivered between 1,800 and 2,400 wrist flexion/extension repetitions. Please note that the electrical stimulation intervention used in this study does not belong to the FET variety of therapies but rather to a muscle strengthening type of interventions. The study results suggest that at discharge, participants who received Bi-Many-Track had improvement in Upper Extremity Fugl-Meyer scores of 16.7 points, while the participants who received electrical stimulation had improvement in Upper Extremity Fugl-Meyer scores of 3.1 points.

We are hopeful that this study will inspire the research community to start comparing equal dose FET and robotic therapy, which are training the same joints and muscle groups, and are delivering equal dose/intensity of intervention.

19.6 Brain-Computer Interface (BCI) Controlled FET

19.6.1 Definition

It is believed that the neuroplasticity induced by FET is mainly due to the involvement of the voluntary intent to perform a task that is supported by FES (see below). During FET, subjects attempt a movement, and only after the therapist gives a patient time to try the task, FES is applied to assist movement completion. When patients attempt the movement, descending motor commands are sent from the brain to the same sensorimotor networks that FES also activates. Repetitive associative stimulation of these same networks is likely responsible for the FET-induced neuroplasticity. However, manual control of FES by the therapist does not ensure that cortical activations are initiated before FES. We propose that brain–computer interface (BCI) systems can be used to synchronize cortical commands and movements generated by FES to elicit neuroplasticity.

19.6.2 Technology

Non-invasive BCI recording and processing advances have aided novel closed-loop applications for controlling FES. While most BCI-FES systems are focused on the idea of restoring movements through control of muscles after SCI [107, 108], recent advances in the field have expanded toward using BCI-controlled FES for improving motor function through FET. In particular, BCI systems can translate brain signals into motor outputs, which can also effectively synchronize cortical commands and movements generated by FES. Synchronized activations of cortical and peripheral networks may also facilitate associative Hebb-type learning. Rehabilitation applications in stroke and SCI patients have demonstrated compelling evidence suggesting cortical neuroplasticity and improved motor function after use of BCI-FET.

Typical BCI-FET systems consist of a "brain switch" with non-invasive brain recordings to detect and trigger a pre-programmed FES sequence. Most BCI-FET applications use binary switch control to detect rest and active (i.e., movement) states since decoding the many different movements required for FET is still challenging using non-invasive recordings. Operation of BCI-FET is typically divided into two steps:

(1) calibration of the BCI system to develop an algorithm for detecting rest and active states, which is performed prior to the intervention and (2) control of BCI-FET in real-time during the intervention. During the calibration, motor imagery-based tasks are presented, or participants attempt to perform movements, while recording synchronous brain activity. Offline analysis is performed to develop a classifier using machine learning approaches or simply by detecting event related desynchronization (ERD) of oscillatory brain activity based on the signal power changes during movements states. The ERD activity represents excitability of the primary motor cortex [109] and spinal motoneurons and has subject-specific characteristics [110–112]. It is therefore important to select the spatial (channel locations) and oscillatory (frequency) characteristics that represent ERD activity, but also to keep in mind the neurobiological constraints such as the affected side of the brain after stroke. After the system is calibrated, FES can be operated in real-time using the BCI-triggered FES. The users typically attempt a movement or perform motor imagery while the algorithm detects the state of the decoder (e.g., rest of active), after which FES is activated. Motor imagery was shown to be effective to elicit muscle-specific excitation in the central nervous system [113], which can be paired with activation of the same muscles during FET. Motor imagery-based two-state threshold decoder was shown to have a success rate of approximately 85% [112], which is comparable to machine learning-based approaches [114, 115]. Our results have also shown that a BCI-FES intervention can elicit rapid neuroplasticity much more effectively compared to FES alone in the muscle groups targeted by the intervention in able-bodied individuals [112]. This supports the notion that BCI-FES systems may also be able to upregulate specific cortical and/or corticospinal connections affected by stroke or SCI through therapeutic interventions.

19.6.3 BCI-FET for Restoration of Upper and Lower Limb Function Following Stroke and SCI

Most BCI-FET applications have been applied in the stroke population, with primarily single-subject case studies [110, 116–119]. These studies served the purpose to provide the proof of concept that BCI-FET technology can improve motor function [116, 119] or investigate the possible mechanisms of improvements, e.g., single-subject crossover design showed marked lateralization of cortical activations after BCI-FET in a stroke patient [117]. Unlike most systems that used BCI-FET to enable single-joint movements, our team showed that a threshold-based "brain switch" BCI-FET can be used to enable a series of functional movements during therapy to enable activity-dependent plasticity and improve motor function in stroke patients after 40 training sessions [110, 111, 118]. The Toronto BCI-FET system and the flow chart that illustrates the system operation were presented by Jovanovic et al. [111] with a summary in Fig. 19.6.

Recently, several more extensive studies have confirmed the efficacy of BCI-FET to improve upper limb motor function after stroke [115, 120, 121]. In a randomized controlled trial, Kim et al. [121] demonstrated superior improvements using BCI-FET compared to FES training, which was applied to enable upper limb functional movements over the course of four weeks that included a total of 20 training sessions. Similarly, in a preliminary study in stroke patients with severe upper extremity paralysis, Li et al. [120] showed superior functional recovery after eight weeks of upper limb BCI-FET, compared to the use of FES alone which was accompanied by increased bilateral and sensorimotor cortex activations. The most comprehensive upper limb trial in stroke patients was performed by Biasiucci et al. [115] over the course of five weeks that included two interventions per week. This trial showed that the BCI-FET stroke group exhibited longer-lasting functional changes compared to the FES group, which lasted 6–12 months after the invention and increased functional connectivity between motor-related cortical areas in the affected hemisphere.

The application of BCI-FET for improving lower limb function in stroke survivors was also investigated recently [122–124], although the clinical evidence is less comprehensive compared to upper limb BCI-FET. Specifically, in a feasibility study in nine chronic stroke survivors, McCrimmon et al. [122] examined the utility of foot dorsiflexion BCI-FET, which was applied over the course of 12 sessions over four weeks. Their results showed that BCI-FET could safely

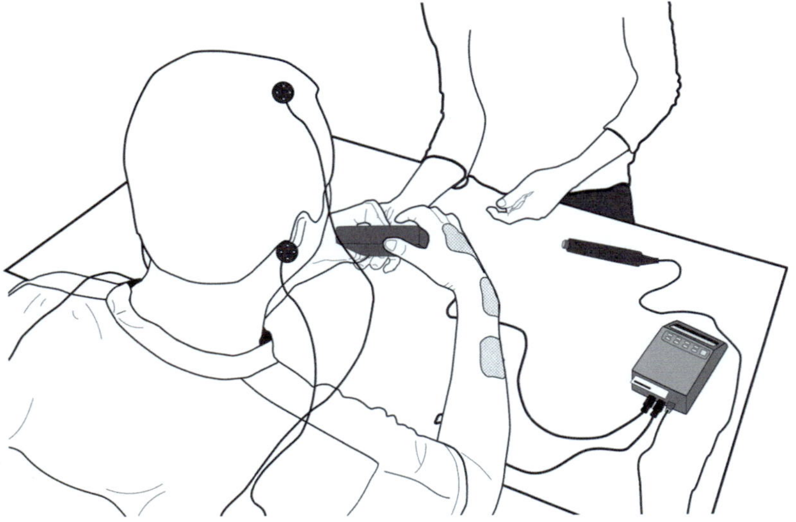

Fig. 19.6 KITE BCI-FET system conceptual illustration showing a participant during the intervention supported by the therapist as shown by Jovanovic et al. [111] © 2021 The Author(s). Published with license by Taylor & Francis Group, LLC on behalf of The Academy of Spinal Cord Injury Professionals, Inc

and effectively deliver therapeutic benefits to increase gait speed and increased ankle dorsiflexion. Moreover, in a pilot randomized controlled trial, Chung et al. [123] compared foot dorsiflexion BCI-FET and FES only during a five-week intervention applied three times per week and showed significantly improved gait velocity and cadence after BCI-FET in stroke patients. Notably, in a randomized intervention aiming to compare BCI-FET against BCI with sub-threshold FES over four weeks of training that included a total of 12 training sessions, Mrachacz-Kersting et al. [124] tested functional improvements and central nervous system excitability in stroke patients. Their results indicated that BCI-FET groups benefited from significantly improved functional recovery as indicated by the lower extremity Fugl–Meyer assessments, which were also accompanied by increased corticospinal excitability after the intervention. However, it should be noted that most of the lower limb BCI-FET interventions were applied to enable ankle dorsiflexion movements and not during walking. Future studies should aim to deliver the intervention during functional walking tasks to enable activity-dependent plasticity.

The applications of BCI-FET in individuals with SCI are fewer and far between, although several recent investigations have shown potentials [111, 114, 125]. Notably, our recent study by Jovanovic et al. [111] applied BCI-FET in five individuals with subacute SCI who completed 12 to 40 training sessions over a few weeks. The results indicated that most patients benefited from improved functional independence scores. Moreover, Osuagwu et al. [114] applied the upper limb BCI-FET intervention in twelve subacute tetraplegic patients with incomplete injuries (C4-C7; ASIA B/C) for a period of 20 sessions, which was compared to the open-loop controlled FES intervention. While the range of motion improved in both groups after the intervention, muscle strength improvements were shown only in the BCI-FET group, suggesting functional improvements accompanied by more focused cortical activations than the FES group. Finally, the application of BCI-FET

for improving lower limb function in individuals with SCI was examined by King et al. [125]. However, no therapeutic measures were quantified in this case study. The results showed that an individual with SCI (T6 AIS B) could operate BCI-FES during overground walking. Further studies are warranted to show the effectiveness of this approach to improve motor function after the intervention.

Taken together, clinical trials in stroke and SCI patients present evidence that functional motor improvements are associated with enhanced cortical activations in the sensorimotor related cortical areas. Moreover, they all agree that BCI-controlled FES/FET is more effective in producing functional and cortical changes compared to FES delivery alone.

19.7 Potential Mechanisms of FET and How BCI-FET Enhances Neuroplasticity

At the present time, the exact mechanisms responsible for the observed FET effect are not known. However, a few hypotheses have been proposed that may provide at least a partial explanation of the FET effect.

Three possible "peripheral" mechanisms might be considered. At first, FET may improve the muscle functions in the remaining motor units through muscle training and strengthening. However, this does not necessarily happen only during FET; other training mechanisms can be used to improve muscle strength and endurance. Second, FET may improve the flexibility and range of motion of the affected limb/joints, and as a consequence, the voluntary function may be improved. However, stretching during physiotherapy should be able to generate similar results. Third, FET reduces the amount of spasticity in the affected limb, and by doing so it may improve the motor function. Although it has been shown in the past that FET does improve the spasticity [32, 33, 126], the FET effect has been observed even in the affected limbs that did not have spasticity. Thus, even though all three above-listed mechanisms may be possible, they alone could not account for the observed FET effect.

It has been reported that cortical reorganization can occur following stroke recovery [127] as well as after FET in traumatic brain injury [36]. As FES activates both motor and sensory nerve fibers, sensory stimulation may be capable of modifying cortical connectivity [128] and spinal reflexes [33] after the stimulating period. Therefore, implications of sensorimotor recruitment through FES go beyond only the contraction of muscles. Our data suggest that FES can induce changes in the central nervous system by modifying the stimulation parameters and protocols of FES delivery (for a review, see Milosevic et al. [129]; Carson et al. [130]). The spinal reflex inhibition [33] may also be related to spasticity reduction after FET. However, previous studies also clearly demonstrated that FES activates the cortical networks during FES through sensory feedback [131–133]. Thus, through repetitive activation, the stimulating pulses that transverse to the brain during FET may promote neuroplasticity in the central nervous system by activating the sensory and motor central nervous system centers [134].

Recently a study by Takeoka et al. [101] demonstrated very elegantly that muscle spindle feedback is critical and probably essential for the functional recovery following SCI. They have shown that if muscle spindles are "removed out of the rehabilitation process" that the animal trained is unable to recover its function. Since FET fully engages muscle spindle feedback system during therapy, it is very likely that the high intensity muscle spindle activation produced by the FET is contributing to the process of recovery of voluntary function. Please note that in the past, it has been frequently suggested that the FES/FET does not activate muscle fibers in physiologically correct manner, i.e., that the fast-twitch muscle fibers are recruited first followed by the slow-twitch muscle fibers [135]. This reverse order of muscle fiber activation could impact the order in which muscle spindle feedback is presented to the central nervous system following FET. However, recent experiments have shown that this notion of reverse

muscle fiber recruitment during FES/FET is incorrect [135], suggesting that the order in which muscle spindle feedback is delivered to the central nervous system should be reasonably close to the natural one. More comprehensive discussion about the sensory feedback systems that may be engaged during FET and how they may contribute to the improvement in the voluntary function following FET can be found in Prochazka's recent article [136].

In addition to the sensory system activation, Rushton [6] suggested a hypothesis that accounts for the neuroplasticity effects that are uniquely due to FES. Electrical stimulation of a motor nerve fiber generates both an orthodromic impulse toward the muscle and an antidromic impulse, which is sent in the opposite direction along the motor axons toward the spine and the brain. When the voluntary, descending command descends from the brain to the spinal motor neuron, they can meet the antidromic impulse at the motor neuron during FES. This coincidence of two impulses can strengthen the synaptic connection via Hebb's rule. The enhancement of the synaptic connection would therefore increase the efficacy of the voluntary, descending command to activate impaired muscle in individuals with stroke and SCI. Recent results that showed the facilitation of motor evoked potential using paired associative stimulation through transcranial magnetic stimulation of the motor cortex after FES support this hypothesis [137–139]. Adherence to spike timing-dependant plasticity, which ensures the timing between cortical and FES activations, was suggested to be relevant for increasing voluntary outputs [139] and as a therapeutic intervention to enhance motor recovery after SCI [138]. However, it should be noted that facilitation of motor evoked potential using paired associative stimulation is not necessarily always guaranteed [140] and that multiple possible pathways could be available to induce neuroplasticity since FET does not strictly adhere to the precise timing of activations. Nonetheless, the above-mentioned Rushton's hypothesis related to FES antidromic activation offers a viable option for FET neuroplasticity that needs to be confirmed.

Another hypothesis that could also explain the mechanisms behind FET is the one proposed by Popovic et al. [29, 36, 61, 67–69, 78, 82, 95, 96, 141]. If a subject, who attempts to execute a motor task, is assisted with the FES to carry out that task, he is effectively voluntarily generating the motor commands through the desire to move the arm, leg, etc. (i.e., command input). At the same time, FET is providing sensory feedback (system's output), indicating that the command was executed successfully. By providing both the command input and system's output to the central nervous system repeatedly over prolonged periods of time, FET facilitates functional reorganization and retraining of intact parts of the central nervous system to allow them to take over the function of the stroke or SCI damaged control centres. It is important to add that during the FET, the subjects perform motor tasks repetitively. Our results clearly indicated that task-specific and repetitive FET could successfully increase cortical activations after 36–40 intervention sessions by integrating voluntary motor commands during diverse and meaningful tasks with high repetition and subject's persistent active engagement (i.e., the subject has to fully devote her/his attention to the tasks performed) and sensorimotor network activation through FES [36]. This may play a critical role in retraining voluntary motor function after stroke and SCI. The mechanisms are also fully in tune with recent findings in the field of neuroplasticity, suggesting that FET is potentially another effective method that can be used to retrain the neuromuscular system.

Taken together, the carry-over effects resulting from FET are probably multifactorial and need to be fully examined. A new explanation proposed by Popovic and his team in Toronto is that the phylogenetically older brain structures (e.g., subcortical structures and/or the brainstem), which are equally able of control limbs, may also be engaged during FET training. Specifically, they hypothesized that FET for reaching and grasping, when it is applied to stroke patients, engages phylogenetically older brain structures and retrains them to perform reaching and grasping tasks, instead of retraining the cortical

structures. Recently, Kawai et al. [142] actually demonstrated in rodents that the motor cortex is required for learning new tasks, but that it is not required for execution of already mastered forelimb motor tasks. This finding suggests that Popovic's hypothesis may be correct, but this still needs to be properly verified. Nonetheless, what is certain is that the FET is an effective method for restoring voluntary upper and lower limb functions in individuals following stroke and SCI.

19.7.1 How BCI Technology Enhances FET

The use of BCI and BCI-FET for improving motor function has gained considerable attention recently, with various comprehensive reviews summarized elsewhere [129, 143]. Evidence from BCI-FES in stroke patients [115, 117] suggests that improved sensorimotor activations in the affected hemisphere may be responsible for the functional improvements. Intact motor areas adjacent to the damaged site within the motor cortex and/or other sensorimotor areas may assume control over the affected muscles as a result of the intervention. Similarly, in individuals with SCI, more focused sensorimotor cortical activations were reported after BCI-FET [114]. In non-human primates after SCI, it was also shown that BCI-FES could induce adaptive cortical changes throughout different sensorimotor cortical sites [144]. Specifically, after using the BCI-FES, cortical activity became localized around an arbitrarily selected cortical area that was used to control FES [144]. Therefore, through associative activation of cortical and FES-activated sensorimotor networks, BCI empowers FET to efficiently create new connections for generating and transmitting neuronal commands from the cortex to the muscles.

The BCI technology could achieve these effects through progressive practice that involves feedback and reward, which likely also includes Hebb-type learning. Specifically, presynaptic inputs in the form of oscillatory cortical activations that are detected by the BCI system can be strengthened as a result of simultaneous or subsequent postsynaptic activation using FES to activate the same sensorimotor networks. In addition to positive reinforcement, which is the likely cause of BCI-FES induced neuroplasticity, some unsuccessful trials also have helpful contributions. Reward-based learning in the brain involves upregulation of dopaminergic excitatory receptors and/or downregulation of GABAergic inhibitory receptors [145] to reinforce these new connections. We propose that presynaptic cortical oscillatory desynchronization detected by the BCI system can make the CNS more susceptible to the subsequent or simultaneous postsynaptic activation of the same sensorimotor networks using FES. The mechanisms of effective neuroplasticity using BCI-controlled FES likely involve Hebb-type plasticity and cortical priming. Thus, BCI-FET training can be used to rapidly elicit neuroplasticity, which can be effective in motor rehabilitation after neurological injuries.

Overall, it is our impression that FET and BCI-FET are very promising interventions that is only now being seriously examined and have the potential to revolutionize the way we rehabilitate individuals with diverse neuromuscular disorders including stroke and SCI.

19.8 Perspectives

This chapter summarizes the research findings regarding the effects of FET in individuals with stroke and SCI. The findings to date clearly show that FET for reaching and grasping is a therapeutic modality that should be implemented in every rehabilitation institution that is treating patients with stroke and SCI. The results obtained in a number of randomized control trials to date clearly point out that FET for upper limb should not be ignored any longer. There is also considerable evidence to support the use of FET as a therapeutic modality to treat drop-foot problem in both stroke and incomplete SCI populations. There are a couple of FES systems on the market that can be used to deliver FET for drop-foot and grasping, and physiotherapists and

occupational therapists should take advantage of this technology. Presently, few teams in the world are investigating use of more complex FES systems (6–16 channels FES systems that stimulate muscles in one of both legs in a physiologically appropriate manner) for retraining voluntary walking function in stroke and incomplete SCI populations. Although comprehensive randomized control trials have not been completed yet with either patient population, preliminary findings are encouraging.

The results obtained to date suggest that FET can be used effectively with both chronic and subacute stroke and SCI patients. However, the results published to date suggest that FET produces better results if it is applied during early rehabilitation, i.e., during subacute phase following injury. Further, the effect of FET has shown good results in individuals with chronic complete and incomplete SCI and stroke subjects. However, to date, statistically significant results have only been obtained with chronic stroke and incomplete SCI patients. It should be noted that FET therapy does not require any voluntary movement in the affected limb as an indication for the therapy. In other words, FET can be applied to individuals who are profoundly paralyzed (i.e., cannot move the limb at all) due to central nervous system injury, and one can expect to see at least partial recovery of the limb function at the end of the FET. However, it should also be noted that peripheral nerve/root paralysis frequently occurs in SCI patients, and in these cases, FET usually has little effect.

As the surface FES technology is continuously improving and delivery methods for FET are evolving due to system's miniaturization, better stimulation electrodes, methods for minimizing fatigue effects, and better stimulation protocols, it is foreseeable that, in next 10–15 years, FET will become one of the dominant interventions for upper and lower limb rehabilitation. Many FET systems are already commercialized, and many more are in the process of being developed and/or commercialized. Clinical utility BCI-FET is yet to be fully demonstrated, but recent advances have provided promising results. Thus, we feel very confident that FET and BCI-FET field is only beginning to evolve, and that, in the future, it may become one of the key therapeutic interventions not only for patients with stroke and SCI but also for patients with other neuromuscular disorders.

Conflict of Interest Disclosure Statement M.R.P. is a shareholder and director in the company MyndTec Inc. M.M. is an Executive Board Member of the International Functional Electrical Stimulation Society (IFESS), a non-profit organization dedicated to promoting the awareness, knowledge, and understanding of electrical stimulation technologies and their applications.

References

1. Loeb GE, Ghez C. The motor unit and muscle action. In: Kandel ER, et al., editors. Principles of Neural Science, Fourth Ed. p. 674–94.
2. Reichel M, Breyer T, Mayr W, Rattay F. Simulation of the three-dimensional electrical field in the course of functional electrical stimulation. Artif Organs. 2002;26:252–5.
3. Kern H, Hofer C, Mödlin M, Forstner C, Raschka-Högler D, Mayr W, Stöhr H. Denervated muscles in humans: limitations and problems of currently used functional electrical stimulation training protocols. Artif Organs. 2002;26:216–8.
4. Nakagawa K, Bergquist AJ, Yamashita T, Yoshida T, Masani K. Motor point stimulation primarily activates motor nerve. Neurosci Lett. 2020;736:135246.
5. Bergquist AJ, Clair JM, Lagerquist O, Mang CS, Okuma Y, Collins DF. Neuromuscular electrical stimulation: implications of the electrically evoked sensory volley. Eur J Appl Physiol. 2011;111:2409–26.
6. Rushton DN. Functional electrical stimulation and rehabilitation—an hypothesis. Med Eng Phys. 2003;25:75–8.
7. Bersch I, Koch-Borner S, Fridén J. Electrical stimulation-a mapping system for hand dysfunction in tetraplegia. Spinal Cord. 2018;56:516–22.
8. Botter A, Oprandi G, Lanfranco F, Allasia S, Maffiuletti NA, Minetto MA. Atlas of the muscle motor points for the lower limb: implications for electrical stimulation procedures and electrode positioning. Eur J Appl Physiol. 2011;111:2461–71.
9. Lim D, Del CM, Bergquist AJ, Milosevic M, Masani K. Contribution of each motor point of quadriceps femoris to knee extension torque during neuromuscular electrical stimulation. IEEE Trans Neural Syst Rehabil Eng. 2021;29:389–96.
10. Garcia-Garcia MG, Jovanovic LI, Popovic MR. Comparing preference related to comfort in torque-matched muscle contractions between two different types of functional electrical stimulation pulses in able-bodied participants. J Spinal Cord Med. 2021;44:S215–24.
11. Mesin L, Merlo E, Merletti R, Orizio C. Investigation of motor unit recruitment during stimulated contractions of tibialis anterior muscle. J Electromyogr Kinesiol. 2010;20:580–9.
12. Doucet BM, Lam A, Griffin L. Neuromuscular electrical stimulation for skeletal muscle function. Yale J Biol Med. 2012;85:201.
13. Bergquist AJ, Babbar V, Ali S, Popovic MR, Masani K. Fatigue reduction during aggregated and distributed sequential stimulation. Muscle Nerve. 2017;56:271–81.
14. Masani K, Popovic MR. Functional electrical stimulation in rehabilitation and neurorehabilitation. In: Springer handbook of medical technology. Berlin, Heidelberg: Springer; 2011. p. 877–96.
15. Barss TS, Ainsley EN, Claveria-Gonzalez FC, Luu MJ, Miller DJ, Wiest MJ, Collins DF. Utilizing physiological principles of motor unit recruitment to reduce fatigability of electrically-evoked contractions: a narrative review. Arch Phys Med Rehabil. 2018;99:779–91.
16. Malešević NM, Popović LZ, Schwirtlich L, Popović DB. Distributed low-frequency functional electrical stimulation delays muscle fatigue compared to conventional stimulation. Muscle Nerve. 2010;42:556–62.
17. Nguyen R, Masani K, Micera S, Morari M, Popovic MR. Spatially distributed sequential stimulation reduces fatigue in paralyzed triceps surae muscles: a case study. Artif Organs. 2011;35:1174–80.
18. Sayenko DG, Nguyen R, Popovic MR, Masani K. Reducing muscle fatigue during transcutaneous neuromuscular electrical stimulation by spatially and sequentially distributing electrical stimulation sources. Eur J Appl Physiol. 2014;114:793.
19. Sayenko DG, Nguyen R, Hirabayashi T, Popovic MR, Masani K. Method to reduce muscle fatigue during transcutaneous neuromuscular electrical stimulation in major knee and ankle muscle groups. Neurorehabil Neural Repair. 2015;29:722–33.
20. Ye G, Ali S, Bergquist AJ, Popovic MR, Masani K. A generic sequential stimulation adapter for reducing muscle fatigue during functional electrical stimulation. Sensors. 2021;21(21):7248.
21. Kuhn A, Keller T, Micera S, Morari M. Array electrode design for transcutaneous electrical stimulation: a simulation study. Med Eng Phys. 2009;31:945–51.
22. Micera S, Keller T, Lawrence M, Morari M, Popovic D. Wearable neural prostheses. IEEE Eng Med Biol Mag. 2010;29:64–9.
23. Popović DB, Popović MB. Automatic determination of the optimal shape of a surface electrode: Selective stimulation. J Neurosci Methods. 2009;178:174–81.
24. Smith B, Tang Z, Johnson MW, Pourmehdi S, Gazdik MM, Buckett JR, Hunter Peckham P. An

externally powered, multichannel, implantable stimulator-telemeter for control of paralyzed muscle. IEEE Trans Biomed Eng. 1998;45:463–75.

25. Davis R, Sparrow O, Cosendai G, Burridge J, Turk R, Wulff C, Schulman J. Post-stroke arm rehabilitation using 5–7 implanted microstimulators: implantation procedures, safety and efficacy. Undefined; 2007.

26. Popovic MB, Popovic DB, Sinkjær T, Stefanovic A, Schwirtlich L. Restitution of reaching and grasping promoted by functional electrical therapy. Artif Organs. 2002;26:271–5.

27. Waters RL. The enigma of "carry-Over." Disabil Rehabil. 1984;6:9–12.

28. Merletti R, Acimovic R, Grobelnik S, Cvilak G. Electrophysiological orthosis for the upper extremity in hemiplegia: feasibility study. Arch Phys Med Rehabil. 1975;56:507–13.

29. Thrasher TA, Zivanovic V, McIlroy W, Popovic MR. Rehabilitation of reaching and grasping function in severe hemiplegic patients using functional electrical stimulation therapy. Neurorehabil Neural Repair. 2008;22:706–14.

30. Kapadia NM, Zivanovic V, Furlan J, Craven BC, Mcgillivray C, Popovic MR. Functional electrical stimulation therapy for grasping in traumatic incomplete spinal cord injury: randomized control trial. Artif Organs. 2011;35:212–6.

31. Goulet C, Arsenault AB, Bourbonnais D, Laramée MT, Lepage Y. Effects of transcutaneous electrical nerve stimulation on H-reflex and spinal spasticity. Scand J Rehabil Med. 1996;28:169–76.

32. Kawashima N, Popovic MR, Zivanovic V. Effect of intensive functional electrical stimulation therapy on upper-limb motor recovery after stroke: case study of a patient with chronic stroke. Physiother Canada. 2013;65:20.

33. Milosevic M, Masugi Y, Obata H, Sasaki A, Popovic MR, Nakazawa K. Short-term inhibition of spinal reflexes in multiple lower limb muscles after neuromuscular electrical stimulation of ankle plantar flexors. Exp Brain Res. 2019;237:467–76.

34. Shin HK, Cho SH, Jeon HS, Lee YH, Song JC, Jang SH, Lee CH, Kwon YH. Cortical effect and functional recovery by the electromyography-triggered neuromuscular stimulation in chronic stroke patients. Neurosci Lett. 2008;442:174–9.

35. Sasaki K, Matsunaga T, Tomite T, Yoshikawa T, Shimada Y. Effect of electrical stimulation therapy on upper extremity functional recovery and cerebral cortical changes in patients with chronic hemiplegia. Biomed Res. 2012;33:89–96.

36. Milosevic M, Nakanishi T, Sasaki A, Yamaguchi A, Nomura T, Popovic MR, Nakazawa K. Cortical re-organization after traumatic brain injury elicited using functional electrical stimulation therapy: a case report. Front Neurosci. 2021;15:1046.

37. Popovic MR, Popovic DB, Keller T. Neuroprostheses for grasping. Neurol Res. 2002;24:443–52.

38. Kapadia N, Moineau B, Popovic MR. Functional electrical stimulation therapy for retraining reaching and grasping after spinal cord injury and stroke. Front Neurosci. 2020;14:718.

39. Taylor PN, Burridge JH, Dunkerley AL, Wood DE, Norton JA, Singleton C, Swain ID. Clinical use of the odstock dropped foot stimulator: its effect on the speed and effort of walking. Arch Phys Med Rehabil. 1999;80:1577–83.

40. Stein RB, Everaert DG, Thompson AK, Chong SL, Whittaker M, Robertson J, Kuether G. Long-term therapeutic and orthotic effects of a foot drop stimulator on walking performance in progressive and nonprogressive neurological disorders. Neurorehabil Neural Repair. 2010;24:152–67.

41. Hausdorff JM, Ring H. Effects of a new radio frequency-controlled neuroprosthesis on gait symmetry and rhythmicity in patients with chronic hemiparesis. Am J Phys Med Rehabil. 2008;87:4–13.

42. Burridge JH, Haugland M, Larsen B, Svaneborg N, Iversen HK, Christensen PB, Pickering RM, Sinkjaer T. Patients' perceptions of the benefits and problems of using the actigait implanted drop-foot stimulator. J Rehabil Med. 2008;40:873–5.

43. Kenney L, Bultstra G, Buschman R, et al. An implantable two channel drop foot stimulator: initial clinical results. Artif Organs. 2002;26:267–70.

44. Van Swigchem R, Vloothuis J, Den Boer J, Weerdesteyn V, Geurts ACH. Is transcutaneous peroneal stimulation beneficial to patients with chronic stroke using an ankle-foot orthosis? A within-subjects study of patients' satisfaction, walking speed and physical activity level. J Rehabil Med. 2010;42:117–21.

45. Daly JJ, Roenigk K, Holcomb J, Rogers JM, Butler K, Gansen J, McCabe J, Fredrickson E, Marsolais EB, Ruff RL. A randomized controlled trial of functional neuromuscular stimulation in chronic stroke subjects. Stroke. 2006;37:172–8.

46. Burridge JH, Taylor PN, Hagan SA, Wood DE, Swain ID. The effects of common peroneal stimulation on the effort and speed of walking: A randomized controlled trial with chronic hemiplegic patients. Clin Rehabil. 1997;11:201–10.

47. Granat MH, Maxwell DJ, Ferguson ACB, Lees KR, Barbenel JC. Peroneal stimulator; evaluation for the correction of spastic drop foot in hemiplegia. Arch Phys Med Rehabil. 1996;77:19–24.

48. Kantrowitz A. Electronic physiological aids: a report of the maimonides Hospital. Maimonides Hospital, Brooklyn; 1960.

49. Strojnik P, Kralj A, Uršič I. Programmed six-channel electrical stimulator for complex stimulation of leg muscles during walking. IEEE Trans Biomed Eng BME. 1979;26:112–6.

50. Kralj A, Bajd T, Turk R. Enhancement of gait restoration in spinal injured patients by functional electrical stimulation. Clin Orthop Relat Res. 1988;233:34–43.

51. Graupe D, Kohn KH. Functional neuromuscular stimulator for short-distance ambulation by certain thoracic-level spinal-cord-injured paraplegics. Surg Neurol. 1998;50:202–7.

52. Graupe D, Davis R, Kordylewski H, Kohn KH. Ambulation by traumatic T4–12 paraplegics using functional neuromuscular stimulation. Crit Rev Neurosurg. 1998;8:221–31.

53. Popovic D, Tomovic R, Schwirtlich L. Hybrid assistive system - the motor neuroprosthesis. IEEE Trans Biomed Eng. 1989;36:729–37.

54. Solomonow M, Baratta R, Hirokawa S, Rightor N, Walker W, Beaudette P, Shoji H, D'Ambrosia R. The RGO generation II: muscle stimulation powered orthosis as a practical walking system for thoracic paraplegics. Orthopedics. 1989;12:1309–15.

55. Bailey SN, Hardin EC, Kobetic R, Boggs LM, Pinault G, Triolo RJ. Neurotherapeutic and neuroprosthetic effects of implanted functional electrical stimulation for ambulation after incomplete spinal cord injury. J Rehabil Res Dev. 2010;47:7–16.

56. Davis JA, Triolo RJ, Uhlir JP, Bhadra N, Lissy DA, Nandurkar S, Marsolais EB. Surgical technique for installing an eight-channel neuroprosthesis for standing. Clin Orthop Relat Res. 2001;237–52.

57. Davis JA, Triolo RJ, Uhlir J, Bieri C, Rohde L, Lissy D, Kukke S. Preliminary performance of a surgically implanted neuroprosthesis for standing and transfers–where do we stand? J Rehabil Res Dev. 38:609–17.

58. Hardin E, Kobetic R, Murray L, Corado-Ahmed M, Pinault G, Sakai J, Bailey SN, Ho C, Triolo RJ. Walking after incomplete spinal cord injury using an implanted FES system: a case report. J Rehabil Res Dev. 2007;44:333–46.

59. Johnston TE, Betz RR, Smith BT, Benda BJ, Mulcahey MJ, Davis R, Houdayer TP, Pontari MA, Barriskill A, Creasey GH. Implantable FES system for upright mobility and bladder and bowel function for individuals with spinal cord injury. Spinal Cord. 2005;43:713–23.

60. Popovic MR, Keller T. Modular transcutaneous functional electrical stimulation system. Med Eng Phys. 2005;27:81–92.

61. Thrasher TA, Flett HM, Popovic MR. Gait training regimen for incomplete spinal cord injury using functional electrical stimulation. Spinal Cord. 2006;44:357–61.

62. Pappas IPI, Popovic MR, Keller T, Dietz V, Morari M. A reliable gait phase detection system. IEEE Trans Neural Syst Rehabil Eng. 2001;9:113–25.

63. Bajd T, Kralj A, Štefančič M, Lavrač N. Use of functional electrical stimulation in the lower extremities of incomplete spinal cord injured patients. Artif Organs. 1999;23:403–9.

64. Wieler M, Stein RB, Ladouceur M, Whittaker M, Smith AW, Naaman S, Barbeau H, Bugaresti J, Aimone E. Multicenter evaluation of electrical stimulation systems for walking. Arch Phys Med Rehabil. 1999;80:495–500.

65. Field-Fote EC, Lindley SD, Sherman AL. Locomotor training approaches for individuals with spinal cord injury: a preliminary report of walking-related outcomes. J Neurol Phys Ther. 2005;29:127–37.

66. Field-Fote EC, Roach KE. Influence of a locomotor training approach on walking speed and distance in people with chronic spinal cord injury: a randomized clinical trial. Phys Ther. 2011;91:48.

67. Giangregorio L, Craven C, Richards K, Kapadia N, Hitzig SL, Masani K, Popovic MR. A randomized trial of functional electrical stimulation for walking in incomplete spinal cord injury: effects on body composition. J Spinal Cord Med. 2012;35:351.

68. Hitzig SL, Craven BC, Panjwani A, Kapadia N, Giangregorio LM, Richards K, Masani K, Popovic MR. Randomized trial of functional electrical stimulation therapy for walking in incomplete spinal cord injury: effects on quality of life and community participation. Top Spinal Cord Inj Rehabil. 2013;19:245–58.

69. Kapadia N, Masani K, Craven BC, Giangregorio LM, Hitzig SL, Richards K, Popovic MR. A randomized trial of functional electrical stimulation for walking in incomplete spinal cord injury: effects on walking competency. J Spinal Cord Med. 2014;37:511.

70. Chae J, Bethoux F, Bohinc T, Dobos L, Davis T, Friedl A. Neuromuscular stimulation for upper extremity motor and functional recovery in acute hemiplegia. Stroke. 1998;29:975–9.

71. Chae J, Harley MY, Hisel TZ, Corrigan CM, Demchak JA, Wong YT, Fang ZP. Intramuscular electrical stimulation for upper limb recovery in chronic hemiparesis: an exploratory randomized clinical trial. Neurorehabil Neural Repair. 2009;23:569–78.

72. Chae J, Hart R. Intramuscular hand neuroprosthesis for chronic stroke survivors. Neurorehabil Neural Repair. 2003;17:109–17.

73. Francisco G, Chae J, Chawla H, Kirshblum S, Zorowitz R, Lewis G, Pang S. Electromyogram-triggered neuromuscular stimulation for improving the arm function of acute stroke survivors: a randomized pilot study. Arch Phys Med Rehabil. 1998;79:570–5.

74. Handricks HT, IJzerman MJ, deKroon JR, Zilvold G. Functional electrical stimulation by means of the "Ness Handmaster Orthosis" in chronic stroke patients: an exploratory study. Clin Rehabil. 2001;15:217–20.

75. Kowalczewski J, Gritsenko V, Ashworth N, Ellaway P, Prochazka A. Upper-extremity functional electric stimulation-assisted exercises on a workstation in the subacute phase of stroke recovery. Arch Phys Med Rehabil. 2007;88:833–9.

76. Sullivan JE, Hedman LD. Effects of home-based sensory and motor amplitude electrical stimulation

on arm dysfunction in chronic stroke. Clin Rehabil. 2007;21:142–50.

77. Sullivan JE, Hedman LD. A home program of sensory and neuromuscular electrical stimulation with upper-limb task practice in a patient 5 years after a stroke. Phys Ther. 2004;84:1045–54.

78. Popovic DB, Popovic MB, Sinkjær T, Stefanovic A, Schwirtlich L. Therapy of paretic arm in hemiplegic subjects augmented with a neural prosthesis: a cross-over study. Can J Physiol Pharmacol. 2004;82:749–56.

79. Popović D, Stojanović A, Pjanović A, Radosavljević S, Popović M, Jović S, Vulović D. Clinical evaluation of the bionic glove. Arch Phys Med Rehabil. 1999;80:299–304.

80. Prochazka A, Gauthier M, Wieler M, Kenwell Z. The bionic glove: an electrical stimulator garment that provides controlled grasp and hand opening in quadriplegia. Arch Phys Med Rehabil. 1997;78:608–14.

81. Rebersek S, Vodovnik L. Proportionally controlled functional electrical stimulation of hand. Arch Phys Med Rehabil. 1973;54:378–82.

82. Popovic MR, Thrasher TA, Zivanovic V, Takaki J, Hajek V. Neuroprosthesis for retraining reaching and grasping functions in severe hemiplegic patients. Neuromodulation. 2005;8:58–72.

83. Gritsenko V, Prochazka A. A functional electric stimulation-assisted exercise therapy system for hemiplegic hand function. Arch Phys Med Rehabil. 2004;85:881–5.

84. Popovic MR, Keller T, Pappas IPI, Dietz V, Morari M. Surface-stimulation technology for grasping and walking neuroprostheses: Improving quality of life in stroke/spinal cord injury subjects with rapid prototyping and portable FES systems. IEEE Eng Med Biol Mag. 2001;20:82–93.

85. Popovic MR, Curt A, Keller T, Dietz V. Functional electrical stimulation for grasping and walking: indications and limitations. Spinal Cord. 2001;39:403–12.

86. Popović MB. Control of neural prostheses for grasping and reaching. Med Eng Phys. 2003;25:41–50.

87. Anderson KD, Wilson JR, Korupolu R, Pierce J, Bowen JM, O'Reilly D, Kapadia N, Popovic MR, Thabane L, Musselman KE. Multicentre, single-blind randomised controlled trial comparing Mynd-Move neuromodulation therapy with conventional therapy in traumatic spinal cord injury: a protocol study. BMJ Open. 2020;10:e039650.

88. Glanz M, Klawansky S, Stason W, Berkey C, Chalmers TC. Functional electrostimulation in poststroke rehabilitation: a meta-analysis of the randomized controlled trials. Arch Phys Med Rehabil. 1996;77:549–53.

89. Cauraugh JH, Kim S. Two coupled motor recovery protocols are better than one: electromyogram-triggered neuromuscular stimulation and bilateral movements. Stroke. 2002;33:1589–94.

90. Ring H, Rosenthal N. Controlled study of neuro-prosthetic functional electrical stimulation in sub-acute post-stroke rehabilitation. J Rehabil Med. 2005;37:32–6.

91. Hebert DA, Bowen JM, Ho C, Antunes I, O'Reilly DJ, Bayley M. Examining a new functional electrical stimulation therapy with people with severe upper extremity hemiparesis and chronic stroke: a feasibility study. Br J Occup Ther. 2017;80:651–9.

92. Kapadia NM, Nagai MK, Zivanovic V, Bernstein J, Woodhouse J, Rumney P, Popovic MR. Functional electrical stimulation therapy for recovery of reaching and grasping in severe chronic pediatric stroke patients. J Child Neurol. 2014;29:493–9.

93. Popovic MR, Thrasher TA. Neuroprostheses, encyclopedia of biomaterials and biomedical engineering. New York: Marcel Dekker Inc.; 2004.

94. Mangold S, Keller T, Curt A, Dietz V. Transcutaneous functional electrical stimulation for grasping in subjects with cervical spinal cord injury. Spinal Cord. 2005;43:1–13.

95. Popovic MR, Thrasher TA, Adams ME, Takes V, Zivanovic V, Tonack MI. Functional electrical therapy: retraining grasping in spinal cord injury. Spinal Cord. 2006;44:143–51.

96. Kapadia N, Zivanovic V, Popovic MR. Restoring voluntary grasping function in individuals with incomplete chronic spinal cord injury: pilot study. Top Spinal Cord Inj Rehabil. 2013;19:279.

97. Kapadia NM, Bagher S, Popovic MR. Influence of different rehabilitation therapy models on patient outcomes: hand function therapy in individuals with incomplete SCI. 2014;37:734–43. https://doi.org/10.1179/2045772314Y0000000203.

98. Tomovic R, Popovic DB, Stein RB. Nonanalytical methods for motor control; 1995. https://doi.org/10.1142/2594.

99. Andrews BJ, Baxendale RH, Barnett R, Phillips GF, Yamazaki T, Paul JP, Freeman PA. Hybrid FES orthosis incorporating closed loop control and sensory feedback. J Biomed Eng. 1988;10:189–95.

100. Solomonow M. Biomechanics and physiology of a practical powered walking orthosis for paraplegics. In: Stein RB, Peckham P, Popovic D, editors. Neural prostheses: replacing motor function after disease or disability. London: Oxford University Press; 1992.

101. Takeoka A, Vollenweider I, Courtine G, Arber S. Muscle spindle feedback directs locomotor recovery and circuit reorganization after spinal cord injury. Cell. 2014;159:1626–39.

102. Freeman CT, Hughes AM, Burridge JH, Chappell PH, Lewin PL, Rogers E. Iterative learning control of FES applied to the upper extremity for rehabilitation. Control Eng Pract. 2009;17:368–81.

103. Stauffer Y, Allemand Y, Bouri M, Fournier J, Clavel R, Metrailler P, Brodard R, Reynard F. The WalkTrainer - a new generation of walking reeducation device combining orthoses and muscle

stimulation. IEEE Trans Neural Syst Rehabil Eng. 2009;17:38–45.

104. Crema A, Mancuso M, Frisoli A, Salsedo F, Raschellà F, Micera S. A hybrid NMES-exoskeleton for real objects interaction. In: International IEEE/EMBS conference on neural engineering NER 2015 July; 2015. p. 663–666.

105. Bergamasco M. An exoskeleton structure for physical interaction with a human being; 2013.

106. Hesse S, Werner C, Pohl M, Rueckriem S, Mehrholz J, Lingnau ML. Computerized arm training improves the motor control of the severely affected arm after stroke: a single-blinded randomized trial in two centers. Stroke. 2005;36:1960–6.

107. Bouton CE, Shaikhouni A, Annetta NV, et al. Restoring cortical control of functional movement in a human with quadriplegia. Nature. 2016;533 (7602):247–50.

108. Ajiboye AB, Willett FR, Young DR, et al. Restoration of reaching and grasping movements through brain-controlled muscle stimulation in a person with tetraplegia: a proof-of-concept demonstration. Lancet. 2017;389:1821–30.

109. Takemi M, Masakado Y, Liu M, Ushiba J. Event-related desynchronization reflects downregulation of intracortical inhibition in human primary motor cortex. J Neurophysiol. 2013;110:1158–66.

110. Marquez-Chin C, Marquis A, Popovic MR. EEG-triggered functional electrical stimulation therapy for restoring upper limb function in chronic stroke with severe hemiplegia. Case Rep Neurol Med. 2016;2016:1–11.

111. Jovanovic LI, Kapadia N, Zivanovic V, Rademeyer HJ, Alavinia M, McGillivray C, Kalsi-Ryan S, Popovic MR, Marquez-Chin C. Brain–computer interface-triggered functional electrical stimulation therapy for rehabilitation of reaching and grasping after spinal cord injury: a feasibility study. Spinal Cord Ser Cases. 2021;71(7):1–11.

112. Suzuki Y, Jovanovic LI, Fadli RA, Yamanouchi Y, Marquez-Chin C, Popovic MR, Nomura T, Milosevic M. Evidence that brain-controlled functional electrical stimulation could elicit targeted corticospinal facilitation of hand muscles in healthy young adults. Neuromodulation Technol. Neural Interface [In press]. 2021.

113. Suzuki Y, Kaneko N, Sasaki A, Tanaka F, Nakazawa K, Nomura T, Milosevic M. Muscle-specific movement-phase-dependent modulation of corticospinal excitability during upper-limb motor execution and motor imagery combined with virtual action observation. Neurosci Lett. 2021;755:135907.

114. Osuagwu BCA, Wallace L, Fraser M, Vuckovic A. Rehabilitation of hand in subacute tetraplegic patients based on brain computer interface and functional electrical stimulation: a randomised pilot study. J Neural Eng. 2016. https://doi.org/10.1088/1741-2560/13/6/065002.

115. Biasiucci A, Leeb R, Iturrate I, et al. (2018) Brain-actuated functional electrical stimulation elicits lasting arm motor recovery after stroke. Nat Commun. 2018;91(9):1–13.

116. Daly JJ, Cheng R, Rogers J, Litinas K, Hrovat K, Dohring M. Feasibility of a new application of noninvasive Brain Computer Interface (BCI): a case study of training for recovery of volitional motor control after stroke. J Neurol Phys Ther. 2009;33:203–11.

117. Mukaino M, Ono T, Shindo K, Fujiwara T, Ota T, Kimura A, Liu M, Ushiba J. Efficacy of brain-computer interface-driven neuromuscular electrical stimulation for chronic paresis after stroke. J Rehabil Med. 2014;46:378–82.

118. Jovanovic LI, Kapadia N, Lo L, Zivanovic V, Popovic MR, Marquez-Chin C. Restoration of upper limb function after chronic severe hemiplegia: a case report on the feasibility of a brain-computer interface-triggered functional electrical stimulation therapy. Am J Phys Med Rehabil. 2020;99:e35–40.

119. McCrimmon CM, King CE, Wang PT, Cramer SC, Nenadic Z, Do AH. Brain-controlled functional electrical stimulation for lower-limb motor recovery in stroke survivors. In: Annual international conference of the IEEE engineering in medicine and biology society. IEEE engineering in medicine and biology society annual international conference 2014; 2014. p.1247–50.

120. Li M, Liu Y, Wu Y, Liu S, Jia J, Zhang L. Neurophysiological substrates of stroke patients with motor imagery-based Brain-Computer Interface training. Int J Neurosci. 2014;124:403–15.

121. Kim T, Kim S, Lee B. Effects of action observational training plus brain-computer interface-based functional electrical stimulation on paretic arm motor recovery in patient with stroke: a randomized controlled trial. Occup Ther Int. 2016;23:39–47.

122. McCrimmon CM, King CE, Wang PT, Cramer SC, Nenadic Z, Do AH. Brain-controlled functional electrical stimulation therapy for gait rehabilitation after stroke: a safety study. J Neuroeng Rehabil. 2015. https://doi.org/10.1186/S12984-015-0050-4.

123. Chung EJ, Park SI, Jang YY, Lee BH. Effects of brain-computer interface-based functional electrical stimulation on balance and gait function in patients with stroke: preliminary results. J Phys Ther Sci. 2015;27:513.

124. Mrachacz-Kersting N, Stevenson AJT, Jørgensen HRM, Severinsen KE, Aliakbaryhosseinabadi S, Jiang N, Farina D. Brain state-dependent stimulation boosts functional recovery following stroke. Ann Neurol. 2019;85:84–95.

125. King CE, Wang PT, McCrimmon CM, Chou CC, Do AH, Nenadic Z. The feasibility of a brain-computer interface functional electrical stimulation system for the restoration of overground walking after paraplegia. J Neuroeng Rehabil. 2015;12:1–11.

126. Granat MH, Ferguson ACB, Andrews BJ, Delargy M. The role of functional electrical stimulation in the rehabilitation of patients with incomplete spinal cord injury–observed benefits during gait studies. Paraplegia. 1993;31:207–15.

127. Willer C, Ramsay SC, Wise RJS, Friston KJ, Frackwiak RSJ. Individual patterns of functional reorganization in the human cerebral cortex after capsular infraction. Ann Neurol. 1993;33:181–9.

128. Ridding MC, Brouwer B, Miles TS, Pitcher JB, Thompson PD. Changes in muscle responses to stimulation of the motor cortex induced by peripheral nerve stimulation in human subjects. Exp brain Res. 2000;131:135–43.

129. Milosevic M, Marquez-Chin C, Masani K, Hirata M, Nomura T, Popovic MR, Nakazawa K. Why brain-controlled neuroprosthetics matter: mechanisms underlying electrical stimulation of muscles and nerves in rehabilitation. Biomed Eng Online. 2020. https://doi.org/10.1186/s12938-020-00824-w.

130. Carson RG, Buick AR. Neuromuscular electrical stimulation-promoted plasticity of the human brain. J Physiol. 2021;599:2375–99.

131. Insausti-Delgado A, López-Larraz E, Omedes J, Ramos-Murguialday A. Intensity and dose of neuromuscular electrical stimulation influence sensorimotor cortical excitability. Front Neurosci. 2021. https://doi.org/10.3389/FNINS.2020.593360.

132. Arendsen LJ, Guggenberger R, Zimmer M, Weigl T, Gharabaghi A. Peripheral electrical stimulation modulates cortical beta-band activity. Front Neurosci. 2021. https://doi.org/10.3389/FNINS.2021.632234.

133. Blickenstorfer A, Kleiser R, Keller T, Keisker B, Meyer M, Riener R, Kollias S. Cortical and subcortical correlates of functional electrical stimulation of wrist extensor and flexor muscles revealed by fMRI. Hum Brain Mapp. 2009;30:963–75.

134. Chae J, Yu D. Neuromuscular stimulation for motor relearning in hemiplegia. Crit Rev Phys Rehabil Med. 1999;11:22.

135. Gregory CM, Bickel CS. Recruitment patterns in human skeletal muscle during electrical stimulation. Phys Ther. 2005;85:358–64.

136. Prochazka A. Sensory control of normal movement and of movement aided by neural prostheses. J Anat. 2015;227:167–77.

137. Suppa A, Quartarone A, Siebner H, Chen R, Di Lazzaro V, Del Giudice P, Paulus W, Rothwell JC, Ziemann U, Classen J. The associative brain at work: evidence from paired associative stimulation studies in humans. Clin Neurophysiol. 2017;128:2140–64.

138. Bunday KL, Perez MA. Motor recovery after spinal cord injury enhanced by strengthening corticospinal synaptic transmission. Curr Biol. 2012;22:2355–61.

139. Taylor JL, Martin PG. Voluntary motor output is altered by spike-timing-dependent changes in the human corticospinal pathway. J Neurosci. 2009;29:11708–16.

140. McGie SC, Masani K, Popovic MR. Failure of spinal paired associative stimulation to induce neuroplasticity in the human corticospinal tract. J Spinal Cord Med. 2014;37:565–74.

141. Popovic MR, Kapadia N, Zivanovic V, Furlan JC, Craven BC, McGillivray C. Functional electrical stimulation therapy of voluntary grasping versus only conventional rehabilitation for patients with sub-acute incomplete tetraplegia: a randomized clinical trial. Neurorehabil Neural Repair. 2011;25:433–42.

142. Kawai R, Markman T, Poddar R, Ko R, Fantana AL, Dhawale AK, Kampff AR, Ölveczky BP. Motor cortex is required for learning but not for executing a motor skill. Neuron. 2015;86:800–12.

143. Mrachacz-Kersting N, Ibáñez J, Farina D. Towards a mechanistic approach for the development of non-invasive brain-computer interfaces for motor rehabilitation. J Physiol. 2021;599:2361–74.

144. Kato K, Sawada M, Nishimura Y. Bypassing stroke-damaged neural pathways via a neural interface induces targeted cortical adaptation. Nat Commun. 2019;101(10):1–13.

145. Nudo RJ. Recovery after brain injury: mechanisms and principles. Front Hum Neurosci. 2013;7:887.

Basis and Clinical Evidence of Virtual Reality-Based Rehabilitation of Sensorimotor Impairments After Stroke

20

Gerard G. Fluet, Devraj Roy, Roberto Llorens,
Sergi Bermúdez i Badia,
and Judith E. Deutsch

Abstract

In the recent years, the use of virtual reality (VR) to enhance motor skills of persons with activity and participation restriction due to disease or injury has become an important area of research and translation to practice. In this chapter, we describe the design of such VR systems and their underlying principles, such as experience-dependent neuroplasticity and motor learning. Further, psychological constructs related to motivation, including salience, goal setting, and rewards are commonly utilized in VR to optimize motivation during rehabilitation activities. Hence, virtually simulated activities are considered to be ideal for [1] the delivery of specific feedback, [2] the ability to perform large volumes of training, and [3] the presentation of precisely calibrated difficulty levels, which maintain a high level of challenge throughout long training sessions. These underlying principles are contrasted with a growing body of research comparing the efficacy of VR with traditionally presented rehabilitation activities in persons with stroke that demonstrate comparable or better outcomes for VR. In addition, a small body of literature has utilized direct assays of neuroplasticity to evaluate the effects of virtual rehabilitation interventions in persons with stroke. Promising developments and findings also arise from the use of off-the-shelf video game systems for virtual rehabilitation purposes and the integration of VR with robots and brain-computer interfaces. Several challenges limiting the translation of virtual rehabilitation into routine rehabilitation practice need to be addressed but the field continues to hold promise to answer key issues faced by modern healthcare.

G. G. Fluet (✉) · D. Roy
Department of Rehabilitation and Movement Sciences, Rutgers–The State University of New Jersey, 65 Bergen Street, Newark, NJ 07101, USA
e-mail: fluetge@shp.rutgers.edu

D. Roy
e-mail: dr756@shp.rutgers.edu

R. Llorens
Instituto de Investigación e Innovación en Bioingeniería, Universitat Politècnica de València, Camino de Vera s/n, 46022 Valencia, Valencia, Spain
e-mail: rllorens@i3b.upv.es

S. Bermúdez i Badia
N-LINCS Madeira and Centro de Competências das Ciências Exatas e da Engenharia, Universidade da Madeira, Funchal, Portugal
e-mail: sergi.bermudez@uma.pt

J. E. Deutsch
Rivers Lab: Department of Rehabilitation and Movement Sciences, Rutgers–The State University of New Jersey, 65 Bergen Street, Newark, NJ 07101, USA
e-mail: deutsch@rutgers.edu

© The Author(s), under exclusive license to Springer Nature Switzerland AG 2022
D. J. Reinkensmeyer et al. (eds.), *Neurorehabilitation Technology*,
https://doi.org/10.1007/978-3-031-08995-4_20

Keywords

Virtual reality · Stroke · Motor rehabilitation ·
Neuroplasticity · Motor learning · Motivation ·
Serious games · Brain-computer interfaces ·
Haptics

20.1 Principles of Virtual Reality in Stroke Sensorimotor Neurorehabilitation

Virtual reality (VR) is an approach to user-computer interface that involves real-time simulation of an environment, scenario, or activity that allows for user interaction via multiple sensory channels [1]. VR is created by using hardware and software (virtual environments- VEs) that allow users to interact with objects and events that appear and sound, and in some cases feel, like those in the real world [2]. VR is used in a rehabilitation context as an approach to improve the sensorimotor and cognitive ability of persons with body function structure, activity, and participation limitations through the use of interactions with VEs [3].

VR aims to substitute the real-world sensations with computer-generated sensory information and to facilitate natural interaction with the virtual world. These characteristics modulate immersion, which is related to the multimodal nature of the perceptual senses. In this chapter, we address how VEs leverage aspects such as immersion and presence to describe the quality of the VE and the user's experience. Further, experience-dependent neuroplasticity and motor learning serve as the basis for modern approaches to the rehabilitation of persons with neurologic dysfunction and inform the design of many virtual rehabilitation systems. Brief orientations to these concepts and examples of virtual rehabilitation applications incorporating them will begin this chapter (Sects. 20.1.1 and 20.1.2). Motivation drives several key attributes of behavior consistent with motor learning, including salience, attention, and repetition. The psychology of motivation as it relates to participation in simulated activities will follow in Sect. 20.1.3, and its importance related to the future of virtual rehabilitation will be underscored

in the conclusion section. Sect. 20.2 reviews the literature describing the role of interfaces and sensory presentations in virtual rehabilitation and their impact on the user experience. Sects. 20.1 and 20.2 can be used by the reader to inform the design or refinement of newer technology-based rehabilitation systems, virtually simulated or otherwise. A review of studies examining the efficacy of a wide variety of virtual rehabilitation systems applied to sensorimotor rehabilitation of persons with stroke will complete the chapter. A majority of these studies compare the relative efficacy of virtual rehabilitation to traditional rehabilitation. This type of evidence can be used to evaluate current approaches to virtual rehabilitation and justify further study. The conclusion section that follows will identify several possible next steps for the efficacy literature, proposing a shift in its focus as well as a discussion of the impact of new technologies.

20.1.1 Immersion, Presence, and Embodiment in Virtual Reality

Immersion, presence, and embodiment are different constructs that are, however, interrelated. The fidelity of the delivered sensory information by VR systems and the extent to which their interaction can support users' sensorimotor contingencies (SCs) modulates immersion [4]. The higher the accuracy of the presentation of sensory stimuli (such as display resolution and field of view, sound, and haptic information) and the more SCs supported (such as head, hand, arm, or full-body tracking), the higher the immersion of a system. Immersion, in turn, affects the sense of presence. Even though there is no standardized definition for presence, it can be understood as the psychological state in which an individual responds to a VE like in the real world [5]. However, there is not a linear relationship between immersion and sense of presence. There is a consensus to characterize presence as a multicomponent construct [6]. According to Slater, the sense of presence relies on the place illusion—the illusion of being there—and the plausibility illusion—the credibility of what is happening [4]. Whereas place illusion is more directly liked to the immersive characteristics of a VR system, plausibility illusion is highly dependent on the implemented VEs. It has been commonly thought that presence is the key mechanism that makes VR work. Presence may be especially relevant in a neurologic population, since the subjective perception when interacting with VEs elicited in persons with CNS dysfunction has been shown to be different to that of healthy subjects [7]. Characteristics of both the user and what and how sensory information are presented by the VE determine the level of presence in VR. With regard to the user, the demographic (age, sex, educational level, etc.), psycho-cultural (social habits, interaction, etc.), and also clinical characteristics (sensorimotor, cognitive, and psychological condition) modulate the perception of the virtual world and the interaction with it [8]. Likewise, a previous experience with VR systems may influence presence [8].

Like presence, embodiment is a multicomponent psychological construct. It has been defined as the sense of one's own body [9], as the bodily self-consciousness [10], or as corporeal awareness [11]. All the existing evidence seems to indicate that presence and embodiment are innately linked since both place illusion and plausibility illusion can support the ownership of a virtual body [4]. This relationship is evidenced by studies showing that the sense of presence can be modulated with avatars that accurately represent the users' actual selves (rather than avatars representing their ideal selves), which can facilitate their embodiment [12]. Although an increasing number of studies investigate the plausibility of physiological indices and behavioral data to evaluate both the subjective sense of presence [13] and embodiment [14], the use of dedicated questionnaires, administered either in the physical or in the virtual world [15], is most frequent in the literature.

Recent research has focused on unifying aspects of the embodied cognition theories and identifying its subcomponents, such as body ownership and agency [16]. Agency refers to the sense that one can move and control one's body [17]. Body ownership can be defined as the sense that the body that one inhabits is one's own [17]. Consequently, body ownership is continuous and omnipresent and is not only elicited during the movement but also during passive mobilization and at rest. Body ownership and agency are key mechanisms to facilitate embodiment in VR, which has traditionally been mediated by avatars representing the user's actions.

Research has shown that specific multisensory stimulation can promote not only illusory ownership of parts of the body, such as rubber hands [18], but of the whole body. Multiple studies report that it is possible to perceive another person's body as one's own [19], but also to induce full-body ownership of a mannequin [20] or a complete virtual body [21]. Embodiment in avatars determines the body ownership and agency of the virtual representation and the user's perception of the world and their behavior. For instance, the illusory ownership of a smaller virtual body (a virtual child) has been shown to cause overestimation of object sizes [22], while the ownership of taller avatars has been shown to promote confidence [23]. In contrast, presence can be elicited by adding emotional valence to the media content, regardless of the media form [6]. In healthy adults the salience of the VE, the hardware used to deliver the VE, and the personal qualities of the participants have been shown to interact in creating a sense of presence and immersion [24]. Complete immersion, however, is not a requirement for presence, as participants post-stroke were shown to be present even in semi-immersive environments [25]. Thus, some characteristics of VR systems such as synchronism of stimuli [21], alignment and continuity of the real and virtual bodies [26], and perspective [20], are determinants for inducing a sense of presence and embodiment and consequently are contributing factors in the effectiveness of VR-mediated therapies. Importantly, these findings have been shown to transfer to individuals with stroke. Borrego and colleagues compared both the sense of embodiment and presence in VR of both healthy subjects and individuals with stroke under different perspectives and levels of immersion [14]. The results of their study showed that, although less intensively, embodiment and presence were similarly experienced by individuals with stroke and by healthy individuals, which could support the vividness of their experience and, consequently, the effectiveness of the VR-based interventions.

20.1.2 Immersion and Cybersickness

Potential users of VR, and practitioners have concerns regarding the use of more immersive VR modalities (head-mounted displays and wide field of view projection screens) and the possibility of developing cyber-sickness, a term used to describe a wide variety of uncomfortable symptoms that include but are not limited to nausea and dizziness, caused by interacting with a VE. Sensory conflict is frequently cited as an important contributor to cyber-sickness. Temporal mismatches between virtual presentations of visual movement and vestibular signals caused by actual patient head movement are the most frequently cited causes of symptoms [27, 28]. Logically, non-immersive displays, typically presented on a television or computer monitor do not eliminate the peripheral visual cues that the brain uses to monitor head movement in space should lead to a lower incidence of symptoms. This said changes in visual information that are temporally matched to head movements would decrease this effect in immersive systems. Improvements in the intuitiveness of virtual world movement and higher levels of user control of navigation within the virtual world have been linked to lower levels of cyber-sickness as well [27].

The literature regarding the impact of immersive VR presentations on virtual rehabilitation interventions is often conflicting. A review by Specht et al. focusing specifically on HMD describes this approach to immersive VR as well tolerated by older adults as well as those with stroke and that treatment with HMD are not hindered by cyber-sickness [29]. Another review by Hoeg et al. describes a slightly higher incidence of reports of cyber-sickness and a higher rate of subject dropouts associated with more immersive

equipment. These authors also pose that slight symptoms of cyber-sickness might be under-reported and could contribute to poor compliance and sub-optimal outcomes [30]. Multiple authors call for better controlled studies of the impact of cyber-sickness on virtual rehabilitation as a priority for future research [27, 29, 30].

20.1.3 Motor Learning Principles

Motor learning principles are defined as the set of processes associated with practice or experience that lead to relatively permanent changes in the ability to perform actions [31]. Different principles have been postulated to modulate motor learning after stroke. Salient, goal-directed, task-specific movement and practice of sufficient intensity are important determinants in motor learning in human skill motor learning [32]. Even though these principles have rarely been analyzed in isolation after VR interventions, the role of motor learning principles has been discussed by authors who described their systems [33], in review papers [3, 34–38], as well as book chapters [39]. One can find motor learning principles embedded in VEs for motor rehabilitation [34, 39]. In the following section, we will discuss a number of principles that have become integral to VEs for promoting skill acquisition in the real world such as enriched environments, augmented feedback, practice dosing, adaptation, motivation, and task-oriented experiences.

20.1.3.1 Enriched Environments

Preclinical research on enriched environments serves as the basis for hypothesizing that enriched VR experiences could serve as rehabilitation tools to promote motor learning [40]. Initial findings with animal models have shown that enriched environments promote sensorimotor functions and learning after stroke [41]. The benefits of enriched environments have also been postulated for human subjects. When persons post-stroke were exposed to enriched environments that motivated exploration, physical training, and social interaction, they increased activity and decreased their alone time [42]. In this context, VR is a promising tool to create synthetic computer-generated environments (VEs) that provide augmented stimulation to stroke survivors.

20.1.3.2 Intrinsic and Extrinsic Feedback

Movement performance is informed by both intrinsic and extrinsic feedback. Intrinsic feedback relates to the sensory-perceptual information that is naturally generated during or after a movement.

Augmented feedback—also known as extrinsic feedback—is an add-on to the intrinsic feedback with the goal of providing further information, in the form of knowledge of performance (KP) and/or knowledge of results (KR), that can facilitate skill learning [42]. Augmented feedback is provided by an external source and not by the movement itself [43]. VEs can provide augmented feedback through different sensory modalities such as visual and auditory information with audiovisual devices and proprioceptive information through specific interfaces such as a haptic apparatus, further described in Sect. 20.2. Consequently, VR systems capitalize on both intrinsic feedback and augmented feedback [42].

There is preliminary evidence supporting that augmented auditory feedback improves the speed and accuracy of virtually simulated activity performance in healthy participants as well as participants with brain injury [44]. Further, because VEs can track the motion of body targets or segments, movement monitoring allows the feedback about movement performance and outcome to be very specific. This fact could be key in the beneficial effect in the recovery of motor function after stroke present in VR approaches [see [45] for review]. In studies comparing real-world performance with comparable VE training, several authors have speculated that the cognitive processing required to process the KP in the VR enhances transfer of training to the real world [46, 47]. It is important to note that feedback from VEs, and in particular from games, can be nonspecific and focus on providing positive feedback to encourage participation. This is especially true with non-

custom commercial video games that have been applied to rehabilitation [35]. To date, little is known about the impact of augmented feedback on the transfer of motor ability improvements from virtual activity to real-world activity [48].

20.1.3.3 Task Specificity

Task specificity has long been a fundamental requirement for designing recovery of function programs. The principle of specificity suggests that motor learning is more effective when practice includes environmental and movement conditions similar to those required for the execution of the movement [49]. This suggests that the benefit of the practice specificity occurs because motor learning is specific to the information available during the learning process. Therefore, removing a source of information that was present during practice (or adding another that was not present) impacts task performance. The specificity of practice hypothesis posits that motor skill learning can be enhanced by practice conditions, especially sensorimotor and perceptual information available, performance context characteristics, and cognitive processes involved [50]. Consistent with this hypothesis, VEs can build on the most appropriate available interfaces and feedback modalities to reproduce the relevant context of tasks, such as haptic feedback to recreate the physics of object manipulation [51], video projections to augment tasks with contextual visual information [52, 53], or combining walking on a treadmill while performing a shopping task [54].

VEs have also been used to recreate meaningful tasks to be performed with the upper limbs. Virtual tasks emulating tasks for independent living have been used for assessing the upper limb motor function after stroke [55], showing correlations with clinical scales. Many different VEs have been successfully used for upper limb rehabilitation with levels of ecological validity that varied widely [56, 57]. Given the multisensory training in VE, there may be essential task requirements, but perfect congruence with the real-world task may not be required [58].

Training walking is characteristically done using simulations in which participants walk on a treadmill as they navigate in parks, cityscapes, or corridors [59–61] (Fig. 20.1) or walk over obstacles [62]. However, several investigators have used pre-gait, balance, and other gait-related activities to train walking [46, 63]. The extent that which the task practiced sensorimotor and perceptual feedback is congruent between the VE and the real-world situation varies greatly based on the VR system. While both Fung and You [60, 63] sought to improve walking post-stroke, each approached it with a different degree of task specificity. For example, in a proof of concept study, Fung had participants post-stroke walking in a virtual scene on an actuated treadmill, which allowed changes in path speed as well as orientation, producing a high degree of vestibular and proprioceptive fidelity with the VE. In contrast, You had participants performing stepping and pre-gait activities on the ground with a level surface, in which a TheraBand™ was placed on the participants' limbs to augment the proprioceptive input. Fung measured and demonstrated participants' ability to adapt their walking based on the environmental demands, while You measured walking performance and demonstrated improvements after training. Their findings suggest that task specificity may be beneficial but not essential in VE constructions in order to demonstrate the transfer of training.

20.1.3.4 Dosing

The dose of the training has been reported as a central factor in motor learning [64]. Dosing depends on three key parameters: training duration and frequency with which the individual performs training and the number of repetitions performed during training. It is known that a sufficient dose of practice needs to be performed in order to produce skilled behavior [65] and neuroplastic changes [66]. VEs are designed to promote repetitive task practice that can be tracked and progressed. The number of lower extremity repetitions in VE training has been reported to be comparable to repetitions in animal studies that successfully induced plasticity [33]. Further, work comparing the number of purposeful movements executed with the upper limb of persons post-stroke during standard of

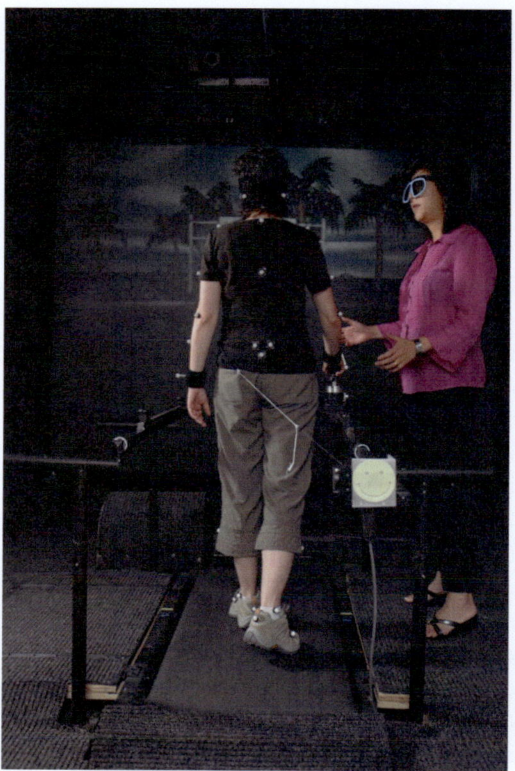

Fig. 20.1 An interactive VR-coupled locomotor system [55] incorporating a self-paced treadmill and dynamic haptics [58] mounted on a six-degree-of-freedom motion platform. Computer-controlled, synchronized animations are rear-projected onto a large screen that can be viewed in 3D with polarized glasses. Such a system can be used to train locomotor adaptation needed to meet demands related to the changing environment (obstruction and surface angle, etc.), tasks (speed requirements, avoiding moving obstacles, dual-tasking, etc.), and cognitive requirements (attention, planning, etc.). Reproduce with permission of Joyce Fung

care was five times lower and slower than when playing Kinect™ [67]. Dose alone, however, is not sufficient for motor learning and neural plasticity (see Sect. 20.3).

20.1.3.5 Adaptability

The repetition of a task is critical for its learning and its refinement. However, the mere repetition of a task has not been shown to induce plastic changes in motor maps. Studies in animals have shown that exposure to a task that requires little or no learning does not produce changes in motor maps or neural morphology [68]. Based on this principle, rehabilitation interventions should involve motor skills with growing difficulty to always pose a motor challenge for post-stroke subjects [69]. The benefits of VEs are, on the one hand, that they can accurately assess the patients' motor condition and, on the other hand, that they can adapt the motor tasks to match this changing condition. Adaptability of the motor tasks has been integrated into several VEs, from the upper limb [56] to balance [70]. VR systems with built-in calibration capabilities or personalization algorithms to autonomously adjust the intensity of training sessions to each patient have been shown to be more effective as compared to conventional therapy [71–73].

20.1.3.6 Motivation

Motivation can be defined as the set of forces that move an individual to act, which may be extrinsic (prompted by an external reward) or intrinsic (propitiated because the task is inherently pleasurable: curiosity, play, etc.). Research has shown that motivation promotes learning [74]. As shown in the following section, motivation plays a major role in VE because it persuades patients to accomplish a task and facilitates presence in the virtual world.

20.1.4 Motivating Through Gaming Elements in Virtual Environments

There are multiple models of motivation, some of which explore intrinsic motivational factors in which the motivation is derived from the act of participation itself or extrinsic factors in which the person is motivated by the purpose of the activity [75]. In the context of sensorimotor rehabilitation, the goal is to facilitate clients to be self-directed and motivated, both because the activity is interesting in itself and because achieving the outcome is important [76]. There is agreement that gaming elements can improve motivation and that, if paired with other activities, they can be harnessed to engage users and achieve desired outcomes. However, there is no consensus regarding the required essential

characteristics of these gaming elements [77], and less than 30% of the studies explicitly reference one or more motivational frameworks or principles [76]. Many elements have been suggested to be important for designing a successful game, such as fun, flow, goals, feedback, game balance, pacing, interesting choices, and narrative structure, among others [78]. In the following sections, we will discuss some of the intrinsic characteristics of games that can affect motivation and learning, and how those are used in the context of motor rehabilitation [79]. While these intrinsic characteristics are discussed as gaming elements in VE, it is important to note that many of them, for example, goal setting, balancing challenge, and reward, overlap with principles of motor learning.

20.1.4.1 Goal Setting

Games generally set multiple goals at different time scales. An appropriate balance of short, medium, and long-term goals has been shown to have a motivating effect in extending gameplay [80]. Further, goals should be achievable, but they should also be attained through a chain of interesting decisions. That is, when players are presented with choices, no one decision should be obviously correct. Most VEs exclusively designed for motor rehabilitation only consider immediate goals (to perform a specific motor task such as reaching or walking) and long-term goals (to collect a sufficiently high amount of rewards). Instead, VEs integrating both cognitive and motor domains seem to be better suited to pose goals at multiple time scales through nontrivial decisions [81–83].

20.1.4.2 Feedback and Rewards

Recent findings suggest that providing appropriate feedback to exercises can stimulate the learning process in rehabilitation therapy [45]. VEs are exceptionally well suited to provide immediate and specific feedback to users, this feature being essential for sustained attention, learning, motivation, and fun [79, 84]. Actions can be rewarded with positive visual and auditory feedback, scores, and specific KP and KR [85, 86]. The simplest way to incorporate KR feedback in VR-based rehabilitation activities is to reinforce successful task completion via general "celebratory" sounds or appropriate sounds when acquiring a target (i.e., explosions during a shooting task). Comparable negative feedback can be provided for unsuccessful performance (collision with an obstacle) [87]. This approach to feedback provides the participant KR, a modality of feedback associated with rapid, effective motor learning [88]. However, rewards can also negatively affect high-interest tasks when rewards are predictable and not associated with performance [89]. More advanced reward systems consider point systems [90, 91], medals [92], bonuses [93], new challenges and tools [94]. Hence, in the ideal scenario, multiple rewards systems need to be selected and manipulated in their number, timing, and quality in order to achieve sustained attention over extended periods. In the case of KP, it does not necessarily require rewards as it can be implemented by providing cues that enable the patient to assess performance, such as the representation of virtual limbs [56, 95], haptic feedback [95], or auditory cues [96].

20.1.4.3 Challenge

VEs for motor rehabilitation should be adjusted in terms of movement demands and dynamics, avoiding situations in which patients lose the ability to control the task directly. It has been suggested that players desire a level of challenge that is neither too easy nor too difficult to perform [97], which is consistent with the early findings of Yerkes and Dodson, when the relation between induced stress and task-learning performance was studied in mice [98] and later replicated in humans in multiple domains [99, 100]. In his flow theory, Csikszentmihalyi describes that user experience during play (anxiety, boredom, and flow) is modulated through the challenge posed and the level of skills required [101]. Flow, defined as the moment of maximum player engagement, is placed at the right balance between user skills and level of challenge. For this reason, the tasks are given as well as the time available to complete them must be calibrated to introduce a controlled challenge

[102]. Therefore, recent developments in VEs for motor training already incorporate transparent and automated modules for the personalization of training by adjusting task difficulty depending on the patient's success rate or by modifying the time available to accomplish a goal [103]. In the cases when VEs are designed to teach complex skills, it is suggested that complex and demanding tasks should be broken down into simpler and more achievable tasks to enhance learning [80]. While simple tasks can be trained by increasing their difficulty in more demanding task settings, complex tasks need to be trained by bringing together previously learned simpler ones, providing a balance of challenge and engagement [104].

20.1.4.4 Sense of Progress

Playing a game entails making decisions and doing actions, with each action influencing the game as a whole. The player must be able to comprehend the immediate effect of their action and how that result was incorporated into the greater context of the game to maintain the motivation to keep playing [105]. Flat and static training tasks can be monotonous and eventually limit the patient's engagement. Malone and Lepper [97] identified curiosity as one of the principal drivers of user engagement in serious games, being it either interest evoked by novel sensations or the desire for knowledge. Narrative elements can be exploited to build an interesting dramatic arc around the training task to increase patients' engagement, facilitate the comprehension of the training objectives, and, most importantly, deliver a clear sense of progress. Multiple elements can be used to shape a narrative curve, such as story events, task difficulty, novel environments, new challenges, or skills. VEs designed to realistically simulate activities, such as navigating a virtual city or shopping in a virtual supermarket, generally provide richer narratives than tasks with simpler cognitive demands [106–109].

20.1.4.5 Socialization

There are multiple ways VEs and games can be used to promote socialization among users.

Thielbar compared a VE for home-based rehabilitation used in multiuser or single-user mode [110]. The multiuser configuration showed a higher compliance rate (10% more), and participants spent more time training when compared to the single-mode version of the system. However, engagement and social involvement do not depend exclusively on VEs being single or multiuser, but also on how user interaction is mediated through the VEs. This can be implemented as a competitive, cooperative, or collaborative interaction [111]. Competitive games have been demonstrated to increase enjoyment [111, 112] and intensity [113]. Collaboration (working together) and cooperation (operating together) have been less studied, with data suggesting that collaboration promotes more behavioral involvement at the expense of having a higher cognitive load [111].

20.1.5 Summary

Motor learning and motivation theories have informed the development of virtual environments and serious games (Table 20.1). Recommendations for the use of augmented feedback or rewards, specifically knowledge of results, are

Table 20.1 Table summarizing some of the key features and their evidence for the design of effective VR systems for motor rehabilitation

		Evidence	References
Motor Learning	Enriched Environments	• Promote activity levels	[41]
	Intrinsic and Extrinsic Feedback	• Knowledge of performance and knowledge of results facilitate skill learning	[42, 45]
		• Knowledge of results has been associated with rapid, effective motor learning	[88]
	Task Specificity	• Virtual tasks emulating ADLs can be used to assess upper limb motor function	[55–57]
		• May be beneficial but not necessary in VR	[144]
	Dosing	• The number of repetitions in VR is comparable to animal studies that induced plasticity	[33]
		• Purposeful movements in VR are performed faster and with higher frequency	[33]
	Adaptability	• VR systems with calibration and/or personalization capabilities are more effective than to conventional therapy	[71, 73, 141]
Motivation	Goal Setting	• An appropriate balance of short, medium and long-term goals has a motivating effect	[80]
		• VEs integrating cognitive and motor domains are better suited to pose goals at multiple time scales	[82, 83, 300]
	Rewards	• Actions should be rewarded with positive visual and auditory feedback, scores and specific knowledge of performance and knowledge of results	[85, 86]
	Challenge	• Task difficulty and time available to complete them should calibrated to control challenge	[102]
		• Complex and demanding tasks should be broken down into simpler and more achievable tasks	[104]
	Sense of progress	• Players must understand the impact of their actions on gameplay	[105]
	Socialization	• Competition increases enjoyment and intensity	[111, 113]
		• Collaboration enhances engagement at the expense of having a higher cognitive load	[111]

consistently found in the VR literature; yet there are few studies to support its use empirically. Instead, the assumption has been made that augmented feedback principles apply in real-world practice and should therefore inform VR design. In contrast, there is modest evidence that VEs promote a high degree of repetition and intensity, and video games deliver higher doses than standard exercises. Until recently, motor learning principles dominated the VR landscape; it is only in more recently that the motivation and game design literature has contributed design principles to guide the appropriate challenge, sense of progress and game modality [79, 114]. Nonetheless, the assumption that motor learning and motivation are essential for the efficacy of virtual rehabilitation is still an open question.

and real-world activities [121–123] as well as differences between two-dimensional and three-dimensional simulated reaching activities [124], and narrow field of view presentations to wide field of view presentations [125]. While there are measurable differences in the movements elicited by comparable activities presented in virtual and veridical worlds, multiple authors describing the training of upper extremity reaching and functional activities by persons with stroke in VEs have shown that comparable real-world improvements in motor abilities can be elicited through repetitive practice in a variety of VEs. Most importantly, upper limb studies show that these improvements are comparable to or better than those elicited by real-world training [36, 126–128].

20.1.6 Visual Presentation

VR systems are frequently classified by the visual presentations they provide to a user and the presence or absence of somatosensory feedback. Visual stimuli are generally grouped by their degree of immersion. Two-dimensional presentations delivered on flat screens are generally considered non-immersive. Three-dimensional presentations utilizing stereoscopic projections or flicker glasses with fixed visual perspectives are considered semi-immersive. Fully immersive systems provide three-dimensional visual information, and perspective is updated with head movements. Full immersion is provided via head-mounted devices or within cave-type environments. Higher levels of immersion are associated with higher levels of agency, presence, and immersion [115–118]

A steadily growing literature has examined the impact of visual presentation on movement kinematics of persons performing reaching movements. Measurable differences in end point and angular measures of upper extremity movement have been noted when comparing two-dimensional simulated movements and comparable real-world activities [119, 120]. Similar differences have been identified in the upper limb when comparing three-dimensional simulated

20.1.7 Point of View

Most immersive and semi-immersive systems, and even some non-immersive systems, present first-person points of view of the workspace during virtual rehabilitation activities. These presentations typically include virtual representations of the participant's limbs or a landscape in which the person might be navigating or acting. However, VR also offers the opportunity to provide users a perspective on movement they may not ordinarily have. For example, video capture-type VR systems present mirror images of the patient as they interact with a VE. These types of augmented reality systems designed for rehabilitation frequently incorporate the ability for the subject to view an image of their own limbs interacting with a VE. One of the reported strengths of this point of view is the high-fidelity feedback regarding patient's posture [129]. This approach presents higher quality information related to limb movement and reduces the need for the brain to rectify differences in somatosensory and visual information associated with the other approaches to VR. One study describes a superior motor performance on a task using an augmented reality system providing a first-person view of the task with the participants' own arms interacting with the VE when compared to a two-dimensional system requiring incongruent motor actions—horizontal forward reaching to elicit vertical movement—in the VE [130]. Two studies suggest that this effect may be enhanced by attaching cameras to a head-mounted device, which improves the fidelity of changes in first-person views of the hands as subtle changes in head position occur [131, 132]. Walking simulations have used both the first- [59] and third-person perspectives [46, 62]. A recent study demonstrated that a first-person point of view enhanced a sense of embodiment in healthy persons and persons with stroke as compared to a third-person view [14]. There are no studies suggesting that an enhanced sense of embodiment might enhance rehabilitation outcomes, but a recent study suggests that an enhanced sense of embodiment might positively affect implicit learning [133].

20.1.8 Auditory Stimuli

Auditory information is a key sensory component of most VEs and has a broad impact on the participant's experience. It is used to enhance immersion in the VE by providing sounds consistent with an activity (i.e., automobile-related sounds for a driving game or the sound of liquid hitting a surface during a pouring activity) [87]. The combination of auditory feedback has also been combined with vibrotactile feedback to enhance collision perception during gait [134], balance [135], and upper extremity training [136]. Spatial sound rendering can also be used to increase the realism of a VE and aid user navigation within a VE (i.e., volume increasing as the virtual representation of the participant approaches the source of a sound in the VE) [87]. The addition of music and specific attributes such as rhythm and cadence has been shown to have a direct impact on the motor performance of healthy and disabled participants [137], particularly when continuous tasks such as gait are simulated [138]. Friedman et al. also found that the addition of music enhanced hand motor performance as well as motivation in the training of functional hand movements [139].

20.1.9 Haptic, Tactile Stimuli and Their Interfaces

Simple or robotic haptic interfaces have allowed for the addition of tactile information and interaction forces into what was previously an essentially visual and auditory experience. Devices of varying complexity are interfaced with more traditional VE presentations to provide haptic feedback that enriches the sensory experience, add physical task parameters, and provide forces that produce biomechanical and neuromuscular interactions with the VE that approximate real-world movement more accurately than visual-only VEs. Simple haptic feedback has been utilized to add the perception of contact to skills like kicking a soccer ball or striking a piano key [140, 141] (Fig. 20.2). Collisions with virtual world obstacles can be used to teach normal movement trajectories such as to place an object on a shelf or the action required to step over a

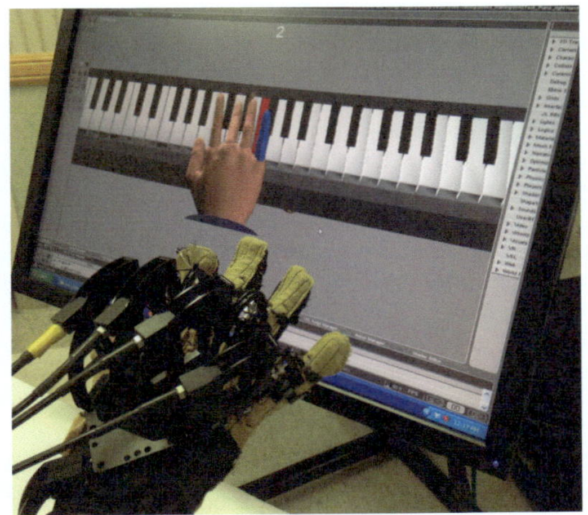

Fig. 20.2 The NJIT-TrackGlove system utilizes a six-degree-of-freedom magnetic tracker, the TrakStar (Ascension Technology Corporation, USA) and a 22-DOF CyberGlove (CyberGlove Systems USA). The simulation pictured also utilizes the CyberGrasp, a cable-actuated robotic exoskeleton. In the pictured simulation, the Virtual Piano Trainer, the magnetic tracker allows the participant to position their hand over the virtual keyboard and the CyberGlove allows them to strike keys with a specific finger. The CyberGrasp can be programmed to provide haptically rendered collisions when keys are pressed or assistance in maintaining extension of non-cued fingers for more impaired subjects [105]

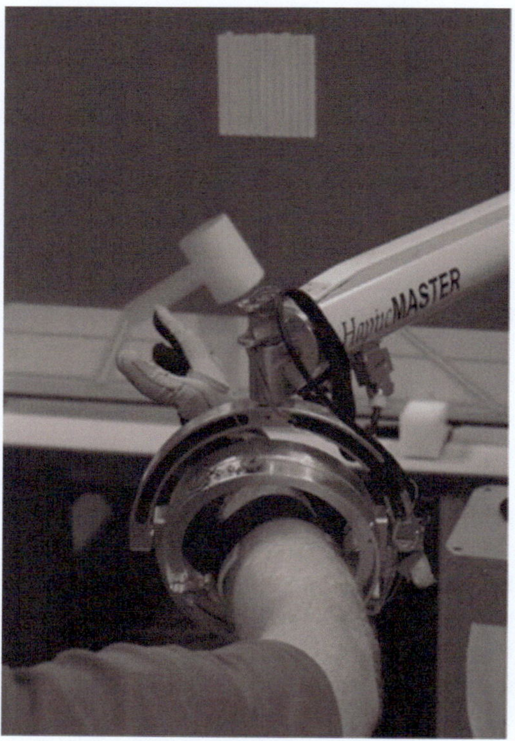

Fig. 20.3 The NJIT-RAVR system utilizes a three-degree-of-freedom robotic (DOF) interface, the Haptic Master (Moog, The Netherlands), three additional passive DOF via a ring-gimbal, and a 22-DOF CyberGlove (CyberGlove Systems USA). The Haptic Master is used to provide haptic rendering of virtual workspaces and add global forces such as gravity to the virtual environments. The ring-gimbal allows for normal positioning of the hand during simulated tasks and the CyberGlove collects data related to finger position. These interfaces are integrated with a suite of complex, virtually simulated tasks to allow for task-based sensorimotor training for persons with upper extremity hemiparesis [67]

curb [62, 72], (Fig. 20.3). Haptic forces can also be synchronized with visual feedback to improve a users' sense of agency in the virtual world. In two small studies involving healthy subjects, this feedback combination was found to be more effective for skill learning than visual-only feedback in healthy subjects [142, 143]. Simulations that aim to shape the behavior of the upper limb have successfully combined haptic feedback with KP to improve upper limb trajectories as post-stroke individuals placed virtual cups on a cupboard [144]. Participants placed their limbs in the haptic master, which

augmented the intrinsic feedback with proprioceptive cues, and the simulation provided information on the trajectory. The coupling of the feedback smoothed out the movement trajectories. Further, haptics has also been used to simulate the interaction forces produced by tools in VEs [117], which increase the sense of immersion and activate neural networks involved with tool manipulation [145]. In a lower extremity application, the addition of haptics improved the accuracy of the limb movement in the VE [33].

20.1.10 Brain-Computer Interfaces

The combination of brain-computer interfaces (BCIs) and VR for stroke rehabilitation has increased in popularity and acceptance during the last decade [146] (Fig. 20.4). BCIs are systems that detect changes in brain signals and translate them into control commands [147]. Such systems exploit the relationships between users' mental states and corresponding electrophysiological signals. In noninvasive BCIs, electroencephalography is commonly used for measuring brain activity. BCIs have gained popularity because evidence relates the mental practice of motor actions with actual movement performance [148]. Motor imagery (MI), the mental practice of motor actions, has been the basis of most BCI approaches to stroke rehabilitation, with a focus on hand and arm training and relying on visual feedback and sometimes combined with Functional Electric Stimulation (FES) or robotic assistance [see [146] for review]. Evidence indicates that the presence of neurofeedback improves MI practice [149]. However, feedback is not the only factor that plays a role. For instance, evidence suggests that motor priming prior to BCI MI can enhance neural activity and improve BCI performance [150]. Avatars in VR and visuo-proprioceptive information can also affect body ownership illusions and modulate the sensorimotor rhythms associated with MI [151, 152]. Also, there are differences between relying on a motor attempt or MI in the underlying neural signals, with evidence suggesting that motor attempt renders

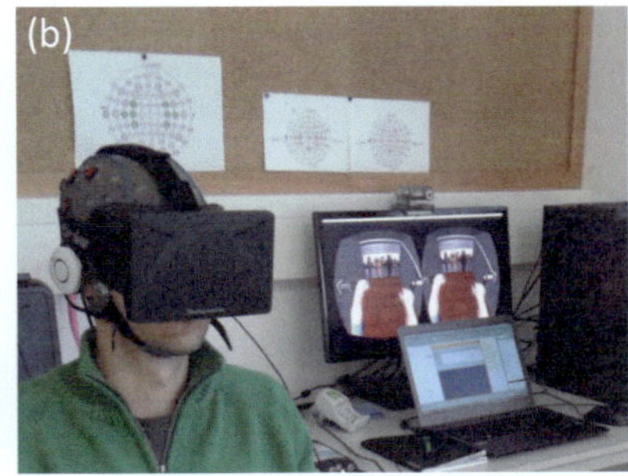

Fig. 20.4 The RehabNet system interfaces a large number of BCI technologies (g.mobiLab, Enobio, Open-BCI, EPOC, Neurosky) and tracking devices (Kinect v1 & v2, Leap Motion, Wii controllers, android phones) with VEs to deliver immersive VR experiences. The RehabNet system is flexible and can work in multiple configurations: **a** MI-BCI neurofeedback training using standard Graz visualization feedback with an 8-channel Enobio acquisition system (Neuroelectrics, Spain); **b** MI-BCI VR training with the virtual representation of upper limbs in a goal-oriented task presented through a head-mounted display and an 8-channel g.mobiLab acquisition system (g.tec, Austria) [114]

better BCI performance [153]. Hence, the lack of standardization on BCI MI methodologies renders BCI studies discrepant and very difficult to compare [154], and consequently, requires significant efforts for the optimization of the settings [155] until standardized protocols are defined [156]. Regardless of the existing difficulties, case studies [157] and RCT findings corroborate that the benefits of MI-based post-stroke rehabilitation are boosted when trained in the context of a BCI paradigm that provides online visual feedback through a VR presentation of the patient's hands [53]. In addition, BCI paradigms allow studying the underlying mechanisms and plastic changes [155, 157], making them a very interesting approach.

20.1.11 Summary

Research into the impact of visual, auditory, and tactile information on virtual rehabilitation activity has started to establish a tentative set of best practices for virtual rehabilitation in terms of the user experience to varying degrees (Table 20.2). The impact of auditory feedback on virtual rehabilitation is at an early stage of development but preliminary work supports the additive effects of rhythm and auditory rendering on the overall effectiveness of the virtual activity. There is a larger body of evidence supporting

that the visual stimulus has a direct, predictable impact on the motor output elicited during simulated activities. However, there is no evidence supporting the notion that higher fidelity visual presentations during virtual rehabilitation translate into larger improvements in the ability of persons with disability to function in the real world. This mismatch between user experience and effectiveness needs to be considered because higher fidelity, fully immersive visual presentations currently require more expensive equipment and more challenging programming to produce. A similar dichotomy exists between VR simulations interfaced with robots to provide tactile feedback and add global forces or with BCIs. Research supports that motor skill learning within the VE is more efficient with these additions. However, this benefit comes at the cost of greater complexity and expense for these integrated systems. These two factors are frequently cited as reasons for the slow adoption of integrated VR-robotic systems into routine clinical practice.

20.2 Neuroscience of Virtual Reality

Knowledge of the neural processes occurring after the central nervous system damage as well as the nervous system's response to activity is necessary to understand the impact of virtual

Table 20.2 Table summarizing key evidence on the role of multisensory information for post-stroke rehabilitation

		Evidence	References
Visual information	2D and 3D simulations	• Exist differences in end point and angular measures with real-world activities	2D: [119, 120] 3D: [121, 122]
		• Improvements are comparable to real-world training	[128, 301]
	Video capture	• Provides high-fidelity feedback on patient's posture	[129]
	1st person view	• Superior task performance	[130]
		• Boosts the effects of motor imagery training supported with online BCI feedback	[53]
Auditory information	Spatial sound	• Increases realism and aids navigation	[87]
	Music	• Rhythm has a direct impact in performance of motor tasks	[137–139]
Haptics and tactile information	Collisions	• Can be used to teach normal movement trajectories	[62, 72, 302]
	Haptic guidance	• Is more effective for skill learning than visual information only	[142, 143]
		• Augments intrinsic feedback with knowledge of performance	[144]
		• Improves accuracy of movements	[144]
	Interaction forces with tools	• Increase immersion and brain activation	[145]

rehabilitation on neural recovery. True recovery is based on behavioral change associated with brain plasticity or neuroplastic changes. After stroke, it is known that perilesional and contralesional brain networks become more excitable, facilitating their reorganization [69, 158]. Research has shown that the recruitment of contralateral or ipsilateral networks largely depends on the integrity of the remaining cortical, subcortical, and corticospinal tracts [159]. As recovery progresses, brain activation patterns of stroke patients become more similar to those of healthy individuals [160, 161], showing that restoration to normal activity patterns correlates with the restoration of motor function.

20.2.1 Brain Plasticity

VR is a particularly interesting research field as it allows creating computer-generated environments that provide customized experiences involving different sensory channels. The motivation of using VR in sensorimotor rehabilitation after a brain lesion is the administration of specific experiences that drive cortical reorganization to support the reacquisition of motor skills. Consequently, neural plasticity is commonly used as an efficacy measure of VR training. Neurophysiological adaptations to training in virtual and real-world environments by people with stroke have been shown to rely on similar neural reorganization processes [117].

An increasing number of studies with many different designs and methodologies have investigated the neural correlates of VR-based interventions focused on sensorimotor rehabilitation (see a recent paper by Hao and colleagues for a review [162]. Interventions included custom and off-the-shelf systems that mostly targeted the upper limb function, followed by lower limb function and balance. Evidences of neural plasticity were explored using functional magnetic resonance imaging (fMRI), electroencephalography (EEG), and transcranial magnetic stimulation. Despite some inconsistent results among studies, fMRI findings support that participation in VR-based sensorimotor interventions increased brain functional connectivity [163–167] and addressed interhemispheric imbalance by increasing cortical activity in the ipsilesional hemisphere [163–165, 168–175]. Interestingly, the increase of the ipsilesional activity ties in with an increase of the cortical representation of the body parts targeted by the VR-based intervention, as derived from the studies that used transcranial magnetic stimulation to explore the plasticity of brain mappings [176–178]. The concomitant manifestation of plastic changes in the brain and improvements in the sensorimotor function after VR-based interventions, as reported by several studies [165, 167, 174, 178–181], could provide evidence of a positive association, although not necessarily causal, between both phenomena.

20.2.2 Visuomotor Representations

It is known that cortical areas involved in the preparation and execution of motor actions undergo plastic changes [182] either due to repeated sessions of proprioceptive stimulation through passive physical training [183] or as a result of task-oriented physical training [184]. Motor deficits do not only arise from the directly damaged tracts by stroke but the networks they disrupt. Hence, its recovery also depends on the intra- and interhemispheric interactions among motor regions [185]. For instance, bilateral recruitment of motor networks can result from unilateral motor movements in hemiparetic stroke patients [185, 186]. Motor training through VE interaction may involve different elements such as object-oriented action planning, action observation, and feedback of the performed action. Unfortunately, there are no standardized protocols for VR motor rehabilitation after stroke, and different interventions have produced distinct effects in both neural reorganization and motor recovery [see [187] for review]. To deliver an optimal rehabilitation process, it becomes essential to identify and understand the neural systems and cerebral processes engaged during motor training mediated by VR.

One of these candidate systems is the human mirror-neuron system (MNS), which is primarily composed of neurons located in the inferior parietal lobe, the ventral premotor cortex, and the caudal part of the inferior frontal gyrus [188]. These are candidate areas for sensory control of action, movement imagery, and imitation [188, 189]. The MNS is of great relevance because it has been shown to be active during the performance of goal-directed actions, their passive observation, and their mental simulation [190]. The MNS has been hypothesized to be involved in action understanding and imitation [191], and, as such, it may represent an important neurophysiological substrate for regaining impaired motor function after stroke [192, 193]. It was suggested that the mere observation of goal-oriented motor actions can be used as a driver [194], and findings corroborate that the use of passive observation of goal-oriented actions can have a positive effect on motor recovery after stroke [195, 196].

From these findings, it is clear that manipulating visual feedback for motor rehabilitation purposes can be an effective ingredient of VR systems. Maeda et al. [197] showed that movement observation can directly enhance and facilitate the motor outcome of the muscles involved in the observed action. In addition, the MNS has been shown to respond to biological as well as robotic effectors [198] and to the manipulation of tools in the real world [199] and VR [200]. Consequently, there is strong evidence supporting that VE interaction can be effective in engaging primary and secondary motor areas for upper extremities [201], locomotion [168], as well as the mirror mechanisms [200, 202]. Consistent with the above findings, the activation of the human MNS has also been documented during the imagination of motor actions [193, 202]. As discussed in Sect. 20.1.10, MI-based BCIs rely on the detection of sensorimotor rhythms, an oscillatory rhythm of synchronized neural brain activity in the alpha and lower beta frequency bands that is measured in sensorimotor brain areas. It has been shown that sensorimotor rhythms can be enhanced utilizing BCI training and that they correlate with motor recovery [53]. Restorative BCIs relying on MI aim at mobilizing neuroplastic changes of the brain in order to achieve reorganization of motor networks and enhance motor recovery [203, 204]. In addition, imaging studies have shown that the combination of first-person observation VR and motor imagery is more effective at recruiting more task-related networks than other conditions for both lower limb [205] and upper limb [206] movements.

The ability to distort visual feedback is an area of inquiry that has been investigated as a possible method to optimize motor adaptations to VR-based rehabilitation activities as well. Preliminary investigations into the visual "augmentation" of small errors during virtual rehabilitation activities performed by persons with stroke have suggested that this approach might enhance motor training outcomes in this population [207]. One possible mechanism for this effect might be an increased level of cortical activity necessary for the brain to rectify virtual movement amplitude that is not scaled to participant movement [208]. One distortion of visual feedback that has been associated with poor responses has been temporal lags between participant movement and corresponding movement within the VE. This may interfere with feed-forward/feedback control of movement, making delayed visual feedback confusing [209]. Recent findings of an RCT also suggest that the visual amplification of upper limb movements can be used to counteract the acquired nonuse of the hemiparetic limb in stroke patients [210].

20.2.3 Summary

After stroke, relearning of motor function is mediated by neuroplasticity. Evidence shows that VR can be a valid tool to drive motor networks, brain plasticity, and functional recovery

Table 20.3 Table summarizing evidence supporting the use of VR to drive neural processes involved in motor recovery

	Evidence	References
Brain plasticity	• Participation in VR-based sensorimotor interventions may increase brain functional connectivity	[164–167]
	• Participation in VR-based sensorimotor interventions may increase cortical activity in the lesioned hemisphere	[165, 173–175]
	• VR-based interventions are associated with increased cortical representation of the body parts targeted by training	[177, 178]
	• Improvements in the sensorimotor function subsequent to VR-based interventions are associated with plastic changes in the brain	[167, 174, 178, 180, 181]
Visuomotor representations	• Bilateral recruitment of motor networks can result from unimanual motor actions	[185, 186]
	• MNS is active during motor action execution, motor observation and mental simulation of motor actions	[190, 193, 202]
	• MNS could be involved in action understanding and imitation	[191]
	• MNS responds to biological, VR, tools and robotic effectors	[198–200]
	• Movement observation facilitates movement of muscles involved in the observed action	[197]
	• Passive observation of motor actions has a positive effect in motor recovery after stroke	[195, 196]
	• Motor imagery BCI training enhances motor recovery	[53, 203, 303]
	• First person VR combined with motor imagery is more effective at recruiting task-related networks	[205, 206]
	• Visual amplification of movements and/or errors in VR might enhance motor training outcomes	[207, 208]

(Table 20.3). Research has shown that after stroke, a window opens when networks become more excitable, and VR has been revealed as an effective tool to engage visuomotor processes such as the ones related to action execution, observation, understanding, and mental simulation. In fact, the manipulation of visual representations has been shown to engage motor networks during passive observation and mental simulation and facilitate the movement of muscles. Thus, the manipulation of these processes through VR cannot only enhance neural activation but also improve motor outcomes.

20.3 Evidence Base: Impact of VR

Virtual reality systems or applications may be divided into custom, those specifically developed for science or rehabilitation and non-custom those that were developed for other purposes (e.g., recreation) but are being adapted for science or recreation. These non-custom systems are often called serious games as they are being applied for science or rehabilitation. We propose that serious games can be further distinguished into rehabilitation or active video games: used to rehabilitate upper limb use, gait and balance, and exergames: used to promote physical activity or exercise. Custom VR systems may include gamification but under these definitions would not be considered a serious game. Defining these terms is an area of ongoing discussion.

Non-custom systems for VR or serious games have included game consoles from Sony, Nintendo, and Microsoft, which were coupled with vision or sensor interfaces. The earliest was the Sony® Eye-Toy®, a camera-based motion capture system designed to be compatible with the PlayStation™ two-entertainment system, which was initially released in 2003. A majority of the initial studies examining rehabilitation applications of this system involved balance activities or gross reaching movements [211]. There were also some upper limb studies that showed evidence of efficacy [212]. Two subsequent systems were released more broadly and have had more substantial impact on the field of rehabilitation, the Wii™ manufactured by Nintendo® and the Kinect™ manufactured by Microsoft®.

The Nintendo® Wii™, which features two accelerometer-based controllers in addition to infrared motion capture capabilities, initially became available in 2006. It was bundled with the Wii-Sports Games and later updated with a more precise controller released with the Wii™ Resort Games. In 2012, the Wii™ Fit game became available. This game was bundled with the Wii™ Balance Board, a force sensor that interfaces with the Wii™ console. These systems have been widely adopted in rehabilitation facilities and nursing homes without modification as a recreation and rehabilitation modality [213]. Surveys of clinicians in Canada and the United States indicate that this system, while discontinued, continues to have the greatest use [214, 215].

The Microsoft® Kinect™, a peripheral for the Xbox series that detects user's movements through a depth-sensing camera, was released to interface with the Xbox 360 in 2010. A substantial body of research related to the validity of measurements of human movement with the Kinect™ has been developed [see [216] for a detailed review]. Analyses of these non-custom games to allow the application to rehabilitation have been conducted for the Wii [35] and the Kinect [217]. These analyses have interpreted the content of the non-custom system's games to include elements of feedback, in particular greater amounts of knowledge of results which may promote game play and engagement, but less knowledge of performance which may lead to poor movement patterns. Therefore, clinicians choosing to incorporate these games into rehabilitation need to carefully observe their clients' movement performance.

The sections on evidence of the impact of VR will be divided by motor control (e.g., upper limb, balance, and gait) and VR system (e.g., custom and non-custom).

20.3.1 Upper Extremities

20.3.1.1 Custom Systems

In 2017, an update was performed on a Cochrane review by Laver et al., which considered the effect of Virtual Reality on upper limb function along with secondary outcomes such as gait, balance, cognitive function, and various QOL measures [128]. This review drew from 72 randomized and quasi-randomized trials and included a sample size of 2470 participants who had experienced a stroke. The results of this review can be broken into two primary categories regarding upper limb function, trials that used VR as the sole treatment strategy for experimental groups, and trials that used VR as a supplementary intervention for experimental subjects. When Virtual Reality was the only intervention, it was found that there was no significant difference in outcomes for intervention versus control groups. However, when Virtual Reality was supplemented to standard therapy it was found that intervention groups had significantly better outcomes when compared to control groups.

It could be argued that the addition of VR as a supplemental form of therapy resulted in more total time spent performing therapeutic interventions, thereby producing significantly better outcomes. Following a stroke, patients are assigned a home exercise program as an adjunct to their regular therapeutic interventions. Normally adherence for such HEPs is low; however, due to VR's effectiveness as a supplemental intervention, it is plausible that the addition of VR interventions could improve adherence, and thereby significantly improve patient outcomes. Studies focused on the relative effects of VR as an adjunct to in-clinic therapy versus traditionally presented exercise as an adjunct are indicated to validate this hypothesis. The balance of this discussion will focus on evidence examining the impact of the effectors trained, interfaces utilized and the severity of the impairment of participants. In addition, some key studies that were not included in these meta-analyses for methodological reasons and papers published following the Cochrane review by Laver [128] will be discussed.

Multiple authors have identified a critical period in which persons in the early subacute period after stroke (less than three months post-stroke) are more able to benefit from motor retraining interventions than persons in the chronic stage of recovery (greater than 6 months) [186, 218]. In order to assess this idea, six studies that evaluated VR's effectiveness in improving upper limb function post-stroke were grouped into two categories. The first category being studies that had sample populations less than 3 months post-stroke, and the second category being studies with sample populations over 6 months post-stroke. In the category of patients less than 3 months post-stroke, two studies were placed; the first study being from Gueye et al. 2020, found a significant difference from the implementation of VR, while the second study from Brunner et al. 2017 found no significant difference [219, 220]. Four different studies examined similar interventions in persons 6 months post-stroke. Two of these studies found a significant difference in upper limb function when VR was implemented [221, 222], but two other studies found no significant difference between control and experimental groups [223, 224]. Taken together, these studies suggest that virtually simulated interventions are not more effective for the delivery of upper extremity therapy during the initial recovery period after stroke. Further studies during the initial recovery period might benefit from refocusing, either on subjects who are too impaired to participate in traditionally presented therapy, or mildly impaired persons with stroke, who are discharged directly to home, without intensive rehabilitation.

Many studies have been written examining the impact of timing and total training volume on the outcomes of relatively short-term interventions utilizing VR (less than 4 weeks). This said, the motivational advantages associated with VR-based interventions and the efficiencies afforded by home-based VR training make the examination of longer intervention periods worthy of attention. To address this question, 12 recent RCT that studied the effectiveness of VR as a treatment for upper limb function post-stroke were examined. Of these 12 articles, 10 fit into the category of being 4 weeks of treatment or less. Of these 10 articles, only 4 showed clearly significant differences between experimental and control group results [219, 221, 225, 226]. Five articles showed no significant difference between VR and control conditions [220, 223, 227–229] and a sixth showed results that differed across outcome measures [230]. In contrast, two studies with treatment lengths of 4 weeks both showed significant differences between experimental and control group outcomes [222, 231]. The mixed results reported by shorter interventions and the consistent group time interactions demonstrated in these two longer studies might imply that treatment length might have some role to play in the effectiveness of VR as an intervention when compared to traditionally presented therapy. Furthermore, an uncontrolled pilot of a twelve-week, home-based intervention in persons with stroke demonstrated excellent adherence and clinically significant improvements in Upper Extremity Fugl Meyer Assessment (UEFMA) score suggests that longer treatment programs are feasible [232]. Clearly, more study of longer interventions is needed.

An important variable of consideration for clinicians designing interventions for patients post-stroke would be frequency. The term "frequency" in this case referring to times per week in which a virtual reality session would take place for a given patient. In order to understand the role of treatment frequency 13 Randomized Control Trials that studied upper limb functional improvement post-stroke when VR was implemented were considered. Nine of these studies utilized treatment protocols with 4 or more treatment sessions per week, and 4 examined protocols with three or less sessions per week. Three of the nine articles with four sessions per week protocols demonstrated statistically significant results [219, 221, 226]. The remaining 5 articles did not demonstrate any significant results [220, 223, 224, 227, 229] and a sixth demonstrated mixed results [230]. Three of the four studies with lower frequencies demonstrated statistically significant differences between VR and control groups [222, 225, 231] and a fourth demonstrated non-significant results [228]. These results suggest that more than two or three VR-based treatment sessions per week might not confer any additional benefits when compared to control therapies. This notion, that Virtual

Reality treatment might elicit significant motor function improvements with a lower treatment frequency is potentially important and warrants further research.

For clinicians who wish to use VR post-stroke it is useful to consider if there are age groups that utilize this family of technology more successfully than others. In order to understand the effects age may have on the effectiveness VR interventions, 13 articles were collected and separated into 2 distinct categories. Ten of these articles examined study populations under the age of sixty. The first category contained all articles with sample populations above the age of 60. Six of these articles reported significant differences between experimental and control groups [221, 222, 225, 227, 231]. One article was found to have mixed results wherein the primary outcome measure, being the UEFMA, was found not to have significant differences between experimental and control trials. However, the secondary outcome measure, the Box and Block Test, did have significant differences. Three of these articles did not demonstrate a difference between VR and controls [220, 224, 227] and a fifth demonstrated mixed results [230]. Interestingly, none of the studies with mean ages above sixty demonstrated better outcomes for VR subjects when compared to controls [223, 228, 229]. This body of evidence suggests that age might play a role in VR therapy effectiveness and that it is plausible that individuals above 60 years of may not benefit from VR-based interventions more than those younger than 60. Alternatively, the differences in effectiveness identified across these studies may be an effect produced by differences in the lived experiences of older subjects, who had less exposure to computer gaming and virtual reality than younger subjects. Large trials with age-stratified samples or smaller studies specifically designed to answer this question are indicated. In addition, previous exposure to technology is a factor that needs to be considered when interpreting the results of technology supported rehabilitation studies. Clinicians should also include an assessment of patient's technology literacy when proposing technology supported interventions.

20.3.1.2 Non-Custom Systems

Several studies of upper extremity rehabilitation have utilized the Wii™ system in patients with stroke. Subjects in several pilot studies of persons with stroke using the Wii™ have demonstrated statistically significant improvements in motor function and activity level clinical tests [233–235]. Even though the Wii™ interface does not collect individual finger movement or grip force data, subjects in another pilot study demonstrated fine motor improvements in persons with stroke following a Wii™-based intervention [236]. Two controlled studies comparing Wii™-based upper extremity interventions and a dose matched traditionally presented upper extremity intervention demonstrated statistically significant improvements at the function and activity levels. Improvements demonstrated by the two groups in both studies did not differ [213, 237]. The Wii™ training group in a third controlled trial made larger improvements on the UEFMA and Box and Blocks test than a dose-matched traditional training group [238]. The Cochrane review by Laver et al. in 2017 identified 7 RCT utilizing an off-the-shelf gaming system compared to 15 RCTs with upper extremity simulated interventions using custom VR systems in persons post-stroke that were methodologically suitable for comparison [128]. Both groups of studies demonstrated significant effects but were not more effective than conventional therapy approaches. A recent systematic review considering 30 studies identified significant benefits for body function and activity measures only for custom VR systems when compared to off-the-shelf VR [239].

A substantial body of research related to the validity of measurements of human movement with the Kinect™ has been developed see [216] for a detailed review as well as a review about translation into practice [240]. However, few studies of the clinical effectiveness of Kinect™-based rehabilitation programs for persons with upper extremity impairments have been published to date. A case/feasibility study with a severely impaired subject demonstrated increased upper extremity active range of motion but no improvements in UEFMA score after a 10-session training program [241]. This subject was severely impaired, which may underestimate the potential of this intervention for less impaired subjects. A case series of five subjects with moderate impairments demonstrated improvements in UEFMA and Wolf Motor Function Test (WMFT) scores that corresponded to increases in cortical activation of the lesioned hemisphere [172]. The changes in clinical test scores and cortical activation demonstrated by subjects in this case series were comparable to those demonstrated by subjects in studies of custom VR systems [164]. Two studies have examined the addition of Kinect™-based upper extremity rehabilitation activities to a program of traditionally presented therapy [57, 242]. Control groups for both of these studies performed the same volume of traditionally presented therapy as the experimental group. As would be expected, the subjects performing the additional Kinect™-based therapy demonstrated larger changes in active range of motion, ADL ability, and larger improvements in UEFMA, WMFT, and Motor Activity Log (MAL) tests. More rigorous testing of Kinect™-based rehabilitation activities will be necessary to evaluate their value relative to custom VR or traditionally presented therapy.

20.3.2 Balance and Gait

20.3.2.1 Custom Systems

Historically, the development and application of VR systems for neurorehabilitation focused on the upper limbs. This may have been motivated by two main factors. First, relative to upper limb use, balance and walking skills are more commonly and extensively recovered after a stroke. Second, building balance and walking VR-based systems require greater technical and space requirements to meet the special physical and safety challenges. In contrast to most upper limb systems, which allow patients to be seated while performing movements with the upper extremities, balance and walking skills, for the most part, require patients to be upright or to walk. There exists a modest yet increasing body of work on the development and use of customized VEs for walking recovery and balance, which is reported in several topic-specific reviews [243–249] as well as in overview reviews [250–252]. In contrast to the 1038 participants who participated in the upper extremity studies included in Laver's Cochrane Review of Stroke Rehabilitation, there were only 139 persons involved in balance and mobility training, with only seven studies where gait speed was measured.

Visual feedback is a common element in evidence-based interventions for balance training post-stroke [253]. It is used to provide participants information about the verticality of their posture, which may be impaired due to sensory and perceptual deficits, as well as their weight distribution. Both of these attributes are incorporated into VEs for balance rehabilitation. The GestureTek® IREX® video capture system based on chroma key technology was first used in studies involving individuals who had sustained a TBI, where slight improvements were detected in balance [254, 255], confidence [256], and reaction time [256], compared to conventional training protocols. The system has also been used with persons post-stroke, providing benefits to the sensory organization, motor function, and balance. In general, training with the system provided benefits that were detected in scales related to balance but not to gait. A randomized controlled trial involving higher functioning persons post-stroke who were inpatients examined the effects of using the system in addition to a conventional rehabilitation program. There was, however, no significant improvement in walking ability and gait speed derived from the use of the system [257].

Force platforms have been used to estimate and visualize participants' center of pressure providing visual feedback during displacements toward the target [253]. The use of force platforms in combination with customized virtual exercises has also been explored. The training of the ankle and hip strategies during weight-shifting exercises adapted to the particular limits of stability of each subject provided benefits to conventional physical therapy interventions in the general balance condition and in the

maximum reachable distance [258] (Fig. 20.5). Interestingly, these effects were retained at follow-up after the intervention [259]. A recent analysis of aggregated data from different studies and unpublished data from Llorens corroborated these results and showed consistent improvements in the Berg Balance Scale and the Functional Reaches Test after an intervention using weight-shifting exercises. The gains were maintained, and even enhanced, one month after the intervention [260]. However, it is important to highlight that the improvement facilitated by these exercises, and more importantly, the maintenance of gains, are severely influenced by time since injury [261]. According to this, fewer gains and more difficulties in maintaining them should be expected with greater chronicity.

Similar to balance platforms, standing frames equipped with gyroscopes can detect postural tilts, enabling interaction with the VE through weight transferences. These systems have been used in home-based interventions with individuals post-stroke, reporting improvements in balance and gait [262, 263]. However, the use of VR did not provide significant benefits to the training with the standing frame alone. Research on the effectiveness of weight-shifting exercises in sitting is very limited. The scant literature about it has focused on training trunk movements through VR-based tasks that required trunk lean and reaching beyond arms' length using Jintronix software (Jintronix, Montreal, Quebec, Canada) interfaced with a pressure mat, showing comparable benefits to conventional physical therapy interventions and variable requirements of trunk stability [264, 265].

Walking on a treadmill interfaced with VE has been used to promote recovery of walking for persons post-stroke. The inclusion of visual and vibrotactile augmentation while stepping over virtual objects during walking on a treadmill improved walking better than stepping over real-world objects. Several studies have reported the combined use of treadmills and VR and its effects on the gait and, to a lesser extent, static balance of stroke survivors. Users commonly walk on a treadmill while the VE is displayed by projectors [59, 266] or TV screens [267–269],

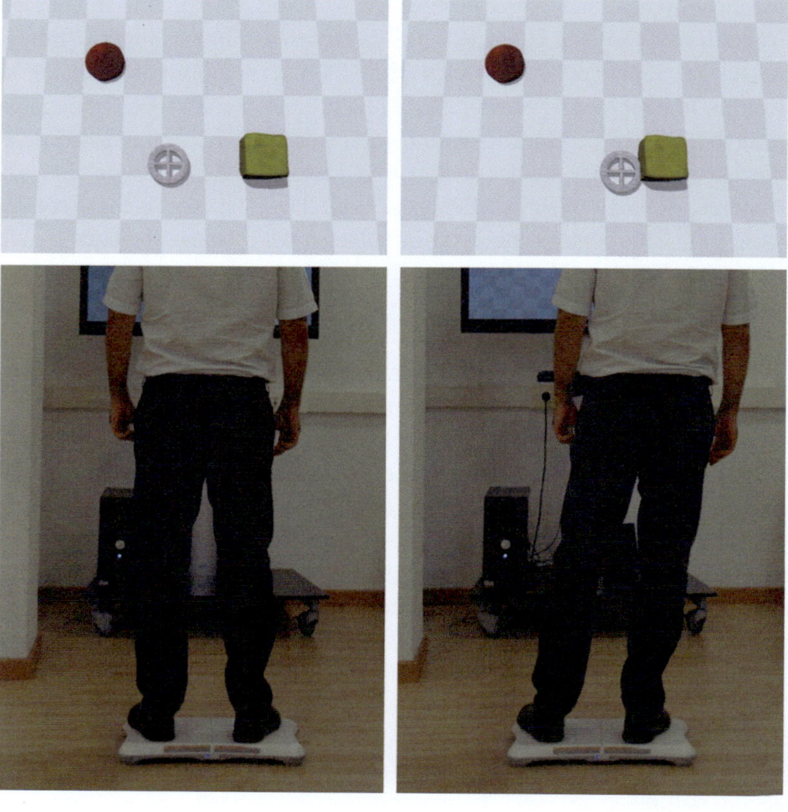

Fig. 20.5 In the system by Llorens et al., after registering their maximum excursion in the medial–lateral and anterior–posterior plane, exercises are adapted to each client's particular motor limitations [186]. Exercises require participants to perform postural adjustments involving the ankle and hip strategies to displace their center of pressure toward different targets

showing real-world video recording [266, 269, 270] or virtual scenarios [267, 268]. Interventions involved tasks of variable difficulty, from walking, dual-task performance, such as remembering and identifying groceries while navigating through a pre-recorded walking scene in a real supermarket [269] or reaching objects with the upper limbs on the SeeMe system (Brontes Processing: Gliwice, Poland) [271], or stepping with either the paretic or nonparetic limb [59]. The use of feedback provided by VR favored not only gait [59, 266, 270–272], but also static balance, sway, sit-to-stand movements, and the use of the paretic limb [266, 267, 269, 271]. The enhanced motor performance after adding VEs to treadmill training could have been promoted by an increased entrainment of brain activity involved in motor planning and learning (maybe through the mirror-neuron system), as suggested by EEG findings on addition of VR to robot-assisted gait training [272].

In addition to treadmill walking simulations, several investigators have used stepping, pre-gait activities, and even training of the lower extremity in sitting to improve walking for persons in the chronic phase post-stroke [46, 70, 168]. Llorens et al. reported that the training through virtual stepping exercises improved balance compared to conventional interventions [273] (Fig. 20.6). Individuals were required to step on items that appeared around a circle with the closest foot while maintaining the other foot inside a circle. This intervention also promoted improvements in gait speed, which could be derived from the training of movements similar to those used in the stance phase of the gait cycle. The system was also used in a home-based intervention with similar results to those obtained in in-clinic interventions. The analysis of aggregated data from 131 individuals with stroke from different studies and unpublished data showed consistent improvements in the Berg Balance

Scale and the 10-m Walk Test after the intervention, which were improved and maintained, respectively, one month after the intervention [260]. Mirelman et al. coupled VR with a robot-based training of the lower extremity, where participants were required to perform movements with the ankle while sitting to navigate a plane or a boat through a VE. When compared to the robot alone, the VR-robot combination was superior in improving walking velocity and distance in laboratory, clinical, and community-based tests [46]. You and colleagues used the IREX® system to promote functional ambulation and waking through the training of stepping movements, side-to-side weight shifting, and sideways navigation. Interestingly, the locomotor recovery was associated with cortical reorganization from aberrant ipsilateral to more normal contralateral activation of the sensorimotor cortex [63]. Recent interventions involving VEs that required similar interaction, and also upper limb movements, provided consistent results with that seminal study, showing improvements in balance that were comparable and almost significantly greater than those provided by conventional physical therapy exercises [274–276].

20.3.2.2 Non-Custom Systems

Studies have reported on outcomes of non-custom systems (e.g., Wii and Kinect) for balance and mobility training of people post-stroke. Early on there were several case reports of people in the chronic phase post-stroke, which reported positive outcomes for balance and mobility interventions [277, 278]. More recently, eight pilot clinical trials using video games to improve balance and mobility have been reported. They have predominantly been conducted with subjects in the chronic phase post-stroke [35, 279–282], but there is now some support for application to persons in the subacute [283] and acute phases of recovery [284, 285].

The quality of the research is improving as more of the trials have active control groups and follow-up measurements [282–286]. However, comparing among studies is complicated based on substantial differences in dose and acuity.

Several studies had unequal doses and did not use active controls [280, 281]. Studies conducted in the acute and subacute care setting using active controls showed a positive effect for balance and functional ambulation tests favoring the games [283, 285]. In contrast, studies with active controls and balanced doses of persons with chronic strokes favored standard of care [35] or showed no difference for balance and mobility measures, but favored the VR group for enjoyment measures [200, 282]. As with the upper limb studies, a better understanding of how acuity modifies the benefits of VE training will guide the future clinical application.

Non-custom systems have used similar technologies as the customized VR systems. PlayStation® two EyeToy: Play™ is similar to the IREX® system [205] and was tested at home in a case study with an individual 2 years post-stroke [277]. The training of postural adaptations during bilateral stance in subjects post-stroke has been mainly facilitated by the Nintendo® Wii™ Balance Board, a force platform peripheral device for the Nintendo® Wii™, which allows interaction through displacements of the center of pressure, it is, through weight shifting [280–283, 285]. Interestingly, some studies have analyzed the combination of static exercises using the Wii™ Balance Board with more dynamic exercises. Deutsch et al. compared standard of care with the Nintendo® Wii™ games and reported no between-group differences, but a greater number of within-group improvements for balance and mobility measures for the standard of care group [35]. Fritz et al. added EyeToy: Play™ games reporting small positive effects of this training compared to traditional therapy [279]. The combined training of weight transferences using the Wii™ Balance Board with dynamic balance exercises with the Microsoft® Kinect™ promoted improvement in the maximum reachable distance in acute subjects post-stroke [284], but were equally effective as conventional physical therapy in maintaining physical function outcomes and ADLs in the chronic population [287].

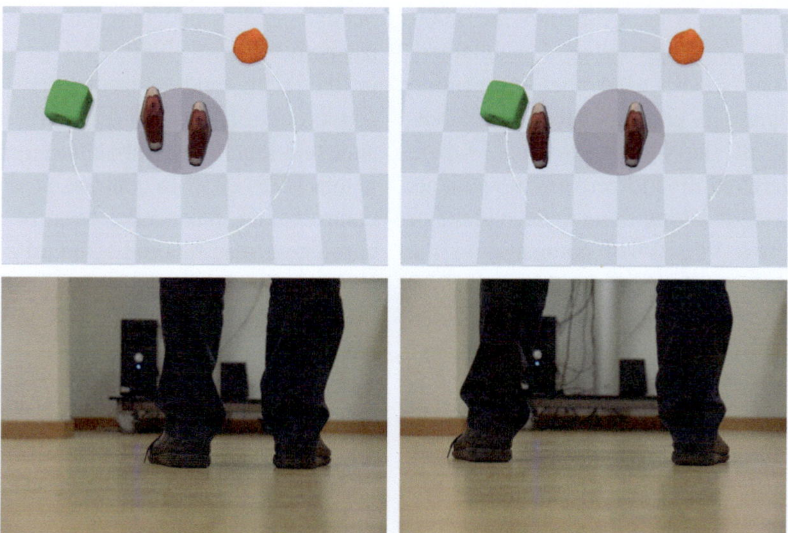

Fig. 20.6 In the system by Llorens et al., the virtual environment consisted of a checkered floor, whose center was indicated by a darkened circle, and jelly items that rose from the ground around the circle [194]. The goal of the exercise was to reach the items with the nearest feet while maintaining the supporting foot within the circle. After reaching the item, the extended extremity had to be recruited to the body within the boundaries of the circle. Otherwise the exercise did not allow new items to be reached

20.3.3 Activity Promotion

Movement-based VR systems have focused on sensorimotor rehabilitation, but there is an emerging application to fitness promotion in persons post-stroke. Given the importance of physical activity [288] and the barriers to exercise encountered by people post-stroke [289], VR is proposed as a facilitator of activity. The VR may be delivered using a custom system coupled with exercise equipment such as a bicycle or a treadmill, or a non-custom system played as an exergame. Custom systems allow the harnessing of heart rate to drive the exercise intensity. A group has developed a VR-augmented cycling system that uses heart rate as an input to the VE [290] (Fig. 20.7). In a pilot study, participants post-stroke who trained on the system had significant improvements in VO2 sub-max bicycle test and mobility outcomes as well as changes in force kinetics during cycling [291].

Non-custom VR systems or exergames have been explored to promote activity and fitness for persons post-stroke. Studies have been either cross-sectional characterizing energy expenditure or clinical trials assessing the cardiovascular benefits of exergames. The ability of persons in the chronic phase post-stroke to increase their exercise intensity using exergames has been reported by three groups [292–294]. Hurkmans et al. characterized two predominantly upper limb Nintendo® Wii™ games (tennis and boxing) and reported that they produced moderate (three to five metabolic equivalents) exercise intensity [293]. Kafri and colleagues in a case-control series compared the energy expenditure and exercise intensity between individuals post-stroke with moderate mobility limitations to semi-active healthy matched controls while playing both Kinect™ and Wii™ games in sitting and standing [292]. The games were categorized as a standing balance task to upper limb predominant (boxing) and lower limb predominant (running). Generally, post-stroke individuals had lower energy expenditure (at the low end of moderate) than the healthy controls (moderate to low end of vigorous), during similar activities. They did, however, exercise in the heart rate intensity recommended for fitness. Silva de Sousa reported similar findings that playing

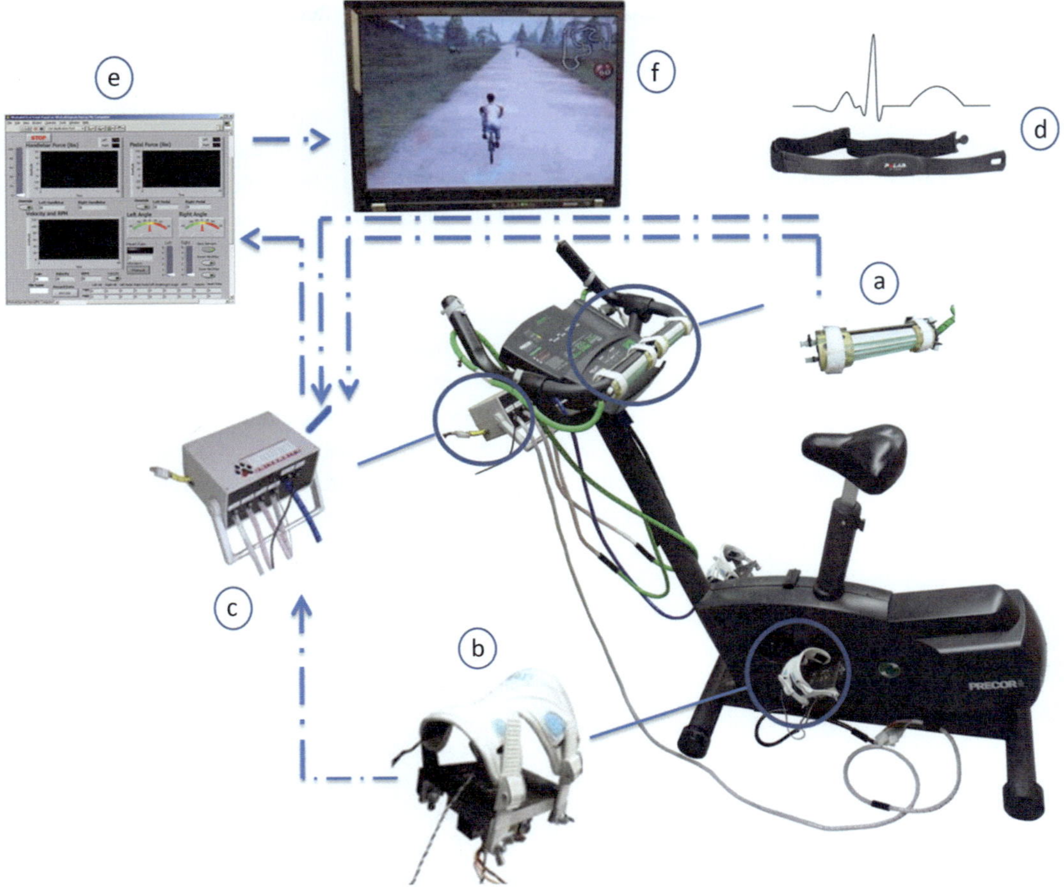

Fig. 20.7 VRACK system complete overview; **a** Handlebar module; **b** Smart pedal; **c** Power supply, preamplifier, and the data acquisition board; **d** Heart rate monitor, **e** Practitioner interface; **f** Virtual reality environment [209]. Reproduced with permission of the Rivers Lab

Kinect™ games of tennis and boxing produced reliable changes in VO_2 which were at a lower aerobic intensity and heart rate responses that were at a higher aerobic intensity [294].

Clinical trials that tested the efficacy of non-custom exergames to improve fitness have been conducted for persons in the chronic as well as subacute phase post-stroke. Game consoles have included the Microsoft with the Kinect™- with dance and Adventure games [295, 296], the Nintendo Wii™ with Sports games [297]. VO_2 and activity improvements were reported for persons in the chronic phase post-stroke who played the Kinect™ Just Dance 3 games [295]. In a large trial (n = 640), Tollar and colleagues reported positive outcomes for persons in the subacute phase (2–4 weeks) post-stroke who played Kinect™ Adventure exergames of Reflex Ridge and Space Pop plus Just Dance 3 \in addition to agility training. They specifically compared a dose of one time a day for five consecutive days over five weeks (25 sessions) to twice a day (for a total of 50 sessions). The higher dose produced significant and clinically meaningful gains both in the six-meter walk test and reductions in systolic blood pressure (interpreted by the authors as anti-hypertensive). These findings are important as they protect against a recurrent stroke. Interestingly in a study that focused on upper limb use comparing Wii-Sports games to modified constraint-induced movement therapy, the Wii™ movement therapy group demonstrated aerobic gains suggesting that upper limb therapy could be combined with aerobic activity [297].

Non-custom gains have been critiqued because they cannot be adjusted for persons post-stroke. A careful comparison between custom and non-custom Kinect™ exergames played in a single session by persons in the chronic phase post-stroke showed that the exercise intensity was statistically greater for the custom game but played in the same intensity bands for METs (moderate) and [298] Heart rate results were similar. Importantly, the participants reported less perceived effort and greater enjoyment with the custom games, and greater symmetry of lower extremity kinematics [299]. It appears the VR in the form of either non-custom or custom exergames may be a valid tool for activity promotion, given their potential to increase motivation for exercise and to promote adherence. Whether custom games are superior to non-custom games remains to be further tested.

20.3.4 Summary

A steady proliferation of studies comparing virtual rehabilitation interventions to traditionally presented rehabilitation in persons with stroke has developed over the past 15–20 years. Comparable outcomes have been reported when comparing virtual and real-world upper extremity training in subjects with more acute strokes. The best-developed area of this literature examines upper extremity interventions in subjects with chronic strokes using customized lab-based systems. These comparisons describe slightly better outcomes for virtual rehabilitation interventions. This advantage is more pronounced in mildly impaired subjects. More, larger, and better controlled studies are required to draw definitive conclusions along these two lines of inquiry.

A smaller literature has examined the relative efficacy of a VR-based rehabilitation on walking ability (as measured by gait speed and distance) in persons with stroke. A non-significant trend toward better outcomes for virtual reality-based training as compared to real-world gait training

has been identified. The balance of studies comparing the impact of these two training approaches considers the kinetics and kinematics of gait. Neither approach to training has been associated with significant advantages across multiple studies. In contrast, balance interventions presented in virtual environments have been associated with significantly better outcomes than traditionally presented balance training across a wide range of balance measures. An expansion of the size and number of studies and a focus on a smaller set of outcome measures will be necessary to identify an additive effect for virtual environments on gait training. Further, VR primarily with non-custom games has some preliminary support as a tool for promoting physical activity.

20.4 Considerations for Future Research

While there is consensus that neuroplasticity is central to the motor recovery process, there is a relatively small literature examining the impact of VR interventions on positive, neuroplastic adaptations in persons with neurologic injuries. Some pioneering investigations utilizing neuroimaging have been conducted. An expansion of this area of inquiry could optimize and accelerate both the design and implementation of VR-based rehabilitation interventions. However, the cost and need for large transdisciplinary teams to perform studies of this type have kept progress in this area slow.

There is also consensus that motor learning is central to the process of neuroplasticity, and VR-based rehabilitation interventions are typically constructed with attention paid to accepted principles of motor learning. Examinations of the motor learning accomplished by virtual interventions have predominantly focused on the transfer of motor skills learned in VEs to veridical world motor skills and performance improvements achieved during virtual interventions to a lesser extent, both with favorable results. A broader implementation of formal motor learning paradigms to the study of virtual rehabilitation might offer a more efficient and cost-effective approach to optimizing virtual rehabilitation. By their nature, interfaces designed for VE-based activities are well suited to collect the necessary data. In addition, simulated activities are easily presented in the systematic, reproducible fashion necessary for studying within and between session learning.

Science related to motivation may, first, enhance the volume of motor practice performed independently by patients in their homes. Home practice is critical in areas with limited access to a therapist due to availability or reimbursement issues, and compliance with home practice schedules is typically poor. Second, motivation science may enhance the frequency and duration of the performance of fitness-oriented activities in persons with disabilities. Motivation and access are primary obstacles to the regular performance of fitness activities with a wide variety of disabilities, both of which can be overcome with well-designed, simulated exercise programs.

20.5   Conclusions

A review of this chapter should leave the reader with the impression that (1) there is a science underpinning virtual rehabilitation, (2) individuals with neurological impairments can effectively use VE, as they feel being as present and bodily represented in them as healthy subjects, and (3) the evidence base related to the efficacy of virtual rehabilitation has confirmed that it can be a viable and, for the upper limb, a superior alternative to traditionally presented activities. While these impressions are validating on the one hand, they also identify a need for continued improvement. This said, trends also emerge, indicating opportunities for optimizing virtual rehabilitation and expanding the populations and areas in which it is practiced.

Early work in virtual reality-based rehabilitation for persons with stroke was informed by concepts of neuroplasticity and motor learning. Simulations incorporated augmented feedback, knowledge of performance, and knowledge of results. An ideal combination of these principles

has not been elucidated. Massed practice was promoted as tool to overcome lack of motivation for repetitive task practice required for behavioral outcomes and neural plasticity. The high number of repetitions per unit time has been robustly supported for both custom and non-custom virtual reality applications.

The user's experience as it is affected by the presentation of information via the user's visual, auditory, kinesthetic, and tactile senses has been another area of study. A small body of literature supports that the presentation approach and quality of sensory information provided to participants with strokes affects the way they move during virtual interventions. A parallel literature describes differences in brain activity during virtual interactions elicited by differing presentations of virtually simulated motor activity. This brain activity has been linked to processes related to the execution, observation, understanding, and mental simulation of real-world movement.

The literature comparing virtual rehabilitation interventions to traditionally presented rehabilitation in persons with stroke has grown slowly but steadily over the past 15–20 years. This literature cites that VR-based interventions produce comparable improvements in upper extremity function and balance when compared to traditionally presented rehabilitation interventions. To date the literature on virtual interventions to improve gait is not developed sufficiently to evaluate its efficacy compared to traditionally presented interventions.

Two important trends will be critical for shaping the future development of virtual reality. One key to the transition of virtual rehabilitation to the home environment has been the development of lower cost, but effective interfaces. The ability to customize the application of Kinect™ like sensors should prove to accelerate this transition, allowing for the use of off-the-shelf equipment to access simulations explicitly designed (custom) for rehabilitation. Clearly, virtual rehabilitation is an expanding area in the field of technology-based rehabilitation and has an evidence base that is growing in terms of size and quality. Several challenges described above need to be addressed but the field continues to hold promise to answer key issues faced by modern healthcare.

References

1. Burdea GC, Coiffet P. Virtual reality technology. Presence. 2003;12(6):663–4.
2. Wilson PN, Foreman N, Stanton D. Virtual reality, disability and rehabilitation. Disabil Rehabil. 1997;19(6):213–20.
3. Adamovich SV, et al. Sensorimotor training in virtual reality: a review. NeuroRehabil. 2009;25(1):29–44.
4. Slater M. Place illusion and plausibility can lead to realistic behaviour in immersive virtual environments. Philos Trans R Soc B Biol Sci. 2009;364(1535):3549–57.
5. Slater M, Wilbur S. A framework for immersive virtual environments (FIVE): Speculations on the role of presence in virtual environments. Presence. 1997;6(6):603–16.
6. Baños RM, et al. Immersion and emotion: their impact on the sense of presence. Cyberpsychol Behav. 2004;7(6):734–41.
7. Llorens R, et al. Tracking systems for virtual rehabilitation: objective performance vs subjective experience: a practical scenario. Sensors. 2015;15(3):6586–606.
8. Stanney KM, Mourant RR, Kennedy RS. Human factors issues in virtual environments: a review of the literature. Presence. 1998;7(4):327–51.
9. Arzy S, et al. Neural mechanisms of embodiment: asomatognosia due to premotor cortex damage. Arch Neurol. 2006;63(7):1022–5.
10. Legrand D. The bodily self: the sensori-motor roots of pre-reflective self-consciousness. Phenom Cog Sci. 2006;5(1):89–118.
11. Berlucchi G, Aglioti S. The body in the brain: neural bases of corporeal awareness. Trends Neurosci. 1997;20(12):560–4.
12. Schultze U. Embodiment and presence in virtual worlds: a review. J Inf technol. 2010;25(4):434–49.
13. Grassini S, Laumann K. Questionnaire measures and physiological correlates of presence: a systematic review. Front Psychol. 2020;11:349.
14. Borrego A, et al. Embodiment and presence in virtual reality after stroke. a comparative study with healthy subjects. Front Neurol, 2019. 10(1061).
15. Schwind V, et al. Using presence questionnaires in virtual reality. In: Proceedings of the 2019 CHI conference on human factors in computing systems. 2019.
16. Longo MR, et al. What is embodiment? A psychometric approach. Cognition. 2008;107(3):978–98.
17. Tsakiris M. My body in the brain: A neurocognitive model of body-ownership. Neuropsychologia. 2010;48(3):703–12.
18. Botvinick M, Cohen J. Rubber hands ′feel/′ touch that eyes see. Nature. 1998;391(6669):756.
19. Petkova VI, Ehrsson HH. If I were you: perceptual illusion of body swapping. PLoS ONE. 2008;3(12):e3832.
20. Petkova VI, Khoshnevis M, Ehrsson HH. The perspective matters! Multisensory integration in ego-centric reference frames determines full body ownership. Front Psychol, 2011. 2.
21. Slater M, et al. Inducing illusory ownership of a virtual body. Front Neurosci. 2009;3(2):214–20.
22. Banakou D, Groten R, Slater M. Illusory ownership of a virtual child body causes overestimation of object sizes and implicit attitude changes. Proc Natl Acad Sci U S A. 2013;110(31):12846–51.
23. Yee N, Bailenson J. The proteus effect: the effect of transformed self-representation on behavior. Human Commun Res. 2007;33(3):271–90.
24. Rand D, et al. Comparison of Two VR platforms for rehabilitation: video capture versus HMD. Presence. 2005;14(2):147–60.
25. Rand D, Kizony R, Weiss PT. The sony playstation II eyetoy: low-cost virtual reality for use in rehabilitation. J Neurol Phys Ther. 2008;32(4):155–63.
26. Perez-Marcos D, Sanchez-Vives M, Slater M. Is my hand connected to my body? The impact of body continuity and arm alignment on the virtual hand illusion. Cog Neurodyn. 2012;6(4):295–305.
27. Weech S, Kenny S, Barnett-Cowan M. Presence and cybersickness in virtual reality are negatively related: a review. Front Psychol. 2019;10:158.
28. Kim J, Luu W, Palmisano S. Multisensory integration and the experience of scene instability, presence and cybersickness in virtual environments. Comput Hum Behav. 2020;113: 106484.
29. Specht J, et al. Acceptance of immersive head-mounted display virtual reality in stroke patients. Comput Human Behavior Reports. 2021;4: 100141.
30. Høeg ER, et al. System immersion in virtual reality-based rehabilitation of motor function in older adults: a systematic review and meta-analysis. Front Virtual Reality. 2021;2:30.
31. Schmidt RA. Motor learning principles for physical therapy. in Contemporary management of motor control problems: Proceedings of the II STEP Conference. Alexandria, VA: Foundation for Physical Therapy. 1991.
32. Wulf G, Shea C, Lewthwaite R. Motor skill learning and performance: a review of influential factors. Med Educ. 2010;44(1):75–84.
33. Deutsch JE, et al. Development and application of virtual reality technology to improve hand use and gait of individuals post-stroke. Restor Neurol Neurosci. 2004;22(3–5):371–86.
34. Holden MK. Virtual environments for motor rehabilitation: Review. Cyberpsychol Behav. 2005;8(3):187–211.
35. Deutsch JE, et al. Nintendo wii sports and wii fit game analysis, validation, and application to stroke

rehabilitation. Top Stroke Rehabil. 2011;18(6):701–19.

36. Fluet GG, Deutsch JE. Virtual reality for sensorimotor rehabilitation post-stroke: the promise and current state of the field. Curr Phys Med Rehabil Rep. 2013;1(1):9–20.

37. Levin MF, Weiss PL, Keshner EA. Emergence of virtual reality as a tool for upper limb rehabilitation: incorporation of motor control and motor learning principles. Phys Ther. 2015;95(3):415–25.

38. Laver KE, et al., Virtual reality for stroke rehabilitation. Cochrane Database Syst Rev, 2015. **2**.

39. Levac, D. and H. Sveistrup, Motor Learning and Virtual Reality, in Virtual Reality for Physical and Motor Rehabilitation, P.L. Weiss, E.A. Keshner, and M.F. Levin, Editors. 2014, Springer: New York.

40. Kleim JA, Barbay S, Nudo RJ. Functional reorganization of the rat motor cortex following motor skill learning. J Neurophysiol. 1998;80(6):3321–5.

41. Janssen H, et al. An enriched environment increases activity in stroke patients undergoing rehabilitation in a mixed rehabilitation unit: a pilot non-randomized controlled trial. Disabil Rehabil. 2014;36(3):255–62.

42. Sigrist R, et al. Augmented visual, auditory, haptic, and multimodal feedback in motor learning: A review. Psychon Bull Rev. 2013;20(1):21–53.

43. Winstein CJ. Knowledge of results and motor learning–implications for physical therapy. Phys Ther. 1991;71(2):140–9.

44. Klinger E, Cherni H, Joseph P-A. Impact of contextual additional stimuli on the performance in a virtual activity of daily living (vADL) among patients with brain injury and controls. Int J Disabil Human Dev. 2014;13(3):377–82.

45. Molier BI, et al. Nature, timing, frequency and type of augmented feedback; does it influence motor relearning of the hemiparetic arm after stroke? A systematic review.. Disabil Rehabil. 2010;32(22):1799–809.

46. Mirelman A, Bonato P, Deutsch JE. Effects of training with a robot-virtual reality system compared with a robot alone on the gait of individuals after stroke. Stroke. 2009;40(1):169–74.

47. Subramanian SK, et al. Arm motor recovery using a virtual reality intervention in chronic stroke: randomized control trial. Neurorehabil Neural Repair. 2013;27(1):13–23.

48. Holden M, et al. Virtual environment training improves motor performance in two patients with stroke: case report. J Neurol Phys Ther. 1999;23(2):57–67.

49. Schmidt, R.A., *Motor learning and performance: a situation-based learning approach*. 2008: Human Kinetics.

50. Magill, R.A., *Motor Learning and Control: Current Concepts and Applications*. 9th ed. 2007: McGraw Hill. 466.

51. Lam P, et al. A haptic-robotic platform for upper-limb reaching stroke therapy: Preliminary design and evaluation results. J NeuroEng Rehabil. 2008;5(1):15.

52. Colomer C, et al. Effect of a mixed reality-based intervention on arm, hand, and finger function on chronic stroke. J Neuroeng Rehabil. 2016;13(1):45.

53. Pichiorri F, et al. Brain–computer interface boosts motor imagery practice during stroke recovery. Ann Neurol. 2015;77(5):851–65.

54. Kizony R, et al. Cognitive load and dual-task performance during locomotion poststroke: a feasibility study using a functional virtual environment. Phys Ther. 2010;90(2):252–60.

55. Adams RJ, et al. Assessing upper extremity motor function in practice of virtual activities of daily living. IEEE Trans Neural Syst Rehabil Eng. 2015;23(2):287–96.

56. Cameirão MS, et al. Neurorehabilitation using the virtual reality based Rehabilitation Gaming System: methodology, design, psychometrics, usability and validation. J Neuroeng Rehabil. 2010;7:48.

57. Shin J-H, Ryu H, Jang SH. A task-specific interactive game-based virtual reality rehabilitation system for patients with stroke: a usability test and two clinical experiments. J Neuroreng Rehabil. 2014;11(1):32.

58. Levac DE, Huber ME, Sternad D. Learning and transfer of complex motor skills in virtual reality: a perspective review. J Neuroeng Rehabil. 2019;16(1):121.

59. Yang YR, et al. Virtual reality-based training improves community ambulation in individuals with stroke: a randomized controlled trial. Gait Posture. 2008;28(2):201–6.

60. Fung J, et al. A treadmill and motion coupled virtual reality system for gait training post-stroke. Cyberpsychol Behav. 2006;9(2):157–62.

61. Fung J, Perez CF. Sensorimotor enhancement with a mixed reality system for balance and mobility rehabilitation. in Engineering in Medicine and Biology Society, EMBC, 2011 Annual International Conference of the IEEE. 2011. IEEE.

62. Jaffe DL, et al. Stepping over obstacles to improve walking in individuals with poststroke hemiplegia. J Rehabil Res Dev. 2004;41(3A):283–92.

63. You SH, et al. Virtual reality–induced cortical reorganization and associated locomotor recovery in chronic stroke an experimenter-blind randomized study. Stroke. 2005;36(6):1166–71.

64. Kwakkel G, et al. Effects of augmented exercise therapy time after stroke: a meta-analysis. Stroke. 2004;35(11):2529–39.

65. Schmidt, R.A. and T. Lee, Motor control and learning. 2011: Human Kinetics Ltd; 5th Revised edition edition (1 Mar. 2011).

66. Kleim JA, Jones TA. Principles of experience-dependent neural plasticity: implications for rehabilitation after brain damage. J Speech Lang Hear Res. 2008;51(1):S225–39.

67. Rand D, et al. Eliciting upper extremity purposeful movements using video games: a comparison with traditional therapy for stroke rehabilitation. Neurorehabil Neural Repair. 2014;28(8):733–9.

68. Remple MS, et al. Sensitivity of cortical movement representations to motor experience: evidence that skill learning but not strength training induces cortical reorganization. Behav Brain Res. 2001;123(2):133–41.

69. Dancause N, Nudo RJ. Shaping plasticity to enhance recovery after injury. Prog Brain Res. 2011;192:273–95.

70. Llorens R, et al. Effectiveness, usability, and cost-benefit of a virtual reality-based telerehabilitation program for balance recovery after stroke: a randomized controlled trial. Arch Phys Med Rehabil. 2015;96(3):418–25.

71. Cameirão M, et al. Virtual reality based rehabilitation speeds up functional recovery of the upper extremities after stroke: A randomized controlled pilot study in the acute phase of stroke using the Rehabilitation Gaming System. Restor Neurol Neurosci. 2011;29(5):287–98.

72. Adamovich SV, et al. Incorporating haptic effects into three-dimensional virtual environments to train the hemiparetic upper extremity. IEEE Trans Neural Syst Rehabil Eng. 2009;17(5):512–20.

73. Fluet GG, et al. Robots integrated with virtual reality simulations for customized motor training in a person with upper extremity hemiparesis: a case report. J Neurol Phys Ther. 2012;36(2):79–86.

74. Homberg V. Evidence based medicine in neurological rehabilitation–a critical review. Acta Neurochir Suppl. 2005;93:3–14.

75. Ryan RM, Deci EL. Intrinsic and extrinsic motivations: Classic definitions and new directions. Contemp Educ Psychol. 2000;25(1):54–67.

76. Swanson LR, Whittinghill DM. Intrinsic or extrinsic? using videogames to motivate stroke survivors: a systematic review. Games Health J. 2015;4(3):253–8.

77. Boyle EA, et al. An update to the systematic literature review of empirical evidence of the impacts and outcomes of computer games and serious games. Comput Educ. 2016;94:178–92.

78. Beristain-Colorado MDP, et al. Standardizing the development of serious games for physical rehabilitation: conceptual framework proposal. JMIR Serious Games. 2021;9(2): e25854.

79. Barrett N, et al. The use and effect of video game design theory in the creation of game-based systems for upper limb stroke rehabilitation. J Rehabilitation Assistive Technol Eng. 2016;3:2055668316643644.

80. Linehan, C., et al. Practical, appropriate, empirically-validated guidelines for designing educational games. in Proceedings of the SIGCHI Conference on Human Factors in Computing Systems. 2011. ACM.

81. Faria AL, et al. Combined cognitive-motor rehabilitation in virtual reality improves motor outcomes in chronic stroke–a pilot study. Front Psychol. 2018;9:854.

82. Kannan L, et al. Cognitive-motor exergaming for reducing fall risk in people with chronic stroke: a randomized controlled trial. NeuroRehabilitation. 2019;44(4):493–510.

83. House G, et al. Integrative rehabilitation of residents chronic post-stroke in skilled nursing facilities: the design and evaluation of the BrightArm Duo. Disabil Rehabil Assist Technol. 2016;11(8):683–94.

84. Parker J, et al. The provision of feedback through computer-based technology to promote self-managed post-stroke rehabilitation in the home. Disabil Rehabil Assist Technol. 2014;9(6):529–38.

85. Cirstea CM, Ptito A, Levin MF. Feedback and cognition in arm motor skill reacquisition after stroke. Stroke. 2006;37(5):1237–42.

86. Kilduski NC, Rice MS. Qualitative and quantitative knowledge of results: effects on motor learning. Am J Occup Ther. 2003;57(3):329–36.

87. Rosati G, et al. On the role of auditory feedback in robot-assisted movement training after stroke: review of the literature. Comp Int Neurosci. 2013;2013:11.

88. Lauber B, Keller M. Improving motor performance: Selected aspects of augmented feedback in exercise and health. Eur J Sport Sci. 2014;14(1):36–43.

89. Cameron J, Banko KM, Pierce WD. Pervasive negative effects of rewards on intrinsic motivation: The myth continues. Behav Analyst. 2001;24(1):1.

90. Paulino, T., A.L. Faria, and S.B. i Badia. Reh@ City v2. 0: a comprehensive virtual reality cognitive training system based on personalized and adaptive simulations of activities of daily living. in 2019 5th Experiment International Conference (exp. at'19). 2019. IEEE.

91. Vourvopoulos, A., A. Ferreira, and S.B. i Badia. *NeuRow: an immersive VR environment for motor-imagery training with the use of brain-computer interfaces and vibrotactile feedback. in International Conference on Physiological Computing Systems. 2016. SCITEPRESS.*

92. Serradilla, J., et al. *Automatic assessment of upper limb function during play of the action video game, circus challenge: validity and sensitivity to change. in 2014 IEEE 3nd International Conference on Serious Games and Applications for Health (SeGAH). 2014. IEEE.*

93. Neil A, et al. Sony PlayStation EyeToy elicits higher levels of movement than the Nintendo Wii: implications for stroke rehabilitation. Eur J Phys Rehabil Med. 2012;49(1):13–21.

94. Burke JW, et al. Optimising engagement for stroke rehabilitation using serious games. Vis Comput. 2009;25(12):1085–99.

95. Koritnik T, et al. Comparison of visual and haptic feedback during training of lower extremities. Gait Posture. 2010;32(4):540–6.

96. Lee, M.H., et al. Towards personalized interaction and corrective feedback of a socially assistive robot for post-stroke rehabilitation therapy. in 2020 29th IEEE International Conference on Robot and Human Interactive Communication (RO-MAN). 2020. IEEE.

97. Malone TW, Lepper MR. Making learning fun: A taxonomy of intrinsic motivations for learning. Aptitude, learning, and instruction. 1987;3:223–53.

98. Yerkes RM, Dodson JD. The relation of strength of stimulus to rapidity of habit-formation. J Comp Neurol and Psychol. 1908;18(5):459–82.

99. Duffy E. Activation and behavior. New York: Wiley; 1962.

100. Malmo RB. Activation: A neuropsychological dimension. Psychol Rev. 1959;66(6):367.

101. Csikszentmihalyi, M., *Beyond Boredom and Anxiety: Experiencing Flow in Work and Play*. Behavioral science series, ed. Jossey-Bass. 1975, San Francisco: Jossey-Bass.

102. Von Ahn L, Dabbish L. Designing games with a purpose. Commun ACM. 2008;51(8):58–67.

103. Zahabi, M. and A.M. Abdul Razak, *Adaptive virtual reality-based training: a systematic literature review and framework*. Virtual Reality, 2020. **24**(4).

104. Bronner S, Pinsker R, Noah JA. Physiological and psychophysiological responses in experienced players while playing different dance exer-games. Comput Human Behav. 2015;51:34–41.

105. Salen, K., K.S. Tekinbaş, and E. Zimmerman, *Rules of play: Game design fundamentals*. 2004: MIT press.

106. Faria AL, et al. Benefits of virtual reality based cognitive rehabilitation through simulated activities of daily living: a randomized controlled trial with stroke patients. J Neuroeng Rehabil. 2016;13(1):1–12.

107. Gonçalves, A., et al., *A virtual reality bus ride as an ecologically valid assessment of balance: a feasibility study*. Virtual Reality, 2021: p. 1–9.

108. Oliveira, J., et al., *Virtual reality-based cognitive stimulation on people with mild to moderate dementia due to alzheimer's disease: a randomized controlled trial*. 2021.

109. Plotnik M, et al. Multimodal immersive trail making-virtual reality paradigm to study cognitive-motor interactions. J Neuroeng Rehabil. 2021;18(1):1–16.

110. Thielbar KO, et al. Home-based upper extremity stroke therapy using a multiuser virtual reality environment: a randomized trial. Arch Phys Med Rehabil. 2020;101(2):196–203.

111. Pereira F, et al. The use of game modes to promote engagement and social involvement in multi-user serious games: a within-person randomized trial with stroke survivors. J Neuroeng Rehabil. 2021;18 (1):1–15.

112. Novak D, et al. Increasing motivation in robot-aided arm rehabilitation with competitive and cooperative gameplay. J Neuroeng Rehabil. 2014;11(1):1–15.

113. Goršič M, Cikajlo I, Novak D. Competitive and cooperative arm rehabilitation games played by a patient and unimpaired person: effects on motivation and exercise intensity. J Neuroeng Rehabil. 2017;14(1):1–18.

114. Verschueren, S., C. Buffel, and G. Vander Stichele, *Developing theory-driven, evidence-based serious games for health: framework based on research community insights*. JMIR serious games, 2019. **7** (2): p. e11565.

115. Max, M.N. and M.N. Sarah, *A Comparative Study of Sense of Presence of Traditional Virtual Reality and Immersive Environments*. Australasian Journal of Information Systems, 2016. **20**(0).

116. Chen X, et al. ImmerTai: Immersive Motion Learning in VR Environments. J Vis Commun Image Represent. 2019;58:416–27.

117. Chandrasekera T, Fernando K, Puig L. Effect of Degrees of Freedom on the Sense of Presence Generated by Virtual Reality (VR) Head-Mounted Display Systems: A Case Study on the Use of VR in Early Design Studios. J Educ Technol Syst. 2019;47 (4):513–22.

118. Shu Y, et al. Do virtual reality head-mounted displays make a difference? A comparison of presence and self-efficacy between head-mounted displays and desktop computer-facilitated virtual environments. Virtual Reality. 2019;23(4):437–46.

119. Ada L, et al. Task-specific training of reaching and manipulation. Adv Psychol. 1994;105:239–65.

120. Viau A, et al. Reaching in reality and virtual reality: a comparison of movement kinematics in healthy subjects and in adults with hemiparesis. J Neuroeng Rehabil. 2004;1(1):11.

121. Subramanian S, et al. Virtual reality environments for post-stroke arm rehabilitation. J Neuroeng Rehabil. 2007;4:20.

122. Knaut LA, et al. Kinematics of pointing movements made in a virtual versus a physical 3-dimensional environment in healthy and stroke subjects. Arch Phys Med Rehabil. 2009;90(5):793–802.

123. Levin, M.F., et al., *Quality of Grasping and the Role of Haptics in a 3D Immersive Virtual Reality Environment in Individuals With Stroke*. IEEE Trans Neural Syst Rehabil Eng, 2015. **PP**(99): p. 1.

124. van den Hoogen W, et al. Visualizing the third dimension in virtual training environments for neurologically impaired persons: beneficial or disruptive? J Neuroeng Rehabil. 2012;9(1):73.

125. Lin C, Tizazu B, Woldegiorgis B. Kinematics of direct reaching in head-mounted and stereoscopic widescreen virtual environments. Virtual Reality. 2021;25:1–14.

126. Erhardsson, M., M. Alt Murphy, and K.S. Sunnerhagen, *Commercial head-mounted display virtual reality for upper extremity rehabilitation in chronic stroke: a single-case design study*. Journal of

NeuroEngineering and Rehabilitation, 2020. **17**(1): p. 154.

127. Fluet, G., et al. *Sensorimotor training in virtual environments produces similar outcomes to real world training with greater efficiency*. in *Virtual Rehabilitation (ICVR), 2013 International Conference on*. 2013. IEEE.

128. Laver, K.E., et al., *Virtual reality for stroke rehabilitation*. Cochrane database of systematic reviews, 2017(11).

129. Weiss PL, et al. Video capture virtual reality as a flexible and effective rehabilitation tool. J Neuroeng Rehabil. 2004;1(1):12.

130. Hondori HM, et al. Utility of Augmented Reality in Relation to Virtual Reality in Stroke Rehabilitation. Stroke. 2014;45:1.

131. Valentini, P.P., *Natural interface for interactive virtual assembly in augmented reality using Leap Motion Controller*. International Journal on Interactive Design and Manufacturing (IJIDeM), 2018. **12**.

132. Bruder, G., et al. *Enhancing Presence in Head-Mounted Display Environments by Visual Body Feedback Using Head-Mounted Cameras*. in *2009 International Conference on CyberWorlds*. 2009.

133. Slater, M., *Implicit Learning Through Embodiment in Immersive Virtual Reality*. 2017. p. 19–33.

134. Held JPO, et al. Augmented Reality-Based Rehabilitation of Gait Impairments: Case Report. JMIR Mhealth Uhealth. 2020;8(5): e17804.

135. Kim, W.S., et al., *Clinical Application of Virtual Reality for Upper Limb Motor Rehabilitation in Stroke: Review of Technologies and Clinical Evidence*. J Clin Med, 2020. **9**(10).

136. Rito, D., et al. *Virtual reality tools for post-stroke balance rehabilitation: a review and a solution proposal*. in *2021 IEEE International Conference on Autonomous Robot Systems and Competitions (ICARSC)*. 2021.

137. Thaut, M.H., G.C. McIntosh, and V. Hoemberg, *Neurobiological foundations of neurologic music therapy: rhythmic entrainment and the motor system*. Front Psychol, 2014. **5**.

138. Powell, W., *Virtually walking: factors influencing walking and perception of walking in treadmill-mediated virtual reality to support rehabilitation*. 2011, University of Portsmouth.

139. Friedman, N., et al. *MusicGlove: Motivating and quantifying hand movement rehabilitation by using functional grips to play music*. in *Engineering in Medicine and Biology Society, EMBC, 2011 Annual International Conference of the IEEE*. 2011. IEEE.

140. Brütsch K, et al. Virtual reality for enhancement of robot-assisted gait training in children with neurological gait disorders. J Rehabil Med. 2011;43 (6):493–9.

141. Adamovich, S., et al., *Design of a complex virtual reality simulation to train finger motion for persons with hemiparesis: a proof of concept study*. J Neuroeng Rehabil, 2009. **6**(28).

142. Huegel, J.C. and M.K. O'Malley. *Progressive haptic and visual guidance for training in a virtual dynamic task*. in *Haptics Symposium, 2010 IEEE*. 2010.

143. Huang FC, Gillespie RB, Kuo AD. Visual and haptic feedback contribute to tuning and online control during object manipulation. J Mot Behav. 2007;39(3):179–93.

144. Merians AS, et al. Robotically facilitated virtual rehabilitation of arm transport integrated with finger movement in persons with hemiparesis. J Neuroeng Rehabil. 2011;8:27.

145. Schweighofer N, et al. Task-Oriented Rehabilitation Robotics. Am J Phys Med Rehabil. 2012;91(11): S270–9.

146. Khan MA, et al. Review on motor imagery based BCI systems for upper limb post-stroke neurorehabilitation: From designing to application. Comput Biol Med. 2020;123: 103843.

147. Wolpaw JR, et al. Brain–computer interfaces for communication and control. Clin Neurophysiol. 2002;113(6):767–91.

148. Poveda-García, A., C. Moret-Tatay, and M. Gómez-Martínez. *The Association between Mental Motor Imagery and Real Movement in Stroke*. in *Healthcare*. 2021. Multidisciplinary Digital Publishing Institute.

149. Stefano Filho, C.A., R. Attux, and G. Castellano, *Actual, sham and no-feedback effects in motor imagery practice*. Biomedical Signal Processing and Control, 2022. **71**: p. 103262.

150. Vourvopoulos, A. and S.B. i Badia, *Motor priming in virtual reality can augment motor-imagery training efficacy in restorative brain-computer interaction: a within-subject analysis*. Journal of neuroengineering and rehabilitation, 2016. **13**(1): p. 1–14.

151. Škola F, Tinková S, Liarokapis F. Progressive training for motor imagery brain-computer interfaces using gamification and virtual reality embodiment. Front Hum Neurosci. 2019;13:329.

152. Le Franc S, et al. Influence of the visuo-proprioceptive illusion of movement and motor imagery of the wrist on EEG cortical excitability among healthy participants. PLoS ONE. 2021;16 (9): e0256723.

153. Chen, S., et al., *The differences between motor attempt and motor imagery in BCI accuracy and event-related desynchronization of hemiplegic patients*. Frontiers in Neurorobotics: p. 147.

154. Aggarwal S, Chugh N. Signal processing techniques for motor imagery brain computer interface: A review. Array. 2019;1: 100003.

155. Blanco-Mora, D.A., et al. *Finding the Optimal Time Window for Increased Classification Accuracy during Motor Imagery*. in *BIODEVICES*. 2021.

156. Mladenović J. Standardization of protocol design for user training in EEG-based brain–computer interface. J Neural Eng. 2021;18(1): 011003.

157. Vourvopoulos A, et al. Efficacy and brain imaging correlates of an immersive motor imagery BCI-driven VR system for upper limb motor rehabilitation: A clinical case report. Front Hum Neurosci. 2019;13:244.

158. Bütefisch CM, et al. Remote changes in cortical excitability after stroke. Brain. 2003;126(2):470–81.

159. Ward NS, et al. Motor system activation after subcortical stroke depends on corticospinal system integrity. Brain. 2006;129(3):809–19.

160. Ward N, et al. Neural correlates of motor recovery after stroke: a longitudinal fMRI study. Brain. 2003;126(11):2476–96.

161. Carey LM, et al. Evolution of brain activation with good and poor motor recovery after stroke. Neurorehabil Neural Repair. 2006;20(1):24–41.

162. Hao, J., et al., *Effects of Virtual Reality Intervention on Neural Plasticity in Stroke Rehabilitation: A Systematic Review*. Arch Phys Med Rehabil, 2021.

163. Saleh S, et al. Mechanisms of neural reorganization in chronic stroke subjects after virtual reality training. Annu Int Conf IEEE Eng Med Biol Soc. 2011;2011:8118–21.

164. Saleh S, Adamovich SV, Tunik E. Resting state functional connectivity and task-related effective connectivity changes after upper extremity rehabilitation: a pilot study. Annu Int Conf IEEE Eng Med Biol Soc. 2012;2012:4559–62.

165. Saleh S, et al. Neural Patterns of Reorganization after Intensive Robot-Assisted Virtual Reality Therapy and Repetitive Task Practice in Patients with Chronic Stroke. Front Neurol. 2017;8:452.

166. Wåhlin A, et al. Rehabilitation in chronic spatial neglect strengthens resting-state connectivity. Acta Neurol Scand. 2019;139(3):254–9.

167. Mekbib DB, et al. Proactive Motor Functional Recovery Following Immersive Virtual Reality-Based Limb Mirroring Therapy in Patients with Subacute Stroke. Neurotherapeutics. 2020;17(4):1919–30.

168. You SH, et al. Virtual reality-induced cortical reorganization and associated locomotor recovery in chronic stroke: an experimenter-blind randomized study. Stroke. 2005;36(6):1166–71.

169. Jang SH, et al. Cortical reorganization and associated functional motor recovery after virtual reality in patients with chronic stroke: an experimenter-blind preliminary study. Arch Phys Med Rehabil. 2005;86(11):2218–23.

170. Tunik E, Adamovich SV. Remapping in the ipsilesional motor cortex after VR-based training: a pilot fMRI study. Annu Int Conf IEEE Eng Med Biol Soc. 2009;2009:1139–42.

171. Turolla, A., et al., *Haptic-based neurorehabilitation in poststroke patients: a feasibility prospective multicentre trial for robotics hand rehabilitation*. Comput Math Methods Med, 2013.

172. Bao X, et al. Mechanism of Kinect-based virtual reality training for motor functional recovery of upper limbs after subacute stroke. Neural Regen Res. 2013;8(31):2904–13.

173. Schuster-Amft C, et al. Intensive virtual reality-based training for upper limb motor function in chronic stroke: a feasibility study using a single case experimental design and fMRI. Disabil Rehabil Assist Technol. 2015;10(5):385–92.

174. Xiao X, et al. Cerebral Reorganization in Subacute Stroke Survivors after Virtual Reality-Based Training: A Preliminary Study. Behav Neurol. 2017;2017:6261479.

175. Wang ZR, et al. Leap Motion-based virtual reality training for improving motor functional recovery of upper limbs and neural reorganization in subacute stroke patients. Neural Regen Res. 2017;12(11):1823–31.

176. Fluet, G.M., Alma ; Patel, Jigna ; Van Wingerden, Anita ; Qiu, Qinyin ; Yarossi, Matthew ; Tunik, Eugene ; Adamovich, Sergei ; Massood, Supriya., *Virtual reality-augmented rehabilitation for patients in sub-acute phase post-stroke: A feasibility study*. Journal of Pain Management, 2016. **9**: p. 227–234.

177. Patel J, et al. Intensive virtual reality and robotic based upper limb training compared to usual care, and associated cortical reorganization, in the acute and early sub-acute periods post-stroke: a feasibility study. J Neuroeng Rehabil. 2019;16(1):92.

178. Yarossi M, et al. The Association Between Reorganization of Bilateral M1 Topography and Function in Response to Early Intensive Hand Focused Upper Limb Rehabilitation Following Stroke Is Dependent on Ipsilesional Corticospinal Tract Integrity. Front Neurol. 2019;10:258.

179. Orihuela-Espina F, et al. Neural reorganization accompanying upper limb motor rehabilitation from stroke with virtual reality-based gesture therapy. Top Stroke Rehabil. 2013;20(3):197–209.

180. Ballester BR, et al. Domiciliary VR-Based Therapy for Functional Recovery and Cortical Reorganization: Randomized Controlled Trial in Participants at the Chronic Stage Post Stroke. JMIR Serious Games. 2017;5(3): e15.

181. De Luca R, et al. Use of virtual reality in improving poststroke neglect: Promising neuropsychological and neurophysiological findings from a case study. Appl Neuropsychol Adult. 2019;26(1):96–100.

182. Richards LG, et al. Movement-dependent stroke recovery: a systematic review and meta-analysis of TMS and fMRI evidence. Neuropsychologia. 2008;46(1):3–11.

183. Carel C, et al. Neural substrate for the effects of passive training on sensorimotor cortical representation: a study with functional magnetic resonance imaging in healthy subjects. J Cereb Blood Flow Metab. 2000;20(3):478–84.

184. Jang SH, et al. Cortical reorganization induced by task-oriented training in chronic hemiplegic stroke patients. NeuroReport. 2003;14(1):137–41.

185. Grefkes, C. and G.R. Fink, *Reorganization of cerebral networks after stroke: new insights from neuroimaging with connectivity approaches*. Brain, 2011.

186. Cramer SC. Repairing the human brain after stroke: I. Mechanisms of spontaneous recovery Ann Neurol. 2008;63(3):272–87.

187. Laver KE, et al. Virtual reality for stroke rehabilitation. Stroke. 2018;49(4):e160–1.

188. Iacoboni M, et al. Cortical mechanisms of human imitation. Science. 1999;286(5449):2526–8.

189. Gallese V, et al. Action recognition in the premotor cortex. Brain. 1996;119(2):593–609.

190. Grezes J, Decety J. Functional anatomy of execution, mental simulation, observation, and verb generation of actions: a meta-analysis. Hum Brain Mapp. 2001;12(1):1–19.

191. Rizzolatti G, Craighero L. The mirror-neuron system. Annu Rev Neurosci. 2004;27:169–92.

192. Buccino G, Solodkin A, Small SL. Functions of the mirror neuron system: implications for neurorehabilitation. Cogn Behav Neurol. 2006;19(1):55–63.

193. Garrison KA, Winstein CJ, Aziz-Zadeh L. The mirror neuron system: a neural substrate for methods in stroke rehabilitation. Neurorehabil Neural Repair. 2010;24(5):404–12.

194. Buccino G, et al. Action observation activates premotor and parietal areas in a somatotopic manner: an fMRI study. Eur J Neurosci. 2001;13 (2):400–4.

195. Sale P, Franceschini M. Action observation and mirror neuron network: a tool for motor stroke rehabilitation. Eur J Phys Rehabil Med. 2012;48 (2):313–8.

196. Ertelt D, et al. Action observation has a positive impact on rehabilitation of motor deficits after stroke. Neuroimage. 2007;36:164–73.

197. Maeda F, Kleiner-Fisman G, Pascual-Leone A. Motor facilitation while observing hand actions: specificity of the effect and role of observer's orientation. J Neurophysiol. 2002;87(3):1329–35.

198. Gazzola V, et al. The anthropomorphic brain: the mirror neuron system responds to human and robotic actions. Neuroimage. 2007;35(4):1674–84.

199. Ferrari PF, Rozzi S, Fogassi L. Mirror neurons responding to observation of actions made with tools in monkey ventral premotor cortex. J Cogn Neurosci. 2005;17(2):212–26.

200. Modrono C, et al. Activation of the human mirror neuron system during the observation of the manipulation of virtual tools in the absence of a visible effector limb. Neurosci Lett. 2013;555: 220–4.

201. August, K., et al. *FMRI analysis of neural mechanisms underlying rehabilitation in virtual reality: activating secondary motor areas*. in *Engineering in Medicine and Biology Society, 2006. EMBS'06. 28th Annual International Conference of the IEEE*. 2006. IEEE.

202. Prochnow D, et al. A functional magnetic resonance imaging study of visuomotor processing in a virtual reality-based paradigm: Rehabilitation Gaming System. Eur J Neurosci. 2013;37(9):1441–7.

203. Dobkin BH. Brain–computer interface technology as a tool to augment plasticity and outcomes for neurological rehabilitation. J Physiol. 2007;579 (3):637–42.

204. Grosse-Wentrup, M., D. Mattia, and K. Oweiss, *Using brain–computer interfaces to induce neural plasticity and restore function*. J Neural Eng, 2011. **8**(2).

205. Villiger M, et al. Enhanced activation of motor execution networks using action observation combined with imagination of lower limb movements. PLoS ONE. 2013;8(8): e72403.

206. Bermúdez i Badia, S., et al., *Using a hybrid brain computer interface and virtual reality system to monitor and promote cortical reorganization through motor activity and motor imagery training*. IEEE Trans Neural Syst Rehabil Eng, 2013. **21**(2): p. 174–181.

207. Abdollahi, F., et al., *Arm control recovery enhanced by error augmentation*, in *IEEE Int Conf Rehabil Robot*. 2011. p. 1–6.

208. Tunik E, Saleh S, Adamovich SV. Visuomotor discordance during visually-guided hand movement in virtual reality modulates sensorimotor cortical activity in healthy and hemiparetic subjects. IEEE Trans Neural Syst Rehabil Eng. 2013;21(2):198–207.

209. Lewis, C.H. and M.J. Griffin, *Human factors consideration in clinical applications of virtual reality*. Stud Health Technol Inform, 1997: p. 35–58.

210. Ballester BR, et al. The visual amplification of goal-oriented movements counteracts acquired non-use in hemiparetic stroke patients. J Neuroeng Rehabil. 2015;12:50.

211. Weiss PL, et al. Video capture virtual reality: A decade of rehabilitation assessment and intervention. Phys Ther Rev. 2009;14(5):307–21.

212. Yavuzer G, et al. "Playstation eyetoy games"improve upper extremity-related motor functioning in subacute stroke: a randomized controlled clinical trial. Eur J Phys Rehabil Med. 2008;44(3):237–44.

213. da Silva Ribeiro NM, et al. Virtual rehabilitation via Nintendo Wii® and conventional physical therapy effectively treat post-stroke hemiparetic patients. Top Stroke Rehabil. 2015;22(4):299–305.

214. Levac, D.E., et al. *A comparison of virtual reality and active video game usage, attitudes and learning needs among therapists in Canada and the US*. in *2019 International Conference on Virtual Rehabilitation (ICVR)*. 2019. IEEE.

215. Levac D, et al. Virtual reality and active videogame-based practice, learning needs, and preferences: a cross-Canada survey of physical therapists and occupational therapists. Games for health journal. 2017;6(4):217–28.

216. Webster D, Celik O. Systematic review of Kinect applications in elderly care and stroke rehabilitation. J Neuroeng Rehabil. 2014;11(1):108.
217. Levac D, et al. "Kinect-ing" with clinicians: a knowledge translation resource to support decision making about video game use in rehabilitation. Phys Ther. 2015;95(3):426–40.
218. Zeiler SR, Krakauer JW. The interaction between training and plasticity in the post-stroke brain. Curr Opin Neurol. 2013;26(6):609.
219. Gueye T, et al. Early post-stroke rehabilitation for upper limb motor function using virtual reality and exoskeleton: equally efficient in older patients. Neurol Neurochir Pol. 2021;55(1):91–6.
220. Brunner I, et al. Virtual reality training for upper extremity in subacute stroke (VIRTUES): a multicenter RCT. Neurology. 2017;89(24):2413–21.
221. Kiper, P., et al., *Virtual reality for upper limb rehabilitation in subacute and chronic stroke: a randomized controlled trial.* Archives of physical medicine and rehabilitation, 2018. **99**(5): p. 834–842. e4.
222. Ögün MN, et al. Effect of leap motion-based 3D immersive virtual reality usage on upper extremity function in ischemic stroke patients. Arq Neuropsiquiatr. 2019;77:681–8.
223. Aşkın A, et al. Effects of Kinect-based virtual reality game training on upper extremity motor recovery in chronic stroke. Somatosens Mot Res. 2018;35(1):25–32.
224. Schuster-Amft C, et al. Effect of a four-week virtual reality-based training versus conventional therapy on upper limb motor function after stroke: A multicenter parallel group randomized trial. PLoS ONE. 2018;13(10): e0204455.
225. Rogers JM, et al. Elements virtual rehabilitation improves motor, cognitive, and functional outcomes in adult stroke: evidence from a randomized controlled pilot study. J Neuroeng Rehabil. 2019;16(1):1–13.
226. Rodríguez-Hernández M, et al. Effects of Specific Virtual Reality-Based Therapy for the Rehabilitation of the Upper Limb Motor Function Post-Ictus: Randomized Controlled Trial. Brain Sci. 2021;11(5):555.
227. Choi, Y.-H. and N.-J. Paik, *Mobile game-based virtual reality program for upper extremity stroke rehabilitation.* Journal of visualized experiments: JoVE, 2018(133).
228. Norouzi-Gheidari N, et al. Feasibility, safety and efficacy of a virtual reality exergame system to supplement upper extremity rehabilitation post-stroke: A pilot randomized clinical trial and proof of principle. Int J Environ Res Public Health. 2020;17(1):113.
229. Park M, et al. Effects of virtual reality-based planar motion exercises on upper extremity function, range of motion, and health-related quality of life: a multicenter, single-blinded, randomized, controlled pilot study. J Neuroeng Rehabil. 2019;16(1):1–13.
230. Afsar SI, et al. Virtual reality in upper extremity rehabilitation of stroke patients: a randomized controlled trial. J Stroke Cerebrovasc Dis. 2018;27(12):3473–8.
231. Lee MM, Lee KJ, Song CH. Game-based virtual reality canoe paddling training to improve postural balance and upper extremity function: A preliminary randomized controlled study of 30 patients with subacute stroke. Medical science monitor: international medical journal of experimental and clinical research. 2018;24:2590.
232. Fluet, G., et al., *Virtual Rehabilitation of the Paretic Hand and Arm in Persons With Stroke: Translation From Laboratory to Rehabilitation Centers and the Patient's Home.* Frontiers in neurology, 2021. **12**.
233. Shiner CT, Byblow WD, McNulty PA. Bilateral Priming Before Wii-based Movement Therapy Enhances Upper Limb Rehabilitation and Its Retention After Stroke A Case-Controlled Study. Neurorehabil Neural Repair. 2014;28(9):828–38.
234. Joo LY, et al. A feasibility study using interactive commercial off-the-shelf computer gaming in upper limb rehabilitation in patients after stroke. J Rehabil Med. 2010;42(5):437–41.
235. Mouawad MR, et al. Wii-based movement therapy to promote improved upper extremity function post-stroke: a pilot study. J Rehabil Med. 2011;43(6):527–33.
236. Paquin, K., et al., *Effectiveness of commercial video gaming on fine motor control in chronic stroke within community-level rehabilitation.* Disabil Rehabil, 2015(0): p. 1–8.
237. Saposnik G, et al. Effectiveness of virtual reality using Wii gaming technology in stroke rehabilitation a pilot randomized clinical trial and proof of principle. Stroke. 2010;41(7):1477–84.
238. Chen, M.-H., et al., *A controlled pilot trial of two commercial video games for rehabilitation of arm function after stroke.* Clin Rehabili, 2014: p. 1–9.
239. Maier M, et al. Effect of specific over nonspecific VR-based rehabilitation on Poststroke motor recovery: a systematic meta-analysis. Neurorehabil Neural Repair. 2019;33(2):112–29.
240. Puh U, Hoehlein B, Deutsch JE. Validity and reliability of the Kinect for assessment of standardized transitional movements and balance: Systematic review and translation into practice. Physical Medicine and Rehabilitation Clinics. 2019;30(2):399–422.
241. Pastor, I., H.A. Hayes, and S.J.M. Bamberg. *A feasibility study of an upper limb rehabilitation system using kinect and computer games.* in *Engineering in Medicine and Biology Society (EMBC), 2012 Annual International Conference of the IEEE.* 2012.
242. Lee G. Effects of training using video games on the muscle strength, muscle tone, and activities of daily living of chronic stroke patients. J Phys Ther Sci. 2013;25(5):595.

243. Mohammadi R, et al. Effects of virtual reality compared to conventional therapy on balance poststroke: a systematic review and meta-analysis. J Stroke Cerebrovasc Dis. 2019;28(7):1787–98.

244. Chen, L., et al., *Effect of virtual reality on postural and balance control in patients with stroke: a systematic literature review.* BioMed Research International, 2016. **2016**.

245. Porras DC, et al. Advantages of virtual reality in the rehabilitation of balance and gait: systematic review. Neurology. 2018;90(22):1017–25.

246. Iruthayarajah J, et al. The use of virtual reality for balance among individuals with chronic stroke: a systematic review and meta-analysis. Top Stroke Rehabil. 2017;24(1):68–79.

247. Garay-Sánchez A, et al. Effects of Immersive and Non-Immersive Virtual Reality on the Static and Dynamic Balance of Stroke Patients: A Systematic Review and Meta-Analysis. J Clin Med. 2021;10 (19):4473.

248. Juras G, et al. Standards of Virtual Reality Application in Balance Training Programs in Clinical Practice: A Systematic Review. Games Health J. 2019;8(2):101–11.

249. Li Z, et al. Virtual reality for improving balance in patients after stroke: a systematic review and meta-analysis. Clin Rehabil. 2016;30(5):432–40.

250. Rutkowski, S., et al., *Use of virtual reality-based training in different fields of rehabilitation: A systematic review and meta-analysis.* Journal of Rehabilitation Medicine, 2020.

251. Wu J, et al. Effects of Virtual Reality Training on Upper Limb Function and Balance in Stroke Patients: Systematic Review and Meta-Meta-Analysis. J Med Internet Res. 2021;23(10): e31051.

252. Zhang B, et al. Virtual reality for limb motor function, balance, gait, cognition and daily function of stroke patients: A systematic review and meta-analysis. J Adv Nurs. 2021;77(8):3255–73.

253. Barclay-Goddard, R., et al., *Force platform feed-back for standing balance training after stroke.* Cochrane Database Syst Rev, 2004(4).

254. Sveistrup H, et al. Experimental studies of virtual reality-delivered compared to conventional exercise programs for rehabilitation. Cyberpsychol Behav. 2003;6(3):245–9.

255. Bisson E, et al. Functional balance and dual-task reaction times in older adults are improved by virtual reality and biofeedback training. Cyberpsychol Behav. 2007;10(1):16–23.

256. Thornton M, et al. Benefits of activity and virtual reality based balance exercise programmes for adults with traumatic brain injury: perceptions of participants and their caregivers. Brain Inj. 2005;19 (12):989–1000.

257. McEwen D, et al. Virtual reality exercise improves mobility after stroke: an inpatient randomized controlled trial. Stroke. 2014;45(6):1853–5.

258. Gil-Gomez JA, et al. Effectiveness of a Wii balance board-based system (eBaViR) for balance rehabilitation: a pilot randomized clinical trial in patients with acquired brain injury. J Neuroeng Rehabil. 2011;8:30.

259. Lloréns, R., et al., *Balance rehabilitation using custom-made Wii Balance Board exercises: clinical effectiveness and maintenance of gains in an acquired brain injury population,* in *Int J Disabil Human Dev* 2014. p. 327.

260. Llorens R. Effect of Virtual Reality-Based Training of the Ankle, Hip, and Stepping Strategies on Balance after Stroke. In: Virtual Reality in Health and Rehabilitation. CRC Press; 2020. p. 85–102.

261. Llorens R, et al. Time since injury limits but does not prevent improvement and maintenance of gains in balance in chronic stroke. Brain Inj. 2018;32(3):303–9.

262. Cikajlo I, et al. Telerehabilitation using virtual reality task can improve balance in patients with stroke. Disabil Rehabil. 2012;34(1):13–8.

263. Krpic A, Savanovic A, Cikajlo I. Telerehabilitation: remote multimedia-supported assistance and mobile monitoring of balance training outcomes can facilitate the clinical staff's effort. Int J Rehabil Res. 2013;36(2):162–71.

264. Sheehy L, et al. Centre of pressure displacements produced in sitting during virtual reality training in younger and older adults and patients who have had a stroke. Disabil Rehabil Assist Technol. 2020;15 (8):924–32.

265. Sheehy L, et al. Sitting balance exercise performed using virtual reality training on a stroke rehabilitation inpatient service: a randomized controlled study. Pm&r. 2020;12(8):754–65.

266. Cho KH, Lee WH. Virtual walking training program using a real-world video recording for patients with chronic stroke: a pilot study. Am J Phys Med Rehabil. 2013;92(5):371–84.

267. Yang S, et al. Improving balance skills in patients who had stroke through virtual reality treadmill training. Am J Phys Med Rehabil. 2011;90 (12):969–78.

268. Walker ML, et al. Virtual reality-enhanced partial body weight-supported treadmill training post-stroke: feasibility and effectiveness in 6 subjects. Arch Phys Med Rehabil. 2010;91(1):115–22.

269. Kim N, Park Y, Lee B-H. Effects of community-based virtual reality treadmill training on balance ability in patients with chronic stroke. J Phys Ther Sci. 2015;27(3):655–8.

270. Kim H, et al. Virtual dual-task treadmill training using video recording for gait of chronic stroke survivors: a randomized controlled trial. J Phys Ther Sci. 2015;27(12):3693–7.

271. Fishbein P, et al. A preliminary study of dual-task training using virtual reality: influence on walking and balance in chronic poststroke survivors. J Stroke Cerebrovasc Dis. 2019;28(11): 104343.

272. Calabrò RS, et al. The role of virtual reality in improving motor performance as revealed by EEG: a randomized clinical trial. J Neuroeng Rehabil. 2017;14(1):1–16.

273. Llorens R, et al. Improvement in balance using a virtual reality-based stepping exercise: a randomized controlled trial involving individuals with chronic stroke. Clin Rehabil. 2015;29(3):261–8.

274. Henrique PP, Colussi EL, De Marchi AC. Effects of exergame on patients' balance and upper limb motor function after stroke: a randomized controlled trial. J Stroke Cerebrovasc Dis. 2019;28(8):2351–7.

275. Junata M, et al. Kinect-based rapid movement training to improve balance recovery for stroke fall prevention: a randomized controlled trial. J Neuroeng Rehabil. 2021;18(1):1–12.

276. Chen S-C, et al. Feasibility and effect of interactive telerehabilitation on balance in individuals with chronic stroke: a pilot study. J Neuroeng Rehabil. 2021;18(1):1–11.

277. Flynn S, Palma P, Bender A. Feasibility of using the Sony PlayStation 2 gaming platform for an individual poststroke: a case report. J Neurol Phys Ther. 2007;31(4):180–9.

278. Deutsch, J.E., et al., *Wii-Based Compared to Standard of Care Balance and Mobility Rehabilitation for Two Individuals Post-Stroke*, in *2009 Virtual Rehabilitation International Conference*. 2009. p. 117–120.

279. Fritz SL, et al. Active video-gaming effects on balance and mobility in individuals with chronic stroke: a randomized controlled trial. Top Stroke Rehabil. 2013;20(3):218–25.

280. Cho KH, Lee KJ, Song CH. Virtual-reality balance training with a video-game system improves dynamic balance in chronic stroke patients. Tohoku J Exp Med. 2012;228(1):69–74.

281. Barcala L, et al. Visual biofeedback balance training using wii fit after stroke: a randomized controlled trial. J Phys Ther Sci. 2013;25(8):1027–32.

282. Hung JW, et al. Randomized comparison trial of balance training by using exergaming and conventional weight-shift therapy in patients with chronic stroke. Arch Phys Med Rehabil. 2014;95(9):1629–37.

283. Morone, G., et al., *The efficacy of balance training with video game-based therapy in subacute stroke patients: a randomized controlled trial*. Biomed Res Int, 2014. **2014**.

284. Rajaratnam, B., et al., *Does the Inclusion of Virtual Reality Games within Conventional Rehabilitation Enhance Balance Retraining after a Recent Episode of Stroke?* Rehabil Research Pract, 2013. **2013**.

285. Bower KJ, et al. Clinical feasibility of the Nintendo Wii for balance training post-stroke: a phase II randomized controlled trial in an inpatient setting. Clin Rehabil. 2014;28(9):912–23.

286. Deutsch, J.E., Bowlby, P. G., & Kafri, M. . *Effects of Optimal Standard of Care Compared with Interactive Video Gaming-Based Balance and Mobility Training for Individuals Post-stroke*. in *World Congress of Physical Therapy*. 2011. Amsterdam, Holland.

287. Singh DK, et al. Effects of substituting a portion of standard physiotherapy time with virtual reality games among community-dwelling stroke survivors. BMC Neurol. 2013;13:199.

288. Gordon NF, et al. Physical activity and exercise recommendations for stroke survivors: an American Heart Association scientific statement from the Council on Clinical Cardiology, Subcommittee on Exercise, Cardiac Rehabilitation, and Prevention; the Council on Cardiovascular Nursing; the Council on Nutrition, Physical Activity, and Metabolism; and the Stroke Council. Circulation. 2004;109(16):2031–41.

289. Rimmer JH, Wang E, Smith D. Barriers associated with exercise and community access for individuals with stroke. J Rehabil Res Dev. 2008;45(2):315–22.

290. Ranky RG, et al. Modular mechatronic system for stationary bicycles interfaced with virtual environment for rehabilitation. J Neuroeng Rehabil. 2014;11:93.

291. Deutsch JE, et al. Feasibility of virtual reality augmented cycling for health promotion of people poststroke. J Neurol Phys Ther. 2013;37(3):118–24.

292. Kafri M, et al. Energy expenditure and exercise intensity of interactive video gaming in individuals poststroke. Neurorehabil Neural Repair. 2014;28(1):56–65.

293. Hurkmans HL, et al. Energy expenditure in chronic stroke patients playing Wii Sports: a pilot study. J Neuroeng Rehabil. 2011;8:38.

294. de Sousa JCS, et al. Aerobic stimulus induced by virtual reality games in stroke survivors. Arch Phys Med Rehabil. 2018;99(5):927–33.

295. Sampaio LMM, et al. Does virtual reality-based kinect dance training paradigm improve autonomic nervous system modulation in individuals with chronic stroke? J Vasc Interv Neurol. 2016;9(2):21.

296. Tollár J, et al. High frequency and intensity rehabilitation in 641 subacute ischemic stroke patients. Arch Phys Med Rehabil. 2021;102(1):9–18.

297. Trinh T, et al. Targeted upper-limb Wii-based Movement Therapy also improves lower-limb muscle activation and functional movement in chronic stroke. Disabil Rehabil. 2017;39(19):1939–49.

298. Deutsch JE, et al. Comparison of neuromuscular and cardiovascular exercise intensity and enjoyment between standard of care, off-the-shelf and custom active video games for promotion of physical activity of persons post-stroke. J Neuroeng Rehabil. 2021;18(1):1–12.

299. James-Palmer, A., et al. Playing self-paced video games requires the same energy expenditure but is more enjoyable and less effortful than standard of care activities. in 2019 International Conference on Virtual Rehabilitation (ICVR). 2019. IEEE.

300. Faria, A., et al. An integrative virtual reality cognitive-motor intervention approach in stroke rehabilitation: a pilot study. in 10th Intl Conf.

Disability, Virtual Reality & Associated Technologies. 2014.

301. Fluet GG, et al. Does training with traditionally presented and virtually simulated tasks elicit differing changes in object interaction kinematics in persons with upper extremity hemiparesis? Top Stroke Rehabil. 2015;22(3):176–84.

302. Wellner M, et al. Obstacle crossing in a virtual environment with the rehabilitation gait robot LOKOMAT. Stud Health Technol Inform. 2006;125:497–9.

303. Cincotti F, et al. EEG-based Brain-Computer Interface to support post-stroke motor rehabilitation of the upper limb. in Engineering in Medicine and Biology Society (EMBC), 2012 Annual International Conference of the IEEE. 2012. IEEE.

Wearable Sensors for Stroke Rehabilitation

21

Catherine P. Adans-Dester, Catherine
E. Lang, David J. Reinkensmeyer,
and Paolo Bonato

Abstract

In this chapter, we provide a review of the current applications of wearable sensors in the field of stroke rehabilitation. Four key points are discussed in this review. First, wearable sensors are a viable solution for monitoring movement during rehabilitation exercises and clinical assessments, but more work needs to be done to derive clinically relevant information from sensor data collected during unstructured activities. Second, wearable technologies provide critical information related to the performance of activities in daily life, information that is not necessarily captured during in-clinic assessments. Third, wearable technologies can provide feedback and motivation to increase movement in the home and community settings. Finally, technologies are rapidly emerging that can complement "traditional" wearable sensors and sometimes replace them as they provide less obtrusive means of monitoring motor function in stroke survivors. These developing technologies, as well as readily available wearable sensors, are transforming stroke rehabilitation, their development is progressing at a fast pace, and their use so far has allowed us to gather important information, that we would have not been able to collect otherwise, which has tremendous potential to further advance stroke rehabilitation.

Keywords

Stroke · Upper limb · Neurorehabilitation · Clinical assessment · Movement tracking · Wearable sensors · Outcome measures

Abbreviations

10 MWT	10-M walk test
ADL	Activity of Daily Living
ARAT	Action Research Arm Test
BBT	Box and Block Test
Cis	Confidence intervals
EMG	Electromyography
FAS	Functional Ability Scale (subscale of the Wolf Motor Function Test)

C. P. Adans-Dester · P. Bonato (✉)
Department of Physical Medicine and Rehabilitation, Harvard Medical School at Spaulding Rehabilitation Hospital, Boston, MA, USA
e-mail: pbonato@mgh.harvard.edu

C. P. Adans-Dester
e-mail: cadans-dester@partners.org

C. E. Lang
Washington University School of Medicine, in St. Louis, St. Louis, MO, USA
e-mail: langc@wustl.edu

D. J. Reinkensmeyer
Department of Mechanical and Aerospace Engineering, Department of Anatomy and Neurobiology, University of California Irvine, University of California Irvine, Irvine, CA, USA
e-mail: dreinken@uci.edu

© The Author(s), under exclusive license to Springer Nature Switzerland AG 2022
D. J. Reinkensmeyer et al. (eds.), *Neurorehabilitation Technology*,
https://doi.org/10.1007/978-3-031-08995-4_21

FMA-UE	Fugl-Meyer Assessment, Upper Extremity subsection
ICF	International Classification of Functioning, Disability and Health
IMU	Inertial Measurement Unit
IoT	Internet of Things
LL	Lower Limb
MAL	Motor Activity Log
MLA	Machine Learning Algorithms
MMG	Mechanomyography
RFID	Radio Frequency IDentification
RMSE	Root Mean Square Error
SARAH	Semi-Automated Rehabilitation at the Home
TIS	Trunk Impairment Scale
TUG	Timed Up-and-Go
UL	Upper Limb
UWB	Ultra-WideBand
WMFT	Wolf Motor Function Test

21.1 Introduction

Wearable sensors could be used in many ways in stroke rehabilitation. They could be used to perform clinical assessments, facilitate the design of patient-specific rehabilitation strategies, enable the delivery of high-dosage interventions, and track clinical outcomes. The use of wearable sensors could help rehabilitation specialists to address the increasing demand and decreasing access to rehabilitation care that the shortage of rehabilitation specialists is expected to cause [1]. Tracking the response of each patient to the prescribed intervention would allow therapists to carefully adjust the intervention strategy throughout the therapy period and achieve optimal clinical outcomes on a patient-by-patient basis. Furthermore, wearable sensors could help rehabilitation specialists to deliver interventions in the home setting and to monitor subjects in the community, thus reducing the therapists' workload and facilitating

the delivery of long-term interventions that may be most effective in maximizing motor gains. Figure 21.1 shows a schematic representation of how we envision that wearable sensors could be applied across the continuum of care.

The material in this chapter is organized in the following four parts:

- Monitoring stroke survivors during the performance of rehabilitation exercises and clinical assessments.
- Monitoring stroke survivors during the performance of activities of daily living (in the home and community settings) with the goal of capturing what patients "do" as opposed to what they "are capable of doing".
- Monitoring stroke survivors to generate feedback and provide motivation to maximize the amount and quality of motor practice.
- Monitoring stroke survivors using emerging technologies that overcome the limitations of "traditional" wearable sensors and systems.

Herein, we will primarily focus on upper-limb (UL) rehabilitation after stroke (i.e., arm and hand movements), though in some sections of the chapter, we will provide insights into the use of wearable technology to monitor and enable lower limb rehabilitation (i.e., balance and mobility training). In each section, we will elaborate on the clinical importance of the applications discussed, provide examples of what has been accomplished so far, and suggest how these technologies should be integrated into the clinical workflow in the future. While this chapter is focused on stroke rehabilitation, many of the applications of wearable technology herein discussed are relevant not only to designing interventions for other neurological diseases, but also to geriatric and musculoskeletal rehabilitative care.

Because we anticipate an interdisciplinary readership, in the box below, we provide the definitions of a few terms utilized throughout this chapter to facilitate a common understanding of the used terminology.

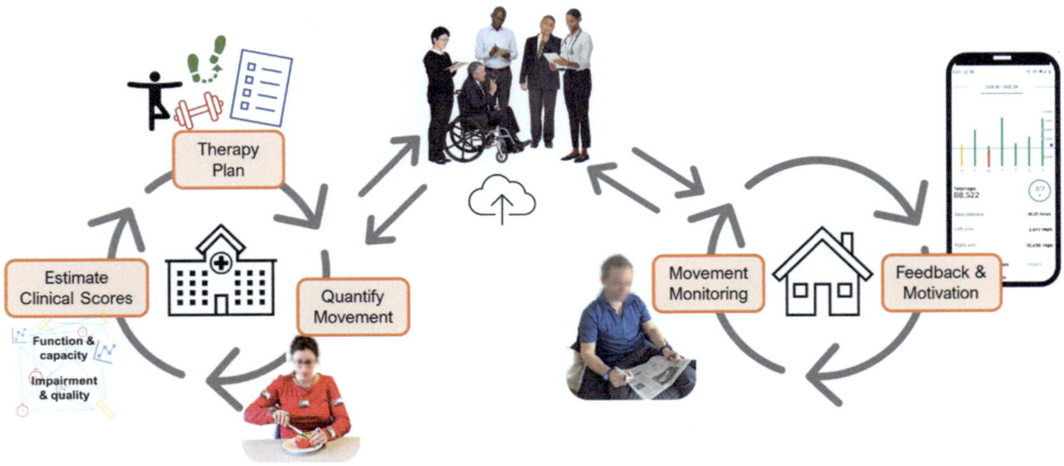

Fig. 21.1 Conceptual representation of the application of wearable sensors across the continuum of rehabilitation care. In the clinic (left), sensors are used to measure movement patterns during the rehabilitation sessions. Sensor data is used to estimate clinical scores and evaluate progression. This information is used to adjust the therapeutic plan. In the home environment (right), sensors can be used to monitor patients' movements as well as provide feedback and motivation to keep practicing in order to improve motor performance. Clinicians and engineers who are part of the care team (top) are given access to the data to make informed decisions

ICF domain definitions and their link with other common terms used across disciplines

Impairment: is a deficit in body structure or function. Example: a loss of muscle strength, or somatosensation in the upper limb post stroke. It is the accumulation of a few or many impairments that lead to limitations in the capacity for and performance of the activity. Common clinical tests to measure impairment in rehabilitation: Fugl-Meyer Upper Extremity test, grip strength, monofilament testing.

Capacity for activity: is the execution of an activity in a structured environment, such as in the clinic or a laboratory. Other common terms used to describe the same idea include "function", "functional capacity", and "capability". Examples: activities such as dressing, typing or walking. Common clinical tests to measure upper limb capacity in rehabilitation: Action Research Arm Test, Wolf Motor Function Test, Box and Block Test.

Performance of activity: is the execution of activity in the unstructured, real-world environment, that is measured in the home and/or community with the existing facilitators and barriers. Examples: activities such as cooking or bathing in the home environment (that might or might not have been modified after the stroke). Common ways to measure upper limb performance: motor Activity Log (self-perceived measure) and accelerometry (direct measure).

Participation: is the fulfillment of life roles and responsibilities. Participation typically requires the performance of multiple activities in the motor, cognitive, and language domains. Examples: caring for a child or working. Common clinical tests to measure participation: stroke impact scale and Neuro-Quality of Life.

21.2 Wearable Sensors for Assessments Performed in the Clinic

21.2.1 Why Would One Want to Use Wearable Sensors for Assessments Performed in the Clinic?

Numerous studies have shown that rehabilitation interventions are beneficial across a number of neurological conditions as they result in a decrease in the severity of disability [2]. However, choosing the most effective intervention among the myriad of available rehabilitation approaches is challenging [3, 4]. High variability in the response to interventions aimed to restore UL function is observed across patients [5, 6], hence pointing to the need for designing "precision rehabilitation" interventions that account for the unique characteristics of each individual. The need to develop patient-specific interventions is paramount in the broad field of medicine [7–9] and is gradually emerging as a topic of great interest in the field of rehabilitation as investigators explore approaches relying on patients' genotype [10–12] and motor phenotype [13–15] to develop subject-specific interventions.

In this context, it is important that rehabilitation specialists be provided with tools to monitor the motor recovery process, assess if the ongoing intervention is leading to the anticipated clinical results, and adjust the intervention if needed. Interventions are typically structured according to the ICF model [16]. Rehabilitation specialists use this framework to evaluate interventions and rely on clinical outcome measures to capture different ICF domains (i.e., *Body Function and Structures*, *Activity*, and *Participation*). Clinical outcome measures are often based on the observation of subjects' motor behaviors (e.g., to capture motor impairments and activity limitations). Unfortunately, these methods suffer from several shortcomings. For instance, only a limited number of rehabilitation specialists undergo the rigorous training needed to properly administer these evaluations. Despite training, substantial inter-rater variability is frequently observed. Besides, oftentimes clinical scales are prone to subjectivity and are marked by low resolution, and hence limited ability to capture change. These assessments are also time-consuming and can be impractical to administer on a regular basis throughout the period of intervention.

Outcome measures are too many times collected only at baseline and at discharge. This is a problem because the lack of longitudinal data tracking progression prevents rehabilitation specialists from examining the potential need to adjust the intervention to maximize motor gains. To address this problem, researchers and clinicians have started to explore the use of wearable sensing technology to collect longitudinal data and derive estimates of clinical outcome measures (i.e., clinical scores). Over the past decade, wearable technology has matured to the extent needed to provide clinicians with an effective tool to monitor outcomes and facilitate delivering interventions [17–20]. This technology has tremendous potential for assessing the benefits of rehabilitation interventions [21]. Wearable sensors are a ubiquitous and unobtrusive tool to quantify movement, gather important data during the administration of clinical assessments, and track motor behaviors during an intervention period to monitor progression.

21.2.2 Assessing Arm and Hand Movements of Stroke Survivors in the Clinic

21.2.2.1 Estimating Movement Kinematics

Most ADLs require the performance of reaching movements, which are marked in stroke survivors by greater trunk movement and limited elbow extension. Nonetheless, clinical tests often fail to measure a range of motion during the performance of motor tasks. Kinematic assessments are considered a gold standard for objective evaluation of movement. In stroke rehabilitation, it is important to capture movement characteristics and deficits in order to refine and evaluate interventions [22]. However, kinematic evaluations in the clinic are limited due to the lack of time, training, cost, and equipment needed (i.e., marker-based optical tracking systems). Over the past decades, quite a few approaches marked by different levels of complexity have used wearable IMUs to track limb movements. For instance, methods have been developed that allow one to reconstruct the kinematics of movement from accelerometer, gyroscope, and magnetometer data recorded using sensors placed on different body segments. Kinematic analysis with wearable sensors has been shown to be an objective, sensitive to change, and quantitative means of measuring motor impairment. A review of all the approaches proposed so far is beyond the scope of this chapter. Herein, we provide instead examples of clinical applications of these technologies.

For example, Schwarz et al. [23] used a portable IMU system to measure UL kinematics. A total of eight IMUs and sensors were placed on the upper body, including a fingertip force-sensing resistor to detect interaction forces between the object and the fingers. Data was collected during the performance of functional reach-to-grasp and object displacement tasks. The authors were able to extract parameters such as trunk compensation, shoulder flexion-extension and abduction-adduction, elbow flexion-extension, forearm supination-pronation, wrist flexion-extension, and flexion-extension of the fingers. In addition, the authors found evidence of joint coupling during the performance of object displacement tasks via the analysis of the correlation between elbow flexion-extension and trunk movements.

Hand function is important for the performance of ADLs, but hand kinematics is difficult to collect. Gloves instrumented with IMUs and magnetic sensors can be used to reconstruct joint motion and provide clinicians with valuable information [24]. Using this type of system in the clinical setting is attractive but has major drawbacks such as the interference with tactile and proprioceptive feedback when manipulating an object, sanitation concerns if the sensors are used by multiple patients and the time required to properly don/doff the glove. Researchers have investigated a novel method of finger movement tracking based on wearable capacitive strain sensors to address some of the glove's limitations [25]. Other emerging technologies will be discussed in Sect. 21.5 of this chapter.

More recently, Nie et al. [26] reported the use of a portable, open-source solution to estimate the position of the wrist during reaching movements with two IMUs. Their method allows one to track the wrist position and average active range of motion during reaching movements with relatively high accuracy (within 1.0–2.5 cm) compared to a marker-based optical tracking system. In addition, a sweeping task allowed the authors to derive two different clinically relevant metrics. The horizontal sweep area "(i.e., reaching workspace)" and the smoothness of the sweeping movement are indicative of movement impairments (smoother movements indicate less impaired UL following a stroke). To improve the clinical implementation of such measures, the authors purposefully decided to make their methods available and transparent for others to use with any sensor capable of estimating limb orientation.

So far, the research findings support the clinical suitability of sensor-based motion analysis to track UL movements in stroke survivors. However, the implementation of such methods outside the research setting remains to be tested.

21.2.2.2 Estimating Clinical Scores

Over the past two decades, researchers have increasingly incorporated the use of wearable sensors into their stroke rehabilitation work [27], both to measure UL activity in the home and to add to the traditional methods of assessments in the clinic. Several research groups have studied the use of IMUs to assess motor function more objectively, some as a way to automate or instrument the assessment of motor function in the clinical setting, others to derive UL motor impairments by estimating various clinical scores from data collected during the performance of predefined motor tasks.

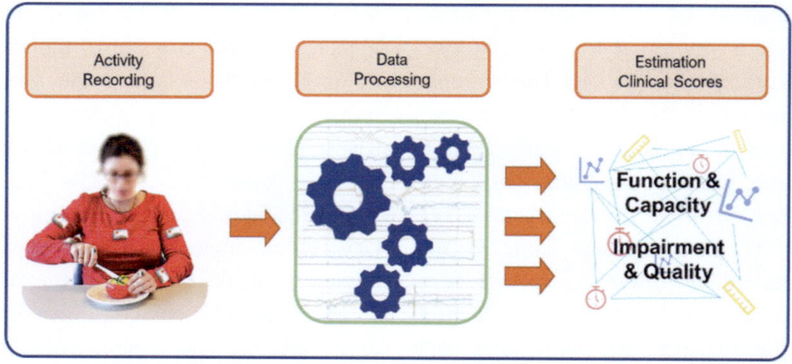

Fig. 21.2 Conceptual representation of methods commonly used to estimate clinical scores. Data is collected with wearable sensors positioned on the upper limbs during the performance of functional tasks. Accelerometer data is fed to a machine learning algorithm to derive estimates of the clinical scores of interest. Reproduced and modified with permission from Adans-Dester et al. (https://doi.org/10.1038/s41746-020-00328-w, licensed under CC BY 4.0)

Figure 21.2 represents a methodology to derive clinically relevant information from wearable sensors. First, different combinations of wearable sensors are used to collect data sometimes during the performance of the clinical test itself, other times during specific tasks or general arm movements. Then, data are processed using machine learning algorithms (MLA) to derive estimates of the outcome of choice. Here, we provide some examples of clinical score estimates relevant to stroke rehabilitation.

Fugl-Meyer Assessment

The Fugl-Meyer Assessment, Upper-Extremity subscale (FMA-UE) is a clinical test designed to evaluate motor impairments that have been tested extensively in the stroke population [28]. A total of 33 items assessing voluntary movement, reflexes, grasp, and coordination are tested; each item is rated on a 3-point ordinal scale.

The first method to derive the FMA-UE scores is the instrumentation of the test with wearable sensors. For example, researchers used two accelerometers and seven flex sensors to monitor the movements of the UL during the performance of 7 movements derived from the FMA-UE [29]. They used MLA to predict the FMA scores based on wearable sensor data and demonstrated the possibility to achieve a coefficient of determination as high as ~ 0.92. Considering that the FMA scale is time-consuming and complicated to perform, using only seven items of the FMA reduces the time to gather data as long as the patient's impairments allow for easy donning and doffing the sensors.

Another method is to use data collected during the performance of functional tasks. Del Din et al. [30] selected a subset of eight tasks from the Wolf Motor Function Test (WMFT) and used six accelerometers placed on the affected arm and the trunk. They used a Random Forest MLA to estimate FMA-UE scores. Their results were marked by a root mean squared error (RMSE) of 4.7 points of the FMA-UE.

Some have tried to combine inertial measurement and mechanomyography (MMG) to better quantify hand and wrist motor function during the estimation of FMA-UE scores. Researchers used 3 IMUs (torso, arm, and forearm) and MMGs placed on finger and wrist flexors to collect data during the performance of FMA-UE tasks. Unfortunately, the detection of the tasks performed by study volunteers was marked by only 75% accuracy for gross movements and 62% accuracy for distal motor tasks (hand and wrist) [31]. These results are not encouraging, not only because of the relatively low accuracy, but also because the data was collected during the performance of the clinical test. Therefore, in this scenario, wearable sensors did not streamline the clinical evaluation.

Functional Ability Scale

The Functional Ability Scale (FAS) is used to assess the quality of movement via observation of the performance of the items of the Wolf Motor Function Test (WMFT). The WMFT is commonly used to quantify UL motor function with timed functional tasks [32]. It consists of 17 items progressing from proximal to distal and from least to most complex UL movements. Each item is used to assess speed and movement quality. The FAS relies on a 6-point ordinal scale to rate the quality of the movement observed by the clinician.

Patel et al. [33] used accelerometers placed on the hand, forearm, upper arm, and trunk to collect data during the performance of a subset of eight motor tasks taken from the WMFT and derive accurate estimates of the total FAS scores provided by a clinician. They showed that it is possible to achieve estimates of the total FAS score marked by a bias of 0.04 points on the scale and a standard deviation of 2.43 points when using as few as three sensors to collect data during the performance of six motor tasks.

Box and Block Test

The Box and Block Test (BBT) is commonly used to measure manual dexterity. The BBT is scored by counting the number of blocks carried over a partition from one compartment to another over one minute [34]. During the investigation of the MusicGlove, a sensorized glove was used to retrain hand function by playing games similar to "Guitar Hero", researchers found that the MusicGlove game scores are strongly correlated with the BBT scores [35].

Action Research Arm Test

The Action Research Arm Test (ARAT) is a common activity level (capacity) measure used in stroke rehabilitation studies. The test has four subscales to evaluate gross motor, grasp, grip,

and pinch. An ordinal scale is used by the clinician to score the observed ability and quality of the task performance. To enhance objectivity and provide additional information on capacity, Resnik et al. instrumented the ARAT test using IMUs and EMG sensors [36]. Five parameters associated with the ARAT were derived (movement time, smoothness, hand trajectories, trunk stability, and grasping muscle activity). They found a strong correlation between the ARAT scores and the movement time and smoothness. While the instrumented ARAT allows one to quantify movement parameters and might provide a better insight into arm motor function, it is quite cumbersome to administer, and the data processing remains lengthy.

To address some of these limitations and to set the preliminary groundwork for evaluating UL outside the clinic, Bochniewicz et al. used a single IMU at the wrist during the performance of four ADL tasks (i.e., laundry and kitchen activities, shopping, and making a bed) [37]. The authors trained a MLA to distinguish between functional (i.e., manipulating an object) and non-functional tasks (i.e., arm swing while walking). The percentage of time spent using the arm to accomplish a functional task was correlated with the ARAT scores. The authors noted the shortcomings of using only one IMU at the wrist for ADLs requiring little to no arm movements.

Estimating More than One Clinical Scale

Adans-Dester et al. estimated two different clinical scores from the same dataset [38]. The authors developed machine learning-based algorithms to estimate FAS and FMA-UE scores via the analysis of accelerometer data collected during the performance of functional motor tasks, that are part of the Wolf Motor Function Test (Fig. 21.3 a). The accelerometer data was segmented to select epochs associated with the performance of specific movement components (e.g., forward arm reaching, pronation-supination movements). Data features were derived from each epoch and fed to a machine learning algorithm based on a regression implementation of a Random Forest. Separate models were built to estimate the FAS

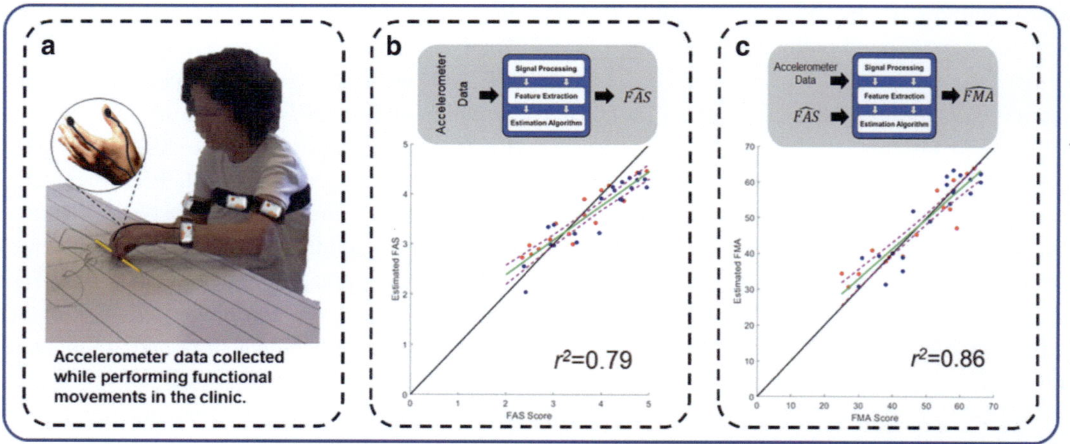

Fig. 21.3 Data collected using accelerometers during the performance of functional tasks (panel A) were used to derive estimates of the FAS (panel B) and the FMA-UE (panel C) clinical scores. Reproduced and modified with permission from Adans-Dester et al. (https://doi.org/10.1038/s41746-020-00328-w, licensed under CC BY 4.0)

and the FMA-UE scores. FAS estimates were marked by an RMSE of 0.38 points and a coefficient of determination (r^2) of 0.79 (Fig. 21.3b). The magnitude of the estimation error was deemed satisfactory, especially given the exploratory nature of the study. For the FMA-UE estimates, the authors used the output of the FAS estimation algorithm as an input to the FMA estimation module. The RMSE was equal to 3.99 points with a coefficient of determination (r^2) equal to 0.86 (Fig. 21.3c). This work is especially relevant to an application in the clinic as with one set of functional tasks, researchers were able to accurately estimate a measure of impairment and one of movement quality.

All the examples discussed above show the feasibility of deriving estimates of clinical scores via the analysis of data collected using wearable sensors. However, these techniques require going through the clinical test items or through a list of predefined motor tasks, which does not help to reduce the burden of administering evaluations. In addition, data processing remains labor-intensive in several of these cases. We hope that, in the future, researchers will find a way to derive clinical scores from wearable sensor data collected during the performance of unstructured activities and to streamline the analysis of such data. As such, using wearable sensors to estimate clinical scores would not only reduce the time needed to perform clinical assessments, but also allow clinicians to evaluate the effects of the intervention more regularly, facilitate the documentation of patients' response to the intervention, and adjust rehabilitation interventions as required to better meet the needs of their patients.

21.2.2.3 Wearable Sensors to Facilitate Upper Limb Training in the Clinic

The *ArmeoSenso* (Hocoma, Switzerland) is an example of a commercially available system for rehabilitation using wearable sensors [39]. Three IMUs are attached to the forearm, upper arm, and trunk to track arm movements in a three-dimensional space. The tracked UL movements serve as input for therapy games. Using such systems can enable group training in the clinic, allowing therapists to treat several patients simultaneously and potentially reduce therapy costs. Although we are not aware of any study using the *ArmeoSenso* for group training, Wittmann et al. [40] provided evidence that the *ArmeoSenso* can be used for self-directed arm therapy and enable high-dosage UL therapy that might result in improvements in arm function.

Also, Widmer et al. [41] used the *ArmeoSenso* in a study in which therapists provided minimum supervision during the training sessions. These studies provide direct evidence of the suitability of the system for self-directed, home-based therapy and indirect evidence of its suitability for group therapy.

The *MusicGlove* is a commercially available instrumented glove that requires the user to practice individual finger and grasping movements to play a music-based video game to retrain hand motor function after stroke (Fig. 21.4). A study comparing conventional UL training to training with the *MusicGlove* in chronic stroke survivors reported improved hand function related to grasping small objects (measured with the Box and Block Test) in the group using the *MusicGlove* [42]. No difference was found between training types for other measures of UL impairment (i.e., FMA-UE, WMTF, force). However, when the device was used for home-based hand therapy and compared to a conventional home exercise program, results showed an improvement in self-reported quality and amount of use (MAL scale) [43]. One of the attractiveness of this device for rehabilitation in the clinic is its low cost ($\sim$\$2500 for the clinic and $\sim$\$350 for the individual version). In addition, the ease of use of the device allows patients to use it by themselves in between therapy sessions or for use in group therapies where a single therapist can oversee numerous patients.

Wearable sensors have also been looked at as a way to provide feedback in the clinic. For example, Arteaga et al. [44] developed and tested a low-cost prototype ($\sim$\$100) of a wearable device to detect undesired postures in stroke survivors. The system consisted of 10 IMUs to track patients' posture and a combination of beeper, vibration, and LED light to provide feedback. While their pilot study showed the ability of the system to detect bad postures, unfortunately, it lacked testing in stroke survivors and seemed cumbersome to use, based on the number of sensors and equipment needed.

Wearable sensors to deliver rehabilitation interventions also provide an objective way to measure arm and hand movements during therapy. The feedback provided on the movement performance can provide much-needed motivation for stroke patients. It is important to note that the cost-benefit ratio of using wearable sensor-based methods to facilitate UL training in a rehabilitation setting needs to be examined.

Fig. 21.4 The *MusicGlove* is a system that integrates wearable technology and an interactive game (e.g., the Guitar Hero) to train hand dexterity. Reproduced with permission from Flint Rehabilitation Devices, LLC

21.2.3 Assessing and Treating Balance and Mobility of Stroke Survivors in the Clinic

Stroke survivors and others living with neurological diseases often present with balance and gait deficits associated with an increased risk of falls which impacts not only the quality of life, but also increased costs of care due to hospitalizations resulting from a fall [45]. It is, therefore, important for clinicians to quantify those deficits and identify patients at risk. For instance, the discharge plan will be different for patients with severe mobility impairments than those with mild ones. These assessments will also guide the rehabilitation plan and choice of assistive devices necessary for safe ambulation.

21.2.3.1 Estimating Clinical Scores

10-m Walk Test

The 10-m walk test (10 MWT) is used in the clinic to assess walking speed and determine the level of gait impairment following a stroke. The test records the time required to ambulate (often at a self-selected pace) 6 m on a 10-m walkway, the distance is then divided by the time to provide the speed in meters per second. However, this test does not provide any information on the gait quality, which may be problematic as one can walk faster but with compensatory strategies or , on the contrary, walk slower but with a better gait quality. Therefore, some researchers tried to complement the traditional gait speed assessment by using wearable sensors. Bergamini et al. [46] used a set of five IMUs to collect 3D linear accelerations and angular velocities from the pelvis, sternum, and head during the 10MWT performance. The amplitude of the accelerations and the gait symmetry measures they derived can provide the clinician with knowledge of the motor strategies and walking abilities of the patients, which complements the traditional speed information. More recently, Garcia et al. [47] tested the use of only one IMU placed at the waist to derive a gait smoothness metric via the estimation of

SPARC (spectral arc length). They identified via the IMU a reduced smoothness (lower SPARC) in stroke survivors, compared to healthy controls. The variability in smoothness during the 10MWT was higher in severely impaired stroke participants. In addition, they found that a smoother gait was correlated with lower limb (LL) spasticity and vice versa. Their results show that IMUs can provide complementary and clinically relevant information to the 10MWT and has the potential to be used in an outdoor environment.

Timed Up-and-Go Test

The Timed Up-and-Go (TUG) test is widely used clinically to evaluate mobility, balance, and fall risks in adults. The instrumented TUG (iTUG) requires patients to walk more than the original, non-instrumented version (7 m vs. 3 m, respectively) but allows one to gather more clinically relevant information than the conventional TUG, which only reports the time to complete the task. Researchers used a set of five IMUs for the iTUG: bilaterally on the wrists, bilaterally on the shanks, and one on the trunk [48, 49]. In addition to the total time, the iTUG can provide a breakdown of the test with the following: sit-to-walk duration and peak velocity, turning duration and peak velocity, and turn-to-sit duration and peak velocity. Gait metrics can also be derived to provide relevant information on the gait quality such as cadence, speed, stride length, and gait asymmetry. Even though it might not be faster than performing instrumented clinical tests, the iTUG allows one to gather more data on movement quality which is not available otherwise with most gait and mobility tests.

Trunk Impairment Scale

Impairments in trunk control often result in decreased balance, increased risk of falls, and can severely affect activities of daily living. In stroke, it can be assessed using clinical outcome measures such as the Trunk Impairment Scale (TIS) [50]. Researchers developed an instrumented version of the TIS with the hope of providing more detailed and clinically relevant information about trunk movement and how it relates to trunk impairments [51]. They used a commercially available system (Valedo, Hocoma, Switzerland) that includes three IMUs to measure trunk movement (in degrees) and velocity of body segments [52]. The system was assessed as a valid and reliable method to estimate trunk movements when compared to using an optoelectronic system in healthy participants [53]. Researchers found a moderate correlation between the instrumented TIS and scores attributed by clinicians. Using the wearable sensor system to instrument the TIS provides more information about trunk movements than the TIS. For instance, the ability to detect small changes in the range of motion that may not be observed clinically [51]. Nonetheless, this system with IMUs only on the trunk cannot account for LL compensatory movements which are commonly used by stroke survivors.

21.2.3.2 Wearable Sensors to Facilitate Gait Training in the Clinic

In the SIRRACT trial, researchers used IMUs bilaterally at the ankles to monitor LL movements performed by stroke survivors during their inpatient stay [54]. The aim of this intervention was to motivate patients and their therapists to engage in more gait practice to obtain improved walking-related outcomes. During this randomized clinical trial, participants followed their conventional therapies while wearing the sensors. Activity summaries (i.e., walking speed, distance, duration) derived from the sensor data was used to provide an augmented feedback intervention that was compared with feedback about walking speed alone. The key findings showed that providing augmented feedback beyond speed alone did not increase the time spent practicing or improve walking outcomes and found that during the inpatient stay, only a modest amount of time was spent walking. The authors pointed out that these results did likely reflect the constraints of inpatient rehabilitation such as space to practice walking and time spent focusing on other aspects of rehabilitation.

Another study by Byl et al. [55] found that providing dynamic visual kinematic biofeedback from pressure sensors and IMUs during gait training had similar effects to verbal feedback provided by the therapist. While these results are not encouraging the use of the system in the clinic when a therapist is available to provide oversight, they demonstrate the potential of using wearable sensors for gait training with limited therapist supervision.

21.2.4 Could Wearable Sensor-Based Evaluations Be Useful to Clinicians? A Possible Future Scenario

If data can be acquired and processed using streamlined procedures, then wearable sensors could enable data to be collected with minimal patients' and clinicians' burdens. These methods could allow clinicians to track the motor recovery trajectory of stroke survivors as schematically represented in Fig. 21.5. The figure shows a hypothetical case in which a patient undergoes a 36-week intervention. During this period of time, wearable sensors are used to monitor the subject. After 18 weeks, clinical score estimates and kinematic parameters derived from the sensor data, are available and define the motor recovery trajectory observed in response to the intervention until that point in time (orange circles in Fig. 21.5). The data can be used by rehabilitation specialists to assess if the patient is responding adequately to the ongoing intervention or if an adjustment to the intervention strategy is needed. Importantly, the information could be used to predict the patient's response to the intervention for the remaining weeks of the intervention period (green circles in Fig. 21.5).

Such models could also account for the patient's clinical phenotype and hence generate predictions based on both the information generated by the wearable sensors and the anticipated response to the intervention based on the patient's clinical characteristics. In this context, the above-described methods could be relied on to assess and predict the effectiveness of a given therapeutic intervention. The approach described in this hypothetical clinical scenario captures the essence

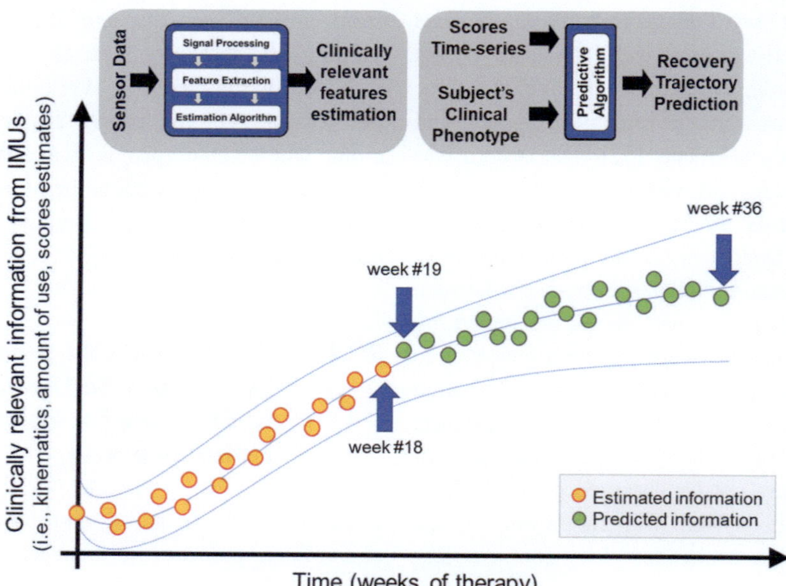

Fig. 21.5 Monitoring the motor recovery trajectory using wearable sensors. The time series represent the recovery trajectory of a hypothetical subject undergoing rehabilitation. The estimated information (orange circles) are clinically relevant measures of arm movements. The predicted information (green circles) is modeled based on the time series of the previously estimated information (orange circles) and the subject's clinical phenotype. Fitting a function (e.g., a polynomial equation) leads to generating a curve that represents the recovery trajectory. In addition, confidence intervals are generated for both the estimated and predicted clinically relevant information. Reproduced and modified with permission from Adans-Dester et al. (https://doi.org/10.1038/s41746-020-00328-w, licensed under CC BY 4.0)

of precision rehabilitation in which clinicians design patient-specific interventions, set clinical objectives, track patient's response using wearable sensors, and periodically evaluate the effectiveness of the ongoing intervention based on the observed recovery trajectory. Future work should fully enable this approach by further improving the unobtrusiveness and ease of use of wearable sensors and by developing fully automated data analysis procedures, for instance, for the segmentation of the sensor data based on detecting data characteristics associated with the performance of motor tasks suitable to derive reliable estimates of clinical scores. Discussing these implementation challenges with patients, clinicians, and engineers during future research and product development will likely result in more widespread and accessible use of wearable sensors in the clinic.

21.3 Wearable Sensors to Measure Movement in the Field

21.3.1 Why Would One Want to Measure Movement in the Field?

The first and simplest answer to this question is because it is movement in the field, i.e., activity performance in everyday life, that persons with stroke care most about. People with stroke are referred to or seek out rehabilitation services to improve the performance of an activity in their home and their community. Indeed, self-identified rehabilitation goals are nearly always (88%) about improving performance in daily life [56]. In contrast, researchers and clinicians rarely place performance of daily activity at the center of their measured treatment goals (Lang et al. unpublished data). Clinicians in the current stroke rehabilitation delivery model focus on measuring impairments and capacity (see Box in the Introduction section for definitions) with the assumption and hope that improvements in these measurement levels will translate to improvements in performance in daily life.

The second answer to this question is that the capacity for movement assessed in the clinic does not necessarily provide accurate and actionable information about the performance of movement in the field. This conflict is illustrated with walking data in Fig. 21.6. The red oval highlights a portion of the data around 0.75 m/s walking speed where some individuals are walking only 2000–4000 steps/day, while others are walking 8000 or even 12,000 steps/day. Without wearable sensors (here attached to the unaffected ankle) quantifying walking performance in the field, neither rehabilitation clinicians nor their patients would know how much walking in the field occurs.

The third answer to why one would want to measure movement in the field is because

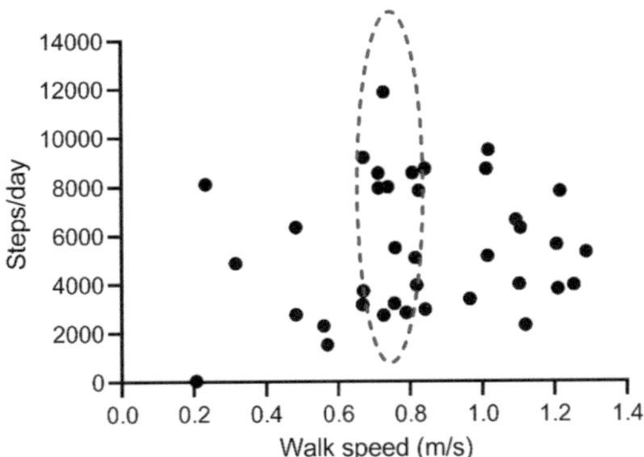

Fig. 21.6 Measures taken in the clinic are not consistently related to measures taken in the field. Scatterplot of people (n = 37) receiving outpatient therapy services post stroke. X-axis: in-clinic measure of walking capacity using the 10 MWT. Y-axis: in the field measurement of walking performance quantified by steps/day. The dashed red oval illustrates how individuals with a small range of walking speeds can have very different amounts of walking in the field. Data from Holleran et al. (https://doi.org/10.1097/NPT.0000000000000327)

improvements in movement assessed at the impairment and capacity levels within the clinic often do not translate to improvements in the performance of activities in daily life. Figure 21.7 shows an example of this, where there is a clear improvement over the course of outpatient therapy services on a common standardized test of UL capacity (Fig. 21.7a) but no change in movement performance in daily life (Fig. 21.7b) as measured with wearable sensors in the field.

Multiple reports have now shown a discrepancy in stroke rehabilitation outcomes in movement capacity assessed in the clinic versus movement performance assessed in the field [57–60]. In a recent analysis (Lang et al., unpublished data, N = 138), the majority (58%) of people receiving outpatient services at five rehabilitation clinics around the United States improved their capacity to complete UL and walking activities, as measured by in-clinic assessments, but failed to improve their movement performance in the field, as measured with wearable sensors. An additional 17% improved both capacity and performance, 24% improved on neither, and 1% improved on performance but not capacity. These data illustrate the point that just because someone can execute actions in a clinic or laboratory does not mean the person will carry over and execute those actions outside the clinic, within an unstructured home and community environment. For example, a person can have the strength and coordination to reach and grasp a cup with the paretic UL and demonstrate that capability on a standardized test, but when at home, may (implicitly) choose to reach and grasp cups with the non-paretic limb due to convenience, efficiency, and/or safety [61, 62]. As implicit choices accumulate across activities, hours, and days in the field, the limited activity of the paretic (or both limbs) can be quantified by numerous wearable sensor variables [63–68] that quantify duration, magnitude, variability, and relative limb activity symmetry. If clinical decisions are based only on the measurement of movement in the clinic, rehabilitation clinicians and patients will be missing information needed to address patient goals and improve movement performance in daily life. Wearable sensors,

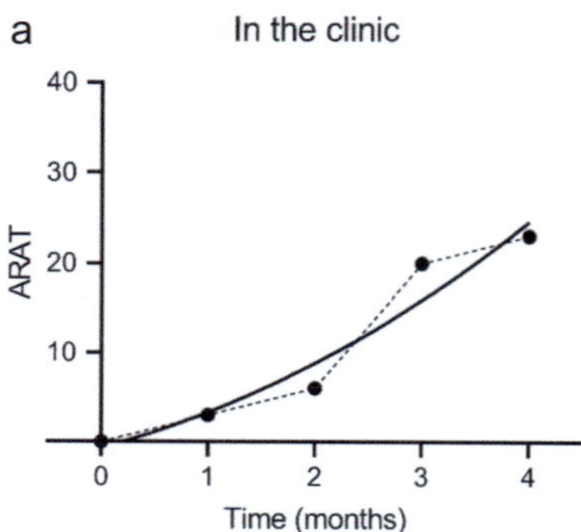

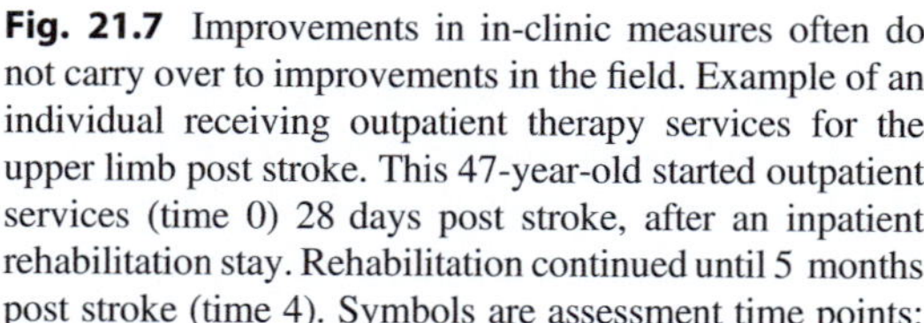

Fig. 21.7 Improvements in in-clinic measures often do not carry over to improvements in the field. Example of an individual receiving outpatient therapy services for the upper limb post stroke. This 47-year-old started outpatient services (time 0) 28 days post stroke, after an inpatient rehabilitation stay. Rehabilitation continued until 5 months post stroke (time 4). Symbols are assessment time points; thick lines represent best fit models. A: Results from in-clinic assessment on the Action Research Arm Test (ARAT, higher = better, 57 = normal). B: Results from monitoring in the field with bilateral, wrist-worn sensors. The use ratio is a ratio of the duration of the paretic limb use to the non-paretic limb use over a 24 h wearing period (higher = better, normative values ∼0.9–1.0)

therefore, provide an important opportunity for future improvement of stroke rehabilitation services and stroke rehabilitation outcomes.

21.3.2 Monitoring Upper Limb Movements in the Field

The most common option to measure UL movement is with tri-axial accelerometers. Many commercially available, research-grade devices also contain gyroscopes, magnetometers, inclinometers, and optical sensors. In patient studies, these devices are typically worn on one or both wrists, with monitoring occurring for at least 24 h [69]. Wrist-worn devices capture movements of the upper arm, forearm, and wrist, but not fine dexterous movement of the fingers. Wrist-worn sensors work well for people who are moderately to severely affected post stroke. Wrist-worn sensors quantify UL movement with reasonable accuracy because these individuals cannot make small, fractionated movements of the fingers without moving the wrist, forearm, and/or upper arm [70]. In persons with very mild stroke, where impairments are relatively isolated to dexterous movement of the fingers, then sensors worn on the fingers in addition to the wrist may be needed to better capture UL movement in the field [71].

If one is to monitor UL movement in the field for adequate durations (e.g., 24 or more hours), then the wearable sensors or system of wearable sensors must meet four practical considerations. *First*, being able to monitor both ULs simultaneously is usually necessary, i.e., the sensors need to be worn on both the paretic and non-paretic limbs post stroke. This is because of the enormous heterogeneity in how much/how often humans move throughout a day, but the tight homogeneity in the relative movement of one limb versus the other in neurologically-intact individuals across the lifespan [63, 72, 73]. *Second*, wearable sensors that are on the wrist or fingers need to be waterproof. Humans wash their hands and encounter water during many activities throughout the day. If the sensors have to be removed every time hand-washing is needed, then the likelihood of the sensors being worn and worn correctly decreases substantially. *Third*, straps or pockets that secure the sensors to the UL need to be comfortable and sufficiently easy for a person with stroke to don and doff (alternatively, sufficiently easy for a caregiver to don/doff). Uncomfortable or too tight sensors on the ULs will be removed, while too loose sensors will not accurately track movement. And *fourth*, the fewer number of sensors can be worn on the limb to get the necessary data, the greater the probability they will be worn for the assigned monitoring period. Wearable systems with multiple sensors [74] are feasible for in-clinic measurement, but often will not be worn, worn correctly, or result in loss of the sensors when monitoring in the field. If wearable sensors have an attractive appearance (e.g., a ring looks like a piece of jewelry), then that will further increase wearing compliance. Developing or adapting sensors and sensor systems that adhere to these practical considerations will further the implementation of wearable sensors into routine stroke rehabilitation care.

One of the major challenges to the widespread adoption of wearable sensors in routine stroke rehabilitation care is the lack of clinical validation [68, 75]. The problem is not in the verification of the sensors themselves, but in the clinical validation of the algorithms developed by researchers to derive metrics of clinical relevance. Clinical validation efforts lag behind the engineering development of sensor hardware and software, perhaps because clinical validation is time-consuming, expensive, and requires interdisciplinary teams. Clinical validation is hampered by four key issues. *First*, a large number of variables have been proposed in various research studies, with many different variable names and often different formulae that may be capturing similar or related constructs of movement [68]. *Second*, variables can be mathematically-complex (e.g., Spectral Arc Length as a quantification for UL movement smoothness [76, 77]) and thus hard to interpret clinically with respect to daily activity in the field. *Third*, there are insufficient validation data to indicate which variables carry clinical meaning and are ready to be deployed widely in clinical practice. Most variables have been evaluated in small samples of control or stroke participants at a single point in time. Only a few variables have been evaluated longitudinally in larger samples but lack data on either responsiveness to change and/or how much change is clinically meaningful to patients. One UL variable, the use ratio, is widely used in research and is close to be ready for

clinical implementation after being proposed 20 years ago [65]. And *fourth*, UL movement performance in daily life is a complex construct that is likely multidimensional [68, 73, 78]. Thus, there is a high probability that UL movement in the field may be most appropriately represented by multiple variables, not any single variable [79, 80]. For example, the use ratio (Fig. 21.7b) provides information about the relative duration and symmetry of UL movement throughout the day, but other variables could be needed to understand the magnitude and variability. Solving these four issues variable-by-variable for the stroke rehabilitation population will require a large investment of engineering and clinical resources if wearable sensors are to become ubiquitous in UL stroke rehabilitation care.

21.3.3 Monitoring Lower Limb Movements in the Field

Monitoring LL movement in the field shares many of the same benefits and challenges as monitoring UL movement. Unlike the UL, walking is the one essential LL movement activity that rises above all the others. Regaining the ability to walk is the number one goal of most persons undergoing stroke rehabilitation [81, 82]. The primary method to quantify walking performance in the field has been with sensors that count steps/day. Clinicians face a dilemma when trying to use wearable sensors to record steps per day in the field for their patients with stroke. On the one hand, consumer-grade devices worn on the wrist can be inexpensive and are readily available, but can be wildly inaccurate for the majority of persons with stroke who walk slowly, asymmetrically, and/or use assistive devices [83–87]. On the other hand, research-grade sensors are expensive and not easy for clinicians to deploy in a busy clinical environment. A collaborative effort to develop a wearable sensor system that is cost-effective, simple to use, and accurately quantifies walking performance in the field across a broad range of walking abilities will be necessary to make monitoring a routine in clinical stroke rehabilitation practice. Study protocols to capture walking performance typically record behavior for more days (e.g., 5–7 days [88]) than are seen in UL studies [1–3, 69], because of the high amount of variability in daily stepping in persons with stroke [89]. Compliance with wearing tends to decrease over time, especially when people have to wear them at multiple time points [90].

As with UL monitoring, walking performance in daily life may eventually be best represented by multiple variables, not a single variable. Steps/day measures the amount but can miss other aspects such as gait asymmetry, the ability to navigate various environments (e.g., outdoor walking, stairs), and potentially falls. Many of the emerging technologies (see below) could present new opportunities for building multivariate feedback regarding walking. If feedback is provided in a simple, compelling interface for clinicians and persons with stroke, there is a greater likelihood of implementation.

21.3.4 Critical Information Learned from Wearable Sensing in the Field that Would not Be Known Otherwise

While there is much work to be done before wearable sensors and systems are perfected, important knowledge for stroke rehabilitation has already been learned by monitoring movement in the field. Here three examples of new knowledge that could only be obtained from wearable sensing are provided. The first two examples are from samples of persons with stroke wearing bilateral, wrist-worn accelerometers for 24 or more hours, while the third is from persons with a stroke wearing a finger/wrist tracking device for 24 h.

Wearable sensors have challenged assumptions about how persons with stroke maintain the overall amount of UL activity in daily life by compensating with their paretic limb. If the overall amount of activity was maintained, then one would expect a negative correlation between the activity of the paretic limb and the activity of the non-paretic limb (i.e., the paretic limb activity would increase as the non-paretic limb decreased

in order to maintain the overall amount of activity). As can be seen in Fig. 21.8, however, there is a strong, positive correlation (r = 0.78, p < 0.01) between the duration of use of the paretic versus non-paretic UL post stroke. This positive correlation indicates that as people move the paretic limb less throughout the day, they move the non-paretic limb less too. They are not compensating as much with the non-paretic limb as assumed, but instead doing less activity overall. Interestingly, this relationship is true both early [91] and later after stroke [92]. Stroke rehabilitation clinicians and researchers would not know about the limited UL movement in daily life without monitoring movement in the field with wearable sensors.

Wearable sensors are also changing perceptions about the recovery of UL movement post stroke. Decades of research on recovery trajectories post stroke indicate that larger, rapid changes occur in the first few weeks, with smaller, slower changes occurring later [93–98]. Changes in impairment generally precede changes in functional capacity by around one week, such that as movement control returns, individuals regain the ability to execute functional tasks

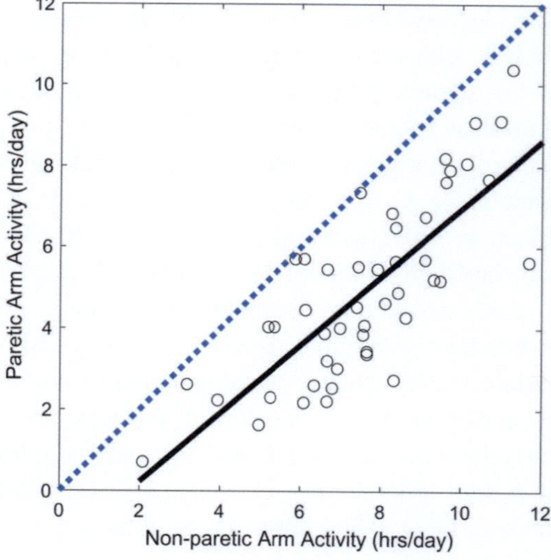

Fig. 21.8 Relationship between movement activity of the paretic (y-axis) vs non-paretic (x-axis) upper limbs in 46 adults with chronic stroke. UL: upper limb. Data from Bailey et al. 2015 (https://doi.org/10.1179/1074935714Z.0000000040)

[96, 97]. The common perception has been that as functional capacity improves in the clinic, then improvements are incorporated into daily life at home and in the community (i.e., activity performance in the field improves). If this perception were correct, one would expect recovery trajectories where a plateau of impairment-level measures occurs first, followed by plateaus in capacity measures, and then finally by plateaus in performance level measures. As can be seen in Fig. 21.9, wearable sensors have discredited that perception [79]. A prospective longitudinal cohort (n = 67) of persons was followed from 2 weeks out to 24 months after first-ever stroke, with bi-weekly measurements of UL impairment (Fugl-Meyer scale [28]), capacity (Action Research Arm test [99]), and performance (use ratio and hours of paretic limb activity [92]). UL performance in daily life (blue line) plateaued surprisingly early after stroke. Plateaus in performance did not lag plateaus in impairment (gray line) and capacity (black line), but instead slightly preceded or occurred at the same time [79]. These data imply that UL movement in the field settles into a stable pattern early and often before neurological and functional recovery is finished. The early plateau in UL performance

strongly suggests that to improve stroke rehabilitation outcomes, interventions that pair motor training *and* intentional health behavioral interventions are needed [100, 101].

Finally, wearable sensors have provided new insights into the detailed nature of the relationship between capacity and performance. As discussed in Sect. 21.3.1 and shown in Fig. 21.9, improvements in movement assessed at the capacity level within the clinic often do not translate into improvements in the performance of activities in daily life. Schweighofer et al. [102] hypothesized that real-world UL performance lags clinically-demonstrated UL capacity until UL capacity reaches a threshold; they generated this "Threshold Hypothesis" based on self-reported use of the amount of hand use at home. Data acquired from a novel wearable sensor (called Manumeter) recently confirmed this hypothesis (Fig. 21.10) [103]. The Manumeter consists of a watch-like sensor and a small permanent magnet worn as a ring. The watch-like sensor uses an array of magnetometers to detect changes in the magnetic field as the ring moves due to finger or wrist movement. A total of 29 stroke survivors wore the Manumeter at home during their daily activities for 6–9 h. Capacity

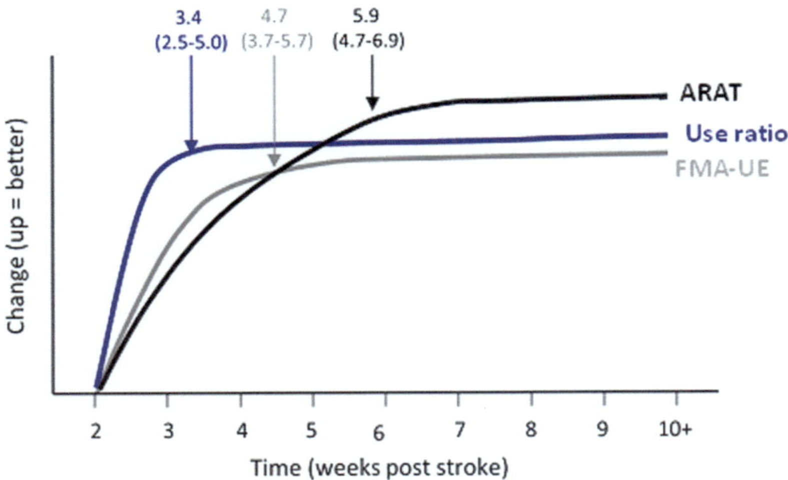

Fig. 21.9 Average trajectories of change over time for impairment (in clinic, gray) capacity (in clinic, black), and performance (in the field, blue). Arrows mark the time of plateau; values are means (95% Cis). Values on the y-axis are theoretical and not intended to be compared across the three trajectories. ARAT: Action Research Arm Test; FMA-UE: Fugl-Meyer Assessment Upper Extremity subsection. Data from Lang et al., 2021 (https://doi.org/10.1177/15459683211041302)

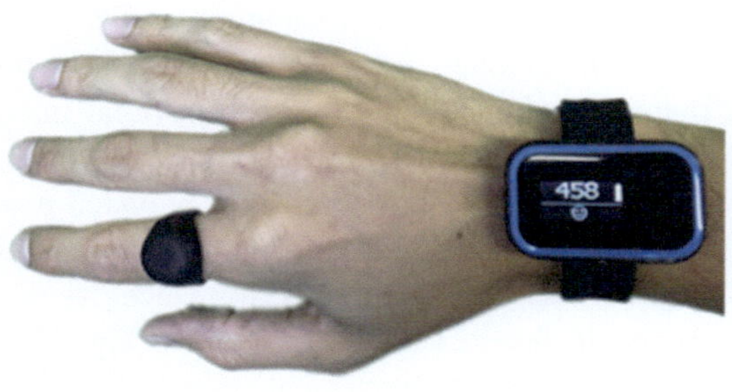

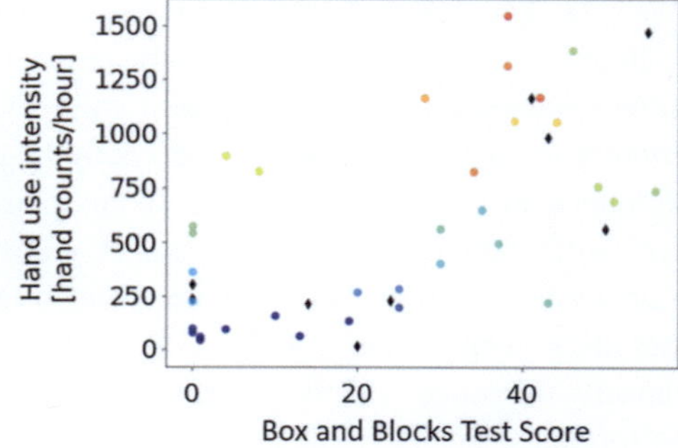

Fig. 21.10 Left: The Manumeter, a device that counts finger/wrist movements by measuring changes in the magnetic field at the wrist sensor produced by the ring. Right: Hand use intensity (i.e., "performance") measured at home for 29 stroke survivors with different levels of hand capacity, as quantified by the Box and Blocks Test (BBT) Score. For the circles, each color represents one subject, and each subject can have one to three samples for up to three different days. Data from Schwerz de Lucena et al. 2021 (https://doi.org/10.3390/s21041502, licensed under CC BY 4.0)

was measured in the laboratory using the Box and Blocks Test (BBT), which requires individuals to pick up and transport as many small blocks as possible in 60 s. Most participants with BBT scores <30 had a low finger/wrist movement count intensity of around 200 counts/h, which is the amount of counts to be expected due to "false positives" from environmental magnetic fields. Then, there was an increase in hand use intensity as participants' BBT scores increased beyond 30, consistent with the Threshold Hypothesis. Thus, achieving a 50% score on a capacity measure predicted the start of use of the hand at home (i.e., increasing performance).

21.4 Wearable Sensors to Motivate Movement and Exercise in the Community

Wearable sensing technologies have proven useful for promoting the activity and health of people without disabilities. For example, a 2007 systematic review in JAMA found that daily pedometer feedback is an effective way to increase walking activity and thereby improve difficult-to-change health outcomes such as body mass index and blood pressure [104]. Recognizing this finding, many companies now sell wearable sensors and phone apps for counting steps. The global fitness tracker market is projected to grow from $36.34 billion in 2020–$114.36 billion in 2028 [105].

Goal setting with feedback is known to be a powerful modulator of performance [106] and indeed it appears to be a key requirement for the successful use of such fitness trackers. Based on an analysis in the systematic review of pedometer feedback referred to above, setting a step goal (e.g., 10,000 steps) was significantly associated with an equivalent one-mile increase in steps/day, while individuals who did not set a goal did not significantly increase their step count. In the context of rehabilitation, a seminal multisite randomized controlled trial on the use of quantitative performance feedback, the SIR-ROWS study, showed that providing individuals post stroke with their completion time in a 10 MWT at regular intervals throughout rehabilitation therapy, along with a simple comment on whether that time exceeded their previous time, significantly improved their gait speed over the course of therapy compared to individuals who did not receive this feedback [107]. Presumably, the quantitative feedback caused patients to set a goal of improving their gait speed at the next test. Goal setting is thought to affect performance through four mechanisms: (1) directing attention toward goal-relevant activities and away from goal-irrelevant activities; (2) energizing greater effort; (3) increasing persistence; and (4) stimulating arousal, discovery, and use of relevant strategies [106], all of which could play a role in rehabilitation.

When considering the motivation for applying wearable sensing in rehabilitation practice, a key issue is increasing amounts of movement practice. While rehabilitation research has not yet been able to precisely define sufficient, patient-specific goals for the amount of practice in a scientific way, there is a broad consensus that patients typically do not practice enough. In-clinic therapy sessions achieve a limited number of practice repetitions [108], and health payors limit the number of reimbursed therapy sessions. Therapists, therefore, create home exercise programs to increase rehabilitation doses. Yet adherence to home programs is low, even if the prescribed exercise program is unambitious [109–111]. Further, as reviewed above, the amount of use of an impaired limb in daily life is often low, even for people with substantial functional capacity. Low daily use of a limb is thought to create a "vicious cycle", contributing to further degradation of movement ability [102]. Thus, a primary goal for wearable sensing after stroke is to create a "virtuous cycle", in which patients move more frequently, whether during exercise sessions or during daily life, in order to promote movement recovery.

21.4.1 Promoting Upper Limb (UL) Movement

Despite the availability of a clear pragmatic goal (i.e., facilitating more movement practice), few wearable sensing studies have attempted to achieve this goal [27]. Initial studies suggest that increasing movement is possible, but likely requires goal setting and coaching along with wearable feedback. Delivering improved UL outcomes may also be possible but may be more difficult than what might have been expected given the pedometer literature.

21.4.1.1 Providing Feedback on UL Movement Amount

Whitford et al. [112] were among the first to study the use of wearable sensors to generate feedback about UL use in stroke survivors. In their study, eight chronic stroke survivors wore accelerometers without feedback screens on both wrists during waking hours for three weeks. Research therapists visited their homes three times per week to collect and process data. They provided feedback on the amount of activity and disparity of activity between arms through verbal discussion and by presenting graphs. At each of these feedback sessions, participants also set two goals related to increasing their paretic UL activity. Their therapist reviewed progress toward these goals with them at the next session. This strategy significantly increased participants' perception of paretic UL activity. Yet no improvements in actual activity of the UL (as measured using the accelerometers) or in functional outcomes were found.

Another recent study provided feedback on the number of wrist and finger movements made throughout the day to try to motivate increased UL use [113]. Twenty chronic stroke participants wore the Manumeter, the wristwatch-like device described above that senses the magnetic field of a small magnet ring worn on the index finger, using a nonlinear detection algorithm to calculate the number of finger movements [103]. Participants in the experimental group received real-time feedback on finger movement counts and a daily goal personalized to their impairment level. Subjects in the control group used the device as a wristwatch, but the device still tracked the number of finger movements. Both groups also were given a home exercise program described in a booklet. After data analysis, it was found that the experimental group chose to wear the Manumeter for approximately one hour more each day, but did not increase their finger movement intensity, measured as counts per hour. Scores on the BBT and MAL did not improve significantly at 3 months, although scores on the FMA-UE and the ARAT improved for both groups.

21.4.1.2 Reminders to Move

A different approach is to use a wearable device to provide reminders (or "nudges") to move the limb, usually in the form of vibrations [18]. In this case, the device can provide the nudge after sensing a period of relative inactivity, or, alternately, the device need not necessarily sense limb movement, but instead can provide reminders based on a timer, similar to a water intake management app. Three pilot studies have shown the feasibility of this approach. Signal et al. [114] studied via an observational methodology conducted when stroke patients were inpatients whether "haptic nudges" caused an increase in probability of moving their arm. They used a Bluetooth-enabled wearable device to provide three consecutive vibratory stimuli of 0.3 s duration at 150 Hz within 1.5 s, with a magnitude similar to a phone vibration. Patients were instructed to "move, try and move, or visualize moving their (affected) arm" following a nudge. Observers discreetly followed stroke inpatients out of their field of view, logging UL movement for one minute every 10 min. They randomly delivered haptic nudges or no intervention just before the observation periods. The odds ratio of moving the UL following a haptic nudge relative to no nudge was 1.44, demonstrating an increase in UL activity in response to the haptic nudge.

A second feasibility study used an acceleration-sensing wristband with seven stroke patients ≤ 28 days post stroke for four weeks [115]. Therapists reviewed movement activity data twice weekly with their patients. The wristband was programmed with a personalized threshold for providing a vibratory prompt (5, 25 or 50% greater than the median activity). Mean activity increased in the hour following a prompt (by 11–29%) compared to the previous hour, as measured by the accelerometer in the wristband. 96% of patients expressed a preference that reminders be delivered once per hour, rather than 2, 3, or 4 times per hour. 75% of patients expressed a preference that the target threshold for triggering vibration be set at the lowest setting (i.e., 5% above the previous median baseline activity). In a follow-up pilot study [116], the same research group studied 33 patients 0–3 months after stroke receiving a four-week, self-directed therapy program with a twice-weekly therapy review. The wristband adjusted the threshold and frequency of delivery of the vibration prompt based on the activity level of the wearer. The wristbands were worn for 79% of the recommended time (between 8 AM and 8 PM). Patients again showed a preference for hourly prompts and not more frequent prompts. While clinical outcome measures were acquired, no statistical comparisons were made with a control group in this pilot feasibility study.

In terms of the therapeutic efficacy of this reminder approach, a vibration-based, remind-to-move sensor was tested in a study with 84 stroke survivors who had the first stroke in the last six months [117]. Participants were randomly allocated to either an experimental group (device worn with vibrations delivered), sham group (device worn with no vibrations), or control group (usual therapy). The patients wore the wrist vibrator for three consecutive hours daily over four weeks. The device emitted a vibration cue similar to the vibration mode of a mobile phone every 10 min. The vibration would not stop until a button on the device was pressed. A small but statistically significant greater improvement in one of the clinical outcomes (the ARAT) was observed. A significant difference in the amount of arm activity between groups (measured by an accelerometer embedded in the wristband) was also observed.

These studies suggest that providing movement reminders through vibratory inputs can increase the amount of UL activity and that this increase may have at least a small therapeutic benefit.

21.4.1.3 Providing Feedback on Exercise Activities

Wearable sensors can also be used to provide users with feedback as they perform exercise activities at home. In this case, other types of non-worn sensors, such as camera-based systems or instrumented objects, can serve similar functions. There is a large and growing literature on clinical trials conducted with a variety of sensor-based exercise systems, and numerous commercial systems are available. Here, we briefly review two important studies that focused on using wearable sensors to provide feedback on exercise activities.

A key concern of rehabilitation therapists in providing home exercise is their inability to provide real-time feedback on the quality of the exercise. Performing poor quality exercise is thought by some to be suboptimal or even detrimental to recovery. Wearable sensors have shown potential to help solve this problem. Lee et al. asked 20 people with stroke to wear IMUs on each wrist as they performed assessments and participated in UL therapy [18]. Using video analysis as the gold standard, they showed that they could distinguish goal-directed movements (such as participating in an UL assessment, ADLs, or therapy) from non-goal-directed movements (such as arm swing, gesturing, and resting periods) with an accuracy of 87%. During the performance of a particular exercise ("arm raise in the sagittal and coronal planes"), they could identify when the therapist provided corrective feedback with an accuracy of 84%.

Chae et al. used a wrist-worn sensor to detect when individuals with chronic stroke were performing UL exercise at home [118]. They assigned patients four UL exercises (Bilateral Flexion, Wall Push, Active Scapula, and Towel Slide). Based on in-clinic data, they showed that they could identify the type of exercise from this small set with up to 98% accuracy. The system also recorded exercise repetition counts and duration of exercise, reporting it via an app to a supervising therapist who then contacted the patients once a week to review their progress. A control group received the exercise program on paper without a sensor and was also contacted once a week. A total of 38 participants were enrolled. All participants in the control group had dropped out after 18 weeks, while 12 of 22 in the wearable sensor group persevered to the end. They observed improvements in the WMFT and range of motion of shoulder flexion and internal rotation in the sensor group, while the control group showed only a significant change in shoulder internal rotation.

These studies outline the potential for wearable sensors to provide movement quality feedback, and to serve as a motivational aid by giving therapists a "window" into their patients' home exercise adherence.

21.4.2 Providing Feedback on Lower Limb (LL) Movement Amount

Research progress with wearable sensors for encouraging walking after stroke is more developed than research to encourage UL activity. A 2018 Cochrane review examined the available evidence regarding the effectiveness of wearable sensors (such as pedometers, Fitbit, and Garmin watches) as well as smartphone activity monitors for increasing physical activity levels for people with stroke [119]. This review found four studies that met its criteria with a total of 245 participants in the subacute or chronic phase post stroke. All studies compared the use of an activity monitor plus another rehabilitation intervention that was focused on walking versus the other intervention alone. The review found no clear effect of the use of activity monitors on step count in a community setting or in an inpatient rehabilitation setting.

More studies have been published since this review with mixed results. Mandigout et al. studied 83 participants at an average time of 2.4 months after stroke [120]. Participants were randomly assigned to receive individualized coaching or standard care for six months. The coaches monitored physical activity with an activity tracker (SenseWear Armband) and conducted home visits and made a weekly phone call to review activity. The difference between the two groups was not significant at any evaluation time point for the primary endpoint, the 6 MWT.

On the other hand, Montserrat randomized 41 chronic stroke survivors to a conventional rehabilitation program or to a Multimodal Rehabilitation Program that monitored adherence to physical activity [121]. The multimodal program combined an app with GPS and accelerometer-based sensing to monitor walking distance and speed, a pedometer, a WhatsApp group, an exercise program with aerobic, task-oriented, balance, and stretching components, and a progressive daily ambulation program that was monitored by the app and pedometer. At the end of the intervention, community ambulation increased more in the intervention group (38.95 vs. 9.47 min), and sitting time decreased more in the intervention group (by 3 vs. 0.5 h/day).

Although it focused on a broader population than just stroke patients, a recent pragmatic clinical trial of 300 mobility-impaired patients (including stroke patients) likewise found a benefit from a multimodal program incorporating digital technology [122]. A physical therapist individually prescribed technology that included virtual reality video games, activity monitors, and handheld computing devices. The technology was used for six months in the hospital and at home; patients used on average four technologies in the hospital and two at home. The most commonly used digital technology in the home was a wearable activity monitor (Fitbit or Garmin, used by 98%), followed by an iPad exercise app (used by 86%). Changes in mobility scores (measured by the performance-based Short Physical Performance Battery) were about 10% higher in the intervention group compared to the control group (p = 0.006). However, there was no evidence of a difference between groups for an upright time at 6 months.

21.4.3　Summary

In summary, the promise of using wearable sensors to encourage UL and LL activity in the community after stroke has not yet been realized. For the UL, at this early stage of research, the reminder paradigm has perhaps shown more potential for increasing activity and reducing impairment than the paradigm of goal setting with performance feedback from a wearable sensor. For the LL, recent studies suggest that programs that incorporate wearable sensors into multimodal therapy programs may be more effective at increasing walking activity than programs that focus on goal setting with performance feedback alone. We suggest that optimizing the programmatic context in which wearable feedback is delivered will be important for realizing the potential of this technology, and will likely include intentional health behavioral interventions, as suggested above. Key factors to consider are the way goals are set, the specific form of the performance feedback (including both quantity and quality feedback), the availability and nature of therapist coaching, and the integration of a diversity of therapeutic activities along with the wearable feedback.

21.5　Emerging Technologies and Their Potential Applications

The previous sections of this chapter have provided an overview of prior work focused on facilitating the implementation of rehabilitation interventions and the assessment of clinical outcomes by relying on wearable sensors consisting of "units" (often relying on wireless technology) that are typically attached to body segments using elastic straps (e.g., wristbands). In this section, we will consider other technologies. Recent advances in e-textiles and materials science have allowed researchers to explore the use of garments with embedded sensors as well as the development of sensors that conform to the anatomy in a way that is similar to an adhesive bandage (often referred to as e-skin sensors). Furthermore, because contextual information is often essential to perform a meaningful analysis of movement patterns, researchers have begun to explore the use of wearable cameras to gather such information. Radio tags and radar-like technologies could be utilized to gather contextual information, but their use has so far received little consideration in the field of rehabilitation. Recent advances in video analysis techniques, largely enabled by the development of deep learning-based algorithms, have generated significant interest among rehabilitation specialists. These techniques provide an unprecedented capability to track movement patterns with low-cost cameras and are likely to replace the use of wearable sensors in systems designed for home-based rehabilitation. Finally, existing and emerging wearable, as well as contactless technologies, provide researchers and clinicians with the ability to monitor the physiology of patients in the home and community settings. Although a thorough discussion of their potential applications to stroke rehabilitation is beyond the scope of this chapter, in this section, we briefly mention a few examples.

21.5.1 E-textiles

E-textiles are fabrics designed to enable embedding electronics in objects and garments, thus allowing researchers and clinicians to monitor patients outside of the laboratory. A recent review by Angelucci et al. [123] provides a summary of the methods (e.g., coating and printing) traditionally used to make conductive yarns and then use them to make e-textile garments by relying on techniques such as knitting, weaving, and embroidery. The development of e-textile systems for patient monitoring was originally motivated by the assumption that providing patients with garments equipped with sensors would have resulted in better compliance than the use of wireless sensors to be strapped to body segments.

The first steps toward developing e-textile systems were marked by major contributions by Jayaraman et al. [124–126] and by De Rossi et al. [127–129]. Seminal work by Jayaraman et al. [124–126] resulted in the development of conductive yarns enabling the connection of sensors embedded in the garment to a data logging unit, hence allowing researchers to monitor patients' physiology. Shortly after the publication of this work, De Rossi et al. [127–129] introduced the use of conductive polymers to print strain sensors on lycra garments. This work was particularly focused on monitoring movement patterns in individuals undergoing rehabilitation. Following their initial work with a focus on developing e-textile garments, De Rossi and colleagues implemented a fully-functioning

platform to monitor stroke survivors and facilitate the performance of rehabilitation exercises [130].

Unfortunately, technical limitations marked these initial prototype e-textile systems. Researchers found it challenging to develop e-textile garments that could be washed multiple times without being damaged. Besides, components such as the connectors between the conductive elements of the garment and the traditional electronics (e.g., data logging units) to be used with the garment turned out to be difficult to manufacture in a way that met the technical specifications of the problem at hand. Nonetheless, this seminal work generated a great deal of interest in the application of e-textiles in the rehabilitation of patients with neurological conditions, including stroke, as summarized in a review paper by McLaren et al. [131]. Interestingly, this review devoted significant attention to e-textile gloves and socks [132]. These are interesting technologies, though e-textile gloves have been found by many researchers to be of limited use in stroke survivors, because these patients have difficulties donning and doffing gloves, particularly on their stroke-affected hand. Similarly, e-textile socks have been seldom utilized in clinical studies, as researchers have often found it more practical to use instrumented insoles to collect proxy measures of ground reaction forces.

New approaches to the development of e-textile garments are currently emerging that appear to have addressed the main limitations of previously developed prototype systems. An example of the techniques used in recently developed e-textile systems is shown in Fig. 21.11. These e-textile garments are based on embedding flexible electronics in pocket-like components typically referred to as "textile channels". The use of traditional integrated circuits allows researchers to take advantage of advances in sensing technology. New materials are used to encapsulate electronic components, thus making them washable and mechanically robust.

It remains to be seen if these new approaches to the development of e-textile garments can

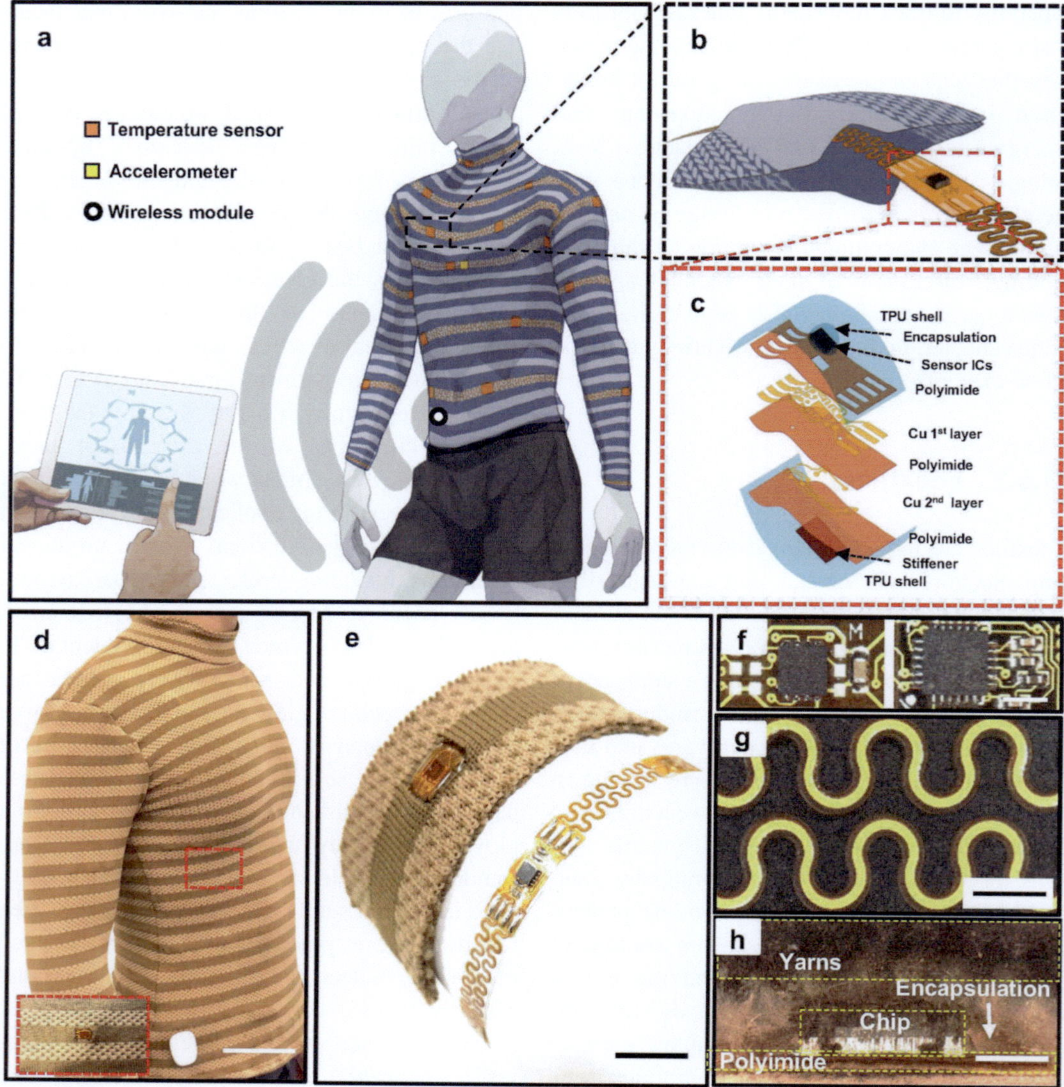

Fig. 21.11 Recent implementations of e-textile garments rely on deploying multiple sensors (panel a) embedded using pocket-like textile channels (panel b) containing sensor islands (panel c) equipped with traditional integrated circuits on a flexible substrate (made of copper and polyimide layers) covered by thermoplastic polyurethane (TPU) and a washable encapsulant. The garment, shown in panel d (scale bar: 10 cm), carries flexible-stretchable electronic strips (right) and woven electronic strips in a knit textile (left) as per the example shown in panel e (scale bar: 1 cm). Examples of temperature and accelerometer integrated circuits are shown in panel f (scale bar: 3 mm). Panel g shows an example of an interconnect module (scale bar: 2 mm). Panel h shows a cross-sectional view of a sensor module embedded in a polydimethylsiloxane (PDMS) layer (scale bar: 2 mm). Reproduced with permission from Wicaksono et al. (https://doi.org/10.1038/s41528-020-0068-y, licensed under CC BY 4.0)

deliver on the promise to achieve higher compliance than wearable sensors that are strapped to body segments. That said, the use of e-textile garments is appealing in clinical applications requiring long-term monitoring as one would anticipate that patients would prefer wearing a garment with embedded sensors rather than having to don and doff multiple elastic straps equipped with sensing technology every day during the monitoring period. Similarly, one would anticipate that patients required to perform vigorous motor activities (e.g., aerobic exercises) would prefer wearing an e-textile garment rather than elastic straps equipped with sensors because elastic straps would be more likely to interfere with the movements to be performed and migrate during the performance of motor tasks.

21.5.2 E-Skin Sensors

The development of stretchable electronics matching the mechanical characteristics of the epidermis, which was pioneered by John Rogers' research group [133, 134], enabled the development of e-skin sensors. These are sensors that, when attached to the skin like an adhesive bandage, stretch in the same way as the skin does in response to the movement of body segments. This technology has recently led to the implementation of movement tracking systems like the

one schematically represented in Fig. 21.12. This figure shows recent work by Kim et al. [135] aimed to detect and estimate the characteristics of movements involving different body segments (panel a) from data collected by relying on e-skin sensors positioned on specific body landmarks. The e-skin sensors allow one to capture skin topographical changes associated with the target movement. For instance, movements of the index finger are detected, and their biomechanical characteristics are estimated using an e-skin sensor positioned at the wrist (panel b of Fig. 21.12). High sensitivity to the movements of the index finger is achieved by relying on laser-induced nanoscale cracking—shown in the inset of panel b (scale bar: 40 μm)—of specific elements of the mesh displayed in panel c (scale bar: 1 mm). Advanced data analysis techniques that rely on deep neural networks are used to estimate the biomechanical characteristics of the index finger movements.

As the technology rapidly evolves and major advances in the field of flexible and printed

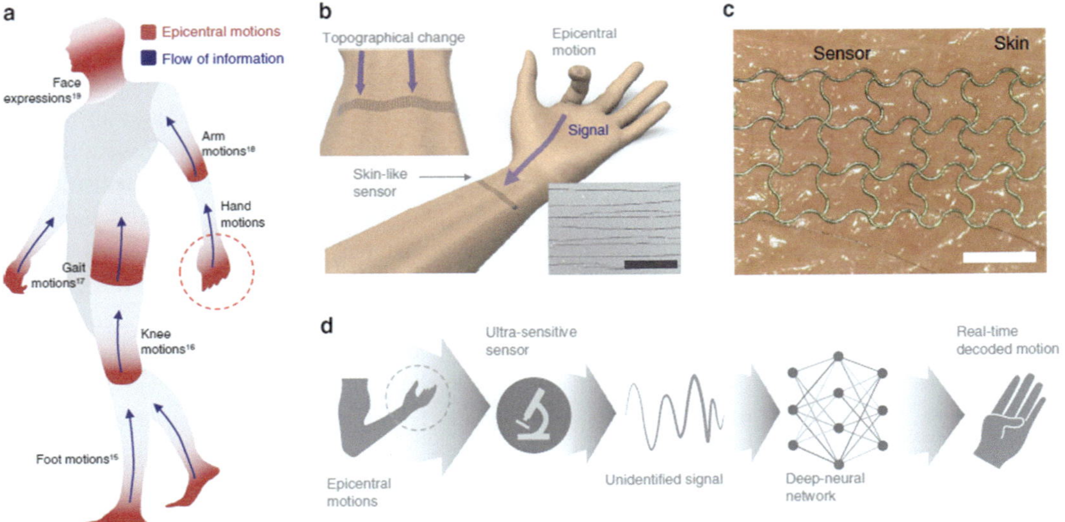

Fig. 21.12 E-skin sensors can be used to monitor movements involving different body segments (panel a). The sensors are positioned on specific anatomical landmarks (panel b) with laser-induced cracking (panel b inset) affecting specific elements of the mesh structure (panel c) used to build the sensor. Data collected using the sensor is processed by deep neural networks (panel d) that generate estimates of the biomechanical characteristics of the movement performed with the monitored body segment. Reproduced with permission from Kim et al. (https://doi.org/10.1038/s41467-020-16040-y, licensed under CC BY 4.0)

electronics are expected over the next few years [136], the interest in potential clinical applications of e-skin sensors, including stroke rehabilitation, is rapidly growing [137]. E-skin sensors are expected not only to facilitate tracking the movement of body segments, but also to enable the detection of the activity of muscles either by electrode arrays mounted on a flexible substrate or by detecting changes in the shape of body segments associated with the contraction of muscles.

The use of e-skin sensors is very appealing for short-term (i.e., a few days) monitoring of motor activities in stroke survivors. When used for longer periods of time, e-skin sensors are likely to cause skin irritation as adhesive components are typically used to secure the sensors to the skin and such materials tend to cause skin irritation when utilized over long periods of time. Nonetheless, e-skin sensors are marked by minimum obtrusiveness and optimal wearability. Hence, a growing interest in this technology is expected over the next years.

21.5.3 Wearable Cameras

The use of wearable cameras (Fig. 21.13, left panel) in the field of rehabilitation was originally proposed to validate the detection of motor activities achieved via the analysis of wearable sensor data and provide contextual information. The manuscripts by Doherty et al. [138] and by Lee et al. [71] are examples of this body of work that relied on egocentric video recordings (i.e., recordings that approximate the visual field of the camera wearer) to capture the environmental conditions in which tasks were performed and hence infer the nature of the tasks. Ad-hoc techniques to analyze egocentric video recordings were developed to facilitate the identification of the environment where motor tasks were performed and the conditions in which they were performed. These techniques allowed researchers to minimize the need to manually annotate lengthy recordings. The manuscript by Yan et al. [139] provides an example of such video analysis techniques.

Wearable cameras have also been used to replace wearable sensors. Seminal work by Zariffa and Popovic [140] explored this application of wearable cameras nearly a decade ago. Subsequently, the research group led by Jose Zariffa further developed this technique in a series of studies carried out first in individuals with spinal cord injury and later in stroke survivors. The right panel of Fig. 21.13 shows examples of the motor tasks analyzed via recordings collected using a wearable camera [141]. In this specific project, the research team used a convolutional neural network to detect the position of the hands in the video frames and used a Random Forest-based classifier to detect when subjects manipulated objects. Subsequent work by the same group explored the combined use of object detectors and trackers [142] as well as the detection of compensatory movement strategies adopted by patients with UL motor impairments [143].

This body of work was focused on the application of wearable cameras to detect and assess the quality of UL movements in individuals with spinal cord injury. Whereas initial work was carried out in the laboratory, recent studies have explored the use of this technology in the field [144]. Importantly in the context of this book chapter, the same research group has started to explore the use of this technology to monitor UL movements performed by stroke survivors [145]. The authors were able to demonstrate the feasibility of tracking hand use and determining if the stroke-affected hand was utilized for the stabilization or the manipulation of objects.

Researchers can now rely on a large body of work focused on the development of techniques for the analysis of egocentric video recordings. Studies relevant to the application of wearable cameras in the field of rehabilitation were recently reviewed by Bandini and Zariffa [146]. The authors surveyed techniques designed to identify the hands or parts of them in the video frames and to detect the task performed by the camera wearer. They also provided a summary of the various applications of these techniques that are currently pursued by researchers, including remote assessment of hand function and gesture recognition. Novel video analysis techniques are

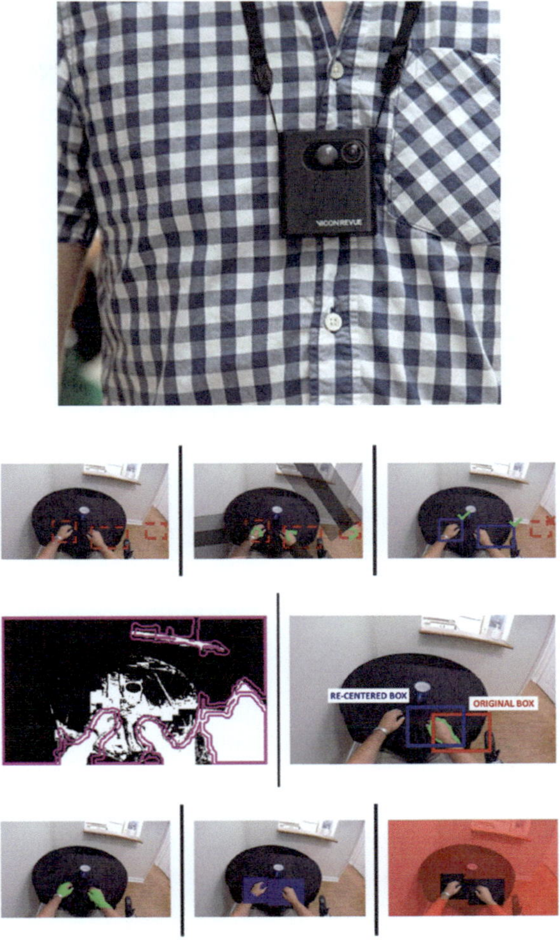

Fig. 21.13 A wearable camera (left panel) can be utilized as an alternative to wearable sensors to detect UL motor activities (right panel). Reproduced with permission (Vicon Revue camera picture courtesy of Oxford Metrics, UK—left panel; Likitlersuang et al. (https://doi.org/10.1186/s12984-019-0557-1, licensed under CC BY 4.0)—right panel)

emerging that are expected to further facilitate the analysis of egocentric video recordings by addressing challenges such as those associated with the continuous change in visual field due the movements of the body segment the wearable camera is attached to [147].

Future developments in this research area are expected to focus on more complex analyses of the contextual information gathered using wearable cameras. For instance, the identification of objects in the video frames could determine the degree of hand dexterity required for their manipulation, which, in turn, could provide a reference to evaluate if the movements of the stroke-affected hand are adequate to meet the task requirements. Furthermore, the development of techniques for quasi-real-time analysis of video recordings gathered using wearable cameras could provide an opportunity to generate stimuli to encourage the use of the stroke-affected hand. Techniques previously developed using wearable sensors positioned bilaterally at the wrist to detect UL activities and deliver stimuli to encourage stroke survivors to use their stroke-affected arm [18] have been shown to be effective in pilot clinical studies (unpublished results). Systems relying on wearable cameras could provide additional information suitable to choose the timing of the stimuli delivered to remind stroke survivors to use their stroke-affected limb in a way that is most likely to lead to a positive behavioral change. For instance, stimuli could be delivered when the patient is engaged in a task that has been identified by the patient—in consultation with the therapist—as a task suitable to increase the use of the stroke-affected arm.

21.5.4 Radio Tags and Radar-Like Technologies to Gather Contextual Information

Whereas the use of wearable cameras could provide useful contextual information, privacy concerns are likely to limit their use and encourage researchers to seek alternative approaches. In a controlled environment (such as the home), systems relying on radio tags and radar-like technologies provide an interesting alternative to the use of wearable cameras. Whereas a review of these technologies is beyond the scope of this chapter, it is important to point out that significant advances have been achieved toward providing accurate data about the position in the home environment of people and objects using a variety of techniques. Herein, we have chosen to briefly comment on the use of ultra-wideband (UWB) radio systems [148] and radar-like technologies [149] as we believe that these techniques are particularly promising.

The localization of UWB radio tags is a well-studied problem. Capra et al. [148] explored the use of this technology to track the position of patients in the home environment and provide immediate assistance when a fall is detected using a wearable sensor connected to the UWB network. Localization algorithms relying on UWB radio tags use sets of transceivers with a minimum of three units utilized as anchor points (i.e., units set in known positions) that serve as a reference to locate radio tags in the environment. Radio signals are exchanged among the anchor point units and the radio tags. Estimates of the time of flight (i.e., the time needed to receive a radio signal) for different anchor points are used as input to a triangulation algorithm that determines the position of the radio tags relative to the

anchor point units. Recent advances in UWB technology include the development of methods for self-calibration of the position of anchor point units [150], thus making the deployment of UWB localization systems both simple and inexpensive.

UWB localization systems can be looked upon as part of a broad category of systems including those that rely on Internet of Things (IoT) and radio frequency identification (RFID) technologies, which are becoming common place and provide the opportunity to track the position of people and objects in the home environment as reviewed by Landaluce et al. [151]. Besides, researchers have developed several techniques to take advantage of and merge the information gathered in the environment using different wireless technologies in ways that are suitable for tracking purpose [152]. Researchers are beginning to envision tracking stroke survivors as they move from room to room in the home environment (e.g., they move to the kitchen at lunchtime) and detect their proximity to objects (e.g., a cutting board on the kitchen counter) that enable inferring that they are engaged in specific activities (e.g., preparing a meal). This contextual information could be utilized to analyze sensor data accounting for the activity that stroke survivors are engaged in. Also, contextual information could be used to generate stimuli to encourage patients to use their stroke-affected arm to perform specific activities.

Radar-like systems [153, 154] designed for deployment in the home setting [155] are rapidly emerging as ideally suited to track people's location. Seminal work by Dina Katabi's group [153, 154] relies on the analysis of how radio signals bounce off the body to track the position of people in the home environment. Figure 21.14 shows a prototype system developed by Dina Katabi's research team at MIT (left panel) and a graphical representation of the radio signals that bounce off the body of the study volunteer as he walks in the room (right panel). The technique developed by this research team can achieve an accuracy of 10–20 cm, which is generally satisfactory in the context of the above-mentioned applications. Among all available technologies for position tracking, this appears to be the most promising. It does not require anything else than positioning in the home a box similar to a WiFi router. Importantly, it does not require that patients wear sensors or radio tags and it does not require a complicated installation. Current implementations are challenged when tracking people in crowded environments. However, that is a situation that seldom occurs in the home of stroke survivors, where typically the system would need to track a few individuals at the most.

21.5.5 Modern Video Analysis Techniques

The development of advanced machine learning techniques that has taken place over the past decade has led to a new generation of video analysis techniques that are dramatically transforming the field of movement science. A number of sophisticated software libraries have been made available to the scientific community including DeepPose [156], DeeperCut [157], OpenPose [158, 159], ArtTrack [160], DeepLabCut [161], Alpha-Pose [162], and MediaPipe [163–165]. These software libraries rely on deep learning algorithms designed to track anatomical landmarks (referred to as "keypoints")—such as the ankle, knee, and hip joint positions—and derive a simplified representation of the body as shown in Fig. 21.15a. This process is referred to as "pose estimation" and can be implemented by using low-cost video cameras. Although the results are not as accurate as those obtained by using traditional, high-cost, camera-based motion capture systems, these modern video analysis techniques provide a valid low-cost alternative when the application at hand does not have stringent accuracy requirements. It turns out that this is the case for many clinical applications such as the assessment of motor patterns to estimate the severity of motor impairments and the generation of feedback during the performance of rehabilitation exercises (e.g., to detect and discourage compensatory movement strategies).

The above-stated considerations have generated tremendous interest among researchers and clinicians for these techniques. Recent reviews have discussed their potential impact on clinical practice [166–168]. However, clinical adoption is still limited. A large number of studies have been focused on the technical validation of these techniques, especially in the analysis of gait patterns [169–172]. Recent publications have started to discuss the possibility of using these techniques to derive proxies for clinical assessment measures [173, 174]. A few studies have explored

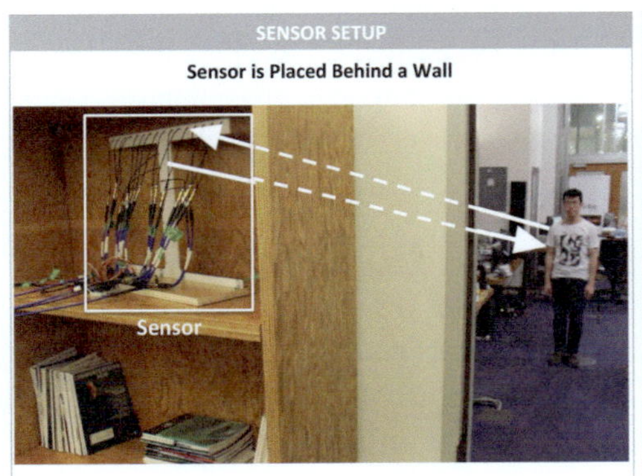

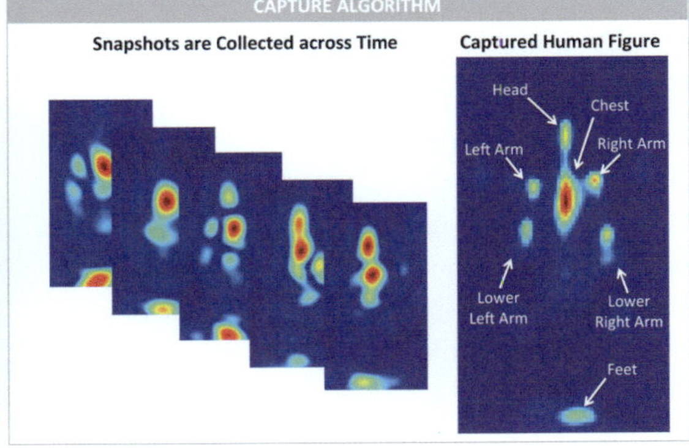

Fig. 21.14 Radar-like systems can provide a totally unobtrusive way to monitor patients' position in the home environment and hence infer contextual information of great use in home-based interventions. The sensor (left panel) consists of a radio transmitter/receiver array. The radio transmission bounces off the person, for instance, while walking and result in a "radio signature" (right panel) from which the position of the patient can be inferred with a 10–20 cm accuracy. Reproduced with permission from Prof Katabi's webpage

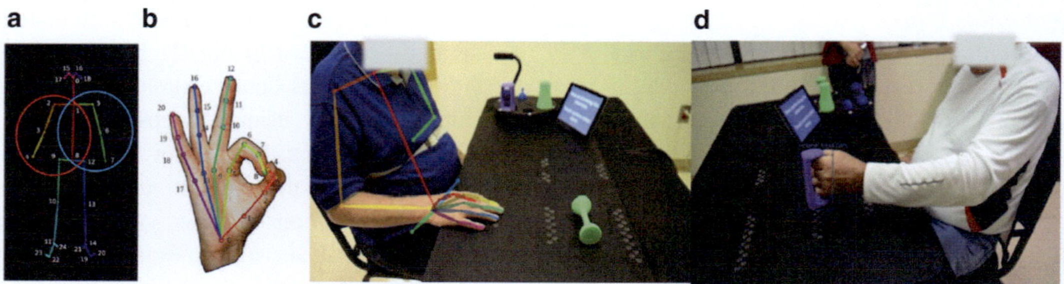

Fig. 21.15 Output of the OpenPose algorithm used to generate a stick figure representation of the patient (**a**). Keypoints used to track hand movements (**b**). Stick figure representation derived using OpenPose overlaid to a video frame (**c**). Output of the algorithm used to track objects (**d**). Reproduced with permission from Ahmed et al.

their use for tracking UL and hand movements [175, 176]. The manuscript by Ahmed et al. is particularly interesting as it explores an important application of modern video analysis techniques, namely tracking movement during the performance of home-based rehabilitation exercises. The manuscript provides details about the work accomplished toward the development of a platform for UL home-based exercises named the Semi-Automated Rehabilitation at the Home (SARAH) system. Figure 21.15 shows some of the key components of the system: the stick figure representation of the body obtained using Open-Pose [158, 159] (panel A), the keypoints used to track hand movements (panel B), the OpenPose stick figure representation overlaid on a video frame (panel C), and the output of the object tracking and recognition algorithm used in the study (panel D) [177].

place in clinical studies [179] and adoption in the clinic is ramping up. Additional applications of wearable technology are emerging. For instance, commercially available wearable systems provide a convenient, unobtrusive way to monitor sleep quality. An example of a sleep report generated by a finger-worn wearable sensor is shown in Fig. 21.16. Sleep quality is important not only as a proxy for wellness and psychological status, but also in the context of motor learning. In fact, motor learning studies have pointed out the important role played by sleep in the processes associated with the consolidation of learned motor patterns [180].

In the future, we envision that metrics of this type will be used routinely in clinical care. However, clinical studies are needed to develop reliable metrics that could inform the design of personalized (i.e., patient-specific) interventions that account for multiple physiological factors and the general well-being of patients.

21.5.6 Collecting Non-motor Data

Although this chapter is devoted to movement tracking-based techniques, wearable sensors can provide additional information that is relevant to stroke rehabilitation. For instance, wearable sensors provide a convenient way to monitor systemic responses associated with vigorous exercise, which should be monitored in stroke survivors [178]. The use of wearable sensors and systems in this context is becoming common

21.5.7 What Emerging Technologies Could Do that "Traditional" Technologies Do not …

E-textiles provide an alternative form factor that patients might prefer over traditional wearable systems for long-term monitoring applications. Traditional wearable sensors are typically

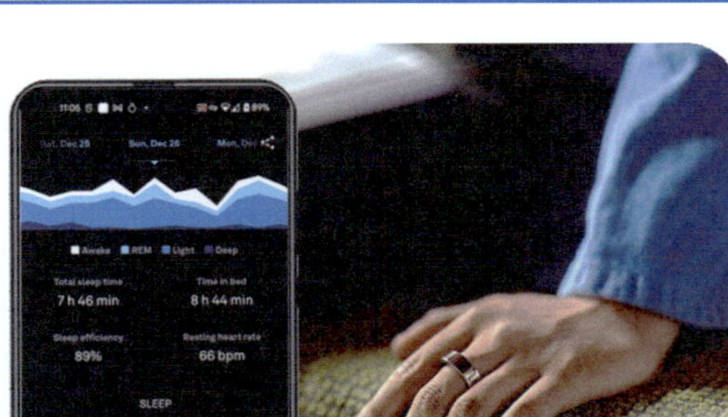

Fig. 21.16 Subject wearing a commercially available ring sensor (right, picture from Oura Health Oy) that provides measures of sleep quality via the companion smartphone application (left). Sleep quality is relevant to stroke rehabilitation in many ways, including the impact on the consolidation of motor gains

attached to the body via elastic straps. When multiple sensors have to be used for a long period of time on a daily basis, they are rapidly perceived by patients as obtrusive. In contrast, an e-textile t-shirt could be used to embed multiple sensors and require donning and doffing a single item.

Nonetheless, e-textiles cannot always conform perfectly to the anatomy and often do not provide a stable contact between the sensing elements and the patient's skin. In these circumstances, researchers can rely on e-skin sensing technology. The quality of the contact with the skin achieved using e-skin sensors is unprecedented. In these specific applications, e-skin sensors deliver high-quality data that would be difficult to achieve with traditional wearable sensing technology as well as with e-textiles.

When the analysis of data collected using wearable sensors requires contextual information, one can rely on wearable cameras to collect egocentric video recordings. Patients would wear traditional sensors or e-textiles or an e-skin sensor or a combination of all of the above. The egocentric video recordings would provide context and hence facilitate the analysis of the data.

However, this approach clearly presents privacy concerns.

When one needs to monitor patients in the home environment, then wearable cameras can be replaced by other technologies such as radio tags and radar-like technology solutions. Whereas further research is needed to develop and test contactless technologies—like the ones mentioned above—to track patients in the home setting with high accuracy, existing radio tag systems and radar-like technology solutions provide sufficient accuracy to enable inferring important contextual information.

Researchers and the rehabilitation technology industry have relied on wearable sensors to generate feedback during the performance of therapeutic exercises (whether in the clinic or at home). The use of modern video analysis techniques is now replacing traditional sensors in this context. In fact, the use of video analysis technology is more convenient as it does not require any donning and doffing of sensor units, which is often problematic for stroke survivors. A preference for video analysis solutions would be expected when one implements rehabilitation interventions using interactive games. This solution would be less attractive when clinicians would like to monitor rehabilitation in an inpatient gym, where a large number of patients would like to be present at the same time.

Finally, it should be emphasized that "traditional" consumer electronics provide data that is highly relevant to stroke rehabilitation. For instance, in this section, we mentioned the capability of several wearable systems of monitoring sleep quality. As it is known that sleep quality affects the consolidation of learned motor patterns, it is expected that—in the near future—we will witness a growing use of wearable sensors to monitor physiological variables such as sleep quality and autonomic dysregulation.

21.6 Conclusions

The body of work discussed in this chapter suggests that the use of wearable sensors will soon become an important tool in rehabilitation, including in the context of home-based monitoring and tele-rehabilitation of stroke survivors. Different clinical applications of wearable sensors are marked by different challenges. For example, setting goals and providing feedback is quite intuitive for users in the case of LL (gait) applications where step counts, distance walked, and stair climbing measures can be used as intuitive metrics to set target levels of activity. Wearable sensors to obtain such metrics can be unobtrusive as typically only one sensor is needed to collect data to derive such metrics, and many commercially available solutions exist. However, translating this approach to UL interventions is challenging. It may require wearing more than one IMU (e.g., wearing sensors to track the movement of multiple fingers) and doing so bilaterally. Consensus on relevant parameters that should be used to set goals and provide feedback to patients has yet to be established. In addition, approaches based on wearable sensors to encourage activity may need to be combined with behavioral interventions to assure that motor gains are sustained over time. Patients need to be engaged and care about the metrics provided by wearable sensors. It is important that patients relate changes in motor behaviors and health outcomes (e.g., decreased stroke risk). In the context of rehabilitation, setting achievable targets that encourage the performance of new activities appears to be an effective strategy to maximize adherence to an exercise program and sustain changes over time. In the context of home-based monitoring and exercises, systems based on wearable sensing technology (examples in Sects. 21.2.2 and 21.2.4) enable the performance of therapeutic exercises, often in a non-immersive virtual environment with gamification, and are a great tool to provide feedback and motivation to patients.

Another factor to consider is accessibility to commercially available wearable sensors for at-home deployment. Currently, about one fifth of the adult population is using a wearable device (i.e., either a smart watch or a wearable fitness tracker) on a regular basis [181]. From a prevention perspective, this is a positive trend as many people are encouraged to develop healthy habits and pay attention to their motor activities, sleep quality, and physiological data. However, reports show that young adults and women from higher-income households and with a college education are among the top users of these technologies. Literacy level and socioeconomic status appear to be highly correlated with the adoption of wearable technology and its use to facilitate adherence to a healthy lifestyle. In addition, the algorithms used in consumer-grade devices often show limited accuracy in patients with motor impairments, which negatively affects the adherence and usefulness of these systems in a rehabilitation context. As activity trackers collect a big amount of data on physical metrics and health, another issue to be carefully considered are the barriers to data sharing that one might encounter either because of patients' preference or because of regulatory requirements. As wearable devices become part of our daily lives and clinical care, ethical considerations about data sharing and use of the data (e.g., for secondary analyses) need to be carefully considered. Also, the use of prompts to change patients' behavior has to be carefully considered from an ethical standpoint.

Nonetheless, wearable sensors are a unique tool that allows to gather data inside and outside the clinic for a longer period than the more classical data snapshots of movement and physiology taken with classical assessment methods used in rehabilitation. In the future, the use of wearable sensors during daily life activities could allow researchers to precisely evaluate the effects of novel therapies (i.e., new rehabilitation approaches or regeneration therapies). This would enable the implementation of precision rehabilitation in which clinicians design patient-specific interventions, set clinical objectives, track patient's response using wearable sensors, and periodically evaluate the effectiveness of the ongoing intervention based on the recovery trajectory defined by the time series of clinical score estimates derived from wearable sensor data.

In conclusion, wearable sensors and their applications in stroke rehabilitation are progressing at a fast pace in research laboratories. Their use in the clinic remains sparse.

Improvements in the adoption of this technology, which has been shown to be clinically useful in many ways, could be achieved by a stronger focus on involving end-users in the early stages of the development of wearable technology solutions. Besides, a stronger focus on developing systems that are very simple to use and require virtually no set-up time would benefit adoption as clinical sites are often extremely busy and every minute of clinicians' schedule is typically fully booked. To achieve a ubiquitous implementation of wearable sensors, researchers, clinicians, stroke survivors, caretakers, and engineers need to work together. It is apparent that more work and research need to be done to improve currently available wearable sensors and systems in terms of their usability and applicability in a clinical setting. However, we should emphasize that research that has relied on wearable technology to collect data from stroke survivors has allowed us to gather important information that we would have not been able to collect without the use of wearable sensors and systems and that such information is reshaping stroke rehabilitation.

References

1. Zimbelman JL, Juraschek SP, Zhang X, Lin VW-H. Physical therapy workforce in the United States: Forecasting Nationwide Shortages. PM&R [Internet]. 2010 Nov;2(11):1021–9. http://doi.wiley.com, https://doi.org/10.1016/j.pmrj.2010.06.015.
2. Langhorne P, Coupar F, Pollock A. Motor recovery after stroke: a systematic review. Lancet Neurol [Internet]. (2009);8(8):741–54. http://www.ncbi.nlm.nih.gov/pubmed/19608100.
3. Teasell RW, Murie Fernandez M, McIntyre A, Mehta S. Rethinking the continuum of stroke rehabilitation. Arch Phys Med Rehabil [Internet]. 2014;95(4):595–6. http://www.ncbi.nlm.nih.gov/pubmed/24529594.
4. Winstein C, Varghese R. Been there, done that, so what's next for arm and hand rehabilitation in stroke? NeuroRehabilitation [Internet]. 2018;43(1):3–18. http://www.ncbi.nlm.nih.gov/pubmed/29991146.
5. Lang CE, Strube MJ, Bland MD, Waddell KJ, Cherry-Allen KM, Nudo RJ, et al. Dose response of task-specific upper limb training in people at least 6 months poststroke: a phase II, single-blind, randomized, controlled trial. Ann Neurol [Internet]. 2016;80(3):342–54. http://www.ncbi.nlm.nih.gov/pubmed/27447365.
6. Ward NS, Brander F, Kelly K. Intensive upper limb neurorehabilitation in chronic stroke: outcomes from the Queen Square programme. J Neurol Neurosurg Psychiatry [Internet]. 2019;90(5):498–506. http://www.ncbi.nlm.nih.gov/pubmed/30770457.
7. Collins FS, Varmus H. A new initiative on precision medicine. N Engl J Med [Internet]. 2015 Feb 26;372(9):793–5. http://www.ncbi.nlm.nih.gov/pubmed/25635347.
8. Hulsen T, Jamuar SS, Moody AR, Karnes JH, Varga O, Hedensted S, et al. From big data to precision medicine. Front Med [Internet]. 2019;6:34. http://www.ncbi.nlm.nih.gov/pubmed/30881956.
9. Niederberger E, Parnham MJ, Maas J, Geisslinger G. 4 Ds in health research-working together toward rapid precision medicine. EMBO Mol Med [Internet]. 2019;11(11):e10917. http://www.ncbi.nlm.nih.gov/pubmed/31531943.
10. Klein TA, Neumann J, Reuter M, Hennig J, von Cramon DY, Ullsperger M. Genetically determined differences in learning from errors. Science [Internet]. 2007 Dec 7;318(5856):1642–5. http://www.ncbi.nlm.nih.gov/pubmed/18063800.
11. Tran DA, Pajaro-Blazquez M, Daneault J-F, Gallegos JG, Pons J, Fregni F, et al. Combining dopaminergic facilitation with robot-assisted upper limb therapy in stroke survivors: a focused review. Am J Phys Med Rehabil [Internet]. 2016;95(6):459–74. http://www.ncbi.nlm.nih.gov/pubmed/26829074.
12. Pearson-Fuhrhop KM, Minton B, Acevedo D, Shahbaba B, Cramer SC. Genetic variation in the human brain dopamine system influences motor learning and its modulation by L-Dopa. PLoS One [Internet]. 2013;8(4):e61197. http://www.ncbi.nlm.nih.gov/pubmed/23613810.
13. Cheung VCK, Turolla A, Agostini M, Silvoni S, Bennis C, Kasi P, et al. Muscle synergy patterns as physiological markers of motor cortical damage. Proc Natl Acad Sci USA [Internet]. 2012 Sep 4;109(36):14652–6. http://www.ncbi.nlm.nih.gov/pubmed/22908288.
14. Miranda JGV, Daneault J-F, Vergara-Diaz G, Torres ÂFS de OE, Quixadá AP, Fonseca M de L, et al. Complex upper-limb movements are generated by combining motor primitives that scale with the movement size. Sci Rep [Internet]. 2018;8(1):12918. http://www.ncbi.nlm.nih.gov/pubmed/30150687.
15. Rosenthal O, Wing AM, Wyatt JL, Punt D, Brownless B, Ko-Ko C, et al. Boosting robot-assisted rehabilitation of stroke hemiparesis by individualized selection of upper limb movements—a pilot study. J Neuroeng Rehabil [Internet]. 2019;16(1):42. http://www.ncbi.nlm.nih.gov/pubmed/30894192.

16. World Health Organization. International classification of functioning, disability and health (ICF). Geneva, Switzerland; 2001

17. Patel S, Park H, Bonato P, Chan L, Rodgers M. A review of wearable sensors and systems with application in rehabilitation. J Neuroeng Rehabil [Internet]. 2012 Apr 20;9:21. http://www.ncbi.nlm.nih.gov/pubmed/22520559.

18. Lee SI, Adans-Dester CP, Grimaldi M, Dowling A V., Horak PC, Black-Schaffer RM, et al. Enabling Stroke Rehabilitation in home and community settings: a wearable sensor-based approach for upper-limb motor training. IEEE J Transl Eng Heal Med [Internet]. 2018;6:1–11. https://ieeexplore.ieee.org/document/8353413/.

19. Maceira-Elvira P, Popa T, Schmid A-C, Hummel FC. Wearable technology in stroke rehabilitation: towards improved diagnosis and treatment of upper-limb motor impairment. J Neuroeng Rehabil [Internet]. 2019;16(1):142. http://www.ncbi.nlm.nih.gov/pubmed/31744553.

20. Dobkin BH, Dorsch A. The promise of mHealth: daily activity monitoring and outcome assessments by wearable sensors. Neurorehabil Neural Repair [Internet]. 25(9):788–98. http://www.ncbi.nlm.nih.gov/pubmed/21989632.

21. Bonato P. Advances in wearable technology and applications in physical medicine and rehabilitation. J Neuroeng Rehabil [Internet]. 2005;2(1):2. http://www.ncbi.nlm.nih.gov/pubmed/15733322.

22. Kwakkel G, Lannin NA, Borschmann K, English C, Ali M, Churilov L, et al. Standardized measurement of sensorimotor recovery in stroke trials: consensus-based core recommendations from the stroke recovery and rehabilitation roundtable. Int J Stroke [Internet]. 2017;12(5):451–61. http://journals.sagepub.com/doi, https://doi.org/10.1177/1747493017711813.

23. Schwarz A, Bhagubai MMC, Wolterink G, Held JPO, Luft AR, Veltink PH. Assessment of upper limb movement impairments after stroke using wearable inertial sensing. sensors [Internet]. 2020;20(17):4770. https://www.mdpi.com/1424-8220/20/17/4770.

24. Kortier HG, Sluiter VI, Roetenberg D, Veltink PH. Assessment of hand kinematics using inertial and magnetic sensors. J Neuroeng Rehabil [Internet]. 2014;11(1):70. http://jneuroengrehab.biomedcentral.com/articles, https://doi.org/10.1186/1743-0003-11-70.

25. Yao S, Vargas L, Hu X, Zhu Y. A Novel Finger Kinematic Tracking Method Based on Skin-Like Wearable Strain Sensors. IEEE Sens J [Internet]. 2018 Apr 1;18(7):3010–5. http://ieeexplore.ieee.org/document/8281089/.

26. Nie JZ, Nie JW, Hung N-T, Cotton RJ, Slutzky MW. Portable, open-source solutions for estimating wrist position during reaching in people with stroke. Sci Rep [Internet]. 2021 Dec 18;11

27. Kim GJ, Parnandi A, Eva S, Schambra H. The use of wearable sensors to assess and treat the upper extremity after stroke: a scoping review. Disabil Rehabil [Internet]. 2021;1–20. http://www.ncbi.nlm.nih.gov/pubmed/34328803.

28. Fugl-Meyer AR, Jääskö L, Leyman I, Olsson S, Steglind S. The post-stroke hemiplegic patient. 1. a method for evaluation of physical performance. Scand J Rehabil Med [Internet]. 1975;7(1):13–31. http://www.ncbi.nlm.nih.gov/pubmed/1135616.

29. Yu L, Xiong D, Guo L, Wang J. A remote quantitative Fugl-Meyer assessment framework for stroke patients based on wearable sensor networks. Comput Methods Programs Biomed [Internet]. 2016;128:100–10. https://linkinghub.elsevier.com/retrieve/pii/S0169260715301905.

30. Del Din S, Patel S, Cobelli C, Bonato P. Estimating Fugl-Meyer clinical scores in stroke survivors using wearable sensors. In: 2011 Annual international conference of the IEEE engineering in medicine and biology society [Internet]. IEEE; 2011. p. 5839–42. http://ieeexplore.ieee.org/document/ss6091444/.

31. Formstone L, Huo W, Wilson S, McGregor A, Bentley P, Vaidyanathan R. Quantification of motor function post-stroke using novel combination of wearable inertial and mechanomyographic sensors. IEEE Trans Neural Syst Rehabil Eng [Internet]. 2021;29:1158–67. https://ieeexplore.ieee.org/document/9455409/.

32. Wolf SL, Catlin PA, Ellis M, Archer AL, Morgan B, Piacentino A. Assessing Wolf motor function test as outcome measure for research in patients after stroke. Stroke. 2001;32(7):1635–9.

33. Patel S, Hughes R, Hester T, Stein J, Akay M, Dy JG, et al. A novel approach to monitor rehabilitation outcomes in stroke survivors using wearable technology. Proc IEEE [Internet]. 2010;98(3):450–61. http://ieeexplore.ieee.org/document/5420034/.

34. Chen H-M, Chen CC, Hsueh I-P, Huang S-L, Hsieh C-L. Test-retest reproducibility and smallest real difference of 5 hand function tests in patients with stroke. Neurorehabil Neural Repair [Internet]. 2009;23(5):435–40. http://www.ncbi.nlm.nih.gov/pubmed/19261767.

35. Friedman N, Chan V, Zondervan D, Bachman M, Reinkensmeyer DJ. MusicGlove: motivating and quantifying hand movement rehabilitation by using functional grips to play music. In: 2011 annual international conference of the IEEE engineering in medicine and biology society [Internet]. IEEE; 2011. p. 2359–63. http://ieeexplore.ieee.org/document/6090659/.

36. Repnik E, Puh U, Goljar N, Munih M, Mihelj M. Using inertial measurement units and electromyography to quantify movement during action research arm test execution. Sensors [Internet]. 2018;18

(1):22491. https://www.nature.com/articles/s41598-021-01805-2.

(9):2767. http://www.mdpi.com/1424-8220/18/9/2767.

37. Bochniewicz EM, Emmer G, McLeod A, Barth J, Dromerick AW, Lum P. Measuring functional arm movement after stroke using a single wrist-worn sensor and machine learning. J Stroke Cerebrovasc Dis [Internet]. 2017;26(12):2880–7. http://www.ncbi.nlm.nih.gov/pubmed/28781056.

38. Adans-Dester C, Hankov N, O'Brien A, Vergara-Diaz G, Black-Schaffer R, Zafonte R, et al. Enabling precision rehabilitation interventions using wearable sensors and machine learning to track motor recovery. NPJ Digit Med [Internet]. 2020;3(1):121. https://www.nature.com/articles/s41746-020-00328-w.

39. Hocoma. ArmeoSenso [Internet]. https://www.hocoma.com/solutions/armeo-senso/.

40. Wittmann F, Held JP, Lambercy O, Starkey ML, Curt A, Höver R, et al. Self-directed arm therapy at home after stroke with a sensor-based virtual reality training system. J Neuroeng Rehabil [Internet]. 2016;13(1):75. http://jneuroengrehab.biomedcentral.com/articles, https://doi.org/10.1186/s12984-016-0182-1.

41. Widmer M, Held JPO, Wittmann F, Valladares B, Lambercy O, Sturzenegger C, et al. Reward during arm training improves impairment and activity after stroke: a randomized controlled trial. Neurorehabil Neural Repair [Internet]. 2022;36(2):140–50. http://journals.sagepub.com/doi, https://doi.org/10.1177/15459683211062898.

42. Friedman N, Chan V, Reinkensmeyer AN, Beroukhim A, Zambrano GJ, Bachman M, et al. Retraining and assessing hand movement after stroke using the MusicGlove: comparison with conventional hand therapy and isometric grip training. J Neuroeng Rehabil [Internet]. 2014;11(1):76. http://jneuroengrehab.biomedcentral.com/articles, https://doi.org/10.1186/1743-0003-11-76.

43. Zondervan DK, Friedman N, Chang E, Zhao X, Augsburger R, Reinkensmeyer DJ, et al. Home-based hand rehabilitation after chronic stroke: randomized, controlled single-blind trial comparing the MusicGlove with a conventional exercise program. J Rehabil Res Dev [Internet]. 2016;53(4):457–72. http://www.ncbi.nlm.nih.gov/pubmed/27532880.

44. Arteaga S, Chevalier J, Coile A, Hill AW, Sali S, Sudhakhrisnan S, et al. Low-cost accelerometry-based posture monitoring system for stroke survivors. In: Proceedings of the 10th international ACM SIGACCESS conference on computers and accessibility. 2008. p. 243–4.

45. Minet LR, Peterson E, von Koch L, Ytterberg C. Occurrence and predictors of falls in people with stroke: six-year prospective study. Stroke [Internet]. 2015;46(9):2688–90. http://www.ncbi.nlm.nih.gov/pubmed/26243230.

46. Bergamini E, Iosa M, Belluscio V, Morone G, Tramontano M, Vannozzi G. Multi-sensor assessment of dynamic balance during gait in patients with subacute stroke. J Biomech [Internet]. 2017;61:208–15. http://www.ncbi.nlm.nih.gov/pubmed/28823468.

47. Garcia F do V, da Cunha MJ, Schuch CP, Schifino GP, Balbinot G, Pagnussat AS. Movement smoothness in chronic post-stroke individuals walking in an outdoor environment—a cross-sectional study using IMU sensors. Najafi B, editor. PLoS One [Internet]. 2021;16(4):e0250100. https://dx.plos.org, https://doi.org/10.1371/journal.pone.0250100.

48. Salarian A, Horak FB, Zampieri C, Carlson-Kuhta P, Nutt JG, Aminian K. iTUG, a sensitive and reliable measure of mobility. IEEE Trans Neural Syst Rehabil Eng [Internet]. 2010;18(3):303–10. http://www.ncbi.nlm.nih.gov/pubmed/20388604.

49. Wüest S, Massé F, Aminian K, Gonzenbach R, de Bruin ED. Reliability and validity of the inertial sensor-based Timed "Up and Go" test in individuals affected by stroke. J Rehabil Res Dev [Internet]. 2016;53(5):599–610. http://www.ncbi.nlm.nih.gov/pubmed/27898161.

50. Verheyden G, Nieuwboer A, Mertin J, Preger R, Kiekens C, De Weerdt W. The trunk impairment Scale: a new tool to measure motor impairment of the trunk after stroke. Clin Rehabil [Internet]. 2004;18(3):326–34. http://journals.sagepub.com/doi, https://doi.org/10.1191/0269215504cr733oa.

51. Alhwoaimel N, Warner M, Hughes A-M, Ferrari F, Burridge J, Wee SK, et al. Concurrent validity of a novel wireless inertial measurement system for assessing trunk impairment in people with stroke. Sensors (Basel) [Internet]. 2020;20(6). Available from: http://www.ncbi.nlm.nih.gov/pubmed/32197493.

52. Hocoma. Valedo [Internet]. http://www.valedotherapy.com/.

53. Bauer CM, Rast FM, Ernst MJ, Kool J, Oetiker S, Rissanen SM, et al. Concurrent validity and reliability of a novel wireless inertial measurement system to assess trunk movement. J Electromyogr Kinesiol [Internet]. 2015;25(5):782–90. https://linkinghub.elsevier.com/retrieve/pii/S1050641115001157.

54. Dorsch AK, Thomas S, Xu X, Kaiser W, Dobkin BH, Emara T, et al. SIRRACT. Neurorehabil neural repair [Internet]. 2015;29(5):407–15. http://journals.sagepub.com/doi, https://doi.org/10.1177/1545968314550369.

55. Byl N, Zhang W, Coo S, Tomizuka M. Clinical impact of gait training enhanced with visual kinematic biofeedback: patients with Parkinson's disease and patients stable post stroke. Neuropsychologia [Internet]. 2015;79(Pt B):332–43. http://www.ncbi.nlm.nih.gov/pubmed/25912760.

56. Waddell KJ, Birkenmeier RL, Bland MD, Lang CE. An exploratory analysis of the self-reported goals of

individuals with chronic upper-extremity paresis following stroke. Disabil Rehabil [Internet]. 2016;38 (9):853–7. http://www.tandfonline.com/doi/full, https://doi.org/10.3109/09638288.2015.1062926.

57. Rand D, Eng JJ. Disparity between functional recovery and daily use of the upper and lower extremities during subacute stroke rehabilitation. Neurorehabil Neural Repair [Internet]. 2012;26 (1):76–84. http://journals.sagepub.com/doi, https:// doi.org/10.1177/1545968311408918.

58. Doman CA, Waddell KJ, Bailey RR, Moore JL, Lang CE. Changes in upper-extremity functional capacity and daily performance during outpatient occupational therapy for people with stroke. Am J Occup Ther [Internet]. 2016;70(3):7003290040p1– 11. https://research.aota.org/ajot/article/70/3/ 7003290040p1/6159/Changes-in-Upper-Extremity-Functional-Capacity-and.

59. Ardestani MM, Henderson CE, Hornby TG. Improved walking function in laboratory does not guarantee increased community walking in stroke survivors: Potential role of gait biomechanics. J Biomech [Internet]. 2019;91:151–9. https:// linkinghub.elsevier.com/retrieve/pii/ S0021929019303434.

60. Waddell KJ, Strube MJ, Bailey RR, Klaesner JW, Birkenmeier RL, Dromerick AW, et al. Does task-specific training improve upper limb performance in daily life poststroke? Neurorehabil Neural Repair [Internet]. 2017;31(3):290–300. http://journals. sagepub.com/doi, https://doi.org/10.1177/ 1545968316680493.

61. Taub E. The behavior-analytic origins of constraint-induced movement therapy: an example of behavioral neurorehabilitation. Behav Anal [Internet]. 2012;35(2):155–78. http://www.ncbi.nlm.nih.gov/ pubmed/23449867.

62. Hirsch T, Barthel M, Aarts P, Chen Y-A, Freivogel S, Johnson MJ, et al. A first step toward the operationalization of the learned non-use phenomenon: a Delphi study. neurorehabil neural repair [Internet]. 2021;35(5):383–92. http://journals. sagepub.com/doi, https://doi.org/10.1177/1545968 321999064.

63. Bailey RR, Lang CE. Upper-limb activity in adults: Referent values using accelerometry. J Rehabil Res Dev [Internet]. 2013;50(9):1213–22. http://www. rehab.research.va.gov/jour/2013/509/pdf/JRRD-2012-12-0222.pdf.

64. Lemmens RJ, Timmermans AA, Janssen-Potten YJ, Smeets RJ, Seelen HA. Valid and reliable instruments for arm-hand assessment at ICF activity level in persons with hemiplegia: a systematic review. BMC Neurol [Internet]. 2012;12(1):21. http:// bmcneurol.biomedcentral.com/articles, https://doi. org/10.1186/1471-2377-12-21.

65. Uswatte G, Miltner WH, Foo B, Varma M, Moran S, Taub E. Objective measurement of functional upper-extremity movement using accelerometer recordings transformed with a threshold filter. Stroke [Internet]. 2000;31(3):662– 7. http://www.ncbi.nlm.nih.gov/pubmed/10700501.

66. Urbin MA, Waddell KJ, Lang CE. Acceleration metrics are responsive to change in upper extremity function of stroke survivors. Arch Phys Med Rehabil [Internet]. 2015;96(5):854–61. https:// linkinghub.elsevier.com/retrieve/pii/S00039993140 12829.

67. Urbin MA, Bailey RR, Lang CE. Validity of Body-Worn Sensor Acceleration Metrics to Index Upper Extremity Function in Hemiparetic Stroke. J Neurol Phys Ther [Internet]. 2015;39(2):111–8. https:// journals.lww.com/01253086-201504000-00006.

68. Lang CE, Barth J, Holleran CL, Konrad JD, Bland MD. Implementation of wearable sensing technology for movement: pushing forward into the routine physical rehabilitation care field. Sensors (Basel) [Internet]. 2020;20(20). http://www.ncbi. nlm.nih.gov/pubmed/33050368.

69. Lang CE, Waddell KJ, Klaesner JW, Bland MD. A method for quantifying upper limb performance in daily life using accelerometers. J Vis Exp [Internet]. 2017;(122). https://www.jove.com/ video/55673/a-method-for-quantifying-upper-limb-performance-daily-life-using.

70. Miller LC, Dewald JPA. Involuntary paretic wrist/finger flexion forces and EMG increase with shoulder abduction load in individuals with chronic stroke. Clin Neurophysiol [Internet]. 2012;123 (6):1216–25. https://linkinghub.elsevier.com/ retrieve/pii/S1388245712000466.

71. Lee SI, Ozsecen MY, Della Toffola L, Daneault J-F, Puiatti A, Patel S, et al. Activity detection in uncontrolled free-living conditions using a single accelerometer. In: 2015 IEEE 12th international conference on wearable and implantable body sensor networks (BSN) [Internet]. IEEE; 2015. p. 1–6. http://ieeexplore.ieee.org/document/7299372/.

72. Hoyt CR, Van AN, Ortega M, Koller JM, Everett EA, Nguyen AL, et al. Detection of pediatric upper extremity motor activity and deficits with accelerometry. JAMA Netw Open [Internet]. 2019;2(4):e192970. http://jamanetworkopen. jamanetwork.com/article.aspx?doi, https://doi.org/ 10.1001/jamanetworkopen.2019.2970.

73. Smith BA, Lang CE. Sensor measures of symmetry quantify upper limb movement in the natural environment across the lifespan. Arch Phys Med Rehabil [Internet]. 2019;100(6):1176–83. https:// linkinghub.elsevier.com/retrieve/pii/ S0003999319300760.

74. Parnandi A, Uddin J, Nilsen DM, Schambra HM. The pragmatic classification of upper extremity motion in neurological patients: a primer. Front Neurol [Internet]. 2019; 18:10. https://www. frontiersin.org/article, https://doi.org/10.3389/fneur. 2019.00996/full.

75. Goldsack JC, Coravos A, Bakker JP, Bent B, Dowling AV., Fitzer-Attas C, et al. Verification, analytical validation, and clinical validation (V3):

the foundation of determining fit-for-purpose for biometric monitoring technologies (BioMeTs). NPJ Digit Med [Internet]. 2020;3(1):55. http://www.nature.com/articles/s41746-020-0260-4.

76. Balasubramanian S, Melendez-Calderon A, Roby-Brami A, Burdet E. On the analysis of movement smoothness. J Neuroeng Rehabil [Internet]. 2015;12(1):112. http://www.jneuroengrehab.com/content/12/1/112.

77. Balasubramanian S, Melendez-Calderon A, Burdet E. A robust and sensitive metric for quantifying movement smoothness. IEEE Trans Biomed Eng [Internet]. 2012;59(8):2126–36. http://ieeexplore.ieee.org/document/6104119/.

78. David A, Subash T, Varadhan SKM, Melendez-Calderon A, Balasubramanian S. A framework for sensor-based assessment of upper-limb functioning in hemiparesis. Front Hum Neurosci [Internet]. 2021;15. https://www.frontiersin.org/articles, https://doi.org/10.3389/fnhum.2021.667509/full.

79. Lang CE, Waddell KJ, Barth J, Holleran CL, Strube MJ, Bland MD. Upper limb performance in daily life approaches plateau around three to six weeks post-stroke. Neurorehabil Neural Repair [Internet]. 2021;35(10):903–14. http://journals.sagepub.com/doi, https://doi.org/10.1177/15459683211041302.

80. Barth J, Lohse KR, Konrad JD, Bland MD, Lang CE. Sensor-based categorization of upper limb performance in daily life of persons with and without neurological upper limb deficits. Front Rehabil Sci [Internet]. 2021;2. https://www.frontiersin.org/articles, https://doi.org/10.3389/fresc.2021.741393/full.

81. Harris JE, Eng JJ. Goal priorities identified through client-centred measurement in individuals with chronic stroke. Physiother Canada [Internet]. 2004;56(03):171. https://journals.bcdecker.com/CrossRef/showText.aspx?path=PTC/volume56%2C2004/issue 03%2CJune/ptc_2004_00017/ptc_2004_00017.xml.

82. Bohannon RW, Andrews AW, Smith MB. Rehabilitation goals of patients with hemiplegia. Int J Rehabil Res. 1988;11(2):181–4.

83. Macko RF, Haeuber E, Shaughnessy M, Coleman KL, Boone DA, Smith GV, et al. Microprocessor-based ambulatory activity monitoring in stroke patients. Med Sci Sport Exerc [Internet]. 2002;34(3):394–9. http://journals.lww.com/00005768-200203000-00002.

84. Fulk GD, Combs SA, Danks KA, Nirider CD, Raja B, Reisman DS. Accuracy of 2 activity monitors in detecting steps in people with stroke and traumatic brain injury. Phys Ther [Internet]. 2014;94(2):222–9. https://academic.oup.com/ptj/article/94/2/222/2735422.

85. Evenson KR, Goto MM, Furberg RD. Systematic review of the validity and reliability of consumer-wearable activity trackers. Int J Behav Nutr Phys Act [Internet]. 2015;12(1):159. http://www.ijbnpa.org/content/12/1/159

86. Larsen RT, Korfitsen CB, Juhl CB, Andersen HB, Langberg H, Christensen J. Criterion validity for step counting in four consumer-grade physical activity monitors among older adults with and without rollators. Eur Rev Aging Phys Act [Internet]. 2020;17(1):1. https://eurapa.biomedcentral.com/articles/https://doi.org/10.1186/s11556-019-0235-0.

87. Rozanski GM, Aqui A, Sivakumaran S, Mansfield A. Consumer wearable devices for activity monitoring among individuals after a stroke: a prospective comparison. JMIR Cardio [Internet]. 2018;2(1):e1. http://cardio.jmir.org/2018/1/e1/.

88. Danks KA, Pohlig RT, Roos M, Wright TR, Reisman DS. Relationship between walking capacity, biopsychosocial factors, self-efficacy, and walking activity in persons poststroke. J Neurol Phys Ther [Internet]. 2016;40(4):232–8. https://journals.lww.com/01253086-201610000-00004.

89. Holleran CL, Bland MD, Reisman DS, Ellis TD, Earhart GM, Lang CE. Day-to-day variability of walking performance measures in individuals post-stroke and individuals with parkinson disease. J Neurol Phys Ther [Internet]. 2020;44(4):241–7. https://journals.lww.com, https://doi.org/10.1097/NPT.0000000000000327.

90. Barak S, Wu SS, Dai Y, Duncan PW, Behrman AL. Adherence to accelerometry measurement of community ambulation poststroke. Phys Ther [Internet]. 2014;94(1):101–10. https://academic.oup.com/ptj/article/94/1/101/2735448.

91. Lang CE, Wagner JM, Edwards DF, Dromerick AW. Upper extremity use in people with hemiparesis in the first few weeks after stroke. J Neurol Phys Ther [Internet]. 2007;31(2):56–63. http://www.ncbi.nlm.nih.gov/pubmed/17558358.

92. Bailey RR, Birkenmeier RL, Lang CE. Real-world affected upper limb activity in chronic stroke: an examination of potential modifying factors. Top Stroke Rehabil [Internet]. 2015;22(1):26–33. http://www.tandfonline.com/doi/full, https://doi.org/10.1179/1074935714Z.0000000040.

93. Duncan PW, Lai SM, Keighley J. Defining post-stroke recovery: implications for design and interpretation of drug trials. Neuropharmacology [Internet]. 2000;39(5):835–41. http://www.ncbi.nlm.nih.gov/pubmed/10699448.

94. Kwakkel G, Kollen B, Lindeman E. Understanding the pattern of functional recovery after stroke: facts and theories. Restor Neurol Neurosci [Internet]. 2004;22(3–5):281–99. http://www.ncbi.nlm.nih.gov/pubmed/15502272.

95. Ramsey LE, Siegel JS, Lang CE, Strube M, Shulman GL, Corbetta M. Behavioural clusters and predictors of performance during recovery from

stroke. Nat Hum Behav [Internet]. 2017;1(3):0038. http://www.nature.com/articles/s41562-016-0038.

96. Cortes JC, Goldsmith J, Harran MD, Xu J, Kim N, Schambra HM, et al. A short and distinct time window for recovery of arm motor control early after stroke revealed with a global measure of trajectory kinematics. Neurorehabil Neural Repair [Internet]. 2017;31(6):552–60. http://journals.sagepub.com/doi, https://doi.org/10.1177/1545968317697034.

97. Jørgensen HS, Nakayama H, Raaschou HO, Vive-Larsen J, Støier M, Olsen TS. Outcome and time course of recovery in stroke. Part II: time course of recovery. The Copenhagen stroke study. Arch Phys Med Rehabil [Internet]. 1995;76(5):406–12. http://www.ncbi.nlm.nih.gov/pubmed/7741609.

98. Vliet R, Selles RW, Andrinopoulou E, Nijland R, Ribbers GM, Frens MA, et al. Predicting upper limb motor impairment recovery after stroke: a mixture model. Ann Neurol [Internet]. 2020;87(3):383–93. https://onlinelibrary.wiley.com/doi, https://doi.org/10.1002/ana.25679.

99. Yozbatiran N, Der-Yeghiaian L, Cramer SC. A Standardized approach to performing the action research arm test. Neurorehabil Neural Repair [Internet]. 2008;22(1):78–90. http://journals.sagepub.com/doi, https://doi.org/10.1177/1545968307305353.

100. Dobkin BH. Behavioral self-management strategies for practice and exercise should be included in neurologic rehabilitation trials and care. Curr Opin Neurol [Internet]. 2016;29(6):693–9. https://journals.lww.com/00019052-201612000-00005.

101. Glasgow RE, Goldstein MG, Ockene JK, Pronk NP. Translating what we have learned into practice. Am J Prev Med [Internet]. 2004;27(2):88–101. https://linkinghub.elsevier.com/retrieve/pii/S0749379704000996.

102. Schweighofer N, Han CE, Wolf SL, Arbib MA, Winstein CJ. A functional threshold for long-term use of hand and arm function can be determined: predictions from a computational model and supporting data from the extremity constraint-induced therapy evaluation (excite) trial. Phys Ther. 2009;89(12):1327–36.

103. Schwerz de Lucena D, Rowe J, Chan V, Reinkensmeyer D. Magnetically counting hand movements: validation of a calibration-free algorithm and application to testing the threshold hypothesis of real-world hand use after stroke. Sensors [Internet]. 2021;21(4):1502. https://www.mdpi.com/1424-8220/21/4/1502.

104. Bravata DM, Smith-Spangler C, Sundaram V, Gienger AL, Lin N, Lewis R, et al. Using pedometers to increase physical activity and improve health: a systematic review. JAMA. 2007;298(19):2296–304.

105. Forbes. Fitness Tracker Market Size, Share & COVID-19 Impact Analysis [Internet]. 2021. https://www.fortunebusinessinsights.com/fitness-tracker-market-103358.

106. Locke EA, Latham GP. Building a practically useful theory of goal setting and task motivation: a 35-year odyssey. Am Psychol. 2002;57(9):705.

107. Dobkin BH, Plummer-D'Amato P, Elashoff R, Lee J. International randomized clinical trial, stroke inpatient rehabilitation with reinforcement of walking speed (SIRROWS), improves outcomes. Neurorehabil Neural Repair [Internet]. 2010;24(3):235–42. http://journals.sagepub.com/doi, https://doi.org/10.1177/1545968309357558.

108. Lang CE, MacDonald JR, Reisman DS, Boyd L, Kimberley TJ, Schindler-Ivens SM, et al. Observation of amounts of movement practice provided during stroke rehabilitation. Arch Phys Med Rehabil. 2009;90(10):1692–8.

109. Jack K, McLean SM, Moffett JK, Gardiner E. Barriers to treatment adherence in physiotherapy outpatient clinics: a systematic review. Man Ther [Internet]. 201015(3):220–8. https://linkinghub.elsevier.com/retrieve/pii/S1356689X09002094.

110. Bassett SF. The assessment of patient adherence to physiotherapy rehabilitation. New Zeal J Physiother. 2003;31(2):60–6.

111. McLean SM, Burton M, Bradley L, Littlewood C. Interventions for enhancing adherence with physiotherapy: A systematic review. Man Ther [Internet]. 2010;15(6):514–21. https://linkinghub.elsevier.com/retrieve/pii/S1356689X10000871.

112. Whitford M, Schearer E, Rowlett M. Effects of in home high dose accelerometer-based feedback on perceived and actual use in participants chronic post-stroke. Physiother Theory Pract. 2018;00(00):1–11.

113. de Lucena DS. New technologies for on-demand hand rehabilitation in the living environment after neurologic injury. Irvine: University of California; 2019.

114. Signal NEJ, McLaren R, Rashid U, Vandal A, King M, Almesfer F, et al. Haptic nudges increase affected upper limb movement during inpatient stroke rehabilitation: multiple-period randomized crossover study. JMIR mHealth uHealth [Internet]. 2020;8(7):e17036. https://mhealth.jmir.org/2020/7/e17036.

115. Da-Silva RH, van Wijck F, Shaw L, Rodgers H, Balaam M, Brkic L, et al. Prompting arm activity after stroke: A clinical proof of concept study of wrist-worn accelerometers with a vibrating alert function. J Rehabil Assist Technol Eng. 2018;5:2055668318761524.

116. Da-Silva RH, Moore SA, Rodgers H, Shaw L, Sutcliffe L, van Wijck F, et al. Wristband accelerometers to motiVate arm Exercises after Stroke (WAVES): a pilot randomized controlled trial. Clin Rehabil. 2019;33(8):1391–403.

117. Wei WXJ, Fong KNK, Chung RCK, Cheung HKY, Chow ESL. Remind-to-move for promoting upper extremity recovery using wearable devices in sub-acute stroke: a multi-center randomized controlled study. IEEE Trans Neural Syst Rehabil Eng

[Internet]. 2019;27(1):51–9. https://ieeexplore.ieee. org/document/8540924/.

118. Chae SH, Kim Y, Lee K-S, Park H-S. Development and Clinical evaluation of a web-based upper limb home rehabilitation system using a smartwatch and machine learning model for chronic stroke survivors: prospective comparative study. JMIR mHealth uHealth [Internet]. 2020;8(7):e17216. http://www.ncbi.nlm.nih.gov/pubmed/32480361.

119. Lynch EA, Jones TM, Simpson DB, Fini NA, Kuys SS, Borschmann K, et al. Activity monitors for increasing physical activity in adult stroke survivors. Cochrane Database Syst Rev. 2018;(7).

120. Mandigout S, Chaparro D, Borel B, Kammoun B, Salle J-Y, Compagnat M, et al. Effect of individualized coaching at home on walking capacity in subacute stroke patients: a randomized controlled trial (Ticaa'dom). Ann Phys Rehabil Med. 2021;64 (4): 101453.

121. Grau-Pellicer M, Lalanza J, Jovell-Fernández E, Capdevila L. Impact of mHealth technology on adherence to healthy PA after stroke: a randomized study. Top Stroke Rehabil [Internet]. 2020;27 (5):354–68. https://www.tandfonline.com/doi/full, https://doi.org/10.1080/10749357.2019.1691816.

122. Hassett L, van den Berg M, Lindley RI, Crotty M, McCluskey A, van der Ploeg HP, et al. Digitally enabled aged care and neurological rehabilitation to enhance outcomes with activity and MObility UsiNg Technology (AMOUNT) in Australia: a randomised controlled trial. Nguyen C, editor. PLOS Med [Internet]. 2020;17(2):e1003029. https://dx.plos.org , https://doi.org/10.1371/journal.pmed.1003029.

123. Angelucci A, Cavicchioli M, Cintorrino IA, Lauricella G, Rossi C, Strati S, et al. Smart textiles and sensorized garments for physiological monitoring: a review of available solutions and techniques. Sensors [Internet]. 2021;21(3):814. https://www.mdpi. com/1424-8220/21/3/814.

124. Gopalsamy C, Park S, Rajamanickam R, Jayaraman S. The wearable motherboard™: the first generation of adaptive and responsive textile structures (ARTS) for medical applications. Virtual Real. 1999;4(3):152–68.

125. Park S, Gopalsamy C, Rajamanickam R, Jayaraman S. The wearable motherboard©: a flexible information infrastructure or sensate liner for medical applications. Stud Health Technol Inform. 1999;62:252–8.

126. Marculescu D, Marculescu R, Zamora NH, Stanley-Marbell P, Khosla PK, Park S, et al. Electronic textiles: a platform for pervasive computing. Proc IEEE [Internet]. 200391(12):1995–2018. http://ieeexplore.ieee.org/document/1246382/.

127. Scilingo EP, Lorussi F, Mazzoldi A, De Rossi D. Strain-sensing fabrics for wearable kinaesthetic-like systems. IEEE Sens J. 2003;3(4):460–7.

128. Lorussi F, Rocchia W, Scilingo EP, Tognetti A, De Rossi D. Wearable, redundant fabric-based sensor arrays for reconstruction of body segment posture. IEEE Sens J. 2004;4(6):807–18.

129. Lorussi F, Scilingo EP, Tesconi M, Tognetti A, De Rossi D. Strain sensing fabric for hand posture and gesture monitoring. IEEE Trans Inf Technol Biomed. 2005;9(3):372–81.

130. Lorussi F, Carbonaro N, De Rossi D, Paradiso R, Veltink P, Tognetti A. Wearable textile platform for assessing stroke patient treatment in daily life conditions. Front Bioeng Biotechnol [Internet]. 2016;4:28. http://www.ncbi.nlm.nih.gov/pubmed/ 27047939.

131. McLaren R, Joseph F, Baguley C, Taylor D. A review of e-textiles in neurological rehabilitation: How close are we? J Neuroeng Rehabil [Internet]. 2016;13(1):59. http://www.ncbi.nlm.nih.gov/ pubmed/27329186.

132. Preece SJ, Kenney LPJ, Major MJ, Dias T, Lay E, Fernandes BT. Automatic identification of gait events using an instrumented sock. J Neuroeng Rehabil. 2011;8:32.

133. Rogers JA, Someya T, Huang Y. Materials and mechanics for stretchable electronics. Science (80-) [Internet]. 2010;327(5973):1603–7. https://www. science.org/doi, https://doi.org/10.1126/science. 1182383.

134. Kim D-H, Lu N, Ma R, Kim Y-S, Kim R-H, Wang S, et al. Epidermal electronics. Science (80-) [Internet]. 2011;333(6044):838–43. https://www. science.org/doi/, https://doi.org/10.1126/science. 1206157.

135. Kim KK, Ha I, Kim M, Choi J, Won P, Jo S, et al. A deep-learned skin sensor decoding the epicentral human motions. Nat Commun [Internet]. 2020;11 (1):2149. http://www.nature.com/articles/s41467-020-16040-y.

136. Bonnassieux Y, Brabec CJ, Cao Y, Carmichael TB, Chabinyc ML, Cheng K-T, et al. The 2021 flexible and printed electronics roadmap. Flex Print Electron [Internet]. 2021;6(2):023001. https://iopscience.iop. org/article, https://doi.org/10.1088/2058-8585/abf986.

137. Khan MA, Saibene M, Das R, Brunner I, Puthusserypady S. Emergence of flexible technology in developing advanced systems for post-stroke rehabilitation: a comprehensive review. J Neural Eng [Internet]. 2021;18(6):061003. https:// iopscience.iop.org/article, https://doi.org/10.1088/ 1741-2552/ac36aa.

138. Doherty AR, Kelly P, Kerr J, Marshall S, Oliver M, Badland H, et al. Using wearable cameras to categorise type and context of accelerometer-identified episodes of physical activity. Int J Behav Nutr Phys Act [Internet]. 2013;10(1):22. http:// ijbnpa.biomedcentral.com/articles, https://doi.org/ 10.1186/1479-5868-10-22.

139. Yan Yan, Ricci E, Gaowen Liu, Sebe N. Egocentric daily activity recognition via multitask clustering. IEEE Trans Image Process [Internet]. 2015;24 (10):2984–95. http://ieeexplore.ieee.org/document/ 7113851/.

140. Zariffa J, Popovic MR. Hand contour detection in wearable camera video using an adaptive histogram region of interest. J Neuroeng Rehabil [Internet]. 2013;10(1):114. http://jneuroengrehab.biomedcentral.com/articles, https://doi.org/10.1186/1743-0003-10-114.

141. Likitlersuang J, Sumitro ER, Cao T, Visée RJ, Kalsi-Ryan S, Zariffa J. Egocentric video: a new tool for capturing hand use of individuals with spinal cord injury at home. J Neuroeng Rehabil [Internet]. 2019;16(1):83. https://jneuroengrehab.biomedcentral.com/articles, https://doi.org/10.1186/s12984-019-0557-1.

142. Visee RJ, Likitlersuang J, Zariffa J. An effective and efficient method for detecting hands in egocentric videos for rehabilitation applications. IEEE Trans Neural Syst Rehabil Eng [Internet]. 2020;28(3):748–55. https://ieeexplore.ieee.org/document/8967132/.

143. Dousty M, Zariffa J. Tenodesis grasp detection in egocentric video. IEEE J Biomed Heal Inform [Internet]. 2021;25(5):1463–70. https://ieeexplore.ieee.org/document/9121713/.

144. Likitlersuang J, Visée RJ, Kalsi-Ryan S, Zariffa J. Capturing hand use of individuals with spinal cord injury at home using egocentric video: a feasibility study. Spinal Cord Ser Cases [Internet]. 2021;7(1):17. http://www.nature.com/articles/s41394-021-00382-w.

145. Tsai M-F, Wang RH, Zariffa J. Identifying hand use and hand roles after stroke using egocentric video. IEEE J Transl Eng Heal Med [Internet]. 2021;1–1. https://ieeexplore.ieee.org/document/9399477/.

146. Bandini A, Zariffa J. Analysis of the hands in egocentric vision: a survey. IEEE Trans Pattern Anal Mach Intell [Internet]. 2020;1–1. https://ieeexplore.ieee.org/document/9064606/.

147. Zhang Y, Sun S, Lei L, Liu H, Xie H. STAC: spatial-temporal attention on compensation information for activity recognition in FPV. Sensors [Internet]. 2021;21(4):1106. https://www.mdpi.com/1424-8220/21/4/1106.

148. Capra M, Sapienza S, Motto Ros P, Serrani A, Martina M, Puiatti A, et al. Assessing the feasibility of augmenting fall detection systems by relying on UWB-based position tracking and a home robot. Sensors (Basel) [Internet]. 2020;20(18). http://www.ncbi.nlm.nih.gov/pubmed/32962142.

149. Vahia I V, Kabelac Z, Hsu C-Y, Forester BP, Monette P, May R, et al. Radio signal sensing and signal processing to monitor behavioral symptoms in dementia: a case study. Am J Geriatr Psychiatry [Internet]. 2020;28(8):820–5. http://www.ncbi.nlm.nih.gov/pubmed/32245677.

150. Ridolfi M, Kaya A, Berkvens R, Weyn M, Joseph W, Poorter E De. Self-calibration and collaborative localization for UWB positioning systems. ACM Comput Surv [Internet]. 2022;54(4):1–27. https://dl.acm.org/doi, https://doi.org/10.1145/3448303.

151. Landaluce H, Arjona L, Perallos A, Falcone F, Angulo I, Muralter F. A review of IoT sensing applications and challenges using RFID and wireless sensor networks. Sensors [Internet]. 2020;20(9):2495. https://www.mdpi.com/1424-8220/20/9/2495.

152. Pascacio P, Casteleyn S, Torres-Sospedra J, Lohan ES, Nurmi J. Collaborative indoor positioning systems: a systematic review. Sensors [Internet]. 2021;21(3):1002. https://www.mdpi.com/1424-8220/21/3/1002

153. Adib F, Kabelac Z, Katabi D, Miller RC. 3d tracking via body radio reflections. In: 11th $\{$USENIX$\}$ Symposium on networked systems design and implementation ($\{$NSDI$\}$ 14). 2014. p. 317–29.

154. Adib F, Katabi D. See through walls with WiFi! In: Proceedings of the ACM SIGCOMM 2013 conference on SIGCOMM [Internet]. New York, NY, USA: ACM; 2013. p. 75–86. https://dl.acm.org/doi, https://doi.org/10.1145/2486001.2486039.

155. Fan L, Li T, Yuan Y, Katabi D. In-home daily-life captioning using radio signals. 2020. p. 105–23. https://link.springer.com, https://doi.org/10.1007/978-3-030-58536-5_7.

156. Toshev A, Szegedy C. DeepPose: human pose estimation via deep neural networks. In: 2014 IEEE conference on computer vision and pattern recognition [Internet]. IEEE; 2014. p. 1653–60. https://ieeexplore.ieee.org/document/6909610.

157. Insafutdinov E, Pishchulin L, Andres B, Andriluka M, Schiele B. DeeperCut: a deeper, stronger, and faster multi-person pose estimation model. 2016. p. 34–50. http://link.springer.com/https://doi.org/10.1007/978-3-319-46466-4_3.

158. Cao Z, Hidalgo G, Simon T, Wei S-E, Sheikh Y. OpenPose: realtime multi-person 2D pose estimation using part affinity fields. IEEE Trans Pattern Anal Mach Intell [Internet]. 2021;43(1):172–86. https://ieeexplore.ieee.org/document/8765346/.

159. Cao Z, Simon T, Wei S-E, Sheikh Y. Realtime multi-person 2D pose estimation using part affinity fields. In: 2017 IEEE conference on computer vision and pattern recognition (CVPR) [Internet]. IEEE; 2017. p. 1302–10. http://ieeexplore.ieee.org/document/8099626/.

160. Insafutdinov E, Andriluka M, Pishchulin L, Tang S, Levinkov E, Andres B, et al. ArtTrack: articulated multi-person tracking in the wild. In: IEEE conference on computer vision and pattern recognition (CVPR) [Internet]. IEEE; 2017. p. 1293–301. http://ieeexplore.ieee.org/document/8099625/.

161. Mathis A, Mamidanna P, Cury KM, Abe T, Murthy VN, Mathis MW, et al. DeepLabCut: markerless pose estimation of user-defined body parts with deep learning. Nat Neurosci [Internet]. 2018;21(9):1281–9. http://www.nature.com/articles/s41593-018-0209-y.

162. Fang H-S, Xie S, Tai Y-W, Lu C. RMPE: regional multi-person pose estimation. 2016; http://arxiv.org/abs/1612.00137.

163. Lugaresi C, Tang J, Nash H, McClanahan C, Uboweja E, Hays M, et al. MediaPipe: a framework for building perception pipelines. 2019.

164. Lugaresi C, Tang J, Nash H, McClanahan C, Uboweja E, Hays M, et al. MediaPipe: a framework for perceiving and processing reality. 2019.

165. Zhang F, Bazarevsky V, Vakunov A, Tkachenka A, Sung G, Chang C-L, et al. MediaPipe hands: on-device real-time hand tracking. 2020. http://arxiv.org/abs/2006.10214.

166. Seethapathi N, Wang S, Saluja R, Blohm G, Kording KP. Movement science needs different pose tracking algorithms. 2019. http://arxiv.org/abs/1907.10226.

167. Arac A. Machine learning for 3D kinematic analysis of movements in neurorehabilitation. Curr Neurol Neurosci Rep [Internet]. 2020;20(8):29. http://www.ncbi.nlm.nih.gov/pubmed/32542455.

168. Cronin NJ. Using deep neural networks for kinematic analysis: challenges and opportunities. J Biomech [Internet]. 2021;123:110460. http://www.ncbi.nlm.nih.gov/pubmed/34029787.

169. Mehdizadeh S, Nabavi H, Sabo A, Arora T, Iaboni A, Taati B. Concurrent validity of human pose tracking in video for measuring gait parameters in older adults: a preliminary analysis with multiple trackers, viewing angles, and walking directions. J Neuroeng Rehabil [Internet]. 2021;18(1):139. http://www.ncbi.nlm.nih.gov/pubmed/34526074.

170. Stenum J, Rossi C, Roemmich RT. Two-dimensional video-based analysis of human gait using pose estimation. PLoS Comput Biol [Internet]. 2021;17(4):e1008935. http://www.ncbi.nlm.nih.gov/pubmed/33891585.

171. Takeda I, Yamada A, Onodera H. Artificial Intelligence-Assisted motion capture for medical applications: a comparative study between markerless and passive marker motion capture. Comput Methods Biomech Biomed Engin [Internet]. 2021;24(8):864–73. https://www.tandfonline.com/doi/full/, https://doi.org/10.1080/10255842.2020.1856372.

172. Viswakumar A, Rajagopalan V, Ray T, Gottipati P, Parimi C. Development of a robust, simple, and affordable human gait analysis system using bottom-up pose estimation with a smartphone camera. Front Physiol [Internet]. 2021;12:784865. http://www.ncbi.nlm.nih.gov/pubmed/35069246.

173. Stenum J, Cherry-Allen KM, Pyles CO, Reetzke RD, Vignos MF, Roemmich RT. Applications of pose estimation in human health and performance across the lifespan. sensors (Basel) [Internet]. 2021;21(21). http://www.ncbi.nlm.nih.gov/pubmed/34770620.

174. Cornman HL, Stenum J, Roemmich RT. Video-based quantification of human movement frequency using pose estimation: a pilot study. PLoS One [Internet]. 2021;16(12):e0261450. http://www.ncbi.nlm.nih.gov/pubmed/34929012.

175. Ahmed T, Thopalli K, Rikakis T, Turaga P, Kelliher A, Huang J-B, et al. Automated movement assessment in stroke rehabilitation. Front Neurol [Internet]. 2021;12:720650. http://www.ncbi.nlm.nih.gov/pubmed/34489855.

176. Zhu Y, Lu W, Gan W, Hou W. A contactless method to measure real-time finger motion using depth-based pose estimation. Comput Biol Med [Internet]. 2021;131:104282. https://linkinghub.elsevier.com/retrieve/pii/S0010482521000767.

177. Ren S, He K, Girshick R, Sun J. Faster R-CNN: towards real-time object detection with region proposal networks. 2015. http://arxiv.org/abs/1506.01497.

178. Regan EW, Handlery R, Stewart JC, Pearson JL, Wilcox S, Fritz S. Integrating survivors of stroke into exercise-based cardiac rehabilitation improves endurance and functional strength. J Am Heart Assoc [Internet]. 2021;10(3):e017907. http://www.ncbi.nlm.nih.gov/pubmed/33499647.

179. Hutchinson K, Sloutsky R, Collimore A, Adams B, Harris B, Ellis TD, et al. A music-based digital therapeutic: proof-of-concept automation of a progressive and individualized rhythm-based walking training program after stroke. Neurorehabil Neural Repair [Internet]. 2020;34(11):986–96. http://www.ncbi.nlm.nih.gov/pubmed/33040685.

180. Backhaus W, Kempe S, Hummel FC. The effect of sleep on motor learning in the aging and stroke population - a systematic review. Restor Neurol Neurosci [Internet]. 2016;34(1):153–64. http://www.ncbi.nlm.nih.gov/pubmed/26835597.

181. Vogels EA. About one-in-five Americans use a smart watch or fitness tracker [Internet]. January 9. 2020 [cited 2021 Dec 11]. https://www.pewresearch.org/fact-tank/2020/01/09/about-one-in-five-americans-use-a-smart-watch-or-fitness-tracker/.

BCI-Based Neuroprostheses and Physiotherapies for Stroke Motor Rehabilitation

Jeffrey Lim, Derrick Lin, Won Joon Sohn, Colin M. McCrimmon, Po T. Wang, Zoran Nenadic, and An H. Do

Abstract

Despite the best available physiotherapy, the stroke survivor population remains affected by significant motor impairment in both upper and lower extremities. Many emerging rehabilitative approaches have ultimately proven to be no better than standard physiotherapy. Hence, there is still a great need for novel methods that can help improve motor outcomes beyond conventional physiotherapy. Brain–computer interfaces (BCIs) may be one such approach. BCIs translate brain signals into control commands for external devices using decoding algorithms. They can be applied to allow those with irreversible paralysis, due to stroke, to directly control prosthetic devices with their brain. Alternatively, they can be applied as novel rehabilitative tools to help improve motor recovery after stroke. However, utilizing BCIs for stroke rehabilitation is a nascent field. While many early clinical studies suggest that BCIs are promising as either neuroprostheses or as rehabilitative tools, there have not been any definitive clinical trials to demonstrate their effectiveness in improving functional or neurological outcomes. Hence, no clinical recommendations can be made for any BCI-based stroke rehabilitation.

Keywords

Brain–computer interface · Bidirectional brain–computer interface · Stroke · Spinal injury · Prosthetic · Neuromodulation · Neurorehabilitation therapies · Neuroplasticity · Clinical trials

22.1 Introduction

There are an estimated 7.6 million stroke survivors in the US alone, with approximately 795,000 new cases annually [1, 2]. Despite spontaneous recovery and intensive physiotherapy [3], 54% of stroke survivors remain affected by significant motor impairment [4], such as upper (21–48% [5, 6]) and lower (50–61% [7, 8]) extremity deficits. Post-stroke motor impairment is directly associated with decreased independence and lost productivity. Gait impairment, in particular, is associated with significant disability

J. Lim (✉) · D. Lin · P. T. Wang · Z. Nenadic
Department of Biomedical Engineering, University of California, Irvine, CA, USA
e-mail: limj4@uci.edu

W. J. Sohn · A. H. Do
Department of Neurology, University of California, Irvine, CA, USA

C. M. McCrimmon
Department of Neurology, University of California, Los Angeles, CA, USA

Z. Nenadic
Department of Electrical Engineering and Computer Science, University of California, Irvine, CA, USA

© The Author(s), under exclusive license to Springer Nature Switzerland AG 2022
D. J. Reinkensmeyer et al. (eds.), *Neurorehabilitation Technology*,
https://doi.org/10.1007/978-3-031-08995-4_22

and reduced physical activity, and is one of the few impairments that is directly linked to poor social re-integration [9, 10]. These problems lead to an increased risk of medical complications and raise a major public health concern in the form of increased healthcare, caregiving, and lost productivity costs. These costs are projected to increase to $184 billion annually by 2030, based on current trends of population aging and increased acute stroke survival rates [11].

For decades clinicians have utilized assistive devices such as orthoses and functional electrical stimulation (FES) systems to mitigate post-stroke motor impairments. However, these devices are cumbersome, may cause discomfort, and their benefits disappear upon removal. Significant effort has been invested to develop new technologies and methodologies to enhance stroke rehabilitation outcomes. However, some of these emerging rehabilitation approaches, such as robotic-assisted therapy, or body weight supported treadmill training have proven to be no

better than conventional physiotherapies [3, 12]. Novel methods are needed to help improve motor outcomes beyond conventional physiotherapy. In a recent example, neuromodulation provided by vagus nerve stimulation [13] was shown to elicit some functional improvements after stroke. In this chapter, we will discuss how brain-computer interface (BCI) technology may also elicit neurological and functional improvements beyond those provided by conventional therapies for stroke by acting as a prosthetic or by inducing neuroplasticity.

BCIs enable direct brain control of assistive devices and prostheses [14]. They employ decoding algorithms to translate electrophysiological signals acquired from the brain (e.g., electroencephalogram [EEG]) into control commands for assistive devices [14]. When integrated with FES [15–17] or robotic orthoses [18, 19], BCIs enable the direct brain control of these assistive devices (see Fig. 22.1). Such integrated systems can be applied as neuroprostheses or as

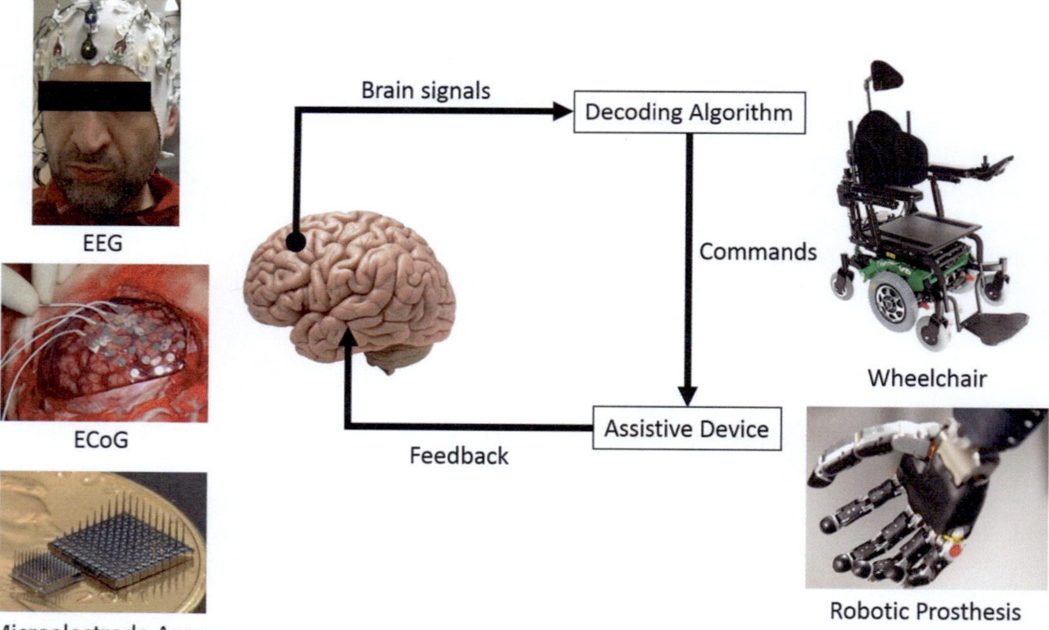

Fig. 22.1 Diagram describing the operation of a typical BCI system. Brain signals are first acquired via EEG, ECoG, or intracortical microelectrodes (local field and action potentials). These signals are analyzed in real time using decoding algorithms. The output of these algorithms is commands for external assistive devices. Their response to the brain-derived commands provides feedback to the user so that adjustments can be made as necessary. This process facilitates learning that may even improve the generation of the underlying neural control signal

novel physiotherapies to restore or improve motor function after stroke. These neuroprosthetic BCI systems are designed to replace motor functions that have been completely lost due to the stroke. More recently, advances in BCI technology have also opened the possibility of feedback using invasive electrocortical stimulation to elicit sensory percepts [20, 21]. A combination of both motor and sensory functions in a BCI constitutes what is referred to as a bidirectional brain–computer interface (BD-BCI). These devices promise to improve BCI control [22] and restore function more holistically, which may lead to improved rehabilitative outcomes. Within the physiotherapy context, BCIs are hypothesized to stimulate a Hebbian plasticity process (where "neurons that fire together, wire together")[23]. This approach could be used synergistically with other rehabilitative options and may ultimately facilitate functional recovery beyond that of conventional physiotherapies. This chapter will explore the emerging use of BCI technology in clinical stroke rehabilitation.

22.2 How BCIs Work

BCIs exploit predictable changes in neural signals to enable brain control of external devices. Brain signals such as EEG, electrocorticogram (ECoG), as well as local field and action potentials, exhibit predictable changes in various motor behaviors. Typically, BCIs use motor imagery (MI) or attempted motor execution (ME) to elicit these changes (see Fig. 22.2). For example, sensorimotor rhythms (SMRs), defined as the 8–30 Hz brain waves from sensory and motor areas [24], are known to be attenuated when an individual initiates movement. This phenomenon is known as event-related desynchronization (ERD). The attenuation of these signals stops with movement cessation, in a process known as event-related synchronization (ERS). In addition to these low-frequency modulations, brain waves in the high-gamma band

(>70 Hz), which can be acquired from subdural ECoG electrodes, also exhibit ERS during movement [25–27]. Lastly, local populations of neurons have been shown to exhibit increased spiking activity for specific movement directions [28–30].

Decoding algorithms powered by machine learning, such as Kalman filtering [31] and more recently deep learning neural networks [32], can utilize a variety of statistical analysis techniques to distinguish an individual's intentions based on changes in the neural signals. Typically, this is performed by classifying neural signals into discrete states, such as a movement class (where the individual intends movement via MI or ME) and an idling class (where no movement is intended). This type of discrete classification can be robustly achieved with EEG signals. To predict limb movement trajectories, higher resolution signals, such as ECoG [33] or neuronal action potentials acquired from microelectrode arrays [34], are required. Upon decoding the movement intention (into either discrete states or continuous trajectories), a computer command is sent to an output device. Common output devices include FES systems, robotic limbs or exoskeletons, wheelchairs, and virtual keyboards/mouses.

The operation of BCI is typically accompanied by simultaneous feedback processes so that users can self-adjust control of the output device. While the feedback for most BCI is visual, feedback can also be provided in the form of sensory percepts, such as those elicited by invasive, direct electrical stimulation of the cortex [20, 21]. This gives rise to a BD-BCI control paradigm that can be implemented by integrating positional or pressure sensors with motor prostheses, allowing for detected changes in the state of a pros thesis to be relayed to the cortex as sensory information. The parameters of the stimulation (e.g., electrode cortical placement, current amplitude, pulse-width modulation) can also be tuned to encode for varying types of sensations [20, 21].

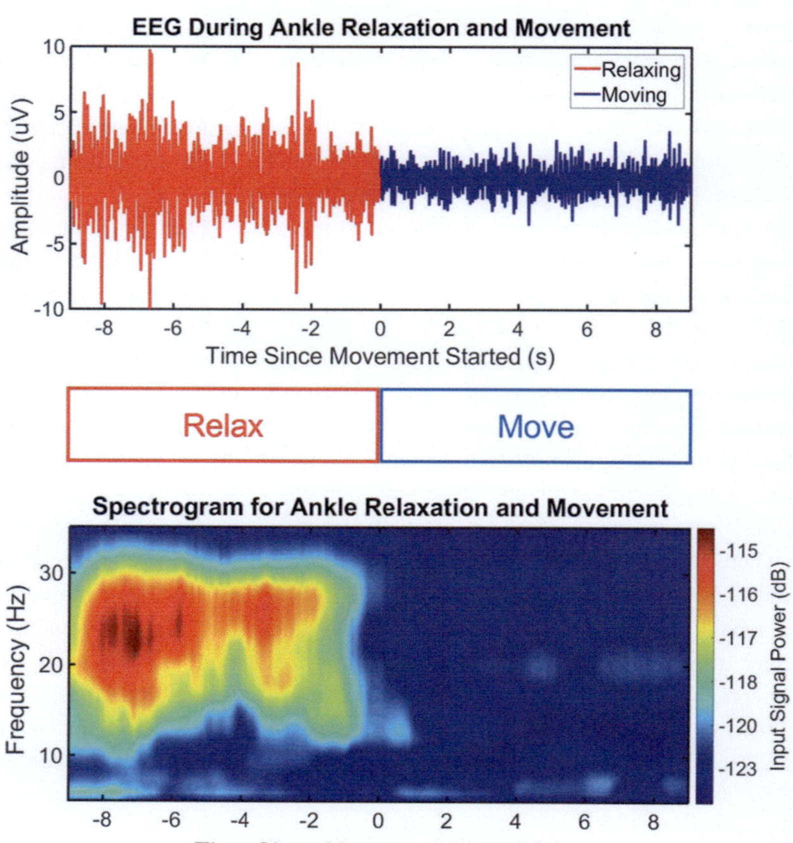

Fig. 22.2 Diagram showing movement-related EEG modulation from the relevant electrode (Cz from the International 10–20 system). When the stroke survivor relaxes the ankle, high amplitude oscillations (5–35 Hz) are seen in the time series data (top) and the time–frequency spectrogram (bottom). When the individual moves his paretic ankle, these waves become desynchronized

22.3 Neuroprosthetic BCI Systems

Historically, BCI systems were developed to target severe forms of motor diseases such as amyotrophic lateral sclerosis (ALS), where they could provide a means of communication with the outside world. Similar BCI systems have since been applied to other neurological diseases, such as high cervical spinal cord injury (SCI), where they are intended to provide the user with volitional and natural upper extremity control. In general, neuroprosthetic BCI systems can be used to acquire movement intention signals from intact brain cortex and subsequently translate them into control commands for devices such as robotic exoskeletons and FES systems, thereby providing a means to restore brain-controlled manipulation of the environment. Even though the original BCI systems were designed for ALS

and SCI, this technology can also be utilized in stroke rehabilitation. More specifically, the ideal post-stroke candidates for such BCI systems would be those with complete or near-complete paralysis due to subcortical strokes (e.g., lesions in the internal capsule) who are unlikely to regain further motor function.

To date, a number of BCI systems have successfully demonstrated that brain signals acquired either non-invasively (typically EEG) or invasively (typically ECoG or intracortical microelectrode arrays) can be exploited to enable brain control of both upper and lower extremity prostheses. For example, Pfurtscheller et al. [35] utilized an EEG-based BCI to enable FES-mediated hand grasping in an individual with C6 SCI. Do et al. [36] also implemented an EEG-based BCI system which enabled an individual with paraplegia to walk using a robotic exoskeleton system. Similarly, King et al.

extended this concept to the restoration of overground walking in a person with paraplegia due to SCI [37]. Invasive BCIs are able to achieve control of more degrees of freedom (DOF) and improved accuracy than existing non-invasive BCI systems and non-BCI technologies (e.g. a mouth joystick). For example, ECoG studies have demonstrated reliable decoding for a number of cortical processes with its superior decoding results attributed to millimeter-spatial and millisecond-temporal resolution [33]. The measurement of high-frequency activity (>70 Hz), such as high- gamma band, and the relative immunity to movement artifacts can provide a potential boost in BCI control accuracy. Wang et al. [38] demonstrated that high-density ECoG signals can be used to decode 6 DOF from the upper extremity. Wang et al. [39] demonstrated that ECoG signals could be used by an individual with tetraplegia to control a robotic arm for a reaching task. Furthermore, Wang et al. [40] demonstrated highly accurate decoding of hand grasping from ECoG signals. In the BrainGate clinical trials, intracortical microelectrode arrays implanted in tetraplegic individuals enabled control of a 6-DOF robotic arm [18, 34]. Collinger et al. [41] utilized two microelectrode arrays implanted into the motor cortex of a patient with severe tetraparesis due to spinocerebellar degeneration to successfully control a 7-DOF robotic arm. Aflalo et al. [42] utilized a microelectrode array to record movement trajectory intention in the posterior parietal cortex and enable a person with tetraplegia to control a robotic arm.

Despite BCI advances enabling many motor functions as above, the means by which feedback is received from these prosthetic limbs are still primitive. Real-life movement invariably involves continuous interaction with external objects and the environment and therefore sensory feedback is critical. For example, this is important in the case of grasping a delicate object, where sensory feedback is necessary to guide the grip strength to hold the object without crushing it. Another example is during ambulation, in which continuous knowledge of feet and leg position is necessary to maintain balance. The loss of somatosensation is known to cause deficits in motor control [43–45]. Moreover, the theory of optimal feedback control [46] corroborates that humans rely on cost and rewards [47, 48], internal models [49, 50], optimal feedback-driven policy [51], and state estimation [51, 52], all of which demand somatosensory feedback as a crucial component of normal motor control. Therefore, the important challenge for BCI development is to realize a BD-BCI system with the capability to convey sensory information back to the brain.

To date, there have been reports of BD-BCIs using either ECoG-based sensory stimulation or intracortical microstimulation (ICMS). Evoked sensory percepts via ECoG electrodes placed over the cortical surface have been evaluated as a viable feedback interface [20, 21, 53] which can potentially guide control of the prostheses.

A number of BD-BCIs utilizing ICMS-driven sensory feedback have demonstrated improved prosthetic arm motor control in grasping and transporting objects [22, 54] compared to when stimulation was disabled. This sheds light on the prospect that mimicking known biological control principles could result in task performance approaching that of the able-bodied human. However, existing BD-BCI systems operate in a constrained laboratory setting, since bulky, non-mobile workstation computers, data acquisition systems, and commercial stimulators prevent the untethered, mobile, everyday use of these systems. Practical implementation of BD-BCIs hinges on the integration of the above components into a special purpose and compact form factor with full programmability. Specifically, an envisioned breakthrough in BD-BCI development would likely come in the form of an embedded system small enough to enable a fully implantable BD-BCI that simultaneously restores limb movement and sensation in persons with neurological injuries.

It is envisioned that the function of the aforementioned systems can be extended to those with hemiplegia due to stroke. For example, distal upper extremity weakness is a common clinical outcome of stroke, and systems such as those by Pfurtscheller et al. [35], may help restore hand movement in stroke survivors. In addition, since as many as 15% of stroke survivors lose their ability to ambulate, BCI-controlled lower extremity prostheses could help restore walking. At the time of publication, there are no BCI-controlled neuroprostheses that have undergone definitive clinical trials for safety and efficacy. In addition, none have been FDA approved for marketing, and hence no clinical recommendations can be made regarding the application of neuroprosthetic BCIs for stroke. Before these devices reach the point where they can be widely used and adopted, several outstanding problems must be addressed. First, these systems are not yet sufficiently accurate to robustly restore movement to paralyzed limbs. While some literature reports that a performance accuracy of 70% may be sufficient for a BCI user to feel that they have reliable control [55–58], even the most accurate BCIs (those that differentiate between moving and idling states) can only reach 95% accuracy, which may still translate into operation errors that are potentially frustrating or dangerous to the user.

Second, they typically require full-sized computers and bulky amplifier arrays, which limit their portability. Significant engineering effort will need to be invested in order to miniaturize the requisite electronic components such that they are wearable, esthetically acceptable, as well as constantly available, and easy to use. Lastly, non-invasive systems require EEG caps which are tedious and time-consuming to don and doff. In order to address these challenges, it may be necessary to develop implantable BCI systems, including electrodes, amplifiers, and special-purpose microcomputers. Ultimately, clinical trials will need to be conducted to determine whether these systems are safe (e.g., do not cause seizures or nervous tissue injuries) and effective (reduce disability).

22.4 BCI Systems for Physiotherapy

22.4.1 Review of Existing BCI Systems for Stroke Rehabilitation and Underlying Mechanisms

The general consensus in stroke motor rehabilitation is that the most effective practices employ repetitive, high-intensity, goal-oriented movement of the impaired limb (such as constraint-induced movement therapy) to overcome learned disuse [59]. Additionally, it is recommended that patients execute these movements as naturally as possible [60, 61]. However, severely disabled individuals may be unable to participate in active movement therapies, and hence BCIs may be applied as novel therapies to facilitate compliance with these rehabilitative guidelines. This is supported by the work of Vouvopoulos et al. [62], which demonstrated that individuals with more severe motor impairments may benefit most from EEG-based neurofeedback compared to EMG-based feedback.

While it can be hypothesized that BCI therapy, when used in conjunction with conventional physiotherapies, may also improve motor function in those with moderate or mild impairment due to stroke, these individuals with largely intact sensorimotor pathways may derive the most benefit from EMG-based feedback [62]. Stroke survivors are still able to modulate EEG without performing any physical movement [60, 63–68], and this can be exploited by BCIs for stroke rehabilitation. Moreover, MI- and ME-based BCI therapies can be carried out in a repetitive [66] and goal-oriented [59] manner that ensures intense focus on the motor function task [60, 64, 69, 70]. These BCIs may facilitate neuroplastic cortical changes similar to repetitive movement practice [61], possibly through operant conditioning. For example, the BCI output could provide feedback to the user about his/her cortical state, and the user could then attempt to modulate his/her SMRs to achieve maximum control of the BCI. This learning process may lead to subsequent beneficial neural changes [63, 67–69, 71–74] and, in turn, to improved motor function (see Fig. 22.3). For example, MI and ME have been shown to strengthen visuospatial [75], primary/associated motor [61, 66, 74–76], and primary somatosensory [60] networks.

Many previous studies have demonstrated that MI and ME generate robust changes in EEG signals that are suitable for BCIs, even in the post-stroke cortex, making BCI-based stroke rehabilitation a possibility. Mohapp et al. [77] used MI and ME on 10 stroke patients with an average BCI classification accuracy between 61.5 and 79.0% (depending on which limb and hemisphere were used). Bai et al. [78] investigated both MI and ME tasks with a BCI that utilized beta-rhythm SMR and found that, without extensive training, the classification accuracy in stroke subjects was comparable to healthy subjects ($\geq 80\%$). Buch et al. [79] reported successful control of a MEG-based BCI by eight stroke patients who utilized mu-rhythm ERD during both ME and MI tasks to control an orthosis attached to the plegic hand. Six out of the eight patients achieved a significant increase in BCI classification accuracy by the 20th session, with a median accuracy of 72.48% across subjects at the final session. Prasad et al. [80] evaluated five chronic stroke patients undergoing combined physical practice and MI-based BCI and found that BCI classification accuracy was 70% on average. McCrimmon et al. [81] investigated an ME-based FES therapy in nine chronic stroke patients and found that the average classification accuracy for these subjects was 80%. From these studies, it is clear that stroke survivors can successfully control MI- and ME-based BCIs. However, the true relationship between the features used in MI- and ME-based BCIs and functional motor recovery remains

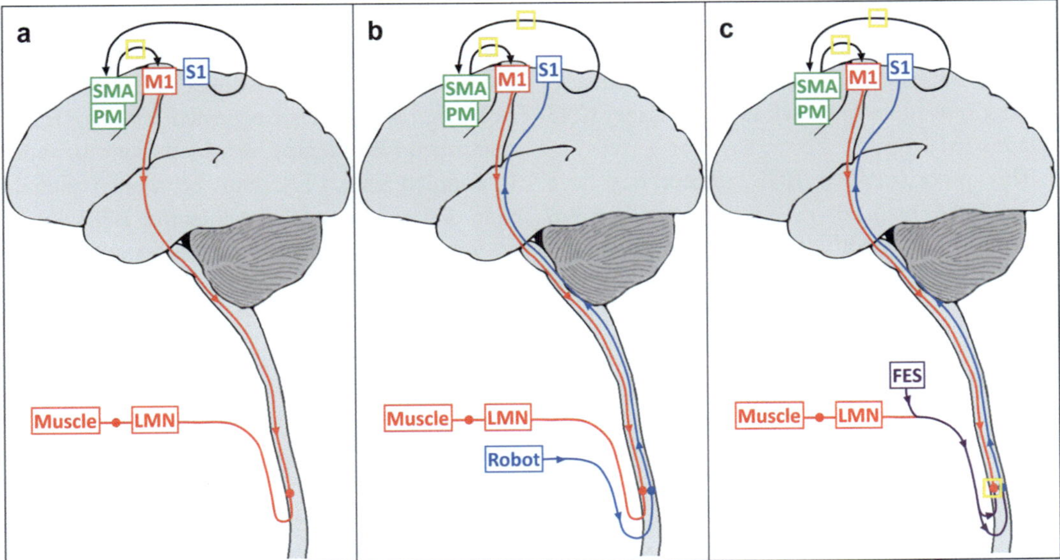

Fig. 22.3 Hypothesized mechanisms of post-stroke motor recovery using BCI systems. Here, the lower motor neuron (LMN) output and subsequently the muscle output are severely impaired. **a** MI- and ME-based BCIs may elicit cortical changes (yellow square) between the primary motor cortex (M1) and the supplementary motor area (SMA), premotor cortex (PM), and even the prefrontal and posterior parietal areas (not shown). **b** MI- and ME-based BCIs that provide robotic assistance may additionally stimulate dorsal sensory pathways and subsequently facilitate neuroplastic changes between the primary somatosensory cortex (S1) and the non-primary motor areas. **c** BCIs that deliver MI- or ME-controlled FES may also promote changes in the anterior horn of the spinal cord at the level of the antidromically activated LMN

elusive, as studies such as [82] suggest that robust motor recovery can occur even without changes in these neurophysiological features.

Adding a proprioceptive feedback mechanism to MI- and ME-based BCIs may further enhance functional recovery in stroke survivors (see Fig. 22.3). More specifically, proprioceptive BCIs pair motor intention (motor and visuomotor activation) with the movement of the paretic limb (e.g., through robotic assistance). This may facilitate Hebbian-like learning and neural reorganization [61, 66, 67, 83] and ultimately improve motor recovery [68, 69, 84]. It is likely that these plastic changes occur at the level of the cortex and primarily affect motor planning and initiation since synaptic changes directly between upper motor neurons and sensory fibers are unlikely to occur. Lau et al. [85] observed that BCI coupled with proprioceptive robotic hand feedback promoted changes in the ipsilesional and contralesional cortices (such as the primary motor, pre-motor, supplementary motor, and parietal areas) and that these changes were endured for at least 6 months. Additionally, any improvement in motor function results in a subsequent increase in proprioceptive feedback, creating a positive feedback loop of further CNS changes [61].

This proprioceptive BCI concept has been successfully realized in several studies. For example, Broetz et al. [72] and Caria et al. [83] trained a hemiplegic patient with no active finger extension with a BCI that drove an orthosis attached to his paralyzed arm. The patient used mu-rhythm modulation to control the orthosis and underwent goal-directed physiotherapy training over the course of one year. Gomez-Rodriguez et al. [84] evaluated BCI-robotic arm-assisted physiotherapy in three chronic stroke patients. Patients attempted either elbow flexion or extension or MI, while the BCI would detect their intention to move and then initiate active robotic assistance. Ang et al. [86] studied the effect of MI-based BCI with haptic feedback in 21 chronic stroke patients in a controlled trial. Subjects participated in 18 therapy sessions in which grasping and knob manipulation tasks were carried out using MI-BCI with robotic

assistance. A recent study by Ramos-Murguialday et al. [87] showed that using an MI-based BCI with sham robotic feedback (that was not based on the user's EEG) led to significantly poorer hand motor recovery after stroke compared to EEG- triggered robotic feedback, both immediately after and 6 months after the intervention. Note that all of the above studies were conducted using non-invasive EEG-based BCIs.

In addition to delivering proprioceptive feedback, BCI can also control functional electrical stimulation (FES) systems as another potential means to drive neuronal plasticity processes (see Fig. 22.3). Using FES with MI- and ME-based BCIs not only activates afferent sensory pathways but also lower motor neurons. Compared to proprioceptive feedback alone, this mechanism may further enhance neural plastic changes, especially in sensorimotor areas [74, 76, 88, 89]. Specifically, the coincident activation of upper and lower motor neurons may induce Hebbian learning via long-term potentiation at their synapse in the spinal cord [89, 90]. Initial evidence from Hara et al. [91] supports the use of therapies that coactivate upper and lower motor neurons. Here, it is suggested that an EMG-controlled FES therapy, in which motor intention is coupled with FES, may be more beneficial than either attempted movement or FES alone.

Several preliminary studies have demonstrated the feasibility of BCI-FES systems for physiotherapy. Daly et al. [60] combined BCI and FES for motor learning in a single chronic stroke patient who was unable to perform isolated finger movements. Visual cues were provided to the subject to relax or move her paretic fingers, and ME- or MI-based motor intention (via the BCI) triggered FES-induced index finger extension. After a small number of training sessions, volitional motor control over the index finger was obtained. McCrimmon et al. [81] utilized a similar paradigm for lower extremity rehabilitation. Nine chronic stroke subjects with foot drop each participated in 12 sessions, in which they followed visual cues and attempted either ankle dorsiflexion or relaxation, while FES was either supplied or withheld, respectively.

Gait function improved in several subjects (details discussed further below). More recently, Lee et al. [92] showed that BCI-FES promoted significant changes in sensorimotor rhythms and functional recovery of the upper extremities after stroke compared to conventional physical therapy. Chung et al. [93] showed that BCI-FES promoted significantly increased functional recovery of gait compared to FES alone (where both therapies were given over an identical duration). In addition, Biasiucci et al. [94] demonstrated that sham-FES, which avoids coactivation of upper and lower motor neurons, is significantly worse than BCI-FES at promoting recovery, and may even promote maladaptive behavior given a decrease in the Modified Ashworth Scale from baseline in these subjects.

22.4.2 BCI-Based Stroke Physiotherapy in Clinical Applications

Recently, there have been an increasing number of interventional clinical studies that examined the feasibility and efficacy of BCI-based therapies for stroke rehabilitation. At the time of this review, clinical trials in various stages of completion have been reported. Studied modalities include pairing EEG-based BCI with robotic orthoses [69, 86, 87, 95–99], exoskeletons [100], FES [81, 93, 95, 101–103], repetitive transcranial magnetic stimulation (rTMS) [104], animations on a computer screen [80, 105], or virtual reality [104], transcranial direct current stimulation (tDCS) [82, 106] or action observation therapy (AOT) [92, 107] for feedback or treatment. Targeted body areas typically are the upper extremities (e.g., finger/wrist extension and upper arm movements) and lower extremities (e.g., ankle dorsiflexion and gait). For these studies, outcome measures typically include Fugl–Meyer Assessment (FMA), Action Research Arm Test (ARAT), Barthel Index (BI, MBI), Wolf Motor Function Test (WMFT), various strength and range of motion (ROM) assessments, various brain connectivity outcomes, and various patient-reported outcome measures.

A representative list of Phases 0 and I (safety and feasibility) trials can be found in Table 22.1. At this stage of a clinical trial, a small number of participants (both stroke and able-bodied) are recruited to ensure that the BCI device and regimen produce no harmful effects and that improvement of a measurable outcome is at least feasible. Results from Phases 0 and I trials are typically used to estimate the effect size and refine the study protocol when designing larger confirmatory clinical trials.

A representative list of Phase II (efficacy) randomized controlled trials (RCT) can be found in Table 22.2. This list only includes studies that have at least 20 participants, and only includes stroke subjects (studies involving able-bodied participants as control were excluded). Studies split participants into experimental and control arms in various ways, such as BCI versus conventional physiotherapy, or BCI-controlled robotic orthosis versus conventionally

Table 22.1 Samples of Phase 0/I (safety and feasibility) and non-controlled studies with at least five participants. Studies already in Table 22.2 are not listed again here. NMES = Neuromuscular electrical stimulation

Study	Feedback	Targeted Limb
Kawakami et al. [95]	Robotic orthosis and NMES	Hand
Cantillo-Negrete et al. [96]	Robotic orthosis	Hand
McCrimmon et al. [81]	FES	Foot
Jang et al. [101]	FES	Shoulder
Sebastian-Romagosa et al. [103]	FES	Hand
Hong et al. [106]	tDCS and MIT-Manus	Hand
Hu et al. [82]	tDCS and MIT-Manus	Hand
Prasad et al. [80]	Computer screen	Hand

Table 22.2 Some published Phase II (efficacy) randomized controlled clinical studies testing the use of closed-loop BCI for stroke rehabilitation. 'nmFU' = n-month Follow-up. FMA = Fugl–Meyer Motor Assessment. '*' denotes a significant efficacious outcome for BCI. '**' denotes significant BCI outperformance over conventional therapy methods. For this claim to be valid, the trial must test the BCI treatment group against a non-BCI treatment group for equal treatment time, such as a 1-h BCI and 1-h non-BCI versus 2-h non-BCI. '†' denotes an ongoing clinical trial with incomplete data analysis. MAL = Motor activity log. MBI = Modified Barthel index. ROM = range of motion. WMFT = Wolf motor function test. MI = Motricity Index of the arm. Dyn = Dynamometer. SIS = Stroke impact scale

Study	Feedback	Targeted Limb	Sample Size	Primary Outcome Measures
Ang et al. [86]	Haptic knob robot	Hand and wrist	21	FMA**
Ang et al. [97]	MIT-Manus robot	Elbow and forearm	26	FMA*
Frolov et al. [98]	Hand exoskeleton	Hand	74	FMA*, ARAT**
Lyukmanov et al. [100]	Hand exoskeleton	Hand	55	FMA*, ARAT**
Mizuno et al. [99]	Robotic orthosis	Hand	40†	FMA
Ramos-Murguialday et al. [69]	Robotic orthosis	Hand	32	FMA*
Ramos-Murguialday et al. [87]	Robotic orthosis	Hand	30 (6mFU)	FMA*
Remsik et al. [102]	FES	Hand	21	ARAT**
Chung et al. [93]	FES	Foot	25	Gait velocity**, Cadence**
Kim et al. [107]	AOT and FES	Hand and wrist	30	FMA*, MAL*, MBI*, ROM*
Lee et al. [92]	AOT and FES	Hand and wrist	26	FMA**, WMFT**, MAL*, MBI*
Sanchez-Cuesta et al. [104]	VR and rTMS	Hand, wrist, arm	42†	MI, FMA, Dyn, SIS
Mattia et al. [105]	Computer screen	Hand	48†	FMA

controlled robotic orthosis. A BCI therapy regimen would be considered efficacious if its primary outcome measure is at least comparable to conventional physiotherapy. All BCIs reported in Table 22.2, other than those in ongoing trials, are efficacious. In addition, BCI was significantly better than conventional physiotherapy for at least one of the outcome measures in six controlled studies [86, 92, 93, 98, 100, 102]. Also, Ramos-Murguialday et al. [87] report that participants treated with BCI-controlled robotic orthosis retained positive outcomes at a similar level to those in the conventional physiotherapy control group after 6 months. Finally, no Phase III (effectiveness) trials or studies involving invasive brain signal acquisition for rehabilitation purposes have been reported at the time of this review. It should be noted that some invasive trials exist for prosthetic applications as discussed in the section above.

Given the absence of any definitive Phase III clinical trials regarding the effectiveness of BCI-based physiotherapy, it is currently not possible to determine whether any of these approaches can be recommended for the stroke population at large. There are several factors that may contribute to why there are still no Phase III studies for BCI therapy despite the multitude of successful Phase II trials (Table 22.2). These include the prohibitive cost of the BCI systems as well as extensive training and setup time. Although not prohibitive, a lack of understanding of the mechanisms underlying BCI therapy may also hinder optimal delivery of the BCI therapy. Several studies have received funding from national agencies such as the National Institute of Health and the National Science Foundation in the United States (Table 22.3) to address these outstanding concerns. Some aim to simplify the hardware and regimen for at-home rehabilitation [108, 109] and long-term use [109]. Some studies aim to elucidate the underlying mechanism of any improvements seen in stroke patients who underwent BCI-based physiotherapies [110].

In summary, the current BCI clinical trial literature provides evidence that BCI-based

Table 22.3 Some ongoing BCI-based stroke rehabilitation studies currently funded by national agencies in the United States. NIH = National Institute of Health. VA = Veteran Affairs. NSF = National Science Foundation

Principal investigator	Targeted areas	Active years and funding agency
K Bhugra (Neurolutions Inc.) [108]	At home, hand rehabilitation	2021–2022 by NIH
DJ Lin (Providence VA) [111]	Arm rehabilitation	2020–2022 by VA
A Do (Univ. of California Irvine) [110]	Gait rehabilitation with FES	2019–2024 by NIH
J Contreras-Vidal (Univ. of Houston) [109]	Long-term, at-home, robot-assisted	2018–2022 by NSF

physiotherapies may be safe and promising enough to warrant large-scale clinical investigations. Simultaneously, fundamental questions still need to be answered, such as understanding the mechanism of action and how brain physiology change over the course of treatment. Furthermore, practical questions will need to be answered. Are there particular characteristics of stroke patients who will respond best to BCI-based physiotherapy? What kind of closed-loop feedback is the best? How can BCI be incorporated into at-home rehabilitation? The answers to these questions have implications for the future justification of BCI-based therapies, particularly if existing dose-matched therapies turn out to be cheaper and just as efficacious.

In addition to the above scientific questions, many practical issues related to the implementation of BCI-based physiotherapies need to be addressed. Since these therapies currently require extensive setup, it is unclear how they will be efficiently and effectively delivered in clinical practice and at home. Are there ways to drastically reduce the setup time of such BCI systems? Additionally, will BCI-based therapies be provided by the physical and occupational therapists in the community? Will they be time and resource-efficient? Will the associated equipment and training costs be acceptable to practicing clinicians? Will patients be interested in such therapies? Will medical insurance providers reimburse or support their clinical use? Since this research field is still in early development, the medical device industry has yet to streamline these systems. As the field matures, it can be expected that BCI-based physiotherapy studies will transition towards large clinical trials. At that time, significant research and development must be performed to understand and address these market issues that may ultimately affect the success of BCI-based physiotherapies.

22.5 Conclusion and Future Directions

In recent years, BCIs have garnered increasing interest as a means of substituting for lost motor functions or for improving post-stroke motor outcomes. Neuroprosthetic BCIs have been designed primarily for SCI, but can be extended to post-stroke paralysis. Significant engineering challenges must still be overcome before these systems can be used in the clinic in a robust and practical manner. BCI systems for physiotherapy may be applied as a novel means of facilitating Hebbian learning mechanisms, which can be elicited by two major strategies. One involves providing BCI-controlled proprioceptive sensory feedback to upregulate the connection between sensory and motor cortices and subsequently cause increased motor output to the lower motor neurons. The second strategy employs BCI-controlled electrical stimulation to simultaneously activate the post-stroke motor areas and the lower motor neurons, thereby increasing their connectivity over time. Both of these strategies can potentially promote motor recovery. In fact, these strategies have already been realized in BCI-controlled robotic and FES therapies. Existing early phase clinical trials suggest that these strategies are promising. However, definitive clinical trials still need to be performed, and many questions still remain regarding the safety and efficacy of BCI-based physiotherapies and whether they can be practically applied in the clinical setting.

References

1. Roger VL, Go AS, Lloyd-Jones DM, Benjamin EJ, Berry JD, Borden WB, et al. Heart disease and stroke statistics–2012 update: a report from the American Heart Association. Circulation. 2012;125 (1):e2–220.

2. Virani SS, Alonso A, Aparicio HJ, Benjamin EJ, Bittencourt MS, Callaway CW, et al. Heart disease and stroke statistics–2021 update: a report from the American Heart Association. Circulation. 2021;143 (8):e254–743.

3. Duncan PW, Sullivan KJ, Behrman AL, Azen SP, Wu SS, Nadeau SE, et al. Body-weight-supported treadmill rehabilitation after stroke. N Engl J Med. 2011;364(21):2026–36.

4. Jørgensen HS, Nakayama H, Raaschou HO, Vive-Larsen J, Støier M, Olsen TS. Outcome and time course of recovery in stroke. Part I: outcome. The Copenhagen stroke study. Arch Phys Med Rehabil. 1995;76(5):399–405.

5. Nakayama H, Jørgensen H, Raaschou H, Olsen T. Recovery of upper extremity function in stroke patients: the copenhagen stroke study. Arch Phys Med Rehabil. 1994;75(4):394–8.

6. Persson HC, Parziali M, Danielsson A, Sunnerhagen KS. Outcome and upper extremity function within 72 hours after first occasion of stroke in an unselected population at a stroke unit. A part of the SALGOT study. BMC Neurol. 2012;12(1):1–6.

7. Hendricks HT, van Limbeek J, Geurts AC, Zwarts MJ. Motor recovery after stroke: a systematic review of the literature. Arch Phys Med Rehabil. 2002;83(11):1629–37.

8. Menezes K, Nascimento LR, Pinheiro MB, Scianni AA, Faria C, Avelino P, et al. Lower limb motor coordination is significantly impaired in ambulatory people with chronic stroke: a cross-sectional study. J Rehabil Med. 2017;49:322–6.

9. Lord SE, McPherson K, McNaughton HK, Rochester L, Weatherall M. Community ambulation after stroke: how important and obtainable is it and what measures appear predictive? Arch Phys Med Rehabil. 2004;85(2):234–9.

10. Perry J, Garrett M, Gronley JK, Mulroy SJ. Classification of walking handicap in the stroke population. Stroke. 1995;26(6):982–9.

11. Ovbiagele B, Goldstein LB, Higashida RT, Howard VJ, Johnston SC, Khavjou OA, et al. Forecasting the future of stroke in the United States: a policy statement from the American Heart Association and American Stroke Association. Stroke. 2013;44(8):2361–75.

12. Lo AC, Guarino PD, Richards LG, Haselkorn JK, Wittenberg GF, Federman DG, et al. Robot-assisted therapy for long-term upper-limb impairment after stroke. N Engl J Med. 2010;362(19):1772–83.

13. Dawson J, Liu CY, Francisco GE, Cramer SC, Wolf SL, Dixit A, et al. Vagus nerve stimulation paired with rehabilitation for upper limb motor function after ischaemic stroke (VNS-REHAB): a randomised, blinded, pivotal, device trial. The Lancet. 2021;397(10284):1545–53.

14. Wolpaw JR, Birbaumer N, McFarland DJ, Pfurtscheller G, Vaughan TM. Brain-computer interfaces for communication and control. Clin Neurophysiol. 2002;113(6):767–91.

15. Lupu RG, Irimia DC, Ungureanu F, Poboroniuc MS, Moldoveanu A. BCI and FES based therapy for stroke rehabilitation using VR facilities. Wirel Commun Mob Comput. 2018;2018.

16. Lupu RG, Ungureanu F, Ferche O, Moldoveanu A. Neuromotor recovery based on BCI, FES, virtual reality and augmented feedback for upper limbs. In: Brain–computer interface research. Springer; 2020. p. 75–85.

17. Pfurtscheller G, Müller-Putz GR, Pfurtscheller J, Rupp R. EEG-based asynchronous BCI controls functional electrical stimulation in a tetraplegic patient. EURASIP J Adv Signal Process. 2005;2005 (19):1–4.

18. Hochberg LR, Bacher D, Jarosiewicz B, Masse NY, Simeral JD, Vogel J, et al. Reach and grasp by people with tetraplegia using a neurally controlled robotic arm. Nature. 2012;485(7398):372–5.

19. Baniqued PDE, Stanyer EC, Awais M, Alazmani A, Jackson AE, Mon-Williams MA, et al. Brain–computer interface robotics for hand rehabilitation after stroke: a systematic review. J NeuroEngineering Rehabil. 2021;18(1):1–25.

20. Hiremath SV, Tyler-Kabara EC, Wheeler JJ, Moran DW, Gaunt RA, Collinger JL, et al. Human perception of electrical stimulation on the surface of somatosensory cortex. PLoS ONE. 2017;12(5): e0176020.

21. Lee B, Kramer D, Armenta Salas M, Kellis S, Brown D, Dobreva T, et al. Engineering artificial somatosensation through cortical stimulation in humans. Front Syst Neurosci. 2018;12:24.

22. Flesher SN, Downey JE, Weiss JM, Hughes CL, Herrera AJ, Tyler-Kabara EC, et al. A brain-computer interface that evokes tactile sensations improves robotic arm control. Science. 2021;372 (6544):831–6.

23. Dimyan MA, Cohen LG. Neuroplasticity in the context of motor rehabilitation after stroke. Nat Rev Neurol. 2011;7(2):76–85.

24. Pfurtscheller G, Neuper C. Future prospects of ERD/ERS in the context of brain-computer interface (BCI) developments. Prog Brain Res. 2006;159:433–7.

25. Miller K, Zanos S, Fetz E, Den Nijs M, Ojemann J. Decoupling the cortical power spectrum

reveals real-time representation of individual finger movements in humans. J Neurosci. 2009;29 (10):3132–7.

26. Schalk G, Kubanek J, Miller K, Anderson N, Leuthardt E, Ojemann J, et al. Decoding two-dimensional movement trajectories using electrocorticographic signals in humans. J Neural Eng. 2007;4(3):264.

27. Wang PT, McCrimmon CM, King CE, Shaw SJ, Millett DE, Gong H, et al. Characterization of electrocorticogram high-gamma signal in response to varying upper extremity movement velocity. Brain Struct Funct. 2017;222(8):3705–48.

28. Georgopoulos AP, Kalaska JF, Caminiti R, Massey JT. On the relations between the direction of two-dimensional arm movements and cell discharge in primate motor cortex. J Neurosci. 1982;2 (11):1527–37.

29. Georgopoulos AP, Schwartz AB, Kettner RE. Neuronal population coding of movement direction. Science. 1986;233(4771):1416–9.

30. Scherberger H, Jarvis MR, Andersen RA. Cortical local field potential encodes movement intentions in the posterior parietal cortex. Neuron. 2005;46 (2):347–54.

31. Wu W, Black MJ, Gao Y, Bienenstock E, Serruya M, Shaikhouni A, et al. Neural decoding of cursor motion using a Kalman filter. Adv Neural Inf Process Syst. 2003;133–40.

32. Zhang X, Yao L, Wang X, Monaghan J, Mcalpine D, Zhang Y. A survey on deep learning-based non-invasive brain signals: recent advances and new frontiers. J Neural Eng. 2021;18 (3):031002.

33. Parvizi J, Kastner S. Promises and limitations of human intracranial electroencephalography. Nat Neurosci. 2018;21(4):474–83.

34. Hochberg LR, Serruya MD, Friehs GM, Mukand JA, Saleh M, Caplan AH, et al. Neuronal ensemble control of prosthetic devices by a human with tetraplegia. Nature. 2006;442(7099):164–71.

35. Pfurtscheller G, Mü ller GR, Pfurtscheller J, Gerner HJ, Rupp R. Thought–control of functional electrical stimulation to restore hand grasp in a patient with tetraplegia. Neurosci Lett. 2003;351 (1):33–6.

36. Do AH, Wang PT, King CE, Chun SN, Nenadic Z. Brain-computer interface controlled robotic gait orthosis. J Neuroeng Rehabil. 2013;10(1):111.

37. King CE, Wang PT, McCrimmon CM, Chou CCY, Do AH, Nenadic Z. The feasibility of a brain-computer interface functional electrical stimulation system for the restoration of overground walking after paraplegia. J Neuroeng Rehabil. 2015;12 (80):1–11.

38. Wang PT, King CE, McCrimmon CM, Lin JJ, Sazgar M, Hsu FP, et al. Comparison of decoding resolution of standard and high-density electrocorticogram electrodes. J Neural Eng. 2016;13 (2):026016.

39. Wang W, Collinger JL, Degenhart AD, Tyler-Kabara EC, Schwartz AB, Moran DW, et al. An electro-corticographic brain interface in an individual with tetraplegia. PLoS ONE. 2013;8(2):e55344.

40. Wang PT, Camacho E, Wang M, Li Y, Shaw SJ, Armacost M, et al. A benchtop system to assess the feasibility of a fully independent and implantable brain-machine interface. J Neural Eng. 2019;16 (6):066043.

41. Collinger JL, Wodlinger B, Downey JE, Wang W, Tyler-Kabara EC, Weber DJ, et al. High-performance neuroprosthetic control by an individual with tetraplegia. Lancet. 2013;381(9866):557–64.

42. Aflalo T, Kellis S, Klaes C, Lee B, Shi Y, Pejsa K, et al. Decoding motor imagery from the posterior parietal cortex of a tetraplegic human. Science. 2015;348(6237):906–10.

43. Rothwell J, Traub M, Day B, Obeso J, Thomas P, Marsden C. Manual motor performance in a deafferented man. Brain. 1982;105(3):515–42.

44. Sainburg RL, Poizner H, Ghez C. Loss of proprioception produces deficits in interjoint coordination. J Neurophysiol. 1993;70(5):2136–47.

45. Gordon J, Ghilardi MF, Ghez C. Impairments of reaching movements in patients without proprioception. I. Spatial errors. J Neurophysiol. 1995;73 (1):347–60.

46. Wolpert DM. Computational approaches to motor control. Trends Cogn Sci. 1997;1(6):209–16.

47. Todorov E, Jordan MI. Optimal feedback control as a theory of motor coordination. Nat Neurosci. 2002;5(11):1226–35.

48. O'Sullivan I, Burdet E, Diedrichsen J. Dissociating variability and effort as determinants of coordination. PLoS Comput Biol. 2009;5(4): e1000345.

49. Kawato M. Internal models for motor control and trajectory planning. Curr Opin Neurobiol. 1999;9 (6):718–27.

50. Shadmehr R, Mussa-Ivaldi FA. Adaptive representation of dynamics during learning of a motor task. J Neurosci. 1994;14(5):3208–24.

51. Körding K. Decision theory: what "should" the nervous system do? Science. 2007;318(5850):606–10.

52. Ernst MO, Banks MS. Humans integrate visual and haptic information in a statistically optimal fashion. Nature. 2002;415(6870):429–33.

53. Cronin JA, Wu J, Collins KL, Sarma D, Rao RP, Ojemann JG, et al. Task-specific somatosensory feedback via cortical stimulation in humans. IEEE Trans Haptics. 2016;9(4):515–22.

54. Raspopovic S, Capogrosso M, Petrini FM, Bonizzato M, Rigosa J, Di Pino G, et al. Restoring natural sensory feedback in real-time bidirectional hand prostheses. Sci Transl Med. 2014;6(222):222ra19–222ra19.

55. Shu X, Chen S, Yao L, Sheng X, Zhang D, Jiang N, et al. Fast recognition of BCI-inefficient users using physiological features from EEG signals: a

screening study of stroke patients. Front Neurosci. 2018;12:93.

56. Kübler A, Neumann N, Wilhelm B, Hinterberger T, Birbaumer N. Predictability of brain- computer communication. J Psychophysiol. 2004;18 (2/3):121–9.

57. McCane LM, Sellers EW, McFarland DJ, Mak JN, Carmack CS, Zeitlin D, et al. Brain-computer interface (BCI) evaluation in people with amyotrophic lateral sclerosis. Amyotroph Lateral Scler Frontotemporal Degener. 2014;15(3–4):207–15.

58. Thompson MC. Critiquing the concept of BCI illiteracy. Sci Eng Ethics. 2019;25(4):1217–33.

59. Langhorne P, Coupar F, Pollock A. Motor recovery after stroke: a systematic review. Lancet Neurol. 2009;8(8):741–54.

60. Daly JJ, Cheng R, Rogers J, Litinas K, Hrovat K, Dohring M. Feasibility of a new application of noninvasive Brain Computer Interface (BCI): a case study of training for recovery of volitional motor control after stroke. J Neurol Phys Ther. 2009;33 (4):203–211. Available from: https://doi.org/10. 1097/NPT.0b013e3181c1fc0b.

61. Daly JJ, Wolpaw JR. Brain–computer interfaces in neurological rehabilitation. Lancet Neurol. 2008;7 (11):1032–43.

62. Vourvopoulos A, Pardo OM, Lefebvre S, Neureither M, Saldana D, Jahng E, et al. Effects of a brain-computer interface with virtual reality (Vr) neurofeedback: a pilot study in chronic stroke patients. Front Hum Neurosci. 2019;13:210. Available from: https://www.frontiersin.org/article/10.3389/fnhum. 2019.00210/full.

63. Pfurtscheller G, Muller-Putz GR, Scherer R, Neuper C. Rehabilitation with brain-computer interface systems. Computer. 2008;41(10):58–65.

64. Ang KK, Guan C, Chua KSG, Ang BT, Kuah C, Wang C, et al. A clinical study of motor imagery-based brain-computer interface for upper limb robotic rehabilitation. In: Proceedings of 31st annual international conference of the engineering in medicine and biology society. IEEE; 2009. p. 5981–4.

65. Teasell RW, Kalra L. Advances in stroke 2003-what's new in stroke rehabilitation. Stroke. 2004;35 (2):383–5.

66. Belda-Lois JM, Mena-del Horno S, Bermejo-Bosch I, Moreno JC, Pons JL, Farina D, et al. Rehabilitation of gait after stroke: a review towards a top-down approach. J Neuroeng Rehabil. 2011;8(1):66.

67. Ramos-Murguialday A, Schürholz M, Caggiano V, Wildgruber M, Caria A, Hammer EM, et al. Proprioceptive feedback and Brain Computer Interface (BCI) based neuroprostheses. PLoS One. 2012;7 (10):e47048.

68. Ang KK, Guan C. Brain-computer interface in stroke rehabilitation. J Comput Sci Eng. 2013;7 (2):139–46.

69. Ramos-Murguialday A, Broetz D, Rea M, Läer L, Yilmaz O, Brasil FL, et al. Brain-machine interface in chronic stroke rehabilitation: a controlled study. Ann Neurol. 2013;74(1):100–8.

70. Wang C, Phua KS, Ang KK, Guan C, Zhang H, Lin R, et al. A feasibility study of non-invasive motor- imagery BCI-based robotic rehabilitation for Stroke patients. In: Proceedings of 31st annual international conference of the engineering in medicine and biology society. IEEE; 2009. p. 271–4.

71. Kaiser V, Kreilinger A, Müller-Putz GR, Neuper C. First steps toward a motor imagery based stroke BCI: new strategy to set up a classifier. Front Neurosci. 2011;5:86.

72. Broetz D, Braun C, Weber C, Soekadar SR, Caria A, Birbaumer N. Combination of brain-computer interface training and goal-directed physical therapy in chronic stroke: a case report. Neurorehab Neural Re. 2010;24:674.

73. Takeuchi N, Izumi SI. Rehabilitation with post-stroke motor recovery: a review with a focus on neural plasticity. Stroke Res Treat. 2013;128641.

74. Meng F, Tong KY, Chan ST, Wong WW, Lui KH, Tang KW, et al. BCI-FES training system design and implementation for rehabilitation of stroke patients. In: Proceedings of international joint conference on neural networks. IEEE; 2008. p. 4103–6.

75. Varkuti B, Guan C, Pan Y, Phua KS, Ang KK, Kuah CWK, et al. Resting state changes in functional connectivity correlate with movement recovery for BCI and robot-assisted upper-extremity training after stroke. Neurorehab Neural Repair. 2013;27(1):53–62.

76. Saito K, Sugawara K, Miyaguchi S, Matsumoto T, Kirimoto H, Tamaki H, et al. The modulatory effect of electrical stimulation on the excitability of the corticospinal tract varies according to the type of muscle contraction being performed. Front Hum Neurosci. 2014;8:835.

77. Mohapp A, Scherer R, Keinrath C, Grieshofer P, Pfurtscheller G, Neuper C. Single-trial EEG classification of executed and imagined hand movements in hemiparetic stroke patients. In: Proceedings of international BCI workshop & training course; 2006.

78. Bai O, Lin P, Vorbach S, Floeter MK, Hattori N, Hallett M. A high performance sensorimotor beta rhythm-based brain-computer interface associated with human natural motor behavior. J Neural Eng. 2007;5(1):24–35.

79. Buch E, Weber C, Cohen LG, Braun C, Dimyan MA, Ard T, et al. Think to move: a neuromagnetic brain-computer interface (BCI) system for chronic stroke. Stroke. 2008;39(3):910–7.

80. Prasad G, Herman P, Coyle D, McDonough S, Crosbie J. Applying a brain-computer interface to support motor imagery practice in people with stroke for upper limb recovery: a feasibility study. J Neuroeng Rehabil. 2010;7:60. Available from: https://doi.org/10.1186/1743-0003-7-60.

81. McCrimmon CM, King CE, Wang PT, Cramer SC, Nenadic Z, Do AH. Brain-controlled functional electrical stimulation therapy for gait rehabilitation after stroke: a safety study. J Neuroeng Rehabil. 2015;12:57.

82. Hu M, Cheng HJ, Ji F, Chong JSX, Lu Z, Huang W, et al. Brain functional changes in stroke following rehabilitation using brain-computer interface-assisted motor imagery with and without TDCS: a pilot study. Front Hum Neurosci. 2021;15:692304. Available from: https://www.frontiersin.org/articles/10.3389/fnhum.2021.692304/full.

83. Caria A, Weber C, Brötz D, Ramos A, Ticini LF, Gharabaghi A, et al. Chronic stroke recovery after combined BCI training and physiotherapy: a case report. Psychophysiology. 2011;48(4):578–82.

84. Gomez-Rodriguez M, Grosse-Wentrup M, Hill J, Gharabaghi A, Schölkopf B, Peters J. Towards brain- robot interfaces in stroke rehabilitation. In: Proceedings of international conference on rehabilation robotics. IEEE; 2011. p. 5975385.

85. Lau CCY, Yuan K, Wong PCM, Chu WCW, Leung TW, Wong W-W, et al. Modulation of functional connectivity and low-frequency fluctuations after brain-computer interface-guided robot hand training in chronic stroke: a 6-month follow-up study. Front Hum Neurosci. 2021;14:611064. Available from: https://www.frontiersin.org/articles/10.3389/fnhum.2020.611064/full.

86. Ang KK, Guan C, Phua KS, Wang C, Zhou L, Tang KY, et al. Brain-computer interface-based robotic end effector system for wrist and hand rehabilitation: results of a three-armed randomized controlled trial for chronic stroke. Front Neuroeng . 2014;7. Available from: http://journal.frontiersin.org/article/10.3389/fneng.2014.00030/abstract.

87. Ramos-Murguialday A, Curado MR, Broetz D, Yilmaz z, Brasil FL, Liberati G, et al. Brain-machine interface in chronic stroke: randomized trial long-term follow-up. Neurorehabilitation Neural Repair. 2019;33(3):188–98. Available from: https://doi.org/10.1177/1545968319827573.

88. Everaert DG, Thompson AK, Chong SL, Stein RB. Does functional electrical stimulation for foot drop strengthen corticospinal connections? Neurorehabil Neural Repair. 2010;24(2):168–77.

89. Takahashi M, Takeda K, Otaka Y, Osu R, Hanakawa T, Gouko M, et al. Event related desynchronization-modulated functional electrical stimulation system for stroke rehabilitation: a feasibility study. J Neuroeng Rehabil. 2012;9:56. Available from: https://doi.org/10.1186/1743-0003-9-56.

90. Rushton D. Functional electrical stimulation and rehabilitation – an hypothesis. Med Eng Phys. 2003;25(1):75–8.

91. Hara Y, Obayashi S, Tsujiuchi K, Muraoka Y. The effects of electromyography-controlled functional electrical stimulation on upper extremity function and cortical perfusion in stroke patients. Clin Neurophysiol. 2013;124(10):2008–15.

92. Lee SH, Kim SS, Lee BH. Action observation training and brain-computer interface controlled functional electrical stimulation enhance upper extremity performance and cortical activation in patients with stroke: a randomized controlled trial. Physiother Theory Pract. 2020;1–9. Available from: https://www.tandfonline.com/doi/full/10.1080/09593985.2020.1831114.

93. Chung E, Lee BH, Hwang S. Therapeutic effects of brain-computer interface-controlled functional electrical stimulation training on balance and gait performance for stroke: a pilot randomized controlled trial. Medicine. 2020;99(51):e22612. Available from: https://journals.lww.com/10.1097/MD.0000000000022612.

94. Biasiucci A, Leeb R, Iturrate I, Perdikis S, Al-Khodairy A, Corbet T, et al. Brain-actuated functional electrical stimulation elicits lasting arm motor recovery after stroke. Nat Commun. 2018;9(1):2421. Available from: http://www.nature.com/articles/s41467-018-04673-z.

95. Kawakami M, Fujiwara T, Ushiba J, Nishimoto A, Abe K, Honaga K, et al. A new therapeutic application of brain-machine interface (BMI) training followed by hybrid assistive neuromuscular dynamic stimulation (HANDS) therapy for patients with severe hemiparetic stroke: a proof of concept study. Restor Neurol Neurosci. 2016;34(5):789–97.

96. Cantillo-Negrete J, Carino-Escobar RI, Carrillo-Mora P, Rodriguez-Barragan MA, Hernandez-Arenas C, Quinzanõs-Fresnedo J, et al. Brain-computer interface coupled to a robotic hand orthosis for stroke patients' neurorehabilitation: a crossover feasibility study. Front Hum Neurosci. 2021;15:293. Available from: https://www.frontiersin.org/article/10.3389/fnhum.2021.656975.

97. Ang KK, Chua KSG, Phua KS, Wang C, Chin ZY, Kuah CWK, et al. A randomized controlled trial of eeg-based motor imagery brain-computer interface robotic rehabilitation for stroke. Clin EEG Neurosci. 2015;46(4):310–20. Available from: http://journals.sagepub.com/doi/10.1177/1550059414522229.

98. Frolov AA, Mokienko O, Lyukmanov R, Biryukova E, Kotov S, Turbina L, et al. Post-stroke rehabilitation training with a motor-imagery-based brain-computer interface (BCI)-controlled hand exoskeleton: a randomized controlled multicenter trial. Front Neurosci. 2017;11:400.

99. Mizuno K, Abe T, Ushiba J, Kawakami M, Ohwa T, Hagimura K, et al. Evaluating the effectiveness and safety of the electroencephalogram-based brain-machine interface rehabilitation system for patients with severe hemiparetic stroke: protocol for a randomized controlled trial (BEST-BRAIN trial). JMIR Res Protoc. 2018;7(12):e12339. Available from: https://www.researchprotocols.org/2018/12/e12339.

100. Lyukmanov RK, Aziatskaya GA, Mokienko OA, Varako NA, Kovyazina MS, Suponeva NA, et al. Post-stroke rehabilitation training with a brain-computer interface: a clinical and neuropsychological study. Zhurnal nevrologii i psikhiatrii im SS Korsakova. 2018;118(8):43. Available from: http://www.mediasphera.ru/issues/zhurnal-nevrologii-i-psikhiatrii-im-s-s-korsakova/2018/8/downloads/ru/1199772982018081043.

101. Jang YY, Kim TH, Lee BH. Effects of brain-computer interface-controlled functional electrical stimulation training on shoulder subluxation for patients with stroke: a randomized controlled trial: bci-controlled fes improved shoulder subluxation in stroke survivors. Occup Ther Int. 2016;23(2):175–85. Available from: https://onlinelibrary.wiley.com/doi/10.1002/oti.1422.

102. Remsik AB, Dodd K, Williams L, Thoma J, Jacobson T, Allen JD, et al. Behavioral outcomes following brain computer interface intervention for upper extremity rehabilitation in stroke: a randomized controlled trial. Front Neurosci. 2018;12:752. Available from: https://www.frontiersin.org/article/10.3389/fnins.2018.00752/full.

103. Sebastia´n-Romagosa M, Cho W, Ortner R, Murovec N, Von Oertzen T, Kamada K, et al. Brain computer interface treatment for motor rehabilitation of upper extremity of stroke patients–a feasibility study. Front Neurosci. 2020;14:591435. Available from: https://www.frontiersin.org/articles/10.3389/fnins.2020.591435/full.

104. Sa´nchez-Cuesta FJ, Arroyo-Ferrer A, Gonza´lez-Zamorano Y, Vourvopoulos A, Badia SBi, Figuereido P, et al. Clinical effects of immersive multimodal BCI-VR training after bilateral neuromodulation with rtms on upper limb motor recovery after stroke. A study protocol for a randomized controlled trial. Medicina. 2021;57(8):736. Available from: https://www.mdpi.com/1648-9144/57/8/736.

105. Mattia D, Pichiorri F, Colamarino E, Masciullo M, Morone G, Toppi J, et al. The promotoer, a brain-computer interface-assisted intervention to promote upper limb functional motor recovery after stroke: a study protocol for a randomized controlled trial to test early and long-term efficacy and to identify determinants of response. BMC Neurol. 2020;20(1):254. Available from: https://bmcneurol.biomedcentral.com/articles/10.1186/s12883-020-01826-w.

106. Hong X, Lu ZK, Teh I, Nasrallah FA, Teo WP, Ang KK, et al. Brain plasticity following MI-BCI training combined with tDCS in a randomized trial in chronic subcortical stroke subjects: a preliminary study. Sci Rep. 2017;7(1):9222. Available from: http://www.nature.com/articles/s41598-017-08928-5.

107. Kim T, Kim S, Lee B. Effects of action observational training plus brain-computer interface-based functional electrical stimulation on paretic arm motor recovery in patient with stroke: a randomized controlled trial: effects of aot plus BCI-FES on arm motor recovery. Occup Ther Int. 2016;23(1):39–47. Available from: https://onlinelibrary.wiley.com/doi/10.1002/oti.1403.

108. Bhugra K. A fully remote telehealth brain computer interface and assessment system for motor rehabilitation of chronic stroke; 2021. https://reporter.nih.gov/.

109. Contreras-Vidal J, Feng J, Shedd B. PFI-RP: brain-controlled upper-limb robot-assisted rehabilitation device for stroke survivors; 2018. https://www.nsf.gov/awardsearch.

110. Do A, Cramer SC, Nenadic Z, Reinkensmeyer DJ. Brain-computer interface-functional electrical stimulation for stroke recovery; 2019. https://reporter.nih.gov/.

111. Lin DJ. Targeting neuroplasticity with brain computer interfaces to maximize motor recovery for veterans with stroke; 2020. https://reporter.nih.gov/.

Passive Devices for Upper Limb Training

23

Marika Demers, Justin Rowe, and Arthur Prochazka

Abstract

Arm and hand motor impairments are frequent after a neurological injury. Motor rehabilitation can improve hand and arm function in many cases, but in the current healthcare climate, the time and resources devoted to physical and occupational therapy after injury are inadequate. This represents an opportunity for technology to be introduced that can complement rehabilitation practices, provide motivating task training and allow remote supervision of exercise training performed in the home. Over the last decades, many research groups have been developing robotic devices for exercise therapy, as well as other methods such as electrical stimulation of muscles or vagus nerve stimulation. Robotic devices tend to be expensive and recent studies have raised some doubt as to whether assistance to movements is always preferable as it can reduce salience and engagement. This chapter reviews the evidence for spontaneous recovery, the means and mechanisms of conventional rehabilitation interventions, the advent of affordable passive devices and other treatment modalities that can be used in combination with passive devices. It is argued that task practice on passive devices, in some cases remotely supervised over the internet or augmented with functional electrical stimulation (FES), is now an affordable and important modality of occupational and physical therapy. Passive devices offer numerous opportunities in the field of neurological rehabilitation to support arm and hand motor recovery.

Keywords

Stroke · Spinal cord injury · Multiple sclerosis · Upper extremity · Tele-rehabilitation · Health technology

23.1 Introduction

Neurological disorders are a leading cause of a disability worldwide [1]. One frequent impairment after a neurological insult is a deficit in movements of the arms and hands. This can range from paresis (weakness) to paralysis (plegia). Arm and hand paresis is characterized by muscle weakness, changed muscle tone, decreased sensation, and impaired voluntary

M. Demers (✉)
Division of Biokinesiology and Physical Therapy, University of Southern California, Los Angeles, CA, USA
e-mail: demers@pt.usc.edu

J. Rowe
Flint Rehabilitation Devices, Irvine, CA, USA
e-mail: jrowe@flintrehab.com

A. Prochazka
University of Alberta, Edmonton, AB, Canada
e-mail: arthur.prochazka@ualberta.ca

© The Author(s), under exclusive license to Springer Nature Switzerland AG 2022
D. J. Reinkensmeyer et al. (eds.), *Neurorehabilitation Technology*,
https://doi.org/10.1007/978-3-031-08995-4_23

movement control resulting in slow, imprecise, and uncoordinated movement [2–4]. Among stroke survivors which are estimated to represent 7.0 million individuals in the United-States (2.5% of the population) [5], approximately 65% experience residual arm motor impairments despite intensive and prolonged rehabilitation [6]. Unilateral or bilateral arm motor impairments impact approximately 30% of individuals with traumatic brain injury [7], 60% of individuals in the first year following the diagnosis of multiple sclerosis [8], and individuals with cervical spinal cord injury, the most common site of spinal cord trauma [9, 10]. Due to the important contribution of the arm and hand to everyday activities, motor impairments often lead to activity limitations and participation restriction [11–14].

Rehabilitation interventions can help remediate arm motor impairments and regain lost function. Common interventions include task-oriented training and repetitive task practice, constraint-induced movement therapy, mental imagery, mirror therapy and virtual reality [15]. However, many health system constraints limit the delivery of neurological arm and hand rehabilitation. The decreasing rehabilitation length of stay, general lack of reimbursement for therapy, disparities in access to rehabilitation care, limited available treatment time, and competing rehabilitation priorities, such as the early focus on improving lower limb mobility and gait, are examples of frequent challenges with neurological rehabilitation [16–18]. Therefore, people with neurological disorders may not receive adequate rehabilitation interventions for arm motor recovery. Moreover, people with neurological disorders are often discharged home with limited opportunities to continue home exercises and engage in evidence-based interventions to drive recovery after therapy has ended. At home, long-term adherence to exercise programs is often low [19], due to low motivation, cognitive impairments, lack of caregiver support, frustration, pain, and musculoskeletal issues [20]. This represents an opportunity for technology to be introduced to address key challenges to neurological rehabilitation.

Passive devices offer an affordable option to continue rehabilitation outside clinical settings. Passive rehabilitation devices are defined by what they do not do. Unlike robotic rehabilitation devices, passive devices do not provide active assistance during motor rehabilitation. Although passive devices do not use actuators such as electric motors or pneumatic cylinders, some may provide postural support using energy storage devices such as springs or moving masses. Others provide no assistance, but rather simplify exercises, making them more approachable by removing degrees of freedom from the task. Still others create a motivational environment to support motor practice. The term 'passive devices' should be distinguished from the term 'passive' by rehabilitation clinicians to describe exercises or movements made without efforts from the patient. This chapter reviews the mechanisms for functional recovery, the means and mechanisms of conventional rehabilitation interventions, the rationale behind passive devices, and current evidence supporting different types of passive devices.

23.2 Mechanisms of Functional Recovery: The Significance of Compensatory Strategies

Spontaneous mechanisms of recovery at the cellular, molecular, and systems levels often follow a neurological insult. However, the degree of spontaneous recovery varies between individuals and neurological conditions, and is generally incomplete [21]. Some of the spontaneous recovery in motor function is evidently a result of the recovery of central nervous structures temporarily inactivated by the injury, or the adaptation of uninjured nervous networks to take over functions of neighboring injured networks, a process called plasticity [22–24]. Various means of early prediction of the extent of recovery have been identified [25–28]. In this regard, the concept of compensation should be distinguished from recovery [29]. Specifically, individuals with neurological disorders may adopt compensatory strategies to accomplish daily tasks, such as the use of alternative motor strategies (i.e., shoulder elevation or trunk flexion) to compensate for lost motor patterns in the elbow and shoulder [3]. At the body function and structure level of the International Classification of Functioning (ICF) [30], compensation is defined as the performance of a movement in a new manner, seen as the appearance of alternative movement patterns during the accomplishment of a task. At the activity level, compensation refers to a successful task completion using different techniques [29]. The adoption of compensatory strategies may be considered maladaptive if compensations limit recovery of independent movements of the most affected arm, contribute to secondary complications such as pain, joint contracture, and discomfort [31], and lead to a pattern of learned maladaptive behavior impeding long-term functional motor recovery [32, 33]. The use of maladaptive compensatory strategies may limit one's ability to generalize movements to a wider array of tasks [34] and contribute to incipient decline after the end of active therapy [35].

Evidence supports the effectiveness of neurological rehabilitation to remediate arm motor impairments and improve function by fostering neuroplastic changes, which often rely on mechanisms similar to those observed during spontaneous recovery [15, 21, 34, 36, 37]. An understanding of the mechanisms through which recovery is achieved after a neurological disorder is important to guide effective rehabilitation interventions. Neuronal plasticity plays a crucial role in neurologic recovery. From the work done in animal models, repetition, intensity, and salience have been identified as critical to drive experience-dependent neuroplasticity [33]. Other factors influencing plasticity include the provision of progressive and optimally adapted rehabilitation interventions tailored to one's capability and the environmental context [38, 39]. In animal stroke models, hundreds of repetitions of motor tasks are needed to induce lasting neural changes [24, 40]. In humans, the critical threshold of rehabilitation intensity needed to engage plastic mechanisms is unknown. A recent review concluded there was a positive relationship between the time scheduled for exercise therapy and the outcome with large doses of exercise therapy leading to clinically meaningful improvements [41]. The authors pointed out that time scheduled did not necessarily equate to the amount of task practice actually performed. They recommended that instead of reporting scheduled time, future studies should report active time in therapy or repetitions of an exercise. The notion that more is better was challenged in recent large, randomized control trials (RCT). Specifically, Winstein et al. [42] compared four dosages of personalized arm task-oriented training in chronic stroke survivors. Higher dosage of training led to greater gains in quality of arm use measured with the Motor Activity Log, but no changes on functional capacity measured with the Wolf Motor Function Test were noted. In another RCT, Lang et al. [43] compared four doses of task-specific training on arm and hand function (i.e., 3,200, 6,400, 9,600, or individualized maximum repetitions). The results showed no evidence of a dose–response effect of task-specific training on arm and hand functional capacity in stroke survivors. However, the number of movement repetitions provided during this trial was far superior to the amount of movement practice normally provided during conventional stroke rehabilitation [44]. Similarly, the most common dosage of home exercises prescribed for adults with neurological conditions is estimated to be 16–30 min per day with a greater focus on fine motor activities, active range of motion, active assistive range of motion, and whole or partial activity of daily living tasks [45]. Despite the lack of consensus on optimal dose, interventions for arm and hand motor impairments may not always be delivered at the most beneficial intensities and may not include enough repetitions to optimize neuroplasticity [46].

23.3 The Role of Rehabilitation to Restore Arm and Hand Function

A variety of rehabilitation interventions can be used to improve arm and hand motor recovery, but the level of evidence available varies depending on the modality and by neurological populations. The Bobath technique and proprioceptive neuromuscular facilitation, two rehabilitation approaches based on neurophysiological principles, were widely adopted in the 1970s with strong adherents in each camp. However, current evidence does not support the superiority of neurodevelopmental techniques over other types of interventions [16]. A RCT that compared these two approaches with conventional rehabilitation concluded that there were no significant between-group differences in improvement of the patients' performance of activities of daily living [47].

Looking specifically at evidence-based interventions for different neurological populations, task-oriented training, resistance and endurance training, constraint-induced movement therapy, and some types of robot-supported training may improve arm and hand function in individuals with multiple sclerosis [36]. However, the evidence for one approach over another is not clear due to large variability between studies and small sample sizes. Similarly, motor training in individuals with cervical spinal cord injury or traumatic brain injury, which can include task practice and functional electrical stimulation (FES), may reduce arm and hand motor impairments and improve function. However, there were wide differences between studies in the types of patients, training, methodology and outcome parameters [37, 48, 49]. For individuals with stroke, a meta-study concluded that sensorimotor training, motor learning training with the use of imagery, electrical stimulation, and the repetitive performance of novel tasks, could all be effective in reducing motor impairment after stroke [16]. Moderate-quality evidence suggests that constraint-induced movement therapy, mental practice, mirror therapy, interventions for sensory impairment, virtual reality and task-specific training may be effective in improving arm and hand function [15, 50].

One of the most investigated interventions to remediate arm and hand motor impairments in neurology is constraint-induced movement therapy (CIMT), a particular form of intensive and supervised task practice [51]. This approach was developed based on experiments in monkeys in which sensory input in one arm was abolished by de-afferentation. Binding of the other, non-affected arm, led to forced use of the de-afferented arm, which was associated with improvements in its motor function [52, 53] and brain reorganization [54, 55]. In humans, CIMT or the modified versions of CIMT include three key components: (1) constraint of the less affected arm for up to 90% of the waking hours to promote the use of the more affected arm, (2) delivery of intensive graded practice of the more affected arm in functionally meaningful tasks (i.e. task shaping), (3) adherence-enhancing behavioral methods designed to transfer the gains to the real-world environment (i.e., transfer package) [56–58]. CIMT and modified CIMT have been shown to have robust, clinically meaningful impacts on stroke survivors' outcomes for arm and hand motor impairment and function, making CIMT one of the most effective interventions for the paretic arm post stroke [59–61]. However, the use of CIMT in clinical practice is limited. In a survey of 92 therapists working in clinical neurorehabilitation in the USA, 75% reported that it would be difficult or very difficult to administer CIMT in their clinics [62]. Challenges with the delivery of CIMT or modified CIMT include the difficulty to achieve the recommended training intensity, stringent inclusion criteria, lack of resources and high delivery costs that may not be reimbursed by third-party payers [62–65].

23.4 Robotic Exercise Devices

Monitoring movement performance and quality is often challenging in clinical practice, as common objects used for task practice (e.g., blocks, cones, therapy putty, peg boards, resistive prehension benches, etc.) do not have sensors to quantify movement kinematics. The supervision by a therapist can prevent the use of maladaptive compensatory movements. However, the supervision by therapists is costly, and in most cases, restricted to clinics, which in turn limits access mainly to subacute patients. Robotic devices have been developed with the aim of delivering high-intensity, repetitive and adaptive training [66]. Robotic devices can be used to provide standardized exercises, take over some supervisory functions and provide quantitative outcome measures to reduce the tedium of conventional rehabilitation. The treatment paradigm for robotics is based on the provision of physical assistance to complete desired motions of the arm and hand in combination with computer games or virtual reality presented on a screen [67]. Robotic assistance is considered advantageous because it (1) allows the challenge level of the task to be adjusted to better suit the needs and abilities of the user, and (2) allows users to move through a larger range of motion—thereby providing a large afferent response that is time correlated with the user's efferent motor intent. Evidence from systematic reviews and meta-analyses support the use of robotic-based training for sensorimotor rehabilitation of stroke survivors and people with spinal cord injury and multiple sclerosis [36, 66, 68–71]. Research on robot-assisted movement therapy has rapidly increased in recent years, as the potential for robotic therapy after a neurological insult remains enormous [67].

Some of the drawbacks of robotic devices are the high cost of the devices and the lack of salience of the tasks (i.e., tasks may not be meaningful or engaging to the users). Robotic devices incorporate actuators and complex control systems, which makes them expensive. Retail prices can vary between $400USD per month for a powered wrist splint (e.g., Hand Mentor, Motus Nova, Atlanta, GA, https://motusnova.com/hand/) to tens of thousands of dollars for exoskeleton robots, such as the KINARM Exoskeleton Robot (Kinarm, Kingston, Ontario, https://kinarm.com/) or the Armeo Power (Hocoma, Volketswil, Switzerland, https://www.hocoma.com/us/solutions/armeo-power/).

The high price point is a barrier to adoption in clinical practice and home use. Another criticism towards robotic devices is that the tasks may not be meaningful to the users, solicit intrinsic motivation or active participation. Ideally, robotic assistance would allow users to practice at their ideal challenge rate, allowing users to be motivated by their success and learn from their occasional failures [72]. However, over-assisting can be counterproductive [73]. Although much work has been devoted to controllers that assist only as needed, the problem of finding the ideal assistance level remains challenging. Since movements are often restricted to a 2D-plane, robotic devices are limited in the variety of movements or tasks practiced, which may not reflect the range of arm and hand movements used for everyday object interaction.

The high-cost and the lack of salience of robotic devices, along with the aforementioned challenges with the delivery of neurorehabilitation stress the need for affordable and effective solutions to harness neuroplasticity. Passive devices are more affordable than robotic devices and unlike CIMT, which has stringent inclusion criteria, they are accessible to individuals with a wide variety of impairments. They can provide motivating task practice to minimize the lack of adherence with home programs and offer the opportunity to deliver salient and intensive task practice outside clinical settings. Because passive devices are defined by what they are not, the remaining field is understandably broad and inclusive. In this chapter, we divide the field of passive devices into 5 groups: (1) passive gravity support systems, (2) tabletop therapy systems, (3) linear rail systems, (4) tone compensating orthoses, and (5) serious game controllers (Fig. 23.1).

23.5 Passive Gravity Support Systems

Mobile arm supports and balanced forearm orthoses have been a part of rehabilitation practice since at least the mid 1960s [74–77]. Although the embodiment has varied, most are either chair-mounted or desk-mounted orthotics

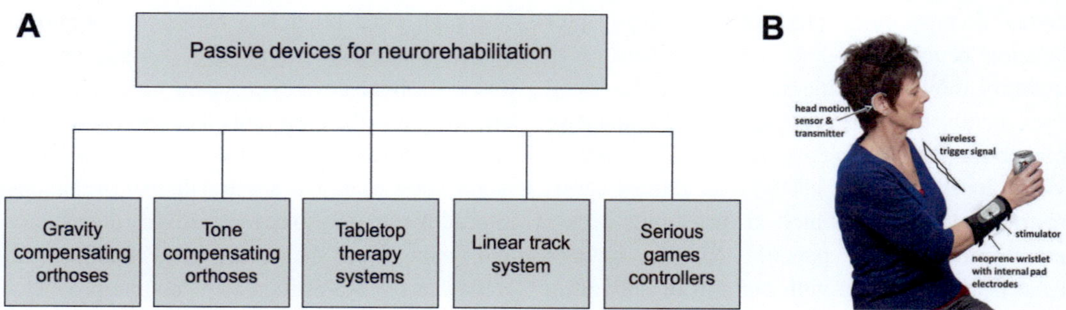

Fig. 23.1 **a** 5 types of passive devices for neurorehabilitation; **b** rehabtronics hand stimulator activated by voluntary head-nods detected by a wireless earpiece

that support the arm at the wrist using either a rigid planar linkage [78, 79], or in many cases, a spring loaded mechanism tuned to balance the weight of the arm. Initially, mobile arm supports and balanced forearm orthoses were used as assistive devices and evaluated based on whether they enabled users with motor impairments to perform activities of daily living that they could not perform otherwise [80]. Early devices were also limited in their degrees of freedom and in the range of motion that they could support [81].

Although chair- or desk-mounted mobile arm support assistive devices have continued to evolve [82], much attention has been given to a newer class of devices designed specifically for rehabilitation. The Armeo Spring (based on the T-WREX, Hocoma, Volketswil, Switzerland) [83], is a counter-balanced multi-segment arm support with six instrumented and lockable degrees of freedom and an instrumented gripper. Positional data from the joint sensor and force data from the gripper are used both to (1) control a suite of games simulating activities of daily living and (2) to quantify features of the user's motor impairment. In early testing, participants with chronic stroke that trained with T-WREX improved their motor capacity (as measured by the Fugl Meyer Assessment) significantly more than participants that trained with tabletop exercises [81]. The gains themselves were modest, but the T-WREX group also showed significantly better retention. Numerous studies have demonstrated that training in T-WREX and Armeo

Spring increases the range of motion—both immediately, while in the orthosis, and to a lesser extent, persistently [81, 82, 84]. Critically, participants prefer training with gravity assistance and assign high value to the exercise [85]. The efficacy of Armeo Spring for subacute recovery is less clear. Recently, a large clinical trial in subacute participants compared therapy in Armeo Spring to dose-matched stretching and basic active exercises. The Fugl-Meyer scores of both groups increased significantly, but the differences between the groups was not significant at either the 4-week assessment or the 12-month follow-up [86].

Although the multi-segment linkage used by Armeo Spring does an impressive job of supporting the weight of the arm without placing unwanted restrictions on its range of motion, adjusting the linkages and the springs for each user can be time consuming. Freebal (sold by Hocoma under the name Armeo Boom, Fig. 23.2), by contrast, is a sling-based gravity support system that does not provide as much freedom of movement as Armeo Spring but is much simpler and requires less adjustment. Like Armeo Spring, Freebal extends the range of motion in which users are able to train [87–90]. Although the effectiveness of Freebal has not been studied as extensively as Armeo Spring, early pilot testing suggested that it had similar and lasting effects on motor capacity and range of motion [90]. The high cost of both devices remains a barrier for clinical or home use.

Fig. 23.2 The
Armeo®Booom, a passive
gravity support system

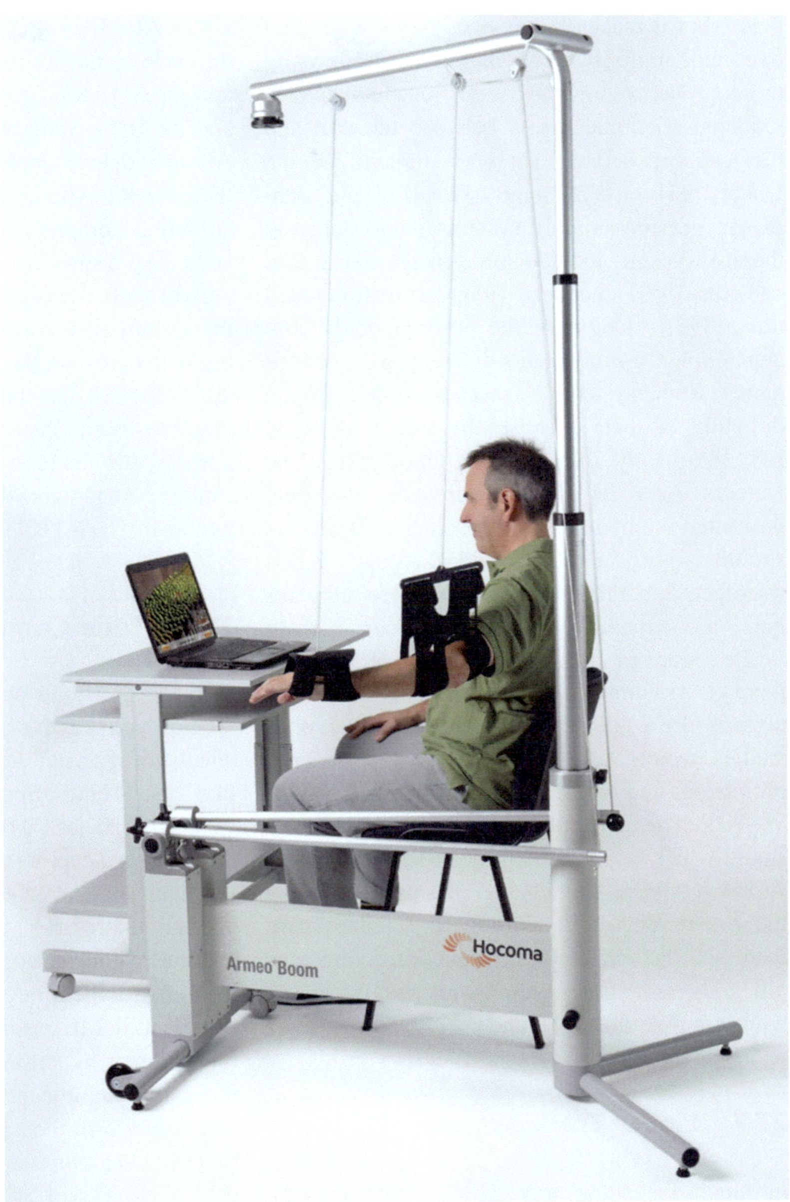

Armeo Spring and Armeo Boom are both stationary devices. There is also a developing class of wearable gravity support exoskeletons that offload the weight of the arm but are not constrained to a particular location [91–93]. The recently developed SpringWear system, for example, is lightweight, and increases the active workspace of its users. However, the exoskeleton did not consistently improve the wearer's ability to complete functional tasks [91, 94].

23.6 Tabletop Therapy Devices

Tabletop therapy systems are similar to passive gravity support systems in that they allow users to practice without supporting the weight of their arm. However, instead of using springs to compensate for gravity, tabletop systems rely on the surface of the table to support the weight. This simplification restricts movements to a single plane, but it makes the devices more cost effective and appropriate for home use. Normally these systems include some mechanism for reducing frictional forces between the arm and the support service such as omnidirectional wheels [95], or a 2D gantry [96]. Like the non-planar gravity support systems, most tabletop therapy systems include some mechanisms (e.g. cameras [95], encoders [97], or instrumented tracks [96]) to monitor the position of the arm and couple the movements with engaging serious games. Some systems allow users to increase the difficulty of their exercises by tilting the table [98, 99] or by introducing friction [97]. The Rutgers Arm II is noteworthy for detecting unwanted compensatory movements from the shoulder [98], and the Rapael Smart Board (Neofect, San Francisco, CA) is noteworthy for being a commercially available device.

Evidence of the effectiveness of tabletop therapy systems is somewhat limited. Most devices are validated using uncontrolled pilot studies, which indicate that the devices are safe, promising, and well received by their users [95, 98, 100]. The main exception is the Rapael Smart Board which was validated by a RCT for chronic stroke survivors. Participants that practiced with the Smart Board in addition to standard care increased their Wolf Motor Function scores significantly more than participants that practiced with a double dose of standard care [96].

23.7 Linear Track System

In much the same way that tabletop therapy devices are a simpler, more constrained alternative to systems like Armeo Spring or Armeo Boom, linear track systems are a simpler and more constrained alternative to tabletop therapy systems. Linear tracks support gravity and reduce unwanted friction, but they also resist unintentional movements caused by imbalanced motor synergies. As long as there is some forward or backward component to the forces that the user applies to the slider, it will progress along the track, making the exercise very forgiving. The three most prominent linear track systems are the

BATRAC [101] which was sold for a time under the trade name "Tailwind", the Reha-Slide [102], and the SMART Arm [103]. Both the BATRAC and the Reha-Slide promote bimanual exercises. All three devices include game-like elements, but the SMART Arm is the only device built around an actual computer gaming system. The Reha-slide has been used as an input to a gaming system with a backend for telerehabilitation, but the commercial version does not yet support these features [104]. Training with any of the three devices has been shown to significantly improve motor capacity in chronic stroke, but none of the devices have proven to be significantly better than dose-matched conventional treatment [105–110].

23.8 Tone Compensating Orthoses

In much the same way that Armeo Spring uses springs and passive mechanisms to cancel the effects of gravity, tone compensating orthoses like the Hand Spring Operated Movement Enhancer (HandSOME), SCRIPT, and EXTEND devices use springs and carefully designed mechanisms to create tunable force profiles that do an impressive job of compensating for unwanted joint torques caused by wrist and finger flexor hypertonia [111–113]. While wearing the HandSOME orthosis, stroke survivors with finger flexor hypertonia were able to both move through a significantly larger range of motion and outperform their unassisted Box and Blocks scores [111]. Similar results were observed for the EXTEND and SCRIPT Orthoses [112, 114]. The stated goal of all three devices is to enable stroke survivors to exercise more and with greater success. However, the efficacy of exercising with a tone-compensating device is far from clear. In an uncontrolled pilot study, the training with the HandSOME orthosis was shown to significantly, but not persistently improve motor capacity [115]. The EXTEND Orthosis has been tested in a controlled, at home clinical trial, but participants in the control group (who performed exercises from a book) improved more than participants that played immersive video games with assistance from the EXTEND Orthosis [116]. Saebo also sells a commercial glove called SaeboFlex, which uses springs and cables to apply forces to resist flexion contractures in the wrist and hand. Unlike the HandSOME system, the springs are not tuned to truly compensate for the unwanted flexor activity, but the device makes up for this in its practicality. It is marketed as an assistive device.

23.9 Serious Games for Home Use

Nearly all the passive devices discussed above are instrumented so that they can be used as inputs to control motivating custom and non-custom video games. The difficulty, repetitiveness, and delayed gratification of motor rehabilitation can make it very challenging for people with a neurological disorder to invest themselves. Serious games can hide the difficulty and repetition inherent to motor rehabilitation by providing a rich and motivating training environment and embedding game-playing elements. The level of difficulty of virtual tasks can often be scaled in ways that purely physical tasks cannot, allowing players to practice at higher success rates. Serious games offer the advantages of providing immediate and enhanced feedback, and can dynamically adjust task difficulty. Many systems offer the opportunity to record and monitor performance and progress, which can be very useful for patients and therapists to monitor performance and track improvements over time. Serious games also offer the opportunity to incorporate motor learning principles, such as motivation, repetitive practice, and enhanced feedback, into rehabilitation interventions [117].

However, the standard input devices (mice, keyboards, and gamepads) used to control commercial video games are poorly suited to games used for rehabilitation because they are designed to require very small, efficient movements, not the types of movements normally prescribed during motor rehabilitation. This has created an interest in unique input devices for rehabilitative serious games. Two of the earliest and most

readily adopted gaming systems used for rehabilitation were the Nintendo Wii (Nintendo, Tokyo, Japan) and the Kinect System (Microsoft, Richmond, WA) [118, 119]. In a survey of 1071 practicing PTs and OTs, 41% reported having access to a Wii in a clinical setting, whereas 10% had the Xbox Kinect [118]. None of the commercial games, such as the Wii and the Kinect, were deliberately designed for rehabilitation, it is not surprising that they have shortcomings. Movement performance and quality may be diminished by the attributes of the virtual environment (e.g., viewing environment, visual, tactile, auditory, and other sensory cues, etc.), which can consequently be detrimental to motor learning [120, 121]. None of the games involve dexterous tasks requiring grasp/release, pronation/supination, pinch-grip/release or picking up and transferring objects. Nevertheless, many of the Wii or Kinect games are considered to be suitable for rehabilitation [122, 123]. Specifically, exercising with the Wii has also proven to be safe, the system is easy to use, and there is some evidence that participants in trials involving the Wii are less likely to drop out [124]. Results from a recent Cochrane review about the effectiveness of virtual rehabilitation after stroke demonstrated that virtual reality and interactive video gaming have a significant but modest effect on improving arm and hand impairments (measured by the Fugl-Meyer Assessment) when used in addition to usual care to increase overall therapy time (standardized mean difference = 0.49, 95% confidence interval 0.21 to 0.77, 210 participants) [125]. The potential benefits of virtual rehabilitation on improving function in everyday activities, quality of life, and reducing participation restrictions were also identified [125–127]. While well adopted, the use of commercially-available serious gaming systems was not found to be superior to usual care [125]. This suggests that these systems should not be viewed as alternatives to usual care, but rather as useful supplements.

More recently, a cohort of low to moderate cost input devices specifically designed for rehabilitation have become available. The most noteworthy of these being the Neofect Smart glove, the AbleX rehabilitation system (AbleX, Auckland, New Zealand), Pablo (Tyromotion, Grz, Austria), the MusicGlove and FitMI systems (Flint Rehabilitation devices, Irvine, CA), the NeuroFenix gameball (NeuroFenix, London, UK), and the Rehabilitation Joystick for Computerized Exercise (ReJoyce) workstation marketed by Saebo (Charlotte, NC).

The Neofect smart glove is an instrumented data glove designed to be easy to don, doff, and clean. It uses inertial sensors to detect movement of the wrist and hand, and resistive bend sensors to detect movement of the fingers. As such, it can facilitate distal grasp-related exercises in addition to some proximal upper limb exercises. In a RCT (N = 13), chronic stroke participants (defined in the study as >4 months post-stroke) who practiced with the smart glove for 15 30-min sessions in addition to 15 30-min sessions of conventional therapy improved significantly more, as measured by the Wolf Motor Function Test, than participants who performed 30 30-min sessions of conventional therapy [128].

The AbleX system consists of two different input devices: the first is an arm skate with a repositionable button that can be used to detect extension of any of the fingers and the second is an inertial measurement unit (IMU)-based controller that can be used both unimanually and bimanually. The AbleX system has not yet been evaluated in a RCT, but early pilot testing suggests that it is safe, motivating, and potentially effective [129].

The Pablo system by Tyromotion includes a Wii-mote-like orientation and force sensor called the "handle", an adapter that allows the handle to be used for bimanual exercises and a ball adapter that allows the handle to be held in a different orientation. The software allows therapists to adjust orientation and force thresholds [130].

The MusicGlove is a distally focused data glove for hand rehabilitation designed and priced to be appropriate for home use. The glove can detect opposition of the thumb to all 5 fingers as well as pincer grip and key pinch grip. It is coupled to a game similar to *Guitar Hero* in which players hit notes by completing the hand grips

indicated by the game [131]. The MusicGlove has been evaluated for both in-clinic [132] and in-home use [131]. In both trials, therapy with the MusicGlove was compared to conventional therapy as a control. The control group performed tabletop exercises in the in-clinic study and therapy guided by a booklet of exercises for the in-home study. In both studies, both groups improved significantly and sustained their improvements (as indicated by changes in Box and Blocks scores), but the MusicGlove groups did not improve significantly more than their corresponding conventional therapy groups [132, 133]. In the in-home study, participants in the MusicGlove group improved significantly more on the Motor Activity Log scores than those of the conventional therapy group. Notably, participants in the MusicGlove group also significantly and voluntarily intensified their dose from an average of 213 ± 301 grips per week during the first week to 466 ± 641 grips per week in weeks 2 and 3 [133].

One of the main limitations of the Music-Glove is that it can only be used for distal exercises. To address this limitation, Flint Rehabilitation Devices released FitMI, a more generic serious game controller with libraries of exercises that support proximal and distal upper limb exercises in addition to lower limb exercises and core strengthening/stretching exercises. The FitMI system consists of two wireless puck-shaped controllers that can detect both forces and movement and can supply both haptic and visual feedback. The FitMI System is currently being evaluated in a RCT.

The NeuroFenix ball is a round ball that can measure movement and orientation changes. It can be strapped to one hand, used bimanually, or secured in a dock that restricts it to orientation changes only. The shape, size, and straps hold the hand in a favorable position and make the device easier to hold than many other comparable devices [134].

The Rehabilitation Joystick for Computerized Exercise (ReJoyce: Rehabtronics.com; Fig. 23.3) comprises a spring-loaded, segmented arm that presents the user with a variety of spring-loaded attachments representing activities of daily

Fig. 23.3 **a** Tele-coaching of an in-home exercise therapy session using the rehabilitation joystick for computerized exercise (ReJoyce) system; **b** participant using ReJoyce workstation to play computer games; **c** movements performed, **d** selected games

living, such as a doorknob, key, gripper, jar lid and peg. Sensors in the arm and the attachments provide signals that are used by the system's software to control serious games that exercise specific types of hand movement.

The system incorporates an automated, quantitative arm and hand function test which takes about 5 min to complete and provides an overall numerical score that correlates well with the Action Research Arm Test and the Upper Extremity Fugl-Meyer Assessment [135]. It also provides scores for specific tasks such as grasp strength, whole-arm range of motion, pronation-supination, pinch-grip and manual dexterity and can be performed in the clinic or remotely. Once a user has done the test, the system automatically suggests games and difficulty levels that match their abilities. This is achieved by an algorithm that considers the user's score on each of the components of the test. If, for example, the user has good ranges of motion but poor pinch-grip strength, games that incorporate pinch-grip are excluded from the suggestion list, and games involving range of motion and grasp-release are included, with difficulty levels corresponding to the relevant test scores.

The ReJoyce system also facilitates remote tele-coaching. A RCT was completed involving 13 tetraplegic participants who had sustained a spinal cord injury more than a year previously [136]. Participants were block-randomized into two groups, both performing exercise therapy at home with tele-coaching for 1 h/day, 5 days/week for 6 weeks. The control group played computer games played with a trackball and 20 min/day with therapeutic electrical stimulation. The treatment group played serious games on a ReJoyce workstation. Voluntary, hand grasp and release were augmented with functional electrical stimulation (FES) triggered by a wireless earpiece that detected small voluntary tooth clicks. The study demonstrated the feasibility of delivering tele-coached functional electrical stimulation-assisted exercises over the Internet. The treatment group showed clinically important improvements in arm and hand function that significantly exceeded those of the control group. Participants commencing with intermediate functional scores improved the most [137]. The ReJoyce system was designed to be affordable for clinics and, through short-term rental, by individual users who could receive tele-supervised treatment in their homes.

23.10　Therapeutic and Functional Electrical Stimulation

The simple and interactive nature of passive devices makes them a natural complement to functional electrical stimulation (FES) and non-motor specific processes like Vagus nerve stimulation (VNS). This is particularly true for passive devices that serve as an input to a computer since inputs from the devices and events from the games could both ostensibly be used to trigger stimulation.

FES refers to intentionally-triggered electrical stimulation of the motor neurons in a targeted muscle group to assist in a functional task. Although often used as a component of an assistive orthosis, it is also used therapeutically; exercising with FES in addition to standard care has been shown to improve motor capacity moderately, but significantly, more than standard care alone [138]. Although FES systems are more commonly controlled by switches [136] or electromyography (EMG) [139], they can also be triggered by passive devices designed to detect movement intent [140]. While useful, FES-enabled passive devices systems are not without their challenges. Although feedback controlled, multi-joint systems do exist [141], creating coordinated multi-joint movements vias FES is challenging, and nearly all systems rely on pre-programmed stimulation profiles targeting one or two muscle groups (e.g. elbow and forearm extensors) [138]. Furthermore, the motor unit recruitment order obtained by FES is difficult to control and leads to premature fatigue [142].

Vagus Nerve Stimulation (VNS) has been proposed as an adjunct to exercise training in neurorehabilitation. The proposed mechanism of VNS is not through muscle activation, as is the case for FES, but rather, through plastic changes in the central nervous system (CNS). Although the vagus nerve, which innervates autonomic organs such as the heart and gastrointestinal tract, might seem like an unlikely player in neurological motor recovery, it is known that vagal afferents project to neuromodulatory networks in the CNS. Neuromodulatory networks are groups of neurons that are influenced by neurotransmitters such as acetylcholine and noradrenaline and that modulate the activity of other CNS centers. It has been suggested that the cholinergic neuromodulatory network affects motor control and that the noradrenergic neuromodulatory network affects awareness and responsiveness to stimuli. There is evidence from animal studies that VNS can promote CNS neuroplasticity [143]. It has been suggested that VNS applied at the right time during motor practice will increase the neuroplastic response to that practice [144–146]. In the only human clinical trials to date, the timing of VNS was controlled manually by a physical therapist. In principle, the timing could be controlled with movement feedback from a passive rehabilitation device.

23.11 Telerehabilitation

From all the above, it is clear that the emerging technologies to deliver task practice have the potential greatly to improve arm and hand function in daily life but providing sufficient support after participants leave rehabilitation clinics is problematic. Although the users may benefit from the devices in the clinic, and initially use them daily at home, in the absence of continuing supervision, usage tends to drop off. This transition is a well-known hurdle in rehabilitation [147]. We reasoned that if participants could only perform regular supervised exercise after discharge, they would benefit much more. However, clinics are not ideal locations for outpatients to perform regular training sessions. Travel is often problematic and stressful, limiting the frequency of attendance. In recent years, telerehabilitation, a form of telemedicine, has slowly gained popularity. One of the early systems used for telerehabilitation was the ReJoyce system used for at-home tele-coaching (Fig. 23.2a). In the study mentioned above, Internet-connected ReJoyce workstations were deployed in the homes of 13 tetraplegic participants, located over a wide geographic region in western Canada. Participants were tele-coached daily by a small team of therapists and students. The logistic challenges that were overcome are detailed in a book chapter [148]. A similar study followed on chronic stroke patients in Canada and the UK [149]. Other studies have also shown that telerehabilitation can be convenient and effective for both therapists and patients [148, 150, 151]. Telerehabilitation promotes flexibility, allows greater access to care and continuity of care, and can help decrease racial and economic disparities in health care [152, 153].

With the recent health care crisis induced by the global COVID-19 pandemic, rapid technological changes have followed and telerehabilitation quickly became widely adopted by rehabilitation services across the world [154]. Key barriers that limited the adoption of telerehabilitation previously, such as reimbursement and clinicians' preference for hands-on interactions, were partly overcome in response to the global pandemic [155]. Resources and guidelines from professional associations were also developed to support clinicians in the delivery of remote rehabilitation (for example, [156, 157]).

A recent trend in stroke rehabilitation is the concept of early supported discharge. Early supported discharge is a multidisciplinary team intervention that facilitates earlier discharge from hospital with rehabilitation care provided in the community [158]. Evidence from meta-analyses supports appropriately-resourced early-supported discharge services delivered by a multidisciplinary team to reduce disability and shorten hospital stays in a selected group of stroke survivors [158, 159]. The use of passive devices is particularly well-suited for home use, early-supported discharge and remote supervision using telerehabilitation.

23.12 Clinical Adoption of Passive Devices

Healthcare professionals play a pivotal role in their patients' access to novel technologies, and they are ultimately the ones who use health technologies, such as passive devices, in their clinical practice or recommend them for home use [160]. Understanding clinicians' perspectives and the factors affecting clinical adoption is crucial to enable clinical changes and better widespread use of passive devices in neurological rehabilitation.

The clinical adoption of passive devices remains low and is not yet commonplace in clinical rehabilitation settings, which is not different from most health technologies [155, 161]. The results from a survey of 1326 healthcare professionals suggest that clinical decisions to acquire and use new technology devices are multifaceted and are based on the benefits for their patients, the technology's appropriateness for the setting and logistical practicality within the service delivery system [162]. Patient characteristics, available financial resources, technology cost, experience with technology and time demands are key factors shown to impact clinical practice patterns [161–163]. The potential benefit of technology to facilitate positive health outcomes, mainly through the provision of meaningful and objective feedback, and repetitive and independent practice, was identified as a main driver to adoption [155]. Clinicians usually have very busy schedules, with little time to deal with new technology. It is therefore vital to provide equipment that is affordable and simple to use, with highly intuitive computer interfaces that do not require procedural memorization from one session to the next [155, 164]. Since barriers to adoption are multifactorial, financial and administrative support from the leadership, and training of clinicians are essential to ensure clinical adoption and use of health technologies [165]. Future technology development should consider following the stepwise approach and conditions for successful implementation of technology in daily clinical practice, as outlined in a recent systematic review [166]. In the later stages of technology development, the incorporation of user-centered design methods and involvement of clinicians and users are important to facilitate technology adoption [167].

23.13 Perspectives and Conclusions

There is general agreement in the field that the time is ripe for physical and occupational therapy to take advantage of new technologies. It is time to move beyond simple equipment currently used in clinics worldwide, to passive devices that provide task-specific, motivating games that can also be performed in the participant's home environment, supervised remotely over the Internet. The advantages of this approach are many: increased compliance, task-specific training on a variety of customized activities, quantification of performance and perhaps most compelling, the ability to provide continuing in-home therapy after acute care in clinics, in a manner that avoids the need for participants to travel, yet retains the important component of one-on-one supervision by enabling therapists to treat participants at times that suit them all. A crucial factor is cost, especially given that costs are generally covered by patients. The cost of passive devices presented in this chapter spans from a few hundred dollars for serious games to a few thousand dollars for tabletop therapy systems, with many options offered below $1,000USD. This chapter has made the case for affordable passive exercise devices that provide entertaining exercises involving full range-of-motion and manual dexterity, with optional tele-coaching and electrical stimulation (summarized in Table 23.1). Task training for arm and hand function on passive devices, with the option of FES-assistance or perhaps VNS, is now an affordable and effective modality of occupational and physical therapy. Passive devices offer numerous opportunities in the field of neurological rehabilitation to support arm and hand motor recovery. Future research could focus on the identification of who might benefit the most from the use of passive devices to guide clinical decision making and maximize the use of scarce health-care resources.

Table 23.1 Summary of passive exercise therapy devices discussed in this chapter

Device	Commercial product	Movement target(s)	Device type	Computer gaming?	Validated upper limb function test?	Integrated Telerehab/Telecoaching	Recent studies
Armeo Boom	Yes	Shoulder, arm, forearm	Gravity support	Yes	No	No	[79, 81–83, 85, 168]
Armeo Spring	Yes	Shoulder, arm, forearm, wrist, hand grasp-release (optional attachment)	Gravity support	Yes	Yes	No	[87–89]
SpringWear	No	Shoulder, arm forearm	Gravity support	Yes	No	No	[91, 94]
Rapael SmartBoard	Yes	Shoulder, arm	Tabletop	Yes	No	No	[96]
Tailwind	Previously	Shoulder, arm	Linear track	No	No	No	[101, 105, 106]
Reha-Slide	Yes	Shoulder, arm	Linear track	Research only	No	Research only	[102, 104]
SMART Arm	No	Shoulder, arm	Linear track	Yes	No	No	[103, 110]
HandSOME	No	Fingers, thumb	Tone Compensation	Yes	No	No	[111, 115]
SCRIPT	No	Wrist, fingers, thumb	Tone Compensation	Yes	No	No	[113, 169]
EXTEND	No	Fingers, thumb	Tone Compensation	Yes	No	No	[112]
SaeboFlex	Yes	Fingers, thumb extension	Tone Compensation	No	No	No	None
Wii	Yes	Shoulder, arm, forearm, wrist, thumb (pushbutton)	Input device	Yes	No	No	[122, 124]
Kinect	Yes	Whole body—software limited	Input device	Yes	No	Software dependent	[123, 170]

(continued)

Table 23.1 (continued)

Device	Commercial product	Movement target(s)	Device type	Computer gaming?	Validated upper limb function test?	Integrated Telerehab/Telecoaching	Recent studies
Neofect SmartGlove	Yes	Wrist individual fingers, thumb	Input device	Yes	No	No	[128]
AbleX	Yes	Arm, wrist, fingers, thumb	Input device/tabletop therapy	Yes	No	No	[129]
MusicGlove	Yes	Individual finger and thumb movement	Input device	Yes	No	No	[131–133]
FitMI	Yes	Ankle, core, shoulder, arm, forearm, wrist, hand, power grip, key pinch	Input device	Yes	No	Research only	[171, 172]
Pablo	Yes	Wrist flexion–extension, pronation-supination, grasp, finger-thumb pinch	Input device	Yes	No	No	None
ReJoyce	Yes	Shoulder, arm, forearm, wrist, hand grasp-release, finger-thumb pinch	Input device	Yes	Yes	Yes	[135–137, 150]

References

1. World Health Organization. Neurological disorders: public health challenges. WHO Press, editor. 2006.
2. Alt Murphy M, Häger CK. Kinematic analysis of the upper extremity after stroke–how far have we reached and what have we grasped? Phys Ther Rev. 2015;
3. Cirstea MC, Levin MF. Compensatory strategies for reaching in stroke. Brain. 2000;123(5):940–53.
4. Lang CE, Wagner JM, Bastian AJ, Hu Q, Edwards DF, Sahrmann SA, et al. Deficits in grasp versus reach during acute hemiparesis. Exp Brain Res. 2005;166(1):126–36.
5. Virani SS, Alonso A, Benjamin EJ, Bittencourt MS, Callaway CW, Carson AP, et al. Heart disease and stroke statistics—2020 update: a report from the American Heart Association. Circulation. 2020.
6. Chen SY, Winstein CJ. A systematic review of voluntary arm recovery in hemiparetic stroke: critical predictors for meaningful outcomes using the international classification of functioning, disability, and health. J Neurol Phys Ther. 2009;33 (1):2–13.
7. Katz DI, Alexander MP, Klein RB. Recovery of arm function in patients with paresis after traumatic brain injury. Arch Phys Med Rehabil. 1998;79 (5):488–93.
8. Kister I, Bacon TE, Chamot E, Salter AR, Cutter GR, Kalina JT, et al. Natural history of multiple sclerosis symptoms. Int J MS Care. 2013;15 (3):146–56.
9. Maynard FM, Bracken MB, Creasey G, Jr JFD, Donovan WH, Ducker TB, et al. International standards for neurological and functional classification of spinal cord injury. Spinal Cord. 1997;35 (5):266–74.
10. Pickett GE, Campos-Benitez M, Keller JL, Duggal N. Epidemiology of traumatic spinal cord injury in Canada. Spine. 2006;31(7):799–805.
11. Desrosiers J, Malouin F, Bourbonnais D, Richards CL, Rochette AA, Bravo G. Arm and leg impairments and disabilities after stroke rehabilitation: relation to handicap. Clin Rehabil. 2003;17(6):666–73.
12. Johansson S, Ytterberg C, Claesson IM, Lindberg J, Hillert J, Andersson M, et al. High concurrent presence of disability in multiple sclerosis: Associations with perceived health. J Neurol. 2007;254 (6):767–73.
13. Mayo NE, Wood-Dauphinee S, Ahmed S, Gordon C, Higgins J, Mcewen S, et al. Disablement following stroke. DisabilRehabil. 1999;21(5–6):258–68.
14. Yozbatiran N, Baskurt F, Baskurt Z, Ozakbas S, Idiman E. Motor assessment of upper extremity function and its relation with fatigue, cognitive function and quality of life in multiple sclerosis patients. J Neurol Sci. 2006;246(1–2):117–22.
15. Pollock A, Farmer SE, Brady MC, Langhorne P, Mead GE, Mehrholz J, et al. Interventions for improving upper limb function after stroke. Cochrane Libr. 2014.
16. Barreca S, Wolf SL, Fasoli S, Bohannon R. Treatment interventions for the paretic upper limb of stroke survivors: a critical review. Neurorehabil Neural Repair. 2003;17(4):220–6.
17. Freburger JK, Holmes GM, Ku L-JE, Cutchin MP, Heatwole-Shank K, Edwards LJ. Disparities in postacute rehabilitation care for stroke: an analysis of the state inpatient databases. Arch Phys Med Rehabil. 2011;92(8):1220–9.
18. Chan L, Wang H, Terdiman J, Hoffman J, Ciol MA, Lattimore BF, et al. Disparities in outpatient and home health service utilization following stroke: results of a 9-year cohort study in Northern California. PM&R. 2009;1(11):997–1003.
19. Miller KK, Porter RE, DeBaun-Sprague E, Van Puymbroeck M, Schmid AA. Exercise after stroke: patient adherence and beliefs after discharge from rehabilitation. Top Stroke Rehabil. 2017;24(2): 142–8.
20. Donoso Brown EV, Fichter R. Home programs for upper extremity recovery post-stroke: a survey of occupational therapy practitioners. Top Stroke Rehabil. 2017;24(8):573–8.
21. Cassidy JM, Cramer SC. Spontaneous and therapeutic-induced mechanisms of functional recovery after stroke. Transl Stroke Res. 2017;8 (1):33–46.
22. Merzenich MM, Nelson RJ, Stryker MP, Cynader MS, Schoppmann A, Zook JM. Somatosensory cortical map changes following digit amputation in adult monkeys. J Comp Neurol. 1984;224(4):591–605.
23. Merzenich MM, Jenkins WM. Reorganization of cortical representations of the hand following alterations of skin inputs induced by nerve injury, skin island transfers, and experience. J Hand Ther. 1993;6(2):89–104.
24. Nudo RJ, Wise BM, SiFuentes F, Milliken GW. Neural substrates for the effects of rehabilitative training on motor recovery after ischemic infarct. Science. 1996;272(5269):1791–4.
25. Curt A, Van Hedel HJA, Klaus D, Dietz V, for the EM-SCI study group. recovery from a spinal cord injury: significance of compensation, neural plasticity, and repair. J Neurotrauma. 2008;25(6):677–85.
26. Kwakkel G, Kollen BJ, Van der Grond JV, Prevo AJH. Probability of regaining dexterity in the flaccid upper limb: Impact of severity of paresis and time since onset in acute stroke. Stroke. 2003;34(9):2181–6.
27. Riley JD, Le V, Der-Yeghiaian L, See J, Newton JM, Ward NS, et al. Anatomy of stroke injury predicts gains from therapy. Stroke. 2011;42 (2):421–6.

28. Stinear C. Prediction of recovery of motor function after stroke. Lancet Neurol. 2010;9(12):1228–32.

29. Levin MF, Kleim JA, Wolf SL. What do motor "recovery" and "compensation" mean in patients following stroke? Neurorehabil Neural Repair. 2009/01/02 ed. 2009;23(4):313–9.

30. World Health Organization. Towards a common language for functionary, disability and health: ICF beginner's guide. Geneva WHO. 2002.

31. Ada L, Canning CG, Carr JH, Kilbreath SL, Shepherd RB. Task-specific training of reaching and manipulation. Adv Psychol. 1994;105:239–65.

32. Alaverdashvili M, Foroud A, Lim DH, Whishaw IQ. "Learned baduse" limits recovery of skilled reaching for food after forelimb motor cortex stroke in rats: a new analysis of the effect of gestures on success. Behav Brain Res. 2008;188(2):281–90.

33. Kleim JA, Jones TA. Principles of experience-dependent neural plasticity: implications for rehabilitation after brain damage. J Speech Lang Hear Res. 2008;51(1):S225–39.

34. Aprile I, Rabuffetti M, Padua L, Di Sipio E, Simbolotti C, Ferrarin M, et al. An evidence-based review of stroke rehabilitation. Stroke. 2014;20(1):289–300.

35. Winstein C, Lewthwaite R, Blanton SR, Wolf LB, Wishart L. Infusing motor learning research into neurorehabilitation practice: a historical perspective with case exemplar from the accelerated skill acquisition program. J Neurol Phys Ther. 2014;38(3):190–200.

36. Lamers I, Maris A, Severijns D, Dielkens W, Geurts S, Van Wijmeersch B, et al. Upper limb rehabilitation in people with multiple sclerosis: a systematic review. Neurorehabil Neural Repair. 2016;30(8):773–93.

37. Lu X, Battistuzzo CR, Zoghi M, Galea MP. Effects of training on upper limb function after cervical spinal cord injury: a systematic review. Clin Rehabil. 2015;29(1):3–13.

38. Lee TD, Wishart LR. Motor learning conundrums (and possible solutions). Quest. 2005.

39. Winstein CJ, Kay DB. Translating the science into practice: shaping rehabilitation practice to enhance recovery after brain damage [Internet]. 1st ed. Vol. 218, Progress in Brain Research. Elsevier B.V.; 2015. pp. 331–360. https://doi.org/10.1016/bs.pbr.2015.01.004.

40. Kleim JA, Barbay S, Nudo RJ. Functional reorganization of the rat motor cortex following motor skill learning. J Neurophysiol. 1998;80(6):3321–5.

41. Lohse KR, Lang CE, Boyd LA. Is more better? Using metadata to explore dose-response relationships in stroke rehabilitation. Stroke. 2014;45(7):2053–8.

42. Winstein C, Kim B, Kim S, Martinez C, Schweighofer N. Dosage matters: a phase IIb randomized controlled trial of motor therapy in the chronic phase after stroke. Stroke. 2019;50(7):1831–7.

43. Lang CE, Strube MJ, Bland MD, Waddell KJ, Cherry-Allen KM, Nudo RJ, et al. Dose response of task-specific upper limb training in people at least 6 months poststroke: a phase II, single-blind, randomized, controlled trial. Ann Neurol. 2016;80(3):342–54.

44. Lang CE, MacDonald JR, Reisman DS, Boyd L, Jacobson Kimberley T, Schindler-Ivens SM, et al. Observation of amounts of movement practice provided during stroke rehabilitation. Arch Phys Med Rehabil. 2009;90(10):1692–8.

45. Proffitt R. Home exercise programs for adults with neurological injuries: a survey. Am J Occup Ther. 2016;70(3):7003290020p1–8.

46. Burke JW, McNeill MDJ, Charles DK, Morrow PJ, Crosbie JH, McDonough SM. Optimising engagement for stroke rehabilitation using serious games. Vis Comput. 2009;25(12):1085–99.

47. Dickstein R, Hocherman S, Pillar T, Shaham R. Stroke rehabilitation: three exercise therapy approaches. Phys Ther. 1986;66(8):1233–8.

48. Subramanian SK, Fountain MA, Hood AF, Verduzco-Gutierrez M. Upper Limb motor improvement after TBI: systematic review of interventions [Internet]. Rehabil Med Phys Ther. 2020 Nov [cited 2021 Oct 5]. http://medrxiv.org/lookup, https://doi.org/10.1101/2020.11.12.20214478.

49. Spooren A, Janssen-Potten YJM, Kerckhofs E, Seelen HAM. Outcome of motor training programmes on arm and hand functioning in patients with cervical spinal cord injury according to different levels of the ICF: a systematic review. J Rehabil Med. 41(7):497–505.

50. Saikaley M, Pauli G, Iruthayarajah J, Mirkowski M, Iliescu A, Caughlin S, et al. Upper extremity motor rehabilitation interventions. 366.

51. Wolf S, Lecraw D, Barton L, Jann B. Forced use of hemiplegic upper extremities to reverse the effect of learned nonuse among chronic stroke and head-injured patients. Exp Neurol. 1989;104(2):125–32.

52. Taub E. Movement in nonhuman primates deprived of somatosensory feedback. Exerc Sport Sci Rev. 1976;4(1):335–74.

53. Taub E, Crago JE, Burgio LD, Groomes TE, Cook EW, DeLuca SC, et al. An operant approach to rehabilitation medicine: overcoming learned nonuse by shaping. J Exp Anal Behav. 1994;61(2):281–93.

54. Pons TP, Garraghty PE, Ommaya AK, Kaas JH, Taub E, Mishkin M. Massive cortical reorganization after sensory deafferentation in adult macaques. Science. 1991;252(5014):1857–60.

55. Richards LG, Stewart KC, Woodbury ML, Senesac C, Cauraugh JH. Movement-dependent stroke recovery: a systematic review and meta-analysis of TMS and fMRI evidence. Neuropsychologia. 2008;46(1):3–11.

56. Taub E, Uswatte G, Mark VW, Morris DMM. The learned nonuse phenomenon: Implications for rehabilitation. Eur Medicophysica. 2006;42(3):241–55.

57. Taub E, Uswatte G, Mark VW, Morris DM, Barman J, Bowman MH, et al. Method for enhancing real-world use of a more affected arm in chronic stroke. Stroke. 2013;44(5):1383–8.

58. Winstein CJ, Wolf SL. Task-oriented training to promote upper extremity recovery. In: Stein J Macko RF, Winstein CJ, Zorowitz RD HRL, editor. Stroke recovery and rehabilitation. 2nd edition. New York: Demos Medical; 2014. pp. 267–90.

59. Corbetta D, Sirtori V, Castellini G, Moja L, Gatti R. Constraint-induced movement therapy for upper extremities in people with stroke. Cochrane Database of Systematic Reviews. 2015.

60. Kwakkel G, Veerbeek JM, van Wegen EEH, Wolf SL. Constraint-induced movement therapy after stroke. Lancet Neurol. 2015;14(2):224–34.

61. Veerbeek JM, van Wegen E, van Peppen R, van der Wees PJ, Hendriks E, Rietberg M, et al. What is the evidence for physical therapy poststroke? A systematic review and meta-analysis. PLoS ONE. 2014;9(2): e87987.

62. Daniel L, Howard W, Braun D, Page SJ. Opinions of constraint-induced movement therapy among therapists in southwestern Ohio. Top Stroke Rehabil. 2012;19(3):268–75.

63. Dobkin BH. Confounders in rehabilitation trials of task-oriented training: lessons from the designs of the excite and scilt multicenter trials. Neurorehabil Neural Repair. 2007;21(1):3–13.

64. Page SJ, Levine P. Modified constraint-induced therapy in patients with chronic stroke exhibiting minimal movement ability in the affected arm. Phys Ther. 2007;87(7):872–8.

65. L. Wolf S, J. Winstein C, Miller JP, Blanton S, Clark PC, Nichols-Larsen D. Looking in the rear view mirror when conversing with back seat drivers: the EXCITE trial revisited. Neurorehabil Neural Repair. 2007;21(5):379–87.

66. Yozbatiran N, Francisco G. Robot-assisted therapy for the upper limb after cervical spinal cord injury. Phys Med Rehabil Clin N Am. 2019;1:30.

67. Reinkensmeyer, David J. Robotic approaches to stroke recovery. In: Brain repair after stroke. Cambridge University Press; 2010. p. 195–205.

68. Bertani R, Melegari C, De Cola MC, Bramanti A, Bramanti P, Calabrò RS. Effects of robot-assisted upper limb rehabilitation in stroke patients: a systematic review with meta-analysis. Neurol Sci. 2017;38(9):1561–9.

69. Holanda LJ, Silva PMM, Amorim TC, Lacerda MO, Simão CR, Morya E. Robotic assisted gait as a tool for rehabilitation of individuals with spinal cord injury: a systematic review. J NeuroEngineering Rehabil. 2017;14(1):126.

70. Khalid S, Alnajjar F, Gochoo M, Renawi A, Shimoda S. Robotic assistive and rehabilitation devices leading to motor recovery in upper limb: a systematic review. Disabil Rehabil Assist Technol. 2021;16:1–15.

71. Veerbeek JM, Langbroek-Amersfoort AC, Van Wegen EEH, Meskers CGM, Kwakkel G. Effects of robot-assisted therapy for the upper limb after stroke. Neurorehabilitation Neural Repair. 2017.

72. Rowe JB, Chan V, Ingemanson ML, Cramer SC, Wolbrecht ET, Reinkensmeyer DJ. Robotic assistance for training finger movement using a Hebbian model: a randomized controlled trial. Neurorehabil Neural Repair. 2017;31(8):769–80.

73. Kahn LE, Lum PS, Rymer WZ, Reinkensmeyer DJ. Robot-assisted movement training for the stroke-impaired arm: does it matter what the robot does? J Rehabil Res Dev. 2006;43(5):619.

74. Chyatte SB, Long C, Vignos PJ. The balanced forearm orthosis in muscular dystrophy. 1965;46(9):633–6.

75. Atkins MS, Baumgarten JM, Yasuda YL, Adkins R, Waters RL, Leung P, et al. Mobile arm supports: evidence-based benefits and criteria for use. J Spinal Cord Med. 2008;31(4):388–93.

76. Haworth R, Dunscombe S, Nichols PJR. Mobile arm supports: an evaluation. Rheumatology. 1978;17(4):240–4.

77. Yasuda YL, Bowman K, Hsu JD. Mobile arm supports: Criteria for successful use in muscle disease patients. Arch Phys Med Rehabil. 1986;67(4):253–6.

78. van Dorsser WD, Barents R, Wisse BM, Herder JL. Gravity-balanced arm support with energy-free adjustment. J Med Devices. 2007;1(2):151–8.

79. Housman SJ, Le V, Rahman T, Sanchez RJ, Reinkensmeyer DJ. Arm-training with T-WREX after chronic stroke: preliminary results of a randomized controlled trial. In: 2007 IEEE 10th international conference on rehabilitation robotics. 2007. pp. 562–8.

80. Mastenbroek B, de Haan E, van den Berg M, Herder JL. Development of a mobile arm support (armon): design evolution and preliminary user experience. In: 2007 IEEE 10th International Conference on Rehabilitation Robotics. 2007. p. 1114–20.

81. Housman SJ, Scott KM, Reinkensmeyer DJ. A randomized controlled trial of gravity-supported, computer-enhanced arm exercise for individuals with severe hemiparesis. Neurorehabil Neural Repair. 2009;23(5):505–14.

82. Iwamuro BT, Cruz EG, Connelly LL, Fischer HC, Kamper DG. Effect of a gravity-compensating orthosis on reaching after stroke: evaluation of the therapy assistant WREX. Arch Phys Med Rehabil. 2008;89(11):2121–8.

83. Sanchez RJ, Liu J, Rao S, Shah P, Smith R, Rahman T, et al. Automating arm movement training following severe stroke: functional exercises with quantitative feedback in a gravity-reduced environment. IEEE Trans Neural Syst Rehabil Eng. 2006;14(3):378–89.

84. Prange GB, Stienen AHA, Jannink MJA, van der Kooij H, IJzerman MJ, Hermens HJ. Increased range of motion and decreased muscle activity during maximal reach with gravity compensation in stroke patients. In: 2007 IEEE 10th international conference on rehabilitation robotics. 2007. pp. 467–71.

85. Reinkensmeyer DJ, Housman SJ. If I can't do it once, why do it a hundred times? Connecting volition to movement success in a virtual environment motivates people to exercise the arm after stroke. In: 2007 Virtual Rehabilitation. 2007. pp. 44–8.

86. Rémy-Néris O, Le Jeannic A, Dion A, Médée B, Nowak E, Poiroux É, et al. Additional, mechanized upper limb self-rehabilitation in patients with sub-acute stroke: the REM-AVC randomized trial. Stroke. 2021;52(6):1938–47.

87. Stienen AHA, Hekman EEG, Prange GB, Jannink MJA, van der Helm FCT, van der Kooij H. Freebal: design of a dedicated weight-support system for upper-extremity rehabilitation. J Med Devices. 2009;3(4): 041009.

88. Stienen AHA, Hekman EEG, Van der Helm FCT, Prange GB, Jannink MJA, Aalsma AMM, et al. Freebal: dedicated gravity compensation for the upper extremities. In: 2007 IEEE 10th International Conference on Rehabilitation Robotics. 2007. p. 804–8.

89. Prange GB, Jannink MJA, Stienen AHA, van der Kooij H, IJzerman MJ, Hermens HJ. Influence of gravity compensation on muscle activation patterns during different temporal phases of arm movements of stroke patients. Neurorehabil Neural Repair. 2009 Jun 1;23(5):478–85.

90. Krabben T, Prange GB, Molier BI, Stienen AH, Jannink MJ, Buurke JH, et al. Influence of gravity compensation training on synergistic movement patterns of the upper extremity after stroke, a pilot study. J NeuroEngineering Rehabil. 2012;9(1):1–12.

91. Chen J, Lum PS. Spring operated wearable enhancer for arm rehabilitation (SpringWear) after stroke. In: 2016 38th annual international conference of the IEEE engineering in medicine and biology society (EMBC). 2016. p. 4893–6.

92. Asgari M, Hall PT, Moore BS, Crouch DL. Wearable shoulder exoskeleton with spring-cam mechanism for customizable, nonlinear gravity compensation. In: 2020 42nd annual international conference of the IEEE engineering in medicine biology society (EMBC). 2020. pp. 4926–9.

93. Hall PT. Effect of continuous, mechanically passive, anti-gravity assistance on kinematics and muscle activity during dynamic shoulder elevation. J Biomech. 2020;8.

94. Chen J, Lum PS. Pilot testing of the spring operated wearable enhancer for arm rehabilitation (SpringWear). J NeuroEngineering Rehabil. 2018;15(1):1–11.

95. Hijmans J, King M, Hale L. Computerised table top exerciser for stroke survivors [Internet]. 2021 [cited 2021 Oct 5]. https://www.researchgate.net/publication/267698027_COMPUTERISED_TABLE_TOP_EXERCISER_FOR_STROKE_SURVIVORS

96. Park J, Jung H-T, Daneault J-F, Park S, Ryu T, Kim Y, et al. Effectiveness of the RAPAEL smart board for upper limb therapy in stroke survivors: a pilot controlled trial. In: 2018 40th annual international conference of the IEEE engineering in medicine and biology society (EMBC). 2018. pp. 2466–9.

97. Peattie A, Korevaar A, Wilson J, S B, Chen X, King M. Automated variable resistance system for upper limb rehabilitation. In: Proceedings of the Australian robotics & automation. 2009. p. 2–4.

98. Burdea GC, Cioi D, Martin J, Fensterheim D, Holenski M. The Rutgers arm II rehabilitation system—a feasibility study. IEEE Trans Neural Syst Rehabil Eng. 2010;18(5):505–14.

99. House G, Burdea G, Grampurohit N, Polistico K, Roll D, Damiani F. Integrative upper-limb rehabilitation with *BrightArm Duo* TM in the early sub-acute phase of recovery post-stroke. J Med Robot Res. 2017;02(02):1740004.

100. Jordan K, Sampson M, King M. Gravity-supported exercise with computer gaming improves arm function in chronic stroke. Arch Phys Med Rehabil. 2014;95(8):1484–9.

101. Whitall J, Waller SM, Silver KHC, Macko RF. Repetitive Bilateral arm training with rhythmic auditory cueing improves motor function in chronic hemiparetic stroke. Stroke. 2000;31(10):2390–5.

102. Hesse S, Schmidt H, Werner C, Rybski C, Puzich U, Bardeleben A. A new mechanical arm trainer to intensify the upper limb rehabilitation of severely affected patients after stroke: design, concept and first case series. Eur Medicophysica. 2007;43(4):463–8.

103. Hayward KS, Barker RN, Brauer SG, Lloyd D, Horsley SA, Carson RG. SMART arm with outcome-triggered electrical stimulation: a pilot randomized clinical trial. Top Stroke Rehabil. 2013;20(4):289–98.

104. Minge M, Ivanova E, Lorenz K, Joost G, Thüring M, Krüger J. BeMobil: Developing a user-friendly and motivating telerehabilitation system for motor relearning after stroke. In: 2017 international conference on rehabilitation robotics (ICORR). 2017. pp. 870–5.

105. Luft AR, McCombe-Waller S, Whitall J, Forrester LW, Macko R, Sorkin JD, et al. Repetitive bilateral arm training and motor cortex activation in chronic StrokeA randomized controlled trial. JAMA. 2004;292(15):1853–61.

106. Richards LG, Senesac CR, Davis SB, Woodbury ML, Nadeau SE. Bilateral arm training with rhythmic auditory cueing in chronic stroke: not always efficacious. Neurorehabil Neural Repair. 2008;22(2):180–4.

107. Whitall J, Waller SM, Sorkin JD, Forrester LW, Macko RF, Hanley DF, et al. Bilateral and Unilateral Arm Training Improve Motor Function Through Differing Neuroplastic Mechanisms: A Single-Blinded Randomized Controlled Trial. Neurorehabil Neural Repair. 2011;25(2):118–29.

108. van Delden A (Lex) EQ, Peper C (Lieke) E, Nienhuys KN, Zijp NI, Beek PJ, Kwakkel G. Unilateral versus bilateral upper limb training after stroke: the upper limb training after stroke clinical trial. Stroke. 2013 Sep;44(9):2613–6.

109. Hesse S, Werner C, Pohl M, Mehrholz J, Puzich U, Krebs HI. Mechanical arm trainer for the treatment of the severely affected arm after a stroke: a single-blinded randomized trial in two centers. Am J Phys Med Rehabil. 2008;87(10):779–88.

110. Barker RN, Hayward KS, Carson RG, Lloyd D, Brauer SG. SMART arm training with outcome-triggered electrical stimulation in subacute stroke survivors with severe arm disability: a randomized controlled trial. Neurorehabil Neural Repair. 2017;31(12):1005–16.

111. Brokaw EB, Black I, Holley RJ, Lum PS. Hand spring operated movement enhancer (HandSOME): a portable, passive hand exoskeleton for stroke rehabilitation. IEEE Trans Neural Syst Rehabil Eng. 2011;19(4):391–9.

112. Haarman CJW, Hekman EEG, Prange GB, van der Kooij H. Joint Stiffness Compensation for Application in the EXTEND Hand Orthosis. In: 2018 7th IEEE International Conference on Biomedical Robotics and Biomechatronics (Biorob). 2018. p. 677–82.

113. Ates S, Haarman CJW, Stienen AHA. SCRIPT passive orthosis: design of interactive hand and wrist exoskeleton for rehabilitation at home after stroke. Auton Robots. 2017;41(3):711–23.

114. S. Nijenhuis. Technology-supported arm and hand training at home after stroke [PhD Thesis]. [Enschede, The Netherlands]: University of Twente; 2018.

115. Chen J, Nichols D, Brokaw EB, Lum PS. Home-based therapy after stroke using the hand spring operated movement enhancer (HandSOME). IEEE Trans Neural Syst Rehabil Eng. 2017;25(12):2305–12.

116. Nijenhuis SM, Prange-Lasonder GB, Stienen AH, Rietman JS, Buurke JH. Effects of training with a passive hand orthosis and games at home in chronic stroke: a pilot randomised controlled trial. Clin Rehabil. 2017;31(2):207–16.

117. Levin MF, Demers M. Motor learning in neurological rehabilitation. Disabil Rehabil. 2020:1–9.

118. Levac D, Glegg S, Colquhoun H, Miller P, Noubary F. Virtual reality and active videogame-based practice, learning needs, and preferences: a cross-Canada survey of physical therapists and occupational therapists. Games Health J. 2017;6(4):217–28.

119. Thomson K, Pollock A, Bugge C, Brady MC. Commercial gaming devices for stroke upper limb rehabilitation: a survey of current practice. Disabil Rehabil Assist Technol. 2016;11(6):454–61.

120. Levin MF, Deutsch JE, Kafri M, Liebermann D. Validity of virtual reality environments for sensorimotor rehabilitation. In: Virtual reality for physical and motor rehabilitation. Springer; 2014. pp. 95–118. (Virtual reality technologies for health and clinical applications).

121. Demers M, Levin MF. Kinematic validity of reaching in a 2D virtual environment for arm rehabilitation after stroke. IEEE Trans Neural Syst Rehabil Eng. 2020;28(3):679–86.

122. Deutsch JE, Brettler A, Smith C, Welsh J, John R, Guarrera-Bowlby P, et al. Nintendo Wii sports and Wii fit game analysis, validation, and application to stroke rehabilitation. Top Stroke Rehabil. 2011;18(6):701–19.

123. Givon Schaham N, Zeilig G, Weingarden H, Rand D. Game analysis and clinical use of the Xbox-Kinect for stroke rehabilitation. Int J Rehabil Res Int Z Rehabil Rev Int Rech Readaptation. 2018;41(4):323–30.

124. Cheok G, Tan D, Low A, Hewitt J. Is Nintendo Wii an Effective Intervention for Individuals With Stroke? A Systematic Review and Meta-Analysis. J Am Med Dir Assoc. 2015;16(11):923–32.

125. Laver KE, Lange B, George S, Deutsch JE, Saposnik G, Crotty M. Virtual reality for stroke rehabilitation. Cochrane Database Syst Rev. 2017;2017(11).

126. Domínguez-Téllez P, Moral-Muñoz JA, Salazar A, Casado-Fernández E, Lucena-Antón D. Game-based virtual reality interventions to improve upper limb motor function and quality of life after stroke: systematic review and meta-analysis. Games Health J. 2020;9(1):1–10.

127. Mekbib DB, Han J, Zhang L, Fang S, Jiang H, Zhu J, et al. Virtual reality therapy for upper limb rehabilitation in patients with stroke: a meta-analysis of randomized clinical trials. Brain Inj. 2020;34(4):456–65.

128. Jung H-T, Kim H, Jeong J, Jeon B, Ryu T, Kim Y. Feasibility of using the RAPAEL Smart Glove in upper limb physical therapy for patients after stroke: a randomized controlled trial. In: 2017 39th Annual International Conference of the IEEE Engineering in Medicine and Biology Society (EMBC). 2017. pp. 3856–9.

129. Jordan K, Sampson M, Hijmans J, King M, Hale L. ImAble system for upper limb stroke rehabilitation.

In: 2011 International Conference on Virtual Rehabilitation. 2011. pp. 1–2.

130. Hartwig M. Fun and evidence - computer-based arm rehabilitation with the Pablo®Plus System [Internet]. Tyromotion. 2011 [cited 2021 Oct 29]. https:// tyromotion.com/wp-content/uploads/2020/11/ HartwigM-2011-Fun-and-evidence-computer-based-arm-rehabiltation-with-the-PabloPlus-System.pdf.

131. Friedman N, Chan V, Zondervan D, Bachman M, Reinkensmeyer DJ. MusicGlove: Motivating and quantifying hand movement rehabilitation by using functional grips to play music. In: 2011 Annual international conference of the IEEE engineering in medicine and biology society. 2011. pp. 2359–63.

132. Friedman N, Chan V, Reinkensmeyer AN, Beroukhim A, Zambrano GJ, Bachman M, et al. Retraining and assessing hand movement after stroke using the MusicGlove: comparison with conventional hand therapy and isometric grip training. J NeuroEngineering Rehabil. 2014;11 (1):1–14.

133. Zondervan DK, Friedman N, Chang E, Zhao X, Augsburger R, Reinkensmeyer DJ, et al. Home-based hand rehabilitation after chronic stroke: Randomized, controlled single-blind trial comparing the MusicGlove with a conventional exercise program. J Rehabil Res Dev. 2016;53(4):457–72.

134. Kilbride C, Scott DJM, Butcher T, Norris M, Ryan JM, Anokye N, et al. Rehabilitation via HOMe based gaming exercise for the upper-limb post Stroke (RHOMBUS): protocol of an intervention feasibility trial. BMJ Open. 2018;8(11): e026620.

135. Prochazka A, Kowalczewski J. A Fully Automated, Quantitative Test of Upper Limb Function. J Mot Behav. 2015;47(1):19–28.

136. Kowalczewski J, Chong SL, Galea M, Prochazka A. In-home tele-rehabilitation improves tetraplegic hand function. Neurorehabil Neural Repair. 2011;25(5):412–22.

137. Kowalczewski J. Upper extremity neurorehabilitation [Internet] [PhD Thesis]. [Edmonton, Canada]: University of Alberta; 2009 [cited 2021 Sep 6]. https://era.library.ualberta.ca/items/a3966268-bf3d-4bd4-9030.

138. Howlett OA, Lannin NA, Ada L, McKinstry C. Functional electrical stimulation improves activity after stroke: a systematic review with meta-analysis. Arch Phys Med Rehabil. 2015;96(5):934–43.

139. Hara Y, Obayashi S, Tsujiuchi K, Muraoka Y. The effects of electromyography-controlled functional electrical stimulation on upper extremity function and cortical perfusion in stroke patients. Clin Neurophysiol. 2013;124(10):2008–15.

140. Chou C-H, Wang T, Sun X, Niu CM, Hao M, Xie Q, et al. Automated functional electrical stimulation training system for upper-limb function recovery in poststroke patients. Med Eng Phys. 2020;84:174–83.

141. Razavian RS, Ghannadi B, McPhee J. Feedback control of functional electrical stimulation for arbitrary upper extremity movements. In: 2017 International Conference on Rehabilitation Robotics (ICORR). 2017. p. 1451–6.

142. Graham GM, Thrasher TA, Popovic MR. The effect of random modulation of functional electrical stimulation parameters on muscle fatigue. IEEE Trans Neural Syst Rehabil Eng. 2006;14(1):38–45.

143. Engineer ND, Kimberley TJ, Prudente CN, Dawson J, Tarver WB, Hays SA. Targeted vagus nerve stimulation for rehabilitation after stroke. Front Neurosci. 2019;13:280.

144. Dawson J, Pierce D, Dixit A, Kimberley TJ, Robertson M, Tarver B, et al. Safety, feasibility, and efficacy of vagus nerve stimulation paired with upper-limb rehabilitation after ischemic stroke. Stroke. 2016;47(1):143–50.

145. Meyers EC, Solorzano BR, James J, Ganzer PD, Lai ES, Rennaker RL, et al. Vagus nerve stimulation enhances stable plasticity and generalization of stroke recovery. Stroke. 2018;49(3):710–7.

146. Hays SA, Rennaker RL, Kilgard MP. Chapter 11—targeting plasticity with vagus nerve stimulation to treat neurological disease. In: Merzenich MM, Nahum M, Van Vleet TM, editors. Progress in brain research. Elsevier; 2013 [cited 2021 Oct 5]. p. 275–99. (Changing Brains; vol. 207). https:// www.sciencedirect.com/science/article/pii/ B9780444633279000102.

147. Andrews K, Stewart J. Stroke recovery: he can but does he? Rheumatology. 1979;18(1):43–8.

148. Kowalczewski J, Prochazka A. Technology improves upper extremity rehabilitation. Prog Brain Res. 2011;192:147–59.

149. Gritsenko V, Prochazka A. A functional electric stimulation—assisted exercise therapy system for hemiplegic hand function11No commercial party having a direct financial interest in the results of the research supporting this article has or will confer a benefit upon the authors(s) or upon any organization with which the author(s) is/are associated. Arch Phys Med Rehabil. 2004;85(6):881–5.

150. Buick AR, Kowalczewski J, Carson RG, Prochazka A. Tele-supervised FES-assisted exercise for hemiplegic upper limb. IEEE Trans Neural Syst Rehabil Eng. 2016;24(1):79–87.

151. Wolf SL, Sahu K, Bay RC, Buchanan S, Reiss A, Linder S, et al. The HAAPI (home arm assistance progression initiative) trial: a novel robotics delivery approach in stroke rehabilitation. Neurorehabil Neural Repair. 2015;29(10):958–68.

152. Brennan DM, Mawson S, Brownsell S. Telerehabilitation: Enabling the remote delivery of healthcare, rehabilitation, and self management. Stud Health Technol Inform. 2009;145:231–48.

153. Soopramanien A, Jamwal S, Thomas PW. Digital health rehabilitation can improve access to care in spinal cord injury in the UK: a proposed solution. Int J Telerehabilitation. 12(1):3–16.

154. Brennan K, Curran J, Barlow A, Jayaraman A. Telerehabilitation in neurorehabilitation: has it passed the COVID test? Expert Rev Neurother. 2021;21(8):833–6.

155. Bower KJ, Verdonck M, Hamilton A, Williams G, Tan D, Clark RA. What factors influence clinicians' use of technology in neurorehabilitation? a multisite qualitative study. Phys Ther. 2021;101(5):pzab031.

156. Covid-19: guide for rapid implementation of remote physiotherapy delivery [Internet]. Chart Soc Physiother. [cited 2021 Sep 22]. https://www.csp.org.uk/publications/covid-19-guide-rapid-implementation-remote-physiotherapy-delivery

157. Occupational Therapy Australia. Telehealth guidelines 2020 [Internet]. 2020 [cited 2021 Oct 6]. https://otaus.com.au/publicassets/553c6eae-ad6c-ea11-9404-005056be13b5/OTA%20Telehealth%20Guidelines%202020.pdf.

158. Langhorne P, Baylan S, Trialists ESD. Early supported discharge services for people with acute stroke. Cochrane Database Syst Rev [Internet]. 2017 [cited 2021 Oct 5];(7). http://www.cochranelibrary.com/cdsr, https://doi.org/10.1002/14651858.CD000443.pub4/full.

159. Langhorne P, Taylor G, Murray G, Dennis M, Anderson C, Bautz-Holter E, et al. Early supported discharge services for stroke patients: a meta-analysis of individual patients' data. The Lancet. 2005;365(9458):501–6.

160. Glegg SMN, Holsti L, Velikonja D, Ansley B, Brum C, Sartor D. Factors influencing therapists' adoption of virtual reality for brain injury rehabilitation. Cyberpsychology Behav Soc Netw. 2013;16(5):385–401.

161. Langan J, Subryan H, Nwogu I, Cavuoto L. Reported use of technology in stroke rehabilitation by physical and occupational therapists. Disabil Rehabil Assist Technol. 2018;13(7):641–7.

162. Chen CC, Bode RK. Factors influencing therapists' decision-making in the acceptance of new technology devices in stroke rehabilitation. Am J Phys Med Rehabil. 2011;90(5):415–25.

163. Hughes AM, Burridge JH, Demain SH, Ellis-Hill C, Meagher C, Tedesco-Triccas L, et al. Translation of evidence-based Assistive Technologies into stroke rehabilitation: Users' perceptions of the barriers and opportunities. BMC Health Serv Res. 2014.

164. Elnady A, Mortenson WB, Menon C. Perceptions of existing wearable robotic devices for upper extremity and suggestions for their development: findings from therapists and people with stroke. JMIR Rehabil Assist Technol. 2018;5(1): e12.

165. Flynn N, Kuys S, Froude E, Cooke D. Introducing robotic upper limb training into routine clinical practice for stroke survivors: Perceptions of occupational therapists and physiotherapists. Aust Occup Ther J. 2019;66(4):530–8.

166. Hochstenbach-Waelen A, Seelen HAM. Embracing change: practical and theoretical considerations for successful implementation of technology assisting upper limb training in stroke. J NeuroEngineering Rehabil. 2012.

167. van Ommeren AL, Smulders LC, Prange-Lasonder GB, Buurke JH, Veltink PH, Rietman JS. Assistive technology for the upper extremities after stroke: systematic review of users' needs. JMIR Rehabil Assist Technol. 2018;5(2): e10510.

168. Chan IHL, Fong KNK, Chan DYL, Wang AQL, Cheng EKN, Chau PHY, et al. Effects of arm weight support training to promote recovery of upper limb function for subacute patients after stroke with different levels of arm impairments. BioMed Res Int. 2016;2016:1–9.

169. Ates S, Leon B, Basteris A, Nijenhuis S, Nasr N, Sale P, et al. Technical evaluation of and clinical experiences with the SCRIPT passive wrist and hand orthosis. In: 2014 7th international conference on human system interactions (HSI). 2014. pp. 188–93.

170. Clark RA, Pua YH, Fortin K, Ritchie C, Webster KE, Denehy L, et al. Validity of the Microsoft Kinect for assessment of postural control. Gait Posture. 2012;36(3):372–7.

171. Swanson VA, Chan V, Cruz-Coble B, Alcantara CM, Scott D, Jones M, et al. A pilot study of a sensor enhanced activity management system for promoting home rehabilitation exercise performed during the COVID-19 pandemic: therapist experience, reimbursement, and recommendations for implementation. Int J Environ Res Public Health. 2021;18(19):10186.

172. Jones M, Collier G, Reinkensmeyer DJ, Deruyter F, Dzivak J, Zondervan D, et al. Big data analytics and sensor-enhanced activity management to improve effectiveness and efficiency of outpatient medical rehabilitation. Int J Environ Res Public Health. 2020;17(3).

Mobile Technology for Cognitive Rehabilitation

24

Amanda R. Rabinowitz,
Shannon B. Juengst,
and Thomas F. Bergquist

Abstract

Many barriers currently prevent persons living with acquired brain injury (ABI) from receiving an adequate "dose" of cognitive rehabilitation therapy via the traditional clinic-based service delivery model. The application of mobile technology to cognitive rehabilitation is an emerging area of great interest. Advances in mobile technology provide new opportunities for cognitive rehabilitation, with the potential to improve access to care, as well as increase patients' opportunities to practice and apply skills in their everyday environments—a notion referred to as *ecologically valid treatment*. The aim of this chapter is to provide an overview of mobile technology in the context of cognitive rehabilitation, with a special focus on applications and interventions targeted towards patients with ABI. We provide a general background on cognitive rehabilitation as a treatment model, discuss opportunities and considerations for developing mobile rehabilitation solutions for users with cognitive impairment, and provide a general overview of the state of the research on mobile cognitive rehab treatments. The vast majority of mobile health apps are not evidence based at present (Ramey et al. in Phys Med Rehabil Clin 30(2):485–97, 2019), however, a small but growing literature is now speaking to the use of smart technologies to both help improve functioning and monitor functioning for persons with ABI. We highlight several areas for future development needed to move towards developing clinical practice guidelines for integrating mobile technology into cognitive rehabilitation.

Keywords

Cognitive rehabilitation · Mobile health (mHealth) · Stroke · Traumatic brain injury (TBI) · Apps · Technology · Rehabilitation

A. R. Rabinowitz (✉)
Moss Rehabilitation Research Institute, 50 Township Line Rd., Elkins Park, PA 19027, USA
e-mail: rabinowa@einstein.edu

S. B. Juengst
TIRR Memorial Hermann Hospital, Houston, TX 77030, USA
e-mail: shannon.juengst@memorialhermann.org

T. F. Bergquist
Mayo Clinic, Rochester, MN, USA
e-mail: bergquist.thomas@mayo.edu

24.1 Introduction

Many barriers currently prevent persons living with acquired brain injury (ABI) from receiving an adequate "dose" of cognitive rehabilitation therapy via the traditional clinic-based service delivery model. This critical service gap exists for a number of reasons, including insufficient clinical resources, poor access to care, and

© The Author(s), under exclusive license to Springer Nature Switzerland AG 2022
D. J. Reinkensmeyer et al. (eds.), *Neurorehabilitation Technology*,
https://doi.org/10.1007/978-3-031-08995-4_24

limited insurance coverage [1]. Some patients with cognitive disabilities are sent home from acute care without any outpatient cognitive rehabilitation at all. Those fortunate enough to access services often receive coverage that is fragmented and front-loaded, with the majority of therapy provided in the weeks and months after a traumatic brain injury (TBI) or stroke. Support dwindles as patients stabilize and return to the community. Studies suggest that the majority of individuals receive inadequate support after discharge from inpatient rehabilitation, at which point caregivers report a progressive decline in both the quality and quantity of services [2]. The short duration of care is a concern. For example, the field increasingly acknowledges that moderate-severe TBI is a chronic, and even progressive disease process [3, 4], for which ongoing rehabilitation is needed to prevent increasing disability. The evidence from the stroke literature strongly suggests that greater intensity therapy over a longer duration, at least within the first year post-stroke, is associated with greater functional improvements [5, 6]. Without sustained access to services many patients are likely to fall short of achieving their optimal level of functioning, and may even be at risk of experiencing further cognitive decline years later [7]. Given all of this, there is a compelling need for remote delivery of cognitive rehabilitation services to address the needs of those living with acquired brain injury (ABI).

The application of mobile technology to cognitive rehabilitation is an emerging area of great interest. Advances in mobile technology provide new opportunities for cognitive rehabilitation, with the potential to improve access to care, as well as increase patients' opportunities to practice and apply skills in their everyday environments—a notion referred to as *ecologically valid treatment*. The aim of this chapter is to provide an overview of mobile technology in the context of cognitive rehabilitation, with a special focus on applications and interventions targeted towards patients with ABI. We provide a general background on cognitive rehabilitation as a treatment model, discuss opportunities and considerations for developing mobile rehabilitation solutions for users with cognitive impairment, and provide a general overview of the state of the research on mobile cognitive rehab treatments. Finally, we provide recommendations for future development and research in this area.

24.2 Cognitive Rehabilitation

Cognitive rehabilitation has been defined as "a systematic, functionally oriented service of therapeutic cognitive activities, based on an assessment and understanding of the person's brain-behavioral deficits. Services are provided by qualified practitioners and are directed to achieve functional changes by [1] reinforcing, strengthening, or reestablishing previously learned patterns of behavior, or [2] establishing new patterns of cognitive activity or compensatory mechanisms for impaired neurological systems" (p. 62) [8]. The process of cognitive rehabilitation can be broadly categorized into interventions which are either restorative or compensatory. **Restorative** interventions, also referred to as direct training or process-specific, are aimed at the impairment level and involve repetitive stimulation of a domain, typically in a systematic, methodical manner. **Compensatory** approaches are targeted at the functional level and are based on the assumption that although damaged neurological functions cannot be restored, one's intact strengths and abilities can be used to circumvent impairments. To this end, compensations can reduce the extent of limitations on activities and participation and promote successful community reintegration in spite of residual physical and cognitive impairments. Cognitive rehabilitation initially focused on the use of restorative approaches. Within the past few decades, the focus has been more on compensation for areas of impairment [9]. Historically, the goal of the restorative approaches has been to reduce or eliminate underlying cognitive impairment [10]. Restorative interventions are focused on exercises targeting specific cognitive impairment after brain injury or disease. Although the notion of directly diminishing cognitive impairment is appealing, many have noted that a program of rehabilitation will only have relevance to patients and their families to the extent that it improves daily functioning. In other words, successful cognitive rehabilitation should produce functional improvements in patients' daily lives, and not just higher scores on laboratory tests of cognition.

The International Classification of Functioning, Disabilities, and Health (ICF) model was developed by the World Health Organization (WHO) to operationalize the functional changes associated with known medical conditions [11]. This model classifies changes associated with brain injury into changes in (1) body functions and structures, (2) activity, and (3) participation. Body functions and structures are measured by assessment of physical or mental functions, and impairment is measured by deviation from expected levels of performance. Activity limitations are an individual's inability to complete basic or instrumental activity of daily living (ADL/IADL) (e.g., inability to recall appointments, follow a recipe while cooking, and follow a medication regimen or balance a checkbook). Participation restrictions are a loss or change in social roles (e.g., loss of a job or inability to parent children). Participation is highly individualized and needs to be assessed by self or family report regarding the degree to which an individual is [1] a productive member of society and/or [2] integrated into family and community life. Participation restrictions reflect, for example, the degree to which individuals are limited in their life roles (e.g., ability to run a household or maintain a network of friends and family, employment, education, and volunteer activities). In this model, there is a dynamic interaction between impairments (both physical and cognitive), activity limitations, and participation restrictions which collectively determine an individual's reintegration into the community. The ICF proposes that ADLs and participation in life roles are the two domains which measure the impact of an injury or illness on daily functioning. It then follows that these same domains best measure treatment effectiveness. Not surprisingly, recent studies of the effectiveness of cognitive rehabilitation largely focus on ADLs and participation as their main measures of treatment outcomes [12].

Cognitive rehabilitation is often time- and labor-intensive and can require multiple sessions to help teach and apply new skills to improve daily functioning. While the restorative approach alone may not be adequate for significant gains in functioning, these approaches are used within the framework of the ICF to rebuild discrete skills in coordination with other approaches tied to functional outcomes. Restorative approaches have a clear and important role in the rehabilitation process by being paired with compensatory approaches towards the goal of improving functioning. Distinct compensation skills, such as those used to increase visual scanning, decrease impulsivity, and/or decrease episodes of forgetfulness will only be effective to the extent that they are reliably and consistently used. While it may seem that once a skill or strategy is learned, it will then be readily applied, that in fact is not necessarily the case [13]. No matter how effective a strategy may be in practice, it will not be effective for the patient until they have the appropriate skills in place to use that strategy in their day-to-day life. This process requires repeated practice, particularly within the patients' usual environments (i.e., their homes and communities).

A compensation strategy designed to enhance functioning can be beneficial only to the extent that its purpose is understood and it is reliably and consistently implemented. Implementation can be accomplished in one of two ways: external cueing or skills training. External cueing can be helpful for giving both reminders to complete a task as well as instructions on how to complete it. This may involve, for instance, setting a timer to remind the patient to use the appropriate strategy, possibly including information on how to use that strategy. A caregiver may also provide a direct cue to the patient to use the correct strategy when needed. While such a reminder from a caregiver may be effective, it also means remaining dependent upon some external agent which reduces independence. It also places an increased burden on caregivers which can add to caregiver burnout. Skills training, in contrast, works to ensure that strategies are used correctly by teaching the patient-specific skills which will help them to directly apply the strategy. In this case, a targeted intervention is used to (re)learn some discrete skills. Training is delivered using trials of mass learning until skill mastery is consistently demonstrated. The patient is then able to independently execute the specific strategy, although some level of cueing may still be necessary for this to happen. Given that this approach involves improving performance on a specific task or skill, it does not always have direct relevance to ADLs. However, these discrete skills are crucial, if not necessary, to using strategies that ultimately lead to improvement in overall functioning. This approach, though often more time-consuming, ultimately results in the individual acting with a greater level of agency.

An example of evidence-based cognitive rehabilitation approach which combines compensatory and restorative approaches is planner acquisition method described by [14]. This has long been a centerpiece of cognitive rehabilitation, into which other compensation strategies are woven. While this strategy is focused on improving memory functioning, the calendar/planner has also been demonstrated as a strategy to address problems including organizational skills, behavioral control, and other areas of cognitive impairment which impact daily functioning. This approach has been supported in multiple subsequent studies as a powerful method to improve functioning in persons with severe memory impairment [12]. The approach works by using both restorative and compensatory approaches in concert, using a three-step process of acquisition application, and adaptation. The **acquisition phase** focuses on learning the names, location, purpose, and use (i.e., acquiring semantic knowledge) for the components of the calendar/planner. Restorative approaches (e.g., errorless learning, spaced retrieval) are used to learn the components of the calendar/planner. Errorless learning is a restorative approach designed to teach specific information to patients by leveraging preserved procedural memory abilities. Learning of the target behavior occurs through the active involvement of the therapist and occurs without their conscious control over its use.

In the **application phase**, the patient practices using the memory notebook in various real-life and/or role-play tasks in the clinic. Working together the patient and therapist choose tasks that are relevant to the patient's life and develop strategies which help to compensate for impairments which interfere with daily functioning. In the final **adaptation phase**, skills learned in the first two stages (acquisition and application phases) are applied within community and naturalistic settings. A task that was previously performed in the clinic can now be performed in the patient's own environment (e.g., remembering to take medications, returning phone calls, and remembering scheduled appointments). The impact on daily functioning in this approach is tied to the use of a specific compensation strategy (in this case, the planner/calendar). However, the use of this strategy is dependent upon the use of restorative approaches that help ensure the strategy is used effectively. Thus, while both the restorative and compensatory components are necessary, neither used in isolation is sufficient to reach goals of greater independence. Success is achieved by using both approaches in concert. The three-stage approach using a combination of both restorative and compensation methods has also been effective to address problems in other cognitive domains, including impairments in attention, executive skills, visual-spatial skills, and communication abilities [15]. Thus, while both restorative and compensatory approaches are important, describing cognitive rehabilitation as making a choice between the two is a false dichotomy and can even be counterproductive. Rather, it is useful to understand the unique contributions of both approaches and how they are used together to achieve goals of greater independence.

24.3 Mobile Technology for Rehabilitation

The WHO defines mobile health (mHealth) technology as "medical and public health practice supported by mobile devices", including (but not limited to) smartphones and tablets and biometric/activity monitors (e.g., wearables). Mobile health (mHealth) has the potential to enhance rehabilitation care in multiple ways, including by providing evidence-based education, supporting compliance to treatment plans, supporting monitoring and management of biometrics and chronic symptoms, facilitating patient–provider communication, and promoting effective long-term self-management [16]. Through mHealth technology, healthcare providers can both collect and share health-related information with their patients. When communication via mHealth occurs in only one direction, as in the case of devices that remotely track biometrics (e.g., heart rate monitors) or apps that collected self-reported data (e.g., symptoms trackers), this is considered a one-way system. Another example of a one-way system would be an app that provided information or guided intervention without collecting any data from the user (e.g., a daily meditation app). A two-way system involves a back-and-forth exchange of information between the user (patient) and healthcare provider. Evidence suggests that two-way systems may be preferred by patients, family members, and clinicians [17].

Mobile technology provides new opportunities for delivering interventions in the "here and now" as patients go about their usual activities—a model referred to as ecological momentary intervention (EMI). EMI has recently garnered widespread interest in healthcare technology development for two principal reasons—its ability to extend an intervention beyond the standard treatment context (e.g., clinician's office) and into individuals' daily lives, and the advantage of providing an intervention in individuals' natural environments, encouraging them to apply new skills and behaviors to their actual experiences [18]. Evidence suggests that EMIs encourage the practice of new behaviors and skills [19–21], which may improve generalization and increase the impact of interventions (e.g., [22, 23]) when combined with face-to-face clinical contacts. For cognitive rehabilitation, this means interventions could be delivered at the very moment that a person with ABI is performing a cognitively demanding task in their home or community, such as planning a shopping list or preparing a meal.

Most adults across all age groups, including adults with disabilities, persons with TBI, use smartphones [24–26]. Beyond being beneficial for using mHealth technology, smartphones are the central hub around which most mHealth technology functions, addressing the frequently voiced need from persons with cognitive disabilities for a single interface to manage a variety of cognitive and health needs [16, 17, 27]. It is therefore not surprising that applications of mHealth technology to health management have grown exponentially in the last decade. In 2017, there were more than 325,000 mHealth apps available, representing a 25% increase over the prior year [28] Apps addressing cognitive issues represent a relatively small share of the market. Most mHealth apps focus on fitness (36%), stress management (17%), and diet (12%), with the remaining 35% focused on disease management [29]. Despite this tremendous growth in the general market, evidence points to a dearth of products that address the particular needs and concerns of users with disability. Disability-focused apps represent less than 2% of all mHealth apps examined [30].

A recently published study evaluated the current state of mobile health care for people with disabilities based on reviews of the scientific literature and web-based resources for locating apps, as well as a survey of 377 users with disabilities [31]. They concluded that mHealth development and application for people with disabilities is in its early stages, with only a handful of mHealth tools targeted towards persons with disability and virtually no scientific evidence of their effectiveness. Furthermore, despite reporting a relatively high rate of adoption of mHealth apps (40%), survey respondents pointed to problems with accessibility and concerns about the accuracy or relevance of the content in mainstream mHealth apps for disabled users. A recent systematic review of mHealth apps specifically within the domain of *rehabilitation* found only seven apps focused on TBI rehabilitation [32]. These apps addressed a variety of issues such as symptom monitoring, wayfinding, memory problems, and emotional and behavioral concerns. Only three apps were evaluated with a quantitative outcome assessment, and only two of these were in randomized controlled trials (RCTs) [32]. The authors concluded that early app design and evaluation efforts show promising results, but clearly, more research is clearly needed.

24.4 Ethical Considerations

The commercially, rather than scientifically, driven proliferation of mHealth apps presents additional challenges to applying mHealth technology in practice. Even as research on effective mHealth technology grows, clinicians may struggle to find evidence-based mHealth apps, particularly those addressing the specific needs of individuals with cognitive disabilities, amid the sea of apps available (less than 1% of currently available mHealth apps are evidence-based) [33, 34]. A survey of over 500 rehabilitation clinicians reported that only 23% felt knowledgeable about available rehabilitation technology and only 51% felt comfortable using this technology in practice [35]. Similarly, in another study of healthcare professionals working with youth with brain injuries, despite 75% of patients using mHealth technology, only 42% of providers discussed or facilitated this mHealth use in their patients [36]. Mirroring this, one study found that only 10% of participants with brain injury reported that a clinician discussed the use of mHealth services with them following their injuries [36, 37]. This highlights a critical issue —individuals want to use mHealth technologies, but they are not receiving informed recommendations from their healthcare providers. This

creates a high risk that these individuals may use mHealth technology that is inappropriate for (or even detrimental to) their health, unique needs, and rehabilitation goals. This concern is heightened for those individuals with cognitive impairments that limit insight and decision-making. A 2018 systematic review of studies on evaluating mHealth apps concluded that clinicians assess the following about mHealth apps to determine if they are appropriate for patients: (1) design, (2) information/content, (3) usability, (4) functionality, (5) ethical issues, (6) security and privacy, and (7) user-perceived value [38]. Inclusion of the patient in this process is essential, to ensure a match between the patient's needs, values, and preferences and the features, potential risks and benefits, and costs of the mHealth technology.

Individuals with cognitive disabilities do report that smartphones generally provide accessible, convenient, effective, and acceptable support. However, they also desire simplicity, better training and support for learning and continuing to use new technology, and better accessibility features (e.g., larger font) [17, 37, 39–42]. Poor accessibility may explain the disparity in mHealth use between those with and without a disability [16]. Inclusion of people with disabilities in the development and evaluation of mHealth technology is essential, though remains a critical gap in current mHealth development that contributes to this ongoing disparity [16]. Even in the growing literature supporting mHealth use for rehabilitation in populations with cognitive impairment, very little work has been done to identify the best approaches to training individuals to use the technology in their daily lives, leading to frustration and technology abandonment. Broadly, practitioners could support the use and adoption of mHealth by their patients by helping them identify appropriate and evidence-based mHealth apps specific to their needs, values, and preferences and providing hands-on opportunities for their patients to practice with the technology itself before moving on to use it independently [27]. The same approaches used in cognitive rehabilitation (e.g., errorless learning, vanishing cues) could also be used to acquire the skills needed to use mHealth technology [43], though instruction should be tailored to the individual learner rather than adopting a one-size-all approach. Further, several strategies should be designed within mHealth platforms themselves, to support ongoing use and reduce abandonment, as we discuss later in this chapter.

Despite its potential, mHealth is not a magic bullet and should not replace—nor be implemented in the absence of—health providers. Careful consideration of privacy, safety, and emotional well-being when using mHealth technology is paramount. Mobile technology is a primary part of the future of health care, presenting opportunities to provide quality care to more individuals, but we need to address the challenges inherent in mHealth—especially for those with cognitive impairment—to maintain an effective and ethical practice.

24.5 Mobile Technology for Cognitive Assessment

Mobile technology has a number of capabilities that hold promise for enriching the assessment of cognitive function. Built-in sensors—such as positional sensors (e.g., accelerometer, gyroscope, and GPS), media sensors (e.g., microphone and camera), inherent sensors (e.g., device timer), and participatory user–device interactions (eg, screen interactions, metadata input, app usage, and device lock and unlock)—could provide helpful information for assessing individuals' behavior for the purposes of training, monitoring, diagnosis, or rehabilitation [44]. For example, sensor-enriched mobile assessment of memory could leverage information from human–device interactions—such as the number of times an individual interacts with the screen to gather necessary information to complete a memory task. Furthermore, mobile technology could serve as a platform for administration, scoring, interpretation, and storage of adapted versions of traditional "paper and pencil" neuropsychological tests. A scoping review published in 2021 found encouraging evidence for

the use of mHealth technology to collect patient-reported outcomes after acquired brain injury, including those capturing neurocognitive functions in day-to-day life [27]. However, despite this great promise, the application of mobile technology for neurocognitive assessment is still in its infancy, and to date, there are no validated mobile assessment tools for neurocognitive assessment in individuals with ABI.

24.6 Mobile Technology for Cognitive Rehabilitative Treatments

There are clear opportunities within the traditional framework of cognitive rehabilitation for mobile technology to enhance treatment. Mobile devices can provide external cueing via alarms and notifications. Using mobile technology in the home and community can support the repeated practice of skills and habits without relying on limited therapist availability for direct supervision in the clinic. Furthermore, mobile technology's ability to deliver instruction and support practice within the home and community allows skill acquisition and practice to take place in the very contexts in which they will be applied. These features have been leveraged by targeted cognitive rehabilitation applications.

In addition to targeted applications, many standard smartphone functions and apps geared towards the general population are now regularly incorporated into cognitive rehabilitation. As indicated earlier, training in the use of external aids—planners, calendars, memory notebooks—has always been a critical aspect of compensatory treatments. Traditional paper and pencil aids however can be bulky and easily misplaced, while smartphones allow users with cognitive impairment to easily and discretely take advantage of electronic alarms, calendars, to-do lists, GPS, and other cognitive aids as they go about their daily lives. A recent study from the TBI Model Systems National Database study on internet use ($N = 337$) of individuals with TBI found that 95% owned a mobile phone and over 75% used their smartphone to access the internet

[24]. A smaller survey of individuals with TBI (N = 29) found that use as a memory and organizational aid was a commonly cited benefit of smartphones for this population [37]. A 2007 review of the literature on efficacy of using external memory aids for managing memory disorders identified critical research gaps in the study of electronic external aids for users with TBI, including, a poor understanding of factors that lead to long-term adoption of external aids [45]. The authors also note that further research is needed on the design and selection of aids and the evaluation, instruction, and ongoing monitoring of people using the devices [45].

The vast majority of mHealth apps are not evidence-based at present [16]. There is however a small but growing literature speaking to the use of smart technologies to both help improve functioning and monitor functioning for persons with ABI. In a systematic review looking at literature through May 2019, Kettlewell and co-authors (2019) were only able to identify four studies that employed a randomized controlled trial methodology and focused on functioning. They concluded that there was insufficient evidence available to support the benefit of personal smart technologies to improve outcomes in this population [46].

In a scoping review, published the same year, Juengst and co-authors (2019), found that mHealth showed some potential for persons with TBI to use as a compensation strategy for cognitive impairment, as well as a method for monitoring reducing symptoms and a way of addressing social educational goals. Looking at literature published between 2012 in 2019, the authors identified 16 papers which met their criteria. These studies used multiple approaches, with most studies focusing on everyday memory functioning and using simple prompts to help complete a specific task at a given time. The studies also collectively show the importance of the involvement of a human component in the intervention and the need for the contact to be meaningful to the person served. Also evident was the importance of understanding the cognitive difficulties experienced by persons with TBI that may cause greater difficulty in learning or using the technology itself. The qualitative evaluation of these interventions indicated encouraging outcomes in terms of better symptom management, decreasing memory problems and mood difficulties, and improving overall well-being than more formal outcome measures showed alone [47].

More recently, Juengst and colleagues (2021) conducted a scoping review of literature from 2015 to 2019 of studies which used mHealth technology for assessment of patient-reported outcomes in community-dwelling individuals with ABI which included ecological momentary assessment for data collection. 12 manuscripts were identified which met inclusion criteria. The interventions were at various stages of development and testing, supporting a range of cognitive skills from everyday memory to executive function (planning, organization, goal setting) to community participation (social, academic/vocational) [27]. Though the evidence remained insufficient to inform clinical guidelines, it did show promise for the future, and overall users with TBI were enthusiastic about using mHealth technology as a compensatory strategy [17].

Given the great variability with respect to what constructs were measured as well as the frequency duration and timing of the exact data collection, the authors were unable to make recommendations regarding either the optimal content or timing of specific mHealth platforms for capturing information. The authors did note that, as with previous reviews, personalization of the technology to account for the unique needs and abilities of each person was crucial to success. This idea has long been one of the main tenants of formulating rehabilitation treatment: to gather data and ask questions to determine which approach has the best "goodness of fit" between technology and a given individual [48].

24.7 Future Directions

Though the extant evidence for use of mHealth technology for cognitive rehabilitation is certainly promising, there are several areas for future development so that we can move towards developing clinical practice guidelines. As discussed in this chapter, mobile technology has the potential to extend the duration and intensity of cognitive rehabilitation (including the delivery of 'booster' sessions or maintenance therapies), provide assessment and treatment in ecologically valid contexts, and permit time-sensitive interventions. Long-term use of mobile technology for cognitive rehabilitation should consider patients' evolving needs as they progress through their individual recovery trajectory, as well as their meaningful life roles and personal treatment goals, which are also subject to change over time [49].

Future design efforts should capitalize on the complimentary expertise of various stakeholder groups from both the "technology domain" (technology developers) and the "healthcare domain" (clinicians, consumers with brain injury, and their caregivers). This co-design approach engages stakeholders representing patients, clinicians, and healthcare administration at the front-end of design to identify clinically meaningful problems and concerns, while still allowing for the opportunistic exploration of "technology-inspired" approaches that envision new healthcare technologies that would be overlooked by strictly "problem-driven" design approaches. That said, we should not forget to pause and ask not just what we *can* do but also what we *should* (or should not) do. Inclusion of relevant stakeholders—from patients to their family members to clinicians to payors—is crucial to ensuring meaningful, feasible, scalable, and ethical development of mHealth to support cognitive rehabilitation [49].

Ongoing work is needed not only in the development and scientific evaluation of specific mHealth technologies and platforms but also in how we operationalize, implement, and reimburse for cognitive rehabilitation services provided via mHealth. For the reasons discussed above, the goal of cognitive rehabilitation is predominantly on compensation and adaptation —that is, long-term behavioral changes to support function—rather than restoration, hence, we also need to carefully consider long-term

consumer engagement and technical support for using these technologies independently and indefinitely. Lastly, the rapid evolution of mHealth often leaves us scrambling to keep up, both with high-quality evidence (especially as technologies rapidly develop, rendering earlier technologies obsolete) and with important ethical considerations. Below we offer suggestions of some specific promising directions for future development.

First, we need to employ a framework for thinking about how to operationalize the specific therapeutic components to include in mHealth technology for different cognitive rehabilitation targets. This is an important, but often overlooked, step for both research and clinical implementation. For research, we cannot test what components are most effective if we do not have a concrete way to measure those components directly. For clinical implementation, we need standardized definitions for documentation and reimbursement. The Rehabilitation Treatment Specification System (RTSS) provides this kind of framework for defining specific treatment components in rehabilitation interventions. The RTSS characterizes rehabilitative treatments according to the targets, ingredients, and mechanisms of action. Thus, RTSS provides a coherent theory-based framework that researchers can use to systematically test the effects of specific ingredients on specific targets [50]. Classifying mHealth treatments for cognitive rehabilitation according to this approach could greatly accelerate development, as it would enable researchers to test the effectiveness of mobile delivery of specific treatment ingredients, and these evidence-based treatment ingredients could then be implemented across various platforms as technology continues to advance.

Second, we need to develop processes for including mHealth technology in the ongoing treatment process and in clinician workflow. This includes processes for adapting mHealth technology use in response to patient progress or decline [49]. Adaptations may be required for the content, timing, and/or frequency of assessments or EMI in response to these changes. Processes may be external to the technology—such as plans for ongoing re-evaluation to ensure the technology is still meeting the patient's and family's needs, and process internal to the technology—such as built-in flexibility and customizability within the apps themselves. Individuals will need feedback from providers as well, so the use of mHealth should continue to be paired with, rather than replace, contact with healthcare providers. Input from clinicians and developing training for clinicians to enable them to identify evidence-based mHealth apps and to provide effective training in the use of mHealth technology to their patients will be critical. Further, clinicians may have different preferences and needs for how they access the data collected from their patients via mHealth, how the data are presented to them (e.g., design of a "clinician portal" in a two-way mHealth system), and how mHealth use would impact, and be incorporated into, their clinical workflow.

Third, we need to develop strategies for engaging individuals with disabilities and their caregivers in the use of mHealth technology, both in the short and long term. This can be done in several ways. One critical approach is to include end-users in the development of mHealth technology. Doing so will promote accessibility, design consistent with patient needs, preferences, and values, and long-term adoption. It will also allow us to identify—and hopefully, circumvent or prevent—barriers to mHealth technology use that result in frustration with technology abandonment. In addition to addressing external factors, like appropriate and accessible tech support, strategies need to be embedded into mHealth technology design to promote engagement and long-term use. Such strategies that have been identified in several past studies include iterative feedback about progress (e.g., not only having individuals track symptoms but also providing feedback and what these symptoms mean), ability to customize mHealth content and notification based on personal needs and preferences (e.g., limiting notifications only to the most salient components for that individuals), and including "gamification" (e.g., use of in-app challenges and rewards based on engagement). Such "gamification" has been successfully employed in other

mHealth technologies, such as those targeting exercise, weight loss, or smoking cessation [51–53], and in one study targeting symptom management after youth concussion [54].

Last, we need to develop better training and early support for learning how to use mHealth technology, to ensure that individuals are able to continue using mHealth technology independently. Training should be individualized based on users' familiarity and comfort with mHealth technology and on their cognitive strengths and limitations.

As technology rapidly develops, so do the future directions that can be explored for using mHealth in cognitive rehabilitation. Technology developers refer to a new era of personal technology—the "post-app era"—wherein the way that users interact with their smartphones is becoming more streamlined. Increasingly users will rely on mobile technology to carry out day-to-day tasks without ever needing to open a traditional mobile application. Cloud computing, advancing in machine learning and artificial intelligence (for example, conversational user interfaces know as virtual assistants or "chatbots"), and the seemingly inevitable transition into the "post-app era" are currently paving the way for previously unthought-of mHealth technology. Exciting as this may be, we cannot rush ahead without considering the ethical implications. These include, but certainly are not limited to the following. First, we must consider what we are tracking or providing to individuals with cognitive disabilities in the absence (or synchronous presence) of a human clinician. Second, we must carefully determine what mHealth technology can replace in current practice versus how it would be best used to complement or supplement current practice. Third, advances in mHealth technology should be made to *close* rather than *widen* health disparities. While high-income countries remain at the forefront of developing the latest mobile technologies used in health care, the rate of penetration of such technologies in low- and middle-income countries has recently exceeded that of their wealthier neighbors [26]. Together, there is much potential for mHealth technology to improve the effectiveness and reach of cognitive rehabilitation interventions, but we must ensure it is done with the same consideration afforded to clinician-delivered care.

References

1. Cullen N. Canadian Healthcare Perspective in traumatic brain injury rehabilitation. J Head Trauma Rehabil. 2007;22(4):214–20.
2. O'Callaghan AM, McAllister L, Wilson L. Experiences of care: perspectives of carers of adults with traumatic brain injury. Int J Speech Lang Pathol. 2011;13(3):218–26.
3. Green R, Colella B, Maller J, Bayley M, Glazer J, Mikulis D. Scale and pattern of atrophy in the chronic stages of moderate-severe TBI. Front Hum Neurosci. 2014;8:67.
4. Masel BE, DeWitt DS. Traumatic brain injury: a disease process, not an event. J Neurotrauma. 2010;27(8):1529–40.
5. Teasell R, Foley N, Salter K, Richardson M. Evidence-based review of stroke rehabilitation (16th edition). Top Stroke Rehabil. 2003;10(1):29–58.
6. Breitenstein C, Grewe T, Flöel A, Ziegler W, Springer L, Martus P, et al. Intensive speech and language therapy in patients with chronic aphasia after stroke: a randomised, open-label, blinded-endpoint, controlled trial in a health-care setting. The Lancet. 2017;389(10078):1528–38.
7. Till C, Colella B, Verwegen J, Green RE. Postrecovery cognitive decline in adults with traumatic brain injury. Arch Phys Med Rehabil. 2008;89(12, Supplement):S25–34.
8. Harley J, Allen C, Braciszewski T, Cicerone K, Dahlberg C, Evans S, et al. Guidelines for cognitive rehabilitation. Neuro Rehabil. 1992;2(3):62–7.
9. Ylvisaker M, Hanks R, Johnson-Greene D. Perspectives on rehabilitation of individuals with cognitive impairment after brain injury: rationale for reconsideration of theoretical paradigms. J Head Trauma Rehabil. 2002;17(3):191–209.
10. Ben-Yishay Y, Silver SM, Piasetsky E, Rattok J. Relationship between employability and vocational outcome after intensive holistic cognitive rehabilitation. J Head Trauma Rehabil. 1987;35–48.
11. Chan F, Gelman JS, Ditchman N, Kim J-H, Chiu C-Y. The World Health Organization ICF model as a conceptual framework of disability. 2009.
12. Cicerone KD, Goldin Y, Ganci K, Rosenbaum A, Wethe JV, Langenbahn DM, et al. Evidence-based cognitive rehabilitation: systematic review of the literature from 2009 through 2014. Arch Phys Med Rehabil. 2019;100(8):1515–33.
13. Bergquist TF, Jackets MP. Awareness and goal setting with the traumatically brain injured. Brain Inj. 1993;7(3):275–82.
14. Sohlberg MM, Mateer CA. Introduction to cognitive rehabilitation: theory and practice. Guilford Press; 1989.
15. Eberle R, Rosenbaum A. Cognitive rehabilitation manual and textbook: translating evidence-based recommendations into practice. 2nd ed. In ACRM Publishing; In Press.
16. Ramey L, Osborne C, Kasitinon D, Juengst S. Apps and mobile health technology in rehabilitation: the good, the bad, and the unknown. Phys Med Rehabil Clin. 2019;30(2):485–97.
17. Osborne CL, Juengst SB, Smith EE. Identifying user-centered content, design, and features for mobile health apps to support long-term assessment, behavioral intervention, and transitions of care in neurological rehabilitation: an exploratory study. Br J Occup Ther. 2021;84(2):101–10.
18. Heron KE, Smyth JM. Ecological momentary interventions: incorporating mobile technology into psychosocial and health behaviour treatments. Br J Health Psychol. 2010;15(1):1–39.
19. Hay CE, Kinnier RT. Homework in counseling. J Ment Health Couns. 1998;20(2):122.
20. Kazantzis N, L'Abate L. Handbook of homework assignments in psychotherapy. Springer; 2007.
21. Newman MG, Consoli A, Taylor CB. Computers in assessment and cognitive behavioral treatment of clinical disorders: anxiety as a case in point. Behav Ther. 1997;28(2):211–35.
22. Kalichman SC, Kalichman MO, Cherry C, Swetzes C, Amaral CM, White D, et al. Brief behavioral self-regulation counseling for HIV treatment adherence delivered by cell phone: an initial test of concept trial. AIDS Patient Care STDS. 2011;25(5):303–10.
23. Malow RM, West JA, Corrigan SA, Pena JM, Cunningham SC. Outcome of psychoeducation for HIV risk reduction. AIDS Education and Prevention. 1994
24. Baker-Sparr C, Hart T, Bergquist T, Bogner J, Dreer L, Juengst S, et al. Internet and social media use after traumatic brain injury: a traumatic brain injury model systems study. J Head Trauma Rehabil. 2018;33(1):E9-17.
25. Morris JT, Sweatman M, Jones ML. Smartphone use and activities by people with disabilities: user survey 2016. J Technol Pers Disabil. 2017;5:50–66.
26. Pew Research Center. Mobile fact sheet. Pew Research Center. 2018;
27. Juengst SB, Terhorst L, Nabasny A, Wallace T, Weaver JA, Osborne CL, et al. Use of mHealth technology for patient-reported outcomes in community-dwelling adults with acquired brain injuries: a scoping review. Int J Environ Res Public Health. 2021;18(4):2173.

28. mHealth Economics 2017 Report: Status and trends in digital health | R2G [Internet]. research2guidance. [cited 2022 Jan 5]. https://research2guidance.com/product/mhealth-economics-2017-current-status-and-future-trends-in-mobile-health/.

29. Aitken M. Patient adoption of mHealth: use, evidence and remaining barriers to mainstream acceptance. IMS Institute of Healthcare Informatics; 2015 Sep.

30. Constantino T. IMS Health Study: patient options expand as mobile healthcare apps address wellness and chronic disease treatment needs [Internet]. IMS Institute for Healthcare Informatics; 2015 Sep [cited 2022 Jan 5]. https://www.businesswire.com/news/home/20150917005044/en/IMS-Health-Study-Patient-Options-Expand-Mobile.

31. Jones M, Morris J, Deruyter F. Mobile healthcare and people with disabilities: current state and future needs. Int J Environ Res Public Health. 2018;15(3):515.

32. Nussbaum R, Kelly C, Quinby E, Mac A, Parmanto B, Dicianno BE. Systematic review of mobile health applications in rehabilitation. Arch Phys Med Rehabil. 2019;100(1):115–27.

33. Anthes E. Mental health: there's an app for that. Nature News. 2016;532(7597):20.

34. Free C, Phillips G, Galli L, Watson L, Felix L, Edwards P, et al. The effectiveness of mobile-health technology-based health behaviour change or disease management interventions for health care consumers: a systematic review. PLoS Med. 2013;10(1): e1001362.

35. Morris J, Jones M, Thompson N, Wallace T, DeRuyter F. Clinician perspectives on mRehab interventions and technologies for people with disabilities in the United States: a national survey. Int J Environ Res Public Health. 2019;16(21):4220.

36. Plackett R, Thomas S, Thomas S. Professionals' views on the use of smartphone technology to support children and adolescents with memory impairment due to acquired brain injury. Disabil Rehabil Assist Technol. 2017;12(3):236–43.

37. Wong D, Sinclair K, Seabrook E, McKay A, Ponsford J. Smartphones as assistive technology following traumatic brain injury: a preliminary study of what helps and what hinders. Disabil Rehabil. 2017;39(23):2387–94.

38. Nouri R, R Niakan Kalhori S, Ghazisaeedi M, Marchand G, Yasini M. Criteria for assessing the quality of mHealth apps: a systematic review. J Am Med Inform Assoc. 2018;25(8):1089–98.

39. Bedell GM, Wade SL, Turkstra LS, Haarbauer-Krupa J, King JA. Informing design of an app-based coaching intervention to promote social participation of teenagers with traumatic brain injury. Dev Neurorehabil. 2017;20(7):408–17.

40. Lannin N, Carr B, Allaous J, Mackenzie B, Falcon A, Tate R. A randomized controlled trial of the effectiveness of handheld computers for improving everyday memory functioning in patients with memory impairments after acquired brain injury. Clin Rehabil. 2014;28(5):470–81.

41. Narad ME, Bedell G, King JA, Johnson J, Turkstra LS, Haarbauer-Krupa J, et al. Social Participation and Navigation (SPAN): description and usability of app-based coaching intervention for adolescents with TBI. Dev Neurorehabil. 2018;21(7):439–48.

42. Powell LE, Wild MR, Glang A, Ibarra S, Gau JM, Perez A, et al. The development and evaluation of a web-based programme to support problem-solving skills following brain injury. Disabil Rehabil Assist Technol. 2019;14(1):21–32.

43. Ehlhardt LA, Sohlberg MM, Kennedy M, Coelho C, Ylvisaker M, Turkstra L, et al. Evidence-based practice guidelines for instructing individuals with neurogenic memory impairments: What have we learned in the past 20 years? Neuropsychol Rehabil. 2008;18(3):300–42.

44. Templeton JM, Poellabauer C, Schneider S. Enhancement of neurocognitive assessments using smartphone capabilities: systematic review. JMIR Mhealth Uhealth. 2020;8(6): e15517.

45. Sohlberg MM, Kennedy M, Avery J, Coelho C, Turkstra L, Ylvisaker M, et al. Evidence-based practice for the use of external aids as a memory compensation technique. J Med Speech-Lang Pathol. 2007;15(1):xv–xv.

46. Kettlewell J, das Nair R, Radford K. A systematic review of personal smart technologies used to improve outcomes in adults with acquired brain injuries. Clin Rehabil. 2019;33(11):1705–12.

47. Juengst SB, Hart T, Sander AM, Nalder EJ, Pappadis MR. Mobile health interventions for traumatic brain injuries. Curr Phys Med Rehabil Rep. 2019;7(4):341–56.

48. Wright BA. Motivating children in the rehabilitation program. 1983;

49. How T-V, Hwang AS, Green REA, Mihailidis A. Envisioning future cognitive telerehabilitation technologies: a co-design process with clinicians. Disabil Rehabil Assist Technol. 2017;12(3):244–61.

50. Hart T, Dijkers MP, Whyte J, Turkstra LS, Zanca JM, Packel A, et al. A theory-driven system for the specification of rehabilitation treatments. Arch Phys Med Rehabil. 2019;100(1):172–80.

51. Burkow TM, Vognild LK, Johnsen E, Bratvold A, Risberg MJ. Promoting exercise training and physical activity in daily life: a feasibility study of a virtual group intervention for behaviour change in COPD. BMC Med Inform Decis Mak. 2018;18(1):136.

52. Rajani NB, Weth D, Mastellos N, Filippidis FT. Use of gamification strategies and tactics in mobile applications for smoking cessation: a review of the UK mobile app market. BMJ Open. 2019;9(6): e027883.

53. Timpel P, Cesena FHY, da Silva CC, Soldatelli MD, Gois E, Castrillon E, et al. Efficacy of gamification-based smartphone application for weight loss in overweight and obese adolescents: study protocol for a phase II randomized controlled trial. Ther Adv Endocrinol Metab. 2018;9(6):167–76.

54. Worthen-Chaudhari L, McGonigal J, Logan K, Bockbrader MA, Yeates KO, Mysiw WJ. Reducing concussion symptoms among teenage youth: Evaluation of a mobile health app. Brain Inj. 2017;31 (10):1279–86.

Telerehabilitation Technology

25

Verena Klamroth-Marganska, Sandra Giovanoli,
Chris Awai Easthope,
and Josef G. Schönhammer

Abstract

Due to physical and cognitive deficits, it is often difficult and costly for individuals who have suffered a stroke to access on-site neurorehabilitation. Telerehabilitation offers the opportunity to improve the rehabilitation process as it can provide intensive supervised rehabilitation in the home environment. The term telerehabilitation refers to the provision of therapeutic services at a distance, enabled by electronic telecommunication and information technologies. Services are provided through a variety of technical systems with different purposes and capabilities. This chapter provides an overview of technical solutions for providing telerehabilitation services to treat the main consequences of stroke, namely paresis of the upper and lower extremities, and communication difficulties. We describe the communication tools, sensor technologies, virtual reality systems, and robots for service delivery and explore the facilitators and barriers to successful implementation. Evidence is summarized in the context of teleassessment, telemonitoring, and teletherapy.

Keywords

Telehealth · Teletherapy · Telemonitoring · Teleassessment · Stroke

25.1 Introduction

In neurorehabilitation, clients face numerous barriers to accessing usual on-site appointments including geographic isolation, limited resources, and shortage of time; all these may lead to the lack of compliance with rehabilitation regimens [1]. Telerehabilitation may help to overcome these barriers and is a suitable supplement or even substitute to usual rehabilitation. The term *telerehabilitation* describes the use of information and communication technologies (ICT) for the delivery of rehabilitation services to people at a distance, for example, in the home environment, in the out-client area, in the in-client sector, or at school [2, 3]. Early attempts to use phones for teletherapy of people with aphasia were made as early as the 1970s. Vaughn et al. designed a device that combined phones "with a

V. Klamroth-Marganska (✉)
Research Unit for Occupational Therapy, School of Health Professions, Zurich University of Applied Sciences ZHAW, Zurich, Switzerland
e-mail: Verena.klamroth@zhaw.ch

S. Giovanoli · C. A. Easthope · J. G. Schönhammer
cereneo Foundation, Center for Interdisciplinary Research CEFIR, Vitznau, Switzerland
e-mail: sandra.giovanoli@cereneo.foundation

C. A. Easthope
e-mail: chris.awai@cereneo.foundation

J. G. Schönhammer
e-mail: josef.schoenhammer@cereneo.foundation

© The Author(s), under exclusive license to Springer Nature Switzerland AG 2022
D. J. Reinkensmeyer et al. (eds.), *Neurorehabilitation Technology*,
https://doi.org/10.1007/978-3-031-08995-4_25

variety of terminal devices, such a speaker phone for speech, handset for other auditory signals, a "Touchtone" keypad for pointing responses, a teletypewriter for typing, a "Telenote" device for handwriting, and computer terminals for more generalized applications" [4, 5]. Standardized assessments such as the modified Barthel Index were also successfully carried out on the phone in the 1980s [6]. In the age of digitalization and the Internet, technologies are advancing, becoming cheaper, more accessible and ubiquitous, and available to a greater number of people. As ICT continues to develop, telerehabilitation is becoming a global treatment option.

25.1.1 Benefits

Telerehabilitation offers many advantages as compared to conventional in-person therapy: By eliminating the need to travel long distances, telerehabilitation helps clients comply with treatment protocols. The remote delivery of service is especially important for those who have difficulties traveling due to physical impairments or who live in rural or remote settings. Telerehabilitation can increase the frequency of services and enhance continuity of care [2].

Perhaps even more importantly, telerehabilitation promotes clients' involvement and empowers them to care for and manage their medical needs and therapeutic interventions [2, 7]. Clients' empowerment and engagement are key elements to enhance neuroplasticity and facilitate functional recovery in neurological disorders [8, 9]. Most health care, perhaps as much as 85%, is self-care [10, 11]. The home environment as an authentic environment for the experiences of functioning encourages clients to develop problem-solving skills and actively engages them in the rehabilitation process [7]. Through the use of new technologies, the client is not alone in his home rehabilitation. As therapists have access to data from home, clients can self-train knowing that their therapist is tracking their progress. Automatized real-time feedback from technologies motivates and engages the client, thus, further promoting neuroplasticity [12, 13].

Educational materials that directly address the current state (as assessed during therapy) can be delivered promptly when needed and improve client knowledge. "The greater the understanding and comprehension that clients have in terms of the rehabilitation process, goals, diagnosis, and healing process, the more invested they will be in their own recovery" [8].

Clients with neurological disorders have often complex healthcare needs that require a multidisciplinary coordination of their rehabilitation. Services are provided by many professionals including physicians, psychologists, rehabilitation engineers, audiologists, nurses, and educators, out-client rehabilitation, and, above all, by occupational and physical therapists as well as speech–language pathologists [3]. Telerehabilitation can be integrated into telemedicine platforms that allow to connect all stakeholders including the client and share all relevant information in real-time.

25.1.2 Interaction from a Distance

In telerehabilitation, therapists and clients communicate from a distance and exchange health data with the use of technologies. These include video conferencing systems, instant messaging platforms, mobile health (mhealth) applications, electronic client portals, and digital client platforms. By incorporating additional technologies, telerehabilitation enables not only audiovisual interaction but also real-time exchange of `hands-on' information. Wearables, for example, monitor specific types of physiological parameters that are readily accessible from outside the human body. Robotic devices can give assistance and transmit information about forces and movements to give therapists a deeper understanding of the client's sensorimotor performance [14, 15]. Environmental sensors provide information about how clients interact with their environment. Taken together, the potential of interaction between therapists and clients is not restricted to audiovisual communication. Innovations in medical devices such as nanosensor technologies and virtual reality may expand the possible information exchanged in the future [16].

25.1.3 Synchronous and Asynchronous Therapy

Telerehabilitation may be delivered synchronously when therapist and client communicate in real-time [17]. Usually, they speak via video conferencing and telemedicine systems. Telerehabilitation may also be performed asynchronously when information for therapy or education is delivered to the client so that he/she can train alone, or when the client trains alone and recorded data is transmitted to the therapist [17]. After a familiarization phase under therapist supervision, clients can continue therapy independently. Thus, telerehabilitation shifts from remote one-to-one therapy to (partially) supervised self-training.

Self-training may involve medical applications (apps) or gaming software run on flat screens, head-mounted devices, or projection systems, that provide clients with visual and auditory feedbacks (but may also include other sensory inputs such as touch, movement, balance, and smell) [18]. The user interacts with the virtual environment by a mouse or joystick, cameras, sensors, or haptic (touch) feedback devices. The data may be further analyzed or processed (e.g., to recognize trends of recovery over days or weeks). Robotics for self-therapy enable clients to practice independently with mechanical assistance [19]. Self-training with integrated feedback may empower and motivate clients to engage in therapy.

25.1.4 Acceptance of Telerehabilitation

Although telerehabilitation is applied regularly in rural regions, its general implementation and acceptance was rather limited among therapists and clients until the Coronavirus disease 2019 (COVID-19) pandemic forced a rapid adoption of telerehabilitation. Because severe acute respiratory syndrome coronavirus 2 (SARS-CoV-2) is highly contagious, vulnerable persons were isolated starting in 2020. Unfortunately, those who require rehabilitation are often vulnerable (e.g., 90% of stroke clients in Switzerland are 65 or above) [3]. Furthermore, the pandemic went along with very early in-client discharge and a suspension of rehabilitation services in out-client settings in 2020, which decreased the access to rehabilitation services and their availability [7, 8]. To aggravate the situation, SARS-CoV-2 infection is not only associated with pneumonia, but also with neurological complications: The hypercoagulable states may lead to stroke and other neurological manifestations. Thus, COVID-19 did not only decrease the supply of rehabilitative services but also additionally increased the number of persons who required these.

Telerehabilitation was a promising way to fill the supply gap of rehabilitative measures during the pandemic. As telerehabilitation reduces the contact between vulnerable persons and the environment, it enables the therapy of persons in strict isolation [15]. Thus, traditional in-person visits were often replaced by tele-services. In Switzerland, for example, about two-thirds of occupational therapists (OT) provided telerehabilitation during the COVID-19 pandemic lockdown although they were often not reimbursed and the pandemic did hardly allow to implement measures associated with the successful implementation of telerehabilitation such as education and training, as well as administrative and technical support [20]. The service relied mainly on audiovisual interaction. The media used most was phone followed by chat services, email, video conferencing systems, and short messages services (SMS). One reason for this rather low use of modern technologies may be clients' preferences. 95% of people aged 65–69 already use the Internet, but only 35% of people aged 85 and over do (numbers from Switzerland 2020 [21]). Three-quarters of the OTs rated their clients' experience of telerehabilitation as positive or rather positive during the lockdown [20]. About two-thirds of OTs described their own experience as broadly positive or rather positive, and about one-quarter as negative or rather negative [22].

Usually, it needs two for out-client telerehabilitation: clinicians and clients. What does it take to improve the telerehabilitation experience for both user groups? The uptake of tele-services is influenced by facilitators and barriers. Almathamy et al. [23] reviewed scientific literature regarding factors that influence clients' adoption of telemedicine. The telemedical counseling they focused on was delivered with synchronous video conferencing systems or software, most studies reported on specially developed systems. The authors distinguished internal and external factors that influence the adoption. While the former describes the users' behaviors and motivations, the latter refers to the system and the surrounding environment. External barriers described were mainly technical (e.g., low internet speed or difficulty to use the system). Internal barriers included resistance to technology, the lack of eye as well as physical and social contacts with telemedicine, and clients' security concerns. Accordingly, understanding of the technology and high internet speed were important facilitators. Family involvement during treatment facilitated the use of systems.

For the service to be effective, clinicians do not only need to find it useful for clients, but they themselves require to be ready for system uptake [24]. To achieve successful implementation and sustainable use of telerehabilitation, it needs preparation, teaching, and support from an organization. Jafni et al. adopted findings from a critical care information system and identified seven significant factors that influence the adoption of telerehabilitation by therapists [25, 26]:

- The system's functions and features should be more useful, faster, and easier compared to existing systems.
- The delivered information must be accurate and understandable for every stakeholder including the client.

- Service quality requires follow-up services and responsiveness to support requests.
- Users should receive regular training that targets technology skills, safety, and security. Implementation plans help avoid problems.
- System performance must be reliable.
- Factors that influence users' decisions on system use include professional hierarchy, age, education, system ease of use, and cutting-edge technology [26]. The system must be effective and useful to meet the expectations even of stakeholders with limited IT skills.
- A strategy is needed to get stakeholders to adopt the new system. This includes stakeholder involvement, interactive communication (e.g., meetings), and a "champion" who can influence and encourage others to use the system.

25.1.5 Effectiveness of Telerehabilitation

The effectiveness of telerehabilitation has not yet been sufficiently proven. There are several studies showing that telerehabilitation interventions have either better or equal salutary effects on motor, higher cortical, and mood disorders as well as quality of life compared with conventional in-person interventions. However, evidence is based mostly on pilot projects that are small in sample size and proof-of-concept in nature, and randomized controlled trials (RCT) are scarce. A meta-analysis in 2020 by Laver et al., involving 22 RCTs, examined the efficacy of telerehabilitation after stroke [12]. The authors concluded that data are not sufficient to draw definitive conclusions as there was variance in interventions and comparators among studies, few adequately powered studies, and several studies with risk of bias. However, at this point telerehabilitation seems not inferior to in-person services or usual care regarding improvement in upper limb function, improvement of health-related quality of life, independence in ADL, or reduction of depressive symptoms [12].

Therapy is an important part, but it is not the only service that is delivered in rehabilitation: Rehabilitation involves also monitoring, assessment, prevention, assistance, supervision, education, consultation, and coaching [3]. We focus on telemonitoring, teleassessments, and teletherapy because current research, especially in these areas, employs new technologies that go beyond audiovisual interaction. Telemonitoring is based on the observation of behavioral, biological, or psychological signals. Standardized teleassessments complement this continuous real-world data in providing metrics that mirror clinical assessments [27]. The term teletherapy describes the delivery of sensory motor and cognitive rehabilitative treatment.

25.1.6 Stroke

We focus on stroke as the manifestations and the recovery process after stroke are good examples of how different technological advances may cover the entire neurorehabilitation spectrum of a client. Stroke is one of the leading causes of long-term disability in the world and survivors often need long-term neurorehabilitation treatment. About 26% remain disabled in basic ADL and 50% suffer from reduced mobility due to hemiparesis [28]. Aphasia and depression are other frequent causes of disability. Thus, stroke is a disease of immense public health importance with growing economic and social consequences [29].

We focus on telerehabilitation for the main complications of stroke, namely paralysis or loss of muscle movement of the upper and lower extremities, and communication difficulties.

25.2 Upper Extremity

A frequent consequence of stroke is a decline in functioning of the upper extremities, like the shoulders, arms, hands, and fingers [30–32]

which, in turn, is often associated with a deterioration in life quality [33, 34].

After a stroke, stroke survivors enter a recovery process. Recovery refers to the improvement in one or more components of an individual's functioning over time [35]. Importantly, at the level of motor behavior, recovery must be distinguished from compensation [36–38]. Behavioral recovery refers to actions that reflect restitution of pre-morbid movements, that is, actions using the same anatomical body parts (i.e., muscles, joints, effectors) for task accomplishment as before the neurological disease, whereas compensation refers to performing a task with different/additional body parts [35]. For example, when an individual reaches for a cup on a table, only shoulder, elbow, wrist, and finger movements may be employed before a stroke. Conversely, after a stroke, a person might make additional trunk movements in direction of reach to support arm and hand movements.

25.2.1 Assessment of Motor Functioning

25.2.1.1 Clinical Assessment

Observation-based Clinical Assessment
Neurorehabilitation research examines the factors of interventions that optimize recovery, like the type of therapy approach and the time of delivery in clients' rehabilitation journey [35]. Evidence for the effectiveness of an intervention is key for generation and accumulation of knowledge, which is fundamental for decision-making in the treatment of individuals [39]. However, the effectiveness of an intervention study is often not comparable to the effectiveness of other studies, due to differences in measured constructs and assessment tools [39]. Therefore, many efforts were made, like systematic reviews and Delphi studies, to determine a set of outcome measures for motor functioning that should be used universally [39–45].

Outcome measures are typically sorted according to several categories [39, 44]. First, they are categorized by the International Classification of Functioning, Disability, and Health (ICF) domains.

The ICF uses the term functioning to refer to an individual's health and disability status. Functioning is regarded as a general term and is defined by three sub-domains [46]:

1. *Body structure and body functions* are anatomical parts of the body and physiological and psychological functions. Problems in body structure or function are defined as impairments.
2. *Activity* is the execution of tasks by an individual. Problems with task accomplishment are referred to as activity limitations.
3. *Participation* is an individual's involvement in life situations. Problems during participation in real-life situations are called participation restrictions.

For example, when an individual experiences a stroke, the incident might result in impairments to the motor cortex (body structure) and the execution of shoulder movements (body function), which might limit reaching movements for objects (activity), which might be a restriction when shopping for groceries (participation).

Clinical outcome measures are often based on the observation of individuals' behavior by a clinical evaluator [44]. Technology-based outcome measures are obtained from sensor data that are usually transformed (e.g., the spatial position of the wrist over time is transformed to movement speed or the number of movement units [45]. Moreover, technology-based outcome measures at the ICF activity level can be classified as capacity (i.e., maximal activity, performed in a controlled environment) and performance measures (i.e., activity in a real-life setting [39, 47, 48]). Outcome measures must fulfill strict criteria regarding validity, reliability, responsiveness, clinical utility, and expert consensus to be recommended [39, 44].

Widely recommended examples of clinical outcome measures of the upper extremity are the Fugl–Meyer Assessment for Upper Extremities (FMA-UE [49, 50]; function level of ICF) and the Action Research Arm Task (ARAT [51, 52], activity level of ICF). The ARAT is an example of a capacity measure.

The FMA-UE covers 33 items that examine reflexes, voluntary movements, and movement synergies of shoulder, elbow, wrist, and fingers. Required materials are a reflex hammer and objects for movement task, like a sheet of paper or a tennis ball. Evaluators rate the degree of function for each item on a three-point scale. Administration time is about 30 minutes [53]. To provide a specific example, in one item of FMA-UE individuals need to abduct their arm. The behavior of interest is the shoulder abduction function (which, after stroke, may be accompanied with undesired elbow movement).

The ARAT comprises of 19 movement tasks, in which clients make reaching, grasp, grip or pinch movements, using standardized objects, like wooden blocks, marbles, pieces of metallic tubes, etc. Evaluators rate clients' ability to perform each task and quality of movement on a four-point scale. Testing time is about 5–20 minutes [52, 54]. One task, for example, is to pick up a wooden ball from a table and put it onto a shelf that is placed within reach. Of interest is the entire sequence of movements.

Clinical outcome measures at the ICF participation level are usually questionnaires that assess participation in daily life as perceived by clients [39, 43]. An example of a recommended questionnaire is the Stroke Impact Scale [43].

Clinical outcome measures, such as the FMA-UE and the ARAT, are well established in clinical practice and have excellent psychometric properties [43]. However, they rely on direct observation, with client and evaluator being present at the same location. Hence, the question arises of how the assessment of motor functioning can be realized in remote settings. One solution is to video-record clinical assessments remotely and to evaluate the videos at a different location [55]. This method has excellent psychometric properties but implies additional time and material costs for cameras and the operation of standardized video-shooting procedures. Thus, for remote assessment, technology-generated outcome measures might be a more actionable solution.

Technology-Based Clinical Assessment

Technology-generated outcome measures are considered as an important complement to clinical outcome measures because of several advantages [39]: They are objective [45] (as they do not depend on the interpretation of an observer) and more sensitive [56] (as they yield continuous scale data). Thus, they can gather more detailed information about recovery which provides better information for the personalization of treatment regimina [56, 57]. Another advantage is that technology-generated outcome measures can distinguish between behavioral recovery and compensation, as they can measure both types of behavior separately, which extends information yielded by the ARAT [57, 58].

For the ICF function level, an expert committee recommended to employ a reaching task in the horizontal plane and capture kinematic outcome measures with an optoelectrical camera system [45, 59]. Optoelectrical systems use multiple cameras with infrared illuminators and triangulation algorithms to reconstruct the three-dimensional (3D) position of reflective markers [60]. Markers are placed at defined anatomical landmarks of individuals' body segments to measure their position and to derive kinematic measures. Position measurement of a marker has less than 1 mm error and measurement of joint angles has an error range of 1–degrees [61]. Alternatives are networks of IMUs and markerless camera systems (multiple high-quality video cameras and computer vision-based analysis). Compared to marker-based systems, these are slightly less [62] or similarly accurate [63], respectively. IMU networks are mobile, whereas markerless camera systems leave individuals' bodies free of markers and sensor straps. However, both systems are similar to optoelectrical systems in price range and complexity of analysis. Hence, these systems imply similar barriers to the widespread use in clinical practice and at home as marker-based systems.

Moreover, various force measurement devices are recommended to measure grip strength (entire hand), precision grip [64] strength (individual fingers), and assess finger individuation [65] (control of one finger independently of the others). An advantage of the assessments for hand and finger strength is that they are quick and easy to administer as compared to other clinical or technology-based assessments, and that muscle strength early after stroke is a good predictor in statistical models of motor improvement capacity [66, 67].

For the ICF activity level, an expert committee recommended a 3D reaching task ("drinking task") and optoelectrical cameras for kinematic measurement [57, 68–70]. For this task, recommendations extend to the analysis of kinematic outcome measures [45], such as the number of movement units of the arm endpoint and angular velocity of the elbow, and trunk displacement and arm abduction angle [58]. Recovery, as opposed to compensation, corresponds to a smaller number of movement units of the arm endpoint, higher angular velocity of the elbow, less trunk inclination, and less arm abduction [57].

While the 3D reaching task assesses capacity at the ICF activity level, experts recommended to assess performance by monitoring arm use in real life, using measures like threshold activity counting [71], which can be obtained with accelerometers or inertial measurement units (IMUs) [39, 72]. However, a standardized way for application and analysis of this assessment is still missing and requires further research [39].

The described technology-based outcome measures are rather established in research than in clinical application. The equipment is expensive (about 10′000–20′000 USD), application is time-intense, and analysis is rather complex, especially in the case of optoelectrical camera systems [39]. Hence, their immediate application in remote settings appears to be impracticable.

25.2.1.2 Tele-Assessment

Low-cost sensor technologies, such as IMUs and depth-sensing cameras, and advances in video-based pose tracking algorithms are expected to enable technology-based assessment in clinical practice and real-life settings, once usability and standardization application and analysis have matured [42, 45].

Accelerometers and IMU sensors are the most frequently used wearable sensors in neurorehabilitation research [73, 74] (see Wang et al. for more types of wearable sensors). Accelerometers measure linear acceleration, gyroscopes measure angular velocity and magnetometers measure magnetic fields; an IMU is an assembly of these three components and captures the mentioned quantities in three orthogonal dimensions, which are usually aligned with the device's housing [75]. Usually, 2–4 sensor units are used and placed at wrists or lower arms, upper arms, or the sternum [74]. Data are frequently used to estimate the orientation of the device in order to estimate orientation of body segments or joint angles [74], with errors lying in the three to eight-degree range [60].

Another technology for clinical practice and home-use are standard digital RGB (Red–Green–Blue) video cameras (e.g., like those of a webcam) combined with infrared depth sensors [76] (RGB-D). Algorithms use these video and depth data to estimate and track 3D joint positions. The position data can be used to estimate joint angles, which have a similar error to IMUs depending on the viewing angle [60, 77].

Pose estimation algorithms [78] are based on computer vision and deep learning, and only use videos of standard digital RGB cameras [79]. The algorithms detect and track joint positions and yield two-dimensional (2D) or 3D position data, depending on the capabilities of the algorithms. Validation studies showed that accuracy can be similar to that of depth-sensing cameras when the plane of body motion and camera viewing angle are aligned (e.g., trunk movement in the frontal plane is captured by a frontal camera [80].

Stereo cameras might be the key technology to provide accurate but low-cost motion sensing [81]. Stereo camera systems consist of two or more RGB video cameras. Depth information is inferred by the difference in an object's positions in the two camera images.

These low-cost technologies can be used to capture motion data and measure motor functioning at each ICF level [73].

Motion data are commonly used to classify the degree of motor function impairment or activity limitation [73]. Observation-based assessments such as the FMU-UE or the ARAT yield ordinal values. Supervised machine learning algorithms can estimate the observation-based scores from the motion data, captured during the clinical assessment tasks [82, 83] or during different functional tasks [84, 85]. For example, FMA-UE scores were estimated from sensor data that were recorded during eight functional motor tasks [85]. Hence, this approach may require fewer movement samples than FMA-UE or ARAT assessments [84]. A drawback of wearable sensors is that they require donning, while camera systems do not have this disadvantage [73].

Estimating clinical scores at the ordinal level has the advantage of linking motion data to validated assessments of motor functioning, but it does not take advantage of the continuous scale data that is provided by the sensors [73]. However, the use of continuous data requires the identification and psychometric validation of the kinematic metrics, as it is the case with the recommended outcome measures obtained with optoelectrical cameras [57].

As mentioned above, motor functioning at the ICF activity performance level has already been quantified by assessing arm motion in real-life, for several hours to several days [73]. Accelerometer data were used, with sensors being placed at one or both wrists [73], and measures as activity counts were employed [71] Activity counts measured whether acceleration magnitude was above a particular threshold [71]. In neurological clients, arm use of the more affected upper limb was compared with the less affected or with data from healthy individuals [71, 73, 86, 87].

Function can be defined as the functionality of a body structure as is the case in the ICF framework. Functional motion can be also defined as the goal-directed and volitional motion of the upper limbs in relation to an object or target during ADL [88]. One approach aims to detect this functional motion and identify its nature during ADL in real-life settings [73]. In these studies, functional motion is defined as the goal-directed and volitional motion of the upper limbs with regard to an object or target in ADL [88]. Functional motion covers functional primitives (e.g., reaching, transporting, repositioning) that can be combined into functional movements (e.g., drinking from a cup) and functional activities (e.g., having dinner). For this approach, usually, IMUs are used and placed at individuals' wrists and/or various other positions [73]. First, several repetitions of each functional task are executed in a standardized setting and categorized by human observers. Then, supervised machine learning is used to estimate the category using the motion data [89–91]. This approach ultimately allows quantification of individuals' functioning on all ICF levels, for example by counting the number of functional activity executions (activity performance) or by measuring movement quality within functional activities [89].

25.2.2 Teletherapy

Neurorehabilitation research and guidelines suggest that recovery of motor functioning is positively influenced by interventions that follow principles of motor learning, such as high intensity and repetitive practice [92–97]. Recovery is positively influenced by the use of technology, as it facilitates the provision of therapy that follows these suggestions [39, 98].

In clinical application, rehabilitation technology may comprise of end-effector and exoskeleton rehabilitation robots linked to multimodal virtual or augmented reality environments [99]. Moreover, low-cost rehabilitation systems may be used that require less direct supervision by health professionals to deliver additional therapy in clinical and home-based settings [12, 98].

Rehabilitation systems for home use can be broadly categorized as feedback systems, virtual reality (VR) systems, and robotic systems [100]. Feedback systems use human motion data to provide feedback about motor performance, primarily in real-life environments [74, 101], whereas VR systems use individuals' motion data to provide interaction with artificially generated environments [102]. The systems may complement each other, as feedback systems are specialized in monitoring and feedback, whereas VR systems focus on training [103]. Robotic systems are usually seen as a separate class [100], even when combined with VR displays, because they comprise of extensive robotic apparatus [104].

25.2.2.1 Feedback Systems

Feedback systems usually consist of wearable sensors (accelerometers or IMUs) on the wrist of the impaired upper limb and sometimes on other body segments (for reviews see [73, 74]). Motion data are used for the estimation of arm use intensity, recognition of specific arm movements, recognition of particular exercises, and movement quality [73]. These analyses yield feedback such as information about upper limb use, the number of completed arm exercises, or numeric values for the degree of compensation [74]. In several studies, feedback was provided in form of vibrotactile signals, as these do not require visual attention. For example, vibrotactile feedback was applied to the affected arm as a reminder to use the arm more frequently [105, 106]. In another study, visual feedback was provided on conventional laptops or tablet screens in form of numbers or graphs which showed summary information of relevant metrics [107–109]. Several studies provided evidence that feedback systems improve motor performance and suggest that feedback systems have the potential to motivate individuals to move, increase adherence, and support motor learning [105, 106, 109]. However, evidence is sparse, especially in form of RCT [74, 110]. Further research is required to explore the full potential of feedback systems, notably with respect to systems that provide feedback during motor performance [73].

25.2.2.2 VR Systems

VR systems use data from motion sensing technologies to allow interaction between individuals and virtual environments [102]. The rationale behind VR systems is that immersion and use of games motivate individuals to practice [111].

VR systems can be classified by the type of motion sensing system and display technology [111]. Motion sensing systems (for lists of systems see [104, 112]) can be grouped into desktop (e.g., joysticks), body-mounted (e.g., IMU systems, sensor gloves, head-mounted displays), and contact-free systems (e.g., camera-based systems). Individual motion sensing systems may contain various motion sensing technologies (e.g., force sensors, pressure sensors, IMUs, depth sensors).

Display technologies of VR systems can be categorized into visual (e.g., 2D flat panel displays), auditory (e.g., two speakers for stereo sound), and haptic displays (e.g., vibrotactile actuators, force feedback [102]).

VR systems differ in the degree of immersion and, hence, can further be classified as non-immersive systems (e.g., systems with conventional tablets) and immersive systems (e.g., 3D stereoscopic vision head-mounted displays with surround sound [104]). Immersion in this definition is a technical quality and refers to the capability of systems to simulate the real world and generate authentic virtual experiences [113, 114].

Individuals may experience the virtual environment from a first- or third-person view. Motion data of individuals' body segments or joints are used to enable synchronous action in the virtual environment.

Examples of commercialized VR rehabilitation of the upper limbs can be found in a recent review [104].

An umbrella review of the meta-analysis concluded that there is evidence of a benefit of VR on motor function, but that evidence is weak as study quality is low or very low [114]. According to the review, important moderating factors for a benefit

of VR are the quality of immersion, interactivity, and customization of VR environments. Immersion and interactivity are important because differences in perceived distance between VR and real life, as well as system delays, might hinder the transference of skills from VR environments to real life [115]. This is critical as clinical evidence for the transfer of skills from VR training to real life is inconclusive [116]. Customization of VR environments is an important capability of VR systems as individuals with motor deficits might also have deficits in perception or cognition (e.g., individuals with attention deficits might require environments with reduced distraction [103]). Thus, VR systems engineered for neurorehabilitation are preferable to commercial VR gaming solutions that rather are designed for able-bodied individuals [114].

Several reviews examined the efficacy of VR-based training in home-based settings [103, 117]. Across reviews, only a few randomized controlled trials were identified. These generally showed a benefit of VR telerehabilitation that is comparable to conventional therapy, but the numbers of participants per study were low, VR Systems and intervention protocols differed strongly, and there was a wide variety in outcome measures. One RCT [118] that observed a benefit of VR-based telerehabilitation (and demonstrated noninferiority to conventional face-to-face therapy) applied several techniques to influence therapy outcomes such as behavioral contracts, stroke education, high ease of use, many input devices to cover a large range arm movement functions, frequent interaction with clients, multiple means of providing client feedback, solutions for creating appointment, and reminders. In several clients, this led to more than 1000 arm movement repetitions per client and day. This points to the conclusions that many aspects of therapy require optimization to obtain the desired doses of practice and beneficial effects on arm movement recovery.

25.2.2.3 Robotic Systems

There is a large and quickly increasing number of rehabilitation robots and robotic devices for the upper extremities (a recent review lists commercial devices for home use [119]). Robotic devices can be classified into grounded end-effectors, grounded exoskeletons, and wearable exoskeletons [120]. End-effector devices are usually only attached to individuals' distal arm parts, such as the hands or fingers. Exoskeletons target one or more joints of a paretic limb and are attached to adjacent segments. Systems can be further divided into passive, active, and interactive systems, depending on whether they merely stabilize, actively control, or react to individuals' movement input (for further classification see [121]). Systems are often equipped with sensors and, hence, can measure kinetics and kinematics that can be used to control VR-based exercises. Robotic devices for home use mainly target wrist, hand, and fingers [120]. Figure 25.1 shows an example of an interactive end-effector robot, the ReHandyBot, a portable robot for hand rehabilitation [122].

In clinical settings, robot-assisted therapy resulted in similar or larger improvements in upper extremity motor function as conventional interventions [123–125], regardless of the type of robotic arm training device [126]. Furthermore, for home-based settings, studies also suggest a benefit of robot-assisted therapy that is similar to conventional therapy, but the number of studies is rather small [127]. Drawbacks of robotic systems are their obtrusiveness, low comfort, and high costs [128, 129]. A recent trend are soft robotic gloves [130]. Soft robotic gloves are wearable and can facilitate hand and finger movements during activities of daily living or while playing VR-based serious games. Many studies report beneficial effects on motor recovery in clinical and home-based settings [130].

25.3 Lower Extremity

While upper limb impairments are common after stroke, the same is true for gait impairments. Over 80% of stroke survivors demonstrate a gait impairment [131] that recovers to some extent in the first two months after stroke [132]. Despite this, community ambulation often remains compromised in most survivors [133–135]. Gait ability specifically has major implications for health; it is an essential predictor for functional independence [136, 137] and long-term survival [137, 138] after stroke, along with participation and quality of life [139, 140]. Regaining gait ability is hence one of the most frequently prioritized goals of stroke survivors [131, 141].

25.3.1 Assessment of Motor Function

25.3.1.1 Clinical Assessment

Observation-Based Clinical Assessment
Gait ability can be separated into two major domains: Functional gait ability and gait quality. Assessments commonly employed in an in-clinic environment typically target the first domain. Best practice guidelines [43, 44] include the 10-m walk test (10MWT) [142], 6-min walk test (6MWT) [143], and timed-up-and-go test (TUG) [144], as they provide a complementary representation of a client's functional gait status and are feasible to implement clinically. These tests are commonly performed throughout stroke rehabilitation, albeit with a variability in terms of protocol and also time/distance criteria [145].

The 10MWT aims to measure maximal gait velocity and reflects the maximal capacity that a client has at a given point in time. For stroke survivors throughout all phases of recovery, the test demonstrates excellent sensitivity (MDC90 = 0.34–0.1 m/s from acute to chronic, respectively), and reliability (ICC = 0.83–0.95 from acute to chronic, respectively [145–147]). The MCID for this test in subacute stroke is estimated between 0.16 and 0.22 m/s [148].

The 6MWT aims to measure gait endurance. Equally, the clinometric properties of this assessment are well understood, and it demonstrates excellent reliability (ICC = 0.97–0.99 from acute to chronic, respectively) and sensitivity (MDC90 = 52 - 28 m from acute to chronic, respectively) [149] in stroke survivors throughout their recovery [150]. In terms of construct validity, the 6MWT correlates significantly with aerobic capacity, mobility, walking speed, strength, balance, and participation [150].

MCID for slow walkers in the subacute phase is estimated at 44m [148].

The TUG represents a compound test that measures the ability to perform sequential motor tasks relative to walking and turning [144, 151]. In this capacity, it can be considered a representation of complex movement [147]—one of the four pillars considered necessary for successful community ambulation [139]. In terms of clinometric properties, the TUG is considered reliable (ICC = 085 - 0.96) [152] and measures with an MDC95 of 2.9–3.5 s [147, 153].

Technology-Based Clinical Assessment

The second domain of gait assessment—gait quality—is not typically breached in clinical routine [154]. Procedures informing on this domain, in order of complexity, include observational gait assessments, pressure mat recordings, monocular video recordings (RGB/RGB-D), inertial-measurement-unit-based recordings, insole-based recordings, multi-camera video recordings, and 3D motion capture with force measurements. The objective and time-efficient quantification of gait quality has long carried the promise of identifying key deficits and guiding therapeutic intervention, however, this has yet to manifest on a large scale in clinical settings [155]. The barriers to large-scale adoption lie mainly in the cost, complexity, training requirements, and unwieldiness of current technology [156]. Advances especially in two domains carry the potential to overcome these limitations and have been frequently evaluated: Improved spatial reconstruction algorithms for IMUs, and improved computer vision pose estimation [157]. Typically, gait quality is described in three domains: Spatio-temporal metrics that describe the spatial and temporal characteristics of the endpoint within the resolution of a single gait cycle (e.g., step length, stance time, double support phase), joint kinematics (time-series of angles between segments that span from one gait event to the next), and kinetics (time-series of the ground reaction force for each limb) [158]. Gait events are considered when the foot strikes the ground (Foot-strike) and when the foot leaves the ground (Foot-off) and are used as anchors to

normalize all cycles to a common timescale [159]. For brevity, this chapter will forgo technologies that are not translatable to the home environment, e.g., force plates. The clinical utility of metrics derived within these domains is not yet well understood and pathology- and deficit-dependent [160, 161].

Camera-based systems are typically best suited for pre-defined walkways or treadmill situations [161], where a large amount of gait cycles can be collected within a small movement volume. For gait, the same optical technologies introduced above are applicable for the calculation of joint angles and segment positions during walking: Multi-camera RGB and RGB-D setups, and single-camera RGB and RGB-D. The benefits and drawbacks are equally largely comparable to the application for the upper extremities, however for walking, the capture volume and movement amplitude are larger, and occlusions are less frequent. Compared to optoelectrical 3D motion capture, the accuracy for multi-camera RGB systems using contemporary pose estimation networks is in the range of 2–3° joint angle deviation [162] and 1–15 mm deviation in 3D joint center position estimation [80]. Multi-camera systems remain complex, as the camera locations must be calibrated relative to each other before measurement. Moving to a single RGB-D camera with an added depth channel results in a reduced accuracy between 5 and 6° for joint angles [163], however, this is at least partially dependent on the camera placement [77]. This is equally applicable when using 2D monocular RGB video to estimate 3D pose [164–166], where out-of-plane errors in the major movement directions can be reduced by strategic camera placement [167]. In synthesis, these values compare favorably to estimated MCIDs for lower limb angles in stroke survivors, which range from between 3 and 9° for the major lower extremity joints and planes [168, 169].

Associated, nascent technologies that may enrich the industry are event-based cameras [170], 2D LiDAR sensors [171–173], or moving camera setups [174–176]. These technologies offer various benefits (high framerate, reduced setup complexity, and increased capture volume, respectively), however, have not yet reached a level of maturity and accuracy for gait analysis that is sufficient for clinical application.

Inertial measurement technology provides a series of additional benefits compared to camera solutions, that are especially relevant to gait. The measurement becomes location-independent and can be performed for long distances overground. Individual measurement units however need to be attached to the subject and calibrated, leading to additional setup time and potential inaccuracies. In a recent review, Poitras et al. report accuracies for joint angle estimation in contemporary commercial systems that are slightly lower than camera-based systems (Hip: < 9.3°; Knee < 11.5°; and Ankle < 18.8°) [177]. However, concerning spatio-temporal parameters [178], IMUs demonstrate accuracy sufficient to describe clinically relevant metrics such as stance duration and asymmetry within the limits of meaningful change after stroke [179–183].

25.3.1.2 Tele-Assessment

Although the advances in technology-based assessments bring the flexibility to perform gait assessments outside clinical environments and without supervision, the link between in-clinic and assessments performed in clients' domestic environments is not yet well understood. The International Classification of Functioning Disability and Health (ICF) suggests that there is a difference between these two settings [46]. The measurements that are carried out in standardized environments such as the clinic reflect the best performance of the clients or their capacity and the assessments that are carried out during daily activities are more representative of the clients' actual performance. Within this thought framework, it can be imagined that a 1:1 translation of clinical fixed-time/distance tests to home rehabilitation may not be trivial, especially concerning functional gait ability. Concerning gait quality, comparisons between clinic and home have been performed using both IMU and camera-based systems [184].

In these comparisons, the accuracy of the technology remains comparable with the clinical setting, however usability and feasibility challenges must be considered.

For cameras, an important aspect is the capture volume: Treadmills are rare in client's typical homes. Furthermore, the calibration of multiple cameras is a time-consuming and painstaking process that is not feasible for clients to perform autonomously. Fixed-camera installations that require minimal calibration are costly to install and require building modifications. Home-recording technologies are thus practically limited to monocular RGB/RGB-D video and stereoscopic cameras that house the sensors in a single discrete unit. In either case, due to the limited field of view camera-based gait analyses are limited to very short walkways or discrete lower limb movements performed on the spot [185]. As a complicating factor, the non-standardized scenes as backgrounds which are common in home applications can confound the tracking algorithms, leading to decreased accuracy [186, 187]. The practical use of cameras for gait analysis in home settings is hence quite limited and confined to a dedicated space.

For IMUs, the limitations concerning the volume of capture do not exist. As these systems are fully portable, they cannot limit the assessment space. However, IMU-based systems also bring challenges that are not yet fully addressed [188]. These are mainly in the usability spectrum, for example, the client-friendly attachment of IMUs to skin/clothing while retaining sufficient data quality, or zero-interaction data transfer and charging solutions [189]. Novel solutions, such as seamlessly including sensing units into shoes with wireless chargers [190], can be expected to provide solace to these usability barriers as costs to components fall and new form factors become available. Finally, many IMU systems require a calibration step in which predefined movements are necessary which may not be possible for a given client or which may lead to recording errors if performed too rapidly or erroneously.

Having scoped the limitations of each technology, there remains the question of which activity to record in a home environment, especially in the context of gait quality. Optimally, the same time/distance tests as performed in the clinic would translate seamlessly to the home environment for both functional gait ability and gait quality. Practical considerations, however, put the use of these tests into question. Specifically, few home environments provide the space necessary for a 6MWT (per clinical definition a 40 m straight walkway), nor even for a 10MWT with acceleration and deceleration phases. These tests are then performed outdoors, which complicates the question of a standardized setting with even surfaces [191]. Furthermore, there remains the question of how to initiate and end recordings for time/distance-limited tests. User interaction at the extrema of the tests invariably leads to a dual-task situation, therefore recording should best be triggered automatically. In the case of time, this is relatively trivial relying on a simple countdown timer. In the case of distance, however, the cumulative error in real-time spatial reconstruction of an IMU can lead to a roughly 5% error in the distance estimation [192]. A 10MWT for instance could be 9.5 m or 10.5 m long, hardly sufficiently reproducible for a reliable clinical outcome. Augmenting IMUs with other technology such as GPS can be helpful in improving the distance estimation [193], however, this is limited to longer tests such as the 6MWT and is strongly sensitive to satellite coverage which can limit its applicability in large cities [194].

The question here arises, whether there is a need for fixed-time/distance tests, or whether the continuous data collection during a portion of a day provides sufficient gait data to either extract representative test segments [195] or predict test performance [196–198]. Continuous data collection or "gait monitoring", holds the promise to also generate novel digital mobility outcomes (DMOs) [199–201] that can enable greater insight into a client's status. A plethora of algorithms are available to extract sequences of interest, such as gait, from the continuous data stream [202, 203]. From the current technology situation and a clinical utility perspective, it is unsurprising that the state of the art is gyrating towards continuous monitoring of gait using IMUs [204–207]. Open questions that remain are the location of the sensors (common are shoe [180], ankles [208], pelvis or pocket, and wrist [209] or a combination of these) and the form factor (dedicated gait monitoring hardware, smartphone sensors, or smartwatch sensors) [210].

25.3.2 Teletherapy

Concerning teletherapy, applications focus either on gait volume and intensity or on gait quality. Volume and intensity are collected throughout the day and presented as summative feedback on demand via web or tablet applications [211], similar to what is present in the lifestyle and wellness segment. Indeed, significant effort has been expended in evaluating lifestyle and wellness wearables for gait monitoring in elderly and clinical populations [212–216].

Applications for gait quality are typically proposed as gait training movements in a stationary setting [13, 217]. This can be performed asynchronously [218] or synchronously [219] with a health professional. There are countless exergaming-like applications that use IMU and/or camera systems coupled to visual displays that provide training programs and movement feedback in various levels of detail [220–224]. This segment is evolving rapidly, accelerated by readily available open-source and lightweight pose estimation algorithms for 2D RGB systems. An overview of commercial offerings would exceed the scope of a single chapter and probably be outdated before the book is printed.

A few research studies [225, 226] have engaged in the challenge of providing continuous real-time feedback on gait quality during ecological activity [227, 228]. While biofeedback is a promising approach in clinical settings [229, 230], challenges remain that limit the transfer to real-world scenarios. Outside of technical and usability limitations such as mounting points and battery capacity, the challenges here entail the appropriate detection of movement context [231] and the careful selection of relevant movement features and feedback modalities and mapping [101, 232, 233]. This area is expected to demonstrate significant innovation in the near future [234–239].

Mixed reality (MR) is a promising pathway for gait training scenarios (Fig. 25.2) [240, 241]. Compared to the previously described VR systems, MR enables the client to mix digital and real worlds through different technologies [242]. One approach for MR environments uses a video stream of RGB video cameras mounted to a headset to display a video of the environment onto a closed head-mounted display (HMD). The environment can be augmented with digital creations that are anchored in physical space, appearing in the HMD along with the video feed. These systems can be highly immersive; however, lag of the video feed and a corresponding mismatch of vestibular and visual information is still a challenge [243, 244]. To compensate for this, video resolution can be scaled down, resulting in a visually jarring imperfect representation of the surroundings. A further issue is the question of video white balance, which is adapted automatically to the lighting of scene and can lead to jumps in brightness of the video stream [245]. Equally open is the question of the matching of light source positioning and environmental situations between virtual and real objects [245]. As technology progresses on the fronts of mobile processing and graphics performance, camera, and display technologies, these issues may be resolved. A second approach uses translucent displays to superimpose a hologram image into the real world [246]. This elegantly circumvents the questions of lag and environment resolution as only the digital portion of the mixed reality needs to be rendered. There is rapid development in this sector, however, the field of view remains limited leading to a low level of immersion. In both cases, the usability of the device for remote deployment to clients [247, 248] and realistic multimodal feedback [249, 250] are remaining challenges that a new generation of rehabilitation-friendly devices should address. The perspective is that MR provides a pathway for highly promising VR-based training interventions from the clinic to extend to a client's home [16, 251, 252].

In synthesis, the technology situation is highly favorable to expand on the solutions currently available for gait in home environments. Improvements in device form factors, usability, measurement accuracy, presentation quality, and visualization of gait data synergize to unlock powerful new insights and provide training over a much longer rehabilitation period than previously feasible.

25.4 Communication

Effective communication in neurological conditions requires intact language, motor speech, and voice production. Aphasia is a language disorder where the use and the comprehension of meaningful speech are affected while motor speech disorder refers to a condition where a person has problems creating or forming speech sounds needed to communicate [253, 254]. In neurological voice disorder, the coordination or strengths of muscles that are needed to produce phonation are affected. All three conditions can occur after stroke, both, independently of each other or in combination, leaving roughly a third of all stroke survivors with some sort of communication impairment [255–257].

These impairments have a great impact on those affected as the majority of labor force depends nowadays on communication skills [258] but also social participation and independent living are dramatically affected by impaired communication. Clients with aphasia after stroke "described intense feelings of frustration, hopelessness, isolation, and depression at not being able to talk" [259].

Telerehabilitation is nowadays attracting broad interest in the field of speech and language therapy. For successful telerehabilitation, there is a need for optimal technology use in both, synchronous and asynchronous therapy delivery. To guide therapists along the recovery path, reliable and meaningful remote measures of voice, speech, and language functions are essential.

Besides the direct need of valid measures for therapy delivery, communication measures might also serve as a general indicator of well-being

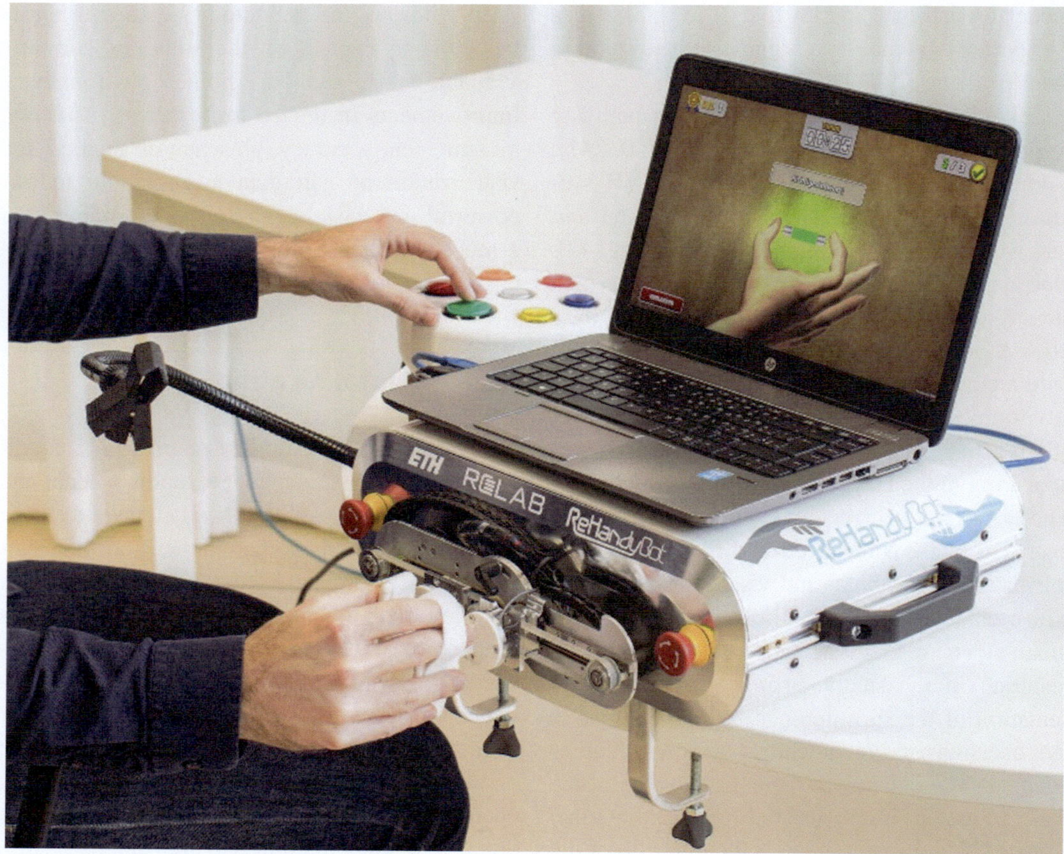

Fig. 25.1 The ReHandyBot (picture: Rehabilitation Engineering Laboratory, ETH Zurich/Stefan Schneller). Fingers and thumb are fixed to separate grippers. Grippers' movement is coupled which allows training hand opening and closing. Moreover, assistive or resistive forces allow haptic feedback and interaction with virtual objects

including emotional load and measure of participation, or early sign of cognitive decline [260, 261].

25.4.1 Assessment of Speech and Language Functions

25.4.1.1 Clinical Assessment

Observation-Based Clinical Assessment
Clinical assessment of functional voice, speech and language impairments consists of a wide range of measurement instruments. They largely depend on the perceptual evaluation by professional speech and language pathologists or task-specific performance measures (dysarthria [262], aphasia [263]). Many of these measurement instruments may not be specific to the stroke population but are used in the respective conditions (i.e., voice, speech, or language impairments) independent of the underlying disease. Most of the functional assessments related to communication are obviously highly language-specific. In the following paragraph, we will give only a few examples of voice, speech, and language scales in English or German to point out the limitations of remote assessments.

The German Bogenhausener Dysarthrieskalen (BoDyS) is a tool to assess voice and speech impairments in different tasks such as picture description [264]. For each task, the speech and

Fig. 25.2 A client after stroke performing a mixed reality gait training parkour with a holographic head-mounted display. Obstacles that require specific gait adaptations are placed throughout a given physical environment, transforming the space into a virtual rehabilitation gym (from [240])

language pathologist rates different symptoms perceived, such as hoarseness or slow talking speed. Even though audio recordings for potential algorithmic analysis are created during the assessment, the analysis requires manual scoring of voice and speech pathologies.

An expert panel has recently defined a core outcome set to assess language impairments in aphasia research [263]. It includes the Western Aphasia Battery-Revised (WAB-R), a diagnostic tool widely used by clinicians to assess language skills [265]. In German-speaking countries, the Aachener Aphasie Test (AAT) is a widely used diagnostic tool for differential diagnosis and follow-up assessment of language impairment [266].

Scales were also developed to assess communication in a more general term. No consensus, however, was reached by Wallace and colleagues on which measures to use in a core outcome set for research, and the discussion on how to best measure communication is ongoing [263].

Measurement instruments related to participation—especially important once a client is at home—are usually complemented by self-reported outcome measures and questionnaires related to emotional well-being and quality of life —measures which seem more meaningful to assess the impact of a disorder on a client's daily living, but may not give much information on the underlying impairments [263, 267].

Taken together, many of the clinically used assessment tools strongly rely on subjective observational methods, i.e., speech and motor behavior are observed by a professional speech and language pathologist and require (rather time-consuming) manual scoring. There is a call for more technology-dependent outcome measures which may help to understand underlying impairment and treatment effects as stated by Elisabeth Armstrong in her commentary article: Let's utilize technology to the maximum and contribute our expertise at the highest level, while at the same time not simply counting what can be counted and neglecting "what counts" [268].

Technology-Based Clinical Assessments

Voice Analysis

Instrumented acoustic voice analysis is a quantitative noninvasive method to complement a multidimensional approach for evaluation of initial impairment and the recovery/therapy progress [269].

Over the past decades, several acoustic measures have been proposed to be sensitive to (overall) voice quality or specific perceptual dimensions such as hoarseness or breathiness ([270] and other reviews). In standard acoustic voice assessments, a voice profile (vocal intensity and pitch measures), frequency and amplitude perturbation (jitter, shimmer), and harmonicity measurements (harmonic to noise ratio) are usually determined that provide information about the underlying functional impairment and recovery progress [270, 271]. The feasibility and validity of such metrics highly depend on the client's task (e.g., sustained vowels vs. reading) and instructions (e.g., comfortable loudness vs. maximal loudness) as well as room setup and technical specifications (e.g., microphone quality, setup, and calibration). An expert panel of the American Speech-Language-Hearing Association has recently approached this issue and published a standardization protocol defining instructions and specifications for qualitative acoustic voice analysis in great detail [269]. Even in a very standardized setup, acoustic parameters are highly variable due to dependence on the vocal intensity used [272]. Most of the traditional perturbation measures depend on the computation of the fundamental frequency which may not be reliable in severe dysphonia. For this reason, Patel et al. further

suggested the calculation of cepstral peak prominence, a measure based on spectral analyses for estimating general dysphonia severity and overall breathiness from connected speech, i.e., during a reading task or a conversation [269, 273–275].

Several computer programs exist which enable instrumented voice analysis also in a clinical setting. One example used in research and clinic is the comprehensive open-source program PRAAT, which was initiated and continuously developed by Jan Boersma and David Weenink from the University of Amsterdam (http://www.PRAAT.org/). Many commercially available voice analysis software exist that come with technical equipment that allows for rapid voice assessment without requiring complex calibration and in-depth technical knowledge.

Speech and Language Analysis

Similar to acoustic voice analysis, certain speech and language parameters can be automatically calculated from audio files using algorithms integrated into voice analysis software such as PRAAT. Such parameters may include speech rate and overall talking time, but also more complex estimates of fluency extracted from connected speech [276, 277].

Fully automized analysis of audio samples to assess language functions is often limited to macroparameters. More relevant [278, 279] linguistic parameters such as lexical diversity, mean utterance lengths, repetitions, and self-corrections can be extracted from transcripts with the aid of specialized speech analysis systems [279]. Examples include the "Aachener Spontansprachanalyse" [280, 281] for German or the "Computerized language analysis" (CLAN) for English [282] which, once the audio sample is manually transcribed, assists with word tagging and linguistic analysis. More content-dependent discourse measures can be assessed using natural language processing by computerized systems [283, 284].

As part of the Aphasia Bank Project, large amounts of standardized speech samples (e.g., the English story-retelling task "Cinderella") from aphasic clients were—and still are—collected, transcribed, and made available to researchers for in-depth discourse analysis [285, 286]. This Big Data approach contributes to the understanding of discourse differences in mildly impaired stroke survivors that may not with scales such as the WAB-R due to ceiling effects [287]. Even though computer-assisted methods facilitate manual transcription, the process is still time-consuming and thus often not feasible in daily clinical practice. Advances in Automated Speech Recognition (ASR)—the automatic *conversion of audio files into written text* [288]—could solve this problem in the future. A recent study suggests that ASR could be used in a clinic setting for aphasic clients. However, the quality of transcripts needs to be investigated separately for each communication disorder [289].

25.4.2 Tele-Assessments

In a remote setting, one way to assess speech and language functions may be to apply standardized assessment batteries or scales via video call in a synchronous session with a speech and language pathologist. This way of applying assessments is limited by the availability of analog equipment on the client's side, and for many tools, a fully computerized adaptation is still lacking. Furthermore, the validity of such teleconference-based application of standardized test batteries needs to be investigated—as has recently been done for the widely used WAB-R [290]—and standards need to be developed [291]. Overall, remote assessments via teleconference systems are feasible, but a revised computerized version rather than ad hoc digital adaptation of valid analog tools would be highly desirable.

Recording audio samples during a tele-assessment via video call for automated analysis is possible, however, signal processing circuits such as automatic gain control or noise cancellation may modify the original microphone signal [269, 271]. Setup and implementation at a client's home may not (easily) be controllable. Audio samples may still be used for synchronous or asynchronous manual scoring of voice, speech, or language impairment or transcription.

Some automated tele-diagnostic instruments are also available for web-based administration or use with smartphone applications. Smartphone applications have been developed for self-administration of acoustic voice assessments in clinical telepractice settings which can produce valid calculations of acoustic parameters [292]. As described above, a highly standardized setup in a clinical setting is crucial for valid acoustic voice assessments. It is thus questionable whether such smartphone-based audio recordings achieve the quality to successfully guide teletherapeutic interventions. For speech and language, there are applications mainly used for research purposes which allow remote application of certain standardized speech and language assessments (e.g., [293]).Tele-monitoring.

Using microphones for monitoring voice, speech, or language functions in daily living is feasible. However, it faces several limitations: Privacy, environmental noise, recording quality, and interference with speech from communication partners. One solution for the privacy issue is to only store information that does not need content recording, such as overall talking time, as done in people with aphasia after stroke [294]. However, this metric does not seem to provide much information about functional recovery as it lacks correlation to impairment level but might be an indicator of participation [294]. Accuracy of measuring talking time is highly dependent on environmental noise. Researchers have started to address this limitation by measuring voice functions via neck surface vibration measures with a single accelerometer placed on the subglottal neck surface [295]. Relevant metrics such as fundamental frequency and sound pressure level as well as spectral parameters can be captured through skin acceleration and correlate for the most part with the parameters received from acoustic measures through a microphone [275, 296, 297].

25.4.3 Teletherapy

25.4.3.1 Synchronous Therapy

Many therapists, clinics or specialized telerehabilitation providers offer speech and language therapy via video conferencing [298], also called telepractice [299, 300] (ASHA 2018). Some use available video call platforms such as Webex or Zoom with standard video call features [301, 302]. Others may use specialized speech and language tele-therapy platforms with integrated digital exercises or social functionality [303]. Such synchronous speech and language therapies were found to have similar effects as in-person therapy (reviewed by [304]). With increasing functionality such as joint editing of digital material and integrated audio recordings, an increased range of therapy content may be provided in the future.

A high-quality microphone and high-speed internet connection are crucial, especially for vocal tele-therapy, and technical hurdles might limit use in the elderly. There are, however, also advantages of video-based synchronous over in-person therapy. Group therapy and group chats for people with communication problems are easily implementable. Group therapy was found to be a valuable and cost-efficient way of delivering therapy. Group chats that socially connect people with similar impairments, medical history, and interests have a positive effect on participation and well-being which in turn can improve specific linguistic effects [305–307]. Another advantage is the high context specificity in which the therapy takes place, allowing speech and language therapy to be tailored to a client's functional environment [308].

25.4.3.2 Asynchronous Therapy

Many commercially available computer programs exist which have digitalized logopedic exercise material and can be used by clients for independent functional training at home. Commercially available programs, such as *Tactus Therapy Solutions, Lingraphica, Constant Therapy,* or *Neolexon* are built to supplement clinician-delivered therapy; the therapist creates a personalized training program and has access to training results or audio recordings via a specialized therapist interface. In many exercises, difficulty levels are automatically adapted based on errors and response time, allowing training without constant evaluation and exercise adaptation by the therapist. Automated detection of errors is mainly available for written input. Thus, any form of automated result or performance feedback is mostly restricted to receptive exercises (i.e., the correct answer can be chosen among suggestions) or to written content. Advances in automatic speech recognition will open the path to more independent verbal training in the future [309]. Feedback during voice training does not rely on speech recognition and helps to adapt vocal behavior—namely pitch and/or loudness—during training [310]. The use of neck vibration monitoring system enables biofeedback during daily living and promotes changes in vocal behavior [311, 312]. Retention of the adapted behavior, however, seems to be strongly dependent on the type and frequency of feedback used [312].

Taken together, such computerized asynchronous therapies are usually well received by stroke survivors and seem a valid option to increase training time, even though efficacy may not exceed in-person therapy effects [313] (see [314] for a review). The field of telerehabilitation is still in its infancy and technologies are developing rapidly. Comprehensive combinations of synchronous and asynchronous options might be more successful in the future. The Australian Tele-CHAT (i.e., Comprehensive High-dose Aphasia Treatment) program is an example of a combination of synchronous, asynchronous, and group therapy and it also supports education and social exchange among stroke survivors. Feasibility and acceptability of this program is still under scientific investigation, but it seems clear that for an efficient and effective therapy delivery, the use of different therapy modes needs to be optimally coordinated.

25.5 Final Remarks

Technological solutions are expected to become an integral part of telerehabilitation. These should be complemented by data platforms that consolidate the diverse streams of information into a coherent and actionable whole. These data can be leveraged by healthcare professionals to gain unprecedented insights into recovery profiles, individual therapeutic needs, and requirements for sustainable technological solutions. AI is already demonstrating some promising performance in specific medical tasks. However, the details of decision-making of the algorithms are often poorly understood, and there is still no robust interpretability and explainability, thus hindering the widespread use of AI in medicine. Ongoing developments of methods for visualization, explanation, and interpretation of AI models aim at making AI transparent and explainable (XAI), and thus robust for application in the medical field [315]. In the future, XAI may allow to target and scale interventions based on a large amount of data precisely and reliably to the needs of the individual.

Clinical trials show promising results with telerehabilitation. The Corona pandemic has promoted the acceptance of telerehabilitation among therapists and clients. To promote widespread and long-term implementation in routine clinical practice, it is up to stakeholders to increase adoption through measures on telecommunications infrastructure, legal certainty around digital health policies and legislation, data security, technical support, teaching, and reimbursement.

The combination of increasing digital literacy, scalability requirements of the neurorehabilitation industry, regulatory changes, and an increased enthusiasm for new technologies are creating a breeding ground for tele-neurorehabilitation to gain wider acceptance and fully deliver on its promise.

In the long term, telerehabilitation could become a ubiquitous tool that enables access to health care and intense, personally optimized neurorehabilitation regardless of geographical location.

Note that this book chapter primarily focused on technology and services for assessment and treatment at the levels of ICF function and activity. Future work will certainly also cover clients' functioning at the ICF participation level, especially because improvements in quality of life and participation in social activities are the higher-order goals of neurorehabilitation. Thus, technology and service design will increasingly need to incorporate clients' social and environmental contexts. Barriers in social contexts can be reduced, for example, through simulation of clients' motor limitations to healthy individuals, using VR and robotic systems [14]. Healthy individuals can thus experience and gain an intuitive understanding of clients' motor limitations [14]. Moreover, caretakers can aid clients to leverage smart homes and smart cities to reduce barriers to social context and increase engagement in these. For example, smart door locks might reduce the stress of clients when entering new social interactions [316] and social games in smart cities can engage to move and interact with others [317].

Hence, as we press on into the future, new technologies and services will become increasingly pervasive in our everyday lives. These technologies can be leveraged for rehabilitation at a distance and might enable a more direct impact on participation than previous approaches.

References

1. Appleby E, et al. Effectiveness of telerehabilitation in the management of adults with stroke: a systematic review. PLoS ONE. 2019;14: e0225150.
2. Brennan DM, Mawson S, Brownsell S. Telerehabilitation: enabling the remote delivery of healthcare, rehabilitation, and self management. Stud Health Technol. 2009;145:231–48.
3. Richmond T, et al. American telemedicine association's principles for delivering telerehabilitation services. Int J Telerehabilitation 2017; 9:63–68.
4. Katz RC. Application of computers to the treatment of US veterans with aphasia. Aphasiology. 2009;23:1116–26.
5. Vaughn GR. Tel-communicology: health-care delivery system for persons with communicative disorders. ASHA. 1976;18:13–7.

6. Shinar D, et al. Reliability of the activities of daily living scale and its use in telephone interview. Arch Phys Med Rehab. 1987;68:723–8.

7. Mayo NE. Stroke Rehabilitation at Home. Stroke. 2018;47:1685–91.

8. Danzl MM, Etter NM, Andreatta RD, Kitzman PH. Facilitating neurorehabilitation through principles of engagement. J Allied Heal. 2012;41:35–41.

9. Isernia S, et al. Efficiency and patient-reported outcome measures from clinic to home: the human empowerment aging and disability program for digital-health rehabilitation. Front Neurol. 2019;10:1206.

10. Vickery DM, et al. Effect of a Self-care Education Program on Medical Visits. JAMA. 1983;250:2952–6.

11. Coulter A, Ellins J. Client-focused interventions: a review of the evidence, vol 1. London: Health Foundation; 2006.

12. Laver KE, et al. Telerehabilitation services for stroke. Cochrane Database Syst Rev. 2020;1: CD010255

13. Deng H, et al. Complex versus simple ankle movement training in stroke using telerehabilitation: a randomized controlled trial. Phys Ther. 2012;92:197–209.

14. Baur K, Rohrbach N, Hermsdörfer J, Riener R, Klamroth-Marganska V. The "Beam-Me-In Strategy"–remote haptic therapist-patient interaction with two exoskeletons for stroke therapy. J Neuroeng Rehabil. 2019;16:85.

15. Mutalib SA, et al. Modernising the dynamometer for grip assessment: comparison between gripable and jamar. 2021. https://doi.org/10.21203/rs.3.rs-720982/v1.

16. Laver KE, et al. Virtual reality for stroke rehabilitation. Cochrane Database Syst Rev. 2017. https://doi.org/10.1002/14651858.cd008349.pub4.

17. Annaswamy TM, et al. Using biometric technology for telehealth and telerehabilitation. Phys Med Rehabil Cli. 2021;32:437–49.

18. Weiss PL, Kizony R, Feintuch U, Katz N. Textbook of neural repair and rehabilitation 2006;182–197. https://doi.org/10.1017/cbo9780511545078.015.

19. Rinne P, et al. Democratizing neurorehabilitation: how accessible are low-cost mobile-gaming technologies for self-rehabilitation of arm disability in stroke? PLoS ONE. 2016;11: e0163413.

20. Klamroth-Marganska V, et al. Does therapy always need touch? A cross-sectional study among Switzerland-based occupational therapists and midwives regarding their experience with health care at a distance during the COVID-19 pandemic in spring 2020. Bmc Health Serv Res. 2021;21:578.

21. Seifert A. Die Nutzung des Internets durch Personen ab 65 Jahren und ihr subjektives Gefühl der Exklusion. Blätter Der Wohlfahrtspflege. 2021;168:149–51.

22. Ballmer T. et al. Behandlungen auf Distanz während des Lockdowns. Ergotherapie. 2021:22–27.

23. Almathami HKY, Win KT, Vlahu-Gjorgievska E. Barriers and facilitators that influence telemedicine-based, real-time, online consultation at patients' homes: systematic literature review. J Med Internet Res. 2020;22: e16407.

24. Bahari M, Jafni TI, Miskon S, Ismail W. A review of success/failure factors influencing healthcare personnel for telerehabilitation. In: 2019 6th International conference res innovation information syst icriis; 2019. p. 1–7.

25. Jafni TI, Bahari M, Ismail W, Radman A. Understanding the implementation of telerehabilitation at pre-implementation stage: a systematic literature review. Procedia Comput Sci. 2017;124:452–60.

26. Yusof M. Mohd. A case study evaluation of a Critical Care Information System adoption using the socio-technical and fit approach. Int J Med Inform. 2015;84:486–499.

27. Creagh AP, et al. Smartphone-based remote assessment of upper extremity function for multiple sclerosis using the Draw a Shape Test. Physiol Meas. 2020;41: 054002.

28. Donkor ES. Stroke in the 21st century: a snapshot of the burden, epidemiology, and quality of life. Stroke Res Treat. 2018;2018:3238165.

29. Feigin VL, et al. Global and regional burden of stroke during 1990–2010: findings from the Global Burden of Disease Study 2010. Lancet. 2014;383:245–55.

30. Broeks JG, Lankhorst GJ, Rumping K, Prevo AJH. The long-term outcome of arm function after stroke: results of a follow-up study. Disabil Rehabil. 2009;21:357–64.

31. Nijland RHM, van Wegen EEH, der Wel BCH, Kwakkel G. Presence of finger extension and shoulder abduction within 72 hours after stroke predicts functional recovery. Stroke. 2010;41:745–50.

32. Kwakkel G, Kollen BJ, van der Grond J, Prevo AJH. Probability of regaining dexterity in the flaccid upper limb. Stroke. 2003;34:2181–6.

33. van Lieshout ECC, van de Port IG, Dijkhuizen RM, Visser-Meily JMA. Does upper limb strength play a prominent role in health-related quality of life in stroke patients discharged from inpatient rehabilitation? Top Stroke Rehabil. 2020;27:1–9.

34. Ytterberg C, et al. A qualitative study of cognitive behavioural therapy in multiple sclerosis: experiences of psychotherapists. Int J Qual Stud Heal. 2017;12:1325673.

35. Bernhardt J, et al. Agreed definitions and a shared vision for new standards in stroke recovery research: The Stroke Recovery and Rehabilitation Roundtable taskforce. Int J Stroke. 2017;12:444–50.

36. Levin MF, Kleim JA, Wolf SL. What do motor "recovery" and "compensation" mean in patients following stroke? Neurorehab Neural Re. 2009;23:313–9.

37. Roby-Brami A, et al. Motor compensation and recovery for reaching in stroke patients. Acta Neurol Scand. 2003;107:369–81.

38. Cirstea MC, Levin MF. Compensatory strategies for reaching in stroke. Brain. 2000;123:940–53.

39. Prange-Lasonder GB, et al. European evidence-based recommendations for clinical assessment of upper limb in neurorehabilitation (CAULIN): data synthesis from systematic reviews, clinical practice guidelines and expert consensus. J Neuroeng Rehabil. 2021;18:162.

40. Burridge J, et al. A systematic review of international clinical guidelines for rehabilitation of people with neurological conditions: what recommendations are made for upper limb assessment? Front Neurol. 2019;10:567.

41. Hughes A-M, et al. Evaluation of upper extremity neurorehabilitation using technology: a European Delphi consensus study within the EU COST Action Network on Robotics for Neurorehabilitation. J Neuroeng Rehabil. 2016;13:86.

42. Schwarz A, Kanzler CM, Lambercy O, Luft AR, Veerbeek JM. Systematic review on kinematic assessments of upper limb movements after stroke. Stroke. 50:718–727.

43. Pohl J, et al. Consensus-based core set of outcome measures for clinical motor rehabilitation after stroke—a Delphi study. Front Neurol. 2020;11:875.

44. Kwakkel G, et al. Standardized measurement of sensorimotor recovery in stroke trials: consensus-based core recommendations from the stroke recovery and rehabilitation roundtable. Int J Stroke. 2017;12:451–61.

45. Kwakkel G, et al. Standardized measurement of quality of upper limb movement after stroke: consensus-based core recommendations from the second stroke recovery and rehabilitation roundtable. Neurorehab Neural Re. 2019;33:951–8.

46. Organization WH. International classification of functioning, disability and health (ICF). 2001.

47. Waddell KJ, Strube MJ, Tabak RG, Haire-Joshu D, Lang CE. Upper limb performance in daily life improves over the first 12 weeks poststroke. Neurorehab Neural Re. 2019;33:836–47.

48. Lee SI, et al. A novel upper-limb function measure derived from finger-worn sensor data collected in a free-living setting. PLoS ONE. 2019;14: e0212484.

49. Fugl-Meyer AR, Jääskö L, Leyman I, Olsson S, Steglind S. The post-stroke hemiplegic patient. 1. a method for evaluation of physical performance. Scand J Rehabil Med. 1975;7:13–31.

50. Sullivan KJ, et al. Fugl-Meyer assessment of sensorimotor function after stroke. Stroke. 2011;42:427–32.

51. Lyle RC. A performance test for assessment of upper limb function in physical rehabilitation treatment and research. Int J Rehabil Res. 1981;4:483–92.

52. Yozbatiran N, Der-Yeghiaian L, Cramer SC. A standardized approach to performing the action research arm test. Neurorehab Neural Re. 2008;22:78–90.

53. Singer B, Garcia-Vega J. The Fugl-Meyer upper extremity scale. j Physiother. 2017;63:53.

54. van der Lee JH, Roorda LD, Beckerman H, Lankhorst GJ, Bouter LM. Improving the action research arm test: a unidimensional hierarchical scale. Clin Rehabil. 2002;16:646–53.

55. Amano S. et al. Reliability of remote evaluation for the Fugl–Meyer assessment and the action research arm test in hemiparetic patients after stroke. Top Stroke Rehabil. 2018:1–6. https://doi.org/10.1080/10749357.2018.1481569.

56. Murphy MA, Häger CK. Kinematic analysis of the upper extremity after stroke–how far have we reached and what have we grasped? Phys Ther Rev. 2015;20:137–55.

57. Thrane G, Sunnerhagen KS, Murphy MA. Upper limb kinematics during the first year after stroke: the stroke arm longitudinal study at the University of Gothenburg (SALGOT). J Neuroeng Rehabil. 2020;17:76.

58. Subramanian SK, Yamanaka J, Chilingaryan G, Levin MF. Validity of movement pattern kinematics as measures of arm motor impairment poststroke. Stroke. 2010;41:2303–8.

59. Ellis MD, Lan Y, Yao J, Dewald JPA. Robotic quantification of upper extremity loss of independent joint control or flexion synergy in individuals with hemiparetic stroke: a review of paradigms addressing the effects of shoulder abduction loading. J Neuroeng Rehabil. 2016;13:95.

60. Milosevic B, Leardini A, Farella E. Kinect and wearable inertial sensors for motor rehabilitation programs at home: state of the art and an experimental comparison. Biomed Eng Online. 2020;19:25.

61. Merriaux P, Dupuis Y, Boutteau R, Vasseur P, Savatier X. A study of vicon system positioning performance. Sensors Basel Switz. 2017;17:1591.

62. Mavor MP, Ross GB, Clouthier AL, Karakolis T, Graham RB. Validation of an IMU suit for military-based tasks. Sensors Basel Switz. 2020;20:4280.

63. Kanko RM, Laende EK, Davis EM, Selbie WS, Deluzio KJ. Concurrent assessment of gait kinematics using marker-based and markerless motion capture. J Biomech. 2021;127: 110665.

64. Mathiowetz V, et al. Grip and pinch strength: normative data for adults. Arch Phys Med Rehab. 1985;66:69–74.

65. Xu J. et al. Separable systems for recovery of finger strength and control after stroke. J Neurophysiol. 2017. https://doi.org/10.1152/jn.00123.2017.

66. Subramanian SK, Cross MK, Hirschhauser CS, Johnson VB, Reistetter TA. Post-stroke upper limb rehabilitation using virtual reality interventions: Do outcome measures assess extent or type of motor improvement? In: 2019 Int Conf Virtual Rehabilitation Icvr. 2019. p. 1–6.

67. Stinear CM, Smith M-C, Byblow WD. Prediction tools for stroke rehabilitation. Stroke. 2019;50:3314–22.

68. Murphy MA, Sunnerhagen KS, Johnels B, Willén C. Three-dimensional kinematic motion analysis of a daily activity drinking from a glass: a pilot study. J Neuroeng Rehabil. 2006;3:18.

69. Murphy MA, Willén C, Sunnerhagen KS. Kinematic variables quantifying upper-extremity performance after stroke during reaching and drinking from a glass. Neurorehab Neural Re. 2011;25:71–80.

70. Frykberg GE, Grip H, Murphy MA. How many trials are needed in kinematic analysis of reach-to-grasp?—A study of the drinking task in persons with stroke and non-disabled controls. J Neuroeng Rehabil. 2021;18:101.

71. David A, Subash T, Varadhan SKM, Melendez-Calderon A, Balasubramanian S. A framework for sensor-based assessment of upper-limb functioning in hemiparesis. Front Hum Neurosci. 2021;15: 667509.

72. Thrane G, Emaus N, Askim T, Anke A. Arm use in patients with subacute stroke monitored by accelerometry: association with motor impairment and influence on self-dependence. J Rehabil Med. 2011;43:299–304.

73. Kim GJ, Parnandi A, Eva S, Schambra H. The use of wearable sensors to assess and treat the upper extremity after stroke: a scoping review. Disabil Rehabil. 2021:1–20. https://doi.org/10.1080/09638288.2021.1957027.

74. Wang Q, Markopoulos P, Yu B, Chen W, Timmermans A. Interactive wearable systems for upper body rehabilitation: a systematic review. J Neuroeng Rehabil. 2017;14:20.

75. Iosa M, Picerno P, Paolucci S, Morone G. Wearable inertial sensors for human movement analysis. Expert Rev Med Devic. 2016;13:1–19.

76. Yang L, Zhang L, Dong H, Alelaiwi A, Saddik AE. Evaluating and improving the depth accuracy of kinect for windows v2. IEEE Sens J. 2015;15:4275–85.

77. Yeung L-F, Yang Z, Cheng KC-C, Du D, Tong RK-Y. Effects of camera viewing angles on tracking kinematic gait patterns using Azure Kinect, Kinect v2 and Orbbec Astra Pro v2. Gait Posture. 2021;87:19–26.

78. Wang J, et al. Deep 3D human pose estimation: a review. Comput Vis Image Und. 2021;210: 103225.

79. Mathis A, Schneider S, Lauer J, Mathis MW. A primer on motion capture with deep learning: principles, pitfalls, and perspectives. Neuron. 2020;108:44–65.

80. Needham L, et al. The accuracy of several pose estimation methods for 3D joint centre localisation. Sci Rep-UK. 2021;11:20673.

81. Arac A. Machine learning for 3D kinematic analysis of movements in neurorehabilitation. Curr Neurol Neurosci. 2020;20:29.

82. Din SD, Patel S, Cobelli C, Bonato P. Estimating Fugl-Meyer clinical scores in stroke survivors using wearable sensors. In: 2011 Annu Int Conf Ieee Eng Medicine Biology Soc. 2011. p. 5839–5842.

83. Kim W-S, Cho S, Baek D, Bang H, Paik N-J. Upper extremity functional evaluation by Fugl-Meyer assessment scoring using depth-sensing camera in hemiplegic stroke patients. PLoS ONE. 2016;11: e0158640.

84. Yu L, Xiong D, Guo L, Wang J. A remote quantitative Fugl-Meyer assessment framework for stroke patients based on wearable sensor networks. Comput Meth Prog Bio. 2016;128:100–10.

85. Adans-Dester C, et al. Enabling precision rehabilitation interventions using wearable sensors and machine learning to track motor recovery. Npj Digital Med. 2020;3:121.

86. Chin LF, Hayward KS, Brauer S. Upper limb use differs among people with varied upper limb impairment levels early post-stroke: a single-site, cross-sectional, observational study. Top Stroke Rehabil. 2019;27:1–12.

87. Strømmen AM, Christensen T, Jensen K. Quantitative measurement of physical activity in acute ischemic stroke and transient ischemic attack. Stroke. 2014;45:3649–55.

88. Schambra HM, et al. A taxonomy of functional upper extremity motion. Front Neurol. 2019;10:857.

89. Lemmens RJM, Janssen-Potten YJM, Timmermans AAA, Smeets RJEM, Seelen HAM. Recognizing complex upper extremity activities using body worn sensors. PLoS ONE. 2015;10: e0118642.

90. Panwar M, et al. Rehab-net: deep learning framework for arm movement classification using wearable sensors for stroke rehabilitation. IEEE T Biomed Eng. 2018;66:3026–37.

91. Biswas D, et al. Recognition of elementary arm movements using orientation of a tri-axial accelerometer located near the wrist. Physiol Meas. 2014;35:1751–68.

92. Ward NS, Brander F, Kelly K. Intensive upper limb neurorehabilitation in chronic stroke: outcomes from the Queen Square programme. J Neurology Neurosurg Psychiatry. 2019;90:498.

93. Ballester BR. et al. Relationship between intensity and recovery in post-stroke rehabilitation: a retrospective analysis. J Neurology Neurosurg Psychiatry jnnp. 2021. https://doi.org/10.1136/jnnp-2021-326948.

94. Krakauer JW, Carmichael ST, Corbett D, Wittenberg GF. Getting neurorehabilitation right. Neurorehab Neural Re. 2012;26:923–31.

95. Maier M, Ballester BR, Verschure PFMJ. Principles of neurorehabilitation after stroke based on motor learning and brain plasticity mechanisms. Front Syst Neurosci. 2019;13:74.

96. Gittler M, Davis AM. Guidelines for adult stroke rehabilitation and recovery. JAMA. 2018;319:820–1.

97. Foundation S. Clinical guidelines for stroke management. 2017. https://informme.org.au/en/Guidelines/Clinical-Guidelines-for-Stroke-Management.

98. Lambercy O, et al. Neurorehabilitation from a distance: can intelligent technology support decentralized access to quality therapy? Front Robot AI. 2021;8: 612415.

99. Lee SH, et al. Comparisons between end-effector and exoskeleton rehabilitation robots regarding upper extremity function among chronic stroke patients with moderate-to-severe upper limb impairment. Sci Rep-UK. 2020;10:1806.

100. Chen Y, et al. Home-based technologies for stroke rehabilitation: a systematic review. Int J Med Inform. 2018;123:11–22.

101. Sigrist R, Rauter G, Riener R, Wolf P. Augmented visual, auditory, haptic, and multimodal feedback in motor learning: a review. Psychon B Rev. 2013;20:21–53.

102. Riener R. Virtual reality for neurorehabilitation. In: Dietz V, Ward N (eds) Oxford textbook of neurorehabilitation. Oxford Medicine Online; 2016. https://doi.org/10.1093/med/9780199673711.003.0034.

103. Amorim P, Santos BS, Dias P, Silva S, Martins H. Serious games for stroke telerehabilitation of upper limb - a review for future research. Int J Telerehabilitation. 2020;12:65–76.

104. Kim W-S, et al. Clinical application of virtual reality for upper limb motor rehabilitation in stroke: review of technologies and clinical evidence. J Clin Medicine. 2020;9:3369.

105. Da-Silva RH, et al. Prompting arm activity after stroke: A clinical proof of concept study of wrist-worn accelerometers with a vibrating alert function. J Rehabil Assistive Technol Eng. 2018;5:2055 668318761524.

106. Da-Silva RH, et al. Wristband Accelerometers to motiVate arm Exercises after Stroke (WAVES): a pilot randomized controlled trial. Clin Rehabil. 2019;33:1391–403.

107. Chae SH, Kim Y, Lee K-S, Park H-S. Development and clinical evaluation of a web-based upper limb home rehabilitation system using a smartwatch and machine learning model for chronic stroke survivors: prospective comparative study. Jmir Mhealth Uhealth. 2020;8: e17216.

108. Lee SI, et al. Enabling stroke rehabilitation in home and community settings: a wearable sensor-based approach for upper-limb motor training. IEEE J Transl Eng Heal Med. 2018;6:2100411.

109. Whitford M, Schearer E, Rowlett M. Effects of in home high dose accelerometer-based feedback on perceived and actual use in participants chronic post-stroke. Physiother Theor Pr. 2018;36:1–11.

110. Parker J, Powell L, Mawson S. Effectiveness of upper limb wearable technology for improving activity and participation in adult stroke survivors: systematic review. J Med Internet Res. 2020;22: e15981.

111. Merians AS, et al. Virtual reality–augmented rehabilitation for patients following stroke. Phys Ther. 2002;82:898–915.

112. Subramanian SK, Cross MK, Hirschhauser CS. Virtual reality interventions to enhance upper limb motor improvement after a stroke: commonly used types of platform and outcomes. Disabil Rehabil Assistive Technol. 2020:1–9. https://doi.org/10.1080/17483107.2020.1765422.

113. Slater M, Wilbur S. A framework for immersive virtual environments (FIVE): speculations on the role of presence in virtual environments. Presence Teleoperators Virtual Environ. 1997;6:603–16.

114. Voinescu A, Sui J, Fraser DS. Virtual reality in neurorehabilitation: an umbrella review of meta-analyses. J Clin Med. 2021;10:1478.

115. Morel M, Bideau B, Lardy J, Kulpa R. Advantages and limitations of virtual reality for balance assessment and rehabilitation. Neurophysiol Clinique Clin Neurophys. 2015;45:315–26.

116. Levac DE, Huber ME, Sternad D. Learning and transfer of complex motor skills in virtual reality: a perspective review. J Neuroeng Rehabil. 2019;16:121.

117. Perrochon A, Borel B, Istrate D, Compagnat M, Daviet J-C. Exercise-based games interventions at home in individuals with a neurological disease: a systematic review and meta-analysis. Ann Phys Rehabil Med. 2019;62:366–78.

118. Cramer SC, et al. Efficacy of home-based telerehabilitation vs in-clinic therapy for adults after stroke: a randomized clinical trial. Jama Neurol. 2019;76:1079.

119. Akbari A, Haghverd F, Behbahani S. Robotic home-based rehabilitation systems design: from a literature review to a conceptual framework for community-based remote therapy during COVID-19 pandemic. Front Robot AI. 2021;8: 612331.

120. Gassert R, Dietz V. Rehabilitation robots for the treatment of sensorimotor deficits: a neurophysiological perspective. J Neuroeng Rehabil. 2018;15:46.

121. Noronha B, Accoto D. Exoskeletal devices for hand assistance and rehabilitation: a comprehensive analysis of state-of-the-art technologies. IEEE Trans Med Robot Bionics. 2021;3:525–38.

122. Ranzani R. Towards minimally-supervised robot-assisted therapy of hand function to increase therapy dose after stroke. 2021.

123. Mehrholz J, Pohl M, Platz T, Kugler J, Elsner B. Electromechanical and robot-assisted arm training for improving activities of daily living, arm function, and arm muscle strength after stroke. Cochrane Db Syst Rev. 2018;2018:CD006876

124. Veerbeek JM, Langbroek-Amersfoort AC, van Wegen EEH, Meskers CGM, Kwakkel G. Effects of robot-assisted therapy for the upper limb after stroke. Neurorehab Neural Re. 2017;31:107–21.

125. Wu J, Cheng H, Zhang J, Yang S, Cai S. Robot-assisted therapy for upper extremity motor impairment after stroke: a systematic review and meta-analysis. Phys Ther. 2021;101.

126. Mehrholz J, Pollock A, Pohl M, Kugler J, Elsner B. Systematic review with network meta-analysis of randomized controlled trials of robotic-assisted arm training for improving activities of daily living and upper limb function after stroke. J Neuroeng Rehabil. 2020;17:83.

127. Linder SM. et al. Improving quality of life and depression after stroke through telerehabilitation. Am J Occup Ther. 2015;69:6902290020p1–6902290020p10.

128. Manjunatha H, et al. Upper limb home-based robotic rehabilitation during covid-19 outbreak. Front Robot AI. 2021;8: 612834.

129. Babič J. et al. Challenges and solutions for application and wider adoption of wearable robots. Wearable Technol. 2021;2.

130. Proulx CE, et al. Review of the effects of soft robotic gloves for activity-based rehabilitation in individuals with reduced hand function and manual dexterity following a neurological event. J Rehabil Assistive Technol Eng. 2020;7:2055668320 918130.

131. Duncan PW, et al. Management of adult stroke rehabilitation care. Stroke. 2005;36:e100–43.

132. Wade DT, Wood VA, Heller A, Maggs J, Hewer RL. Walking after stroke. measurement and recovery over the first 3 months. Scand J Rehabil Med. 1987;19:25–30.

133. Combs SA, et al. Is walking faster or walking farther more important to persons with chronic stroke? Disabil Rehabil. 2012;35:860–7.

134. Blennerhassett JM, Levy CE, Mackintosh A, Yong A, McGinley JL. One-quarter of people leave inpatient stroke rehabilitation with physical capacity for community ambulation. J Stroke Cerebrovasc Dis. 2018;27:3404–10.

135. de Graaf JA, et al. Long-term restrictions in participation in stroke survivors under and over 70 years of age. Disabil Rehabil. 2017;40:1–9.

136. Portelli R, et al. Institutionalization after stroke. Clin Rehabil. 2005;19:97–108.

137. Newman AB, et al. Association of long-distance corridor walk performance with mortality, cardiovascular disease, mobility limitation, and disability. JAMA. 2006;295:2018–26.

138. Wade DT, Skilbeck CE, Wood VA, Hewer RL. Long-term survival after stroke. Age Ageing. 1984;13:76–82.

139. Grau-Pellicer M, Chamarro-Lusar A, Medina-Casanovas J, Ferrer B-CS. Walking speed as a predictor of community mobility and quality of life after stroke. Top Stroke Rehabil. 2019;26:349–58.

140. Zahuranec DB, Skolarus LE, Feng C, Freedman VA, Burke JF. Activity limitations and subjective well-being after stroke. Neurology. 2017;89:944–50.

141. Harris JE, Eng JJ. Goal priorities Identified through client-centred measurement in individuals with chronic stroke. Physiother Can. 2004;56:171.

142. Wade DT. Measurement in neurological rehabilitation. Curr Opin Neurol Neu. 1992;5:682–6.

143. Butland RJ, Pang J, Gross ER, Woodcock AA, Geddes DM. Two-, six-, and 12-minute walking tests in respiratory disease. Br Med J Clin Res Ed. 1982;284:1607.

144. Podsiadlo D, Richardson S. The timed "Up & Go": a test of basic functional mobility for frail elderly persons. J Am Geriatr Soc. 1991;39:142–8.

145. Cheng DK-Y, et al. Distance-limited walk tests post-stroke: A systematic review of measurement properties. NeuroRehabilitation. 2021;48:413–39.

146. Cheng DK, Nelson M, Brooks D, Salbach NM. Validation of stroke-specific protocols for the 10-meter walk test and 6-minute walk test conducted using 15-meter and 30-meter walkways. Top Stroke Rehabil. 2019;27:1–11.

147. Lexell J, Flansbjer U-B, Holmbäck AM, Downham D, Patten C. Reliability of gait performance tests in men and women with hemiparesis after stroke. J Rehabil Med. 2005;37:75–82.

148. Tilson JK, et al. Meaningful gait speed improvement during the first 60 days poststroke: minimal clinically important difference. Phys Ther. 2010;90:196–208.

149. Fulk GD, Echternach JL, Nof L, O'Sullivan S. Clinometric properties of the six-minute walk test in individuals undergoing rehabilitation poststroke. Physiother Theor Pr. 2009;24:195–204.

150. Salbach NM, et al. Considerations for the selection of time-limited walk tests poststroke. J Neurol Phys Ther. 2017;41:3–17.

151. Morris S, Morris ME, Iansek R. Reliability of measurements obtained with the timed "Up & Go" test in people with Parkinson Disease. Phys Ther. 2001;81:810–8.

152. Hafsteinsdóttir TB, Rensink M, Schuurmans M. Clinimetric properties of the timed up and go test for patients with stroke: a systematic review. Top Stroke Rehabil. 2014;21:197–210.

153. Chan PP, Tou JIS, Tse MM, Ng SS. Reliability and validity of the timed up and go test with a motor task in people with chronic stroke. Arch Phys Med Rehab. 2017;98:2213–20.

154. Coutts F. Gait analysis in the therapeutic environment. Manual Ther. 1999;4:2–10.

155. Mohan DM, et al. Assessment methods of post-stroke gait: a scoping review of technology-driven approaches to gait characterization and analysis. Front Neurol. 2021;12: 650024.

156. Burridge JH, Hughes A-M. Potential for new technologies in clinical practice. Curr Opin Neurol. 2010;23:671–7.

157. Parks MT, Wang Z, Siu K-C. Current low-cost video-based motion analysis options for clinical rehabilitation: a systematic review. Phys Ther. 2019;99:1405–25.

158. Kirtley C. Clinical gait analysis: theory and practice. 2006.

159. Perry J, Burnfield J. Gait analysis: normal and pathological function. 2010.

160. Simon SR. Quantification of human motion: gait analysis—benefits and limitations to its application to clinical problems. J Biomech. 2004;37:1869–80.

161. Muro-de-la-Herran A, Garcia-Zapirain B, Mendez-Zorrilla A. Gait analysis methods: an overview of wearable and non-wearable systems highlighting clinical applications. Sensors. 2014;14:3362–94.

162. Colyer SL, Evans M, Cosker DP, Salo AIT. A review of the evolution of vision-based motion analysis and the integration of advanced computer vision methods towards developing a markerless system. Sports Med Open. 2018;4:24.

163. Balta D. et al. A two-dimensional clinical gait analysis protocol based on markerless recordings from a single RGB-Depth camera. In: 2020 IEEE Int symposium medical meas appl memea. 2020. p. 1–6.

164. Mehta D, et al. VNect. Acm Trans Graph Tog. 2017;36:1–14.

165. Bartol K, Bojanic D, Petkovic T, D'Apuzzo N, Pribanic T. A review of 3D human pose estimation from 2D images. In: 11th International conference and exhibition on 3D body scanning and processing technologies. 2020.

166. Sarafianos N, Boteanu B, Ionescu B, Kakadiaris IA. 3D Human pose estimation: a review of the literature and analysis of covariates. Comput Vis Image Und. 2016;152:1–20.

167. Stamm O, Heimann-Steinert A. Accuracy of monocular two-dimensional pose estimation compared with a reference standard for kinematic multiview analysis: validation study. Jmir Mhealth Uhealth. 2020;8: e19608.

168. Guzik A, et al. Establishing the minimal clinically important differences for sagittal hip range of motion in chronic stroke patients. Front Neurol. 2021;12: 700190.

169. Guzik A, Drużbicki M, Wolan-Nieroda A, Turolla A, Kiper P. Estimating minimal clinically important differences for knee range of motion after stroke. J Clin Medicine. 2020;9:3305.

170. Calabrese E. et al. DHP19: dynamic vision sensor 3D human pose dataset. In: 2019 IEEE Cvf conf comput vis pattern recognit work cvprw. 2019. p. 1695–1704.

171. Yoon S, et al. Development and validation of 2D-LiDAR-based gait analysis instrument and algorithm. Sensors. 2021;21:414.

172. Sakdarakse S, Somboon P. Development of LIDAR based gait training system with gait assessment. In: 2020 17th Int conf electr eng electron comput telecommun information technology ecti-con. 2020. p. 263–266.

173. Glandon A. et al. 3D skeleton estimation and human identity recognition using lidar full motion video. In: 2019 Int Jt Conf Neural Networks Ijcnn. 2019. p. 1–8.

174. von Marcard T, Henschel R, Black MJ, Rosenhahn B, Pons-Moll G. Recovering accurate 3D human pose in the wild using IMUs and a moving camera. In: Computer vision–ECCV 2018, 15th European conference. 2018. p. 614–631. https://doi.org/10.1007/978-3-030-01249-6_37.

175. Basu CK, Deacon F, Hutchinson JR, Wilson AM. The running kinematics of free-roaming giraffes, measured using a low cost unmanned aerial vehicle (UAV). PeerJ. 2019;7: e6312.

176. Lafferty L. et al. Clinical indoor running gait analysis may not approximate outdoor running gait based on novel drone technology. Sports Heal Multidiscip Approach. 2021:194173812110509. https://doi.org/10.1177/19417381211050931.

177. Poitras I, et al. Validity and reliability of wearable sensors for joint angle estimation: a systematic review. Sensors Basel Switz. 2019;19:1555.

178. Petraglia F. et al. Inertial sensors versus standard systems in gait analysis: a systematic review and meta-analysis. Eur J Phys Rehab Med. 2019;55.

179. Soulard J, Vaillant J, Balaguier R, Vuillerme N. Spatio-temporal gait parameters obtained from foot-worn inertial sensors are reliable in healthy adults in single- and dual-task conditions. Sci Rep-UK. 2021;11:10229.

180. Lefeber N, Degelaen M, Truyers C, Safin I, Beckwée D. Validity and reproducibility of inertial physilog sensors for spatiotemporal gait analysis in patients with stroke. IEEE T Neur Sys Reh. 2019;27:1865–74.

181. Wüest S, Massé F, Aminian K, Gonzenbach R, de Bruin ED. Reliability and validity of the inertial sensor-based Timed "Up and Go" test in individuals affected by stroke. J Rehabil Res Dev. 2016;53:599–610.

182. Yang S, Zhang J-T, Novak AC, Brouwer B, Li Q. Estimation of spatio-temporal parameters for post-stroke hemiparetic gait using inertial sensors. Gait Posture. 2013;37:354–8.

183. Zhang W, et al. Gait symmetry assessment with a low back 3D accelerometer in post-stroke patients. Sensors. 2018;18:3322.

184. Peters DM, et al. Utilization of wearable technology to assess gait and mobility post-stroke: a systematic review. J Neuroeng Rehabil. 2021;18:67.

185. Ferraris C, et al. Monitoring of gait parameters in post-stroke individuals: a feasibility study using RGB-D sensors. Sensors. 2021;21:5945.

186. Cao Z, Simon T, Wei SE, Sheikh Y. Realtime multi-person 2D pose estimation using part affinity fields. In: 2017 IEEE Conf comput vis pattern recognit cvpr. 2017. p. 1302–1310. https://doi.org/10.1109/cvpr.2017.143.

187. Pedersen JCL. et al. Improving the accuracy of intelligent pose estimation systems through low level image processing operations. In: International conference on digital image & signal processing (DISP'19). 2019.

188. Routhier F, et al. Clinicians' perspectives on inertial measurement units in clinical practice. PLoS ONE. 2020;15: e0241922.

189. Block VAJ, et al. Remote physical activity monitoring in neurological disease: a systematic review. PLoS ONE. 2016;11: e0154335.

190. Wu J, et al. An intelligent in-shoe system for gait monitoring and analysis with optimized sampling and real-time visualization capabilities. Sensors. 2021;21:2869.

191. MacLellan MJ, Patla AE. Adaptations of walking pattern on a compliant surface to regulate dynamic stability. Exp Brain Res. 2006;173:521–30.

192. Truong PH, Lee J, Kwon A-R, Jeong G-M. Stride counting in human walking and walking distance estimation using insole sensors. Sensors. 2016;16:823.

193. van de Port I, Wevers L, Kwakkel G. Is outdoor use of the six-minute walk test with a global positioning system in stroke patients' own neighbourhoods reproducible and valid? J Rehabil Med. 2011;43:1027–31.

194. Wing MG, Eklund A, Kellogg LD. Consumer-grade global positioning system (GPS) accuracy and reliability. J Forest. 2005;103:169–73.

195. Ullrich M, et al. Detection of unsupervised standardized gait tests from real-world inertial sensor data in Parkinson's Disease. IEEE T Neur Sys Reh. 2021;29:2103–11.

196. Kawai H, et al. Association between daily living walking speed and walking speed in laboratory settings in healthy older adults. Int J Environ Res Pu. 2020;17:2707.

197. Takayanagi N, et al. Relationship between daily and in-laboratory gait speed among healthy community-dwelling older adults. Sci Rep-uk. 2019;9:3496.

198. Toosizadeh N, et al. Motor performance assessment in Parkinson's Disease: association between objective in-clinic, objective in-home, and subjective/semi-objective measures. PLoS ONE. 2015;10: e0124763.

199. Polhemus A, et al. Walking on common ground: a cross-disciplinary scoping review on the clinical utility of digital mobility outcomes. Npj Digital Med. 2021;4:149.

200. Warmerdam E, et al. Long-term unsupervised mobility assessment in movement disorders. Lancet Neurol. 2020;19:462–70.

201. Atrsaei A, et al. Gait speed in clinical and daily living assessments in Parkinson's disease patients: performance versus capacity. Npj Park Dis. 2021;7:24.

202. Martindale CF, Christlein V, Klumpp P, Eskofier BM. Wearables-based multi-task gait and activity segmentation using recurrent neural networks. Neurocomputing. 2021;432:250–61.

203. Dehzangi O, Taherisadr M, ChangalVala R. IMU-based gait recognition using convolutional neural networks and multi-sensor fusion. Sensors. 2017;17:2735.

204. Rast FM, Labruyère R. Systematic review on the application of wearable inertial sensors to quantify everyday life motor activity in people with mobility impairments. J Neuroeng Rehabil. 2020;17:148.

205. Jourdan T, Debs N, Frindel C. The contribution of machine learning in the validation of commercial wearable sensors for gait monitoring in patients: a systematic review. Sensors. 2021;21:4808.

206. Porciuncula F, et al. Wearable movement sensors for rehabilitation: a focused review of technological and clinical advances. Pm&r. 2018;10:S220–32.

207. Chen S, Lach J, Lo B, Yang G-Z. Toward pervasive gait analysis with wearable sensors: a systematic review. IEEE J Biomed Health. 2016;20:1521–37.

208. Werner C, Easthope CA, Curt A, Demkó L. Towards a mobile gait analysis for patients with a spinal cord injury: a robust algorithm validated for slow walking speeds. Sensors. 2021;21:7381.

209. Tietsch M, et al. Robust step detection from different waist-worn sensor positions: implications for clinical studies. Digital Biomarkers. 2020;4:50–8.

210. Dorsey ER, Glidden AM, Holloway MR, Birbeck GL, Schwamm LH. Teleneurology and mobile technologies: the future of neurological care. Nat Rev Neurol. 2018;14:285–97.

211. Bravata DM, et al. Using pedometers to increase physical activity and improve health: a systematic review. JAMA. 2007;298:2296–304.

212. Rozanski GM, Aqui A, Sivakumaran S, Mansfield A. Consumer wearable devices for activity monitoring among individuals after a stroke: a prospective comparison. Jmir Cardio. 2018;2: e1.

213. Degroote L, Bourdeaudhuij ID, Verloigne M, Poppe L, Crombez G. The accuracy of smart devices for measuring physical activity in daily life: validation study. Jmir Mhealth Uhealth. 2018;6: e10972.

214. Tedesco S, et al. Accuracy of consumer-level and research-grade activity trackers in ambulatory settings in older adults. PLoS ONE. 2019;14: e0216891.

215. Henriksen A, et al. Using fitness trackers and smartwatches to measure physical activity in research: analysis of consumer wrist-worn wearables. J Med Internet Res. 2018;20: e110.

216. Huang Y, Xu J, Yu B, Shull PB. Validity of FitBit, Jawbone UP, Nike+ and other wearable devices for level and stair walking. Gait Posture. 2016;48:36–41.

217. Altilio R, Liparulo L, Panella M, Proietti A, Paoloni M. Multimedia and gaming technologies for telerehabilitation of motor disabilities [Leading Edge]. IEEE Technol Soc Mag. 2015;34:23–30.

218. Held JP, et al. Autonomous rehabilitation at stroke patients home for balance and gait: safety, usability and compliance of a virtual reality system. Eur J Phys Rehab Med. 2017. https://doi.org/10.23736/s1973-9087.17.04802-x.

219. Cikajlo I, Rudolf M, Goljar N, Matjačić Z. Virtual reality task for telerehabilitation dynamic balance training in stroke subjects. In: 2009 Virtual rehabilitation international conference. 2009. p.121–125. https://doi.org/10.1109/icvr.2009.5174217.

220. Trombetta M, et al. Motion rehab AVE 3D: a VR-based exergame for post-stroke rehabilitation. Comput Meth Prog Bio. 2017;151:15–20.

221. Henrique PPB, Colussi EL, Marchi ACBD. Effects of exergame on patients' balance and upper limb motor function after stroke: a randomized controlled trial. J Stroke Cerebrovasc Dis. 2019;28:2351–7.

222. Kannan L, Vora J, Bhatt T, Hughes SL. Cognitive-motor exergaming for reducing fall risk in people with chronic stroke: a randomized controlled trial. NeuroRehabilitation. 2019;44:493–510.

223. Lupo A, et al. Effects on balance skills and patient compliance of biofeedback training with inertial measurement units and exergaming in subacute stroke: a pilot randomized controlled trial. Funct Neurol. 2018;33:131–6.

224. Huber SK, Held JPO, de Bruin ED, Knols RH. Personalized motor-cognitive exergame training in chronic stroke patients—a feasibility study. Front Aging Neurosci. 2021;13: 730801.

225. Shull PB, Damian DD. Haptic wearables as sensory replacement, sensory augmentation and trainer–a review. J Neuroeng Rehabil. 2015;12:59.

226. Shull PB, Jirattigalachote W, Zhu X. An overview of wearable sensing and wearable feedback for gait retraining. In: Proceedings of the 6th international conference, Intelligent Robotics and Applications, ICIRA 2013, Busan, South Korea, 25–28 September 2013, p. 434–443. https://doi.org/10.1007/978-3-642-40852-6_44.

227. van Meulen FB, et al. Objective evaluation of the quality of movement in daily life after stroke. Front Bioeng Biotechnol. 2016;3:210.

228. Schließmann D, et al. Trainer in a pocket - proof-of-concept of mobile, real-time, foot kinematics feedback for gait pattern normalization in individuals after stroke, incomplete spinal cord injury and elderly patients. J Neuroeng Rehabil. 2018;15:44.

229. Mikolajczyk T, et al. Advanced technology for gait rehabilitation: an overview. Adv Mech Eng. 2018;10:1687814018783627.

230. Genthe K, et al. Effects of real-time gait biofeedback on paretic propulsion and gait biomechanics in individuals post-stroke. Top Stroke Rehabil. 2018;25:1–8.

231. Massé F, et al. Improving activity recognition using a wearable barometric pressure sensor in mobility-impaired stroke patients. J Neuroeng Rehabil. 2015;12:1–15.

232. Tate JJ, Milner CE. Real-time kinematic, temporospatial, and kinetic biofeedback during gait retraining in patients: a systematic review. Phys Ther. 2010;90:1123–34.

233. van Gelder LMA, Barnes A, Wheat JS, Heller BW. The use of biofeedback for gait retraining: a mapping review. Clin Biomech. 2018;59:159–66.

234. Spencer J, Wolf SL, Kesar TM. Biofeedback for post-stroke gait retraining: a review of current evidence and future research directions in the context of emerging technologies. Front Neurol. 2021;12: 637199.

235. Xu J, et al. Configurable, wearable sensing and vibrotactile feedback system for real-time postural balance and gait training: proof-of-concept. J Neuroeng Rehabil. 2017;14:102.

236. Biesmans S, Markopoulos P. Design and evaluation of SONIS, a wearable biofeedback system for gait retraining. Multimodal Technol Interact. 2020;4:60.

237. Ma CZ-H, Zheng Y-P, Lee WC-C. Changes in gait and plantar foot loading upon using vibrotactile wearable biofeedback system in patients with stroke. Top Stroke Rehabil. 2017;25:20–7.

238. Zhang H, et al. Wearable biofeedback system to induce desired walking speed in overground gait training. Sensors. 2020;20:4002.

239. Afzal MR, Pyo S, Oh M-K, Park YS, Yoon J. Evaluating the effects of delivering integrated kinesthetic and tactile cues to individuals with unilateral hemiparetic stroke during overground walking. J Neuroeng Rehabil. 2018;15:33.

240. Held JPO, et al. Augmented reality-based rehabilitation of gait impairments: case report. Jmir Mhealth Uhealth. 2020;8: e17804.

241. Massetti T, et al. The clinical utility of virtual reality in neurorehabilitation: a systematic review. J Central Nerv Syst Dis. 2018;10:1179573518813541.

242. Brewster S. et al. What is mixed reality? In: Proceedings of the 2019 Chi conf hum factors comput syst. 2019, p. 1–15. https://doi.org/10.1145/3290605.3300767.

243. Morishima S. et al. Effects of head-display lag on presence in the oculus rift. In: Proceedings of the 24th Acm symposium virtual real softw technology 83. 2018. https://doi.org/10.1145/3281505.3281607.

244. Chan ZYS, et al. Walking with head-mounted virtual and augmented reality devices: effects on position control and gait biomechanics. PLoS ONE. 2019;14: e0225972.

245. Kolivand H, Sunar MS. Realistic real-time outdoor rendering in augmented reality. PLoS ONE. 2014;9: e108334.

246. Kress BC, Cummings WJ. Optical architecture of HoloLens mixed reality headset. Digital Opt Technol 2017;2017:103350K-1–103350K-10. https://doi.org/10.1117/12.2270017.

247. Tuena C, et al. Usability issues of clinical and research applications of virtual reality in older people: a systematic review. Front Hum Neurosci. 2020;14:93.

248. Lloréns R, Noé E, Colomer C, Alcañiz M. Effectiveness, usability, and cost-benefit of a virtual reality-based telerehabilitation program for balance recovery after stroke: a randomized controlled trial. Arch Phys Med Rehab. 2015;96:418-425.e2.

249. Vaquero-Melchor D, Bernardos AM. Enhancing interaction with augmented reality through mid-air haptic feedback: architecture design and user feedback. Appl Sci. 2019;9:5123.

250. Ott R, Thalmann D, Vexo F. Haptic feedback in mixed-reality environment. Vis Comput. 2007;23:843–9.

251. Corbetta D, Imeri F, Gatti R. Rehabilitation that incorporates virtual reality is more effective than

standard rehabilitation for improving walking speed, balance and mobility after stroke: a systematic review. J Physiother. 2015;61:117–24.

252. de Rooij IJM, van de Port IGL, Meijer J-WG. Effect of virtual reality training on balance and gait ability in patients with stroke: systematic review and meta-analysis. Phys Ther. 2016;96:1905–18.

253. Berthier ML. Poststroke aphasia. Drug Aging. 2005;22:163–82.

254. Duffy J. Motor speech disorders. (2019).

255. Flowers HL, Silver FL, Fang J, Rochon E, Martino R. The incidence, co-occurrence, and predictors of dysphagia, dysarthria, and aphasia after first-ever acute ischemic stroke. J Commun Disord. 2013;46:238–48.

256. Wray F, Clarke D. Longer-term needs of stroke survivors with communication difficulties living in the community: a systematic review and thematic synthesis of qualitative studies. BMJ Open. 2017;7: e017944.

257. Cock ED, et al. Dysphagia, dysarthria and aphasia following a first acute ischaemic stroke: incidence and associated factors. Eur J Neurol. 2020;27:2014–21.

258. Ruben RJ. Redefining the survival of the fittest: communication disorders in the 21st century. Laryngoscope. 2000;110:241–241.

259. Worrall L, et al. What people with aphasia want: Their goals according to the ICF. Aphasiology. 2011;25:309–22.

260. Puyvelde MV, Neyt X, McGlone F, Pattyn N. Voice stress analysis: a new framework for voice and effort in human performance. Front Psychol. 2018;09:1994.

261. Themistocleous C, Eckerström M, Kokkinakis D. Voice quality and speech fluency distinguish individuals with mild cognitive impairment from healthy controls. PLoS ONE. 2020;15: e0236009.

262. Chiaramonte R, Vecchio M. A systematic review of measures of dysarthria severity in stroke patients. Pm&r. 2021;13:314–24.

263. Wallace SJ. et al. Many ways of measuring: a scoping review of measurement instruments for use with people with aphasia. Aphasiology. 2020:1–66. https://doi.org/10.1080/02687038.2020.1836318.

264. Ziegler W, Staiger A, Schölderle T, Vogel M. Gauging the auditory dimensions of dysarthric impairment: reliability and construct validity of the Bogenhausen Dysarthria Scales (BoDyS). J Speech Lang Hear Res. 2017;60:1516–34.

265. Kertesz A, Raven JC. The Western Aphasia Battery-Revised (WAB-R). (PsychCorp, 2007).

266. Huber W, Poeck K, Weniger D, Willmes K. Der Aachener Aphasie Test (AAT). (Hogrefe, 1983).

267. Brandenburg C, Worrall L, Rodriguez A, Bagraith K. Crosswalk of participation self-report measures for aphasia to the ICF: what content is being measured? Disabil Rehabil. 2014;37:1113–24.

268. Armstrong E. The challenges of consensus and validity in establishing core outcome sets. Aphasiology. 2017:1–4. https://doi.org/10.1080/02687038.2017.1398804.

269. Patel RR, et al. Recommended protocols for instrumental assessment of voice: american speech-language-hearing association expert panel to develop a protocol for instrumental assessment of vocal function. Am J Speech-Lang Pat. 2018;27:887–905.

270. Maryn Y, Roy N, Bodt MD, Cauwenberge PV, Corthals P. Acoustic measurement of overall voice quality: a meta-analysisa). J Acoust Soc Am. 2009;126:2619–34.

271. Reetz S, Bohlender JE, Brockmann-Bauser M. Do standard instrumental acoustic, perceptual, and subjective voice outcomes indicate therapy success in patients with functional dysphonia? J Voice. 2019;33:317–24.

272. Brockmann-Bauser M, Bohlender JE, Mehta DD. Acoustic perturbation measures improve with increasing vocal intensity in individuals with and without voice disorders. J Voice. 2018;32:162–8.

273. Bogert B, Healy MJR, Tukey JW. The quefrency alanysis of time series for echoes ; Cepstrum, pseudo-autocovariance, cross-cepstrum and saphe cracking. In: Rosenblatt M (ed) Proceedings of the symposium on time series analysis. p. 209–243 (Wiley; 1963).

274. Fraile R, Godino-Llorente JI. Cepstral peak prominence: a comprehensive analysis. Biomed Signal Proces. 2014;14:42–54.

275. Murton O, Hillman R, Mehta D. Cepstral peak prominence values for clinical voice evaluation. Am J Speech-Lang Pat. 2020;29:1596–607.

276. de Jong NH, Wempe T. Praat script to detect syllable nuclei and measure speech rate automatically. Behav Res Methods. 2009;41:385–90.

277. de Jong NH, Pacilly J, Heeren W. PRAAT scripts to measure speed fluency and breakdown fluency in speech automatically. Assess Educ Princ Policy Pract. 2021:1–21. https://doi.org/10.1080/0969594x.2021.1951162.

278. Grande M, et al. Basic parameters of spontaneous speech as a sensitive method for measuring change during the course of aphasia. Int J Lang Comm Dis. 2008;43:408–26.

279. Hussmann K, et al. Computer-assisted analysis of spontaneous speech: quantification of basic parameters in aphasic and unimpaired language. Clin Linguist Phonet. 2012;26:661–80.

280. Barthel G, Djundja D, Meinzer M, Rockstroh B, Eulitz C. Aachener Sprachanalyse (ASPA): evaluation bei Patienten mit chronischer Aphasie. Sprache Stimme Gehoer. 2006;30:103–10.

281. Grande M, Springer L, Huber W. Richtlinien fuer die Transkription mit dem Programm ASPA (Aachener Sprachanalyse). Sprache Stimme Geh R. 2006;30:179–85.

282. MacWhinney B. The Childes Project. 2000. https://doi.org/10.4324/9781315805641.

283. Brown C, Snodgrass T, Kemper SJ, Herman R, Covington MA. Automatic measurement of propositional idea density from part-of-speech tagging. Behav Res Methods. 2008;40:540–5.

284. Ferguson A, Spencer E, Craig H, Colyvas K. Propositional idea density in women's written language over the lifespan: computerized analysis. Cortex. 2014;55:107–21.

285. MacWhinney B, Fromm D. AphasiaBank as big-data. Semin Speech Lang. 2016;37:010–22.

286. MacWhinney B, Fromm D, Forbes M, Holland A. AphasiaBank: Methods for studying discourse. Aphasiology. 2011;25:1286–307.

287. Fromm D, et al. Discourse characteristics in aphasia beyond the western aphasia battery cutoff. Am J Speech-Lang Pat. 2017;26:762–8.

288. Carrier M. Automated speech recognition in language learning: potential models, benefits and impact. Train Lang Cult. 2017;1:46–61.

289. Jacks A, Haley KL, Bishop G, Harmon TG. Automated speech recognition in adult stroke survivors: comparing human and computer transcriptions. Folia Phoniatr Logo. 2019;71:286–96.

290. Dekhtyar M, Braun EJ, Billot A, Foo L, Kiran S. Videoconference administration of the western aphasia battery–revised: feasibility and validity. Am J Speech-Lang Pat. 2020;29:673–87.

291. Wood E, Bhalloo I, McCaig B, Feraru C, Molnar M. Towards development of guidelines for virtual administration of paediatric standardized language and literacy assessments: considerations for clinicians and researchers. Sage Open Med. 2021;9:205031212110505.

292. Grillo EU, Wolfberg J. An assessment of different praat versions for acoustic measures analyzed automatically by voice value and manually by two raters. J Voice. 2020. https://doi.org/10.1016/j.jvoice.2020.12.003.

293. Mueller A, et al. Digital endpoints for self-administered home-based functional assessment in pediatric Friedreich's ataxia. Ann Clin Transl Neur. 2021;8:1845–56.

294. Brandenburg C, Worrall L, Copland D, Rodriguez A. An exploratory investigation of the daily talk time of people with non-fluent aphasia and non-aphasic peers. Int J Speech-lang Pa. 2016;19:1–12.

295. Hillman RE, Heaton JT, Masaki A, Zeitels SM, Cheyne HA. Ambulatory monitoring of disordered voices. Ann Otology Rhinol Laryngol. 2006;115:795–801.

296. Mehta DD, Stan JHV, Hillman RE. Relationships between vocal function measures derived from an acoustic microphone and a subglottal neck-surface accelerometer. IEEE ACM Trans Audio Speech Lang Process. 2016;24:659–68.

297. Mehta DD, et al. Using ambulatory voice monitoring to investigate common voice disorders: research update. Front Bioeng Biotechnol. 2015;3:155.

298. Association ASLH. Telepractice: overview. https://www.asha.org/PRPSpecificTopic.aspx?folderid=8589934956§ion=Overview (2018).

299. Association ASLH. Scope of practice in audiology. https://www.asha.org/policy/sp2018-00353/ (2018).

300. Association ASLH. Scope of practice in speech-language pathology. https://www.asha.org/policy/sp2016-00343/ (2016).

301. Grillo EU. Building a successful voice telepractice program. Perspect Asha Spec Interest Groups. 2019;4:100–10.

302. Grillo EU. Results of a survey offering clinical insights into speech-language pathology telepractice methods. Int J Telerehabil. 2017;9:25–30.

303. Pitt R, Theodoros D, Hill AJ, Rodriguez AD, Russell T. The feasibility of delivering constraint-induced language therapy via the Internet. Digital Heal. 2017;3:2055207617718767.

304. Weidner K, Lowman J. Telepractice for adult speech-language pathology services: a systematic review. Perspect Asha Spec Interest Groups. 2020;5:326–38.

305. Walker JP, Price K, Watson J. Promoting social connections in a synchronous telepractice, aphasia communication group. Perspect Asha Spec Interest Groups. 2018;3:32–42.

306. Pitt R, Theodoros D, Hill AJ, Russell T. The impact of the telerehabilitation group aphasia intervention and networking programme on communication, participation, and quality of life in people with aphasia. Int J Speech-lang. 2018:1–11. https://doi.org/10.1080/17549507.2018.1488990.

307. Marshall J, et al. A randomised trial of social support group intervention for people with aphasia: a Novel application of virtual reality. PLoS ONE. 2020;15: e0239715.

308. Hinckley JJ, Patterson JP, Carr TH. Differential effects of context- and skill-based treatment approaches: preliminary findings. Aphasiology. 2001;15:463–76.

309. Jamal N, Shanta S, Mahmud F, Sha'abani M. Automatic speech recognition (ASR) based approach for speech therapy of aphasic patients: a review. Aip Conf Proc. 2017;1883:020028.

310. Maryn Y, Bodt MD, Cauwenberge PV. Effects of biofeedback in phonatory disorders and phonatory performance: a systematic literature review. Appl Psychophys Biof. 2006;31:65–83.

311. Stan JHV, Mehta DD, Hillman RE. The effect of voice ambulatory biofeedback on the daily performance and retention of a modified vocal motor behavior in participants with normal voices. J Speech Lang Hear Res. 2015;58:713–21.

312. Stan JHV, Mehta DD, Sternad D, Petit R, Hillman RE. Ambulatory voice biofeedback: relative frequency and summary feedback effects on performance and retention of reduced vocal intensity in the daily lives of participants with normal voices. J Speech Lang Hear Res. 2017;60:853–64.

313. Palmer R, et al. Self-managed, computerised speech and language therapy for patients with chronic aphasia post-stroke compared with usual care or attention control (Big CACTUS): a multicentre, single-blinded, randomised controlled trial. Lancet Neurol. 2019;18:821–33.

314. Zheng C, Lynch L, Taylor N. Effect of computer therapy in aphasia: a systematic review. Aphasiology. 2015:1–34. https://doi.org/10.1080/02687038.2014.996521.

315. Tjoa E, Guan C. A survey on explainable artificial intelligence (XAI): toward medical XAI. IEEE T Neur Net Lear. 2021;32:4793–813.

316. Willems EMG, Vermeulen J, van Haastregt JCM, Zijlstra GAR. Technologies to improve the participation of stroke patients in their home environment. Disabil Rehabil. 2021:1–11. https://doi.org/10.1080/09638288.2021.1983041.

317. Cities P. The city as a digital playground. Gaming Media Soc Eff. 2017. https://doi.org/10.1007/978-981-10-1962-3.

Part VI

Robotic Technologies for Neurorehabilitation: Upper Extremity

Forging Mens et Manus: The MIT Experience in Upper Extremity Robotic Therapy

26

Hermano Igo Krebs, Dylan J. Edwards, and Bruce T. Volpe

Abstract

MIT's motto is "Mens et Manus" (Mind and Hand) and we have adopted it as the guiding rule (principle) for our line of research: using robotics and information technology to re-connect the brain to the hand. Training and treatment protocols enhance this re-connection phenomenon, reduce impairment, increase function, and improve the quality of life beyond natural recovery. This chapter describes our efforts towards attaining this goal since the initial development of the MIT-MANUS in 1989. Numerous clinical trials involving thousands of participants working with (receiving therapy using) different versions of the MIT-Manus have been conducted since then and we have created a complete robotic gym for the upper extremity. In fact, for over 10 years, the American Heart Association and the Veterans Affairs/Department of Defense endorsed the use of robot-assisted therapy in stroke rehabilitation for upper extremities, and we have been focusing on how to tailor and augment therapy to a particular patient's need and in determining who is a responder (and non-responder) to this kind of intervention.

Keywords

Rehabilitation robotics · Robot-assisted therapy · Robotic therapy · MIT-Manus · Upper extremity · Neuro-modulation · Stroke

26.1 Introduction

The use of robotic technology to assist recovery after a neurological injury has proven to be safe, feasible and effective—at least to reduce impairment—for the upper extremity of stroke participants. Nevertheless, there is vast room for improvement. But what is the best way to pursue further improvement? Ultimately, we would like to prescribe customized therapy to optimize and augment a patient's recovery. In this chapter, we review our experience developing upper extremity robotic therapy and applying it in clinical practice. Based on that experience, we propose the most productive way to refine and optimize this technology and its application. Needless to say, this personal viewpoint will almost certainly neglect or under-emphasize important developments; however, that should not

H. I. Krebs (✉)
Massachusetts Institute of Technology, 77 Massachusetts Avenue, Office 3-155, Cambridge, MA 02139, USA
e-mail: hikrebs@mit.edu

D. J. Edwards
Moss Rehabilitation Research Institute, Elkins Park, USA
e-mail: EdwardDy@einstein.edu

B. T. Volpe
Feinstein Institutes for Medical Research, Manhasset, NY, USA
e-mail: Bvolpe1@northwell.edu

© The Author(s), under exclusive license to Springer Nature Switzerland AG 2022
D. J. Reinkensmeyer et al. (eds.), *Neurorehabilitation Technology*,
https://doi.org/10.1007/978-3-031-08995-4_26

be construed as a dismissal of other work but more as a symptom of the explosive growth of research in this field. Despite its inevitable limitations, we trust our perspective may have value.

This chapter is organized as follows. We first discuss the expert consensus expressed in the American Heart Association guidelines (AHA) and introduce the MIT-Manus robotic gym (Sects. 26.1.1 and 26.1.2). We then discuss some of our evidence with over 1,000 stroke participants in both sub-acute and chronic phases of recovery that are in line with AHA's guidelines (Sects. 26.2.1 and 26.2) and what we believe to be the best way to employ robotic technology (Sect. 26.2.3). We also discuss a few misconceptions that have been experimentally debunked including some approaches that failed to augment outcomes on robot-mediated therapy (Sects. 26.2.4. thru 26.2.6). We review some of the robot control schemes employed in our studies and which one promotes the best outcomes (Sect. 26.2.7), as well as the potential of robot-mediated assay to evaluate recovery (Sect. 26.2.8). We wrap-up this chapter with a short discussion and some closing remarks (Sect. 26.3).

26.1.1 The State of the Art

The 2010 American Heart Association (AHA) guidelines for stroke care recommended that: "Robot-assisted therapy offers the amount of motor practice needed to relearn motor skills with less therapist assistance. Most robots for motor rehabilitation not only allow for robot assistance in movement initiation and guidance but also provide accurate feedback; some robots additionally provide movement resistance. Most trials of robot-assisted motor rehabilitation concern the upper extremity (UE), with robotics for the lower extremity (LE) still in its infancy... Robot-assisted UE therapy, however, can improve motor function during the inpatient period after stroke." In 2010 AHA suggested that robot-assisted therapy for the UE has already achieved Class I, Level of Evidence A for Stroke Care in the Outpatient Setting and Care in Chronic Care Settings. It suggested that robot-

assisted therapy for UE has achieved Class IIa, Level of Evidence A for stroke care in the inpatient setting. Class I is defined as: "Benefit >>> Risk. Procedure/Treatment SHOULD be performed/administered;" Class IIa is defined as: "Benefit >> Risk, IT IS REASONABLE to perform procedure/administer treatment;" Level A is defined as "Multiple populations evaluated: Data derived from multiple randomized clinical trials or meta-analysis" [1]. In 2016 AHA did not differentiate between inpatient and outpatient stroke populations and it suggested Class IIa across different time points following a stroke.

This is not an isolated opinion. The 2010 Veterans Administration/Department of Defense guidelines for stroke care came to the same conclusion endorsing the use of rehabilitation robots for the upper extremity. More specifically, the VA/DOD 2010 guidelines for stroke care "Recommend robot-assisted movement therapy as an adjunct to conventional therapy in patients with deficits in arm function to improve motor skill at the joints trained." The VA/DOD suggested that robot-assisted therapy for the UE has already achieved rating level B, "A recommendation that clinicians provide (the service) to eligible patients. At least fair evidence was found that the intervention improves health outcomes and concludes that benefits outweigh harm" [2].

These endorsements came on the 21st anniversary of our initial efforts begun in 1989 that led to what became known as "MIT-Manus." It would be difficult to deny the impact of this work on neuro-rehabilitation, described by a senior clinical colleague as "perhaps one of the most important developments in neuro-recovery in the last 75 years" (the late Dr. Fletcher H. McDowell, Director, The Burke Rehabilitation Center, Cornell-NY Hospital, White Plains, NY). Creating this level of trust required decades of perseverance. The enormity of the challenge cannot be understated. This type of research is the antithesis of the rapid-fire breakthroughs expected in, say, information technologies. It requires slow and painstaking experimental trials and the creation of a large body of experimental evidence to demonstrate progress, but that is essential. Neuro-rehabilitation depends on neural plasticity and its potential to augment recovery. The central challenge of rehabilitation robotics is to provide tools to manage and harness positive plastic changes. It is not simply to automate conventional practices. Primarily due to a lack of tools for measurement and experimental control, many conventional practices lack the support of scientific evidence. As a result, there is no clear design target for the technology nor any reliable "gold standard" against which to gauge its effectiveness. The message seems clear: We must study the process of neuro-recovery as well as the technologies that might augment this process.

26.1.2 An Upper Extremity Gym of Robots

To begin with, we had to invent the technology since the available technologies were inadequate. We developed interactive robots to work with the shoulder-and-elbow (with and without gravity compensation), the wrist and the hand, as well as combinations of these modules. We further developed exoskeletal robots for neuroscience research (see Fig. 26.1).

26.1.2.1 Modularity
We chose to pursue a modular approach for several reasons. The foremost was entirely pragmatic: As we intended to introduce new technology to a clinical environment, it needed to be minimally disruptive—i.e., not too big, complex or intimidating. A secondary reason was our recognition that engineers were unlikely to create optimal technology on the first pass. Though a design to address over 200 DOFs of the human skeleton was technically feasible, it would have been large, complex and—most important—difficult to revise or modify. With a modular approach, individual modules could be refined and optimized without re-design of other modules.

26.1.2.2 Gravity-Compensated Shoulder-And-Elbow Robot
The centerpiece of our effort for the upper extremity became known as MIT-Manus, from MIT's Motto "Mens et Manus" (Mind and Hand). Unlike most industrial robots, MIT-Manus was configured for safe, stable, and highly compliant operation in close physical contact with humans. This was achieved using impedance control, a key feature of the robot control system. Its computer control system modulated the way the robot reacted to mechanical perturbation from a patient or clinician and ensured a gentle compliant behavior. The machine was designed to have a low intrinsic end-point impedance (i.e., be backdrivable) to allow weak patients to express movements without constraint and to offer minimal resistance at speeds up to 2 m/s (the approximate upper limit of unimpaired human performance, hence the target of therapy, and the maximum speed observed in some pathologies, e.g., the shock-like movements of myoclonus). MIT-Manus had 2 active degrees of freedom (DOF) and one passive DOF. It consisted of a semi-direct-drive, five-bar-linkage SCARA mechanism (Selective Compliance Assembly Robot Arm) driven by brushless motors [3, 4]. Since then, several variants were deployed commercially.

26.1.2.3 Gravity-Compensated Shoulder-Elbow-And-Wrist Exoskeletal Robot
Based on their mechanical interface with a human, robots can be classified as end-effector or exoskeletal designs. End-effector robots interact with the human via a handshake, i.e., the interaction takes place through a single point of contact. In other words, there is power exchange only at the tip of the robot. Exoskeletal robots are mounted on distinct human limb segments with more than one point of contact. End-effector robot designs like the MIT-Manus are simpler, and afford significantly faster "don" and "doff" (set-up time much smaller) than exoskeleton designs, but typically occupy a larger volume. We employ a "rule of thumb" to guide us in the selection of configuration based on the target range of motion. For limb segment movements requiring joint angles to change by 45 degrees or less, end-effector designs appear to offer better

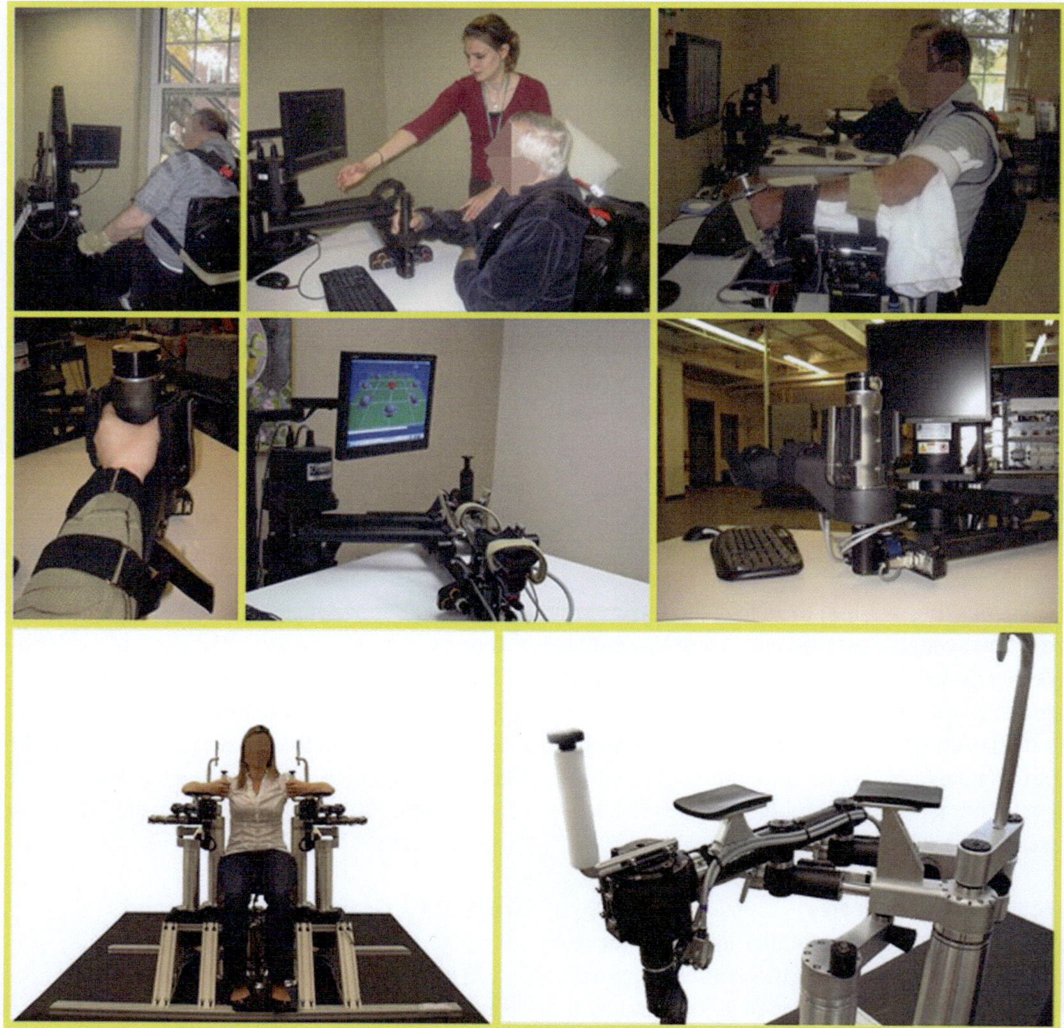

Fig. 26.1 A Gym of Upper Extremity Robots. The top row, left shows a person with chronic stroke working with the anti-gravity shoulder-and-elbow robot. The top row, middle panel shows a person working with the planar shoulder-and-elbow robot. The top row, right panel shows the wrist robot during therapy at the Burke Rehabilitation Hospital. The middle row, left panel shows the hand module for grasp and release. The middle row, middle panel shows reconfigurable robots. The robotic therapy shoulder-and-elbow and wrist modules can operate in stand-alone mode or be integrated into a coordinated functional unit. The middle row, right panel shows the shoulder-and-elbow and hand module integrated into a coordinated functional unit. The bottom row shows an exoskeletal robot with 3 active DOFs designed for psychophysical studies of the shoulder, elbow and wrist. For this exoskeletal robot, the links must be adjusted to the person's limb segments (using laser pointers). Once the arm, forearm, and wrist are properly adjusted, psychophysical experiments can assist or selectively apply perturbation force fields to the shoulder, elbow, and wrist (either flexion/extension or abduction/adduction)

compromises. Conversely, exoskeletal designs appear to offer better choices for larger ranges of motion. That said, in some circumstances, the application dictates the configuration. One such case occurs during psychophysical experiments in which we may want to carefully apply and control perturbations to one, but not another, joint. Hence we designed a highly-backdrivable,

3 active DOF, gravity-compensated shoulder-elbow-and-wrist exoskeletal robot as shown in Fig. 26.1. Two Exos can be configured for bimanual use (see other examples in [5–7]).

26.1.2.4 Gravity Non-compensated Shoulder-And-Elbow Robot

A 1-DOF module was conceived to extend the benefits of planar robotic therapy to spatial arm movements, including movements against gravity. Incorporated in the design are therapists' suggestions that functional reaching movements often occur in a range of motion close to shoulder scaption. That is, this robotic module was designed for therapy to focus on movements within the 45° to 65° range of shoulder abduction and from 30° to 90° of shoulder elevation or flexion [8]. The module can permit free motion of the patient's arm, or can provide partial or full assistance or resistance as the patient moves against gravity. As with MIT-Manus, the system is highly-backdrivable.

26.1.2.5 Wrist Robot

To extend treatment beyond the shoulder-and-elbow, we designed and built a wrist module for robotic therapy [9]. The device accommodates the range of motion of a normal wrist in everyday tasks, i.e., flexion/extension 60°/60°, abduction/adduction 30°/45°, and pronation/supination 70°/70°. The torque output from the device is capable of lifting the patient's hand against gravity, accelerating the inertia, and overcoming most forms of hypertonicity. As with all of our exoskeletal designs, we purposely under-actuated the wrist robot with fewer degrees of freedom than are anatomically present. Not only does this simplify the mechanical design, but it also allows the device to be installed quickly without problems of misalignment with the patient's joint axes. In this case, the axes of the wrist's ulnar-radial and flexion-extension joints do not intersect and the degree of non-intersection varies between individuals [10]. If robots and humans had the same number of degrees of freedom but these were not co-aligned, the motion might evoke excessive forces or torques. By allowing the human joint more degrees of freedom than the robot, excessive loads are avoided. Ease of use is another critical consideration in all our designs. We consider it a major determinant of success or failure in the clinical rehabilitation environment. The wrist robot must be attached to or removed from the patient (donned or doffed) within 2 min. Finally, the wrist robot module can be operated in isolation or mounted at the tip of the shoulder-and-elbow, gravity-compensated robot. Hence it enables a combination of translating the hand (with the shoulder-and-elbow robot) to a location in space and orienting the hand (with the wrist robot) to facilitate object manipulation.

26.1.2.6 Hand Robot

Moving a patient's hand (for purposes of post-stroke training focusing on the fingers) is not a simple task since the human hand has 15 joints with a total of 22 DOF (27 including the wrist); therefore, it was prudent to determine how many DOF are necessary for a patient to perform the majority of everyday functional tasks. Here our clinical experience with over 2,000 stroke patients was invaluable in that it allowed us to identify what was most likely to work in the clinic (and what probably would not). Though individual digit opposition (e.g., thumb to pinkie) may be important for the unimpaired human hand, it is clearly beyond the realistic expectations of most of our patients whose impairment level falls between severe and moderate; a device to manipulate 22 DOF is unnecessary (or at least premature). Our hand therapy module is a novel design that converts rotary into linear movement using a single brushless DC electrical motor as a free-base mechanism with what is traditionally called the stator being allowed to rotate freely [11]. The stator (strictly, the "second rotor") is connected to a set of arms, while the rotor is connected to another set of arms. When commanded to rotate, the rotor and stator work like a double crank and slider mechanism, in opposing configuration, where the crank is represented by a single arm and the slider is the shell or panel that interacts with the hand of the patient (see Fig. 26.1). The hand robot is used to simulate grasp and release with its impedance determined by the torque evoked by relative movement between stator and rotor. A torsional spring (connected in geometric parallel) is available to compensate for a patient's hypertonicity (inability to relax). The hand robot is capable of providing continuous passive motion, strength, sensory, and sensorimotor training for grasp and release; it can be employed in stand-alone operation or mounted at the tip of the planar robot (see Fig. 26.1).

26.2 Harnessing Plasticity to Augment Recovery

26.2.1 Clinical Evidence for Inpatient Care

The earliest pilot experiments using the MIT-Manus device to alter the outcome of post-stroke upper extremity impairment demonstrated remarkable safety and significant effectiveness whether the patients were treated within a month, at the inpatient rehabilitation setting, or 6 months after the acute stroke, in an outpatient setting [12, 13]. Volpe et al. reported composite results of the controlled tests of robotic therapy compared to appropriate sham conditions in 96 stroke inpatients admitted to Burke Rehabilitation Hospital in White Plains, NY [14]. All participants received conventional neurological rehabilitation during their participation in the study. The goal of these trials was to build on the positive pilot data and test whether movement therapy had a measurable impact on recovery. Consequently, we provided one group of patients with as much movement therapy as possible to address a fundamental question: Does goal-oriented movement therapy have a positive effect on neuromotor recovery after stroke? In retrospect, at the time of these studies, the answer to this question was far from clear.

Placement of subjects in an experimental (robot-trained) or control (robot-exposure) group was done in a random fashion. Individuals in the experimental group received no less than 25 sessions of sensorimotor robotic therapy for the paretic arm (one-hour session daily every weekday). Patients were asked to perform goal-directed, visually-guided and visually-evoked reaching movements with their paretic arm. MIT-Manus' low impedance and low-friction assured that the robot would not suppress the patient's attempts to move. The robot afforded gentle guidance and assistance only when a patient could not move or deviated from the desired path [15]. We named this intervention "sensorimotor" therapy and it was similar to the "hand-over-hand" assistance that a therapist often provides during usual care. It is interesting to note that this form of "assistance as needed," which has been a central feature of our approach from the outset, has been adopted by other groups [12, 16, 17].

Individuals assigned to the robot-exposure (control) group were asked to perform the same planar reaching tasks as the robot therapy group. However, the robot did not actively assist the patient's movement attempts. When the subject was unable to reach toward a target, he or she could assist with the unimpaired arm or the technician in attendance could help to complete the movement. The robot supported the weight of the limb while offering negligible impedance to motion. For this control group, the task, the visual display, the audio environment (e.g., noise from the motor amplifiers) and the therapy context (e.g., the novelty of a technology-based treatment) were all the same as for the experimental group, so this served as a form of "placebo" of robotic movement therapy. Patients in this group were seen for only 1 h per week during their inpatient hospitalization.

The study was "double blinded" in that patients were not informed of their group assignment and therapists who evaluated their motor status did not know to which group patients belonged. Standard clinical evaluations included the upper extremity sub-test of the Fugl-Meyer Assessment (FM, maximum score = 66); the MRC Motor Power score for four shoulder-and-elbow movements (MP, maximum score = 20); and the Motor Status Score (MSS, maximum score = 82) [18–20]. The Fugl-Meyer test is a widely accepted measure of impairment in sensorimotor and functional grasp abilities. To complement the Fugl-Meyer scale, Burke Rehabilitation Hospital developed the Motor Status Scale to further quantify discrete and functional movements in the upper limb. The MSS scale expands the FM and has met standards for inter-rater reliability, significant intra-class correlation coefficients and internal item consistency for inpatients [21].

Although the robot-exposure (control) and robot-treated (experimental) groups were comparable on admission, based on sensory and motor evaluation and on clinical and demographic scales, and both groups were in-patients in the same stroke recovery unit and received the same standard care and therapy for comparable lengths of stay, the robot-trained group demonstrated significantly greater motor improvement (higher mean interval change ± sem) than the control group on the MSS/E and MP scores (see Table 26.1). In fact, the robot-trained group improved twice as much as the control group in these measures. Though this was a modest beginning, it provided unequivocal evidence that movement therapy of the kind that might be delivered by a robot had a significant positive impact on recovery.

26.2.2 Clinical Evidence for Chronic Care

The natural history of motor recovery of the paretic upper limb after stroke reveals a dynamic process that has traditionally been described by a period of flaccidity that is followed by changes in tone and reflex, as well as the frequent development of synkinesis or associated movement disorders. This synkinesis is characterized by involuntary, composite movement patterns that accompany an intended motor act [22]. Complete motor recovery, when it occurs, will unfold rapidly in hours or days. The more commonly observed partial recovery, with broad variability in final motor outcomes, unfolds over longer periods [23, 24]. At the time of our initial studies, the state of knowledge regarding motor recovery post-stroke indicated that the majority of gains in motor abilities occurred within the first three months after stroke onset, and that over 90% of motor recovery was complete within the first five months [25]. However, we were able to recall one third of the 96 stroke inpatients mentioned earlier three years after discharge. We observed that both groups continued to improve after discharge from the hospital and after five months post-stroke. Our data suggested that previous results limiting the potential of chronic patients' recovery were based on the effects of general rather than task-specific treatments during the recovery period post-stroke [23, 24]. In 2010 the Veterans Affairs completed the VA-ROBOTICS study (CSP-558), a landmark multi-site, randomized clinical trial in a

Table 26.1 Burke rehabilitation hospital inpatient studies

Between group comparisons: Final evaluation minus initial evaluation	Robot trained (N = 55)	Control (N = 41)	P-Value
Impairment measures (±sem)			
Fugl-Meyer shoulder/elbow (FM-se)	6.7 ± 1.0	4.5 ± 0.7	NS
Motor power (MP)	4.1 ± 0.4	2.2 ± 0.3	<0.01
Motor status shoulder/elbow (MS-se)	8.6 ± 0.8	3.8 ± 0.5	<0.01
Motor status wrist/hand (MS/wh)	4.1 ± 1.1	2.6 ± 0.8	NS
Disability evaluation			
Function Independence Measure (FIM)	32.0 ± 5	25.5 ± 6.5	NS

(n = 96) Mean interval change in impairment and disability (significance p < 0.05)

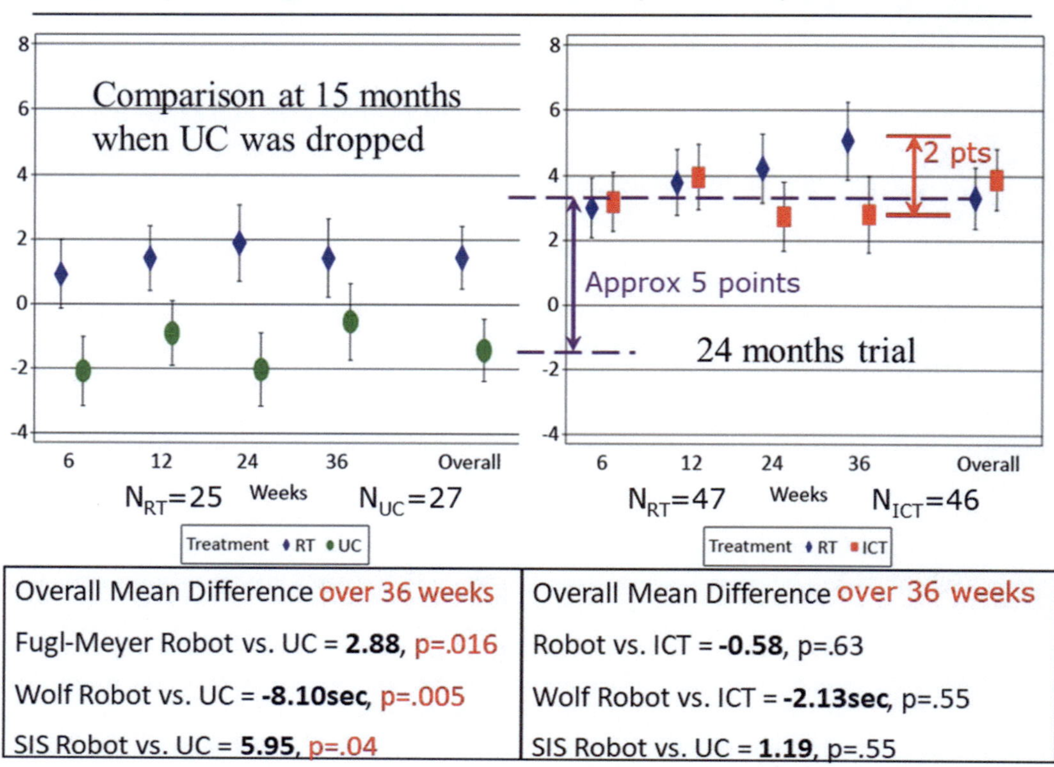

Fig. 26.2 Changes Over Time in the VA-ROBOTICS. The training lasted for 12 weeks with an additional six-month follow-up after completion of the intervention. The left panel shows the comparison of the 1st half of the robot group with the usual care (1st half as therapists learned how to employ the system). The right panel shows the comparison of the complete robot group with the Intensive Comparison Training (both groups executed 1,024 reaching movements with the paretic arm in an hour session). Arrows indicate the changes between usual care and robot group and between robot group and ICT at 36 weeks of evaluation

chronic stroke of upper extremity rehabilitation robotics employing our gym of robots (planar shoulder-and-elbow, anti-gravity, wrist, and hand robots) [26].

The VA-ROBOTICS study vanquished for good the old dogma that an adult brain was hardwired and static. It demonstrated that even for persons with multiple strokes, severe strokes, and many years post-stroke there is a real opportunity for meaningful improvement. At follow-up, 6 months after completing the intervention, the robot group demonstrated sustainable and significant improvement over the usual care group on impairment, disability, and quality of life. The results are even more impressive if we consider the results of the complete program of robotic treatment rather than an analysis that focused on the first half of the study (see Fig. 26.2). In a nutshell, while the results at 12 weeks show on the left plot that the difference between the first half of the robotic treatment group and usual care was slightly over 2 Fugl-Meyer points (as the therapists were learning how to use the robots), once the therapists were proficient in using the technology, the difference between the second half of the robotic treatment group and usual care was almost 8 points in the Fugl-Meyer assessment (note on the right plot that the total robotic group versus the total usual care showed around a 5-point change which corresponds to the MCID threshold—Minimum Clinically Important Difference for chronic stroke and allow us to estimate participants 8 points improvement in the second half of the study [27, 28]).

It is quite important to stress that VA-ROBOTICS enrolled moderately to severely impaired chronic stroke patients and over 30% of these patients had multiple strokes. As such, the group represented a spectrum of disability burdens that many studies have avoided, and that, in our research, represented the majority of the cases (65% of the volunteers that were enrolled). Thus, even if the positive changes in the robotic therapy group might appear modest, the persistent statistically significant improvement at the 6 month follow-up evaluation suggests improved robustness and perhaps an incremental advantage that prompted further improvement even without intervention.

In this era of cost containment, cost-benefit analysis is essential and in this case, it provided an important result [29]. As expected, active interventions added costs beyond the usual care offered in the VA; for example, the extra cost of the robotic equipment plus an additional therapist cost the VA $5,152 per patient. However, when we compared the total cost, which included the clinical care needed to take care of these Veterans, there were no significant differences between active intervention and usual care. In fact, the robotic group was less costly to the VA. The total healthcare utilization cost of the usual care group was $19,098 per patient, compared to $17,831 total healthcare cost for the robotic group (including the additional cost of robotic therapy). To check the possibility that a Hawthorne-like effect may have biased the cost analysis, we requested that the VA examine whether the total healthcare costs increased for the robotic therapy group after the cessation of the intervention. It did not. In fact, the total healthcare cost for the robotic group continued to decrease, perhaps because patients continued to improve even without intervention (see Fig. 26.2) [29]. These results strongly suggest better care for the same or lower total cost.

The impact of rehabilitation robotics in terms of impairment was confirmed in the UK National Health Service (NHS) and its Health Technology Assessment (HTA) Program [30, 31]. RATULS was the largest ever randomized clinical trial in robotic therapy. It enrolled 770 stroke participants. Like the VA study, RATULS included a positive control group with 257 subjects enrolled in the robot therapy (RT) group, 259 in the enhanced upper limb therapy (EULT) group, and 254 in the usual care group (UC) [32]. We had introduced the concept of a positive control treatment to test whether there was something inherent in either intensive therapy delivered by a therapist or by a robot; essentially the key factor was the intensive physical training [26, 33, 34]. Participants were evaluated at the completion of the intervention at 12 weeks, and again at 6 months from the beginning of the intervention. This was a pragmatic trial and its primary outcome was the ARAT-Success (functional outcome), secondary outcomes included the Fugl-Meyer Assessment (FMA) and other scales. All groups improved significantly from admission to completion beyond the MCID threshold for most scales (see for example the table on the bottom left of the figure that shows the FMA at baseline and completion of the intervention—the difference is larger than the MCID for all groups), but no statistically significant difference between groups was observed in the primary outcome. Of notice, in terms of impairment reduction both the RT and EULT showed a significant and clinically meaningful advantage over the UC in the Fugl-Meyer assessment with the RT group (but not EULT) maintaining this significant advantage at 6 months (see Fig. 26.3 panel B—as defined in RATULS' Statistical Analysis Plan the odds-ratio of the adjusted mean difference must be positive and include the MCID for the RT vs UC and for the EULT vs UC at completion and only for TR vs UC after 6 months).

RATULS also afforded a unique opportunity to compare the cost with the VA-ROBOTICS. In RATULS at 6 months, the total healthcare utilization cost of the usual care group was £3,785 per participant, compared to £4,451 total healthcare cost for the EULT group and £5,387

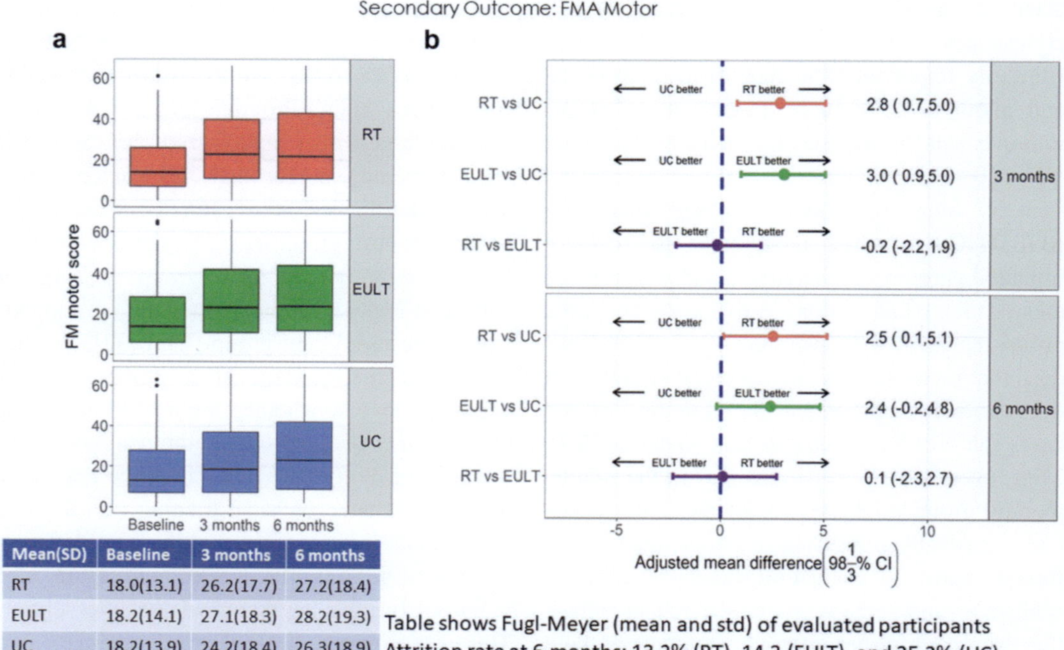

Mean(SD)	Baseline	3 months	6 months
RT	18.0(13.1)	26.2(17.7)	27.2(18.4)
EULT	18.2(14.1)	27.1(18.3)	28.2(19.3)
UC	18.2(13.9)	24.2(18.4)	26.3(18.9)

Table shows Fugl-Meyer (mean and std) of evaluated participants
Attrition rate at 6 months: 13.2% (RT), 14.3 (EULT), and 25.2% (UC)

Fig. 26.3 Fugl-Meyer motor score at baseline, 3 months, and 6 months **a** Fugl–Meyer motor score. **b** Pair-wise comparison of Fugl–Meyer motor score. In (**a**), the horizontal black line is the median, the box is the IQR, and whiskers extend to the closest value within the upper or lower quartile ±1·5 multiplied by the IQR; the black dots are any values outside of this range. EULT = enhanced upper limb therapy. RT = robot-assisted training. UC = usual care

for the robotic group which included the acquisition cost of 2 robots per site [35].

Comparing the VA-ROBOTICS and RATULS study leads to interesting insights for future RCTs. First, the VA study offered a very important enticement to participants randomized to usual care: At the completion of the protocol, they could select to receive either RT or the intensive matching therapy (therapist delivered). We speculate that this led to much lower attrition in the VA-ROBOTICS usual care group (10.7%) as compared to RATULS were at 6 months, there was an attrition rate at a remarkable 25.2% for the UC group (planned attrition rate was 10%). It is quite understandable that even stoic participants would feel pretty distraught by their luck-of-the-draw. Indeed, RATULS' UC group performance is quite puzzling and unexpected in many ways.

Participants in RATULS were evaluated using standardized clinical assessments spanning WHO ICF domains including the Action Research Arm Test (ARAT—primary outcome) and Fugl-Meyer (FMA). The FMA and ARAT were recommended as the preferred scales by the Stroke Recovery and Rehabilitation Roundtable for body structure and function, and activity [13]. We disagree with the Roundtable recommendation and strongly opposed from the onset the selection of the ARAT in RATULS. In our opinion, the consensus of the Stroke Recovery and Rehabilitation Roundtable did a disservice to the field by recommending a scale that poorly characterizes and measures function among stroke survivors with severe-to-moderate impairment (which is the focus group of most rehabilitation robotics interventions). The ARAT is most appropriate for participants who have active movement in the wrist and hand after stroke (essentially mild stroke survivors who have greater function than those included in RATULS). Specifically, the ARAT has four subtests that separately assess grasp, grip, pinch and gross movement during hierarchically arranged tasks, such as picking up wooden cubes or ball bearings of different sizes, pouring water from glass to glass, or placing the paretic hand behind the head [36]. Not surprisingly, the post-hoc analysis led the RATULS core team to concur with our opinion (see "Analysing the Action Research Arm Test (ARAT): a cautionary tale from the RATULS trial") [37]. We reproduced below one of the paper's main findings showing the ARAT distribution for 769 stroke participants.

Except for gross movement, the ARAT is not sensitive to the enrolled participants. It is L- or U-shaped with most participants unable to perform and some performing very well. However, few participants fall in the middle range demonstrating the inadequacy and insensitivity of this scale to characterize and track improvement among severe-to-moderate strokes (see Fig. 26.4).

From the economic perspective, the RATULS trial was designed to deliver robot-assisted

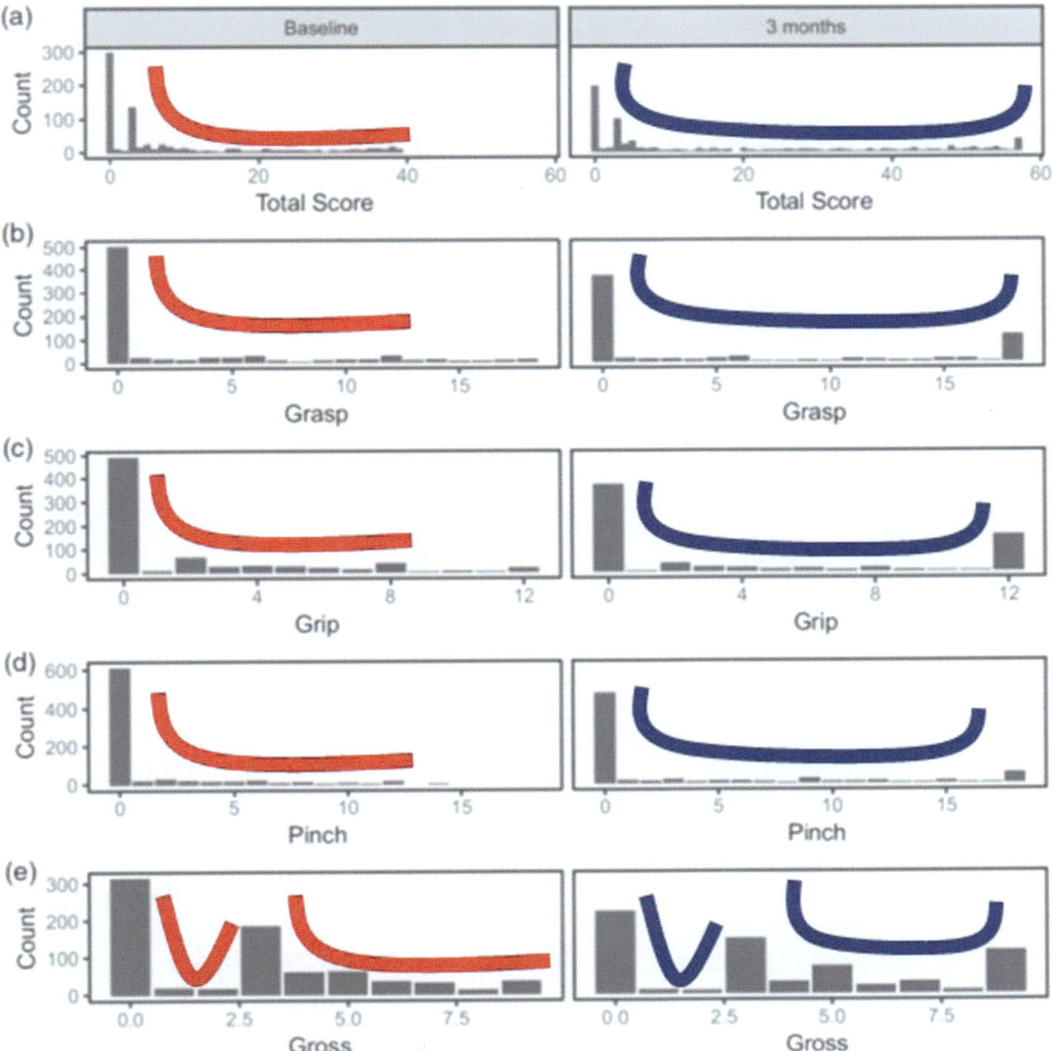

Fig. 26.4 Distribution of the ARAT total score and subscales at baseline ($n = 769$) and 3 months ($n = 669$ except gross where $n = 668$) for RATULS in **a** Total, with **b** Grasp, **c** Grip, **d** Pinch, and **e** Gross representing different components of the test. ARAT = Action Research Arm Test. Note that the distribution is essentially binomial. Note the initial L-shaped and then U-shaped distribution

Table 26.2 Primary outcome measures

Fugl-Meyer	RTT (N = 11)			RTI (N = 10)			Mixed Anova Group x Time		
Mean (SD)	pre	post	f/u	Pre	post	f/u	F	p	n^2
Overall	29.9(6.1)	38.0(7.0)*	37.6(9.1)*	31.6(7.1)	34.3(6.8)†	34.0(7.8)	6.77	0.01	0.43
Proximal	23.7(2.5)	28.5(3.1)*	28.0(4.0)*	25.4(3.9)	27.0(3.8)	27.3(4.0)	3.72	0.04	0.29
Distal	6.2(4.0)	9.6(4.6)*	9.6(6.0)*	6.2(5.0)	7.3(4.7)	6.7(5.7)	5.43	0.01	0.38

training on a one-to-one basis, i.e., the robotic intervention consisted of a therapist interacting with a single patient training with a robotic gym that included two distinct robots [35]. As implemented in RATULS, the total additional cost of the robotic intervention that included two types of robots per site is the difference between the cost of the RT and usual care group (£5,387 − £3,785 = £1,602). Considering the 36 therapy sessions, this leads to a cost of £44.5 per session. However, if we consider the alternative arrangement of the VA-ROBOTICS trial, which assumed a robotic gym that includes the two robots working simultaneously with two patients under the oversight of a single therapist, the cost per session is approximately £35.25 per session per participant. This information is quite relevant within the UK setting that might offer this type of therapy for private pay.

Nevertheless, while robotics is no panacea, there is now objective evidence that in the "real" therapy world away from the research environment, robotic therapy that involves an interactive high intensity, intention-driven therapy based on "assist-as-needed" principles leads to better outcomes than usual care in stroke at distinct recovery phases (acute to chronic), at least in terms of impairment reduction.

26.2.3 Impairment-Based or Functionally-Based Robotic Therapy: Transition-to-Task

The selection of the appropriate measurement tool is quite important as discussed in the example of the selection of the ARAT as the primary outcome to attempt to identify functional differences among participants with severe-to-moderate impairment in the RATULS study. Nevertheless, Valerie Pomeroy, an experienced UK-based clinician argued, 20 years ago, that most participants of robotic interventions were within the severe-to-moderate range and the robotic interventions should indeed aim at impairment reduction with the therapist assisting in the translation of impairment gains to function. This insight is borne out in the table below which shows the results of Chung-shan Hung and colleagues who demonstrated the advantages of this approach (see Table 26.2 and notice that the RTT group improved by roughly 8.1 points in the Fugl-Meyer while the RTI group improved 2.7 points) [38]. We obtained a similar result in a larger study and recommend this approach in clinical practice [39].

26.2.4 Clinical Evidence Contrary to Common Clinical Perceptions

While appropriate robotic therapy has been demonstrated to augment recovery, we still don't know how to tailor therapy to meet a particular patient's needs. We do not know the optimal dosage. What is the minimum intensity to promote actual change? Is too much therapy detrimental? To whom: severe, moderate, mild stroke patients? Should therapy progress from proximal to distal or the other way around? Should we train subcomponents of a movement, such as reaching in a compensated environment and raising the arm against gravity, or train the complete spatial movement against gravity?

Should our training require bilateral coordination or does unilateral training suffice? Should we assist-as-needed, resist, or perturb and augment error? Who might be the responders who benefit most from these interventions? How should we integrate robotic gyms with therapy practice?

Our ignorance was never more evident than when we tested a common perception among clinicians that training must involve spatial movement. While Lo and colleagues demonstrated that a combination of planar, vertical, wrist, and hand robot training improves both arm impairment and functional recovery, as well as the quality of life [26], the added value of anti-gravity/spatial training was not addressed in that study. Though therapists long held the belief that training must be spatial, investigations comparing training in gravity-compensated and non-compensated environments had not been performed. To address this question, we compared in a randomized clinical trial a combination of anti-gravity and planar robot training with planar training alone and compared its effectiveness to a control group who received intensive conventional arm exercise (ICAE) [40]. We hypothesized that planar robot training combined with robot-assisted reaching outside the constrained gravity-compensated horizontal plane would be superior to gravity-compensated planar robot therapy alone (see Fig. 26.5). We also hypothesized that a six-week program of robot- assisted motor training would be more efficacious than

ICAE across impairment, function and activity measures. Training for this study was half (6 weeks) of that employed in the VA-ROBOTICS or RATULS study).

All interventions were provided by the same therapist for 6 weeks: 1 h, 3 times a week for a total of 18 sessions. Robot therapy included the use of 2 different robots employed in the VA-ROBOTICS study. Robot-assisted planar reaching was performed with a 2 active degrees of freedom (DOF) InMotion2 shoulder-elbow robot. The combined robot group (planar + vertical) used the planar shoulder-elbow robot for gravity-compensated horizontal reaching followed by the 1-DOF InMotion-linear robot in its vertical position for reaching against gravity. The robots' compliant and backdrivable behavior allowed for the expression of movement outside a rigid trajectory and provided assistance with a performance-based algorithm, adapting forces as needed to challenge or assist movement. This algorithm and its variations, first introduced over 20 years ago and described further below, continuously challenges the patient by modifying (a) the time allotted for the patient to make the move and (b) the primary stiffness of the impedance controller that guides the movement [15, 41, 42]. The controller updates its characteristics after each group of 5 games; the better the patient performs, the less guidance is provided and the more s/he is challenged to move quickly [15]. The intensive conventional arm exercise (ICAE)

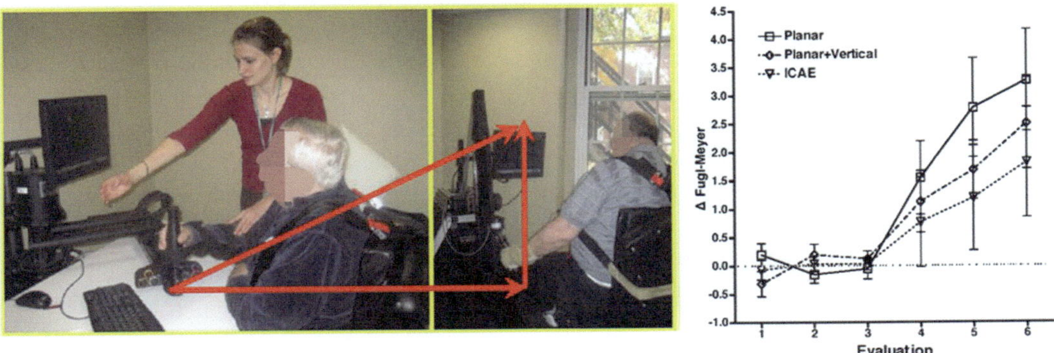

Fig. 26.5 Component Training and Spatial Composition. The FMA change at each point (mean, STD) with ICAE standing for intensive conventional arm exercise. The baseline demonstrates stability and no difference among groups. Changes from baseline to final and follow-up showed a significant benefit for both robotic groups

sessions were time-matched with the robotic sessions. The rate of movement repetition was not precisely matched to the robot, but the overall intensity was much greater than with a conventional exercise program.

In the primary outcome, all 3 groups showed modest gains from baseline to final training without significant differences. The 2 robotic groups, however, showed significant within-group changes not seen in the ICAE control group, both at the end of treatment and after a retention period. Remarkably, contrary to expectations, the combined-training group was not superior to the gravity-compensated robot training group; in fact, it improved less. Moreover, the planar (gravity-compensated) robot training subjects showed the greatest change [40].

Independence in everyday living activities includes the ability to execute reaching motions at any given moment despite the opposition of gravity. In this investigation, the robot interventions were primarily differentiated by the presentation of 2 different types of reaching in a horizontal and in a vertical plane (gravity-compensated and non-compensated) versus reaching in a single (gravity-compensated) horizontal plane. It was hypothesized that a combined robotic training program would enhance recovery by increasing task challenge and generalization of reaching in more than one context. However, the successive presentation of arm activities with different environmental and motor demands did not lead to better overall group outcomes.

One interpretation of these results is that the motor system may use two distinct modules for anti-gravity compensation and gravity-compensated reaching, and training each movement type in close succession interfered with motor consolidation [43, 44]. This interpretation is supported by a prior robotic study which found that non-gravity-compensated vertical reaching promoted further recovery in chronic stroke beyond that resulting from gravity-compensated planar reaching if it followed, rather than abutted, gravity-compensated planar reaching, i.e., 6 weeks of planar reaching training followed by 6 weeks of anti-gravity training [8]. Whether

motor memories require an interval to consolidate [45, 46] or whether practicing the whole arm movement is necessary to promote optimal recovery [47] is a complex question that this study design cannot answer. But more recently Joon-Ho Shin and colleagues did perform a direct comparison and found that whole arm movement did not lead to superior results when compared to more modular focused training. On the contrary, Joon-Ho Shin and colleagues debunked the dogma that whole arm training is required [48].

That said, preliminary studies using the anti-gravity function of the linear shoulder robot demonstrated successful treatment of shoulder subluxation after stroke [49].

26.2.5 Bilateral Versus Unilateral Motor Learning and Rehabilitation Interventions

Two main theoretical motor learning frameworks have dominated the research on bimanual control: the information-processing and dynamic framework [50]. In the first framework, bimanual movement is a special case of dual-task performance and is limited by interference. Hence, studies have focused on the interference effects when each limb performs a different task such as "patting the head while rubbing the stomach" [51]. When interference is present, one limb's movement gets biased towards the other [52]. In both discrete and cyclical bimanual tasks interference appears to be absent when limbs are moved in a mirror-symmetric fashion. However, such bias increases dramatically when the required trajectories deviate from symmetry [53]. It appears that inter-hemispheric communication plays a key role in mediating interference because non mirror-symmetric bilateral interference is not present when the corpus-callosum is transected [54, 55]. In the second framework, the system organizes itself into attractor states that afford elegant mathematical modeling. For rhythmic bimanual coordination, a system of coupled nonlinear oscillators, the Haken-Kelso-

Bunz model, has successfully modeled many observed kinematic features [56]. Under this framework, learning is a dynamic process whereby new patterns emerge corresponding to the stabilization of a dynamic system around a new attractor [57].

Several experimentalists have studied uni- and bimanual control using force field perturbations. For example, Nozaki suggested that there is a partial but not complete overlap in the learning processes of uni- and bimanual skills [58]. Bays demonstrated that subjects can learn to stabilize coupled fields [59]. Howard applied coupled and uncoupled force fields with results suggesting that the representation of coupled and uncoupled fields were independent [60]. Tcheang compared learning of a force field in one arm when the other arm made movements in a null field or in a force field showing that the learning was the same regardless of whether the other arm moved in a force field or in a null field [61]. Moreover, learning bimanual tasks showed no significant differences compared to uni-manual learning, when one arm experienced a force field and the other was at rest. They concluded that during bimanual movements the application of a force field to one arm neither interferes with nor facilitates learning of a force field applied to the other arm. Opposing force fields can be learned bimanually, without interference. This is in sharp contrast to results that suggested interference. Casadio and Morasso run a similar experiment but obtained a distinct result with directional dependency [62]. Burgess compared the skill transfer between limbs during reaching and bimanual transfer (both hands clasp a common handle) finding mild transfer from bimanual grip to the dominant and non-dominant hand that was comparable to the amount of transfer from non-dominant to the dominant hand [63].

The effectiveness of bilateral versus unilateral rehabilitation training is also a contentious issue. Many argued that bimanual might lead to superior outcomes as compared to unilateral ones. In rehab, the concept of bilateral training emanates from clinical observations that many patients while attempting to move the paretic arm also move the unimpaired arm and support strategies that promote concurrent activation of homologous homunculus, corpus-callosum inter-limb coupling, and timely sensory input [64]. Comparisons pre- and-post bilateral therapy had shown bilateral training clinical benefits [40, 65–68] and identified changes in the rubrospinal tract [69]. While benefits are clearly present, a direct longitudinal comparison of the approaches did not demonstrate that bimanual therapy was superior to unilateral therapy, quite the opposite [70, 71]. In a direct comparison of bimanual versus unilateral training, Lum and colleagues held treatment duration constant across groups, with all participants receiving 24 one-hour sessions of therapy over a 2-month period. Participants randomly assigned to the robot bilateral group practiced reaching using the MIME (mirror image movement enabler) robot to 12 spatial targets. Participants in the robot unilateral group practiced reaching with the paretic arm only, while participants assigned to the third group of combined uni- and bilateral training received a combination of both. Contrary to Lum's expectation, the bilateral robot group improved less than the unilateral or the combined approach [70]. Chung-shan Hung and colleagues also compared bimanual vs unilateral training employing distinct robotic tools and likewise found that unilateral training led to a significantly larger change in terms of impairment as reflected in the Fugl-Meyer Assessment as compared to bimanual [72–75].

26.2.6 Augmenting Robotic-Mediated Therapy: Neuro-modulation

We have been investigating modes to increase the impact of robotic therapy. In particular, we have been investigating the potential of combining robotic-mediated therapy with transcranial direct current stimulation (tDCS) [76–78]. We have been focusing particularly on anodal tDCS that can transiently increase corticomotor excitability of intrinsic hand muscles and improve upper limb function in patients with chronic stroke.

We tested whether the increased corticomotor excitability might extend to muscles acting about the wrist in patients with a residual motor deficit due to chronic stroke, and remain present during robotic training involving active wrist movements. We employed TMS and measured the motor evoked potentials (MEPs) in the flexor carpi radialis (FCR). In particular, we measured corticomotor excitability and short-interval cortical inhibition (SICI) before and immediately after a period of tDCS (1 mA, 20 min, anode and TMS on the same affected hemicranium) and robotic wrist training (1 h). We observed following tDCS an escalation in MEP amplitude increase (mean 168 ± 22%SEM; p < 0.05), that remained increased after robot training (163 ± 25%; p < 0.05). Conditioned MEPs were of significantly higher amplitude after tDCS or robotic training (62 ± 6% pre-TDCS, p < 0.05; 89 ± 14% post-tDCS, p = 0.40; 91 ± 8% post-robot; p < 0.28), suggesting that the increased corticomotor excitability is associated with reduced intracortical inhibition [76]. These effects continued during an expanded period of robotic motor training, demonstrating that a motor retraining program can co-exist with tDCS-induced changes in cortical motor excitability (see Fig. 26.6). This result supports the concept of employing brain stimulation to potentially augment robotic therapy outcomes.

In addition, we investigated whether these combinatorial therapeutic approaches further augment clinical outcomes. We conducted a randomized controlled trial in 82 chronic ischemic stroke patients (inclusion >6 months post-injury, dominant hemisphere, first stroke; residual hemiparesis) who were split into two groups to receive tDCS (M1-SO montage, anode ipsilesional, 5 × 7 cm electrodes, 2 mA, 20 min) or sham tDCS, prior to robotic upper limb training (12 weeks; 36 sessions; shoulder-

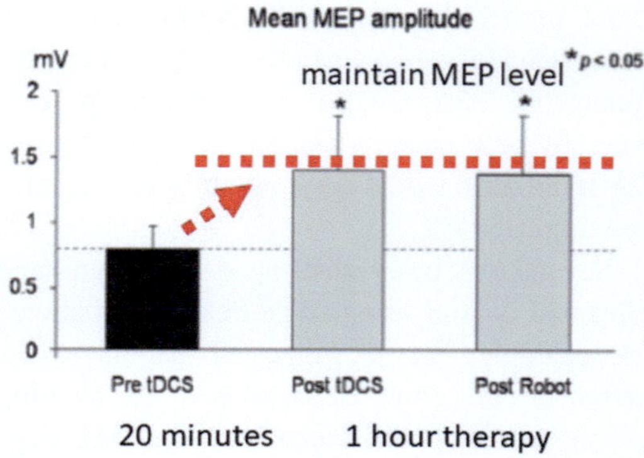

Fig. 26.6 Mean (±SEM) MEP amplitude from across subjects. MEPs were recorded from the FCR muscle during a low-level isometric wrist flexion, before and immediately following 20 min anodal brain stimulation (tDCS), then again after 1 h of robotic wrist therapy. Following tDCS, MEP amplitude was significantly elevated, and remained significantly elevated after robotic therapy

elbow robot or wrist robot on alternating sessions). The primary end-point was taken after 12 weeks of training, and assessed with the Upper Extremity Fugl-Meyer assessment (FMA). For the combined group (n = 82; post-training) robotic training increased the FMA by 7.36 points compared to baseline (p < 0.0001). But there was no difference in the FMA increase between the tDCS and sham groups (6.97 and 7.73 respectively, p = 0.46). In both groups, clinically meaningful improvement (≥ 5 points) from baseline was evident in the majority of patients (56/77), was sustained six months later (54/72), and could be attained in severe, moderate and mild baseline hemiparesis (see Fig. 26.7). This study confirms again the benefit of intensive robot-assisted training in stroke recovery. Participants improved an average of 7.36 points in the FMA (typical improvement in line with for example RATULS study [30, 31]). However, conventional tDCS did not confer a further advantage to robotic training, although our study cannot rule out that tDCS changes the rate of recovery (our study did not include intermediate evaluations) [78, 79].

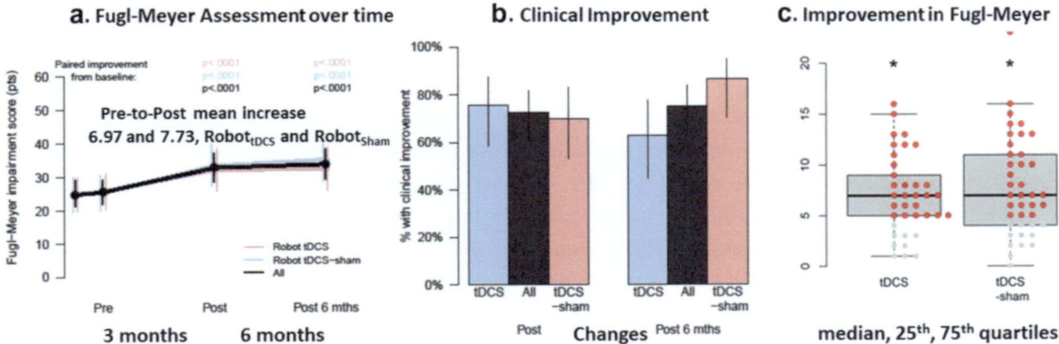

Fig. 26.7 Fugl-Meyer impairment score before and after intervention within Robot$_{tDCS}$ (n = 40) and Robot$_{Sham}$ (n = 37). Mean raw scores and 95% confidence intervals are presented in figure A, with p-values corresponding to within-person paired comparisons of improvement from baseline. A significant increase in FM score was observed for each group post intervention, and was sustained 6 months later. There was no significant difference between intervention groups at each time point. Figure B shows the proportion of patients in each group who improved at least the minimum clinically important difference from baseline. >70% of all patients demonstrated a clinically meaningful improvement, and this proportion remained at 6 months. At the primary end-point post intervention, there was no difference in the proportion of clinical responders between Robot$_{tDCS}$ and Robot$_{Sham}$ groups, while at 6 months post intervention there was a significant separation in favor of Robot$_{Sham}$. Individual data are shown in Figure B and illustrate that the majority of participants in each group achieved at least 5 points. Red dots indicate individual participant scores ≥ 5 points. * indicates significant improvement from baseline (t test p < 0.05)

26.2.7 Which Processes Underlie Neuro-Recovery?

A common assumption is that sensory-motor therapy works by helping patients "re-learn" motor control [80]. Though intuitively sensible, this notion may need to be refined. In the first place, normal motor learning does not have to contend with the neuromuscular abnormalities that are common sequelae of neurological injury, including focal spasticity, the abnormal tone throughout supporting musculo-skeleton, disrupted or unbalanced sensory pathways and muscular weakness. Thus recovery faces more obstacles to smoothly executed movements than motor learning in the developing brain. Also, it is abundantly clear that normal motor learning is not fully understood. For example: What variables or parameters of action does the brain command and control? How are these encoded and represented in the brain? How are these encodings or representations acquired and retained? These questions have practical relevance for therapy. For example, if the brain represents an action as a sequence of muscle activations, it would seem profitable to focus sensory-motor therapy on muscles. However, a large and growing body of evidence indicates that under many circumstances the brain does not directly control muscles; instead, in the case of control of the upper extremity, there is a primary requirement to meet kinematic specifications (such as the simple motion of the hand in a visually-relevant coordinate frame), so that muscle activity adjusts to compensate for movement-by-movement variation of mechanical loads. That would suggest it may be more productive to focus sensory-motor therapy on motions rather than muscles and on motor learning rather than muscle strengthening. In our research on robotic stroke rehabilitation, we have attempted to assess some of these possibilities and have developed adaptive treatment algorithms to incorporate such ideas.

Our performance-based adaptive algorithm uses nonlinear impedance control to implement a "virtual slot" extending between the start and goal positions during reaching movements [15]. Lateral deviation from the desired trajectory was discouraged by the stiffness and damping of the slot sidewalls. The desired motion was assisted by moving the back wall of the slot along a minimum-jerk virtual trajectory so that the slot progressively "collapsed" to a "virtual spring" centered on the reaching movement goal position. However, motion along with the "virtual slot" (well-aimed and faster than the nominal desired trajectory) was unimpeded. Tests of this algorithm proved highly successful [81].

A request to move was signalled by a target in the visual display changing color. If the patient failed to trigger the robot within two seconds, the robot began to act (i.e., the back wall of the "virtual slot" closed on the goal position). To trigger the robot, the patient had to move the handle (in any direction) at a speed above a modest threshold value. Even severely impaired patients with a paretic arm could trigger the robot —although trunk motion was discouraged by restraining seatbelts, in practice sufficient trunk motion was possible to move the handle and trigger the robot; no particular instruction was given other than to try to reach the target. Though ultimately inappropriate trunk motion is to be discouraged, this mode of triggering the robot encouraged severely impaired patients to participate actively rather than passively allow the robot to drive the arm.

Secondly, the revised algorithm continuously monitored the patient's performance. By combining records of the kinematics of actual patient motion and the kinetics of mechanical interaction between robot and patient, five performance measures were computed: (a) patients' ability to initiate movement, (b) patients' movement range or an extension towards the reaching movement target goal, (c) amount of mechanical power that the robot exerted to assist the hand towards the target, (d) the smoothness of the movement, and (e) the aiming/deviation from a straight line connecting the start point to the reaching goal. These measures were used to adjust the parameters of the controller during a therapy session. For the first five cycles through the eight goal positions, the time allotted for a movement (the duration of the nominal minimum-jerk trajectory) and the stiffness (impedance) of the "virtual slot" sidewalls were adjusted to approximately track the patient's current performance and need for guidance. This was important as patient performance typically declined between the end of one

therapy session and the beginning of the next as commonly seen in motor learning (acquisition of a skill and its retention). For every subsequent five cycles of the game, the controller parameters were adjusted based on the patient's performance and its variability during the previous batch of moves. The intent here was not just to track patients' performance but also to challenge them to improve. As patients aimed better, the stiffness of the "virtual slot" sidewalls was decreased, requiring better accuracy (and vice versa). As patients moved faster, the time allotted for the movement was decreased, requiring faster movements (and vice versa). The speed threshold to trigger the robot was also adjusted to 10% of the peak speed of a minimum-jerk trajectory of that duration. Consequently, if nominal movement duration increased, the speed of motion required to trigger the robot decreased (and vice versa). Thus the motor ability required to trigger the robot and move to the target was less demanding for more impaired patients and more demanding as performance improved. Again, this was intended to encourage active participation of even the most impaired patients and yet continuously challenge patients as they recovered.

Thirdly, to provide motivation, positive reinforcement and knowledge of results, the revised algorithm provided specific, movement-related feedback in the form of a simple graphical display consisting of five indicators reflecting the patient's performance in the last batch of five repetitions [82]. Each read-out was determined by the five performance measures discussed earlier. The therapist could elect to hide displays that were not meant for a patient to avoid discouraging patients who could not yet move well without boring patients who could.

This performance-based progressive therapy algorithm provided support for patients to progress from complete hemiplegia to normal arm movement, and the clinical studies that first implemented them were encouraging and successful (see Table 26.3 for typical results which were confirmed in multiple subsequent studies) [81]. The ability to initiate a movement was stressed for severely impaired patients, helping to ensure appropriate timing of afferent and efferent

Table 26.3 Motor impairment outcomes of performance-based progressive robotic therapy

Severity	Impairment measure (Mean±Sem)	FM SEC (Max = 42)	% Change	MP (Max = 70)	% Change
Moderate	Before treatment	17.0 ± 1.3		37.2 ± 2.5	
N = 12	After treatment	$22.5 \pm 1.3^*$	32%	$45.4 \pm 1.7^*$	22%
CNS > 4; NIHSS < 15	Follow up (3 months)	$24.5 \pm 0.9^*$	44%	$46.5 \pm 1.9^*$	25%
Severe	Before treatment	8.2 ± 0.7		17.3 ± 1.8	
N = 16	After treatment	$10.9 \pm 0.9^*$	33%	$23.7 \pm 2.0^*$	52%
CNS < 4 NIHSS > 15	Follow up (3 months)	$12.5 \pm 0.9^*$	37%	$26.3 \pm 2.2^*$	52%

FM SEC: Fugl-Meyer, Shoulder-Elbow Component; MP: Motor Power; CNS: Canadian Neurological Scale; NIHSS: National Institutes of Health Stroke Scale; * denotes significant change, $P < 0.001$

signals. Movement range is an important clinical measure of function but also rewards hypertonic patients for relaxing their arms, allowing the impedance controller to move their hands closer to the target. The amount of power that the robot exerted encourages a patient to attempt to do more of the movement (i.e., the robot brings the participant's limb close to the target but requires the participant to complete the attempt to hit the target). Finally, smoothness and aiming (deviation from a straight path) quantify the tradeoff between speed and accuracy that is characteristic of unimpaired movement and probably most important for patients with moderate impairment.

This adaptive algorithm was evaluated in multiple studies including VA-ROBOTICS and RATULS. Here we recount the typical changes observed in chronic stroke patients as reported elsewhere [81]. All patients were evaluated six times: three times in a two-month period prior to the start of therapy to assess baseline stability (phase-in phase), then at the midpoint and at the discharge from robotic therapy (18 one-hour sessions of robotic training, three times a week for six weeks) and finally at a follow-up evaluation session three months after training. Evaluators were blinded to the protocol used for treatment.

The first three evaluations showed no significant changes on any of the impairment scales, verifying that subjects were indeed at the chronic phase of their recovery in which no spontaneous improvement was observed. Subsequent evaluations showed that the adaptive protocol evoked a statistically significant improvement in motor performance which was maintained at the three-month follow-up. More important for our understanding of recovery, the magnitude of the improvement achieved with this adaptive algorithm was greater than that achieved with our previous robotic therapy. The only change was the robot control scheme; the same robot-assisted with the same set of reaching movements during the same number of sessions. A treatment protocol that is adapted to the patient in order to present a continuous challenge substantially enhanced recovery.

An important and informative detail is that like others we found that this enhancement of recovery was achieved with fewer repetitions [83]. Because the adaptive protocol adjusted the time allotted for a movement and allowed long movement durations as needed, fewer repetitions could be accomplished in a one-hour therapy session. Under this adaptive protocol, patients typically made just over 12,000 movements over the course of treatment. Under the previous hand-over-hand sensory-motor protocol, patients made just over 18,000 movements in the same number of sessions.

This confirms that, although the process of recovery may share some features of motor learning (such as specificity), the relationship between learning and recovery may be subtle. Though the movement is beneficial, movement alone is not sufficient; active involvement of the patient is essential. Though repetition may be beneficial, repetition alone is not sufficient; these results strongly suggest that the benefits of robotic therapy do not exclusively derive from the high "dosage" of movement delivered but also from the interactive nature of the therapy protocol.

26.2.8   Robot-Mediated Assay

First proposed over thirty years ago, robot-aided neuro-rehabilitation is increasingly being incorporated into everyday clinical practices. In addition to delivering high intensity and reproducible sensorimotor therapy, these devices are precise and reliable "measuring" tools. These measurements are objective and repeatable. Reducing the time to evaluate a patient's movement ability may offer new opportunities for designing therapeutic programs and for providing superior biomarkers [5, 84–91]. Clinical scales and robotic devices were used at two clinical sites on 208 patients with moderate to severe acute ischemic stroke to measure (determine) the range of arm movement 7, 14, 21, 30, and 90 days after the event.

Kinematic and kinetic parameters were compared to clinical assessments. Robot measures accurately forecast the clinical outcomes (cross-validated R2 of modified Rankin scale = 0.60; NIH Stroke Scale = 0.63; Fugl-Meyer = 0.73; Motor Power = 0.75). The robotic measures revealed greater sensitivity in measuring the recovery of patients (increased standardized effect = 1.47—see Fig. 26.8), demonstrating that robotic measures will more than adequately capture outcome and the altered effect size will reduce noticeably the required sample size by close to 70%. Reducing sample size will substantially improve study efficiency [92].

The reliability of human-administered clinical scales has often been questioned; for example,

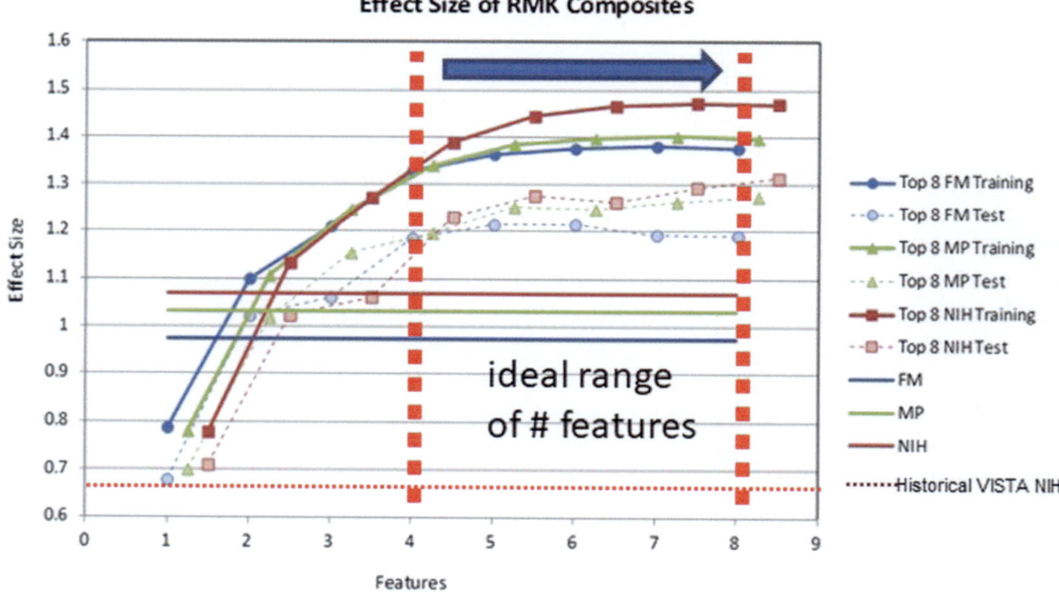

Fig. 26.8 Optimization of effect size for robot-derived robot-assisted measurement of kinematic and kinetic (RMK2) metrics. The horizontal lines show the day 7 to day 90 effect size for comparable patients of the historical Virtual International Stroke Trials Archive (VISTA) data for the NIHSS, as well as the effect sizes for the NIHSS, FM and MP assessments scales for our completers cohort. The figure also shows the performance of the robot-derived RMK2 composites optimized for the effect size for the trained (solid lines) and cross-validated sets (dashed lines). Note the increase of over 20% in cross-validated effect size for the RMK2 composites over the clinical scales with 4-features for this study (and over 70% over the historical data).

Sanford reported an inter-rater variability of ±1.8 points (95% confidence interval) for the total Fugl-Meyer scale, pointing out that small patient improvements will not be identified by this score [93, 94]. Krebs found up to a 15% discrepancy between therapists when evaluating the same patient for the upper extremity FMA scale [95, 96] Gregson estimated an inter-rater agreement of 59% for the Modified Ashworth Scale (MAS) [97]. The MAS is considered a reliable clinical scale by some [98], but totally unreliable by others [99, 100]. Besides having questionable reliability, human-administered clinical scales are also time-consuming. In contrast, robot measurements can potentially provide therapists and patients with immediate feedback. Real-time scoring cannot only greatly reduce the amount of time required to evaluate patients' motor progress, but it is also becoming a key requisite for the new robot-aided neuro-rehabilitation scenarios. These include systems that continuously adapt the amount and type of delivered therapy based on the patient's motor abilities [15, 42].

26.3 Discussion

To briefly summarize the points we have attempted to make, the available evidence demonstrates unequivocally that some forms of robotic therapy can be significantly effective and clinically efficacious as compared to usual care across the continuum of care for patients who have sustained a stroke. We submit that the technology we have deployed to date for upper extremity therapy is straightforward in that it is simple, efficient and easy to administer. However it is non-trivial, derived from decades of neuroscience research, it is firmly based on features of the neural control of upper extremity behavior, foremost the concept of movement organization with kinematics on top and subservient dynamics. We observed that in the recovery of the normal pattern of movement, kinematic coordination is preeminent and that an adaptive treatment that continuously challenges and assists a patient to improve coordination can yield

substantial advantages, while passive motion does not. These advantages can be observed via properly selected clinical scales (or missed otherwise). The same holds for aggregate kinematics measurements of speed and/or movement duration with measures of coordination such as smoothness being the most informative. We anchored this adaptive training protocol on our working model of a behavioral recovery that proceeds by re-acquiring an elementary "alphabet" of primitive movements, then over a longer period, re-developing the means to modulate and smoothly combine these elements. These primitives consist of motions—discrete and rhythmic movement primitives—and mechanical impedances—important in posture and interacting with the objects [101–103]. Of course, it is important to consider the training attributes that might speed up, or not, mastering these primitives during recovery such as: salience of the task and attention, intensity, frequency, duration, specificity, generalization, transference, and interference. While the complete understanding of the neural control of human movement is incomplete, we believe the need for continual revision will prompt further experiments and promote further clinical benefits to so many.

Acknowledgements and Disclosures This work was supported in part by a grant from NIH R21 HD060999 and R01 HD069776. H. I. Krebs is a co-inventor in several MIT-held patents for the robotic technology. He was one of the founders and the Chairman of the Board of Directors of Interactive Motion Technologies, Watertown, MA, USA from 1998 to 2016. He sold it to Bionik Laboratories, where he served as its Chief Science Officer and on the Board of Directors until July 2017. He later founded 4Motion Robotics.

References

1. Miller EL, et al. Comprehensive overview of nursing and interdisciplinary rehabilitation care of the stroke patient: a scientific statement from the American Heart Association (in eng). Stroke Practice Guideline. 2010;41(10):2402–48.
2. G. Management of Stroke Rehabilitation Working, VA/DOD Clinical practice guideline for the management of stroke rehabilitation (in eng). J Rehabil Res Dev, Practice Guideline. 2010;47(9):1–43.

3. Hogan N, Krebs HI, Charnnarong J, Srikrishna P, Sharon A. MIT-MANUS: a workstation for manual therapy and training. I. In: 1992 Proceedings of IEEE International Workshop on Robot and Human Communication. 1992 (pp. 161–65). IEEE.

4. Krebs HI, Hogan N, Aisen ML, Volpe BT. Robot-aided neurorehabilitation, (in eng). IEEE Trans Rehabil Eng, Clinical Trial Controlled Clinical Trial Research Support, Non-U.S. Gov't Research Support, U.S. Gov't, Non-P.H.S. 1998;6(1):75–87.

5. Scott SH, Dukelow SP. Potential of robots as next-generation technology for clinical assessment of neurological disorders and upper-limb therapy, (in eng). J Rehabil Res Dev, Research Support, Non-U. S. Gov't Review. 2011;48(4):335–53.

6. Dukelow SP, Herter TM, Bagg SD, Scott SH. The independence of deficits in position sense and visually guided reaching following stroke, (in eng). J Neuroeng Rehabil, Research Support, Non-U.S. Gov't.2012;9:72.

7. Casadio M, Sanguineti V, Morasso PG, Arrichiello V. Braccio di Ferro: a new haptic workstation for neuromotor rehabilitation. Technol Health Care. 2006;14(3):123–42.

8. Krebs HI et al. Rehabilitation robotics: pilot trial of a spatial extension for MIT-Manus, (in Eng). J Neuroeng Rehabil. 2004;1(1):5.

9. Krebs HI et al. Robot-aided neurorehabilitation: a robot for wrist rehabilitation, (in eng). IEEE Trans Neural Syst Rehabil Eng, Research Support, N.I.H., Extramural Research Support, Non-U.S. Gov't Review. 2007;15(3):327–35.

10. Charles SK, Hogan N. Dynamics of wrist rotations, (in eng). J Biomech, Research Support, Non-U.S. Gov't. 2011;44(4):614–21.

11. Masia L, Krebs HI, Cappa P, Hogan N. Design and characterization of hand module for whole-arm rehabilitation following stroke (in Eng). IEEE ASME Trans Mechatron. 2007;12(4):399–407.

12. Riener R, Lunenburger L, Jezernik S, Anderschitz M, Colombo G, Dietz V. Patient-cooperative strategies for robot-aided treadmill training: first experimental results, (in eng). IEEE Trans Neural Syst Rehabil Eng, Clinical Trial Research Support, Non-U.S. Gov't. 2005;13(3):380–94.

13. Kwakkel G, et al. Standardized measurement of sensorimotor recovery in stroke trials: consensus-based core recommendations from the stroke recovery and rehabilitation roundtable. Int J Stroke. 2017;12(5):451–61.

14. Volpe BT, Krebs HI, Hogan N. Is robot-aided sensorimotor training in stroke rehabilitation a realistic option? (in eng). Curr Opin Neurol, Research Support, Non-U.S. Gov't Research Support, U.S. Gov't, P.H.S. Review.2001;14(6):745–52.

15. Krebs HI, et al. Rehabilitation robotics: performance-based progressive robot-assisted therapy. Auton Robot. 2003;15(1):7–20.

16. Emken JL, Benitez R, Sideris A, Bobrow JE, Reinkensmeyer DJ. Motor adaptation as a greedy optimization of error and effort, (in eng). J Neurophysiol, Research Support, N.I.H., Extramural Research Support, Non-U.S. Gov't Research Support, U.S. Gov't, Non-P.H.S. 2007;97(6):3997–4006.

17. Wolbrecht ET, Chan V, Reinkensmeyer DJ, Bobrow JE. Optimizing compliant, model-based robotic assistance to promote neurorehabilitation, (in eng). IEEE Trans Neural Syst Rehabil Eng, Research Support, N.I.H., Extramural. 2008;16 (3):286–97.

18. Aisen ML, Krebs HI, Hogan N, McDowell F, Volpe BT. The effect of robot-assisted therapy and rehabilitative training on motor recovery following stroke, (in eng), Arch Neurol, Research Support, Non-U.S. Gov't. 1997;54(4):443–6.

19. Krebs HI, Volpe BT, Aisen ML, Hogan N. Increasing productivity and quality of care: robot-aided neuro-rehabilitation, (in eng). J Rehabil Res Dev, Research Support, Non-U.S. Gov't Research Support, U.S. Gov't, Non-P.H.S. Research Support, U. S. Gov't, P.H.S. 2000;37(6):639–52.

20. Volpe BT, Krebs HI, Hogan N, Edelsteinn L, Diels CM, Aisen ML. Robot training enhanced motor outcome in patients with stroke maintained over 3 years," (in eng), Neurology, Clinical Trial Randomized Controlled Trial Research Support, Non-U.S. Gov't Research Support, U.S. Gov't, Non-P.H.S. 1999;53(8):1874–6.

21. Ferraro M et al. Assessing the motor status score: a scale for the evaluation of upper limb motor outcomes in patients after stroke, (in eng). Neurorehabil Neural Repair, Clinical Trial Comparative Study Research Support, Non-U.S. Gov't Research Support, U.S. Gov't, P.H.S. 2002;16(3):283–9.

22. Twitchell TE. The restoration of motor function following hemiplegia in man, (in eng). Brain. 1951;74(4):443–80.

23. Nakayama H, Jørgensen HS, Raaschou HO, Olsen TS. Recovery of upper extremity function in stroke patients: the Copenhagen Stroke Study," (in eng), Arch Phys Med Rehabil, Research Support, Non-U.S. Gov't.1994;75(4):394–8.

24. Jørgensen HS, Nakayama H, Raaschou HO, Vive-Larsen J, Støier M, Olsen TS. Outcome and time course of recovery in stroke. Part I: outcome. The Copenhagen stroke study, (in eng). Arch Phys Med Rehabil, Research Support, Non-U.S. Gov't. 1995;76(5):399–405.

25. Jørgensen HS, Nakayama H, Raaschou HO, Vive-Larsen J, Støier M, Olsen TS. Outcome and time course of recovery in stroke. Part II: Time course of recovery. The Copenhagen stroke study, (in eng). Arch Phys Med Rehabil, Research Support, Non-U. S. Gov't. 1995;76(5):406–12.

26. Lo AC et al. Robot-assisted therapy for long-term upper-limb impairment after stroke, (in eng). N Engl J Med, Comparative Study Multicenter Study Randomized Controlled Trial Research Support, U. S. Gov't, Non-P.H.S. 2010;362(19):1772–83.

27. Page SJ, Fulk GD, Boyne P. Clinically important differences for the upper-extremity Fugl-Meyer Scale in people with minimal to moderate impairment due to chronic stroke," (in eng). Phys Ther, Research Support, Non-U.S. Gov't. 2012;92 (6):791–8.

28. Wu X, Guarino P, Lo AC, Peduzzi P, Wininger M. Long-term effectiveness of intensive therapy in chronic stroke. Neurorehabil Neural Repair. 2016;30(6):583–90.

29. Wagner TH et al. An economic analysis of robot-assisted therapy for long-term upper-limb impairment after stroke, (in eng). Stroke. 2011.

30. Rodgers H et al. Robot assisted training for the upper limb after stroke (RATULS): study protocol for a randomised controlled trial. Trials. 2017;18 (1):340.

31. Rodgers H, et al. Robot assisted training for the upper limb after stroke (RATULS): a multicentre randomised controlled trial. Lancet. 2019;394 (10192):51–62.

32. Bosomworth H, et al. Evaluation of the enhanced upper limb therapy programme within the Robot-Assisted Training for the Upper Limb after Stroke trial: descriptive analysis of intervention fidelity, goal selection and goal achievement. Clin Rehabil. 2021;35(1):119–34.

33. Volpe BT et al. Intensive sensorimotor arm training mediated by therapist or robot improves hemiparesis in patients with chronic stroke, (in eng). Neurorehabil Neural Repair, Comparative Study Randomized Controlled Trial Research Support, N. I.H., Extramural Research Support, Non-U.S. Gov't. 2008;22(3):305–10.

34. Lo AC et al. Multicenter randomized trial of robot-assisted rehabilitation for chronic stroke: methods and entry characteristics for VA ROBOTICS, (in eng). Neurorehabil Neural Repair, Multicenter Study Randomized Controlled Trial Research Support, U.S. Gov't, Non-P.H.S. 2009;23(8):775–83.

35. Fernandez-Garcia C et al. Economic evaluation of robot-assisted training versus an enhanced upper limb therapy programme or usual care for patients with moderate or severe upper limb functional limitation due to stroke: results from the RATULS randomised controlled trial. BMJ Open;11(5): e042081.

36. Lyle RC. A performance test for assessment of upper limb function in physical rehabilitation treatment and research. Int J Rehabil Res. 1981;4 (4):483–92.

37. Wilson N, Howel D, Bosomworth H, Shaw L, Rodgers H. Analysing the Action Research Arm Test (ARAT): a cautionary tale from the RATULS trial. Int J Rehabil Res. 2021;44(2):166–9.

38. Hung CS, Hsieh YW, Wu CY, Lin YT, Lin KC, Chen CL. The effects of combination of robot-assisted therapy with task-specific or impairment-oriented training on motor function and quality of life in chronic stroke. PM R. 2016;8(8):721–9.

39. Conroy SS, Wittenberg GF, Krebs HI, Zhan M, Bever CT, Whitall J. Robot-assisted arm training in chronic stroke: addition of transition-to-task practice. Neurorehabil Neural Repair. 2019;33(9):751–61.

40. Conroy SS et al. Effect of gravity on robot-assisted motor training after chronic stroke: a randomized trial, (in Eng). Arch Phys Med Rehabil. 2011.

41. Perez-Ibarra JC, Siqueira AAG, Silva-Couto MA, de Russo TL, Krebs HI. Adaptive impedance control applied to robot-aided neuro-rehabilitation of the ankle, (in English). IEEE Robot Autom Lett. 2019;4(2):185–92.

42. Michmizos KP, Rossi S, Castelli E, Cappa P, Krebs HI. Robot-aided neurorehabilitation: a pediatric robot for ankle rehabilitation, (in Eng). IEEE Trans Neural Syst Rehabil Eng. 2015.

43. Brashers-Krug T, Shadmehr R, Bizzi E. Consolidation in human motor memory, (in eng). Nature, Research Support, Non-U.S. Gov't Research Support, U.S. Gov't, P.H.S. 1996;382(6588):252–5.

44. Shadmehr R. Generalization as a behavioral window to the neural mechanisms of learning internal models, (in eng). Hum Mov Sci, Research Support, N.I.H., Extramural Research Support, U.S. Gov't, P. H.S. 2004;23(5):543–68.

45. Caithness G et al. Failure to consolidate the consolidation theory of learning for sensorimotor adaptation tasks, (in eng). J Neurosci, Research Support, Non-U.S. Gov't. 2004;24(40):8662–71.

46. Stickgold R. Sleep-dependent memory consolidation, (in eng). Nature, Research Support, N.I.H., Extramural Research Support, U.S. Gov't, P.H.S. Review. 2005;437(7063):1272–8.

47. Overduin SA, Richardson AG, Lane CE, Bizzi E, Press DZ. Intermittent practice facilitates stable motor memories, (in eng). J Neurosci, Research Support, N.I.H., Extramural. 2006;26(46):11888–92.

48. Lee SH et al. Comparisons between end-effector and exoskeleton rehabilitation robots regarding upper extremity function among chronic stroke patients with moderate-to-severe upper limb impairment. Sci Rep. 2020;10(1):1806.

49. Dohle CI, Rykman A, Chang J, Volpe BT. Pilot study of a robotic protocol to treat shoulder subluxation in patients with chronic stroke, (in eng). J Neuroeng Rehabil, Clinical Trial Research Support, N.I.H., Extramural. 2013;10:88.

50. Swinnen SP, Wenderoth N. Two hands, one brain: cognitive neuroscience of bimanual skill. Trends Cogn Sci. 2004;8(1):18–25.

51. Swinnen SP, Walter CB, Lee TD, Serrien DJ. Acquiring bimanual skills: contrasting forms of information feedback for interlimb decoupling. J Exp Psychol Learn Mem Cogn. 1993;19(6):1328–44.

52. Wenderoth N, Debaere F, Sunaert S, van Hecke P, Swinnen SP. Parieto-premotor areas mediate directional interference during bimanual movements. Cereb Cortex. 2004;14(10):1153–63.

53. Swinnen SP. Intermanual coordination: from behavioural principles to neural-network interactions. Nat Rev Neurosci. 2002;3(5):348–59.

54. Kennerley SW, Diedrichsen J, Hazeltine E, Semjen A, Ivry RB. Callosotomy patients exhibit temporal uncoupling during continuous bimanual movements. Nat Neurosci. 2002;5(4):376–81.

55. Sternad D, Wei K, Diedrichsen J, Ivry RB. Intermanual interactions during initiation and production of rhythmic and discrete movements in individuals lacking a corpus callosum, (in eng). Exp Brain Res, Research Support, N.I.H., Extramural Research Support, Non-U.S. Gov't Research Support, U.S. Gov't, Non-P.H.S. 2007;176(4):559–74.

56. Haken H, Kelso JA, Bunz H. A theoretical model of phase transitions in human hand movements. Biol Cybern. 1985;51(5):347–56.

57. Zanone PG, Kelso JA. Evolution of behavioral attractors with learning: nonequilibrium phase transitions. J Exp Psychol Hum Percept Perform. 1992;18(2):403–21.

58. Nozaki D, Kurtzer I, Scott SH. Limited transfer of learning between unimanual and bimanual skills within the same limb. Nat Neurosci. 2006;9 (11):1364–6.

59. Bays PM, Wolpert DM. Actions and consequences in bimanual interaction are represented in different coordinate systems. J Neurosci. 2006;26(26):7121–6.

60. Howard IS, Ingram JN, Wolpert DM. Composition and decomposition in bimanual dynamic learning. J Neurosci. 2008;28(42):10531–40.

61. Tcheang L, Bays PM, Ingram JN, Wolpert DM. Simultaneous bimanual dynamics are learned without interference. Exp Brain Res. 2007;183(1):17–25.

62. Casadio M, Squeri V, Masia L, Morasso P, Sanguineti V, Sandini G. Characterizing "direct" bimanual coordination while adapting to a dynamic environment. In: Proceedings of the 2nd Biennial IEEE/RAS-EMBS International Conference on Biomedical Robotics and Biomechatronics, 2008.

63. Burgess JK, Bareither R, Patton JL. Single limb performance following contralateral bimanual limb training. IEEE Trans Neural Syst Rehabil Eng. 2007;15(3):347–54.

64. Lewis GN, Perreault EJ. An assessment of robot-assisted bimanual movements on upper limb motor coordination following stroke. IEEE Trans Neural Syst Rehabil Eng. 2009;17(6):595–604.

65. Burgar CG, Lum PS, Shor PC, Van der Loos HM. Development of robots for rehabilitation therapy: the Palo Alto VA/Stanford experience. J Rehabil Res Dev. 2000;37(6):663–73.

66. Whitall J, Waller SM, Silver KH, Macko RF. Repetitive bilateral arm training with rhythmic auditory cueing improves motor function in chronic hemiparetic stroke. Stroke. 2000;31(10):2390–5.

67. Lum PS, Burgar CG, Shor PC. Evidence for improved muscle activation patterns after retraining of reaching movements with the MIME robotic system in subjects with post-stroke hemiparesis. IEEE Trans Neural Syst Rehabil Eng. 2004;12 (2):186–94.

68. Hesse S, Werner C, Pohl M, Mehrholz J, Puzich U, Krebs HI Mechanical arm trainer for the treatment of the severely affected arm after a stroke: a single-blinded randomized trial in two centers, (in eng). Am J Phys Med Rehabil, Multicenter Study Randomized Controlled Trial Research Support, Non-U.S. Gov't.2008;87(10):779–88.

69. Luft AR et al. Repetitive bilateral arm training and motor cortex activation in chronic stroke: a randomized controlled trial, (in eng). JAMA, Clinical Trial Randomized Controlled Trial Research Support, Non-U.S. Gov't Research Support, U.S. Gov't, Non-P.H.S. Research Support, U.S. Gov't, P.H.S. 2004;292(15):1853–61.

70. Lum PS, Burgar CG, Van der Loos M, Shor PC, Majmundar M, Yap R. MIME robotic device for upper-limb neurorehabilitation in subacute stroke subjects: A follow-up study. J Rehabil Res Dev. 2006;43(5):631–42.

71. Richards LG, Senesac CR, Davis SB, Woodbury ML, Nadeau SE. Bilateral arm training with rhythmic auditory cueing in chronic stroke: not always efficacious. Neurorehabil Neural Repair. 2008;22(2):180–4.

72. Hung CS, et al. Unilateral vs bilateral hybrid approaches for upper limb rehabilitation in chronic stroke: a randomized controlled trial. Arch Phys Med Rehabil. 2019;100(12):2225–32.

73. Hung CS et al. Comparative assessment of two robot-assisted therapies for the upper extremity in people with chronic stroke. Am J Occup Ther. 2019;73(1):7301205010p1–7301205010p9.

74. Hung CS, et al. Hybrid rehabilitation therapies on upper-limb function and goal attainment in chronic stroke. OTJR (Thorofare N J). 2019;39(2):116–23.

75. Yang CL, Lin KC, Chen HC, Wu CY, Chen CL. Pilot comparative study of unilateral and bilateral robot-assisted training on upper-extremity performance in patients with stroke. Am J Occup Ther 2012;66(2):198–206.

76. Edwards DJ et al. Raised corticomotor excitability of M1 forearm area following anodal tDCS is sustained during robotic wrist therapy in chronic stroke, (in eng). Restor Neurol Neurosci, Clinical Trial Research Support, N.I.H., Extramural. 2009;27(3):199–207.

77. Volpe BT et al. Robotic devices as therapeutic and diagnostic tools for stroke recovery, (in eng). Arch Neurol, Research Support, N.I.H., Extramural Research Support, Non-U.S. Gov't Review.2009;66(9):1086–90.

78. Moretti CB et al. Robotic Kinematic measures of the arm in chronic Stroke: part 1—Motor Recovery patterns from tDCS preceding intensive training. Bioelectron Med. 2021;7(1):20.

79. Edwards DJ, et al. Clinical improvement with intensive robot-assisted arm training in chronic

stroke is unchanged by supplementary tDCS. Restor Neurol Neurosci. 2019;37(2):167–80.

80. Hogan N et al. Motions or muscles? Some behavioral factors underlying robotic assistance of motor recovery, (in eng). J Rehabil Res Dev, Research Support, N.I.H., Extramural Research Support, Non-U.S. Gov't Research Support, U.S. Gov't, Non-P.H.S. Review. 2006;43(5):605–18.

81. Ferraro M, Palazzolo JJ, Krol J, Krebs HI, Hogan N, Volpe BT. Robot-aided sensorimotor arm training improves outcome in patients with chronic stroke, (in eng). Neurology, Clinical Trial Research Support, Non-U.S. Gov't Research Support, U.S. Gov't, P.H.S. 2003;61(11):1604–7.

82. Schmidt R, Lee T. Motor Control and learning: a behavioral emphasis Champaign, Il: Human Kinetics. 2005.

83. Duarte JE, Gebrekristos B, Perez S, Rowe JB, Sharp K, Reinkensmeyer DJ. Effort, performance, and motivation: Insights from robot-assisted training of human golf putting and rat grip strength, (in eng). IEEE Int Conf Rehabil Robot. 2013;2013:1–6.

84. Bosecker C, Dipietro L, Volpe B, Igo Krebs H. Kinematic robot-based evaluation scales and clinical counterparts to measure upper limb motor performance in patients with chronic stroke, (in eng), Neurorehabil Neural Repair, Research Support, N.I.H., Extramural Research Support, U.S. Gov't, Non-P.H.S. 2010;24(1):62–9.

85. Agrafiotis DK, et al. Accurate prediction of clinical stroke scales and improved biomarkers of motor impairment from robotic measurements. PLoS ONE. 2021;16(1): e0245874.

86. Colombo R et al. Robotic techniques for upper limb evaluation and rehabilitation of stroke patients, (in eng). IEEE Trans Neural Syst Rehabil Eng, Clinical Trial Controlled Clinical Trial Research Support, Non-U.S. Gov't. 2005;13(3):311–24.

87. Colombo R et al. Assessing mechanisms of recovery during robot-aided neurorehabilitation of the upper limb, (in eng). Neurorehabil Neural Repair, Research Support, Non-U.S. Gov't. 2008;22(1):50–63.

88. Alt Murphy M, Sunnerhagen KS, Johnels B, Willen C. Three-dimensional kinematic motion analysis of a daily activity drinking from a glass: a pilot study," (in eng), J Neuroeng Rehabil, vol. 3, p. 18, 2006.

89. Alt Murphy M, Willen C, Sunnerhagen KS. Kinematic variables quantifying upper-extremity performance after stroke during reaching and drinking from a glass, (in eng), Neurorehabil Neural Repair, Research Support, Non-U.S. Gov't. 2011;25(1): 71–80.

90. Alt Murphy M, Willen C, Sunnerhagen KS. Responsiveness of upper extremity kinematic measures and clinical improvement during the first three months after stroke, (in eng). Neurorehabil Neural Repair, Research Support, Non-U.S. Gov't. 2013;27 (9):844–53.

91. Moretti CB et al. Robotic kinematic measures of the arm in chronic Stroke: part 2 - strong correlation with clinical outcome measures. Bioelectron Med. 2021;7(1):21.

92. Krebs HI, et al. Robotic measurement of arm movements after stroke establishes biomarkers of motor recovery, (in eng). Stroke. 2014;45(1):200–4.

93. Fugel-Meyer AR, Jaasko L, Leyman I, Ollson S, Steglind S. The post-stroke hemiplegic patient. 1. a method for evaluation of physical performance, (in eng). Scand J Rehabil Med. 1975;7(1):13–31.

94. Duncan PW, Propst M, Nelson SG. Reliability of the Fugl-Meyer assessment of sensorimotor recovery following cerebrovascular accident, (in eng). Phys Ther. 1983;63(10):1606–10.

95. Krebs HI et al. Robot-aided neurorehabilitation: from evidence-based to science-based rehabilitation, (in eng). Top Stroke Rehabil. 2002;8(4):54–70.

96. See J et al. A standardized approach to the Fugl-Meyer assessment and its implications for clinical trials, (in eng). Neurorehabil Neural Repair, Research Support, N.I.H., Extramural. 2013;27 (8):732–41.

97. Gregson JM, Leathley MJ, Moore AP, Smith TL, Sharma AK, Watkins CL. Reliability of measurements of muscle tone and muscle power in stroke patients, (in eng). Age Ageing. 2000;29(3):223–8.

98. Bohannon RW, Smith MB. Interrater reliability of a modified Ashworth scale of muscle spasticity, (in eng). Phys Ther. 1987;67(2):206–7.

99. Gregson JM, Leathley MJ, Moore AP, Smith TL, Sharma AK, Watkins CL. Clinical measures: reliable or not? (in eng). Age Ageing Comment Lett. 2001;30(1):86–7.

100. Pomeroy VM et al. The unreliability of clinical measures of muscle tone: implications for stroke therapy, (in eng). Age Ageing, Research Support, Non-U.S. Gov't. 2000;29(3):229–33.

101. Levy-Tzedek S, Krebs HI, Song D, Hogan N, Poizner H. Non-monotonicity on a spatiotemporally defined cyclic task: evidence of two movement types? (in eng). Exp Brain Res, Research Support, N.I.H., Extramural Research Support, Non-U.S. Gov't Research Support, U.S. Gov't, Non-P.H.S. 2010;202(4):733–46.

102. Krebs HI, Aisen ML, Volpe BT, Hogan N. Quantization of continuous arm movements in humans with brain injury, (in eng). Proc Natl Acad Sci U S A, Research Support, Non-U.S. Gov't Research Support, U.S. Gov't, Non-P.H.S. 1999;96(8):4645–9.

103. Hogan N, Sternad D. Dynamic primitives of motor behavior. Biol Cybern. 2012;106:727–39.

Three-Dimensional Multi-Degree-of-Freedom Arm Therapy Robot (ARMin)

27

Tobias Nef, Verena Klamroth-Marganska, Urs Keller, and Robert Riener

Abstract

Rehabilitation robots have become an important tool to complement rehabilitation training in patients with neurological disorders such as stroke and spinal cord injury. Arm rehabilitation robots can create a motivational, activity-based environment supporting an intensive rehabilitation training with frequent and numerous repetitions. Therefore, robots have the potential to improve the rehabilitation process in patients with lesions of the central nervous system. In this chapter, the three-dimensional, multi-degree-of-freedom ARMin arm robot, and the related ChARMin and Armeo Power robots, are presented. The devices have an exoskeleton structure that enables the training of activities of daily living. Patient-responsive control strategies assist the patient only as much as needed and stimulate patient activity. This chapter covers the mechanical setup, the therapy modes, and the clinical evaluation of the exoskeleton robots. It concludes with an outlook on ongoing developments.

T. Nef (✉)
Gerontechnologoy and Rehabilitation Group, ARTORG Center for Biomedical Engineering Research, University of Bern, Murtenstrasse 50, 3010 Bern, Switzerland
e-mail: tobias.nef@unibe.ch

Department of Neurology, University Hospital Inselspital, Bern, Switzerland

V. Klamroth-Marganska
Research Unit for Occupational Therapy, School of Health Professions, Zurich University of Applied Sciences ZHAW, Zurich, Switzerland
e-mail: verena.klamroth@zhaw.ch

U. Keller
Swiss Children's Rehab, University Children's Hospital Zurich, Affoltern am Albis, Switzerland
e-mail: Urs.Keller@kispi.uzh.ch

R. Riener
Sensory-Motor Systems Lab, Institute of Robotics and Intelligent Systems, ETH Zurich, and University Hospital Balgrist, Tannenstrasse 1, 8092 Zurich, Switzerland
e-mail: riener@hest.ethz.ch

Keywords

Exoskeleton · Rehabilitation · Stroke · Upper extremity · Virtual reality

27.1 State of the Art

27.1.1 Rationale for Application of Current Technology

Stroke remains the leading cause of permanent disability: Recent studies estimate that it affects more than one million people in the European Union [1, 2] and more than 0.7 million in the United States each year [3]. The major symptom of a stroke is severe sensory and motor

© The Author(s), under exclusive license to Springer Nature Switzerland AG 2022
D. J. Reinkensmeyer et al. (eds.), *Neurorehabilitation Technology*,
https://doi.org/10.1007/978-3-031-08995-4_27

"

hemiparesis of the contralesional side of the body [4]. The degree of recovery depends on the location and the severity of the lesion [5]. However, only 18% of stroke survivors regain full motor function after 6 months [6]. Restoration of arm and hand function is essential to resuming daily living tasks and regaining independence in life. Several studies show that sensorimotor arm therapy has positive effects on the rehabilitation progress of stroke patients [7–9].

The goal is to induce brain plasticity and improve functional outcomes. Relevant factors for successful therapy are training intensity [10–12] including frequency, duration [13, 14], and repetition [15]. With respect to these criteria, one-to-one manually assisted training has several limitations. It is labor-intensive, time-consuming, and expensive. The disadvantageous consequence is that the training sessions are often shorter than required for an optimal therapeutic outcome. Finally, manually assisted movement training lacks repeatability and objective measures of patient performance and progress.

Some shortcomings can be overcome by the use of robotics. With robot-assisted arm therapy, the number and duration of training sessions can be increased while reducing the number of therapists required per patient can potentially be reduced. Thus, it is expected that personnel costs can be reduced. Furthermore, robotic devices can provide quantitative measures and they support the objective observation and evaluation of the rehabilitation progress.

27.1.2 Therapeutic Actions and Mechanism

Numerous groups have been working on arm rehabilitation robots, and several different types of rehabilitation robots have been developed and tested with stroke patients. In this article, we discuss different types of robotic arm therapy by analyzing several arm robots. This is not an exhaustive analysis of arm therapy robots, and the interested reader is referred to appropriate review articles [16–21].

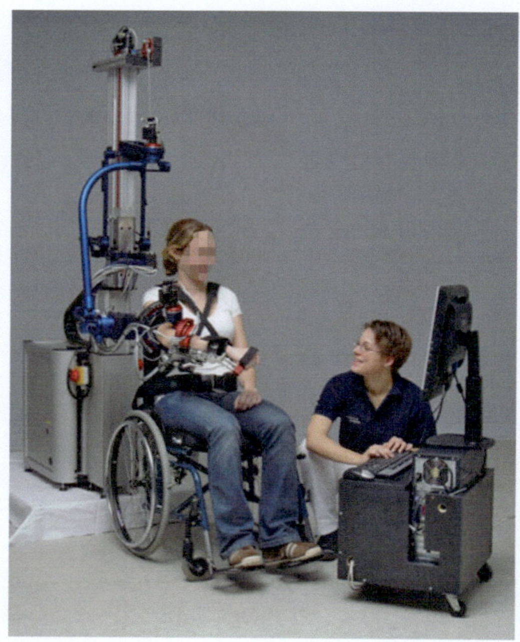

Fig. 27.1 Typical setup for a robot-supported arm therapy system

The typical setup for robot-supported arm therapy consists of the seated stroke patient with the most affected arm connected to the robotic device (Fig. 27.1). In most applications, the patient looks at a graphical display—either a large, immersive 3D projection or a standard computer screen. The robotic device is characterized by its mechanical structure, the number and type of actuated joints, and the actuation principle. This section discusses these three key characteristics and their influence on rehabilitation training.

27.1.2.1 Mechanical Structure: End-Effector-Based Robots and Exoskeleton Robots

End-effector-based robots are connected to the patient's hand or forearm at a single point (Fig. 27.2). Depending on the number of links of the robot, the human arm can be positioned and/or oriented in space. The robot's axes generally do not correspond with the human-joint rotation axes. That is why, from a mechanical point of view, these robots are easier to build and use.

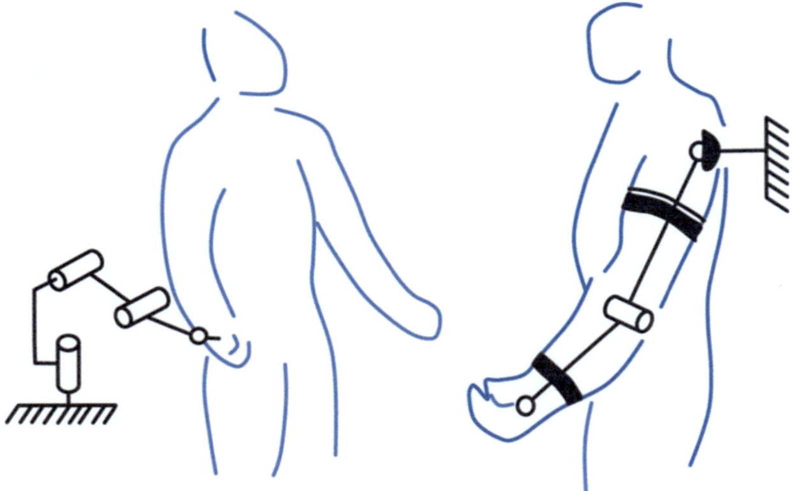

Fig. 27.2 Schematic view of end-effector-based (left) and exoskeleton (right) robots

Many researchers have developed and evaluated end-effector-based robots. The MIT Manus [22], the Mirror Image Motion Enabler [23], the Bi-Manu-Track [24], the GENTLE/s [25], and the Arm Coordination Training Robot [26] are examples of end-effector-based robotic devices. An important advantage of these robots is that they are easy to adjust to different arm lengths. A disadvantage is that, in general, the arm posture and/or the individual joint interaction torques are not fully determined by the robot because the patient and the robot interact just through one point—the robot's end-effector.

The mechanical structure of the exoskeleton robot resembles the human arm anatomy, and the robot's links correspond with human joints. Consequently, the human arm can be attached to the exoskeleton at several points. Adaptation to different body sizes is, therefore, more difficult than in end-effector-based systems because the length of each robot segment must be adjusted to the patient's arm length. Since the human shoulder girdle is a complex joint, this is challenging and requires advanced mechanical solutions for the robot's shoulder actuation [27]. However, with an exoskeleton robot, the arm posture is fully determined, and the applied torques to each joint of the human arm can be controlled separately. The ability to separately control the interacting torques in each joint is essential, such as when the subject's elbow flexors are spastic. The mobilization of the elbow joint must not induce reaction torques and forces in the shoulder joint, which can be guaranteed by an exoskeleton robot, but not by an end-effector-based one. That is also why therapists use both hands to mobilize a spastic elbow joint. To avoid exercising forces to the shoulder, one hand holds the lower arm while the other hand holds the upper arm. This is comparable to an exoskeleton robot with a cuff affixed to the lower arm and another cuff affixed to the upper arm. Some examples of arm rehabilitation exoskeletons include the Dampace [28], the Armeo Spring (former T-Wrex) [29], the MGA-Exoskeleton [30], the L-Exos [31], the Caden-7 [32], the Intelligent Robotic Arm [33], as well as the ARMin I, II, and III devices [27, 34].

While it seems clear that end-effector-based robots have practical advantages (usability, simplicity, and cost-effectiveness) and exoskeleton robots have biomechanical advantages (better guidance), it remains an open research question whether and how this disparity influences therapeutic outcomes.

27.1.2.2 Number and Type of Actuated Joints

Apart from the mechanical structure, the number and type of actuated joints are another point of differentiation among robotic devices. Some groups focus on functional training that includes the entire arm and hand (proximal and distal joints). This functional training can be based on activities of daily living (ADL) and requires sophisticated and complex robotic devices such as the GENTLE/s, the Dampace, the Armeo Spring, or the ARMin robot. The reason for ADL training is that there is evidence that functional and task-oriented training shows good results in stroke patients [9, 35]. This confirms previous observations made with the constraint-induced movement therapy. Interventional studies have shown that forcing the affected limb to perform ADLs yields functional gains, allowing the stroke patient to increase the use of the affected arm in the "real-world" environment [36–39].

Other groups have developed robots that focus on the training of distal parts of the human arm such as the hand [40], the wrist, and the lower arm [41, 42]. One may speculate that the distal approach results in a more powerful activation of the sensorimotor cortex, given their larger cortical representation [43]. The suggested competition between proximal and distal arm segments for plastic brain territory after stroke [44] would imply shifting treatment emphasis from the shoulder to the forearm, hand, and fingers. Other devices work proximal to the elbow and shoulder [26, 45]. Namely, the Act3D robot implements an impairment-based, 3D robotic intervention that specifically targets abnormal joint torque coupling between the elbow and shoulder joint [45].

An interesting research question is whether robotic training should focus on whole-arm/hand functional movements, simply in a distal fashion or by combining distal and proximal modes. There is evidence from Krebs et al. [46] that training both the transport of the arm and manipulation of an object did not confer any advantage over solely training transport of the arm. This calls for further investigation with other robotic devices—especially with whole-arm exoskeletons.

27.1.2.3 Actuation Principle: Nonmotorized Robots and Motorized Robots

Most motorized rehabilitation robots are powered by electric motors. Depending on the underlying control paradigm, the motors can either control the interaction force/torque between the patient and the robot or the position of the robot. This allows the robotic device to support the human arm against gravity, canceling gravitational forces and making it easier for the patient to move his or her arm. Also, motorized robots can support the patient in the movement toward a target, such as an object within an ADL training scenario. If required, electric motors can also resist the patient in the movement, making the patient's arm heavier or making the patient feel that he is carrying an object with a given mass. Motorized robots can be used as an evaluation tool to objectively measure voluntary force, range of motion, and level of spasticity [47–49]. Another important application is having the robot introduce force fields onto the endpoint of the human. The adaptation of human to different force fields is expected to trigger plasticity changes in the brain and enhance rehabilitation.

Some recent rehabilitation devices have been developed to work without motors [28, 29]. The commercially available Armeo Spring device is based on the former T-Wrex device [49] and works without any motors. In this exoskeleton device, springs support the human arm against gravity. The mechanical design allows the therapist to adjust the spring length and select the proper amount of support. Sensors measure the position and orientation of the human arm, which is transmitted to the graphical display where the patient can see his or her own movement on the computer screen. Compared to motorized robots, this approach has the great advantage of significantly lower costs and weight. Moreover, the device is easier to use and intrinsically safe. The disadvantage is that it is not possible to support the patient other than against gravity, so, for instance, the device cannot support the patient in directed reaching movements, nor can it challenge the patient by resisting movement. Some devices overcome this by adding brakes to the robot that dissipate energy and challenge the patient's movements [28]. Current evidence suggests that nonmotorized devices might be very well suited for the training of mildly impaired stroke patients who do not need as much support as heavily impaired subjects [49].

27.2 Review of Experience and Evidence for the Application of the Armin Robot System

27.2.1 Technical Evaluation of the ARMin Robot System

The first version of the arm therapy robot, ARMin I, was designed and tested from 2003 to 2006 at the ETH Zurich in close collaboration with therapists and physicians from the University Hospital Balgrist, Zürich [34, 50, 51]. This version is characterized by 4 degrees of freedom (DOF) actuating the shoulder in 3D and flex/extend the elbow (Fig. 27.3). The upper arm is connected to the robot by an end-effector-based structure. Like later versions of the ARMin, the device could be operated in three modes: passive mobilization, active game-supported arm therapy, and active training of activities of daily living (ADL). The improved version, ARMin II, was characterized by a complete exoskeletal structure with two additional DOF (six altogether) allowing also pronation/supination of the lower arm and wrist flexion/extension (Fig. 27.1). Particular efforts were undertaken to optimize shoulder actuation: a sophisticated coupling mechanism enables the center of rotation of the shoulder to move in a vertical direction when the arm is lifted [52, 53]. This function is required to provide an anatomically correct shoulder movement that avoids shoulder stress from misalignment of the robot and anatomical joint axes when lifting the upper arm above face level.

ARMin III (Fig. 27.4) was further improved with respect to mechanical robustness, complexity, user operation, and reliability [27]. Five ARMin III devices have been developed for a multicenter clinical trial. The next section describes the mechanics of the ARMin III robot in more detail.

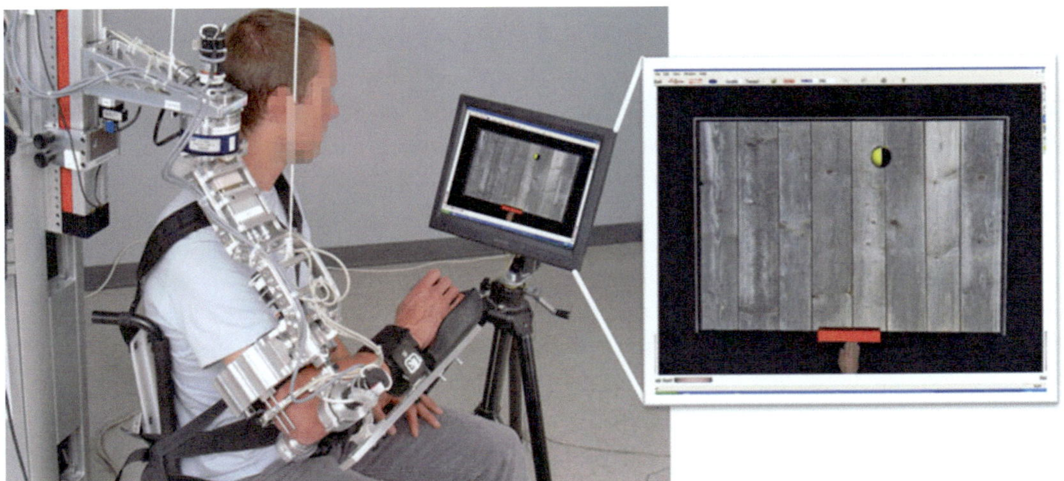

Fig. 27.3 ARMin I robot with a healthy test person (left). The person is looking at a computer monitor showing the movement task (right)

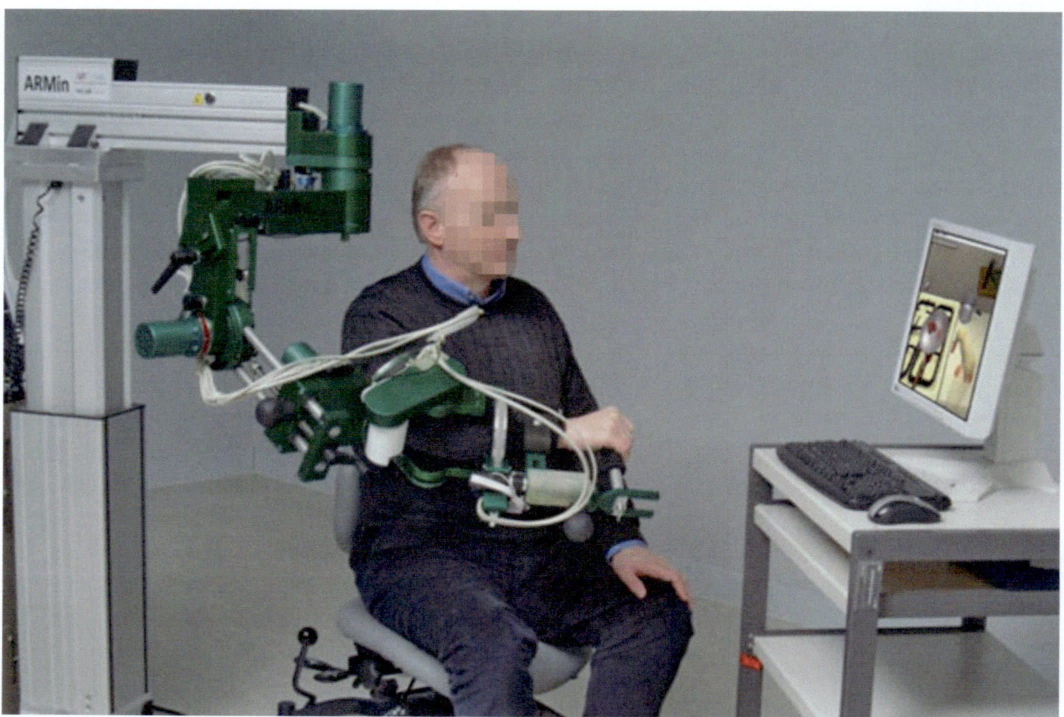

Fig. 27.4 ARMin III setup

27.2.2 Mechanical Setup of the ARMin III Robot

The ARMin III robot (Fig. 27.4) has an exoskeleton structure with six electric motors allowing it to move the human arm in all possible directions. Three motors actuate the shoulder joint for shoulder flexion/extension, horizontal abduction/adduction, and internal/external rotation. The elbow joint has two motors that actuate elbow flexion/extension and forearm pronation/supination. The last motor actuates wrist flexion/extension [27]. An optional module to support hand opening and closing can be attached to the ARMin III robot. All motors are equipped with two position sensors for redundant measurements. The motor and gears are carefully selected so that the friction is small and the back-drivability is good which is an important require-ment for sensorless force-control [52] and impedance-control strategies.

The patient's arm is affixed to the exoskeleton via two adjustable cuffs, one for the upper arm and one for the lower arm. To accommodate patients of varying body plans, the shoulder height can be adjusted via an electric lifting column, and the lengths of the upper and lower arms are adjustable. Laser pointers indicating the center of the glenohumeral joint help the thera-pist position the patient in the ARMin III device. The ARMin III robot can be configured to accommodate either the left or the right-arm. The transition between the two configurations does not require tools and takes less than 15 s.

A spring in the uppermost horizontal robotic link compensates for part of the weight of the exoskeleton. This lessens the load of the electric motor and has the desired effect of balancing the robotic arm when the power is off. Experience has shown that this is crucial for safety and for easy handling of the patient. The robotic shoul-der actuation compensates for scapula motion during the arm-elevation movement, resulting in a comfortable and ergonomic shoulder motion [27].

27.2.3 Therapy Modes

The motorized ARMin robots work in three training modes: mobilization, game training, and ADL training. We found it was beneficial to start a typical 1-h training session with a slow and gentle mobilization exercise. Chronic stroke patients in particular seemed to profit from the passive mobilization that reduced spasms and "loosened" the arm and hand. After 10–15 min of passive mobilization, active training followed, including games, reaching exercises, and ADL training scenarios [54, 55].

27.2.3.1 Passive and Active Mobilization

In the mobilization-training mode, the robot moves the patient's arm on a predefined trajectory. The robot is position-controlled, and the feedback loops help the motors compensate for any resistance that the patient produces. This means that, regardless of what the patient is doing, the robot will follow the predefined trajectory. If the patient moves together with the robot in the desired direction (active mobilization), the motors have less work than if the patient remains passive (passive mobilization). However, in both cases, the resulting movement will look the same. Since it is often desirable for the patient to actively contribute to the movement, the motor torque can be recorded and used as a performance measure to monitor how actively the patient contributes to the movement. In this case, the audiovisual display is used as a feedback modality to let the patient and therapist know how actively the patient is contributing to the movement [50]. Note that, from a technical point of view, this position-controlled training is based on industry-standard position control and is straightforward to implement.

The mobilization requires predefined trajectories that fit the patient's needs in terms of velocity and range of motion. The therapist can either input the data via a computer graphical user interface (GUI) or—more conveniently—use a teach-and-repeat procedure that enables the robot to directly learn a desired trajectory from the therapist. To do this, the therapist moves the robotic arm together with the human arm in the desired way, and the robot records and stores the position data that enable the robot to repeat the movement as shown by the therapist.

27.2.3.2 Game Therapy

Computer games are a good way to motivate the patient to participate actively in the training and contribute as much as possible to a particular movement task. For example, in the ball game, a virtual ball is presented on a computer monitor. It rolls down on an inclined table (Fig. 27.5). The patient can catch the ball with a virtual handle that replicates the movement of the human hand. Thus, the patient "catches" the virtual ball by moving his or her hand to the appropriate position. An assist-as-much-as-needed control paradigm has been implemented to support the patient in this task: If the patient can catch the ball on his or her own, the robot does not deliver any support. If the patient cannot catch the ball, the robot supports the patient with an adjustable force that pushes or pulls the hand to the ball position and helps the patient to initiate and execute the appropriate movement.

Whenever the robotic device supports the patient, the color of the handle changes from green to red, and an unpleasant sound is produced to alert both patient and therapist that the robot has supported the movement. The goal for the patient is to perform the task with as little support as possible. The therapist selects the supporting force, typically scaled so that the patient can successfully catch 80% of the balls. Several options enable the therapist to select the therapy mode that best fits the patient's needs. For instance, the incline angle of the virtual table can be modified, resulting in faster or slower rolling. The size of the handle and the ball can be changed, and the behavior of the ball (multiple reflections with the wall and the handle) can be changed to challenge the patient further. For advanced patients, disturbing forces and force fields can be introduced by the robot to make the task harder and to challenge the patient even more. Also, the number and kind of joints, as well as the range of motion of the involved joints, can be adjusted to the patient's needs.

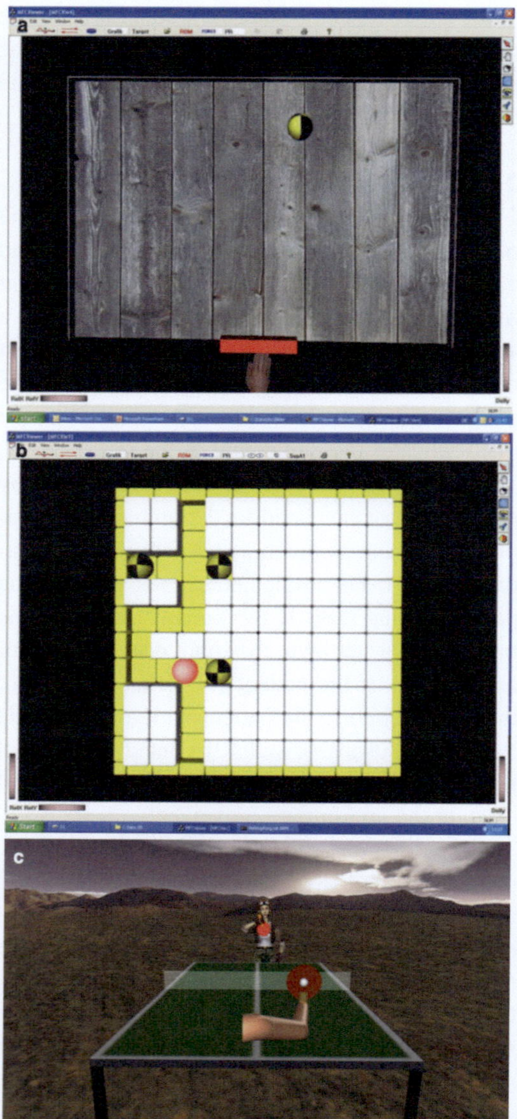

Fig. 27.5 Virtual reality scenarios for arm training. Ball game (**a**), labyrinth (**b**), and ping-pong game (**c**)

opponent. At the highest level of difficulty, the patient must control the position, orientation, and impulse of the virtual racket to hit the incoming ball so that it lands on the computer-opponent's side of the table. At easier levels, the robot takes care of the orientation and velocity of the racket, and the patient need only move the racket to a position where it will hit the incoming ball.

If required, the robot can also support the patient's arm and provide a force that pulls the hand to the desired spot. To increase the patient's motivation and engagement, a multiplayer application—where the patient plays virtual ping-pong against another patient instead of a virtual opponent—has been implemented and tested. This application allowed remote patients from different hospitals to meet virtually for a ping-pong game.

Another therapeutic computer game is the labyrinth game, where the patient navigates his or her hand through a virtual labyrinth. A red dot on the screen indicates the actual position of the human hand. The patient must move the red dot through the labyrinth. Virtual walls block the red dot and robot motors produce resistance that prevents the hand from passing through the walls. Force-feedback technology delivers a realistic impression of the virtual wall to the patient.

We found the labyrinth game particularly useful for patient therapy since the patient can use the walls for guidance. By following the walls, his or her movements remain free in three movement directions and are restricted only in the direction of the wall. This seemed to help patients move their hands in straight lines [55]. If required, the patient can be supported by the robot in completing the labyrinth task. In these instances, the labyrinth task is selected in the way that the patient must elevate his or her arm in the course of the exercise. This means that the starting point is at the bottom of the labyrinth and the goal is on top of the labyrinth. The therapist can choose from two supporting strategies. One compensates for the weight of the human arm, thus supporting the patient in lifting the arm. In the case of 100% weight support, the patient's

A prerequisite for this assist-as-needed control strategy is that the intended movement of the patient (i.e., where the patient wants to move his or her hand) is known. For the ball game, this is the position where the ball falls.

A similar supporting strategy has been implemented for a ping-pong game (Fig. 27.5). Here, the patient holds a virtual ping-pong racket and plays a ping-pong match against a virtual

arm floats somewhat, and it is very easy for the patient to lift his or her arm. In the second supporting scheme, the robot allows upward arm movements but resists downward movements. With this strategy, the patient must lift his or her arm by him- or herself, but whenever he or she gets tired, he or she can rest, and the arm will stay at the current position without any effort. Both strategies can also be combined [56]. To increase patient motivation, scoring is used based on the time, intensity, number, and time of collisions with the wall as well as the number of objects (positioned along the course of the labyrinth) that are collected by the patient.

27.2.3.3 Training of Activities of Daily Living

The purpose of ADL training is to support the patient in relearning ADL tasks, make the training a better simulation of real-life tasks, and further motivate the patient. An ADL task is presented on the computer screen, and the patient tries to complete the task. As with game therapy, the robot supports the patient as much as needed and only interferes if necessary. Current research focuses on the implementation and evaluation of appropriate ADL tasks for robotic therapy. To

date, implemented ADL tasks and used within ARMin therapy include:

- Setting a table
- Cooking potatoes
- Filling a cup
- Cleaning a table
- Washing hands
- Playing the piano
- Manipulating an automatic ticketing machine.

For the kitchen scenario (Fig. 27.6), a virtual arm is presented on the computer screen. The arm reflects the movement of the patient's arm, including shoulder, elbow, wrist, and hand opening and closing movements. A cooking stove, a kitchen table, and a shelf are fixed elements of the scenario. Cooking ingredients include several potatoes, black pepper, salt, and oregano. Available cooking tools include a pan and a dipper. Spoken instructions guide the patient through the cooking process. For instance, the patient must position the pan on the stove, turn on the heat, wait until the pan is hot, grasp the potatoes with his or her hand and put them into the pan, and wait until he or she hears the sound of roasting, add pepper and salt, and stir the pan.

Fig. 27.6 Kitchen scenario

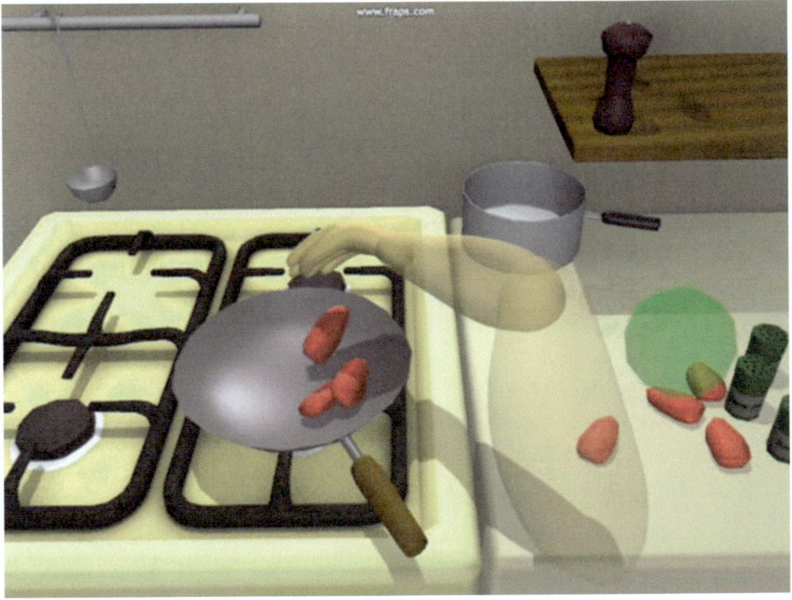

For this training scenario, the robot supports the patient only as much as needed, the patient has enough freedom to select his or her own movement trajectory, and the patient always sees feedback on how much he or she is currently supported by the robotic device. This is technically challenging because the cooking scenario involves several different movements [56, 57]. One possible solution that has been implemented with the ARMin system is to use virtual tunnels spanning from the start point to the goal point [58].

For instance, with the subtask of positioning potatoes in the pan, an invisible virtual tunnel starts at the initial location of the potatoes and ends above the pan. The robot lets the patient move freely within this tunnel. But once the patient hits the walls of the tunnels, the robot resists movement (similar to the labyrinth). Thus, the patient must follow the predefined path and not deviate from it. The diameter of the tunnel defines the amount of freedom the patient has. Furthermore, the patient is also free to select the timing and velocity of the movement. In addition, if required, the robot can also compensate for part of the arm weight and make the movement easier. Similar support strategies are implemented for the other ADL tasks [56].

27.2.4 Measurement Functionality of the ARMin Robot

The ability to objectively assess patient performance is one of the key benefits of robot-supported arm rehabilitation and allows the therapist to quantify therapy effects and patient progress. With the ARMin robot, the following parameters can be measured:

- Active range of motion
- Passive range of motion
- Muscle strength
- Abnormal joint synergies
- Spatial precision of hand positioning.

The active and passive range of motion (ROM) is measured for each joint individually.

When measuring, for example, the ROM of the elbow joint, all other joints are locked in a predefined position. The joint under investigation is controlled so that the patient can move it without resistance from the robot. The motor is only used to compensate for friction and gravity. The patient is instructed to extend the elbow as much as possible, and the robot measures the position of the elbow and stores the maximum values. When the passive range of motion is determined, the patient remains passive, and the joint is moved by the therapist while the robot records the maximum values of the joint position.

Muscle strength is measured with all joints locked in a predefined position. The motors are position-controlled with a fixed-reference position. Each joint is tested individually. For example, if the muscle strength of the abduction movement is tested, the patient is asked to abduct his or her arm as much as possible. Since the robot is position-controlled, and—in almost all cases—stronger than the human, the arm will not move. But the electric motor will need more current to work against the abduction torque. By measuring the motor current, the abduction torque can be determined using a model of the ARMin robot. The model describes the effects of gravity, friction, and the current-torque relationship in the electric motor.

Abnormal synergies result from abnormal muscle coactivation and loss of interjoint coordination. This means that, if a patient tries to abduct his or her arm, this goes together with an elbow flexion, forearm supination, and wrist and finger flexion [59]. To quantify abnormal synergies, all joints are locked in a predefined position. The patient abducts his or her arm as much as possible, and during the abduction torque, the joint torques produced by the patient in the shoulder, elbow, lower arm, and wrist are measured and recorded by the robotic device.

Moreover, a procedure to assess the resistance to the passive movement was developed. This measurement allows us to draw conclusions about the spasticity present in the affected arm. Here, the robot moves the human limb at different velocities and measures the required force. This technique has been implemented and evaluated for the lower limb within the Lokomat gait training robot [60].

The different ARMin assessments were evaluated in twenty-four healthy subjects and five patients with a spinal cord injury. The assessment was shown to be applicable and safe and that the measurements are widely reliable and comparable to clinical scales for arm motor function [61].

27.2.5 From ARMin for Adults to ChARMin for Children

As mentioned above an intensive, task-oriented rehabilitation training with active participation is crucial for the recovery of arm motor functions in adult stroke patients. These key features can be addressed using robotic support during arm training. That is why robots are increasingly used to complement rehabilitation training in stroke (e.g. ARMin III) and SCI patients (e.g. ARMin IV) or patients suffering from other neurological or motor impairments.

For children who suffer from cerebral palsy (CP) and other motor deficits, it is also known, that an intensive training [62] with active participation [63] is important to maintain and improve arm motor function. A small number of robots are available, that were tested with young patients (i.e., InMotion2 [64], NJIT-RAVR [65], REAPlan [66] and ArmeoSpring Pediatric [67]). The first results suggest that children profit from the intense training provided by the robot.

Based on the knowledge acquired with the adult arm robot ARMin and in close collaboration with the Rehabilitation Center for Children and Adolescents, Affoltern a. A., Switzerland, a new prototype—ChARMin—was developed for the use with children with neurological diagnoses including congenital or acquired brain lesions [68]. To the best of our knowledge, ChARMin is the first active robotic platform able to support single-joint and spatial movements and was built specifically for the needs of the pediatric target group.

Multiple aspects had to be changed in the new pediatric robot to achieve a design that covers the requirements of children. The robot needs to cover the target group of 5–18-years-old children and adolescents. The anthropometric ranges that need to be covered are too large to have it realized in a single system. Therefore, a modular design was chosen for ChARMin consisting of a proximal module that covers the entire range from 5 to 18-year-old children and a distal module that covers children aged 5–13 and 13–18 years. With this modular design and adjustable length settings for the shoulder height, the upper arm, the forearm and the hand length, the robot is applicable to all the children within the target group.

The kinematic shoulder structure of ARMin could not be transferred to the ChARMin concept as miniaturization would lead to robotic parts very close to the patient's head. The new mechanical structure uses a parallel remote center

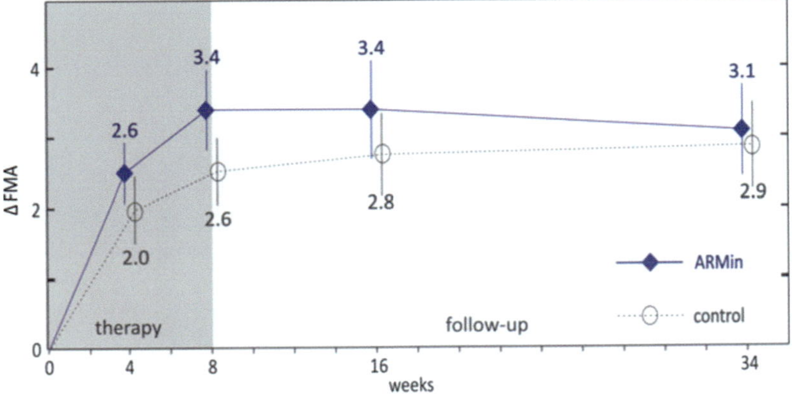

Fig. 27.7 Change in FMA over 8 weeks of therapy and during follow-up for ARMin and control groups; error bars are SE

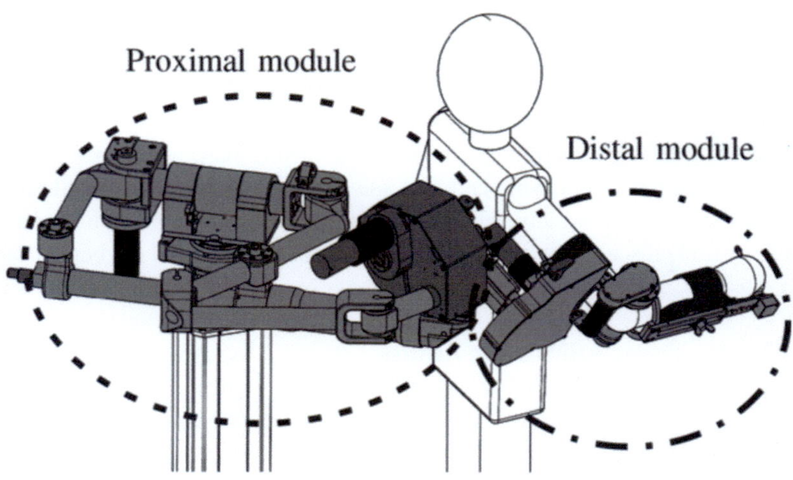

Fig. 27.8 Modular design of the ChARMin exoskeleton. The distal module is exchangeable according to the size of the child being trained. The robot is shown with a 13-year-old avatar. (Copyright IEEE, used with permission)

of rotation mechanism (Fig. 27.7, proximal module) to actuate the horizontal shoulder rotation and another parallel structure for the shoulder internal/external rotation. This combination of serial and parallel kinematics provides the safety distance needed between the robot and the child. The two robotic concepts for ARMin and ChARMin are shown in Fig. 27.8.

Similar to the adult ARMin version, the pediatric version has six DOF (three DOF for the shoulder and a single DOF for elbow, pro-/supination and wrist). Instead of an actuated hand module, ChARMin has an instrumented rubber bulb that detects the grip pressure, which can be used as an input for the software. The robot can be used for the right and left-arm sides and is mobile for transportation and positioning according to the patient. A passive gravity compensation mechanism and backdrivable joints allow for safe conditions even in the case of power loss.

An audiovisual interface with game-like scenarios is used to motivate the child to actively participate during the therapy session (Fig. 27.9).

While the passive mobilization and parts of the active game-supported arm therapy were transferred to the ChARMin robot, the ADL tasks were replaced with more child-friendly gaming scenarios. Different game scenarios were implemented that allow for a diversified training

Fig. 27.9 a ARMin IV robot for rehabilitation of stroke and SCI patients compared with **b** the ChARMin robot for pediatric arm rehabilitation (same scale). (Copyright IEEE, used with permission)

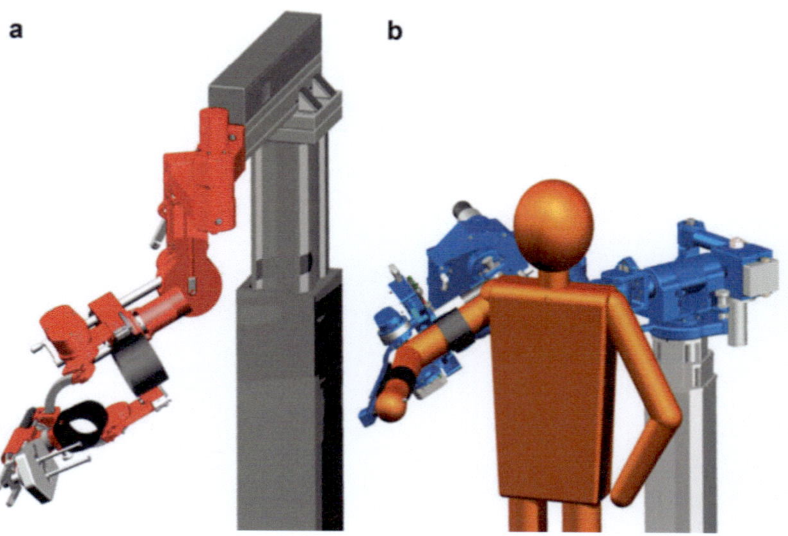

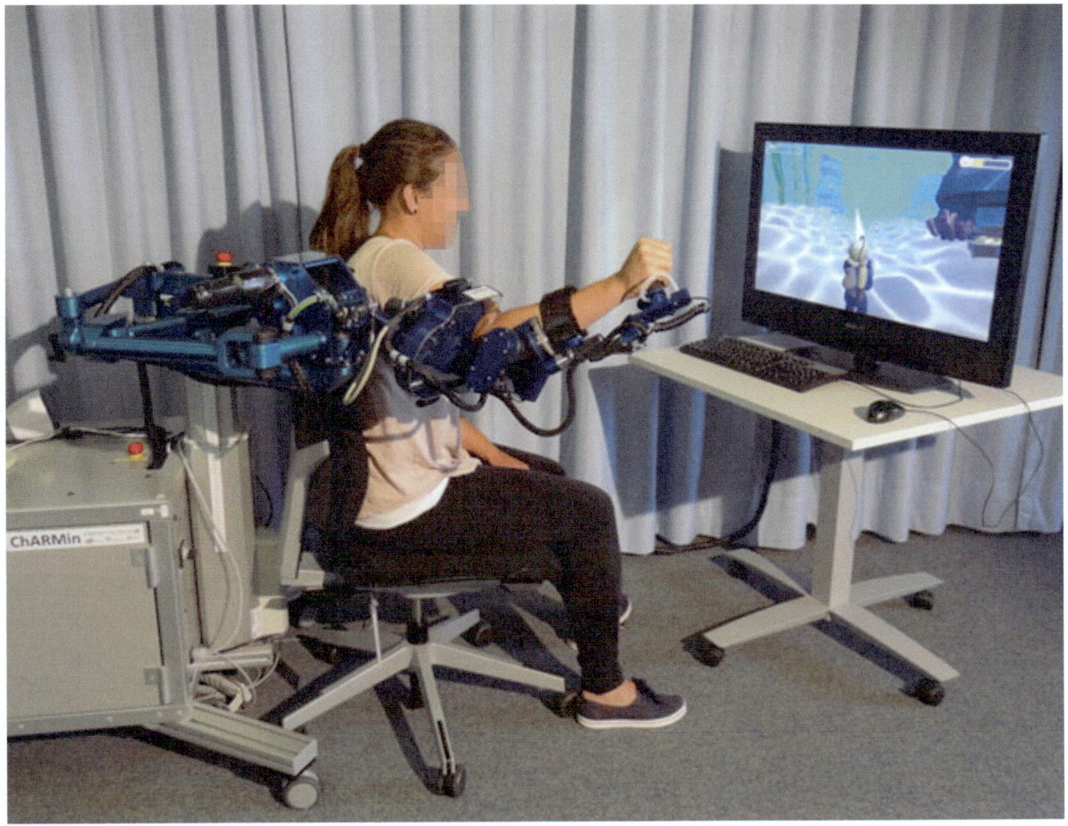

Fig. 27.10 Visualization of a possible setup of ChARMin and the visual interface shown with a healthy subject

(Fig. 27.10). While some games are played with single joints (joint-based) others allow to perform multi-joint movements (end-effector based) in a workspace that is previously defined by the therapist (Fig. 27.11).

Different support strategies are used to support the patient when needed. The support can be changed continuously from free non-supported movements to completely guided movements, where the patient can stay passive. Between these extreme conditions, the support can be changed to optimally support the patient such that he or she is challenged but not bored or over-challenged.

Moreover, the interface supports robot-assisted assessments. Five different assessment packages, which were previously evaluated in SCI patients with ARMin IV [61], can be used to assess the active and passive joint range of motion, the cubic workspace of the hand, the quality of point-to-point movements, the resistance to passive movements and the isometric joint torques for the six different joints.

A first ChARMin feasibility study is planned in the Rehabilitation Center for Children and Adolescents, Affoltern a. A., Switzerland, after receiving ethical approval. The study will investigate the applicability of the robot to children with cerebral palsy or other neurological diagnoses. Furthermore, the different support modes will be evaluated and the psychometric properties of the robot-assisted assessments determined.

Fig. 27.11 Various games are available for ChARMin that can be played on a joint- or end-effector level. **a** Airplane multi-joint, **b** diver multi-joint, **c** whack-a-mole single-joint and multi-joint, **d** tennis multi-joint, **e** ball single-joint, **f** spaceship multi-joint

27.2.6 Armeo Power®—Commercial Version of the ARMin Robot

The ARMin III robot also serves as a model for the prototype of the commercial version of the ARMin device, which is being developed and sold by Hocoma AG (Volketswil, Switzerland). The commercial version of the ARMin robot, named Armeo Power, was further optimized with respect to reliability, mechatronic robustness, user-friendliness, ergonomic function, and design, as well as optimized manufacturing processes and costs. The Armeo therapy concept presented by Hocoma consists of three Armeo

products that are all driven from the same software platform. Each product is optimized for a specific phase of the rehabilitation process. Shortly after the injury, a patient with no or very little voluntary activation of arm muscles trains with the motorized robotic device Armeo Power (former ARMin III). Once his or her motor function improves and some active movements are possible, the patient continues arm training with the nonmotorized, weight-supported exoskeleton Armeo Spring (former T-Wrex) [29]. After further improvements, the patient might continue training with the Armeo Boom, which consists of an overhead sling suspension system. This training seems suitable for patients who can actively move the arm but still exhibit reduced workspace and poor motor control [69].

Further distribution of the commercialized products would allow selling companies such as Hocoma AG to increase the body of clinical data of specific rehabilitation robots since a large number of rehabilitation facilities would use the same device for clinical practice and for research (Fig. 27.12).

27.2.7 Evaluation of the ARMin Technology

Three different versions of the ARMin device (I–III) were used to evaluate the ARMin technology. Evaluation of the ARMin technology was carried out with different versions of the ARMin.

27.2.7.1 Technical Tests with Healthy Subjects

Before the robotic device can be used with test subjects, it must be tested without a person in it. The appropriate test procedure verifies device safety and tests all situations defined as critical in the risk-management document. After testing, the technical specifications of the robot were validated by measurement. Table 27.1 shows the measured technical data for the ARMin III robot [27].

The next step was to evaluate the robot with healthy subjects. After appropriate approval by an independent ethics committee (internal review board), a thorough technical evaluation was

Fig. 27.12 The Armeo Product line, with the commercial ARMin device Armeo®Power (**a**), Armeo®Spring (**b**), and Armeo®Boom (**c**). (Copyright Hocoma AG, Switzerland, www.hocoma.com; used with version of the permission)

Table 27.1 Measured technical data for the ARMin III robot

Maximal endpoint load[a,b]	4.6 kg
Weight (excl. controller, hardware, frame)[b]	18.755 kg
Repeatability (endpoint)[b]	±0.5 mm
Stiffness (endpoint)[a,c]	0.364 mm/M
Force (endpoints)[a,b]	$F_{max} = (451\ N,\ 804\ N,\ 706\ N)^{T}$ with $G = (-g,0,0)^{T}$
Bandwidth for small endpoint movements (±1.5 cm)[d]	1.28 Hz

[a] Worst-case exoskeleton position
[b] Measured without subject (exoskeleton only)
[c] Stiffness measured at the endpoint by applying 20 N, while the motors are position-controlled
[d] Measured with healthy subject

performed on healthy subjects before the robot was used with patients. After providing written informed consent, the test subjects were exposed to the robotic device. The purposes of this evaluation included:

- Testing the handling of the robotic device. This includes positioning the test subject, adapting the robotic device for different body plans, changing from left-arm use to right-arm use, and comfort evaluation.
- Functional testing of the software. The questions were whether the test subject understood the instructions, whether he or she could successfully perform the exercises, and whether he or she liked the exercises. Special attention was also given to unwanted side effects, i.e., motion sickness and others.

Questionnaires validated the comfort and subjective feelings of the test subjects. One important side effect of this technical testing was that the therapist learned how to manipulate and use the robotic device before being exposed to patients.

27.2.7.2 Technical Tests with Stroke Patients

After the tests with healthy subjects concluded, technical tests with stroke patients were performed. After written informed consent was obtained, chronic stroke patients tested the device in one to five therapy sessions. The purpose of these tests was not to measure possible improvements in the patient's health status but to evaluate the technical ergonomic functionality of the ARMin robot. Specific goals included:

- Testing the handling of the ARMin device with stroke patients. Assessing the subjective feelings regarding comfort and ergonomics. Evaluating all training modes, including passive and active mobilization, game-supported therapy, and ADL training.
- Testing the level of difficulty of the tasks and the level of assistance that the robot provides to support the patients.
- Assessing patient motivation.

More than 20 stroke subjects participated in these preliminary tests [34].

27.2.7.3 Clinical Pilot Studies with Stroke Patients

A pilot study with three chronic stroke subjects (at least 14 months post-stroke) was performed with the ARMin I robot to investigate whether arm training with the ARMin I improves motor function of the paretic upper extremity [55]. The study had an A–B design with 2 weeks of multiple baseline measurements (A) and 8 weeks of training (B) with repetitive measurement and follow-up measurements 8 weeks after training. The training included shoulder and elbow movements induced by ARMin I. Two subjects had three 1-h sessions per week, and one subject received five 1-h sessions per week. The main outcome measurement was the upper limb motor portion of the Fugl-Meyer Assessment (FMA). It showed moderate, but significant improvements in all three subjects ($p < 0.05$): Starting with 14, 26 and 15 out of a maximum score of 66 points, the gains were 3.1, 3.0, and 4.2 points, respectively. Most improvements were maintained 8 weeks after discharge. However, patients stated that the daily use of their paretic arm in the real-world did not change. This finding was supported by constant ARAT and Barthel Index scores. This could be explained by the fact that, due to limitations of the ARMin I device, primarily non-ADL-related proximal joint movements were trained.

Therefore, another study was performed to investigate the effects of intensive arm training on motor performance using the ARMin II robot, where distal joints and ADL tasks were also incorporated into the training [54]. The study was conducted with four chronic stroke subjects (at least 12 months post-stroke). The subjects received robot-assisted therapy over a period of 8 weeks, 3–4 days per week, 1-h per day. Two patients had four 1-h training sessions per week, and the other two patients had three 1-h training sessions per week.

The primary outcome measurement was again the upper extremity portion of the FMA. The secondary outcome measures were the Wolf Motor Function Test (WMFT), maximum voluntary joint torques, and additional scores to assess transfer effects. Three out of four patients showed significant improvements ($p < 0.05$) in the primary outcome. Starting with 21, 24, 11, and 10 out of a maximum score of 66 points, the gains at the end of therapy were 17.6, 3.1, 6.8, and 2.1, and at six month follow-up 29, 5, 8, and 3 points, respectively. Improvements in FMA scores aligned with the torque measurements.

Most improvements were maintained, and some even further increased, between discharge and a 6-month follow-up. The data clearly indicate that intensive arm therapy with the robot ARMin II can significantly improve motor function of the paretic arm in some stroke patients. Even those who are in a chronic state achieve sustainable improvements. Care must be taken in analyzing the results of this pilot study. Participants were selected outpatients, there was no control group, and there were only four participants. Thus, one cannot generalize these results. However, the result justified the start of a subsequent controlled, randomized, multicenter clinical trial.

27.2.7.4 Clinical Trials with Stroke Patients

In order to investigate the effectiveness of arm treatment with ARMin, a clinical study with subjects in the chronic phase post-stroke was performed [70]. It was the first large-scale clinical study to offer neurorehabilitative therapy of the arm with an exoskeleton robot. A key aspect was to investigate the effects of ADL training based on reaching and grasping movements. ARMin III provides the required functions: audiovisual ADL tasks, large movement ranges in the three-dimensional space, actuation of proximal and distal joints including hand opening and closing, and a patient-responsive control.

Four hospitals participated in the trial. Seventy-seven patients in the chronic phase (i.e., more than six months) post-stroke with moderate to severe impairment of an arm (as tested with FMA: 8–38/max 66 points) were randomly assigned to either ARMin training or conventional, physical or occupational therapy. During therapy with ARMin, each of three therapy modes (mobilization, games, and ADL training)

had to be performed for at least ten minutes. Conventional therapy resembled the regular therapy given in outpatient clinics. Both groups were trained for eight weeks, three times per week, with one hour for each training session (total of 24 sessions). Outcome measures were obtained at five time points: prior to, during (after four weeks), directly after and two and six months after the training phase. The primary outcome measure was the FMA, a well-established clinical test that measures impairment of the arm. Further outcome measures were performed to evaluate task-oriented function (by means of the Wolf Motor Function Test and the Motor Activity Log). Furthermore, participation in life was assessed (with the Stroke Impact Scale). With ARMin, isometric strength in the arm (i.e., of shoulder abduction, adduction, anteversion, and retroversion, and of elbow flexion and extension) was measured.

Results confirmed the hypothesis: after eight weeks of training, ARMin therapy was not only as successful as conventional therapy but the improvements in motor function significantly exceeded those of conventional therapy (FMA, mean difference: 0.78 points, 95% CI 0.03–1.53) (Fig. 27.7). Especially the most severely affected profited from robotic therapy (mean difference 1.91 points, 95% CI 1.00–2·82). Of note, the robotic group gained significantly less strength than the conventional group. We speculate that the variables for the path assistance chosen during ARMin therapy might have been too supportive, tempting patients to diminish their own effort and therefore restricting strength training. A future focus for chronic patients would be to integrate specific strength training tasks in the robot. The other tests showed no significant difference between the two groups.

The higher motor functional gains in the ARMin group were still too small to be clinically meaningful for the single subject, but promising taking into consideration that the patients were in the chronic phase when a plateau of recovery is approached and gains in most cases are only limited.

Palermo and colleagues [71] tested the translational effects of robotic therapy in subacute stroke patients using the Armeo Power robot in addition to conventional rehabilitation therapy. In the study, ten subacute stroke survivors underwent a robotic training program of 20 sessions, each lasting 50 min, five sessions per week in addition to usual conventional rehabilitation therapies. Besides clinical scales, a sophisticated kinematic assessment of the upper limb, both pre-and post-treatment, was performed. The authors report remarkable differences in most parameters and significant correlations between the kinematic parameters and clinical scales. The data, although from a rather small sample, suggests that 3D robot-mediated rehabilitation, in addition to conventional therapy, could represent an effective method for the recovery of upper limb disability and that kinematic assessment may represent a valid tool for objectively evaluating treatment efficacy.

Calabro et al. [72] conducted a very interesting study with 35 patients with a first-ever ischemic supratentorial stroke at least two months before enrollment and unilateral hemiplegia. The study was designed to identify potential neurophysiologic markers to predict the responsiveness of stroke patients to upper limb robotic treatment. All patients underwent 40 Armeo Power training sessions that lasted one hour each (five times a week, for eight weeks). Spasticity and motor function of the upper limbs were assessed by means of the Modified Ashworth scale and the Fugl-Meyer assessment, respectively. The cortical excitability of the bilateral primary motor areas was assessed in response to the repetitive paired associative stimulation paradigm using transcranial magnetic stimulation. The results showed that patients with significant repetitive paired associative stimulation after-effects at baseline exhibited an evident increase in cortical plasticity in the affected hemisphere, and a decrease in interhemispheric inhibition. These findings were paralleled by clinical improvements (Fugl-Meyer assessment) and Armeo Power kinematic improvement, suggesting that the use of Armeo power may improve upper limb motor function recovery as predicted by baseline cortical excitability. The same team [73] conducted a pilot randomized controlled trial to investigate whether robotic rehabilitation combined with muscle vibration improves upper limb spasticity and function. Twenty patients suffering from unilateral post-stroke upper limb spasticity were included and they received 40 daily sessions of Armeo Power training (1-h/session, 5 sessions/week, for 8 weeks) with or without muscle vibration. The group with muscle vibration showed a greater reduction of spasticity measured with the modified Ashworth Scale and greater functional outcome measured with the Fugl-Meyer Assessment of the Upper Extremity. The authors write that this combined rehabilitative approach could be a promising option for improving upper limb spasticity and motor function.

The updated Cochrane Review on upper limb rehabilitation robotics [74] includes 45 trials involving 1619 participants and 24 different devices. The quality of the evidence was rated as high and the authors conclude that "robot-assisted arm training improved activities of daily living in people after stroke, and function and muscle strength of the affected arm. As adverse events, such as injuries and pain, were seldom described, these devices can be applied as a rehabilitation tool, but we still do not know when or how often they should be used". Studies with the following devices account for the largest numbers of patients for the review: 25% MIT Manus/InMotion2 [75], 14% Bi-manu-track [76], 7% Hand Master [77], 7% MIME [78], 6% T-Wrex/ArmeoSpring [49], 6% ARMin/Armeo Power, 5% ReoGo [79] and 4% Amadeus [80].

27.2.7.5 Clinical Trials with Spinal Cord Injured Patients

A pilot randomized controlled trial to evaluate the clinical efficacy of upper limb robotic therapy in people with tetraplegia with the Armeo power was conducted in an inpatient hospital in Seoul, Korea [61]. Participants were randomly allocated to a robotic therapy or an occupational therapy group, both groups receiving usual care plus 30 min additional therapy per day for four weeks. Primary outcomes were the Medical Research Council scale of each key muscle and Upper Extremity Motor Score (UEMS) for the trained arm. A total of 34 individuals with tetraplegia were included (17 in each group). At four weeks, the median change in the UEMS in the robotic group was 1/25 (0–3) points compared with 0/24 (−1 to 1) points in the occupational therapy group. The differences were not statistically significant, and the authors conclude that further studies are required for a better understanding of the effects of robotic therapy on people with tetraplegia.

In a concept study [81], 24 healthy subjects and five patients after spinal cord injury underwent robot-based assessments using the ARMin robot. Five different tasks were performed with aid of a visual display. Ten kinematic, kinetic, and timing assessment parameters were extracted on both joint- and end-effector levels including active and passive range of motion, cubic reaching volume, movement time, distance-path ratio, precision, smoothness, reaction time, joint toques, and joint stiffness. A subsequent comparison with clinical scores revealed good correlations between robot-based joint torques and the Manual Muscle Test. Reaction time and distance-path ratio showed a good correlation with the "Graded and redefined assessment of Strength, Sensibility and Prehension" (GRASP) and the Van Lieshout Test (VLT) for movements towards a predefined target in the center of the frontal plane. The authors conclude that these preliminary results suggest that the measurements are widely reliable and comparable to clinical scales for arm motor function.

27.2.7.6 Perspectives for Future Clinical Testing

We believe that objective device-based measurements are a relevant part of standardized clinical outcomes and should be integrated into clinical evaluation studies. Future studies on patients should be performed in the first days to weeks after stroke, when the potential for real recovery, rather than compensation, is highest. Here, an exoskeleton robot should be the ideal tool as it enables to train purposeful movements with control of the whole arm from the shoulder to the hand. It is, thus, capable of guiding the arm in an almost physiological manner during task training. Different learning strategies that have been proven to be successful can be implemented in the software. Through the measurement functionality of ARMin, the VR tasks can be adapted continuously to the subjects' abilities to achieve a patient-tailored, intensified therapy.

27.3 Current Developments and Ongoing Testing

27.3.1 Technical Developments for Improving the Human-Robot Interaction

A common problem in actuated arm exoskeletons, namely in the ARMin robot and the Armeo Power robot, is that the serial kinematic structure results in a system that suffers from high inertia and friction altering the effective haptic rendering properties of the virtual training environment [82]. This can result in increased patient fatigue, limiting the potential use of active training paradigms. Indeed, a transparent robot is needed to support patients to perform motor tasks. One possibility to solve this problem is presented by Özen and colleagues [83]. They propose to equip the exoskeleton with force sensors measuring the interaction force between the robot and the human arm (Fig. 27.13). The authors demonstrate that high control loop rates and advanced motion control techniques in combination with disturbance observers allowed to achieve high transparency even for fast movements. Because of the force sensors, this could be achieved without the need for precise modeling of the robot.

In a perfectly transparent robot, the patient would have to carry the weight of his own arm. This is not possible for extended training duration and arm weight compensation needs to be implemented. In a recent study, Just et al. [84] introduce new methods for human arm weight compensation. Arm weight compensation is an important requirement for stroke rehabilitation because it allows to increase the active range of motion and to reduce the effects of pathological muscle synergies. As the authors emphasize, it is, however, hard to effectively assess and compare

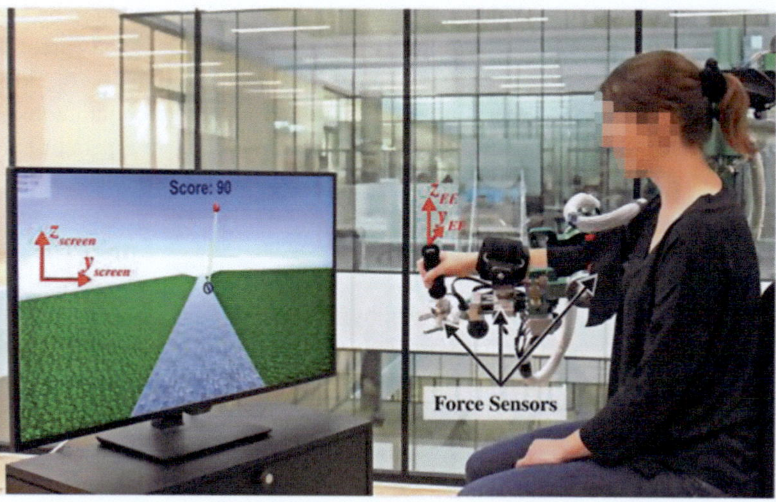

Fig. 27.13 Improved haptic rendering performance thanks to additional force sensors measuring the interaction force between the user's am and the robotic device. The picture shows an inverted pendulum task using the ARMin exoskeleton as a visuo-haptic interface (Copyright IEEE, with permission).

human arm weight relief because of the differences in structure, performance, and control algorithms among the existing robotic platforms. They introduce criteria for ideal am weight compensation, and they propose and analyze three distinct arm weight compensation methods. They could show that all methods reduce EMG activity by at least 49%.

Both the optimized transparency methods and the improved weight compensation are important elements to further improve the human-robot interaction and the therapeutic modes of motorized arm exoskeletons. The disadvantage of these approaches is that they require rather expensive six degree of freedom force sensors and it is unclear if the industry is willing to upgrade the motorized exoskeletons with this feature.

In a validation study with 40 healthy participants [85] it was shown that the improved haptic rendering significantly increased participant's movement variability during the training and the ability to synchronize with the task. Furthermore, the weight support enhanced the participant's movement variability during the training and reduces the participant's physical effort. The haptic rendering enhanced motor learning and skill transfer. On the other side, the authors found, contrary to the expectations, that the weight support hampers motor learning compared to training without weight support.

A possible explanation could be that the weight support disrupts motor learning because participants rely on the assistance during the training and fail to learn the motor commands required to perform the desired task. It becomes evident that further work is needed to better understand the interactions between robotic assistance, haptic rendering and the effects on motor learning, especially in brain-injured patients.

27.3.2 When Music Meets Robotics—An Innovative Approach to Increase Training Motivation

In a study with sixteen healthy subjects, Baur et al. [86] developed an audio-haptic task where participants could generate sounds by moving their arms. As music is known to improve motivation in neurorehabilitation training, the authors aimed at integrating music creation into robotic assisted motor therapy. The task was designed so that it could be performed either with or without a graphical display as an audio-haptic environment only. The game environment was developed to target horizontal movements at table height as this type of arm motion is required for activities of daily living, such as cleaning a table or moving objects on a table. The arm of the participants was supported by means of a virtual table-top allowing participants to move the arm within a horizontal plane. The horizontal left-right movement served as game input allowing participants to move into different sound zones consisting of fourteen different pairs of sound samples and two pairs of sound effects. The sound samples consisted of synthetic piano, mallets, marimba, vibraphone, pads, hi-hats, and claps. After the training, participants were invited to collect and keep the produced sound files for future listening. While the study did not show statistically significant motivational differences between the tested conditions, the authors conclude that the combination of music and activities promoting creativity in motor training promotes enjoyment, and thus, intrinsic motivation of subjects performing robot-assisted training. They found that the audio-haptic environment is sufficient to create a meaningful gameplay and that music tasks can be performed without a visual display. The study demonstrated the feasibility of playing an audio-haptic music game and the authors suggest a follow-up study on stroke survivors.

27.3.3 Multiplayer Games—How to Increase Training Motivation

Multiplayer environments increase training intensity in robot-assisted therapy after stroke [87]. Compared to single-player modes they improve the game experience and game performance. Baur et al. [88] tested two multiplayer games with the ARMin robot. The Air Hockey game is a competitive game in which the different abilities of the players were compensated by individual haptic guidance or damping forces. Forty patients in the subacute phase post-stroke played the game in single-player and multiplayer modes. Sixteen of them preferred the multiplayer mode. The competitive form was more motivating and increased exercise intensity more than the cooperative mode. In the Haptic Kitchen game, a healthy person (e.g. therapist or spouse) could assist the patient's movements in the ARMin with

a virtual force field applied by moving an HTC Vive hand controller©. Both the force field and the location of the application were visualized. In a single caste study with two patients post-stroke and their spouses, patients showed a tendency to be more motivated in the multiplayer mode as compared to the single-player mode.

27.3.4 A Novel Neuro-Animation Experience to Facilitate High-Dosage and High-Intensity Training

Krakauer and colleagues [89] implemented a custom-designed immersive animation-based audiovisual scenario named "I am Dolphin" (KATA John Hopkins University). In this setting, the patient's paretic arm was unweighted using the Armeo Power exoskeleton device. This allowed the practice of multi-joint 3D arm movements despite weakness without requiring a therapist to activity lift the paretic arm. The 3D movements of the paretic arm controlled the movement of a virtual dolphin, swimming through different ocean scenes with various task goals including chasing and eating fish, eluding attacks, and performing jumps. The tasks were designed to promote movement in all planes throughout the active ranges of motion, and titrated based on successful completion of progressive levels of difficulties. A total of 24 patients (within 6 weeks post-stroke) were randomized to the experimental group (Dolphin scenario) and to conventional occupational therapy and underwent 30 sessions of 60 min in addition to standard care. Both groups were also matched to a historical cohort, which received only 30 min of upper limb therapy per day. There were no significant between-group differences in Fugl-Meyer Upper Extremity motor score (primary outcome), Action Research Arm Test (ARAT) or other secondary outcomes at any timepoint. Both high-dose groups showed greater recovery on the ARAT, but not on the Fugl-Meyer score when compared with the historical cohort. The authors conclude that neuro-animation may offer an enjoyable and scalable way to deliver high-dose and intensive upper limb therapy.

27.4 Conclusions

Current studies, including the updated Cochrane Review on upper limb rehabilitation robotics [74], indicate that stroke and spinal cord injury patients benefit from robot-assisted upper limb training. The functional gains of robotic training are in the same range as those of manual therapy. This contrasts with the training of the lower extremities, where robotic therapy is more efficient than conventional training (Cochrane). Several points could contribute to this finding: (a) Post-stroke recovery of ambulatory function is generally better than the recovery of arm/hand function [73]. (b) Conventional upper limb training might better reflect daily activities with all its facets (bimanual, manipulations of real objects) than robotic training with VR simulated activities. In other words, robot-supported gait training more closely matches real-life activity than robot-supported arm/hand therapy. (c) The technological level of clinically available upper limb rehabilitation robots does not fully explore the technological potential (e.g. training strategies; hand actuation; (bimanual) object manipulation) of upper limb robotics. (d) Clinical findings pertaining to the upper extremity include a larger number of different devices which complicates a coherent analysis of findings across these studies.

Based on current findings, we can neither advocate nor condemn the clinical use of upper limb rehabilitation robots. There are, however, some indications that severely affected patients might indeed benefit most from robot-supported upper limb training [70]. An interesting possibility is to establish so called "robot studios" where a therapist supervises several patients working with different devices as an add-on to existing conventional therapy [90].

Current and future research to better understand the mechanisms of action, as well as which patients benefit most from robotic therapy, is extremely important to clarify future clinical use. Clinical research must go hand in hand with research into technological aspects. This includes the question about the patient-specific optimal audiovisual input, optimal training and support strategies, and improved control strategies.

Acknowledgements We thank all people who contributed to the development and clinical application of ARMin, including Prof. Dr. med. V. Dietz, M. Guidali, A. Brunschweiler, A. Rotta, and A. Kollmar. Furthermore, we want to thank all participating patients and our clinical partners contributing to the multicenter study. The research was and is still funded in part by NCCR Neuro, Swiss National Science Foundation, Hans-Eggenberger Foundation, Bangerter-Rhyner Foundation, and ETH Foundation.

References

1. Brainin M, Bornstein N, Boysen G, Demarin V. Acute neurological stroke care in Europe: results of the European stroke care inventory. Eur J Neurol. 2000;7:5–10.
2. Thorvaldsen P, Asplund K, Kuulasmaa K, Rajakangas AM, Schroll M. Stroke incidence, case fatality, and mortality in the WHO MONICA project. World health organization monitoring trends and determinants in cardiovascular disease. Stroke. 1995;26 (3):361–7.
3. Rosamond W, Flegal K, Friday G, et al. Heart disease and stroke statistics update. A report from the American Heart Association Statistics Committee and Stroke Statistics Subcommittee. Circulation. 2007;115:69–171.
4. Maeurer HC, Diener HC. Der Schlaganfall. Stuttgart: Georg Thieme Verlag; 1996.
5. Rossini PM, Calautti C, Pauri F, Baron JC. Post-stroke plastic reorganisation in the adult brain. Lancet Neurol. 2003;2:493–502.
6. Nakayama H, Jrgensen HS, Raaschou HO, Olsen TS. Recovery of upper extremity function in stroke patients: the Copenhagen stroke study. Arch Phys Med Rehabil. 1994;75:394–8.
7. Barreca S, Wolf SL, Fasoli S, Bohannon R. Treatment interventions for the paretic upper limb of stroke survivors: a critical review. Neurorehabil Neural Repair. 2003;17(4):220–6.
8. Platz T. Evidence-based arm rehabilitation—a systematic review of the literature. Nervenarzt. 2003;74 (10):841–9.
9. Dobkin BH. Strategies for stroke rehabilitation. Lancet Neurol. 2004;3(9):528–36.
10. Ottenbacher KJ, Jannell S. The results of clinical trials in stroke rehabilitation research. Arch Neurol. 1993;50:37–44.
11. Kwakkel G, Wagenaar RC, Koelman TW, Lankhorst GJ, Koetsier JC. Effects of intensity of rehabilitation after stroke. A research synthesis. Stroke. 1997;28(8):1550–6.

12. Nelles G. Cortical reorganization-effects of intensive therapy. Arch Phys Med Rehabil. 2004;22:239–44.
13. Sunderland A, Tinson DJ, Bradley EL, Fletcher D, Langton Hewer R, Wade DT. Enhanced physical therapy improves recovery of arm function after stroke. A randomised controlled trial. J Neurol Neurosurg Psychiatry. 1992;55(7):530–5.
14. Kwakkel G, Kollen BJ, Wagenaar RC. Long term effects of intensity of upper and lower limb training after stroke: a randomised trial. J Neurol Neurosurg Psychiatry. 2002;72:473–9.
15. Butefisch C, Hummelsheim H, Denzler P, Mauritz KH. Repetitive training of isolated movements improves the outcome of motor rehabilitation of the centrally paretic hand. J Neurol Sci. 1995;130:59–68.
16. Prange GB, Jannink MJA, Groothuis-Oudshoorn CGM, Hermens HJ, MJ Ijzerman. Systematic review of the effect of robot-aided therapy on recovery of the hemiparetic arm after stroke. J Rehabil Res Dev. 2006;43:171–84.
17. Riener R, Nef T, Colombo G. Robot-aided neurorehabilitation for the upper extremities. Med Biol Eng Comput. 2005;43:2–10.
18. Kwakkel G, Kollen BJ, Krebs HI. Effects of robotassisted therapy on upper limb recovery after stroke: a systematic review. Neurorehabil Neural Repair. 2008;22(2):111–21.
19. Singh H, Unger J, Zariffa J, Pakosh M, Jaglal S, Craven BC, Musselman KE. Robot-assisted upper extremity rehabilitation for cervical spinal cord injuries: a systematic scoping review. Disabil Rehabil Assist Technol. 2018;13(7):704–15. https://doi.org/10.1080/17483107.2018.1425747 Epub 2018 Jan 15 PMID: 29334467.
20. Monardo G, Pavese C, Giorgi I, Godi M, Colombo R. Evaluation of patient motivation and satisfaction during technology-assisted rehabilitation: an experiential review. Games Health J. 2021;10(1):13–27. https://doi.org/10.1089/g4h.2020.0024 Epub 2020 Jul 2 PMID: 32614618.
21. Palermo E, Hayes DR, Russo EF, Calabrò RS, Pacilli A, Filoni S. Translational effects of robot-mediated therapy in subacute stroke patients: an experimental evaluation of upper limb motor recovery. PeerJ. 2018;4(6): e5544. https://doi.org/10.7717/peerj.5544.PMID:30202655;PMCID:PMC6128258.
22. Krebs HI, Ferraro M, Buerger SP, et al. Rehabilitation robotics: pilot trial of a spatial extension for MIT-Manus. J Neuroeng Rehabil. 2004;1:5–9.
23. Lum PS, Burgar CG, Shor PC, Majmundar M, Van der Loos M. Robot-assisted movement training compared with conventional therapy techniques for the rehabilitation of upper-limb motor function after stroke. Arch Phys Med Rehabil. 2002;83(7):952–9.
24. Hesse S, Werner C, Pohl M, Mehrholz J, Puzich U, Krebs HI. Mechanical arm trainer for the treatment of the severely affected arm after a stroke: a single-blinded randomized trial in two centers. Am J Phys Med Rehabil. 2008;87(10):779–88.
25. Coote S, Murphy B, Harwin W, Stokes E. The effect of the GENTLE/s robot-mediated therapy system on arm function after stroke. Clin Rehabil. 2008;22(5):395–405.
26. Dewald J, Ellis MD, Holubar BG, Sukal T, Acosta AM. The robot application in the rehabilitation of stroke patients. Neurol Rehabil. 2004;4:S7.
27. Nef T, Guidali M, Riener R. ARMin III—arm therapy exoskeleton with an ergonomic shoulder actuation. Appl Bionics Biomech. 2009;6(2):127–42.
28. Stienen AHA, Hekman EEG, Van der Helm FCT, et al. Dampace: dynamic force-coordination trainer for the upper extremities. Proc IEEE. 2007;10:13–5.
29. Sanchez RJ, Liu J, Rao S, et al. Automating arm movement training following severe stroke: functional exercise with quantitative feedback in a gravityreduced environment. IEEE Trans Neural Syst Rehabil Eng. 2006;14:378–89.
30. Roderick S, Liszka M, Carignan C.Design of an arm exoskeleton with scapula motion for shoulder rehabilitation ICAR '05. In: Proceedings of the 12th international conference on advanced robotics; 2005. p. 524–31. https://doi.org/10.1109/ICAR.2005.1507459.
31. Frisoli A, Borelli L, Montagner A, et al. Arm rehabilitation with a robotic exoskeleleton in virtual reality. In: IEEE 10th international conference on rehabilitation robotics, vol. 1. and 2. Noordwijk; 2007. p. 631–42.
32. Rosen J, Perry JC, Manning N, Burns S, Hannaford B.The human arm kinematics and dynamics during daily activities - toward a 7 DOF upper limb powered exoskeleton ICAR '05. In: Proceedings of the 12th international conference on advanced robotics; 2005. p. 532–9. https://doi.org/10.1109/ICAR.2005.1507460.
33. Zhang LQ, Park FS, Ren YP. Developing an intelligent robotic arm for stroke rehabilitation. In: 2007 IEEE 10th international conference on rehabilitation robotics, vol. 1 and 2. Noordwijk; 2007, p. 984–93.
34. Nef T, Mihelj M, Riener R. ARMin: a robot for patient-cooperative arm therapy. Med Biol Eng Comput. 2007;45:887–900.
35. Bayona NA, Bitensky J, Salter K, Teasell R. The role of task-specific training in rehabilitation therapies. Top Stroke Rehabil. 2005;12:58–65.
36. Wolf SL, Lecraw DE, Barton LA, Jann BB. Forced use of hemiplegic upper extremity to reverse the effect of learned nonuse among chronic stroke and head-injured patients. Exp Neurol. 1989;104:125–32.
37. Taub E, Uswatte G, Pidikiti R. Constraint-induced movement therapy: a new family of techniques with broad application to physical rehabilitation—a clinical review. J Rehabil Res Dev. 1999;36:237–51.
38. Miltner WHR, Bauder H, Sommer M, Dettmers C, Taub E. Effects of constraint-induced movement therapy on patients with chronic motor deficits after stroke. A replication. Stroke. 1999;30:586–92.

39. Dromerick AW, Edwards DF, Hahn M. Does the application of constraint-induced movement therapy during acute rehabilitation reduce arm impairment after ischemic stroke? Stroke. 2000;31:2984–8.

40. Lambercy O, Dovat L, Gassert R, Burdet E, Teo CL, Milner T. A haptic knob for rehabilitation of hand function. IEEE Trans Neural Syst Rehabil Eng. 2007;15(3):356–66.

41. Hesse S, Schulte-Tigges G, Konrad M, Bardeleben A, Werner C. Robot-assisted arm trainer for the passive and active practice of bilateral forearm and wrist movements in hemiparetic subjects. Arch Phys Med Rehabil. 2003;84:915–20.

42. Krebs HI, Volpe BT, Williams D, et al. Robot-aided neurorehabilitation: a robot for wrist rehabilitation. IEEE Trans Neural Syst Rehabil Eng. 2007;15(3):327–35.

43. Hesse S, Werner C, Pohl M, Rueckriem S, Mehrholz J, Lingnau ML. Computerized arm training improves the motor control of the severely affected arm after stroke: a single-blinded randomized trial in two centers. Stroke. 2005;36(9):1960–6.

44. Muellbacher W, Richards C, Ziemann U, et al. Improving hand function in chronic stroke. Arch Neurol. 2002;59(8):1278–82. 42. Krebs HI, Hogan N, Aisen ML, Volpe BT. Robot-aided neurorehabilitation. IEEE Trans Rehabil Eng. 1998;6:75–87.

45. Ellis MD, Sukal-Moulton TM, Dewald JP. Impairment-based 3-D robotic intervention improves upper extremity work area in chronic stroke: targeting abnormal joint torque coupling with progressive shoulder abduction loading. IEEE Trans Robot. 2009;25(3):549–55.

46. Krebs HI, Mernoff S, Fasoli SE, Hughes R, Stein J, Hogan N. A comparison of functional and impairment-based robotic training in severe to moderate chronic stroke: a pilot study. NeuroRehabilitation. 2008;23(1):81–7. PMID: 18356591; PMCID: PMC4692808.

47. Bolliger M, Banz R, Dietz V, Lunenburger L. Standardized voluntary force measurement in a lower extremity rehabilitation robot. J Neuroeng Rehabil. 2008;5:23.

48. Lunenburger L, Colombo G, Riener R. Biofeedback for robotic gait rehabilitation. J Neuroeng Rehabil. 2007;4:1.

49. Housman SJ, Scott KM, Reinkensmeyer DJ. A randomized controlled trial of gravity-supported, computer-enhanced arm exercise for individuals with severe hemiparesis. Neurorehabil Neural Repair. 2009;23(5):505–14.

50. Nef T, Mihelj M, Colombo G, Riener R. ARMin robot for rehabilitation of the upper extremities. In: IEEE international conference on robotics and automation, Orlando; 2006. p. 3152–7.

51. Mihelj M, Nef T, Riener R. ARMin II—7 DoF rehabilitation robot: mechanics and kinematics. In: Proceedings of the 2007 IEEE international conference on robotics and automation, vol. 1–10, Rome; 2007. p. 4120–5.

52. Nef T, Lum P. Improving backdrivability in geared rehabilitation robots. Med Biol Eng Comput. 2009;47(4):441–7.

53. Staubli P, Nef T, Klamroth-Marganska V, Riener R. Effects of intensive arm training with the rehabilitation robot ARMin II in chronic stroke patients: four single-cases. J Neuroeng Rehabil. 2009;6:46.

54. Nef T, Quinter G, Muller R, Riener R. Effects of arm training with the robotic device ARMin I in chronic stroke: three single cases. Neurodegener Dis. 2009;6(5–6):240–51.

55. Nef T, Mihelj M, Kiefer G, Perndl C, Mueller R, Riener R. ARMin—exoskeleton for arm therapy in stroke patients. In: 2007 IEEE 10th international conference on rehabilitation robotics, vol. 1 and 2, Noordwijk; 2007. p. 68–74.

56. Guidali M, Duschau-Wicke A, Broggi S, Klamroth-Marganska V, Nef T, Riener R. Med Biol Eng Comput. 2011;49(10):1213–23 Epub 2011 Jul28 PMID:21796422.

57. Mihelj M, Nef T, Riener R. A novel paradigm for patient-cooperative control of upper-limb rehabilitation robots. Adv Robot. 2007;21(8):843–67.

58. Duschau-Wicke A, von Zitzewitz J, Caprez A, Lunenburger L, Riener R. Path control: a method for patient-cooperative robot-aided gait rehabilitation. IEEE Trans Neural Syst Rehabil Eng. 2010;18(1):38–48.

59. Dewald JP, Beer RF. Abnormal joint torque patterns in the paretic upper limb of subjects with hemiparesis. Muscle Nerve. 2001;24:273–83.

60. Schmartz AC, Meyer-Heim AD, Muller R, Bolliger M. Measurement of muscle stiffness using robotic assisted gait orthosis in children with cerebral palsy: a proof of concept. Disabil Rehabil Assist Technol. 2011;6(1):29–37.

61. Keller U, Schölch S, Albisser U, Rudhe C, Curt A, Riener R, Klamroth-Marganska V. Robot-assisted arm assessments in spinal cord injured patients: a consideration of concept study. PLoS ONE. 2015;10(5): e0126948. https://doi.org/10.1371/journal.pone.0126948.PMID:25996374;PMCID:PMC4440615.

62. Sakzewski L, Gordon A, Eliasson A-C. The state of the evidence for intensive upper limb therapy approaches for children with unilateral cerebral palsy. J Child Neurol. 2014;29(8):1077–90.

63. Damiano D. Activity, activity, activity: rethinking our physical therapy approach to cerebral palsy. Phys Therap. 2006;86(11):1534–40.

64. Fasoli S, Fragala-Pinkham M, Hughes R, Hogan N, Krebs H, Stein J. Upper limb robotic therapy for children with hemiplegia. Am J Phys Med Rehabil. 2008;87(11):929.

65. Fluet G, Qiu Q, Kelly DParikh H, Ramirez D, Saleh S, Adamovich S. Interfacing a haptic robotic system with complex virtual environments to treat impaired upper extremity motor function in children

with cerebral palsy. Develop Neurorehabil 2010; 13 (5):335–45.

66. Gilliaux M, Renders A, Dispa D, Holvoet D, Sapin J, Dehez B, Detrembleur C, Lejeune TM, Stoquart G. Upper limb robot-assisted therapy in cerebral palsy a single-blind randomized controlled trial. Neurorehabil Neural Repair. 2014;29(2):183–92.

67. http://www.hocoma.com/products/armeo/armeospring-pediatric/.

68. Keller H, Riener R. Design of the pediatric arm rehabilitation robot ChARMin. In: IEEE international conference on biomedical robotics and biomechatronics (BioRob). IEEE; 2014. p. 530–535

69. Gassert R, Dietz V. Rehabilitation robots for the treatment of sensorimotor deficits: a neurophysiological perspective. J Neuroeng Rehabil. 2018;15(1):46. https://doi.org/10.1186/s12984-018-0383-x.PMID:29866106;PMCID:PMC5987585.

70. Klamroth-Marganska V, Blanco J, Campen K, et al. Three-dimensional, task-specific robot therapy of the arm after stroke: a multicentre, parallel-group randomised trial. Lancet Neurol. 2014;13(2):159–66.

71. Mehrholz J, Thomas S, Kugler J, Pohl M, Elsner B. Electromechanical-assisted training for walking after stroke. Cochrane Database Syst Rev. 2020;10(10): CD006185. https://doi.org/10.1002/14651858.CD006185.pub5. PMID: 33091160; PMCID: PMC8189995.

72. Calabrò RS, Naro A, Russo M, Milardi D, Leo A, Filoni S, Trinchera A, Bramanti P. Is two better than one? Muscle vibration plus robotic rehabilitation to improve upper limb spasticity and function: a pilot randomized controlled trial. PLoS ONE. 2017;12 (10): e0185936. https://doi.org/10.1371/journal.pone.0185936.PMID:28973024;PMCID:PMC5626518.

73. Paci M, Nannetti L, Casavola D, Lombardi B. Differences in motor recovery between upper and lower limbs: does stroke subtype make the difference? Int J Rehabil Res. 2016;39:185.

74. Lo AC, et al. Robot-assisted therapy for long-term upper-limb impairment after stroke. New Engl J Med. 2010;362:1772–83.

75. Hesse S, Gotthard S-T, Konrad M, Bardeleben A, Werner C. Robot-assisted arm trainer for the passive and active practice of bilateral forearm and wrist movements in hemiparetic. Arch Phys Med Rehab. 2003;84:915–20.

76. Wolf SL, et al. The HAAPI (home Arm assistance progression initiative) trial. Neurorehabil Neural Repair. 2015;29:958–68.

77. Burgar C, Lum P, Shor P, der Loos MH. Development of robots for rehabilitation therapy: the Palo Alto VA/Stanford experience. J Rehabil Res Dev. 2000;37:663–73.

78. Takahashi K, et al. Efficacy of upper extremity robotic therapy in subacute poststroke hemiplegia. Stroke. 2018;47:1385–8.

79. Hwang C, Seong J, Son D-S. Individual finger synchronized robot-assisted hand rehabilitation in subacute to chronic stroke: a prospective randomized

80. Kim J, Lee BS, Lee HJ, Kim HR, Cho DY, Lim JE, Kim JJ, Kim HY, Han ZA. Clinical efficacy of upper limb robotic therapy in people with tetraplegia: a pilot randomized controlled trial. Spinal Cord. 2019;57(1):49–57. https://doi.org/10.1038/s41393-018-0190-z. Epub 2018 Sep 11. Erratum in: Spinal Cord. 2019 Feb 4;: PMID: 30206423.

81. Metzger J, Lambercy O, Gassert R. Performance comparison of interaction control strategies on a hand rehabilitation robot. In: IEEE international conference on rehabilitation robotics (ICORR); 2015. p. 846–51. https://doi.org/10.1109/ICORR.2015.7281308.

82. Özen Ö, Penalver-Andres J, Ortega EV, Buetler KA, Marchal-Crespo L.Haptic rendering modulates task performance, physical effort and movement strategy during robot-assisted training. In: 8th IEEE RAS/EMBS international conference for biomedical robotics and biomechatronics (BioRob); 2020. p. 1223–1228. https://doi.org/10.1109/BioRob49111.2020.9224317.

83. Just F, Özen Ö, Tortora S, Klamroth-Marganska V, Riener R, Rauter G. Human arm weight compensation in rehabilitation robotics: efficacy of three distinct methods. J Neuroeng Rehabil. 2020;17 (1):13. https://doi.org/10.1186/s12984-020-0644-3.PMID:32024528;PMCID:PMC7003349.

84. Özen Ö, Buetler KA, Marchal-Crespo L. Towards functional robotic training: motor learning of dynamic tasks is enhanced by haptic rendering but hampered by arm weight support. J Neuroeng Rehabil. 2022;19(1):19. https://doi.org/10.1186/s12984-022-00993-w PMID: 35152897.

85. Baur K, Speth F, Nagle A, Riener R, Klamroth-Marganska V. Music meets robotics: a prospective randomized study on motivation during robot aided therapy. J Neuroeng Rehabil. 2018;15(1):79. https://doi.org/10.1186/s12984-018-0413-8.PMID:30115082;PMCID:PMC6097420.

86. Krakauer JW, Kitago T, Goldsmith J, Ahmad O, Roy P, Stein J, Bishop L, Casey K, Valladares B, Harran MD, Cortés JC, Forrence A, Xu J, DeLuzio S, Held JP, Schwarz A, Steiner L, Widmer M, Jordan K, Ludwig D, Moore M, Barbera M, Vora I, Stockley R, Celnik P, Zeiler S, Branscheidt M, Kwakkel G, Luft AR. Comparing a novel neuroanimation experience to conventional therapy for high-dose intensive upper-limb training in subacute stroke: the SMARTS2 randomized trial. Neurorehabil Neural Repair. 2021;35(5):393–405. https://doi.org/10.1177/15459683211000730 Epub 2021 Mar 20 PMID: 33745372.

87. Baur K, Schättin A, de Bruin ED, Riener R, Duarte JE, Wolf P. Trends in robot-assisted and virtual reality-assisted neuromuscular therapy: a systematic review of health-related multiplayer games. J Neuroeng Rehabil. 2018;15(1):1–19.

88. Baur K, Wolf P, Klamroth-Marganska V, Bierbauer W, Scholz U, Riener R, Duarte JE. Robot-

supported multiplayer rehabilitation: feasibility study of haptically linked patient-spouse training. In: IEEE/RSJ international conference on intelligent robots and systems (IROS). IEEE; 2018. p. 4679–84.

89. Calabrò RS, Russo M, Naro A, Milardi D, Balletta T, Leo A, Filoni S, Bramanti P. Who may benefit from armeo power treatment? a neurophysiological approach to predict neurorehabilitation outcomes. PM R. 2016;8(10):971–8. https://doi.org/10.1016/j.pmrj.2016.02.004 Epub 2016 Feb 20 PMID: 26902866.

90. Buschfort R, et al. Arm studio to intensify the upper limb rehabilitation after stroke: concept, acceptance, utilization and preliminary clinical results. J Rehabil Med. 2010;42:310–4.

Upper-Extremity Movement Training with Mechanically Assistive Devices 28

David J. Reinkensmeyer,
Daniel K. Zondervan,
and Martí Comellas Andrés

Abstract

This chapter describes the development of mechanically assistive devices to enhance upper-extremity movement training after neurologic injury. We use the term "mechanically assistive devices" to refer to non-powered devices that incorporate springs, guides, pulleys, ramps, and/or levers to assist a patient in moving his or her weakened arm primarily by reducing the effect of gravity. As a case study of this approach, we first describe the development of the T-WREX exoskeletal training device, which was then commercialized and further tested as ArmeoSpring. Next, we provide a summary of clinical evidence for the effectiveness of mechanically assistive devices. We discuss why training with mechanically assistive devices reduces arm impairment, highlighting motivational, strengthening, and proprioceptive effects. We conclude by describing our recent efforts to democratize mechanically assistive devices for arm training by incorporating them directly onto wheelchairs as armrests.

Keywords

Movement rehabilitation · Motor learning · Rehabilitation technology · Stroke · Computer games · Upper-extremity exercise

D. J. Reinkensmeyer (✉)
Department of Anatomy and Neurobiology, University of California at Irvine, 4200 Engineering Gateway, Irvine, CA 92617, USA
e-mail: dreinken@uci.edu

Department of Mechanical and Aerospace Engineering, University of California at Irvine, 4200 Engineering Gateway, Irvine, CA 92617, USA

Department of Biomedical Engineering, University of California at Irvine, 4200 Engineering Gateway, Irvine, CA 92617, USA

D. K. Zondervan
Flint Rehabilitation Devices, 18023 Sky Park Circle, Suite H2, Irvine, CA 92614, USA
e-mail: dzondervan@flintrehab.com

M. C. Andrés
Department of Mechanical Engineering, Escuela Politécnica Superior de La Universidad de Lleida, Lleida, Spain
e-mail: marti.comellas@udl.cat

28.1 Introduction: A Case Study of the Development of a Mechanically Assistive Device

We begin this chapter by reviewing the motivation behind and development of an exemplar mechanically assisted device, T-WREX, which eventually was commercialized and became one of the most widely used devices for arm training after stroke, ArmeoSpring.

© The Author(s), under exclusive license to Springer Nature Switzerland AG 2022
D. J. Reinkensmeyer et al. (eds.), *Neurorehabilitation Technology*,
https://doi.org/10.1007/978-3-031-08995-4_28

"""

28.1.1 From Traditional Mechanically Assistive Devices to Robotic Rehabilitation

Prior to the late 1980s, several pieces of rehabilitation equipment took a mechanically assistive approach to allow people with arm weakness to practice arm movement. These included overhead slings, mobile arm supports, or simply a towel on a tabletop. However, despite their presence in rehabilitation facilities, there was little clinical evidence on the effectiveness of these approaches in reducing arm impairment.

A key realization in the late 1980s was that such rehabilitation technology might be improved by adding powered actuators to improve adjustability/assistance and sensors to provide feedback. Out of this rationale came several new robotic devices, including the MIT-Manus [1], the MIME [2], the ARM-Guide [3], and the Bi-Manu-Trac [4]. Each device took the approach of providing powered assistance to arm movements as users played simple computer games. It was these robotic devices that laid the scientific groundwork for the observation that mechanical assistance can be beneficial for arm training after a stroke.

Specifically, thousands of persons with a stroke have now participated in randomized controlled trials (RCTs) with these devices, two of which are commercially available (MIT-Manus as InMotion ARM Interactive Therapy System and Bi-Manu-Trac). The studies indicate that people with an acute or chronic stroke can recover a modest amount of additional movement ability if they exercise for tens of hours with these assistive devices; the transfer to functional movement is typically small [5–12]. Exercise with a robotic device has also been found to be as effective or, in some cases, more effective than a matched amount of exercise performed with a therapist [8–10, 13–15], or a matched amount of exercise performed with other rehabilitation technologies, such as electromyogram-triggered functional electrical stimulation [16] or sensor-based approaches [17].

28.1.2 From Robotics Back to Mechanically Assistive Devices

Robotic rehabilitation devices are at the high end of complexity in the spectrum of therapeutic technology. While these devices have proven to be useful tools for studying rehabilitative movement training, it is still unclear whether their modest therapeutic benefit justifies their cost, and, indeed, clinical uptake of robotic therapy devices is still sporadic. In the 1990s, we asked whether it would be possible to gain the benefits of robotic assistance without powered motors—i.e. with a mechanically assistive device—but with better adjustability and feedback compared to the "old school" devices.

With National Institute of Disability and Rehabilitation Research (NIDRR) support, we began developing a new device called T-WREX (or "Therapy-Wilmington Robotic Exoskeleton") (Fig. 28.1a), which was described in the doctoral dissertation research of Dr. Robert Sanchez [18]. We used a spring orthosis as the basic platform, allowing T-WREX to be nonrobotic but still capable of assisting severely weakened patients in moving by providing gradable assistance against gravity with elastic bands. To achieve this, we collaborated with Dr. Tariq Rahman of the A.I. duPont Institute for Children, who also with NIDRR support had developed the innovative arm support called WREX to assist children with weakened arms in moving their arms [19]. We scaled up the WREX design to be large enough and strong enough to support movements by adults with a stroke.

We also designed T-WREX to support functional upper-extremity movements by integrating a grip sensor that allowed detection of even trace amounts of hand grasp, thus allowing people with weakened, essentially "useless" hands to practice using their hands in a meaningful way for simulated activities of daily living in a virtual world, in coordination with their arms. We developed a suite of computer games that were easy to learn yet engaging and which approximated the movements needed for cooking, shopping, bathing, and cleaning.

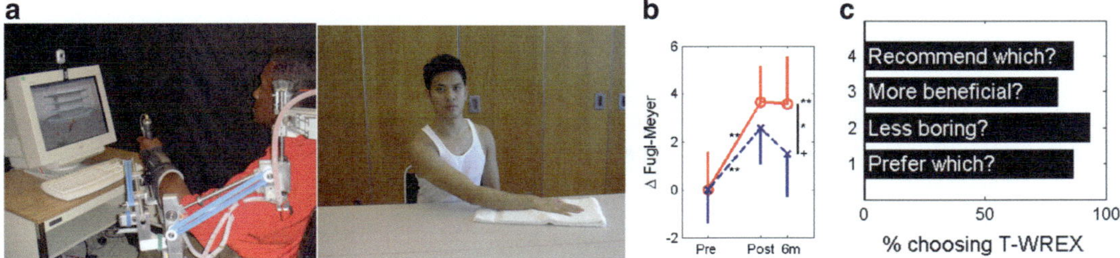

Fig. 28.1 **a** The T-WREX arm support exoskeleton was based on WREX and relieves the weight of the arm while allowing a wide range of motion of the arm. In a single-blind randomized controlled trial of T-WREX, we compared training with T-WREX to training of the arm on a tabletop with a towel. **b** Improvements in upper-extremity (UE) movement ability as measured with the UE Fugl–Meyer (FM) scale following chronic stroke with 2 months of T-WREX therapy ($n = 14$) and conventional tabletop exercise ($n = 14$) were significantly different at 6-month follow-up ($p = 0.05$). **c** Percentage of subjects preferring T-WREX therapy, compared to conventional, self-directed tabletop exercise, measured in our study. Subjects in both groups were given a chance to try each therapy and then select which one they preferred in ten categories, of which four are summarized here. (From Housman et al. [21] © 2009; reprinted with permission from SAGE Publications)

The first study of T-WREX quantified the effect of the gravity balance provided by T-WREX on voluntary arm movements. We measured how well volunteers with moderate-to-severe stroke (mean Fugl–Meyer upper-extremity score 25 out of 66, $n = 9$) could perform various arm movements while they wore the orthosis with and without arm gravity balance [18]. The most dramatic results came when the volunteers attempted to trace the outline of a large plastic disk placed in the frontal plane about 20 cm in front of their torso. The gravity balancing provided by T-WREX significantly improved the accuracy of the drawn circles for those who were able to draw a circle. Most strikingly, some participants who were unable to draw circles without support could draw them with support. Subsequent testing with T-WREX showed that the device improved the quality of movements of people with stroke, as measured by both the smoothness and timing of the movements [20].

As described next, we proceeded to test the therapeutic effects of providing assistance with T-WREX, which was eventually commercialized as ArmeoSpring by Hocoma A.G. and then further tested.

28.2   Summary of Clinical Evidence for the Effectiveness of Mechanically Assistive Devices

In this section, we discuss the evidence for the therapeutic effectiveness of training with mechanical assistance. We start with studies with T-WREX, progressing to studies with ArmeoSpring and other prominent mechanically assistive devices.

28.2.1   Effect of Movement Training Provided by T-WREX

We performed a pilot therapeutic test of T-WREX at UC Irvine [18]. Volunteers (n = 5) with moderate-to-severe arm impairment after chronic stroke (mean starting FM score 22) practiced moving with T-WREX three times per week, 45 min per session, over an 8-week period. They improved their movement ability as quantified by an average change in Fugl–Meyer score of 20% compared to baseline, hand grasp strength by 50%, as well as unsupported and supported reaching range of motion by 10%.

They achieved these improvements with approximately 6 min of direct contact with a rehabilitation therapist per 45 min of training.

Encouraged by these results, we refined T-WREX and performed a single-blind, randomized controlled trial of it at the Rehabilitation Institute of Chicago, under the supervision of the occupational therapist Sarah Housman [21]. We compared movement training with T-WREX against the standard approach for semiautonomous exercise at RIC, which was to train the weakened arm by using a tabletop to support the arm and a towel to remove the friction between the arm and the table (Fig. 28.1a). Twenty-eight chronic stroke survivors were randomly assigned to the experimental (T-WREX) or control (tabletop exercise) treatment. A blinded evaluator rated upper-extremity movement before and after 24 1-h treatment sessions and at a 6-month follow-up. The volunteers were also asked to rate their preference for T-WREX versus tabletop exercise after a single-session crossover treatment. The volunteers significantly improved upper-extremity motor control (Fugl–Meyer [22]), active reaching range of motion (ROM), and self-reported quality and amount of arm use (Motor Activity Log [23]). Improvements in the T-WREX group were better sustained at 6 months (Fugl–Meyer score improvement of 3.6 ± 3.9 versus 1.5 ± 2.7 points, mean $\pm$ SD, $p = 0.05$, Fig. 28.1b). The volunteers reported a strong preference for the T-WREX training compared to the tabletop training (Fig. 28.1c). The amount of supervision time required for both groups was about 3 min, following an initial training period of three sessions.

These results were encouraging: training with T-WREX produced detectably better results than a matched duration of the tabletop towel exercise and was substantially preferred by patients. It also required minimal direct supervision time.

28.2.2 Further Clinical Validation of the Mechanically Assistive Approach with ArmeoSpring

Hocoma AG licensed the intellectual property for T-WREX from the University of California at Irvine and then improved the mechanical, electrical, and software design of T-WREX for usability and manufacturability. The resulting ArmeoSpring device (Fig. 28.2) is as of 2021 being used in over 1000 rehabilitation facilities around the world. Multiple research studies have been conducted with ArmeoSpring measuring its therapeutic effects and expanding its use by other patient populations as we briefly review here.

Training with ArmeoSpring improved impairment and activity measures in chronic stroke patients with more mild hemiparesis than had been tested in previous studies with T-WREX (average starting Fugl–Meyer Upper-Extremity Score 45.7/66) [24]. Training with ArmeoSpring by individuals in the acute phase after stroke, as opposed to the chronic stage, was found to be about as effective as conventional one-on-one training with a therapist [25, 26]. In one of these studies, the group that trained with ArmeoSpring significantly improved shoulder range of motion and movement smoothness, while the control group did not [26]. The ArmeoSpring group also expressed higher satisfaction with the therapy [26]. ArmeoSpring was also combined with an iterative electrical stimulation system, allowing an improvement in UEFM score of almost 10 points in individuals with chronic stroke [27].

Another study used ArmeoSpring to investigate if the weight support provided by the device was in and of itself therapeutically advantageous [28]. This study compared the therapeutic effects of a single computer game, played alone, or with haptic input from a haptic robot, or with arm support from ArmeoSpring. All three groups

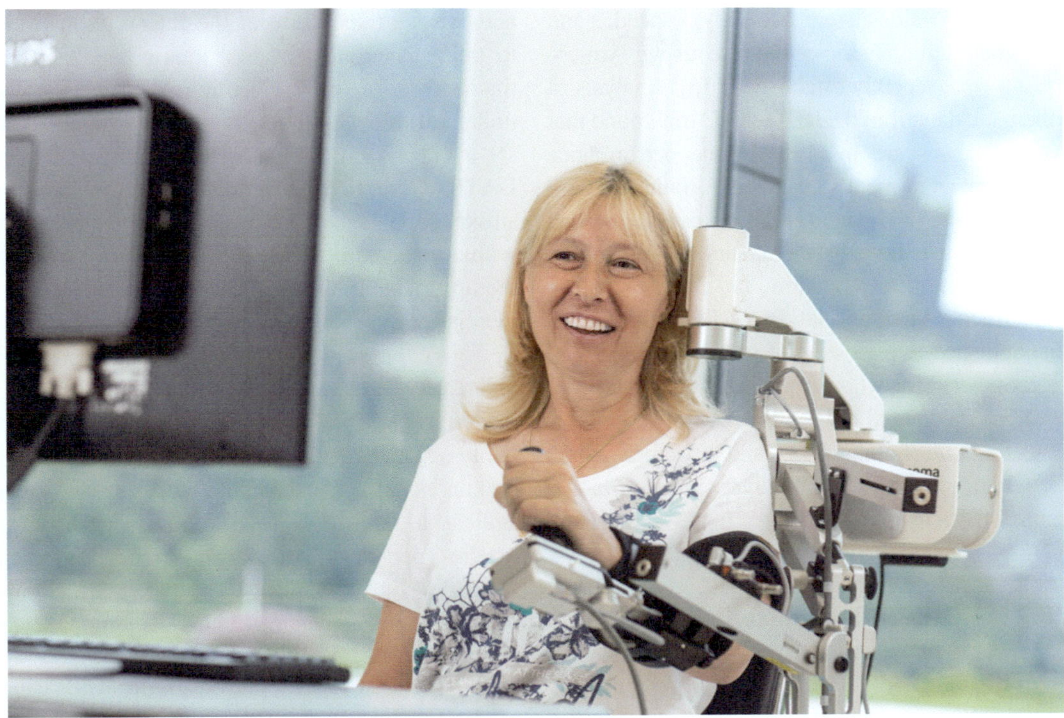

Fig. 28.2 ArmeoSpring, developed by Hocoma AG based on T-WREX, is designed to be more quickly adjustable than T-WREX for easier clinical use (Picture: Hocoma, Switzerland)

improved a comparable amount, although the haptic group improved more on the Box and Blocks score. The mechanical constraints inherent to ArmeoSpring (it doesn't allow shoulder internal/external rotation) appeared to prevent learning of some compensatory movements.

The largest study to date of ArmeoSpring is the REM-AVC trial, which was a multicenter RCT with a 12-month follow-up that enrolled 215 persons in the subacute phase after stroke [29]. The Exo group performed games-based exercises using ArmeoSpring. The control group performed stretching plus basic active exercises. The primary outcome was a change in upper-extremity Fugl–Meyer Assessment score at 4 weeks. The Fugl–Meyer score improved by 13.3 points in the Exo group and 11.8 in the control group (P = 0.22). The improvement in the primary functional measure, the ARAT, favored the Exo group (15.2 vs. 11.7 points), but only approached significance (p = 0.07). Participants in the Exo group rated the ease of learning and performing the self-rehabilitation significantly higher.

ArmeoSpring has also now been tested with other patient populations besides individuals with stroke. ArmeoSpring was found to increase the amount of training while reducing the amount of active therapist time required and to have a small therapeutic benefit for individuals with subacute cervical spinal cord injury, but only for individuals with partial hand function at baseline [30]. Training with ArmeoSpring benefited individuals with multiple sclerosis in a pilot study with ten individuals with a high level of disability [31], as well as individuals with proximal humeral fractures [32].

In terms of assessment, ArmeoSpring was shown to provide reliable measurement of active arm workspace for people with cervical spinal cord injury [33]. A variety of kinematic measurements obtained from ArmeoSpring during therapeutic game play accurately predicted clinical scores of upper-extremity movement ability after SCI [34].

Normative values for accuracy, speed, and smoothness for a single exercise using ArmeoSpring were recently established [35]. Analysis of kinematic data from the REM-AVC trial found that two processes are involved in the performance improvements measured when training with ArmeoSpring [36]. There is a fast process related to learning to use the exoskeleton and a slow process that reflects the reduction in upper-extremity impairment. Another analysis of REM-AVC data distinguished two clusters of persons with stroke: "Recoverers" for whom shoulder/elbow joint correlations converged toward the respective correlations for control participants, and "Compensators" for whom joint correlations diverged from that of control participants [37].

28.2.3 Other Mechanically Assistive Approaches

Other types of mechanically assistive devices have been developed and clinically tested, also demonstrating therapeutic benefits. We highlight three prominent devices here.

The FreeBal device [38–40] uses an overhead sling and cable/spring system to assist in three-dimensional movement and incorporates sensors and computer games. This device was commercialized as ArmeoBoom by Hocoma. In a multi-site study with 70 subacute stroke patients, training with ArmeoBoom produced comparable results to conventional training, although the patients rated the therapy as having higher interest and enjoyment than the conventional training [41].

The BATRAC [42] features two linear slides with hand grips positioned shoulder-width apart on a table with a joint allowing the linear slides to be raised or lowered to create an inclined plane (i.e. to allow forward motion of the hand to mechanically assist in raising the arm). This device was used to test a novel form of repetitive bilateral arm training with rhythmic auditory cueing (resulting in the acronym BATRAC). Training with BATRAC showed promise in an initial pilot study [43], with a follow-up randomized trial [44] indicating that the bilateral

and/or rhythmic nature of the intervention leads to unique neural reorganizations compared to a matched dose of conventional (i.e. typically unilateral) arm training. A larger follow-up RCT (N = 111; [45]) found that training with BATRAC reduced arm impairment in chronic stroke patients by a modest amount compared with a matched dose of conventional one-on-one therapy. The BATRAC device was later commercialized as Tailwind.

Feys et al. used a rocking chair and arm splint to create a mechanically assistive arm training device [46]. They had patients with subacute stroke (N = 100) rock themselves backwards in rocking chairs by reaching forward to push against a rail for a total of 15 h (500–1000 reaches per day) with their extended elbows supported by the splint. These patients had significantly greater increases in UE Fugl–Meyer (FM) score of 17 points at a five-year follow-up [46] compared to a control group who were passively rocked. The Feys study supported the concept that early and repetitive practice of relatively simple arm movements can translate into clinically meaningful benefits, particularly if delivered early after a stroke at a high intensity.

28.3 Why is Mechanical Assistance Beneficial for Promoting Motor Recovery?

In this section, we discuss three plasticity-related mechanisms that appear to play a role in producing the therapeutic effect associated with training with mechanical assistance: motivation, neural strengthening, and proprioceptive effects.

28.3.1 The Motivational Effect

Mechanical assistance allows weakened people to practice movements that are normally impossible or difficult to practice. This has the effect of improving the motivation for training. In the words of a volunteer in a T-WREX study, "If I can't do something once, why would I do it a hundred times?" [47].

A recent study of robotic hand training rigorously tested the motivational effect of mechanical assistance [48]. Participants (n = 30) at least six months after stroke and with some residual hand movement ability (minimum Box and Blocks Test score = 3, average Box and Blocks Test score = 32 ± 18 SD, and upper-extremity Fugl–Meyer score = 46 ± 12 SD) actively moved their index and middle fingers to targets while playing the musical game similar to Guitar Hero 3 h/wk for 3 weeks, achieving about 8000 movements during the nine training sessions. The participants were randomized to receive high assistance (causing 82% success at hitting targets) or low assistance (55% success) using the FINGER robotic device [49]. Note that without assistance the participants in both groups could only achieve about 20% success on average. High assistance boosted motivation, as measured with the Intrinsic Motivation Inventory after every training session (Fig. 28.3 left). High assistance also boosted self-efficacy, measured as the self-predicted improvement in BBT each week (Fig. 28.3 middle).

Motivated patients will presumably practice with more engagement and at a greater frequency, particularly if left unsupervised. However, the effect of improved motivation may go beyond encouraging more and better practice by helping cement motor learning. In the FINGER robotic study described above, high assistance boosted the change in Upper-Extremity Fugl–Meyer score, particularly for individuals with more severe baseline motor impairment (Fig. 28.3 right). A potential explanation is that higher assistance improves success, which in turn promotes better motor retention through dopaminergic mechanisms, a known effect in motor learning studies [50].

28.3.2 The Strengthening Effect

Even if a user is more motivated when practicing with mechanical assistance, the motivation won't be of benefit unless there is a neural plasticity mechanism in play that improves sensory motor control through practice. One such mechanism relevant to rehabilitative movement training with mechanically assistive devices is neural strengthening.

Weakness is a major culprit in reducing functional ability after stroke [51–54]. Weakness following stroke primarily has a neurologic rather than muscular origin, as, for example, electrical stimulation can produce near-normal muscle forces [55]. Strength for unimpaired people also has a large neurologic component, as, for example, the initial increases in force production caused by strength training cannot be explained by muscle hypertrophy, which requires time-delayed protein synthesis [56]. Further,

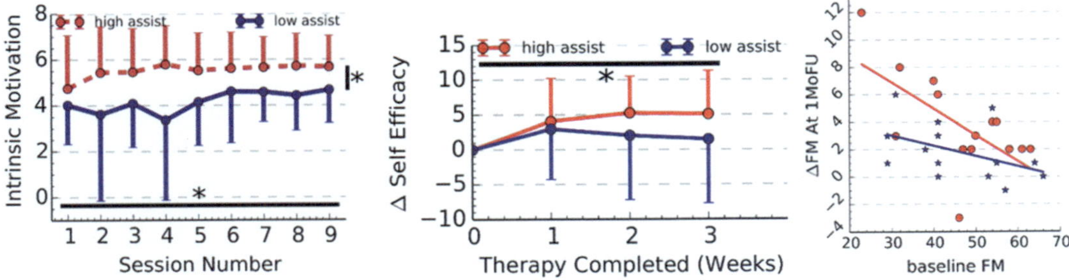

Fig. 28.3 Left: High mechanical assist provided by the FINGER robotic training device boosted self-reported motivation across the nine finger movement training sessions, judged by the Intrinsic Motivation Inventory. Middle: High assist also boosted self-efficacy across the three weeks of training. Self-efficacy was measured by asking participants to estimate how many more blocks they thought they could move in the Box and Blocks Test each week. Right: High robot assist (red) benefited Upper-Extremity Fugl–Meyer score at the one-month follow-up (1MoFU) more than low assist (blue), especially for more severely impaired participants. At the 1-month follow-up assessment, the intercept of the high-assistance group was significantly higher (Fig. 28.3E), P = 0.03, and the difference in slopes trended toward significance, P = 0.13. The figure was used with permission from [48]

imagined contractions alone can improve maximum force output [56].

Active assist movement training has been found to be beneficial for neural strengthening, almost certainly because such training requires an efferent contribution from the patient [57, 58]. Mechanically assistive devices are mechanically passive (i.e. unpowered) devices, so they will not move unless the patient initiates and drives the movement. Thus, when a patient practices with a mechanically assistive device, one should expect improvements in the neurally mediated component of strength due to repetitive efferent activation. Strength improvements, in turn, should translate into better movement ability, particularly for very weak patients, since weakness is a major culprit in reducing functional ability after stroke, as noted above.

28.3.3 The Proprioceptive Effect

A third, more speculative explanation for why movement training with mechanical assistance is therapeutic relates to proprioception. When a person moves poorly with reduced speed and range, they generate a paucity of proprioceptive input to their motor system. Mechanical assistance likely enhances the diversity of proprioceptive information delivered to the brain during training by allowing people to move with a greater variety of speeds, ranges, and directions. This may in turn promote beneficial sensory motor plasticity, through use-dependent or Hebbian-like mechanisms.

If this proprioceptive hypothesis is true, then we should expect the outcomes of movement training to depend on the integrity of proprioception at baseline. In the FINGER robotic hand training study mentioned above [48], the primary outcome was a functional measure of fine manipulation ability—the BBT score at one-month follow-up. There was no difference between groups in the primary endpoint. However, individuals with more impaired finger proprioception at baseline (measured with the FINGER Crisscross Assessment) benefited less from the training in terms of BBT (R2 = 0.335,

p = 0.002) regardless of the assist mode applied. Thus, the efficacy of robotic finger therapy in promoting hand function depended on finger proprioception at baseline.

This result was mapped to a neuroanatomical basis by examining over 60 measures of clinical characteristics, sensor motor behavior, neural injury (via MRI-based analysis of both sensory and motor structures), and neural function (via activation and connectivity analysis using fMRI and resting-state EEG) to explain the observed variability in treatment response [59]. Proprioceptive ability and measures of somatosensory network injury and function best explained inter-subject differences in treatment-related hand function gains.

Impaired proprioception was also found to be one of the strongest predictors of therapeutic benefit in one of the largest, most successful upper-extremity RCTs for persons with a stroke, the EXCITE clinical trial [60]. This study reported that "Patients with impaired proprioception had a 20% probability of achieving a clinically meaningful outcome compared with those with intact proprioception". From a computational neuroscience perspective, these results make sense: proprioception likely provides the teaching signal in both supervised learning and reinforcement-learning processes that shape neural activity after stroke [61]. Further, a more diverse stimulation of proprioceptive pathways, such as occurs when gravity assistance is provided, likely stimulates plasticity of those pathways better than a more stereotyped stimulation.

28.4 Democratizing Mechanically Assistive Devices

While mechanically assistive devices such as ArmeoSpring have proven useful, access to such devices is limited because of cost. This fact was highlighted by a visit to our laboratory by Dr. Don Schoendorfer, the founder of Free Wheelchair Mission, a non-profit organization that seeks to provide low-cost wheelchairs to more than 100 million individuals in developing nations who cannot afford a wheelchair [62].

While Dr. Schoendorfer was enthusiastic about robotic rehabilitation technologies, he challenged us to develop simpler devices. We review our attempts here, which are focused on using a ubiquitous mechanically assistive device for mobility—the manual wheelchair—as a platform for a more accessible mechanically assistive arm training device.

28.4.1 Resonating Arm Exerciser (RAE)

We first developed a lever-based device that provides mechanical assistance for arm training but in a much simpler way than T-WREX or ArmeoSpring. The Resonating Arm Exerciser (RAE) is comprised of a lever and arm support that attaches to a wheelchair wheel and enables users to practice something like the stationary "rocking therapy" used in the Feys study described above while seated in their wheelchair (RAE, see [63, 64]) (Fig. 28.4 left).

In a home-based randomized controlled trial with persons with severe arm impairment after chronic stroke, we found that the use of RAE reduced UE impairment more than a conventional home exercise program [64]. We also found that individuals with subacute stroke could safely perform hundreds of reaching movements per day with RAE [64]. We also tested RAE in a rehabilitation facility in Vietnam with physical therapist collaborators from Cal State Northridge, resulting in positive clinical results [65].

However, the key feedback from patients and clinicians in these studies was that, although RAE was simple and effective, it rendered the user's wheelchair immobile, and it was too troublesome to keep attaching and removing RAE when they desired to use the wheelchair in the normal fashion.

28.4.2 Lever-Assisted Rehabilitation for the Arm (LARA)

Taking this feedback into account, we next developed a novel lever-driven wheelchair (LARA, see [66–72]) that was designed to be used as a user's primary wheelchair, thus enabling UE rehabilitation without sacrificing mobility, requiring a transfer, or requiring a separate device to be attached to the wheelchair before use (Fig. 28.4 middle). Specifically, LARA allowed people to perform stationary UE rehabilitation in their wheelchair by moving attached levers back and forth with their impaired arm or to propel their wheelchair bimanually with the levers using a hand clutching system to repeatedly engage and disengage the lever from the wheel. By timing the hand clutching with arm movement, the user can steer, ambulate, and even turn in place.

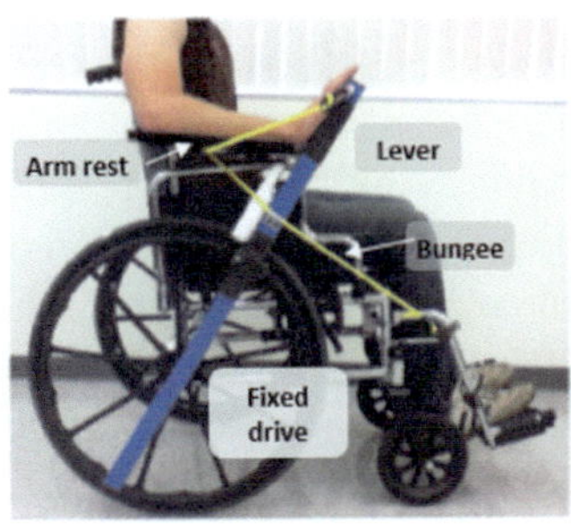

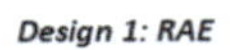

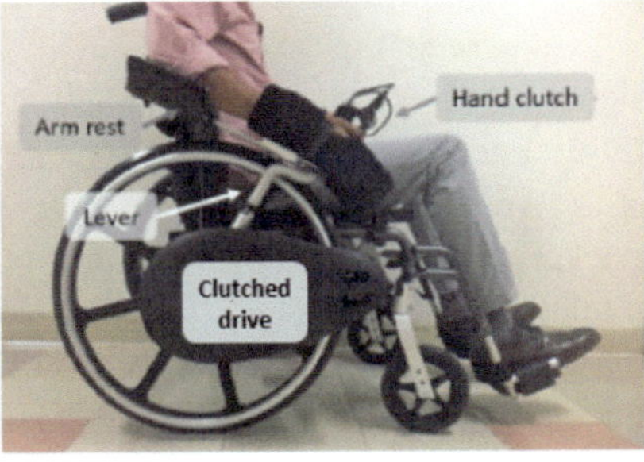

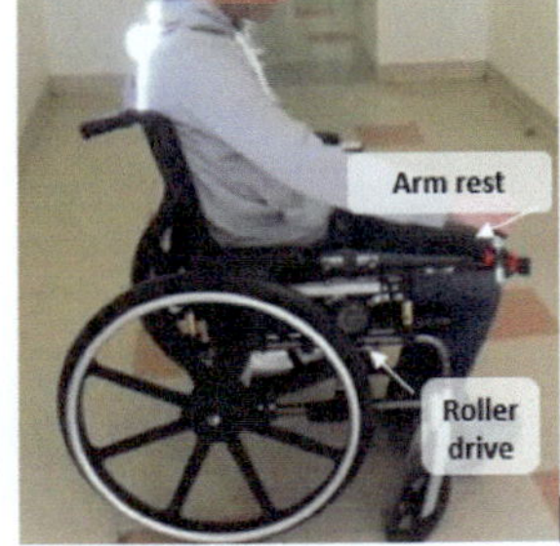

Design 1: RAE
- Detachable lever with bungees
- Positive therapeutic effect
- Cumbersome to attach/detach preventing normal wheelchair use

Design 2: LARA
- Custom lever drive wheelchair allowing paretic arm participation in ambulation & stationary exercise
- Positive therapeutic effect
- Requires a transfer/too large & bulky/difficult to learn overground ambulation due to hand clutching

Design 3: Boost
- Custom armrest quickly attaches and detaches to standard wheelchair
- No transfer needed; wheelchair can be used like a normal wheelchair
- Easy to learn to use; no "hand clutching"

Fig. 28.4 Three design cycles toward a wheelchair-based, mechanically assistive arm training device

Using motion capture and EMG, we confirmed that persons with stroke achieved wheelchair propulsion with LARA by moving their impaired arm with normative biomechanics while activating elbow extension muscles [70]. We then tested LARA in a pilot, two-site randomized controlled trial with individuals with subacute stroke. We found that both stationary and overground exercise with LARA led to a significantly greater reduction in arm impairment than conventional treatment at a one-month follow-up [73], conforming Fey's initial observation that early, repetitive stimulation of the arm is beneficial.

However, clinicians were still resistant to the idea of using LARA in routine clinical practice because it required too much cognitive demand for patients to learn to use the hand clutching and it was too bulky to be used as a patient's primary wheelchair. Therapists also did not like the idea of having to transfer patients to another large piece of rehabilitation equipment (i.e. a standalone LARA chair) and worried about where they would physically store LARA when not in use. They stated they would use a device like LARA if it were smaller and could be quickly attached to a patient's conventional manual wheelchair without impeding normal use/mobility.

28.4.3 Boost

In a third design iteration, we synthesized the lessons learned from the experiences with RAE and LARA and invented a method to deliver in-wheelchair UE rehabilitation in a clinically and commercially viable hardware package, resulting in Boost (Fig. 28.4 right). Boost replaces a conventional manual wheelchair's existing armrest by quickly and easily "clicking in" to the existing armrest slots. Once attached, Boost safely supports the arm in an ergonomic posture while enabling individuals to practice a full range-of-motion forward reaching task in two modes: (1) against low resistance and with an adjustable range of motion to achieve a "rep", with the chair remaining stationary ("Stationary Mode"), or

(2) against moderate resistance provided by the wheelchair wheel itself through an innovative one-way reel-drive that translates forward pushing into rotation of the wheel ("Overground Mode"). In Overground Mode, the user contributes to propelling their wheelchair with their impaired arm.

The reel-drive is comprised of a cable attached to the armrest that is then wrapped around a reel, which is in turn coupled to a friction disk via a one-way bearing. When the armrest is pushed forward, the cable spins the reel. When the reel is engaged with a mechanical switch for Overground Mode, it drives the wheelchair tire via the friction disk. After completing a push, a torsional spring inside the reel pulls the cable back, assisting the user in returning their arm to its initial position.

Critically, Boost's small and lightweight design does not interfere with the practice of the "good arm + good leg" propulsion technique currently taught to stroke survivors, which is essential for timely discharge from the hospital. Rather, it transforms this compensatory propulsion technique into a "good arm + good leg + impaired arm" therapeutic technique, encouraging the use of the paretic limb. That is, the patient can choose to try to incorporate their impaired arm as they ambulate in their wheelchair, thus stimulating their arm motor system.

In unpublished pilot testing of Boost with five subacute stroke patients with arm impairment, all were able to exercise the arm with Boost in stationary mode. Three ambulated overground exceeding 2 m/s after 2–5 practice trials. Two of these three were unable to push the rim to propel the wheelchair. Thus, this dynamic armrest provides a way to train arm movement, right on the wheelchair.

We recently solicited feedback on Boost from 16 physical and occupational therapists from two different hospitals. They strongly agreed that Boost was easy to set up, intuitive for patients to use, may improve their patients' motor recovery, and may improve their patients' wheelchair mobility. In addition, 100% reported that they would use Boost during one-on-one therapy sessions with moderately impaired patients, and

88% with severely impaired patients. 88% said they would allow moderately impaired patients to use Boost in the clinic between therapy sessions, and 94% would want patients to use Boost on their own at home. We are now proceeding to a randomized controlled trial to test whether we can provide early, repetitive arm stimulation with Boost and whether that stimulation is both pragmatic and therapeutic.

28.5 Conclusion

In this chapter, we traced the evolution of mechanically assistive devices for upper-extremity arm therapy after stroke, starting with a case study of T-WREX/ArmeoSpring. We briefly reviewed the large body of evidence that indicates that repetitive movement training with such devices is therapeutic, resulting in modest reductions in arm impairment that are comparable in magnitude to other forms of intense movement training. We focused mainly on arm therapy, as providing mechanical assistance for hand movement is challenging because the hand changes orientation with respect to gravity, and hand forces are often dominated by passive and spastic restraint. However, the mechanical assistance approach has been applied with success to the hand as well (see for example [74, 75]).

Importantly, mechanical assistance has also now been proven to improve motivation for training. We discussed how training with mechanical assistance also likely causes improvements in the neural component of strength, particularly for weak individuals, and may also improve the diversity of proprioceptive input to the brain, with therapeutic benefit. An important direction for future research is to precisely define the relative advantages and disadvantages of non-powered, mechanically assistive devices compared with powered ones.

Finally, we discussed a user-centric, iterative design approach that makes use of a patient's wheelchair armrest to provide mechanical assistance for arm training. We are hopeful that this approach can democratize mechanical assistance, making it accessible to a large number of people.

Acknowledgements Supported by NIDILRR grant 90REGE005-01 and NIH grant R44HD106850.

Disclosure David Reinkensmeyer has a financial interest in Hocoma AG, a maker of rehabilitation equipment. David Reinkensmeyer and Daniel Zondervan have a financial interest in Flint Rehabilitation Devices, a company that develops technologies for home-based rehabilitation after stroke and other neurologic injuries. The terms of these arrangements have been reviewed and approved by the University of California, Irvine, in accordance with its conflict of interest policies.

References

1. Hogan N, Krebs HI, Charnarong J, Sharon A. Interactive robot therapist. 1995.
2. Lum PS, Burgar CG, Shor PC, Majmundar M, Van der Loos M, P.C S. Robot-assisted movement training compared with conventional therapy techniques for the rehabilitation of upper-limb motor function after stroke. Arch Phys Med Rehabil. 2002;83(7):952–9.
3. Reinkensmeyer DJ, Dewald JPA, Rymer WZ. Guidance-based quantification of arm impairment following brain injury: a pilot study. IEEE Trans Rehabil Eng. 1999;vol.7(no.1):1–11.
4. Hesse S, Schulte- G, Konrad M, Bardeleben A, Werner C. Robot-assisted arm trainer for the passive and active practice of bilateral forearm and wrist movements in hemiparetic subjects. Arch Phys Med Rehabil. 2003;84(6):915–20.
5. Prange GB, Jannink MJA, Groothuis CGM, Hermens HJ, Ijzerman MJ. Systematic review of the effect of robot-aided therapy on recovery of the hemiparetic arm after stroke. J Rehabil Res Dev. 2006;43(2):171–84.
6. Brewer BR, McDowell SK, Worthen-Chaudhari LC. Poststroke upper extremity rehabilitation: a review of robotic systems and clinical results. Top Stroke Rehabil. 2007;14(6):22–44.
7. Mehrholz J, Platz T, Kugler J, Pohl M. Electromechanical and robot-assisted arm training for improving arm function and activities of daily living after stroke. Cochrane Database Syst Rev Online. 2008;(4)(4):CD006876.
8. Lo AC, Guarino PD, Richards LG, Haselkorn JK, Wittenberg GF, Federman DG, et al. Robot-assisted therapy for long-term upper-limb impairment after stroke. N Engl J Med. 2010;362(19):1772–83.
9. Reinkensmeyer DJ, Wolbrecht ET, Chan V, Chou C, Cramer SC, Bobrow JE. Comparison of 3D, assist-as-needed robotic arm/hand movement training provided with Pneu-WREX to conventional table top therapy following chronic stroke. Am J Phys Med Rehabil. 91(11 Suppl 3):S232–41.
10. Klamroth-Marganska V, Blanco J, Campen K, Curt A, Dietz V, Ettlin T, et al. Three-dimensional,

task-specific robot therapy of the arm after stroke: a multicentre, parallel-group randomised trial. Lancet Neurol. 2014 Feb;13(2):159–66.

11. Veerbeek JM, Langbroek-Amersfoort AC, van Wegen EEH, Meskers CGM, Kwakkel G. Effects of Robot-Assisted Therapy for the Upper Limb After Stroke: A Systematic Review and Meta-analysis. Neurorehabil Neural Repair. 2017 Feb 1;31(2):107–21.

12. Mehrholz J, Pohl M, Platz T, Kugler J, Elsner B. Electromechanical and robot-assisted arm training for improving activities of daily living , arm function , and arm muscle strength after stroke (Review). Cochrane Database Syst Rev. 2018;(9).

13. Bertani R, Melegari C, De Cola MC, Bramanti A, Bramanti P, Calabrò RS. Effects of robot-assisted upper limb rehabilitation in stroke patients: a systematic review with meta-analysis. Neurol Sci Off J Ital Neurol Soc Ital Soc Clin Neurophysiol. 2017 Sep;38(9):1561–9.

14. Ferreira FMRM, Chaves MEA, Oliveira VC, Petten AMVNV, Vimieiro CBS. Effectiveness of robot therapy on body function and structure in people with limited upper limb function: A systematic review and meta-analysis. PLOS ONE. 2018 Jul 12;13(7): e0200330.

15. Rodgers H, Bosomworth H, Krebs HI, van Wijck F, Howel D, Wilson N, et al. Robot assisted training for the upper limb after stroke (RATULS): a multicentre randomised controlled trial. The Lancet. 2019 Jul 6;394(10192):51–62.

16. Hesse S, Werner C, Pohl M, Rueckriem S, Mehrholz J, Lingnau ML. Computerized arm training improves the motor control of the severely affected arm after stroke: a single-blinded randomized trial in two centers. Stroke. 2005;36(9):1960–6 Epub.

17. Kahn LE, Zygman ML, Rymer WZ, Reinkensmeyer DJ. Robot-assisted reaching exercise promotes arm movement recovery in chronic hemiparetic stroke: A randomized controlled pilot study. J Neuroengineering Neurorehabilitation. 2006;3:12.

18. Sanchez RJ, Liu J, Rao S, Shah P, Smith R, Rahman T, et al. Automating arm movement training following severe stroke: functional exercises with quantitative feedback in a gravity-reduced environment. IEEE Trans Neural Syst Rehabil Eng Publ IEEE Eng Med Biol Soc. 2006;14(3):378–89.

19. Rahman T, Sample W, Seliktar R, Alexander M, Scavina M. A body-powered functional upper limb orthosis. J Rehabil Res Dev. 2000;37(6):675–80.

20. Iwamuro BT, Cruz EG, Connelly LL, Fischer HC, Kamper DG. Effect of a gravity-compensating orthosis on reaching after stroke: evaluation of the Therapy Assistant WREX. Arch Phys Med Rehabil. 2008;89(11):2121–8.

21. Housman SJ, Scott KM, Reinkensmeyer DJ. A Randomized Controlled Trial of Gravity-Supported, Computer-Enhanced Arm Exercise for Individuals With Severe Hemiparesis. Neurorehabil Neural Repair. 2009;23(5):505–14.

22. Fugl-Meyer AR, Jaasko L, Leyman I, Olsson S, Steglind S. The post-stroke hemiplegic patient. 1. a method for evaluation of physical performance. Scand J Rehabil Med. 1975;7(1):13–31.

23. van der Lee JH, Beckerman H, Knol DL, de Vet HCW, Bouter LM. Clinimetric properties of the Motor Activity Log for the assessment of arm use in hemiparetic patients. Stroke. 2004;35(6):1404–10.

24. Colomer C, Baldoví A, Torromé S, Navarro MD, Moliner B, Ferri J, et al. Efficacy of Armeo® Spring during the chronic phase of stroke. Study in mild to moderate cases of hemiparesis. Neurol Barc Spain. 2013 Jun;28(5):261–7.

25. Joo M, Park H, Noh S, Kim J, Kim H, Jang C. Effects of Robot-assisted Arm Training in Patients with Subacute Stroke. Brain NeuroRehabilitation. 2014;7 (2):111–7.

26. Bartolo M. Arm weight support training improves functional motor outcome and movement smoothness after stroke. Funct Neurol. 2014;Vol. XXIX (No. 1):15–22.

27. Meadmore KL, Hughes A-M, Freeman CT, Cai Z, Tong D, Burridge JH, et al. Function electrical stimulation mediated by iterative learning control and 3D robotics reduces motor impairment in chronic stroke. J Neuroengineering Rehabil. 2012;9(1):32.

28. Cameirão MS, Badia SB i, Duarte E, Frisoli A, Verschure PFMJ. The combined impact of virtual reality neurorehabilitation and its interfaces on upper extremity functional recovery in patients with chronic stroke. Stroke J Cereb Circ. 2012 Oct;43 (10):2720–8.

29. Rémy- O, Le Jeannic A, Dion A, Médée B, Nowak E, Poiroux É, et al. Additional, Mechanized Upper Limb Self-Rehabilitation in Patients With Subacute Stroke: The REM-AVC Randomized Trial. Stroke. 2021;52(6):1938–47.

30. Zariffa J, Kapadia N, Kramer JL, Taylor P, Alizadeh-M, Zivanovic V, et al. Feasibility and efficacy of upper limb robotic rehabilitation in a subacute cervical spinal cord injury population. Spinal Cord. 2012;50(3):220–6.

31. Gijbels D, Lamers I, Kerkhofs L, Alders G, Knippenberg E, Feys P. The Armeo Spring as training tool to improve upper limb functionality in multiple sclerosis: a pilot study. J Neuroeng Rehabil. 2011;8(5):5.

32. Schwickert L, Klenk J, Stähler A, Becker C, Lindemann U. Robotic-assisted rehabilitation of proximal humerus fractures in virtual environments: a pilot study. Z Gerontol Geriatr. 2011;44(6):387–92.

33. Rudhe C, Albisser U, Starkey ML, Curt A, Bolliger M. Reliability of movement workspace measurements in a passive arm orthosis used in spinal cord injury rehabilitation. J Neuroengineering Rehabil. 2012 Jan;9:37.

34. Zariffa J, Kapadia N, Kramer JL, Taylor P, Alizadeh-M, Zivanovic V, et al. Relationship between clinical assessments of function and measurements from an upper-limb robotic rehabilitation device in cervical

spinal cord injury. IEEE Trans Neural Syst Rehabil Eng. 2012;20(3):341–50.

35. Merlo A, Longhi M, Giannotti E, Prati P, Giacobbi M, Ruscelli E, et al. Upper limb evaluation with robotic exoskeleton. Normative values for indices of accuracy, speed and smoothness. NeuroRehabilitation. 2013 Jan;33(4):523–30.

36. Schweighofer N, Wang C, Mottet D, Laffont I, Bakhti K, Reinkensmeyer DJ, et al. Dissociating motor learning from recovery in exoskeleton training post-stroke. J NeuroEngineering Rehabil. 2018 Oct 5;15(1):89.

37. Nibras N, Liu C, Mottet D, Wang C, Reinkensmeyer D, Remy- O, et al. Dissociating Sensorimotor Recovery and Compensation During Exoskeleton Training Following Stroke. Front Hum Neurosci. 2021 Apr;30(15): 645021.

38. Stienen AHA, Hekman EEG, Van der Helm FCT, Prange GB, Jannink MJA, Aalsma AMM, et al. Freebal: dedicated gravity compensation for the upper extremities. IEEE 10th Int Conf Rehabil Robot. 2007;804–8.

39. Prange GB, Jannink MJ, Stienen AH, van der Kooij H, Ijzerman MJ, Hermens HJ. Influence of gravity compensation on muscle activation patterns during different temporal phases of arm movements of stroke patients. Neurorehabil Neural Repair. 2009;23(5):478–85.

40. Stienen A, Hekman E, van der Helm F, van der Kooij H. Self-aligning exoskeleton axes through decoupling of joint rotations and translations. IEEE T Robot. 2009;25(3):628–33.

41. Prange GB, Kottink AIR, Buurke JH, Eckhardt MMEM, van Keulen-Rouweler BJ, Ribbers GM, et al. The effect of arm support combined with rehabilitation games on upper-extremity function in subacute stroke: a randomized controlled trial. Neurorehabil Neural Repair. 2015 Feb;29(2):174–82.

42. McCombe S, Whitall J. Fine motor control in adults with and without chronic hemiparesis: baseline comparison to nondisabled adults and effects of bilateral arm training. Arch Phys Med Rehabil. 2004 Jul;85(7):1076–83.

43. Whitall J, McCombe S, Silver KH, Macko RF. Repetitive bilateral arm training with rhythmic auditory cueing improves motor function in chronic hemiparetic stroke. Stroke. 2000 Oct;31(10):2390–5.

44. Luft AR, McCombe S, Whitall J, Forrester LW, Macko R, Sorkin JD, et al. Repetitive bilateral arm training and motor cortex activation in chronic stroke: a randomized controlled trial. JAMA. 2004 Oct 20;292(15):1853–61.

45. Whitall J, Waller SM, Sorkin JD, Forrester LW, Macko RF, Hanley DF, et al. Bilateral and unilateral arm training improve motor function through differing neuroplastic mechanisms: a single-blinded randomized controlled trial. Neurorehabil Neural Repair. 2011 Feb;25(2):118–29.

46. Feys H, De Weerdt W, Verbeke G, Steck GC, Capiau C, Kiekens C, et al. Early and repetitive stimulation of the arm can substantially improve the long-term outcome after stroke: a 5-year follow-up study of a randomized trial. Stroke. 2004;35(4):924–9.

47. Reinkensmeyer DJ, Housman SJ. "If i can't do it once, why do it a hundred times?": Connecting volition to movement success in a virtual environment motivates people to exercise the arm after stroke. In 2007.

48. Rowe JB, Chan V, Ingemanson ML, Cramer SC, Wolbrecht ET, Reinkensmeyer DJ. Robotic Assistance for Training Finger Movement Using a Hebbian Model: A Randomized Controlled Trial. Neurorehabil Neural Repair. 2017 Aug 14;31(8):769–80.

49. Taheri H, Rowe JBJB, Gardner D, Chan V, Gray K, Bower C, et al. Design and preliminary evaluation of the FINGER rehabilitation robot: controlling challenge and quantifying finger individuation during musical computer game play. J Neuroengineering Rehabil. 2014;11(1):10.

50. Abe M, Schambra H, Wassermann EM, Luckenbaugh D, Schweighofer N, Cohen LG. Reward Improves Long-Term Retention of a Motor Memory through Induction of Offline Memory Gains. Curr Biol. 2011 Apr 12;21(7):557–62.

51. Bohannon RW, Warren ME, Cogman KA. Motor variables correlated with the hand-to-mouth maneuver in stroke patients. Arch Phys Med Rehabil. 1991;7:682–4.

52. Canning CG, Ada L, Adams R, O'Dwyer NJ. Loss of strength contributes more to physical disability after stroke than loss of dexterity. Clin Rehabil. 2004;18 (3):300–8.

53. Lang CE, Wagner JM, Edwards DF, Dromerick AW. Upper Extremity Use in People with Hemiparesis in the First Few Weeks After Stroke. J Neurol Phys Ther JNPT. 2007;31(2):56–63.

54. Harris JE, Eng JJ. Paretic upper-limb strength best explains arm activity in people with stroke. Phys Ther. 2007;87(1):88–97.

55. Landau WM, Sahrmann SA. Preservation of directly stimulated muscle strength in hemiplegia due to stroke. Arch Neurol. 2002;59(9):1453–7.

56. Enoka RM. Neural adaptations with chronic physical activity. J Biomech. 1997;30(5):447–55.

57. Reinkensmeyer DJ, Maier MA, Guigon E, Chan V, Akoner OM, Wolbrecht ET, et al. Do robotic and non-robotic arm movement training drive motor recovery after stroke by a common neural mechanism? Experimental evidence and a computational model. Proc. of IEEE EMBC 2009; 2439–2441.

58. Milot M-H, Spencer SJ, Chan V, Allington JP, Klein J, Chou C, et al. A crossover pilot study evaluating the functional outcomes of two different types of robotic movement training in chronic stroke survivors using the arm exoskeleton BONES. J NeuroEngineering Rehabil. 2013;10(1).

59. Ingemanson ML, Rowe JR, Chan V, Wolbrecht ET, Reinkensmeyer DJ, Cramer SC. Somatosensory system integrity explains differences in treatment response after stroke. Neurology. 2019;92(10).

60. Park S-W, Wolf SL, Blanton S, Winstein C, Nichols-Larsen DS. The EXCITE Trial: Predicting a Clinically Meaningful Motor Activity Log Outcome. Neurorehabil Neural Repair. 2008 Sep 1;22(5):486–93.

61. Reinkensmeyer DJ, Guigon E, Maier MA. A computational model of use-dependent motor recovery following a stroke: Optimizing corticospinal activations via reinforcement learning can explain residual capacity and other strength recovery dynamics. Neural Netw. 2012;29–30.

62. Shore SL. Use of an economical wheelchair in India and Peru: Impact on health and function. Med Sci Monit. 2008;14(12):71–9.

63. Zondervan DK, Palafox L, Hernandez J, Reinkensmeyer DJ. The Resonating Arm Exerciser: design and pilot testing of a mechanically passive rehabilitation device that mimics robotic active assistance. J NeuroEngineering Rehabil. 2013;10(39).

64. Zondervan DK, Augsburger R, Bodenhoefer B, Friedman N, Reinkensmeyer DJ, Cramer SC. Machine-Based, Self-guided Home Therapy for Individuals With Severe Arm Impairment After Stroke: A Randomized Controlled Trial. Neurorehabil Neural Repair. 2015 Jun;29(5):395–406.

65. Beling J, Zondervan D, Snyder B, Jiggs G, Reinkensmeyer D. Use of a mechanically passive rehabilitation device as a training tool in Vietnam: impact on upper extremity rehabilitation after stroke. In: Physiotherapy, Volume 101. Singapore: World Confederation for Physical Therapy Congress; 2015. p. e138–139.

66. Zondervan DK, Smith B, Reinkensmeyer DJ. Lever-actuated resonance assistance (LARA): A wheelchair-based method for upper extremity therapy and overground ambulation for people with severe arm impairment. In: IEEE International Conference on Rehabilitation Robotics. 2013. p. 1–6.

67. Smith B, Zondervan D, Lord TJ, Chan V, Reinkensmeyer D. Feasibility of a bimanual, lever-driven wheelchair for people with severe arm impairment after stroke. In: Intl Conf of the IEEE Engineering in Medicine and Biology Society. 2014.

68. Sarigul-Klijn Y, Lobo-Prat J, Smith BW, Thayer S, Zondervan D, Chan V, et al. There is plenty of room for motor learning at the bottom of the Fugl-Meyer: Acquisition of a novel bimanual wheelchair skill after chronic stroke using an unmasking technology. IEEE Int Conf Rehabil Robot Proc. 2017 Jul;2017:50–5.

69. Sarigul-Klijn Y, Smith BW, Reinkensmeyer DJ. Design and experimental evaluation of yoked hand-clutching for a lever drive chair. Assist Technol Off J RESNA. 2018 May 16;30(5):281–8.

70. Smith BW, Bueno DR, Zondervan DK, Montano L, Reinkensmeyer DJ. Bimanual wheelchair propulsion by people with severe hemiparesis after stroke. Disabil Rehabil Assist Technol. 2019 Jun;1–14.

71. Smith B, Lobo-Prat J, Zondervan D, Lew C, Chan V, Chou C, et al. Using a bimanual lever-driven wheelchair for arm movement practice early after stroke: a randomized, assessor-blind, controlled trial. Clin Rehabil. 2020;In Review.

72. Lobo-Prat J, Zondervan D, Lew C, Smith B, Chan V, Chou C, et al. Using a Novel Lever-Drive Wheelchair to Increase Arm Movement Practice Early After Stroke: Preliminary Results of a Randomized Controlled Trial. Neurorehabil Neural Repair. 2018; Abstracts(Annual Meeting):F37.

73. Smith BW, Lobo-Prat J, Zondervan DK, Lew C, Chan V, Chou C, et al. Using a bimanual lever-driven wheelchair for arm movement practice early after stroke: A pilot, randomized, controlled, single-blind trial. Clin Rehabil. 2021 Nov 1;35(11):1577–89.

74. Chen J, Nichols D, Brokaw EB, Lum PS. Home-Based Therapy After Stroke Using the Hand Spring Operated Movement Enhancer (HandSOME). IEEE Trans Neural Syst Rehabil Eng Publ IEEE Eng Med Biol Soc. 2017 Dec;25(12):2305–12.

75. Casas R, Sandison M, Nichols D, Martin K, Phan K, Chen T, et al. Home-Based Therapy After Stroke Using the Hand Spring Operated Movement Enhancer (HandSOME II). Front Neurorobotics. 2021;15: 773477.

Part VII

Robotic Technologies
for Neurorehabilitation:
Gait and Balance

Laura Marchal-Crespo and Robert Riener

Abstract

Rehabilitation robots allow for longer and more intensive locomotor training than that achieved by conventional therapies. Robot-assisted gait training also offers the possibility to provide objective haptic, visual, and auditory feedback to the patients and/or therapists within one training session and to monitor functional improvements over time. This chapter provides an overview of the technical approach for one of the most widely used systems known as "Lokomat" including features such as hip abduction/adduction actuation, cooperative control strategies, assessment tools, and augmented feedback.

L. Marchal-Crespo (✉)
Department of Cognitive Robotics, Delft University of Technology, Mekelweg 2, 2628 CD Delft, Netherlands
e-mail: L.MarchalCrespo@tudelft.nl

L. Marchal-Crespo
Motor Learning and Neurorehabilitation Laboratory, ARTORG Center for Biomedical Engineering Research, University of Bern, Bern, Switzerland

R. Riener
Sensory-Motor Systems Lab, Institute of Robotics and Intelligent Systems, Department of Health Sciences and Technology, ETH Zurich, Tannenstrasse 1, 8092 Zurich, Switzerland

R. Riener
Spinal Cord Injury Center, Balgrist Campus, Lengghalde 5, 8008 Zurich, Switzerland

These special technical functions may be capable of further enhancing training quality, training intensity, and patient participation.

Keywords

Exoskeleton · Actuated gait orthosis · Gait rehabilitation · Cooperative control · Augmented feedback · Lokomat

29.1 Introduction

A major limitation of manual-assisted, body weight-supported treadmill therapy (BWSTT) is that a training session relies upon the ability and availability of physical therapists to appropriately assist the patient's leg movement through the gait cycle. Robotic devices can eliminate this problem through the use of a mechatronic system that automates the assistance of the leg movement [1, 2]. This chapter presents the technological steps in the evolution of the design and development of Lokomat, an internationally well-established robot for gait therapy.

Manually assisted BWSTT involves therapists' assistance, while the patient practices stepping movements on a motorized treadmill with simultaneous unloading of a certain percentage of the body weight. Manual assistance is provided as necessary (and as far as possible) to enable upright posture and to induce physiological leg movements associated with physiological

© The Author(s), under exclusive license to Springer Nature Switzerland AG 2022
D. J. Reinkensmeyer et al. (eds.), *Neurorehabilitation Technology*,
https://doi.org/10.1007/978-3-031-08995-4_29

human gait. Over the last decades, there has been growing supporting evidence for the use of this technique in neurorehabilitation programs for stroke survivors and subjects with spinal cord injury (SCI). A large randomized clinical trial, known as the LEAPS study, has confirmed that walking training on a treadmill using body weight support and practice overground at clinics was superior to usual care in improving walking, regardless of the severity of initial impairment [3]. Yet, it has to be noted that in the LEAPS study, body-weight-supported treadmill training did not lead to superior results when compared to a home program of flexibility, range of motion, strength training, and balance of the same duration.

Whereas evidence demonstrates improvement in locomotor function following manually assisted treadmill training, its practical implementation in the clinical setting is limited by the labor-intensive nature of the method. Specifically, training sessions tend to be short because of the non-ergonomic physical demands and time costs placed upon the therapists' resources. This resource constraint yields significant limitations upon access to the therapy and, ultimately, to the effectiveness of the therapeutic approach with patients, as it limits the intensity of the training, resulting in a hindrance of functional gains in the lower limbs [4–6]. Particularly, in individuals with limb paralysis and/or a high degree of spasticity, appropriate manual assistance is difficult to provide; these patients require more than two therapists, which increases the already high cost and further limits training time [7]. The success and promise of BWSTT and the limitations and resource constraints in the therapeutic environment have inspired the design and development of robotic devices to assist in the rehabilitation of ambulation in patients following a stroke or SCI.

The research team of the Spinal Cord Injury Center of the University Hospital Balgrist in Zurich, Switzerland, an interdisciplinary group of physicians, therapists, and engineers, began to work on a driven gait orthosis in 1995 that would essentially replace the cumbersome and exhausting physical labor of therapists in the administration of locomotor training [1]. The "Lokomat" (commercially available from Hocoma AG, Volketswil, Switzerland) consists of a computer-controlled robotic exoskeleton that moves the legs of the patient in adjustable conjunction with a body weight support system (Fig. 29.1). It is the most widely used rehabilitation robot worldwide, with about 1000 installed devices until February 2020.

Later on, other exoskeletal systems were developed including the "AutoAmbulator" by Healthsouth Inc. (USA). Like the Lokomat, the AutoAmbulator is a four degrees-of-freedom (DOF) treadmill-based rehabilitation device, which consists of actuated robotic orthoses that guide the patient's knee and hip joints within the sagittal plane. In Europe, the device is sold as "ReoAmbulator" (www.motorica.com). Another treadmill-based robotic exoskeleton is LOPES (lower-extremity powered exoskeleton) [8]. It combines an actuated pelvis segment with a leg exoskeleton. The pelvis can move in translational directions, whereas the legs have two active rotary DOF at the hip (flexion/extension and abduction/adduction) and one active DOF at the knee (flexion/extension). The leg joints of the robot are actuated with Bowden cable-driven series elastic actuators resulting in a lightweight and compliant robotic system. The lateral pelvis translation is equipped also with the same actuation principle, whereas the anterior/posterior motion is driven by a linear actuator. A new version, LOPES II, uses an end-effector structure approach with parallel actuation that facilitates the alignment of human-robot joints [9]. Another gait rehabilitation robot is the active leg exoskeleton (ALEX) [10]. The so-called walker supports the weight of the device, and the orthosis incorporates several passive and actuated DOF with respect to the walker. The trunk of the orthosis (connected to the walker) has three DOF, namely, vertical and lateral translations and rotation about the vertical axis. All the DOF in the trunk are passive and held in position by springs. The hip joint of the orthosis has two DOF with respect to the trunk of the orthosis allowing actuated hip flexion/extension and passive abduction/adduction movements. A final

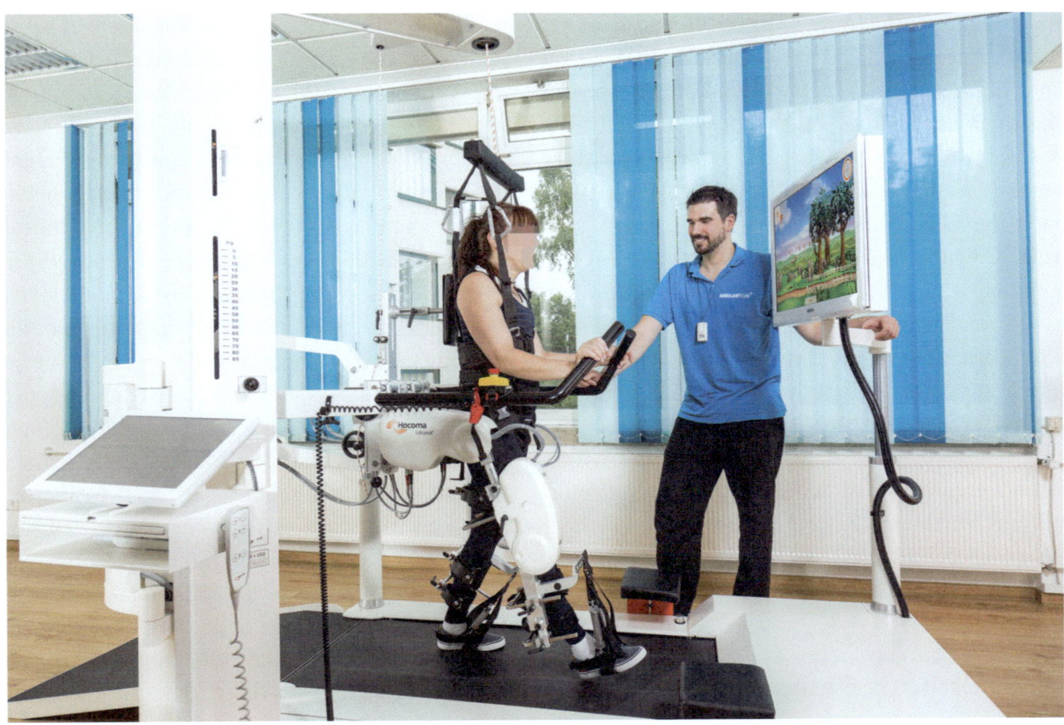

Fig. 29.1 Current version of the Lokomat system with a spinal cord-injured patient (Printed with permission of Hocoma AG, Volketswil)

example of a pioneer robotic device exoskeleton is the pelvic assist manipulator (PAM), a six DOF pneumatically operated device developed at the University of California Irvine that assists the pelvic motion during human gait training on a treadmill [11] and "POGO" (pneumatically operated gait orthosis), which moves the patient's legs with linear actuators attached to a frame placed around the subject [12].

Recent efforts have been made into developing wearable exoskeletons that allow overground walking. Although initially designed to assist SCI subjects in their daily ambulation, recent research highlights their potential also for gait rehabilitation [13]. Wearable robotic exoskeletons promote active participation of the user—crucial to driving brain plasticity and recovery [14]—as they require the user's active participation for both swing initiation and foot placement [15]. Several powered exoskeletons are already commercially available, such as the Ekso (Ekso Bionics, USA), the ReWalk (ReWalk Robotics, Israel), and the hybrid assistive leg (HAL) (Cyberdyne, Japan).

The interest in robot-assisted gait training has increased exponentially in the last years. This is reflected in the considerable number of reviews published within the last decade. Here we only presented some examples of hardware solutions. A detailed comparison between different gait-assisted devices can be found in e.g., [13, 16, 17].

An alternative to exoskeletal systems is end-effector-based systems, which connect the patients' leg (usually at the foot level) to the robot end-effector. One of the first footplate-based systems was the Gait Trainer [2], currently commercialized by Reha-Stim, Switzerland. The Gait Trainer operates like a conventional elliptical trainer, where the subject's feet are strapped into two footplates, moving the feet along a trajectory that is similar to a gait trajectory. As the Gait Trainer moves each leg only in one degree of freedom (DOF), Hesse and colleagues from the Fraunhofer Institute IPK developed a more complex device, called the "HapticWalker" [18]. The device comprises two end-effector-based platforms that move each foot in three DOF. Based on the knowledge gained with Gait Trainer and HapticWalker, Hesse et al. [19] developed the G-EO robot (EO is Latin meaning "I walk"), which is commercially available by the company Reha Technology AG in Switzerland (www.rehatechnology.com). As in the HapticWalker, the G-EO consists of two footplates, which move each foot with three DOF in the sagittal plane and enable the training of freely programmable tasks such as stair climbing. More recent examples of commercial end-effector systems include, e.g., Lokohelp (Woodway, USA), and THERA-Trainer lyra (medica Medizintechnik GmbH, Germany).

29.2 Orthosis Design

29.2.1 Mechanical Aspects

The Lokomat® is a bilaterally driven gait orthosis that is used in conjunction with an active body weight support system [1]. The Lokomat moves the patient's legs through the gait cycle in the sagittal plane (Fig. 29.1). The device's hip and knee joints are actuated by linear drives integrated into an exoskeletal structure. Passive foot lifters support ankle dorsiflexion during the swing phase. The orthosis is fixed to the rigid frame of the body weight support system via a parallelogram construction that, originally, only allowed passive vertical translations of the orthosis while keeping the orientation of the robotic pelvis segment constant, and thus, restricting the gait pattern to a two-dimensional trajectory in the body sagittal plane. Newer versions of the Lokomat (from 2014) also incorporate lateral translation and transverse rotation of the pelvis (Fig. 29.2; FreeD Module, Hocoma, Switzerland). These two movements are mechanically coupled and actuated through a linear actuation on the new pelvis module. With the FreeD addition, the pelvis is now movable in

the frontal plane to a lateral translation of up to 4 cm (per side) and in the transversal plane to a pelvic rotation of up to 4° (per side). Additionally, the legs cuffs can passively move laterally. This promotes a more natural gait pattern as the new module allows the natural lateral pelvis displacement as well as weight shifting during walking, and thus the excitation of more physiological sensory information from cutaneous, muscular, and joint mechanoreceptors and the possibility to train balance.

The linear drives on the orthoses are equipped with redundant position sensors as well as force sensors. The angular positions of each leg are measured by potentiometers attached to the lateral sides of the hip and knee joints of the orthosis. The hip and knee joint trajectories can be manually adjusted to the individual patient by changing the amplitude and offsets of a predefined gait trajectory. Knee and hip joint torques of the orthosis and pelvis module are measured by force sensors integrated in series with the linear drives. The signals from the force sensors may be used to determine the interaction torques between the patient and the device, which allows the estimation of the voluntary physical effort produced by the patient. This important information may be optimally used for various control strategies as well as for specific biofeedback and assessment functions.

The patient is fixed to the orthosis with straps around the waist, thighs, and shanks. The Lokomat geometry can be adjusted to the subject's individual anthropometry. The lengths of the thighs and shanks of the robot are adjustable via telescopic bars, so that the orthosis may be used by subjects with different femur lengths ranging between 35 and 47 cm. A special version of the Lokomat was designed and developed in 2006 to accommodate pediatric patients with shorter femur lengths between 21 and 35 cm (equivalent to body heights between approximately 1.00 and 1.50 m). The width of the hip orthosis can also be adjusted by changing the distance between the two lower limbs. The FreeD module accommodates pelvic widths between 29 and 51 cm (between 17 and 28 cm in the

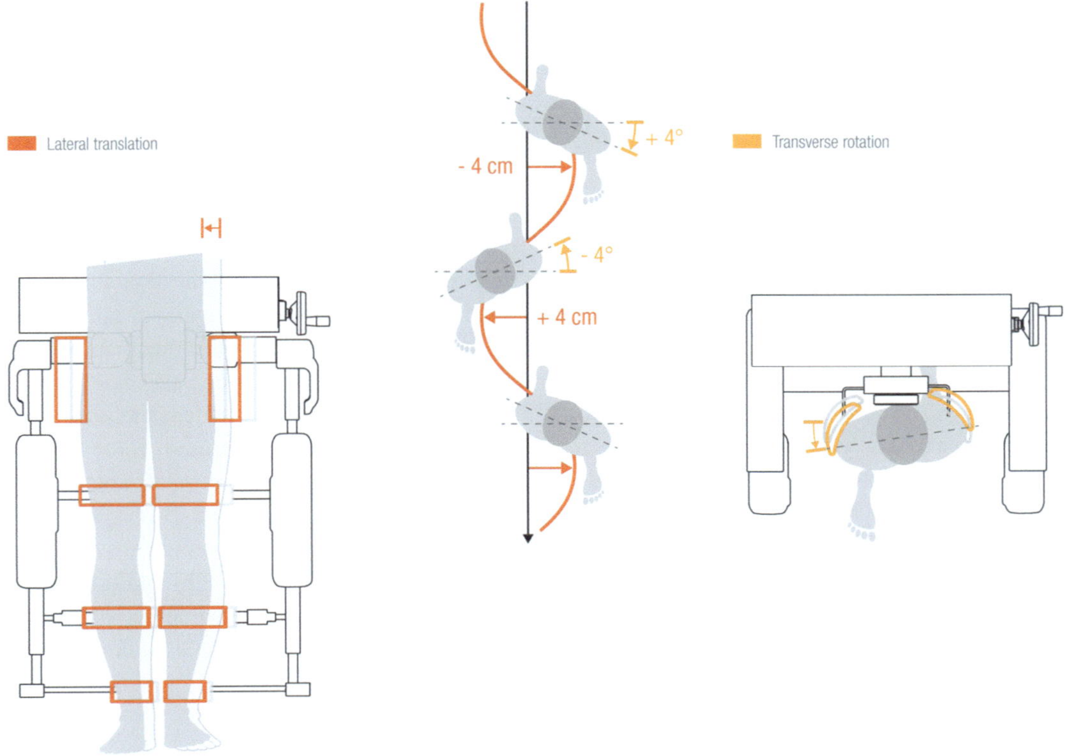

Fig. 29.2 Lateral translation and transverse rotation of FreeD. Left: Coupled lateral pelvis movement and rotation during physiological walking. Middle: Possibility of lateral pelvis and leg translation with the new FreeD. Right: Possibility of pelvis rotation with the new FreeD. Images with courtesy of Hocoma AG

pediatric version). The fixation straps, available in different sizes, are used to safely and comfortably hold the patient's limbs to the orthosis.

29.2.2 Drives

Ruthenberg and coworkers [20] reported the maximal hip torque during gait to be approximately 1 Nm per kilogram of body weight and an estimated average torque of approximately 35 Nm. In the Lokomat, hip and knee joints are actuated by custom-designed drives with a precision ball screw. The nut on the ball screw is driven by a toothed belt, which is in turn driven by a DC motor. The nominal mechanical power of the motors is 150 W. This yields an average torque of approximately 30 and 50 Nm at the knee and hip, respectively. Maximum peak torques are 120 and 200 Nm, respectively. This design has been demonstrated to be sufficient to move the legs against gravitational and inertial loads and, thus, to generate a functional gait pattern required in a clinical environment and suitable for most patients, even those with severe spasticity.

29.2.3 Safety

Whereas the mentioned peak torques are required to move the patient's joints in the presence of considerable interaction forces produced at the joints (e.g., due to spasticity) or between the patient's feet and treadmill (e.g., due to minor deviations of robot and treadmill speed), they can pose an inherent risk to the musculoskeletal system of the patient. To minimize this risk, various measures of safety were implemented into electronics, mechanics, and software. The electronic and mechanical safety measures follow principles of medical device safety regulations and standards (e.g., galvanic insulation). Additionally, passive back drivability and mechanical end stops avoid human joints getting overstressed or blocked in case of actuator malfunction. The software safety measures manage the proper operation of the device through the monitorization of nominal ranges of force sensors and the use of redundant position sensors. The software safety layer also checks the plausibility of movement and stops the device as soon as the movement deviates too much from the predefined desired gait trajectory. Another important safety feature is realized by the existence of the body weight support system, where the patient can be brought to a safe state when all drives have to be deactivated, e.g., when stumbling, or when spasticity causes the interaction forces to exceed the given threshold values. A wireless sensor system tracks the therapist's presence and regularly prompts input from the therapist to ensure the therapist's attention, and thus, improve the patient's safety. Furthermore, several manual emergency stops enable the therapist and/or patient to cause a sudden stop of movement whenever desired.

29.3 Body Weight Support System

Body weight support systems enable patients with leg paresis to participate in functional gait therapy, both on the treadmill and overground walking [21, 22]. The simplest system consists of a harness worn by the patient, ropes and pulleys, and a counterweight used to partially unload the patient. However, these simple systems do not ideally accommodate the wide range of conditions a patient with sensorimotor deficits encounters in gait therapy. The supporting vertical force varies mainly because of the effect of inertia that is induced by the vertical movement components performed during gait [23]. The lack of transparency of simple BWS solutions, the support force vector direction, and attachment to the harness influence the quality of the gait patterns during gait neurorehabilitation [24].

A mechatronic body weight support system called "Lokolift" has been developed to allow more precise unloading during treadmill walking. The Lokolift combines the key principles of both passive elastic and active dynamic systems [23]. In this system, at unloading levels of up to 85 kg and walking speeds of up to 3.2 km/h, the mean unloading error was less than 1 kg, and the maximum unloading error was less than 3 kg. This system can perform changes of up to 20 kg in the desired unloading within less than 100 ms. With this feature, not only constant body weight support but also gait cycle-dependent or time-variant changes of the desired force can be achieved with a high degree of accuracy. More recently, a spring-based (passive) system has been developed that allows similar results to the Lokolift system [25]. With the addition of the FreeD module, a new actuated degree of freedom was included in the Lokolift to allow for the pelvis lateral movement.

29.4 Control Strategies

In early clinical applications, the Lokomat was only used in a position control mode, where the measured hip and knee joint angles were fed into a conventional Position-Derivative (PD) controller. In the position control mode, the Lokomat does not systematically allow for deviation from the predefined gait pattern. However, rigid execution and repetition of the same pattern are not optimal for learning [26] and might lead to a reduction of patients' effort [27], and therefore, limit the therapeutic efficacy of the training [28]. In contrast, movement variability and the possibility to make errors are considered essential components of practice for motor learning [29]. Bernstein's demand that training should be "repetition without repetition" [30] is considered to be a crucial requirement. This is supported by recent advances in computational models of plasticity and motor learning to predict recovery [31]. More specifically, the study by Lewek et al. [32] demonstrated that intralimb coordination after stroke was improved by manual training that enabled kinematic variability, but was not improved by position-controlled Lokomat training, which reduced kinematic variability to a minimum. Another study performed with transected spinal rats also showed that kinematic variability facilitates spinal learning [33].

In response to this important finding, "patient-cooperative" control strategies were developed that "recognize" the patient's movement intention and motor abilities by estimating the patient's physical effort and adapting the robotic assistance to the patient's contribution, thus giving the patient more movement freedom and promoting more movement variability than during position control [34, 35]. It is recommended that the robotic control and feedback strategies should behave as qualified human therapists, i.e., they assist the patient's movement only as much as needed and inform the patient about how to optimize voluntary muscle effort and coordination to achieve and/or improve a particular movement.

The first step towards more movement freedom is to allow a variable deviation from a predefined leg trajectory by means of compliant control, such as impedance control. Impedance control allows a variable deviation from the predefined joint trajectories while an adjustable assisting torque is applied depending on the deviation between the current and desired joint trajectories, which depends on the patient's effort and behavior (e.g., measured using force/torque sensors). This assistive torque is usually defined as a function of angular position and its derivatives and is more generally called mechanical impedance [36]. More recent compliant controllers (e.g., path control) also include a deadband—i.e., a volume around the desired trajectory in which no assistance is provided—or the amount of impedance force varies with space and time [10, 35]. Figure 29.3 depicts a block diagram of an impedance controller [34].

An impedance controller was initially tested with the Lokomat in several subjects without neurological disorders and several subjects with incomplete paraplegia [34]. In the impedance control mode, angular deviations increased with increasing robot compliance (decreasing impedance) as the robot applied a smaller amount of force to guide the human legs along a given trajectory. It was found that inappropriate muscle activation produced by high muscle tone, spasms, or reflexes could affect the movement quality and yield a physiologically incorrect gait pattern, depending on the magnitude of the impedance chosen. In contrast, subjects with minor to moderate motor deficits stated that the gentle behavior of the robot feels good and comfortable.

The disadvantage of a standard impedance controller is that the patient needs to retain sufficient voluntary effort to move along a physiologically correct trajectory, which limits the range of application to patients with only mild lesions. Furthermore, the underlying desired gait trajectory allows no flexibility in time, i.e., leg position can deviate only orthogonally but not tangentially to the given time-dependent trajectory. Therefore, the original impedance controller

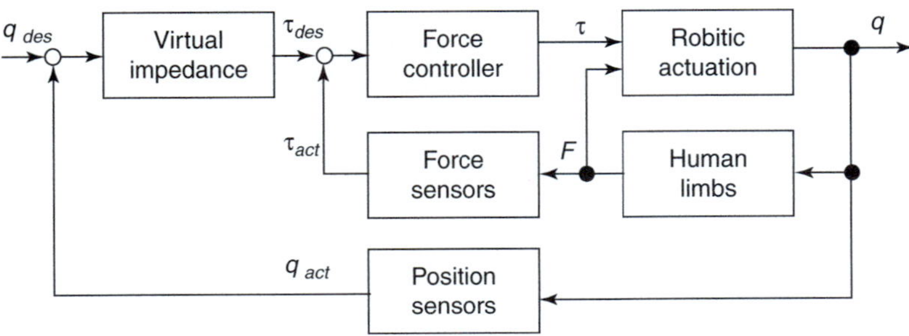

Fig. 29.3 Example of an impedance control architecture for the compliance of rehabilitation robot [34]. Symbols: q is the vector of generalized positions or joint angles; τ is the vector of generalized joint torques; F is the interaction force between robot and human; index "des" refers to the desired reference signal; index "act" refers to the actual, measured signal

has been extended to a so-called path controller [35], in which the time-dependent walking trajectories are converted to walking paths with free timing. Furthermore, the impedance along the path can vary to obtain physiological satisfactory movements, especially at critical phases of gait (e.g., before heel contact) [35]. This is comparable to fixing the patient's feet to soft rails, thus limiting the accessible domain of foot positions calculated as functions of hip and knee angles. Along these "virtual rails," the patients are free to move. Supplementary to these *corrective* actions of the Lokomat, a *supportive* force field of adjustable magnitude can be added to provide extra support to patients along the path. Depending on the actual position of the patient's legs, the supportive force act in the direction of the desired path. The support is derived from the desired angular velocities of the predefined trajectory at the current path location. Compared to the more simple impedance controller, the path controller gives the patient more freedom in timing, while she or he can still be guided through critical phases of the gait. The path controller has been evaluated in several single-case studies [37–39]. Most stroke patients improved their gait performance after several weeks of training with the path controller.

New trends in robot-based motor learning and neurorehabilitation suggest that challenge-based controllers—i.e., controllers that make movement tasks more difficult or challenging [40]—might enhance recovery by, e.g., strengthening the muscles by opposing the movement (e.g., resistive methods [41]), facilitating the detection of tracking errors (e.g., error augmentation methods [42–44]), and increasing movement variability (e.g., force perturbations [45]). These challenge-based control strategies might lead to improvements in recovery, especially in people in the late stages of neurorehabilitation or with mild impairments [41, 46–48].

Finally, adaptive controllers that provide tailored assistance or resistance based on real-time measurements of the users' performance (e.g., tracking error [49]) during gait training might be more effective in promoting motor learning than those that do not adapt the assistance to the patients' especial needs [50]. The controller adaptation might be done by modifying the parameters of a reference trajectory or the dynamics of a virtual compliant controller [9, 49, 51]. The Hocoma 2020 software release for the Lokomat, the LokomatPro Sensation, provides new therapy options that include intelligent algorithms which create a maximum challenge by personalizing the assistance based on patients' performance.

29.5 Assessment Tools

Using robotic devices in locomotor training can have more advantages than just supporting the movement, thus, increasing the intensity of training. Data recorded by the position and force transducers can also be used to assess the clinical state of the patients throughout the therapy [52]. The following clinical measures can be assessed by the Lokomat.

29.5.1 Mechanical Stiffness

Spasticity is an alteration in muscle activation with increased tone and reflexes. It is a common side effect of neurological disorders and injuries affecting the upper motor neuron, e.g., after brain or spinal cord injuries. Formally, spasticity is usually considered as "a motor disorder characterized by a velocity-dependent increase of tonic stretch reflexes (muscle tone) with exaggerated tendon jerks, resulting from hyperexcitability of stretch reflexes" [53]. It appears as an increased joint resistance during passive movements. Sanger et al. [54] used a more functional rather than physiological definition describing spasticity as "a velocity-dependent resistance of a muscle to stretch." Most commonly, spasticity is evaluated by the Ashworth Test [55] or Modified Ashworth Test [56]. In both tests, an examiner moves the limb of the patient, while the patient tries to remain passive. The examiner rates the encountered mechanical resistance to passive movement on a scale between 0 and 4. However, such an evaluation is subject to variable factors, such as the speed of the movement applied during the examination and the experience of the examiner and interrater variability.

The mechanical resistance can also be measured with the Lokomat [57, 58], which is capable of simultaneously recording joint movement and torques. The actuation principle allows for the assessment of the hip and knee flexion and extension movements in the sagittal plane. The stiffness measurement can be performed immediately before and following the usual robotic movement training without changing the setup. To measure the mechanical stiffness with the Lokomat, the subject is lifted from the treadmill by the attached body weight support system so that the feet can move freely without touching the ground. The Lokomat then performs controlled flexion and extension movements of each of the four actuated joints subsequently at different velocities. The joint angular trajectories are squared sinusoidal functions of time replicating the movements applied by an examiner performing a manual Ashworth Test. Measured joint torques and joint angles are used to calculate the elastic stiffness as slopes of the linear regression of the torque-position plots. As the recorded torques also include passive physical effects of the Lokomat and the human leg, the measured torque is offline-compensated for inertial, gravitational, Coriolis, and frictional effects obtained from an identified segmental model of the orthosis including the human leg. Patient data comparisons with manual assessments of spasticity based on the Modified Ashworth Scale demonstrated that higher stiffness values measured by Lokomat corresponded with higher ratings of spasticity [57, 58]. A feasibility study with ten children with CP showed that the Lokomat is a feasible tool to measure stiffness, but it is not sensitive enough to detect small changes in muscle tone [59]. Assessment of spasticity is still in experimental status and needs further validation in future studies.

29.5.2 Voluntary Force

For some patients, maximum voluntary force is a measure of limiting factor for walking. In order to assess the maximum voluntary force in the Lokomat [57], the examiner instructs the patient to generate force in each joint, first in flexion and then in extension directions. The force is generated against the Lokomat, which is position-controlled to a predefined static posture, thus providing a quasi-isometric measurement condition. Simultaneously, the joint moments are measured by the built-in force transducers and displayed to the patient and the therapist. The maximum moments for flexion and extension are used as outcome variables. An improved version standardizes the computerized sequence and instructions and uses a time-windowed calculation for the output values [60]. It was shown that this measurement method has a high inter- and intratester reliability and can be used to assess the strength of the lower extremities [61]. It has also been shown that the recorded peak forces during isometric contractions could be employed as an outcome measure to monitor changes in muscle strength in incomplete SCI subjects following robot-aided gait training [62].

29.5.3 Range of Motion and Lower Limb Proprioception

In a manner similar to the conventional clinical range of motion assessments, the therapist moves the leg of the patient until the passive torque produced by the patient's joint reaches a certain threshold that is qualitatively predefined by the therapist based on his or her expertise. As the patient's legs are attached to the device with the anatomical and technical joint axes in alignment with each other, and the recorded joint angles correspond with the patient's joint angles, the passive range of motion is determined by the maximum and minimum joint angles measured. This parameter can be used for further assessments and training. The Lokomat measures the joint range of motion within values typical for human gait and may represent only a fraction of the patient's physiological range. This test provides important additional measures of the patient relevant to the gait and further conditions making contractures and other joint limitations (e.g., due to shortened tendons) quantifiable. These measures are directly relevant to activities of daily living.

Lower limb proprioception has also been listed as a relevant assessment in clinical practice [63]. Some attempts have been made to use the position sensors of the Lokomat for evaluating hip and knee joint proprioception. Joint position reproduction (JPR) was tested in healthy subjects and 23 incomplete SCI subjects [64]. The participants' legs were positioned at predetermined hip and knee angles and then displaced to a different (distractor) position. Using a joystick to control the robot, participants were requested to place their limb back at the remembered initial position and the error between the initial and remembered position was measured. Although the test-retest reliability in SCI participants was found to be between fair and substantial at the hip and knee respectively, the JPR score correlated well with the clinical assessment of proprioception. The ability to sense movement, i.e., the threshold to detection of passive motion (TTDPM), was also incorporated in a different study with healthy participants and 17 individuals with SCI [65]. The Lokomat was used to passively move the hip and knee joints at four different speeds (0.5, 1.0, 2.0, and 4.0 deg/s) in both flexion and extension. Participants were requested to press a button whenever a movement was felt. The measures showed high test-retest reliability and high correlation with several assessments of lower limb joint kinesthesia.

29.6 Biofeedback

Compared to manual treadmill therapy, robotic gait retraining changes the nature of the physical interaction between the therapist and the patient. Therefore, it is important to incorporate the features into the Lokomat system to assess the patient's contribution and performance during training and to provide necessary real-time feedback and instructions derived from precise measurements taken by the system. The patient may have deficits in sensory perception and cognition interfering with her/his ability to objectively assess movement performance and making it difficult to engage the patient and encourage active participation in the movement and training. With the feature of Lokomat, the technology of multisensory biofeedback has the potential to challenge and engage the patient in order to increase the benefit of motor recovery and neurological rehabilitation [66, 67].

The built-in force transducers can estimate the muscular efforts contributed by the patient's knee and hip joints. Incorporating this information into an audiovisual display can simulate the "feedback" the therapist usually gives to the patient during manual training, where the therapist estimates the patient's activity based on the effort required to guide the patient's legs.

The goal of the biofeedback function is to derive and display performance values that quantify the patient's activity and performance in relation to the target gait function such that the patient can improve muscle activity toward a more functional gait pattern. An early implementation of a force-biofeedback strategy for the Lokomat has been described [34, 68, 69].

To obtain relevant biofeedback values, the gait cycle is divided into the stance phase and swing phase. For each phase, weighted averages of the forces are calculated at each joint independently, thus yielding two values per stride per joint. Eight biofeedback values are available for each gait cycle from all four joints of the two lower limbs. Because of the bilateral symmetry, four weighting functions are required for the averaging procedure (hip stance, hip swing, knee stance, knee swing). The weighting functions were selected heuristically to provide positive biofeedback values when the patient performs therapeutically reasonable activities (e.g., active weight bearing during stance, sufficient foot clearance during the swing, active hip flexion during swing, active knee flexion during early swing, knee extension during late swing). The graphical display of these values has been positively rated by the patients and leads to an increased instantaneous activity by the patients [70, 71] and improvements in cognitive functioning and psychological well-being in patients with chronic stroke [72] and traumatic brain injury [73]. However, there is no direct clinical evidence showing that this training with computerized feedback leads to better rehabilitation outcomes or faster recovery compared to Lokomat training without feedback. In the Lokomat 2020 software release, the LokomatPro Sensation, a new assessment tool was incorporated to evaluate in real-time the patients' ability to walk. Using adaptive algorithms [74], the assistance provided by the device can be automatically adjusted at each gait step (i.e., the impedance of the joints and the unloading of the body weight) based on the patient's ability to follow a predefined gait trajectory [49]. Although not systematically evaluated in SCI and/or brain-injured patients, a first experimental evaluation of the assessment algorithm in eight healthy participants suggests that this method can be a promising tool to objectively assess walking function during training in clinical practice.

To further increase patients' engagement and motivation, virtual reality and computer game techniques may be used to provide virtual environments that encourage active participation during training (Fig. 29.4). A first feasibility study showed that the majority of subjects could navigate through a virtual environment by appropriately controlling and increasing their activity of left and right legs while walking through a forest scenario and other scenarios [75]. Wagner et al. showed how such kind of VR-enhanced Lokomat training activates premotor and parietal areas [76]. Calabrò et al. further showed that combining robotic-based rehabilitation with avatars animated in a 2D VR in chronic stroke patients may entrain several brain areas involved in motor planning and learning, resulting in enhanced motor performance [77].

Fig. 29.4 Walking through a virtual environment. Lokomat in combination with a virtual reality back-projection display system

More immersive VR visualization displays, such as large projections and off-the-shelf head-mounted displays (HMD), could also be employed to modulate motor behavior through the manipulation of the virtual environment [78]. The multisensory integration of visual and non-visual information (e.g., auditory, vestibular, and somatosensory information) allows the manipulation of the perception of the environment. This in turn can be employed to modulate the planning and execution of movements. For example, it has been shown that the walking speed is affected by changing the speed of the optic flow [79] and that visualizing avatars as self-representations of the users' own bodies can be exploited to induce changes in the gait pattern in healthy participants [80]. However, further research with patients is needed to unveil the real potential of more immersive VR in robot-aided rehabilitation.

29.7 Clinical Outcomes

Robotic technology is still very much in development, and there are a lot of new devices and technical features that might further enhance the potential of therapeutic training. Nevertheless, there have already been more than 200 clinical investigations applying the Lokomat technology to different patient groups. It was applied for the therapy of patients with SCI, hemiplegia after stroke, traumatic brain injuries, multiple sclerosis, Parkinson's disease, cerebral palsy, and other pathologies (see [81]). Most of these studies show positive outcomes with the Lokomat compared to conventional therapies or usual care.

Although robot-aided gait training with Lokomat has proved to be feasible in a number of pathologies such as iSCI [6], multiple sclerosis [82], and cerebral palsy [83], the majority of clinical studies have been done with stroke subjects. Often cited are the ones from Hidler et al. [28] and Hornby et al. [84], who applied the Lokomat on subacute and chronic stroke patients, respectively, and compared it with conventional gait therapy. Both studies showed that participants who received conventional training experienced greater gains in gait parameters such as walking speed, walking distance, or single limb stance than those trained on the Lokomat. Hidler et al. and Hornby et al. concluded that for stroke participants, conventional gait training interventions appear to be more effective than robot-assisted gait training. However, both studies included only ambulatory patients, although the Lokomat is recommended to be used primarily for nonambulatory patients. Furthermore, the Lokomat was used in most simple control modes (position controller or impedance controller with reduced guidance force), without any other features such as augmented feedback or biofeedback functions. Of course, this kind of mode cannot compete with the quality and gentleness of a trained therapist or more advanced robotic features, such as cooperative and self-adaptive control strategies. A more recent study is the one by Dundar et al. [85] who compared conventional physiotherapy and robotic training combined with conventional therapy on 107 subacute and chronic stroke patients. They found that robotic training combined with conventional therapy produced better improvement in a large number of different stroke scales.

A recent Cochrane report [72] analyzing 62 trials involving 2440 stroke patients revealed that people who receive electromechanical-assisted gait training, such as provided by the Lokomat or the Gait Trainer, in combination with physiotherapy after stroke are more likely to achieve independent walking than people who receive gait training without these devices. Specifically, people in the first three months after stroke and those who are not able to walk seem to benefit most from this type of intervention. The role of the type of device is still not clear.

Fewer studies have focused on evaluating the effectiveness of electromechanical-assisted gait training on cognitive function and quality of life in patients with walking disabilities, despite their importance in rehabilitation. There is initial evidence that training with the Lokomat together with virtual reality positively affects cognitive recovery and psychological well-being in patients with chronic stroke [73] and traumatic brain injury [86].

The current evidence suggests that future research should go into performing a large definitive, pragmatic, phase III trial to address specific questions such as: "What frequency or duration of electromechanical-assisted gait training might be most effective?" and "how long does the benefit last?" Importantly, due to the heterogeneity of the studied stroke population, future clinical trials should consider time post-stroke in their trial design.

29.8 Conclusion

Robotic rehabilitation devices such as the Lokomat become increasingly important and popular in clinical and rehabilitation environments to facilitate prolonged duration of training, increased number of repetitions of movements, improved patient safety, and less strenuous operation by therapists. Novel sensor, displays, and control technologies improved the function, usability, and accessibility of the robots, thus increasing patient participation and improving performance. Improved and standardized assessment tools provided by the robotic system can be an important prerequisite for the intra- and intersubject comparison that the researcher and the therapist require to evaluate the rehabilitation process of individual patients and entire patient groups. Some rehabilitation robots offer an open platform for the implementation of advanced technologies, which will provide new forms of training for patients with movement disorders.

With the use of different cooperative control strategies and particular virtual reality technologies, patients can be encouraged not only to increase engagement during walking training but also to improve motivation to participate in therapy sessions. Especially promising are adaptive controllers that can accommodate the patient's specific pathology and level of disability by identifying and reducing hindrances that could impede the recovery process, or by challenging more skilled patients. However, to date, there is a lack of standardized comparisons among control strategies to analyze the relation between control strategies and clinical outcomes [87, 88]. More immersive VR visualization displays that provide a realistic representation of virtual environments and avatars that mimic the patients' movements could potentially further enhance robot-aided rehabilitation by leveraging psychological factors such as motivation, presence, and embodiment [89]. Yet, more research is needed to unveil the real potential of immersive VR in robot-aided rehabilitation.

Several clinical trials have been performed showing that the application of rehabilitation devices is at least as effective as the application of conventional therapies. Further clinical studies are required to find predictors for the success of a Lokomat treatment in order to distinguish therapy responders from nonresponders. From such investigations, it is expected to figure out which choice of technical Lokomat features (controller complexity, number of actuated joints, kind of feedback, etc.) have to be applied to which kind of patient characteristics (kind of pathology, severity, and time since lesion, anthropometry, etc.) in order to obtain the best therapeutic outcome. New sensitive assessment methods, which include non-sensorimotor assessments such as cognitive ability and quality of life, will be required to better distinguish among the different patient characteristics and detect already small changes in the therapeutic outcomes.

Acknowledgements We would like to thank Lars Luenenburger and Daniele Munari from Hocoma, Switzerland, for their enriching feedback. Further, we would like to thank Katherine G. August for her support in the organization of the structure and language of this manuscript. This work was supported in parts by the Swiss National Science Foundation (SNF) under the grant PP00P2 163800, the Dutch Research Council (NWO) Talent Program Vidi TTW 2020, the United States National Institute on Disability and Rehabilitation Research (NIDRR) under Grant H133E070013, the National Center of Competence in Research (NCCR) of the Swiss National Science Foundation (SNF) on Neural Plasticity and Repair and on Robotics, the European Union Project MIMICS funded by the European Community's Seventh Framework Program (FP7/2007–2013) under grant agreement n 215756, the Swiss Commission for Technology and Innovation (CTI) projects 6199.1-MTS and 7497.1 LSPP-LS, the Olga Mayenfisch Foundation, and the Bangerter-Rhyner Foundation.

References

1. Colombo G, Joerg M, Schreier R, Dietz V. Treadmill training of paraplegic patients using a robotic orthosis. J Rehabil Res Dev. 2000;37:693–700.
2. Hesse S, Uhlenbrock D. A mechanized gait trainer for restoration of gait. J Rehabil Res Dev. 2000;37:701–8.
3. Nadeau SE, Wu SS, Dobkin BH, Azen SP, Rose DK, Tilson JK, et al. Effects of task-specific and impairment-based training compared with usual care on functional walking ability after inpatient stroke rehabilitation: LEAPS trial. Neurorehabil Neural Repair. 2013;27:370–80. https://doi.org/10.1177/1545968313481284.
4. Kwakkel G, van Peppen R, Wagenaar RC, Wood Dauphinee S, Richards C, Ashburn A, et al. Effects of augmented exercise therapy time after stroke: a meta-analysis. Stroke. 2004;35:2529–39. https://doi.org/10.1161/01.STR.0000143153.76460.7d.
5. Globas C, Becker C, Cerny J, Lam JM, Lindemann U, Forrester LW, et al. Chronic stroke survivors benefit from high-intensity aerobic treadmill exercise: a randomized control trial. Neurorehabil Neural Repair. 2012;26:85–95. https://doi.org/10.1177/1545968311418675.
6. Wirz M, Zemon DH, Rupp R, Scheel A, Colombo G, Dietz V, et al. Effectiveness of automated locomotor training in patients with chronic incomplete spinal cord injury: a multicenter trial. Arch Phys Med Rehabil. 2005;86:672–80. https://doi.org/10.1016/j.apmr.2004.08.004.
7. Morrison SA, Backus D. Locomotor training: is translating evidence into practice financially feasible? J Neurol Phys Ther. 2007;31:50–4. https://doi.org/10.1097/NPT.0b013e3180690679.
8. Veneman JF, Kruidhof R, Hekman EEG, Ekkelenkamp R, Asseldonk EHFV, van der Kooij H. Design and evaluation of the LOPES exoskeleton robot for interactive gait rehabilitation. IEEE Trans Neural Syst Rehabil Eng. 2007;15:379–86. https://doi.org/10.1109/TNSRE.2007.903919.
9. Meuleman J, van Asseldonk E, van Oort G, Rietman H, van der Kooij H. LOPES II–design and evaluation of an admittance controlled gait training robot with shadow-leg approach. IEEE Trans Neural Syst Rehabil Eng. 2016;24:352–63. https://doi.org/10.1109/TNSRE.2015.2511448.
10. Banala SK, Kim SH, Agrawal SK, Scholz JP. Robot assisted gait training with active leg exoskeleton (ALEX). IEEE Trans Neural Syst Rehabil Eng. 2009;17:2–8. https://doi.org/10.1109/TNSRE.2008.2008280.
11. Aoyagi D, Ichinose WE, Harkema SJ, Reinkensmeyer DJ, Bobrow JE. A robot and control algorithm that can synchronously assist in naturalistic motion during body-weight-supported gait training following neurologic injury. IEEE Trans Neural Syst Rehabil Eng. 2007;15:387–400. https://doi.org/10.1109/TNSRE.2007.903922.
12. Reinkensmeyer DJ, Aoyagi D, Emken JL, Galvez JA, Ichinose W, Kerdanyan G, et al. Tools for understanding and optimizing robotic gait training. J Rehabil Res Dev. 2006;43:657–70. https://doi.org/10.1682/jrrd.2005.04.0073.
13. Rodríguez-Fernández A, Lobo-Prat J, Font-Llagunes JM. Systematic review on wearable lower-limb exoskeletons for gait training in neuromuscular impairments. J Neuroeng Rehabil. 2021;18:22. https://doi.org/10.1186/s12984-021-00815-5.
14. Cramer SC, Sur M, Dobkin BH, O'Brien C, Sanger TD, Trojanowski JQ, et al. Harnessing neuroplasticity for clinical applications. Brain. 2011;134:1591–609. https://doi.org/10.1093/brain/awr039.
15. Chen G, Chan CK, Guo Z, Yu H. A review of lower extremity assistive robotic exoskeletons in rehabilitation therapy. Crit Rev Biomed Eng. 2013;41:343–63. https://doi.org/10.1615/critrevbiomedeng.2014010453.
16. Hobbs B, Artemiadis P. A review of robot-assisted lower-limb stroke therapy: unexplored paths and future directions in gait rehabilitation. Front Neurorobotic. 2020;14. https://doi.org/10.3389/fnbot.2020.00019.
17. Tucker MR, Olivier J, Pagel A, Bleuler H, Bouri M, Lambercy O, et al. Control strategies for active lower extremity prosthetics and orthotics: a review. J NeuroEngineering Rehabil. 2015;12:1. https://doi.org/10.1186/1743-0003-12-1.
18. Schmidt H, Hesse S, Bernhardt R, Krüger J. HapticWalker—a novel haptic foot device. ACM Trans Appl Percept. 2005;2:166–80. https://doi.org/10.1145/1060581.1060589.
19. Hesse S, Waldner A, Tomelleri C. Innovative gait robot for the repetitive practice of floor walking and stair climbing up and down in stroke patients. J Neuroeng Rehabil. 2010;7:30. https://doi.org/10.1186/1743-0003-7-30.
20. Ruthenberg BJ, Wasylewski NA, Beard JE. An experimental device for investigating the force and power requirements of a powered gait orthosis. J Rehabil Res Dev. 1997;34:203–13.
21. Apte S, Plooij M, Vallery H. Influence of body weight unloading on human gait characteristics: a systematic review. J Neuroeng Rehabil. 2018;15:53. https://doi.org/10.1186/s12984-018-0380-0.
22. Lünenburger L, Lam T, Riener R, Colombo G. Gait retraining after neurological disorders. Wiley Encyclopedia of Biomedical Engineering, Wiley; 2006. https://doi.org/10.1002/9780471740360.ebs1393.

23. Frey M, Colombo G, Vaglio M, Bucher R, Jörg M, Riener R. A novel mechatronic body weight support system. IEEE Trans Neural Syst Rehabil Eng. 2006;14:311–21. https://doi.org/10.1109/TNSRE.2006.881556.

24. Plooij M, Apte S, Keller U, Baines P, Sterke B, Asboth L, et al. Neglected physical human-robot interaction may explain variable outcomes in gait neurorehabilitation research. Sci Robot. 2021. https://doi.org/10.1126/scirobotics.abf1888.

25. Colombo G, Bucher R. Device for adjusting the prestress of an elastic means around a predetermined tension or position, 2008.

26. Basalp E, Wolf P, Marchal-Crespo L. Haptic training: which types facilitate (re)learning of which motor task and for whom answers by a review. IEEE Trans Haptics 2021:1–1. https://doi.org/10.1109/TOH.2021.3104518.

27. Israel JF, Campbell DD, Kahn JH, Hornby TG. Metabolic costs and muscle activity patterns during robotic- and therapist-assisted treadmill walking in individuals with incomplete spinal cord injury. Phys Ther. 2006;86:1466–78. https://doi.org/10.2522/ptj.20050266.

28. Hidler J, Nichols D, Pelliccio M, Brady K, Campbell DD, Kahn JH, et al. Multicenter randomized clinical trial evaluating the effectiveness of the Lokomat in subacute stroke. Neurorehabil Neural Repair. 2009;23:5–13. https://doi.org/10.1177/1545968308326632.

29. Wu HG, Miyamoto YR, Gonzalez Castro LN, Ölveczky BP, Smith MA. Temporal structure of motor variability is dynamically regulated and predicts motor learning ability. Nat Neurosci. 2014;17:312–21. https://doi.org/10.1038/nn.3616.

30. Bernshtein NA. The co-ordination and regulation of movements. Oxford; New York: Pergamon Press; 1967.

31. Reinkensmeyer DJ, Burdet E, Casadio M, Krakauer JW, Kwakkel G, Lang CE, et al. Computational neurorehabilitation: modeling plasticity and learning to predict recovery. J Neuroeng Rehabil. 2016;13:42. https://doi.org/10.1186/s12984-016-0148-3.

32. Lewek MD, Cruz TH, Moore JL, Roth HR, Dhaher YY, Hornby TG. Allowing intralimb kinematic variability during locomotor training poststroke improves kinematic consistency: a subgroup analysis from a randomized clinical trial. Phys Ther. 2009;89:829–39. https://doi.org/10.2522/ptj.20080180.

33. Ziegler MD, Zhong H, Roy RR, Edgerton VR. Why variability facilitates spinal learning. J Neurosci. 2010;30:10720–6. https://doi.org/10.1523/JNEUROSCI.1938-10.2010.

34. Riener R, Lünenburger L, Jezernik S, Anderschitz M, Colombo G, Dietz V. Patient-cooperative strategies for robot-aided treadmill training: first experimental results. IEEE Trans Neural Syst Rehabil Eng. 2005;13:380–94. https://doi.org/10.1109/TNSRE.2005.848628.

35. Duschau-Wicke A, von Zitzewitz J, Caprez A, Lunenburger L, Riener R. Path control: a method for patient-cooperative robot-aided gait rehabilitation. IEEE Trans Neural Syst Rehabil Eng. 2010;18:38–48. https://doi.org/10.1109/TNSRE.2009.2033061.

36. Hogan N. Impedance control: an approach to manipulation. American Control Conference. 1984;1984:304–13.

37. Krishnan C, Kotsapouikis D, Dhaher YY, Rymer WZ. Reducing robotic guidance during robot-assisted gait training improves gait function: a case report on a stroke survivor. Arch Phys Med Rehabil. 2013;94:1202–6. https://doi.org/10.1016/j.apmr.2012.11.016.

38. Krishnan C, Ranganathan R, Kantak SS, Dhaher YY, Rymer WZ. Active robotic training improves locomotor function in a stroke survivor. J Neuroeng Rehabil. 2012;9:57. https://doi.org/10.1186/1743-0003-9-57.

39. Schück A, Labruyère R, Vallery H, Riener R, Duschau-Wicke A. Feasibility and effects of patient-cooperative robot-aided gait training applied in a 4-week pilot trial. J Neuroeng Rehabil. 2012;9:31. https://doi.org/10.1186/1743-0003-9-31.

40. Marchal-Crespo L, Reinkensmeyer DJ. Review of control strategies for robotic movement training after neurologic injury. J Neuroeng Rehabil. 2009;6:20. https://doi.org/10.1186/1743-0003-6-20.

41. Veldema J, Jansen P. Resistance training in stroke rehabilitation: systematic review and meta-analysis. Clin Rehabil. 2020;34:1173–97. https://doi.org/10.1177/0269215520932964.

42. Wei Y, Patton J, Bajaj P, Scheidt R. A real-time haptic/graphic demonstration of how error augmentation can enhance learning. In: Proceedings of the 2005 IEEE international conference on robotics and automation. 2005. p. 4406–11. https://doi.org/10.1109/ROBOT.2005.1570798.

43. Marchal-Crespo L, Michels L, Jaeger L, Lopez-Oloriz J, Riener R. Effect of error augmentation on brain activation and motor learning of a complex locomotor task. Front Neurosci. 2017;11. https://doi.org/10.3389/fnins.2017.00526.

44. Marchal-Crespo L, Tsangaridis P, Obwegeser D, Maggioni S, Riener R. Haptic error modulation outperforms visual error amplification when learning a modified gait pattern. Front Neurosci. 2019;13. https://doi.org/10.3389/fnins.2019.00061.

45. Blanchette AK, Noël M, Richards CL, Nadeau S, Bouyer LJ. Modifications in ankle dorsiflexor activation by applying a torque perturbation during walking in persons post-stroke: a case series. J Neuroeng Rehabil. 2014;11:98. https://doi.org/10.1186/1743-0003-11-98.

46. Ouellette MM, LeBrasseur NK, Bean JF, Phillips E, Stein J, Frontera WR, et al. High-intensity resistance training improves muscle strength, self-reported function, and disability in long-term stroke survivors. Stroke. 2004;35:1404–9. https://doi.org/10.1161/01.STR.0000127785.73065.34.

47. Lamberti N, Straudi S, Malagoni AM, Argirò M, Felisatti M, Nardini E, et al. Effects of low-intensity endurance and resistance training on mobility in chronic stroke survivors: a pilot randomized controlled study. Eur J Phys Rehabil Med. 2017;53:228–39. https://doi.org/10.23736/S1973-9087.16.04322-7.

48. Liu LY, Li Y, Lamontagne A. The effects of error-augmentation versus error-reduction paradigms in robotic therapy to enhance upper extremity performance and recovery post-stroke: a systematic review. J Neuroeng Rehabil. 2018;15:65. https://doi.org/10.1186/s12984-018-0408-5.

49. Maggioni S, Reinert N, Lünenburger L, Melendez-Calderon A. An adaptive and hybrid end-point/joint impedance controller for lower limb exoskeletons. Front Robot AI. 2018;5:104. https://doi.org/10.3389/frobt.2018.00104.

50. Lotze M, Braun C, Birbaumer N, Anders S, Cohen LG. Motor learning elicited by voluntary drive. Brain. 2003;126:866–72. https://doi.org/10.1093/brain/awg079.

51. Fricke SS, Bayón C, der Kooij H van, F. van Asseldonk EH. Automatic versus manual tuning of robot-assisted gait training in people with neurological disorders. J NeuroEng Rehabil 2020;17:9. https://doi.org/10.1186/s12984-019-0630-9.

52. Chaparro-Rico BDM, Cafolla D, Tortola P, Galardi G. Assessing stiffness, joint torque and rom for paretic and non-paretic lower limbs during the subacute phase of stroke using lokomat tools. Appl Sci. 2020;10:6168. https://doi.org/10.3390/app10186168.

53. Lance J. Pathophysiology of spasticity and clinical experience with Baclofen. Spasticity: disordered motor control. Lance JW, Feldmann RG, Young RR, Koella WP, editors. Chicago: Year Book; 1980, p. 184–204.

54. Sanger TD, Delgado MR, Gaebler-Spira D, Hallett M, Mink JW. Task force on childhood motor disorders. Classification and definition of disorders causing hypertonia in childhood. Pediatrics 2003;111:e89–97. https://doi.org/10.1542/peds.111.1.e89.

55. Ashworth B. Preliminary trial of carisoprodol in multiple sclerosis. The Practitioner 1964.

56. Bohannon RW, Smith MB. Interrater reliability of a modified Ashworth scale of muscle spasticity. Phys Ther. 1987;67:206–7.

57. Lunenburger L, Colombo G, Riener R, Dietz V. Clinical assessments performed during robotic rehabilitation by the gait training robot Lokomat. In: 9th International conference on rehabilitation robotics, ICORR; 2005. p. 345–8. https://doi.org/10.1109/ICORR.2005.1501116.

58. Riener R, Lünenburger L, Colombo G. Human-centered robotics applied to gait training and assessment. J Rehabil Res Dev. 2006;43:679–94. https://doi.org/10.1682/jrrd.2005.02.0046.

59. Schmartz AC, Meyer-Heim AD, Müller R, Bolliger M. Measurement of muscle stiffness using robotic assisted gait orthosis in children with cerebral palsy: a proof of concept. Disabil Rehabil Assist Technol. 2011;6:29–37. https://doi.org/10.3109/17483107.2010.509884.

60. Bolliger M, Lunenburger L, Bircher, Colombo G, Dietz D. Reliability of measuring isometric peak torque in the driven gait orthosis Lokomat. Hong Kong: 2006.

61. Bolliger M, Banz R, Dietz V, Lünenburger L. Standardized voluntary force measurement in a lower extremity rehabilitation robot. J Neuroeng Rehabil. 2008;5:23. https://doi.org/10.1186/1743-0003-5-23.

62. Galen SS, Clarke CJ, Mclean AN, Allan DB, Conway BA. Isometric hip and knee torque measurements as an outcome measure in robot assisted gait training. NeuroRehabilitation. 2014;34:287–95. https://doi.org/10.3233/NRE-131042.

63. Maggioni S, Melendez-Calderon A, van Asseldonk E, Klamroth-Marganska V, Lünenburger L, Riener R, et al. Robot-aided assessment of lower extremity functions: a review. J Neuroeng Rehabil. 2016;13:72. https://doi.org/10.1186/s12984-016-0180-3.

64. Domingo A, Lam T. Reliability and validity of using the Lokomat to assess lower limb joint position sense in people with incomplete spinal cord injury. J Neuroeng Rehabil. 2014;11:167. https://doi.org/10.1186/1743-0003-11-167.

65. Chisholm AE, Domingo A, Jeyasurya J, Lam T. Quantification of lower extremity kinesthesia deficits using a robotic exoskeleton in people with a spinal cord injury. Neurorehabil Neural Repair. 2016;30:199–208. https://doi.org/10.1177/1545968315591703.

66. Basmajian J. Muscles Alive—their functions revealed by electromyography. 4th ed. Baltimore: Williams and Wilkins; 1978.

67. Schmidt RA, Wrisberg CA. Motor learning and performance: a situation-based learning approach, 4th ed. Champaign, IL, US: Human Kinetics; 2008.

68. Lünenburger L, Colombo G, Riener R, Dietz V. Biofeedback in gait training with the robotic orthosis Lokomat. Conf Proc IEEE Eng Med Biol Soc. 2004;2004:4888–91. https://doi.org/10.1109/IEMBS.2004.1404352.

69. Lünenburger L, Colombo G, Riener R. Biofeedback for robotic gait rehabilitation. J Neuroeng Rehabil. 2007;4:1. https://doi.org/10.1186/1743-0003-4-1.

70. Banz R, Bolliger M, Colombo G, Dietz V, Lünenburger L. Computerized visual feedback: an adjunct to robotic-assisted gait training. Phys Ther. 2008;88:1135–45. https://doi.org/10.2522/ptj.20070203.

71. Banz R, Bolliger M, Muller S, Santelli C, Riener R. A Method of estimating the degree of active participation during stepping in a driven gait orthosis based on actuator force profile matching. IEEE Trans Neural Syst Rehabil Eng. 2009;17:15–22. https://doi.org/10.1109/TNSRE.2008.2008281.

72. Mehrholz J, Thomas S, Kugler J, Pohl M, Elsner B. Electromechanical-assisted training for walking after

stroke. Cochrane Database Syst Rev. 2020. https://doi.org/10.1002/14651858.CD006185.pub5.

73. Manuli A, Maggio MG, Latella D, Cannavò A, Balletta T, De Luca R, et al. Can robotic gait rehabilitation plus virtual reality affect cognitive and behavioural outcomes in patients with chronic stroke? A randomized controlled trial involving three different protocols. J Stroke Cerebrovasc Dis. 2020;29: 104994. https://doi.org/10.1016/j.jstrokecerebrovasdis.2020.104994.

74. Emken JL, Harkema SJ, Beres-Jones JA, Ferreira CK, Reinkensmeyer DJ. Feasibility of manual teach-and-replay and continuous impedance shaping for robotic locomotor training following spinal cord injury. IEEE Trans Biomed Eng. 2008;55:322–34. https://doi.org/10.1109/TBME.2007.910683.

75. Lunenburger L, Wellner M, Banz R, Colombo G, riener r. combining immersive virtual environments with robot-aided gait training. in: ieee 10th international conference on rehabilitation robotics, 2007. p. 421–4. https://doi.org/10.1109/ICORR.2007.4428459.

76. Wagner J, Solis-Escalante T, Scherer R, Neuper C, Müller-Putz G. It's how you get there: walking down a virtual alley activates premotor and parietal areas. Front Hum Neurosci. 2014;8:93. https://doi.org/10.3389/fnhum.2014.00093.

77. Calabrò RS, Naro A, Russo M, et al. The role of virtual reality in improving motor performance as revealed by EEG: arandomized clinical trial. J Neuroeng Rehabil. 2017;14(1):53. Published 2017 Jun 7. https://doi.org/10.1186/s12984-017-0268-4.

78. Keshner EA, Lamontagne A. The untapped potential of virtual reality in rehabilitation of balance and gait in neurological disorders. Front Virtual Rity. 2021;2:6. https://doi.org/10.3389/frvir.2021.641650.

79. Aburub AS, Lamontagne A. Altered steering strategies for goal-directed locomotion in stroke. J Neuroeng Rehabil. 2013;10:80. https://doi.org/10.1186/1743-0003-10-80.

80. Willaert I, Aissaoui R, Nadeau S, Duclos C, Labbe DR. Modulating the gait of a real-time self-avatar to induce changes in stride length during treadmill walking. In: IEEE conference on virtual reality and 3D user interfaces abstracts and workshops (VRW). 2020. p. 718–9. https://doi.org/10.1109/VRW50115.2020.00210.

81. Riener R, Lünenburger L, Maier I, Colombo G, Dietz V. Locomotor training in subjects with sensorimotor deficits: an overview of the robotic gait orthosis lokomat. J Healthc Eng. 2010;1:197–216. https://doi.org/10.1260/2040-2295.1.2.197.

82. Lo AC, Triche EW. Improving gait in multiple sclerosis using robot-assisted, body weight supported treadmill training. Neurorehabil Neural Repair. 2008;22:661–71. https://doi.org/10.1177/1545968308318473.

83. Borggraefe I, Schaefer JS, Klaiber M, Dabrowski E, Ammann-Reiffer C, Knecht B, et al. Robotic-assisted treadmill therapy improves walking and standing performance in children and adolescents with cerebral palsy. Eur J Paediatr Neurol. 2010;14:496–502. https://doi.org/10.1016/j.ejpn.2010.01.002.

84. Hornby TG, Campbell DD, Kahn JH, Demott T, Moore JL, Roth HR. Enhanced gait-related improvements after therapist- versus robotic-assisted locomotor training in subjects with chronic stroke: a randomized controlled study. Stroke. 2008;39:1786–92. https://doi.org/10.1161/STROKEAHA.107.504779.

85. Dundar U, Toktas H, Solak O, Ulasli AM, Eroglu S. A comparative study of conventional physiotherapy versus robotic training combined with physiotherapy in patients with stroke. Top Stroke Rehabil. 2014;21:453–61. https://doi.org/10.1310/tsr2106-453.

86. Maggio MG, Torrisi M, Buda A, De Luca R, Piazzitta D, Cannavò A, et al. Effects of robotic neurorehabilitation through lokomat plus virtual reality on cognitive function in patients with traumatic brain injury: a retrospective case-control study. Int J Neurosci. 2020;130:117–23. https://doi.org/10.1080/00207454.2019.1664519.

87. Campagnini S, Liuzzi P, Mannini A, et al. Effects of control strategies on gait in robot-assisted poststroke lower limb rehabilitation: a systematic review. J NeuroEngineering Rehabil. 2022;19(52). https://doi.org/10.1186/s12984-022-01031-5.

88. de Miguel-Fernandez J, Lobo-Prat J, Prinsen E, Font-Llagunes JM, Marchal-Crespo L. Control strategies used in lower limb exoskeletons for gait rehabilitation after brain injury: a systematic review and analysis of clinical effectiveness. Research Square; 2022. https://doi.org/10.21203/rs.3.rs-1195778/v1.

89. Wenk N, Penalver-Andres J, Buetler KA, et al. Effect of immersive visualization technologies on cognitive load, motivation, usability, and embodiment. Virtual Reality (2021). https://doi.org/10.1007/s10055-021-00565-8.

Using Robotic Exoskeletons for Overground Locomotor Training

30

Arun Jayaraman, William Z. Rymer, Matt Giffhorn, and Megan K. O'Brien

Abstract

Over the past decade, overground robotic exoskeletons have emerged as promising technologies that can be integrated into the rehabilitation process to help individuals maintain or regain neuromuscular health following neurological injury. Early studies suggest that individuals recovering from stroke, spinal cord injury, and other neurological conditions can benefit from the use of exoskeletons, either alone or as a complement to traditional rehabilitation strategies, to improve mobility and independence. Due to the broad range of impairments observed after neurological injury, clinicians should consider different types of exoskeletons to best suit the goals of each patient. The use of these exoskeletons as clinical tools also requires clinicians to understand how to operate and monitor the device, to identify which patient population(s) are appropriate and how they may benefit from the device in rehabilitation, and the limitations and safety measures required for each device. More research in this field, including large-scale clinical trials to assess the therapeutic benefits and limitations of exoskeletons, is required to achieve a greater understanding of how to optimize the use of these devices in the clinic and for personal mobility.

A. Jayaraman (✉) · W. Z. Rymer · M. Giffhorn · M. K. O'Brien
Shirley Ryan AbilityLab, Chicago, IL, USA
e-mail: ajayaraman@sralab.org;
a-jayaraman@northwestern.edu

M. Giffhorn
e-mail: mgiffhorn@sralab.org

M. K. O'Brien
e-mail: mobrien02@sralab.org

A. Jayaraman · W. Z. Rymer · M. K. O'Brien
Department of Physical Medicine and Rehabilitation,
Northwestern University Feinberg School of
Medicine, Chicago, IL, USA

W. Z. Rymer
Department of Neuroscience, Northwestern
University Feinberg School of Medicine, Chicago,
IL, USA

Department of Biomedical Engineering,
Northwestern University McCormick School of
Engineering, Evanston, IL, USA
e-mail: zrymer@sralab.org

Keywords

Exoskeletons · Neurological injury · Assistive robots · Robotics · Rehabilitation

30.1 Introduction and Brief History of Exoskeletons

Exoskeletons are assistive technologies designed to augment human capabilities. Some of the earliest concepts of exoskeleton-like devices date back to the late nineteenth century [1, 2]. More

© The Author(s), under exclusive license to Springer Nature Switzerland AG 2022
D. J. Reinkensmeyer et al. (eds.), *Neurorehabilitation Technology*,
https://doi.org/10.1007/978-3-031-08995-4_30

recent attempts at developing powered exoskeletons began in the 1960s and continued through the 1980s and 1990s, spurred by military funding and a desire to enhance soldier performance. Unfortunately, many of these early exoskeleton prototypes never came to fruition due to their size, weight, and power-supply issues [3].

In the past two decades, research on exoskeletons has significantly advanced because of continued U.S. military funding and support, as well as global interest in developing robotic machines that augment human capabilities, such as for individuals with weakened muscles due to aging, disease, or injury. In the early 2000s, the Defense Advanced Research Projects Agency (DARPA) initiated the Exoskeleton for Human Performance Augmentation (EHPA) project, which led to the development of additional exoskeleton designs, including the BLEEX (Berkley Lower Extremity Exoskeleton) designed at U.C. Berkeley. Powered by hydraulic actuators at the hip, knee, and ankle, the BLEEX was designed to allow a user to carry heavy loads with substantially reduced effort. It was the first successfully demonstrated exoskeleton to accomplish this goal and maintained maneuverability of the user while walking, running, squatting, bending, or twisting [4, 5].

More recently, exoskeletons have been developed and commercialized specifically for their potential as therapeutic tools and personal mobility devices. The most common populations served are persons with functional gait deficits, such as individuals with spinal cord injury (SCI), stroke, and the elderly. The therapeutic use of exoskeletons may fill the gap in effective treatments for restoring upper and lower limb function in these individuals. For persons who are confined to wheelchairs and experience secondary medical complications due to immobility, exoskeletons may help build endurance and strengthen muscle groups related to walking, prevent secondary adverse effects associated with chronic injury, and regain mobility if they choose to use the device for ambulation in the home or community. For persons with considerable gait deficits, exoskeletons may offer therapeutic benefits such as overground stepping practice, loading of the limbs, and balance and posture control, while reducing the need for therapist assistance.

30.2 Currently Available Devices

The landscape of available exoskeletons is continuously developing. Each of these devices can differ in its structure, weight, hardware, and control mechanisms. As evidence unfolds regarding the efficacy of these devices, so do the recommended locomotor training protocols. The breadth of use with varied clinical populations also continues to expand.

There are currently hundreds of exoskeletons being developed and tested in academic and industry settings. In this chapter, we discuss examples of available, powered, overground exoskeletons for locomotor training, categorized by type of device (rigid lower-body, modular, soft) and their potential clinical applications. These devices are summarized in Table 30.1, all of which are FDA/CE approved or in the process of obtaining approval. Additional literature is available for more extensive overviews of various exoskeleton designs [6, 7], guidelines [8], and efficacy [9–11].

30.3 Rigid Lower-Body Exoskeletons

This class of exoskeletons provides the highest level of assistance to the user. Designed for individuals with severe mobility deficits, these devices are comprised of a rigid, over-body frame connecting all segments of the lower limbs with actuation at multiple joints. Due to their size, weight, and complex control schema, clinician support and supervision are essential during use to maximize safety and effectiveness. For individuals who attain sufficient recovery after training with these exoskeletons, alternative, less-constrained devices may be considered for downstream training to promote functional progression.

Table 30.1 Summary of powered, overground exoskeletons discussed in this chapter

Device name	Type	Target population (s)	Powered joints	Unilateral (U) or Bilateral (B)
ReWalk	Rigid lower-body	SCI	Hip and knee	U/B
Ekso	Rigid lower-body	SCI, stroke, brain injury	Hip and knee	U/B
Indego	Rigid lower-body	SCI, stroke	Hip and knee	U/B
Atalante	Rigid lower-body	SCI	Hip and knee	B
Honda walking assist device	Modular	Stroke, aging	Hip	B
Samsung GEMS-H	Modular	Stroke, aging	Hip	B
Keeogo	Modular	Stroke	Knee	B
ReWalk restore	Soft	Stroke	Ankle	U
MyoSuit	Soft	General weakness	Knee	B
BiOMOTUM Spark	Soft	Cerebral palsy	Ankle	U/B

***ReWalk*™ (ReWalk Robotics, Inc., Marlboro, MA, USA and Yokneam Illit, Israel):** ReWalk Robotics, Inc., formerly Argo Medical Technologies, was founded in 2001 by Amit Goffer, PhD, an electrical and computer engineer whose own experience with SCI inspired him to develop the ReWalk™ [12]. The ReWalk™, a lower limb exoskeleton, provides powered hip and knee motion to assist individuals with SCI in standing upright and walking, as well as ascending and descending stairs.

ReWalk Inc. produces two exoskeleton products that differ slightly—The ReWalk™ Rehabilitation, for therapeutic purposes, and the ReWalk™ Personal, designed to provide personal mobility in the home and community. The ReWalk™ Personal was the first exoskeleton to have received FDA clearance for personal use in the United States; this device is also available in parts of Europe. In this chapter, we will focus on the ReWalk™ Rehabilitation unit (Fig. 30.1). This exoskeleton includes bilateral hip and knee joint actuator motors powered by rechargeable batteries, as well as sensors that measure upper-body tilt angle, joint angles, and ground contact [13, 14]. Users operate the ReWalk™ through minor trunk movements; for example, a user shifts the trunk forward, which is detected by the tilt sensors, and initiates the device's leg swing to begin walking. The user must then return the

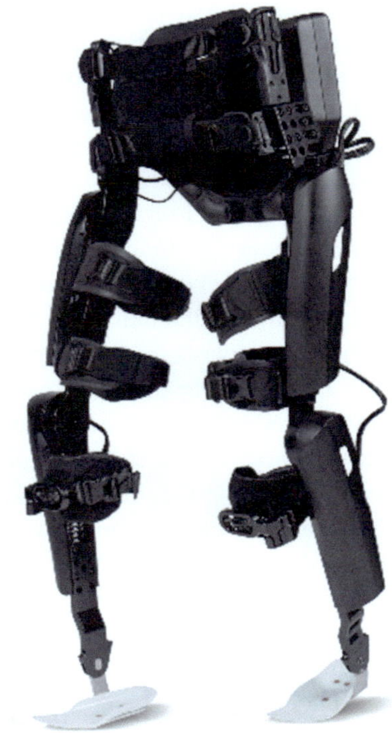

Fig. 30.1 The ReWalk 6.0 (Photo courtesy of ReWalk Robotics, Inc.)

trunk to an upright position after step initiation to complete the swing and ensure that the leg clears the ground. Arm crutches are used to maintain stability during activity with the exoskeleton.

Clinical Applications: The ReWalk™ is intended for persons with lower extremity paralysis or paresis due to SCI (level T7 to L5 for home use when accompanied by a trained caregiver, and levels T4 to T6 when used in a rehabilitation setting) [15]. Individuals who use the ReWalk™ must have adequate bilateral upper limb strength (for stabilizing the body on crutches), adequate trunk control, adequate femur and lower limb length, and sufficient lower extremity range of motion that allows for ambulation. They must also have adequate bone density (no fractures), adequate blood pressure tolerance for upright positioning, and be able to tolerate standing and participating in a walking program.

Mobility Training: Multiple studies have evaluated the safety and efficacy of the ReWalk™ to help restore locomotor abilities in people with SCI at the motor complete thoracic level [13, 16, 17]. Additional studies are underway to assess training strategies (for ambulation over level surfaces, non-level surfaces, stairs, ramps, and curbs) for using the ReWalk™ in persons with SCI.

At present, it is not clear how many training sessions are necessary for individuals to become proficient with the ReWalk™. During therapy sessions, the goal is for the individual to use the device with as little assistance from the therapist as possible. Early training sessions should focus on specific balance exercises, how to move with the crutches, how to weight shift, and finding a center standing position [14, 16]. The user must be proficient in these tasks before they begin to walk with the device. Users must then master the process of triggering the first step, followed by learning proper weight shifts and timing with the crutches. Device settings can be adjusted as users' walking ability improves. Specific training guidelines are outlined in a training manual provided by ReWalk Robotics Inc. Additionally, Esquenazi et al. outlined a gait training schedule involving 18 sessions. In sessions 1–4, therapists would perform all necessary measurements and begin helping the user with sit-to-stand and stand-to-sit transitions within parallel bars; progressively, the user begins walking with the crutches (manual trigger mode followed by using the tilt

sensor) and starts performing sit-to-stand and stand-to-sit transitions using the crutches [13].

In a two-year clinical trial, Esquenazi et al. examined the safety and effectiveness of the ReWalk™ for 12 individuals with paraplegia due to SCI when performing routine ambulatory functions, such as sit-to-stand transitions or walking [13]. Training included 2–3 sessions per week for 8 to 10 weeks (i.e., 13 to 24 sessions), with each session lasting 1–2 h. The authors concluded that participants could walk independently using the ReWalk™ for at least 50 to 100 m continuously, for a period of at least 5–10 min, and perform transfers without therapist assistance. Walking velocities ranged from 0.03 to 0.45 m/s (mean, 0.25 m/s). Additionally, some subjects reported enhanced physical benefits, such as reduced spasticity and pain, as well as improved bowel and bladder function. Other studies have identified improvements in walking speed and endurance [14, 17].

Some adverse effects have been reported with using the ReWalk™, including minor fall-like events, skin breakdown issues, and hairline fractures; however, no serious adverse events have been reported [16]. Benson et al. evaluated the neurological and functional effects of using the ReWalk™ in individuals with chronic SCI [16]. They observed that walking speeds and distances improved in ReWalk™ users compared to patients who did not use the device; however, perceived benefits of using the exoskeleton did not always meet subjects' expectations, and the authors also noted a relatively high number of device-related skin aberrations.

To date, there have been few evaluations of rigid exoskeletons outside of the clinic. One such study examined the usability of the ReWalk™ Personal in the home and community [18]. Following an 8-week training, 14 individuals with complete SCI were given the exoskeleton for 2–3 weeks for personal use without restrictions (under the supervision of a companion). Their device usage—including amount, purpose, and location of use—was recorded. Median use of the exoskeleton occurred on 9 out of 16 days for 49 min per day. Participants predominantly used the exoskeleton for exercise and social events, and 48% of sessions occurred exclusively outdoors. Participants reported overall satisfaction with the exoskeleton, but on average were dissatisfied with its weight, effectiveness, ease of use, and safety, and they perceived that the exoskeleton had limited usability in home and for indoor daily activities. Overall, this study demonstrates the potential of the ReWalk™ to facilitate exercise and social interaction for individuals with complete SCI in the home and community. Improving transportability and ease of use may increase its potential as an assistive device for other daily activities.

EksoNR™ (Ekso Bionics, Berkeley CA, USA): Researchers from the University of California Berkeley Robotics & Human Engineering Laboratory formed Ekso Bionics—formerly known as Berkeley Bionics—in 2005 with grant support from the Department of Defense; the company also has partnerships with UC Berkeley and licensing technology agreements with Lockheed Martin Corporation [19]. Initially named eLEGS, the EksoNR™ is a lower extremity exoskeleton intended for use as a gait training tool, designed for individuals with lower extremity weakness following SCI, stroke, or acquired brain injury. The EksoNR™ features a variable assist program that allows therapists to adjust how much assistance the device provides based on the user's needs and current ability; reducing the amount of assistance allows the user to do more on their own. Initially, patients must learn to balance and shift their weight when wearing the EksoNR™, so that they can safely and effectively ambulate with the device.

Clinical Applications: The EksoNR™ has been FDA approved for individuals who have lower extremity weakness or paralysis as a result of complete SCI (T4 or below), incomplete SCI (C7 or below) with functional bilateral upper extremity strength or functional strength of one upper extremity and one lower extremity, and stroke (hemiparesis or hemiplegia). EksoNR™ is currently the only FDA-cleared exoskeleton for use in acquired brain injury (ABI). During patient evaluation and selection, precautions should be considered for persons with open wounds, uncontrolled orthostatic hypotension, active heterotopic ossification, or cognitive impairments that interfere with their ability to communicate. Ekso Bionics provides detailed recommendations for inclusion and exclusion criteria in their training manual.

Patients should receive medical clearance from a physician and be evaluated by a physical therapist prior to training with the device. Training will differ for each population and their specific needs and goals; however, for all populations, the first step should focus on teaching the patient how to balance and weight shift properly in order to operate the exoskeleton. Although the EksoNR™ provides external stability in order to keep the patient upright, the patient must practice static and dynamic standing balance activities to ensure they can maintain their balance without strenuous physical effort. The patient should then practice forward, backward, and lateral weight shifting using their own balance reactions (if they are able to use their legs) or their upper extremities (if they do not have lower extremity control) to return to the balanced start position. Once the patient can adequately find balance, the user can begin walking in the EskoNR™ using a walker, with the goal of achieving a natural heel-strike pattern. Users then progress through the variable assist walk modes, with supervision and assistance from a physical therapist.

Mobility Training: Clinical trials with the EskoNR™ began in 2012 and have primarily focused on the safety and feasibility of the device for individuals with SCI and stroke, and on evaluating its potential as an effective therapeutic training device. Training sessions have ranged from 6 to 24 sessions, from one to three times a week [20–25]. Outcome measures have included energy expenditure, walking speeds and distances, balance, exercise conditioning effects, changes in spasticity, as well as blood pressure and pain levels [20]. Decreasing the amount of robotic assistance during training sessions allows patients to utilize emerging muscle activation during ambulation. Thus, for individuals affected by moderate to severe stroke, the EksoNR™ has the potential to increase step length, stride length, gait speed, walking endurance, overall balance, and confidence with ambulation [26]. With practice, some patients with SCI demonstrated improvements in walking speed and balance, though the authors noted that larger clinical trials are needed before widespread use [22, 27]. In a prospective pilot study, Kolakowsky-Hayner et al. evaluated the feasibility and safety of using the Ekso to aid individuals with SCI (complete T1 SCI or below) with ambulation and found the device safe for use in a controlled environment with a trained professional [21]. A recent 12-week exoskeleton-based gait training regimen with the EksoNR™ led to a clinically meaningful improvement in independent gait speed, in 25 community-dwelling participants with chronic incomplete spinal cord injury (iSCI) [28].

Indego **(Parker Hannifin Corporation, Macedonia, OH)**: The Indego—formerly known as the Vanderbilt exoskeleton or Parker Hannifin exoskeleton—is a lower extremity exoskeleton device designed at Vanderbilt University to allow individuals with lower extremity weakness following SCI to stand and walk. The Indego has been available since 2014 in Europe and received FDA clearance for use in SCI and stroke in the United States in 2017.

The Indego weighs approximately 26 pounds and assists users in sit-to-stand transitions, stand-to-sit transitions, standing, and walking (Fig. 30.2). It is intended to be used with platform walkers, rolling walkers, forearm crutches, or other devices that assist with stability, and can be donned in a user's wheelchair if there is adequate space. The Indego features five modular components—a hip segment, right and left upper leg segments, as well as right and left lower leg segments—that each come in three sizes that can be mixed to fit different body types. A wireless controller allows therapists to control the operation of the exoskeleton, modify settings, and capture and export data. Standing, sitting, and walking movements of the exoskeleton are based on the user's shift in body weight and change in body position.

Clinical Applications: The device is currently intended for users with complete or incomplete

Fig. 30.2 The Indego (Photo courtesy of Parker Hannifin Corporation)

SCI level C5 or lower. It can be used as a clinical tool for gait training, as well as a personal mobility device in the home and community. It may also be used as a gait training tool for individuals who have had a stroke. In addition to meeting the specific height and weight criteria, users must have an adequate passive range of motion at their shoulders, hips, knees, and ankles, and must demonstrate sufficient upper-body strength. Contraindications include insufficient upper extremity strength, uncontrolled spasticity, spinal instability, and conditions that prevent proper fit of the device (e.g., excessive soft tissue).

Mobility Training: Training has included a series of approximately 5 to 28 sessions, lasting about an hour and a half each, that involve evaluating users' walking distance, endurance, community ambulation skills (going through automatic doors, getting on/off elevators), as well as assessments when they are not using the device and follow-up phone calls. The first session focused on achieving the correct fit and balance when upright and practicing sit-to-stand transitions. Further sessions focus on standing, taking steps, and assessing changes to parameters. Outcome measures have included walking speed, endurance, as well as measures of a person's independence, muscle strength, stability, ability to walk on various surfaces, and ease of donning and doffing the device [29, 30]. One study of 16 individuals with tetraplegia and paraplegia reported that these individuals learned to use the Indego on a variety of indoor and outdoor surfaces at the end of five sessions that lasted 1.5 h each; some participants also achieved walking speeds and distances while wearing the device that would indicate the functional ability for limited community ambulation [30]. Another study found that when using an exoskeleton during assisted overground walking, cardiorespiratory and metabolic demands were "consistent with physical activities performed at a moderate intensity" [29].

Atalante (**Wandercraft, Paris France**): Atalante is designed to enable individuals with disabilities to perform ambulatory functions and mobility exercises, hands-free, under the supervision of a trained operator. In Europe, this device is CE-marked for patients at least 18 years of age with complete motor paraplegia. The application of the device for individuals with other disabilities (e.g., incomplete paraplegia, hemiplegia) is currently under review. The device is also in the investigational stage in the United States.

The Atalante is a motorized exoskeleton with 12 actuators that drive the hip, knee, and ankle (Fig. 30.3). The thigh and leg segments of the device are adjustable in length to allow adaption to different patient morphologies. The motion of

Fig. 30.3 Atalante (Photo courtesy of Wandercraft)

the exoskeleton is based on a control interface and postural cues to trigger transitions during walking and other activities. A fabric vest is equipped with an inertial motion unit (IMU) which detects the intention of the patient inside the exoskeleton based on chest motion.

30.4 Modular Exoskeletons

This class of exoskeletons is intended for individuals with mild to moderate mobility deficits. Their modularity is characterized by the ability to change sections of the device to better accommodate a wide range of body sizes and anthropometrics. Actuation typically occurs at a single joint, such as the hip or knee. Overall, these devices are lighter weight and less constrained than rigid, lower-body exoskeletons. As such, they usually do not deliver as much support or assistance to the user, but they can often be deployed more readily to the home or community setting to improve functional mobility.

Honda Walking Assist Device (**Honda Research and Development Company, Ltd., Wako, Japan**): The Honda Walking Assist

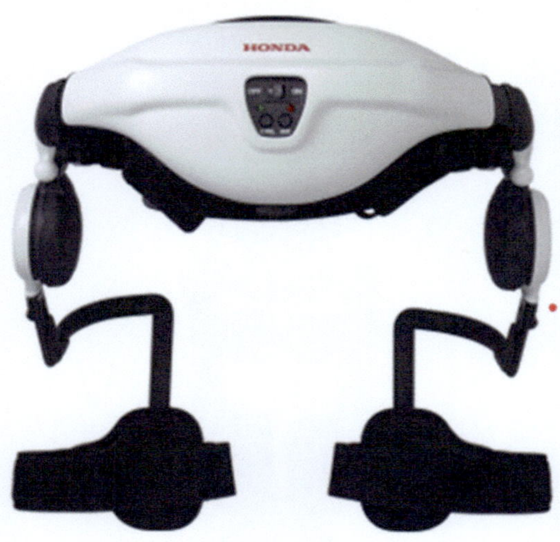

Fig. 30.4 Honda Walking Assist Device (Photo courtesy of Honda Research and Development Company, Ltd.)

Device (Fig. 30.4) is an assistive exoskeleton designed to regulate walking pace in individuals who can walk but have mild gait deficits due to aging or medical conditions such as stroke or osteoarthritis [31]. Honda initiated research into developing an assistive walking device in 1999 and began conducting collaborative testing of the device in 2008 with Shinseikai Medical Group at Kasumigaseki-Minami Hospital in Kawagoe, Japan. In 2013, the company began leasing the device to hospitals in Japan in order to study its usability and applicability; clinical research in the United States evaluating the device for therapeutic purposes in individuals who have experienced stroke also began in 2013 at the Rehabilitation Institute of Chicago. It was FDA approved for use by individuals with stroke in 2018.

The Honda Walking Assist Device uses angle sensors embedded in the actuators that detect the wearer's hip joint angles throughout the gait cycle. The hip actuators then produce assist torques at specific instances during the gait cycle to regulate walking patterns. This torque is transmitted to the thighs via flexible thigh frames. The goal is to adjust a user's stride and walking rhythm within a preprogrammed range, by assisting hip flexion and extension for each side independently, as needed.

Clinical Applications: The Honda Walking Assist Device has been tested in healthy young adults and elderly people in Japan as well as individuals with stroke in the U.S. All potential subjects must be assessed for strength, flexibility, balance, sensation, endurance, transfers, and gait. Contraindications to the use of the Honda Walking Assist Device include symptomatic cardiovascular disease, hypertension, heart failure, or severe pain.

Initial gait training with the Honda Walking Assist Device emphasizes safety and balance, to minimize the risk of falls while wearing the device. In studies involving elderly adults, subjects walked on a treadmill or outside anywhere from 30 to 90 min [32, 33]. These studies found that the device improved subjects' walk ratio, walking speed, and step length, indicating that using the Honda Walking Assist Device alongside a walking intervention program may improve the walking ability of the elderly. In a randomized clinical trial in the U.S., 50 individuals with chronic stroke ($\geq$ 30 days poststroke, 18–85 years old) completed a gait training protocol with or without the Honda Walking Assist Device over 18 sessions of outpatient physical therapy lasting 45 to 60 min. The post-training evaluation revealed greater improvements in walking speed, endurance, step count during training, and excitability of the paretic lower limb for device users compared to the control group [34]. Another study involving 10 young adults (aged 21–32 years) demonstrated reduced energy expenditure and improved endurance while walking with the device [35].

Samsung Gait Enhancing and Motivating System—Hip (GEMS) (Samsung Electronics Ltd, Suwon-si, South Korea): The Gait Enhancing and Motivating System (GEMS-H) was developed by Samsung Electronics Co. Ltd. It is a hip-based robotic exoskeleton worn around the waist and fastened to the thighs to provide assistance to hip flexion and extension. The GEMS-H device has a pair of actuators that generate assistive forces at each hip joint. The device currently comes in three sizes, and the width of each version can be adjusted to fit individual body size. The device is controlled through a custom-built application on a tablet. Through the application, the therapist is able to turn on/off torque applied at the hip, and modify the assistance and timing for the torque, referred to as gain and delay. "Gain" increases or decreases the amplitude of assistance provided and the maximum value for gain is 15 (up to $\sim$ 12 Nm). "Delay" allows the assistance to be applied earlier or later in the gait cycle. The range of delay is between 0.15–0.25 s.

Clinical Applications: The GEMS-H has been tested in healthy, elderly people of Korea, and is currently undergoing clinical evaluation in the United States with the stroke population. All potential subjects must be assessed for strength, flexibility, balance, sensation, endurance, transfers, and gait. Contraindications to the use of the GEMS-H include symptomatic cardiovascular disease, hypertension, heart failure, or severe pain. Initial gait training with the GEMS-H emphasizes safety and balance, to minimize the risk of falls. In a study involving 15 elderly adults, the use of this device decreased cardiopulmonary metabolic cost during stair climbing when compared to not using the device [36].

Mobility Training: This device is new to the U.S. market, and research is limited on the exact number of training sessions required for improvements in walking speed, endurance, or balance; however, similar principals of other modular hip-based exoskeletons can be applied.

***Keeogo* (B-Temia Inc., Saint-Augustin-de-Desmaures, Quebec, CA)**: The Keeogo is a computer-controlled lower extremity motorized orthosis worn over the user's hips and legs. The Keeogo's controller box contains sensors that supply information about the kinematics and the kinetics of the user's lower extremities and includes software that recognizes the user's mobility intentions. A lithium-polymer battery powers the system. The leg brace assembly is mainly comprised of the actuator, the electronic boards, hip joint, and soft goods (cuffs, belts) for affixing the assembly to the user's legs. The waist belt comes in various sizes adapted to each

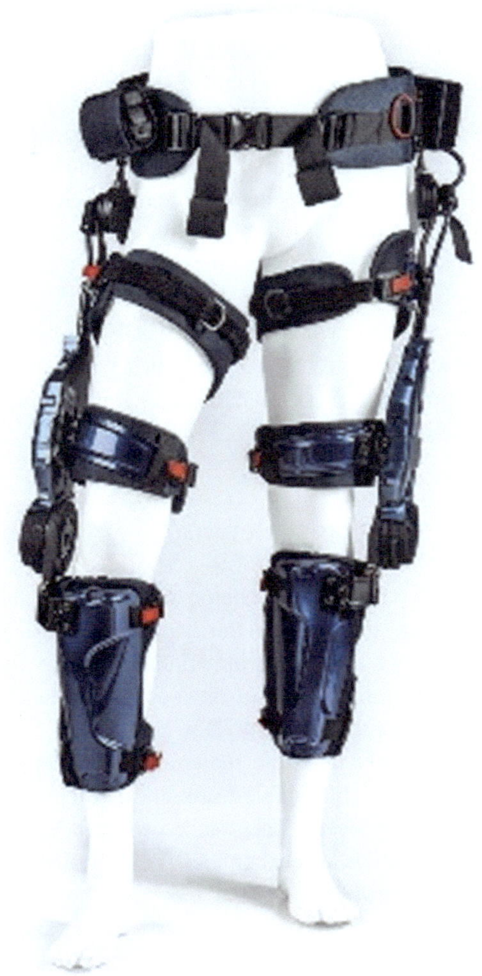

Fig. 30.5 Keeogo (Photo courtesy of B-Temia Inc.)

wearer and adds additional support to the device (Fig. 30.5).

A tablet is used to customize the assistance parameters of the device for the specific needs of each patient. Symmetrical or asymmetrical assistance may be programmed to match the user's needs. Additionally, assistance can be applied to a specific phase of the gait cycle.

Clinical Applications: The Keeogo received FDA clearance in 2020, with intended applications for patients with stroke who are undergoing rehabilitation. To date, published research about this device has examined its usability and user experience for individuals with multiple sclerosis and osteoarthritis [37–39]. Investigations regarding its effectiveness for improving gait function and quality, as well as its recommended clinical applications, are still ongoing.

Mobility Considerations: Keeogo does not initiate any movement but waits for the user's lead. Once the user makes the first move, Keeogo assists according to the activity. For example, Keeogo will help push the user up when the user leans forward to stand up, or Keeogo will slowly support the user when the user does the motion to sit down. Keeogo can be used during functional activities such as sit-to-stand, gait training, and stair climbing.

30.5 Soft Exoskeletons

A notable shift in exoskeleton development and design has centered on the use of fabrics and other soft or compliant materials to reduce the overall size and weight of devices. Assistance is delivered to the user by Bowden cables, soft tubes, or other dynamic materials. These materials typically still enable users and clinicians to modulate the level of assistance provided by the device. Textiles often connect these active and passive control elements, or they may be incorporated into garments or even worn under clothing. In addition to their lightweight, low profiles, and easier fit for a user, soft exoskeletons can be advantageous because natural joint motions are not impeded by a rigid device structure. The trade-off, however, is the lightweight, low-profile designs may not provide enough support or torque for individuals with severe impairments. For these individuals, a rigid or modular exoskeleton may be a better choice.

Recent work in soft exoskeletons for lower limb impairments has been propelled by the work of Conor Walsh, Ph.D. at the Wyss Institute for Biologically Inspired Engineering at Harvard University. Though this class of exoskeletons is still in the early stages, they are promising technologies to improve gait function in individuals with a variety of lower limb impairments.

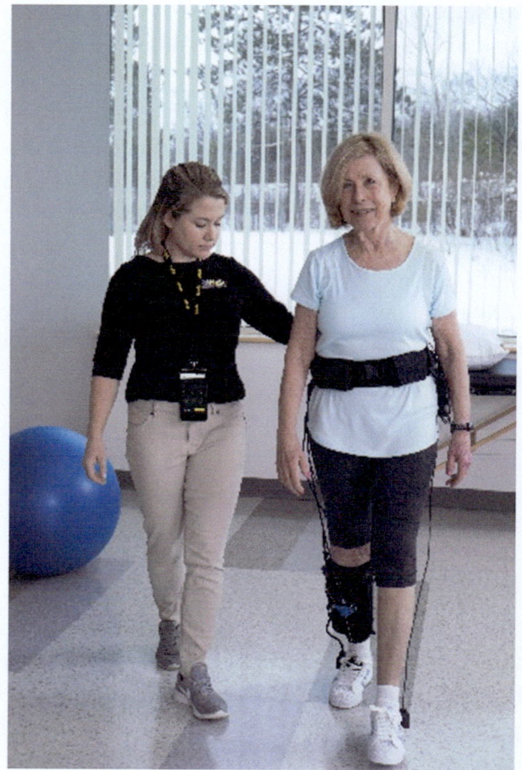

Fig. 30.6 ReWalk Restore (Photo courtesy of ReWalk Robotics)

ReWalk Restore **(ReWalk Robotics, Marlborough, MA, USA):** The Wyss Institute collaborated with ReWalk Robotics, Ltd. in 2016 to accelerate the development of soft exoskeleton technology [40]. The commercialized product that emerged from this collaboration, the ReWalk Restore, is a lightweight, wearable, soft exosuit intended to assist with ambulatory function in rehabilitation settings for people with hemiplegia/hemiparesis due to stroke (Fig. 30.6). The device is comprised of textile, mechanical, and electrical components, as well as Bowden cables, sensors, and a user interface. It is designed to provide plantarflexion and dorsiflexion assistance to the paretic ankle during walking by transmitting mechanical forces from motors at the waist, through contraction and relaxation of Bowden cables, to attachment points worn on the calf and insole. IMUs are attached to both of the user's shoes, which informs the timing of assistance from the device.

Mobility Training: There are three basic modes of operation for the ReWalk Restore. "Assist Mode" provides dynamic assistance for ankle plantarflexion and dorsiflexion during forward walking on level ground or on a treadmill. The amount of assistance provided by the device is modulated through the user interface by a trained clinician. "Brace Mode" provides static support by restricting the movement of the Bowden cables to maintain a neutral ankle position while walking. "Slack Mode" releases all tension in the cables to eliminate resistance and enable a full range of ankle motion. The ReWalk Restore is not intended for sports or stair climbing.

Clinical Applications: In 2019, the ReWalk Restore received FDA clearance for use in stroke rehabilitation. A recent multi-site clinical trial with 44 individuals demonstrated safety and reliability of the ReWalk Restore during post-stroke gait rehabilitation [41]. The trial found that patients improved their walking speed, both with and without the device, following five days of device training (including up to 20 min of treadmill walking and up to 20 min of overground walking). Other work has noted improvements in walking speed and distance [42, 43], increased ground clearance [44, 45], reduced interlimb propulsion asymmetry [45], reduced metabolic effort [45], and reduced compensatory gait patterns [46] while walking with the ReWalk Restore.

***MyoSuit* (MyoSwiss AG, Zurich, Switzerland)**: The MyoSuit provides assistance at the knee during anti-gravity movements while providing passive support during transitional movements with gravity, locomotion, and ADLs. A single actuator powers extension while flexion is controlled by a passive spring-like material. These elements are connected by a garment layer, which can be adjusted to fit a variety of body types. In this way, the Myosuit design mimics the function of the muscles, bones, and ligaments in the human body for activity-

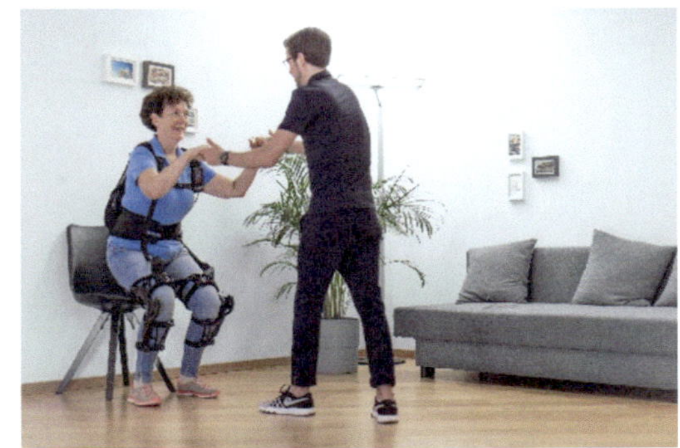

Fig. 30.7 Myosuit (Photo courtesy of Myoswiss AG)

dependent support during a variety of activities (Fig. 30.7). The device has two modes of operation: Transparency Mode minimizes resistance felt by the user while Assistance Mode provides support to the user when moving against or with gravity. Inertial Measurement Units (IMUs) located on both shanks, thighs, and in the tendon driver unit measure linear accelerations and rates of rotation.

Clinical Applications: In recent studies, the Myosuit was found to be safe, feasible, and well tolerated by individuals with a variety of gait impairments [47], with five out of eight participants demonstrating improved walking speed after training with the device, compared to their baseline. In another study with a single participant, the Myosuit was shown to assist with sit-to-stand and stand-to-sit transitions by leveraging anti-gravity control [48].

***BiOMOTUM Spark* (BiOMOTUM, Flagstaff, AZ, USA)**: The BiOMOTUM Spark is an intelligent, powered ankle device designed to increase independence, improve mobility, and deliver gait training to children with movement disorders, such as cerebral palsy.

It utilizes carbon fiber materials to create a lightweight yet durable design. The heaviest components, including the motors and battery, are located near the waist, which minimizes the metabolic energy (Fig. 30.8). Following growth spurts, lightweight components, like the cables, carbon fiber footplates, and calf cuffs can be exchanged. Each Spark device comes with the

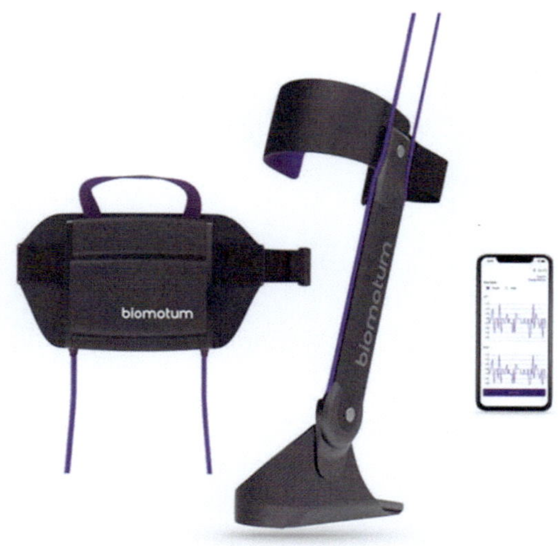

Fig. 30.8 BiOMOTUM Spark (Photo courtesy of BiOMOTUM)

necessary components to fit a child for an entire year. Motors at the ankle influence plantarflexion and dorsiflexion throughout the gait cycle. The Spark can be operated in two modes: "Assist-and-Go", which provides assistance in plantarflexion and dorsiflexion or "Resist-to-Restore", which provides resistance to plantarflexion. Once a profile has been created in the application, the system automatically calibrates the foot sensors during the first three steps of walking, after which the torque gradually builds to the peak, set by the therapist. While this torque can be predetermined by a trained clinician, the app allows control of the torque throughout training for a customizable profile to meet individual users' needs.

30.6 Considerations for Clinical Use

General Assessment

Before an exoskeleton is prescribed or used in a therapy session, a trained professional (physiatrist or physical therapist) should assess each potential candidate. Considering that an overground robotic exoskeleton can be used for therapeutic purposes or as an assistive device for at-home personal mobility, the therapist or physiatrist should consider the user's overall function and prognosis to determine which device will best suit their rehabilitation goals. In general, for most exoskeleton devices, the ideal user should be:

- Between the age group of 18–70
- Able to fit into the device and have a joint motion to allow safe ambulation in the device
- Able to tolerate upright standing for 15–30 min
- Exhibiting sufficient balance (or ability to use stabilizing tools) to allow ambulation with the device

Since exoskeleton devices require the use of assistive devices such as crutches, walkers, or canes to provide additional stability to the user, an individual's upper extremity function, as well as their overall cardiovascular health and bone density, must be assessed prior to using the device. Persons with joint contractures that limit the range of motion of any extremity, and those with any medical issues that prevent them from fully bearing their weight, are also not ideal candidates. Individuals should also be evaluated for skin issues, as the use of exoskeletons may cause additional skin irritation/breakdown and lead to further complications in areas that are in contact with the device. For these reasons, potential users should receive medical clearance from their physician before initiating therapeutic training or a personal mobility regimen with an exoskeleton.

One of the most important clinical assessments involves optimizing the fit of the exoskeleton, which requires obtaining detailed measurements of the person in order to match the joints/motors of the device with the joints of the person. If the leg or hip length is mismatched, it can change how the user's body moves within the device or lead to safety issues. For the stroke population, therapists should aim to have the movement of the device match the user's normal walking pattern as much as possible. For individuals with spinal cord injuries, poor fit may lead to pressure injuries, which are not easily detected by an individual who has lost sensory feedback.

Documentation

Thorough documentation is an essential component of rehabilitative treatment to monitor patient safety and progression and to ensure proper billing. The same is also true for gait training with robotic exoskeletons, which can benefit from additional documentation specific to the device use. Furthermore, because exoskeleton research is still in its early stages, anecdotal clinical findings can continue to guide the practical implementation of the device across various patient populations.

Initial therapy notes should include a patient's blood pressure, heart rate, spasticity, strength, range of motion, and overall skin condition to gauge whether they can safely use the device. During the session, therapists should record the number of minutes spent walking, steps per session, device settings if applicable, as well as how much assistance they needed to perform certain skills (walking on ramps or steps, going in and out of an elevator), and any instances of the device overheating or malfunctioning. Of particular interest is a therapist's practical experience using an exoskeleton to target specific impairments or functions that wouldn't have been achievable with traditional training strategies. Finally, skin irritation, bruises, or other adverse events should also be documented in detail, and the device use should be modified appropriately as needed to ensure patient safety and well-being.

General Locomotor Training Strategies

Training for personal mobility focuses on the user becoming proficient with specific tasks, such as getting in and out of the device, overground walking, going up and down ramps or curbs, and being able to stop. Training strategies for rehabilitation will differ for each device and for each individual depending on their goals, the type or level of injury, or the severity of impairment. In recent years, locomotor clinical practice guidelines suggest that gait training should occur at moderate to high intensities in order to improve walking function following neurologic injury [49]. If a clinician is able to match patient impairments with the proper device and control mechanism, it is possible to implement this recommendation in a clinical setting using an exoskeleton. That same guideline recommends that body weight supported treadmill training (BWSTT) should not be used in ambulatory individuals with chronic neurologic injury, as there seems to be a little benefit compared with overground walking or other interventions. In this case, the use of overground robotic exoskeletons may be an excellent choice to facilitate locomotor training. By reducing energy expenditure, robot-driven assistance can enable patients to participate in higher dosages of stepping than they could independently or with therapist assistance alone. For individuals in an acute/subacute recovery phase or those who require support to ambulate, an overground robotic exoskeleton may be used as a supplement to BWSTT and standard overground training. Though the optimal number of training sessions with a certain device is still unknown, many of the devices listed above can be used in conjunction with traditional rehabilitation tools in order to best address gait-related goals. Psychological factors such as motivation and time since injury may also play a role in how quickly a person acclimates to using an exoskeleton.

Safety Considerations

Externally worn and controlled robotic devices can pose a safety risk beyond that of traditional rehabilitation tools; however, with proper safety protocols in place, risk can be mitigated, and these technologies can be implemented without harm to the patient or therapist. A thorough assessment of the patient's sensation and skin integrity, especially for neurologic populations, should be completed prior to device use. Skin checks before and after device use should also be completed and adjustments (such as fit, cushioning, or cessation of use) should be made if issues arise. While there is no evidence that overground robotic exoskeletons increase fall risk, therapists should take typical precautions to avoid loss of balance or falls. Many overground robotic exoskeletons can be used with overhead harness systems as an additional safety measure. Each device has an individualized protocol for assisting patients in the event of a fall.

Current Limitations

Several limitations must be addressed before exoskeletons become more prevalent in the home and clinical setting. For use in the home and community, individuals must be able to achieve speed levels that are practical for activities of daily living, such as crossing a busy intersection in an urban setting. Additionally, exoskeletons are expensive investments, costing anywhere from $50,000 to $250,000. This high cost makes the devices out of reach for most individuals, or for researchers without grant support. Currently, the lack of extensive clinical evidence on the efficacy of these devices has limited insurance reimbursement, and therefore, implementation in a clinical setting.

30.7 Regulatory Status and Future Expectations

In the United States, all exoskeletons must receive approval from the U.S. Food and Drug Administration (FDA) before they can be commercialized and used in clinics and rehabilitation settings. In February 2015, the FDA announced that exoskeletons will be classified as Class II devices (special controls) [50]. The report cited falls, bruising, skin abrasions, changes in blood pressure, adverse tissue reaction, premature battery failure, burns, and device malfunction as potential risks. In the European Union, exoskeletons undergo a clinical evaluation to obtain a CE mark, which identifies a device that complies with safety, health, and environmental regulations.

30.8 Conclusions

Exoskeletons are rapidly evolving technologies that have the potential to reduce impairments, restore function, and improve overall recovery in persons who have experienced stroke, spinal cord injury, or other neuromuscular injuries/diseases.

Recent studies have demonstrated the effectiveness of overground locomotor training with exoskeletons to improve gait, balance, and mobility in individuals with lower limb deficits due to neurological injury or aging. Moving forward, proper clinical validation of exoskeletons will require more randomized control trials to test the safety and effectiveness of training with these devices compared to traditional gait therapy.

Designs are shifting toward modularity and lightweight materials to improve the fit, comfort, and portability of these devices. More exoskeletons are also featuring variable assistance controls, which can better accommodate the diverse impairments of different patient populations while also allowing therapists to implement progression strategies that promote functional change. However, additional technological advances are likely needed before widespread clinical use and deployment to the home and community, such as making the devices lighter and easier for patients to use independently. Simplified designs and control interfaces may make exoskeletons more approachable to clinicians and patients, thereby facilitating implementation. Furthermore, larger-scale clinical trials are needed at the clinical and home/community levels to establish whether robotic exoskeletons should be widely adopted in evidence-based practices for a continuum of care overground locomotor training. Finally, the cost of these devices needs to go down substantially to make these devices affordable across an economic spectrum.

Although exoskeletons are promising technologies that may be integrated into the rehabilitation process, exoskeletons are still in the early stages of development and implementation and require experienced and trained physical therapists and caregivers to monitor and assist users at all times. Further research is needed to identify which patients derive the most benefit from these devices, as well as to establish the optimal dosage of exoskeleton training to achieve meaningful functional gains. Finally, this chapter only covers a subset of devices that are commercially available or in the process of commercialization; additional devices are currently under development in the industry and research domains.

References

1. Yagn N, Inventor apparatus for facilitating walking. USA patent US420179 A. 1890.
2. Dollar AM, Herr H. Lower extremity exoskeletons and active orthoses: challenges and state of-the-Art. IEEE Trans Rob. 2008;24(1):144–58.
3. The story behind the real 'Iron Man' suit GE reports [Internet]. 2010 Nov 23 [cited 2015 May 1, 2015]. Available from: http://www.gereports.com/post/7857 4114995/the-story-behind-the-real-iron-man-suit.
4. Zoss A, Kazerooni H, Chu A. On the mechanical design of the Berkeley lower extremity exoskeleton (BLEEX). IEEE/RSJ Int Conf Intellifent Robot Syst. 2005;3132–3139.
5. UC Berkeley News. UC Berkeley researchers developing robotic exoskeleton that can enhance human strength and endurance [Press release]; 2004, March 3. Retrieved from https://www.berkeley.edu/news/media/releases/2004/03/03_exo.shtml
6. Lajeunesse V, Vincent C, Routhier F, Careau E, Michaud F. Exoskeletons' design and usefulness evidence according to a systematic review of lower limb exoskeletons used for functional mobility by people with spinal cord injury. Disabil Rehabil Assis Technol. 2015;11(7):535–47.
7. Meng W, Liu Q, Zhou Z, Ai Q, Sheng B, Xie SS. Recent development of mechanisms and control strategies for robot-assisted lower limb rehabilitation. Mechatronics. 2015;31:132–45.
8. Rupal BS, Rafique S, Singla A, Singla E, Isaksson M, Virk GS. Lower-limb exoskeletons: research trends and regulatory guidelines in medical and non-medical applications. Int J Adv Robot Syst. 2017;14:6.
9. Rodriguez-Fernandez A, Lobo-Prat J, Font-Llagunes JM. Systematic review on wearable lower-limb exoskeletons for gait training in neuromuscular impairments. J NeuroEngineering Rehabil. 2021;18:22.
10. Federici S, Meloni F, Bracalenti M, De Filippis ML. The effectiveness of powered, active lower limb exoskeletons in neurorehabilitation: a systematic review. NeuroRehabilitation. 2015;37(3):321–40.
11. Bruni MF, Melegari C, De Cola MC, Bramanti A, Bramanti P, Calabrò RS. What does best evidence tell us about robotic gait rehabilitation in stroke patients: a systematic review and meta-analysis. J Clin Neurosci. 2018;48:11–7.
12. ReWalk: Company [Internet]. 2015 [cited 2015 April 29, 2015]. www.rewalk.com/company/
13. Esquenazi A, Talaty M, Packel A, Saulino M. The ReWalk powered exoskeleton to restore ambulatory function to individuals with thoracic-level motor-complete spinal cord injury. Am J Phys Med Rehabil/Assoc Acad Physiatr. 2012;91(11):911–21.
14. Talaty M, Esquenazi A, Briceno JE. Differentiating ability in users of the ReWalk(TM) powered exoskeleton: an analysis of walking kinematics. IEEE Int Conf Rehabil Robot: [Proc]. 2013;2013:6650469.
15. FDA allows marketing of first wearable, motorized device that helps people with certain spinal cord injuries to walk [Internet press release]. 2014 [cited May 6, 2015]. Available from: http://www.fda.gov/NewsEvents/Newsroom/PressAnnouncements/ucm402970.htm
16. Benson I, Hart K, Tussler D, van Middendorp JJ. Lower-limb exoskeletons for individuals with chronic spinal cord injury: findings from a feasibility study. Clin Rehabil. 2015.
17. Zeilig G, Weingarden H, Zwecker M, Dudkiewicz I, Bloch A, Esquenazi A. Safety and tolerance of the ReWalk exoskeleton suit for ambulation by people with complete spinal cord injury: a pilot study. J Spinal Cord Med. 2012;35(2):96–101.
18. van Dijsseldonk RB, van Nes IJW, Geurts ACH. et al. Exoskeleton home and community use in people with complete spinal cord injury. Sci Rep. 2020; 10:15600.
19. Ekso Bionics—Who We Are [Internet]. 2015 [cited 2015 May 6, 2015]. Available from: http://www.eksobionics.com/ourstory.
20. Kressler J, Thomas CK, Field-Fote EC, Sanchez J, Widerstrom-Noga E, Cilien DC, et al. Understanding therapeutic benefits of overground bionic ambulation: exploratory case series in persons with chronic, complete spinal cord injury. Arch Phys Med Rehabil. 2014;95(10):1878–87 e4.
21. Kolakowsky-Hayner SA, Crew J, Moran S, Shah A. Safety and feasibility of using the Ekso Bionic Exoskeleton to aid ambulation after spinal cord injury. J Spine. 2013.
22. Jayaraman A, editor Evaluation of the clinical criteria for safe and efficent use of exoskeletons in individuals with SCI. In: American Spinal Injury Association (ASIA) Conference, 2013; 2013; Chicago, Illinois.
23. Jayaraman AH, Esquenazi A, Franceschini, A., editors. Effectiveness of robotic exoskeletons in everyday rehabilitation. In: Symposium, ISPRM Conference; 2014 June 2014; Cancun, Mexico.
24. Jayaraman AR, WZ. Exoskeletons for Neurologically Impaired Individuals. Academy of Spinal Cord Injury Professionals. St. Louis2014.
25. Kozlowski A; Spungen, AM; Jayaraman A; Tefertiller C; Forrest GF; Hartigan C; Evans N. Assisted walking for persons with neurological conditions: the state of science. In: Proceedings from the American congress of rehabilitation medicine; 2014; Toronto, Canada.
26. Nolan KJ. Gait training of stroke patients using a robotic exoskeleton during inpatient rehabilitation: feasibility study. Int Work Wearable Robot. 2014; 2014; Baiona, Pontevedra, Spain.
27. Forrest G, et al., Kessler Foundation, editor. The potential of the Ekso Exoskeleton for affecting long-term health and well-being in the SCI population. In: Proceedings from the academy of spinal cord injury professionals; September 2012; Las Vegas, Nevada.

28. Edwards DJ, Forrest G, Cortes M, et al. Walking improvement in chronic incomplete spinal cord injury with exoskeleton robotic training (WISE): a randomized controlled trial. Spinal Cord. 2022. https://doi.org/10.1038/s41393-022-00751-8.

29. Evans NH, Kandilakis C, Pharo E, Clesson I. Acute cardiorespiratory and metabolic responses during exoskeleton-assisted walking overground among persons with chronic spinal cord injury. Top Spinal Cord Inj Rehabil. 2015;21(2):122–32.

30. Hartigan C, Kandilakis C, Dalley S, Clausen M, Wilson E, Morrison S, Etheridge S, Farris R. Mobility outcomes following five training sessions with a powered exoskeleton. Top Spinal Cord Inj Rehabil. 2015;21(2):93–9.

31. Walking Assist Device with Stride Management System http://corporate.honda.com/innovation/walk-assist/2015 [Internet]. 2015 [cited 2015 June 9, 2015]. Available from: http://corporate.honda.com/innovation/walk-assist/

32. Shimada H, Suzuki T, Kimura Y, Hirata T, Sugiura M, Endo Y, et al. Effects of an automated stride assistance system on walking parameters and muscular glucose metabolism in elderly adults. Br J Sports Med. 2008;42(11):922–9.

33. Shimada H, Hirata T, Kimura Y, Naka T, Kikuchi K, Oda K, et al. Effects of a robotic walking exercise on walking performance in community-dwelling elderly adults. Geriatr Gerontol Int. 2009;9(4):372–81.

34. Jayaraman A, O'Brien MK, Madhavan S, Mummidisetty CK, Roth HR, Hohl K, et al. Stride management assist exoskeleton vs functional gait training in stroke: arandomized trial. Neurology. 2019;92(3):e263–73.

35. Kitatani R, Ohata K, Takahashi H, Shibuta S, Hashiguchi Y, Yamakami N. Reduction in energy expenditure during walking using an automated stride assistance device in healthy young adults. Arch Phys Med Rehabil. 2014;95(11):2128–33.

36. Kim D-S, Lee H-J, Lee S-H, Chang WH, Jang J, Choi B-O, et al. A wearable hip-assist robot reduces the cardiopulmonary metabolic energy expenditure during stair ascent in elderly adults: a pilot cross-sectional study. BMC Geriatr. 2018;18:230.

37. McGibbon CA, Sexton A, Jayaraman A, Deems-Dluhy S, Gryfe P, Novak A, et al. Evaluation of the Keeogo exoskeleton for assisting ambulatory activities in people with multiple sclerosis: an open-label, randomized, cross-over trial. J Neuroeng Rehabil. 2018;15(1):117.

38. McGibbon C, Sexton A, Gryfe P, Dutta T, Jayaraman A, Deems-Dluhy S, et al. Effect of using of a lower-extremity exoskeleton on disability of people with multiple sclerosis. Disabil Rehabil Assist Technol. 2021;27:1–8.

39. McGibbon C, Sexton A, Jayaraman A, Deems-Dluhy S, Fabara E, Adans-Dester C, et al. Evaluation of a lower-extremity robotic exoskeleton for people with knee osteoarthritis. Assist Technol. 2021;20:1–14.

40. Wyss Institute for Biologically Inspired Engineering at Harvard University. (2016, May 17). Wyss Institute collaborates with ReWalk Robotics to develop wearable exosuits for patients with limited mobility [Press release]. Retrieved from https://wyss.harvard.edu/news/wyss-institute-collaborates-with-rewalk-robotics-to-develop-wearable-exosuits-for-patients-with-limited-mobility/

41. Awad LN, Esquenazi A, Francisco GE, Nolan KJ, Jayaraman A. The ReWalk ReStore™ soft robotic exosuit: a multi-site clinical trial of the safety, reliability, and feasibility of exosuit-augmented post-stroke gait rehabilitation. J Neuroeng Rehabil. 2020;17(1):80.

42. Awad LN, Lewek MD, Kesar TM, Franz JR, Bowden MG. These legs were made for propulsion: advancing the diagnosis and treatment of post-stroke propulsion deficits. J Neuroeng Rehabil. 2020;17(1):139.

43. Awad LN, Kudzia P, Revi DA, Ellis TD, Walsh CJ. Walking faster and farther with a soft robotic exosuit: implications for post-stroke gait assistance and rehabilitation. IEEE Open J Eng Med Biol. 2020;1:108–15.

44. Sloot L, Bae J, Baker L, O'Donnell K, Menard N, Porciuncula F, Choe D, Ellis T, Awad L, Walsh C. A soft robotic exosuit assisting the paretic ankle in patients post-stroke: effect on muscle activation during overground walking. Gait Posture. 2018; S0966–6362(18):30881–6.

45. Awad LN, Bae J, O'Donnell K, De Rossi SMM, Hendron K, Sloot LH, et al. A soft robotic exosuit improves walking in patients after stroke. Sci Transl Med. 2017;9(400):eaai9084.

46. Awad LN, Bae J, Kudzia P, et al. Reducing circumduction and hip hiking during hemiparetic walking through targeted assistance of the paretic limb using a soft robotic exosuit. Am J Phys Med Rehabil. 2017;96(10 Suppl 1):S157–64.

47. Haufe FL, Schmidt K, Duarte JE, Wolf P, Riener R, Xiloyannis M. Activity-based training with the Myosuit: a safety and feasibility study across diverse gait disorders. J Neuroeng Rehabil. 2020;17(1):135.

48. Schmidt K, Duarte JE, Grimmer M, Sancho-Puchades A, Wei H, Easthope CS, Riener R. The Myosuit: bi-articular anti-gravity exosuit that reduces hip extensor activity in sitting transfers. Front Neurorobot. 2017;11:57.

49. Hornby TG, Reisman DS, Ward IG, Scheets PL, Miller A, Haddad D, Fox EJ, Fritz NE, Hawkins K, Henderson CE, Hendron KL, Holleran CL, Lynskey JE, Walter A, the Locomotor CPG Appraisal Team. Clinical practice guideline to improve locomotor function following chronic stroke, incomplete spinal cord injury, and brain injury. J Neurol Phys Ther. 2020 Jan;44(1):49–100.

50. Medical devices; physical medicine devices; classification of the powered exoskeleton (2015). Docket no. FDA-2014-N-1903. Company [cited 2015 May 27, 2015]. Available from: https://federalregister.gov/a/2015-03692

Arun Jayaraman PT, Ph.D. is Executive Director of the Technology & Innovation Hub (tiHUB) and Director of the Max Näder Center for Rehabilitation Technologies & Outcomes Research at the Shirley Ryan AbilityLab (SRAlab). He also is an Associate Professor in the Departments of Physical Medicine and Rehabilitation, Medical Social Sciences, and Physical Therapy and Human Movement Sciences at the Northwestern University Feinberg School of Medicine. Dr. Jayaraman has been involved in rehabilitation research for the past ten years with a focus on outcomes research in the field of prosthetics, orthotics, rehabilitation robotics, and in other assistive and adaptive technologies to treat physical disability. A large part of his research has been focused on SCI, specifically to understand mechanisms of recovery. His research has included using both human and animal models; studying the muscle physiology and neurophysiology of SCI recovery; assessing therapeutic approaches, including rehabilitation robotics, neuroprosthetics, orthotics, and imaging; and performing clinical trials in rehabilitation for SCI. His current research funding investigates the impact of new technologies and therapeutic interventions on social interaction, community mobility, and overall quality of life for people with a disability. Dr. Jayaraman is the Principal Investigator on several federal and industrial grants, including the Department of Defense, The National Institutes of Health, and a NIDILRR-funded RERC on Rehabilitation Robotics, Timing & Dosage, and Manipulation & Mobility.

William Z. Rymer MD, Ph.D., is Director of the Single Motor Unit (SMU) Laboratory at the Shirley Ryan AbilityLab. In addition to his research roles at SRAlab, he holds appointments as Professor of Physiology and Physical Medicine and Rehabilitation at the Northwestern University Feinberg School of Medicine, as well as a Professor of Biomedical Engineering at Northwestern University's McCormick School of Engineering. As a physician and research scientist specializing in neurophysiology, Dr. Rymer has dedicated his career toward investigating regulation of movement in persons with neurological impairments; physiological effects of spinal cord injury; sources of altered motoneuronal and inter-neuronal responses in spinal segments below a partial or complete spinal cord transaction using electro-physiological; pharmacological and biomechanical techniques; and rehabilitation robotics. He has published more than 250 papers in the fields of biomechanics and control of movement, and has served as the Project Director of the NIDILRR-funded RERC (MARS) for the past 3 years.

Matt Giffhorn PT, DPT, NCS, is a research physical therapist in the Max Näder Center for Rehabilitation Technologies & Outcomes Research at the Shirley Ryan AbilityLab (SRAlab). He also holds an associated faculty appointment at Northwestern University in the Department of Physical Therapy and Human Movement Sciences. Matt has a clinical background in acute rehabilitation of neurologic populations, with emphasis on spinal cord injury and stroke. His research involves investigation on the impact of lower extremity robotics on functional outcomes in aging and neurologic populations.

Megan K. O'Brien Ph.D., is a Research Scientist within the Technology & Innovation Hub (tiHUB) at the Shirley Ryan AbilityLab. She also holds an appointment as Research Assistant Professor in the Department of Physical Medicine and Rehabilitation at the Northwestern University Feinberg School of Medicine. Dr. O'Brien's work focuses on utilizing the capabilities of mobile technologies to measure, monitor, and improve rehabilitation outcomes. Her research has spanned applications for stroke, spinal cord injury, Parkinson's disease, and pediatrics.

Beyond Human or Robot Administered Treadmill Training

31

Hermano Igo Krebs, Conor J. Walsh,
Tyler Susko, Lou Awad, Konstantinos Michmizos,
Arturo Forner-Cordero, and Eiichi Saitoh

Abstract

The demand for rehabilitation services is growing apace with the graying of the population. This situation creates both a need and an opportunity to deploy technologies such as rehabilitation robotics, and in the last two decades many research groups have deployed variations of this technology for gait rehabilitation. While gait robotic technology is elegant and sophisticated, results so far are mixed. We argue here that much of this technology may be misguided in its focus, providing highly repeatable control of rhythmic movement but ultimately overfocusing on this one aspect of gait. Our approach to lower extremity therapeutic robots is guided by our model of dynamic primitives in locomotion, which posits that walking is a composite of three dynamic primitives including oscillations (rhythmic movements), but also submovements (discrete movements), and mechanical impedances (balance). We developed devices based on the principle that the machine should allow the patient to express those dynamic primitives as much as (s)he can, while accommodating a large spectrum of pathological gaits. In the following, we review four innovative solutions for lower extremity (LE) rehabilitation based on this approach: Anklebot, MIT-Skywalker, Soft Exosuit, and Variable-Friction Cadense Shoes.

H. I. Krebs (✉)
Massachusetts Institute of Technology, Cambridge, MA, USA
e-mail: hikrebs@mit.edu

C. J. Walsh
Harvard University, Boston, MA, USA
e-mail: walsh@seas.harvard.edu

T. Susko
University of California, Santa Barbara, CA, USA
e-mail: susko@ucsb.edu

L. Awad
Boston University, Boston, MA, USA
e-mail: louawad@bu.edu

K. Michmizos
Rutgers University, New Brunswick, NJ, USA
e-mail: km1078@cs.rutgers.edu

A. Forner-Cordero
Escola Politécnica da Universidade de Sao Paulo, Sao Paulo, Brasil
e-mail: aforner@usp.br

E. Saitoh
Fujita Health University, Nagoya, Japan
e-mail: esaitoh@fujita-hu.ac.jp

Keywords

Rehabilitation robotics · Robot-assisted therapy · Robotic therapy · Assistive technology · Lower extremity · Stroke · Cerebral palsy

© The Author(s), under exclusive license to Springer Nature Switzerland AG 2022
D. J. Reinkensmeyer et al. (eds.), *Neurorehabilitation Technology*,
https://doi.org/10.1007/978-3-031-08995-4_31

31.1　Introduction

In the following, we review four innovative solutions for lower extremity (LE) rehabilitation ranging from rehabilitation robotics to assistive technology for participants with LE impairment: Anklebot [1], MIT-Skywalker [2], Soft Exosuit [3, 4], and Variable-Friction Cadense Shoes [5]. These designs depart from the most common LE robotic therapy, and we highlight here some of the initial results while investigating what might constitute best practice. Our approach to lower extremity therapeutic robots is guided by our model of dynamic primitives in locomotion (see next section); by the principle that the machine should allow the patient to express those dynamic primitives as much as (s)he can (i.e., it should be able to "get out of the way"); and by the need to accommodate a vast spectrum of pathological gaits and impairment levels as defined in [6]. The Anklebot and the MIT-Skywalker exemplify our approach to train at least three independent training modes (rhythmic, discrete, and balance training) that can be added or subtracted depending on the patient's needs as showcased later. The Exosuit expands the training to a wearable technology that can be employed outside the clinical setting and train or assists during walking. Last, but not least, the Variable-Friction Cadence shoes embody the concept behind the MIT-Skywalker on a wearable solution that can also train or assist.

31.2　A Competent Model for Walking

We propose a competent model of human walking (as well as arm movement) based on dynamic primitives [7]. By "competent model" we mean that it may only be a first approximation of a fundamental theory, but it is good enough to improve the design of robots and regimens for both UE and LE therapy. The theory of dynamic motor primitives is succinctly outlined by Hogan and Sternad [8]. To accommodate real-life walking with all its variations, we propose that walking is a composite of three dynamic primitives, specifically submovements (discrete movements) [9], oscillations (rhythmic movements) [10, 11], and mechanical impedances (balance) [12–16]. The three primitives are related via the concept of a virtual trajectory, which in a nutshell operates like a reference trajectory to standard motion controller with no assumption that dynamics are meaningful or fast [8]. To render precision, a discrete movement is defined as one with a clear start and stop posture. Because the term "rhythmic" has numerous confusing variations of meaning, the corresponding dynamic primitive is defined as an almost-periodic oscillation [17]. Mechanical impedance is defined as the operator that determines the force or torque evoked by imposed displacement [18].

These dynamic primitives have different neural substrates. In a functional MRI study, Schaal et al. demonstrated that a discrete wrist movement recruited more regions of the brain than did the same movement performed rhythmically [10]. Perhaps more important, they influence learning in different ways. It has been shown that motor learning of discrete arm movements has a positive transfer to rhythmic movements but not vice-versa [19]. To the extent that recovery after neural injury resembles motor learning, this suggests that discrete training as in pointing with the ankle may be more effective as it appears to have a positive transfer to rhythmic training of locomotion than vice-versa [19]. Discrete locomotor therapy would consist of patients working on self-directed, visually guided, discrete steps to initiate movement or pointing movements to targets with the lower limb [20].

Upright walking requires active balance mechanisms that often include modulating mechanical impedance. The posture or configuration of the limbs profoundly affects the response to perturbations, i.e., mechanical impedance. Challenges to balance commonly evoke changes of lower limb posture, for example, a wider stance. Impaired balance is a common symptom in most neurological injuries such as stroke and cerebral palsy [21–24]. Balance training has been shown to reduce postural asymmetry associated with hemiparesis and was a part of the home-based protocol in the LEAPS study which resulted in walking benefits similar to those achieved with body-weight supported treadmill training BWSTT [25].

A similar combination of dynamic primitives has been proposed to underlie upper extremity actions [26]. This suggests that the differences between upper extremity and lower extremity control may be smaller than previously considered in the literature.

31.2.1 Anklebot

We focused our initial LE robotics development efforts on the ankle because it is critical for propulsion, shock absorption, and balance during walking. Following stroke, "drop foot" is a common impairment. It is caused by a weakness in the dorsiflexor muscles that lift the foot. Two major complications of drop foot are "slapping" of the foot after heel strike in the early stance (foot slap) and dragging of the toe during swing, making it difficult to clear the ground (toe drag). In addition to inadequate dorsiflexion ("toe up"), the paretic ankle also suffers from excessive inversion (sole towards midline). Both begin in the swing phase and result in toe contact (as opposed to heel contact) and lateral instability during stance, a major cause of ankle injuries. Lack of proper control during these phases increases the likelihood of trips and falls. In fact, deficits of swing clearance, propulsion, and balance contribute to more than 70% of stroke survivors sustaining a fall within six months [21], leading to higher risks for hip and wrist fractures in the first year [22–24]. The ankle is also the largest source of mechanical power during terminal stance [27]. The plantarflexors contribute as much as 50% of positive mechanical work in a single stride to enable forward propulsion [28–31]. In pre-swing plantarflexors also act to advance the leg into swing phase while promoting knee flexion at toe-off [32]. Additionally, the ankle helps maintain body-weight support during gait [33–35] and balance. Finally, the ankle musculature helps absorb impact forces during

foot strike to enable controlled landing. In summary, given its importance in overground foot-floor swing clearance, propulsion, shock absorption, and balance, we elected to focus first on the ankle. The Anklebot has the potential to address both swing clearance and propulsion, as well as balance problems since it is actuated in both the sagittal and frontal planes [1].

The design, characterization, donning procedure, and safety features of the adult and pediatric version of the Anklebot have been previously described [36, 37]. Here, we will briefly summarize the salient design features and measurement capabilities of the two versions of the robot. It is a portable, tethered wearable exoskeletal ankle robot that allows normal range of motion in all three degrees of freedom of the ankle and shank during walking overground, on a treadmill, or while sitting (25° of dorsiflexion, 45° of plantar flexion, 25° of inversion, 20° of eversion, and 15° of internal or external rotation). It also provides independent assistance or resistance in two of those degrees of freedom (dorsi-plantarflexion and eversion/inversion) via two linear actuators mounted substantially in parallel. Anatomically, internal–external rotation is limited at the ankle, the orientation of the foot in the transverse plane being controlled primarily by rotation of the leg at the hip. Under-actuation, i.e., actuating fewer degrees of freedom than are anatomically present, affords one key advantage: it allows the device to be installed without requiring precise alignment with the patient's joint axes (ankle and subtalar joints). This is actually an important characteristic of all our robotic devices. In this configuration, if both actuators push or pull in the same direction, a dorsi-plantarflexion torque is produced. Similarly, if the two links push or pull in opposite directions, an inversion–eversion torque results.

The Anklebot is a backdriveable robot with low intrinsic mechanical impedance, weighs less than 3.6 kg (2.5 kg for the pediatric version) can deliver a continuous net torque of approximately 23 N m in dorsi-plantarflexion and 15 N m in eversion–inversion (7.21 and 4.38 N m for the pediatric version). The robot can estimate ankle angles with an error less than 1° in both planes of movement over a wide range of movement (60° in dorsi-plantarflexion and 40° in eversion–inversion), and can measure ankle torques with an error less than 1 N m. It has low friction (0.74 N m) and inertia (0.8 kg per actuator for a total of 1.6 kg at the foot) to maximize back-driveability. Of course, the Anklebot torque capability does not allow lifting the weight of a patient. At best, we can cue the subject to use his/her voluntary plantarflexor function by providing supplemental support to the paretic ankle plantarflexors during the stance phase. Our design is aimed at supporting foot clearance during swing phase assisting a controlled landing at foot contact. The torque generated by the Anklebot can compensate for drop foot during early and final stance phases of gait and insufficient muscle activity during push-off. We can also generate torque during the mid-swing phase to evoke concentric activity in the dorsiflexor muscles. In this respect, the Anklebot can provide continuous torques up to $\sim$23 N m in the sagittal plane ($\sim$7 N m for the pediatric version), which is higher than required to position the foot in dorsiflexion during mid-swing.

We conclude this description of the salient features of the Anklebot by noting that we showed that unilaterally loading the impaired leg with an unpowered adult or pediatric Anklebot's additional mass had no detrimental effect on the gait pattern of subjects with chronic hemiparesis or children with cerebral palsy [38, 39].

31.2.2 Translating to Practice: Training in Seated Position

Results with stroke survivors with chronic hemiparetic gait and children with cerebral palsy who underwent a 6 week interactive seated anklebot training program were quite promising [1, 20, 36]. Follow-up studies confirmed the potential benefits of paretic ankle training on impairment and that reducing impairment would translate into functional improvement in overground walking speed. We used a visually guided, visually evoked, training paradigm in which the amount of assistance changed and challenged participants to improve performance. In these trials, we trained subjects in a seated position ("open chain") and not in task-specific gait training (see Fig. 31.1). Task difficulty (i.e., target locations on the screen) was initially set proportional to baseline deficit severity (i.e., paretic ankle active range of motion). Training parameters (i.e., target locations, speed) were adjusted every 2 weeks based on individual subject performance and included discrete and rhythmic pointing movements with the ankle.

For example, Chang and colleagues reported a study with participants with chronic stroke (>6 month) and hemiparetic gait (N = 29) who received 18 sessions of isolated robot-assisted motor training of the ankle (3x/week for 6 weeks). All participants had stable clinical baseline scores across three admission measures, and no participant was receiving simultaneous outpatient rehabilitation. Baseline gait speed defined three impairment groups: high, >0.8 m/s; medium, 0.4–0.8 m/s; low, <0.4 m/s. Outcome measures included the Berg Balance Scale, the 6 min Walk Test, and the 10 m Walk Test, and were recorded upon admission, discharge, and 3 months following intervention [40].

Three distinct and significant between-group patterns of recovery emerged for gait speed. The within-group analysis showed that the medium and high group exhibited significant improvements in gait speed and endurance upon discharge, that were maintained at 3 months. Gait speed improvements were clinically significant (>0.16 m/s) for the high function group across all gait speed and endurance measures at discharge and at 3 months. The moderate group also exhibited clinically significant improvements at follow-up on the 10 m Walk Test, fast pace (0.16 m/s), and approached clinical significance for the 10 m Walk comfortable pace (0.12 m/s). The low group had small but significant improvements, at discharge on two of the three gait measures, and these improvements were

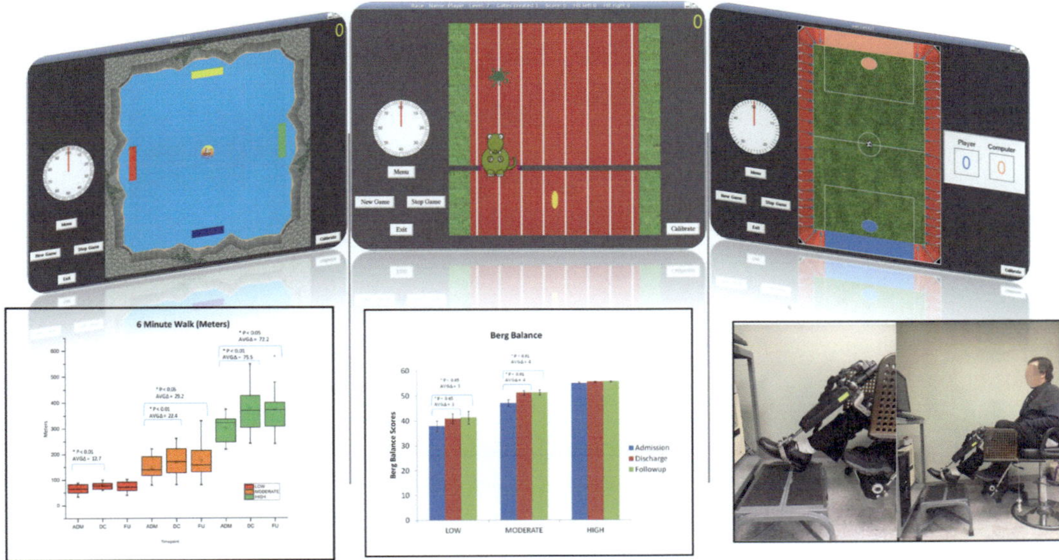

Fig. 31.1 Training in seated position. Top row shows some of the serious games developed for the anklebot. Left bottom row panel shows the endurance test (6 min walk test) in which patients walked continuously for 6 min, and total meters walked were measured. Low, moderate, and high groups showed significant differences at discharge (Low: mean change = 12.7 m, p < 0.01. Moderate: mean change = 22.4 m, p < 0.01. High: mean change = 75.5 m, p < 0.01). At follow-up, low group maintained small but non-significant change (mean change = 6.6 m, p > 0.05). Moderate group showed further improvement (mean change = 29.2 m, p < 0.05) and high group maintained significant changes (mean change = 72.2 m, p < 0.05). Middle bottom panel shows balance scores at admission, discharge, and follow-up (x/56). Higher scores indicate better functioning. Impairment groups: low, moderate, and high were based on average admission gait speed (low, <0.4 m/s; moderate, 0.4–0.8 m/s; high function, >0.8 m/s). Low and moderate groups showed significant changes at discharge (Low: mean change from admission to discharge = 3, p < 0.05. Moderate: mean change from admission to discharge = 4, p < 0.01) and maintained improvements at follow-up (Low: mean from admission to follow up = 3, p < 0.05. Moderate: mean change = 4, p < 0.01). High group showed non-significant changes at discharge and follow-up; admission score for high group approached ceiling (mean = 55 out of maximum 56 points) and plateaued at discharge and follow-up. I bars indicate standard error. Right bottom panel shows the side view of patient wearing ankle robot in a seated position (right) and close up of robotic training device (left)

maintained at 3 months. For balance measures, the low and moderate impairment groups had significant improvements at discharge that were robust on follow-up measure. The high function group demonstrated no significant change in balance.

Joint-specific robotic training of the paretic ankle provided the most benefit to individuals with moderate or mild gait speed impairments after stroke. Baseline gait speed function (low, moderate, high) was associated with three distinct recovery profiles. This suggests that severity-specific intervention may be critical to improving efficiency of stroke recovery. Of course, we must take the results in these small studies with the appropriate caveats as the number of subjects is small, the intensities and duration of the interventions are different, the patient populations are distinct, and they are non-controlled studies. However, it is important to highlight that initially we did not expect that training while seated to be successful as load receptor input is essential for a physiological leg muscle activation during stance and gait [41]. Yet our initial and subsequent experimental results told a different story. We speculate that the observed overground changes with training while seated are related to changes of ankle mechanical impedance leading to a more ecological foot landing during gait [12, 42–45].

31.2.3 MIT-Skywalker

The MIT-Skywalker robot is inspired by the concept of passive dynamic walkers [46]. In conventional gait physiotherapy, the therapist pushes or slides the patient's swing leg forward, either on the ground or on a treadmill. In kinematically-based robot-assisted gait therapy, the leg is propelled forward by the robotic orthosis acting on the patient's leg (e.g., in Lokomat or Autoambulator). Instead of lifting the patient's leg manually or mechanically, we achieve forward propulsion during swing in MIT-Skywalker using the concepts of the passive walker by lowering the walking surface at maximum hip extension. This provides swing clearance and takes advantage of gravity and the pendular dynamics of the leg to propel the leg forward, while allowing proper neural inputs due to hip extension near swing onset and ecological heel strike at swing termination. Moreover, since the working principle takes advantage of the natural dynamics of the leg, no mechanism attached to the patient's leg is needed. This maximizes safety by eliminating the possibility of exerting unwanted forces on the leg due to mismatch between the artificial (robot) and natural (human) degrees of freedom. Equally important, it significantly reduces the don and doff time required—a significant consideration for clinically practical designs. Preliminary tests demonstrated its ability to provide therapeutic assistance without restricting the movement to any pre-determined kinematic profile, providing ecological heel strike and hip extension to maximize patient participation during therapy [2]. More details on the hardware architecture and characteristics of MIT-Skywalker can be found elsewhere [2, 47], as well as details of our control algorithm used to track the patient's gait abilities and challenge them to increase participation and improve speed and symmetry [48, 49].

31.2.4 Translating to Practice: MIT-Skywalker

Here we report on our initial feasibility study in which the MIT-Skywalker was employed to deliver three distinct modes of training in line with our model of walking: rhythmic, discrete, and balance.

31.2.4.1 Rhythmic Training Mode

The timing of the track drops is determined by the vision system estimating the position of the heel on the track. When a minimum x-position is found (indicating the onset of patient-directed swing phase), a signal is sent to drop the track. In the interest of a quick but soft drop, the sagittal plane drive was programmed to drop 2.5° (approx. 3.3 cm below the horizontal plane at the mid frontal plane) and back to horizontal in 0.7 s. Acceleration of the initial drop was four times the deceleration at the end of the perturbation, resulting in a soft feel on heel strike. Our initial target of 0.4 s for swing was based on healthy gait at 2 m/s. Training speeds for study participants were mostly done below or at 1 m/s resulting in longer swing times of the paretic limb. The soft feel of the final track movement was comfortable for subjects even if the foot hit the track early. When delivering the rhythmic program, three additional goals were implemented for some participants.

31.2.4.2 Speed Enhancing Programs

On top of the standard rhythmic protocol described above, the speed-enhancing programs focused on raising participant's training speed.

31.2.4.3 Asymmetric Speed Programs

The asymmetric speed programs focused on altering the step-length asymmetry via speed distortion (asymmetric split-belt speeds).

31.2.4.4 Vision Distortion Programs

A visual display presented in front of patients distorted the perceived length of each step while instructing participants to equalize the distorted steps to induce changes in step-length symmetry as seen in [49].

31.2.4.5 Discrete Training Mode

The MIT-Skywalker is the first rehabilitation robot to introduce discrete training for post-stroke lower extremity training. In this mode of training, the treadmill tracks operate in position mode. A random target is projected onto the treadmill track from an overhead projector. The patient is instructed to land the heel on the target. Once the vision system recognizes that the patient's heel has landed, the algorithm compares the x-position of the heel with the x-position of the target to determine if the target was hit. The treadmill track gently moves the heel back to a neutral position underneath the body. A half second later, a new target is displayed. The number of successfully hit targets and the success rate is displayed at the front end of the treadmill and the level of difficulty (target size) and location can be adjusted. Patients considered this simple game very engaging.

31.2.4.6 Balance Training

The MIT-Skywalker system is capable of imposing perturbations in both the frontal and sagittal planes. This is achieved by lowering or raising the walking surface or rotating the whole system in the frontal plane. In this feasibility study, only frontal plane perturbations were used with a stereotyped sinusoidal profile ranging from ($0–2.5°$ $2.5°$ $0°$) in 1.4 s. This is a fairly gentle profile for a healthy subject but challenging for our patients. The initial rotational direction was presented randomly and perturbation timing was randomized between 2 and 4 s. For stroke and cerebral palsy adult participants with a moderate impairment, the frontal plane perturbations were used in concert with the rhythmic program. For our most severe participant, the frontal plane perturbations were used exclusively to develop balance during standing alone. We employed a video game in the form of a surfer to indicate the frontal plane rotation.

Before and after each session, participants in this feasibility study were asked to walk for approximately 30 s to 1 min while the MIT-Skywalker vision system recorded hip and knee kinematics. During training, kinematics and heart rate were also recorded. Clinical Evaluations were performed by a physical therapist before and after the 1 month-long study at least one day removed from therapy. Subjects underwent clinical evaluations that included a 6 min walk test, self-selected and maximum walking velocity tests (measured as the average velocity of the middle 6 m of a 10 m walk test), the Berg balance scale, the Tardieu scale, and sagittal plane kinematic analysis using a 3D Guidance Trak-STAR system (Ascension Technology Co. Milton, VT). Furthermore, we monitored heart rate. We observed an average increase in heart rate between the standing and training periods of 14.7 bpm for rhythmic training sessions. Each training block lasted approximately 5 min and each rest period was between 1 and 5 min depending on the state of the participant (Table 31.1).

This initial study marks the first time the MIT-Skywalker system has been tested with persons with neurological impairments. This initial study demonstrated the feasibility of the three different training routines and showed their promise for the rehabilitation therapy of various disabilities (stroke and cerebral palsy) at three impairment levels. MIT-Skywalker showed its versatility to accommodate each. Further, each participant was able to make substantial gains in one or more of

Table 31.1 Clinical evaluations before and after 1-month training

	Participant 1		Participant 2		Participant 3	
	Initial	Final	Initial	Final	Initial	Final
6 min walk test (m)	478	546	200	209	213	204
SSV (m/s)	0.89	1.17	0.50	0.50	0.24	0.22
MSV (m/s)	1.50	1.65	0.59	0.63	0.26	0.26
Berg balance test	54	55	10	37	52	55

the tested parameters even though the injury onset was more than 5 years in the past (in the case of our CP patients, the injury was over 25 and 56 years prior).

That said, these are just a feasibility study, and proper clinical controlled studies must be performed to better understand how to tailor lower extremity therapy and how move robotics for the lower extremity beyond its "infancy" [50].

31.2.5 From Traditional Anklebots to Soft Exosuits for Restoration of Walking for Individuals Post Stroke

Post-stroke hemiparesis results in asymmetric and slow walking. Unfortunately, the current rehabilitation environment emphasizes the rapid attainment of walking independence over gait restoration. Although walking independence is an important short-term goal for survivors of stroke, independence is often achieved via compensatory mechanisms that limit recovery. Indeed, gait compensations are associated with a reduced fitness reserve, increased risk of falls, reduced endurance, and reduced speed [51, 52]. Although assistive devices such as canes, walkers, and ankle-foot orthoses are highly utilized after stroke, persisting gait deficits (such as impaired paretic propulsion [53, 54] result in a high energy cost of walking and walking disability.

Interventions that can reduce the high energy cost of walking after stroke have the potential to facilitate improved long-distance walking capacity and reduce walking-related disability [55, 56]. Indeed, a high energy cost of walking is a primary contributor to physical inactivity across neurological diagnostic groups. In people post-stroke, recent work has shown that gait interventions that facilitate faster walking only have a positive effect on the energy cost of walking if they concurrently facilitate more symmetric walking [57]. This finding may account for why 76% of individuals in the chronic phase after stroke identify deficits in their ability to walk farther distances as limiting engagement at home and in the community, whereas only 18% identify deficits in walking speed as a limiting factor [58]. That is, walking faster may not be sufficient to improve everyday walking behavior if it is not also economical.

Next-generation soft wearable robots, called exosuits, assist paretic dorsiflexion during swing phase to facilitate ground clearance and paretic plantarflexion during stance phase to enhance propulsion [59]. The development of these systems was guided by a human-in-the-loop approach where iterative development helped uncover user needs and system requirements in conjunction with new concept and technology development [60]. The result was new approaches to attaching and anchoring to the body through the use of functional apparel components that combine extensible (e.g., knits) with inextensible (e.g., woven) textile materials, placed at strategic anatomical locations. Integrated lightweight laminates provide reinforcement and create force transmission paths that distribute pressure and enhance anchoring and enable the possibility of assisting multiple joints with a single actuator through the use of multi-articular textile architectures [61]. An important aspect of their control approach for the ankle and hip is that active assistance is triggered coincidently with key biomechanical events (detected with wearable sensors), thus making it suitable for adapting to different walking speeds or step lengths [62, 63]. Combined with lightweight and efficient actuators, these innovations have enabled the demonstration of lightweight, autonomous wearable systems that can assist the ankle and hip joints for healthy individuals [64, 65].

Preliminary research on exosuits for individuals poststroke that focused on device development [3, 66] (see Fig. 31.2) demonstrated immediate, within-session improvements in both paretic ground clearance and forward propulsion [3], interlimb symmetry, energy cost of walking

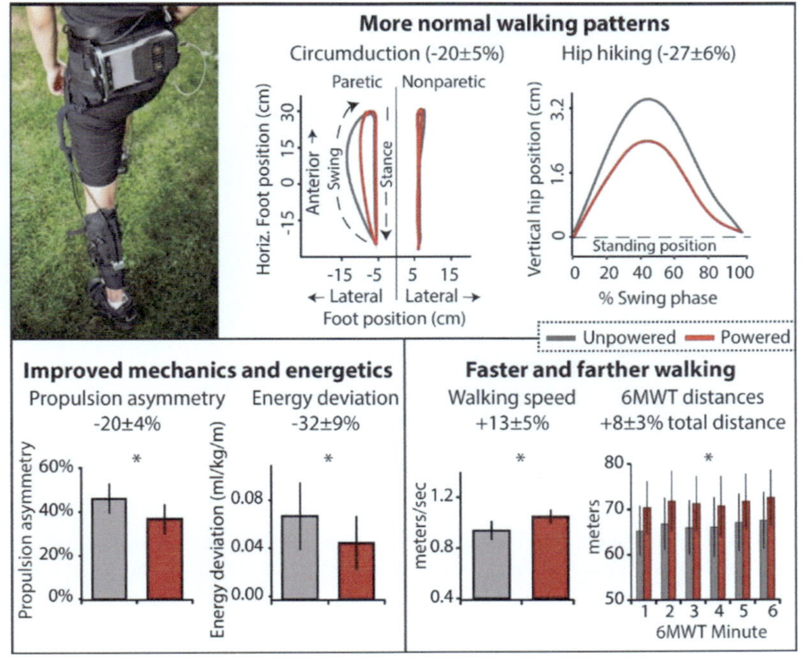

Fig. 31.2 Soft robotic exosuit technology that has been shown to improve post-stroke walking patterns, improve the mechanics and energetics of hemiparetic walking, facilitate faster and farther post-stroke walking. See references for primary sources and additional detail

[59], and reduced gait compensations [66]. The level of assistance applied was relatively low ($\sim 12\%$ of biological joint torques), yet the exosuit assistance was able to facilitate an immediate 5.33° increase in the paretic ankle's swing phase dorsiflexion and 11% increase in the paretic limb's generation of forward propulsion. These improvements in paretic limb function contributed to a 20% reduction in forward propulsion interlimb asymmetry and a 10% reduction in the energy cost of walking, compared to walking with the exosuit unpowered, which is equivalent to a 32% reduction in the metabolic burden associated with poststroke walking [3]. In [66], it was shown that the same soft exosuit targeting the paretic ankle could reduce common poststroke gait compensations. Specifically, compared to walking with the exosuit unpowered, walking with the exosuit powered resulted in significant reductions in hip hiking (27%) and circumduction (20%). Together, these immediate biomechanical benefits enabled clinically meaningful increases in both short- and long-distance walking speeds [4].

31.2.6 Translating to Practice: The Robotic Exosuit Augmented Locomotion (REAL)

Though promising, the value of exosuits in the context of gait rehabilitation is unknown; the potential for training-related effects that are retained beyond the use of exosuits is not known. Building on our previous findings of immediate improvements in speed and propulsion when walking with a soft robotic exosuit [3, 53], we designed the Robotic Exosuit Augmented Locomotion (REAL) gait training program (Fig. 31.3). REAL training merges the exosuit technology with contemporary motor learning concepts to provide an individualized and progressive gait training protocol designed to therapeutically retrain faster walking by way of increased paretic propulsion. More specifically, REAL training combines (i) paretic propulsion augmentation, (ii) progressive speed training, and (iii) goal-based strategic feedback in an algorithm-based therapeutic program centered on high intensity, task-specific, and progressively

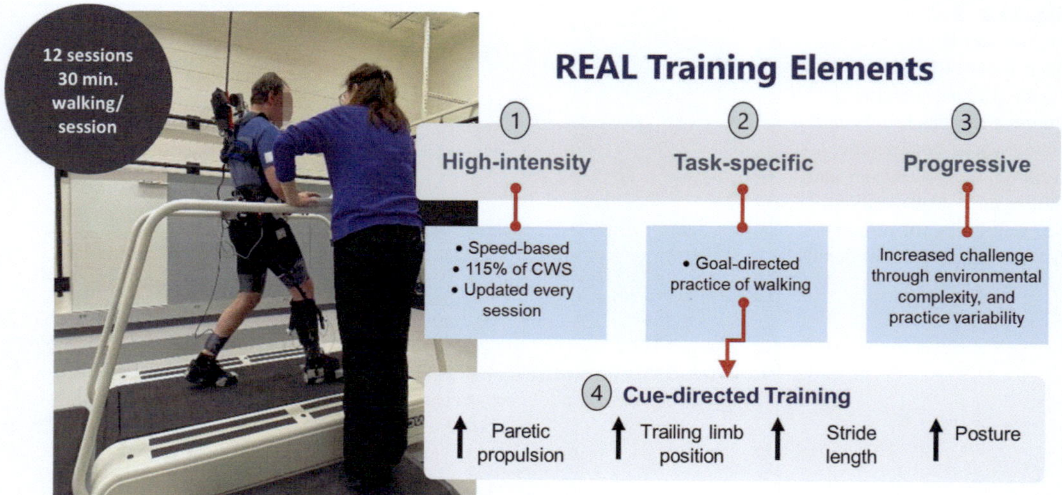

Fig. 31.3 Illustration of participant in REAL protocol and overview of different elements that are part of training. Walking begins on treadmill but transitions to overground [67]

challenging walking practice—principles which are known to be important in motor learning, and relevant for contemporary robot augmented rehabilitation interventions.

The REAL training program is currently undergoing clinical trials. A recent consideration of concept trial with a single stroke survivor demonstrated the feasibility and therapeutic potential of the REAL program [67]. The subject underwent gait training over five daily sessions. Each session consisted of 30 min of total walking practice, divided into five 6 min training bouts. The first two bouts were conducted on the treadmill, followed by three bouts overground. Data from the trial showed that comfortable walking speed was stable at 0.96 m/s prior to training and increased by 0.30 m/s after training. Clinically meaningful increases in maximum walking speed (change of 0.30 m/s) and 6 min walk test distance (change of 59 m) were similarly observed. Improvements in paretic peak propulsion (change of 2.80% BW), propulsive power (change of 0.41 W/kg), and trailing limb angle (change of 6.2°) were observed at comfortable walking speed (p's < 0.05). Likewise, improvements in paretic peak propulsion (change of 4.63% BW) and trailing limb angle (change of 4.30°) were observed at maximum walking speed (p's < 0.05). These results demonstrate that the

REAL training program is feasible to implement after stroke and capable of facilitating rapid and meaningful improvements in paretic propulsion, walking speed, and walking distance. This early-stage clinical investigation provides several design considerations and insights that can inform subsequent clinical trials of the soft robotic exosuit technology and next generation robot-assisted gait rehabilitation.

As we consider transitioning the exosuit technology and REAL training paradigms to the community, we can leverage the exosuit sensors for remote monitoring and assessment. In an early proof of concept study, it has been shown that inertial measurements on the feet can capture changes in clinically relevant variables during walking in free-living settings [68]. Moreover, these sensor measurements can facilitate automatic adjustments to the exosuit's assistance profiles to better adapt to the changing needs of the patient across varying task demands and environmental contexts. The vision underlying the application of soft robotic exosuit technology as a long-term neurorehabilitation intervention spanning both clinical and community settings is the gradual reduction of gait asymmetries and undesirable compensatory motions such as hip hiking and circumduction, in favor of more physiological gait mechanics. The exosuit technology has the potential to influence post-stroke rehabilitation from the very early stages of recovery. When combined with adjuvant therapies such as body weight support, the gait-restorative effects of the exosuit can be used even in those who do not have independent ambulatory ability. As patients progress, the exosuit can provide the combined ability to apply gait-restorative forces and provide quantitative feedback during community walking. This will extend the abilities of clinicians to the real world, providing a unique tool to retrain gait through the design and progression of personalized community-based walking rehabilitation programs.

31.3 Extending the MIT-Skywalker to Variable-Friction Cadense Shoes: An Accessible New Technology for Disabled Gait

As discussed earlier, one of the most common impairments following a neurological injury is drop foot which leaves patients with difficulty advancing the foot during the swing phase of gait. A common compensatory strategy is circumduction, which involves moving the leg outward in a circle to advance the foot during the swing phase and is a natural response to the challenge of clearing the floor. Circumduction is energetically inefficient and taxing on hip adductors and flexors, which leads to a decrease in stamina, walking speed, gait symmetry, and rhythmicity.

As discussed earlier the MIT-Skywalker introduced the concept of "removing the floor constraint" during the swing phase of gait. The MIT-Skywalker employs parallel treadmill tracks that independently drop under the foot when the patient initiates swing, thereby restoring rhythmicity and symmetry [69]. The track returns to the horizontal position to meet the foot at heel strike [70]. This work showed promise in a month-long feasibility study [7] and led to the development of the Cadense shoe (Fig. 31.4). The Cadense shoe works by providing a low friction surface between the floor and shoe during swing and a high friction surface between the floor and shoe during stance, thereby reducing the penalty for failure to clear the floor during the swing phase. The shoe is constructed with low friction plastic *pucks* arranged below soft foam. The pucks protrude from the shoe outsole and are tuned to remain exposed under the load of a foot scuff but to depress into the midsole under the weight of stance. When the pucks are depressed, the high friction rubber material is exposed to the floor creating a high friction surface between the shoe and floor.

A small pilot study with the Cadense shoe showed a 9–56% increase in maximum speed and comfortable gait speed in the 10 m walk test with a 41–66% decrease in the frontal plane hip angle for three study participants that otherwise exhibiting exaggerated circumduction [5].

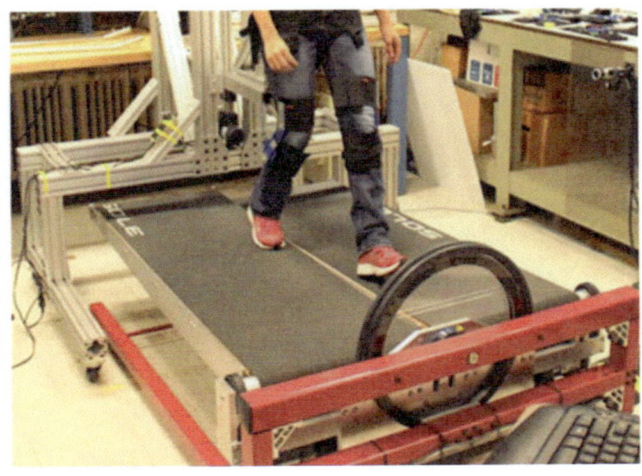

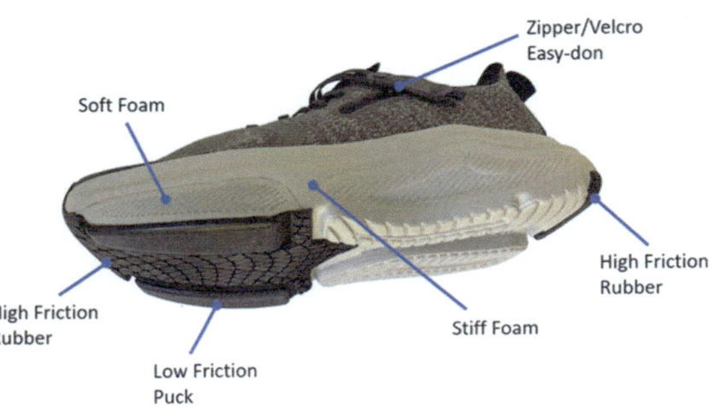

Fig. 31.4 MIT-Skywalker shown with a left track drop and the Cadense shoe

Interestingly, these changes occurred after only two minutes of warming up with the shoe without any instruction. The Cadense shoe extends the concepts incorporated in the MIT-Skywalker and has the potential to provide comparable therapy at a steep cost reduction, improving global accessibility.

31.4 Conclusion

An NIH-sponsored randomized controlled trial (RCT) demonstrated that contrary to expectations of its clinical proponents, body-weight-supported treadmill training administered by 2 or 3 therapists did not lead to superior results when compared with a home program of strength training and balance (LEAPS Study). This is a remarkable and extremely important result, one that must be acknowledged and explored further by roboticists: The goal of rehabilitation robotics is to optimize care and augment the potential of individual recovery. It is not simply to automate current rehabilitation practices, which for the most part lack a sound basis of scientific evidence. This is not a criticism of clinical practitioners, who must provide treatment as best they know how, but is primarily due to a lack of tools suitable to properly assess clinical practices themselves. To move LE robotics beyond its infancy, we have to determine what constitutes "best practice." Here robotics offers tools to carefully and methodically build evidence- and science-based approaches that allow a patient to harness plasticity and recover within only the limitations of biology. In this chapter, we examined two pairs of alternatives: (a) the Anklebot and the Soft Exosuit, and (b) the MIT-Skywalker and Cadense shoes, discussing our working model for gait and locomotion, which suggested the need to engage the supraspinal network explicitly—much like we do in upper extremity robotic therapy and, we suspect, as occurs in usual-care gait training approaches.

Of course, these are only the initial, faltering steps towards our goal. We recognize the present conclusion of the American Heart Association's statement in its guidelines: "… robotics for the lower extremity (LE) still in its infancy…" We still don't know how to tailor therapy for a particular patient's needs. We do not know the optimal dose, or in cost–benefit terms: What is the minimum intensity to promote actual change? Should we deliver impairment-based approaches (as in seated "open-chain" ankle training, i.e., joint-based, non-task specific) or functionally-based approaches (as in the soft exosuit, task specific) and to whom: those who had suffered severe, moderate, mild strokes? How can we predict potential responders versus non-responders based on stratification of impairments and deficit severities? What types of serious games should be designed and which patients' behavioral metrics should be used to drive these games? If impairment-based approaches, should therapy focus on each joint one at a time? If so, should therapy progress proximal to distal restricting all but a few limited degrees of freedom and then expand to additional degrees of freedom? Should we assist-as-needed, resist, or perturb and augment error? Who might be the responders who benefit most from these interventions? How should we integrate the robotic gyms in therapy practices? Should we consider "dual use" technology approaches as the Cadense shoes that are assistive technology in nature but may also promote long-term impairment reduction.

Acknowledgements and Disclosures H. I. Krebs is a co-inventor in several MIT-held patents for the robotic technology. He was one of the founders and the Chairman of the Board of Directors of Interactive Motion Technologies, Watertown, MA, USA from 1998 to 2016. He sold it to Bionik Laboratories, where he served as its Chief Science Officer and on the Board of Directors until July 2017. He later founded 4Motion Robotics.

C. J. Walsh declares that Harvard University has entered into a licensing and collaboration agreement with ReWalk Robotics to commercialization the soft exosuit. C. J. W. is a paid consultant for ReWalk Robotics.

T. Susko holds an equity position in BrainE Labs, the company that manufactures the Cadense Shoe.

L. Awad serves a scientific advisor to ReWalk Robotics and a paid consultant with MedRhythms.

A. Forner-Cordero acknowledges the research grant 312236/2019-0 from the Brazilian National Council for Scientific and Technological Development (CNPq) and acknowledges CAPES-PRINT for Visiting Professorship at the SCUOLA SUPERIORE SANT'ANNA (processo #. 88887.573215/2020-00).

E. Saitoh has research grants with Toyota Motor Co, Aska Co, Imasen Co, Brother Co, NTT Co, NTT Docomo Inc, Tomei Brace, Moritoh Co, Kissei-Comtec Co.

References

1. Roy A, et al. Robot-aided neurorehabilitation: a novel robot for ankle rehabilitation. IEEE Trans Rob. 2009;25(3):569–82.
2. Susko TG, Massachusetts Institute of Technology. MIT Skywalker: a novel robot for gait rehabilitation of stroke and cerebral palsy patients. Department of Mechanical Engineering. p. 322.
3. Awad LN, et al. A soft robotic exosuit improves walking in patients after stroke. Sci Transl Med. 2017;9(400).
4. Awad LN, Kudzia P, Revi DA, Ellis TD, Walsh CJ. Walking faster and farther with a soft robotic exosuit: implications for post-stroke gait assistance and rehabilitation. IEEE Open J Eng Med Biol. 2020;1:108–15.
5. Evora A, Sloan E, Castellino S, Hawkes E, Susko T. Pilot study of cadence, a novel shoe for patients with foot drop. In: 2019 41st annual international conference of the IEEE engineering in medicine and biology society (EMBC). 2019. p. 5291–6.
6. Cai LL, et al. Implications of assist-as-needed robotic step training after a complete spinal cord injury on intrinsic strategies of motor learning. J Neurosci. 2006;26(41):10564–8.
7. Susko T, Swaminathan K, Krebs HI. MIT-Skywalker: a novel gait neurorehabilitation robot for stroke and cerebral palsy. IEEE Trans Neural Syst Rehabil Eng. 2016;24(10):1089–99.
8. Hogan N, Sternad D. Dynamic primitives in the control of locomotion. Front Comput Neurosci. 2013;7:71.
9. Krebs HI, Aisen ML, Volpe BT, Hogan N. Quantization of continuous arm movements in humans with brain injury. Proc Natl Acad Sci U S A. 1999;96(8):4645–9.
10. Schaal S, Sternad D, Osu R, Kawato M. Rhythmic arm movement is not discrete. Nat Neurosci. 2004;7(10):1136–43.
11. Levy-Tzedek S, Krebs HI, Song D, Hogan N, Poizner H. Non-monotonicity on a spatio-temporally defined cyclic task: evidence of two movement types? Exp Brain Res. 2010;202(4):733–46.
12. Lee H, Rouse EJ, Krebs HI. Summary of human ankle mechanical impedance during walking. IEEE J Transl Eng Health Med. 2016;4:2100407.
13. Hyunglae L, Ho P, Rastgaar M, Krebs HI, Hogan N. Multivariable static ankle mechanical impedance with active muscles. IEEE Trans Neural Syst Rehabil Eng. 2014;22(1):44–52.
14. Lee H, Ho P, Rastgaar M, Krebs HI, Hogan N. Multivariable static ankle mechanical impedance with active muscles. IEEE Trans Neural Syst Rehabil Eng. 2014;22(1):44–52.
15. Lee H, Ho P, Rastgaar MA, Krebs HI, Hogan N. Multivariable static ankle mechanical impedance with relaxed muscles. J Biomech. 2011;44(10):1901–8.
16. Lee H, Hogan N. Time-varying ankle mechanical impedance during human locomotion. IEEE Trans Neural Syst Rehabil Eng. 2015;23(5):755–64.
17. Hogan N, Sternad D. On rhythmic and discrete movements: reflections, definitions and implications for motor control. Exp Brain Res. 2007;181(1):13–30.
18. Mussa-Ivaldi FA, Hogan N, Bizzi E. Neural, mechanical, and geometric factors subserving arm posture in humans. J Neurosci. 1985;5(10):2732–43.
19. Ikegami T, Hirashima M, Taga G, Nozaki D. Asymmetric transfer of visuomotor learning between discrete and rhythmic movements. J Neurosci. 2010;30(12):4515–21.
20. Forrester LW, Roy A, Krebs HI, Macko RF. Ankle training with a robotic device improves hemiparetic gait after a stroke. Neurorehabil Neural Repair. 2010.
21. Forster A, Young J. Incidence and consequences of falls due to stroke: a systematic inquiry. BMJ. 1995;311(6997):83–6.
22. Ramnemark A, Nyberg L, Borssen B, Olsson T, Gustafson Y. Fractures after stroke. Osteoporos Int. 1998;8(1):92–5.
23. Dennis MS, Lo KM, McDowall M, West T. Fractures after stroke: frequency, types, and associations. Stroke. 2002;33(3):728–34.
24. Kanis J, Oden A, Johnell O. Acute and long-term increase in fracture risk after hospitalization for stroke. Stroke. 2001;32(3):702–6.
25. Duncan PW, et al. Protocol for the locomotor experience applied post-stroke (LEAPS) trial: a randomized controlled trial. BMC Neurol. 2007;7:39.
26. Hogan N, Sternad D. Dynamic primitives of motor behavior. Biol Cybern. 2012;106(11–12):727–39.
27. Robertson DG, Winter DA. Mechanical energy generation, absorption and transfer amongst segments during walking. J Biomech. 1980;13(10):845–54.
28. Sawicki GS, Ferris DP. A pneumatically powered knee-ankle-foot orthosis (KAFO) with myoelectric activation and inhibition. J Neuroeng Rehabil. 2009;6:23.
29. Eng JJ, Winter DA. Kinetic analysis of the lower limbs during walking: what information can be gained from a three-dimensional model? J Biomech. 1995;28(6):753–8.
30. Teixeira-Salmela LF, Nadeau S, Milot MH, Gravel D, Requiao LF. Effects of cadence on energy generation and absorption at lower extremity joints during gait. Clin Biomech (Bristol, Avon). 2008;23(6):769–78.
31. Umberger BR, Martin PE. Mechanical power and efficiency of level walking with different stride rates. J Exp Biol. 2007;210(Pt 18):3255–65.

32. Perry J, STk, Davids JR. Gait analysis: normal and pathological function. J Pediatr Orthop. 1992;12(6):815.

33. Neptune RR, Kautz SA, Zajac FE. Contributions of the individual ankle plantar flexors to support, forward progression and swing initiation during walking. J Biomech. 2001;34(11):1387–98.

34. Meinders M, Gitter A, Czerniecki JM. The role of ankle plantar flexor muscle work during walking. Scand J Rehabil Med. 1998;30(1):39–46.

35. Gottschall JS, Kram R. Energy cost and muscular activity required for propulsion during walking. J Appl Physiol (1985). 2003;94(5):1766–72.

36. Michmizos K, Rossi S, Castelli E, Cappa P, Krebs H. Robot-aided neurorehabilitation: a pediatric robot for ankle rehabilitation. IEEE Trans Neural Syst Rehabil Eng. 2015.

37. Roy A, Krebs HI, Williams DJ, Bever CT, Forrester LW, Macko RM, Hogan N. Robot-aided neurorehabilitation: a novel robot for ankle rehabilitation. IEEE Trans Robot. 2009;25(3):569–82.

38. Khanna I, Roy A, Rodgers MM, Krebs HI, Macko RM, Forrester LW. Effects of unilateral robotic limb loading on gait characteristics in subjects with chronic stroke. J Neuroeng Rehabil. 2010;7:23.

39. Rossi S, Colazza A, Petrarca M, Castelli E, Cappa P, Krebs HI. Feasibility study of a wearable exoskeleton for children: is the gait altered by adding masses on lower limbs? PLoS ONE. 2013;8(9):e73139.

40. Chang JL, Lin RY, Saul M, Koch PJ, Krebs HI, Volpe BT. Intensive seated robotic training of the ankle in patients with chronic stroke differentially improves gait. NeuroRehabilitation. 2017;41(1):61–8.

41. Ivanenko YP, Grasso R, Macellari V, Lacquaniti F. Control of foot trajectory in human locomotion: role of ground contact forces in simulated reduced gravity. J Neurophysiol. 2002;87(6):3070–89.

42. Forrester LW, Roy A, Krebs HI, Macko RF. Ankle training with a robotic device improves hemiparetic gait after a stroke. Neurorehabil Neural Repair. 2011;25(4):369–77.

43. Roy A, Krebs HI, Bever CT, Forrester LW, Macko RF, Hogan N. Measurement of passive ankle stiffness in subjects with chronic hemiparesis using a novel ankle robot. J Neurophysiol. 2011;105(5):2132–49.

44. Roy A, Forrester LW, Macko RF, Krebs HI. Changes in passive ankle stiffness and its effects on gait function in people with chronic stroke. J Rehabil Res Dev. 2013;50(4):555–72.

45. Coelho R, Durand S, Martins J, Krebs HI. Multivariable passive ankle impedance in stroke patients: a preliminary study. J Biomech. 2022;130(110829).

46. Collins S, Ruina A, Tedrake R, Wisse M. Efficient bipedal robots based on passive-dynamic walkers. Science. 2005;307(5712):1082–5.

47. Bosecker C, Krebs HI. MIT-Skywalker. In: Proc IEEE 11th international conference on rehabilitation robotics. 2009. p. 542–9.

48. Artemiadis PK, Krebs HI. On the control of the MIT-Skywalker. Conf Proc IEEE Eng Med Biol Soc. 2010;2010:1287–91.

49. Kim SJ, Krebs HI. Effects of implicit visual feedback distortion on human gait. Exp Brain Res. 2012;218(3):495–502.

50. Miller EL, et al. Comprehensive overview of nursing and interdisciplinary rehabilitation care of the stroke patient: a scientific statement from the American Heart Association. Stroke, Practice Guideline. 2010;41(10):2402–48.

51. Bowden MG, Behrman AL, Neptune RR, Gregory CM, Kautz SA. Locomotor rehabilitation of individuals with chronic stroke: difference between responders and nonresponders. Arch Phys Med Rehabil. 2013;94(5):856–62.

52. Mayo NE, et al. Disablement following stroke. Disabil Rehabil. 1999;21(5–6):258–68.

53. Awad LN, Esquenazi A, Francisco GE, Nolan KJ, Jayaraman A. The ReWalk ReStore soft robotic exosuit: a multi-site clinical trial of the safety, reliability, and feasibility of exosuit-augmented post-stroke gait rehabilitation. J Neuroeng Rehabil. 2020;17(1):80.

54. Awad LN, Hsiao H, Binder-Macleod SA. Central drive to the paretic ankle plantarflexors affects the relationship between propulsion and walking speed after stroke. J Neurol Phys Ther. 2020;44(1):42–8.

55. Reisman DS, Binder-MacLeod S, Farquhar WB. Changes in metabolic cost of transport following locomotor training poststroke. Top Stroke Rehabil. 2013;20(2):161–70.

56. Reisman D, et al. Time course of functional and biomechanical improvements during a gait training intervention in persons with chronic stroke. J Neurol Phys Ther. 2013;37(4):159–65.

57. Awad LN, Binder-Macleod SA, Pohlig RT, Reisman DS. Paretic propulsion and trailing limb angle are key determinants of long-distance walking function after stroke. Neurorehabil Neural Repair. 2015;29(6):499–508.

58. Combs SA, Van Puymbroeck M, Altenburger PA, Miller KK, Dierks TA, Schmid AA. Is walking faster or walking farther more important to persons with chronic stroke? Disabil Rehabil. 2013;35(10):860–7.

59. Bae J, et al. Biomechanical mechanisms underlying exosuit-induced improvements in walking economy after stroke. J Exp Biol. 2018;221(Pt 5).

60. Walsh C. Human-in-the-loop development of soft wearable robots. Nat Rev Mater. 2018;3:78–80.

61. Asbeck AT, De Rossi SMM, Galiana I, Ding Y, Walsh CJ. Stronger, smarter, softer: next-generation wearable robots. IEEE Robot Autom Mag. 2014;21(4):22–33.

62. Perez-Ibarra JC, Siqueira AAG, Krebs HI. Identification of gait events in healthy subjects and with Parkinson's disease using inertial sensors: an adaptive unsupervised learning approach. IEEE Trans Neural Syst Rehabil Eng. 2020;28(12):2933–43.

63. Perez-Ibarra JC, Siqueira AAG, Krebs HI. Real-time identification of gait events in impaired subjects using a single-IMU foot-mounted device. IEEE Sens J. 2020;20(5):2616–24.
64. Kim J, et al. Reducing the metabolic rate of walking and running with a versatile, portable exosuit. Science. 2019;365(6454):668–72.
65. Quinlivan BT, et al. Assistance magnitude versus metabolic cost reductions for a tethered multiarticular soft exosuit. Sci Robot. 2017;2(2).
66. Awad LN, et al. Reducing circumduction and hip hiking during hemiparetic walking through targeted assistance of the paretic limb using a soft robotic exosuit. Am J Phys Med Rehabil. 2017;96(10 Suppl 1):S157-64.
67. Porciuncula F, et al. Targeting paretic propulsion and walking speed with a soft robotic exosuit: a consideration-of-concept trial. Front Neurorobot. 2021;15: 689577.
68. Arens P, et al. Real-time gait metric estimation for everyday gait training with wearable devices in people poststroke. Wearable Technol. 2021;2.
69. Susko T, Krebs H. Cycle variance: proposing a novel ensemble-based approach to assess the gait rhythmicity on the MIT-Skywalker. In: 2015 IEEE international conference on rehabilitation robotics (ICORR). 2015. p. 101–6.
70. Susko T, Krebs H. MIT-Skywalker: a novel environment for neural gait rehabilitation In: 5th IEEE RAS/EMBS international conference on biomedical robotics and biomechatronics. 2014. p. 677–82.

A Flexible Cable-Driven Robotic System: Design and Its Clinical Application for Improving Walking Function in Adults with Stroke, SCI, and Children with CP

Ming Wu

Abstract

A cable-driven locomotor training system (3DCaLT) has been developed to understand locomotor adaptation and improve walking function in adults following a hemispheric stroke or spinal cord injury (SCI), and children with cerebral palsy (CP). A key component of this system is that it is highly back drivable and allows for variation in the trajectory of the gait pattern. In addition, this robotic system can provide controlled forces in both the sagittal and frontal planes at targeted phases of gait. The robotic trainer uses a light-weight cable-driven with controlled forces applied to the pelvis and leg. The 3DCaLT is compliant, and gives patients the freedom to voluntarily move their pelvis and legs in a natural gait pattern while providing controlled assistance/resistance forces during body weight supported treadmill training (BWSTT). Results from these randomized controlled studies suggest that applying targeted lateral assistance force to the pelvis during treadmill training seems more effective than treadmill only training in improving endurance in adults with SCI and in children with CP. In addition, applying a targeted resistance force to both legs during treadmill training is more effective than applying assistance for improving walking function in children with CP. Applying a resistance or assistance force to the paretic leg during treadmill training may induce improvement in walking function in individuals post-stroke, but applying resistance force was not greater than assistance force for improving walking function, which may be due to the compensatory movement from the non-paretic leg. Thus, the flexible cable-driven robotic system, i.e., 3DCaLT, may be used to improve the locomotor function in adults post-stroke or with SCI, and children with CP. Further studies with a large sample size of subjects are warranted.

Keywords

Locomotion · Cable-driven · Robot · Motor adaptation · Spinal cord injury · Stroke · Children with cerebral palsy · Constraint induced movement therapy

M. Wu (✉)
Department of Biomedical Engineering, University of Illinois at Chicago, Chicago, IL 60607, USA
e-mail: w-ming@northwestern.edu; mingwu9@uic.edu

M. Wu
Legs and Walking Lab, Shirley Ryan AbilityLab, Chicago, IL 60611, USA

M. Wu
Department of Physical Medicine and Rehabilitation, Northwestern University, Chicago, IL 60611, USA

© The Author(s), under exclusive license to Springer Nature Switzerland AG 2022

D. J. Reinkensmeyer et al. (eds.), *Neurorehabilitation Technology*,
https://doi.org/10.1007/978-3-031-08995-4_32

32.1 Introduction

Body weight supported treadmill training has been used for improving locomotor function in humans with spinal cord injury (SCI) [1–3], stroke [4–6], and children with cerebral palsy [7, 8]. One limitation of this technique is the requirement of greater involvement of physical therapists during locomotor training. In addition, it can be a labor intensive work for the physical therapist who conducts the training, particularly for those patients who require substantial walking assistance. As a consequence, several robotic gait training systems have been developed. While these robotic gait training systems are effective in reducing the labor intensity of physical therapists, they showed limited functional gains for some patients. Thus, there is clear need for the development of new robotic gait training systems and the examination of the motor learning mechanisms during locomotor training. In order to fit the need, we developed a 3D cable-driven robotic gait training system. In this chapter, we focused on the development and clinical application of this cable-driven robotic gait training system.

32.1.1 Relevant Pathophysiology Background

32.1.1.1 Stroke

Stroke is currently the leading cause of serious, long-term disability in the U.S. [9]. Impaired mobility is an important factor in determining the degree of physical disability after stroke [10]. While up to 80% of individuals with stroke may ultimately recover the ability to walk a short distance [11], most of them do not achieve the locomotor capacity necessary for community ambulation [12]. Limited community walking ability reduces the probability of a successful return to work and decreases participation in community activities [13].

Walking ability post-stroke is characterized primarily by reduced walking speed [14] and endurance [15], residual spatial and temporal left-right asymmetry [16], and impaired postural stability [17]. Patients suffer a greatly reduced knee flexion at toe-off and during the swing of the paretic leg, as compared to the non-paretic leg, which is usually associated with compensatory movements such as pelvic hiking and leg circumduction [18]. The impaired hip and knee flexion during swing may result in a decreased forward progression and gait velocity, shortened step length and toe drag at the initial swing [19]. These impairments restrict independent mobility and severely impact quality of life of individuals post-stroke.

32.1.1.2 Spinal Cord Injury

The estimated prevalence of spinal cord injury (SCI) in the United States is approximately 296,000, with an incidence of approximately 17,900 new cases every year [20]. One of the major goals of patients with SCI is to regain walking ability [21], as limitations in mobility can adversely affect most activities of daily living [22, 23]. Following SCI, descending spinal motor pathways are usually damaged. The loss of descending input to spinal neurons may reduce the synaptic drive to locomotor networks, and also compromise the ability to produce voluntary movements of the limbs. In addition, there is often impaired control of balance, and this impairment, together with associated weakness of lower extremity muscles may adversely impact walking. As a consequence, patients with SCI walk with reduced speed and shorter stride length [24], require assistive devices, such as rolling walkers, and spend more of the gait cycle in double limb support [25]. In addition, subjects with SCI may demonstrate excessive pelvis and trunk motion to compensate for the lower limb deficits due to the spinal cord lesion [26], resulting in an abnormal gait pattern.

32.1.1.3 Children with Cerebral Palsy (CP)

CP is the most prevalent physical disability originating in childhood with an incidence of 2–3 per 1,000 live births [27, 28]. Of the children who are diagnosed with CP, as much as 90% of children with CP have difficulty in walking [29, 30]. Reduced waking speed and endurance are two of the main functional problems, particularly in children with more severe disabilities [31].

Attaining functional walking ability is often an important functional goal for children with CP. Ambulation plays a central role in healthy bone development [32] and cardiopulmonary endurance [33] and children who are able to ambulate are more accomplished in activities of daily living and social roles, such as participation in the community, than children who use a wheelchair [34]. The development of independent gait and efficiency of walking are often the focus of gait rehabilitation for children with CP.

32.1.2 Rationale for Application of Current Technology (The Role of Neural plasticity)

32.1.2.1 Neuroplasticity of Adults with Stroke and SCI, and Children with CP

Although the loci of neuraxis lesions obviously differ between stroke, SCI and CP, the extent of injury to the motor system and to motor-related cognitive networks often overlaps. In particular, the mechanisms of the neural adaptations that accompany training and learning are not dependent on the disease (i.e., stroke, SCI, or CP) as much as they rely on the available plasticity in relevant neural networks [35]. The neural reorganization achieved during rehabilitation is highly dependent on the magnitude and specificity of neural activity. Thus, increasing the intensity of neural activity during locomotor training should improve the training effect, consistent with use-dependent synaptic plasticity, as expressed in "Hebb's Rule" [36]. Observations in spinalized cats in which targeted standing training or locomotor training produced only task-specific improvements in motor function demonstrates that practice is more effective when it is task-specific [37, 38]. Furthermore, motor training paradigms that emphasize active movements are more effective in producing plasticity

in spinal circuits and should increase volitional locomotor performance when compared to passive movement training [39, 40]. Thus, to maximize locomotor recovery, rehabilitation for adults after stroke and SCI, and children with CP should emphasize active, repetitive and task-specific practice that maximizes neuromuscular activity.

32.1.3 Therapeutic Action/Mechanisms and Efficacy

32.1.3.1 Task-Oriented Practice in Individuals Post-Stroke

To improve gait performance and functional outcomes following a neurological injury, rehabilitation efforts have been focused on re-establishing normal walking patterns [41]. Towards this end, the use of body weight supported treadmill training (BWSTT) has demonstrated significant improvements in walking capability in individuals post-stroke and SCI [42] and is becoming increasingly popular. Actually, the use of treadmill training for people with neurological disorders has its roots in previous animal studies where spinal cats were able to regain locomotor functions of the hindlimbs with weight support through treadmill training [43]. The underlying mechanism of the effectiveness of this technique is thought to be the reorganization capacity of the central system when task-specific motor practice is provided through treadmill training [44]. In clinics, the use of a treadmill (with or without body weight support) permits a greater number of steps to be performed within a training session. That is, it increases the amount of task-specific walking practice [45]. By providing partial body weight support over a treadmill and manual facilitation from therapists, previous research has demonstrated improvements in waling function and temporal-spatial gait patterns, including gait velocity [4–6], endurance [46], balance [5], and symmetry [47]. For instance, previous studies in non-ambulatory hemiparetic subjects revealed that BWSTT was superior to conventional physiotherapy with regard to restoration of gait ability and improvement of overground walking velocity [4]. Changes in impairments and functional limitations observed with intensive BWSTT are often greater than that achieved during conventional or lower intensity physical therapy [6, 48].

However, two randomized, controlled trials in acute stroke survivors failed to show a superiority of BWSTT compared with conventional physical therapy focusing on overground training [49, 50]. For instance, results from a multicenter trial in hemiparetic patients (n = 73) indicated that there was no significant difference between the BWSTT and the control group (who completed overground walking training) with regard to Functional Independence Measures (FIM), walking velocity, Fugl-Meyer Stroke Assessment, and balance assessments [49]. However, in a subgroup of severe stroke subjects, the BWSTT group demonstrated a greater improvement in walking speed and endurance compared to the control group [50]. In addition, in studies that have employed high intensity walking regimens in individuals with chronic stroke (i.e., those without presumed spontaneous recovery), the average increase in walking speed ranges from 0.09 to 0.13 m/s following 1–6 months of training [6, 46]. While significant statistically, these changes are relatively small considering the effort required to perform such training.

32.1.3.2 Task-Oriented Practice in Humans With SCI

BWSTT with manual assistance given to the legs and the pelvis has also been used as a promising rehabilitation method designed to improve locomotor function in people with SCI [2, 51–54]. For instance, BWSTT has been shown to provide significant improvements in locomotor ability and motor function in humans with SCI [53]. Specifically, 89 patients with incomplete SCI underwent BWSTT and were compared with 64 patients treated conventionally. The results indicated that the BWSTT group improved their mobility more than the control group (i.e., conventional treatment group). For the acute patients, 92% of those initially wheelchair-bound became independent walkers following BWSTT, while only 50% were able to walk independently following conventional therapy. For chronic patients, 76% of those initially wheelchair-bound learned to walk independently following BWSTT, while only 7% returned to walking following conventional therapy [53].

Conversely, results from a recent large multicenter randomized clinical trial with acute incomplete SCI patients indicated that both groups improved their outcome measurements related to walking performance, but no significant differences were found between the BWSTT and the conventionally trained groups [3].

Even though BWSTT may only be as effective as conventional training, it is still a valuable technique for locomotor training in humans with SCI. The technique may be safer and more convenient for assisting ASIA A and B subjects to stand and step when compared with conventional physical therapy [3]. Also, it may allow for earlier gait training in patients with limited locomotor capabilities, allowing them to repeat a gait-like motion and alternative loading of the lower limbs [51, 53]. Despite this, BWSTT often requires the effort of multiple physical therapists (generally up to 3) to assist the legs and control trunk movement. It can be a labor intensive work for physical therapists, particularly for those patients who require substantial walking assistance following SCI. This suggests that there is a need to improve the current BWSTT system.

32.1.3.3 Task-Oriented Practice in Children with CP

BWSTT has also been used to improve the locomotor function in children with CP [8]. While statistically significant improvements in walking capacity with BWSTT have been shown, the function gains are relatively small (increased only 0.07 m/s in walking speed) [7]. In particular, recent randomized controlled studies indicated that BWSTT is not more effective than overground walking for improving walking speed and endurance for children with CP [55, 56], although another randomized controlled study indicated that BWSTT is more effective than overground gait training in improving walking function in children with CP [57]. Thus, there is still insufficient evidence about the effect of BWSTT in improving locomotor function in children with CP [58–60]. In addition, BWSTT requires greater involvement of the physical therapist [61].

32.1.4 Review of Experience and Evidence for the Application of Specific Technology

Due to the high effort level required by therapists to assist patients during BWSTT, several robotic systems have been developed for automating locomotor training of individuals post-stroke or SCI, and children with CP, including the Lokomat [62], the Gait Trainer (GT) [63] and the AutoAmbulator [64]. The Lokomat is a motorized exoskeleton that drives hip and knee motion with a fixed trajectory [62]. The GT rigidly drives the patient's feet through a stepping motion using a crank-and-rocker mechanism attached to foot platforms [63]. The AutoAmbulator is a body-weight supported treadmill robot system with robotic arms strapped to the patient's leg at the thigh and ankle, which move the legs in a quasi-normal walking pattern. These robotic systems had at their initiation the basic design goal of firmly assisting patients in producing correctly shaped and timed locomotor movements. This approach is effective in reducing therapist labor in locomotor training and increasing the total duration of training. Also, the number of therapists required to provide robotic BWSTT is significantly less than that required for manually assisted treadmill training [65].

32.1.5 Robotic Gait Training in Individuals Post-Stroke

While robotic gait training relieves the strenuous effort of the therapists, the functional gains are limited for some patients [66, 67]. For instance,

results from a study using the Lokomat with 30 acute stroke patients indicated that there was only 0.06 m/s gait speed improvement following 4 weeks of training, and there was no significant difference between the therapy on the Lokomat and gait training overground [66]. In particular, in a study with 63 subacute stroke patients, results indicated that participants who received conventional gait training experienced significantly greater gains in walking speed and distance than those trained on the Lokomat [68]. In addition, results from a study with 48 chronic ambulatory stroke survivors indicated that robotic-assisted BWSTT with a fixed trajectory control strategy is less effective in improving walking ability in individuals post-stroke than physical therapist-assisted locomotor training [69]. In contrast, results from a study with 155 non-ambulatory subacute stroke patients show that robotic-assisted gait training (using the Gait Trainer) plus conventional physiotherapy resulted in a significantly better gait ability compared with conventional physiotherapy only [70]. Recent literature reviews suggest that robotic gait training in combination with physiotherapy increased the odds of participants becoming independent in walking, although did not significantly increase walking speed (mean difference = 0.04 m/s) and endurance (i.e., mean difference = 3 m walked in 6 min) [67, 71, 72]. In particular, the type of robotic systems (i.e., exoskeleton robotic system, such as the Lokomat versus end-effector, such as the Gait Trainer) might have an impact on the outcome measures of gait rehabilitation of individuals post-stroke [71]. For instance, a meta-analysis indicated that the use of end-effector robotic gait training systems significantly increased the walking velocity with the pooled mean difference for walking velocity being 0.12 m/s [67]. In contrast, a meta-analysis indicated that the use of exoskeleton robotic systems for gait rehabilitation even significantly decreased the walking velocity with the pooled mean difference for walking velocity being −0.05 m/s [73], although direct empirical comparisons between the two types of robotic gait training systems are still lacking.

32.1.6 Robotic Gait Training in Humans with SCI

Similar results have been observed in humans with SCI [74]. For instance, results from a randomized study with 27 chronic SCI patients indicated that all modalities of locomotor training were associated with improved walking speed, and there were no significant differences between the group with robotic gait training using the Lokomat and other groups [54]. Similarly, in a study with 30 acute SCI patients randomly assigned to three groups: robotic-assisted BWSTT using the Lokomat, therapist-assisted BWSTT, and overground ambulation with a mobile suspension system used for safety and support as necessary, results indicated that there were no significant differences in the rate and extent of motor and functional recovery among the three groups [75], although the total distance ambulated during robotic BWSTT was significantly greater than that with overground training. Such results suggest that current robotic-assisted BWSTT methods may reduce the requirements and labor effort for the physical therapist, but do not necessarily offer an advantage in terms of regaining locomotor function in humans with SCI.

32.1.7 Robotic BWSTT in Children with CP

The Pediatric Lokomat (Hocoma AG, Volketswil, Switzerland) has been developed to provide robotic assistance to children with CP during treadmill training [76]. While the current Pediatric Lokomat is effective in reducing therapist labor intensity during locomotor training and increasing the total duration of the training, it shows relatively limited functional gains for some children with CP [76]. For instance, a recent randomized controlled study indicated that robotic treadmill training using the Pediatric Lokomat was not more effective than conventional physical therapy for improving walking function in children with CP [77]. In contrast, results from other studies indicated that robotic treadmill training induced significant improvements in walking speed [76],

and gross motor function in children with CP [78], but these studies did not have a control group, which may preclude a firm conclusion about the efficacy of robotic treadmill training in children with CP. Thus, there is still insufficient evidence for determining the effect of robotic treadmill training on walking function in children with CP. As a consequence, there is a need for the development of novel robotic training paradigms and/or systems.

32.1.7.1 Limitations of Current Robotic Systems

While these first generation robotic systems are effective in reducing therapist labor in locomotor training, they do have some limitations [79]. For instance, a fixed trajectory control strategy that was used in previous robotic systems may encourage passive rather than active training. During robotic BWSTT, the driven gait orthosis passively moves the legs in a kinematically correct pattern. The robot essentially takes over the movement task, sharply reducing the patient's participation level [80]. In addition, a fixed trajectory training eliminates the variability in the kinematics of the lower limbs, which may be crucial for successful motor learning as demonstrated in animal studies [81].

Another limitation of current robotic gait training systems is the relatively expensive cost, which may be a significant barrier to widespread clinical application and use. For instance, many rehabilitation settings will be unable to deliver this type of therapeutic intervention to a larger patient population. As a consequence, there is a need to develop new cost-effective techniques of robotic BWSTT in order to produce greater functional improvements in individuals post-stroke, SCI, or children with CP.

In an attempt to improve the efficacy of robotic BWSTT, we have developed a cable-driven gait training system (CaLT) [82]. This robotic trainer uses a light-weight cable-driven with controlled forces applied to legs. A key component of this robotic system is that it is highly back drivable, which means that the patient can readily overcome the forces generated by the robot. This unique feature offers key advantages over both the ball-screw mechanisms used in the Lokomat [62], and the crank-and-rocker mechanism, as used in the Gait Trainer [63], in that it allows for variation in the lower limb kinematics and increases active participation of the patient during training.

Recently, this cable-driven gait training system has been further developed by the integration of the pelvis component [83]. Specifically, two motor and pulley systems have been attached at the side of the treadmill to provide controlled assistance force to the pelvis during the stance phase of gait (for assisting weight shift) while the subject walks on a treadmill. As suggested in previous studies, these components of gait training are critical to maximize motor learning and functional improvements in adults with stroke and SCI, and children with CP [81].

In the current design, four nylon-coated stainless-steel cables (1.6 mm diameter), driven by four motors (AKM33H, Kollmorgen) through 4 cable spools and pulleys, are affixed to custom cuffs that are strapped to the legs (routinely around the ankles) to produce an assistance/resistance force of up to 45 N (see Fig. 32.1). Additional two cables, driven by two motors (AKM33H, Kollmorgen), are affixed to custom braces that are strapped to the pelvis and provide controlled assistance forces for facilitating weight shifting in the mediolateral direction during treadmill walking. Ankle kinematics of both legs are recorded using two custom, 3-dimensional position sensors. The ankle position signals are used by the operator to control the timing and magnitude of applied forces, at targeted phases of gait.

Control is implemented through a custom LabVIEW program, which sends control signals to the motor drives through an analog output to set the applied forces. The controller automatically adjusts the load provided by the cables based on the kinematic performance of the subject. The load is applied to legs starting at pre-swing ($\sim$ 10% gait cycle prior to toe-off) through mid-swing of gait [84]. In addition, the pelvis load is applied from heel strike to mid-stance of the ipsilateral leg for facilitating weight shifting.

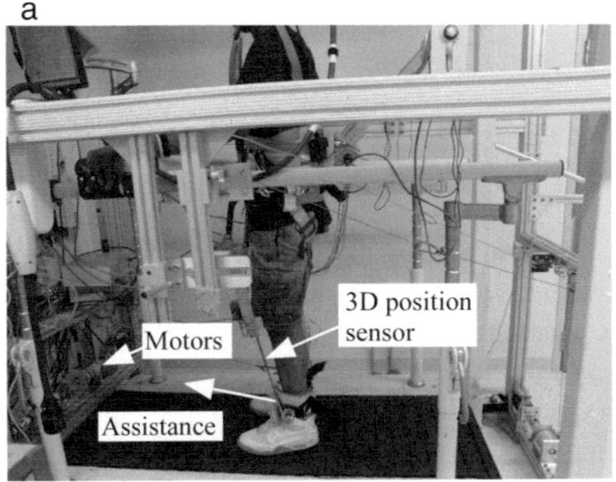

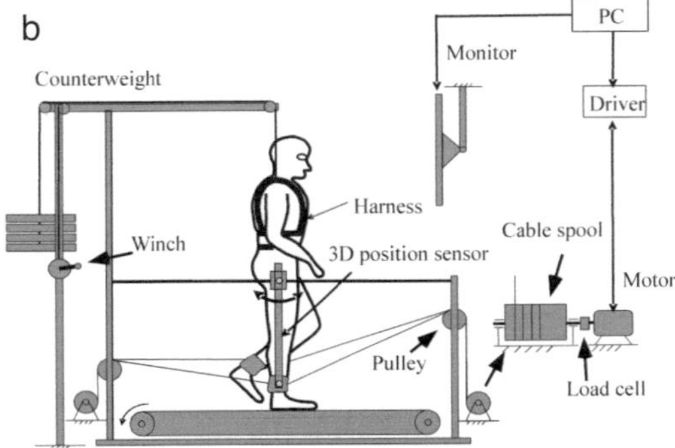

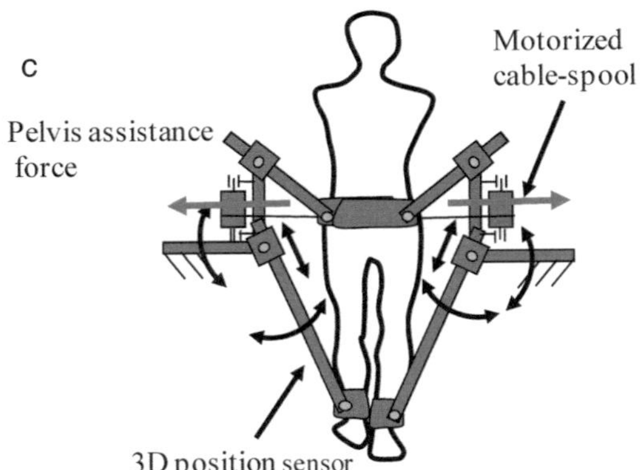

Fig. 32.1 This figure illustrates the cable robot, a motor-driven cable apparatus that was used with a treadmill and body weight support system. Six cables driven by six motors, pulleys, and cable spools were used to apply resistance/assistance loads to the legs during the swing phase of gait, and assistance loads to the pelvis during the stance phase of gait. A personal computer was used to control the load produced by the six motors, applying targeted assistance or resistance loads

Two control algorithms were designed for either an assistance or resistance strategy. For the assistance paradigm, the force applied to the legs was determined in real time using the following equation:

$$F_a(t) = -k_P(x(t) - x_d(t)) - k_D(x(t) - x_d(t))$$
(32.1)

where t is time; k_P and k_D are the position and velocity gains (e.g., k_P and k_D are adjustable depending on the tolerance of the subject); $x(t)$, and $x_d(t)$ are measured and desired ankle horizontal position and velocity during the swing phase. For the resistance paradigm, a similar equation was used to determine the amount of force, but a resistance load was applied. For the pelvis assistance paradigm, a control algorithm similar to leg assistance was used.

32.2 Current Developments and Ongoing Testing

32.2.1 Locomotor Training in Individuals Post-Stroke

32.2.1.1 Introduction

Previous studies demonstrated that active motor training is more effective than passive training in eliciting performance improvement [40]. Further, data from hemiparetic subjects practicing upper limb movements with forces that provide passive guidance versus error enhancement indicate that greater improvements in performance are achieved when errors are magnified [85]. These results suggest that error-augmentation training may also be used as an effective way to improve locomotor function in individuals post-stroke. We postulated that by applying a controlled resistance load to increase kinematic errors of the paretic leg during treadmill walking, motor learning would be accelerated during treadmill training in individuals post-stroke.

On the other hand, providing a controlled assistance load to the paretic leg may facilitate leg swing and induce a longer step length, which mimics the way that clinical therapists provide assistance to the paretic leg during treadmill training. We postulated that providing an assistance load to the paretic leg might also improve locomotor function in individuals post-stroke through a use-dependent motor learning mechanism [86]. However, it remains unclear whether leg resistance versus assistance is more effective in improving locomotor function in individuals post-stroke. The purpose of this study was to assess locomotor function (i.e., walking speed, endurance, balance) after resistance versus assistance (applied to the paretic leg) treadmill training in individuals post-stroke. The hypothesis was that subjects from both groups would show improvements in locomotor function, although there would be greater improvements in subjects who underwent resistance training in comparison with those who underwent assistance training [87]. The 3DCaLT robotic gait training system was used to provide controlled resistance or assistance load to the paretic leg during treadmill training.

Thirty individuals with chronic hemiparetic stroke were recruited to participate in this study. Subjects were randomly assigned to assistance or resistance groups after the first evaluation and 28 participants finished all the training and test sessions (n = 14 in each group).

A significant improvement in walking speed was observed for subjects from the resistance group after 6 weeks of treadmill training. Specifically, self-selected and fast walking speeds significantly increased from 0.53 ± 0.25 m/s to 0.61 ± 0.28 m/s (P = 0.002, ANOVA), and from 0.72 ± 0.36 m/s to 0.82 ± 0.39 m/s (P = 0.001), respectively, after resistance training. Further, improvements in walking speed were partially retained at follow up (P = 0.03 and P = 0.002 for self-selected and fast walk speeds, respectively). The 6-min walking distance slightly increased but this was not significant (P = 0.18). Berg Balance Scale (BBS) score also slightly increased although this was not significant (P = 0.11).

On the other hand, a significant improvement in walking function was obtained for subjects from the assistance training group after training. Specifically, self-selected and fast walk speeds significantly increased from 0.47 ± 0.24 m/s to 0.56 ± 0.32 m/s (P = 0.01), and from 0.65 ± 0.38 m/s to 0.76 ± 0.45 m/s (P = 0.002), respectively, after training. Further, the improvements in walking speeds were partially retained at follow up (P = 0.01 and P = 0.004 for self-selected and fast walking speeds, respectively). Also, the 6-min walk distance significantly increased from 177.4 ± 99.9 m to 197.5 ± 109.5 m (P = 0.002), and was partially retained at follow up (191.1 ± 108.5 m, P = 0.02). The BBS score significantly increased from 43.6 ± 9.0 to 45.5 ± 8.8 (P = 0.02) after assistance training but was not significant at follow up (P = 0.41).

There was no significant difference in improvements in walking speed between subjects who underwent resistance versus assistance training. The improvement in the 6-min walk distance tended to be greater for the assistance group than for the resistance group, although this was not significant (p = 0.06). In addition, the improvement in the BBS score was comparable, with no significant difference between the two groups (P = 0.63).

32.2.1.2 Constraint Induced Forced Use of the Paretic Leg During Walking

One possible reason that the robotic resistance training was not more effective than assistance training may be due to the compensatory movement of the non-paretic leg that often occurs during treadmill training [88]. For instance, individuals post-stroke often rely more on the non-paretic leg to generate more power to properly move the body forward during treadmill walking [89]. Repetitive practice in this manner may actually lead to reinforcing this compensatory strategy that relies more on the paretic leg and allows for restricted use of the paretic leg, which may result in modest improvement in motor control of the paretic leg after treadmill training. As a consequence, many individuals post-stroke still walk with asymmetrical gait patterns after treadmill training [90, 91]. These spatiotemporal asymmetries are associated with an increase in energetic cost [92], and are often negatively associated with slow walking speed [93]. The asymmetrical gait pattern was primarily due to the weakness of the paretic leg,

particularly the ankle joint, which was significantly impaired in many individuals post-stroke [94].Thus, interventions that target the deficits of the ankle plantarflexors and/or hip extensors of the paretic leg have been advocated to improve locomotor function in individuals post-stroke [95].

Constraint induced movement therapy (CIMT) is a promising paradigm designed to restore the motor function of the affected arm of individuals post-stroke [96, 97]. CIMT was developed based on previous studies in non-human primates from which the concept of learned nonuse of the affected arm was proposed [98]. Previous randomized studies indicated that CIMT is more effective than conventional interventions in improving motor function of the affected arm in individuals post-stroke [97, 99]. However, it is challenging to transfer this paradigm to lower limb training in individuals post-stroke because both the paretic and non-paretic legs need to be engaged for walking practice [100].

Previous studies in healthy adults and infants have demonstrated that perturbation applied to one leg during the swing phase of gait induced enhanced muscle activity of the ankle plantarflexors of the contralateral leg (i.e., standing leg) during treadmill walking [101–103]. Thus, we postulated that when a constraint force (i.e., a backward resistance force) was applied to the non-paretic leg during the swing phase would result in enhanced muscle activity of the paretic leg during stance (i.e., the standing leg) in individuals post-stroke during treadmill walking. Further, we postulated that repetitive practice in this pattern would induce a retention of enhanced muscle activity of the paretic leg even when the constraint force was removed, which might be achieved through a use-dependent motor learning mechanism [104]. In addition, we postulated that the enhanced use of the paretic leg during treadmill walking would be transferred to overground walking, resulting in an improvement in the walking pattern during overground walking after treadmill walking.

Sixteen individuals (8 female) post chronic (>6 months) stroke was recruited in this study.

Each subject participated in two test conditions, i.e., with constraint force and treadmill only, with a 10-min sitting break in between. The order of the two test conditions was randomized across subjects.

Each subject walked on a treadmill without force for 1-min (i.e., baseline), followed by the application of the constraint force for an additional 7 min (adaptation). Then, the constraint force was removed without warning and the subjects continued walking on the treadmill for another 1-min (post-adaptation). Afterwards, subjects were allowed to take a 1-min standing break and walked on the treadmill for another 5 min with the application of constraint force (re-adaptation). Overground walking gait speed was tested before treadmill walking, immediately post treadmill walking, and 10-min after the end of treadmill walking. For the treadmill only condition, a protocol that was similar to the constraint force condition was used but no force was applied.

Averages of EMG activity (during stance phase) of the paretic leg for the constraint force and treadmill only conditions are shown in see Fig. 32.2 [105]. For the constraint force condition, the EMG integral of TA, MG, SOL, VM, RF, MH, ADD, and ABD increased during the early adaptation period, and still remained higher during the late adaptation period. Further, the enhanced muscle activity of TA, MG, RF and MH was partially retained following the release of the constraint force during the post-adaptation period. In contrast, for the treadmill only condition, the EMG integral of all muscles showed modest change during the course of the adaptation period and post-adaptation period, Fig. 32.2.

For the constraint condition, the symmetry of step length during overground walking significantly improved after treadmill walking ($F_{2,13}$, = 3.56, p = 0.043, n = 14, data from two subjects were excluded due to outlier, ANOVA). The post-hoc test indicated that the symmetry index of step length significantly decreased from 0.13 $\pm$ 0.06 at baseline to 0.09 $\pm$ 0.07 at 10-min after the end of treadmill walking, p = 0.03, (a smaller number means a more symmetrical step length). Overground walking speed had no

significant changes (p > 0.05). For the treadmill only condition, there were no significant effects on step length symmetry, and overground walking speed (all p > 0.05).

Between conditions, the comparison showed more improvement in the symmetry of step length after treadmill walking with constraint force than after treadmill only (p < 0.04) suggesting that subjects walked with a more symmetrical gait pattern after treadmill walking with the application of the leg constraint force than after treadmill only condition.

32.2.2 Locomotor Training in Human with Incomplete SCI

32.2.2.1 Introduction

Recent reviews of clinical studies on the effectiveness of current robotic training in humans with SCI suggest that robot-assisted gait training is not superior to other gait training modalities [74, 106]. One possibility is that these robotic training modalities do not provide adequate challenges to drive motor learning in humans with SCI during locomotor training [107]. For instance, muscle activities are significantly lower during passive guided, robotic locomotor training than with physical therapist-assisted treadmill training in humans with SCI [108]. Previous studies have shown that an error-augmentation training paradigm may enhance arm recovery in individuals post-stroke [85]. Thus, we postulated that error-augmentation also would facilitate motor learning during locomotor training in humans with SCI.

By applying a resistance load to the leg during treadmill walking, which may increase leg kinematic errors [109], recent studies have indicated that humans with SCI adapt to the resistance load applied and demonstrate an aftereffect consisting of an increase in step length following load release [110]. The presence of aftereffects suggests the formation of anticipatory locomotor commands in response to the resistance load. In particular, a previous study indicated that this aftereffect could be transferred from treadmill training to overground walking in humans with

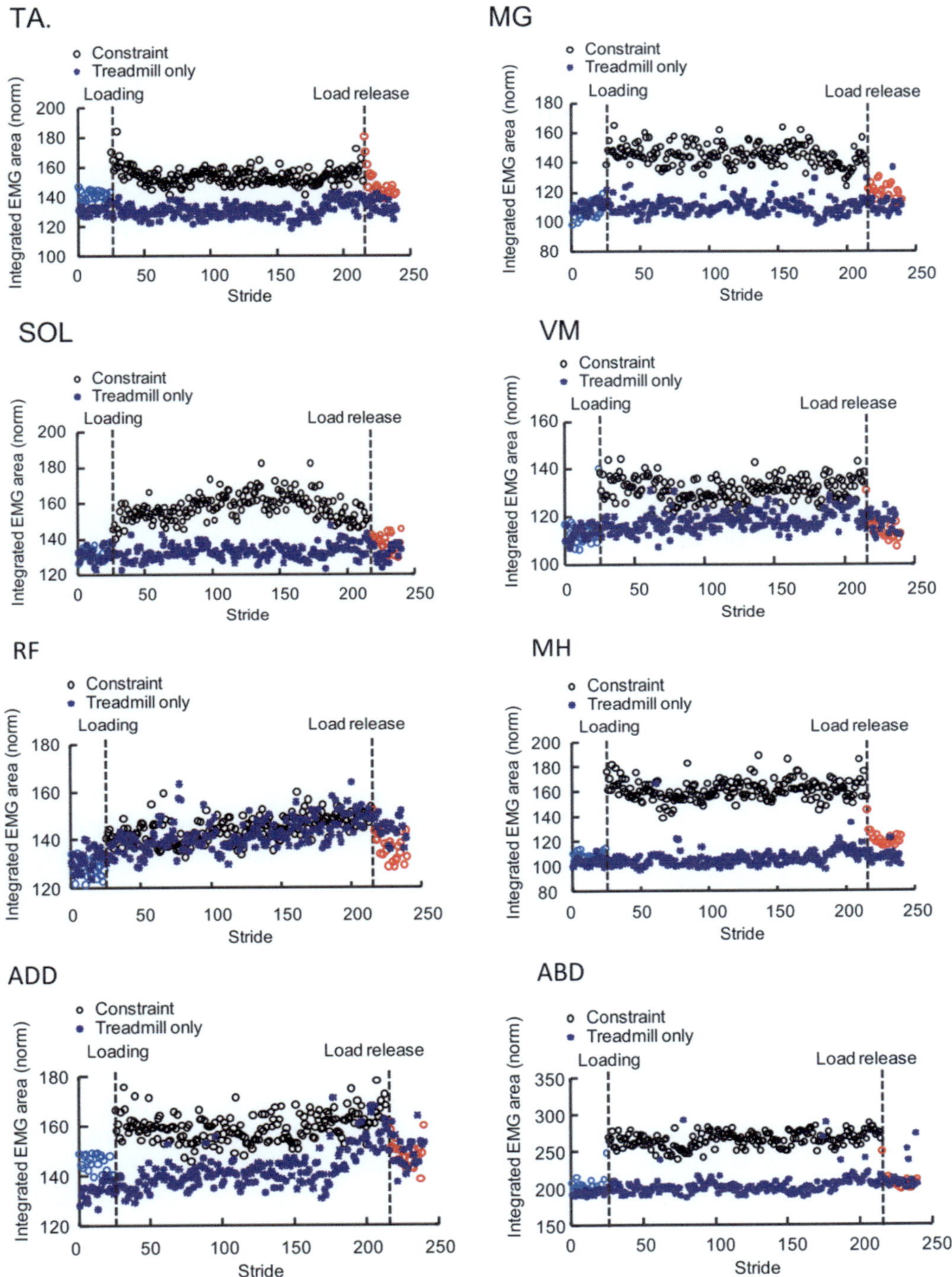

Fig. 32.2 Step-by-step integrated EMG of the paretic leg during the course of treadmill walking for the conditions with a constraint force applied to the non-paretic leg and treadmill only (control). Data shown are average across 15 subjects post-stroke (data from one subject were excluded due to large artifacts). The step numbers were different across subjects during the adaptation period because the treadmill speeds were different. Thus, the data from the first 160 steps and the last 30 steps during the adaptation period were used to calculate the average of integrated EMG. (Adopted from Wu et al. [105]). Abbreviation: tibialis anterior (TA), medial gastrocnemius (MG), soleus (SOL), vastus medialis (VM), rectus femoris (RF), medial hamstrings (MH), adductors (ADD) and abductors (ABD)

SCI [111]. However, locomotor adaptation and the aftereffects are generally short-lived, i.e., the increase in step length returns back to baseline within tens of steps during the post-adaptation period, after one session of force perturbation training, which may have a limited clinical impact on walking function in humans with SCI. A previous study using a split-belt treadmill paradigm indicated that prolonged repeated exposure to split-belt perturbation induces a long-term retention of improved step length symmetry in individuals post-stroke [112]. Thus, we postulated that a prolonged repeated exposure to swing resistance perturbations during treadmill training might also induce long-term retention of improved step length, resulting in improvements in the walking function of humans with SCI. The purpose of this study was to determine whether robotic resistance or assistance treadmill training by using a cable-driven robotic system would be effective in improving locomotor function in humans with chronic incomplete SCI [113].

Ten individuals with chronic incomplete SCI (i.e., >12 months post injury) with an injury level from C2 to T10 were recruited to participate in this study. All subjects were classified by the American Spinal Injury Association (ASIA) as ASIA grade D.

In order to test the locomotor training effect of the cable-driven robot in the SCI population, an 8 weeks training trial was conducted using a randomized crossover schedule. Specifically, subjects were blocked by gait speed into slow (<0.5 m/s) or fast (>0.5 m/s) subgroups and then randomized to either the assistance or resistance training first. After the first 4 weeks of training, subjects from both groups were switched from assistance to resistance or from resistance to assistance training, and completed another 4 weeks of training. Three assessments of gait were used to determine the training effects.

32.2.2.2 Results

In this pilot study, 8 out of 10 subjects finished 8 weeks of robotic treadmill training, with 2 subjects dropping out of the study. For the 8 patients that finished 8 weeks of robotic gait training, we found a significant improvement in self-selected overground walking speed (p = 0.03, one-way repeated measures ANOVA), i.e., the gait speed improved from 0.67 ± 0.20 m/s to 0.76 ± 0.23 m/s (see Fig. 32.3a). Fast walking speed also improved from 0.96 ± 0.31 m/s to 1.06 ± 0.32 m/s, although no significant difference was obtained due to the small sample size (p = 0.19, one-way repeated measures ANOVA). In addition, scores on the Berg Balance Scale significantly improved from 42 ± 12 at pre-training to 45 ± 12 post 8 weeks of robotic gait training (see Fig. 32.4). There were no significant changes in walking distance at the pre and post robotic training evaluation sessions (p = 0.12), although averaged 6-min walk distance increased from 223 ± 81 m at pre-training to 247 ± 88 m at post training (see Fig. 32.3b).

In order to compare the effect of 3DCaLT robotic treadmill training versus conventional treadmill training on walking function in people with SCI, we conducted a randomized controlled study. In particular, one limitation of previous robotic gait training systems is the constraint of the pelvis movement in the mediolateral direction [3], which may limit the weight shifting that occurs during the training. The ability to initiate and control weight shifting is a prerequisite for independent walking [63]. The weight shifting ability of many patients with SCI is impaired. Insufficient weight shifting to the ipsilateral leg may limit the unloading of the contralateral leg, which may affect the leg swing of the contralateral leg because the load afferent is a key input that modulates the translation from stance to swing during locomotion [114]. Thus, improving the weight shifting ability of the patients with SCI may improve their walking function. The goal of this study was to determine whether 3DCaLT robotic treadmill training would improve walking function in people with SCI. We hypothesized that the integration of the weight shifting component into the treadmill training paradigm would improve the weight

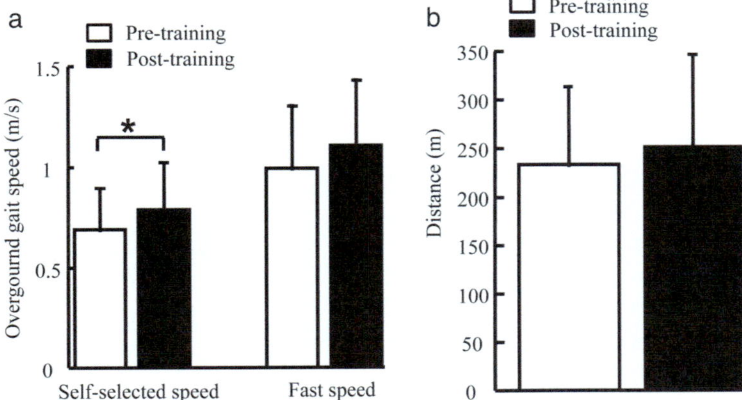

Fig. 32.3 Self-selected overground gait speed (**a**) and 6 min walk distance (**b**) at pre and post 8 weeks of robotic treadmill training in human SCI. The bar and error bar indicates the mean and standard deviation of the gait speed and walking distance across 8 subjects for pre and post training. Asterisk (*) indicates a significant effect of treatment

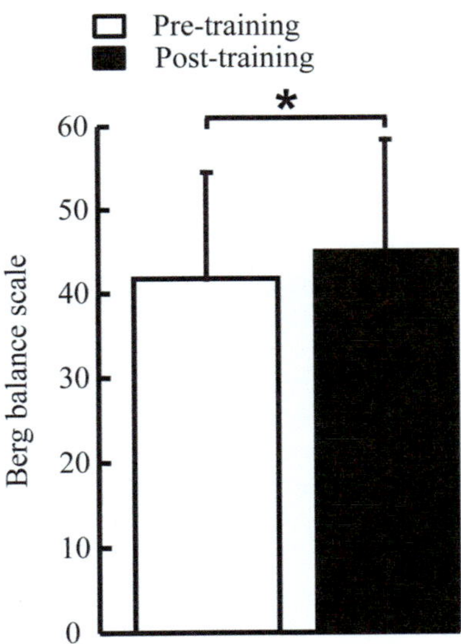

Fig. 32.4 Berg balance scale at pre and post 8 weeks robotic treadmill training in humans with SCI. The bar and error bar indicates the mean and standard deviation across 8 subjects for pre and post training. Asterisk (*) indicates a significant effect of treatment of robotic gait training

shifting ability and/or balance of humans with SCI, resulting in improvements in walking speed and endurance after robotic treadmill training.

Sixteen humans with motor incomplete SCI (i.e., ASIA Impairment Scale Level of C or D) were recruited in this study and were randomly assigned to receive robotic or treadmill only training, and underwent 6 weeks of locomotor training. The 3DCaLT cable-driven robotic system was used to provide controlled forces to the pelvis and legs during treadmill walking.

Outcome measures consisted of overground walking speed, 6-min walking distance, and other clinical measures, and were assessed pre and post 6 weeks of training, and 8 weeks after the end of training (Fig. 32.4).

32.2.2.3 Results

Fourteen participants completed all the training and evaluation sessions and 2 participants dropped out of the study. Robotic treadmill training with weight shift induced significant improvement in the endurance of patients with SCI. Specifically, 6-min walking distance significantly increased after robotic treadmill training ($p = 0.02$) [115]. Post-hoc tests indicated that 6-min walking distance significantly increased from 120 ± 37 m to 157 ± 59 m (29% increase) after robotic training ($p = 0.04$), and remained to be significantly greater than baseline at the follow up test, i.e., 151 ± 60 m (24% increase, $p = 0.04$). In addition, self-selected overground walking speed was intended to increase after robotic treadmill training, i.e., from 0.33 ± 0.15 m/s to 0.39 ± 0.20 m/s (15% increase), although this was not significant (p = 0.07), and was 0.38 ± 0.19 m/s (13% increase) at the follow up test. There were no significant changes in fast walking speed (p = 0.16) after robotic treadmill training. In contrast, treadmill only training induced no significant improvement in the endurance of patients with SCI. Specifically, the 6-min walking distance had no significant change after treadmill only training (p = 0.65). In addition, self-selected walking speed (p = 0.89) and fast walking speed (p = 0.43) showed no significant change pre and post treadmill only training.

Between group comparisons indicated that gains in 6-min walking distance were significantly greater for the robotic training group than that for treadmill only training group (p = 0.03), but gains in self-selected and fast walking speeds were not significantly different between the two groups (p = 0.06 and p = 0.12 for self-selected walking speed and fast walking speed, respectively), Fig. 32.5.

32.2.3 Locomotor Training in Children With CP

32.2.3.1 Introduction

As mentioned previously, weight shifting in the mediolateral direction is one of the key components during human locomotion [116]. However, the weight shifting ability is often impaired in children with CP compared to children who are typically developed [117]. For instance, children with CP performed weight shifting less efficiently as demonstrated by a shorter range of motion of the center of pressure (COP) and slower velocity of COP displacement during standing compared to children who are typically developed. It was suggested that weight shift training might improve dynamic balance during walking in children with CP [118]. Thus, we postulated that applying a mediolateral assistance force at the pelvis during the stance phase of gait might facilitate weight shifting in children with CP during treadmill walking, and repeat practice

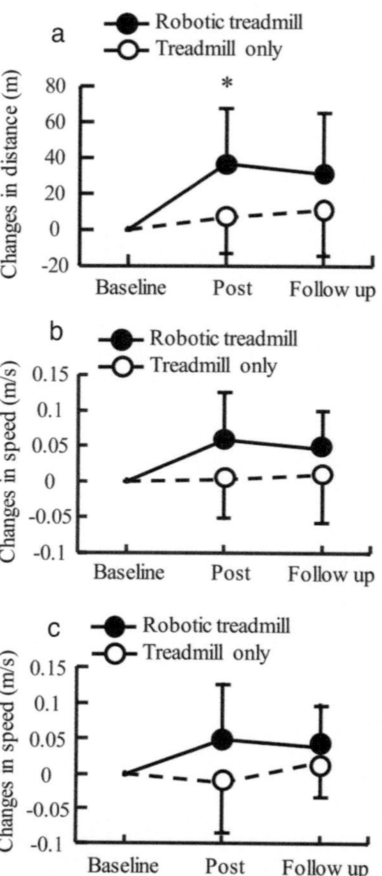

Fig. 32.5 Changes in 6 min walking, **a**, self-selected overground walking speed, **b**, fast walking speed, **c**, pre and post 6 weeks of robotic treadmill training or treadmill only training, and 8 weeks after the end of training. Data were averaged across subjects in each group. * indicates a significant difference, p < 0.05. (Adopted from Wu et al. [115])

of weight shifting during treadmill training may improve dynamic balance and improve walking speed in children with CP.

Evidence from spinalized mice indicates that motor learning is more effective with assistance as needed than with a fixed trajectory paradigm [81]. Thus, a robotic system that allows for variability in the stepping pattern during treadmill training will be effective in improving walking speed in children with CP. In addition, results from motor learning studies indicate that when there are more similarities between learning tasks and the application of those tasks, a greater transfer of motor skills will take place

[118]. Thus, a robotic BWSTT technique that provides less constraints and allows for a natural dynamic gait pattern during treadmill walking will be an effective method for transferring motor skills from treadmill training to overground walking in children with CP as measured by increased overground walking speed after robotic BWSTT. The purpose of this study is to assess improvements in the locomotor function of children with CP after robotic treadmill training with the application of applying controlled forces to both the pelvis, for facilitating weight shift, and the leg at the ankle, for facilitating leg swing, through the 3DCaLT robotic gait training system [83]. We hypothesized that robotic treadmill training that applies assistance at the pelvis for facilitating weight shifting would be more effective than treadmill only training in improving walking function in children with CP.

Twenty three children with CP were recruited (14 boys and 9 girls, the average age were 10.9 ± 3.2 years old, Gross Motor Function Classification System (GMFCS) levels [119] were I–IV). Each subject was randomly assigned to either a robotic training group (n = 11) or a treadmill only training group (n = 12).

Treadmill training was performed 3 times per week for 6 weeks. Gait assessment was made pre, post 6 weeks of robotic treadmill training, and at 8 weeks after the end of the training, using gait speed, endurance, and clinical measures of motor function (the dimensions D (standing) and E (walking, running, jumping) of the Gross Motor Function Measure (GMFM-66), [120]).

Results.

Eleven participants from the robotic training group and 10 participants from the treadmill only group completed all the training and evaluation sessions (the dropout rate was 8.7%). The walking function of children with CP improved after 3DCaLT robotic treadmill training. Specifically, self-selected walking speed significantly increased after robotic training (p = 0.03), see Fig. 32.6 [121]. The post-hoc test indicated a significant difference between pre versus post training tests (15.4% increase, p = 0.04), although there were no significant differences

between pre versus follow up tests (p = 0.08). Six-minute walking distance significantly increased after robotic training (p = 0.048), Fig. 32.6. Post-hoc test indicated a significant difference between pre versus post training tests (12.8% increase, p = 0.04), although there were no significant differences between pre versus follow up tests (p = 0.28). The GMFM scores had no significant changes after robotic training (p > 0.06).

Walking speed and endurance had no significant change after treadmill only training (p > 0.05), see Fig. 32.6. In addition, treadmill only training induced no significant change in dimension E of GMFM (p = 0.34), but induced a significant increase in the dimension D of GMFM (p = 0.01). Post-hoc tests indicated a significant difference between pre versus post training tests (p = 0.03), and pre versus follow up test (p = 0.02).

A greater gain in 6-min walking distance was obtained for the participants from the robotic group than that from the treadmill only group (p = 0.01), but the gain in self-selected walking speed had no significant difference between the two groups (p = 0.12). Specifically, changes in 6-min walking distance were 42.2 ± 57.4 m (post training) and 25.1 ± 52.0 m (follow up) after robotic training, and were -3.8 ± 35.9 m (post training) and -8.2 ± 46.8 m (follow up) after treadmill only training. Changes in self-selected walking speed were 0.10 ± 0.15 m/s (post training) and 0.09 ± 0.09 m/s (follow up) after robotic training, and were 0.04 ± 0.11 m/s (post training) and 0.04 ± 0.11 m/s (follow up) after treadmill only training.

In order to determine whether robotic resistance versus assistance training was more effective in improving walking function in children with CP, we conducted another randomized controlled study. Specifically, results from motor learning studies indicated that active training is more effective than passive training in improving the efficacy of motor training [40]. Thus, active engagement from children with CP might improve the efficacy of locomotor training. We proposed that applying a resistance force to leg

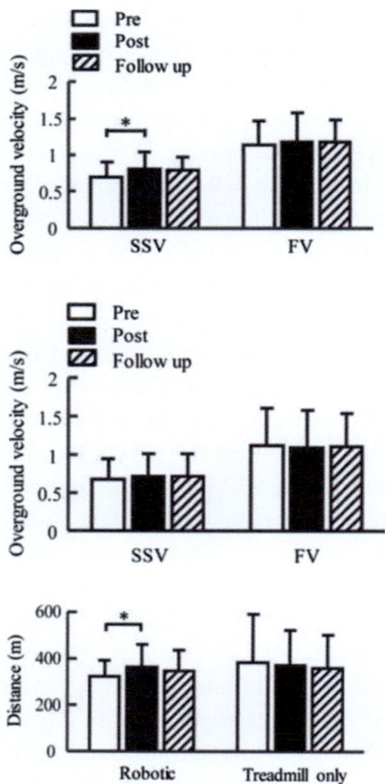

Fig. 32.6 Average of self-selected, fast walking velocities pre, post 6 weeks of robotic treadmill training, **a**, or treadmill only training, **b**, and 8 weeks after the end of treadmill training, i.e., follow up. Three trials were tested and averaged across each test session and averaged across participants for each group. **c**. average of 6-min walking distance pre, post 6 weeks of robotic treadmill training or treadmill only training, and 8 weeks after the end of training. Error bars indicate the standard deviation of each gait parameter (n = 10 for the robotic treadmill training group, data from one subject was excluded for the 6-min walking distance test because this subject was sick immediately before post test, which significantly impacted his endurance performance based on subject's self-report). Error bars indicate the standard deviation of each gait parameters. SSV, self-selected velocity; FV, fast velocity. * indicates a significant difference, p < 0.05. (Adopted from Wu et al. [121])

swing may force them to be more actively involved. In addition, applying a resistance force to leg swing may produce a deviation in step kinematics, i.e., an increase in the kinematic errors, which is supported by previous studies in individuals post-stroke [122, 123] and spinal cord injury [109]. Error augmentation may accelerate motor learning during treadmill training [87, 124], resulting in an improvement in the efficacy of locomotor training in children with CP.

On the other hand, providing leg assistance force may facilitate leg swing, which imitates the way that physical therapist provides leg assistance during treadmill training, and improve locomotor function in children with CP through use-dependent motor learning mechanisms [86]. Thus, the purpose of this study was to assess functional changes after robotic resistance versus assistance treadmill training in children with CP. We hypothesized that children with CP from both groups would show improvements in locomotor function, although greater improvements were expected after resistance than that after assistance training.

In this study, 23 children with CP (11 males and 12 females) were recruited. The average age of participants was 10.6 years old (ranged from 6 to 14). Their Gross Motor Function Classification System level ranged from I–IV (I (1), II (11), III (8), IV (3)).

Participants were randomly assigned to receive assistance or resistance force applied to both legs at the ankle during treadmill walking. Training sessions took place 3 times a week for 6 weeks.

32.2.3.2 Results

Twenty participants completed all the training and assessment sessions with 3 participants dropping out. Robotic resistance treadmill training improved walking function in children with CP. Specifically, fast walking speed significantly increased after robotic resistance treadmill training (F (2, 9) = 4.12, p = 0.03), [125]. Post-hoc tests indicated a significant increase from baseline to post testing (18% increase, from 0.98 ± 0.39 m/s to 1.13 ± 0.38 m/s, p = 0.01), although there was no significant difference between baseline and follow up testing (17% increase, p = 0.13). Self-selected walking speed tended to increase (32% increase), but this was not significant (F (2, 9) = 1.66, from 0.63 ± 0.30 m/s to 0.72 ± 0.24 m/s, p = 0.21).

In addition, 6 min walking distance significantly increased after resistance training (F (2, 9) = 10.04, p = 0.001). Post-hoc tests indicated a significant increase from baseline to post testing (30% increase, from 272.7 ± 113.0 m to 336.3 ± 104.9 m, p = 0.01), and from baseline to follow up testing (35% increase, from 272.7 ± 113.0 m to 353.9 ± 125.8 m, p = 0.001).

In contrast, robotic assistance treadmill training induced modest changes in walking function in children with CP. Specifically, both fast and self-selected walking speeds had no significant changes after assistance training (F (2, 9) = 1.85, from 0.84 ± 0.34 m/s to 0.84 ± 0.37 m/s, p = 0.19, and F (2, 9) = 0.51, from 0.54 ± 0.22 m/s to 0.52 ± 0.18 m/s, p = 0.6, for fast and self-selected walking speeds, respectively). In addition, 6 min walking distance had no significant change after assistance training (F (2, 9) = 0.48, from 216.3 ± 116.8 m to 230.1 ± 119.2 m, p = 0.63).

Greater functional gains were observed for participants who underwent robotic resistance training than for those who underwent robotic assistance training. Specifically, changes in self-selected walking speed, fast walking speed, and 6 min walking distance after treadmill training were significantly greater for participants from the resistance group than that from the assistance group (F (1,1) = 5.36, p = 0.03, F (1,1) = 10, p = 0.003, and F (1,1) = 12.23, p = 0.001, for self-selected walking speed, fast walking speed, and 6 min walking distance, respectively), Fig. 32.7. Functional gains were partially retained or even slightly increased during the follow up period.

GMFM scores significantly changed after robotic resistance training (F (2, 9) = 5.21, p = 0.02). Post-hoc test indicated significant differences between post testing and follow up testing scores (p = 0.02), although there was no significant difference between pre testing versus post testing scores (p = 0.89), and pre testing versus follow up testing scores (p = 0.05). In contrast, GMFM scores had no significant changes after robotic assistance training (F (2, 9) = 0.49, p = 0.69).

32.2.3.3 Discussion

The purpose of these studies was to determine the effect of robotic treadmill training using the 3DCaLT on walking function in adults with chronic stroke and motor incomplete SCI, and children with CP. We found that locomotor training using the 3D cable-driven robotic system may induce improvements in locomotor function in adults with stroke and SCI, and children with CP. In particular, results show that the functional gains after robotic treadmill training were greater than that after treadmill only training in adults

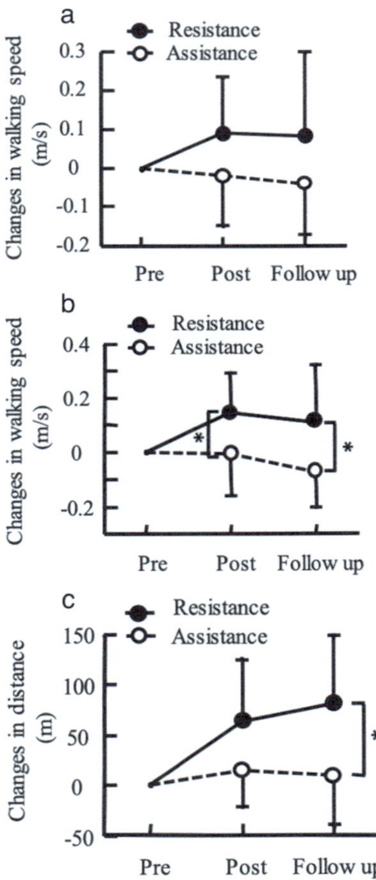

Fig. 32.7 Changes in walking self-selected walking speeds, (**a**), fast walking speed, (**b**), and 6-min walking distance, **c**, after robotic resistance/assistance treadmill training, and 8 weeks after the end of the training, i.e., follow up test. Data shown in the figure are the mean and standard deviation of walking speeds and distance across participants. * indicates a significant difference, p < 0.05. (Adopted from Wu et al. [126])

with SCI and children with CP. Further, the improvements in walking function were partially retained at 8 weeks after the end of the training, indicating the clinical significance of such robotic treadmill training.

32.2.4 Improved Walking Function in Individuals Post-Stroke

Applying a controlled resistance or assistance load to the paretic leg during treadmill training using the cable-driven robotic system significantly improved walking function in individuals post-stroke. However, robotic resistance training, which was applied to the paretic leg, was not superior to assistance training in improving endurance, balance in individuals post-stroke [87].

The increase in kinematic errors produced by the resistance load may elicit an error correction process that accelerates motor learning during locomotor training in individuals post-stroke. For instance, for the subjects who were assigned to the resistance training group, the resistance applied to the paretic leg produced a deviation in leg kinematics, that is, increased kinematic errors. An enhanced error has been shown to be more effective than passive guidance in improving arm performance in individuals post-stroke [85]. For the lower limb, previous studies indicated that individuals post-stroke adapted to the resistance load applied to the paretic leg and showed an aftereffect consisting of increased step length of the paretic leg after load release [122, 123].

Further, repeated exposure to resistance load during treadmill training may induce a prolonged retention of aftereffects of the paretic leg in individuals post-stroke. In this study, repeated exposure to a resistance load was applied to the paretic leg during 6 weeks of treadmill training. As a result, the step length of the paretic leg during overground walking increased after resistance training, suggesting that the aftereffect of an increased step length may be accumulated

and transferred from one context (i.e., treadmill walking) to another context (i.e., overground walking) in individuals post-stroke. In particular, we observed a partial retention of the increased step length of the paretic leg at follow up.

On the other hand, for subjects who were assigned to the assistance training group, an assistance force provided to the paretic leg may facilitate the leg swing to induce a longer step length on the paretic side during treadmill training. The increased step length of the paretic leg may be accumulated and transferred to overground walking through 6 weeks of locomotor training, resulting in an improvement in walking function after assistance treadmill training in individuals post-stroke. However, because the assistance force was applied at the paretic leg facilitating the leg to swing forward, instead of resisting the leg to induce kinematic deviation, we postulated that the motor learning mechanisms involved in robotic assistance training would be different from those involved in resistance training. A use-dependent motor learning mechanism may be involved during robotic assistance treadmill training [86]. The synaptic efficacy of sensorimotor pathways involved in the leg swing of the paretic leg may be enhanced by repetitive stepping assisted by the cable-driven robot [127].

Maintaining variation in kinematics during BWSTT is considered to be critical in improving the locomotor function in individuals post-stroke. For instance, results from animal experiments show that motor learning is more effective with a robotic algorithm that allows variability in the stepping pattern than with a fixed trajectory paradigm [81]. In addition, results from the human study have shown that intralimb coordination after stroke was improved by physical therapist-assisted BWSTT, which allowed for kinematic variability, but not robotic gait training with fixed trajectory, which reduces kinematic variability [128].

In the current study, the 3DCaLT robotic system, which is highly back drivable, has limited constraints on the leg kinematics during treadmill training [82]. The cable- driven system can be moved by the patient with the smallest possible resistance as opposed by the robot. Thus, the cable system allows the patients greater flexibility in controlling their gait patterns. The cable-driven robotic did not significantly affect the variability in ankle trajectory while the controlled load was applied to the legs during treadmill walking, see Fig. 32.8a. For instance, a previous study indicated that there were no significant changes in the variability of ankle trajectory of humans with SCI for different loading conditions (i.e., at baseline, with cable attached, and with assistance load applied (ANOVA, $p = 0.6$), see Fig. 32.8b [82]. This type of training seems more effective than fixed trajectory training in improving locomotor function in individuals post-stroke.

Results from this study also indicated that resistance training was not superior to assistance training in improving speed, endurance, balance in individuals post-stroke. One possible reason may be due to the compensatory movement of the non-paretic leg that often occurs during treadmill training when the resistance force was applied to the paretic leg during the swing phase [88]. For instance, in this study, we found enhanced muscle activities in hip extensors, and ankle plantarflexors when the constraint (i.e., resistance) force was applied to the non-paretic leg during the swing phase (which is corresponding to the time of period of the stance phase of the paretic leg). The enhanced muscle activities in hip extensors and ankle plantarflexors of the paretic leg may be because individuals post-stroke need to counteract the backward resistance force and move the body forward over the standing leg (i.e., the paretic leg) by extending the hip and ankle joints. These results suggest that the application of targeted constraint force to the non-paretic leg during the swing phase may induce forced use of the paretic leg. This is also consistent with previous CIMT, a strategy that has been extensively used in arms training in individuals post-stroke to improve motor function and daily use of the paretic arm [129].

Further, repetitive forced use of the paretic leg may result in an enhanced muscle activity of the

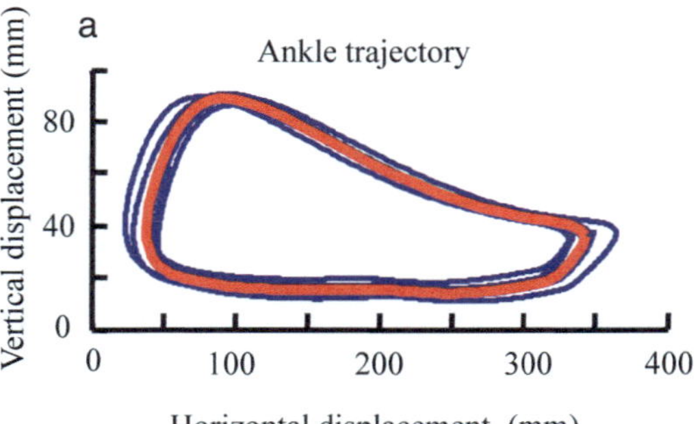

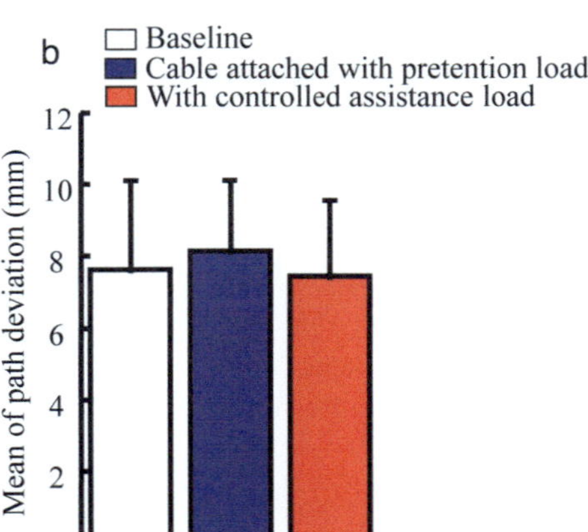

Fig. 32.8 **a** Ankle trajectories in the sagittal plane are shown from one patient with incomplete SCI during treadmill walking without the attachment of a cable robot. The solid thick line shown is the ensemble average trajectory across 7 step cycles. **b** Variability of ankle trajectory for 3 different loading conditions, i.e., baseline without attachment of the cable-driven robot; cable robot attached with 4 N pretension load, and cable robot attached with controlled assistance load. Path deviation of ankle trajectory in the sagittal plane for each condition was used to quantify the variability of ankle movement of each subject during treadmill walking. The bar and error bar indicate the mean and SD of the RMS error of ankle trajectory across subjects. (Modified from Fig. 3 of Wu et al. [82])

paretic leg even when the constraint force was removed during the post-adaptation period, which may be achieved through a use-dependent motor learning mechanism [104]. The repetitive movement or neural activity in a particular pattern over time, such as enhanced muscle activity of the paretic leg, in this case, may influence future movement or neural activity patterns even when the external perturbation is removed [130].

32.2.5 Improved Walking Function in Humans with SCI

The locomotor functional gains obtained using the 3DCaLT robotic gait training system are comparable to or even greater than that of using treadmill only training or robotic systems with a fixed trajectory control strategy. For instance, in a randomized trial involving 27 participants with SCI, the use of robotic-assisted BWSTT with a fixed trajectory did not significantly increase walking velocity (mean difference was −0.05 m/s) [54, 106, 131]. However, in another study with 20 subjects with chronic SCI, results indicated that the use of robotic-assisted treadmill training with a fixed trajectory may significantly improve walking speed in the SCI population [2]. The functional gains were 0.11 ± 0.11 m/s following robotic gait training, which is comparable to the results obtained in the current study.

In addition, results from the current study indicate an improvement in balance control in human SCI following cable-driven robotic gait training, i.e., Berg Balance Scale scores increased 3.3 ± 2.3. This is a functional gain not previously seen in studies with the Lokomat. The unnecessary medial-lateral support may reduce the potential functional gains in balance control following robotic gait training using the Lokomat. Recent studies indicate that there is a strong relationship between balance and walking capacity in patients with SCI [132, 133]. Thus, training stereotypical gait patterns in human SCI without challenging balance control may squander training time by focusing the training on the impairment that is not the bottleneck for achieving a greater walking speed [134].

In addition, the intensive task-specific walking practice may be delivered through a cable-driven robotic-assisted BWSTT system with the help of only 1 therapist, and can be performed for a longer duration (dependent upon the tolerance of the patient) thereby increasing the amount of practice of stepping behaviors. While the sample size is small, our results indicate that the improvements in locomotor function in our ambulatory subject population were statistically significant, with self-selected gait speed and Berg Balance Scale scores increasing by 0.09 ± 0.10 m/s (13%) and 3.3 ± 2.3 (8%), respectively, post robotic training. These improvements were qualitatively similar to those achieved by people with a similar diagnosis and chronicity of injury who performed therapist-assisted BWSTT [54]. Thus, 3DCaLT robotic BWSTT may achieve comparable functional gains when compared to therapist-assisted BWSTT, but can substantially reduce the labor effort and personnel cost of physical therapists.

Applying pelvis assistance could facilitate weight shift and induce additional challenges in maintaining lateral balance control and stability of the hip of the standing leg of patients with SCI. Thus, we speculate that patients with SCI may have to recruit additional muscle activation of the hip abductors/adductors to maintain lateral balance during walking. Repeated exposure to a pelvis assistance force during long-term treadmill training may improve motor control of hip abductors/adductors through use-dependent motor learning mechanisms [36], resulting in an improvement in lateral balance control, particularly of the standing leg of patients with SCI after robotic training. The improvement in the lateral balance of the standing leg may allow more time for patients with SCI to take a longer step length with the swing leg, which is supported by the trend of improvements in single leg support time after robotic training. In addition, given the importance of lateral balance control during walking, the improvements in lateral balance may reduce the energy cost [135], resulting in a more efficient gait pattern. This may be one reason why we observed a significant improvement in the endurance of patients with SCI after robotic treadmill training.

In contrast, results from this study indicated that treadmill only training did not induce significant improvements in the walking function of patients with SCI, which is consistent with several previous systematic reviews [106, 136]. For instance, the average gain in 6-min walking distance obtained after treadmill only training was 11.0 ± 24.4 m, which is comparable to gains obtained from previous randomized controlled studies using treadmill training [136], but is only approximately equal to 30% of the gain obtained after robotic treadmill training, i.e., 36.8 ± 30.5 m (>minimal clinical important difference, [137], and > clinical meaningful difference, which was suggested to be 31 m) [136]. One possible reason why treadmill only training seems less effective than robotic-assisted treadmill training may be that the challenge inherent in the task of treadmill only training for patients with SCI was not strong enough to induce improvement.

32.2.6 Improved Walking Function in Children with CP

We observed significant improvements in self-selected walking speed and endurance after robotic training, but not after treadmill only training. The participants from the 3DCaLT robotic training group showed a greater increase in 6-min walking distance that those from the treadmill only training group. In addition, greater functional gains in walking were obtained for participants from the resistance group than those from the assistance group, and the functional gains obtained after resistance training were partially retained during the follow up period. Results from this study suggest that treadmill training in conjunction with the application of controlled forces to the pelvis and/or resistance force to the legs while allowing for a natural stepping pattern is effective in improving walking function in children with CP.

Applying assistance load to the pelvis during treadmill training may improve weight shifting ability in children with CP. In this study, the 3DCaLT robotic system provided assistance force for facilitating weight shift at the targeted phase of gait while children with CP walk on a treadmill. The repeat practice of weight shifting during treadmill training may improve the weight shifting ability of children with CP through a use-dependent motor learning mechanism [86].

In addition, applying a mediolateral assistance force at the pelvis may enhance muscle activation of hip abductors/adductors, key muscles for balance control in the frontal plane during walking [138]. Further, repeat activation of these sensorimotor pathways induced by repeat pelvis assistance load during treadmill training may reinforce circuits and synapses used for lateral balance control during walking through use-dependent neuronal plasticity mechanisms [36], leading to long-term improvements in lateral balance control. The improvement in balance control may lead to improved lateral stability on the stance leg, allowing for the contralateral leg to move forward, resulting in improvements in walking speed and endurance in children with CP after training.

In addition, greater improvements in walking function were observed for children who underwent resistance training than those who underwent assistance training. One possible reason for the differences in functional gains may be due to children undergoing resistance training were more actively involved in the locomotor training session than those who underwent assistance training. Specifically, for children who were assigned the resistance training group, the resistance load applied to both legs during the swing phase may force participants to generate additional joint torque to counteract the load and move the leg forward during treadmill training, which may require participants to increase voluntary activation through enhanced supraspinal input to the motoneuron pool and/or increase motoneuron excitability [139].

In contrast, for children who underwent assistance training, the central nervous system may adapt to the assistance force applied to the leg(s) during the swing phase of gait by reducing the motor output of the leg muscles [140], probably due to optimization of the energy cost [141]. As a consequence, the training effect could be suboptimal. This is also consistent with a previous study, which indicated that only a modest increase in the gait speed of children with CP was observed after robotic training in which a passive guidance force was applied to both legs [77].

The functional gains obtained after robotic resistance treadmill training are comparable to or even greater than that from previous studies using conventional treadmill training. For instance, the functional gain in walking speed (i.e., 0.15 m/s, although <0.17 m/s, minimal clinical importance difference, MCID) [142] after robotic resistance training was greater/comparable with that after treadmill training with applied manual assistance force (0.01–0.07 m/s) [7, 55], or with robotic assistance using the pediatric Lokomat (i.e., 0.02 m/s) [77], which is comparable with functional gains obtained after assistance training (i.e., −0.02 m/s), and after robotic training using the GaitTrainer (i.e., 0.12 m/s) [143]. Similarly, the functional gain in 6 min walking distance after resistance training (81.2 m > 61.5 m, MCID) was greater than that after treadmill training with manual assistance (i.e., −25.0 − 19.8 m) [7, 55], which is comparable to functional gains obtained after assistance training (i.e., 13.9 m), and after robotic training using the GaitTrainer (i.e., 69 m) [143].

This study has several limitations. For instance, the sample size is small, which warrants further studies involving a larger cohort. In addition, while participants were randomly assigned into two different groups (i.e., 3D robotic and treadmill only training, or robotic resistance and assistance), physical therapists who conducted intervention and outcome assessments were not blinded, which might potentially bias the results. Further studies are needed to optimize this resistance training paradigm.

32.2.7 Other Advantages of the Cable-Driven Robotic System

The cable-driven robot system can apply for compliant assistance as needed or even resistance as tolerated to the paretic leg (s) during treadmill training. The cable-driven robot system is easy to setup compared to an exoskeleton robot system, such as the Lokomat, which requires the rotation center of robotic arms to be aligned with the patient's hip and knee joints [62]. The setup time of the cable-driven system is shorter than that of the exoskeleton systems, which is critical for the long-term treadmill training. In addition, the cost of the cable-driven robot system is less expensive than the current robotic systems, such as the Lokomat or AutoAmbulator. Thus, the cable-driven robotic system has multiple potential advantages to allow for the delivery of this type of therapy to a larger patient population.

Robotic-assisted treadmill systems provide for the training of a repetitive walking pattern that is critical for locomotor recovery in individuals post-stroke or with SCI. However, the sensory feedback provided to the patients who are trained on the treadmill is distinct from overground walking. For instance, the optical flows are different for these two walking conditions. Visual cues are in conflict with proprioceptive signals from the legs during treadmill walking, which is not experienced during overground walking [144]. Such factors may limit the transfer of the motor skill learned on the treadmill to overground walking. For instance, a previous study showed a partial transfer of motor adaptation obtained from split-belt treadmill training to overground walking [145].

32.3 Conclusion

The cable-driven locomotor training system proposed in this study provides a promising adjunct for the treatment of patients post-stroke, patients with incomplete SCI, and children with CP through robotic-assisted treadmill training. This system is highly back drivable, complaint, and allows patients to voluntarily move their legs during BWSTT. The 3DCaLT can apply controlled assistance/resistance forces to the pelvis (in the frontal plane) and legs (in the sagittal plane) at the targeted phase of gait while subjects walk on a treadmill. Results from these randomized controlled studies suggest that applying targeted lateral assistance force to the pelvis during treadmill training seems more effective than treadmill only training in improving endurance in adults with SCI and in children with

CP. In addition, applying a targeted resistance force to both legs during treadmill training to induce active involvement of children with CP is crucial for improving the effect of treadmill training. However, applying a resistance force to the paretic leg during treadmill training was not greater than applying assistance force for improving walking function in individuals post-stroke, which may be due to the compensatory movement from the non-paretic leg. In addition, the cable-driven robot is easy to set up and cost-effective to allow for delivery of this type of therapeutic intervention to a larger patient population.

Acknowledgements These studies were supported by NIH/NICHD, R01HD083314, R01HD082216, R21HD058267, R21HD066261, and PVA, #2552, NIDRR/RERC, H133E100007. We thank Dr. Schmit BD, Dr. Hornby TG, Dr. Gaebler-Spira DJ, Dr. Rymer WZ, Dr. Roth E, Dr. Yen SC, Dr. Wei F, Dr. Hsu CJ, Ms. Kim J, Mrs. Dee W, Ms. Arora P, Mrs. Rafferty M, Ms. Landry JM, and Mrs. Zhang YH for their assistance.

References

1. Dietz V, Muller R, Colombo G. Locomotor activity in spinal man: significance of afferent input from joint and load receptors. Brain. 2002;125(Pt 12):2626–34.
2. Wirz M, et al. Effectiveness of automated locomotor training in patients with chronic incomplete spinal cord injury: a multicenter trial. Arch Phys Med Rehabil. 2005;86(4):672–80.
3. Dobkin B, et al. Weight-supported treadmill vs over-ground training for walking after acute incomplete SCI. Neurology. 2006;66(4):484–93.
4. Hesse S, et al. Treadmill training with partial body weight support compared with physiotherapy in nonambulatory hemiparetic patients. Stroke. 1995;26(6):976–81.
5. Visintin M, et al. A new approach to retrain gait in stroke patients through body weight support and treadmill stimulation. Stroke. 1998;29(6):1122–8.
6. Sullivan KJ, Knowlton BJ, Dobkin BH. Step training with body weight support: effect of treadmill speed and practice paradigms on poststroke locomotor recovery. Arch Phys Med Rehabil. 2002;83(5):683–91.
7. Dodd KJ, Foley S. Partial body-weight-supported treadmill training can improve walking in children with cerebral palsy: a clinical controlled trial. Dev Med Child Neurol. 2007;49(2):101–5.
8. Willoughby KL, Dodd KJ, Shields N. A systematic review of the effectiveness of treadmill training for children with cerebral palsy. Disabil Rehabil. 2009;31(24):1971–9.
9. Mozaffarian D, et al. Heart disease and stroke statistics–2015 update: a report from the American Heart Association. Circulation. 2015;131(4):e29-322.
10. Perry J, et al. Classification of walking handicap in the stroke population. Stroke. 1995;26(6):982–9.
11. Jorgensen HS, et al. Recovery of walking function in stroke patients: the Copenhagen Stroke Study. Arch Phys Med Rehabil. 1995;76(1):27–32.
12. Goldie PA, Matyas TA, Evans OM. Deficit and change in gait velocity during rehabilitation after stroke. Arch Phys Med Rehabil. 1996;77(10):1074–82.
13. Treger I, et al. Return to work in stroke patients. Disabil Rehabil. 2007;29(17):1397–403.
14. von Schroeder HP, et al. Gait parameters following stroke: a practical assessment. J Rehabil Res Dev. 1995;32(1):25–31.
15. Dean CM, Richards CL, Malouin F. Walking speed over 10 metres overestimates locomotor capacity after stroke. Clin Rehabil. 2001;15(4):415–21.
16. Hsu AL, Tang PF, Jan MH. Analysis of impairments influencing gait velocity and asymmetry of hemiplegic patients after mild to moderate stroke. Arch Phys Med Rehabil. 2003;84(8):1185–93.
17. de Haart M, et al. Recovery of standing balance in postacute stroke patients: a rehabilitation cohort study. Arch Phys Med Rehabil. 2004;85(6):886–95.
18. Chen G, et al. Gait differences between individuals with post-stroke hemiparesis and non-disabled controls at matched speeds. Gait Posture. 2005;22(1):51–6.
19. Dobkin B. The clinical science of neurologic rehabilitation. 2nd ed. New York: Oxford University Press; 2003.
20. NSCISC. *Spinal cord injury: facts and figures at a glance.* 2021 [cited 2021].
21. Anderson KD. Targeting recovery: priorities of the spinal cord-injured population. J Neurotrauma. 2004;21(10):1371–83.
22. Post M, Noreau L. Quality of life after spinal cord injury. J Neurol Phys Ther. 2005;29(3):139–46.
23. Noreau L, et al. Participation after spinal cord injury: the evolution of conceptualization and measurement. J Neurol Phys Ther. 2005;29(3):147–56.
24. Barbeau H, et al. Tapping into spinal circuits to restore motor function. Brain Res Brain Res Rev. 1999;30(1):27–51.
25. Pepin A, Ladouceur M, Barbeau H. Treadmill walking in incomplete spinal-cord-injured subjects: 2. Factors limiting the maximal speed. Spinal Cord. 2003;41(5):271–9.
26. Leroux A, Fung J, Barbeau H. Postural adaptation to walking on inclined surfaces: II. Strategies

following spinal cord injury. Clin Neurophysiol. 2006;117(6):1273–82.

27. Odding E, Roebroeck ME, Stam HJ. The epidemiology of cerebral palsy: incidence, impairments and risk factors. Disabil Rehabil. 2006;28(4):183–91.

28. Rosen MG, Dickinson JC. The incidence of cerebral palsy. Am J Obstet Gynecol. 1992;167(2):417–23.

29. Pharoah PO, et al. Epidemiology of cerebral palsy in England and Scotland, 1984–9. Arch Dis Child Fetal Neonatal Ed. 1998;79(1):F21–5.

30. Hutton JL, Pharoah PO. Effects of cognitive, motor, and sensory disabilities on survival in cerebral palsy. Arch Dis Child. 2002;86(2):84–9.

31. Duffy CM, et al. Energy consumption in children with spina bifida and cerebral palsy: a comparative study. Dev Med Child Neurol. 1996;38(3):238–43.

32. Wilmshurst S, et al. Mobility status and bone density in cerebral palsy. Arch Dis Child. 1996;75 (2):164–5.

33. Chien F, Knutson DS, Fowler EL. The use of the 600 yard walk run test to assess walking endurance and speed in children with cerebral palsy. Pediatr Phys Ther. 2006;18:86–87.

34. Lepage C, Noreau L, Bernard PM. Association between characteristics of locomotion and accomplishment of life habits in children with cerebral palsy. Phys Ther. 1998;78(5):458–69.

35. Dobkin BH. Motor rehabilitation after stroke, traumatic brain, and spinal cord injury: common denominators within recent clinical trials. Curr Opin Neurol. 2009;22(6):563–9.

36. Hebb D. The organization of behavior: a neuropsychological theory. New York: Wiley; 1949.

37. Edgerton VR, et al. Use-dependent plasticity in spinal stepping and standing. Adv Neurol. 1997;72:233–47.

38. Henry F. Specificity vs. generality in learning motor skill. In: Brown RAK, GS editors. Classical studies on physical activity; 1968, Prentice-Hall: Englewood Cliffs, NJ.

39. Kaelin-Lang A, Sawaki L, Cohen LG. Role of voluntary drive in encoding an elementary motor memory. J Neurophysiol. 2005;93(2):1099–103.

40. Lotze M, et al. Motor learning elicited by voluntary drive. Brain. 2003;126(Pt 4):866–72.

41. Field-Fote EC. Spinal cord control of movement: implications for locomotor rehabilitation following spinal cord injury. Phys Ther. 2000;80(5):477–84.

42. Barbeau H, Fung J. The role of rehabilitation in the recovery of walking in the neurological population. Curr Opin Neurol. 2001;14(6):735–40.

43. Barbeau H, Rossignol S. Recovery of locomotion after chronic spinalization in the adult cat. Brain Res. 1987;412(1):84–95.

44. Rossignol S, et al. The "beneficial" effects of locomotor training after various types of spinal lesions in cats and rats. Prog Brain Res. 2015;218:173–98.

45. Hesse S, et al. Treadmill training with partial body weight support after stroke. Phys Med Rehabil Clin N Am. 2003;14(1 Suppl):S111–23.

46. Macko RF, et al. Treadmill exercise rehabilitation improves ambulatory function and cardiovascular fitness in patients with chronic stroke: a randomized, controlled trial. Stroke. 2005;36(10):2206–11.

47. Trueblood PR. Partial body weight treadmill training in persons with chronic stroke. NeuroRehabilitation. 2001;16(3):141–53.

48. Pohl M, et al. Speed-dependent treadmill training in ambulatory hemiparetic stroke patients: a randomized controlled trial. Stroke. 2002;33(2):553–8.

49. Nilsson L, et al. Walking training of patients with hemiparesis at an early stage after stroke: a comparison of walking training on a treadmill with body weight support and walking training on the ground. Clin Rehabil. 2001;15(5):515–27.

50. Kosak MC, Reding MJ. Comparison of partial body weight-supported treadmill gait training versus aggressive bracing assisted walking post stroke. Neurorehabil Neural Repair. 2000;14(1):13–9.

51. Behrman AL, Harkema SJ. Locomotor training after human spinal cord injury: a series of case studies. Phys Ther. 2000;80(7):688–700.

52. Dietz V, et al. Locomotor capacity of spinal cord in paraplegic patients. Ann Neurol. 1995;37(5):574–82.

53. Wernig A, et al. Laufband therapy based on "rules of spinal locomotion" is effective in spinal cord injured persons. Eur J Neurosci. 1995;7(4):823–9.

54. Field-Fote EC, Lindley SD, Sherman AL. Locomotor training approaches for individuals with spinal cord injury: a preliminary report of walking-related outcomes. J Neurol Phys Ther. 2005;29(3):127–37.

55. Willoughby KL, et al. Efficacy of partial body weight-supported treadmill training compared with overground walking practice for children with cerebral palsy: a randomized controlled trial. Arch Phys Med Rehabil. 2010;91(3):333–9.

56. Swe NN, et al. Over ground walking and body weight supported walking improve mobility equally in cerebral palsy: a randomised controlled trial. Clin Rehabil. 2015;29(11):1108–16.

57. Grecco LA, et al. A comparison of treadmill training and overground walking in ambulant children with cerebral palsy: randomized controlled clinical trial. Clin Rehabil. 2013;27(8):686–96.

58. Valentin-Gudiol M, et al. Treadmill interventions with partial body weight support in children under six years of age at risk of neuromotor delay: a report of a Cochrane systematic review and meta-analysis. Eur J Phys Rehabil Med. 2013;49(1):67–91.

59. Damiano DL, DeJong SL. A systematic review of the effectiveness of treadmill training and body weight support in pediatric rehabilitation. J Neurol Phys Ther. 2009;33(1):27–44.

60. Mutlu A, Krosschell K, Spira DG. Treadmill training with partial body-weight support in

children with cerebral palsy: a systematic review. Dev Med Child Neurol. 2009;51(4):268–75.

61. Schindl MR, et al. Treadmill training with partial body weight support in nonambulatory patients with cerebral palsy. Arch Phys Med Rehabil. 2000;81 (3):301–6.

62. Colombo G, et al. Treadmill training of paraplegic patients using a robotic orthosis. J Rehabil Res Dev. 2000;37(6):693–700.

63. Hesse S, Uhlenbrock D. A mechanized gait trainer for restoration of gait. J Rehabil Res Dev. 2000;37 (6):701–8.

64. HealthSouth®. http://www.healthsouth.com. 2010 [cited 2010].

65. Hornby TG, Zemon DH, Campbell D. Robotic-assisted, body-weight-supported treadmill training in individuals following motor incomplete spinal cord injury. Phys Ther. 2005;85(1):52–66.

66. Husemann B, et al. Effects of locomotion training with assistance of a robot-driven gait orthosis in hemiparetic patients after stroke: a randomized controlled pilot study. Stroke. 2007;38(2):349–54.

67. Mehrholz J et al. Electromechanical-assisted training for walking after stroke. Cochrane Database Syst Rev. 2020;10:CD006185.

68. Hidler J, et al. Multicenter randomized clinical trial evaluating the effectiveness of the Lokomat in subacute stroke. Neurorehabil Neural Repair. 2009;23(1):5–13.

69. Hornby TG, et al. Enhanced gait-related improvements after therapist-versus robotic-assisted locomotor training in subjects with chronic stroke: a randomized controlled study. Stroke. 2008;39 (6):1786–92.

70. Pohl M, et al. Repetitive locomotor training and physiotherapy improve walking and basic activities of daily living after stroke: a single-blind, randomized multicentre trial (DEutsche GAngtrainerStudie, DEGAS). Clin Rehabil. 2007;21(1):17–27.

71. Mehrholz J, Pohl M. Electromechanical-assisted gait training after stroke: a systematic review comparing end-effector and exoskeleton devices. J Rehabil Med. 2012;44(3):193–9.

72. Alashram AR, Annino G, Padua E. Robot-assisted gait training in individuals with spinal cord injury: a systematic review for the clinical effectiveness of Lokomat. J Clin Neurosci. 2021;91:260–9.

73. Mehrholz J et al. Electromechanical-assisted training for walking after stroke. Cochrane Database Syst Rev. 2013;7):CD006185.

74. Swinnen E, et al. Effectiveness of robot-assisted gait training in persons with spinal cord injury: a systematic review. J Rehabil Med. 2010;42 (6):520–6.

75. Hornby T, et al. Clinical and quantitative evaluation of robotic-assisted treadmill walking to retrain ambulation after spinal cord injury. Top Spinal Cord Inj Rehabil. 2005;11:1–17.

76. Meyer-Heim A, et al. Improvement of walking abilities after robotic-assisted locomotion training in

children with cerebral palsy. Arch Dis Child. 2009;94(8):615–20.

77. Druzbicki M, et al. Functional effects of robotic-assisted locomotor treadmill therapy in children with cerebral palsy. J Rehabil Med. 2013;45 (4):358–63.

78. Meyer-Heim A, et al. Feasibility of robotic-assisted locomotor training in children with central gait impairment. Dev Med Child Neurol. 2007;49 (12):900–6.

79. Edgerton VR, Roy RR. Robotic training and spinal cord plasticity. Brain Res Bull. 2009;78(1):4–12.

80. Wolbrecht ET, et al. Optimizing compliant, model-based robotic assistance to promote neurorehabilitation. IEEE Trans Neural Syst Rehabil Eng. 2008;16(3):286–97.

81. Cai LL, et al. Implications of assist-as-needed robotic step training after a complete spinal cord injury on intrinsic strategies of motor learning. J Neurosci. 2006;26(41):10564–8.

82. Wu M, et al. A cable-driven locomotor training system for restoration of gait in human SCI. Gait Posture. 2011;33(2):256–60.

83. Wu M, et al. Locomotor training through a 3D cable-driven robotic system for walking function in children with cerebral palsy: a pilot study. Annu Int Conf IEEE Eng Med Biol Soc. 2014;2014:3529–32.

84. Perry J, Burnfield JM. Gait analysis. Normal and pathological function. Vol. Second. 2010, Pomona, California: SLACK.

85. Patton JL, et al. Evaluation of robotic training forces that either enhance or reduce error in chronic hemiparetic stroke survivors. Exp Brain Res. 2006;168(3):368–83.

86. Sawaki L, et al. Effects of somatosensory stimulation on use-dependent plasticity in chronic stroke. Stroke. 2006;37(1):246–7.

87. Wu M, et al. Robotic resistance/assistance training improves locomotor function in individuals post-stroke: a randomized controlled study. Arch Phys Med Rehabil. 2014;95(5):799–806.

88. Raja B, Neptune RR, Kautz SA. Coordination of the non-paretic leg during hemiparetic gait: expected and novel compensatory patterns. Clin Biomech (Bristol, Avon). 2012;27(10):1023–30.

89. Souissi H, et al. Muscle force strategies for poststroke hemiparetic patients during gait. Top Stroke Rehabil. 2019;26(1):58–65.

90. Patterson SL, et al. Effect of treadmill exercise training on spatial and temporal gait parameters in subjects with chronic stroke: a preliminary report. J Rehabil Res Dev. 2008;45(2):221–8.

91. Gama GL, et al. Effects of gait training with body weight support on a treadmill versus overground in individuals with stroke. Arch Phys Med Rehabil. 2017;98(4):738–45.

92. Sanchez N, Finley JM. Individual differences in locomotor function predict the capacity to reduce asymmetry and modify the energetic cost of

walking poststroke. Neurorehabil Neural Repair. 2018;32(8):701–13.

93. Gonzalez-Suarez CB, et al. Motor impairment and its influence in gait velocity and asymmetry in community ambulating hemiplegic individuals. Arch Rehabil Res Clin Transl. 2021;3(1): 100093.

94. Nadeau S, et al. Plantarflexor weakness as a limiting factor of gait speed in stroke subjects and the compensating role of hip flexors. Clin Biomech (Bristol, Avon). 1999;14(2):125–35.

95. Roelker SA, et al. Paretic propulsion as a measure of walking performance and functional motor recovery post-stroke: a review. Gait Posture. 2019;68:6–14.

96. Taub E, Uswatte G, Elbert T. New treatments in neurorehabilitation founded on basic research. Nat Rev Neurosci. 2002;3(3):228–36.

97. Wolf SL, et al. Effect of constraint-induced movement therapy on upper extremity function 3 to 9 months after stroke: the EXCITE randomized clinical trial. JAMA. 2006;296(17):2095–104.

98. Taub E, Heitmann RD, Barro G. Alertness, level of activity, and purposive movement following somatosensory deafferentation in monkeys. Ann N Y Acad Sci. 1977;290:348–65.

99. Stevenson T, et al. Constraint-induced movement therapy compared to dose-matched interventions for upper-limb dysfunction in adult survivors of stroke: a systematic review with meta-analysis. Physiother Can. 2012;64(4):397–413.

100. Hase K, et al. Effects of therapeutic gait training using a prosthesis and a treadmill for ambulatory patients with hemiparesis. Arch Phys Med Rehabil. 2011;92(12):1961–6.

101. Dietz V, et al. Obstruction of the swing phase during gait: phase-dependent bilateral leg muscle coordination. Brain Res. 1986;384(1):166–9.

102. Yang JF, Stephens MJ, Vishram R. Transient disturbances to one limb produce coordinated, bilateral responses during infant stepping. J Neurophysiol. 1998;79(5):2329–37.

103. Savin DN, Tseng SC, Morton SM. Bilateral adaptation during locomotion following a unilaterally applied resistance to swing in nondisabled adults. J Neurophysiol. 2010;104(6):3600–11.

104. Diedrichsen J, et al. Use-dependent and error-based learning of motor behaviors. J Neurosci. 2010;30 (15):5159–66.

105. Wu M, Hsu CJ, Kim J. Forced use of paretic leg induced by constraining the non-paretic leg leads to motor learning in individuals post-stroke. Exp Brain Res. 2019;237(10):2691–703.

106. Mehrholz J, Kugler J, Pohl M, *Locomotor training for walking after spinal cord injury.* Cochrane Database Syst Rev, 2012. **11**: p. CD006676.

107. Lam T, et al. Training with robot-applied resistance in people with motor-incomplete spinal cord injury: Pilot study. J Rehabil Res Dev. 2015;52(1):113–29.

108. Israel JF, et al. Metabolic costs and muscle activity patterns during robotic- and therapist-assisted treadmill walking in individuals with incomplete spinal cord injury. Phys Ther. 2006;86(11): 1466–78.

109. Yen SC, Landry JM, Wu M. Size of kinematic error affects retention of locomotor adaptation in human spinal cord injury. J Rehabil Res Dev. 2013;50 (9):1187–200.

110. Houldin A, Luttin K, Lam T. Locomotor adaptations and aftereffects to resistance during walking in individuals with spinal cord injury. J Neurophysiol. 2011;106(1):247–58.

111. Yen SC, et al. Locomotor adaptation to resistance during treadmill training transfers to overground walking in human SCI. Exp Brain Res. 2012;216 (3):473–82.

112. Reisman DS, et al. Repeated split-belt treadmill training improves poststroke step length asymmetry. Neurorehabil Neural Repair. 2013;27(5):460–8.

113. Wu M, et al. Robotic resistance treadmill training improves locomotor function in human spinal cord injury: a pilot study. Arch Phys Med Rehabil. 2012;93(5):782–9.

114. Pearson KG. Proprioceptive regulation of locomotion. Curr Opin Neurobiol. 1995;5(6):786–91.

115. Wu M, Kim J, Wei F. Facilitating weight shifting during treadmill training improves walking function in humans with spinal cord injury: a randomized controlled pilot study. Am J Phys Med Rehabil. 2018;97(8):585–92.

116. Inman V, Ralston HJ, Todd F. Human walking. Baltimore: Williams and Wilkins; 1981.

117. Ballaz L, et al. Impaired visually guided weight-shifting ability in children with cerebral palsy. Res Dev Disabil. 2014;35(9):1970–7.

118. Liao HF, et al. The relation between standing balance and walking function in children with spastic diplegic cerebral palsy. Dev Med Child Neurol. 1997;39(2):106–12.

119. Palisano R, et al. Development and reliability of a system to classify gross motor function in children with cerebral palsy. Dev Med Child Neurol. 1997;39(4):214–23.

120. Russell DJ, et al. The gross motor function measure: a means to evaluate the effects of physical therapy. Dev Med Child Neurol. 1989;31(3):341–52.

121. Wu M, et al. Effects of the integration of dynamic weight shifting training into treadmill training on walking function of children with cerebral palsy: a randomized controlled study. Am J Phys Med Rehabil. 2017;96(11):765–72.

122. Savin DN, et al. Poststroke hemiparesis impairs the rate but not magnitude of adaptation of spatial and temporal locomotor features. Neurorehabil Neural Repair. 2013;27(1):24–34.

123. Yen SC, Schmit BD, Wu M. Using swing resistance and assistance to improve gait symmetry in individuals post-stroke. Hum Mov Sci. 2015;42: 212–24.

124. Wu M, et al. Repeat exposure to leg swing perturbations during treadmill training induces

long-term retention of increased step length in human SCI: a pilot randomized controlled study. Am J Phys Med Rehabil. 2016;95(12):911–20.

125. Wu M, et al. Robotic resistance treadmill training improves locomotor function in children with cerebral palsy: a randomized controlled pilot study. Arch Phys Med Rehabil. 2017;98(11):2126–33.

126. Wu M et al. Robotic resistance treadmill training improves locomotor function in children with Cerebral Palsy: a randomized controlled pilot study. Arch Phys Med Rehabil. 2017.

127. Edgerton VR, et al. Training locomotor networks. Brain Res Rev. 2008;57(1):241–54.

128. Lewek MD, et al. Allowing intralimb kinematic variability during locomotor training poststroke improves kinematic consistency: a subgroup analysis from a randomized clinical trial. Phys Ther. 2009;89(8):829–39.

129. Wolf SL, et al. Retention of upper limb function in stroke survivors who have received constraint-induced movement therapy: the EXCITE randomised trial. Lancet Neurol. 2008;7(1):33–40.

130. Classen J, et al. Rapid plasticity of human cortical movement representation induced by practice. J Neurophysiol. 1998;79(2):1117–23.

131. Mehrholz J, Kugler J, Pohl M. Locomotor training for walking after spinal cord injury. Cochrane Database Syst Rev. 2008(2):CD006676.

132. Scivoletto G et al. Clinical factors that affect walking level and performance in chronic spinal cord lesion patients. Spine (Phila Pa 1976). 2008;33 (3):259–64.

133. Lemay JF, Nadeau S. Standing balance assessment in ASIA D paraplegic and tetraplegic participants: concurrent validity of the Berg Balance Scale. Spinal Cord. 2010;48(3):245–50.

134. Hornby TG, Reinkensmeyer DJ, Chen D. Manually-assisted versus robotic-assisted body weight-supported treadmill training in spinal cord injury: what is the role of each? PM R. 2010;2 (3):214–21.

135. Donelan JM, et al. Mechanical and metabolic requirements for active lateral stabilization in human walking. J Biomech. 2004;37(6):827–35.

136. Mehrholz J et al. Is body-weight-supported treadmill training or robotic-assisted gait training superior to overground gait training and other forms of physiotherapy in people with spinal cord injury? A systematic review. Spinal Cord. 2017.

137. Norman GR, Sloan JA, Wyrwich KW. Interpretation of changes in health-related quality of life: the remarkable universality of half a standard deviation. Med Care. 2003;41(5):582–92.

138. Winter DA, et al. An integrated EMG/biomechanical model of upper body balance and posture during human gait. Prog Brain Res. 1993;97:359–67.

139. Ekblom MM. Improvements in dynamic plantar flexor strength after resistance training are associated with increased voluntary activation and V-to-M ratio. J Appl Physiol (1985). 2010;109(1):19–26.

140. Wu M, et al. Kinematic and EMG responses to pelvis and leg assistance force during treadmill walking in children with Cerebral Palsy. Neural Plast. 2016;2016:5020348.

141. Reinkensmeyer DJ, et al. Slacking by the human motor system: computational models and implications for robotic orthoses. Conf Proc IEEE Eng Med Biol Soc. 2009;2009:2129–32.

142. Norman G. The effectiveness and effects of effect sizes. Adv Health Sci Educ Theory Pract. 2003;8 (3):183–7.

143. Smania N, et al. Improved gait after repetitive locomotor training in children with cerebral palsy. Am J Phys Med Rehabil. 2011;90(2):137–49.

144. Torres-Oviedo G, Bastian AJ. Seeing is believing: effects of visual contextual cues on learning and transfer of locomotor adaptation. J Neurosci. 2010;30(50):17015–22.

145. Reisman DS, et al. Split-belt treadmill adaptation transfers to overground walking in persons poststroke. Neurorehabil Neural Repair. 2009;23 (7):735–44.

Body Weight Support Devices for Overground Gait and Balance Training

33

Andrew Pennycott and Heike Vallery

Abstract

Regaining the ability to walk overground, to climb stairs and to perform other functional tasks such as standing up and sitting down are important rehabilitation goals following neurological injury or disease. However, these activities are often difficult to practice safely for patients with severe impairments due to the risk of injury, not only to the patient but also to therapists. The emergence of various technologies that provide a degree of body weight support can play a role in rehabilitation focused on recovering overground gait and balance functions. These can greatly reduce the risk of falls and thus allow more intense and longer training sessions. Therefore, the systems empower individuals with the ability to practice the types of activities and functions they need in order to return home and to be reintegrated into the community as much as possible. This chapter explores the origin of body weight supported devices and considers which groups could derive benefit from the training. An overview of the main training platforms available today—which comprise both robotic and non-robotic technologies—is then provided, followed by a discussion regarding outcomes of the devices thus far and possible future directions of the technology.

Keywords

Robotics · Rehabilitation · Body weight support · Gait · Walking · Stroke · Spinal cord injury

33.1 Clinical Rationale for Body Weight Supported Training

33.1.1 Origins and Evolution

In body weight supported (BWS) gait training, a harness is placed around a person's torso and / or pelvis and connected to an unloading system in order to provide a variable degree of unloading as the subject walks. Originally motivated by a rich literature of studies with felines [6] and rodents [18] which demonstrated that stepping patterns could be restored through treadmill-based

A. Pennycott · H. Vallery(✉)
Faculty of Mechanical, Maritime and Materials Engineering, Delft University of Technology, Delft, The Netherlands
e-mail: h.vallery@tudelft.nl

A. Pennycott
e-mail: A.Pennycott@tudelft.nl

H. Vallery
Department for Rehabilitation Medicine, Erasmus MC, Rotterdam, The Netherlands

© The Author(s), under exclusive license to Springer Nature Switzerland AG 2022
D. J. Reinkensmeyer et al. (eds.), *Neurorehabilitation Technology*,
https://doi.org/10.1007/978-3-031-08995-4_33

training with BWS, this training method enables patients with a high degree of weakness and/or poor coordination to undergo gait training following neurological and musculoskeletal injuries.

One major advantage of body weight support devices is that they reduce the fear of falling and thereby allow individuals to concentrate more freely on the main training tasks. It has been shown that fear itself can have effects on gait characteristics; for example, older adults who reported being more fearful during walking tended to walk more slowly with a greater step width and shorter step length in one study [14].

There are low-cost solutions available for providing assistance from therapists during transfers and gait training such as the gait and walking belt devices [46], which allow therapists to apply supportive forces close to a patient's centre of gravity and provide handles for assistance. It has been shown that the risk of falling was lower in rehabilitation programs incorporating gait belts and similar devices, and that using a gait belt during assisted falling reduced the risk of injury [50]. However, due to the exertion levels required and sometimes awkward postures required for performing assistance, manual gait training has been associated with increased risk of injury to therapists [12].

Moreover, research has shown that increased fall efficacy—a term representing an individual's degree of confidence in conducting activities of daily living without falling—is a predictor of gait and indeed other functional rehabilitation outcomes [8]. Fear of falling may have contributed to the findings of some earlier research, for instance with incomplete spinal cord injured subjects [17], which pointed to greater gains being achieved through treadmill-based as opposed to overground walking, since in the latter studies, the subjects were supported only by simple ambulatory assistive devices such as crutches and canes and hence likely had a more pronounced fear of falling during the training.

Another possible contributor may be that walking with body weight support could reduce fatigue and hence increase the feasible training duration. Indeed, it has previously been shown that body weight support reduces the net metabolic rate

during gait [16]. Furthermore, 'verticalisation' of individuals, which can be promoted through BWS, is an important goal in rehabilitation of different groups such as stroke patients, since it can have benefits regarding circulation, prevention of pneumonia and clots, while also providing stimulation for the autonomic and sensory nervous systems [26].

The first generation of body weight support systems were mounted over a treadmill with therapists providing assistance to the patient. The observation that the training duration and consistency were frequently limited by therapist fatigue led to the development of robot-assisted gait training platforms such as the Lokomat [11]. Beneficial effects on kinematic, kinetic and spatiotemporal parameters of robot-assisted training have been noted, for instance, for chronic stroke patients [7].

However, other studies have suggested that for certain groups such as subacute stroke participants with moderate to severe walking impairments, the variability and diversity afforded by overground walking may be more beneficial than treadmill-based training with robotic assistance [19]. This could be partly due to the many changes in gait biomechanics that occur when walking on a treadmill [30]. It has also been argued that the treadmill belt could make the training less intense and challenging, and that the treadmill-based systems cannot provide task-specific overground training [25].

Several new body weight support systems now offer a safe environment that is not dependent on treadmills, which help individuals to overcome their fear of falling and thereby to practice overground gait. This technology could lead to improved rehabilitation outcomes compared with previous treadmill-centred methods. Researchers have argued that overground training could prove beneficial due to greater kinematic variability being permitted in addition to offering gait training in a more real-world environment [56].

33.1.2 Target Groups

There are various groups of people who can benefit from body weight supported gait and balance training such as individuals who have suffered from an injury or neurological pathology. Though the various patient groups are affected in different ways and have different patterns of gait and balance impairment, the main task of the training devices remains the same: to provide a degree of body weight support so that the people can perform gait and balance training with a reduced risk of falling.

The incidence of spinal cord injury (SCI) has been estimated to lie between 10.4 and 83 per million people per year [57]. In addition to the detrimental effects on sensory and motor function, which can manifest in gait as reductions in gait speed and alterations in walking pattern [28], SCI can lead to secondary health conditions such as ulcers, amputations and major depressive disorder [27]. In addition to potential benefits with regard to ambulation, robot-based rehabilitation has shown positive results concerning posture, intestinal, cardio-respiratory and metabolic function [20].

Over 50 million people per year worldwide have a traumatic brain injury (TBI) [32], which is characterised as a change in brain function and other pathology due to an external force [36]. As well as the impacts on cognitive, behavioral, and emotional functioning, the effects on motor skills can lead to considerable alterations in gait, including decreased walking speed and step length, with exaggerated knee flexion on initial ground contact with the foot being evident [54]. The impact on walking also leads to increased incidence of falls in this group [35]. Independent walking is a common discharge rehabilitation goal for TBI patients [24].

Stroke is the second leading cause of death globally and a major cause of long-term disability in adults [45]. Furthermore, as stroke is associated with increased age, the prevalence of stroke could well be increasing due to the ageing of the population [15]. It can lead to impaired ambulatory function, which is in turn associated with a decreased quality of life [39].

Multiple sclerosis (MS) is a disease that causes the myelin sheath of nerve cells in both the brain and spinal cord to be damaged, leading to sensory and motor impairment and physical and mental issues. The incidence of multiple sclerosis is over 35 per 100,000 people and appears to be increasing with time [52]. MS manifests in gait by decreased gait speed, endurance, step length, cadence and joint kinematics [9]. For multiple sclerosis patients, in common with the other patient groups, gait is strongly associated with wider participation in society [23].

Cerebral palsy (CP) represents the most prevalent physical disability in children, affecting from 2 to 2.5 per 1,000 children in the United States [29]. CP can affect gait development in various ways, and CP patients typically have stiff knee action in the swing phase, crouched gait, excessive hip flexion, intoeing, and ankle equinus (limited dorsiflexion of the foot due to a lack of flexibility in the ankle joint) [55]. As noted for the other groups in this section, lower limb and gait dysfunction can have wider impacts on the overall quality of life for CP patients [22].

33.2 Overground Training Devices

33.2.1 Robotic Devices

Motivation Robotic body weight support devices use actuation to control the forces acting on an individual, possibly in multiple directions and varying over the gait cycle. They can actively track the movement of a subject in both the vertical and horizontal directions during gait, thus avoiding undesired interference between the individual and the device itself.

Adjusting the level of unloading is usually conveniently done through user interfaces that are operated by therapists. Moreover, the devices, which are summarised in the following sections, can be broadly categorised as being ceiling-mounted or mobile frames.

Ceiling-Mounted Systems The ZeroG® gait and balance training system (Fig. 33.1a) is commercially available through Aretech, LLC (Ashburn, Virginia, US). The system can provide around 200 kg and 90 kg of static and dynamic of body-weight support, respectively. Mounted to an overhead track, a small motor propels a trolley to track a patient's movements. More than one ZeroG trolley can be placed on the same track, thereby providing the opportunity for multiple patients to train simultaneously. Patients can practice walking overground, walk up and down steps, perform sit-to-stand movements and practise other balance-centred tasks. These activities are important since the patients will encounter such challenges in their everyday lives.

The SafeGait 360° (Gorbel Medical, Victor, NY, U.S.) and the Vector Gait & Safety System® (Bioness Inc., Valencia, CA, U.S.) are further examples of ceiling-mounted body weight support systems that actively track a subject's movement during gait in the longitudinal direction.

A drawback of systems based on single rails is that they restrict the user to a specific path. Another limitation is that the cable will unavoidably transmit lateral force components whenever the individual moves laterally. Even if these lateral movements are small during gait, the 'pendulum effect' due to the mounting below a pivot could potentially disturb balance or support it more than intended or needed [13, 41]. This stabilising effect makes it more difficult to purposefully train lateral balance. However, there can be benefits from stabilising a patient in the lateral plane in some cases, particularly those with lateral propulsion syndrome [38].

Therefore, multiple systems have been proposed that enable training in a 3D workspace. The Free Levitation for Overground Active Training (FLOAT) system, commercially available through Reha-Stim (Schlieren, Switzerland) and shown in Fig. 33.1b), is a 3D body weight support system that capitalises on cable robot technology developed at ETH Zürich [48]. The system transmits forces to a person via wires that are actuated by motorised winches positioned at the ceiling in the four corners of the desired workspace.

Conventional cable robots tend to have a limited workspace because the angle of the cables with respect to the horizontal plane determines how much force needs to be transmitted for a given vertical unloading level. To enable a large workspace without incurring excessive cable forces, the FLOAT uses a mechanical configuration of moving cable deflection units (pulleys) [51]. The deflection units are not actu-

ated but rather are moved by the tension in the cables they deflect. The design reduces moving masses to a minimum and it enables control of a three-dimensional (3D) force vector, including relieving patients of a percentage of their body weight and providing longitudinal assistance or resistance.

Nevertheless, a design drawback of the FLOAT is that, as in conventional cable robots, all the winches must act in combination to actuate the different degrees of freedom, rather than in a decoupled fashion. This requires high-power actuators because they need to serve both the low-speed, high-force vertical degree of freedom and the high-speed, low-force horizontal degrees of freedom.

The RYSEN™ (Motek Medical B.V., Houten, the Netherlands) is also a cable robot but has lower-power motors. This 3D body weight support system (Fig. 33.1c) mechanically decouples the degrees of freedom across the motors such that these can make different speed-torque trade-offs [42]. The RYSEN uses moving cable deflection units on rails and posterior-anterior motion is actuated by motors. Springs are applied in series with two main motors being used for predominantly vertical actuation and a double-sided variable-radius winch for lateral actuation. The low-power motors limit the bandwidth of closed-loop control in the vertical direction but the design does achieve very good force tracking in the horizontal direction [42, 43].

Besides cable robot technology, another option to provide 3D body weight support is to use gantry systems. The Active Response Gravity Offload System (ARGOS) is used by NASA for astronaut training in simulated reduced gravity environments and is based on an active overhead gantry crane system. Motion in the horizontal plane is controlled by electric motors, while the degree of body weight support is controlled by a crane connected to the user via a steel cable and shock absorber.

The NaviGAITor is a similar example of a multidirectional body weight support system [44]. A movable bridge is mounted on a pair of rails to enable movement in the longitudinal direction, while lateral motion is permitted by movement

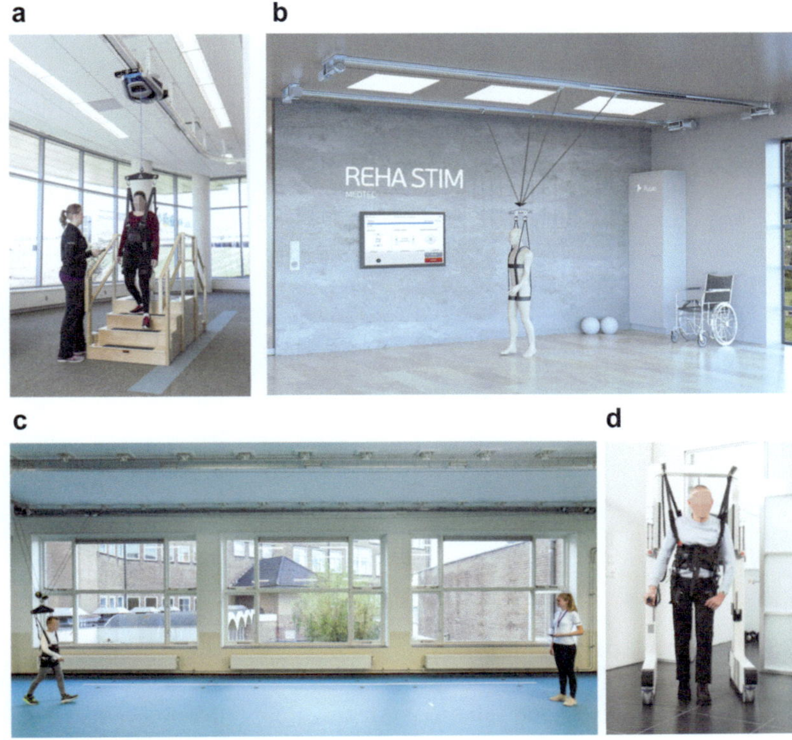

Fig. 33.1 Robotic body-weight support devices. **a** Zero-G (courtesy of Aretech, US), **b** THE FLOAT (courtesy of REHA-STIM MEDTEC AG, CH), **c** Motek Medical's RYSEN™ at the Rehabilitation Center Heliomare, Wijk aan Zee, NL (courtesy of Motek Medical), **d** ANDAGO (courtesy of Hocoma AG, CH)

of a trolley along the bridge. Actuation is realised in these two directions via electric motors, and a further motor drives a hoist to provide vertical forces in order to realise different degrees of body weight support.

A major advantage of 3D systems over systems based on single rails is that, in principle, they do not restrict the user to a specific path, and, if they are sufficiently transparent, do not cause restoring horizontal forces, thereby preventing the aforementioned pendulum effect. However, a 3D setup may impose different limitations on the practice space; for example, it fits less easily into narrow or curved rooms or corridors. Furthermore, such a device can typically only support one patient at at time, the practice walking length is limited, and the design may require high ceilings for installation.

An inherent limitation of the gantry-based systems is that large masses need to be moved when tracking a walking user. Even if closed-loop force control is applied, the user will feel some remaining inertia because the reduction in apparent inertia is limited in causal control schemes [10]. This limits the devices' ability to render purely vertical forces on a user, potentially disturbing their gait by imposing undesired horizontal force components.

Mobile Frames The Andago® (Fig. 33.1d), developed by Hocoma AG (Volketswil Switzerland), comprises a mobile frame mounted on wheels and a BWS [34]. Patient trunk movements are tracked, and hence he or she can practice walking without being confined to one specific training room. The training platform can be used in patient-following mode in which the person's movements are followed not only in the forward and backward directions but also in turning movements, straight-line mode in which turning inputs are not followed, and finally in manual mode in which the device is controlled by a therapist via a joystick.

A further system which was originally mobile is the KineAssist [40], which interacts with users through a pelvis and torso harness. The device senses interaction forces at the pelvis and thereby controls the movement of a robotic platform. Today, however, the KineAssist is only available as a treadmill-mounted system from Woodway (Waukesha, WI, U.S.).

 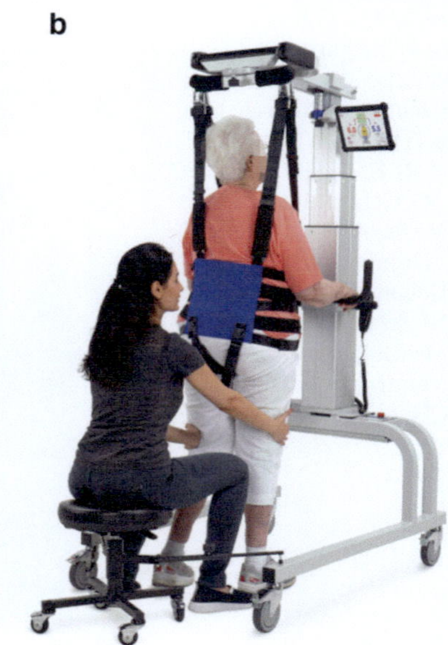

Fig. 33.2 Passive body weight support devices. **a** Zero-G passive (Courtesy of Aretech, US), **b** LiteGait (courtesy of Mobility Research, US)

33.2.2 Non-Robotic Devices

Motivation Various non-robotic systems are available on the market that provide static body weight support during overground gait and balance training. The principal advantage of these systems is clearly their cost: they do not require the expensive sensing and actuation hardware of their robotic counterparts. The disadvantage is that the vertical support cannot be precisely controlled and the options for applying forces in the horizontal plane—for instance perturbation forces for balance training—are more limited. Nevertheless, the devices do allow people who cannot support their entire body weight to practice overground walking.

Ceiling-Mounted Systems A large number of commercial solutions are available that are based on passive trolleys mounted on rails in order to provide support during overground ambulation. For example, the design of the ZeroG-Passive system (Aretech, LLC, Ashburn, VA, US) closely follows the robotic ZeroG platform, but rather than actuating the trolley position via a motor, it is simply pulled along by the patients as they ambulate (Fig. 33.2a). The FreeStep SAS (Biodex, Inc, Shirley, New York, US) operates similarly to the ZeroG-Passive and also uses small and lightweight trolleys.

In light of the findings concerning how BWS reduces the risk and fear of falling, a particularly innovative building design is the Shirley Ryan AbilityLab in Chicago (formerly known as the Rehabilitation Institute of Chicago, RIC). Many areas of this hospital are equipped with rails mounted to the ceiling such that training with a safety harness and passive trolleys is not limited to dedicated gym areas. Instead, users can walk with a safety harness along 'patient highways' in many locations in and close by their own rooms. This reduces the risk of falls and reduces the transfer time between therapy sessions. If these aspects are considered when planning the construction of a hospital or rehabilitation center, the installation of such rails is much less expensive than retrofitting. However, the noise levels of passive carts on the rails should also be taken into consideration.

Mobile Frames The LiteGait system by Mobility Research (Tempe, AZ, U.S.) comprises a mobile cart mounted on castors and an overhead bracket for attachment to a harness (Fig. 33.2b). Owing to a locking system, the device can be used for either treadmill or overground training.

Other castor-mounted systems include the NxStep[TM] Unweighing system by Biodex (Shirley, New York, US) and the PhysioGait

(HealthCare International, Langley, US). Like their mobile robotic support counterparts, mobile frames are not tied to being used in a specific training room. However, a disadvantage is the relatively large mass of the device that must be moved by the patient, which potentially affects the gait biomechanics.

33.3 Device Characteristics

33.3.1 Transparency

To maximise motor learning outcomes and functional gains, training devices should only provide just enough support to create a safe environment; the devices should be able to 'hide' their presence in terms of interaction forces between the device itself and the user. The degree to which this is possible is often referred to as transparency [5]. This can be promoted through hardware design that emphasises low inertia and friction as these factors can lead to higher interaction forces and distort the mechanics of gait.

In robotic body weight support devices, the ability of a system to track the movement of an individual will govern the realisable degree of transparency to a large extent. As mentioned in Sect. 33.2.1, even with closed-loop force control, the device dynamics cannot be 'hidden' completely, and the physical mass of a device is a governing factor of transparency, with friction between static and moving parts also playing a role. Again, systems mounted on mobile platforms and 3D gantry-based systems tend to have limited transparency since relatively large masses must be moved by the user.

33.3.2 Vertical Support Forces

For body weight support systems, levels of body weight support above 30% appear to significantly alter some gait characteristics such as the kinematics and kinetics of the hip and knee joints [3]. Indeed, modelling has shown that there could be changes in the sign of joint moments above this level of unloading [43]. Therefore, an excessive level of body weight unloading potentially causes large kinematic and kinetic changes from physiological gait and therapists should aim to apply levels of vertical support below this apparent threshold during training sessions.

Besides the mean value of support over the gait cycle, variations in vertical unloading should also be considered. Some systems intrinsically exhibit variable forces, especially when using counterweights or springs. A prevalent paradigm is to keep vertical unloading forces as constant as possible, with the argument in favour of this approach being that gravity itself is constant. This presumed requirement increases device complexity to keep force levels constant despite the oscillatory vertical movement of the user during gait. This goal is often pursued by closed-loop force-controlled actuation. However, although a constant unloading vertical force can relieve the body of a portion of its weight, its inertia in the vertical direction remains unchanged, which disturbs the relationship between gravitational and inertial forces and may thus alter the natural frequency of gait.

In fact, simulation results suggest that a simple passive spring suspension—in which a spring stiffness is chosen to favour natural gait cadence—distorts gait mechanics to a lesser extent than more complex approaches that actively control the vertical force to a constant level [4]. During gait, acceleration-dependent inertial forces can largely be compensated by position-dependent spring forces and the relationship between weight and inertia is thereby maintained constant despite the unloading, and therefore, the gait frequency remains closer to its physiological value (without unloading). Hence, a passive elastic suspension is not only easier and less expensive to realise, but also may disturb gait to a lesser degree than an actively controlled constant force suspension.

This insight can, furthermore, influence the design of robotic systems, namely the inclusion of series-elastic elements in combination with low-bandwidth actuators. For example, the RYSEN employs series springs that cover oscillations in the vertical direction during gait where the low-power motors lack bandwidth. Only for movements with lower frequency need the motors move to change the setpoint of the springs (mainly in order to manoeuvre over stairs or to sit down).

33.3.3 Longitudinal Forces

The notion that individuals would prefer zero or positive forces in the longitudinal direction (forces acting in the forward direction of ambulation) due to energy efficiency considerations has been challenged by research using the RYSEN [43]. Though resistive longitudinal forces applied by the devices would require greater energy consumption by the subject during gait, healthy subjects in this study tended to favour small negative forces when walking in the robotic platform. The reason suggested was that unlike a backward fall, a forward fall can be recovered through a swift movement of the swing leg and, therefore, a small negative force encouraging a slight forward lean during gait could allow subjects to feel safer and have less fear of dangerous falls.

Overground gait training devices can also include perturbation forces in various directions, including along the longitudinal axis. For example, an additional module has been integrated into the FLOAT platform to apply perturbation forces in different directions [37]. It has been argued that applying perturbations in this way, as opposed to a treadmill, presents less risk of falling and hence less risk of injury and a reduced sense of fear during the training.

33.3.4 Lateral Forces

As mentioned in Sect. 33.2.1, research has pointed to a pendulum effect from the body weight support mechanism itself: the systems can produce lateral restoring forces which decrease the challenge of maintaining balance in the frontal plane. Therefore, systems that reduce these restoring forces could be useful for gait training that incorporates active balance control. Systems that do not permit lateral movement are likely to lead to greater restoring forces and hence may limit the ability

to train balance in the frontal plane, which is an important element of gait [33].

However, systems that, in principle, do allow lateral movement may still generate undesired lateral forces, depending on their degree of transparency. Moreover, the point of application of the resultant unloading force needs to be considered to know which stabilising or destabilising moments are applied about the center of mass as discussed in the next section.

33.3.5 Harness and Attachment

Ease of attachment to the harness is important as long set-up times will detract from the time available for actual training. Wheeled mobile platforms should offer sufficient space so as to allow wheelchair access and to avoid necessitating additional transfers. Comfort while exercising in the harness should also be considered in the design as discomfort may limit the training duration tolerable by the subjects and, similarly to the fear of falling, may detract from their concentration on the active gait task. This has been considered in some designs for which different harness versions are available for male and female users.

The attachment of the harness also determines the line of action of the resultant force acting on the user and how the unloading forces are distributed along the body. The location of the line of action with respect to the body's center of mass defines the resultant moments of the unloading force. Inmoments caused by ground reaction forces, this governs the rate of change of centroidal angular momentum, which is a key variable for bipedal stability and balance control.

Results with able-bodied subjects indeed indicate that the self-selected walking speed is affected by the attachment mechanism [43]. Furthermore, the optimal type of attachment may vary according to the degree of impairment of the subject. For less impaired subjects, attachment near the pelvis could be better since this will not produce stabilising moments about the hip joint, while for individuals who are less able to maintain their balance independently, attachment higher up on the torso would be better due to the stabilising moments this affords.

For treadmill walking, attachment via a harness has been shown to reduce vertical acceleration. This is due to restrictions in both linear and rotational movements imposed by the harness that lead to the trunk being less able to absorb shocks than in normal gait [1].

33.4 Outcomes of Overground Gait Training

There have been a limited number of studies thus far which have investigated the outcomes of applying the various devices described here to overground gait and balance for different patient groups. In contrast, there has been a greater volume of research focused on the outcomes of treadmill-based training and rehabilitation as summarised, for instance, in the review by Wessels et al. [53]. Although research has suggested that conventional therapy incorporating overground walking with manual assistance seems to yield broadly comparable outcomes as compared to treadmill training with BWS [31], there have not been comprehensive comparisons between overground training using BWS and treadmill-centred training. Nevertheless, the results from studies using BWS devices in rehabilitation have generally been encouraging so far; some of the main findings from the studies with overground devices are summarised below.

Huber and Sawaki compared overground gait training for non-traumatic SCI using the Zero-G system, finding better outcomes in terms of sphincter control for the robot-assigned group compared to standard-of-care therapy [21]. Both groups achieved significant gains in Functional Independence Measure (FIM) scores but no significant cross-group differences were apparent.

Anggelis et al. compared the outcomes of training with and without dynamic body weight support using the Zero-G platform in traumatic brain injury patients [2]. The Zero-G group demonstrated significantly greater improvements in functional independence measures (FIM) as

well as cognitive improvements than the control group. The authors suggest that this is due to the greater intensity facilitated by dynamic body weight support systems. Furthermore, the importance of reducing the fear of falling and thereby allowing the individuals to concentrate more fully on the training tasks was highlighted.

Brunelli et al. conducted a study with subacute stroke patients and compared body weight supported overground training with the LiteGait to conventional physiotherapy [8]. While both groups showed improvements in all the outcome measures—which included the Rivermead Mobility Index, Barthel Index, and the Six Minute Walk Test—greater gains were shown concerning the Functional Ambulation Classification for the individuals who participated in the BWS training.

Tay et al. compared outcomes for patients undergoing training with the Andago system in addition to conventional therapy to other individuals who only had conventional therapy [47]. Statistically significant FIM gains were made in both groups; though the robotic training group showed greater improvement in functional ambulation scores, there were no statistically significant differences in the other outcomes.

Van Hedel et al. assessed the use of the Andago platform for children and youths with gait impairments [49]. They observed that the device prevented several falls and also that variability in stride duration and the degree of inter-joint coordination were higher with the overground training than treadmill walking.

33.5 Future Directions

The cost of the various overground robotic devices remains an obstacle to their widespread adoption in rehabilitation programs. Notably, the actuators used to track gait movement lead to additional safety requirements and hence higher certification costs. Non-robotic systems are, therefore, much less expensive than robotic body weight support platforms. Since the former category may provide most of the potential benefits of the body weight support systems at much lower cost, research comparing usability and clinical outcomes—for example in terms of functional gait—between passive and robotic systems is needed in order to evaluate whether the additional costs of actuation are actually justified.

More generally, there remains a paucity of results concerning clinical outcomes from using the various devices for different patient groups. Additional evidence for the efficacy of the devices from studies with larger subject groups is needed to justify funding of the devices by health care systems. Though there are promising results showing improvements achieved through overground robotic training, whether these lead to improved ambulation in the long term remains uncertain. Therefore, studies including longer-term follow-ups for the various different types of training are needed, along with investigations as to whether or not training in the clinical setting can really translate into increased community participation and integration.

Acknowledgements This chapter is an update of a previous edition that was co-authored by Joe Hidler (Aretech, US) and Arno Stienen (Motek Medical B.V., Houten, NL).

Disclosure of Financial Interests H. Vallery is an inventor on multiple patents and patent applications related to the FLOAT and to the RYSEN and may financially benefit from sales of either device.

References

1. Aaslund MK, Moe-Nilssen R. Treadmill walking with body weight support: Effect of treadmill, harness and body weight support systems. Gait & Posture. 2008;28(2):303–8.
2. Anggelis E, Powell ES, Westgate PM, Glueck AC, Sawaki L. Impact of motor therapy with dynamic body-weight support on functional independence measures in traumatic brain injury: an exploratory study. NeuroRehabilitation. 2019;45(4):519–24.
3. Apte S, Plooij M, Vallery H. Influence of body weight unloading on human gait characteristics: a systematic review. J Neuroeng Rehabil. 2018;15(1):1–18.
4. Apte S, Plooij M, Vallery H. Simulation of human gait with body weight support: benchmarking models and unloading strategies. J Neuroeng Rehabil. 2020;17(1):1–16.
5. Bannwart M, Bolliger M, Lutz P, Gantner M, Rauter G. Systematic analysis of transparency in the gait rehabilitation device the float. In: 2016 14th international conference on control, automation, robotics and vision (ICARCV). IEEE; 2016. p. 1–6.
6. Barbeau H, Rossignol S. Recovery of locomotion after chronic spinalization in the adult cat. Brain Res. 1987;412(1):84–95.
7. Bonnyaud C, Pradon D, Boudarham J, Robertson J, Vuillerme N, Roche N. Effects of gait training using a robotic constraint (lokomat®) on gait kinematics and kinetics in chronic stroke patients. J Rehabil Med. 2014;46(2):132–8.
8. Brunelli S, Iosa M, Fusco FR, Pirri C, Di Giunta C, Foti C, Traballesi M. Early body weight-supported overground walking training in patients with stroke in subacute phase compared to conventional physiotherapy: a randomized controlled pilot study. Int J Rehabil Res. 2019;42(4):309–15.
9. Cameron MH, Wagner JM. Gait abnormalities in multiple sclerosis: pathogenesis, evaluation, and advances in treatment. Curr Neurol Neurosci Rep. 2011;11(5):507–15.
10. Colgate E, Hogan N. An analysis of contact instability in terms of passive physical equivalents. In: Proceedings of the IEEE international conference on robotics and automation (ICRA). Scottsdale, AZ, USA; 1989. p. 404–9.
11. Colombo G, Joerg M, Schreier R, Dietz V, et al. Treadmill training of paraplegic patients using a robotic orthosis. J Rehabil Res Dev. 2000;37(6):693–700.
12. Darragh AR, Campo M, King P. Work-related activities associated with injury in occupational and physical therapists. Work. 2012;42(3):373–84.
13. Dragunas AC, Gordon KE. Body weight support impacts lateral stability during treadmill walking. J Biomech. 2016;49(13):2662–8.
14. Dunlap P, Perera S, VanSwearingen JM, Wert D, Brach JS. Transitioning to a narrow path: the impact of fear of falling in older adults. Gait & Posture. 2012;35(1):92–5.
15. Giles MF, Rothwell PM. Measuring the prevalence of stroke. Neuroepidemiology. 2008;30(4):205.
16. Grabowski A, Farley CT, Kram R. Independent metabolic costs of supporting body weight and accelerating body mass during walking. J Appl Physiol. 2005;98(2):579–83.
17. Gupta N, Mehta R. Comparison of gait performance of spinal cord injury subjects: body weight supported treadmill training versus over ground gait training. Apollo Med. 2009;6(1):21–7.
18. Heng C, de Leon R. Treadmill training enhances the recovery of normal stepping patterns in spinal cord contused rats. Exp Neurol. 2009;216(1):139–47.
19. Hidler J, Nichols D, Pelliccio M, Brady K, Campbell DD, Kahn JH, Hornby TG. Multicenter randomized clinical trial evaluating the effectiveness of the lokomat in subacute stroke. Neurorehabil Neural Repair. 2009;23(1):5–13.
20. Holanda LJ, Silva PMM, Amorim TC, Lacerda MO, Simão CR, Morya E. Robotic assisted gait as a tool for rehabilitation of individuals with spinal cord injury: a systematic review. J Neuroeng Rehabil. 2017;14(1):1–7.

21. Huber JP, Sawaki L. Dynamic body-weight support to boost rehabilitation outcomes in patients with non-traumatic spinal cord injury: an observational study. J NeuroEng Rehabil. 2020;17(1):1–9.

22. Jaspers E, Verhaegen A, Geens F, Van Campenhout A, Desloovere K, Molenaers G. Lower limb functioning and its impact on quality of life in ambulatory children with cerebral palsy. Eur J Paediatr Neurol. 2013;17(6):561–7.

23. Johansson S, Ytterberg C, Gottberg K, Holmqvist LW, von Koch L, Conradsson D. Participation in social/lifestyle activities in people with multiple sclerosis: changes across 10 years and predictors of sustained participation. Mult Scler J. 2020;26(13):1775–84.

24. Katz DI, White DK, Alexander MP, Klein RB. Recovery of ambulation after traumatic brain injury. Arch Phys Med Rehabil. 2004;85(6):865–9.

25. Kim SK, Park D, Yoo B, Shim D, Choi J-O, Choi TY, Park ES. Overground robot-assisted gait training for pediatric cerebral palsy. Sensors. 2021;21(6):2087.

26. Knecht S, Hesse S, Oster P. Rehabilitation after stroke. Deutsches Ärzteblatt Int. 2011;108(36):600.

27. Krause JS, Saunders LL. Health, secondary conditions, and life expectancy after spinal cord injury. Arch Phys Med Rehabil. 2011;92(11):1770–5.

28. Krawetz P, Nance P. Gait analysis of spinal cord injured subjects: effects of injury level and spasticity. Arch Phys Med Rehabil. 1996;77(7):635–8.

29. Krigger KW. Cerebral palsy: an overview. Am Fam Physician. 2006;73(1):91–100.

30. Lee SJ, Hidler J. Biomechanics of overground vs. treadmill walking in healthy individuals. J Appl Physiol. 2008;104(3):747–55.

31. Lura DJ, Venglar MC, van Duijn AJ, Csavina KR. Body weight supported treadmill vs. overground gait training for acute stroke gait rehabilitation. Int J Rehabil Res. 2019;42(3):270–4.

32. Maas AIR, Menon DK, Adelson PD, Andelic N, Bell MJ, Belli A, Bragge P, Brazinova A, Büki A, Chesnut RM, et al. Traumatic brain injury: integrated approaches to improve prevention, clinical care, and research. Lancet Neurol. 2017;16(12):987–1048.

33. MacKinnon CD, Winter DA. Control of whole body balance in the frontal plane during human walking. J Biomech. 1993;26(6):633–44.

34. Marks D, Schweinfurther R, Dewor A, Huster T, Paredes LP, Zutter D, Möller JC. The andago for overground gait training in patients with gait disorders after stroke-results from a usability study. Physiother Res Rep. 2019;2:1–8.

35. Medley A, Thompson M, French J. Predicting the probability of falls in community dwelling persons with brain injury: a pilot study. Brain Inj. 2006;20(13–14):1403–8.

36. Menon DK, Schwab K, Wright DW, Maas AI, et al. Position statement: definition of traumatic brain injury. Arch Phys Med Rehabil. 2010;91(11):1637–40.

37. Meyer A, Cutler E, Hellstrand J, Meise E, Rudolf K, Hrdlicka HC, Grevelding P, Nankin M. Inducing body weight supported postural perturbations during gait and balance exercises to improve balance after stroke: a pilot study; 2021.

38. Ness D. Dynamic over-ground body weight support training in patients with pusher syndrome after stroke: case series. APTA combined sections meeting. In Case series. APTA combined sections meeting; 2014.

39. Park J, Kim T-H. The effects of balance and gait function on quality of life of stroke patients. NeuroRehabilitation. 2019;44(1):37–41.

40. Patton J, Brown DA, Peshkin M, Santos-Munné JJ, Makhlin A, Lewis E, Colgate EJ, Schwandt D. KineAssist: design and development of a robotic overground gait and balance therapy device. Top Stroke Rehabil. 2008;15(2):131–9.

41. Pennycott A, Wyss D, Vallery H, Riener R. Effects of added inertia and body weight support on lateral balance control during walking. In: 2011 IEEE international conference on rehabilitation robotics. IEEE; 2011. p. 1–5.

42. Plooij M, Keller U, Sterke B, Komi S, Vallery H, Von Zitzewitz J. Design of RYSEN: an intrinsically safe and low-power three-dimensional overground body weight support. IEEE Robot Autom Lett. 2018;3(3):2253–60.

43. Plooij M, Apte S, Keller U, Baines P, Sterke B, Asboth L, Courtine G, von Zitzewitz J, Vallery H. Neglected physical human-robot interaction may explain variable outcomes in gait neurorehabilitation research. Sci Robot. 2021;6(58):eabf1888.

44. Shetty D, Fast A, Campana C. Ambulatory suspension and rehabilitation apparatus. US Patent 7,462,138, 9 Dec 2008.

45. Strong K, Mathers C, Bonita R. Preventing stroke: saving lives around the world. Lancet Neurol. 2007;6(2):182–7.

46. Tang R, Holland M, Milbauer M, Olson E, Skora J, Kapellusch JM, Garg A. Biomechanical evaluations of bed-to-wheelchair transfer: Gait belt versus walking belt. Workplace Health & Saf. 2018;66(8):384–92.

47. Tay SS, Visperas CA, Abideen ABZ, Tan MMJ, Zaw EM, Lai H, Neo EJR. Effectiveness of adjunct robotic therapy with a patient-guided suspension system for stroke rehabilitation using a 7-days-a-week model of care: a comparison with conventional rehabilitation. Arch Rehabil Res Clin Transl. 2021;3(3):100144.

48. Vallery H, Lutz P, von Zitzewitz J, Rauter G, Fritschi M, Everarts C, Ronsse R, Curt A, Bolliger M. Multidirectional transparent support for overground gait training. In: 2013 IEEE 13th international conference on rehabilitation robotics (ICORR). IEEE; 2013. p. 1–7.

49. van Hedel HJA, Rosselli I, Baumgartner-Ricklin S. Clinical utility of the over-ground bodyweight-supporting walking system andago in children and youths with gait impairments. J Neuroeng Rehabil. 2021;18(1):1–20.

50. Venema DM, Skinner AM, Nailon R, Conley D, High R, Jones KJ. Patient and system factors associated with unassisted and injurious falls in hospitals: an observational study. BMC Geriatr. 2019;19(1):1–10.

51. Von Zitzewitz J, Fehlberg L, Bruckmann T, Vallery H. Use of passively guided deflection units and energy-storing elements to increase the application range of wire robots. In: Cable-driven parallel robots. Berlin: Springer; 2013. p. 167–84.

52. Walton C, King R, Rechtman L, Kaye W, Leray E, Marrie RA, Robertson N, La Rocca N, Uitdehaag B, van der Mei I, et al. Rising prevalence of multiple sclerosis worldwide: Insights from the atlas of MS. Mult Scler J. 2020;26(14):1816–21.

53. Wessels M, Lucas C, Eriks I, de Groot S. Body weight-supported gait training for restoration of walking in people with an incomplete spinal cord injury: a systematic review. J Rehabil Med. 2010;42(6):513–9.

54. Williams G, Morris ME, Schache A, McCrory PR. Incidence of gait abnormalities after traumatic brain injury. Arch Phys Med Rehabil. 2009;90(4):587–93.

55. Wren TAL, Rethlefsen S, Kay RM. Prevalence of specific gait abnormalities in children with cerebral palsy: influence of cerebral palsy subtype, age, and previous surgery. J Pediatr Orthop. 2005;25(1):79–83.

56. Wright A, Stone K, Martinelli L, Fryer S, Smith G, Lambrick D, Stoner L, Jobson S, Faulkner J. Effect of combined home-based, overground robotic-assisted gait training and usual physiotherapy on clinical functional outcomes in people with chronic stroke: A randomized controlled trial. Clin Rehabil. 2021;35(6):882–93.

57. Wyndaele M, Wyndaele J-J. Incidence, prevalence and epidemiology of spinal cord injury: what learns a worldwide literature survey? Spinal Cord. 2006;44(9):523–9.

Epilogue: Robots for Neurorehabilitation—The Debate

34

John W. Krakauer and David J. Reinkensmeyer

Abstract

There is an ongoing debate over the value of robotic movement training devices for clinical rehabilitation practice and their promise for the future. In this Epilogue, we break down the debate into a series of specific propositions and then provide commentary from two perspectives. JK is a neurologist and neuroscientist who has developed novel location-based immersive training environments combined with proprietary gaming software that encourages active exploration to reduce upper limb impairment after stroke. DR is an engineer and neuroscientist who has developed robotic and sensor-based rehabilitation training systems and computational models of motor recovery.

Keywords

Robots · Rehabilitation · Stroke · Movement training · Sensory-motor recovery · Plasticity · Motor learning · Somatosensation

Proposition 1: Robotic movement training devices are useful scientific tools.

DR: Agree. Robotic movement training devices such as those shown in Fig. 34.1 have facilitated key contributions to rehabilitation science. In terms of scientific method, they have made scientists and clinicians *define* the content of rehabilitation movement training because they must be specifically designed and precisely programmed before they can be used to do anything. And then, when they are used to assist in movement training, they maintain a record of every force and movement the patient experiences, generating an abundance of information that has only just begun to be tapped for scientific purposes. These twin "superpowers" of codification and quantification have led to more precise hypotheses—about the effects of different types of training paradigms and about how we might detect and predict movement recovery.

J. W. Krakauer
Departments of Neurology, Neuroscience and Physical Medicine & Rehabilitation, Baltimore, MD, USA
e-mail: jkrakau1@jhmi.edu

The John Hopkins University School of Medicine, Baltimore, MD, USA

The Santa Fe Institute, Santa Fe, NM, USA

D. J. Reinkensmeyer (✉)
Department of Mechanical and Aerospace Engineering, University of California at Irvine, 4200 Engineering Gateway, 92617 Irvine, CA, USA
e-mail: dreinken@uci.edu

Department of Anatomy and Neurobiology, University of California at Irvine, 4200 Engineering Gateway, 92617 Irvine, CA, USA

Department of Biomedical Engineering, University of California at Irvine, 4200 Engineering Gateway, 92617 Irvine, CA, USA

© The Author(s), under exclusive license to Springer Nature Switzerland AG 2022
D. J. Reinkensmeyer et al. (eds.), *Neurorehabilitation Technology*,
https://doi.org/10.1007/978-3-031-08995-4_34

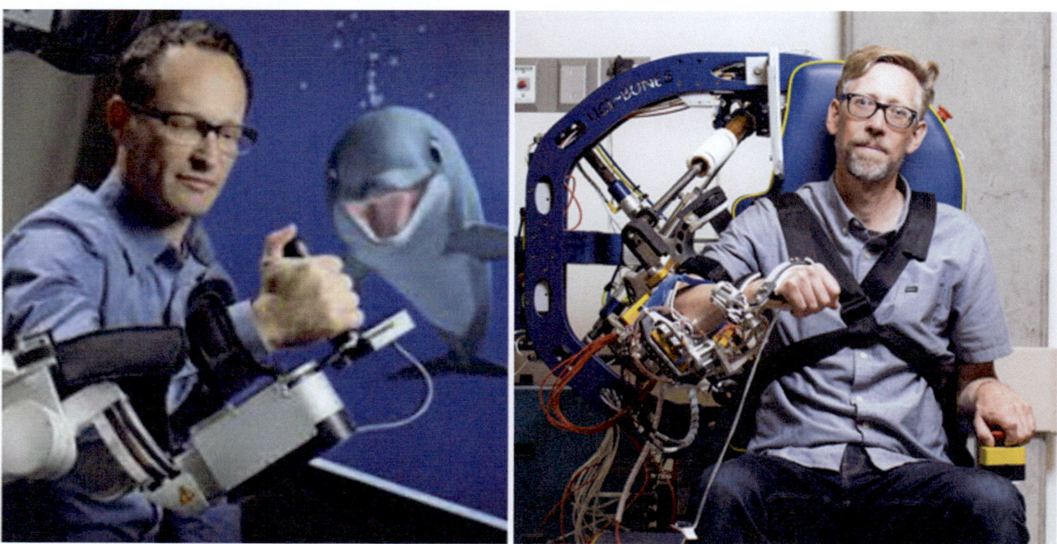

Fig. 34.1 The authors in robotic movement training devices

Some key scientific findings enabled by robotic devices include the importance of efference (i.e., active patient effort) in movement training [1–3], and relatedly, the algorithmic tendency of the motor system to "slack" when physical assistance is provided [4–7]; the interplay of error, challenge, and motivation in stimulating motor learning [8–12]; the presence of multiple learning/adaptation process during recovery [13, 14]; the importance of proprioception for the effectiveness of motor training [15]; and the diagnostic power of movement kinematics such as movement smoothness [16–18]. There have also been many equivocal therapeutic findings in studies conducted with robotic therapy devices. This is likely due to underpowering related to the high variability and modest size of treatment response, even with the improved quantification ability. But at least robots have forced scientists to ask well-defined questions, which is an improvement over the field from when I entered it 30 years ago.

JK. Agree. I think that robots broadly defined can be thought about and used in three ways. The first is as scientific tool to study motor control and motor learning. The seminal work on force-field adaptation by Reza Shadmehr and colleagues is a key example. Second, robots can be used to quantify motor deficits in patients, the studies by Stephen Scott and colleagues, and Amy Bastian and colleagues serve as good examples of this. Third, and most questionably in my view, is as therapeutic devices.

Proposition 2: The treatment effect due to robotic movement training is so modest that the technology is not a clinically useful tool.

DR: Agree with caveats. Yes, the *mean* treatment effect due to robotic therapy is modest, below what can be considered clinically important. But note that this isn't a result unique to robotic therapy—it's a problem with all neurorehabilitation movement training and probably reflects fundamental limits imposed by damaged neural tissue. Also note that when you consider the statistical distribution of patient responses, some percentage (I estimate 10–30%) experience treatment effects due to robotic therapy that are above the minimally clinically important difference (MCID). So, while the statement is true on average, it's not accurate for all individuals. Much of the challenge of robotic therapy (and neurorehabilitation in general) now lies in figuring out who responds, to what protocol, and why, in order to better shape patient and treatment selection. Finally, robotic movement training devices typically remain relatively difficult to use and thus are not clinically viable because of uptake issues, not effectiveness issues (they mostly match and sometimes exceed the results achievable with conventional training—see below). Thus, the dream of providing therapists and patients with a tool that allows them to automate key aspects of movement training with less cost has only been achieved sporadically by certain therapists and facilities.

JK: Agree with caveats. The reason why the treatment effect of robots has been modest (to say the least) is because of conceptual confusion rather than a failure of engineering. This confusion is a sad echo of a similar conceptual confusion that lies at the heart of rehabilitation in general. It is a strange move to defend robotics by saying that the whole field is doing poorly! What is the confusion? It takes two forms—the first is the assumption that assisted repetition of movements is an effective form of training—it isn't. The second is the assumption that motor learning is the way to fix a motor control deficit—it isn't. With regard to the first false assumption—completing the proprioceptive feeling of the correct kinematics of a movement is no more likely to help you find the motor commands required than watching someone serve will turn you into a tennis player. Unfortunately, this haptic guidance fallacy will not die and it is not rescued by the fact that patients make some initial contribution to the movement. With regard to the second assumption, motor learning capacity is normal in patients with hemiparesis from corticospinal lesions—if learning is all that was required for repair then patients who have had years to learn after stroke should all have recovered!

Proposition 3: Continuing to refine the device design and training algorithms won't improve on the treatment effect size enough to matter.

DR: Agree with caveats. I must admit that I am skeptical that continuing to tweak device designs or movement training algorithms can have a major effect unless coupled with plasticity-enhancing interventions. However, can science

really ever confidently assert an absolute negative? Recall dogmas such as "There are no physical laws other than Newton's" or "The adult brain can't create new neurons"? Somewhere out there, there is an ingenious Ph.D. student who just might knock it out of the park with a new approach. And, in the meanwhile, I like to keep in mind that science is often Edisonian, in the sense that it relies on extensive trial-and-error, but also, and perhaps more relevantly, in the sense that it requires protracted effort and the careful assembling of incremental improvements to finally generate meaningful effects. Take as one inspiring example the evolution of the treatment of childhood leukemia [19].

JK: Agree with caveats. Again what is needed is clearer biological thinking not better engineering. As is so often the case these days, the methodological tail wags the conceptual dog. It is an unfortunate historical fact that rehabilitation of late has been driven by an unholy alliance between engineers and clinicians with a relative lack of neuroscience in between. This techno-utopianism badly needs tempering. What it has led to is a fuzzy thinking and confusion. For example, is a robot meant to mimic, assist, or ultimately substitute for a therapist? Or is a robot meant to be a complementary approach? The lack of clear answers to these questions is exasperating. That said, I do think that a next generation of robots could be used to enhance recovery of the upper limb after stroke and other forms of neurological injury. For example, to provide varying degrees of weight support, allow bilateral training, and restrict compensatory strategies. What we must stop doing is seeing robots primarily as devices to provide triggered assistance along the direction of movement. Indeed, a recent study showed that robotic weight support was sufficient to improve 3D movements in a circle-drawing task with no additional improvements obtained when path assistance was added [20]. Thus, to the degree that some therapeutic gains have been seen for assistive modes, they are small and probably attributable to a second-order effect acting through motivation and effort, rather than any direct effect on either learning or repair.

Repetition of this kind is not synonymous with deliberative practice. As to motivation, there are other ways to promote it that do not require expensive actuated robots [21].

Proposition 4: Refining patient selection criteria won't improve on the treatment effect size enough to matter.

DR: Disagree. I think the evidence is already pretty clear for the upper extremity after stroke that people with some residual corticospinal tract will benefit more than people with totally disrupted CST, and the size of the additional benefit is above the MCID for a fraction of patients [22, 23]. Timing matters enough to be meaningful as well, as the recent study by Dromerick et al. convincingly showed (training in the subacute phase was significantly more beneficial than training in the chronic phase) [24]. Baseline proprioception is a powerful predictor of benefit as well [15, 25]. The challenge comes in the logistical and ethical implementations of these criteria. But I do believe that, if they could be implemented appropriately, the result would be to bump up the mean treatment effect of clinical rehabilitation practice.

JK. Agree. It is always a last resort to say that there is a subset of patients that will respond to a treatment that is not showing impressive results overall. I of course agree that we should stratify intelligently, especially when investigating new approaches. It would have been odd if the first heart transplants had been tried on octogenarians with severe medical comorbidities. That said you can't rescue a new approach that was ill-conceived conceptually from the get-go by holding out hope that there are some patients out there that will show large effect sizes. It is indeed the case that there are some chronic patients that respond well to larger doses and intensities of rehabilitation of any kind. Whether this is true recovery/restoration versus peripheral effects, strengthening or compensation remains unclear. Again, what is the specific biological hypothesis about robotics over other forms of approach that make it the ideal choice for a subset of latent recoverers? I have yet to hear a clear answer.

Proposition 5: Robotic therapy treatment effects would not be significantly amplified even if patients were given orders of magnitude more training time with the devices.

DR: Disagree with caveats. The amount of movement practice typically achieved by neurologic patients is woefully small—in-clinic therapy is limited and home adherence is low, even if the prescribed exercise program is unambitious. Historically, even studies that have set out to study "high dose" have probably undershot the target as well. Jeffers et al.'s meta-study in a rodent model of stroke suggested the dose–response curve is a nonlinear, "hockey stick" function—flat at first, up to a relatively high threshold of training activity, and then correlated thereafter [26]. For walking after stroke, a recent study convincingly showed the benefit of a large dose of walking compared to a small dose [27]. But in that study, which had three dose levels, the therapeutic effects of the two higher dose levels were about the same, and those effects, while clinically significant, were *just* clinically significant. So, maybe it's a "broken hockey stick". Variations in low doses (which clinicians might sometimes mistakenly think of as relatively "high" because of historic practice habits) don't matter (the blade); variations that get past the threshold matter (the shaft), but there is no zone of "hyperdose" that can ultimately overcome the effects of severe neuroanatomical damage (the broken, top shaft). See Chap. 3 for an extended discussion of these issues.

JK Agree. We can all agree that patients need more behavioral intervention at higher dose and intensity focused on impairment reduction. The question is what should this behavioral intervention be? At first blush, robots would seem to be ideal—high-dose and high-intensity movement repetition machines! Alas, this superficial impression is wrong but still, as I have said above, holds powerful sway over the minds of many in the field. Motor learning is never about simple repetition and recovery is not about motor learning. What we need to do is take a step back and think. The success of treadmill training coupled with epidural stimulation for spinal cord injury is a reason for optimism and points to a potential analogous use of upper limb robots as "treadmills" for the arm. As I said above, for a robot to be effective it must help patients find their own commands not just provide them with kinematic sensory experiences. I am just not convinced that assistive robots are the best way to promote exploration of residual command space.

Proposition 6: There are no mechanistic reasons to believe that providing active assistance with a robotic therapy device should be more effective than conventional rehabilitation exercise that does not provide physical assistance.

DR: Disagree with caveats. In the case of gait training, if some physical assistance isn't provided, the walking task can't be safely practiced. In that case, it is difficult for me to believe that providing no assistance can achieve similar results to providing some assistance (robotic or not). But even when there aren't safety issues, providing active assistance with a robot can provide benefits in at least three ways. First, providing active assistance improves motivation to engage in training [10]. Improved motivation can in turn improve not only dose achieved (by encouraging a greater willingness to engage in practice) but also the amount of motor learning achieved (possibly through dopaminergic mechanisms associated with experiences of success) [28, 29]. Yet, it may also be possible to generate similar motivational effects without active assistance provided by robots, for example, through sensor-based systems with clever feedback about success, or by sociopsychological effects—such as collaboration, cooperation, or simply practicing in a group [30]—it remains to be seen. Second, active assistance promotes neural strengthening. Weakness is a strong predictor of limb function after stroke but cannot be attributed to muscle atrophy. Repeated stimulation of descending pathways improves force output by multiple mechanisms. For example, repeated efferent stimulation that is accompanied by movement performance feedback likely helps the motor system sort out which residual CST neurons are most useful to activate desired muscles with adequate force. We have previously laid out a model of this concept of neural strengthening with mathematical detail [31–33]. The point for this discussion is that a robotic device that appropriately titrates mechanical assistance will help a person repetitively engage damaged descending pathways (in a motivated way!—see above). This will cause improvements in the neurally mediated component of strength, and, indeed, this is a well-documented outcome of robotic movement training [34, 35]. Improvements in neural strength should be expected to translate, in turn, to some amount of impairment reduction on a variety of scales (and indeed, neural strengthening is impairment reduction by definition!). For some subsets of activities, this will translate to modest improvements in function. Third, movement with active assistance promotes somatosensory stimulation compared to what happens during self-driven movement that is very impaired. And by stimulation I don't mean kinematic demonstration, which I agree has limited benefit not only in movement rehabilitation but also in movement training with unimpaired participants [36]. Rather, I mean generation of a greater diversity of somatosensory input. Such somatosensory stimulation may then work through use-dependent, Hebbian, or reward-based learning mechanisms to drive beneficial cortical plasticity. If the somatosensory story turns out to be true, then the role of robots may increase, as stimulating somatosensation seems like it will require active technologies that impose movements.

In summary, then, what do assistive robots bring that can serve as a useful tool for therapists and patients? Imagine you are a therapist who would like your very weak patient to practice a large dose of functionally relevant movement in a motivating way. You also want to make sure that the patient is motivated and continuously puts out the effort needed to rebuild neural strength. Then, as a bonus, you would like them to experience a diversified portfolio of somatosensory input. Ideally, you'd like for them to do this without you being constantly present, as you are incredibly busy and the amount of reimbursed time you have with them is limited.

JK. Agree with caveats. At the risk of repeating myself: What is the extra thing that a robot does compared to conventional therapy? Is there a qualitative difference or a quantitative one? "Proprioceptive stimulation" is a slippery notion. As David agrees, it can't just be proprioceptive "demonstration", analogous to watching the correct movement. So what is it? Is the proprioceptive stimulation instructing cortex or is it increasing the receptivity of segmental circuits to residual descending commands? Not yet knowing the answer would be ok if the effect sizes of robotic intervention were large, but sadly they are not. In the recent RATULS trial [37], the largest robotic rehabilitation trial to date, robot-assisted training of the upper limb was no better than usual care on the primary outcome measure, the ARAT at 3 months. Additionally, robotics was no better than enhanced upper limb training (EULT) on the Fugl–Meyer score and was inferior to UELT on ADLs at 3 months. These results are, to say the least, hardly a ringing endorsement for robotic rehabilitation no matter how badly David wants to put a positive spin on RATULS (see below). My belief is that assistive robotics falls between two chairs—it is neither optimal for impairment reduction nor functional training. Robots have just not worked out. If they are to have a future in rehabilitation then movement-path assistance needs to go to the back of the line. There are other ways to "help a person repetitively engage damaged descending pathways (in a motivated way!)" as David states, without path assistance.

Proposition 7: Robotic training holds no advantages compared to other non-robotic technologies for training, such as sensor-based training.

DR: Agree with caveats. In RATULS, as in the vast majority of previous trials of robotic movement training, robot-assisted training was no worse than some form of enhanced upper limb training for the primary outcome (ARAT success) and it was better than usual care for impairment reduction (measured by UEFM change, which, by the way, is meaningful to patients at around 4 points [38, 39]—RATULS found the mean benefit over usual care was 2.8 points, indicating that a substantial fraction of robot participants exceeded by 4 points what usual care participants achieved). But, surprisingly to me at this mature stage of clinical testing, the robotic movement training in RATULS was delivered "face-to-face" "with a one-to-one patient-to-therapist ratio" (quotes from the study). The study thus did not test what is perhaps the most obvious potential advantage of robotic movement training: motivating real-world patients to engage in further beneficial training outside of formal therapy times. That said, I have to acknowledge that my ideas about the uniquely, positive package of influence that active assistance might exert over and above sensor-only approaches still needs to be further experimentally verified. I could fall back on the robot superpowers of codification and quantification mentioned above, but sensors have these superpowers to some degree as well and are missing only the power of being able to precisely control patterns of force application. So, pragmatically, yes, I agree—use of sensor-based systems for routine clinical practice seems warranted for the most part at this time. By way of full disclosure, I'd like to note that the two systems that I have worked on that have been commercialized with some success—as Armeo-Spring and MusicGlove—are essentially sensor-based systems with some functionally targeted. non-motorized, mechanical design. ArmeoSpring foregoes path assistance, as John suggests, providing gradable weight support for 3D hand movement.

JK: Agree with caveats. Actuated robots are not going to be required in general—can use far cheaper devices for weight support and sensing. That said, I do think that a next generation of robots could occupy a niche for specific forms of intervention and measurement, especially of forces, that address the problem of 3D movements in general, and the complexities of the paretic shoulder in particular.

Acknowledgements DJR supported by NIH grant R01HD062744 and NIDILRR grant 90REGE0005-01.

Disclosure David Reinkensmeyer has a financial interest in Hocoma AG and Flint Rehabilitation devices, makers of rehabilitation equipment. The terms of these arrangements have been reviewed and approved by the University of California, Irvine in accordance with its conflict of interest policies.

References

1. Hu XL, Tong K-Y, Song R, Zheng XJ, Leung WWF. A comparison between electromyography-driven robot and passive motion device on wrist rehabilitation for chronic stroke. Neurorehabil Neural Repair. 2009;23(8):837–46.
2. Hidler J, Nichols D, Pelliccio M, Brady K, Campbell DD, Kahn JH, et al. Multicenter randomized clinical trial evaluating the effectiveness of the Lokomat in subacute stroke. Neurorehabil Neural Repair. 2009;23(1):5–13.
3. Hicks AL. Locomotor training in people with spinal cord injury: is this exercise? Spinal Cord. 2021;59(1):9–16.
4. Wolbrecht ET, Chan V, Reinkensmeyer DJ, Bobrow JE. Optimizing compliant, model-based robotic assistance to promote neurorehabilitation. IEEE Trans Neural Syst Rehabil Eng. 2008;16(3).
5. Emken JL, Benitez R, Sideris A, Bobrow JE, Reinkensmeyer DJ. Motor adaptation as a greedy optimization of error and effort. J Neurophysiol. 2007;97(6).
6. Smith BW, Rowe JB, Reinkensmeyer DJ. Real-time slacking as a default mode of grip force control: implications for force minimization and personal grip force variation. J Neurophysiol. 2018 Oct 1;120(4):2107–20.
7. Israel JF, Campbell DD, Kahn JH, Hornby TG. Metabolic costs and muscle activity patterns during robotic- and therapist-assisted treadmill walking in individuals with incomplete spinal cord injury. Phys Ther. 2006;86(11):1466–78.
8. Patton JL, Stoykov ME, Kovic M, Mussa-Ivaldi FA. Evaluation of robotic training forces that either enhance or reduce error in chronic hemiparetic stroke survivors. Exp Brain Res. 2006;168(3):368–83.
9. Rosenthal O, Wing AM, Wyatt JL, Punt D, Brownless B, Ko-Ko C, et al. Boosting robot-assisted rehabilitation of stroke hemiparesis by individualized selection of upper limb movements - A pilot study. J NeuroEngineering Rehabil. 2019;16(1).
10. Rowe JB, Chan V, Ingemanson ML, Cramer SC, Wolbrecht ET, Reinkensmeyer DJ. Robotic Assistance for Training Finger Movement Using a Hebbian Model: A Randomized Controlled Trial. Neurorehabil Neural Repair. 2017;31(8).
11. Brown DA, Lee TD, Reinkensmeyer DJ, Duarte JE. Designing robots that challenge to optimize motor learning. 2016. (Neurorehabilitation Technology, Second Edition).
12. Duarte JE, Reinkensmeyer DJ. Effects of robotically modulating kinematic variability on motor skill learning and motivation. J Neurophysiol. 2015;113(7).
13. Schweighofer N, Wang C, Mottet D, Laffont I, Bakhti K, Reinkensmeyer DJ, et al. Dissociating motor learning from recovery in exoskeleton training post-stroke. J NeuroEngineering Rehabil. 2018;15(1):89.
14. Nibras N, Liu C, Mottet D, Wang C, Reinkensmeyer D, Remy- O, et al. Dissociating sensorimotor recovery and compensation during exoskeleton training following stroke. Front Hum Neurosci. 2021;30(15): 645021.
15. Ingemanson ML, Rowe JR, Chan V, Wolbrecht ET, Reinkensmeyer DJ, Cramer SC. Somatosensory system integrity explains differences in treatment response after stroke. Neurology. 2019. https://doi.org/10.1212/WNL.0000000000007041.
16. Krebs HI, Krams M, Agrafiotis DK, DiBernardo A, Chavez JC, Littman GS, et al. Robotic measurement of arm movements after stroke establishes biomarkers of motor recovery. Stroke. 2014;45(1):200–4.
17. Bosecker C, Dipietro L, Volpe B, Krebs HI. Kinematic robot-based evaluation scales and clinical counterparts to measure upper limb motor performance in patients with chronic stroke. Neurorehabil Neural Repair. 2010 Jan;24(1):62–9.
18. Rohrer B, Fasoli S, Krebs HI, Hughes R, Volpe B, Frontera WR, et al. Movement smoothness changes during stroke recovery. J Neurosci Off J Soc Neurosci. 2002;22(18):8297–304.
19. Madhusoodhan PP, Carroll WL, Bhatla T. Progress and prospects in Pediatric Leukemia. Curr Probl Pediatr Adolesc Health Care. 2016;46(7):229–41.
20. Raghavan P, Bilaloglu S, Ali SZ, Jin X, Aluru V, Buckley MC, et al. The Role of robotic path assistance and weight support in facilitating 3D movements in individuals with Poststroke hemiparesis. Neurorehabil Neural Repair. 2020;34(2):134–47.
21. Krakauer JW, Kitago T, Goldsmith J, Ahmad O, Roy P, Stein J, et al. Comparing a novel neuroanimation experience to conventional therapy for high-dose intensive upper-limb training in subacute stroke: the SMARTS2 randomized trial. Neurorehabil Neural Repair. 2021;35(5):393–405.
22. Stinear CM. Prediction of motor recovery after stroke: advances in biomarkers. Lancet Neurol. 2017;16(10):826–36.
23. Senesh MR, Reinkensmeyer DJ. Breaking Proportional Recovery After Stroke. Neurorehabil Neural Repair. 2019;33(11).

24. Dromerick AW, Geed S, Barth J, Brady K, Giannetti ML, Mitchell A, et al. Critical period after stroke study (CPASS): a phase II clinical trial testing an optimal time for motor recovery after stroke in humans. Proc Natl Acad Sci U S A. 2021;118(39): e2026676118.

25. Park S-W, Wolf SL, Blanton S, Winstein C, Nichols-Larsen DS. The EXCITE trial: predicting a clinically meaningful motor activity log outcome. Neurorehabil Neural Repair. 2008;22(5):486–93.

26. Jeffers MS, Karthikeyan S, Gomez-Smith M, Gasinzigwa S, Achenbach J, Feiten A, et al. Does stroke rehabilitation really matter? part B: an algorithm for prescribing an effective intensity of rehabilitation. Neurorehabil Neural Repair. 2018;32(1):73–83.

27. Klassen TD, Dukelow SP, Bayley MT, Benavente O, Hill MD, Krassioukov A, et al. Higher doses improve walking recovery during stroke inpatient rehabilitation. Stroke. 2020;51(9):2639–48.

28. Wong AL, Lindquist MA, Haith AM, Krakauer JW. Explicit knowledge enhances motor vigor and performance: motivation versus practice in sequence tasks. J Neurophysiol. 2015;114(1):219–32.

29. Abe M, Schambra H, Wassermann EM, Luckenbaugh D, Schweighofer N, Cohen LG. Reward improves long-term retention of a motor memory through induction of offline memory gains. Curr Biol. 2011;21(7):557–62.

30. Wuennemann MJ, Mackenzie SW, Lane HP, Peltz AR, Ma X, Gerber LM, et al. Dose and staffing comparison study of upper limb device-assisted therapy. NeuroRehabilitation. 2020;46(3):287–97.

31. Reinkensmeyer DJ, Guigon E, Maier MA. A computational model of use-dependent motor recovery following a stroke: Optimizing corticospinal activations via reinforcement learning can explain residual capacity and other strength recovery dynamics. Neural Netw. 2012;29–30.

32. Norman SL, Lobo-Prat J, Reinkensmeyer DJ. How do strength and coordination recovery interact after stroke? A computational model for informing robotic training. IEEE Int Conf Rehabil Robot; 2017. p. 181–6.

33. Norman SL, Wolpaw JR, Reinkensmeyer DJ. Targeting Neuroplasticity to Improve Motor Recovery after Stroke [Internet]. 2020 [cited 2021 Dec 4]. Available from: https://resolver.caltech.edu/CaltechAUTHORS: 20200911-133136049.

34. Ferreira FMRM, Chaves MEA, Oliveira VC, Petten AMVNV, Vimieiro CBS. Effectiveness of robot therapy on body function and structure in people with limited upper limb function: A systematic review and meta-analysis. PLOS ONE. 2018 Jul 12;13(7): e0200330.

35. Reinkensmeyer DJ, Maier MA, Guigon E, Chan V, Akoner O, Wolbrecht ET, et al. Do robotic and non-robotic arm movement training drive motor recovery after stroke by a common neural mechanism? Experimental evidence and a computational model. Conf Proc Annu Int Conf IEEE Eng Med Biol Soc IEEE Eng Med Biol Soc Conf. 2009.

36. Heuer H, Lüttgen J. Robot assistance of motor learning: a neuro-cognitive perspective. Neurosci Biobehav Rev. 2015;56:222–40.

37. Rodgers H, Bosomworth H, Krebs HI, van Wijck F, Howel D, Wilson N, et al. Robot assisted training for the upper limb after stroke (RATULS): a multicentre randomised controlled trial. The Lancet. 2019;394 (10192):51–62.

38. Page P, Fulk S, Boyne G. Clinically important differences for the upper-extremity fugl-meyer scale in people with minimal to moderate impairment due to chronic stroke. Phys Ther. 2012;92(6):791–8.

39. Lundquist CB, Maribo T. The Fugl-Meyer assessment of the upper extremity: reliability, responsiveness and validity of the Danish version. Disabil Rehabil. 2017;39(9):934–9.

Correction to: Spinal Cord Stimulation to Enable Leg Motor Control and Walking in People with Spinal Cord Injury

Ismael Seáñez, Marco Capogrosso,
Karen Minassian, and Fabien B. Wagner

Correction to:
Chapter 18 in: D. J. Reinkensmeyer et al. (eds.), *Neurorehabilitation Technology*, https://doi.org/10.1007/978-3-031-08995-4_18

Chapter 18 (Spinal Cord Stimulation to Enable Leg Motor Control and Walking in People with Spinal Cord Injury) was previously published non-open access. It has now been changed to open access under a CC BY 4.0 license and the copyright holder updated to 'The Author(s)'. The book has also been updated with this change.

The updated version of this chapter can be found at
https://doi.org/10.1007/978-3-031-08995-4_18

© The Author(s) 2023
D. J. Reinkensmeyer et al. (eds.), *Neurorehabilitation Technology*,
https://doi.org/10.1007/978-3-031-08995-4_35

Index

© The Editor(s) (if applicable) and The Author(s), under exclusive license to Springer
Nature Switzerland AG 2022
D. J. Reinkensmeyer et al. (eds.), *Neurorehabilitation Technology*,
https://doi.org/10.1007/978-3-031-08995-4

Index

MIX
Papier aus verantwortungsvollen Quellen
Paper from responsible sources
FSC® C105338

If you have any concerns about our products,
you can contact us on
ProductSafety@springernature.com

In case Publisher is established outside the EU,
the EU authorized representative is:
**Springer Nature Customer Service Center GmbH
Europaplatz 3, 69115 Heidelberg, Germany**

Printed by Libri Plureos GmbH
in Hamburg, Germany